STUDIES IN MATHEMATICS

Probability

T V M McKirgan

Ginn London
Cheshire Melbourne

077009 ISBN 0 602 21387 8

Published in Great Britain by
Ginn & Company Ltd, 18 Bedford Row, London WC1
and in Australia by
F W Cheshire Publishing Pty Ltd, 346 St Kilda Road, Melbourne 3004

Made and printed in Great Britain by R & R Clark Ltd, Edinburgh

PREFACE

This book is intended for students in sixth forms, colleges of education and the first year of universities. The presentation of the subject is different from that adopted by the more traditional books: it is hoped that by the use of set theory a much clearer understanding of the subject will ensue.

The book can be used by a student with minimal supervision, and to help such a student, there are many worked examples to illustrate the various points in the text. Supervised students, too, should find these examples help their understanding of the subject.

The standard of the exercises varies from elementary to scholarship level. The exercises at the ends of the chapters are not in order of presentation in the text. I feel that too often students can do questions simply because they know that those at the beginning of a revision set deal with the points mentioned at the beginning of the chapter.

I thank Professor C D Kemp of the School of Mathematics, University of Bradford, for reading the manuscript and for his very helpful suggestions, my wife for typing the manuscript and the staff of Ginn and Company for all their help. I also acknowledge my debt to the Cambridge Local Examinations Syndicate for permission to use questions which have been set in their papers, and which are indicated by the letter C.

Belfast
June 1970

T V M McKirgan

CONTENTS

1 | THE CHANCES ARE . . .

1.1 Introduction

Phrases such as 'It is not likely to rain to-day', 'He's got a fifty–fifty chance of passing the examination', 'It is almost certain that . . .' are part and parcel of our everyday speech. They are used when we want to give some indication of the likelihood of something going to, or not going to, happen. It is not by any means a reliable scale, nor is it meant to be. Generally we consider three cases: unlikely, fifty–fifty, almost certain, and into these three categories we place our forecasts of what is going to happen. Now and again we may even stretch a point and introduce a 'sixty–forty' category or when something is almost certain we use the even more general phrase 'It is odds on that . . .'.

Gamblers are obviously very interested in the likelihood of winning money in games of chance, and it is mainly due to them that an interest was founded in that branch of mathematics which came to be known as the theory of probability.

It should be emphasized at this stage that probability concerns itself with the long-term result. In other words, if we say that in the tossing of a fair coin the probability of a head is $\frac{1}{2}$, it does not follow that in 10 throws the result will be 5 heads and 5 tails. Nor does it follow that every woman's fourth child will be Chinese, although 1 in 4 children born in the world is Chinese.

If someone tosses a fair coin, *i.e.*, a head or tail is equally likely, and obtains 10 heads in a row, what would be your reaction to the chance of a tail on the next throw? Most people would say that by the 'law of averages' there is a very good chance of a tail. This is, in fact, wrong. The probability of a tail is still $\frac{1}{2}$, since probability only takes into account what is going to happen and not what has happened. On the other hand, if you are asked what is the chance of obtaining 10 heads in 10 throws of a fair coin, you would be right in saying that it is very unlikely.

There are two aspects of probability, one found by observation and the other by analysis. Suppose we want to know the probability of a rainy day in June. From the records of previous years we can find the number of days on which rain fell for more than a given time, say 2 hours. Suppose we find that, for the month of June over the past 10 years (300 days), 25 days had rain for more than 2 hours, then we can say that 25 out of 300, or 1 day out of 12, should have rain. Again it should be pointed out that if, in a given month of June, the first 3 days have rain it does not necessarily mean that the rest of the month will be dry.

This is finding the probability by observation. We shall now proceed to finding probability by analysis.

1.2 Sample spaces and events

When we toss a coin there are two possible results—head or tail. We can denote this using set notation by saying that the set $\{H, T\}$ gives all possible results of throwing a coin. In the case of throwing a die the possible results would be given by $\{1, 2, 3, 4, 5, 6\}$. These two sets are known as *sample spaces* for each of the two experiments.

In working out a question on probability it is always necessary to know the sample space, though obviously it is not always possible or feasible to write it out in full. For example, if we were considering a problem where the sample space was all the children in a school, then it would only be necessary to write {All the children in the school} rather than put down all the names. And as we shall see later, it is only necessary to know the number of elements (or members) in the set.

We shall define an *event* as a subset of the sample space.

A *single event* is a subset having one element.

The set $\{H, T\}$ has four subsets, and so we have four events. Two obvious ones are $\{H\}$ and $\{T\}$ but we must remember that $\{H, T\}$ and $\emptyset$ are also subsets. The symbol $\emptyset$ denotes the empty set, that is the set which has no elements (or members).

The set $\{1, 2, 3, 4, 5, 6\}$ has 64 subsets. In fact if there are n elements in the sample space there will be 2^n subsets. A proof of this and a fuller discussion of sets is given in 'Sets and Symbolic Logic' in this series.

Exercise 1a

For each of the following experiments list, or describe, the sample space, and write down the subset describing the event.

1. Obtaining two heads in the throwing of two coins.
2. In tossing a coin and throwing a die, obtaining a head and an even number.
3. Throwing two dice, adding the results together and finding the answer to be 2 or 6.
4. Drawing a white ball from a box containing one white, one red and one black ball.
5. Selecting a left-handed child from your school.
6. Selecting an ace from a well shuffled pack of cards.
7. Selecting two aces from a well shuffled pack of cards.
8. Selecting a three-digit even number from the digits 1, 2, 3, 4, no repetitions being allowed.
9. Selecting a three-digit even number from the digits 0, 1, 2, 3, no repetitions being allowed.
10. Finding that the result of adding the three numbers obtained when three dice are tossed is divisible by 3.

1.3 Probability

We define the ***probability*** of an event A as:

$$P(A) = \frac{n(E)}{n(S)}$$

where $n(E)$, $n(S)$ are the numbers of elements in the event and sample space, respectively.

Example 1. What is the probability of a head in the tossing of a fair coin?

In this case the sample space is $\{H, T\}$ and the event is $\{H\}$, and so the probability is given by $P(\text{Head}) = \frac{1}{2}$.

Example 2. What is the probability of an odd number in the throwing of a fair die?

The sample space here is given by $\{1, 2, 3, 4, 5, 6\}$ and the event by $\{1, 3, 5\}$. Thus the probability is given by $P(\text{Odd number}) = \frac{3}{6}$.

Example 3. One ball is chosen at random from a box containing two red, one green and two black balls. What is the probability that it is red?

The sample space is $\{R_1, R_2, G, B_1, B_2\}$, where R_1, R_2, G, B_1, B_2 represent the separate red, green and black balls, respectively, and the event is $\{R_1, R_2\}$. Thus the probability is $P(\text{Red ball}) = \frac{2}{5}$.

Example 4. If three fair coins are tossed, what is the probability of obtaining two heads and one tail?

The sample space is:

$$\{HHH, HHT, HTH, THH, HTT, THT, TTH, TTT\}$$

and the event is:

$$\{HHT, HTH, THH\}$$

Thus the probability is given by $P(2 \text{ heads}, 1 \text{ tail}) = \frac{3}{8}$. Note that in this case the event is not just $\{HHT\}$.

Example 5. Suppose in Example 3 two balls are drawn out at random. What is the probability that (*a*) they are the same colour, and (*b*) at least one is black?

The sample space will be all the ways of selecting two balls from the five in the box. If we let R_1, R_2, G, B_1, B_2 again represent the five balls then the sample space is:

$$\{R_1R_2, R_1G, R_1B_1, R_1B_2, R_2G, R_2B_1, R_2B_2, GB_1, GB_2, B_1B_2\}$$

For part (*a*) the event is:

$$\{R_1R_2, B_1B_2\}$$

and for part (*b*):

$$\{R_1B_1, R_1B_2, R_2B_1, R_2B_2, GB_1, GB_2, B_1B_2\}$$

Note that 'at least one black' includes both being black. Therefore $P(\text{Both same colour}) = \frac{2}{10}$ and $P(\text{At least 1 black}) = \frac{7}{10}$.

Note. If we were asked to write down the set the elements of which are the letters of the word *calculus*, we should write:

$$\{c, a, l, u, s\}$$

since elements are not repeated within the brackets. But, if we were asked the probability of selecting a letter at random and that letter turning out to be c, then, since there are two cs in the word, there is twice the chance of selecting c than, say, s. We should have to take this into account in our sample space so that the number of elements corresponds to the number of letters in the word. We could, if we wished, discriminate between the like letters by using subscripts. Our sample space would then be:

$$\{c_1, a, l_1, c_2, u_1, l_2, u_2, s\}$$

Example 6. Two dice are thrown and the resulting numbers multiplied together. What is the probability that this product is (*a*) less than 4 or greater than 24, and (*b*) a prime number?

First of all we write down all the possible results, from throwing the two dice, in the form (x, y) where x is the number on the first die and y the number on the second:

(1, 1)	(2, 1)	(3, 1)	(4, 1)	(5, 1)	(6, 1)
(1, 2)	(2, 2)	(3, 2)	(4, 2)	(5, 2)	(6, 2)
(1, 3)	(2, 3)	(3, 3)	(4, 3)	(5, 3)	(6, 3)
(1, 4)	(2, 4)	(3, 4)	(4, 4)	(5, 4)	(6, 4)
(1, 5)	(2, 5)	(3, 5)	(4, 5)	(5, 5)	(6, 5)
(1, 6)	(2, 6)	(3, 6)	(4, 6)	(5, 6)	(6, 6)

Our sample space is the corresponding products for each of these:

$$\left\{\begin{matrix} 1, & 2, & 3, & 4, & 5, & 6 \\ 2, & 4, & 6, & 8, & 10, & 12 \\ 3, & 6, & 9, & 12, & 15, & 18 \\ 4, & 8, & 12, & 16, & 20, & 24 \\ 5, & 10, & 15, & 20, & 25, & 30 \\ 6, & 12, & 18, & 24, & 30, & 36 \end{matrix}\right\}$$

Though some elements are repeated, they come from different throws of the dice. For example, 4 appears three times, since it can come from either (1, 4) (2, 2) or (4, 1).

For part (*a*) our subset describing the event is:

$$\{1, 2(1, 2), 2(2, 1), 3(1, 3), 3(3, 1), 25, 30(5, 6), 30(6, 5), 36\}$$

and for part (*b*):

$$\{2(1, 2), 2(2, 1), 3(1, 3), 3(3, 1), 5(1, 5), 5(5, 1)\}$$

where 2(1, 2), for example, means a product of 2 from 1 on the first die and 2 on the second. 1 is not considered a prime number.

Therefore, the probability for each part is given by:

$$P(\text{Number less than 4 or greater than 24}) = \tfrac{9}{36} = \tfrac{1}{4}$$

and

$$P(\text{Prime number}) = \tfrac{6}{36} = \tfrac{1}{6}$$

Example 7. A card is drawn at random from a well shuffled pack of playing cards. What is the probability that it is (*a*) an ace and, (*b*) not of a red suit?

In this example it would be too cumbersome to list the sample space, so we can just describe it as:

{All 52 playing cards in the pack}

For part (*a*) the subset will be:

$$\{A_H, A_D, A_C, A_S\}$$

where A_H, A_D, A_C, A_S represent the aces of hearts, diamonds, clubs and spades, respectively. For part (*b*) the subset will be all the cards of a black suit. It is again rather cumbersome to list in full so we write:

{26 playing cards of the black suits}

Therefore the probability that the card drawn is an ace is:

$$P(\text{Ace}) = \tfrac{4}{52} = \tfrac{1}{13}$$

and

$$P(\text{Not of a red suit}) = P(\text{A card of a black suit}) = \tfrac{26}{52} = \tfrac{1}{2}$$

Exercise 1b

In each of the following, list or describe the sample space, write down the subset describing the event and then find the probability.

1. What is the probability that a prime number will appear on the throw of an unbiased die? [Note that 1 is not considered to be a prime number.]
2. A coin is tossed and a die thrown. Find the probability that:
   (*a*) a head and a 6 appear
   (*b*) a head and an odd number appear
   (*c*) a tail and a number less than 5 appear
   (*d*) a head or a tail and an even number appear
3. What is the probability of obtaining two heads in three tosses of a fair coin?

4. Two unbiased dice are thrown. If the two numbers that come up are added together, what is the probability that the result will be:
(*a*) 7
(*b*) greater than 4
(*c*) an odd number
(*d*) less than 3 or greater than 9
(*e*) a prime number
5. Four identical pieces of paper have the letters *A*, *B*, *C*, *D* printed on them, one letter to each piece of paper. Three pieces are picked at random, one at a time. What is the probability that they will, in order, spell out a recognizable three-letter word?
6. Out of a pack of well shuffled playing cards one card is drawn at random. Find the probability that it is:
(*a*) the 3 of hearts
(*b*) the 3 of any suit
(*c*) a king or a queen
(*d*) the ace of diamonds or a jack
(*e*) an even-numbered card of a red suit
(*f*) not a diamond
7. A bag contains three discs marked 1, 2 and 3. Two discs are drawn out at random. What is the probability that the two-digit number that could be formed with them is:
(*a*) even (*b*) prime (*c*) greater than 13
8. There are two white socks and two black socks in a drawer. If I lift out two socks at random, what is the probability that they will be of the same colour?
9. If the digits 1, 2, 3, 4 are written down in a random order what is the probability that:
(*a*) they will be in numerical order
(*b*) only two digits will be in their proper place
(*c*) at least two digits will be in their proper place
(*d*) at most two digits will be in their proper place
10. A red and a black die are tossed. What is the probability that:
(*a*) the number on the red die exceeds that on the black die by more than 3
(*b*) the number on the red die is a multiple of the number on the black die

(*c*) the number on the red die is either a multiple or a factor of the number on the black die. (Consider that a number is a factor of itself.)

(*d*) if the two numbers shown are multiplied together they form a perfect square

11. The rule for divisibility by 3 is that the sum of the digits of the number should be divisible by 3. From the digits 1, 2, 3, 4, 5 I write down three digits, in numerical order. What is the probability that the number so formed is divisible by 3?

1.4 Scale of probability

The maximum number of elements in an event occurs when we consider the sample space as a subset of itself, and gives a maximum value for the probability of 1. The minimum number of elements in an event occurs when we have the empty set $\emptyset$ as a subset of the sample space, and gives a minimum value of 0 for the probability. So, for any event A we now have a scale of probability:

$$0 \leqslant P(A) \leqslant 1$$

In describing this scale we can say that if $P(A) = 0$ the event is impossible, and if $P(A) = 1$ the event is a certainty.

Example 8. What is the probability of getting a number greater than or equal to 1 and less than or equal to 6, in the throwing of a fair die?

The sample space is $\{1, 2, 3, 4, 5, 6\}$, and the event is $\{1, 2, 3, 4, 5, 6\}$. Thus, the probability is given by:

$$P(1 \leqslant n \leqslant 6 \mid n \text{ is a single-digit number}) = \tfrac{6}{6} = 1$$

Example 9. What is the probability of getting a number greater than 6 in the throwing of a fair die?

The sample space is $\{1, 2, 3, 4, 5, 6\}$, and the event is $\emptyset$. Thus, the probability is given by:

$$P(\text{Number greater than } 6) = \tfrac{0}{6} = 0$$

Exercises on Chapter 1

In questions marked with an asterisk, you will find it particularly useful to write down the sample space and the subset describing the event before working out the probability: in other questions you may not find it necessary.

*1. Two dice are thrown. What is the probability of:
(*a*) getting a double
(*b*) one number being twice the other
(*c*) the difference between the numbers being more than 2

2. A class consists of 30 pupils of whom 18 are boys. Their names are written on identical pieces of paper which are put into a tin and the tin is shaken. What is the probability that the first name drawn out belongs to:
(*a*) a boy (*b*) a girl

3. A box contains 20 red, 10 green and 20 blue marbles. One marble is drawn from the box. What is the probability that it is:
(*a*) blue (*c*) green or red
(*b*) green (*d*) not red

*4. There are 100 discs numbered from 1 to 100 in a bag. One disc is drawn at random. What is the probability that it has:
(*a*) an even number
(*b*) a multiple of 10
(*c*) a perfect square
(*d*) a prime number
(*e*) a perfect cube
(*f*) a two-digit number the sum of whose digits is greater than 15
(*g*) either an odd number or a factor of 36
(*h*) a number of two digits whose ten's digit exceeds its unit's digit by 5
(*i*) a number which has two digits the same
(*j*) a prime number whose digits add up to 5
[Note that 1 is not considered a prime number and that a number is a factor of itself.]

5. The following data gives the number of men, in a sample of 100, who wear a particular size of shoe:

Size of shoe	7	8	9	10	11	12
Number of men	3	12	35	27	16	7

If one man is picked at random out of the 100, what is the probability that he will be wearing a shoe of size:

(*a*) 9
(*b*) less than 10
(*c*) 10 or 11
(*d*) 7 or 12
(*e*) greater than the average

*6. Four coins are tossed. What is the probability of obtaining:
(*a*) 4 heads
(*b*) no tails
(*c*) at least 1 head
(*d*) at most 2 tails
(*e*) either 3 heads or 3 tails

7. An urn contains 100 balls of which 15 are white, 20 are green, 25 are black, 10 are orange, 5 are blue and the rest are red. One ball is selected at random. What is the probability that it is:
(*a*) green
(*b*) red
(*c*) orange or black
(*d*) not blue
(*e*) white or red or green
(*f*) neither black nor white

8. I have dealt the ace, king and queen of clubs. What is the probability that the next card will be:
(*a*) the jack of clubs
(*b*) a jack
(*c*) the jack of a red suit

9. If a letter is chosen at random from the word *mathematical*, what is the probability that it is:
(*a*) a vowel
(*b*) an *m* or a *c*
(*c*) not a *t*

10. In a class of 40 there are two sets of twins. What is the probability that a person selected at random will have a twin?

*11. Mr and Mrs Jones, Mr Smith, Mr Green and Miss Brown form the Social Committee of a tennis club. It is decided to make a sub-committee of three. If the names of the five people are put in a hat and three names drawn out, what is the probability that the sub-committee so formed will have:
(*a*) all men on it
(*b*) Mr and Mrs Jones on it
(*c*) at least one woman on it

12. At the Hockey Club dance there are 200 people, of whom 120 are women. There are 40 married couples and the rest are single. All 50 playing members of the Ladies' Hockey Club are there.
(*a*) If a man selects a woman at random, what is the probability that she is:
(*i*) single
(*ii*) not a member of the Ladies' Hockey Club
(*b*) In the 'Ladies' Choice' what is the probability that a man, selected by a particular woman, is single?

*13. From three positive and two negative integers, three integers are chosen at random. What is the probability that the square root of their product will not be a real number?

14. The local Rugby Club has been given 54 prizes for its Tombola Stall. The drawing of a marked disc wins a prize. How many unmarked discs must there be if the first person to draw is to have a probability of $\frac{3}{94}$ of winning a prize?

2 | MORE THAN ONE EVENT

2.1 Combination of events

In Chapter 1 we were concerned with one event; here we must consider more than one event. We must first discuss the implication of the words 'or' and 'and.'

The word 'or' can be taken to mean:

(*i*) either . . . or . . . or both

(*ii*) either . . . or . . . but not both.

For example, if I want to know how many people are wearing a hat or gloves, then I should include those wearing both a hat and gloves. However, if I want to know how many people are wearing a pair of brown or a pair of black shoes, I should not expect to meet someone wearing both a pair of brown and a pair of black shoes! For our work on probability we shall take the first meaning for 'or', that is the union of the sets. And so, $P(A \cup B)$ will be taken to mean the probability of either a hat (event A) or gloves (event B) or both.

When we use the word 'and' we shall be thinking in terms of the intersection of the sets. For example, if we wish to know the probability of the number of people wearing a hat and gloves, we shall only be concerned with those people wearing both articles. We shall interpret $P(A \cap B)$ as being the probability of the event A and the event B.

2.2 Addition theorem

Example 1. Suppose that out of 20 people in a room, 10 are wearing hats, 12 are wearing gloves and 4 are wearing both, what is the probability that a person chosen at random will be wearing either a hat or gloves? (Remember that this will include people wearing both a hat and gloves.)

Let A be the set of people wearing hats and B the set of people wearing gloves. What we are interested in finding is $P(A \cup B)$.

We can illustrate the situation with a Venn diagram (Fig. 2.1).

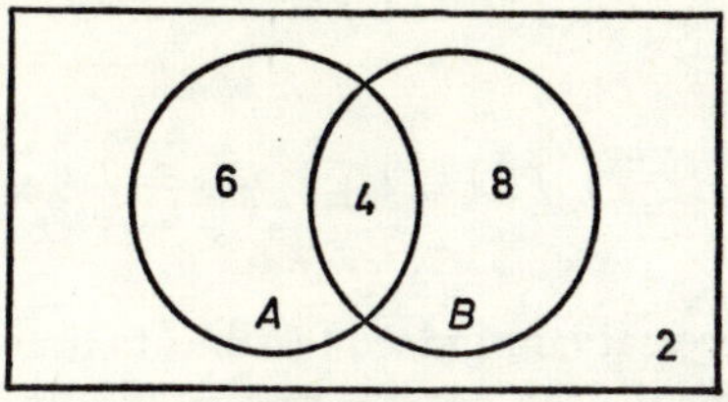

FIGURE 2.1

By our definition of probability on p. 3:

$$P(A \cup B) = \frac{n(A \cup B)}{n(S)} = \tfrac{18}{20}$$

The number of elements in the union of A and B is not the number in A added to the number of elements in B, because the union contains elements common to both sets. If we did this then we would be adding the number of elements in the intersection of the two sets twice. It is clear that:

$$n(A \cup B) = n(A) + n(B) - n(A \cap B)$$

Dividing both sides of this equation by $n(S)$ gives:

$$\frac{n(A \cup B)}{n(S)} = \frac{n(A)}{n(S)} + \frac{n(B)}{n(S)} - \frac{n(A \cap B)}{n(S)}$$

and this, from the definition of probability, gives:

$$P(A \cup B) = P(A) + P(B) - P(A \cap B)$$

This is known as the *addition theorem.*

Example 2. Let $E = \{1, 2, 3, 4, 5, 6, 7, 8, 9, 10\}$, $A = \{1, 3, 5, 7, 9\}$, $B = \{1, 2, 3, 8, 9, 10\}$ and $C = \{2, 3, 8, 10\}$. What is the probability that, if you ask someone to give you a number from E, it will be a member of:

(*i*) A or B (*ii*) B or C (*iii*) A or C

(*i*) We know that $P(A \cup B) = P(A) + P(B) - P(A \cap B)$. Since $(A \cap B) = \{1, 3, 9\}$, $P(A \cap B) = \tfrac{3}{10}$. Hence:

$$P(A \cup B) = \tfrac{5}{10} + \tfrac{6}{10} - \tfrac{3}{10} = \tfrac{4}{5}$$

(*ii*) Since $(B \cap C) = \{2, 3, 8, 10\}$, $P(B \cap C) = \frac{4}{10}$ and so:

$$P(B \cup C) = \tfrac{6}{10} + \tfrac{4}{10} - \tfrac{4}{10}$$
$$= \tfrac{3}{5}$$

(*iii*) Similarly:

$$P(A \cup C) = \tfrac{5}{10} + \tfrac{4}{10} - \tfrac{1}{10}$$
$$= \tfrac{4}{5}$$

It is not always necessary, or indeed feasible, to draw a Venn diagram, but sometimes this is the easiest way of finding the number of elements in each subset.

Example 3. Out of a group of 80 students studying at least one of physics, chemistry or mathematics, 50 study physics, 40 study chemistry and 33 study mathematics. There are 5 students studying all three subjects. Of those who study chemistry, 13 study mathematics as well. There are 10 studying mathematics only and 12 chemistry only. If a student is selected at random what is the probability that he is studying:

(*i*) physics or chemistry
(*ii*) mathematics or physics
(*iii*) chemistry or mathematics

First we draw a Venn diagram to illustrate the information, letting p be the number taking physics, c chemistry and m mathematics.

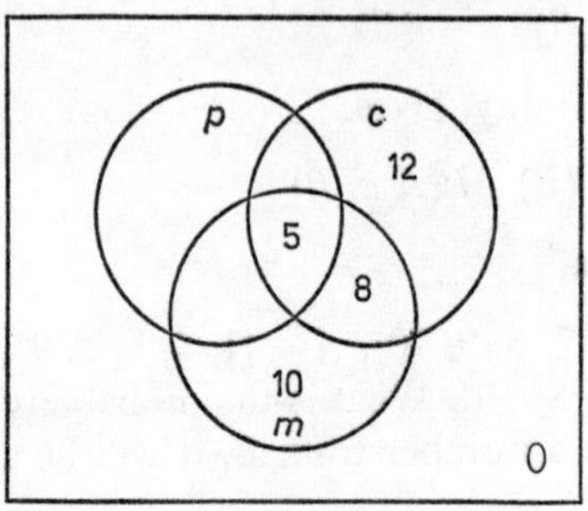

FIGURE 2.2

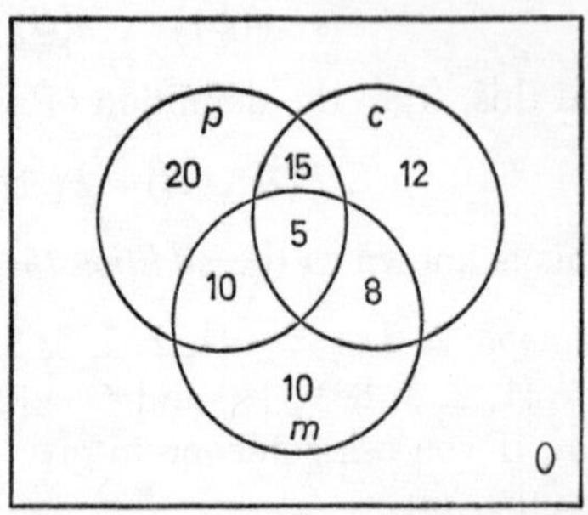

FIGURE 2.3

Figure 2.2 shows the number of students taking all three subjects, the number studying chemistry and mathematics, the number studying chemistry only and the number studying mathematics only. Figure 2.3 is the completed diagram.

(*i*) $n(p \cup c) = 70$, and so:

$$P(p \cup c) = \tfrac{70}{80} = \tfrac{7}{8}$$

(*ii*) $n(m \cup p) = 68$, giving:

$$P(m \cup p) = \tfrac{68}{80} = \tfrac{17}{20}$$

(*iii*) $n(c \cup m) = 60$, and thus:

$$P(c \cup m) = \tfrac{60}{80} = \tfrac{3}{4}$$

Exercise 2a

1. Let $E = \{1, 2, 3, 4, 5, 6, 7, 8, 9, 10\}$, $A = \{1, 2, 3, 4, 5\}$ and $B = \{3, 4, 5, 6, 7\}$. What is the probability that a number selected at random from 1 to 10 will belong to either A or B?
2. In a particular class of 40 pupils there are some studying either French or German or both. Of the 27 who study French, 12 study German as well. There are 22 studying German. What is the probability that a pupil, selected at random, will be studying French or German?
3. Twelve office girls go out for lunch. With the main course there is a choice of creamed or roast potatoes, or both. Eight of the girls ask for roast potatoes and, of these, two ask for both varieties. Four of the 12 have creamed potatoes on their plates. What is the probability that any one of the girls has roast or creamed potatoes?
4. Figure 2.4 represents three subsets L, M and G out of a universal set of 100 people. L is the set of left-handed people, M the set of men and G the set of golfers. What is the probability that a person selected at random is:
   (*a*) either a man or left-handed
   (*b*) either left-handed or a golfer
   (*c*) either a man or a golfer

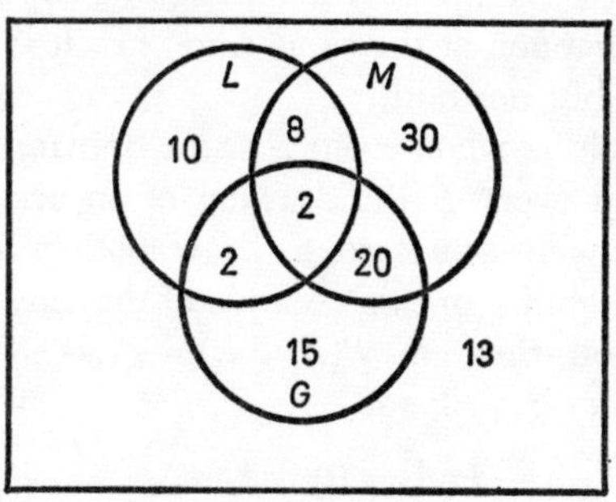

FIGURE 2.4

5. A swimming team consists of 18 members. Each member performs in at least one of the three events: freestyle, backstroke, diving. Only one member performs in all three events. Every diver also races. Of the 11 who take part in the freestyle race, 5 take part in this event only. Of the 10 who take part in the backstroke race, 7 do not dive and, of these 7, 2 also swim in the freestyle race. What is the probability that if one person is selected at random he will have entered for either the freestyle or the diving events?

2.3 Mutually exclusive events

It is quite conceivable that in a particular situation $A \cap B = \emptyset$, and so $P(A \cap B) = 0$. The Venn diagram would look like Fig. 2.5.

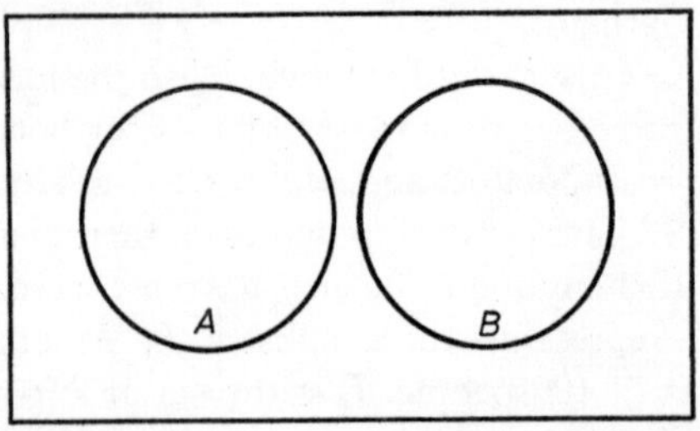

FIGURE 2.5

When we have a situation like this the events are said to be *mutually exclusive*. This means that the events A and B cannot happen simultaneously. A simple example is the tossing of a coin. The two events $\{H\}$, $\{T\}$ cannot happen at the same time, *i.e.*, it is possible to have one or the other, but not both.

Another example is when event A is the drawing of a king from a pack of cards and event B the drawing of an ace. Obviously it is impossible in drawing one card to draw both a king and an ace. Consider the probability in this example of drawing a king or an ace. From the addition theorem $P(A \cup B) = P(A) + P(B) - P(A \cap B)$. In this case, $P(A \cap B) = 0$, so:

$$\begin{aligned} P(A \cup B) &= \tfrac{4}{52} + \tfrac{4}{52} \\ &= \tfrac{2}{13} \end{aligned}$$

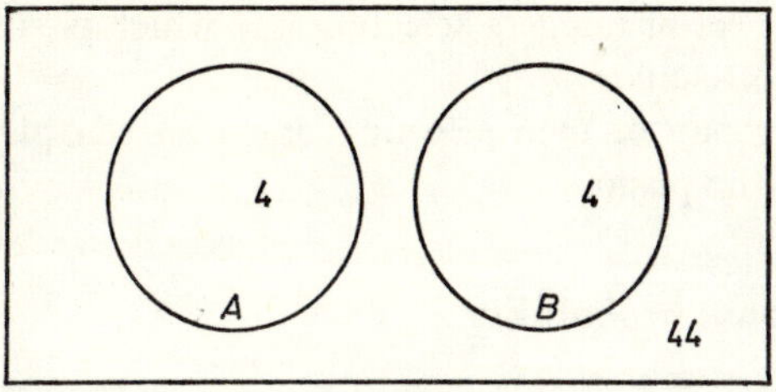

FIGURE 2.6

Figure 2.6 is a Venn diagram showing this. Note again that when the events are mutually exclusive the sets are disjoint.

A particular case of mutually exclusive events is when one event is the complement of the other. In this case we have:

$$P(A \cup \bar{A}) = P(A) + P(\bar{A})$$

since $P(A \cap \bar{A}) = 0$. [In most books on sets the complement of A is written A', but it is general practice in probability to write $\bar{A}$.] Now $n(A \cup \bar{A}) = n(S)$, and so, $P(A \cup \bar{A}) = 1$, giving:

$$1 = P(A) + P(\bar{A}) \text{ or } P(\bar{A}) = 1 - P(A)$$

For example, if the probability of winning a game is $\frac{2}{3}$, then the probability of losing is $\frac{1}{3}$.

Exercise 2b

Which of the following situations describe mutually exclusive events?

1. Throwing a die and obtaining a 1 or a 6.
2. Obtaining 2 heads or 2 tails from the tossing of three coins.
3. Drawing an ace or a heart from a well shuffled pack of cards.
4. From the numbers 1 to 100 selecting one which is either a perfect square or an odd number.
5. From the numbers 1 to 100 selecting one which is either prime or a multiple of 2.
6. From a box containing red, white, blue and green balls, selecting one which is either red or white.
7. Selecting a person from a crowd and finding whether he takes milk or sugar with his tea.
8. Throwing two dice and obtaining a double or a total of 6.

9. From the set of integers selecting one which is either negative or a perfect square.
10. From a group of men selecting one who can play either the violin or the piano.

2.4 Conditional probability

Let us go back to Example 1 (p. 12) involving the hats and gloves and ask: 'What is the probability that a person chosen at random is wearing a hat, given that he is wearing gloves?'

Here we have placed a condition on the question, in that we are only concerned with those people wearing gloves. Look again at Fig. 2.1. A was the set of people wearing hats and B the set of people wearing gloves. In the previous case our sample space consisted of 20 people. Now we only consider the 12 people wearing gloves. This is the new sample space and the subset of people wearing hats has 4 elements in it. The probability of an event A occurring when it is known that event B has occurred is said to be the 'probability of A given B'. In symbolic form this is written $P(A \mid B)$, where the bar is read as 'given'.

And so we have:

$$P(A \mid B) = \frac{n(A \cap B)}{n(B)} = \tfrac{4}{12}$$

If we divide both the numerator and the denominator of the fraction $n(A \cap B)/n(B)$ by $n(S)$, where $n(S)$ is the number of elements in the original sample space, we have:

$$P(A \mid B) = \frac{n(A \cap B)/n(S)}{n(B)/n(S)} = \frac{P(A \cap B)}{P(B)}$$

This gives:

$$P(A \cap B) = P(A \mid B).P(B)$$

which can also be written as:

$$P(A \cap B) = P(A).P(B \mid A)$$

Example 4. What is the probability, in drawing two cards from a pack of 52 playing cards without replacement, that the cards obtained will both be aces?

Let the event A be the drawing of the first ace and event B the drawing of the second ace. By 'both be aces' we mean 'the first card an ace and the second card an ace'. So what we have to find is $P(A \cap B)$. Now $P(A \cap B) = P(A).P(B|A)$, where $P(B|A)$ means the probability of the second card being an ace given that the first card drawn is an ace. If the first card drawn is an ace this leaves three aces in the remaining 51 cards. Therefore, $P(B|A) = \frac{3}{51}$, and so $P(A \cap B) = \frac{4}{52} \times \frac{3}{51} = \frac{1}{221}$.

Example 5. What is the probability that in the drawing of a card from a pack of 52 playing cards it will be either an ace or a heart?

By the addition theorem, $P(A \cup B) = P(A) + P(B) - P(A \cap B)$, where the event A is the drawing of an ace and event B the drawing of a heart. These two events are not mutually exclusive since we could draw the ace of hearts. Now, $P(A \cap B) = P(A|B).P(B)$, where $P(A|B)$ is the probability of drawing an ace given that the card is a heart. There are 13 hearts and one ace of hearts, and therefore $P(A|B) = \frac{1}{13}$, and so $P(A \cap B) = \frac{1}{13} \times \frac{13}{52} = \frac{1}{52}$. (Note that this part of the problem could have been simplified a little by just thinking in terms of the probability of drawing the ace of hearts from a pack of cards.)

Substituting into the addition theorem, we have:

$$P(A \cup B) = \frac{4}{52} + \frac{13}{52} - \frac{1}{52} = \frac{16}{52}$$

Just to emphasize the point let us illustrate the situation with a Venn diagram (Fig. 2.7). Since there is only one ace of hearts, $n(A \cap B) = 1$, which means that:

$$n(\bar{A} \cap B) = 12 \qquad [(13-1) \text{ hearts}]$$

and $$n(A \cap \bar{B}) = 3 \qquad [(4-1) \text{ aces}]$$

Hence $n(A \cup B) = 16$ and $P(A \cup B) = \frac{16}{52}$.

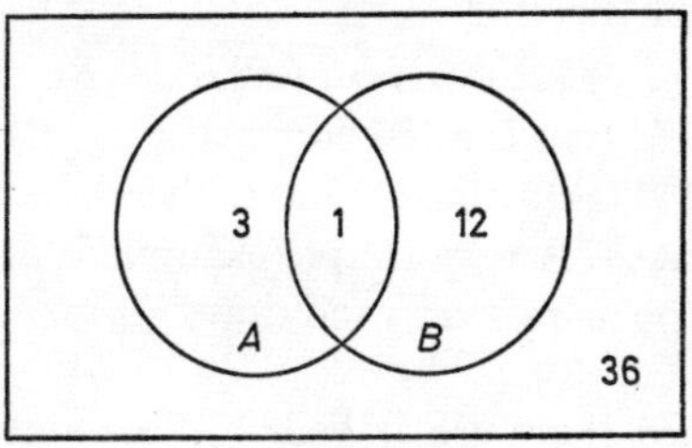

FIGURE 2.7

Exercise 2c

1. If I draw two cards from a well shuffled pack, what is the probability that both will be kings?
2. A box contains 4 red, 2 white and 6 green balls. I draw out two, without replacement. What is the probability that they are:
   (*a*) both red
   (*b*) both white
   (*c*) both green
   (*d*) the first one is red and the second is white
   (*e*) the first is red and the second green
3. What would the answers to question 2 be if the first ball was replaced before the second is drawn?
4. Two dice are thrown. What is the probability of obtaining two sixes?
5. Two cards are drawn successively without replacement from a well shuffled pack. What is the probability that:
   (*a*) they are both aces
   (*b*) the first is an ace and the second is a king
   (*c*) they are both red cards
   (*d*) the first card is the seven of clubs and the second is the seven of spades
   (*e*) the first is a queen and the second is not
6. Repeat question 5, but now with the condition that replacement is made after the first draw.

2.5 Independent events

Two events A and B are said to be *independent* when:

$$P(A \cap B) = P(A).P(B)$$

By comparing this with:

$$P(A \cap B) = P(A|B).P(B)$$

we see that this condition is satisfied when $P(A|B) = P(A)$. In other words, when event B takes place, it has no bearing on the result of the probability of event A. A very simple example is any sort of draw in which replacement occurs after each draw.

Example 6. In a box there are 10 red, 15 blue and 5 green marbles. I draw out 1 marble, note its colour and replace it. Then I draw another marble and note its colour. What is the probability that they will both be green?

Let the event A be the drawing of the first marble and event B the drawing of the second marble. The probability of drawing a green marble will be the same for both events, since the sample space and the events contain the same number of elements before and after the first draw. Hence the events are independent and:

$$P(A \cap B) = P(A).P(B)$$
$$= \frac{5}{30} \cdot \frac{5}{30} = \frac{1}{36}$$

The reader should be warned not to confuse the two notions of mutually exclusive events and independent events. In the case of mutually exclusive events, $P(A \cap B) = 0$, and so such events cannot be independent, unless of course the events were themselves null events and then no problem would arise. For non-null, independent events $P(A \cap B) \neq 0$, and so they cannot be mutually exclusive.

Just remember that for mutually exclusive events, if one takes place the other cannot; whereas for independent events they can both take place but the first has no effect on the probability of the second.

Exercise 2d

1. Figure 2.8 illustrates two events A and B: are the events independent?

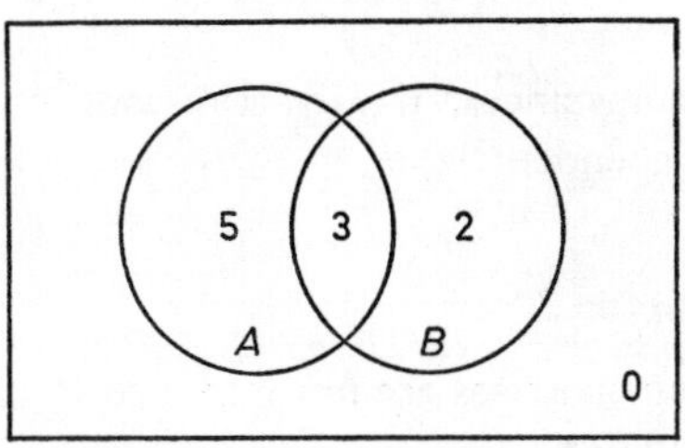

FIGURE 2.8

2. What would be the value of x in Fig. 2.9 so that the events A and B are independent?

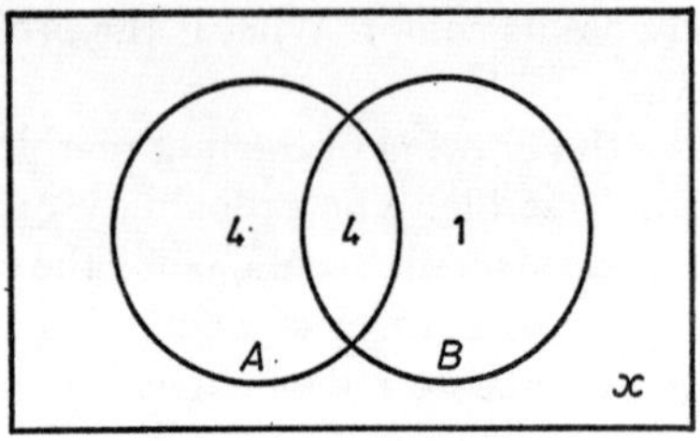

FIGURE 2.9

3. If the probability of an event A is $\frac{1}{4}$, and the probability of A and another independent event B is $\frac{1}{8}$, what is the probability of the event B?
4. From the information in Fig. 2.10 which of the following events are independent of each other:
   (*a*) A and B
   (*b*) A and C
   (*c*) B and C
   (*d*) A, B and C?

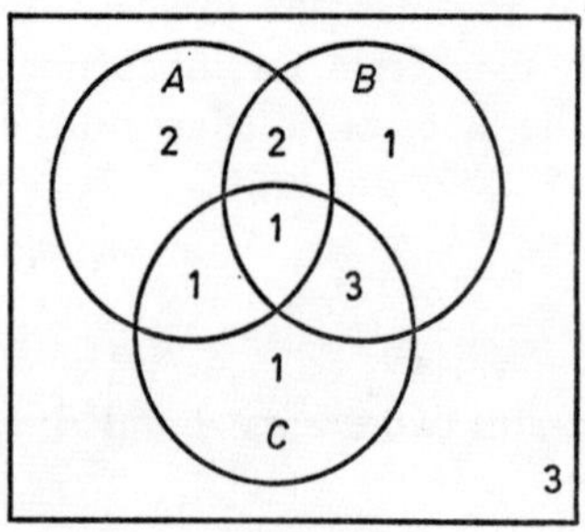

FIGURE 2.10

5. In what circumstances, if any, can two mutually exclusive events be independent?

Exercises on Chapter 2

Note. The following exercises are not intended to be in any order of difficulty or presentation.

1. (*a*) Suppose A, B and C represent three events which are not mutually exclusive. Show that:

$$P(A \cup B \cup C) = P(A) + P(B) + P(C) - P(A \cap B) - P(B \cap C) - P(A \cap C) + P(A \cap B \cap C)$$

(*b*) What does $P(A \cup B \cup C)$ become when:
(*i*) the three events are mutually exclusive
(*ii*) when only A and B are mutually exclusive
Illustrate these two cases on a Venn diagram.

2. Out of a class of 30 students, 10 study Greek, 6 study ancient history, 4 study Greek and Latin only, 1 studies Greek and ancient history only, 4 study Greek only, 2 study ancient history and Latin only, and 11 do not study any of these three subjects. If a student is selected at random and it is known he studies Latin what is the probability that:
(*a*) he studies Greek as well
(*b*) he also studies Greek and ancient history
(*c*) he studies only one of the other two subjects
(*d*) he studies at least one of the other two subjects
3. If the probability that player A will win a tennis singles tournament is 0·4 and the probability that either he or player B will win is 0·6, what is the probability that B will win?
4. A coin is tossed and a card drawn from a pack of well shuffled cards. What is the probability that there will be:
(*a*) a head or a king
(*b*) a head and a king
(*c*) a tail and the ace of diamonds
(*d*) a tail or the ace of diamonds
5. In what circumstances, if any, can two independent events be mutually exclusive?
6. If three events A, B and C are not independent of each other, how would you determine $P(A \cap B \cap C)$?
7. The number of elements in the sets A, B and C are shown in Fig. 2.11.

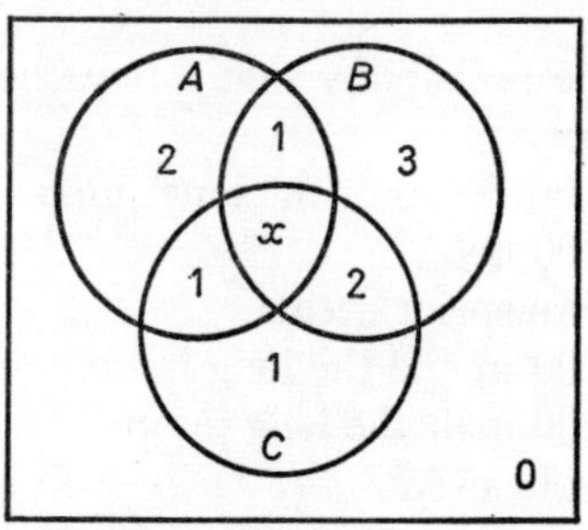

FIGURE 2.11

(*a*) What would be the value of x so that the events A, B and C are independent?

(*b*) For this value of x which of the following events are independent:

(*i*) A and B (*ii*) B and C (*iii*) A and C

8. At a recent examination given to a class of 30 students, it was found that in the subjects mathematics and biology 25 passed either one or other or both, 8 passed in both subjects and 15 passed in biology. If a student is selected at random, what is the probability that he will have passed mathematics?

9. A box contains a number of red, blue and yellow coloured balls. If the probability of drawing a red ball is $\frac{3}{8}$ and that of drawing a red or yellow one is $\frac{2}{3}$, what is the probability of drawing a yellow ball?

10. Out of a class of 40 boys a number enter for four events in the school sports. A, B, C, D in Fig. 2.12 represent the sets of boys of this class who entered for the 100 m, 200 m, long jump and high jump, respectively.

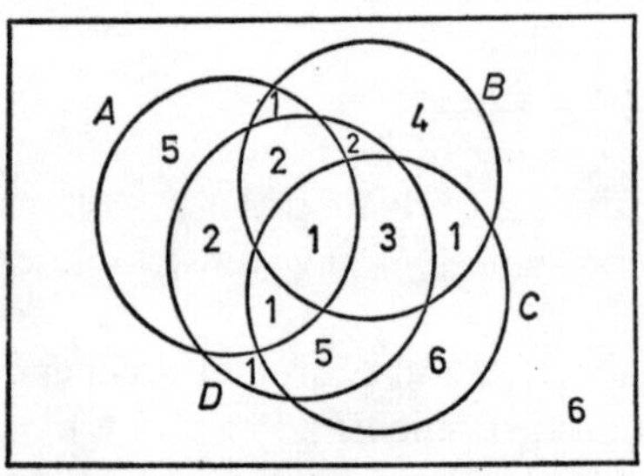

FIGURE 2.12

(*a*) What is the probability that a boy chosen at random has entered for:

(*i*) the high jump or the long jump
(*ii*) the 100 m
(*iii*) the two sprint events
(*iv*) the 200 m and the long jump
(*v*) the 200 m or the long jump
(*vi*) only one event
(*vii*) only two events

(*viii*) at least one event
(*ix*) at most two events

(*b*) Given that he has entered for the 100 m, what is the probability that he has:
(*i*) also entered for the 200 m
(*ii*) not entered for anything else
(*iii*) entered for only one other event

11. One thousand tickets are sold for a raffle for which there are three prizes of equal value. If you bought only three tickets show that your chances of winning all three prizes is less than 1 in 150 000 000 000.

12. In successive throws of an unbiased coin what is the probability of obtaining:
(*a*) 3 heads (*b*) 10 heads (*c*) *n* heads

13. A box holds 20 red, 10 white and 15 blue marbles. If two are selected at random, one at a time without replacement, what is the probability that:
(*a*) they are both red
(*b*) they are not both blue
(*c*) they are red or white
(*d*) the first is red and the second red or white
(*e*) the second is not red
(*f*) at least one is white
(*g*) at most one is white
(*h*) only one is white
(*i*) the first is not red and the second is not blue
(*j*) they are the same colour
(*k*) they are not the same colour

14. Repeat question 13, with the condition that replacement is made after the first draw.

15. If in question 13 three marbles are drawn out, what is the probability that:
(*a*) they are all red
(*b*) they are all white
(*c*) they are all of the same colour
(*d*) there are two red and one white
(*e*) they are of different colours
(*f*) the first two are of the same colour and the third is of a different colour

16. What would be the answers to question 15 if:
   (*i*) replacement was made after each draw
   (*ii*) replacement was only made after the first draw
   (*iii*) the first two marbles were drawn out without replacement, then both of these replaced before the third draw was made
17. A group of 60 people were asked whether they read *The Times*, *The Daily Telegraph* or *The Guardian*. The following table gives the results:

Times	*Telegraph*	*Guardian*	*Times* & *Telegraph*	*Times* & *Guardian*	*Telegraph* & *Guardian*	All 3 papers
23	27	31	9	11	13	5

   What is the probability that a person, selected at random from this group, reads:
   (*a*) *The Times* or *The Guardian*
   (*b*) *The Guardian*, given that he already reads *The Times*
   (*c*) only one of the papers
   (*d*) at least one of the papers
   (*e*) at the most, two of the papers
   (*f*) none of the papers
18. Two men A and B toss a coin and the first to obtain a head wins. What is the probability that B will win on his first throw if A throws first?
19. If the game in question 18 is continued until an eventual winner is found, show that A's chance of winning is $\frac{2}{3}$ and B's is $\frac{1}{3}$.
20. (*a*) If 3 men were to toss a coin in succession until an eventual winner (*i.e.*, the first one to obtain a head) is found, what chance does each have of being the winner?
   (*b*) If n men play this game, show that the first man's probability of winning is $2^{n-1}/(2^n - 1)$, and then find the probability that:
      (*i*) the second man is the winner
      (*ii*) the nth man is the winner
   Find also the probability that the kth man wins, and check the results with (*i*) and (*ii*) above, by putting $k = 2$ and n.

3 | PICKING AND CHOOSING

3.1 Tree measure

When we have two or more events, it is possible to illustrate them by use of what is known as a *tree diagram*. We shall see that this can be particularly useful for calculating probabilities.

Example 1. In three tosses of a fair coin, what is the probability of obtaining two heads and a tail?

There are two equally likely outcomes, *i.e.*, $P(H) = P(T) = \frac{1}{2}$ on the first toss. We can illustrate these with two branches, by putting

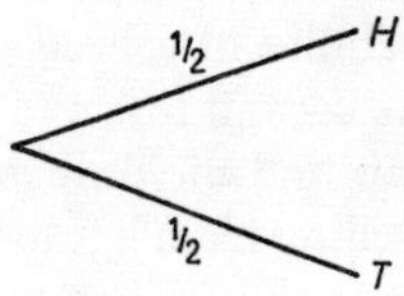

FIGURE 3.1

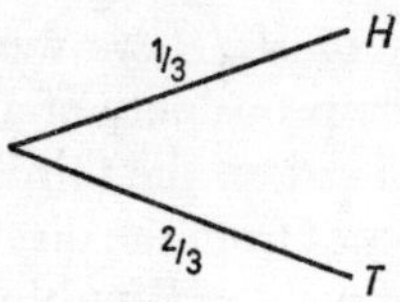

FIGURE 3.2

the probability of a $\frac{1}{2}$ onto each branch (see Fig. 3.1.). (Note that if the coin had been biased so that $P(H) = \frac{1}{3}$, say, then our diagram would have looked like Fig. 3.2.) For each of the two outcomes on the first toss there are two more on the second toss, as shown in Fig. 3.3.

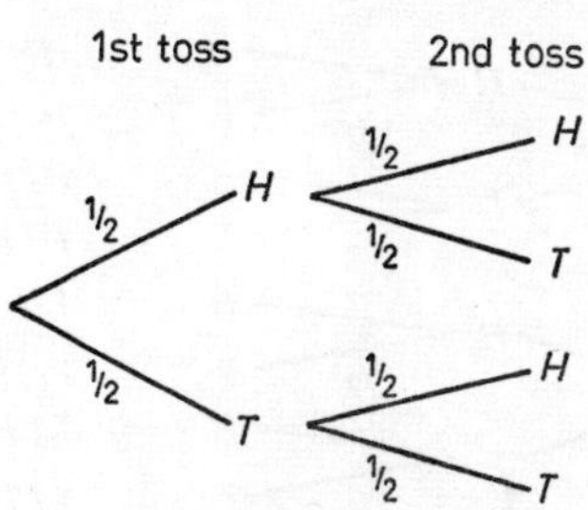

FIGURE 3.3

Similarly, on the third toss there are two outcomes for each one in the second toss (see Fig. 3.4). The last column gives the final outcome of

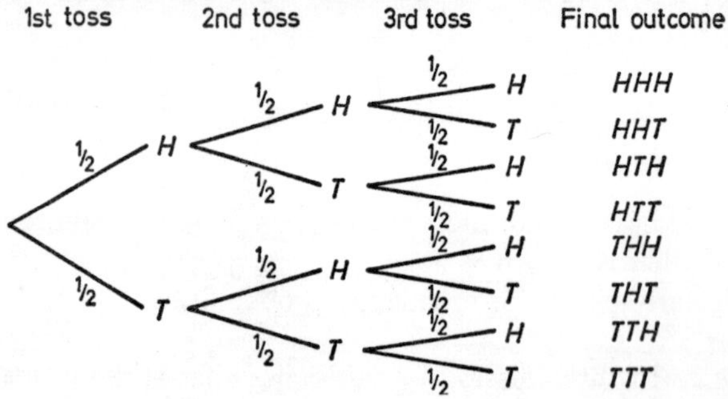

FIGURE 3.4

travelling from left to right along any one of the branches. To find the probability of *HHH* we want the probability of a head on the first toss and the probability of a head on the second toss and the probability of a head on the third toss. We can find the probability of the final outcomes from the diagram by travelling from left to right along any one branch and multiplying together the probabilities at each stage.

The completed diagram is given in Fig. 3.5. Note that the sum of the probabilities in the final column is 1.

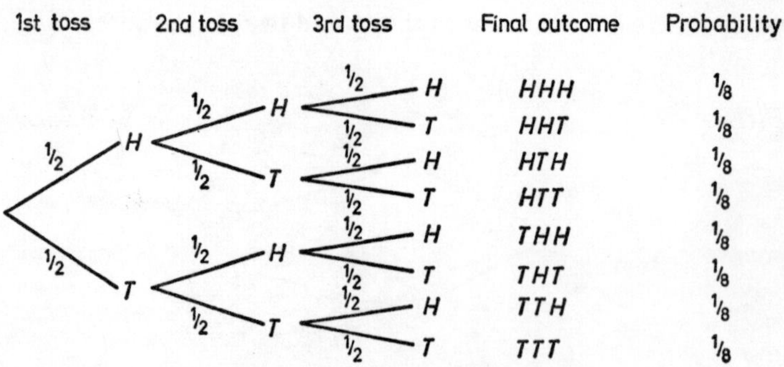

FIGURE 3.5

We have been asked for the probability of obtaining two heads and a tail, *i.e.*, *HHT* or *HTH* or *THH*. Since these events are mutually exclusive, we have, by the addition theorem (page 13):

$$P\{(HHT) \cup (HTH) \cup (THH)\} = P(HHT) + P(HTH) + P(THH)$$
$$= \tfrac{1}{8} + \tfrac{1}{8} + \tfrac{1}{8} = \tfrac{3}{8}$$

In general on a tree diagram, we multiply the probabilities as we move along the branches from left to right, and add the probabilities in the final column as we move vertically. Again it is worth pointing out that the sum of these probabilities in the final column should be 1, and this can be used as a check on the work.

Example 2. Suppose we are using a coin which is biased so that the probability of a head is $\frac{2}{3}$ and, hence, the probability of a tail is $\frac{1}{3}$. What is now the probability of two heads and a tail after three tosses?

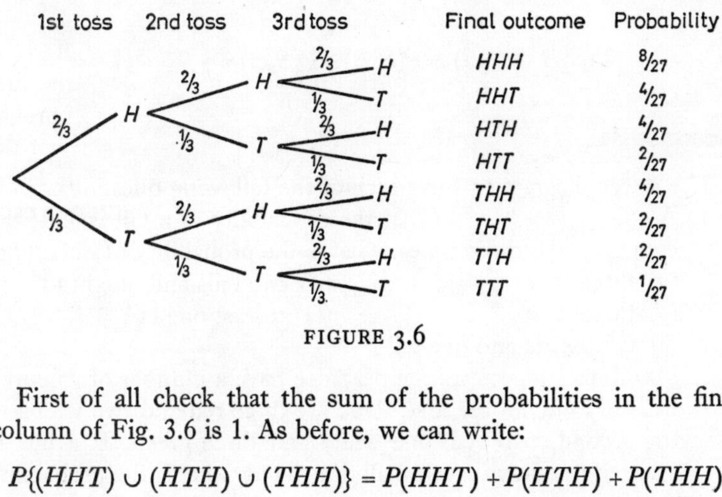

FIGURE 3.6

First of all check that the sum of the probabilities in the final column of Fig. 3.6 is 1. As before, we can write:

$$P\{(HHT) \cup (HTH) \cup (THH)\} = P(HHT) + P(HTH) + P(THH)$$
$$= \tfrac{4}{27} + \tfrac{4}{27} + \tfrac{4}{27}$$
$$= \tfrac{12}{27} = \tfrac{4}{9}$$

Example 3. A sack contains two black balls and five white balls. A second identical sack contains two black balls and one white ball. A sack is chosen at random and a ball taken out. What is the probability that it is black?

The first stage of our tree diagram is concerned with the choice of sack (S_1 or S_2) and the second with the choice of ball (B, black, or W, white). Since the sacks are identical in appearance, the probability of selecting either one is $\frac{1}{2}$. Figure 3.7 is the complete tree diagram,

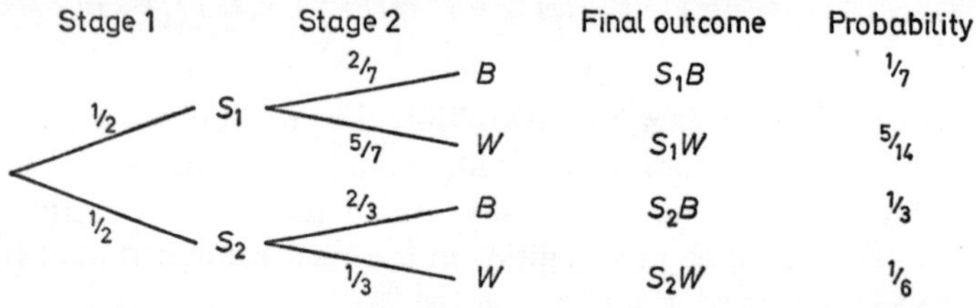

FIGURE 3.7

where, for example, S_1B means choosing sack 1 and drawing out a black ball. First check that the sum of the probabilities in the final column is one. The probability of selecting a black ball will be given by:

$$P(S_1B \cup S_2B) = P(S_1B) + P(S_2B) = \tfrac{1}{7} + \tfrac{1}{3} = \tfrac{10}{21}$$

Exercise 3a

Use a tree diagram to solve each of the following questions.

1. A coin is weighted so that the probability of a tail is $\frac{3}{5}$. If the coin is tossed three times, what is the probability of obtaining:
   (*a*) three tails
   (*b*) three heads
   (*c*) two heads and one tail
   (*d*) two tails and one head
   (*e*) at least one tail
2. Two boxes identical in appearance have a number of coloured balls in each. In the first, there are three red and two white; in the second, two red, one white and three green. If a box is selected at random and a ball drawn out, what is the probability that it is:
   (*a*) green
   (*b*) red
   (*c*) white
   (*d*) red or green
   (*e*) green or white
3. A fair die is rolled and on the result of this one of six boxes is selected. Each box contains three numbered discs. In the first the numbers are 1, 2, and 3, in the second they are 4, 5 and 6, and so on until the sixth which has the numbers 16, 17 and 18.

Out of the selected box a disc is drawn. What is the probability that the number on it is:
(*a*) divisible by 5 (*b*) a prime number (*c*) a perfect square
[Note that 1 is not a prime number]

4. In each of three drawers there is a coin. The coins in the first two drawers are fair but the coin in the third drawer is weighted so that the probability of a head is $\frac{3}{4}$. If a drawer is selected at random and the coin in that drawer tossed twice, what is the probability of obtaining:
(*a*) two heads
(*b*) two tails
(*c*) a head on the first toss and a tail on the second

3.2 *A posteriori* probability

The following example will explain what we mean by the term *a posteriori* probability.

Example 4. Three identical boxes each have two drawers. In one of the boxes there is a gold coin in each drawer, in the second there is a gold coin in one drawer and a silver coin in the other, and in the third a silver coin in each drawer. If we select a box at random, open one of the drawers and find a gold coin, what is the probability that the other drawer has a silver coin?

There are two stages in the tree diagram, the choosing of a box and the choosing of a coin. The tree diagram is shown in Fig. 3.8, where B_1, B_2 and B_3 represent the three boxes and G, S the gold and silver coins. It should be noted that if B_1 is chosen, the selection of a gold coin is a certainty, and the probability is 1. Similarly, the probability of a silver coin from B_3 is 1.

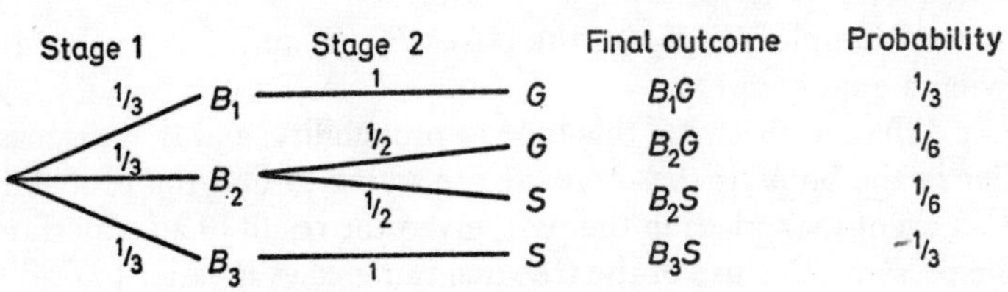

FIGURE 3.8

The experiment resulted in a gold coin so the sample space is restricted to $\{B_1G, B_2G\}$, and the particular subset we are concerned with is $\{B_2G\}$, since B_2 is the box with the silver coin in the other drawer. Since $P(A|B) = P(A \cap B)/P(B)$ (see p. 18), the probability that we choose the second box is given by:

$$P(B_2|G) = \frac{P(B_2G)}{P(B_1G) + P(B_2G)}$$

where $P(B_2|G)$ means the probability that box 2 was chosen, given that a gold coin was in one of the drawers.

This is a simplification of a theorem known as *Bayes Theorem* which states:

> If $E_1, E_2, E_3, \ldots E_n$ are mutually exclusive and exhaustive events (*exhaustive* means all possible cases have been taken into account), and if A is the result of an experiment such that $P(A) \neq 0$, then:

$$P(E_1|A) = \frac{P(E_1A)}{P(E_1A) + P(E_2A) + \ldots + P(E_nA)} \tag{3.1}$$

If we use the sigma notation, where $\sum_{i=1}^{n} X_i$ represents the sum $X_1 + X_2 + \ldots X_n$, then equation (3.1) may be written in the form:

$$P(E_1|A) = \frac{P(E_1A)}{\sum_{i=1}^{n} P(E_iA)}$$

Another form, which you can easily verify, is:

$$P(E_1|A) = \frac{P(E_1 \cap A)}{\sum_{i=1}^{n} P(E_i \cap A)}$$

Similar results hold for $E_2, E_3, \ldots E_n$.

In our example, E_1, E_2 are the boxes B_1, B_2 and A is the event of drawing a gold coin.

The difference between this type of probability, and that discussed earlier in the book, is that here we are trying to find the probability that an event took place in the past, given the result of an experiment in the present. The use of the tree diagram makes it easier to find the probabilities and to tabulate them for each outcome.

Example 5. On a particular Saturday in November, the local football team has to play a match at home. In dry conditions they win 4 out of 5 home matches, but in wet conditions they only have a 50–50 chance of winning at home. Suppose that the weather records show that in November an average of 10 out of the 30 days are wet. You read in the paper the following Monday that the team won. What is the probability that it was wet that Saturday?

There are two stages: the weather conditions, and the result of the match. Note that $P(\text{Wet day}) = \frac{1}{3}$; $P(\text{Dry day}) = \frac{2}{3}$; $P(\text{Winning on a wet day}) = \frac{1}{2}$; $P(\text{Winning on a dry day}) = \frac{4}{5}$. The probabilities of the various outcomes are shown in Fig. 3.9.

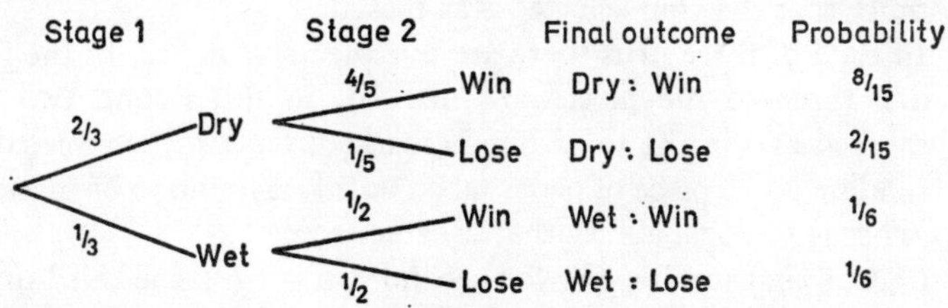

FIGURE 3.9

Our sample space is {Dry : Win, Wet : Win}, and the required subset is {Wet : Win}. Therefore, from Fig. 3.9:

$$P(\text{Wet} \mid \text{Win}) = \frac{P(\text{Wet : Win})}{P(\text{Dry : Win}) + P(\text{Wet : Win})}$$

$$= \frac{\frac{1}{6}}{\frac{8}{15} + \frac{1}{6}} = \frac{\frac{1}{6}}{\frac{21}{30}} = \frac{5}{21}$$

where $P(\text{Wet} \mid \text{Win})$ means the probability that the day was wet given that the team won.

Exercise 3b

1. Two boxes are identical in appearance. In the first box there are three red and four blue marbles, and in the second box five red and two blue marbles. One box is selected at random and a marble taken out. It turns out to be blue. What is the probability that it came from:
   (*a*) the first box (*b*) the second box

2. Two machines A and B produce, respectively, 60% and 30% of the total output of a factory. For machine A it has been found from past experience that 2% of the goods produced are defective, and that for machine B 5% are defective. One item is tested and found to be defective. What is the probability that it came from:
(*a*) machine A (*b*) machine B
3. Two coins of identical appearance are placed on a table. One of them is fair but the other is weighted so that a head comes up twice as often as a tail. One of the coins is selected at random and tossed twice. The result is two heads. What is the probability that the coin selected was biased?
4. In each of three urns there are five pieces of paper. In the first urn three of the pieces are marked, in the second two are marked and in the third one is marked. One urn is selected at random and a piece of paper taken out. It is found to be marked. What is the probability that it came from:
(*a*) the first urn (*b*) the second urn (*c*) the third urn
5. Of the number of people attending the out-patients department of a certain hospital, 10% come without a note from their own doctor. For those with a note the probability of seeing the hospital doctor within 30 min of their arrival is 0·8; for those without a note, the probability is 0·3. Someone comes out complaining that he had to wait nearly an hour before being seen. What is the probability that he had no note from his own doctor?
6. Tom and Jim play a game of singles in the local tennis club's tournament. They toss a coin to decide who will serve first. From their past play it is known that if Tom is serving he will win a game 8 out of 10 times, but if Jim is serving Tom has only a 50–50 chance of winning the game. If you hear that Tom won the first game, what is the probability that he served first?

3.3 Permutations

In evaluating probabilities it is sometimes more convenient to make use of permutations and combinations. First we consider permutations, which involve the placing of things in a particular order.

Example 6. In how many ways can I write down the letters A, B, C? The tree diagram in Fig. 3.10 can help us to see the solution to this problem.

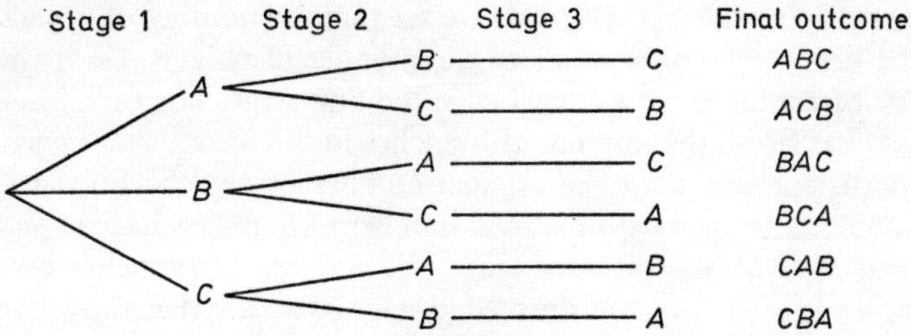

FIGURE 3.10

The first stage is the choice of any one of the three letters; the second stage is the choice of one of the remaining two letters; and the third stage is the addition of the last remaining letter. For this example, there are six possible outcomes. Figure 3.11 shows the outcome for four letters. As before, the first stage is the choice of any one of the three letters, and the subsequent stages the choice of the remaining letters. In this case there are 24 ways of arranging the four letters.

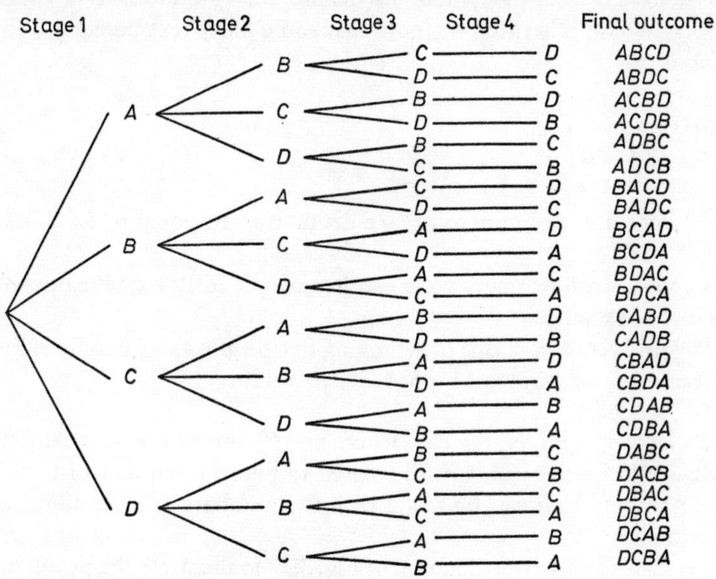

FIGURE 3.11

Obviously as the number of objects increases a tree diagram becomes much too cumbersome, and we need a general rule for finding the number of ways of arranging n unlike objects.

By studying the tree diagrams we see that the number of branches in the first stage corresponds to the number of objects. So if there are n objects there are n branches in the first stage. For each branch in the first stage, the number of branches in the second stage corresponds to one less than the original number of objects. So if there are n objects to start with the total number of branches in the second stage is $n(n-1)$. At the third stage, the number of branches corresponding to each branch of the second stage is two less than the original number of objects. In the case of n objects, then, the total number of branches at the third stage is $n(n-1)(n-2)$.

This procedure is carried on until, at the final stage, there is only one branch corresponding to each branch of the previous stage. For n objects the total number of branches at this final stage is, thus:

$$n(n-1)(n-2)\ldots 3.2.1$$

where the dots indicate a continuation of the series.

We have a special symbol to denote this product. It is called *factorial n* and is written $n!$ (note that some older text books use the symbol $\underline{n|}$).

Example 7.
$3! = 3.2.1 = 6$
$8! = 8.7.6.5.4.3.2.1 = 40\,320$
As $0!$ does not mean anything, we define it to be equal to 1.

Example 8. In how many ways can five people form a queue outside a theatre box office?

The problem is just one of arranging five people and, therefore, the total number of ways is $5! = 5.4.3.2.1 = 120$.

We have considered the case where no repetitions are allowed, but what would be the result if we wanted to find the number of three-digit numbers that can be made with the numbers 1, 2, 3, allowing repetitions?

We shall use the tree diagram in Fig. 3.12 to illustrate the problem. As before, the number of branches in the first stage corresponds to the

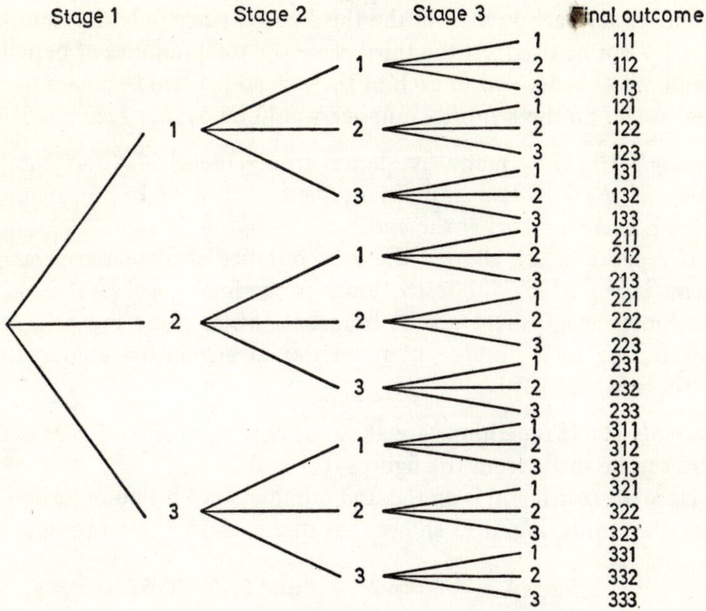

FIGURE 3.12

number of items—in this case 3. But at the second and subsequent stages, instead of the number of branches, corresponding to each branch of the preceding stage, being reduced in number, they in fact remain the same; in this case 3. So at the end of the second stage there are 3.3 branches and at the end of the third stage there are 3.3.3 branches.

Therefore, if we are arranging n unlike objects where repetitions are allowed, the number of branches at the end of the second stage would be n^2 and at the end of the third stage n^3. Thus at the rth stage the total number of branches is n^r and at the final stage n^n.

Example 9. If repetitions are allowed, how many four-digit numbers can be formed from the numbers 1, 2, 3 and 4?

Applying our theory there are 4^4 (*i.e.*, 256) four-digit numbers.

It is possible that a restriction may be imposed on the question in Example 9. For example, the required number may have to be even, which would mean that at the final stage there would be only two branches for each branch of the third stage, since only the numbers 2 and 4 can be used. At the third stage the total number of branches would be $4^3 = 64$, and to each of these is added two branches in the final stage. So the required number would be $64.2 = 128$.

Example 10. How many five-letter arrangements are there of the letters *A*, *B*, *C*, *D* and *E*, if repetitions are allowed? In how many of these is there a vowel at the end?

If repetitions are allowed the total number of five-letter arrangements is $5^5 = 3125$. The restriction is in the final stage. At the end of the fourth stage there are 5^4 branches. Since there are only two vowels, the total number of five-letter arrangements ending in a vowel is $5^4.2 = 1250$.

Example 11. If repetitions are allowed, how many three-figure numbers can be made from the figures 0, 1 and 2?

Here the restriction is on the first number since it cannot be 0. The tree diagram in Fig. 3.13 shows that there are 18 different numbers.

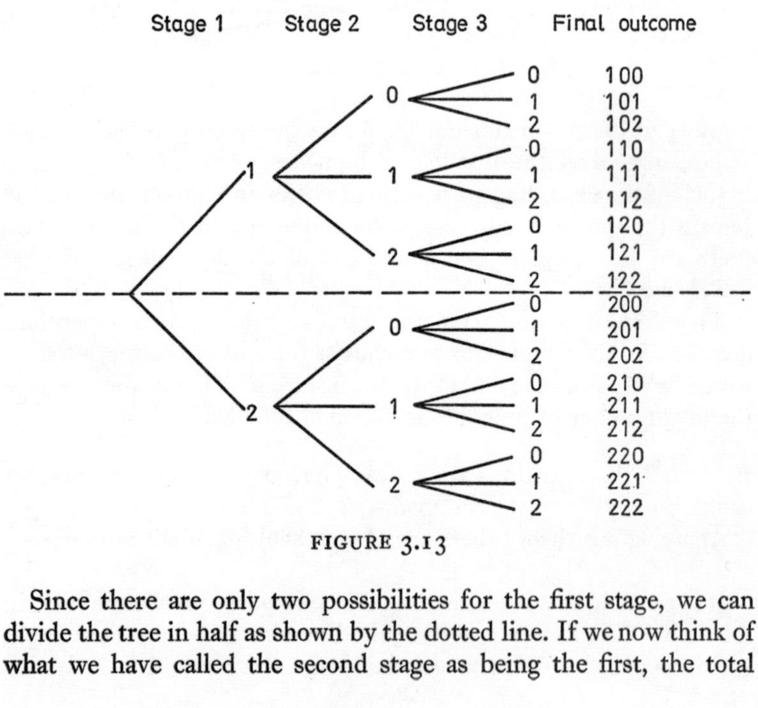

FIGURE 3.13

Since there are only two possibilities for the first stage, we can divide the tree in half as shown by the dotted line. If we now think of what we have called the second stage as being the first, the total number of branches in the top half of the final stage is $3^2 = 9$. (Since the second stage is considered as the first then the final stage, *i.e.*, stage 3, is considered as the second.) Now all we do is multiply by two to include the other half of the tree.

Example 12. Allowing repetitions how many five-letter arrangements are there of the letters A, B, E, F and O if:

(*i*) the first letter must be a vowel
(*ii*) the first and last letters must be vowels

(*i*) The first letter has to be a vowel so the first stage has three branches. If we take one of these branches there are four further stages after it. Since there are five letters, there are $5^4 = 625$ ways of arranging the letters in these stages. So, including the other two branches in the first stage, the total number of five-letter arrangements satisfying the restriction that the first letter is a vowel is $625.3 = 1875$.

(*ii*) At the end of the fourth stage the total number of branches is $5^3.3 = 375$. In the fifth and final stage there are three branches for each branch of the fourth stage, since the last letter has to be A, E or O. So the total number of five-letter arrangements beginning and ending with a vowel is $375.3 = 1125$.

3.4 Selective arrangements

Let us now consider the problem of what happens when we have to arrange, say, three objects out of five.

Example 13. If repetitions are not allowed, how many ways are there of arranging three letters out of the five letters A, B, C, D and E?

In this problem we only have three stages. The first stage has five branches, and for each of these there are four branches in the second stage and, corresponding to each branch in the second stage, there are three branches in the third. So the total number of branches is $5.4.3 = 60$.

When we have to arrange n unlike objects r at a time, we represent the number of ways by the symbol ${}_nP_r$. There are only r stages, and the first has n branches. For each of these branches there are $(n-1)$ branches in the second stage and, corresponding to each of these, there are $(n-2)$ branches in the third. [Note: $(n-2)$ in the 3rd stage.] It follows that in the rth stage there are $[n-(r-1)]$ branches corresponding to each branch in the preceding stage. The total number of arrangements is:

$$n(n-1)(n-2)\ldots[n-(r-1)] = n(n-1)(n-2)\ldots(n-r+1)$$

where the dots denote the continuation of the series. Now:

$$n(n-1)(n-2)\ldots(n-r+1) = \frac{n(n-1)(n-2)\ldots(n-r+1)(n-r)(n-r-1)\ldots3.2.1}{(n-r)(n-r-1)\ldots3.2.1}$$

$$= \frac{n!}{(n-r)!} = {}_nP_r$$

The number of ways of arranging n objects n at a time is, therefore, ${}_nP_n = n!/(n-n)! = n!/0! = n!$, which is the result we have already obtained.

If we arrange the n objects r at a time and repetitions are allowed, the number of arrangements is simply n^r, since there are r stages.

Example 14. There are ten applicants for six vacancies. In how many ways can the vacancies be filled?

The number of ways is given by:

$${}_{10}P_6 = \frac{10!}{(10-6)!} = \frac{10!}{4!} = 10.9.8.7.6.5 = 151\ 200$$

Let us now consider what happens when we have to arrange a number of objects, some of which are identical in appearance. For example, how many four-letter arrangements can be made of the word *book*? Here we have two *o*s, which we shall call o_1 and o_2. The number of ways of arranging $b\ o_1\ o_2\ k$ is $4! = 24$. For each of these arrangements involving o_1 and o_2, in that order, but not necessarily together, there is a corresponding one involving o_2 and o_1, since o_1 and o_2 can be arranged in 2! ways. When we remove the subscripts these *o*s are identical. So to find the number of four-letter arrangements of the letters of the word *book* we divide the number of arrangements of b, o_1, o_2, and k by 2!; *i.e.*, $4!/2! = 24/2 = 12$.

Example 15. How many five-letter words can be made from the word *eerie*? The words need not have any meaning but are just an arrangement of letters.

If we denote the three *e*s by e_1, e_2 and e_3, the number of ways of arranging the letters $e_1e_2rie_3$ is $5! = 120$. But the number of ways of arranging e_1, e_2 and e_3 is $3! = 6$, and so for each arrangement involving $e_1e_2e_3$ in that order, but not necessarily together, there are five others with an e in the same position. Consider, for example, the arrangement $e_1rie_2e_3$. With the *es* in the same position, we can also have $e_1rie_3e_2$, $e_2rie_1e_3$, $e_2rie_3e_1$, $e_3rie_1e_2$ and $e_3rie_2e_1$. Hence for this position there is a total of six arrangements of the *es*. Similarly for each other position of the *es*. When the subscripts are removed, the *es* are of course, identical. And so, to find the number of five-letter words obtainable from *eerie*, we divide the number of ways of arranging the five letters (5!) by the number of ways of arranging the *es* (3!). That is, the number is $5!/3! = 120/6 = 20$.

Example 16. How many arrangements can be made of the letters of the word *average*?

If we use subscripts to differentiate between like letters the word becomes $a_1ve_1ra_2ge_2$, and the number of arrangements of these letters is $7! = 5040$. If we remove the subscripts from the *as* the number of arrangements is $7!/2! = 2520$. If we now remove the subscripts from the *es* the number of arrangements is $7!/(2!\,2!) = 1260$.

In general, if we have to arrange n objects of which there are n_1 alike of one kind, n_2 alike of another kind, n_3 alike of a third kind, and so on for k kinds of objects, then the total number of arrangements is:

$$\frac{n!}{n_1!\,n_2!\,n_3!\ldots n_k!}$$

where $n_1 + n_2 + n_3 + \ldots + n_k = n$. This general formula is still true in the particular cases where some of the n_is are equal to 1.

Example 17. How many arrangements can be made of the letters of the word *proportion*?

In this case
$n_1 = 2$(the number of *ps*)
$n_2 = 2$(„ „ *rs*)
$n_3 = 3$(„ „ *os*)
$n_4 = 1$(„ „ *ts*)
$n_5 = 1$(„ „ *is*)
$n_6 = 1$(„ „ *ns*)

and $n = n_1 + n_2 + n_3 + n_4 + n_5 + n_6 = 10$

The number of arrangements is $10!/(2!\,2!\,3!\,1!\,1!\,1!) = 37\,800$. Obviously there is no need to put in the 1! but it is only added here to complete the picture.

Example 18. In how many ways can the digits 1, 2, 3, 4, 5, 6, 7, 8, 9 be written down if the odd numbers have to remain in the order 1, 3, 5, 7, 9?

Since we are not allowed to arrange the odd numbers other than one way, we can regard them as being identical. So the total number of arrangements is $9!/5! = 3024$.

Note that, if it had been required that the odd numbers should remain in the same positions and order as at the start, then the answer would have been found simply by arranging the other four numbers, *i.e.*, $4! = 24$.

To finish this section consider the following example.

Example 19. In how many ways can six people be seated at a round table if (*i*) they can sit anywhere and (*ii*) two particular people must sit together?

(*i*) The method of solving this type of problem, where people are to be seated at a round table, is to seat one person and then find the number of ways of arranging the remainder. In this case the total number of arrangements is $(6-1)! = 5! = 120$.

(*ii*) When the restriction is that of a number of people having to sit together, we consider them as one person, work out the arrangement of this new number of people around the table and then multiply this answer by the number of ways of arranging the people who have to sit together. In this case two people have to sit together, and we consider the number of ways of arranging five people at a round table and multiply our result by 2! The number of ways, then, is $4!2! = 24.2 = 48$.

If the problem involved two people who must not sit together, we solve it by finding the number of ways when they do sit together and subtract from the total number of ways of seating the original six. So the number of ways in which these two people are not sitting together is $120 - 48 = 72$.

Exercise 3c

1. Evaluate:

(*a*) $\dfrac{8!}{4!}$ (*b*) ${}_7P_5$ (*c*) $\dfrac{n!}{n(n-2)!}$

2. In how many ways can four people be seated on a bench?

3. How many ways are there of setting out seven different books on a shelf?
4. If no repetitions are allowed, how many six-digit numbers can be made from the numbers 1, 2, 3, 4, 5 and 6?
5. If repetitions are allowed, how many four-digit numbers can be made from the numbers 1, 2, 3 and 4?
6. What would the answer to question 5 be if the numbers are 0, 1, 2 and 3?
7. What would the answers be to questions 5 and 6 if no repetitions are allowed?
8. How many ways are there of arranging four 3s and six 5s?
9. What is the answer to question 8 if the number has to be divisible by 5?
10. (*a*) How many arrangements are there of the letters of the word *theory*?
    (*b*) How many arrangements are there in which:
    (*i*) the *t* is at the beginning
    (*ii*) the vowels remain in the same order
    (*iii*) the vowels remain in the same position
    (*iv*) the *t* is at the beginning and the *y* is at the end
11. Mr and Mrs Jones, Mr and Mrs Smith, Mr Brown and Miss Green are seated at a round table. How many ways are there of arranging them if:
    (*a*) they can sit anywhere
    (*b*) Mr Brown and Miss Green are to sit together
    (*c*) the three men are not to sit together
12. How many ways are there of arranging the letters of the word *barometric* if:
    (*a*) the vowels are to be together
    (*b*) the consonants are to be together
    (*c*) the vowels are to be in alphabetical order
13. (*a*) How many three-digit numbers can be made from the digits 1, 3, 4, 5, 7 and 8, no repetitions being allowed?
    (*b*) If the numbers in part (*a*) have to be greater than 500, how many will there be?
    (*c*) How many of the arrangements in part (*a*) are even numbers?
14. How many five-different-letter code words can be made from

the letters A, B, C, D, E, F and G? How many of these contain at least one vowel?

15. How many arrangements of three letters can be made from the word *algebra*?

3.5 Combinations

In the previous section the word 'arrangement', rather than 'permutation', was used deliberately to emphasize the idea of order. We shall now consider 'selections' or 'combinations', where the order does not matter.

Example 20. How many ways are there of selecting three letters from the letters A, B, C, D and E?

In this case it is not too difficult to write down the selections:

ABC ABE ACE BCD BDE ABD ACD ADE BCE CDE

Altogether there are 10, and in each case the letters themselves can be arranged in 3! ways. For example, taking *ABC* we can also have *ACB, BAC, BCA, CAB, CBA*. So for this example we can write:

Total number of arrangements = number of selections × 3!

And so:

Number of selections = number of arrangements ÷ 3!

We have already found the total number of arrangements of 3 letters from 5 is given by ${}_5P_3 = 5!/2!$. Therefore, the number of selections of 3 letters from 5 is $5!/(3!\,2!) = 10$.

If we select r objects from n, then any one selection can itself be arranged in $r!$ ways. And so:

$r! \times$ number of ways of selecting r objects from n

= number of ways of arranging r objects from n

giving:

number of ways of selecting r objects from n

$$= {}_nP_r/r!$$

$$= \frac{n!}{r!(n-r)!}$$

In symbols, the number of selections can be represented by ${}_nC_r$.

A more modern notation is, $\binom{n}{r}$; *i.e.*, $\binom{n}{r} = \dfrac{n!}{r!(n-r)!}$.

Example 21. In how many ways can a football eleven be chosen from 15 boys?

Here we want the number of ways of selecting 11 boys from 15. The required number is:

$$\binom{15}{11} = \frac{15!}{4!\,11!} = 1365$$

Example 22. Find the value of n for which $\binom{n}{6} = \binom{n}{4}$.

$$\binom{n}{6} = \frac{n!}{6!\,(n-6)!} \quad \text{and} \quad \binom{n}{4} = \frac{n!}{4!\,(n-4)!}$$

So, if $\binom{n}{6} = \binom{n}{4}$, we have:

$$\frac{n!}{6!\,(n-6)!} = \frac{n!}{4!\,(n-4)!}$$

$$\frac{(n-4)!}{(n-6)!} = \frac{6!}{4!}$$

$$(n-4)(n-5) = 30$$

$$n^2 - 9n + 20 = 30$$

$$n^2 - 9n - 10 = 0$$

$$(n+1)(n-10) = 0$$

$$\therefore n = -1, 10$$

Since n cannot be negative, the required value is 10.

Example 23. In how many ways can a committee of five be chosen from six men and six women if it must comprise three men and two women?

The number of ways of selecting three men from six is:

$$\binom{6}{3} = \frac{6!}{3!\,3!} = 20$$

The number of ways of selecting two women from six is:

$$\binom{6}{2} = \frac{6!}{2!\,4!} = 15$$

Now for each selection of two women there will be 20 ways of selecting 3 men. So the total number of ways of selecting 3 men and 2 women is:

$$\binom{6}{3}\cdot\binom{6}{2} = 20.15 = 300$$

Example 24. A boy has four different coins in his pocket. How many different sums of money can he form with the coins?

He can select either 1 out of 4, 2 out of 4, 3 out of 4 or 4 out of 4 coins, that is:

$$\binom{4}{1} + \binom{4}{2} + \binom{4}{3} + \binom{4}{4}$$

$$= \frac{4!}{1!\,3!} + \frac{4!}{2!\,2!} + \frac{4!}{3!\,1!} + \frac{4!}{4!\,0!}$$

$$= 4 + 6 + 4 + 1$$

$$= 15$$

We can illustrate the solution to this type of problem another way by using the tree diagram in Fig. 3.14, where S, $\bar{S}$ stand for the selection or non-selection of a coin. Altogether there are 16 final outcomes, though this includes the non-selection of all coins ($\bar{S}\ \bar{S}\ \bar{S}\ \bar{S}$). If at least one coin has to be chosen, the number of different sums formed will be 15.

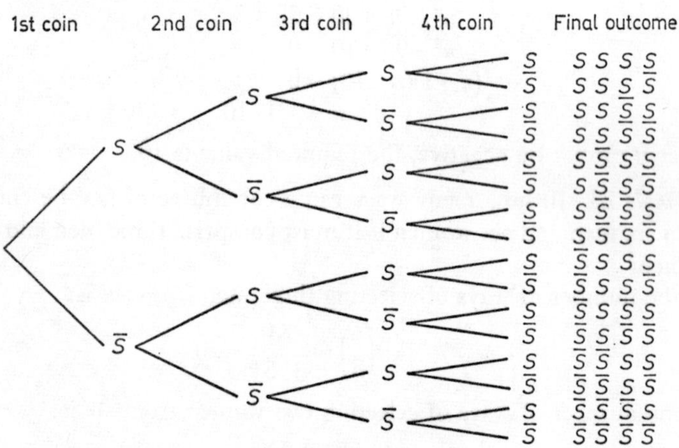

FIGURE 3.14

By looking at the formation of the tree diagram we can arrive at a general result for the selection of one or more objects from *n* unlike objects. At the first stage there are 2 branches—selection or non-selection. For each branch of the first stage there are 2 branches in the second stage, and so the total number of branches at the end of the second stage is 2.2 or 2^2. Similarly the number of branches in the third stage, corresponding to each branch in the second stage, is 2, giving a total of $2^2.2$ or 2^3 for the number of branches at the end of the third stage. Continuing the process it can be easily seen that at the end of n stages the total number of branches is 2^n. But this includes the case where no selection of any of the objects is made. Thus, if we want at least one of the objects to be selected the total number is $2^n - 1$.

Example 25. A boy has three 50p coins and four 5p coins. How many different sums of money can he form?

The difference in this problem is that we have three things alike of one kind and four things alike of another. Consider first the 50p coins. We can select 0, 1, 2 or 3 coins, *i.e.*, there are four ways of dealing with them. For the 5p coins we can select 0, 1, 2, 3 or 4, *i.e.*, five ways of dealing with them. Since, for each way of dealing with the 50p coins, there are five ways of dealing with the 5p coins, the total number of ways of dealing with all the coins is $4.5 = 20$.

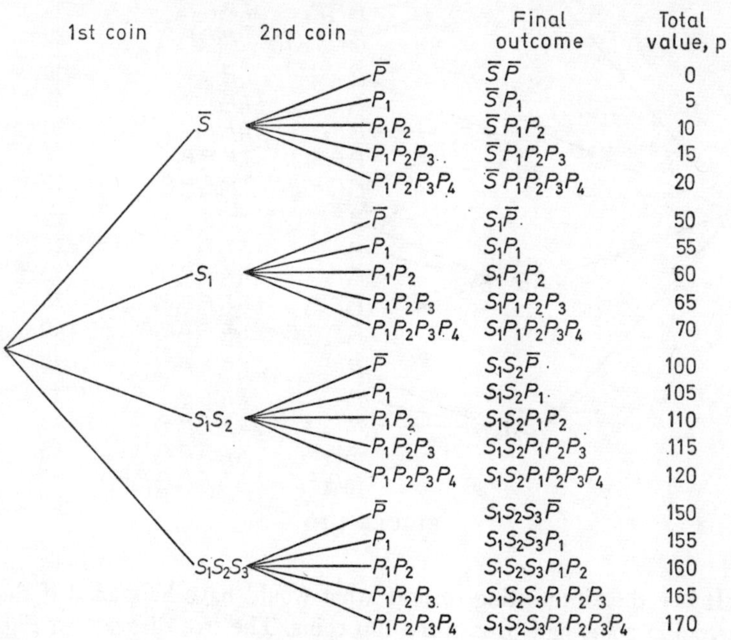

FIGURE 3.15

But this includes rejection of all the coins and, if we have to select at least one coin, the total number of ways of dealing with the coins is then $20 - 1 = 19$. This is illustrated in Fig. 3.15, where S_1, S_2, S_3 represent the 50p coins, P_1, P_2, P_3, P_4 the 5p coins and $\bar{S}$, $\bar{P}$ the non-selection of the coins. In general, the number of branches in the first stage of the tree diagram will be one more than the number of objects. So if there are p objects alike of one kind there will be $(p+1)$ ways of dealing with them. At each subsequent stage the number of branches corresponding to each branch of the preceding stage will be one more than the number of objects. So if we have p objects alike of one kind, q alike of another and r alike of a third kind the number of ways of selecting at least one of them is $(p+1)(q+1)(r+1) - 1$.

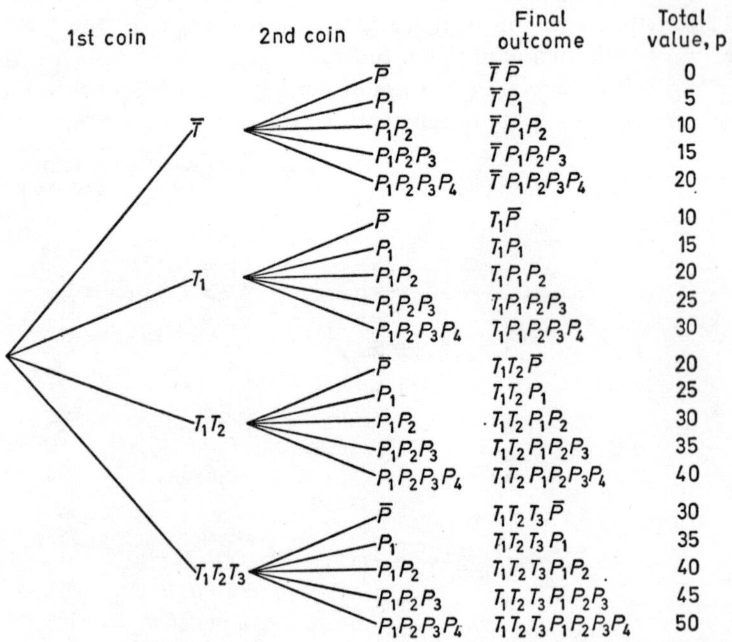

FIGURE 3.16

It is rather interesting to note what would have happened if the three 50p coins had been three 10p coins. The tree diagram in Fig. 3.16 is similar to the previous one, except that I have used T with subscripts and $\bar{T}$ to represent the selection and non-selection, respectively, of the 10p coins. If you look at the column on the right-hand side you will see that some values are repeated. In fact the number of different sums of money in this case is only 9. It is worth noting how to recognize this type of problem. It happens when a number of the smaller-valued coins taken together exceed, or equal, the value of one of the other higher-valued coins. Here, for example, two of the 5p coins are equivalent to one 10p coin.

Exercise 3d

1. Evaluate:
   (*a*) $\binom{5}{3}$ (*b*) $\binom{8}{6}$ (*c*) $\binom{99}{97}$
2. Find the value of n for which:
   (*a*) $\binom{n}{2} = 28$ (*b*) $\binom{n}{11} = \binom{n}{9}$
3. How many ways are there of selecting three people from a group of seven?
4. A box contains eight balls. In how many ways can I select three of them?
5. In how many ways can I select 2 cards from a pack of 52?
6. A soccer team of 11 has to be selected from 14 players.
   (*a*) How many ways are there of selecting the team?
   (*b*) If there is only one person capable of playing in goal, how many ways are there of selecting the team?
7. On the finance committee of a particular organization there are eight men and six women. A sub-committee of five is to be formed. How many ways are there of doing this if the sub-committee is to have:
   (*a*) three men and two women on it
   (*b*) two men and three women on it
   (*c*) at least one woman on it
   (*d*) all men on it
   (*e*) at the most two women on it
8. A badminton team consists of four men and four women. There are seven men and eight women from which the team can be selected. How many ways are there of selecting the team?

9. A boy has five coins each of a different denomination. How many different sums of money can he form?
10. A box holds nine red, eleven white, four blue and nineteen orange marbles. How many ways are there of selecting at least one marble?
11. (*a*) In her purse a woman has two £1 notes, one 50p coin, four 10p coins and one 5p coin. How many different sums of money can she form?
    (*b*) After having made a number of purchases she has one 50p coin, three 10p coins and three 5p coins left. How many different sums of money could be made from these?
12. Suppose, as well as having p objects alike of one kind, q alike of another kind and r alike of a third kind, there are k unlike objects, what would be the number of ways of selecting at least one of them?

3.6 Applications

Example 26. There are eight people from whom a committee of five is to be formed. What is the probability that a particular person will be on the committee?

The sample space will be the number of ways of selecting five people from eight, that is $\binom{8}{5}$. The subset will be the number of ways of selecting five people from eight with this particular person included in the five. To do this we put this person on the committee and find the numbers of ways of selecting four others from the remaining seven. The number of elements in the subset will be $\binom{7}{4}$. Therefore, the probability that a particular person will be on the committee is given by $\binom{7}{4}\Big/\binom{8}{5}$. But $\binom{7}{4} = \dfrac{7!}{3!\,4!} = 35$ and $\binom{8}{5} = \dfrac{8!}{3!\,5!} = 56$. Therefore the required probability is $\frac{35}{56} = \frac{5}{8}$.

Example 27. Suppose the eight people in the previous example comprise five men and three women. In the committee of five what is the probability that there will be three men and two women on it?

As before the number of elements in the sample space is $\binom{8}{5}$. The subset in this case is the number of ways of selecting three men from five and, with each of these, the number of ways of selecting two women from three. That is:

$$\binom{5}{3} \times \binom{3}{2} = \frac{5!}{2!\,3!} \cdot \frac{3!}{1!\,2!} = 30$$

The probability, then, of the committee comprising three men and two women is given by $\binom{5}{3} \times \binom{3}{2} \Big/ \binom{8}{5} = \frac{30}{56} = \frac{15}{28}$.

Example 28. A bag contains three red marbles, two white marbles and four green marbles. If two marbles are drawn at random, what is the probability that they are (*i*) both white, (*ii*) red and green?

(*i*) The sample space is the number of ways of selecting two marbles from nine, *i.e.*:

$$\binom{9}{2} = \frac{9!}{7!\,2!} = 36$$

The number of elements in the subset is the number of ways of selecting two white marbles from two, *i.e.*:

$$\binom{2}{2} = \frac{2!}{0!\,2!} = 1$$

Therefore, the probability is given by:

$$P(\text{Both marbles are white}) = \binom{2}{2} \Big/ \binom{9}{2} = \tfrac{1}{36}$$

(*ii*) The number of elements in the sample space is again 36. The number of elements in the subset is the number of ways of selecting one red marble from three multiplied by the number of ways of selecting one green marble from four. The probability of the marbles being red and green is:

$$\binom{3}{1} \times \binom{4}{1} \Big/ \binom{9}{2} = \frac{3.4}{36} = \tfrac{1}{3}$$

Instead of using permutations and combinations in this example, we could have applied the theory of Chapter 2 and found the probability where there is no replacement. Thus, letting R, W and G represent a red, white and green marble, respectively:

(*i*) $P(WW) = \frac{2}{9}.\frac{1}{8} = \frac{1}{36}$

(*ii*) $P\{(RG) \cup (GR)\} = P(RG) + P(GR)$

$= (\frac{3}{9}.\frac{4}{8}) + (\frac{4}{9}.\frac{3}{8}) = \frac{1}{6} + \frac{1}{6} = \frac{1}{3}$

Example 29. Five discs in a bag are numbered 1 to 5. If I draw out three discs at random, what is the probability that I draw them out in numerical order?

The first thing to note is that order is involved. The sample space is the number of ways of arranging 3 numbers from 5. The number of elements is ${}_5P_3 = 5!/2! = 60$. For each arrangement of 3 numbers, there is only one in which the discs are in numerical order. Therefore the subset is the number of ways of selecting 3 numbers from 5. Remember that all six arrangements of 3 objects count as one selection. So the number of elements in our subset is:

$$\binom{5}{3} = \frac{5!}{2!\,3!} = 10$$

Thus the probability of drawing out 3 discs in numerical order is:

$$\frac{\binom{5}{3}}{{}_5P_3} = \frac{10}{60} = \frac{1}{6}$$

Exercise 3e

1. A sub-committee of five has to be formed from the main committee of seven. What is the probability that a person selected at random will be on this sub-committee?
2. A rugby team is made up of 15 players. If there are 18 people available for selection, what is the probability that a person selected at random will be on the team?
3. Out of a group of eight men and four women we want to form a smaller group of seven people. What is the probability that:
   (*a*) a particular person is in this smaller group
   (*b*) a particular man is in this smaller group
   (*c*) a particular woman is in this smaller group
   (*d*) the smaller group consists of five men and two women
   (*e*) the smaller group has at least one woman on it
   (*f*) the smaller group has at least four men on it

4. A box contains five green, five yellow, four blue and three orange marbles. If two marbles are drawn at random, what is the probability that they are:
(*a*) both green
(*b*) both yellow
(*c*) both blue
(*d*) both orange
5. If, in question 4, three marbles are drawn out at random, what is the probability that they are:
(*a*) all green
(*b*) all yellow
(*c*) all blue
(*d*) all orange
(*e*) two blue and one green
(*f*) two orange and one yellow
(*g*) all the same colour
(*h*) two green and the other not blue
6. From the numbers 1, 2, 3, 4, 5, 6, 7, 8, 9 and 10, three are selected at random. What is the probability that:
(*a*) they are in numerical order
(*b*) they are even numbers
(*c*) they are prime numbers (NB 1 is not a prime number)
(*d*) they are less than 5 or greater than 6
7. Four cards are selected at random from a well shuffled pack of 52 playing cards. What is the probability that:
(*a*) they are all aces
(*b*) there are 2 kings and 2 aces
(*c*) 3 of them are hearts and the other is a diamond
(*d*) 3 of them are of one suit and the other is not of that suit
(*e*) at least one of them is an ace
(*f*) they will each be of different suits
8. What is the probability that if I select 3 letters from the word *nought*, at least one of them is a vowel?

3.7 Mathematical expectation

Example 30. Suppose there are 100 discs in a bag and one of them is marked in a particular way, so that if you draw it you win £5. The probability of winning is $\frac{1}{100}$.

In Chapter 1, it was shown that this does not necessarily mean that if you make 100 draws (assuming replacement) you will at some stage draw the marked disc. What it does mean is that if you continue making draws for a long time the *average* number of wins tends towards 1 out of every 100. (Average is here taken as the arithmetic average, usually called the mean. Other interpretations will be discussed in 'Statistics' in this series.)

What then are your expected winnings? If you take 1000 draws you expect to win 10 times and lose 990 times. So in terms of money you win (£5 × 10) + (£0 × 990) = £50. Your average winnings on 1 draw are then £$\frac{50}{1000}$ = £0·05. The £50 comes from (£5 × 10) and, to find the average winnings for 1 draw, we calculate £5.10/1000. This can be written £5.$\frac{1}{100}$, that is, multiplying the £5 by the probability of winning it.

The technical name for this £0·05 is the *mathematical expectation* or *expected value* of the amount of winnings.

In other words, if instead of having one marked disc you decide that every disc is to be marked, then the amount of money paid out for each disc will be £$\frac{5}{100}$ = £0·05, *i.e.*, the £5 averaged over the 100 discs.

Let us now give a more formal definition of mathematical expectation.

If some variable X takes values $x_1, x_2, x_3, \ldots$ with probabilities $p_1, p_2, p_3, \ldots$ then we write the *mathematical expectation* or *expected value* of X as:

$$E(X) = p_1x_1 + p_2x_2 + p_3x_3 + \ldots$$
$$= \sum_{r=1}^{r=n} p_rx_r$$

Note that $E(X)$ need not necessarily take a value of x_i nor need it be integral (*i.e.*, a whole number) if the values of x_i are integral. You may have met this situation before, when, for example, people say that the average number of children in a family is 2·3.

Example 31. What is the expected number when a fair die is thrown?

In this case the variable X is the number thrown and so:

$$x_1 = 1,\ x_2 = 2,\ x_3 = 3,\ x_4 = 4,\ x_5 = 5,\ x_6 = 6$$

and

$$p_1 = p_2 = p_3 = p_4 = p_5 = p_6 = \tfrac{1}{6}$$

Therefore:

$$E(X) = p_1x_1 + p_2x_2 + p_3x_3 + p_4x_4 + p_5x_5 + p_6x_6$$
$$= \tfrac{1}{6}.1 + \tfrac{1}{6}.2 + \tfrac{1}{6}.3 + \tfrac{1}{6}.4 + \tfrac{1}{6}.5 + \tfrac{1}{6}.6$$
$$= \tfrac{1}{6}(1 + 2 + 3 + 4 + 5 + 6) = \tfrac{21}{6} = 3{\cdot}5$$

Look again at Example 30. Obviously in games of chance the owner of the game is not a philanthropist and is out to make money. If he charges £0·10 for each draw he will make a profit. From the owner's point of view the beauty of this arrangement is that, although somebody will pay for 1 draw and win £5, the player's expected winnings for each draw is only £0·05 and, as he is paying £0·10, the owner of the game makes a tidy profit.

Example 32. A 'one-armed bandit' has 3 wheels, each with 10 symbols on it made up of 1 bell, 2 oranges, 2 lemons and 5 cherries. A 2p coin is needed to play, and the machine pays out as follows:

3 bells	50p	2 bells	5p
3 oranges	20p	2 oranges	2p
3 lemons	10p	2 lemons	2p
3 cherries	5p	anything else pays nothing	

What is a player's expected profit?

In this case X is the profit made in each situation. Note that it is the profit in which we are interested, *i.e.*, the amount of money the machine pays out less the 2p needed to play the machine. Therefore:

$$x_1 = 48,\ \ x_2 = 18,\ \ x_3 = 8,\ \ x_4 = 3,\ \ x_5 = 3,\ \ x_6 = 0,\ \ x_7 = 0,\ \ x_8 = -2$$

x_1, x_2, x_3, x_4, x_5, x_6, x_7 and x_8 are, respectively, the turning up of 3 bells, 3 oranges, 3 lemons, 3 cherries, 2 bells, 2 oranges, 2 lemons and anything else.

If p_1 is the probability of 3 bells, then:

$$p_1 = P(BBB)$$
$$= P(B \text{ on 1st wheel}).P(B \text{ on 2nd wheel}).P(B \text{ on 3rd wheel})$$
$$= \tfrac{1}{10}\cdot\tfrac{1}{10}\cdot\tfrac{1}{10} = \tfrac{1}{1000}$$

Similarly, if p_2, p_3 and p_4 are the probabilities of 3 oranges, 3 lemons and 3 cherries, we have:

$$p_2 = P(OOO) = \tfrac{2}{10}\cdot\tfrac{2}{10}\cdot\tfrac{2}{10} = \tfrac{8}{1000}$$
$$p_3 = P(LLL) = \tfrac{2}{10}\cdot\tfrac{2}{10}\cdot\tfrac{2}{10} = \tfrac{8}{1000}$$
$$p_4 = P(CCC) = \tfrac{5}{10}\cdot\tfrac{5}{10}\cdot\tfrac{5}{10} = \tfrac{125}{1000}$$

If p_5, p_6 and p_7 are the probabilities of 2 bells, 2 oranges and 2 lemons, respectively, then:

$$p_5 = P(BB\bar{B}) + P(B\bar{B}B) + P(\bar{B}BB)$$

where $\bar{B}$ means not a bell. Now:

$$P(BB\bar{B}) = \tfrac{1}{10}.\tfrac{1}{10}.\tfrac{9}{10} = \tfrac{9}{1000}$$

and since the three wheels are independent of each other:

$$P(BB\bar{B}) = P(B\bar{B}B) = P(\bar{B}BB) = \tfrac{9}{1000}$$

Therefore:

$$p_5 = \tfrac{9}{1000} + \tfrac{9}{1000} + \tfrac{9}{1000} = \tfrac{27}{1000}$$

Similarly:

$$p_6 = P(OO\bar{O}) + P(O\bar{O}O) + P(\bar{O}OO)$$

But $\quad P(OO\bar{O}) = P(O\bar{O}O) = P(\bar{O}OO) = \tfrac{2}{10}.\tfrac{2}{10}.\tfrac{8}{10} = \tfrac{32}{1000}$

Hence $p_6 = \tfrac{96}{1000}$. Similarly, $p_7 = \tfrac{96}{1000}$. This means that:

$$p_1 + p_2 + p_3 + p_4 + p_5 + p_6 + p_7 = \tfrac{361}{1000}$$

and hence $p_8 = 1 - \tfrac{361}{1000} = \tfrac{639}{1000}$. We can now write:

$$\begin{aligned} E(X) &= p_1x_1 + p_2x_2 + p_3x_3 + p_4x_4 + p_5x_5 + p_6x_6 + p_7x_7 + p_8x_8 \\ &= \tfrac{1}{1000}(48) + \tfrac{8}{1000}(18) + \tfrac{8}{1000}(8) + \tfrac{125}{1000}(3) + \tfrac{27}{1000}(3) + \tfrac{96}{1000}(0) \\ &\quad + \tfrac{96}{1000}(0) + \tfrac{639}{1000}(-2) \\ &= \tfrac{1}{1000}[48 + 144 + 64 + 375 + 81 + 0 + 0 - 1278] = -\tfrac{566}{1000} \end{aligned}$$

The minus sign indicates a loss, and so in playing this machine once the player should expect to lose £5·66/1000. Obviously this is very little but it means that in, say, 1000 plays (which is not very many) the profit to the machine is just over £5·50.

Example 33. In a bag there are three white marbles and two black ones. If I draw out one marble at a time (no replacement), what is the number of draws I expect to make in order to remove the two black marbles?

If X is the number of draws required to remove the two black marbles, it can take values $x_1 = 2$, $x_2 = 3$, $x_3 = 4$, $x_4 = 5$. Let B, W represent a black and white marble, respectively. Then, p_1, the probability of drawing out the 2 black marbles in 2 draws is given by:

$$p_1 = P(BB) = \tfrac{2}{5}.\tfrac{1}{4} = \tfrac{1}{10}$$

Note that since there is no replacement the events are dependent. If p_2 is the probability of it taking 3 draws to draw out the 2 black marbles, then:

$$\begin{aligned} p_2 &= P(BWB) + P(WBB) \\ &= (\tfrac{2}{5}\cdot\tfrac{3}{4}\cdot\tfrac{1}{3}) + (\tfrac{3}{5}\cdot\tfrac{2}{4}\cdot\tfrac{1}{3}) \\ &= \tfrac{1}{10} + \tfrac{1}{10} = \tfrac{1}{5} \end{aligned}$$

Similarly:

$$\begin{aligned} p_3 &= P(BWWB) + P(WBWB) + P(WWBB) \\ &= (\tfrac{2}{5}\cdot\tfrac{3}{4}\cdot\tfrac{2}{3}\cdot\tfrac{1}{2}) + (\tfrac{3}{5}\cdot\tfrac{2}{4}\cdot\tfrac{2}{3}\cdot\tfrac{1}{2}) + (\tfrac{3}{5}\cdot\tfrac{2}{4}\cdot\tfrac{2}{3}\cdot\tfrac{1}{2}) \\ &= \tfrac{1}{10} + \tfrac{1}{10} + \tfrac{1}{10} = \tfrac{3}{10} \end{aligned}$$

and:

$$\begin{aligned} p_4 &= P(BWWWB) + P(WBWWB) + P(WWBWB) + P(WWWBB) \\ &= (\tfrac{2}{5}\cdot\tfrac{3}{4}\cdot\tfrac{2}{3}\cdot\tfrac{1}{2}\cdot\tfrac{1}{1}) + (\tfrac{3}{5}\cdot\tfrac{2}{4}\cdot\tfrac{2}{3}\cdot\tfrac{1}{2}\cdot\tfrac{1}{1}) + (\tfrac{3}{5}\cdot\tfrac{2}{4}\cdot\tfrac{2}{3}\cdot\tfrac{1}{2}\cdot\tfrac{1}{1}) + (\tfrac{3}{5}\cdot\tfrac{2}{4}\cdot\tfrac{1}{3}\cdot\tfrac{2}{2}\cdot\tfrac{1}{1}) \\ &= \tfrac{1}{10} + \tfrac{1}{10} + \tfrac{1}{10} + \tfrac{1}{10} = \tfrac{2}{5} \end{aligned}$$

Hence:

$$\begin{aligned} E(X) &= p_1x_1 + p_2x_2 + p_3x_3 + p_4x_4 = \tfrac{80}{20} = 4 \\ &= \tfrac{1}{10}(2) + \tfrac{1}{5}(3) + \tfrac{3}{10}(4) + \tfrac{2}{5}(5) \end{aligned}$$

In this example it is relatively easy to write out all the possible cases, but there may be examples where this is not so. Suppose, for the sake of argument, the number of white marbles is 18 and the number of black remains at 2. What is the probability of taking 20 draws to remove the 2 black marbles?

Consider first the probability that the last two marbles to be drawn are black:

$$\begin{aligned} &P(WWWWWWWWWWWWWWWWWWBB) \\ &= \tfrac{18}{20}\cdot\tfrac{17}{19}\cdot\tfrac{16}{18}\cdot\tfrac{15}{17}\cdot\tfrac{14}{16}\cdot\tfrac{13}{15}\cdot\tfrac{12}{14}\cdot\tfrac{11}{13}\cdot\tfrac{10}{12}\cdot\tfrac{9}{11}\cdot\tfrac{8}{10}\cdot\tfrac{7}{9}\cdot\tfrac{6}{8}\cdot\tfrac{5}{7}\cdot\tfrac{4}{6}\cdot\tfrac{3}{5}\cdot\tfrac{2}{4}\cdot\tfrac{1}{3}\cdot\tfrac{2}{2}\cdot\tfrac{1}{1} \\ &= \frac{18!\times 2!}{20!} = \frac{1}{190} \end{aligned}$$

The question concerns the number of arrangements of the marbles. Since we have to take 20 draws, the last marble to be drawn must be black, and what we want is the number of ways of arranging the remaining 18 white and 1 black. It is fairly obvious that this is 19 since there are 19 positions for the black marble.

You can also use the formula for the arrangement of n things of which p are alike, *i.e.*, $n!/p!$ In this case, $n = 19$ and $p = 18$ and therefore $n!/p! = 19!/18! = 19$. Hence, the probability of taking 20 draws to remove the 2 black marbles is $\frac{1}{190}\cdot 19 = \frac{1}{10}$.

But repeating this for each draw is rather tedious. Let us see if there is a general rule for finding the expected number of draws to remove two black marbles from a box containing n marbles of which two are black and the rest white. To do this we shall consider the probability that it would take k draws, multiply this probability by k (the number of draws) and sum over k from 2 to n.

Suppose the last two marbles, in k draws, to be removed are the two black ones. We shall then have removed $(k-2)$ white marbles. The probability of removing $(k-2)$ white marbles from a total of n marbles is given by:

$$\frac{(n-2)}{n}\cdot\frac{(n-3)}{(n-1)}\cdot\frac{(n-4)}{(n-2)}\cdots\frac{(n-k-1)}{(n-k+1)}$$

Note that the last term is $\dfrac{n-(k+1)}{n-(k-1)}$ which, when the inside brackets are removed becomes $(n-k-1)/(n-k+1)$.

$$(n-2).(n-3).(n-4)\ldots(n-k-1)=\frac{(n-2)!}{(n-k-2)!}$$

and $$n.(n-1).(n-2)\ldots(n-k+1)=\frac{n!}{(n-k)!}$$

Therefore:

$$P(\text{Removing } k-2 \text{ white marbles})=\frac{(n-2)!}{(n-k-2)!}\cdot\frac{(n-k)!}{n!}$$
$$=\frac{(n-k)(n-k-1)}{n(n-1)}$$

The probability of removing the two black marbles next is given by:

$$\frac{2}{(n-k)}\cdot\frac{1}{(n-k-1)}$$

And so the probability of the last 2 marbles drawn in k draws being black is given by:

$$\frac{(n-k).(n-k-1)}{n.(n-1)}\cdot\frac{2}{(n-k)(n-k-1)}=\frac{2}{n(n-1)}$$

We have, in fact, shown here that the probability of the last two marbles being black is independent of the draw. If the last marble drawn in k draws is black, there are $(k-1)$ places in which the other black marble can be. Therefore the probability of it taking k draws to remove the two black marbles is given by $\dfrac{2(k-1)}{n(n-1)}$.

Therefore
$$E(X) = \sum_{k=2}^{n} \frac{2k(k-1)}{n(n-1)}$$
$$= \frac{2}{n(n-1)} \sum_{k=2}^{n} (k^2 - k)$$

Now
$$\sum_{k=1}^{n} (k^2 - k) = \tfrac{1}{6}n(n+1)(2n+1) - \tfrac{1}{2}n(n+1)$$
$$= \frac{n(n+1)(n-1)}{3}$$

and
$$(k^2 - k) = 0, \text{ when } k = 1$$

Therefore
$$\sum_{k=2}^{n} (k^2 - k) = \tfrac{1}{3}n(n+1)(n-1)$$

And so
$$E(X) = \frac{2}{n(n-1)}\left[\frac{n(n+1)(n-1)}{3}\right] = \frac{2(n+1)}{3}$$

Example 34. Sometimes the probability of a series of events can be tabulated as follows:

X	3	5	8	10
$p(X)$	0·1	0·2	0·3	0·4

From these results:

$$E(X) = (0{\cdot}1)(3) + (0{\cdot}2)(5) + (0{\cdot}3)(8) + (0{\cdot}4)(10)$$
$$= 0{\cdot}3 + 1{\cdot}0 + 2{\cdot}4 + 4{\cdot}0 = 7{\cdot}7$$

It is also possible to calculate other expected values from this table. We may, for example, compute $E(X^2)$, remembering that $P(X^2 = 9)$ is the same as the $P(X = 3)$.

$$E(X^2) = (0{\cdot}1)(3^2) + (0{\cdot}2)(5^2) + (0{\cdot}3)(8^2) + (0{\cdot}4)(10^2)$$
$$= 0{\cdot}9 + 5{\cdot}0 + 19{\cdot}2 + 40{\cdot}0 = 65{\cdot}1$$

The interesting thing here is that $E(X^2) \neq [E(X)]^2$.

We can also compute $E(3X)$:

$$E(3X) = (0{\cdot}1)(3.3) + (0{\cdot}2)(3.5) + (0{\cdot}3)(3.8) + (0{\cdot}4)(3.10)$$
$$= 0{\cdot}9 + 3{\cdot}0 + 7{\cdot}2 + 12 = 23{\cdot}1$$

In this case, $E(3X) = 3E(X)$. In other words, trebling every number trebles the mean; note that squaring every number does not square the mean.

$$\begin{aligned}&E[(X-7{\cdot}7)^2]\\&\quad=(0{\cdot}1)(3-7{\cdot}7)^2+(0{\cdot}2)(5-7{\cdot}7)^2+(0{\cdot}3)(8-7{\cdot}7)^2+(0{\cdot}4)(10-7{\cdot}7)^2\\&\quad=2{\cdot}209+1{\cdot}458+0{\cdot}027+2{\cdot}116=5{\cdot}810\end{aligned}$$

But $[E(X)]^2=7{\cdot}7^2=59{\cdot}29$. Hence:

$$E(X^2)-[E(X)]^2=65{\cdot}1-59{\cdot}29=5{\cdot}81=E[(X-7{\cdot}7)^2]$$

7·7 is the mean of this particular distribution. We can write the mean of X as $\bar{X}$. It is also common practice to use the Greek letter μ to denote the mean. Consider another example:

$$\begin{aligned}E(X^2+X)&=(0{\cdot}1)(3^2+3)+(0{\cdot}2)(5^2+5)+(0{\cdot}3)(8^2+8)+(0{\cdot}4)(10^2+10)\\&=1{\cdot}2+6{\cdot}0+21{\cdot}6+44{\cdot}0=72{\cdot}8\end{aligned}$$

But $$E(X^2)+E(X)=65{\cdot}1+7{\cdot}7=72{\cdot}8$$
$$\therefore E(X^2+X)=E(X^2)+E(X)$$

If this is so we have:

$$\begin{aligned}E[(X-\bar{X})]^2&=E(X^2-2\bar{X}.X+\bar{X}^2)\\&=E(X^2)-2\bar{X}E(X)+\bar{X}^2\\&=E(X^2)-2[E(X)]^2+[E(X)]^2\\&=E(X^2)-[E(X)]^2\end{aligned}$$

as we have already found. Note that the expected value of a constant is the constant itself and so $E(\bar{X}^2)=\bar{X}^2$.

This particular function $E[(X-\bar{X})]^2$ is known as the *variance* of the distribution and is the mean of the sum of the squares of the deviations of the values of X_i from the mean $\bar{X}$. But this is discussed fully in 'Statistics', in this series. Let it suffice at the moment to recognize the name and know that the positive square root of the variance is known as the *standard deviation.*

Exercise 3f

1. (*a*) If the probability of winning £5 in a ballot is $\frac{1}{100}$, what is a fair price for a ticket?
   (*b*) Suppose that the chance of winning a second prize of £1 in the same ballot is $\frac{7}{100}$, what is now a fair price for a ticket?
2. The probability in a particular venture of a businessman making

a profit of £5000 is $\frac{1}{50}$, but at the same time the probability of his losing £1000 is $\frac{1}{12}$. What is his expected gain or loss?

3. How many heads would you expect to appear in the tossing of 3 fair coins?
4. From the following table compute $E(X)$, $E(X^2)$, $E(4X)$, $E(2X+3)$ and show that $E[X-E(X)]^2 = E(X^2) - [E(X)]^2$.

X	-3	-1	0	1	3
$p(X)$	0·2	0·1	0·3	0·2	0·2

5. A particular sample of eight items is known to have three defective ones among them. If the items are drawn at random, one at a time and without replacement, what is the expected number of draws to remove all the defective items?

Exercises on Chapter 3

These exercises are not in any order of difficulty or presentation.

1. In one of two identical boxes there are two red and three white balls and in the other three red and two white. A box is selected at random and two balls are drawn from it, without replacement. They are found to be both white. What is the probability that the remaining three balls in the box are red?
2. Prove Pascal's Rule:

$$\binom{n}{r} + \binom{n}{r-1} = \binom{n+1}{r}$$

3. How many ways are there of forming a committee of 3 from 20 people, if 2 particular people will not serve together?
4. Two boxes are marked A and B. In box A there are four red balls and one white ball and in box B there are two red balls and two white balls. A coin is tossed. If it comes up heads a ball is drawn from box A, and if it comes up tails a ball is drawn from box B. What is the probability that a red ball will be drawn?
5. Four men toss a die, which is loaded so that the probability of a six is $\frac{1}{5}$. They decide each man will throw once, and the first one to throw a six will win a prize of £100. They also decide that

if no one wins on this throw the £100 will be divided equally among them. What are their individual expectations?

6. In a competition eight things have to be placed in order of merit. If it costs 1p for each entry, how much would you have to pay to be certain of having a winning entry?
7. If in the competition mentioned in question 6 the eight things have to be selected from twelve and then arranged, how much would it cost to be sure of a winning line?
8. How many ways are there of selecting 3 cards from a pack of 52 playing cards if:
   (*a*) at least one of them is a king
   (*b*) at most one of them is a king
9. The map in Fig. 3.17 shows a section of roadway. A person sets

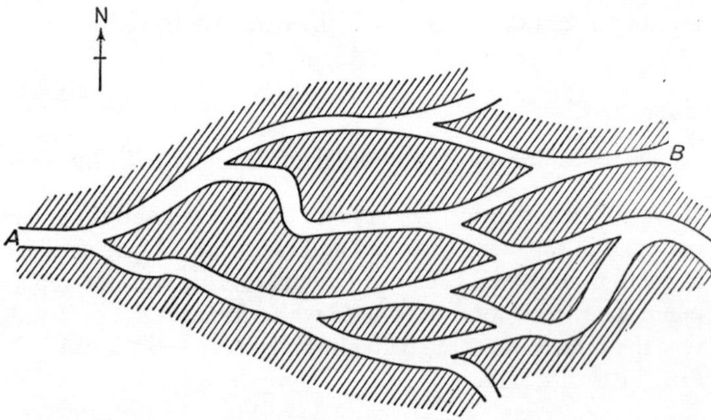

FIGURE 3.17

out from *A* and travels in an easterly direction. At each fork in the road which he meets, he tosses a coin to decide whether he goes left or right. What is the probability that he reaches:
(*a*) *B* (*b*) *C*

10. If I ask five people each to give me a letter of the alphabet, what is the probability that the letters will be in alphabetical order?
11. A coin is tossed until two heads have been obtained. What is the probability that:
    (*a*) this will have happened in two tosses
    (*b*) it will take three tosses
    (*c*) it will take at the most four tosses

Calculate the probability that it would take $n \geqslant (n2)$ tosses to obtain two heads.

12. A man has a pound note, a 50p coin, three 10p coins and one 5p coin. What is the probability that a sum of money formed from these is greater than £1·50?
13. If two dice are tossed what is the expected sum when the two numbers shown are added together?
14. Find the total number of arrangements of the letters of the word *statistics*.
15. Eight people are waiting for a bus. When the bus arrives there are only five seats available, the rest have to stand. How many ways are there of filling the five seats?
16. Four English, three French and two German books are to be put on a shelf.
    (*a*) If the books are all distinctive in appearance, how many ways are there of arranging them if:
    (*i*) they can be arranged randomly
    (*ii*) the English books are to stay together
    (*iii*) the French books are to stay together
    (*iv*) the German books are to stay together
    (*v*) they are placed so that the English books are together, the French books are together and the German books are together
    (*vi*) one particular English book is to be at the left-hand end of the row
    (*vii*) one particular English book is to be at one end and one particular German book at the other

    (*b*) If the English books are identical in appearance, how many ways are there of arranging them?

    (*c*) If the English books are identical, the French books are identical and the German books are identical, how many ways are there of arranging them?
17. From five positive integers and six negative integers, three are chosen at random. What is the probability that their product will be positive?
18. In one of two identical boxes there are two red and three white marbles and in the other three red and two white. After having selected a box at random, the following rule is applied: a ball is

drawn and, if it is red, another is drawn without replacing the first, but, if the first drawn ball is white, it is replaced and a second ball drawn. What is the probability that:
(*a*) both balls will be red
(*b*) both balls will be white
(*c*) the first will be red and the second white
(*d*) one will be red and the other white

19. Four dice are identical in appearance, but while three of them are unbiased, the fourth is weighted so that the probability of a six is $\frac{1}{4}$. One of the dice is selected at random and rolled twice. The result is two sixes. What is the probability that this was the loaded die?
20. Ten points in a plane are such that no three are collinear. How many triangles is it possible to make with these points as vertices?
21. Suppose in question 20, six of the points lie in a straight line, how many triangles can now be formed?
22. Show that, if there are n points in a plane and m of them are collinear, then the greatest number of triangles which can be formed using these points as vertices is:

$$\frac{(n-m)}{3!}\{(n+m)(n+m-3)-nm+2\}$$

23. If in question 22 the remaining $(n-m)$ points are also in a straight line and no point is common to both lines, show that the greatest number of triangles that could now be formed with these points as vertices is $\frac{1}{2}m(n-m)(n-2)$.
24. Out of a batch of 10 items, two are selected at random. If it is known that there are 4 defective items in the batch, what is the probability that:
(*a*) both selected items are defective
(*b*) at least one of them is defective
(*c*) at most one of them is defective
25. A coin is tossed. If the result is heads, a box marked A is chosen and a ball taken out. If the result is tails then a second box B is chosen and a ball taken out. In box A there are two white and three red balls and in box B there are four white balls and one red one. If the drawn ball is red, what is the probability that the tossing of the coin resulted in tails?

26. What would be the probability of a chimpanzee typing out eight letters on a typewriter and making the word *equation*?
27. Six cards are selected at random from a well shuffled pack of 52 playing cards. What is the probability that:
    (*a*) 4 aces are selected
    (*b*) 4 aces and 2 kings are selected
    (*c*) they are all hearts
    (*d*) they are all of the same suit
    (*e*) they are not court cards (*i.e.*, not king, queen or jack)
28. Three identical bags each contain the same number of coins. In the first bag there are among the coins three which are gold, in the second there are two gold and in the third there is one which is gold. Having selected a bag at random a coin is drawn out and found to be gold. Show that the probability that the first bag was selected is $\frac{1}{2}$.
29. A bag contains five discs numbered 1, 2, 3, 4 and 5. Four people *A*, *B*, *C*, *D* in that order each draw a disc out of the bag and does not replace it. The first to draw an even number wins £10. What are their respective expectations?
30. A box contains five red and three green balls. Another identical box has four red and four green balls. A box is selected at random and a ball taken out. It turns out to be green. What is the probability that the boxes now each hold the same number of green balls?
31. Suppose in a tub for Tombola, there are *n* discs of which *r* are marked. Show that the expected number of draws to take out all the marked discs is:

$$\frac{r(n-r)!}{n!}\sum_{k=r}^{n}\frac{k(k-1)!}{(k-r)!}$$

    where *k* is the number of the draw.
32. In a particular community 60% of the voters are Conservative, the other 40% are Labour. On a certain issue which concerns all in the community, of those who would vote Conservative, 75% are for it, the rest against it. But for those who would vote Labour only 10% are in favour, the rest against it. A man is stopped in the street and asked about this issue. He is in favour of it. What is the probability that he would vote Conservative?

4 | DISTRIBUTION AND FREQUENCY

4.1 Binomial distribution

Suppose we have an experiment which results in one of two outcomes, *e.g.*, the tossing of a coin (head or tail) or the obtaining of a six in the throw of a die (six or not a six). Let us further suppose in the tossing of the coin that we are interested in the number of heads appearing in repetitive tosses of the coin. For the purpose of illustration let us further suppose that the coin is biased in such a way that the probability of a head is p ($p \neq \frac{1}{2}$). The probability of a tail will then be q where $q = 1 - p$. A tree diagram will help to illustrate this. But this time we shall draw the tree diagram with the branches spreading

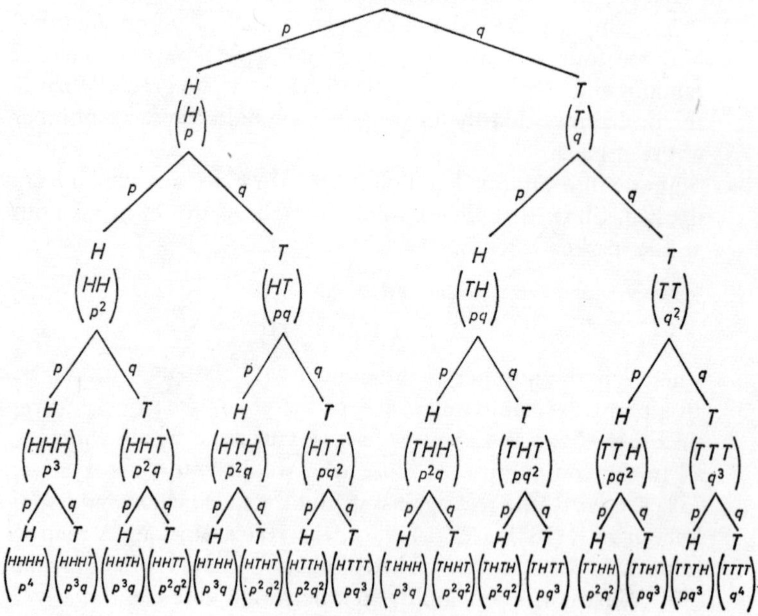

FIGURE 4.1

downwards (Fig. 4.1) and insert the various outcomes with their respective probabilities at each stage. If we tabulate the results as in Table 4.1, we can see what happens to the probability of a number of heads appearing in a number of tosses of a biased coin.

TABLE 4.1

Number of tosses	Number of heads 0	1	2	3	4
1	q	p	–	–	–
2	q^2	$2pq$	p^2	–	–
3	q^3	$3pq^2$	$3p^2q$	p^3	–
4	q^4	$4pq^3$	$6p^2q^2$	$4p^3q$	p^4

To arrive at the result $6p^2q^2$, for example, in the 4th toss we would want the probability of 2 heads appearing in 4 tosses. This would be given by:

$$P(HHTT)+P(HTHT)+P(HTTH)+P(TTHH)+P(THTH)+P(THHT)$$
$$=p^2q^2+p^2q^2+p^2q^2+p^2q^2+p^2q^2+p^2q^2=6p^2q^2$$

The probabilities in Table 4.1 are easily recognizable as being the terms in the expansion of $(q+p)^n$ where $n = 1, 2, 3, 4$.

According to the binomial theorem (see 'Sets and Symbolic Logic' in this series):

$$(a+x)^n = a^n + na^{n-1}x + \frac{n(n-1)}{2!}a^{n-2}x^2 + \frac{n(n-1)(n-2)}{3!}a^{n-3}x^3 + \ldots + x^n$$

Using this theorem to expand $(q+pt)^n$ when $n = 4$ gives:

$$(q+pt)^4 = q^4 + 4pq^3t + 6p^2q^2t^2 + 4p^3qt^3 + p^4t^4$$

The probability of a head was p and so the probability of 0, 1, 2, 3 or 4 heads appearing is the coefficient of t^0, t, t^2, t^3 or t^4.

For each value of the variable—in this case the number of heads appearing in successive tosses of the coin—we have a corresponding probability. This can be tabulated and is known as a *probability distribution.* For example the probability distribution for the sum of the scores when two fair dice are tossed is given as follows:

Total	2	3	4	5	6	7	8	9	10	11	12
Probability	$\frac{1}{36}$	$\frac{2}{36}$	$\frac{3}{36}$	$\frac{4}{36}$	$\frac{5}{36}$	$\frac{6}{36}$	$\frac{5}{36}$	$\frac{4}{36}$	$\frac{3}{36}$	$\frac{2}{36}$	$\frac{1}{36}$

And this distribution can be graphed as in Fig. 4.2. Since the distribution that we are concerned with in such examples has only two outcomes, it is known as the *binomial distribution*.

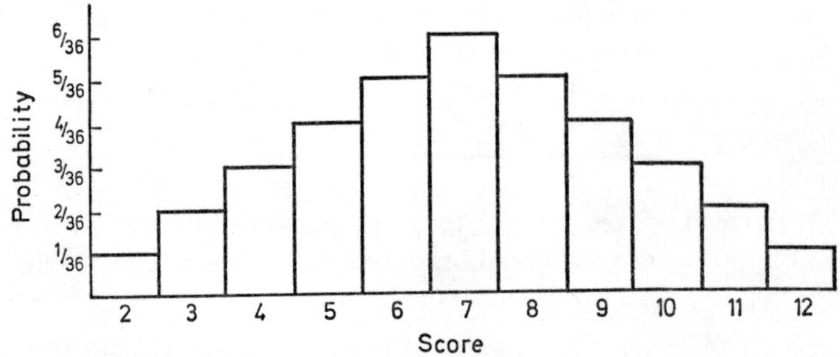

FIGURE 4.2

It is usual practice to label the two outcomes 'success' when the event happens and 'failure' when it does not. Further we say that the probability of a success is p and of a failure q where $p+q=1$. That being so we can now say that the probability of r successes in n trials is the coefficient of t^r in the expansion of $(q+pt)^n$:

$$P(r \text{ successes}) = \binom{n}{r} p^r q^{n-r} \tag{4.1}$$

The terms 'success' and 'failure' are rather unfortunate since in problems a 'success' may be the selection of a defective item. It is best to think of a 'success' as the taking place of the event in which you are interested.

We can use a tree diagram, or at least the idea of it, to arrive at equation (4.1). In a tree diagram with n stages one way of finding the probability of r successes would be to travel along r branches each ending in a success and then travel along the remaining $n-r$ branches each ending in a failure. The probability of this is simply $p^r q^{n-r}$.

The question now is to find how many ways there are of travelling along n branches of which r end in success and the rest in failure. This is simply a question of the number of ways of selecting r from n and, referring to Chapter 3, this is $\binom{n}{r}$. So the probability of r successes in n trials is given by:

$$P(r \text{ successes}) = \binom{n}{r} p^r q^{n-r}$$

where p is the probability of a success, q the probability of a failure and $p + q = 1$.

Let us now apply this to the tossing of the biased coin mentioned earlier. We shall consider four tosses and find the probability distribution for the appearance of heads, that is, we shall calculate the separate probabilities for the obtaining of 0, 1, 2, 3 and 4 heads. In this case the event in which we are interested is the obtaining of head and this will be considered a success whose probability is p. For this example $n = 4$ and $r = 0, 1, 2, 3, 4$:

$$P(0 \text{ heads}) = \binom{4}{0} p^0 q^4 = \frac{4!}{0!\,4!} q^4 = q^4$$

$$P(1 \text{ head}) = \binom{4}{1} p^1 q^3 = \frac{4!}{1!\,3!} pq^3 = 4pq^3$$

$$P(2 \text{ heads}) = \binom{4}{2} p^2 q^2 = \frac{4!}{2!\,2!} p^2 q^2 = 6p^2 q^2$$

$$P(3 \text{ heads}) = \binom{4}{3} p^3 q^1 = \frac{4!}{3!\,1!} p^3 q = 4p^3 q$$

$$P(4 \text{ heads}) = \binom{4}{4} p^4 q^0 = \frac{4!}{4!\,0!} p^4 = p^4$$

And these results agree with those we obtained at the beginning of the section.

Example 1. If 10% of the items produced by a particular machine are defective, what is the probability that out of 5 items chosen at random:

(*i*) none will be defective
(*ii*) 1 will be defective
(*iii*) at most 3 will be defective

In this example we are interested in the choosing of defective items and so this will be our 'success'. Since 10% of the items are defective the probability (p) of a 'success' is 0·1 and hence $q = 0{\cdot}9$. The number (n) of items is 5 and so the probability of choosing r defective ones in the 5 is given by:

$$P(r\text{ defective}) = \binom{5}{r}p^r q^{5-r} = \binom{5}{r}(0{\cdot}1)^r(0{\cdot}9)^{5-r}$$

(*i*) $$P(0\text{ defective}) = \binom{5}{0}(0{\cdot}1)^0(0{\cdot}9)^5 = 0{\cdot}59049$$

(*ii*) $$P(1\text{ defective}) = \binom{5}{1}(0{\cdot}1)(0{\cdot}9)^4 = 0{\cdot}32805$$

(*iii*) $$P(\text{at most 3 defective}) = P(0\text{ defective}) + P(1\text{ defective}) + P(2\text{ defective} + P(3\text{ defective})$$

We have calculated P(0 defective) and P(1 defective).

$$P(2\text{ defective}) = \binom{5}{2}(0{\cdot}1)^2(0{\cdot}9)^3 = 0{\cdot}0729$$

$$P(3\text{ defective}) = \binom{5}{3}(0{\cdot}1)^3(0{\cdot}9)^2 = 0{\cdot}0081$$

Therefore:

$$P\text{ (at most 3 defective)} = 0{\cdot}59049 + 0{\cdot}32805 + 0{\cdot}0729 + 0{\cdot}0081 = 0{\cdot}99954$$

If the question had only asked for this third part it would have saved quite a bit of calculation to have made use of the fact that:

$$P(\text{at most 3 defective}) = 1 - [P(4\text{ defective}) + P(5\text{ defective})]$$

$$P(4\text{ defective}) = \binom{5}{4}(0{\cdot}1)^4(0{\cdot}9) = 0{\cdot}00045$$

$$P(5\text{ defective}) = \binom{5}{5}(0{\cdot}1)^5(0{\cdot}9)^0 = 0{\cdot}00001$$

Therefore:

$$
\begin{aligned}
P(\text{at most 3 defective}) &= 1 - [0{\cdot}00045 + 0{\cdot}00001] \\
&= 1 - 0{\cdot}00046 \\
&= 0{\cdot}99954
\end{aligned}
$$

Example 2. A fair die is tossed four times. Find the probability distribution for the appearance of a six.

The question is to find the probability of 0, 1, 2, 3 or 4 sixes in 4 tosses of a fair die. The event we are interested in is a six and so this is a success, and $p = \frac{1}{6}$ and $q = \frac{5}{6}$:

$$
\begin{aligned}
P(r \text{ sixes}) &= \binom{4}{r} p^r q^{4-r} \\
&= \binom{4}{r} (\tfrac{1}{6})^r (\tfrac{5}{6})^{4-r}
\end{aligned}
$$

$$
\begin{aligned}
P(0 \text{ sixes}) &= \binom{4}{0} (\tfrac{1}{6})^0 (\tfrac{5}{6})^4 \\
&= \tfrac{625}{1296}
\end{aligned}
$$

$$
\begin{aligned}
P(1 \text{ six}) &= \binom{4}{1} (\tfrac{1}{6}) (\tfrac{5}{6})^3 \\
&= \tfrac{500}{1296}
\end{aligned}
$$

$$
\begin{aligned}
P(2 \text{ sixes}) &= \binom{4}{2} (\tfrac{1}{6})^2 (\tfrac{5}{6})^2 \\
&= \tfrac{150}{1296}
\end{aligned}
$$

$$
\begin{aligned}
P(3 \text{ sixes}) &= \binom{4}{3} (\tfrac{1}{6})^3 (\tfrac{5}{6}) \\
&= \tfrac{20}{1296}
\end{aligned}
$$

$$
\begin{aligned}
P(4 \text{ sixes}) &= \binom{4}{4} (\tfrac{1}{6})^4 (\tfrac{5}{6})^0 \\
&= \tfrac{1}{1296}
\end{aligned}
$$

Example 3. Evaluate $E(X)$ for the binomial distribution.

You will remember from the previous chapter that $E(X)$ is the mean of the distribution and is defined by:

$$E(X) = \sum_{x=0}^{n} Xp(X)$$

For the binomial distribution:

$$P(X) = \binom{n}{X} p^X q^{n-X}$$

and so:

$$E(X) = \sum_{X=0}^{n} X.\binom{n}{X} p^X q^{n-X}$$

$$= \sum_{X=0}^{n} X.\frac{n!}{X!\,(n-X)!} p^X q^{n-X}$$

$$= np \sum_{X=1}^{n} \frac{(n-1)!}{(X-1)!\,(n-X)!} p^{X-1} q^{n-X}$$

The reason for taking out the factor np is that in the X numerator divides into $X!$ in the denominator $(X-1)!$ times and so we now have a term:

$$\sum_{X=1}^{n} \frac{(n-1)!}{(X-1)!\,(n-X)!} p^{X-1} q^{n-X}$$

which is equal to:

$$\sum_{X=1}^{n} \binom{n-1}{X-1} p^{X-1} q^{n-X}$$

which in turn is equal to $(q+p)^{n-1}$. Now $(q+p) = 1$ and so:

$$\sum_{X=1}^{n} \frac{(n-1)!}{(X-1)!\,(n-X)!} p^{X-1} q^{n-X} = 1$$

Therefore, $E(X) = np$.

[Note: Limits of summation have been changed from '$X = 0$ to $X = n$' to '$X = 1$ to $X = n$', so that the power of p is non-negative and to ensure that the denominator has non-negative factorials.]

Example 4. Evaluate $E(X-\bar{X})^2$ for the binomial distribution.

$$E(X-\bar{X})^2 = E(X^2 - 2X.\bar{X} + \bar{X}^2)$$
$$= E(X^2) - 2E(X).\bar{X} + \bar{X}^2$$
$$= E(X^2) - 2[E(X)]^2 + [E(X)]^2 \qquad \text{since } E(X) = \bar{X}$$

Therefore:

$$E(X-\bar{X})^2 = E(X^2) - [E(X)]^2$$

We know $E(X) = np$. Therefore:

$$
\begin{aligned}
E(X^2) &= \sum_{X=0}^{n} X^2 p(X) \\
&= \sum_{X=0}^{n} [X(X-1)+X]p(X) \\
&= \sum_{X=0}^{n} X(X-1)p(X) + \sum_{X=0}^{n} Xp(X) \\
&= \left[\sum_{X=0}^{n} X(X-1)\binom{n}{X} p^X q^{n-X}\right] + np \text{ since } \sum_{X=0}^{n} Xp(X) = E(X) \\
&= \left[\sum_{X=0}^{n} X(X-1)\frac{n!}{X!\,(n-X)!} p^X q^{n-X}\right] + np \\
&= \left[\sum_{X=2}^{n} \frac{n!}{(X-2)!\,(n-X)!} p^X q^{n-X}\right] + np \\
&= \left[n(n-1)p^2 \sum_{X=2}^{n} \frac{(n-2)!}{(X-2)!\,(n-X)!} p^{X-2} q^{n-X}\right] + np \\
&= n(n-1)p^2 + np
\end{aligned}
$$

since:

$$
\sum_{X=2}^{n} \frac{(n-2)!}{(X-2)!(n-X)!} p^{X-2} q^{n-X} = 1
$$

the reasoning as in Example 3. Therefore:

$$
\begin{aligned}
E(X-\bar{X})^2 &= n(n-1)p^2 + np - n^2p^2 \\
&= np - np^2 \\
&= np(1-p) \\
&= npq
\end{aligned}
$$

$E(X-\bar{X})^2$ is known as the variance, and the square root of this is the *standard deviation*. The two results for the binomial distribution [mean $= np$ and standard deviation $= \sqrt{(npq)}$] are very important. We shall see how they can be used in a later section of this chapter.

[Note: Limits of summation have been changed from '$X = 0$ to $X = n$' to '$X = 2$ to $X = n$', to avoid a non-negative power p and to ensure that the denominator has non-negative factorials.]

Standard deviation

Consider the following two sets of numbers:

$$A = \{0, 10, 15, 28, 37\} \qquad B = \{16, 17, 18, 19, 20\}$$

If you calculate the mean for each you will find that in both cases it is 18. However in B the numbers are close to the mean whereas in A they are spread out. There are several ways of measuring this spread, but one of the most useful as far as statistics is concerned is the standard deviation, which is denoted by the Greek letter σ (sigma). We have seen that the variance [*i.e.*, σ^2] is defined by $E(X^2) - [E(X)]^2$, which, in words, is 'the mean of the squares minus the square of the mean'. To apply it here we form Table 4.2.

TABLE 4.2

Set A x	x^2	Set B x	x^2
0	0	16	256
10	100	17	289
15	225	18	324
28	784	19	361
37	1369	20	400
	$\Sigma x^2 = 2478$		$\Sigma x^2 = 1630$

Therefore, for set A $\quad \sigma^2 = \frac{2478}{5} - (18)^2 = \frac{858}{5}$

and, for set B $\quad \sigma^2 = \frac{1630}{5} - (18)^2 = 2$

And so $\quad \sigma_A = \sqrt{\frac{858}{5}} \approx 13{\cdot}10$

and $\quad \sigma_B = \sqrt{2} \approx 1{\cdot}41$

It can be seen from this that the higher the value of σ the bigger the spread of the variable about the mean.

It should be pointed out that there are other ways of calculating σ and there are other ways of measuring the spread of the variable about the mean, but it is sufficient here to understand what standard deviation means.

Example 5. If the probability of a six appearing in the throw of a fair die is $\frac{1}{6}$, find the mean and standard deviation for the distribution of sixes in 720 throws of the die.

Mean $= np = 720(\frac{1}{6}) = 120$

Standard deviation $= \sqrt{(npq)} = \sqrt{720(\frac{1}{6})(\frac{5}{6})} = 10$

Summary

In using the binomial distribution a 'success' is the event looked for, and the probability of this event is p. The probability of a 'failure' or of the event not taking place is q where $q = 1 - p$. Sometimes it is easier to calculate the probability of the event not taking place, and then subtract the result from 1.

Example 6. What is the probability of obtaining 2 or more heads in 5 tosses of a fair coin?

In this case, $p = q = \frac{1}{2}$, and the answer is given by:

$$P(2 \text{ or more heads}) = P(2 \text{ heads}) + P(3 \text{ heads}) + P(4 \text{ heads}) + P(5 \text{ heads})$$

However, it would be easier to use:

$$P(2 \text{ or more heads}) = 1 - [P(0 \text{ heads}) + P(1 \text{ head})]$$

$$P(0 \text{ heads}) = \binom{5}{0} p^0 q^5 = \tfrac{1}{32}$$

$$P(1 \text{ head}) = \binom{5}{1} p q^4 = \tfrac{5}{32}$$

Therefore:

$$P(2 \text{ or more heads}) = 1 - [\tfrac{1}{32} + \tfrac{5}{32}] = \tfrac{26}{32}$$

Example 7. A test paper consists of 8 questions. Each question has 5 answers given; only one of which is correct. What is the probability that a student relying on sheer guesswork will get at least 5 questions right?

Since the student is relying on guesswork his probability of giving the correct answer to any one question is $\frac{1}{5}$. In this example a 'success' is getting the correct answer so $p = \frac{1}{5}$, $q = \frac{4}{5}$ and $n = 8$.

At least 5 right means 5, 6, 7 or 8 right. It is no advantage in this example to work with the complement, since there would be more terms and, as can be seen from the working, terms with $(\frac{4}{5})^4$ $(\frac{4}{5})^5$ $(\frac{4}{5})^6$ $(\frac{4}{5})^7$ and $(\frac{4}{5})^8$ would be involved.

$$P(8 \text{ right}) = \binom{8}{8} (\tfrac{1}{5})^8 (\tfrac{4}{5})^0 = \tfrac{1}{390\,625}$$

$$P(7 \text{ right}) = \binom{8}{7} (\tfrac{1}{5})^7 (\tfrac{4}{5}) = \tfrac{32}{390\,625}$$

$$P(6 \text{ right}) = \binom{8}{6} (\tfrac{1}{5})^6 (\tfrac{4}{5})^2 = \tfrac{448}{390\,625}$$

$$P(5 \text{ right}) = \binom{8}{5}(\tfrac{1}{5})^5(\tfrac{4}{5})^3 = \tfrac{3584}{390\,625}$$

Therefore:

$$P(\text{at least 5 right}) = \tfrac{3584}{390\,625} + \tfrac{448}{390\,625} + \tfrac{32}{390\,625} + \tfrac{1}{390\,625}$$
$$\approx 0{\cdot}01$$

Exercise 4a

1. Using the binomial distribution find the probability of obtaining 4 heads in 6 tosses of a fair coin.
2. An unbiased die is tossed 5 times. What is the probability that:
   (*a*) no sixes appear
   (*b*) no even numbers appear
   (*c*) at most 2 threes appear
   (*d*) at least 1 six appears
3. Obtain the probability distribution for the number of tails appearing in 5 tosses of a fair coin.
4. Repeat question 3, given that the coin is biased so that the probability of a tail is $\frac{2}{3}$.
5. Assuming equal probabilities for boys and girls, out of 10 000 families each with 4 children how many would you expect to have:
   (*a*) 0 boys
   (*b*) 2 boys and 2 girls
   (*c*) at least 1 girl
   (*d*) at most 1 girl
6. In a particular experiment a card is drawn at random from a well shuffled pack of 52 playing cards and note is taken of whether it is a heart or not. The card is replaced, the pack shuffled and the experiment repeated. The experiment is done six times. Determine the probability distribution for the number of hearts drawn.
7. A football team playing at home has a probability of $\frac{3}{4}$ of winning the match. Out of the next 5 home matches what is the probability, assuming there will be no matches drawn, that:
   (*a*) they will win them all
   (*b*) they will win 3
   (*c*) they will lose at least one
   (*d*) they will either win them all or lose them all

4.2 Normal distribution

Up to the moment we have only concerned ourselves with what are known as discrete values of the variable, that is, between certain limits the variable takes integral values only. If the variable can take any value between certain limits it is said to be continuous. For example the heights or weights of people are continuous. Saying that someone is 1 m 77 cm tall can be taken as saying that that person's height lies between 1 m 75 cm and 1 m 80 cm. This is because measurement of this kind is not exact and is only as accurate as the instrument used allows. If the heights of a number of people are recorded and a histogram drawn, showing the distribution of these heights, the lines joining the mid-points of the top of the rectangles have a distinctive bell shape (Fig. 4.3).

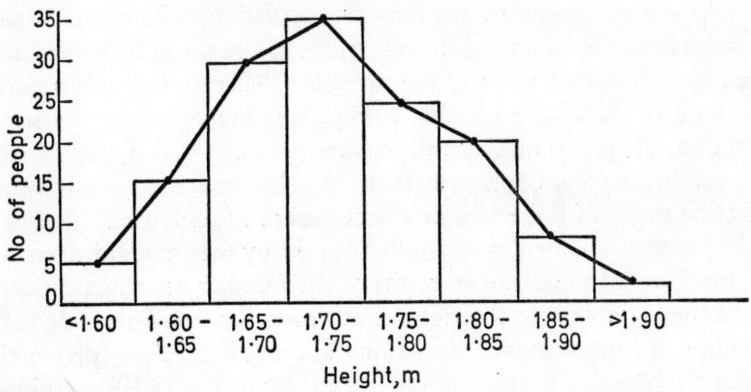

FIGURE 4.3

As the number in the group increases, and the group width is decreased, the lines forming the mid-points of the rectangles approximate to a bell-shaped curve until eventually the shape appears as in

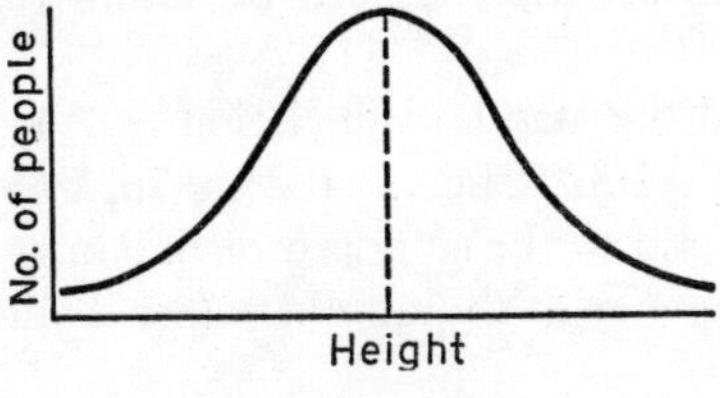

FIGURE 4.4

Fig. 4.4. This curve is known as the *normal curve* and the distribution as the *normal distribution.* Note that the curve is symmetrical about a value which is, in fact, the mean of the distribution. This curve is very common in statistics and its equation is:

$$y = \frac{1}{\sigma\sqrt{2\pi}} \exp\left[-\tfrac{1}{2}\frac{(X-\mu)^2}{\sigma^2}\right] \tag{4.2}$$

where μ is the mean and σ the standard deviation. [Note that in complicated exponential expressions, e^x is often written exp $[x]$: both versions have the same meaning. The derivation of the equation is beyond the scope of this book, but forms part of an undergraduate course in mathematics.] It is easy to see that different values of μ and σ will produce different graphs. It would be particularly useful if we could have one graph to use for our calculations. We can eliminate the variability caused by different values of the mean by subtracting the mean from each value of the variable. This results in the mean of the new values of the variable being zero. For example, the set A (Section 4.1, p. 73) now has the values $\{-18, -8, -3, 10, 19\}$ and the set B the values $\{-2, -1, 0, 1, 2\}$. The reader can easily verify that the mean of these two new sets is zero in each case.

We now eliminate the variability caused by the standard deviation by marking off the x-axis in terms of the number of standard deviations the variable is to the right or left of the mean. This is done by dividing by the standard deviation. For example, if we divide the different values of $\{-18, -8, -3, 10, 19\}$ by σ_A $(=13{\cdot}10)$, we obtain the set $\{-1{\cdot}37, -0{\cdot}61, -0{\cdot}23, 0{\cdot}76, 1{\cdot}45\}$. Similarly, if we divide the elements of $\{-2, -1, 0, 1, 2\}$ by σ_B $(=1{\cdot}41)$, we obtain the set $\{-1{\cdot}42, -0{\cdot}71, 0{\cdot}00, 0{\cdot}71, 1{\cdot}42\}$. As you can see, these sets are now almost the same.

Note that if a constant is added or subtracted to each and every member of a given set the standard deviation remains unchanged.

Let us now find the standard deviation of:

$$\{-1{\cdot}37, -0{\cdot}61, -0{\cdot}23, 0{\cdot}76, 1{\cdot}45\}$$

The mean is 0, and so the standard deviation is the square root of the mean of the squares of the numbers (see Table 4.3).

TABLE 4·3

X	X^2
−1·37	1·8769
−0·61	0·3721
−0·23	0·0529
0·76	0·5776
1·45	2·1025
	$\Sigma X^2 = 4{\cdot}9820$

$$\sigma = \sqrt{\frac{4{\cdot}9820}{5}} = \sqrt{0{\cdot}9964}$$
$$= 1{\cdot}00 \text{ (to 2 decimal places)}$$

Verify that the standard deviation of {−1·42, −0·71, 0·00, 0·71, 1·42} is also 1·00.

We now have two sets of numbers whose mean in each case is 0 and whose standard deviation in each case is 1.

This procedure of subtracting the mean from each value of the variable and dividing by the standard deviation gives what is known as the *standard score* or *Z score* and can be written more formally as:

$$Z = \frac{X - \mu}{\sigma} \tag{4.3}$$

The graph of equation (4.2) extends from $-\infty$ to $+\infty$ and the area under the curve is given by:

$$A = \int_{-}^{\infty} \frac{1}{\sigma\sqrt{(2\pi)}} \exp\left[-\tfrac{1}{2}\frac{(X-\mu)^2}{\sigma^2}\right] dX$$

We can evaluate this by using equation (4.3) as a substitution:

Since $$Z = \frac{X - \mu}{\sigma}$$

we have $$dZ = \frac{1}{\sigma}\, dX$$

The limits remain the same, and so:

$$A = \int_{-\infty}^{\infty} \frac{1}{\sqrt{(2\pi)}} \exp\,[-\tfrac{1}{2}Z^2]dZ$$
$$= \frac{1}{\sqrt{(2\pi)}} \int_{-\infty}^{\infty} \exp\,[-\tfrac{1}{2}Z^2]dZ$$

But $\int_{-\infty}^{\infty} \exp\left[-\tfrac{1}{2}Z^2\right]dZ = \sqrt{(2\pi)}$

Hence: $A = 1$

The graph of $y = \dfrac{1}{\sqrt{(2\pi)}} \exp\left[-\tfrac{1}{2}Z^2\right]$ is shown in Fig. 4.5. It is symmetrical about the line $Z = 0$. The mean is 0 and the standard deviation is 1.

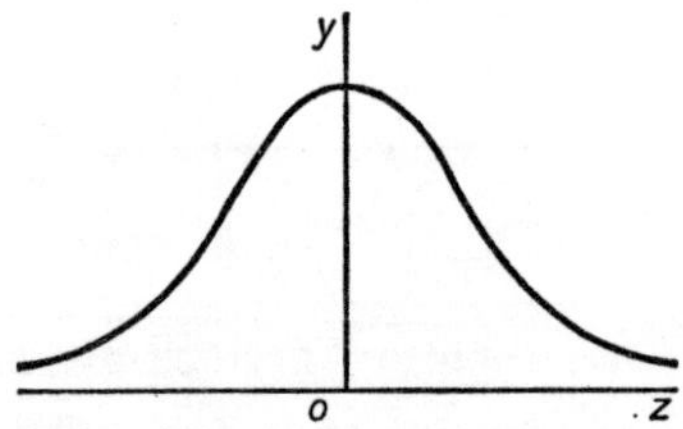

FIGURE 4.5

There are tables which give the values of the areas under the normal curves between given values of Z. It is usual practice to give the area from $Z = 0$ to some particular value of Z, as, for example, the shaded area in Fig. 4.6. Remember that, since the curve is symmetrical about the mean which is now zero, the value from $Z = 0$ to a positive Z is the same as from $Z = 0$ to a corresponding negative Z.

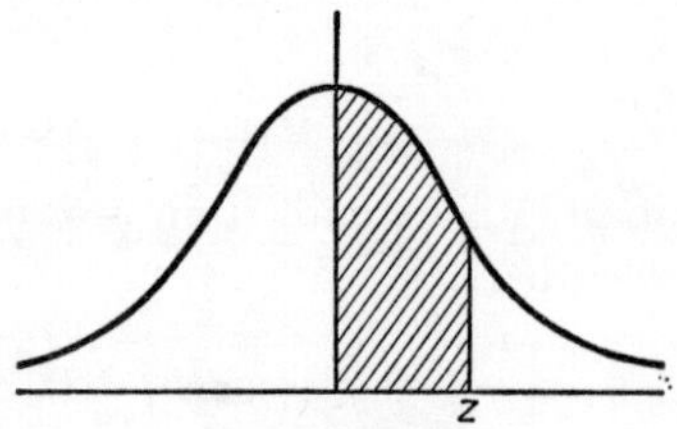

FIGURE 4.6

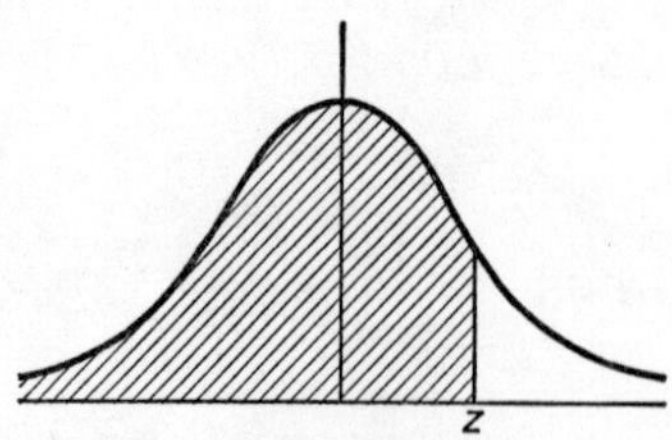

FIGURE 4.7

There are some tables which give the area from $-\infty$ to Z, such as that shown in Fig. 4.7. There are tables also giving the value of the area to the right of the value of Z (see Fig. 4.8).

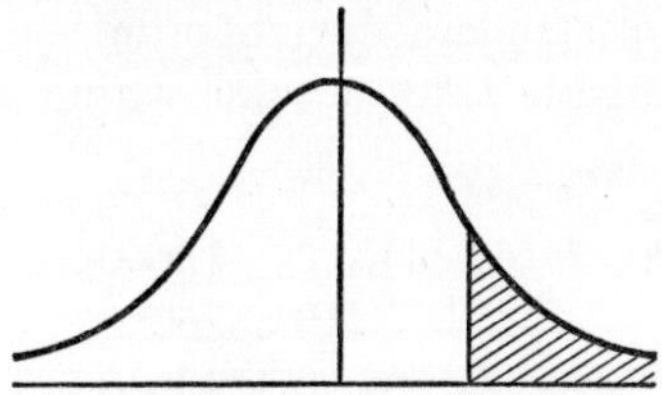

FIGURE 4.8

You must be sure which type of table you are using. If the value at $Z = 0$ is 0, then it is the first type, *i.e.*, the one given in the Appendix on pp. 124–125. If the value at $Z = 0$ is 0·5, then it is either the second or third type; the third type is easily detected, since the values decrease as Z increases.

Example 8. Using the table given on pp. 124–125, we can find the following areas under the normal curve:

(*a*) From $Z = 0$ to $Z = 2{\cdot}5$: area $= 0{\cdot}4938$
(*b*) From $Z = 0$ to $Z = 1{\cdot}54$: area $= 0{\cdot}4382$
(*c*) From $Z = -1{\cdot}05$ to $Z = 0$: this is the same as the area from $Z = 0$ to $Z = 1{\cdot}05$, *i.e.*, area $= 0{\cdot}3531$
(*d*) From $Z = -0{\cdot}57$ to $Z = 1{\cdot}21$: splitting this into two parts, from $Z = -0{\cdot}57$ to $Z = 0$, area $= 0{\cdot}2157$, and from $Z = 0$ to $Z = 1{\cdot}21$, area $= 0{\cdot}3869$. Therefore, required area $= 0{\cdot}6016$
(*e*) $Z > 1{\cdot}35$: from $Z = 0$ to $Z = 1{\cdot}35$, area $= 0{\cdot}4115$; therefore for $Z > 1{\cdot}35$, area $= 0{\cdot}5000 - 0{\cdot}4115 = 0{\cdot}0885$
(*f*) $Z < -2{\cdot}01$: from $Z = -2{\cdot}01$ to $Z = 0$, area $= 0{\cdot}4778$; therefore for $Z < -2{\cdot}01$, area $= 0{\cdot}5000 - 0{\cdot}4778 = 0{\cdot}0222$
(*g*) $Z < 0{\cdot}94$: from $Z = 0$ to $Z = 0{\cdot}94$, area $= 0{\cdot}3264$; therefore required area $= 0{\cdot}5000 + 0{\cdot}3264 = 0{\cdot}8264$
(*h*) $Z > -1{\cdot}23$: from $Z = -1{\cdot}23$ to $Z = 0$, area $= 0{\cdot}3907$; therefore required area $= 0{\cdot}5000 + 0{\cdot}3907 = 0{\cdot}8907$
(*i*) From $Z = 1{\cdot}73$ to $Z = 2{\cdot}01$: from $Z = 0$ to $Z = 2{\cdot}01$, we know area $= 0{\cdot}4778$ and from $Z = 0$ to $Z = 1{\cdot}73$, area $= 0{\cdot}4582$; therefore required area $= 0{\cdot}4778 - 0{\cdot}4582 = 0{\cdot}0196$

Since the area under the normal distribution curve is 1 and Z must lie between $-\infty$ and $+\infty$, this area can be used to represent the probability that a variable lies between particular values of Z. The probability that Z lies between 0 and 2·5 is $P(0<Z<2{\cdot}5)=0{\cdot}4938$. Since Z is in units of standard deviation, let us see what the probability is that a variable takes a value within 1, 2 or 3 standard deviations of the mean:

$$\begin{aligned} P(-1<Z<1) &= 0{\cdot}3413 + 0{\cdot}3413 \\ &= 0{\cdot}6826 \\ P(-2<Z<2) &= 0{\cdot}4772 + 0{\cdot}4772 \\ &= 0{\cdot}9544 \\ P(-3<Z<3) &= 0{\cdot}4987 + 0{\cdot}4987 \\ &= 0{\cdot}9974 \end{aligned}$$

Hence about 68·3%, 95·4% and 99·7% of values lie within 1, 2 and 3 standard deviations of the mean, respectively, where the distribution is normal. This is a very important result in that branch of statistics known as inferential statistics, which concerns decision-making based on the result of samples. For example, the fact that 99·7% of values lie within 3 standard deviations from the mean, would make you suspicious of any results beyond 3 standard deviations; suspicious, that is, that they did not belong to the same population for which the graph was drawn.

Example 9. One thousand students sit an examination the total marks for which are 100. When the results are known they are found to be normally distributed with a mean of 55 and a standard deviation of 15. How many students would you expect to pass if the pass mark is 45? [Note that, since we are dealing with what is in effect a continuous distribution, the probability of a student passing is the probability of his scoring more than 44·5. The choice of score we use in the calculation depends on the question. Thus in this example, since a pupil passes if he scores 45, we are concerned with pupils scoring 44·5 or more. If, however, a pass was awarded to those scoring more than 45, we would use the value 45·5 in the calculation.]

We now find the Z score corresponding to the score of 44·5:

$$Z=\frac{4{\cdot}45-55}{15}=-0{\cdot}7$$

And so $P(Z > -0{\cdot}7) = 0{\cdot}5000 + 0{\cdot}2580$
$= 0{\cdot}7580$

That is, 75·8% of the students would be expected to pass. Therefore, out of 1000 students you would expect 758 to pass.

Example 10. In this examination it was decided that the top 10% should be awarded a distinction. What is the least score a student would need to obtain such an award?

Let X_1 be the least score. We now want the score corresponding to $X_1 - 0{\cdot}5$:

$$Z = \frac{(X_1 - 0{\cdot}5) - 55}{15}$$

We now have to find the value of Z such that 10% of the area lies to the right of this value. That is, we want the value of Z from our table which gives an area of $0{\cdot}5000 - 0{\cdot}1000 = 0{\cdot}4000$. The nearest value to this is $Z = 1{\cdot}28$. Therefore:

$$\frac{X_1 - 55{\cdot}5}{15} > 1{\cdot}28$$

that is $X_1 > 74{\cdot}7$

So a student would have to score at least 75 to be sure of being in the top 10%.

Example 11. What is the probability of selecting a student at random and finding he has scored 90 or more, given that he has gained a distinction?

First we find the Z scores corresponding to 89·5 and 74·5:

$$Z_{89{\cdot}5} = \frac{89{\cdot}5 - 55}{15} = 2{\cdot}3 \qquad Z_{74{\cdot}5} = \frac{74{\cdot}5 - 55}{15} = 1{\cdot}3$$

Referring back to our work on conditional probability in Section 2.4 the probability of Z being greater than 2·3, given that it is greater than 1·3, can be written:

$$P(Z > 2{\cdot}3 \mid Z > 1{\cdot}3) = \frac{P(Z > 2{\cdot}3)}{P(Z > 1{\cdot}3)} = \frac{0{\cdot}0107}{0{\cdot}0968} \approx 0{\cdot}11$$

Example 12. How many students would be expected to score 60 in this same test?

Previously the questions have been of a type where you were asked the probability that the variable would fall within (or outside) certain limits. This question is no exception. We wish to find the probability that the variable will lie between 59·5 and 60·5:

$$Z_{60\cdot5} = \frac{60\cdot5 - 55}{15} = 0\cdot37 \qquad Z_{59\cdot5} = \frac{59\cdot5 - 55}{15} = 0\cdot30$$

$$P(Z<0\cdot37) = 0\cdot6443 \qquad P(Z<0\cdot30) = 0\cdot6179$$

Therefore:

$$P(0\cdot30<Z<0\cdot37) = 0\cdot6443 - 0\cdot6179 = 0\cdot0264$$

And so, out of 1000 students, we should expect 26 to score 60.

Exercise 4b

1. Find the area under the normal curve in each of the following cases:
   (*a*) $Z = 0$ to $Z = 1\cdot31$
   (*b*) $Z = 0$ to $Z = 2\cdot51$
   (*c*) $Z = -0\cdot9$ to $Z = 0$
   (*d*) $Z = -1\cdot76$ to $Z = 0$
   (*e*) $Z = -1\cdot5$ to $Z = 1\cdot5$
   (*f*) $Z = -2\cdot01$ to $Z = 1\cdot97$
   (*g*) $Z>1\cdot11$
   (*h*) $Z< -1\cdot22$
   (*i*) $Z> -0\cdot34$
   (*j*) $Z<2\cdot0$
2. In a test the mean of the marks scored was 60 and the standard deviation 10. Convert the following raw scores into Z scores:
   (*a*) 80 (*b*) 90 (*c*) 47 (*d*) 62 (*e*) 35
3. What is the probability that in the test in question 2 a candidate scored:
   (*a*) more than 80
   (*b*) less than 30
   (*c*) between 50 and 70 (both scores inclusive)
   (*d*) exactly 75
   (*e*) more than 70 given that he has scored over 65
   The top 5% are to move into another form. What is the least mark that will put a student into this category?
4. The mean weight of 1000 men is 75 kg and the standard devia-

tion is 4 kg. Assuming the weights to be normally distributed, how many men would you expect to weigh:
(*a*) 80 kg and over
(*b*) 60 kg and under
(*c*) between 65 kg and 85 kg inclusive

5. If the heights of students are normally distributed about a mean of 1 m 75 cm and with a standard deviation of 7·5 cm, what percentage of students are:
(*a*) greater than 1 m 85 cm (*b*) less than 1 m 72 cm

4.3 Standard scores

We have discussed standard scores known as *Z*-scores. These are not the only type of standard score. *Z*-scores often result in negative numbers. A way to avoid this is to use what are known as *T*-scores. We first multiply the *Z* score by 10 and then add 50. This means we now have a set of scores whose mean is 50 and standard deviation 10. Since 99·7% of all results, which are normally distributed, will fall within 3 standard deviations of the mean then practically all the results, written as *T*-scores, will fall within the range 20 to 80. The formula for computing the *T*-score is:

$$\frac{(X-\mu)10}{\sigma}+50$$

For the set $A = \{0, 10, 15, 28, 37\}$ the *T*-scores would be:

$$\{36{\cdot}3,\ 43{\cdot}9,\ 47{\cdot}7,\ 57{\cdot}6,\ 64{\cdot}5\}$$

The mean of this set is $\frac{250}{5} = 50$ as expected. The standard deviation is given by:

$$\begin{aligned}\sigma &= \sqrt{\{E(X^2)-[E(X)]^2\}}\\ &= \sqrt{\left[\frac{12998{\cdot}20}{5}-(50)^2\right]}\\ &= \sqrt{99{\cdot}64}\\ &= 9{\cdot}982 = 10{\cdot}0 \text{ (to 1 decimal place)}\end{aligned}$$

T-scores for set $B = \{16, 17, 18, 19, 20\}$ are $\{35{\cdot}8, 42{\cdot}9, 50, 57{\cdot}1, 64{\cdot}2\}$
Verify that the mean is 50 and the standard deviation 10.

In fact any scores can be standardized to a particular mean and standard deviation. For example another form of standard score is one used in intelligence testing, where the mean is taken to be 100 and the standard deviation 15. This 'IQ score' is found by computing $15Z + 100$ where, as before, $Z = \frac{X - \mu}{\sigma}$, μ is the mean of the results and σ is the standard deviation of these results.

One use of standard scores is to compare marks in different subjects. For example suppose a student obtained the marks shown below in the named subjects:

TABLE 4.4

Mathematics	Physics	Chemistry	English	French
$X = 80$	$X = 75$	$X = 60$	$X = 45$	$X = 55$
$\mu = 60$	$\mu = 49$	$\mu = 55$	$\mu = 51$	$\mu = 55$
$\sigma = 8$	$\sigma = 10$	$\sigma = 5$	$\sigma = 12$	$\sigma = 7$
$Z = 2{\cdot}5$	$Z = 2{\cdot}6$	$Z = 1{\cdot}0$	$Z = -0{\cdot}5$	$Z = 0{\cdot}0$
$T = 75$	$T = 76$	$T = 60$	$T = 45$	$T = 50$

The mean and standard deviation for each subject are given and the Z-score and T-score computed for each. As can be seen from the Z-scores he has done relatively better in physics than in mathematics, since in physics he obtained a mark 2·6 standard deviations above the mean, whereas in mathematics his mark of 80 only corresponded to 2·5 standard deviations above the mean.

4.4 Relation between binomial and normal distribution

As n increases in value it becomes more and more cumbersome to evaluate the binomial coefficients. But if we proceed with the calculations and draw graphs for each value of n for a given value of p we find that on joining the mid-points of the tops of the rectangles we have the characteristic shape of the normal curve. Figure 4.9 shows the values of $\binom{25}{r} p^r q^{n-r}$ for $p = q = \frac{1}{2}$ and $r = 0, 1, 2 \ldots 25$. So for large values of n ($\geqslant 10$) we can say that the binomial distribution approximates very closely to the normal distribution provided that neither p nor q is close to zero.

The mean is np and the standard deviation is $\sqrt{(npq)}$.

And so the Z-score is calculated from $\frac{X - np}{\sqrt{(npq)}}$.

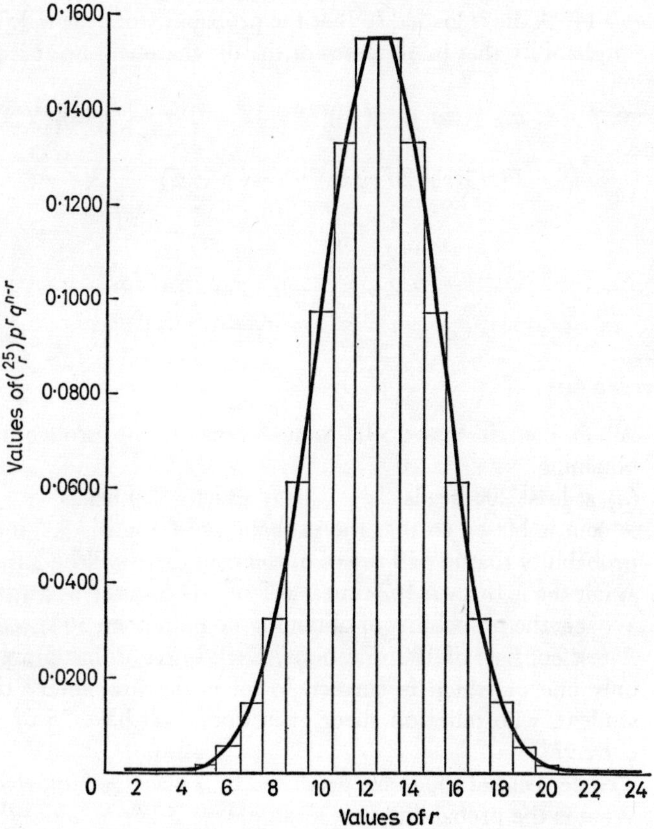

FIGURE 4.9

Example 13. Find the probability of obtaining between 45 and 55 heads (inclusive) in 100 tosses of a fair coin.

Since the coin is fair $p = q = \frac{1}{2}$, therefore $\mu = (100)(\frac{1}{2}) = 50$ and $\sigma = \surd[(100)(\frac{1}{2})(\frac{1}{2})] = 5$. Therefore:

$$Z_{55\cdot5} = \frac{55\cdot5 - 50}{5} = 1\cdot1 \qquad Z_{44\cdot5} = \frac{44\cdot5 - 50}{5} = -1\cdot1$$

and so:

$$P(-1\cdot1 < Z < 1\cdot1) = 0\cdot3643 + 0\cdot3643 = 0\cdot7286$$

Example 14. A die is loaded so that the probability of a six is $\frac{1}{4}$. What is the probability that in 48 tosses of the die there will be at least 20 sixes?

Here, $p = \frac{1}{4}$, $q = \frac{3}{4}$, so $\mu = (48)(\frac{1}{4}) = 12$ and $\sigma = \sqrt{[(48)(\frac{1}{4})(\frac{3}{4})]} = 3$. Thus:

$$P(\text{at least 20 sixes}) = P(X > 19{\cdot}5)$$

But
$$Z_{19\cdot5} = \frac{19{\cdot}5 - 12}{3} = 2{\cdot}5$$

$$\therefore P(Z > 2{\cdot}5) = 0{\cdot}5000 - 0{\cdot}4938 = 0{\cdot}0062$$

Exercise 4c

1. A fair coin is tossed 400 times. What is the probability of obtaining:
   (*a*) at least 200 heads (*b*) exactly 200 heads
2. A coin is biased so that the probability of a tail is $\frac{2}{3}$. Find the probability that in 450 tosses of the coin there will be 325 tails.
3. A fair die is thrown 162 times. A 5 or 6 is counted as a success. What is the probability of obtaining no more than 50 successes?
4. A test consists of 100 questions. Each question has 5 answers only one of which is correct. What is the probability that a student, who relies on sheer guesswork, will have 25 or more correct?
5. Ten per cent of the items produced by a machine are defective. What is the probability that a sample of 25 has no more than 1 defective item?

4.5 Poisson distribution

In the preceding section mention was made that provided neither p nor q is close to zero the normal curve is a good approximation to the binomial distribution. When an event is rare, that is, p is small, then provided the number n is large enough for np to remain finite, the binomial distribution approximates to another distribution known as the Poisson distribution after the man who discovered it.

For the binomial distribution, the probability of r successes is given by the coefficient of t^r in the expansion of $(q + pt)^n$. We can write:

$$(q+pt)^n = (1-p+pt)^n$$
$$= [1+p(t-1)]^n$$

Since p is now very small and n large enough for np to be a relatively small number, we can substitute $np = m$. We then have:

$$(q+pt)^n = \left[1+\frac{m(t-1)}{n}\right]^n$$

We now let $n \to \infty$ giving:

$$\lim_{n\to\infty} (q+pt)^n = \lim_{n\to\infty} \left[1+\frac{m(t-1)}{n}\right]^n$$
$$= e^{m(t-1)}$$

since $\lim_{n\to\infty} \left(1+\frac{x}{n}\right)^n = e^x$

Now $\qquad e^{m(t-1)} = e^{-m}.e^{mt}$

$$= e^{-m}\left[1+mt+\frac{m^2t^2}{2!}+\ldots+\frac{m^rt^r}{r!}+\ldots\right]$$

(Full details of the mathematics of this derivation are given in 'Relations and Functions' in this series.)

The probability of r successes is the coefficient of t^r, *i.e.*,

$$P(r \text{ successes}) = \frac{e^{-m}m^r}{r!}$$

The most famous example involving the Poisson distribution is its application by Quetelet and von Bortkiewicz to the number of deaths over 20 years in 10 Prussian Army Corps from kicks by a horse.

TABLE 4.5

Actual deaths per Corps	Observed frequency
0	109
1	65
2	22
3	3
4 and over	1
Total	200

The results they collected were as shown in Table 4.5. The mean of the observed frequency is given by:

$$\tfrac{1}{200}[(0\times 109)+(1\times 65)+(2\times 22)+(3\times 3)+(4\times 1)]$$
$$=\tfrac{122}{200}=0{\cdot}61$$

Therefore $\quad m = np = 0{\cdot}61$

Now $\quad e^{-0\cdot 61} = (2{\cdot}718)^{-0\cdot 61}$

$$=\frac{1}{(2{\cdot}718)^{0\cdot 61}}$$
$$=0{\cdot}5434$$

This gives us:

$$P(0 \text{ deaths}) = 0{\cdot}5434$$
$$P(1 \text{ death}) = (0{\cdot}5434)(0{\cdot}61) = 0{\cdot}3315$$
$$P(2 \text{ deaths}) = \frac{(0{\cdot}5434)(0{\cdot}61)^2}{2!} = 0{\cdot}1011$$
$$P(3 \text{ deaths}) = \frac{(0{\cdot}5434)(0{\cdot}61)^3}{3!} = 0{\cdot}0206$$

Therefore:

$$P(4 \text{ deaths and over}) = 1 - [0{\cdot}5434 + 0{\cdot}3315 + 0{\cdot}1011 + 0{\cdot}0206]$$
$$= 0{\cdot}0034$$

We now add to our table a column of calculated frequencies (each of the above multiplied by 200), as shown in Table 4.6.

TABLE 4.6

Actual deaths per Corps	Observed frequency	Calculated frequency
0	109	108·7
1	65	66·3
2	22	20·2
3	3	4·1
4 and over	1	0·7
Total	200	200·0

As you can see the agreement is very good. I should, however, point out that we have assumed that the value of p did not change over 20 years! Remember that the Poisson distribution is an approximation, and that you are assuming that the probability remains unchanged. Neither of these points, of course, affects our work in this book.

Example 15. In any given year in a certain colliery the probability of a man being killed is 0·002. What is the probability that this colliery, which employs 1000 men will have:

(*a*) a year free of fatal accidents

(*b*) no more than 2 fatalities in a given year

(*a*) $m = np = (1000)(0{\cdot}002) = 2$ and $e^{-2} = 0{\cdot}1353$. So

$$P(0 \text{ deaths}) = 0{\cdot}1353$$

(*b*) $P(\text{no more than 2 deaths})$

$$= P(0 \text{ deaths}) + P(1 \text{ death}) + P(2 \text{ deaths})$$
$$= e^{-2} + e^{-2}.(2) + \frac{e^{-2}(2)^2}{2!}$$
$$= (0{\cdot}1353)(1 + 2 + 2)$$
$$= 0{\cdot}6765$$

This second part is merely an academic exercise. Deaths in colliery accidents are usually due to explosions or cave-ins, and though the probability of a person being killed may only be 0·002 when an accident does happen it is more than likely that the number killed would be in excess of 2.

Example 16. Fit a Poisson distribution to the following data:

X	0	1	2	3	4	5	6 and over
Observed frequency	200	325	257	138	55	20	5

It is useful if we can fit a distribution to data since we can then forecast the probability for other values of the variable.

In this example the sum of the frequencies is 1000, and therefore the mean of the distribution is:

$$\tfrac{1}{1000}[(0 \times 200) + (1 \times 325) + (2 \times 257) + (3 \times 138) + (4 \times 55)$$
$$+ (5 \times 20) + (6 \times 5)]$$
$$= 1{\cdot}60$$

Therefore:

$$m = 1{\cdot}60 \text{ and } e^{-1{\cdot}60} = 0{\cdot}2019$$

and so:

$$P(X = 0) = e^{-1{\cdot}60} = 0{\cdot}2019$$
$$P(X = 1) = (0{\cdot}2019)(1{\cdot}60) = 0{\cdot}3230$$
$$P(X = 2) = \frac{(0{\cdot}2019)(1{\cdot}60)^2}{2!} = 0{\cdot}2584$$
$$P(X = 3) = \frac{(0{\cdot}2019)(1{\cdot}60)^3}{3!} = 0{\cdot}1378$$
$$P(X = 4) = \frac{(0{\cdot}2019)(1{\cdot}60)^4}{4!} = 0{\cdot}0551$$
$$P(X = 5) = \frac{(0{\cdot}2019)(1{\cdot}60)^5}{5!} = 0{\cdot}0176$$
$$P(X = 6 \text{ and over}) = 1 - [0{\cdot}2019 + 0{\cdot}3230 + 0{\cdot}2584 + 0{\cdot}1378 + 0{\cdot}0551 + 0{\cdot}0176] = 0{\cdot}0062$$

We now multiply each of these probabilities by 1000:

X	0	1	2	3	4	5	6 and over
Calculated frequency	201·9	323·0	258·4	137·8	55·1	17·6	6·2

As you can see the agreement between calculated and observed frequencies is good.

Exercise 4d

1. Given that $P(r \text{ successes})$ is given by:

$$\frac{e^{-4{\cdot}6}(4{\cdot}6)^r}{r!}$$

find $P(0)$, $P(1)$, $P(2)$.

2. The probability of a defective bolt being made by a particular machine is 0·002. What is the probability that out of 1000 bolts inspected at random there will be:

(*a*) none defective
(*b*) 1 defective
(*c*) no more than 2 defective

3. It has been found that 5% of the electric light bulbs manufactured by a certain company are useless. Assuming a Poisson distribution, what is the probability that in a sample of 100 there are no more than 5 useless bulbs? Compare your result with that obtained by using a normal distribution.
4. Fit a Poisson distribution to the following data:

X	0	1	2	3	4 and over
Observed frequency	63	30	6	1	0

5. In a factory the probability of an employee having a particular accident in a given week is 0·0015. The factory employs 2000 workers. What is the probability that the factory will have:
(*a*) a week free of this type of accident
(*b*) no more than 2 accidents of this type in a given week

4.6 Relation between Poisson and normal distributions

It can be shown that the variance of the Poisson distribution is np and hence the standard deviation is $\sqrt{(np)}$.

If we consider p to be very small then q is very nearly equal to one. For the binomial distribution the standard deviation is $\sqrt{(npq)}$ and so in this limiting form $\sqrt{(npq)} \approx \sqrt{(np)}$.

As n increases indefinitely we can use the normal distribution with $Z = \dfrac{X - m}{\sqrt{m}}$, where $m = np$.

Example 17. Between the hours of 10.00 am and 12 noon, the average number of calls going through a particular telephone exchange is 100/min. If the maximum number of calls which can be handled by this exchange is 125, and if the calls are Poisson distributed, find the probability that on a given morning between these hours the switchboard will be jammed.

In this example $m = 100$, and we can approximate using the normal distribution with $Z = \dfrac{X - 100}{\sqrt{100}}$. In this case X is 125·5 and so:

$$Z = \frac{125{\cdot}5 - 100}{10} = 2{\cdot}55$$

$$\begin{aligned} P(Z > 2{\cdot}55) &= 0{\cdot}5000 - 0{\cdot}4946 \\ &= 0{\cdot}0054 \end{aligned}$$

Exercise 4e

1. Suppose the probability of a person suffering from a particular disease is 0·001. What is the probability that in a population of 225 000 there are less than 185 people suffering from this disease?
2. Assuming that the number of calls a switchboard operator receives between 2.30 pm and 2.40 pm follows a Poisson distribution with $\mu = 9$, and that the switchboard can handle 15 calls at the most, find the probability that in this time interval on a given day the switchboard will be overloaded.
3. A car-hire firm finds that the hiring of cars during a week follows a Poisson distribution with mean 100. How many cars should they have so as to be able with 99% probability to satisfy a demand?

Exercises on Chapter 4

1. The probability that a student in a particular university will complete his course satisfactorily is 0·6. What is the probability that out of a group of 6 students, chosen at random, at least one will graduate?
2. The probability that a marksman will hit a target is 0·4. What is the probability of his hitting the target 80, or more, times in 150 shots?
3. In intelligence testing of school children the mean is considered to be 100 and the standard deviation 15. Assuming normal distribution, what is the probability that a child will have an IQ:
   (*a*) greater than 140
   (*b*) less than 90
   (*c*) between 60 and 130 inclusive

4. The lengths of metal rods are normally distributed with mean 13·20 cm and standard deviation 0·025 cm. For a particular job the maximum tolerance allowable is between 13·14 cm and 13·26 cm inclusive. Calculate the percentage of rods acceptable for this job.
5. A serum has been developed and it has been found that 0·1% of the population suffer a bad reaction from it. In a sample of 1000 people what is the probability that not more than 2 of them will suffer a bad reaction?
6. Two unbiased dice are tossed 100 times. What is the probability that when the two numbers shown are added together there will be exactly 50 odd numbers?
7. A coin is weighted so that tails appear, on the average, 2 out of every 3 tosses. The coin is tossed 4 times. Determine the probability distribution for the number of tails appearing.
8. Fit a binomial distribution to the following data:

X	0	1	2	3	4
Observed frequency	2401	4121	2640	753	85

9. Telephone calls coming into a switchboard average 3/min. Assuming they follow a Poisson distribution find the probability that in a particular minute there are:
   (*a*) no calls
   (*b*) 1 call
   (*c*) 2 calls
   (*d*) 3 calls
   (*e*) more than 3 calls
10. Fit a binomial distribution to the following data:

X	0	1	2	3
Observed frequency	214	436	286	64

11. Fit a Poisson distribution to the following data:

X	0	1	2	3	4
Observed frequency	55	33	9	3	0

12. A card is drawn from a well shuffled pack of 52 playing cards. The suit is noted and the card returned to the pack, which is again shuffled. The process is continued until the total number of draws is 48. What is the probability that:
(*a*) 13 or more hearts are drawn
(*b*) 7 or less hearts are drawn
13. The life-hours of electric light bulbs of a certain type are normally distributed with mean 200 h and standard deviation 30 h. What is the probability that:
(*a*) a bulb will last for 250 h or more
(*b*) if the bulb has been burning for 235 h it will continue to burn for another 30 h
14. Rubber washers made by a certain machine have a mean inside diameter of 0·450 cm with a standard deviation of 0·005 cm. Any of these washers which have diameters less than, or equal to, 0·444 cm and greater than, or equal to, 0·456 cm are rejected. If these measurements are normally distributed what is the probability of a particular washer being rejected? If the machine makes 10 000 per day, how many would you expect to be rejected?
15. A firm, which specializes in a delivery service, has found from past records that in any one month the probability of being more than 2 days late with a delivery is 0·001. During a given month they have to deliver 2000 items. What is the probability that:
(*a*) none of these will arrive more than 2 days late
(*b*) only in one case will the delivery be more than 2 days late
(*c*) in no more than 3 cases will deliveries be over the guaranteed time
16. For every 10 000 vehicles on the road it has been estimated that in a particular area there are 4·5 accidents per day. What is the probability that a firm owning 100 vehicles will:
(*a*) have no accidents on a given day
(*b*) have no more than 1 accident on a given day
Assume a Poisson distribution.
17. Letting $f(r)$ be the function describing the probability of r successes in a binomial distribution, we can write $f(r) = \binom{n}{r} p^r q^{n-r}$.

Show that:

$$f(r+1) = \frac{n-r}{r+1} \cdot \frac{p}{q} f(r)$$

[This formula is useful when n is relatively large. $f(0)$ can first be calculated and then the other values can be obtained by relatively simple calculation.]

18. Given that $f(r) = \binom{12}{r}(\frac{2}{3})^r(\frac{1}{3})^{12-r}$ and $(\frac{1}{3})^{12} \approx 0{\cdot}000\ 002$, use the formula in question 17 to find the probability distribution for the various values of r.
19. A box contains 10 coloured marbles of which 4 are red. A marble is taken out, the colour noted, and then replaced. This procedure is carried out 8 times altogether. What is the probability that at most 3 of the red marbles are drawn out of the box?
20. Suppose that 4% of the items produced by a machine are considered to be defective. Samples of 24 are taken at regular intervals for the purposes of inspection. If there are 2 or more defective items in a sample, all items made since the preceding inspection are scrapped. What is the probability that this will happen in a given inspection?
21. If the probability that a six will appear on the throw of a biased die is $\frac{1}{4}$, find the mean and standard deviation for the distribution of sixes in 108 throws of this die.
22. In an examination the mean was found to be 52 and the standard deviation 12. The top 10% are to receive scholarships. If the examination was marked out of 100 and the marks were normally distributed, what is the least mark that a candidate needs to win a scholarship?
23. Suppose a particular machine produces items of which 10% are defective. How many items should be taken in a sample to ensure that the probability of obtaining no more than one defective item is less than 0·9?
24. A class is given a test which consists of 6 questions. Each question has 4 answers given of which only one is correct. To pass the test a student must get 4 questions correct. A particular student knows the answers to 3 of the questions and guesses the answers for the remainder. What is the probability that he will pass the test?

25. The probability that a certain football team will win a match is $\frac{2}{3}$ the probability it will lose. Find the probability of this team winning at least 2 out of the next 5 games. Assume no draws.
26. In an experiment the probability of a 'success' is p and of a 'failure' q, where $q = 1 - p$. Show that in n trials of this experiment the probability of no more than one 'success' is given by $q^{n-1}[n - (n - 1)q]$.

5 | SPECIAL TOPICS

5.1 Multinomial distribution

In Chapter 4, working with the binomial distribution, we were concerned with an experiment which resulted in one of two outcomes. Let us now consider an experiment where there are k outcomes. For convenience we shall label these $E_1, E_2, \ldots E_k$ and their respective probabilities $p_1, p_2, \ldots p_k$. We wish to find the probability that in n trials of this experiment $E_1, E_2, \ldots E_k$ will occur $n_1, n_2, \ldots n_k$ times, respectively, where $n_1 + n_2 + \ldots + n_k = n$.

We shall use a method similar to that used in Chapter 4. If we move along n_1 branches ending in the event E_1 and then n_2 branches ending in E_2, n_3 ending in E_3 and so on until the last n_k branches ending in E_k, then the probability of this will be $p_1^{n_1} . p_2^{n_2} \ldots p_k^{n_k}$. The question now arises as to how many ways there are of doing this. It resolves itself into the problem of finding how many ways there are of arranging n objects of which $n_1, n_2, \ldots n_k$ are alike. In Section 3.3 we saw the answer was:

$$\frac{n!}{n_1!\, n_2! \ldots n_k!}$$

And so the required probability is given by:

$$P(E_1, E_2, \ldots E_k \text{ occurring } n_1, n_2, \ldots n_k \text{ times, respectively}) = \frac{n!}{n_1!\, n_2! \ldots n_k!} p_1^{n_1} . p_2^{n_2} \ldots p_k^{n_k} \quad (5.1)$$

It is easy to show that if we put $n_3 = n_4 = \ldots = n_k = 0$, equation (5.1) reduces to the familiar binomial distribution.

As the binomial distribution was concerned with experiments involving two outcomes, this distribution, being concerned with more than two outcomes, is known as the *multinomial distribution*. And as the general term for the binomial distribution was the general term of $(q+p)^n$, so in this case the general term is the general term of the expansion of $(p_1+p_2+\ldots+p_k)^n$.

Example 1. If a fair die is tossed 6 times what is the probability that 3 ones and 3 sixes will occur?

Let E_1 be the occurrence of a one and E_2 the occurrence of a six. $p_1=p_2=\frac{1}{6}$ and $n_1=n_2=3$. Therefore:

$$P(3 \text{ ones and } 3 \text{ sixes}) = \frac{6!}{3!\,3!}(\tfrac{1}{6})^3(\tfrac{1}{6})^3 = \tfrac{5}{11664}$$

Example 2. A box contains 4 red, 3 green and 3 blue marbles. A marble is drawn at random, its colour noted, and returned to the box. Find the probability that in 6 draws there will be 2 of each colour.

If E_1 is the drawing of a red marble, E_2 is the drawing of a green marble and E_3 is the drawing of a blue marble, then:

$$n_1=n_2=n_3=2, \text{ and } p_1=\tfrac{4}{10},\ p_2=\tfrac{3}{10} \text{ and } p_3=\tfrac{3}{10}$$

Therefore:

$$P(2 \text{ of each colour in } 6 \text{ draws}) = \frac{6!}{2!\,2!\,2!}(\tfrac{4}{10})^2(\tfrac{3}{10})^2(\tfrac{3}{10})^2 = \tfrac{729}{6250}$$

Remember that the formula $\dfrac{n!}{n_1!\,n_2!\ldots n_k!}\,p_1^{n_1}\,p_2^{n_2}\ldots p_k^{n_k}$ is for convenience. Most problems on probability can be worked out from first principles. To illustrate this point let us consider the example of the box containing the red, green and blue marbles. Here $P(\text{red marble})=\frac{4}{10}$, $P(\text{green})=\frac{3}{10}$ and $P(\text{blue})=\frac{3}{10}$, giving:

P(drawing 2 red marbles with replacement) $=\frac{4}{10}\times\frac{4}{10}=(\frac{4}{10})^2$

P(drawing 2 green marbles with replacement) $=\frac{3}{10}\times\frac{3}{10}=(\frac{3}{10})^2$

P(drawing 2 blue marbles with replacement) $=\frac{3}{10}\times\frac{3}{10}=(\frac{3}{10})^2$

And so the probability of drawing 2 red, 2 green and 2 blue marbles, in that order, with replacement is $(\frac{4}{10})^2.(\frac{3}{10})^2.(\frac{3}{10})^2$. The number of ways of arranging 2 red, 2 green and 2 blue marbles is:

$$\frac{6!}{2!\,2!\,2!}$$

So P(2 of each colour in 6 draws) $= \dfrac{6!}{2!2!\,2!}\left(\frac{4}{10}\right)^2\left(\frac{3}{10}\right)^2\left(\frac{3}{10}\right)^2$

which, of course, comes from substitution in equation (5.1).

Exercise 5a

1. What is the probability of each number appearing once in 6 throws of a fair die?
2. A bag contains 10 balls of which 5 are black, 2 are white, 2 are green and 1 is blue. A ball is selected at random from the bag, its colour noted and returned to the bag. This is done 5 times altogether. What is the probability that the balls drawn out are made up of:
   (*a*) 2 black, 2 white and 1 green
   (*b*) 3 black and 2 green
   (*c*) 2 white, 2 green and 1 blue
   What is the probability that all 4 colours appear?
3. Five cards are drawn with replacement from a well shuffled pack of 52 playing cards. What is this probability that there are:
   (*a*) 3 aces and 2 kings (*b*) 2 aces, 2 kings and a queen

5.2 Geometric distribution

In the binomial distribution we were concerned with the probability of r successes in n trials of an experiment where p was the probability of a success and q the probability of a failure ($p + q = 1$).

Suppose instead we are interested in finding out how many trials of the experiment are needed before a success occurs.

Let the variable X be the number of the trial in which a success occurs. Therefore $X = 1, 2, \ldots$ and is, in fact, an infinite series.

We now wish to find the probability that $X = k$. This can be found by calculating the probability of $k-1$ failures and 1 success:

$$P(X = k) = q^{k-1}p \qquad k = 1, 2, \ldots$$

This is known as the *geometric distribution.*

Example 3. Show that $P(X = k) = q^{k-1}p$ represents a probability distribution.

For this to be so $P(X = k)$ must be non-negative, which it is, since $p > 0$ and $q > 0$. As well as this, $\sum_{k=1}^{\infty} P(X = k)$ must equal 1, since:

$$\sum_{k=1}^{\infty} P(X = k) = \sum_{k=1}^{\infty} q^{k-1}p$$

$$= p(1 + q + q^2 + \ldots)$$

$$= p\left(\frac{1}{1-q}\right) = 1$$

Example 4. What is the probability that there will be 4 throws of a fair die before a six appears?

In this example the six is to appear on the 5th throw; therefore $k = 5$. This means:

$$P(X = 5) = q^4p = (\tfrac{5}{6})^4(\tfrac{1}{6}) = \tfrac{625}{7776}$$

This particular distribution is just an application of the theory of independent events developed in Section 2.5.

The probability in this example could have been calculated thus:

$$P(X = 5) = P(\overline{6}).P(\overline{6}).P(\overline{6}).P(\overline{6}).P(6)$$

where $P(\overline{6})$ means the probability of 'not a six.'

5.3 Markov chains

Consider the following example.

Example 5. A professor is either early or late for his lectures. He is never certain to be early two days running. In fact, if he is early one day the probability of his being early the next is only 0·3, but if he is late on any one day the probability of his being early the next day is 0·8. Suppose he is early on Monday morning, what is the probability of his being early on Wednesday morning?

The significant difference here is that the probability of the event of his being early can take two values—0·3 if he was early the preceding day and 0·8 if he was late the preceding day.

A tree diagram helps to solve this problem. From Fig. 5.1, the probability of his being early on Wednesday is $0{\cdot}09 + 0{\cdot}56 = 0{\cdot}65$.

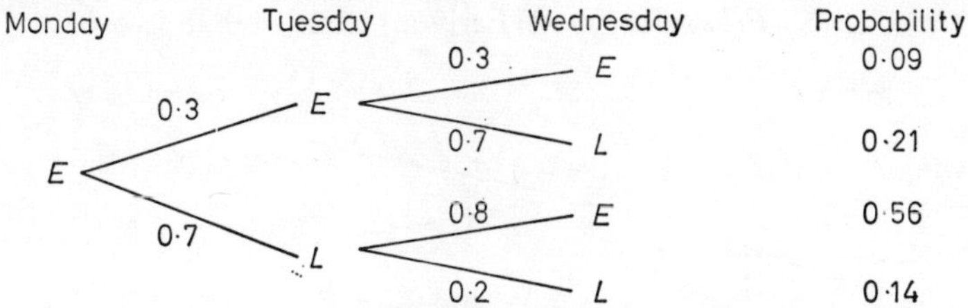

FIGURE 5.1

This problem could be extended for any number of days. A process like this where the probability of an event depends on the preceding event is known as a *Markov chain*. The information given in the question can be put in matrix form:

$$\begin{array}{cc} & \begin{array}{cc} E & L \end{array} \\ \begin{array}{c} E \\ L \end{array} & \begin{pmatrix} 0{\cdot}3 & 0{\cdot}7 \\ 0{\cdot}8 & 0{\cdot}2 \end{pmatrix} \end{array}$$

Let us call this matrix P. It gives the probabilities as we move from one state (E or L) to another (E or L), and so we can write:

$$P = \begin{pmatrix} 0{\cdot}3 & 0{\cdot}7 \\ 0{\cdot}8 & 0{\cdot}2 \end{pmatrix}$$

The first thing to notice about this matrix is that it has no negative entries. The second, that the sum of the entries of each row is 1. A matrix having these two properties is known as a *stochastic matrix*.

There are only two states (E, L) in this problem and so the matrix P is of order 2×2. We shall consider an example later which has three states and P is of order 3×3. Obviously there is no restriction to the number of states other than that imposed by involved computation.

We shall use P_n to denote the matrix giving the probabilities at the nth stage of the process.

Let us see what P_2 is for our example. The second stage of the process gives the probabilities of his being early or late on Wednesday morning given that he was early or late on Monday (see Fig. 5.2).

$$P(\text{Early on Wed} \mid \text{Early on Mon}) = 0{\cdot}65$$
$$P(\text{Late on Wed} \mid \text{Early on Mon}) = 0{\cdot}35$$

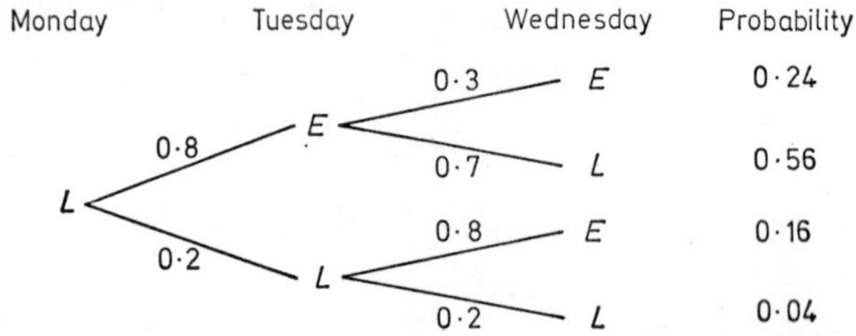

FIGURE 5.2

And so:

$$P(\text{Early on Wed} \mid \text{Late on Mon}) = 0{\cdot}40$$
$$P(\text{Late on Wed} \mid \text{Late on Mon}) = 0{\cdot}60$$

Therefore:

$$P_2 = \begin{pmatrix} 0{\cdot}65 & 0{\cdot}35 \\ 0{\cdot}40 & 0{\cdot}60 \end{pmatrix}$$

As we would expect, this too is a stochastic matrix. Note that the first entry in the first row gives the probability of his being early on Wednesday given that he was early on Monday and the second entry in the first row the probability of his being late on Wednesday given that he was early on Monday. The second row gives the probabilities of his being early or late on Wednesday given that he was late on Monday.

A rather interesting thing happens if we square the matrix P:

$$P^2 = \begin{pmatrix} 0{\cdot}3 & 0{\cdot}7 \\ 0{\cdot}8 & 0{\cdot}2 \end{pmatrix} \begin{pmatrix} 0{\cdot}3 & 0{\cdot}7 \\ 0{\cdot}8 & 0{\cdot}2 \end{pmatrix}$$
$$= \begin{pmatrix} 0{\cdot}65 & 0{\cdot}35 \\ 0{\cdot}40 & 0{\cdot}60 \end{pmatrix}$$

And so

$$P^2 = P_2$$
$$P^3 = P^2 P$$
$$= \begin{pmatrix} 0{\cdot}65 & 0{\cdot}35 \\ 0{\cdot}40 & 0{\cdot}60 \end{pmatrix} \begin{pmatrix} 0{\cdot}3 & 0{\cdot}7 \\ 0{\cdot}8 & 0{\cdot}2 \end{pmatrix}$$
$$= \begin{pmatrix} 0{\cdot}475 & 0{\cdot}525 \\ 0{\cdot}600 & 0{\cdot}400 \end{pmatrix}$$

(For a full treatment of matrices and matrix operations, see 'Matrices' in this series.)

Let us now find P_3 and see if it is equal to P^3. P_3 is the probabilities at the third stage, *i.e.*, Thursday (see Fig. 5.3). And so:

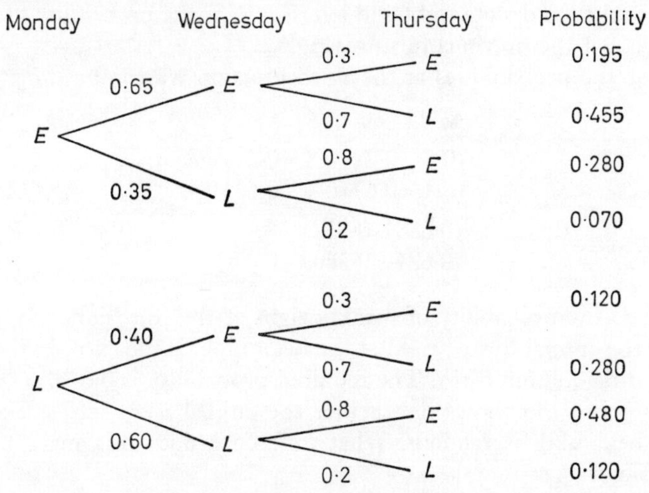

FIGURE 5·3

$$P(\text{Early on Thurs} \mid \text{Early on Mon}) = 0{\cdot}195 + 0{\cdot}280 = 0{\cdot}475$$
$$P(\text{Late on Thurs} \mid \text{Early on Mon}) = 0{\cdot}455 + 0{\cdot}070 = 0{\cdot}525$$
$$P(\text{Early on Thurs} \mid \text{Late on Mon}) = 0{\cdot}120 + 0{\cdot}480 = 0{\cdot}600$$
$$P(\text{Late on Thurs} \mid \text{Late on Mon}) = 0{\cdot}280 + 0{\cdot}120 = 0{\cdot}400$$

Therefore:

$$P_3 = \begin{pmatrix} 0{\cdot}475 & 0{\cdot}525 \\ 0{\cdot}600 & 0{\cdot}400 \end{pmatrix}$$

which is the same as P^3. In fact, it can be shown that $P^n = P_n$.

Example 6. In a laboratory experiment, an animal is sent through a maze. At each junction it has the choice of going left or right. If it goes left it receives a mild electric shock but if it goes right it receives food. On receiving an electric shock the probability of the animal turning right at the next junction is 0·4, but on receiving food the probability of turning right is 0·8. If it turns right at the first junction what is the probability of it turning right at the third junction?

For this example:

$$P = \begin{matrix} & r & l \\ r & 0{\cdot}8 & 0{\cdot}2 \\ l & 0{\cdot}4 & 0{\cdot}6 \end{matrix}$$

where the r, l denote right and left. So 0·2 is the probability of turning left, if the previous turn was right.

For the probabilities at the third junction we compute P_3:

$$\begin{aligned} P_3 &= P^3 \\ &= \begin{pmatrix} 0{\cdot}8 & 0{\cdot}2 \\ 0{\cdot}4 & 0{\cdot}6 \end{pmatrix}\begin{pmatrix} 0{\cdot}8 & 0{\cdot}2 \\ 0{\cdot}4 & 0{\cdot}6 \end{pmatrix}\begin{pmatrix} 0{\cdot}8 & 0{\cdot}2 \\ 0{\cdot}4 & 0{\cdot}6 \end{pmatrix} \\ &= \begin{pmatrix} 0{\cdot}688 & 0{\cdot}312 \\ 0{\cdot}624 & 0{\cdot}376 \end{pmatrix} \end{aligned}$$

And so the probability of turning right at the third junction, given that the animal turned right at the beginning, is the entry in the first row, first column of P_3. The required probability is 0·688.

Suppose the choice of whether the animal went left or right at the beginning is random. What difference does this make to our problem?

The probability of going right at the beginning is the same as the probability of going left—$\frac{1}{2}$ in each case.

As we represented the other probabilities by stochastic matrices we shall do the same here and denote it by P_0, where $P_0 = (0{\cdot}5 \quad 0{\cdot}5)$. Since we have added one more stage to the beginning of the Markov chain, we can find the probability of going right at the third junction by evaluating $P_0 P_3$:

$$\begin{aligned} P_0 P_3 &= P_0 P^3 \\ &= (0{\cdot}5 \quad 0{\cdot}5)\begin{pmatrix} 0{\cdot}688 & 0{\cdot}312 \\ 0{\cdot}624 & 0{\cdot}376 \end{pmatrix} \\ &= (0{\cdot}656 \quad 0{\cdot}344) \end{aligned}$$

This matrix tells us that if the animal has an equal probability of turning right or left at the beginning, then at the third junction the animal has a probability of 0·656 of turning right and one of 0·344 of turning left.

Suppose the probability of turning right at the beginning was p then the probability of turning left was $1 - p$. Let us denote by X_0

the matrix $(p \quad 1-p)$. Is there such a matrix X_0 such that $X_0 = X_0P$? If this is true, then:

$$(p \quad 1-p) = (p \quad 1-p)\begin{pmatrix} 0{\cdot}8 & 0{\cdot}2 \\ 0{\cdot}4 & 0{\cdot}6 \end{pmatrix}$$

$$= (0{\cdot}4 + 0{\cdot}4p \quad 0{\cdot}6 - 0{\cdot}4p)$$

Since the matrices are equal, corresponding entries are equal. Hence $p = 0{\cdot}4 + 0{\cdot}4p$ and $1 - p = 0{\cdot}6 - 0{\cdot}4p$, which gives $p = \frac{2}{3}$ and $1 - p = \frac{1}{3}$. Thus $X_0 = (\frac{2}{3} \quad \frac{1}{3})$. Now $X_0 = X_0P$, and so:

$$X_0P_2 = X_0P^2 = (X_0P)P = X_0P = X_0$$

Similarly: $$X_0P_3 = X_0P^3 = (X_0P^2)P = X_0P = X_0$$

Continuing this we eventually arrive at $X_0P_n = X_0$. This means that given this matrix $X_0 = (p \quad 1-p)$, then the probability of going right or left at every junction is p or $1-p$. Let us test it for P_3:

$$X_0P_3 = (\tfrac{2}{3} \quad \tfrac{1}{3})\begin{pmatrix} 0{\cdot}688 & 0{\cdot}312 \\ 0{\cdot}624 & 0{\cdot}376 \end{pmatrix}$$

$$= (\tfrac{2}{3} \quad \tfrac{1}{3})\begin{pmatrix} \frac{688}{1000} & \frac{312}{1000} \\ \frac{624}{1000} & \frac{376}{1000} \end{pmatrix}$$

$$= (\tfrac{2000}{3000} \quad \tfrac{1000}{3000}) = (\tfrac{2}{3} \quad \tfrac{1}{3})$$

It can also be shown that if some power of P has no zero entries, then, as n increases, P_n approaches a matrix A which has as each of its rows the matrix X_0. A matrix with the property that some power of the matrix has no zero entries is said to be *regular*.

Example 7. Which, if either, of the following two matrices is regular?

(*i*) $\begin{pmatrix} 0 & 1 \\ 1 & 0 \end{pmatrix}$ (*ii*) $\begin{pmatrix} 0 & 1 \\ 1 & 1 \end{pmatrix}$

(*i*) $\begin{pmatrix} 0 & 1 \\ 1 & 0 \end{pmatrix}^2 = \begin{pmatrix} 0 & 1 \\ 1 & 0 \end{pmatrix}\begin{pmatrix} 0 & 1 \\ 1 & 0 \end{pmatrix} = \begin{pmatrix} 1 & 0 \\ 0 & 1 \end{pmatrix}$

This is the identity matrix, which means that all odd powers of this matrix are the matrix itself and all even powers are the identity matrix. So every power has two zero entries and, therefore, this matrix is not regular.

(*ii*) $\begin{pmatrix} 0 & 1 \\ 1 & 1 \end{pmatrix}^2 = \begin{pmatrix} 0 & 1 \\ 1 & 1 \end{pmatrix}\begin{pmatrix} 0 & 1 \\ 1 & 1 \end{pmatrix} = \begin{pmatrix} 1 & 1 \\ 1 & 2 \end{pmatrix}$

which has no zero entries and, therefore, this matrix is regular.

Example 8. In the long run, how often will the professor, mentioned at the beginning of this section, be early for his class?

$$P = \begin{pmatrix} 0\cdot3 & 0\cdot7 \\ 0\cdot8 & 0\cdot2 \end{pmatrix}$$

which is a regular matrix. We wish to find the matrix $X_0 = (p \quad 1-p)$ such that $X_0 = X_0P$:

$$(p \quad 1-p) = (p \quad 1-p)\begin{pmatrix} 0\cdot3 & 0\cdot7 \\ 0\cdot8 & 0\cdot2 \end{pmatrix}$$

$$= (0\cdot8 - 0\cdot5p \quad 0\cdot2 + 0\cdot5p)$$

Therefore $p = 0\cdot8 - 0\cdot5p$ and $1-p = 0\cdot2 + 0\cdot5p$, which gives $p = \frac{8}{15}$ and $1-p = \frac{7}{15}$. So as n increases, P_n approaches matrix A where:

$$A = \begin{pmatrix} \frac{8}{15} & \frac{7}{15} \\ \frac{8}{15} & \frac{7}{15} \end{pmatrix}$$

This means that the professor will be early $\frac{8}{15}$ of the time.

Example 9. A machine is designed to produce the digits 1, 2, 3. If the digit 1 is put in, the probability of a 1 appearing is 0·5, the probability of a 2 is 0·4, and of a 3 is 0·1. If the digit 2 is put in the probabilities of 1, 2 and 3 appearing are 0·3, 0·2 and 0·5, respectively. If the digit 3 is put in the respective probabilities are 0·4, 0·4 and 0·2. Any digit which appears is fed back into the machine and the process repeated. If the digits 1, 2 and 3 stand an equal chance of being selected to start the process, what is the probability of a 2 appearing at the second stage? Suppose the machine is allowed to run for a long time, what is then the probability of a 2 appearing?

This example involves three states and so the matrix P, which shows the transition probabilities, is of the order 3×3:

$$P = \begin{pmatrix} 0\cdot5 & 0\cdot4 & 0\cdot1 \\ 0\cdot3 & 0\cdot2 & 0\cdot5 \\ 0\cdot4 & 0\cdot4 & 0\cdot2 \end{pmatrix}$$

Remember the first row gives the probabilities of 1, 2 or 3 appearing with an input of 1, the second row with an input of 2, and the third row shows the respective probabilities with an input of 3.

At the beginning of the process the digits 1, 2, 3 stand an equal chance of being selected; therefore $P_0 = (\frac{1}{3} \;\; \frac{1}{3} \;\; \frac{1}{3})$. And so the probability of a 2 appearing at the second stage is found by evaluating P_0P_2:

$$\begin{aligned} P_0P_2 &= P_0P^2 \\ &= (\tfrac{1}{3} \;\; \tfrac{1}{3} \;\; \tfrac{1}{3})\begin{pmatrix} 0{\cdot}5 & 0{\cdot}4 & 0{\cdot}1 \\ 0{\cdot}3 & 0{\cdot}2 & 0{\cdot}5 \\ 0{\cdot}4 & 0{\cdot}4 & 0{\cdot}2 \end{pmatrix}\begin{pmatrix} 0{\cdot}5 & 0{\cdot}4 & 0{\cdot}1 \\ 0{\cdot}3 & 0{\cdot}2 & 0{\cdot}5 \\ 0{\cdot}4 & 0{\cdot}4 & 0{\cdot}2 \end{pmatrix} \\ &= (\tfrac{1}{3} \;\; \tfrac{1}{3} \;\; \tfrac{1}{3})\begin{pmatrix} 0{\cdot}41 & 0{\cdot}32 & 0{\cdot}27 \\ 0{\cdot}41 & 0{\cdot}36 & 0{\cdot}23 \\ 0{\cdot}40 & 0{\cdot}32 & 0{\cdot}28 \end{pmatrix} \\ &= (\tfrac{122}{300} \;\; \tfrac{100}{300} \;\; \tfrac{78}{300}) \end{aligned}$$

And so the probability of a 2 appearing is $\frac{100}{300} = \frac{1}{3}$.

Since P has no zero entries it is a regular matrix, and we can find the matrix A, which P_n approaches as n gets very large, by finding the matrix X_0 which satisfies $X_0 = X_0P$.

Now since we are involved with 3 states, the matrix X_0 will be of the form $(p \;\; r \;\; 1-p-r)$. Note that the three elements in this row matrix must add up to 1. If $X_0 = X_0P$:

$$\begin{aligned} (p \;\; r \;\; 1-p-r) &= (p \;\; r \;\; 1-r)\begin{pmatrix} 0{\cdot}5 & 0{\cdot}4 & 0{\cdot}1 \\ 0{\cdot}3 & 0{\cdot}2 & 0{\cdot}5 \\ 0{\cdot}4 & 0{\cdot}4 & 0{\cdot}2 \end{pmatrix} \\ &= (0{\cdot}1p - 0{\cdot}1r + 0{\cdot}4 \quad -0{\cdot}2r + 0{\cdot}4 \quad -0{\cdot}1p + 0{\cdot}3r + 0{\cdot}2) \end{aligned}$$

Equating elements we have:

$$\begin{aligned} p &= 0{\cdot}1p - 0{\cdot}1r + 0{\cdot}4 \\ r &= -0{\cdot}2r + 0{\cdot}4 \\ 1-p-r &= -0{\cdot}1p + 0{\cdot}3r + 0{\cdot}2 \end{aligned}$$

and solving these equations simultaneously we have:

$$p = \tfrac{11}{27} \quad r = \tfrac{1}{3} \quad 1-p-r = \tfrac{7}{27}$$

Therefore $X_0 = (\frac{11}{27} \;\; \frac{1}{3} \;\; \frac{7}{27})$, and hence:

$$A = \begin{pmatrix} \frac{11}{27} & \frac{1}{3} & \frac{7}{27} \\ \frac{11}{27} & \frac{1}{3} & \frac{7}{27} \\ \frac{11}{27} & \frac{1}{3} & \frac{7}{27} \end{pmatrix}$$

And so no matter which digit is put in at the beginning of the process, the probability of a 2 appearing after a long run is $\frac{1}{3}$. We can also add that, in these circumstances, the probability of 1 appearing is $\frac{11}{27}$ and of a 3 appearing, $\frac{7}{27}$.

5.4 Application of probability to genetics

The various characteristics—height, build, colour of hair, *etc.*—of people, plants and animals are determined by factors known as genes. In simple Mendelian heredity a gene may be either dominant (A) or recessive (a). They exist in pairs, and so we can have AA, Aa or aa. (The letters A and a are only used here for convenience and have no special significance.) AA and aa are said to be homozygous and Aa is said to be heterozygous. An individual with AA genes is said to be dominant, one with aa genes is said to be recessive and a hybrid is an individual with Aa genes.

Note that the characteristic of AA and Aa will be the same, the difference only becomes apparent in mating.

The basic assumption of genetics is that the genes are selected at random and independently of each other. This being so we can apply our knowledge of probability to the various situations after successive matings.

If we let X_1, X_2 represent the two parents, then we can construct Table 5.1 to show the probabilities of the various offspring.

TABLE 5.1

	X_1					
X_2	AA		Aa		aa	
AA	AA	1	AA	$\frac{1}{2}$	AA	0
	Aa	0	Aa	$\frac{1}{2}$	Aa	1
	aa	0	aa	0	aa	0
Aa	AA	$\frac{1}{2}$	AA	$\frac{1}{4}$	AA	0
	Aa	$\frac{1}{2}$	Aa	$\frac{1}{2}$	Aa	$\frac{1}{2}$
	aa	0	aa	$\frac{1}{4}$	aa	$\frac{1}{2}$
aa	AA	0	AA	0	AA	0
	Aa	1	Aa	$\frac{1}{2}$	Aa	0
	aa	0	aa	$\frac{1}{2}$	aa	1

We can show how the table was constructed by the use of two examples.

First consider both parents to be hybrid (Aa).

Then there are four possibilities for the offspring:

(*i*) A from X_1, A from X_2, giving AA
(*ii*) A from X_1, a from X_2, giving Aa
(*iii*) a from X_1, A from X_2, giving aA
(*iv*) a from X_1, a from X_2, giving aa

Since Aa and aA are the same, the various probabilities are

$$P(AA) = \tfrac{1}{4} \quad P(Aa) = \tfrac{1}{4} + \tfrac{1}{4} = \tfrac{1}{2} \quad P(aa) = \tfrac{1}{4}$$

Secondly let us consider the case where $X_1 = AA$ and $X_2 = aa$. Since the offspring takes one gene from each parent the only possible pairing will be Aa, and so:

$$P(Aa) = 1 \quad P(AA) = P(aa) = 0$$

Let us now find out what happens when we cross a dominant parent with one of unknown genetic character. Suppose $X_1 = AA$ then X_2 will take the values AA, Aa and aa. In general then the offspring can be AA, Aa or aa. We can construct Table 5.2 to show the various probabilities.

TABLE 5.2

X_2	Offspring AA	Aa	aa
AA	1	0	0
Aa	$\frac{1}{2}$	$\frac{1}{2}$	0
aa	0	1	0

Table 5.2 could be written:

$$\begin{array}{c} \\ AA \\ Aa \\ aa \end{array} \begin{array}{c} \begin{array}{ccc} AA & Aa & aa \end{array} \\ \begin{pmatrix} 1 & 0 & 0 \\ \frac{1}{2} & \frac{1}{2} & 0 \\ 0 & 1 & 0 \end{pmatrix} \end{array}$$

which is a stochastic matrix describing a Markov chain. Here:

$$P_2 = P^2 = \begin{pmatrix} 1 & 0 & 0 \\ \frac{1}{2} & \frac{1}{2} & 0 \\ 0 & 1 & 0 \end{pmatrix} \begin{pmatrix} 1 & 0 & 0 \\ \frac{1}{2} & \frac{1}{2} & 0 \\ 0 & 1 & 0 \end{pmatrix}$$

$$= \begin{pmatrix} 1 & 0 & 0 \\ \frac{3}{4} & \frac{1}{4} & 0 \\ \frac{1}{2} & \frac{1}{2} & 0 \end{pmatrix}$$

Writing this matrix in the form:

$$\begin{array}{c} \\ AA \\ Aa \\ aa \end{array} \begin{array}{c} \begin{array}{ccc} AA & Aa & aa \end{array} \\ \begin{pmatrix} 1 & 0 & 0 \\ \frac{3}{4} & \frac{1}{4} & 0 \\ \frac{1}{2} & \frac{1}{2} & 0 \end{pmatrix} \end{array}$$

and remembering that one parent is dominant and that each row gives the probabilities of the different genetic types of offspring corresponding to the different genetic types of the second parent, we can say that the probability of a hybrid offspring in the second generation is 0, $\frac{1}{4}$, $\frac{1}{2}$, depending on whether the second parent is *AA*, *Aa* or *aa*.

Example 10. Suppose an individual who is hybrid is crossed with another of unknown genetic character. What is the probability after successive crossings of having a dominant offspring?

For this example, $X_1 = Aa$ and $X_2 = AA$, Aa or aa; again we draw up a probability table (Table 5.3).

TABLE 5.3

	Offspring		
X_2	*AA*	*Aa*	*aa*
AA	$\frac{1}{2}$	$\frac{1}{2}$	0
Aa	$\frac{1}{4}$	$\frac{1}{2}$	$\frac{1}{4}$
aa	0	$\frac{1}{2}$	$\frac{1}{2}$

$$P = \begin{pmatrix} \frac{1}{2} & \frac{1}{2} & 0 \\ \frac{1}{4} & \frac{1}{2} & \frac{1}{4} \\ 0 & \frac{1}{2} & \frac{1}{4} \end{pmatrix}$$

$$P^2 = \begin{pmatrix} \frac{1}{2} & \frac{1}{2} & 0 \\ \frac{1}{4} & \frac{1}{2} & \frac{1}{4} \\ 0 & \frac{1}{2} & \frac{1}{2} \end{pmatrix} \begin{pmatrix} \frac{1}{2} & \frac{1}{2} & 0 \\ \frac{1}{4} & \frac{1}{2} & \frac{1}{4} \\ 0 & \frac{1}{2} & \frac{1}{2} \end{pmatrix}$$

$$= \begin{pmatrix} \frac{3}{8} & \frac{1}{2} & \frac{1}{8} \\ \frac{1}{4} & \frac{1}{2} & \frac{1}{4} \\ \frac{1}{8} & \frac{1}{2} & \frac{3}{8} \end{pmatrix}$$

This matrix has no zero entries; therefore P is regular. Then as n increases P_n approaches a matrix A whose rows are formed by the matrix $X_0 = (p \quad r \quad 1 - p - r)$, where $X_0 = X_0 P$:

$$(p \quad r \quad 1-p-r) = (p \quad r \quad 1-p-r)\begin{pmatrix} \frac{1}{2} & \frac{1}{2} & 0 \\ \frac{1}{4} & \frac{1}{2} & \frac{1}{4} \\ 0 & \frac{1}{2} & \frac{1}{2} \end{pmatrix}$$

$$= (\tfrac{1}{2}p + \tfrac{1}{4}r \quad \tfrac{1}{2} \quad -\tfrac{1}{2}p - \tfrac{1}{4}r + \tfrac{1}{2})$$

Equating elements we have:

$$\begin{aligned} p &= \tfrac{1}{2}p + \tfrac{1}{4}r \\ r &= \tfrac{1}{2} \\ 1-p-r &= -\tfrac{1}{2}p - \tfrac{1}{4}r + \tfrac{1}{2} \end{aligned}$$

Solving these simultaneously we have:

$$p = \tfrac{1}{4} \quad r = \tfrac{1}{2} \quad 1-p-r = \tfrac{1}{4}$$

Therefore $X_0 = (\frac{1}{4} \ \frac{1}{2} \ \frac{1}{4})$ and so as n increases P_n approaches the matrix A where:

$$A = \begin{pmatrix} \frac{1}{4} & \frac{1}{2} & \frac{1}{4} \\ \frac{1}{4} & \frac{1}{2} & \frac{1}{4} \\ \frac{1}{4} & \frac{1}{2} & \frac{1}{4} \end{pmatrix}$$

The probability then of having a dominant offspring after successive crossings is $\frac{1}{4}$. Note that this does not depend on the genetic character of the second parent.

Exercise 5b

1. Determine which of the following matrices are stochastic:

(a) $\begin{pmatrix} 1 & 0 \\ \frac{1}{2} & \frac{1}{2} \end{pmatrix}$

(b) $\begin{pmatrix} 2 & -1 \\ \frac{3}{4} & \frac{1}{4} \end{pmatrix}$

(c) $\begin{pmatrix} \frac{1}{2} & \frac{1}{4} & \frac{1}{4} \\ 1 & 0 & 0 \\ 0 & \frac{1}{2} & \frac{1}{2} \end{pmatrix}$

(d) $\begin{pmatrix} \frac{1}{3} & \frac{1}{3} & \frac{1}{6} & \frac{1}{6} \\ \frac{1}{2} & 0 & 0 & \frac{1}{2} \\ \frac{1}{4} & 0 & \frac{3}{4} & 0 \\ \frac{1}{8} & \frac{1}{2} & \frac{1}{4} & \frac{1}{8} \end{pmatrix}$

(e) $\begin{pmatrix} \frac{1}{3} & \frac{1}{3} & \frac{1}{3} \\ \frac{1}{2} & \frac{1}{4} & \frac{1}{4} \\ \frac{1}{8} & \frac{1}{4} & \frac{9}{16} \end{pmatrix}$

2. Determine which of the following matrices are regular:

(a) $\begin{pmatrix} 1 & 0 \\ 0 & 1 \end{pmatrix}$

(b) $\begin{pmatrix} \frac{1}{2} & \frac{1}{2} \\ \frac{1}{4} & \frac{3}{4} \end{pmatrix}$

(c) $\begin{pmatrix} 0 & 1 \\ \frac{1}{2} & \frac{1}{2} \end{pmatrix}$

(d) $\begin{pmatrix} 1 & 0 \\ \frac{1}{2} & \frac{1}{2} \end{pmatrix}$

(e) $\begin{pmatrix} 0 & 1 & 0 \\ 0 & 0 & 1 \\ \frac{1}{2} & \frac{1}{2} & 0 \end{pmatrix}$

3. A machine is designed to produce the digits 0 and 1 on the input of either 0 or 1. If a 0 is put in, the probability of a 0 appearing is 0·5 and, if a 1 is put in, the probability of a 1 appearing is 0·3.
   (a) Write down the matrix P whose elements are the transition probabilities.
   (b) Evaluate P_2 and P_3.
   (c) If the choice of which digit to start the process is decided by the toss of a coin, find the probability of a 0 appearing at the second stage.
   (d) In the long run, what is the probability of a 1 appearing?
4. A man changes his car each year. He restricts himself to two types of car, A and B. If he owns a car of type A the probability of his buying another one is 0·8, but if he owns a car of type B the probability of his buying the same type again is 0·5. In 1966 he owned a car of type A. What is the probability that in 1968 he owned a car of:
   (a) type A (b) type B
   Over a long period of time, how much of the time will he have driven a car of type A?
5. A pupil is either early, late or on time for school. No two days running are ever the same. If he is early one day he is just as likely to be late or on time the next day, and if he is on time he is just as likely to be early or late the next day. However, if he is late he does make some effort to be early the next day, and, in fact, the probability of his being early is $\frac{3}{4}$.
   (a) If he is early on Monday what is the probability of his being on time on Wednesday?
   (b) Over the school year, how often is he early?

6. An individual, who is recessive, mates with another individual of unknown genetic character. What are the probabilities of a hybrid offspring in the second generation?

Exercises on Chapter 1–5

1. A three-man delegation is to be chosen from 4 Americans and 5 Canadians. The delegates are chosen by putting their names into a hat and picking out 3 at random. Find the probability that all three delegates will be Canadian. (C)
2. A manufacturer states that 50% of the blades of his cricket bats will be sound at the end of a full season's use.
   (*a*) What is the probability that at least 4 out of 6 bats would be sound after a full season?
   (*b*) If it is found that handles are equally likely to break (but not both blade and handle) what would be the revised probability? (C)
3. A housewife buys a dozen eggs of which two turn out to be bad. She chooses four eggs to scramble for breakfast. Find the chance that she chooses:
   (*a*) all good eggs
   (*b*) three good and one bad
   (*c*) two good and two bad
   (*d*) at least one bad egg (C)
4. *A* and *B*, in that order, draw one ball at a time from a bag originally containing 2 white and 3 black balls and do not replace the balls. What are their respective chances of being the first to draw a white ball? (C)
5. In a shooting competition a marksman can score a bull 4 times out of 5. If he has three shots, find the chance of his scoring a bull:
   (*a*) not at all
   (*b*) exactly once
   (*c*) at least twice (C)
6. In the manufacture of a certain article, one in every ten is found to be faulty. Calculate the probabilities that in a random sample of three, the number of faulty articles is 0, 1, 2, 3. (C)
7. A bag contains 7 white and 3 black balls. A drawing of three

balls is to be made, each ball being replaced after it is drawn. Calculate the probabilities that the drawing will give:

(*a*) 2 white and 1 black (in any order)

(*b*) all three of the same colour (c)

8. In a selected group of 100 people, 60 have brown hair and 45 have blue eyes. Find the chance that a particular member of the group has:

(*a*) brown hair and blue eyes

(*b*) brown hair and eyes which are not blue

(*c*) neither brown hair nor blue eyes (c)

9. Figure 5.4 shows a map of six towns $A, B, \ldots, F$, and the roads connecting them. A man starts from O to visit the towns. He leaves each town by a road other than the one by which he enters and there are equal chances of his leaving by any of these other roads. Calculate the probability that the third town he visits is:

(*a*) F (*b*) C

Note that the first town he visits must be A. (c)

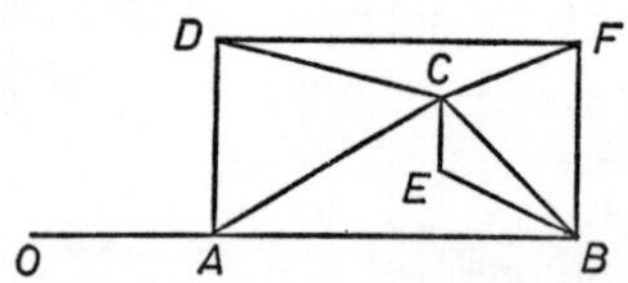

FIGURE 5.4

10. Each of the two boxes A and B contains 2 white balls and 2 black balls. One ball is taken at random from box A and placed into box B. The 5 balls in box B are now mixed and one of them is picked at random and placed into box A. What is the chance that each box again contains 2 white balls and 2 black balls? (c)

11. In a large collection of bulbs, 3 out of 4 are daffodils and the rest are tulips. If they are planted at random, find the chance that in a row of 5 plants:

(*a*) all are daffodils (*b*) at least 4 are daffodils (c)

12. If the mean of a binomial distribution is 24 and its standard deviation is 4, calculate the number of observations and the frequency of occurrence of the event. (c)

13. If the 'crude' mark x of each of the n pupils in a class is replaced

by a 'standardized' mark $(x - m)/\sigma$, where m is the mean mark and σ the standard deviation of the marks in that particular subject, calculate, for the standardized marks:
(*a*) the mean (*b*) the standard deviation
By standardizing the crude marks given in the following Table, place the four boys *A*, *B*, *C* and *D* in an order of merit for tests in the three subjects, English, Mathematics and French, taken in the same examination:

	m	σ	*A*	*B*	*C*	*D*
English	55	10	60	50	60	70
Mathematics	70	4	80	60	70	80
French	60	12	50	80	60	40

(c)

14. In a game for two players a turn consists of throwing a die either once or twice, once if the score obtained is less than 6, twice if the score at the first throw is 6. The score for the turn is in the first case the score of the single throw, and in the second case the total score of the two throws. What are the probabilities of a player:
(*a*) scoring more than 9 in a single turn
(*b*) scoring a total of more than 20 in two succeeding turns
(*c*) obtaining equal scores in two succeeding turns (c)
15. In a certain competition for teams of six the qualifying event consists of each member of the team being allowed up to three attempts at a particular trial, the team as a whole qualifying if at least four members out of the six succeed in the trial. Assuming that the chance of failure in any one attempt by any member of the team is q, evaluate the chance of the team qualifying. (c)
16. (*a*) The probability that *A* can solve a certain problem is $\frac{3}{4}$. The probability that *B* can solve it is $\frac{2}{5}$. If both *A* and *B* attempt the problem, what is the probability that it will be solved?
(*b*) In a set at tennis, *A* is leading *B* by 5 games to 3. What is the probability that *A* will win the set, if the probability of his winning any game is $\frac{1}{2}$? The set is won when either player has won at least six games and leads by two games. (c)

17. Certain mass-produced articles, of which 1 in 200 is defective, are packed in cartons each containing 100 articles. Assuming that the number of defective articles per carton follows Poisson's distribution, find the probability that a carton selected at random will contain:
(*a*) no defective article
(*b*) two or more defective articles (C)

18. In a television panel game each player is asked a series of questions until either he answers one incorrectly or he has answered three correctly; his score for the turn is the number of correct answers he has given, with a bonus mark if he gives three correct answers. Assume that the questions are independent and that the chance of giving a correct answer to any question is p. Obtain the mean score for a turn.
Show that for $p = \frac{1}{2}$ the chance that a team of three players obtain a total score over 3 after one turn each is $\frac{25}{64}$. (C)

19. An experiment consists in drawing balls one at a time from a bag containing 8 white and 2 black balls initially. Balls drawn from the bag are not replaced. A note is kept of the number $r (1 \leqslant r \leqslant 9)$ of the draw which first yields a black ball. If the experiment is repeated many times, what is:
(*a*) the most probable value of r
(*b*) the probability that r exceeds 5 (C)

20. The number of bacteria in 1 ml of inoculum may be assumed to follow a Poisson distribution with mean 2. Assuming further that at least 3 bacteria are needed for a dose to be infective, calculate the probability that out of 10 doses each of 1 ml. exactly 4 will produce infection. (C)

21. In a certain examination the candidates are divided into 4 classes on the written papers and those in the top 3 classes are interviewed. After the interview the candidates are classified A, B or C, the C class including those placed in the fourth class on the written papers. The chance of a candidate obtaining class i ($i = 1, 2, 3, 4$) on the written paper is p_i, and the chance of his then being finally classified as A, B, C is p_{iA}, p_{iB}, p_{iC} ($i = 1, 2, 3$); thus p_{2C} is the chance of a candidate who has been placed in the second class on the written papers being finally classified as C, so that $p_{2A} + p_{2B} + p_{2C} = 1$. Write down the chance:

(*a*) that of 8 given candidates at least 6 will be interviewed
(*b*) that a candidate chosen at random will finish in:
(*i*) class *A* (*ii*) class *C*
(*c*) that of 4 given candidates none will finish in class *A*. (c)

22. The fluoride content (in parts per million) of water which has passed through a certain filtering plant is a normal variable with mean 46 ppm and standard deviation 7 ppm. It is desired that the fluoride content shall only rarely rise above 58 ppm; evaluate what proportion of the output of the plant does in fact exceed this level. (c)

23. On my journey to work I have to pass through 12 sets of traffic lights which operate independently; the chance that any set is at green when I reach it is $\frac{1}{2}$. If I am stopped at fewer than 3 sets I have time for a cup of coffee before work, but if I am stopped at more than 8 sets I am late; if I am late more than twice in a five-day week I have to see the Office Manager. What is the probability that:
(*a*) I have time for coffee on two given consecutive mornings
(*b*) I have to see the Office Manager in the week after my holiday (c)

24. A number is drawn at random from the set of 37 numbers 0, 1, 2, 3, . . ., 34, 35, 36, and is then replaced. What is the chance that:
(*a*) the number drawn will be odd
(*b*) in two consecutive draws the numbers drawn will be the same
(*c*) in 100 consecutive draws the number 0 will not be drawn
(*d*) in five consecutive draws none of the numbers drawn will be odd

Give the answers to (*a*) and (*b*) as vulgar fractions, and evaluate the answers to (*c*) and (*d*) approximately as decimals. (c)

25. In the so-called St Petersburg game the player throws a coin until a head is obtained and then receives from the bank £2^n, where n is the number of throws. What is the chance that he receives:
(*a*) £16 (*b*) more than £32

What is the chance that, if the bank holds £10^6, the player will break the bank in a single play? What can be said about the player's average receipt if the bank's resources are unlimited? (c)

26. In lawn tennis a set is won by the first player to win 6 games, except that if the score reaches 5 – 5 the set is won by the first player to lead by 2 games. Two players have chances respectively p and q of winning in any game ($p+q=1$); games may be treated as independent. Find the chance that a set lasts exactly $2n+2$ games ($n \geqslant 5$). (C)

27. An encyclopaedia consisting of $n(\geqslant 4)$ similar volumes is kept on a shelf with the volumes in correct numerical order; that is, with volume one on the left, volume two next and so on. The volumes are all taken down for cleaning and replaced on the shelf in random order. Prove that the probabilities of finding exactly n, $(n-1)$ $(n-2)$, $(n-3)$ volumes in their correct positions on the shelf are, respectively:

$$\frac{1}{n!}, 0, \frac{1}{(n-2)!\,2}, \frac{1}{(n-3)!\,3}$$ (C)

28. A bag contains m oranges and n lemons: sampling is random without replacement. What are the probabilities of the following events:
    (*a*) the first fruit drawn is an orange
    (*b*) the first two fruits drawn are oranges
    (*c*) the second fruit is a lemon, given that the first was an orange
    (*d*) the third fruit is an orange (C)

29. At a seed-testing station it is found that a proportion 0·4 of a certain type of seed is fertile. By accident the remaining stock of this seed (whose total amount is very large) is completely mixed with an equal quantity of a second type of seed which is believed to be completely infertile. If this latter assumption is true, what is the probability that a seed taken at random from the mixture will germinate? (C)

30. A draw is held each calendar month to allot prizes to holders of Premium Savings Bonds. For a person who holds a certain number of bonds, the chance of winning at least one prize in a single draw is 0·08. What is the chance that the holder:
    (*a*) will not receive a prize in a particular month
    (*b*) will not receive a prize in a particular year
    (*c*) will receive at least one prize in a particular year
    Give the answers to (*b*) and (*c*) correct to two decimal places. (C)

31. A certain number, r, of persons sit down in a random arrangement on chairs labelled with their names. If u_r denotes the number of ways that all r chairs can be filled wrongly, write down the values of u_2 and u_3 and calculate u_4 and u_5.
Show that, when $r = 5$, the chance that more than one person is in his right chair slightly exceeds 0·25. (C)

32. The following Table shows the results of recording the telephone calls handled at a village telephone exchange between 1 pm and 2 pm on each of 100 weekdays (*e.g.*, on 36 days no such calls were made):

Calls	0	1	2	3	4 or more
Days	36	35	22	7	0

Assuming that calls arrive independently and at random, estimate:
(*a*) the mean m of the corresponding Poisson probability distribution
(*b*) the probability that if the operator is absent for 10 minutes no call will be missed
(*c*) the probability that if the operator is absent for 10 minutes two or more calls will be missed
(Answers need only be given to an accuracy of three significant figures.) (C)

33. A farmer keeps two breeds (A, B) of chicken; 70% of the egg production is from birds of breed A. Of the eggs laid by the A hens, 30% are large, 50% standard and the remainder small: for the B hens the corresponding proportions are 40%, 30% and 30%. Egg colour (brown or white) is manifested independently of size in each breed; 30% of A eggs and 40% of B eggs are brown. Find the probability that:
(*a*) an egg laid by an A hen is large and brown
(*b*) an egg is large and brown
(*c*) a brown egg is large (C)

34. An ordinary unbiased cubical die with faces marked 1 to 6 is thrown twice. Tabulate the probability distribution of the sum of the two throws.
In a certain game a player makes three throws. At each throw he scores according to the face uppermost, if it is a six or if he

has previously thrown a six, but not otherwise. Tabulate the probability distribution of his total score.

Show that his chance of scoring more than 11 is less than 1 in 6, but that his chance of scoring more than 10 exceeds 1 in 6. (c)

35. An electronic audio apparatus is liable to two faults T and R (respectively failure of a transistor and failure of a resistor), which occur independently with respective probabilities 0·2 and 0·1. Verify that the prior probabilities of the four states N (fault-free) T, R, and TR (*i.e.*, T and R simultaneously) are as given in the Table below.

The Table also gives the probabilities, conditional on its state, of the apparatus emitting a hum H, a whine W or a squeal S. Draw up a table giving (either as fractions or as decimals) the posterior probabilities of the four states conditional on H, W or S:

State	Prior probability	Conditional probability of H	W	S	Total
N	0·72	1·0	0	0	1·0
T	0·18	0·5	0·4	0·1	1·0
R	0·08	0·7	0·2	0·1	1·0
TR	0·02	0·2	0·3	0·5	1·0

(c)

36. The diameter of a mass-produced rod is a normal variable with mean 0·85 cm and standard deviation 0·03 cm. The internal diameter of the socket which holds the rod is a normal variable with mean and standard deviation, respectively 0·94 cm and 0·04 cm. Rods and sockets are paired randomly during assembly. What proportion of assemblies are rejected because:

(*a*) the socket is too small for the rod

(*b*) the rod is too loose in the socket, which is the case if the diameter of the socket exceeds that of the rod by more than 0·17 cm

If the diameter of the socket exceeds that of the rod by more than 0·12 cm, the assembly requires a spring washer. What proportion of accepted assemblies require a spring washer? (c)

37. Half the population of the city of Ekron are Philistines and the other half are Canaanites. A Philistine never tells the truth but

a Canaanite speaks truthfully, with a probability $\frac{2}{5}$ and falsely with a probability $\frac{3}{5}$. What is the probability that a citizen encountered at random will give a correct answer to a question? Tabulate (as fractions, decimals or percentages) the probabilities that 0, 1, 2 or 3 men out of a sample of 3 citizens taken at random from Ekron will affirm a proposition:

(*a*) when it is true (*b*) when it is false

The proposition has a prior probability $\frac{3}{4}$ of being true. What is the (posterior) probability of it being true conditional on it being affirmed by only 1 out of the 3? (c)

APPENDIX

Areas under the standard normal curves from 0 to *Z*

Z	0	1	2	3	4	5	6	7	8	9
0·0	0·0000	0·0040	0·0080	0·0120	0·0160	0·0199	0·0239	0·0279	0·0319	0·0359
0·1	0·0398	0·0438	0·0478	0·0517	0·0557	0·0596	0·0636	0·0675	0·0714	0·0754
0·2	0·0793	0·0832	0·0871	0·0910	0·0948	0·0987	0·1026	0·1064	0·1103	0·1141
0·3	0·1179	0·1217	0·1255	0·1293	0·1331	0·1368	0·1406	0·1443	0·1480	0·1517
0·4	0·1554	0·1591	0·1628	0·1664	0·1700	0·1736	0·1772	0·1808	0·1844	0·1879
0·5	0·1915	0·1950	0·1985	0·2019	0·2054	0·2088	0·2123	0·2157	0·2190	0·2224
0·6	0·2258	0·2291	0·2324	0·2357	0·2389	0·2422	0·2454	0·2486	0·2518	0·2549
0·7	0·2580	0·2612	0·2642	0·2673	0·2704	0·2734	0·2764	0·2794	0·2823	0·2852
0·8	0·2881	0·2910	0·2939	0·2967	0·2996	0·3023	0·3051	0·3078	0·3106	0·3133
0·9	0·3159	0·3186	0·3212	0·3238	0·3264	0·3289	0·3315	0·3340	0·3365	0·3389
1·0	0·3413	0·3438	0·3461	0·3485	0·3508	0·3531	0·3554	0·3577	0·3599	0·3621
1·1	0·3643	0·3665	0·3686	0·3708	0·3729	0·3749	0·3770	0·3790	0·3810	0·3830
1·2	0·3849	0·3869	0·3888	0·3907	0·3925	0·3944	0·3962	0·3980	0·3997	0·4015
1·3	0·4032	0·4049	0·4066	0·4082	0·4099	0·4115	0·4131	0·4147	0·4162	0·4177
1·4	0·4192	0·4207	0·4222	0·4236	0·4251	0·4265	0·4279	0·4292	0·4306	0·4319
1·5	0·4332	0·4345	0·4357	0·4370	0·4382	0·4394	0·4406	0·4418	0·4429	0·4441
1·6	0·4452	0·4463	0·4474	0·4484	0·4495	0·4505	0·4515	0·4525	0·4535	0·4545
1·7	0·4554	0·4564	0·4573	0·4582	0·4591	0·4599	0·4608	0·4616	0·4625	0·4633
1·8	0·4641	0·4649	0·4656	0·4664	0·4671	0·4678	0·4686	0·4693	0·4699	0·4706
1·9	0·4713	0·4719	0·4726	0·4732	0·4738	0·4744	0·4750	0·4756	0·4761	0·4767

Z	0	1	2	3	4	5	6	7	8	9
2·0	0·4772	0·4778	0·4783	0·4788	0·4793	0·4798	0·4803	0·4808	0·4812	0·4817
2·1	0·4821	0·4826	0·4830	0·4834	0·4838	0·4842	0·4846	0·4850	0·4854	0·4857
2·2	0·4861	0·4864	0·4868	0·4871	0·4875	0·4878	0·4881	0·4884	0·4887	0·4890
2·3	0·4893	0·4896	0·4898	0·4901	0·4904	0·4906	0·4909	0·4911	0·4913	0·4916
2·4	0·4918	0·4920	0·4922	0·4925	0·4927	0·4929	0·4931	0·4932	0·4934	0·4936
2·5	0·4938	0·4940	0·4941	0·4943	0·4945	0·4946	0·4948	0·4949	0·4951	0·4952
2·6	0·4953	0·4955	0·4956	0·4957	0·4959	0·4960	0·4961	0·4962	0·4963	0·4964
2·7	0·4965	0·4966	0·4967	0·4968	0·4969	0·4970	0·4971	0·4972	0·4973	0·4974
2·8	0·4974	0·4975	0·4976	0·4977	0·4977	0·4978	0·4979	0·4979	0·4980	0·4981
2·9	0·4981	0·4982	0·4982	0·4983	0·4984	0·4984	0·4985	0·4985	0·4986	0·4986
3·0	0·4987	0·4987	0·4987	0·4988	0·4988	0·4989	0·4989	0·4989	0·4990	0·4990
3·1	0·4990	0·4991	0·4991	0·4991	0·4992	0·4992	0·4992	0·4992	0·4993	0·4993
3·2	0·4993	0·4993	0·4994	0·4994	0·4994	0·4994	0·4994	0·4995	0·4995	0·4995
3·3	0·4995	0·4995	0·4995	0·4996	0·4996	0·4996	0·4996	0·4996	0·4996	0·4997
3·4	0·4997	0·4997	0·4997	0·4997	0·4997	0·4997	0·4997	0·4997	0·4997	0·4998
3·5	0·4998	0·4998	0·4998	0·4998	0·4998	0·4998	0·4998	0·4998	0·4998	0·4998
3·6	0·4998	0·4998	0·4999	0·4999	0·4999	0·4999	0·4999	0·4999	0·4999	0·4999
3·7	0·4999	0·4999	0·4999	0·4999	0·4999	0·4999	0·4999	0·4999	0·4999	0·4999
3·8	0·4999	0·4999	0·4999	0·4999	0·4999	0·4999	0·4999	0·4999	0·4999	0·4999
3·9	0·5000	0·5000	0·5000	0·5000	0·5000	0·5000	0·5000	0·5000	0·5000	0·5000

ANSWERS

Exercise 1a

$S \equiv$ sample space, $E \equiv$ event, $A \equiv$ ace, $S \equiv$ spades, $H \equiv$ hearts, $D \equiv$ diamonds, $C \equiv$ clubs

1. $S = \{HH, HT, TH, TT\}$ $\quad E = \{HH\}$
2. $$S = \left\{\begin{array}{l} (H, 1); (H, 2); (H, 3); (H, 4); (H, 5); (H, 6) \\ (T, 1); (T, 2); (T, 3); (T, 4); (T, 5); (T, 6) \end{array}\right\}$$
   $E = \{(H, 2); (H, 4); (H, 6)\}$
3. $$S = \left\{\begin{array}{llllll} (1,1)=2 & (1,2)=3 & (1,3)=4 & (1,4)=5 & (1,5)=6 & (1,6)=7 \\ (2,1)=3 & (2,2)=4 & (2,3)=5 & (2,4)=6 & (2,5)=7 & (2,6)=8 \\ (3,1)=4 & (3,2)=5 & (3,3)=6 & (3,4)=7 & (3,5)=8 & (3,6)=9 \\ (4,1)=5 & (4,2)=6 & (4,3)=7 & (4,4)=8 & (4,5)=9 & (4,6)=10 \\ (5,1)=6 & (5,2)=7 & (5,3)=8 & (5,4)=9 & (5,5)=10 & (5,6)=11 \\ (6,1)=7 & (6,2)=8 & (6,3)=9 & (6,4)=10 & (6,5)=11 & (6,6)=12 \end{array}\right\}$$
   $E = \{(1, 1) = 2 \quad (1, 5) = 6 \quad (2, 4) = 6 \quad (3, 3) = 6 \quad (4, 2) = 6 \quad (5, 1) = 6\}$
4. $S = \{$White ball, red ball, black ball$\}$ $\quad E = \{$White ball$\}$
5. $S = \{$All the children in your school$\}$
   $E = \{$All the left-handed children in your school$\}$
6. $S = \{$All 52 playing cards in a pack$\}$ $\quad E = \{A_H, A_D, A_C, A_S\}$
7. $S = \{$All 1326 ways of selecting 2 cards from a pack of 52$\}$
   $E = \{A_H$ and A_D; A_H and A_C; A_H and A_S; A_D and A_C; A_D and A_S; A_C and $A_S\}$
8. $$S = \left\{\begin{array}{l} 123, 124, 132, 134, 142, 143 \\ 213, 214, 231, 234, 241, 243 \\ 312, 314, 321, 324, 341, 342 \\ 412, 413, 421, 423, 431, 432 \end{array}\right\} \qquad E = \left\{\begin{array}{l} 124, 132, 134, 142 \\ 214, 234, 312, 314 \\ 324, 342, 412, 432 \end{array}\right\}$$
9. $$S = \left\{\begin{array}{l} 102, 103, 120, 123, 130, 132 \\ 201, 203, 210, 213, 230, 231 \\ 301, 302, 310, 312, 320, 321 \end{array}\right\} \qquad E = \left\{\begin{array}{l} 102, 120, 130, 132, 210 \\ 230, 302, 310, 312, 320 \end{array}\right\}$$
10. $S = \{$All 216 possible triples$\}$
    $$E = \left\{\begin{array}{lll} (1,1,1)=3 & (2,1,3)=6 & (3,1,2)=6 \\ (1,1,4)=6 & (2,1,6)=9 & (3,1,5)=9 \\ (1,2,3)=6 & (2,2,2)=6 & (3,2,1)=6 \\ (1,2,6)=9 & (2,2,5)=9 & (3,2,4)=9 \\ (1,3,2)=6 & (2,3,1)=6 & (3,3,3)=9 \\ (1,3,5)=9 & (2,3,4)=9 & (3,3,6)=12 \\ (1,4,1)=6 & (2,4,3)=9 & (3,4,2)=9 \\ (1,4,4)=9 & (2,4,6)=12 & (3,4,5)=12 \\ (1,5,3)=9 & (2,5,2)=9 & (3,5,1)=9 \\ (1,5,6)=12 & (2,5,5)=12 & (3,5,4)=12 \\ (1,6,2)=9 & (2,6,1)=9 & (3,6,3)=12 \\ (1,6,5)=12 & (2,6,4)=12 & (3,6,6)=15 \end{array}\right\}$$
    & $\{6,6,6\} = 18$

Exercise 1b

$S \equiv$ sample space, $E \equiv$ event, $P \equiv$ probability of the event

1. $S=\{1, 2, 3, 4, 5, 6\}$ $\quad E=\{2, 3, 5\}$ $\quad P=\frac{3}{6}=\frac{1}{2}$
2. $$S=\left\{\begin{array}{l}(H, 1); (H, 2); (H, 3); (H, 4); (H, 5); (H, 6)\\ (T, 1); (T, 2); (T, 3); (T, 4); (T, 5); (T, 6)\end{array}\right\}$$
   (*a*) $E=\{H, 6\}$ $\quad P=\frac{1}{12}$
   (*b*) $E=\{(H, 1); (H, 3); (H, 5)\}$ $\quad P=\frac{3}{12}=\frac{1}{4}$
   (*c*) $E=\{(T, 1); (T, 2); (T, 3); (T, 4)\}$ $\quad P=\frac{4}{12}=\frac{1}{3}$
   (*d*) $E=\{(H, 2); (H, 4); (H, 6); (T, 2); (T, 4); (T, 6)\}$ $\quad P=\frac{6}{12}=\frac{1}{2}$
3. $S=\{HHH, HHT, HTH, THH, HTT, THT, TTH, TTT\}$
   $E=\{HHT, HTH, THH\}$ $\quad P=\frac{3}{8}$
4. $$S=\left\{\begin{array}{llllll}(1, 1)=2; & (2, 1)=3; & (3, 1)=4; & (4, 1)=5; & (5, 1)=6; & (6, 1)=7\\ (1, 2)=3; & (2, 2)=4; & (3, 2)=5; & (4, 2)=6; & (5, 2)=7; & (6, 2)=8\\ (1, 3)=4; & (2, 3)=5; & (3, 3)=6; & (4, 3)=7; & (5, 3)=8; & (6, 3)=9\\ (1, 4)=5; & (2, 4)=6; & (3, 4)=7; & (4, 4)=8; & (5, 4)=9; & (6, 4)=10\\ (1, 5)=6; & (2, 5)=7; & (3, 5)=8; & (4, 5)=9; & (5, 5)=10; & (6, 5)=11\\ (1, 6)=7; & (2, 6)=8; & (3, 6)=9; & (4, 6)=10; & (5, 6)=11; & (6, 6)=12\end{array}\right\}$$
   (*a*) $E=\{(1, 6)=7; (2, 5)=7; (3, 4)=7; (4, 3)=7; (5, 2)=7; (6, 1)=7\}$
   $P=\frac{6}{36}=\frac{1}{6}$
   (*b*) $$E=\left\{\begin{array}{llllll}(1, 4)=5; & (1, 5)=6; & (1, 6)=7; & (2, 3)=5; & (2, 4)=6; & (2, 5)=7\\ (2, 6)=8; & (3, 2)=5; & (3, 3)=6; & (3, 4)=7; & (3, 5)=8; & (3, 6)=9\\ (4, 1)=5; & (4, 2)=6; & (4, 3)=7; & (4, 4)=8; & (4, 5)=9; & (4, 6)=10\\ (5, 1)=6; & (5, 2)=7; & (5, 3)=8; & (5, 4)=9; & (5, 5)=10; & (5, 6)=11\\ (6, 1)=7; & (6, 2)=8; & (6, 3)=9; & (6, 4)=10; & (6, 5)=11; & (6, 6)=12\end{array}\right\}$$
   $P=\frac{30}{36}=\frac{5}{6}$
   (*c*) $$E=\left\{\begin{array}{llllll}(1, 2)=3; & (1, 4)=5; & (1, 6)=7; & (2, 1)=3; & (2, 3)=5; & (2, 5)=7\\ (3, 2)=5; & (3, 4)=7; & (3, 6)=9; & (4, 1)=5; & (4, 3)=7; & (4, 5)=9\\ (5, 2)=7; & (5, 4)=9; & (5, 6)=11; & (6, 1)=7; & (6, 3)=9; & (6, 5)=11\end{array}\right\}$$
   $P=\frac{18}{36}=\frac{1}{2}$
   (*d*) $$E=\left\{\begin{array}{l}(1, 1)=2; (4, 6)=10; (5, 5)=10; (6, 4)=10;\\ (5, 6)=11; (6, 5)=11; (6, 6)=12\end{array}\right\} P=\frac{7}{36}$$
   (*e*) $$E=\left\{\begin{array}{lllll}(1, 1) = 2 & (1, 2) = 3 & (1, 4) = 5 & (1, 6) = 7 & (2, 1) = 3\\ (2, 3) = 5 & (2, 5) = 7 & (3, 2) = 5 & (3, 4) = 7 & (4, 1) = 5\\ (4, 3) = 7 & (5, 2) = 7 & (5, 6) = 11 & (6, 1) = 7 & (6, 5) = 11\end{array}\right\}$$
   $P=\frac{15}{36}=\frac{5}{12}$
5. $$S=\left\{\begin{array}{l}ABC, ACB, BAC, BCA, CAB, CBA, ABD, ADB\\ BAD, BDA, DAB, DBA, ACD, ADC, CAD, CDA\\ DAC, DCA, BCD, BDC, CBD, CDB, DBC, DCB\end{array}\right\}$$
   $E=\{BAD, CAB, CAD, DAB\}$ $\quad P=\frac{4}{24}=\frac{1}{6}$
6. $S=\{52$ playing cards$\}$
   (*a*) $E=\{3_H\}$ $\quad P=\frac{1}{52}$
   (*b*) $E=\{3_H, 3_D, 3_S, 3_C\}$ $\quad P=\frac{4}{52}=\frac{1}{13}$
   (*c*) $E=\{K_H, K_D, K_S, K_C, Q_H, Q_D, Q_S, Q_C\}$ $\quad P=\frac{8}{52}=\frac{2}{13}$
   (*d*) $E=\{A_D, J_H, J_D, J_S, J_C\}$ $\quad P=\frac{5}{52}$
   (*e*) $E=\{2_H, 4_H, 6_H, 8_H, 10_H, 2_D, 4_D, 6_D, 8_D, 10_D\}$ $\quad P=\frac{10}{52}=\frac{5}{26}$
   (*f*) P (Not a diamond) $= P$ (heart, spade or club) $=\frac{39}{52}=\frac{3}{4}$

7. $S=\{12, 21, 13, 31, 23\ P, 32\}$
   (*a*) $E=\{12, 32\}=\frac{2}{6}=\frac{1}{3}$
   (*b*) $E=\{13, 23, 31\}$ $P=\frac{3}{6}=\frac{1}{2}$
   (*c*) $E=\{21, 23, 31, 32\}$ $P=\frac{4}{6}=\frac{2}{3}$
8. $W\equiv$ white sock, $B\equiv$ black sock
   $S=\{W_1W_2, W_1B_1, W_1B_2, W_2B_1, W_2B_2, B_1B_2\}$
   $E=\{W_1W_2, B_1B_2\}$ $P=\frac{2}{6}=\frac{1}{3}$
9. $$S=\begin{Bmatrix}1234, 1243, 1324, 1342, 1423, 1432, 2134, 2143\\ 2314, 2341, 2413, 2431, 3124, 3142, 3214, 3241\\ 3412, 3421, 4123, 4132, 4213, 4231, 4312, 4321\end{Bmatrix}$$
   (*a*) $E=\{1234\}$ $P=\frac{1}{24}$
   (*b*) $E=\{1243, 1324, 1432, 2134, 3214, 4231\}$ $P=\frac{6}{24}=\frac{1}{4}$
   (*c*) $E=\{1234, 1243, 1324, 1432, 2134, 3214, 4231\}$ $P=\frac{7}{24}$
   (NB 'At least 2' includes 3 or 4. 3 is out since if 3 are in order so must the 4th)
   (*d*) $E=\{\text{All except } 1234\}$ $P=\frac{23}{24}$
   (NB 'At most 2' includes 0 or 1)
10. $R\equiv$ red die, $B\equiv$ black die
   $$S=\begin{Bmatrix}R_1B_1, R_2B_1, R_3B_1, R_4B_1, R_5B_1, R_6B_1\\ R_1B_2, R_2B_2, R_3B_2, R_4B_2, R_5B_2, R_6B_2\\ R_1B_3, R_2B_3, R_3B_3, R_4B_3, R_5B_3, R_6B_3\\ R_1B_4, R_2B_4, R_3B_4, R_4B_4, R_5B_4, R_6B_4\\ R_1B_5, R_2B_5, R_3B_5, R_4B_5, R_5B_5, R_6B_5\\ R_1B_6, R_2B_6, R_3B_6, R_4B_6, R_5B_6, R_6B_6\end{Bmatrix}$$
   (*a*) $E=\{R_5B_1, R_6B_1, R_6B_2\}$ $P=\frac{3}{36}=\frac{1}{12}$
   (*b*) $E=\{R_2B_1, R_3B_1, R_4B_1, R_5B_1, R_6B_1, R_4B_2, R_6B_2, R_6B_3\}$
   $P=\frac{8}{36}=\frac{2}{9}$
   (*c*) $$E=\begin{Bmatrix}R_1B_1, R_2B_1, R_3B_1, R_4B_1, R_5B_1, R_6B_1, R_1B_2\\ R_2B_2, R_4B_2, R_6B_2, R_1B_3, R_3B_3, R_6B_3, R_1B_4\\ R_2B_4, R_4B_4, R_1B_5, R_5B_5, R_1B_6, R_2B_6, R_3B_6, R_6B_6\end{Bmatrix}$$
   $P=\frac{22}{36}=\frac{11}{18}$
   (*d*) $$E=\begin{Bmatrix}(R_1B_1)=1 & (R_1B_4)=4 & (R_2B_2)=4 & (R_3B_3)=9\\ (R_4B_1)=4 & (R_4B_4)=16 & (R_5B_5)=25 & (R_6B_6)=36\end{Bmatrix}$$
   $P=\frac{8}{36}=\frac{2}{9}$
11. $$S=\begin{Bmatrix}123, 124, 125, 134, 135\\ 145, 234, 235, 245, 345\end{Bmatrix}$$
   $E=\{123, 135, 234, 345\}$
   $P=\frac{4}{10}=\frac{2}{5}$

Exercises on Chapter 1

1. (*a*) $\frac{1}{6}$ (*b*) $\frac{1}{6}$ (*c*) $\frac{1}{3}$
2. (*a*) $\frac{3}{5}$ (*b*) $\frac{2}{5}$
3. (*a*) $\frac{2}{5}$ (*b*) $\frac{1}{5}$ (*c*) $\frac{3}{5}$ (*d*) $\frac{3}{5}$
4. (*a*) $\frac{1}{2}$ (*b*) $\frac{9}{100}$ (*c*) $\frac{1}{10}$ (*d*) $\frac{1}{4}$
   (*e*) $\frac{1}{25}$ (*f*) $\frac{3}{50}$ (*g*) $\frac{14}{25}$ (*h*) $\frac{1}{20}$
   (*i*) $\frac{1}{10}$ (*j*) $\frac{1}{50}$ (NB in (*i*) 100 has 2 digits the same)
5. (*a*) $\frac{7}{20}$ (*b*) $\frac{1}{2}$ (*c*) $\frac{43}{100}$ (*d*) $\frac{1}{10}$
   (*e*) $\frac{1}{2}$ (NB in (*e*) average is between 9 and 10)

6. (*a*) $\frac{1}{16}$ (*b*) $\frac{1}{16}$ (*c*) $\frac{15}{16}$ (*d*) $\frac{11}{16}$
(*e*) $\frac{1}{2}$
(NB in (*b*) 'no tails' is the same as 'all heads'; in (*c*) 'at least 1 head' includes 2, 3 or 4; in (*d*) 'at most 2 tails' includes 0 or 1)
7. (*a*) $\frac{1}{5}$ (*b*) $\frac{1}{4}$ (*c*) $\frac{7}{20}$ (*d*) $\frac{19}{20}$
(*e*) $\frac{3}{5}$ (*f*) $\frac{3}{5}$
8. (*a*) $\frac{1}{49}$ (*b*) $\frac{4}{49}$ (*c*) $\frac{2}{49}$
9. (*a*) $\frac{5}{12}$ (*b*) $\frac{1}{4}$ (*c*) $\frac{5}{6}$
10. $\frac{1}{10}$
11. (*a*) $\frac{1}{10}$ (*b*) $\frac{3}{10}$ (*c*) $\frac{9}{10}$
12. (*a*) (*i*) $\frac{2}{3}$ (*ii*) $\frac{7}{12}$ (*b*) $\frac{1}{2}$
13. $\frac{3}{5}$
14. $\frac{54}{N}=\frac{3}{94}$: $N=1692$

Exercise 2a

1. $\frac{7}{10}$
2. $\frac{37}{40}$
3. $\frac{5}{6}$
4. (*a*) $\frac{18}{25}$ (*b*) $\frac{57}{100}$ (*c*) $\frac{77}{100}$
5. 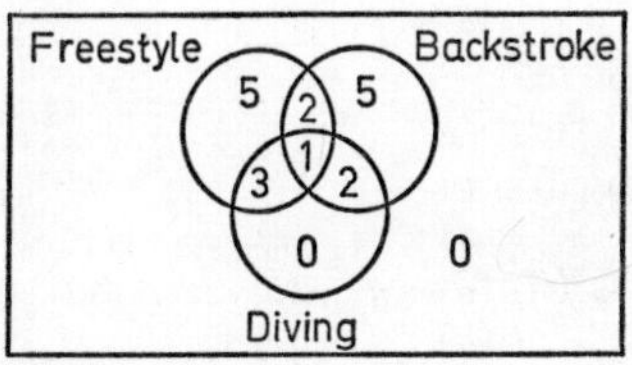

$P=\frac{13}{18}$

Exercise 2b

Nos. 1, 2, 5, 6 and 9 are mutually exclusive events.

Exercise 2c

1. $\frac{4}{52}\cdot\frac{3}{51}=\frac{1}{221}$
2. (*a*) $\frac{4}{12}\cdot\frac{3}{11}=\frac{1}{11}$ (*b*) $\frac{2}{12}\cdot\frac{1}{11}=\frac{1}{66}$
(*c*) $\frac{6}{12}\cdot\frac{5}{11}=\frac{5}{22}$ (*d*) $\frac{4}{12}\cdot\frac{2}{11}=\frac{2}{33}$
(*e*) $\frac{4}{12}\cdot\frac{6}{11}=\frac{2}{11}$
3. (*a*) $\frac{4}{12}\cdot\frac{4}{12}=\frac{1}{9}$ (*b*) $\frac{2}{12}\cdot\frac{2}{12}=\frac{1}{36}$
(*c*) $\frac{6}{12}\cdot\frac{6}{12}=\frac{1}{4}$ (*d*) $\frac{4}{12}\cdot\frac{2}{12}=\frac{1}{18}$
(*e*) $\frac{4}{12}\cdot\frac{6}{12}=\frac{1}{6}$
4. $\frac{1}{6}\cdot\frac{1}{6}=\frac{1}{36}$
5. (*a*) $\frac{4}{52}\cdot\frac{3}{51}=\frac{1}{221}$ (*b*) $\frac{4}{52}\cdot\frac{4}{51}=\frac{4}{663}$
(*c*) $\frac{26}{52}\cdot\frac{25}{51}=\frac{25}{102}$ (*d*) $\frac{1}{52}\cdot\frac{1}{51}=\frac{1}{2652}$
(*e*) $\frac{4}{52}\cdot\frac{48}{51}=\frac{16}{221}$
6. (*a*) $\frac{4}{52}\cdot\frac{4}{52}=\frac{1}{169}$ (*b*) $\frac{4}{52}\cdot\frac{4}{52}=\frac{1}{169}$
(*c*) $\frac{26}{52}\cdot\frac{26}{52}=\frac{1}{4}$ (*d*) $\frac{1}{52}\cdot\frac{1}{52}=\frac{1}{2704}$
(*e*) $\frac{4}{52}\cdot\frac{48}{52}=\frac{12}{169}$

Exercise 2d

1. $P(A)=\frac{4}{5}$; $P(B)=\frac{1}{2}$; $P(A\cap B)=\frac{3}{10}$
$P(A\cap B)\neq P(A).P(B)$; Not indep.
2. $P(A)=\dfrac{8}{x+9}$; $P(B)=\dfrac{5}{x+9}$

$P(A\cap B)=\dfrac{4}{x+9}$

$\dfrac{4}{x+9}=\dfrac{8}{x+9}\cdot\dfrac{5}{x+9}$, whence $x=1$
3. $P(A)=\frac{1}{4}$; $P(A\cap B)=\frac{1}{8}$; $P(B)=\frac{1}{2}$
4. (*a*) $P(A)=\frac{3}{7}$, $P(B)=\frac{1}{2}$,
$P(A\cap B)=\frac{3}{14}$, Independent
(*b*) $P(A)=\frac{3}{7}$, $P(C)=\frac{3}{7}$,
$P(A\cap C)=\frac{1}{7}$, Not independent
(*c*) $P(B)=\frac{1}{2}$, $P(C)=\frac{3}{7}$,
$P(B\cap C)=\frac{2}{7}$, Not independent
(*d*) $P(A)=\frac{3}{7}$, $P(B)=\frac{1}{2}$, $P(C)=\frac{3}{7}$,
$P(A\cap B\cap C)=\frac{1}{14}$, Not independent

Exercises on Chapter 2

1. (*b*) (*i*) $P(A \cup B \cup C) = P(A) + P(B) + P(C)$

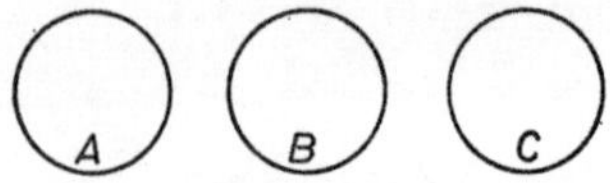

(*ii*) $P(A \cup B \cup C) = P(A) + P(B) + P(C) - P(B \cap C) - P(A \cap C)$

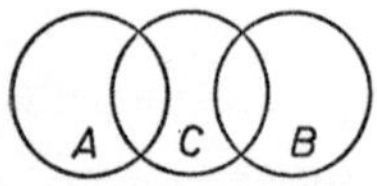

2. (*a*) $\frac{5}{12}$ (*b*) $\frac{1}{12}$ (*c*) $\frac{1}{2}$ (*d*) $\frac{7}{12}$

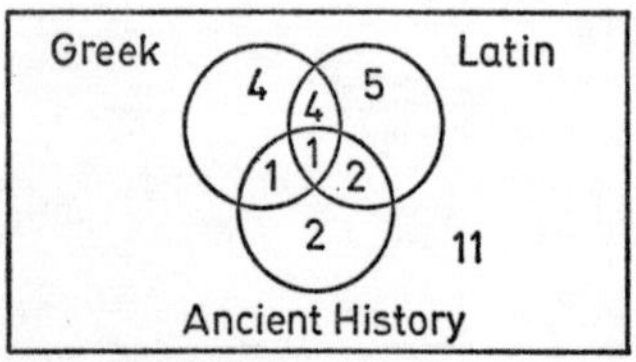

3. 0·2 (NB these events are not mutually exclusive)
4. (*a*) $\frac{15}{26}$ (*b*) $\frac{1}{26}$ (*c*) $\frac{1}{104}$ (*d*) $\frac{53}{104}$ (NB these events are independent but not mutually exclusive)
5. Where one, or other, or both events are null.
6. $P(A \cap B \cap C) = P(A).P(B|A).P(C|A \cap B)$
7. (*a*) 2 (*b*) (*i*) No (*ii*) Yes (*iii*) Yes
8.

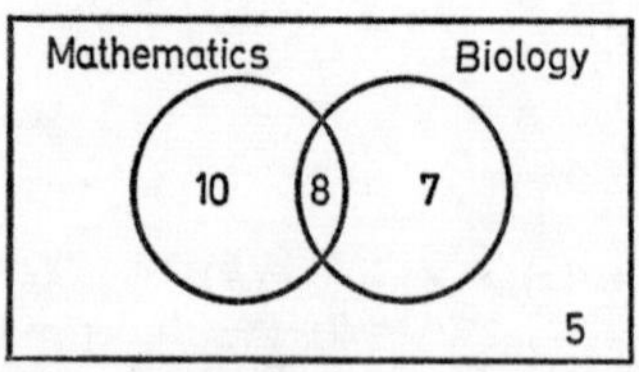

$P = \frac{3}{5}$

9. $\frac{7}{24}$. (NB these events are mutually exclusive)
10. (*a*) (*i*) $\frac{3}{5}$ (*ii*) $\frac{3}{10}$ (*iii*) $\frac{11}{20}$ (*iv*) $\frac{1}{8}$ (*v*) $\frac{23}{40}$ (*vi*) $\frac{2}{5}$ (*vii*) $\frac{11}{40}$ (*viii*) $\frac{17}{20}$ (ix) $\frac{33}{40}$
    (*b*) (*i*) $\frac{1}{3}$ (*ii*) $\frac{5}{12}$ (*iii*) $\frac{1}{4}$
11. P (winning 3 prizes) $= \frac{3}{1000} \cdot \frac{2}{999} \cdot \frac{1}{998} = \frac{1}{166\,167\,000}$
12. (*a*) $\frac{1}{8}$ (*b*) $\frac{1}{1024}$ (*c*) $\frac{1}{2^n}$
13. (*a*) $\frac{19}{99}$ (*b*) $\frac{59}{66}$ (*c*) $\frac{29}{66}$ (*d*) $\frac{29}{99}$ (*e*) $\frac{5}{9}$ (*f*) $\frac{79}{198}$ (*g*) $\frac{21}{22}$ (*h*) $\frac{35}{99}$ (*i*) $\frac{37}{99}$ (*j*) $\frac{34}{99}$ (*k*) $\frac{65}{99}$
    (NB in (*i*) first not red and second not blue means either white and white, or white and red or blue and white or blue and red.)
14. (*a*) $\frac{16}{81}$ (*b*) $\frac{8}{9}$ (*c*) $\frac{4}{9}$ (*d*) $\frac{8}{27}$ (*e*) $\frac{5}{9}$ (*f*) $\frac{32}{81}$ (*g*) $\frac{77}{81}$ (*h*) $\frac{28}{81}$ (*i*) $\frac{10}{27}$ (*j*) $\frac{29}{81}$ (*k*) $\frac{52}{81}$
15. (*a*) $\frac{38}{473}$ (*b*) $\frac{4}{473}$ (*c*) $\frac{91}{2838}$ (*d*) $\frac{190}{1419}$ (*e*) $\frac{100}{473}$ (*f*) $\frac{1895}{8514}$
16. (*i*) (*a*) $\frac{64}{729}$ (*b*) $\frac{8}{729}$ (*c*) $\frac{11}{81}$ (*d*) $\frac{32}{243}$ (*e*) $\frac{16}{81}$ (*f*) $\frac{2}{9}$
    (*ii*) (*a*) $\frac{76}{891}$ (*b*) $\frac{1}{99}$ (*c*) $\frac{233}{1782}$ (*d*) $\frac{118}{891}$ (*e*) $\frac{20}{99}$ (*f*) $\frac{725}{3564}$
    (*iii*) (*a*) $\frac{76}{891}$ (*b*) $\frac{1}{99}$ (*c*) $\frac{233}{1782}$ (*d*) $\frac{118}{891}$ (*e*) $\frac{20}{99}$ (*f*) $\frac{170}{891}$
    (NB in (*ii*)(*d*) $RRW + RWR + WRR = \frac{20}{45} \cdot \frac{20}{45} \cdot \frac{10}{44} + \frac{20}{45} \cdot \frac{10}{45} \cdot \frac{20}{44} + \frac{10}{45} \cdot \frac{20}{45} \cdot \frac{19}{44}$)
17.

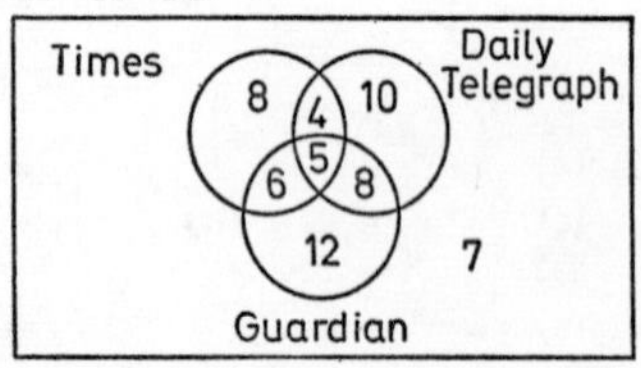

(*a*) $\frac{43}{60}$ (*b*) $\frac{11}{23}$ (*c*) $\frac{1}{2}$ (*d*) $\frac{53}{60}$ (*e*) $\frac{11}{12}$ (*f*) $\frac{7}{60}$

18. $\frac{1}{4}$
    $P(B \text{ win}) = P(A \text{ not win}).P(B \text{ win})$
19. $P(A$ winning) is the sum of the infinite series $\frac{1}{2} + \frac{1}{8} + \frac{1}{32} + \ldots$
    $P(B$ winning) is the sum of the infinite series $\frac{1}{4} + \frac{1}{16} + \frac{1}{64} + \ldots$
20. (*a*) $\frac{4}{7}$, $\frac{2}{7}$ and $\frac{1}{7}$
    (*b*) (*i*) $\dfrac{2^{n-2}}{2^n - 1}$ (*ii*) $\dfrac{1}{2^n - 1}$
    $P(k\text{th man would win}) = \dfrac{2^{n-k}}{2^n - 1}$

Exercise 3a

1. (*a*) $\frac{27}{125}$ (*b*) $\frac{8}{125}$ (*c*) $\frac{36}{125}$
(*d*) $\frac{54}{125}$ (*e*) $\frac{117}{125}$
2. (*a*) $\frac{1}{4}$ (*b*) $\frac{7}{15}$ (*c*) $\frac{17}{60}$
(*d*) $\frac{43}{60}$ (*e*) $\frac{8}{15}$
3. (*a*) $\frac{1}{6}$ (*b*) $\frac{7}{18}$ (*c*) $\frac{2}{9}$
4. (*a*) $\frac{17}{48}$ (*b*) $\frac{3}{16}$ (*c*) $\frac{11}{48}$

Exercise 3b

1. (*a*) $\frac{2}{3}$ (*b*) $\frac{1}{3}$
2. (*a*) $\frac{4}{9}$ (*b*) $\frac{5}{9}$
3. $\frac{16}{25}$
4. (*a*) $\frac{1}{2}$ (*b*) $\frac{1}{3}$ (*c*) $\frac{1}{6}$
5. $\frac{7}{25}$
6. $\frac{8}{13}$

Exercise 3c

1. (*a*) 1680 (*b*) 2520 (*c*) $n-1$
2. 24
3. 5040
4. 720
5. 256
6. 192
7. 24, 18
8. 210
9. 126
10. (*a*) 720 (*b*) (*i*) 120
(*ii*) 360 if vowels can be separated; 120 if vowels kept together
(*iii*) 24 (*iv*) 24
11. (*a*) 120 (*b*) 48 (*c*) 84
12. (*a*) 60 480 (*b*) 43 200 (*c*) 2520
13. (*a*) 120 (*b*) 60 (*c*) 40
14. 2520, 2400
15. 105

Exercise 3d

1. (*a*) 10 (*b*) 28 (*c*) 4851
2. (*a*) 8 (*b*) 20
3. 35
4. 56
5. 1326
6. (*a*) 364 (*b*) 286
7. (*a*) 840 (*b*) 560 (*c*) 1946
(*d*) 56 (*e*) 1316
8. 2450
9. 31
10. 11 999
11. (*a*) 59 (*b*) 31
12. $2^k(p+1)(q+1)(r+1)-1$

Exercise 3e

1. $\frac{5}{7}$
2. $\frac{5}{6}$
3. (*a*) $\frac{7}{12}$ (*b*) $\frac{7}{12}$ (*c*) $\frac{7}{12}$
(*d*) $\frac{14}{33}$ (*e*) $\frac{98}{99}$ (*f*) $\frac{92}{99}$
4. (*a*) $\frac{5}{68}$ (*b*) $\frac{5}{68}$ (*c*) $\frac{3}{68}$
(*d*) $\frac{3}{136}$
5. (*a*) $\frac{1}{68}$ (*b*) $\frac{1}{68}$ (*c*) $\frac{1}{170}$
(*d*) $\frac{1}{680}$ (*e*) $\frac{3}{68}$ (*f*) $\frac{3}{136}$
(*g*) $\frac{5}{136}$ (*h*) $\frac{11}{68}$
6. (*a*) $\frac{1}{6}$ (*b*) $\frac{1}{12}$ (*c*) 30
(*d*) $\frac{7}{15}$
7. (*a*) $\frac{1}{270\,725}$ (*b*) $\frac{36}{270\,725}$
(*c*) $\frac{286}{20\,825}$ (*d*) $\frac{3432}{20\,825}$
(*e*) $\frac{15\,229}{54\,145}$ (*f*) $\frac{2197}{20\,825}$
8. $\frac{4}{5}$

Exercise 3f

1. 5p
2. 12p
3. Expected gain of £16·67
4. 1·5
5. $E(X) = 0{\cdot}1$ $E(X^2) = 3{\cdot}9$
   $E(4X) = 0{\cdot}4$ $E(2X+3) = 3{\cdot}2$
6. 6·75

Exercises on Chapter 3

1. $\frac{1}{4}$
3. 969
4. $\frac{13}{20}$
5. £30·24; £26·24; £23·04; £20·48
6. £403·20
7. £199 584
8. (*a*) 4804 (*b*) 21 808
9. (*a*) $\frac{1}{4}$ (*b*) $\frac{3}{8}$
10. $\frac{1}{120}$
11. (*a*) $\frac{1}{4}$ (*b*) $\frac{1}{4}$ (*c*) $\frac{11}{16}$
    $\frac{n-1}{2^n}$
12. $\frac{7}{31}$
13. 7
14. 50 400
15. 56
16. (*a*) (*i*) 362 880 (*ii*) 17 280
    (*iii*) 30 240 (*iv*) 80 640
    (*v*) 1728 (*vi*) 40 320
    (*vii*) 5040 if ends are named;
    10 080 if ends unnamed
    (*b*) 15 120
    (*c*) 1260
17. $\frac{17}{33}$
18. (*a*) $\frac{1}{5}$ (*b*) $\frac{13}{50}$ (*c*) $\frac{3}{10}$
    (*d*) $\frac{27}{50}$
19. $\frac{3}{7}$
20. 120
21. 100
24. (*a*) $\frac{2}{15}$ (*b*) $\frac{2}{3}$ (*c*) $\frac{13}{15}$
25. $\frac{1}{4}$
26. $\frac{1}{26^8}$
27. (*a*) $\frac{3}{54\,145}$ (*b*) $\frac{3}{10\,179\,260}$
    (*c*) $\frac{33}{391\,510}$ (*d*) $\frac{66}{195\,755}$
    (*e*) $\frac{124\,741}{124\,774}$
29. £4; £3; £2; £1
30. $\frac{4}{7}$
32. $\frac{45}{49}$

Exercise 4a

1. $\frac{15}{64}$
2. (*a*) $\frac{3125}{7776}$ (*b*) $\frac{1}{32}$ (*c*) $\frac{625}{648}$
   (*d*) $\frac{4651}{7776}$
3. $\frac{1}{32}$ $\frac{5}{32}$ $\frac{10}{32}$ $\frac{10}{32}$ $\frac{5}{32}$ $\frac{1}{32}$
4. $\frac{1}{125}$ $\frac{10}{125}$ $\frac{40}{125}$ $\frac{80}{125}$ $\frac{80}{125}$ $\frac{32}{125}$
5. (*a*) 625 (*b*) 3750 (*c*) 9375
   (*d*) 3125
6. $\frac{729}{4096}$ $\frac{1458}{4096}$ $\frac{1215}{4096}$ $\frac{540}{4096}$ $\frac{135}{4096}$
   $\frac{18}{4096}$ $\frac{1}{4096}$
7. (*a*) $\frac{243}{1024}$ (*b*) $\frac{135}{512}$ (*c*) $\frac{781}{1024}$
   (*d*) $\frac{61}{256}$

Exercise 4b

1. (*a*) 0·4049 (*b*) 0·4940
   (*c*) 0·3159 (*d*) 0·4608
   (*e*) 0·8664 (*f*) 0·9534
   (*g*) 0·1335 (*h*) 0·1112
   (*i*) 0·6331 (*j*) 0·9772
2. (*a*) 2 (*b*) 3 (*c*) −1·3
   (*d*) 0·2 (*e*) −2·5

3. (*a*) 0·0202 (*b*) 0·0011
(*c*) 0·7062 (*d*) 0·0129
(*e*) $\frac{0{\cdot}1469}{0{\cdot}2912} = 0{\cdot}5045$ (*f*) 77
4. (*a*) $Z > 1{\cdot}73$; 129
(*b*) $Z < 3{\cdot}63$ None
(*c*) $-2{\cdot}63 < Z < 2{\cdot}63$; 991
5. (*a*) $Z > 1.34$ 9%
(*b*) $Z < -0{\cdot}41$ 34%
(NB standard deviation given to 1 decimal place. Change measurements to millimetres)

Exercise 4c

1. (*a*) 0·5199 (*b*) 0·0398
2. 0·0017
3. 0·2810
4. 0·1292
5. 0·2514

Exercise 4d

1. $P(0) = 0{\cdot}0101$ $P(1) = 0{\cdot}0465$
$P(2) = 0{\cdot}1069$
2. (*a*) 0·1353 (*b*) 0·2706
(*c*) 0·6765
3. 0·6124
0·6026 (by normal distribution)
4. Calc. freq. 64 29 6 1 0
5. (*a*) 0·0498 (*b*) 0·4233

Exercise 4e

1. 0·0035
2. $Z > 2{\cdot}17$; 0·0150
3. 123

Exercises on Chapter 4

1. 0·9959
2. 0·0006
3. (*a*) 0·0035 (*b*) 0·2420
(*c*) 0·9753
4. 99%
5. 0·9198
6. 0·0796
7. $\frac{1}{81}$ $\frac{8}{81}$ $\frac{24}{81}$ $\frac{32}{81}$ $\frac{16}{81}$
8. 2401 4116 2646 756 81
(NB $n = 4$, $\mu = 1{\cdot}2$
$\therefore 4p = 1{\cdot}2$, $p = 0{\cdot}3$)
9. (*a*) 0·0498 (*b*) 0·1494
(*c*) 0·2241 (*d*) 0·2241
(*e*) 0·3526
10. Calc. freq. 216 432 288 64
11. Calc. freq. 55 33 10 2 0
12. (*a*) 0·4325 (*b*) 0·0668
13. (*a*) 0·0495 (*b*) 0·1263
14. 0·2714, 2714
15. (*a*) 0·1353 (*b*) 0·2706
(*c*) 0·8569
16. (*a*) 0·9560 (*b*) 0·9990
18. $f(0) = 0{\cdot}000002$; $f(1) = 0{\cdot}000048$;
$f(2) = 0{\cdot}000528$; $f(3) = 0{\cdot}00352$;
$f(4) = 0{\cdot}01584$; $f(5) = 0{\cdot}050688$;
$f(6) = 0{\cdot}118272$; $f(7) = 0{\cdot}202752$;
$f(8) = 0{\cdot}253440$; $f(9) = 0{\cdot}22528$;
$f(10) = 0{\cdot}135168$; $f(11) = 0{\cdot}049152$;
$f(12) = 0{\cdot}008192$
(NB the sum of the probabilities is not 1 because of the slight inaccuracy of taking $\left(\frac{1}{3}\right)^{12}$ as 0·000002)

19. $\left(\frac{3}{5}\right)^8 + 8\left(\frac{3}{5}\right)^7\left(\frac{2}{5}\right) + \binom{8}{2}\left(\frac{3}{5}\right)^6\left(\frac{2}{5}\right)^2 + \binom{8}{3}\left(\frac{3}{5}\right)^5\left(\frac{2}{5}\right)^3 = \frac{46\,413}{78\,125}$
20. 0·2877
21. Mean = 27 stand dev. = 4·5
22. 68
23. $n \geqslant 6$
24. $\frac{37}{64}$
25. $\frac{857}{3125}$

Exercise 5a

1. $\frac{5}{324}$
2. (*a*) $\frac{3}{50}$ (*b*) $\frac{1}{20}$ (*c*) $\frac{3}{625}$; $P = \frac{3}{25}$
3. (*a*) $\frac{5}{512}$ (*b*) $\frac{15}{512}$

Exercise 5b

1. (*a*), (*c*), (*d*) are stochastic
2. (*b*), (*c*) are regular matrices
3. (*a*) $P = \begin{pmatrix} 0{\cdot}5 & 0{\cdot}5 \\ 0{\cdot}7 & 0{\cdot}3 \end{pmatrix}$

   (*b*) $P_2 = \begin{pmatrix} 0{\cdot}60 & 0{\cdot}40 \\ 0{\cdot}56 & 0{\cdot}44 \end{pmatrix}$

   $P_3 = \begin{pmatrix} 0{\cdot}580 & 0{\cdot}420 \\ 0{\cdot}588 & 0{\cdot}412 \end{pmatrix}$

   (*c*) 0·580
   (*d*) $\frac{5}{12}$
4. (*a*) 0·74 (*b*) 0·26; $P = \frac{5}{7}$
5. (*a*) $\frac{1}{8}$ (*b*) $\frac{7}{18}$
6. $\frac{1}{2}$, $\frac{1}{4}$, 0, depending on whether second parent is *AA*, *Aa* or *aa*

Exercises on Chapters 1–5

1. $\frac{5}{42}$
2. (*a*) $\frac{11}{32}$ (*b*) $\frac{77}{2048}$
3. (*a*) $\frac{14}{33}$ (*b*) $\frac{16}{33}$ (*c*) $\frac{1}{11}$
   (*d*) $\frac{19}{33}$
4. *A* first, $\frac{3}{5}$; *B* first, $\frac{2}{5}$
5. (*a*) $\frac{1}{125}$ (*b*) $\frac{12}{125}$ (*c*) $\frac{112}{125}$
6. 0·729, 0·243, 0·027, 0·001
7. (*a*) $\frac{441}{1000}$ (*b*) $\frac{370}{1000}$
8. (*a*) 0·27 (*b*) 0·33 (*c*) 0·22
9. (*a*) $\frac{13}{36}$ (*b*) $\frac{5}{18}$
10. $\frac{3}{5}$
11. (*a*) $\frac{243}{1024}$ (*b*) $\frac{81}{128}$
12. 72, $\frac{1}{3}$
13. (*a*) 0 (*b*) 1 : *D*, *A*, *C*, *B*
14. (*a*) $\frac{1}{12}$ (*b*) $\frac{5}{648}$ (*c*) $\frac{31}{216}$
15. $(1-q^3)^4\,(1-4q^3+10q^6)$
16. (*a*) $\frac{17}{20}$ (*b*) $\frac{7}{8}$
17. (*a*) 0·6065 (*b*) 0·0903
18. $p(1+p+2p^2)$
19. (*a*) $r=1$ (*b*) $\frac{2}{9}$
20. 0·22
21. (*a*) $(1-p_4)^6(1+6p_4+21p_4^2)$
    (*b*) (*i*) $p_1p_{1A}+p_2p_{2A}+p_3p_{3A}$
    (*ii*) $p_4+p_1p_{1C}+p_2p_{2C}+p_3p_{3C}$
    (*c*) $(1-p_1p_{1A}-p_2p_{2A}-p_3p_{3A})^4$
22. 4·3%
23. (*a*) $\left(\frac{79}{2^{12}}\right)^2$ (*b*) 0·003476
24. (*a*) $\frac{18}{37}$ (*b*) $\frac{1}{37}$ (*c*) 0·065
    (*d*) 0·036
25. (*a*) $\frac{1}{16}$ (*b*) $\frac{1}{32}$
    $1{\cdot}91.10^{-2}$. Player's average receipt is also unlimited
26. $\frac{63}{8}(2pq)^n(p^2+q^2)$
28. (*a*) $m/(m+n)$
    (*b*) $m(m-1)/(m+n)(m+n-1)$
    (*c*) $n/(m+n-1)$ (*d*) $m/(m+n)$
29. 0·2
30. (*a*) 0·92 (*b*) 0·37 (*c*) 0·63
31. 1; 2; 9; 44
32. (*a*) 1 (*b*) 0·846 (*c*) 0·013
33. (*a*) 0·09 (*b*) 0·111 (*c*) 0·336

34.

Sum	2	3	4	5	6	7	8	9	10	11	12
Prob.*	1	2	3	4	5	6	5	4	3	2	1

* As a multiple of $\frac{1}{36}$

Score	0	6	7	8	9	10	11
Prob.†	125	25	5	6	7	8	9

Score	12	13	14	15	16	17	18
Prob.†	10	6	5	4	3	2	1

† As a multiple of $\frac{1}{216}$

35.

State	Posterior probability conditional on *H*	*W*	*S*
N	0·828	0	0
T	0·103	0·766	0·500
R	0·064	0·170	0·222
TR	0·005	0·064	0·278

36. (*a*) 0·036 (*b*) 0·055 0·241

37. $P=\frac{1}{5}$

	True	False
0	$\frac{64}{125}$	$\frac{1}{125}$
1	$\frac{48}{125}$	$\frac{12}{125}$
2	$\frac{12}{125}$	$\frac{48}{125}$
3	$\frac{1}{125}$	$\frac{64}{125}$

$P=\frac{12}{13}$

INDEX

INTRODUCTION

There are many works, written in several European languages, on general, animal, and more especially vegetational ecology, but the space devoted to the study and description of the various ecosystems – both terrestrial and aquatic – of Europe is very limited, and this even in the books of European workers. Indeed, today we are in the strange position that, while some extra-European continents or parts of continents such as South Africa, the west part of South American, Australia, the arid zones of the Old World, not to mention Antarctica (that recently has attracted the attention of a great number of students) have been very thoroughly studied and illustrated – for European ecosystems we have very scant information, if we make allowance for the many analytical works on single natural or artificial environments (generally belonging to Central or Northern Europe). We have a beautiful book on the animal ecology of Palestine (today comprising Israel plus Jordan) written in the mid-thirties by Bodenheimer. However, Palestine does not belong to Europe, although some of the ecological and geographical peculiarities of the western part of the country do correspond very closely to those of Southern Europe.

This is the reason why I have felt it opportune, even necessary to gather the most original, recent and serious data we have about European natural and artificial ecosystem, both terrestrial and aquatic, including both freshwater and inland seas, and to write a book which would give the most complete and clear, though synthetic, idea possible about these environments. It is unnecessary to say that for Europe – more so than for other continents, where industrialization is much less developed – this subject has a theme of urgency and immediacy due to the rapid human development of practically every ecosystem, which, without any exaggeration, in a short time is going to transform the face of a part of our continent and to alter greatly the existing relationships between man and his environment.

Of course, the main problem was how to treat the great number of European ecosystems. I thought it wise to preserve the classic subdivision of 'biomes' that we find in most ecological works, i.e. tundra, taiga, deciduous forest, mixed forest, sclerophyllous forest and steppes, making allowance of course for all the intermediate or 'transitional' formations. In the book we will start with the southernmost biome of Europe, the sclerophyllous forest, and we will finish with the tundra or the arctic frigorideserta.

It is a pleasure to acknowledge the help and advice I have had from my colleague, the Reverend Professor Cleto Corrain of the Institute of Anthropology of the University of Padova, for all parts inherent to pre-history and the impact of man on European ecosystems in pre- and proto-historic times. He also put his personal library at my disposition. Dr Zoltan Kaszab of the National Museum of Natural History, Budapest, presented me with many Tenebrionid beetles, which have been used for several pictures. Furthermore he gave me several good pictures of the Hungarian puszta. Prof Dr med Georg Pilleri, of the Brain Institute of the University of Bern, gave me many good pictures of sandy shores (Tuscany) and rocky shores (Färoer). Dr Sandro Minelli, Institute of Animal Biology of the University of Padova, lent me some insects to be photographed for use in this book.

My thanks are also due to the technicians of the Institute of Animal Biology of the University of Padova, Mr Renzo Mazzaro, who prepared all original drawings, and Mr Umberto Arezzini, who made all the original photographs of animals. Both of them have contributed very positively to the technical part of this work.

Padova, July 27, 1973

THE SCLEROPHYLLOUS FOREST

'Behold, the Lord thy God giveth thee a good land,
a land of water, brooks and fountains that spring out
of the valley and depths,
a land of wheat and barley, of vines, figs and pome-
granates, of olive oil and honey,
a land in which thou shalt eat bread without scarceness,
thou shalt not lack anything in it'

(Deuteron., 8, 7-9)

A. GENERAL PART

The climate

The Mediterranean vegetation, which originally was mainly represented by the sclerophyllous forest, is present throughout the southernmost part of Europe (Fig. 1, 2).

The climate throughout this region belongs to the category of Mediterranean and modified subdesert climates. A classification of the different climates prevailing in the different areas can be based on the so-called xerothermic index x, i.e. an index of hot weather drought, which for the dry season is the sum of the calculated monthly indices. This index gives the number of 'biologically dry' days during the dry season. As well as the xerothermic index, x_m is also taken into consideration, i.e. the index of a given dry month, based on the number of days with rain, days with fog and dew and average atmospheric humidity of the month considered (Fig. 3-7).

On this basis, 5 climatic types can be recognized in the European part of the Mediterranean basin, namely: xerothermo-mediterranean (dry hot), with an x between 150 and 200; modified thermo-mediterranean climate, with x between 125 and 150; modified thermo-mediterranean climate, with x from 100 to 125; modified mesomediterranean climate, with x from 75 to 100, and finally modified mesomediterranean climate, with x between 40 and 75. Below 45, we find the so-called submediterranean climate, characterized by a transitional temperate deciduous forest, composed mainly of *Ostrya, Fraximus ornus*, etc., widely distributed in Italy and the Balkans, where several elements of the sclerophyllous forest – both plants and animals – can thrive.

The xerothermo-mediterranean climate is found almost exclusively in Almeria (Spain) and on the western shores of the Caspian Sea, from Koura up to Makhatch-Kala.

The modified thermo-mediterranean climate is present in the Gulf of Cadiz (Spain), from Faro up to Jerez de la Frontera; in western Sicily; in the Gulf of Taranto (Puglie); in Greece (Athens, most of Euboea), in the Cyclades and Dodecanneses, which are the most arid parts of Greece, on Crete, with the exception of the western part, where the climate is modified thermo-mediterranean, and the southern part of the island, where the climate belongs to the mesomediterranean type. Finally, this climate prevails at the mouth of the Koura river.

The attenuated thermomediterranean climate can be found along the Guadalquivir and near Cordoba; in Puglia, Calabria and Sicily; in Greece (a coastal belt of Peloponnesus), on the Ionian islands (Cephalonia and Zante), near the Turkish border, in the Koura valley and finally in north-eastern Crimea.

The accentuated mesomediterranean climate prevails in the Iberian Peninsula, on the Thyrrenian part of Italy, from the Tuscan Archipelago southwards, on the Sporades, Lemma and Tharos (Aegean islands) and in the Koura valley to Tiflis.

Finally, the attenuated mesomediterranean climate is present in the Iberian Peninsula, France, Liguria, part of the Thyrrenian coast and on the Adriatic coast of Italy as far north as Ravenna and on the Adriatic islands (Dalmatia, etc.) with the exception of the northernmost ones; in Greece, from the Albanian coast up to the Peloponnesus, on some Ionian islands (Korkyra), on the southern and north-

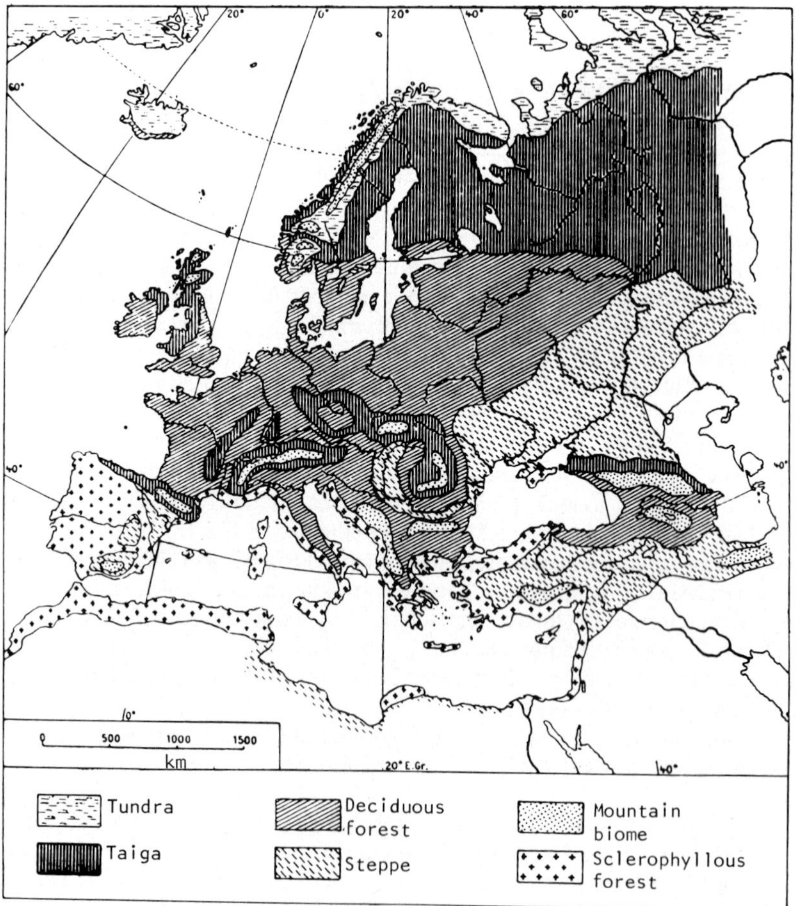

Fig. 1. The main European biomes according to 'Enciclopedia Italiana', Treccani, Roma. (According to Dizionario Enciclopedico Italiano, Istituto della Enciclopedia Italiana fondata da Giovanni Treccani).

eastern coastal belt in the Crimea, in the Kerch Peninsula and on the Russian continent to Krasnodar.

The soil

The most common soils in the European Mediterranean area belong to the following categories: 1. Red earths s.l., ocra earths, etc., both calcareous and non-calcareous, with various amounts of iron and aluminium sesquioxides. They vary according to the geological substratum, as we can see from the following scheme: a) on limestone: terra fusca terrosa; terra rossa; terra rossa sialitica, terra rossa alitica; b) on non-calcareous rocks: clayey soils or 'plastosols' (Kubiena): brown loam; decoloured brown loam; red loam. 2. Mediterranean brown earth. 3. Xerorendzina. 4. Lithosols, or skeletal soils (French authors), or also Rohböden (German authors). 5. Regosols. 6. Alluvial soils (Fig. 8).

The so-called terra fusca terrosa, belonging to the brown loams s. lato, can be found in the arid region of Southern Europe. It varies from loose to slightly aggregated, with good clod structure. The colour is always a dark, dull brown. Generally the vegetation growing on this soil is steppe, or dry maquis, or a thicket

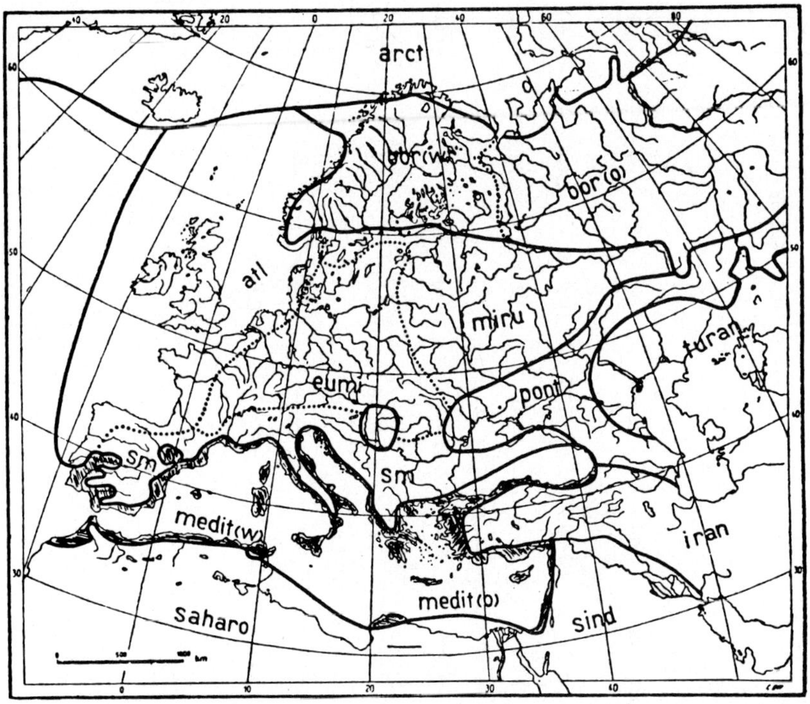

Fig. 2. Biogeographical 'provinces' of Europe according to Freitag (Reproduced with the permission of Fischer, Stuttgart). The sclerophyllous forests correspond to the black areas surrounding the Mediterranean Sea.

I. Terra rossa

skeleton (% of soil)	soil proper (% of soil)	sand %	silt %	clay %	pH	$CaCO_3$ %	C %	N %	C/N
0.6-37	63-99.4	30.6-77	10.5-29	8-46.1	6.6-8	0-56.7	1.35-7.1	0.13-0.6	8.5-14
average 11.4	88.6	50	21.4	28.6	7.1	14.1	3.2	0.28	10.8

II. Sandy calcareous soils

skeleton	soil proper	sand	silt	clay	pH	$CaCO_3$	C	N	C/N
2.5-56.6	57.6-97.5	39-92.6	3.5-40.7	3.9-24.4	8-8.3	50-92	0.9-3	0.07-0.33	8.7-12.8
25	77	78.36	13.71	8.93	8.15	71.47	1.51	0.15	9.78

III. Humiferous soils

skeleton	soil proper	sand	silt	clay	pH	$CaCO_3$	C	N	C/N
7.65	92.35	68.9-74	12.6-16	92.35	6.8-7.2	0.53-2.85	5.5-6.5	0.52-0.6	9.1-12.5

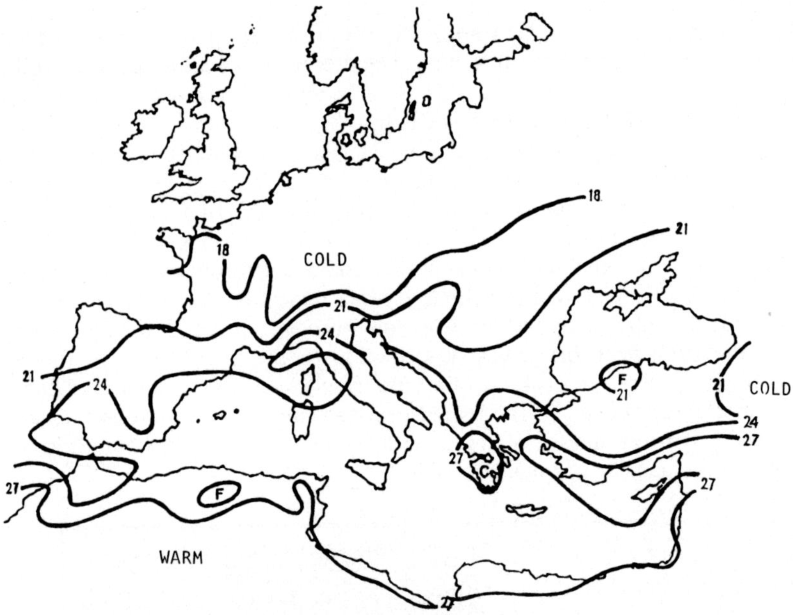

Fig. 3a. Mean temperature in the Mediterranean Basin in August. The sclerophyllous forests are practically limited by the isotherm 24°C (from: Polunin & Huxley).

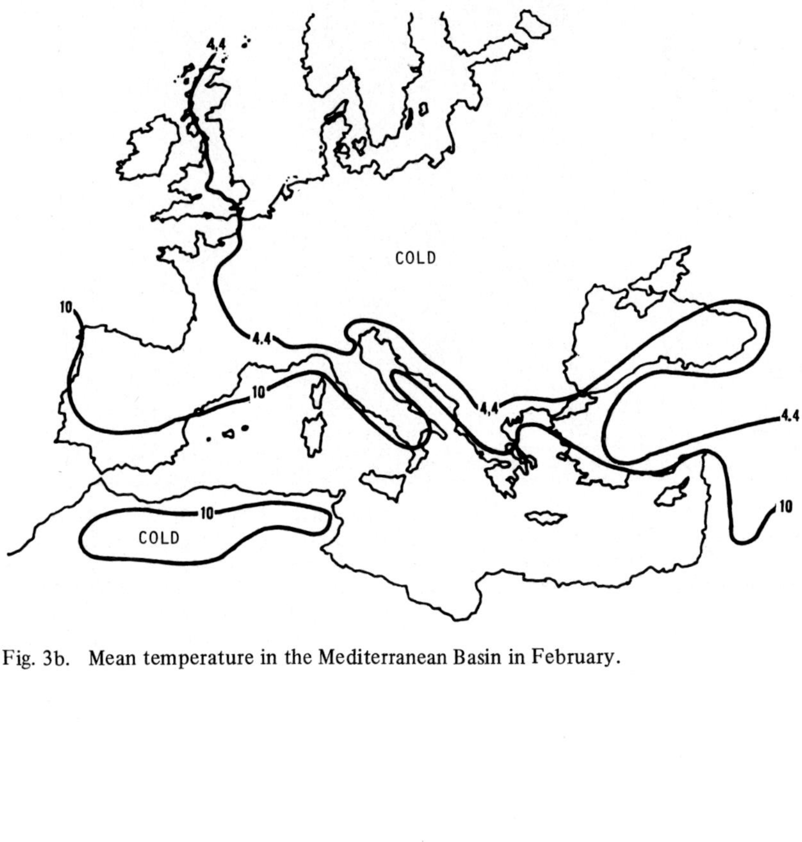

Fig. 3b. Mean temperature in the Mediterranean Basin in February.

of oaks. The decomposition and humification are relatively poor, as is revealed by the poorly decomposed plant substances included within the humus. This soil is found in the arid parts of Europe side by side with terra rossa, on calcareous rocks, particularly in Spain, Greece and Crete.

The 'sialitic' terra rossa is rich in colloidal substances; when wet, it shows high plasticity and tenacity; it is compact and rich in silicic acid. This soil expands strongly when it is wet. Since most of the terra rossa are relict formations, their biology depends greatly on the new environmental conditions, maybe with the exception of that of some endogeous or edaphobious arthropods of very remote origin. In the humid regions life is limited more or less to the surface, and does not penetrate the inner layer, which is of low porosity. In the arid region, on the contrary, the drought brings about a retraction of the upper layer of the soil which is therefore very accessible to life; nevertheless, this is rather scarce due to the lack of water, although in the humid regions a mull is always present, decomposition and humification are not complete while in the arid regions humus is very scarce due to the paucity of organic remains.

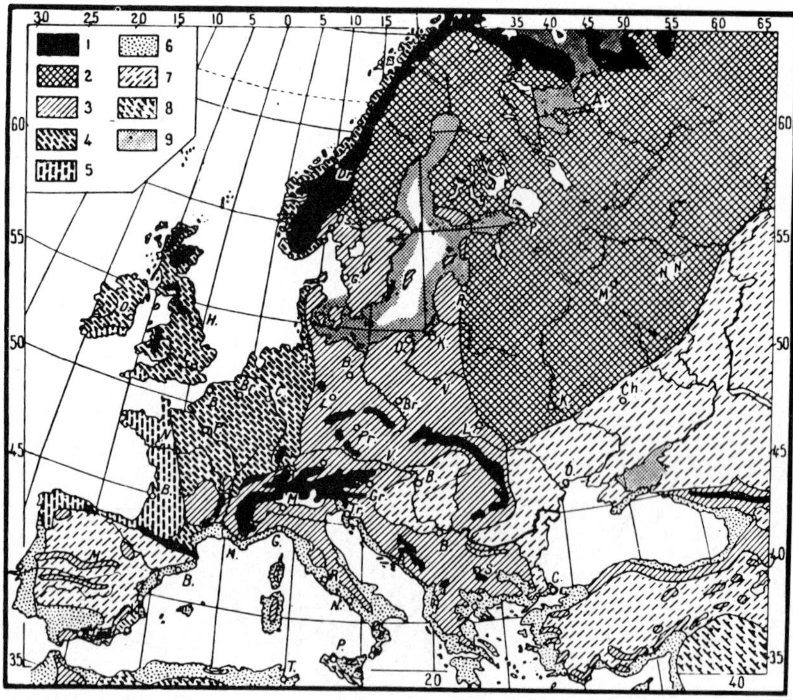

Fig. 4. Climatological map of Europe according to 'Enciclopedia Treccani'. 1. Arctic climate of Northern Europe and of the mountains of Central Europe. 2. Continental climate of Eastern Europe. 3. Transitional climate of Central Europe. 4. Attenuated oceanic climate. 5. True oceanic climate. 6. Mediterranean climate. 7. Continental climate of Southern Russia. 8. Subtropical continental climate. 9. Presence of ice near the shores in winter.
(According to Dizionario Enciclopedico Italiano, Istituto della Enciclopedia Italiana fondata da Giovanni Treccani).

This type of soil is confined mainly to the karstic regions of Southern Europe, in Spain, Southern France, Southern Italy, Corsica, Sardinia, Sicily, Jugoslavia, in Central-Western and Southern Greece, Turkey and Crete. The classic regions of 'terra rossa' are Istria and Dalmatia.

Alitic terra rossa is a terra rossa deficient in silicic acid, rich in aluminium and iron hydroxides; it is loose, with strong clodding, but with a very feeble plasticity. The soil is loose, clodded, high in porosity, with an extraordinary spongey and strongly fragmented texture. Also, in presence of water, the particles show no affinity one for another. The main deposits are known from Istria and Croatia.

Brown loam (decoloured). Due to permeability, there is a decoloured horizon light yellow, light grey to whitish, with strong impoverishment of colloidal substances in the upper layer. The B horizon is often minutely fragmented. There is a constant emigration of all colloidal substances, so that the decoloured brown loam is sandy to silty-sandy. Only when it has lost iron is it silty to clayey.

These soils are sometimes present as relict soils, sometimes with elements of the relict vegetation, as in Southern Andalusia and more especially on the Sierra de Algeciras.

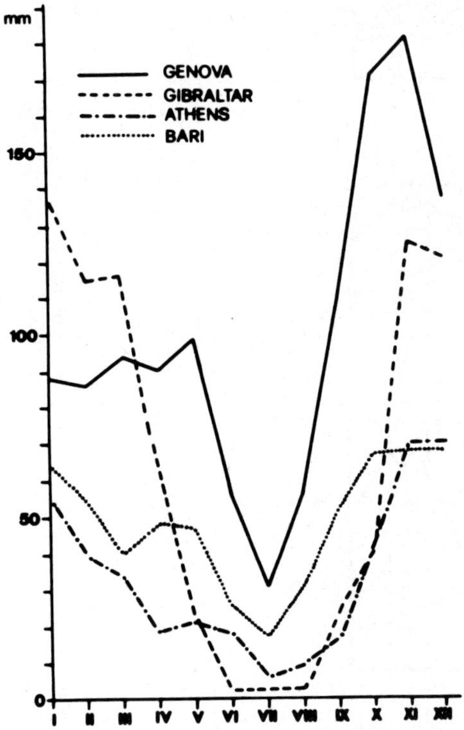

Fig. 5. Annual rainfall for various localities belonging to the sclerophyllous forest (from: Polunin).

Red loam is almost always very compact and heavy, with some colloidal substances, rich in silicic acid (very mobile) with iron hydroxide predominantly in colloidal form, easily diffusible, which can only partially be irreversibly separated as red concretions, deposits and patches.

This soil is very compact, low in porosity and with a slightly permeable texture. When wet, it expands and upon drying it disaggregates into a number of small pieces of stoney consistency. Life, despite the presence of a mull, is relatively poor due to the compact texture. It is very widely distributed in Southern Europe, generally as a relict soil (Greece, Spain) or buried by more recent soils of different kinds.

Southern brown earth. The colour is a dull light desert-brown; since this colour is never present in Central Europe, it may be utilized for the recognition of this soil type. The parent rock is generally a compact siliceous rock, i.e. granite and similar eruptive rocks, gneiss, micaceous schists, chlorite-schists, phyllites and clayish slates.

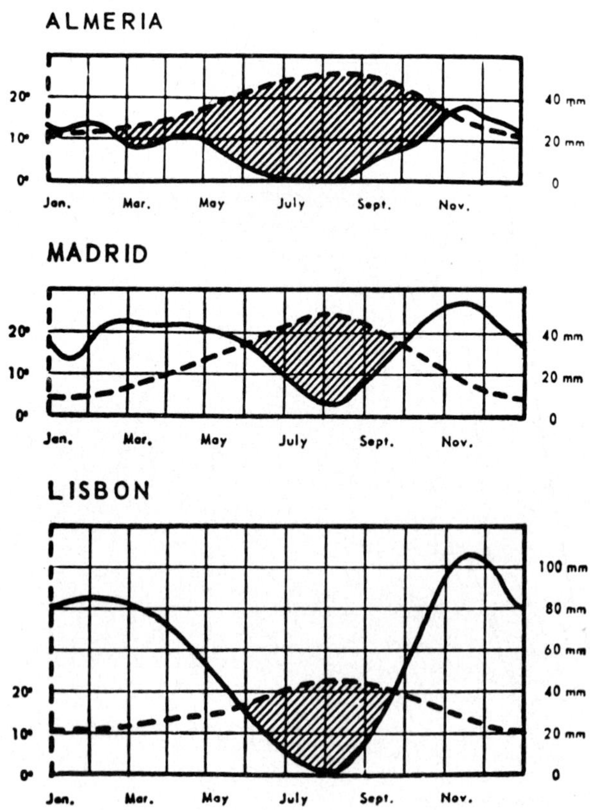

Fig. 6. Ombrothermic diagrams for three localities belonging to the dominion of the sclerophyllous forest.

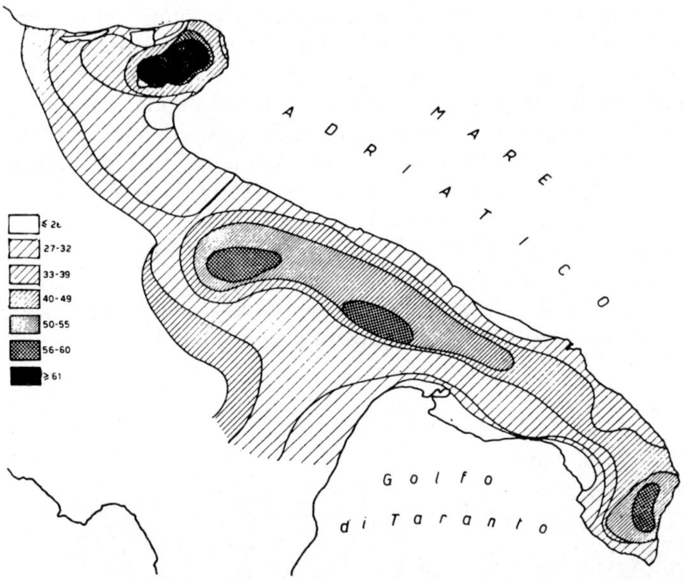

Fig. 7. The value of the rain factor (Lang) for a region belonging mainly to the sclerophyllous forest, Puglie (Southern Italy). Values can reach 39; above this value, the mesophilous, deciduous forest is present.

Fig. 8. Four examples of profile of soils present in the sclerophyllous forest (Puglie, Southern Italy).

These soils are almost always covered by a very sparse vegetation, represented by shrubs such as *Quercus ilex, Juniperus oxycedrus, Thymus vulgaris*, etc., secondarily with a dry forest of *Pinus pinea* and *P. halepensis*. The biology of the southern brown earth is very intense during the short humid period, but on the whole it is inhibited by drought. Notwithstanding, decomposition and humification are relatively good, but the humus is rather scarce due to the paucity of plant remains. The southern brown earth is found in the arid regions of Southern Europe, especially in Greece, Southern Italy and the Iberian Peninsula.

Xerorendzina. This is a rendzina of the arid regions, grey-ash coloured, almost always very loose, when dry like a powder, extremely rich in calcium and sometimes also in soluble salts, especially gypsum. This soil is very similar to serosem: indeed, xerorendzina is a soil very variable in construction and composition, devel-

Fig. 9. Aspects of the vegetation of the sclerophyllous forest: a. shrub of *Quercus calliprinos* in an olive grove on the Ionian coast of Salento (Southern Puglie). b. low maquis with *Q. calliprinos* and *Phlomis fruticosa*; c. very evoluted aspect of 'maquis' of coastal dunes; d. *Juniperus phoenicea*; e. *Cistus monspeliensis* (Salento); f. *Phlomis fruticosa* (Salento) (Photos Prof. G.G. Lorenzoni).

oped above compact parent rocks, whereas serosem is a soil of arid steppes, developed above a relatively loose parent material. The vegetation is generally scattered, so that areas of barren soil are always visible. Many plants show a tendency to lignification and formation of thorny branches. Though dry, the soil shows good decomposition and humification, but the humus is very scarce due to the paucity of organic material. When the soil is clayey, there are some characteristic earthworms, whose presence is revealed by rather small coproliths (3-5 mm long) of cylindrical shape. This soil is widely distributed in mountains where karstic phenomena are present, in the dry regions of Southern Europe, above limestone, marl, or gypsum. The soil generally has stones of different sizes scattered in it.

The lithosols present in the mountains and hills of all Mediterranean areas, are characteristic of rocky substrates where the local conditions, i.e. relief, rainfall, wind etc. have not yet allowed the formation of a true 'soil'. Within this category we may perhaps also place sand obtained from the weathering of granite.

The regosols come from soft or loose substrates, like clay and sands; sometimes they assume the aspect of a ranker (on acid rocks), a pararendzina (on calcareous rocks) or a stagnogley (in a badly drained region). In the case of the so-called 'calanchi' they are very similar to lithosols.

Alluvial soils are formed above recent alluvium; they also include the gley-hydromorphous soils (where there is the influence of the water table). They may evolve either towards brown earths (both leached and typical) or towards black soils, according to the prevailing climate. Within the hydromorphous soils proper we find, in the Mediterranean area, gleys (badly drained soils, poor in oxygen), but sometimes also some well drained soils, like sand dunes or salty clays, covered with halophilous vegetation (see the chapter on humid environments).

The vegetation

μύρτα τε καὶ ἴα καὶ ἑλίχρυσος
μᾶλά τε καὶ ῥόδα καὶ τέρεινα δάφνα

(the myrtle and the violets, and the helichrysos,
and apples and roses and delicate laurels)

Ibycus

---- μελαμφύλλοι δάφνηι χλωρῆι
τ' ἐλαίηι ταντελίζει

(the dark leaves of the laurel quiver,
and the pale ones of the olive tree)

Anacreon

The vegetation of the Mediterranean area corresponds entirely to the mediterranean and attenuated subdesert climates, and more precisely to the following vegetation types:

Shrub or tree pseudosteppes and open forests in very dry climates: these are present only in Spain and more precisely are represented by a pseudosteppe in the very dry part of the eastern Andalusian coast. Some *Tetraclinis* and *Anabasis* are also found there and man has successfully introduced the date palm (*Phoenix dactylifera*) and several tropical crops which require irrigation in summer, such as sugar cane, pineapple (*Annona squamosa*) and avocado pear (*Persea gratissima*).

Shrub or tree pseudosteppes and open forests in less dry climates: In the islands of the Aegean we find *Pinus halepensis* or *P. brutia, Cypressus sempervirens* (on mountains such as on Rhodes); in Greece and Crete we find *Pinus halepensis* and *P. brutia* formations; here the degraded formations with *Paliurus australis* are called 'phryganes'. In France we have *Pinus halepensis* in Provence and eastern Languedoc, the species having spread widely as a result of fires which have destroyed the other trees. There is often a garrigue undergrowth with *Rosmarinus, Globularia alypum, Thymus vulgaris*, and some degenerated garrigues with *Brachypodium ramosum, Lavandula latifolia* and several species of *Asphodelus.*

Fig. 10. Aspect of the sclerophyllous forest in which *Quercus ilex* is the dominant tree (Salento, Puglie).

Quercus coccifera is often an important element. In Italy *Pinus halepensis* is scattered throughout the Oleo-lentiscetum; a few Aleppo pine areas are also found in Sicily, on the east coast. In Spain the Aleppo pine is very widely distributed in the eastern part. The flora of this region is not dissimilar to that of Provence, with the addition of *Chamaerops humilis*; the carob tree (*Ceratonia siliqua*), both self-sown and cultivated, grows near the coasts (Fig. 9-16).

Mainly high steppe with or without trees, or shrubs in cold and cold temperate climates: this vegetation type is mainly represented in the Crimea and in Spain, with *Festuca* and *Stipa*-steppes. Here *Juniperus thurifera* is the characteristic species of the hard winter-summer drought type of Mediterranean vegetation. It is present in the middle Ebro valley; associated with *Juniperus* are *Rhamnus lycioides, Genista scorpius, Thymus vulgaris, Rosmarinus vulgaris*. A product of degeneration of this vegetation is the so-called 'tomillar', composed of white absinth, *Salsola vermiculata, Thymus vulgaris* and *Asphodelus fistulosus*.

Formations of the western Mediterranean evergreen oak stage, or true sclerophyllous forest: this is distributed all along the European mediterranean coasts. In Cyprus we find a forest and shrub vegetation formed by *Pistacia lentiscus, Ceratonia siliqua, Quercus infectoria* and *Q. calliprinos*; in Dalmatia we find mainly *Quercus ilex* (holm-oak) mixed with *Pinus halepensis*; this formation is very fre-

Fig. 11. Aspect of highly degraded, anthropized maqui, growing on some 'calanchi' in Basilicata (Southern Italy).

quently substituted by a degradation product, the so-called 'maquis'. In Albania maquis is present from 0 to 300 m, with *Quercus ilex, Pistacia lentiscus* and *Erica arborea*; from 300 to 600 m the holm oak is dominant; in Greece and Crete there is a very widely distributed *Quercus coccifera* and *Q. aegilops* 'phrygane' formation. In France two main types of vegetation are recognizable according to geological substratum, namely a *Quercus suber* series on siliceous rocks, and a *Quercus ilex* series on limestone. The former is present in Provence and in the Albères mountains; the products of degeneration of this forest are a scrub with *Erica arborea, Arbutus unedo, Phyllirea angustifolia, Smilax aspera*, and a garrigue with *Lavandula stoechas, Ulex parviflorus, Calycotome spinosa, Pistacia lentiscus, Myrtus communis, Olea* and *Cistus monspeliensis. Quercus ilex* is widely distributed in Provence, Languedoc and Rousillon; often it is substituted by a garrigue with *Genista scorpius, Lavandula latifolia, Quercus coccifera, Rosmarinus, Pistacia lentiscus, Phlomis lychnitis, Brachypodium ramosum*, etc. In Corsica we find, in the eastern part, the cork oak *Quercus suber*, and in the west the holm oak *Quercus ilex*. The Corsican maquis is similar to that of Provence, and *Helianthemum halimifolium* and *Juniperus phoenicea* are frequently found; a calcareous garrigue occurs only on the southern tip of the island.

In Italy the *Quercus ilex* formation is common on the coast south of Rome (from Ostia onwards) and is present on all the western Thyrrenian coasts, and the eastern Ionian and Adriatic coasts as far north as Ravenna. On the main islands, Sicily and Sardinia, the holm oak is also dominant. The degenerate formations are scrub and garrigue as in France. On the west coast there are populations of umbrella pine (*Pinus pinea*) whereas *Quercus suber* is found in a few localities in Calabria, Silicy and Sardinia, where it is more widespread. In Sicily, *Quercus coccifera* is widespread, but *Q. calliprinos* is very localised. Extremely localised is *Quercus aegilops* in Puglie.

The majority of the Spanish plains show a natural Mediterranean evergreen oak vegetation, often altered by the action of man. The *Quercus suber* series is present mainly in Catalonia and in the extreme south of the Peninsula. *Pinus pinea* is particularly well developed in Old Castile and in the Huelva region. The degraded sclerophyllous forest is called 'tomillares' here and is represented mainly by *Thymus vulgaris*. Further degradation results in a steppe-like facies, known locally as 'caveros'. The *Quercus ilex* (subsp. *cordifolia*) forests are particularly widespread in the Sierra Morena, accompanied by *Cistus ladaniferus*.

On the Balearic islands *Pinus halepensis* and *Quercus ilex* prevail. In Portugal the south central part is covered by *Quercus suber*, whereas in the Algarve, on calcareous soils, the palm *Chamaerops humilis* is to be found. *Quercus lusitanica* is common; the garrigues are represented by *Genista triacanthos, Cistus* and *Erica*; a further degree of degradation is represented by an asphodel dominated dry grassland.

Finally we come to the formations of the sub-humid Mediterranean stage. On the extreme eastern limit of the Mediterranean area as we have delimited it, in Azerbaijan, we find *Quercus longipes, Q. infectoria* and *Q. iberica*; in the south-east Caucasus there is a caspian and pontic deciduous forest, which has practically nothing to do with Mediterranean vegetation, with *Quercus castaneaefolia, Parrotis persica* and *Carpinus caucasica*. On the island of Cyprus this vegetational type is represented by *Pinus brutia, Quercus calliprinos* and *Olea europaea* in open forest. On Crete *Cypressus sempervirens, Ceratonia siliqua* and *Acer orientale* are to be found. In Italy the Mediterranean vegetation is represented mainly by the so-called xerothermic 'oasis' of Northern Italy (insubrian climate restricted to insubrian lakes). Finally, in Spain there are large forests of *Pinus salzmanni* in the Serrania de Cuenca and on the Pyrenees, which constitute a transitional forest.

As an example of the main biological forms (Lebensformen) present in the Mediterranean vegetation and their percentage occurrence in a given flora, we represent here Raunkjaer's spectrum compiled for the flora of Monte Argentario (Tuscany, near Orbetello, adjacent to Island Giglio; from Susmel, 1971-2):

No. of species	Ph phanerophytes	Ch chamaeph.	H Hemicryptophytes	G Geophytes	Th (in %) Therophytes
866	12	6	29	11	42

Fig. 12. *Quercetum ilicis* in very good state of preservation, at the Villa Reale of Portici (near Naples).

Fig. 13. Distribution of the olive tree in the Mediterranean Basin according to Polunin & Huxley.

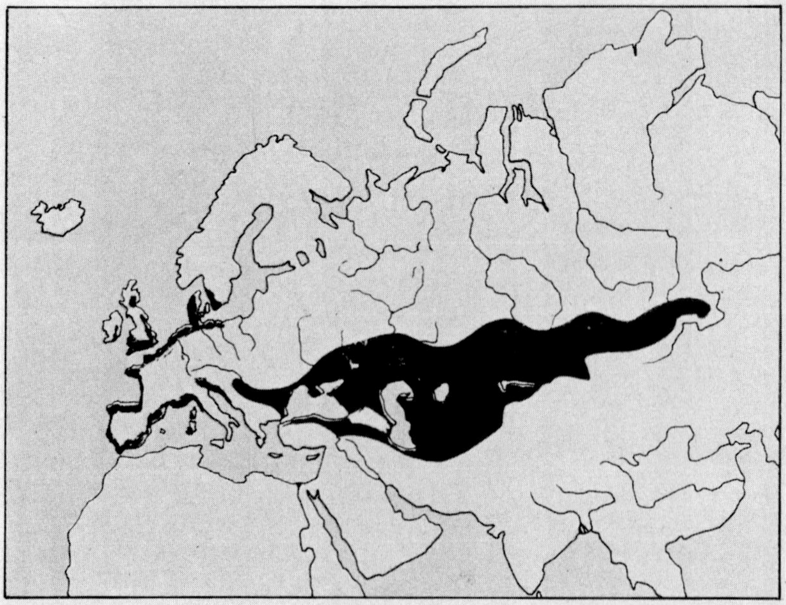

Fig. 14. *Artemisia maritima*, an element coming to Europe from the Central Asian steppes (according to Walter, in Zangheri) (Reproduced with the permission of Ist. Geogr. De Agostini).

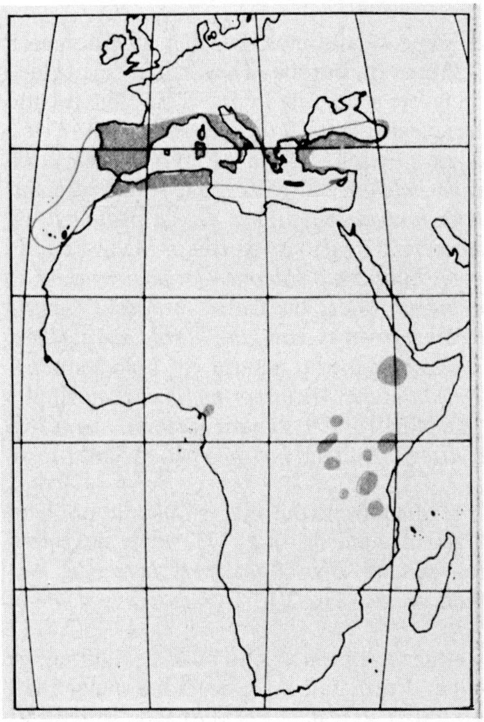

Fig. 15. Distribution of the broom *Erica arborea* (according to Vester, in Zangheri).

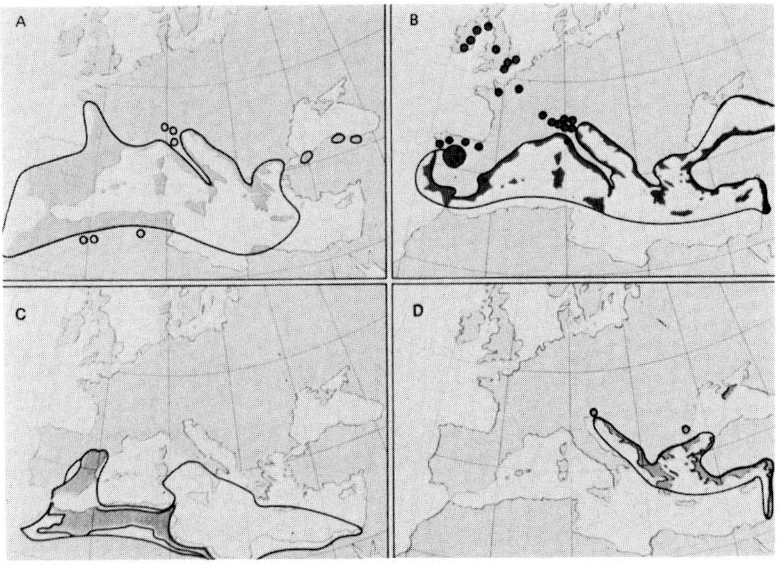

Fig. 16. A. Distribution of *Quercus ilex* in Europe. B. Distribution of *Pinus pinea* in Europe: black line – typical areal of the species, dots and circles – localities where it has been introduced by man. C. *Stipa tenacissima* (dark) and *Lygeum spartum* (light). D. *Cercis siliquastrum*. (According to Rikli & Meusel, in Zangheri).

It is worthwhile to mention here some of the most peculiar Mediterranean plants which thrive on the southern shores of Europe. They can be subdivided according to their distribution (and therefore according to prevailing climate) into several categories: a) Eastern Mediterranean plants: *Poterium spinosum; Cercis siliquastrum, Calycotome villosa* (see later, present also in Algeria), *Genista acanthoclados, Medicago arborea; Trigonella balansae, T. graeca* and *T. coerulescens, Euphorbia acanthommos, E. veneta, Asphodeline liburnica, Gagea peduncularis, Tulipa* sp. plur., *Lloydia graeca, Sternbergia sicula*. b) Northern Mediterranean shores (European): *Pyracantha coccinea, Euphorbia spinosa, Cotinus coggyria*. c) Western Mediterranean: *Calycotome infesta* (vicariant with *C. villosa*!), *Genista cinerea, G. equisetifolia, G. hirsuta, Retama monosperma, Euphorbia biumbellata, Chamaerops humilis, Iris xiphium*. d) Circum-mediterranean (or holomediterranean): *Crataegus monogyna, Anagyris phoetida, Medicago marina* (psammophilous species, thence limited more by edaphic than by climatic factors), *Euphorbia dendroides, E. characias, Asphodeline lutea, Fritillaria messanensis*. e) Sardo-corsican species: *Pancratium illyricum*.

Several plants have recently been studied from the cytotaxonomic point of view, for instance *Arum maculatum* and *A. italicum* (Fig. 17), where hexaploid races are found in the Mediterranean climate; *Blackstonia perfoliata* (Fig. 18) and *B. acuminata* (Fig. 19); *Centaurium minus* (Fig. 20), *C. majus* (Fig. 21) and several other species of this genus.

For *Blackstonia* and *Centaurium* Zeltner has been able to build a phylogenetic tree on the basis of chromosome number. Furthermore, Favarger has shown that, in several plants such as *Petrorhagia saxifraga, Amelanchier ovalis* and *Cotoneaster*

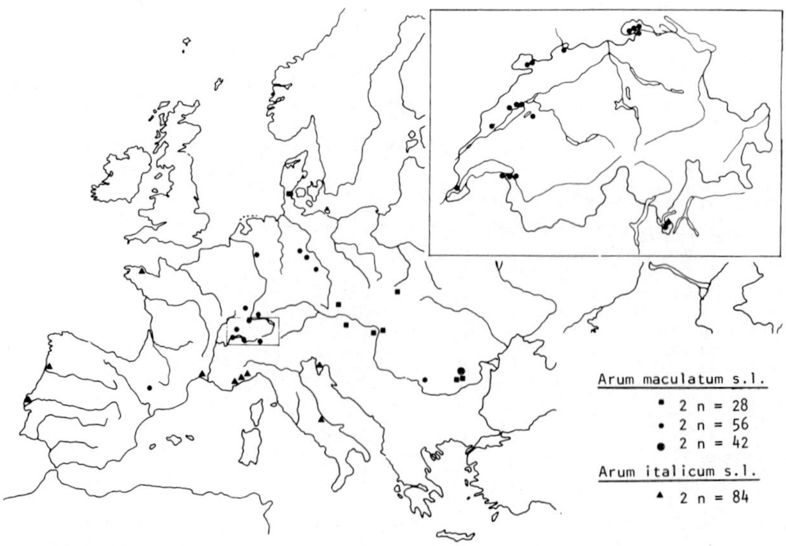

Fig. 17. Chromosome numbers of *Arum maculatum* and *A. italicum* in Europe and (inset) limited to Switzerland (from Beuret).

integerrima, there are diploid populations only in the Mediterranean climate (i.e. in the dominium of the sclerophyllous forest), whereas in the deciduous forests of Central Europe only tetraploid races occur. The diploid race area corresponds sometimes to the original habitat of the species, sometimes only to a refuge where the species retreated during glacial periods.

Effects of man on the sclerophyllous forest

The dominion of the sclerophyllous forest in the Mediterranean, and more particularly in the European mediterranean area, is one of the most altered in the world, as a result of human intervention which has had its effects here since pre- and protohistoric times. Indeed, the Mediterranean Basin has, since very remote times, seen the development of several cultures, which have alternated and subdued each other, or which, in some instances, have been preserved, in particularly isolated (districts), for a long time. Mainly the climate – relatively uniform all

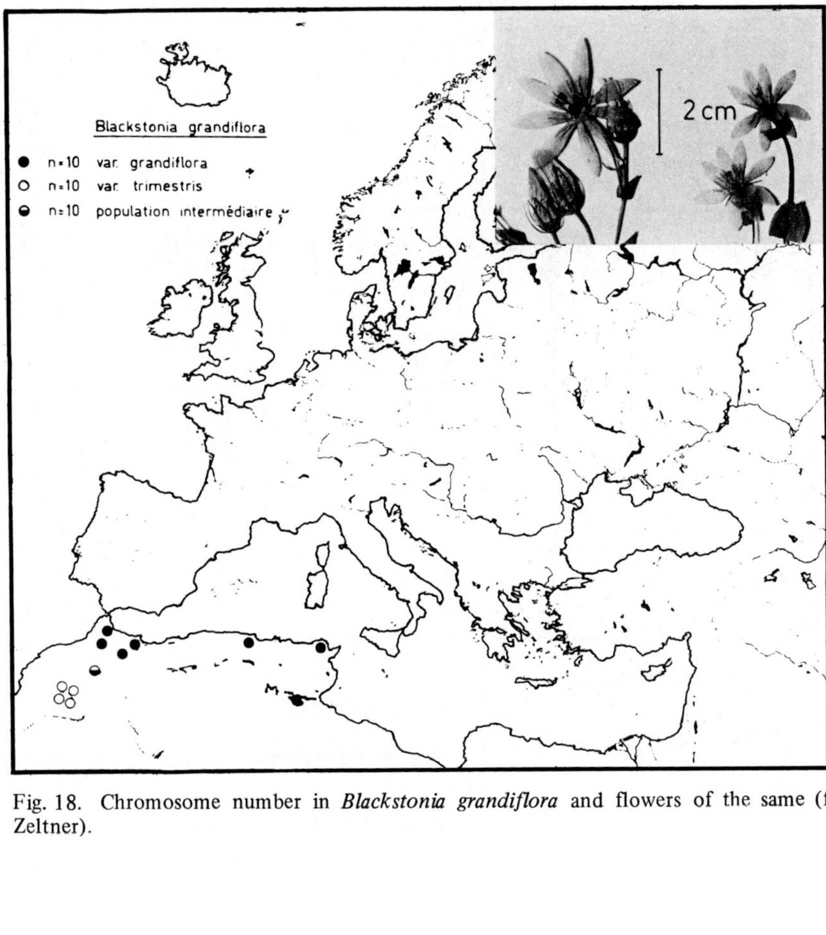

Fig. 18. Chromosome number in *Blackstonia grandiflora* and flowers of the same (from Zeltner).

along the shores of the Mediterranean Sea, except of course in a few desert districts such as Libya and Egypt – has permitted the origin and successive spreading of types of construction common to all the shores of the Mediterranean Sea, and in some instances they have spread as far as the shores of the Atlantic Ocean (Fig. 22-23).

An example is given by megalithic constructions, originating probably in Egypt and the Middle East (rock of Tiryns, cyclopic walls of Mycene) and thence spreading all over North Africa, from where they successively reached Spain, France, Corsica and Sardinia. Examples of these constructions are given by the 'mastaba'

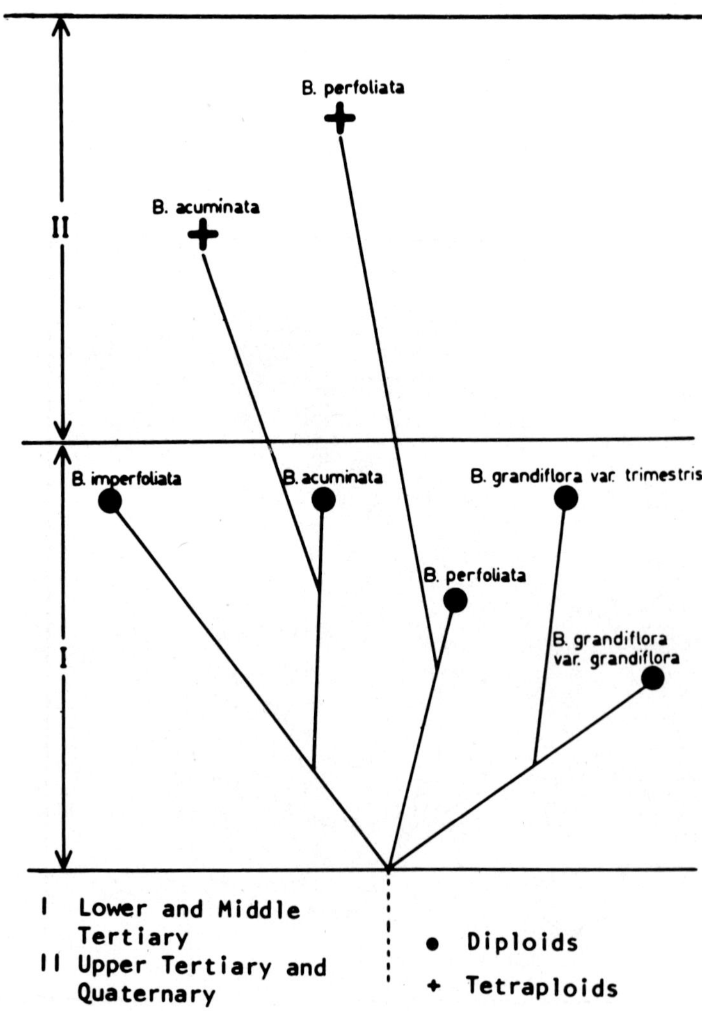

Fig. 19. Dendrogram of the genus *Blackstonia* (from Zeltner).

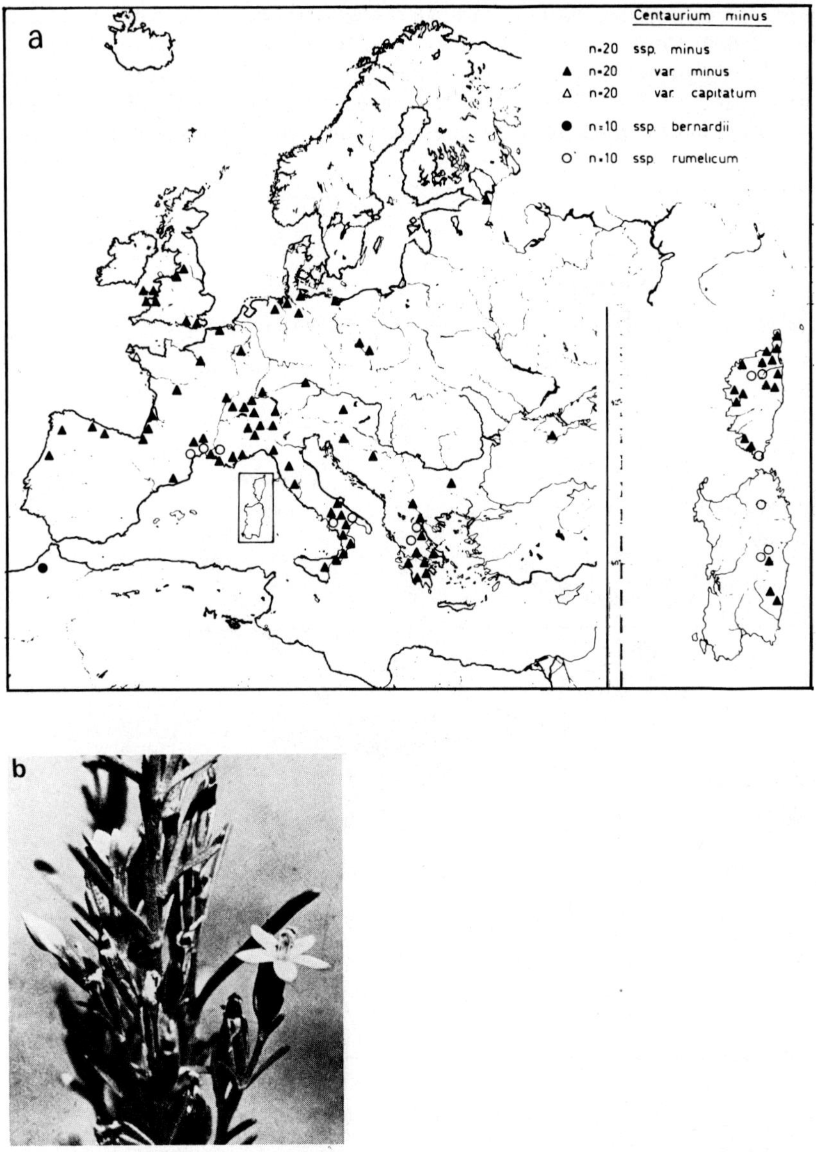

Fig. 20. Evolution of the genus *Centaurium*. a. Chromosome number of *Centaurium minus*. b. Flowers of *Centaurium favargeri*.

of Egypt, the dolmens[1] present all through the Western mediterranean region, the menhirs[2] and the 'betils', also very widespread. We should also mention here the so-called 'rocky graves', originating in the East (Cyprus), from whence they spread to several parts of the Western Mediterranean, from the Iberian Peninsula to France, Sicily and Sardinia. In some cases an entire mountain has been transform-

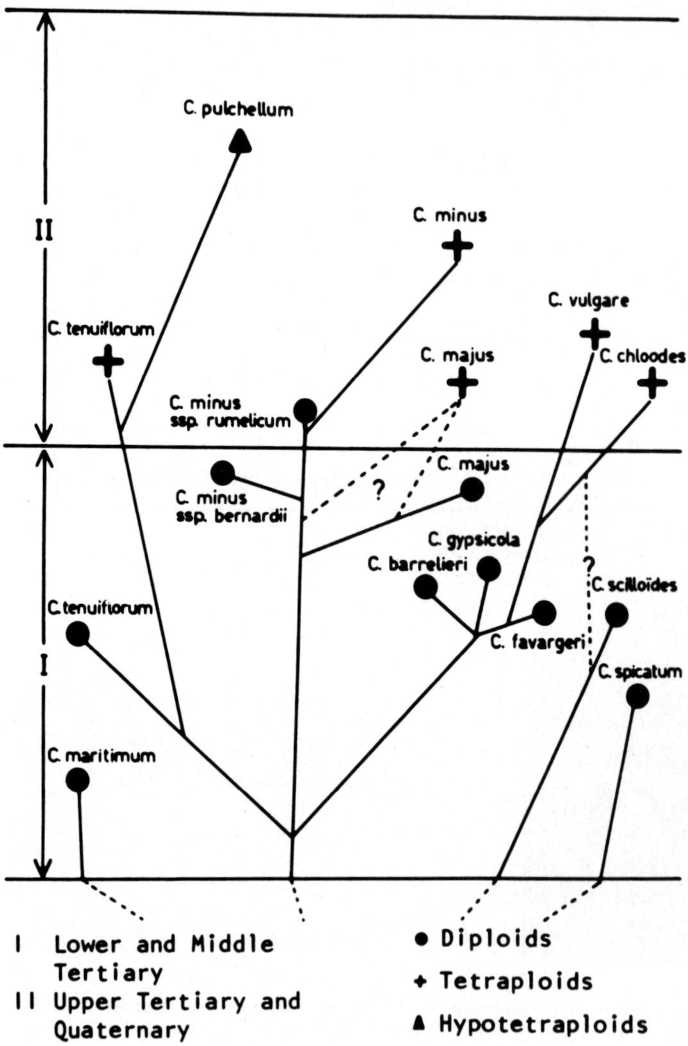

Fig. 21. Dendrogram of the genus *Centaurium* (from Zeltner).

[1] From daul (table) and moen (stone) in Celtic. They are also known as 'tumulus sepulchres'.
[2] From men (stone) and hir (high or long) in Breton.

ed into a grave-yard as at Pantalica (Siracuse, Sicily), Palmella (Portugal) or Petit-Morin (Marne, France). In Sardinia we find rocky graves dating from 1700-1500 years B.C., side by side with sepulchres of the megalithic-dolmen type, and with the so-called 'megalithic cists' (Arzachena, Gallura). The latter resemble somewhat the Corsican 'cromlechs': indeed, the migratory currents coming from the ibero-french coasts, must have reached Corsica about 2000-1800 years B.C.

At the same time, corresponding to the alteration of the natural environment in the Mediterranean Basin, as a direct response to climate, another completely different style originated in the Aegean islands and Eastern Mediterranean, the so-called jut system or false-dome system. This system was initially present in the large Minoan-Mycenae pseudo-domes of Cyprus (IV millenium B.C.), in the villages of the Near-East (Mesopotamia), in Greece (tholos) and thence (2000 years B.C.) in Spain, the Balearic islands and Sardinia. Here, in a relatively similar hot and dry climate, this system gave rise to the 'nuraghe' (plur. nuraghi) and this is a confirmation that the climate in the Mediterranean area has not changed in the last millenia, and that the vegetation, though much richer than the present one, was always xerophilous (Fig. 24).

Whereas in the East the jut system (or pseudo-dome system) was very soon replaced by new architectural types, similar to the present dwellings of these

Fig. 22. Typical garrigue in Puglia (Martina Franca). The soil is partially denuded, and the rock (cretaceous limestone) is visible, together with terra rossa.

lands, in some peripheral and isolated places – such as the Balearic islands and Sardinia – the jut system (in Sardinia the nuraghe, on the Balearic islands the 'talayots' and the 'navetas') must have remained for a long period of time and undergone a completely autochtonous evolution, until the IX-VIII centuries B.C. Of the same epoch are the so-called 'well-temples' where the rain water was kept for religious functions and as a reserve for dry periods. In the temple the stair reached the water level and only the elect were allowed there, where they officiated in front of the precious liquid. The climate/religion relationship (or if we prefer ritual) is very close. Water was, as it is today, so scarce and valuable that man felt the necessity to erect temples to it. Here again we have a confirmation that the climate was definitely no more humid than now as people have often affirmed. The cult of water is incidentally quite accepted in arid lands: in Tunisia for instance there are the so-called 'sacred wells', still venerated today, and maybe they are, with some substitutions, the same as in antiquity. According to Corrain,

Fig. 23. Effect of man on the original vegetation of the sclerophyllous forest: the grape was cultivated in Greece earlier than in Italy. In the picture some grapes in the Zoological Station at Naples.

Zampini & Fabbri (1972) those traditions we find today in the moslem, North African world, have deep roots in antiquity: rain water in North Africa has its own sacred properties, from which the great number of magic ceremonies performed to promote rain take their origin.

Another proof of the close relationship between climate and type – or types – of human dwellings is the existence in present-day Sardinia of some constructions inspired by 'nuraghi', the 'pinneta' for farmers, and the 'pinneta' for shepherds.

Fig. 24a. Betils with some mammilarly hollows (Paulitatino, Sardinia) (from Stacul).

Fig. 24b. A 'nuraghe' near Silanus, Sardinia (from Stacul, with permission of Mondadori).

Incidentally, in Puglia, with a basically similar climate, we find the 'trulli' (plur. of trullo) from which some rough, rural constructions of stones, with a pseudo-dome now take inspiration. These constructions resemble the Sardinian nuraghe a little. Now, according to Battaglia (1952), there is no fundamental difference between the monocellular hut, with a false dome or jut roof, built up with stones, which is used as a repository for agricultural tool or as a temporary refuge for peasants and shepherds, and the true trullo, a permanent dwelling. According to Battaglia, the constructions we find in Puglia and the nuraghe can be dated back to a single origin, i.e. a circular hut of stones, with a false dome roof. 'The architectural system of trulli – says this author – reveals itself as a produce of more evolved cultures and of a particular geographical environment, characterized by an arid climate with scanty vegetation' (transl. from Italian) (Fig. 25).

The neolithic culture (to which both megalithic sepulchres and jut-constructions belong), originated in the East and then slowly spread to the West, demonstrating that even in these remote epochs there was a certain density of population along the shores of the Mediterranean Sea, and that journeys were relatively easy. To the Neolithic belongs the tholos of Haghia Triada: according to some the Neolithic in Crete can be dated to 5800 years ago, to others (Mosso) to 9000 years ago. The Neolithic of Phaestos lasted 2500 years. Some date the original palaces of Phaestos and Knossos to the Middle Minoic period[1], apparently contemporary with the XII dynasty in Egypt (i.e. 1991-1778 B.C.) and therefore with the tholos of Haghia Triada. During the XII-XVIII (1991-1318 B.C.) dynasties of Egypt there were very close relationships between Crete and Egypt, so that the culture of the two countries must have been very similar.

In the Mediterranean basin, i.e. in the area of the sclerophyllous forests, Neolithic man must have been very numerous. The great abundance of weapons

Fig. 25a. Present-day 'pinneta' for peasants near Sassari, Sardinia (from Stacul).
Fig. 25b. Some 'trulli' in Puglia.

[1] The term minoic has its origin from the mythical king of Crete, Minos, and is a synonym of Cretan.

constructed of copper in Tuscany and Umbria shows that agriculture must have been flourishing and people rich, and that they were able to obtain large amounts of food from peoples that still lived in the Stone Age in exchange for the metals they could give them. Probably the population was dense in Etruria.

Following nuraghi and other jut-constructions we find in the East a type of house similar to the present-day one, for instance at Knossos about 2000 years B.C. The wood utilized for these constructions came from Southern Italy, which shows 1) that in that epoch the climate of the Aegean islands, Crete and Greece was the same as today and that no arboreal vegetation was present (that is confirmed by the distribution of some planticolous insects such as Tenebrionids); 2) that in Southern Italy there were very flourishing sclerophyllous forests, some of which have been preserved until today, and that deforestation had started even in Neolithic times. In Southern Italy this deforestation caused the spread of malaria, which has been known in the Mediterranean area since very remote times: according to Bel, malaria is mentioned in orphic poems dating from 1000 years B.C. Homer (between the X and VIII centuries B.C.) also talked about it. In the V century Hippocrates gave a very neat description of this disease from the medical point of view. Incidentally, in North Africa there were no forests, contrary to what some would like us believe today. During the I dynasty of Egypt (i.e. 2850 years B.C.) they used reed instead of wood for the construction of boats, and the same holds good for the III dynasty (2650-2600 B.C.). Only in 1500 B.C. did the Egyptians utilize the wood of Lebanese forests for their craft, where evidently deforestation had already begun. Incidentally, Solomon was obliged to import cedars from Lebanon for the construction of his temple, and wood workers were also imported from that country.

As well as human influence, in the area of the sclerophyllic forest some bradyseismic or even orogenic movements have possibly occurred. For instance, on the island of Malta, 100 km from Sicily, some very large rivers must have been present to allow the existence of a species of *Hippopotamus*, and the erosion valleys present on this island show that once there must have been large water courses. Mosso thinks that during Neolithic times in Southern Italy there were some modifications of environment due solely to seismic phenomena, greater than those that occurred during historic times. Also, eustatismus must have played its role in the changes in the sea level.

The development of agriculture and the alteration of forests at the end of the Neolithic were rather conspicuous: during the Neolithic both South and North of the Alps, the local populations cultivated wheat, barley, beans, rye, millet, lentils, chickpeas and other crops. The climate was colder than the present one. On the Italian pre-Alps there were *Pinus*, *Larix* and *Abies*, whereas in the lower parts oaks, elms and beeches could thrive. In the palisades of that period some chestnut conkers have been found which show that during the Neolithic this tree was already growing in Italy. In the palisades of the Bronze Age at Peschiera (not far from the Lake of Garda, where today olive trees are growing) some kernels of olive have been found. The same is true for Mentone (today situated in the true sclerophyllous forest) where kernels of olive have been attributed to a period before the Neolithic, which shows that either this plant is indigenous (as affirmed by someone) or it was introduced in the Mesolithic. Several kinds of lint were even cultivated in the Neolithic. Grapes were cultivated in Greece far earlier than in Italy.

What is certain is that, in Southern Italy, primary forest had been destroyed and the soil cultivated well before the Neolithic, since in this period around Etna there were fields and meadows: Homer indeed says in his Odyssey (XII Chapt., verse 164 fll.) 'then the beautiful shores of Trinacria will come to you, where the cattle of the Sun is grazing; 7 herds of oxen, and 7 of lambs, each of 50 heads. ... The immortal Neera gave them to care after the father's cows with the moon-shaped forehead, and the father's shining sheeps ...' and again (verse 342 fll.) 'after got through Scylla and Charybdis, in front of us the fertile, agreeable island appeared, where the cattle of the Sun is grazing, and from the far stables the bleatings and the bellowings came to us, ...'.[1]

Finally, we must not forget that, around 1000 B.C., Indoeuropeans reached Europe, and their pastoral habits must have contributed decisively to the destruction of the original vegetation, not only in the sclerophyllous forest, but also in the deciduous one (Fig. 26).

The fauna of the sclerophyllous forest

Vertebrates

In the vast dry areas of the Peninsula of Peljesac and in its magnificent, almost impenetrable forests of *Pinus halepensis*, a very rare European animal, the jackal or *Canis aureus* may be found. This animal, social like its kin the wolf, always lives in small packs. The peninsula of Peljesac is probably the northernmost point of distribution of this mammal, which is very common, on the contrary, in the Near East, e.g. in Palestine, where it has been known since biblical times (in the Bible it

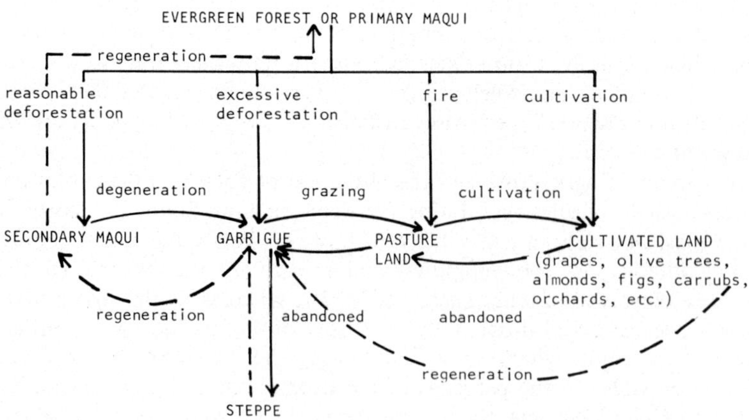

Fig. 26. The various stages of degeneration and regeneration of vegetation communities and the influence of man on the vegetation of the Mediterranean region (sclerophyllous forest) (from Polunin & Huxley).

[1] Free translation from Greek.

is called 'shual'), and where its nocturnal howling, prolonged and mournful, is sometimes unbearable. Very harmful to vineyards, which in Dalmatia are among the most important agricultural resources, it also eats corn, water melons, pumpkins, and it is said that one jackal can destroy a whole field of melons in one night. When hungry, it does not hesitate to penetrate villages, and if driven by hunger it will also assault man. In this case it is absolutely necessary to undergo anti-rabies treatment. It will also eat small mammals and birds, and – if nothing else is available – it will eat beetles.

The wild cat of Mediterranean Europe, *Felis lybica*, thrives all through Africa, Asia, Palestine, North India and Turkestan. In Europe it is limited to Sardinia and probably Corsica and the Balearic Islands, where it is represented by a peculiar race, *F.l. sarda*. It is the normal prey of the screech owl, eagle and other carnivores (Fig. 27).

The porcupine *Hystrix cristata* is very harmful to trees, because it gnaws the bark and trunk. Introduced into Italy by man, we find it today in the central-

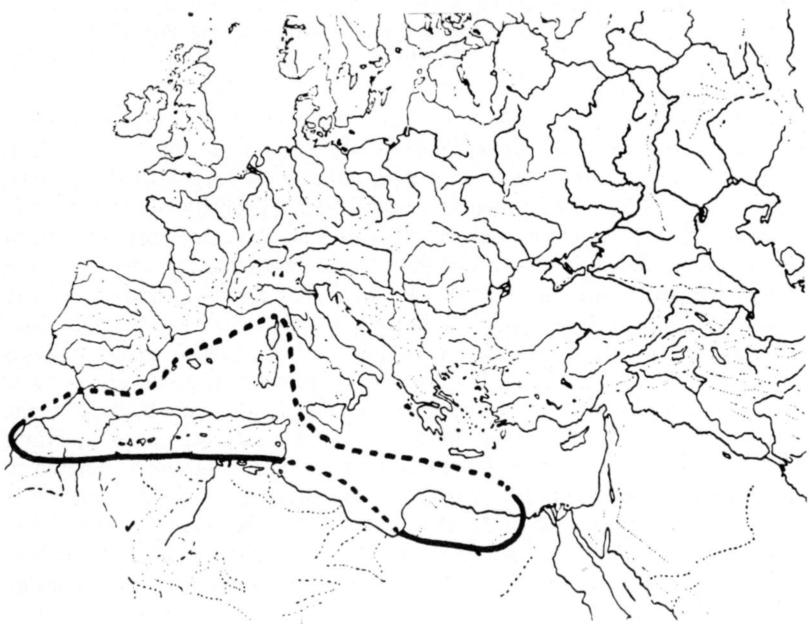

Fig. 27. The Sardinian wild-cat *Felis lybica*, and its distribution (from Bacetti).

southern parts of the peninsula, and in Sicily. It is widespread in Africa, Albania and central-northern parts of the Balkans. It has nocturnal habits, and eats predominantly tubers, roots, ears of maize, fruits and cereals.

Another beautiful carnivore, typical of the sclerophyllous forest is the southern lynx (*Lynx pardina*), nowadays extremely rare. It is limited to the Balkan and the Iberian peninsula, while in the Carpathians it lives together with the northern lynx, *Lynx lynx*. Its slender and elegant body is covered with beautiful red brown fur, black spotted or striped. It is not properly established whether this form is a true species or only a subspecies of *Lynx lynx*.

The deer (*Dama dama*), the goat (*Capra hircus*) and the mouflon (*Ovis musimon*) are the characteristic, though not the only, ungulates of the sclerophyllous forest both in plains and in mountains, at the altitudes at which it is still growing. The deer was endemic to Central Europe before the Glacial Epoch, but gradually retreated into the Mediterranean area, where it formed two subspecies, a western and an eastern one. The western race was introduced in Roman times to other countries. It loves the silence and solitude of the well-forested areas of Europe (except the northernmost parts) and Asia Minor. It seems that in Sardinia it has been preserved in a wild state even outside the reserves. In Italy it can be found only in some territories belonging to the State and in some reserves (Mesola, S. Rossore, Capallio, Montalto di Castro, Cansiglio). It thrives furthermore on an islet off the western Istrian coast (Brioni) and until World War II it also lived on the Island of Rhodos. It is a resident mammal, with a home range only a few kilometres wide, though in some particular conditions it can wander. It prefers forested areas.

The wild goat seems to belong to the 'species' *Capra hircus*, closely related to the ibex (*Capra ibex* or *Capra hircus ibex*). It inhabits some of the Aegean islands (Cyclades, Sporades, Crete), the Caucasus, Southern Turkey, Asia Minor, Iran, Beluchistan, Western Sind and India. In Europe we are probably confronted with some feral forms (Scotland, England, Island of Montecristo, Dalmatia etc.). These populations should belong to the geographical race *aegagrus*, whereas the eastern ones should belong to the race *blythi*. The wild goat thrives in arid, rocky and inhospitable localities, covered with a shrub vegetation, belonging to the maquis, or an arboreal vegetation, belonging to the sclerophyllus forest. It feeds on grasses and branches, and can also live without water. On some Mediterranean islands its food is composed exclusively of rosemary (*Rosmarinus vulgaris*). Because of the scarcity of vegetation in this environment, the wild goat represents a serious danger to the natural vegetation. The mouflon *Ovis musimon*, closely related to the domestic sheep, which according to some descended from the former, is also an inhabitant of Mediterranean areas. Originally from Corsica and Sardinia, it has been successfully introduced in several parts of Europe (as far north as the Netherlands), in Italy and on the Island of Zannone. The mouflon of Cyprus, according to some, belongs to a different, eastern species, *Ovis orientalis*. Like the wild goat, the mouflon jumps and climbs very actively; it thrives in inaccessible mountains or hills, covered with the typical elements of the maquis (*Quercus suber, Q. ilex*) but also in clearings and fields. It is the normal prey of wolves, foxes, wild cats, martins, eagles and screech-owls. In Southern Europe and in Asia Minor it is found both on the plain and in the mountains. It is a typical forest animal, and comes out to graze early in the morning or in the evening. Generally it lives in flocks, with an old male heading the flock. It can be easily tamed.

The only monkey of Europe is *Pithecus inuus*, last reminder of a climate now disappeared from Europe, living only on the Rock of Gibraltar, witness to the old (Pliocene) connections between the Iberian Peninsula and Morocco, or better still the Atlas. Indeed some specimens are still introduced from the latter into Gibraltar, in order to avoid the disappearance of this monkey from Europe.

The genet *Genetta genetta* is a small carnivore present in the Iberian Peninsula: it lives on the Spanish slope of the Pyrenees, on the right bank of the Ebro valley, in the southernmost portion of the Peninsula and on Mallorca (Balearic Islands) which perhaps it reached before the 'Villafranchian', when the Balearic islands separated from the Iberian Peninsula.

Finally, the mongoose *Mungos widdringtoni* or *Herpestes ichneumon* is restricted to Sierra Morena and to some districts of Andalusia (Fig. 28).

Among Mediterranean birds we can recognize a series of species present throughout the European Mediterranean area (and often in the whole of the Mediterranean Basin), and some present exclusively, or more commonly, in the eastern part of it. Very few species are exclusive to the western part. Among the widely distributed birds we can say a little about Cory's shearwater, *Puffinus kuhlii kuhlii*, a relatively big bird with the upper parts brownish-grey on the back with a tendency to sooty. It is present from Gibraltar to Asia Minor and Palestine. In Italy it is abundant some 15-20 km off the coast, both on the Adriatic and the Mediterranean sides. It nests on very small islands; rarely nearing the coast or going inland along rivers. It lands practically only to nest. *Puffinus kuhlii* is a good swimmer and a good diver, remaining under water looking for food for more than 20 seconds, but generally it fishes on the surface. *Puffinus kuhlii* lives in small flocks; it has more nocturnal than diurnal habits. When it is in its nest, it is very easy to catch with the hands. It does not attempt to escape, but it defends itself

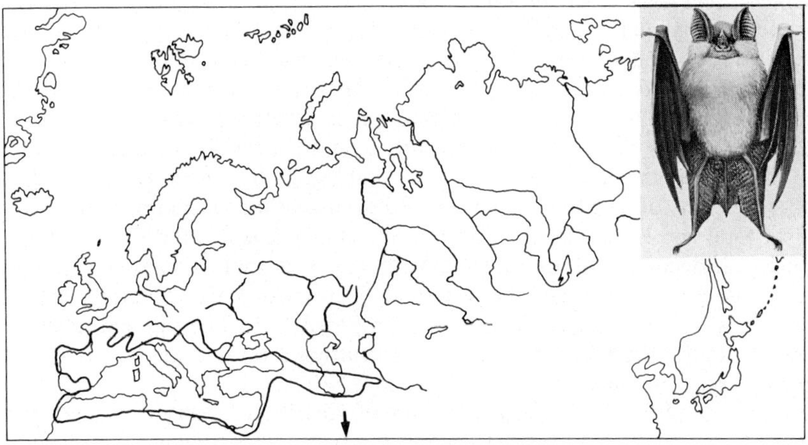

Fig. 28. Rhinolophus euryale and its distribution (modified, according to Lange).

pecking and spitting. It is generally a silent bird, but during the night it is possible now and then to hear its melancholy cry.

It is worthwhile recalling here another kind of procellariiform, *Puffinus kuhlii yelkouan*, which on some islands such as the Tremiti, may live along with the former (see later).

P. kuhlii kuhlii nests in colonies, often in a great number, on grassey, inaccessible cliffs, vertical to the sea, sometimes digging small holes of various shapes and sizes in the barren earth or inside small cavities or also under stones. It feeds on molluscs, crustaceans, small fishes, fatty substances, and also on vegetable debris.

The anseriform *Oxyura leucocephala*, or white-headed duck, is relatively small, with the upper parts of the body (male) reddish brown, crossed by many thin, black, zig-zag stripes. At any age and in both sexes it is recognizable thanks to the long and stiff tail feathers not covering each other, so that they look separate and independent of one another. The species is a regular spring and autumn migrant. It thrives in freshwater ponds, lakes and rivers, rich in aquatic and swampy vegetation, amongst which it generally hides. It is a very good diver, and when it is pursued it does not take flight, but submerges itself and remains under water for a long time. It nests among the vegetation of fresh- and brackish waters. It builds its nest in a depression of swampy soil, covered with stalks of plants, and sparsely lined with down. It lays up to 12 white-bluish eggs, with a rough shell. It feeds mainly on molluscs, crustaceans and insects.

The ruddy shelduck, *Casarca ferruginea*, prefers freshwater to brackish. It is a typical gregarious animal during migration and in the wintertime, but for the rest of the year it lives isolated or in couples. It nests in holes in any kind of tree, in rocks, in crevices, and it can also utilize old, abandoned, nests of magpies or large falcons, as high as 1000 m altitude. The nest is made of feathers, moss and grasses. This bird feeds on water insects, molluscs, crustaceans, and also on plant remains, and it goes – like geese – to the fields or to wheat cultivations to look for food. During winter time it is abundant on the sandy shores of rivers and the sea, in fields and also on dry steppes.

The Egyptian vulture, *Neophron percnopterus*, lives in rocky mountains as well as on open plains, and in eastern countries – such as Turkey, Egypt, etc. – also inside the towns and villages, where it is considered a sacred bird and protected because of the good it does for hygiene by destroying carrion. In Europe it is generally a solitary bird, or lives in couples, but sometimes it lives and migrates in big groups. It nests on rocks, even at quite low altitudes. Its nest consists of a rough area which the bird occupies for several years and which is composed of branches, remains of small mammals and fishes and other substances which have served as food and have a terrible, almost unbearable, smell. It feeds on carcasses, particularly if decomposed, of lambs, kids, dogs, foxes, rabbits, hares, hedgehogs, moles, voles, etc., but it can also eat small living mammals and birds, snakes, tortoises, toads, crustaceans and fishes, decomposed fruits like figs and dates, or even faeces. In Southern Europe the Egyptian vulture is both resident and migrant (double passage), but very rare.

The griffon vulture, *Gyps fulvus*, is a widely distributed creature, common in Sardinia, rare in Sicily and Corsica, where it is resident. It is frequently mentioned in the Bible. The Talmud says that 'even when the griffon vulture is flying upon

Babylon, it can discover carrion in Palestine.' A few years ago, in Palestine (today Israel) the slaughter-houses were visited daily by specimens of this species, from nests on the shores of the Dead Sea. In Palestine they say that it sometimes makes sheep or goats fall into crevasses, in order to eat them.

The bearded vulture, or *Gypaetus barbatus*, known also as 'the Lammergyr', is easily recognized thanks to a kind of 'beard' composed of hairy-like feathers under the chin, a characteristic present in no other bird of prey. It is resident and nests in Sardinia, but is very rare in Corsica and Sicily. It lives solitary on the highest mountains, flying here and there looking for food, and staying in one area only to breed. People say that sometimes it attacks animals as big as chamois, wild goats (*Capra ibex*), goats and lambs, pushing them with its wings in order to make them fall on to the rocks and die, after which it can eat them. It also has the habit, as do other birds of prey, of taking bones to a very great height and letting them fall on to the rocks in order to break them and to eat the marrow. The same is done with living tortoises. It nests on high mountains in the wildest and most inaccessible places, anywhere it can find an area suitable for a range. The nest is a big, rough mass of branches and dry roots, plus remains of wool, pieces of skin, etc.

The imperial eagle, *Aquila heliaca*, is limited (in Europe) to Spain (except the northernmost parts) and the Balkans (except the westernmost portion) but reaches as far east as Mongolia and China. Very rarely it can be seen in Italy. It lives in the plains, steppes and swamps. It builds a huge nest on high, isolated trees.

The so-called Bonelli's eagle, *Hieraaëtus fasciatus*, is strong as a true eagle, but appears more like a goshawk. It is a circummediterranean element, living on mountains and in wooded localities, not far from swamps or lakes, rich in game, which it hunts very actively. It nests in crevices in rocks and more rarely in trees, and the nest consists of a mass of branches, stems, etc. lined with leaves and other vegetative substances. In some places, such as Sardinia, it is very harmful to hares, partridges, quails, and in general to water fowls. In Italy it is resident in Sardinia, and rare in Sicily and Corsica.

Another *Hieraaëtus, H. pennatus* or the booted eagle, is a widely distributed species, present on the mountains and in the forests of Southern Europe; in Italy it is a common migrant (double passage, but very irregular). It inhabits the large forests of pines and broad leaved trees. It nests in trees such as elms, oaks, chestnuts, etc. It is extremely dangerous to game, particularly to pheasants on whom it inflicts veritable massacre if it enters a game reserve.

The pratincole, *Glareola pratincola*, lives in swamps and on the shores of the sea, rivers, lakes and ponds. It builds no nest, but lays eggs in small colonies in the salty swamps, on dry islets, in a small depression in the mud, often among the halophilous vegetation. It feeds mainly on insects, particularly beetles, hoppers, crickets, etc., which, like terns, it catches in flight.

Among the passerine birds of the sclerophyllous forest there is the calandra lark, *Melanocorypha calandra*, typical of Southern Europe: in Italy it is resident in the central, southern and insular districts. It lives in dry, stoney places, in fields and on steppes; the nest is built on the ground. *Cettia cetti*, Cetti's warbler, is relatively abundant and resident in Southern Europe and the Mediterranean islands. It lives isolated in the dense *Phragmites* stands, so that it may easily escape enemies, man included. It sings both during the day and night. It builds its nest in bushes of *Tamarix gallica* or other species, half a metre from the ground and near water. It feeds on insects, small molluscs and green seeds.

Another passerine bird is *Hippolais pallida*, the olivaceous warbler, present in cultivated areas and in gardens. It nests in shrubs, hedges, etc. or often at a very great height in palms. In Western Italy it is extremely rare. The warblers (*Sylvia*) are among the most characteristic birds of the sclerophyllous forest. They are typical arboreal creatures, also living in vineyards, olive groves, wooded areas, shrubs etc. During winter they prefer thick heaths, the dense maquis with myrtle, *Arbutus*, ivy, holly and other evergreen plants, among which they find a sure and ready refuge against enemies. They do not travel very much. The nest is built among shrubs and in trees, generally scarcely off the ground. Their feeding habits are mainly insectivorous, so that they are extremely useful to agriculture: they eat

Fig. 29. Distribution of the two Galliforms *Alectoris graeca* and *A. barbara* (modified according to Baccetti).

beetles, hoppers, caterpillars, worms and aphids (for example the pea aphid). During summer and autumn they eat the fruit of holly, ivy, honeysuckle, other berries, *Ligustrum, Sorbus, Sambucus* and sometimes also oranges and figs.

The rufous warbler, *Erythropygia galactotes*, is restricted to the southernmost parts of the Iberian Peninsula and the Balkans, living in gardens, vineyards, in olive trees, palms, etc. It nests in hedges of *Opuntia ficus-indica* and in palms, so that it can be considered as one of the most thermophilous birds of Europe.

The fan tailed warbler, *Cisticola juncidis*, is present in Southern Spain, Southern France, Italy (except the Alps), the large islands (Balearic, Corsica, Sardinia and Sicily) and the southern part of Greece (Peloponnesus). It thrives in dry and humid places, in corn fields, in uncultivated areas and in swamps. It builds a deep, bottle-shaped nest, hanging from high herbs and shrubs. Despite its geographical distribution, it cannot be considered as an inhabitant of European sclerophyllous forest.

The last passerine bird we want to mention here is the Spanish sparrow, *Passer hispaniolensis*, confined to the three main peninsulae of the Mediterranean. In Italy it is limited to central-southern parts, and lives also in Sicily and Sardinia. Unlike the house sparrow, this small bird does not live near houses, but prefers wooded areas, shrubs or trees growing near the banks of rivers. It makes its nest in colonies or sometimes isolated, often on the branches of trees. Finally, another kind of bird more or less restricted to the sclerophyllous forest is the pallid swift, *Apus pallidus*, on the wing undistinguishable from the related *A. apus*. It lives in the air, it can thrive everywhere, but prefers places in which it can nest. Its nest is situated on buildings, under the gutters, occasionally among rocks. It is present in the southern part of France and of the Iberian Peninsula, on the Balearic islands, Corsica, central-southern Italy and Southern Dalmatia. Its area of distribution corresponds very closely with that of the sclerophyllous forest.

As we said, many birds are limited to the eastern Mediterranean basin. Restricting our studies to European species, let us mention here the pigmy cormorant, *Phalacrocorax pygmaeus*, an essentially aquatic bird, frequent on rocky seashores, but also on those of lakes and rivers, particularly if covered with high trees. It is a good swimmer; if frightened, it swims practically completely submerged, with only the head and neck out of the water. It nests in colonies sometimes together with herons, shovelers, glossy ibis, etc. in submerged, swampy localities, on the top of small shrubs, nearly 2 m above water. The nest is rough and large, constituted of stalks and swamp grasses. It feeds on fishes, both freshwater and marine, such as trout, salmon, eels, mullet, herring, but also on crustaceans and molluscs. It can be harmful to fish-culture.

Puffinus kuhli yelkouan, or 'diomedea' has a call which closely resembles a child's cry (Moltoni, 1957). It is called by this name because it lives on the Tremiti Islands off the Adriatic coast of Puglie, and the Tremiti are called also 'Diomedeae Insulae' since people once thought that Diomede's sepulchre was here, and they said that Diomede's companions had been transformed into birds, namely 'Diomedeae Aves'.

The white pelican, *Pelecanus onocrotalus*, is characterized by the presence of a huge fold of skin, hanging from the mandible, which, when extended, can contain several litres of water but which generally is contracted so that it is almost invisible. This 'sack' is used like a net to catch fish; when it is full, the bird closes the beak and raises it, contracting the skin in order to make the water flow out of the sides and then swallows the fishes. The pelican feeds its nestlings with small fishes, partially macerated, which the parent ejects from the crop. It thrives in the shallow waters along the shores of the sea and lagoons, of rivers and lakes. It is an essentially gregarious creature, living in flocks of thousands individuals, together with cormorants and other aquatic birds. It swims very well, but cannot dive, so that it is restricted to shallow waters. Its irregular migrations also seem to be caused by the necessity to find shallow waters. The pelican nests in large colonies on islets and in areas covered with aquatic vegetation and *Phragmites.* The nest is composed of a simple depression of the swampy ground, on which the animal accumulates aquatic and swampy vegetation, together with the remains of its food, which, after decomposition, have a terrible smell. The pelican fishes together with cormorants, sea gulls, terns etc. In a sense pelicans and cormorants are synergic, possibly symbionts, since the latter make the fish swim hither and thither frightened, so that they can be easily preyed upon by the former.

The Lanner Falcon, *Falco biarmicus*, restricted in Europe to central-southern Italy, Sardinia, Sicily and western and southern parts of the Balkans, lives on rocks, the slopes of rocky mountains, and also in stoney and semi-desert plains. It nests on rocks or occasionally in trees.

The black-headed bunting, *Emberiza melanocephala*, a passeriform bird, thrives in plains and hills, even in arid areas, though it prefers cultivated areas, rich in trees, either deciduous or evergreen, on which it often sits. It can nevertheless also live in swampy places rich in *Phragmites*, birches, and small willows. It nests in the higher area of the swamp, among the *Phragmites* and *Carex*. The nest is a rough, oval construction, consisting of moss, stalks of herbaceous plants, roots, etc. lined with hairs, feathers, down and woolly pappus of swamp plants. It feeds not only on the grains and seeds of several plants, on the fruits of wild species, young buds, but also on small molluscs and crustaceans. The parents feed the nestlings on insects (beetles, butterflies, moths, bugs, larvae).

Finally, the olive tree warbler, *Hippolais olivetorum*, inhabits the southernmost part of the Balkans: it thrives in the dense canopy of olive trees and oaks, where it hides easily. It nests on the branches of olive trees. It can be therefore considered a typical inhabitant of the European sclerophyllous forest.

One of the few birds restricted to the central portion of the European Mediterranean is the Italian sparrow, *Passer italiae*, exclusive to Italy (except the Alps), Corsica and the north-eastern part of Sicily. It is therefore a vicariant species (both in the geographical and ecological sense) of the common house sparrow, *Passer domesticus*, with which it shares habits. According to one source (Bauer & Rokitansky) the Italian sparrow can occasionally reach some northern regions such as the Inn valley and Southern Carinthia (Austria).

Among the few birds restricted (in Europe) to the Western Mediterranean, we can mention two warblers, *Sylvia undata* and *S. conspicillata*. The latter, called in English the spectacled warbler, is present in southern Spain, southern France, central-southern Italy, Corsica, Sardinia and Sicily. It thrives mainly on sea shores covered with *Salicornia*, or – together with *S. undata* – on low shrubs, where it builds its nest.

The melodious warbler, *Hippolais polyglotta*, lives in the Iberian peninsula, most of France, the whole of Italy, Corsica and Sicily, and the Balearic islands. It is found in gardens, parks and cultivated areas, and in forests, hedges, or among rich vegetation near water. It nests in shrubs or even trees. We want to finish this small section by mentioning the black wheatear, *Oenanthe leucura*, a small creature present in the Iberian Peninsula, southern France, Liguria, Sardinia and Sicily. It thrives on inaccessible rocks, on cliffs near the sea, on stoney hill slopes, and in arid, stoney areas within meadows. Its food consists of insects (beetles, flies, *Hymenoptera*), small fruits and seeds. It nests in crevices or cracks in rocks. The nest is built of dry grasses, hairs and feathers, but sometimes also moss and remains of animal fur.

If we bear in mind the climatic characteristics of the sclerophyllous forest – still more if compared with those of the other European biomes – we realize that a particularly favoured vertebrate group is the reptiles. Indeed, they are poikilothermous animals (commonly called 'cold blooded animals') in general widely distributed and very abundant in the tropics, and relatively euryhygrous, since they can easily live without water. There is only one other European biome where reptiles are so favoured, the steppe.

Snakes, lizards and secondarily tortoises are practically omnipresent in the sclerophyllous forest, from the sea shores to the upper limit of the *Quercetum ilicis*, in the dense forest as well as in the driest garrigue, on rocks emerging from the sea with a surface of a few square metres (for instance the 'faraglioni').

The smallest European snake is *Typhlops vermicularis*, belonging to the family *Typhlopidae*, which includes the smallest species of the world. It can be found in the Balkans south of 42°N and in a zone from Southern Egypt to Tagikistan, Western India and Afghanistan, including Transcaucasia. Not very different from an earthworm, no longer than 36 cm, it thrives in sandy biotypes with a scanty xerophilois vegetation, characterized by a short vegetative period. It does not live above 1600 m altitude. Its subterranean habits make it a very rare species. It feeds on ants and their larvae which it searches for in their nests.

Contia modesta, related to some steppe species, is a small snake, olive grey with the head bearing three dark stripes. It lives on the Aegean islands and the European part of the Caucasus (Azerbaidjan, Georgia, Armenia and Dagestan on the shores of the Caspian Sea). It lives in dry places covered with herbaceous or shrub vegetation, never above 1800 m, and feeds mainly on insects.

A poisonous snake is *Coelopeltis* (= *Malpolon*) *monspessulanus*, belonging to the *Opistoglypha* insomuch that it shows the poison groove only in the posterior teeth. This species is greenish with some small dark spots, and may reach a length of 170 cm. It thrives in relatively dry places all through the sclerophyllous forest dominion. Its bite, if prolonged for some minutes, can be lethal to small mammals. Adult specimens feed on lizards, small snakes and small mammals, younger specimens on insects. Restricted to the eastern part of the Mediterranean basin is *Tarbophis fallax*, grey with blackish spots, whose bite is harmless to man. In Europe it is found in the Balkans, on some islands of the eastern Mediterranean, Azerbaidjan, Armenia and Georgia. It prefers dry, sandy biotopes, sometimes covered with shrubs. It does not avoid human settlements and in Azerbaidjan it climbs on roofs covered with earth. It feeds on lizards.

Lizards are omnipresent: in sunny or shady places, on the mountains up to the limit of the evergreen vegetation and in the deciduous forest which succeeds it (submediterranean forest), on small rocks only a few metres from the sea shore, in subdesert, steppe, biotopes, they are always present.

A genus more specialized to life in deserts than in forests is *Acanthodactylus*, furnished with a fringe of long squamulae all along the toes, which resembles somewhat the hairs ornamenting the tarsi of some desert beetles (*Pimelia*). This fringe would help the animal not to sink into the sand. In Europe we find *A. vulgaris* in southern France, Spain and Portugal, and *A. schreiberi* is restricted to Cyprus. Probably these lizards are witnesses to a warmer, drier climate all through the Mediterranean area.

Many species are of the genus *Lacerta*. We can mention *L. oxycephala*, olive green with small black spots, living on the rocks near the sea shore in Dalmatia and on the nearby continent; *L. graeca* present in the Peloponnesus; *L. hispanica* of the Iberian Peninsula and North Africa, *L. sicula* (the commonest lizard in Italy, together with the wall lizard *Lacerta muralis*), subdivided into a great number of subspecies or, in some instances, 'demes', in Dalmatia replaced by *L. melisellensis*,

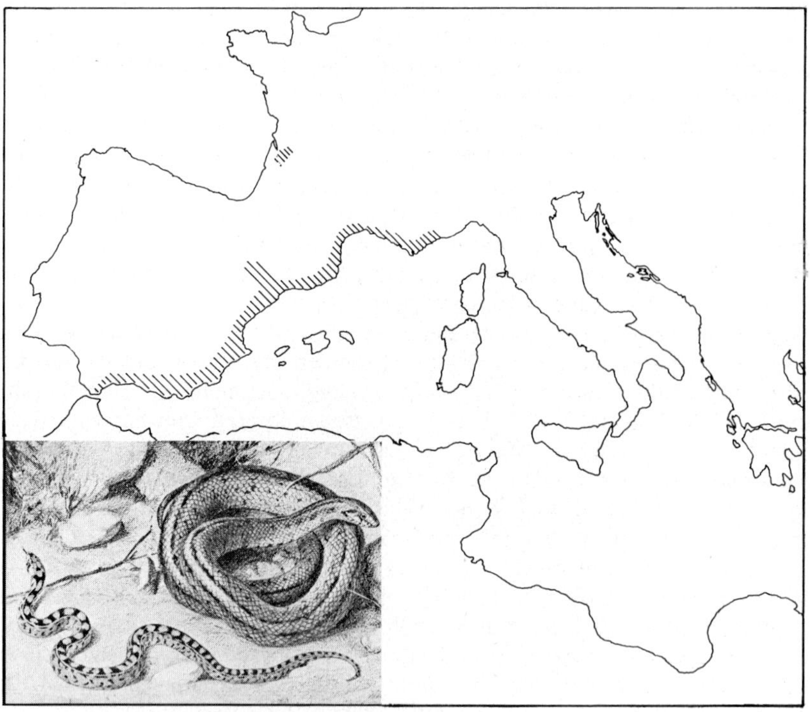

Fig. 30. *Coluber scalaris* (from Brehm).

a true vicariant species in the geographical and maybe also in the ecological sense. Some species, *L. bedriagae, L. filfolensis, L. wagleriana* and *L. tiliguerta* (Fig. 31) furthermore are endemic to some of the Italian islands. The evolution of all these species, subspecies and local races into which they are often subdivided must be attributed to the geographical isolation of ancient populations after the separation of the islands from the mainland and from each other. Tectonic and orogenetic movements, together with eustacy, are responsible for these fragmentations of former land-masses and therefore for speciation or subspeciation (even incipient subspeciation).

A very interesting species is *Ophiops elegans*, represented by two geographical races, one restricted to Cyprus (*O. e. schlüteri*) and another extending from Transcaucasia and Asia Minor to the Punjab (*O. e. elegans*). It thrives at the surface of the soil – particularly in steppe environments – and can easily be captured, thanks to its slow movements.

Very different are the lizards belonging to *Scincidae*, with a short cylindrical body, shining, smooth skin, a short, conical tail and more-or-less reduced limbs. *Scincidae* have rather subterranean habits. Some of them are typical of the sclerophyllous forest: *Chalcides chalcides* is restricted to the western Mediterranean (including Italy), *C. ocellatus*, on the contrary, is found in the south-eastern Mediterranean, as far west as Sardinia, Sicily and the southernmost part of the Balkans. It is divided into a number of subspecies, a sign of antiquity. A complete-

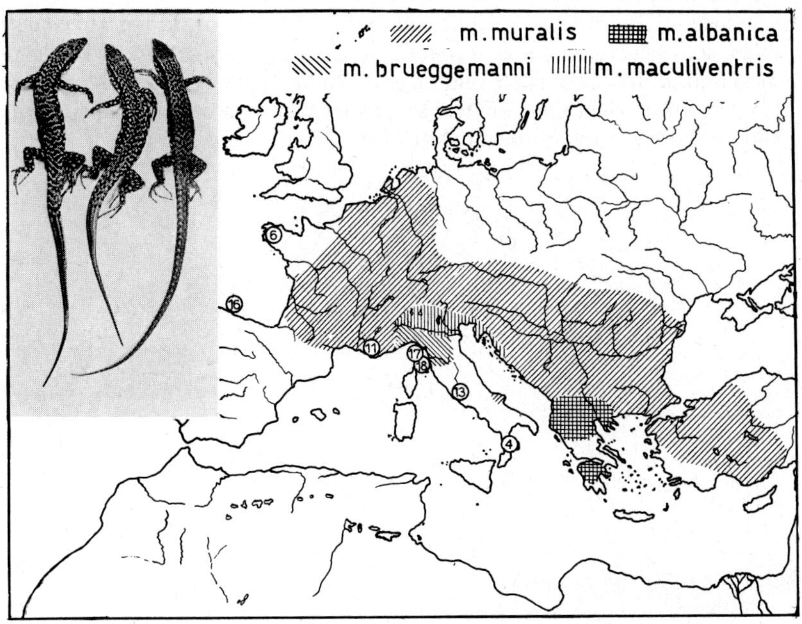

Fig. 31. Distribution of the main subspecies of the 'Rassenkreis' *Lacerta muralis*. The numbers within circles correspond to the following subspecies: 4. *breviceps*; 6. *calbis*; 11. *oyensis*; 13. *nigriventris*; 16. *rasquinettii*; 17. *vinciguerrai*; 18. *beccarii*.

ly blind and limbless lizard is *Ophiomorus punctatissimus*. It is a yellow, small, earthworm-like creature, living under stones in Greece, Asia Minor and Galilea (Fig. 32a).

The species which most resembles an earthworm is nevertheless *Blanus strauchi*, quite limbless and with anterior and posterior extremities of the body quite indistinguishable. The eyes are covered with tegument and therefore without any function; there is no tympanic membrane, and the animal can be considered the best adapted species for subterranean life. *Blanus strauchi* belongs to a taxon, *Amphisbaenidae* (from ἀμφί at both extremities, and βαινειν to go, to walk, since the animal can proceed either forwards or backwards), widely distributed in the tropics, but not necessarily restricted to dry situations, which probably reached the Mediterranean Basin during a warmer period and left – as a reminder of it – only one species. (The same is true for some beetles, for instance *Strongylium* or *Doliema*). *Blanus strauchi* is present in Greece, Rhodes, Asia Minor, Syria and the region around Istanbul.

Ophisaurus apus (from α and πους i.e. limbless) related to the common blind-worm *Anguis fragilis*, is another lizard typical of sclerophyllous forest. Much larger than the blind-worm (it can reach more than 1 m in length) it is gradually tapered from the head towards the tail. It thrives in several kinds of biotopes, including gardens and kitchen gardens. It feeds on snails, small mammals, reptiles and insects. It lives in all South-Eastern Europe, from Dalmatia to Russia (in the Crimea it is known as 'sheltopuz'), the Caucasus, Transcaucasia, and Turkestan, where it is called 'saradzhilyan', which means 'yellow snake'.

The chameleon *Chamaeleon vulgaris*, the only European lizard able to change colour under some causal stimulus, is present all along the Mediterranean coasts of Africa, Asia Minor, Southern Spain and Cyprus. Probably it can be considered as a relict of older times, when the climate was warmer than now and very likely more humid, due to the typically arboreal habits of the genus.

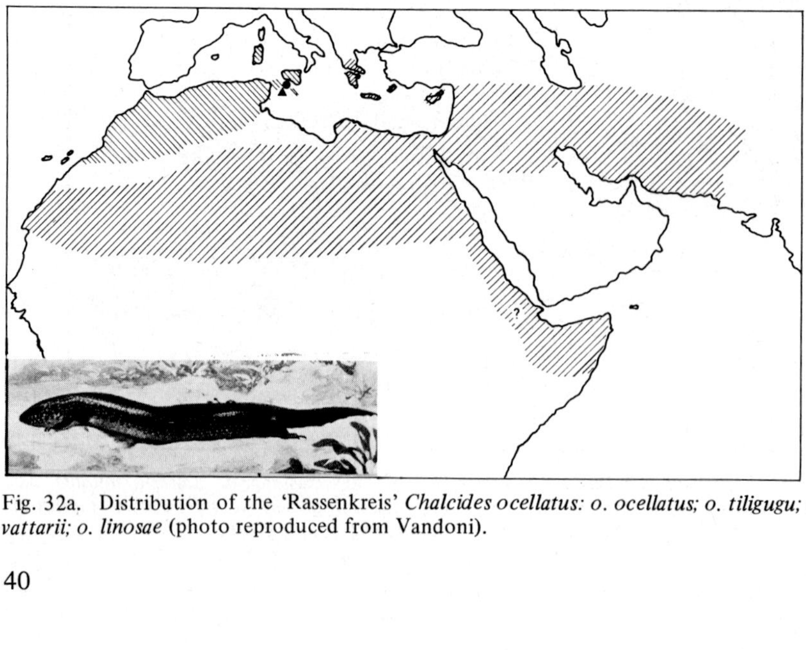

Fig. 32a. Distribution of the 'Rassenkreis' *Chalcides ocellatus: o. ocellatus; o. tiligugu; o. zavattarii; o. linosae* (photo reproduced from Vandoni).

Another group very widespread in the Mediterranean area and almost omnipresent is the *Gekonidae* or 'geckos', whose origin is probably tropical, due to the enormous number of species of this family living in the tropics. They are characterized by an adhesive disk on the under-surface of the toes, which permits the animal to climb on completely vertical, smooth surfaces, as well as to walk on the ceiling of a room, equalled only by flies, which incidentally geckos are catching.

They are completely without eyelids, which can be related to the nocturnal habits of the family. The commonest genera are *Hemidactylus*, spread all along the shores of the Mediterranean Sea (*H. turcicus*), *Gymnodactylus*, represented by several species, the best known of which is *G. kotschyi*, present in all the Mediterranean area, Italy included. *G. colchicus* is endemic to the south-western part of Transcaucasia; *G. danilewskii* is endemic to the Crimea (where it is called 'akhreb') and finally *G. caspius* and *G. fedtschenkoi* are oriental species, reaching the south-western limit of their distribution in Transcaucasia.

The common tarentola, *Tarentola mauritanica*, is known in all the Mediterranean basin, where it continuously catches flies and harmful insects, but where it is the object of superstitions and dreadful stories. It is considered a very poisonous animal and everywhere persecuted to death (Fig. 32b).

Before finishing this small section it is necessary to mention some tortoises which can be often seen in the area of the maquis. *Testudo graeca* extends into southern Europe (including the Pitiusas Islands), south-western Asia and North Africa. It has been introduced in several Italian localities where it is now naturalized. It has vegetation feeding habits, but can also eat earthworms and snails. *Testudo hermanni* is present in southern Europe, including the Balkans, Corsica, Pianosa, Sardinia and central-southern Italy. Finally, we must mention here *Testudo marginata*, endemic to southern Greece, introduced into Sardinia and other countries where it is now perfectly naturalized.

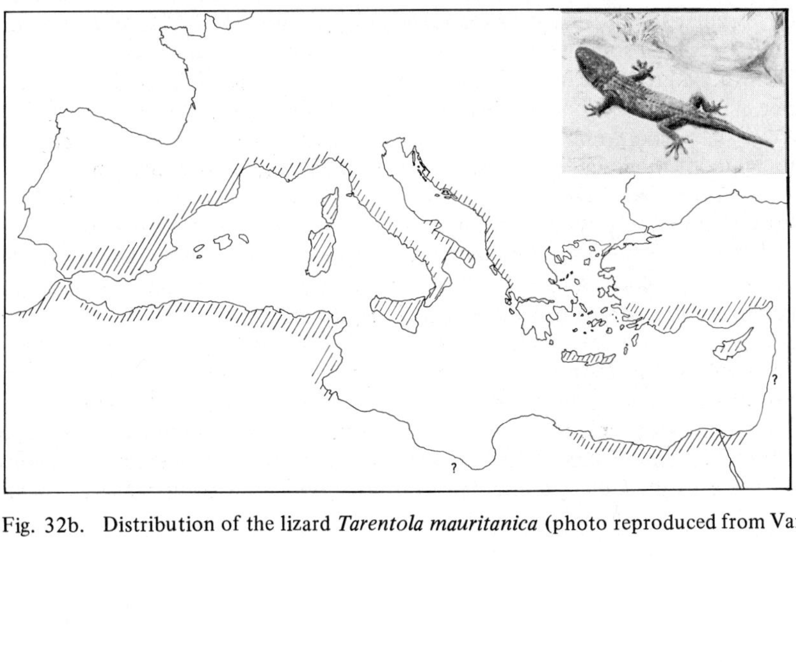

Fig. 32b. Distribution of the lizard *Tarentola mauritanica* (photo reproduced from Vandoni).

Invertebrates

The invertebrates are much less well-known than the vertebrates.

Many invertebrates are harmful to vegetation within the sclerophyllous forest. We will just mention a few beetles as an example. Within the Coleopterorous family *Cerambycidae* we find *Macrotoma scutellaris* on the trunks of sclerophyllous evergreen oaks; *Criocephalus syriacus* in Italy and Cyprus, probably inside the dead trunks of *Pinus; Cerambyx nodulosus*, in south-eastern Europe, also on living fruit trees; *C. carinatus, C. velutinus*; *Hesperophanes* (name which in Greek means 'which appears in the evening', since the insect flies only at twilight) *sericeus*, characteristic of the maquis, where its larva developes inside living or dead tree trunks, *Pistacia lentiscus*, oaks, vines etc., *H. griseus*, in dead branches of figs and also *H. fasciculatus*, in living branches of the holm oak and other trees; *Stenopterus ater*, in *Pistacia lentiscus* and *P. terebinthus; Phymatodes glabratus*, on *Juniperus* and cypress; *P. alni* in the dead branches of evergree oaks; *Plagionotus arcuatus*, similarly on living oaks; *P. scalaris; Clytus clavicornis; C. nigripes; C. pelletieri; C. rhamni* (present also in the deciduous forest); *C. pilosus* (on holm oak); *Purpuricenus dalmatinus*, only on evergreen oaks; *P. budensis*, on the diseased twigs of holm oaks; *P. koehleri*, also on fruit trees, often on *Paliurus* and *Lonicera.* One of the smallest *Cerambycidae* of the sclerophyllous forest is *Parmena pubescens*, whose larva feeds on *Euphorbia* and other plants near the sea shore. Other *Cerambycidae* are *Dorcatypus tristis*, polyphagous on figs, mulberry, cypress, vines and sometimes also carnations; *Monochamus galloprovincialis*, on *Pinus halepensis; Pogonochaerus perroudi*, on the dead twigs of *Pinus; Agapanthia irrorata*, on *Onopordon illyricum; A. kirbyi*, on *Verbascum*, etc.; *Oberea erythrocephala*, on *Euphorbia; Phytoecia melanocephala, P. coerulea* and other congeneric species.

Among Chrysomelids of the sclerophyllous forest we can mention the beautiful *Chrysomela americana*, a shining, red and green coloured small insect, living only on rosemary; *Crioceris paracenthesis*, living exclusively on *Asparagus acutifolius* and present also in some xerothermic oases of Northern Italy and at the Adige mouth, where it has been found by the present author (Fig. 33-34 & 36).

Other phytophagous beetles belong to the *Scarabaeidae*, like the beautiful *Polyphylla fullo*, the *Buprestidae*, e.g. *Capnodis* and *Julodis*, the *Elateridae*, weevils, etc. Many of them will be recorded in the special part of the Chapter.

The evergreen oaks, *Quercus coccifera* and *Q. ilex*, are often attacked by a scale insect, *Kermococcus vermilio*, belonging to a group of animals known since biblical times; according to Bodenheimer the Palestinian species named in the Bible are two, *Kermes nahalali* and *K. greeni. Hermococcus vermilio* is mentioned as '*Kermes vermilio*' in one of Charlemagne's capitularies.

A real pest in the sclerophyllous forest can be the hopper *Dociostaurus maroccanus*, known as the 'Moroccan locust'. Interesting work has been done on its parasites, which of course are a means of attack against this harmful insect.

As far as predatory insects, carnivores, are concerned, we will limit ourselves to mentioning Carabids, Staphylinids, Cantharids, etc. Most of the saprophagous insects belong to the Coleopterous family *Tenebrionidae.* Among parasitic insects there is the genus *Meloë*. Coprophagous insects are also sometimes frequent: we can mention here the elegant *Copris hispanus*.

There is very scant information about the soil fauna of the sclerophyllous forest, not only in Europe but also in North America, especially when compared to that available for other biomes (deciduous forest, taiga, etc.). We want to mention here the data obtained by the present writer on the soil fauna of a sclerophyllous forest of Abruzzo (Italy) and more precisely a *Quercetum ilicis* near Torino di Sangro. The locality belongs to the almost typical Mediterranean forest, with *Quercus ilex, Q. pubescens, Fraxinus ornus, Acer* sp. etc. In the

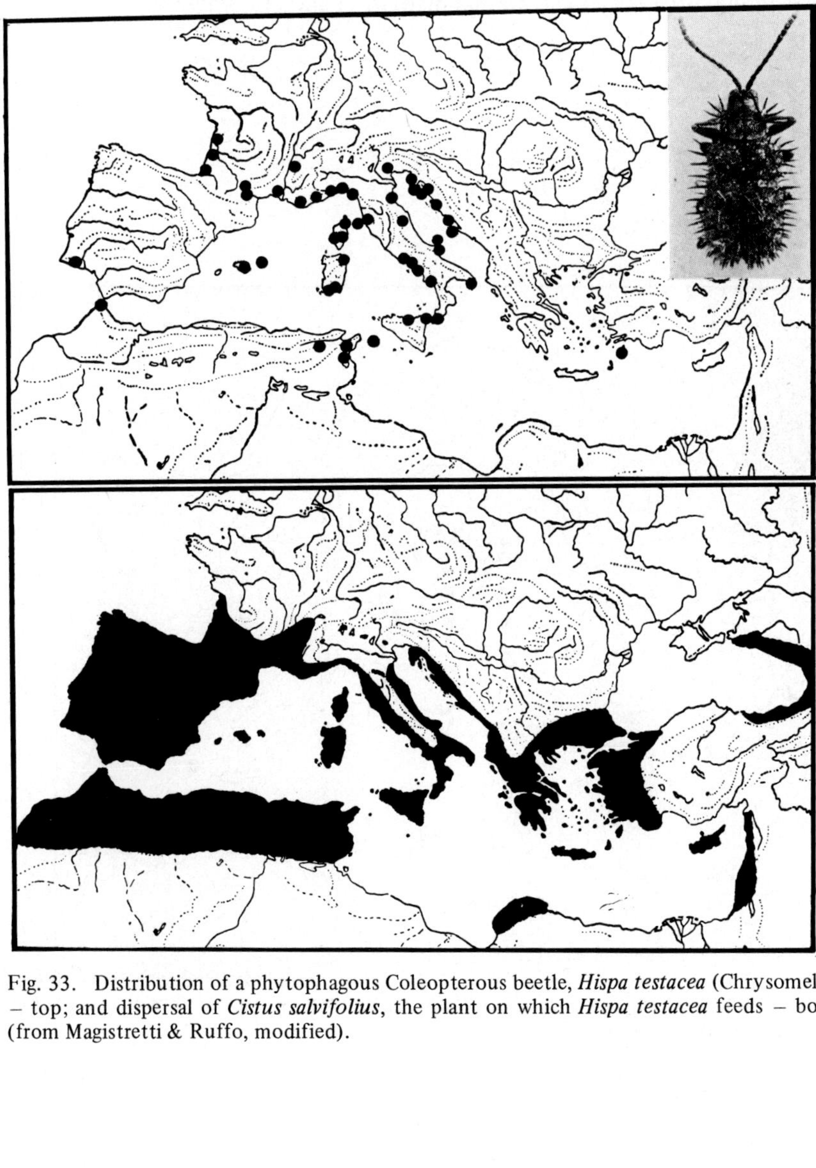

Fig. 33. Distribution of a phytophagous Coleopterous beetle, *Hispa testacea* (Chrysomelidae) – top; and dispersal of *Cistus salvifolius*, the plant on which *Hispa testacea* feeds – bottom (from Magistretti & Ruffo, modified).

undergrowth ivy and *Ruscus aculeatus* (the so-called butcher's broom) predominate. The ivy is practically sterile. On the floor of the forest, where light never penetrates, a thick layer of litter (mainly leaves of holm oak) accumulated. The soil belongs to Mediterranean brown earth ('Braunerde'). Near the holm-oak stand several perennials as *Cistus, Pistacia* and *Smilax aspera*, grow.

These are the main chemico-physical characteristics of the soil:

clay	silt	sand	$CaCO_3$	C	N	C/N	pH
13.1%	42.4%	44.5%	–	6.4%	0.32%	20	6.53

The average number of soil animals in 500 cm^3 of soil and their approximate biomass in the different seasons are summarized in the following table:

Fig. 34. The distribution of the xerophil Orthopterous insect *Arachnocephalus vestitus,* a small cricket-like creature. In Northern Italy it is limited to the xerothermic oasis (from Magistretti & Ruffo, modified).

	spring				summer				autumn				winter			
	litter		soil		litter		soil		litter		soil		litter		soil	
	No	b.	No	b.	No	b.	No	b.	No	b.	No	b.	No	b.	No	b.
Nematoda									1	1	3	3			1,5	1,5
Enchytraeidae	8	4,16							3,5	1,82	8,5	4,42			1	0,52
Isopoda	2,5	16			1	6,4			1,5	9,6	1	6,4			0,5	3,2
Araneae	0,5	3,2	1	6,4	1	6,4	1	6,4	0,5	3,2	0,5	3,2	1,5	9,6		
Pseudoscorpiones	3,5	22,4	0,5	3,2	2	12,8	1,5	9,6	0,5	3,2	1	6,4	0,5	3,2		
Acari	1825	730	991,5	396,6	421	168,4	810	324	1435	574	1046,5	418,6	639,5	255,8	335	134
Pauropoda	4,5	28,8	24,5	156,8	5	32	71,5	457,6	12	76,8	18	115	2	12,8	4	25,6
Diplopoda	10	64	2,5	16	3,5	22,4	0,5	3,2			1	6,4	1	6,4	0,5	3,2
Symphyla			11	70,4			14	89,6	0,5	3,2	11	70,4			11,5	73,6
Chilopoda	1	6,4	1	6,4			2,5	16	1,5	9,6	2,5	16			0,5	3,2
Protura			68,5	27,4			9,5	3,8			8,5	3,4			2	0,8
Collembola	777,5	311	521	208,4	99	39,6	765,5	306,2	592	236,8	737,5	295	84	33,6	107	42,8
Diplura					0,5	0,2	0,5	0,2					0,5	0,2		
Psocoptera			0,5	3,2							0,5	3,2				
Thysanoptera	0,5	3,2											0,5	3,2		
Homoptera											1					
Diptera (larvae)	17	108,8	5,5	35,2	1,5	9,6	5	32	8,5	54,4	4	25,6	1	6,4		
Coleoptera (l.)			4,5	28,8	1,5	9,6	8	51,2	1,5	9,6	4,5	28,8			2,5	16
Lepidoptera (l.)							1,5	9,6								
Diptera (ad.)					0,5	3,2							0,5	3,2		
Coleoptera (ad.)	0,5	3,2	0,5	3,2	0,5	3,2			0,5	3,2	2,5	16	0,5	3,2		
Formicidae	0,5		1				0,5		1		1,5		0,5			
Hym. not Form.	1,5	9,6			0,5	3,2										
Gastropoda	1	6,4							1	6,4			0,5	3,2		
TOTAL	2654		1633		537		1691		2060		1853		732		466	

Note. The biomass is expressed for Nematodes in μg for all other groups in mg. No is number of specimens, b = biomass

The fauna of litter fluctuates much more than that of soil (Fig. 35).

In the soil some Nematodes are present such as *Aporcelaimellus obtusicaudatus, Rotylenchus* sp. (a genus which attacks the roots of several plants), many species of springtails, among which are some limited to the sclerophyllous forest such as the endemic species *Tullbergia italica* and *Heteromurus major* (found as far north as Austria). The fauna (both macro- and mesofauna) of litter and soil includes woodlice such as *Haplophthalmus capricornis, Philoscia affinis, Cylisticus convexus*, whip-scorpions such as *Chthonius tetrachelatus, C. ischnocheles, C. jonicus, C. austriacus, Hysterochelifer tuberculatus, Roncus lubricus*, the Opilionid *Trogulus coriziformis*, the millipedes *Brachydesmus* sp., *Polydesmoidea, Chilognathi* and *Pselaphognathi, Lophoproctus lucidus* and *Polyxenus* sp., the centipedes *Monotarsobius microps, Chaetechelyne vesuviana, Clinopodes flavidus*; the *Psocoptera Lachesilla pedicularia* (a cosmopolitan species) and an unclassified *Liposcelis.* Furthermore, there are some *Thysanoptera* (*Haplothrips* sp.), some parasitic micro-hymenoptera such as *Aphanogmus* and *Platygaster*, the beetles *Arcopagus puncticollis*(?), *Arcopagus marcuzzii* (an endemic species), *Cartodere separanda* (present only in western, central and southern Europe), *Cryptophagus*

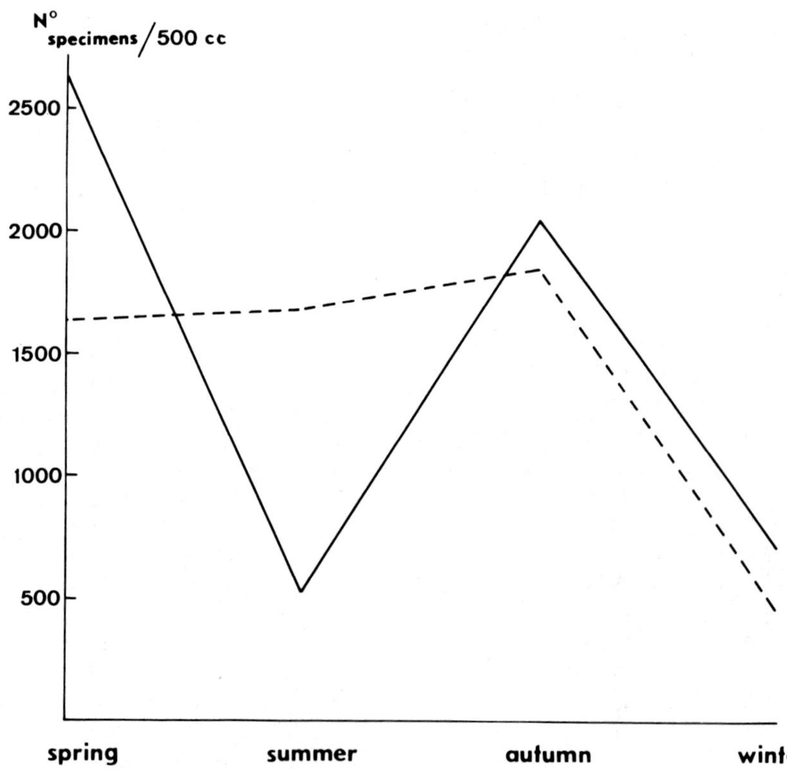

Fig. 35. Number of species of the soil fauna at Torino di Sangro (Abruzzo, Central Italy) during the year. ——— litter fauna; ------ soil fauna.

distinguendus, Coxelus pictus, Dermestes undulatus (under carrion), *Silpha obscura,* and finally the weevils *Acalles lemur* and *Torneuma grouvellei*. The planticolous fauna is represented by *Chrysomela lutea, Oxythyrea funesta* and the Neuropteran *Hemerobius micans*, whose larvae are more or less bound to the soil.

Parasites and poisonous animals in Southern Europe

Because of the mild climate of the sclerophyllous forest, terrestrial parasites and poisonous animals are very frequent, much more so than in other European biomes, where the winter cold acts as a limiting factor for many of them, especially for haematophagous arthropods.

Diptera are responsible for the transmission of several diseases in the European Mediterranean area. First of all the *Psychodidae* genus *Phlebotomus,* within which *P. papatasii, P. perfilievi*, and other species transmit *Leishmania tropica*, a Flagellate Protozoon responsible for a cutaneous leishmaniasis called 'bouton d'Orient', present in tropical countries but also in some European districts such as Italy

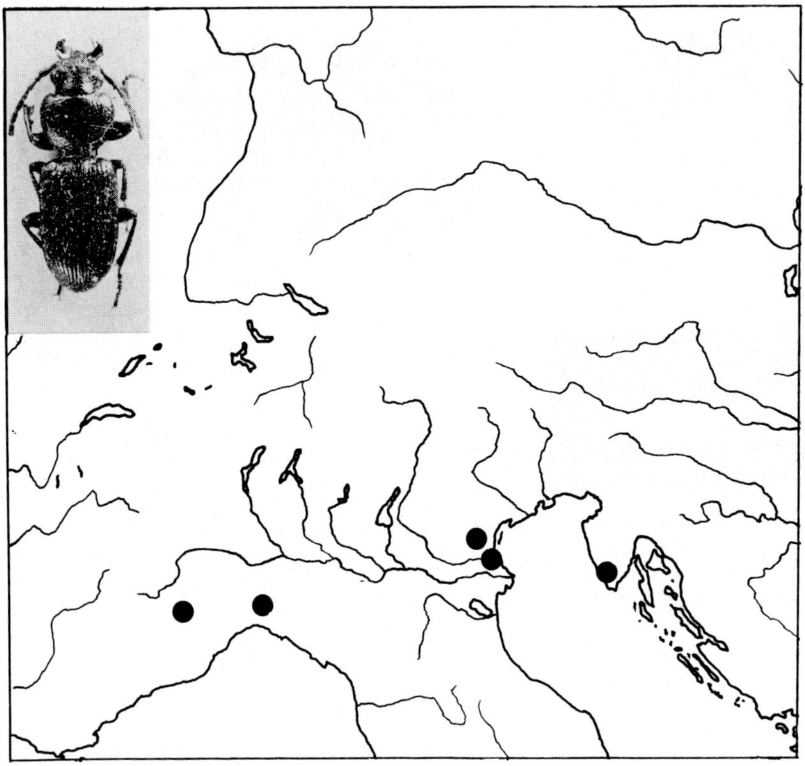

Fig. 36. Example of a distribution bound to the so-called xerothermic oasis in Northern Italy – *Carterus dama* (Carabidae, Coleoptera).

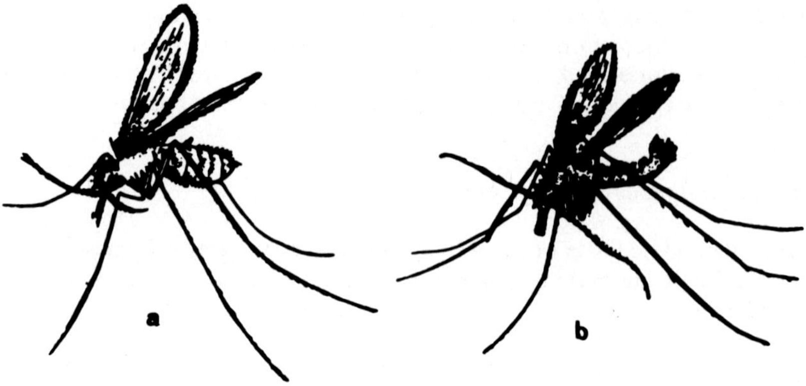

Fig. 37. *Phlebothomus papatasi* (Psichodidae, Diptera), transmitter of *Leishmania tropica.* Furthermore it is the transmitter of kala-azar in India, of a virus in Turkestan, of the virus of the so-called papataci-fever (three-days-fever, or – in German – 'Drei Tage-Fieber'). In Palestine it causes the so-called harara, a kind of cutaneous lesion similar to stropholus (from Brumpt) (reproduced with permission of Masson).

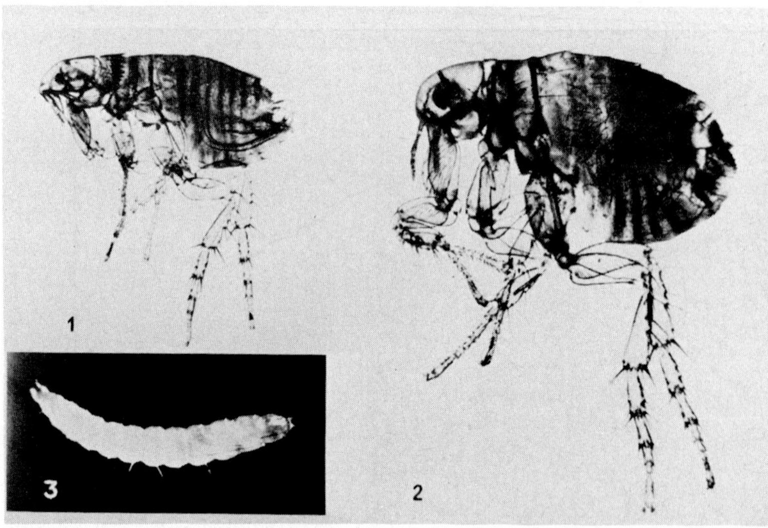

Fig. 38. Some fleas of Southern Europe. 1. *Ctenocephalides felis*, the common cat-flea, a potential transmitter of plague. 2. *Pulex irritans*, the common human flea, potential transmitter of plague only from man to man. 3. *Xenopsylla cheopis*, the normal transmitter of plague (from Geigy & Herbig) (with the permission of Verl. Recht & Gesells, Basel).

(Sicily, Calabria, Abruzzo, Romagna and Veneto), Greece, Crete and Spain. *Phlebotomus perniciosus* and some other related species transmit *Leishmania donovani* instead (according to some also *L. infantum*), the agent of a visceral disease known as Mediterranean kala-azar, present throughout the Mediterranean basin. In Italy it can be found in Silicy, Sardinia, Calabria, Puglie and Campania; recently (1972) a case has been notified in Emilia (Bologna). The *Phlebotomus* larvae live in humid debris, in the soil and in the cavities of dead and rotten trunks, so that prophylaxis of this disease, in part the use of mosquito-nets, is very difficult. Another disease transmitted by *Phlebotomus* is the so-called Three day fever, or also papatasi fever (drei Tage Fieber), a feverish acute disease, generally non-lethal, whose specific agent seems to be a virus. It is characterized by a fever lasting three days, followed by periods without fever (Fig. 37).

Other arthropods which transmit human diseases in the area of the sclerophyllous forest are ticks, which are very common in wooded areas, in fields, uncultivated areas, etc. A widely spread species is the dog tick, *Rhipicephalus sanguineus*, which transmits the so-called 'fièvre boutouneuse' or Marseille fever, in Rome described as 'eruptive fever' because of a hexanthema which resembles a little the hexanthematic typhus. Like the latter, it is due to a *Rickettsia*, but is very rarely lethal (1-2% of cases). In the fields there is a small mite, *Trombicula autumnalis* (= *Microtrombidium pusillum*), c. 2 mm long, with a hairy body (different from ticks), living in the soil. Its larva, 0.2 mm long and red, is a parasite of moles, several rodents and bats, and can attach itself to human legs, particularly those of peasants who work barefoot sometimes in Southern Europe. It causes the so-called autumnal erythema, dangerous only because of the infections which generally accompany erythema due to scratching the intense itch. On the contrary, the cattle tick, *Ixodes ricinus*, is very dangerous, because it can transmit a disease due to a Sporozoon, *Babesia* (= *Pyroplasma*) *bovis*, which causes relatively high mortality in adult cattle, although it spares calves. It is characterized by a high fever, haemoglobinuria (i.e. presence of haemoglobin in the urine) and often diarrhoea.

Very abundant throughout the area of the sclerophyllous forest are poisonous animals, though they are generally not lethal to man: spiders, scorpions, centipedes, insects, but particularly snakes, which thrive on the soil, under stones, along the lanes and in the sunny fields, where they can easily attack the man who heedlessly touches or tramples them.

We want to mention here only two kinds of spiders, the Volterra spider (in Italian 'malmignatta') *Lathrodectes 13-guttatus*, congeneric with the black widow of the USA, whose bite can be lethal even to man. It is black with some elegant white spots, the female much bigger than the male. The other very poisonous spider, although never lethal to man, is *Lycosa tarentula*, in Italian 'tarantola' (from the name of the town Taranto), frequent under stones in central-southern Italy, Spain and Dalmatia. Its bite causes various nervous phenomena known as 'tarantolism' or 'tarantism'. The poison of this spider enters the central and peripheral nervous systems and near the bitten region, and only secondarily into the blood, lungs, heart, liver and spleen. It is constituted by several protein fractions and three non-protein fractions. It is toxic to mammals; horses and camels being particularly susceptible, dogs to a much lesser extent. Lizards are resistant, frogs on the other hand are very sensitive (Fig. 39).

Both species have been the object of superstitions, which flourish particularly in southern Italy and the Italian islands due to the poor education of the inhabitants. In Sardinia, the *Lathrodectes 13-guttatus*, often associated by people with *Mutilla*, a Hymenopterous insect which stings but is not dangerous to man, is called argia, arza or even alza (from the Latin term varius, by the loss of the consonant). Until recently, on this island people used some practices of expelling evil spirits against the bite of this spider with a complex of rites consisting of music, dances and songs, whilst the patient was buried up to the neck in a dunghill, covered only with a sack. The old inhabitants of Sardinia practically put all their hopes of healing in the dances and songs, as if these were to appease the anger of some demon. The songs generally consisted of a sequence of cursings and imprecations against the spider, repeating the name of the person who had been bitten. Since the civil status of the spider was unknown, the dancers belonged to three categories: unmarried women ('bagadias'), married ones ('cojuadas') and widows ('fiviudas') (Fiori).

As far as *Lycosa tarentula* is concerned, the treatment consists of a special dance, the so-called 'tarantella', a dance whose extremely quick rhythm can be compared with some African initiation dances, true transfer rites, by means of which the unrestrained running compels 'spirits' to go out of the body, panting and perspiring. In the specific case of tarantella, the movements, according to the magic of 'transfer', should permit the elimination of the poison introduced into the body by the *Lycosa*-bite. According to Bouisson, these beliefs have a basin in

Fig. 39. Parasites or poisonous animals: left, *Mutilla europaea*, right, *Lathrodectes 13-guttatus*.

an observation of which modern medicine takes note, when it sometimes prescribes treatments associated with hot drinks in order to cause perspiration, which favours the elimination of toxins. The poison contains some neurotoxins, basic proteins of low molecular weight (11-18,000). The LΔ/50 for a mouse is 3 μgm.

There are several species of scorpions in southern Europe: as well as the genus *Euscorpius* (*E. italicus, E. carpathicus*, etc.), which is never dangerous to man, there is also *Buthus occitanus*, of southern France and the Alpes Maritimes, whose bite can cause general poisoning, sometimes lethal to children or ill, weak people. The centipede *Scolopendra cingulata* can reach a length of 13 cm: it is frequent under stones in all dry, sunny biotopes of Mediterranean Europe but its bite is not very dangerous. The arachnids *Solifuga* (very rare in Europe) are generally non-venomous, but the bite of *Galeodes* is painful out of proportion to the real damage it does.

Among poisonous insects there are several *Hymenoptera*, i.e. wasps, bees and ants, particularly the former (sufficient to mention the common hornet, *Vespa crabro*). Another group of *Hymenoptera* whose sting is particularly feared, though not at all dangerous, is the family *Mutillidae*, winged or sometimes wingless, so to resemble an ant. They are remarkable because of their vivid colours (black and red, black and yellow, etc., with metallic bluish wings) so that we can speak of true aposematic coloration, possible only among animals actively protected against enemies by means of poisons or repellent substances. The sting of the *Mutillidae* is very painful but not dangerous, even if the common people have fantastic ideas about the effects of the sting and the necessary therapy, which in some countries such as Sardinia and Cyprus consists of real practices for expelling evil spirits! The venom of these elegant insects contains little or no 5-hydroxytryptamine: it appears to be a protein or a substance attached to protein.

The only poisonous snakes of Europe lethal to man are vipers (adders) among which *V. aspis, V. berus* (which can be found also on mountains at very high altitudes), *V. ammodytes*, limited to the south-eastern parts of Europe and finally to Abruzzo, in Jugoslavia present on the west coast, on the island of Krk, and also in European Turkey.

Two other poisonous snakes belonging not to the *Viperidae* (Proteroglypha) but to the *Colubridae* (Opistoglypha) are *Tarbophis fallax* and *Coelopeltis monspessulanus* but their bite is not at all dangerous to man.

B. SPECIAL PART

Dalmatia

As an example of a region covered, at least originally, with a sclerophyllous forest, we take first into consideration Dalmatia, i.e. the coastal region of the western Balkans going from Privlaka (near Nin) up to Budva (42° 10′ 12″). The northern limit would be represented not by the mainland but by an island, Rab (or Arbe), whose latitude is 44°45′21″ (cf. Müller, 1921; Novak, 1952; Marcuzzi, 1968). This region has an area of about 17,830 km^2, of which 3500 are islands. The climate belongs generally to the Mediterranean type (Fig. 40-41).

As far as soils are concerned, we can distinguish (Marcuzzi, 1968b) terra rossa sensu lato, sandy calcareous soils, and very rarely, humiferous soils.

As far as the vegetation is concerned, there are very small differences between that of the mainland and that characteristic of the enormous number of islands present all along the coast of Dalmatia. On the mainland we mainly find the following species: *Quercus ilex, Pinus halepensis, Pistacia, Smilax aspera* (which sometimes makes the forest or maquis quite impenetrable to man), *Asparagus, Myrtus communis, Erica arborea, Juniperus phoenicea, Juniperus communis, Euphorbia spinosa, E. dendroides, Coronilla emerus, Salvia officinalis, Cistus* spp. plur., *Arbutus unedo, Lonicera* sp. and *Phyllirea* sp. Very common on the islands are *Quercus ilex, Pinus halepensis* (for instance on the islands Jakljan, Šipan, Mljet, etc.), *Euphorbia dendroides, E. spinosa, Salvia officinalis, Helichrysum italicum, Pistacia lentiscus, Inula viscosa, Inula candidissima* (very common on Mljet), *Arbutus unedo, Juniperus macrocarpa, J. phoenicea, Cistus monspeliensis* and *Cistus incanus, Smilax aspera, Rubus, Carduus, Thymus capitatus, Erica arborea, Laurus nobilis* (the common laurel), *Phyllirea, Rosmarinus* (or rosemary), *Limonium cancellatum, Myrtus communis* (the common myrtle), *Viburnum tinus, Asparagus* sp. plur., *Coronilla emerus*, etc. Near the seashore some halophil species are to be found, such as *Crythmum maritimum* (Fig. 42).

Vertebrate life is not very abundant except reptiles, most of which are common to southern Italy (see later). One peculiar species is *Ophisaurus apus*, similar to a snake, but belonging to the same family as the common blind snake, *Anguis fragilis* (*Anguidae*). The commonest lizard by far is *Lacerta melisellensis*, which in a sense is vicariant to both *Lacerta muralis* and *L. sicula* of Italy (see later). This species is subdivided into a number of small races, each limited to a small islet (Fig. 43). On the rocks by the sea shore there is an elegant lizard, *Lacerta oxycephala*, almost as big as the common green lizard, *Lacerta viridis*, extremely swift and ready to hide at the slightest sign of danger, so that it is extremely difficult to catch. It is restricted to a small part of Western Jugoslavia and to adjacent islands (Fig. 44). A characteristic terrestrial tortoise is *Testudo hermanni*. The most interesting snakes are *Coluber leopardinus*, the conspicuous *Coluber quatuor-lineatus*, the Opistoglyphon *Coelopeltis monspessulana*, poisonous but not dangerous to man, and the elegant *Tarbophis fallax*. The Viperids are represented chiefly by the horn viper, *Vipera ammodytes*.

The invertebrates are very similar to those of Puglia, as a consequence of the geographical position and similarity of environments.

The beetles predominate among insects, and more especially the Tenebrionids. There are *Tentyria italica, Stenosis intermedia* and *S. angustata, Asida fascicularis* and *A. acuticollis* (two Dalmatian endemic species), *Dendarus dalmatinus* (a trans-Adriatic species), *Pedinus fallax*, one of the northernmost species of the genus, *Pedinus helopioides, Allophylax picipes* (also a trans-Adriatic species) and *Opatrum verrucosum*. According to previous research (Marcuzzi & Dalla Venezia, 1968) some Tenebrionids prefer alkaline soils, i.e. *Stenosis angustata, Allophylax picipes* and *Gonocephalum rusticum*, while others prefer relatively acid soils ($pH \leqslant 8$), i.e. *Stenosis intermedia, Pedinus helopioides, Gonocephalum pusillum* and *Opatrum verrucosum*. As far as clay is concerned some Tenebrionids prefer clay soils (*Dendarus dalmatinus, Pedinus helopioides* and *Opatrum verruco-*

Fig. 40. Dalmatia and its main islands.

sum) whereas others are found in soils deficient in clay (*Stenosis angustata, Allophylax picipes, Gonocephalum pusillum* and *G. rusticum*). Some species seem to prefer soils with a high C/N ratio, i.e. *Stenosis angustata, Asida fascicularis, Allophylax picipes* and *Gonocephalum rusticum*; others, on the contrary, prefer soils with a low C/N ratio (*Tentyria italica, Stenosis intermedia, Pedinus helopioides* and *Gonocephalum pusillum*).

Other Coleopterous families very common in the sclerophyllous forest of Dalmatia or in one of its degenerate forms (maquis, garrigue, etc.) are Carabids, Curculionids (weevils), Scarabeids, Chrysomelids and Cerambicids.

The Carabids are represented by many species of *Harpalus* such as *H. punticeps* (euro-anatolic species), *H. subquadratus* (Mediterranean part of Europe), *H. tenebrosus* (western European, Mediterranean, Caucasus and Turkestan), *H. attenuatus* (western European-Mediterranean; also present in Istria at Pola), *H. incisus* (southern Europe from Spain to Greece, often on flowers of *Umbelliferae*), *Zabrus incisus*, a phytophagous species feeding on grains of cereals, limited to the Balkans and Asia Minor, extending as far north as Veglia (Krk), *Olistopus fuscatus*, a Mediterranean species present also in Istria (Pola), *Cymindis axillaris* subsp. *adriatica*, present in Central Europe and the Mediterranean region, but a race limited to Dalmatia and southern Istria (Pola), *Licinus silphoides* (very frequent), *Pterostichus melas* subsp. *depressus*, a race limited to Dalmatia and Istria, *Acinopus picipes*, in southern Europe and the Pontic region as far east as the Caspian Sea, *Calathus fuscipes graecus*, a subspecies vicariant with the Italian *fuscipes latus*.

Among the Curculionids we find several species of the genus *Otiorrhynchus*, occurring on many plants, some of which are cultivated, such as *O. cardiniger*, a balkan species extending as far north as Venezia Giulia and Arbe (Rab), *O. villosus*, also a Balkan species, occurring as far north as Lussin Island, *O. corruptor*, common in Italy and Dalmatia, *O. lugens*, a trans-Adriatic species, sometimes very abundant, *O. lasioscelis*, limited to Dalmatia, going as far north as Arbe (Rab),

Fig. 41. Dalmatia. Luka Sipan, seen from the harbour. The original vegetation has been almost completly replaced by introduced plants, such as olive trees, cypresses and palms.

Fig. 42a. Dalmatia: Peninsula of Peliesac, looking towards Mljet. In the foreground a typical scerophyllous forest with *Pinus halepensis* and *Quercus ilex*, very well preserved. The region is one of the least inhabited ones of all Dalmatia.

Fig. 42b. Lastovo Island: Ubli, a *Pinus halepensis* stand. In the foreground some plants of *Asphodelus.*

Cherso (Kres) and Lussin Island, a rare inconspicuous species, living under small stones during the daytime, sometimes near the sea shore together with the interesting Cerambicid beetle *Parmena pubescens, O. alutaceus*, also a trans-Adriatic species, *O. crinipes*, present in Dalmatia and only in a very few Italian localities, such as the Foresta Umbra on the Gargano (Puglie), *O. heydeni*, living under stones at the base of some trees such as olive trees, present in Dalmatia, Istria and

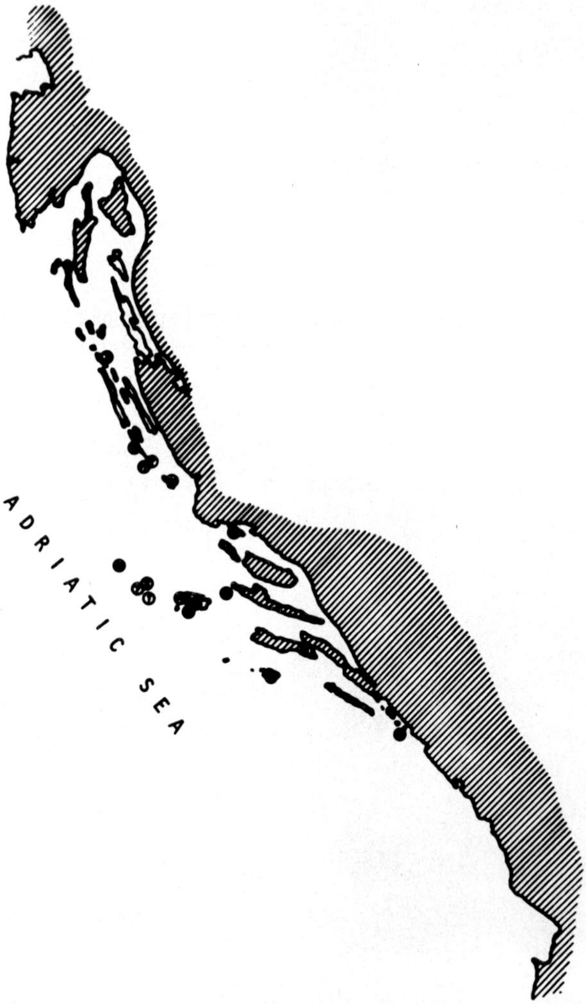

Fig. 43. Distribution in Dalmatia and Istria of the lizard *Lacerta melisellensis.* The hatched part is inhabited by the subspecies *fiumana.* The numbers in the circles show the different subspecies or 'subspecies-populations' into which the species has split, probably as a consequence of the isolation of the different islands following eustacy and isostacy, by which the eastern part of the Adriatic Sea has been uplifted (and the western part has been sunk). (Constructed on the basis of the data of Mertens & Wermuth).

Gargano, and finally *O. dalmatinus*, also a trans-Adriatic species. Very little or nothing is known about the life history of these animals, which are among the most interesting creatures from the biogeographical standpoint (Fig. 45-46).

Other Curculionids are the elegant, slender *Lixus anguinus*, present in southern Europe, Algeria and Morocco, living on *Cheiranthus* (e.g. *C. excelsior*) on which the larva develops; several species of *Larinus*, generally to be found on the flowers of several thistles (*Cirsium, Carduus*, etc.), *Coniocleonus nigrosuturatus*, a holo-Mediterranean species, rather rare in Dalmatia, living on hot, well-exposed soils, with unknown feeding habits; and finally *Pachycerus scabrosus*, limited to Southern Europe, living on several *Borraginaceae* in which the larva develops, such as *Echium vulgare, Cynoglossum pictum, C. cheirifolium, Anchusa italica*, etc. The adult is to be found on the same species.

Fig. 44. Distribution in Istria (sclerophyllous forest of the western part of the peninsula) of the common lizard, *Lacerta sicula*. Each islet or group of islets has its own 'subspecies-population' or population. (Constructed on the basis of the data of Mertens & Wermuth).

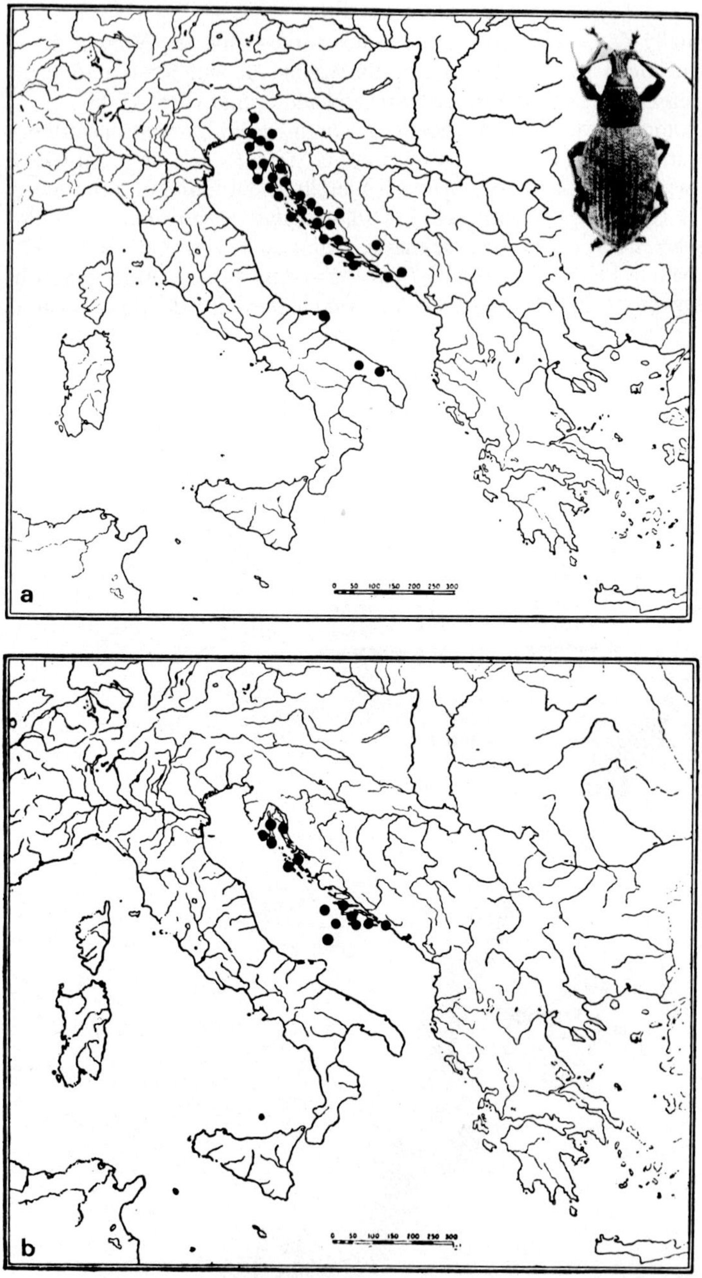

Fig. 45. Distribution of some weevils characteristic of the sclerophyllous forests, with the present transadriatic distribution which points to an ancient land connection between the Balkans (Dalmatia) and Southern Italy (Puglia). a. *Otiorrhynchus alutaceus.* b. *O. lasioscelis.* c. *O. crinipes.* d. *O. lugens* (maps from Gridelli, 1950). To the Puglie localities some others should be added according to personal research of the author.

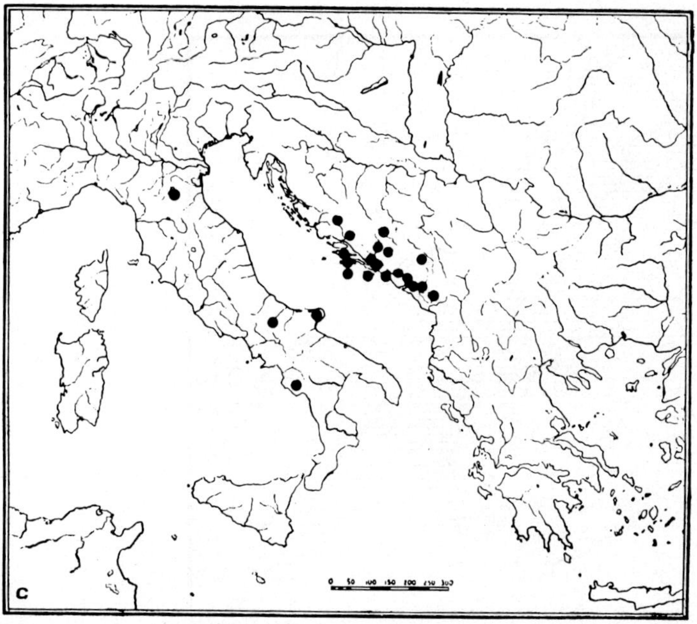
0 50 100 150 200 250 300
c

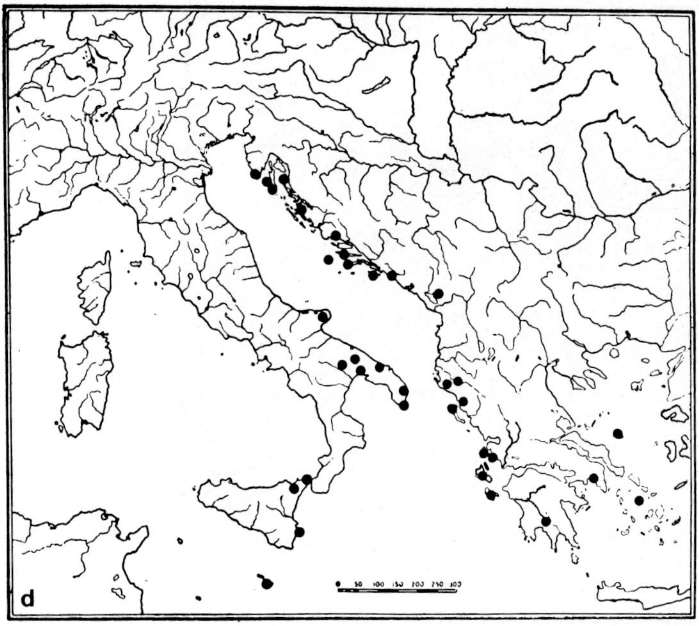
0 50 100 150 200 250 300
d

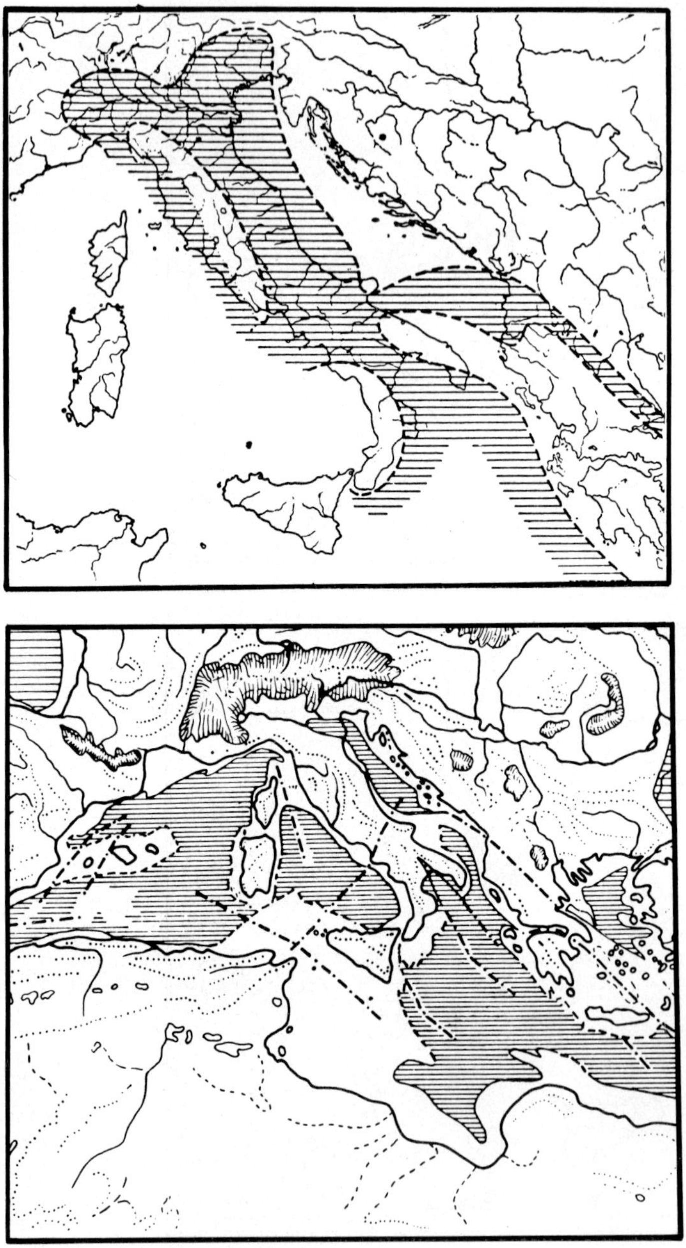

Fig. 46. The paleogeography of the Mediterranean area in the eras in which the transadriatic and the transionic types of distribution originated. Top – Middle Miocene. Bottom – regression (Rissian or Milazzian II) (Top, from Gridelli 1950, bottom from Pasa, 1953).

The Scarabeid beetles are represented by some coprophagous species such as *Scarabaeus variolosus, Oniticellus* sp., *Caccobius schreibersi*, etc. and by some phytophagous ones, such as *Oxythyrea funesta, Potosia cuprea* and *Polyphylla fullo.*

Very frequent and abundant are the Chrysomelids: we can list here *Lachnaea longipes*, a pontic species living mainly on *Quercus* (especially *Q. pubescens*), *Cryptocephalus bipunctatus*, a widely distributed species, in Dalmatia represented by the typical form (*C.b. bipunctatus*), present often on *Quercus pubescens, Cryptocephalus moreai*, living mainly on the flowers of *Hypericum, C. fulvus signaticollis*, a subspecies limited to the Mediterranean basin, *Chrysomela herbacea*, living on various species of *Mentha* and other *Labiatae, C. vernalis*, exclusive to Dalmatia, Albania, Greece, Central and Southern Italy, *C. americana*, a holo-Mediterranean species feeding on rosemary and *Lavandula, C. grossa*, limited to the Western Mediterranean, reaching as far east as the western part of Yugoslavia, living on several *Labiatae, C. geminata*, living on species of the genus *Hypericum, C. cerealis mixta*, a subspecies limited to southern Europe, on *Labiatae*, very often on rosemary, *Macrolenes bimaculata*, a holo-Mediterranean species feeding on the leaves of *Pistacia* and sometimes on those of *Quercus ilex*, to which it can be very harmful, and finally *Emarhopa rufa*, found generally near sea shores, a ponto-Mediterranean element, living on *Convolvulus sepium* and *C. arvensis.*

Among the Cerambicids there are the red and black coloured *Leptura cordigera* (var. *illyrica*), present in Central and Southern Europe, and Western Asia, where it lives on flowers such as *Daucus, Orlaya, Rubus* and *Achillea, Phytoecia caerulescens*, a Central European-Mediterranean species living on *Borraginaceae, Dorcatypus tristis*, limited to Southern Europe; its larva feeds on figs, *Morus*, willows, *Cupressus* and sometimes grapes, to which it can be very harmful. *Parmena pubescens* is very rare; it is a Mediterranean element, whose larvae will feed on several plants (polyphagous). *Hesperandrius griseus*, finally, is also a Mediterranean element, whose larva feeds on figs and therefore is a potential pest, especially in areas such as Dalmatia, where the fig is a very important cultivated plant.

Together with beetles, bugs are very abundant throughout all the Mediterranean, xerophytic, parts of Dalmatia. We will mention only a few here, such as *Rhinocoris iracundus*, a stinging bug very common in Dalmatia, *Lygaeus equestris, Deraeocoris schach*, holo-Mediterranean, and *D. punctulatus*, a widely distributed species, *Dicranocephalus agilis*, also a widely distributed element, the rare *Arenocoris waltli*, an Atlantic-Mediterranean species, *Aphanus rolandri*, widely distributed, *Ventocoris rusticus*, Southern European, present both on the mainland and on the islands, *Ancyrosoma leucogramma*, a Euro-Mediterranean species very common on the islands, the extremely common *Cydnus aterrimus*, a palaearctic species present also in the Italian sclerophyllous forest, *Stagonomus amoenus*, a Mediterranean-Iranian species, found by the writer only on the island of Hvar (Lesina), *Emblethis verbasci*, widely distributed, but in Dalmatia apparently rather rare, the conspicuous *Carpocoris m. mediterraneus*, a Mediterranean-Iranian taxon, *Odontoscelis fuliginosus* (palaearctic) and *O. dubius* (Mediterranean), *Rhyparochromus alboacuminatus*, the psammo-xerophilous *Aethus flavicornis*, holo-Mediterranean, found by the writer only on the small islands of Biševo (Busi), *Geotomus elongatus, Psacasta hexanthematica*, a holo-Mediterranean species, and finally *Himacerus myrmecoides*, a Euro-Mediterranean element, similar to an ant (whence its name – *myrmecoides*) and very common in Dalmatia.

Other invertebrates to be found are Molluscs, Isopods (or woodlice), Diplopods (or millipedes such as the interesting *Glomeris pulchra*), Chilopods or Centipedes, scorpions, spiders, etc. Very little is known about these inconspicuous creatures.

Puglia and the Tremiti Islands

A second example of the environment prevailing in the parts of Southern Europe belonging to the sclerophyllous forest is provided by Puglia, an Italian region today terribly threatened by development (tourism, house building, roads, or,

Fig. 47. Vegetation on the Tremiti Islands. a. Sclerophyllous forest (*Pinus halepensis*). b. Maquis, with rosemary (*Rosmarinus officinalis*) and *Cistus* sp. plur. c and d. Garrigue with *Helichrysum italicum, Pistacia lentiscus*, etc., near the sea at the western point of S. Domino island.

more generally, by the speculations of local and foreign economists). Of course, only the parts of Puglia covered by sclerophyllous forest or the products of its degeneration are taken into consideration. The climate is the typical Mediterranean one, except in the innermost parts of the region (Murge, Gargano). It can be visualized, at least partly, by means of the distribution of the values of Lang's rain factor (Fig. 47, 48). The soils mainly belong to the following categories 1) calcareous sandy soils (Principi), characterized by a low level of organic matter, high pH (8.2-8.6), high amount of $CaCO_3$ and sand (in some cases 84% of $CaCO_3$ and 88.1% of sand). This soil has developed on recent limestone, i.e. Quaternary or Pleistocene deposits, relatively near the sea; 2) red soils, rich in clay and poor in humus, corresponding to the so-called terra rossa (Principi, Mancini) and to Kubiena's category of 'terrae calxis', and more precisely to 'terra fusca terrosa'. In some cases, under sclerophyllous forest, the colour is brown, so that we can think of a 'brunified red soil' or even a typical brown loam. In other cases the soil is very rich in clay and corresponds to typical brown loam and to 'earthy brown loam', respectively.[1] 3) Organic soils, present mainly under *Pinetum halepensis*; they could perhaps be attributed to the category of rendzinas sensu lato.

The vegetation is rather varied corresponding to the local climatic conditions (ecoclimates) and still more to the degree of human development to which it has been subjected since remote times. We will list here only some of the commonest plants of Puglia: *Quercus ilex, Q. coccifera, Pirus piraster, Olea oleaster, Pinus halepensis, Pistacia lentiscus, Inula viscosa, Scilla maritima, Rosmarinus officinalis, Cistus salvifolius* and *C. incanus*, the introduced *Opuntia ficus-indica, Alyssum* sp., *Sideritis sicula* (at some 600-750 m altitude), *Asphodelus* spp., *Asphodeline lutea* (also at some 600-750 m), *Euphorbia spinosa, Prunus spinosa, Fraxinus ornus* (typical of the cold belt of Lauretum), *Helichrysum italicum* (typical of garri-

Fig. 48. Tremiti Islands, off Gargano (Puglie), Adriatic Sea: right, Cretaccio, the smallest island of the archipelago: very remarkable signs of erosion in Miocenic limestone and marl; left, Capparara, an unhabited island of the archipelago: some rocks composed of conglomerates emerge from the sea near the southern shore.

[1] For more detailed information about Italian terra rossa see also Comel, 1933-1957.

gues), *Juniperus phoenicea, Anthemis maritima* (near sea shores), *Limonium cancellatum* (idem), *Thymus capitatus* and *Thymelaea hirsuta*, both characteristic of the garrigues now so extensive throughout Puglia.

Among the vertebrates, reptiles are definitely the most interesting, because of their abundance and variety. There are the 'gekkos', such as *Gymnophthalmus kotschyi, Hemidactylus turcicus*, extremely common, *Tarentola mauritanica*; the blind snake *Anguis fragilis*, the green lizard *Lacerta viridis*, the common lizard *Lacerta sicula* and the wall lizard, *Lacerta muralis.* Another very interesting and rare species of saurian is *Chalcides c. chalcides.* Among the snakes, it is enough to list *Zamenis gemonensis, Coluber longissimus, Coluber leopardinus, Coluber quatuor-lineatus* and *Coronella austriaca.*

Among the invertebrates of Puglia, beetles as a whole predominate throughout the sclerophyllous forest. The most represented family is *Tenebrionidae*, followed by *Carabidae, Staphylinidae, Chrysomelidae, Curculionidae* (the so-called weevils), *Scarabaeidae* and several other minor families. We will name only a few species, particularly abundant and frequent in all the Mediterranean parts of Puglie, such as the Tenebrionids *Pimelia rugulosa, Tentyria italica, Scaurus striatus* (polysaprophagous species), *Allophylax picipes, Pedinus meridianus, Dendarus caelatus* and *D. dalmatinus* (phytosaprophagous species, which take part in the break-down of litter and therefore in humification); the Chrysomelids *Chrysomela vernalis, C. viridana, Timarcha niceaensis* and *T. pimelioides, Galeruca* sp. (phytophagous), the weevils *Brachycerus undatus, Otiorrhynchus* spp. plur., some of which are of great biogeographical interest, *Donus crinitus* (phytophagous like the Chrysomelids), the Carabids (carnivorous except some genera such as *Carterus) Harpalus* spp. plur., *Licinus silphoides, Pterostichus melas, Calathus fuscipes, Odontonyx* sp., *Cymindis* sp. The most frequent Scarabeids are *Oxythyrea funesta* and *Tropinota hirta, Aphodius, Copris hispanus, Bubas bison, Geotrupes, Scarabaeus affinis, Gymnopleurus* spp. plur. (coprophagous). Particularly striking because of their showy colour are some *Meloidae* such as *Mylabris variabilis* and other congeneric species, which generally display aposematic colours due to their toxicity to all insectivorous vertebrates (birds, mammals, lizards). Particularly abundant and frequent are the Ladybirds, generally aphidivorous, and also furnished with aposematic colours.

The Tenebrionids of this region have been compared with the main chemico-physical characteristics of the soils, and it has been demonstrated (Marcuzzi, 1964) that there are some relationships, limited to some species. More precisely, a low pH is sought by *Dendarus dalmatinus*, a high pH by *Allophylax picipes* and *Gonocephalum pusillum*; a high sand content is sought by *Tentyria italica, Stenosis intermedia, Gonocephalum pusillum*, whereas a low content is sought by *Dendarus dalmatinus.* Practically the côntrary holds for clay. As far as the C/N ratio is concerned, *Dendarus dalmatinus* and *Gonocephalum pusillum* prefer a high value, *Tentyria italica* and *Stenosis intermedia* show no clear preference.

As far as the other terrestrial invertebrates are concerned, we will just mention Isopods, Diplopods (particularly some Julidae and *Glomeris pulchra*), Chilopods (among which the conspicuous *Scolopendra cingulata* can reach 13 cm length), spiders, scorpions (*Euscorpius carpathicus*) and whip-scorpions (several species, some of which are endemic to Puglia or to Italy). Extremely widespread all through the sclerophyllous forest are Molluscs (cf. Paget, Forcart, Marcuzzi & Lo Casto).

Off the northern coast of Gargano, the northernmost part of Puglia, there is a small archipelago, the Tremiti, composed of 4 principal islands surrounded by a series of inconspicuous rocks (among them the so-called 'Pagliai'). The climate is a typical Mediterranean one. As far as soils are concerned, we can refer to the following data:

Locality	Main chemico-physical characteristics						soil type according to Kubiena's classification tion
	pH	$CaCO_3$ %	C %	clay %	silt %	sand %	
S. Nicola	8.1	33.14	5.4	4.25	24.00	71.75	xerorendzina
S. Domino	7.8	9.83	4.56	14.25	27.25	48.50	reddish brown loam (rich in calcium)
S. Domino (garrigue at west-point)	8.1	3.03	3.06	31.10	18.55	50.35	reddish brown loam (or red loam)
S. Domino	8.4	0.59	0.12	3.50	5.65	90.85	

As far as vegetation and flora are concerned, for S. Domino we can list *Pinus halepensis, Cistus incanus* and *C. monspeliensis, Pistacia lentiscus*, the myrtle or *Myrtus communis, Olea oleaster* and *Thymelaea hirsuta.* In Spring the ground of the sclerophyllous forest or the garrigue is covered with many flowers, such as *Anemone hortensis* and *Ophrys* spp. plur. At S. Nicola an impoverished flora has replaced the sclerophyllous forest or the maquis: as well as *Pistacia lentiscus, Euphorbia dendroides* and *Urginea maritima* (known also as *Scilla maritima*) dominate on the flat, sun-dried plateau which occupies most of the island. The vegetation of Caprara (or Capparara) is the product of human influence, so that

Fig. 49. Some plants characteristic of the Tremiti Islands: a. *Matthiola* sp., a Crucifera; b. *Cistus salviaefolius.*

today we find a secondary vegetation following first the cultivation of cereals and secondly cattle breeding. The most common plants today are *Marrubium, Carduus pycnocephalus, C. corymbosus* (thistles). Only the northern and southern edges of the island are covered with *Pistacia lentiscus*, mixed here and there with *Atriplex halimus* and *Artemisia* sp (Fig. 49).

The vegetation of Cretaccio (the smallest of the main islands of the archipelago) gives a clear sign of the struggle between the inorganic kingdom – the barren rocks, marls or conglomerates – and plants. A part, a small flat surface on the top of the hill, is covered with a relatively rich herbaceous vegetation (*Lotus, Muscari, Teucrium polium, Daucus gingidium, Helichrysum italicum, Brachypodium ramosum*), on the remaining part of the island there are only a few sterile *Gramineae* and the halophil *Limonium cancellatum.* Only near the sea shore can some definitely halophil plants thrive such as *Crithmum maritimum* and *Atriplex halimus.*

Animal life is represented by many birds. Apart from the marine species (*Puffinus, Larus, Sterna, Chlidonias*, etc.) there are the peregrine or *Falco peregrinus brookei*, Eleonora's Falcon *Falco eleonorae*, the kestrel *Falco t. tinnunculus*, the kingfisher *Alcedo atthis*, many passerine birds, the hooded crow *Corvus c. cornix*, the woodcock *Scolopax rusticola*, the common sandpiper *Actitis hypoleucos*, the pallid swift *Apus pallidus* and the white wagtail, *Motacilla a. alba.* Finally, there are other migrants such as the quail *Coturnix coturnix*, the turtle dove *Streptopelia turtur*, the song thrush, *Turdus ericetorum* and the wood pigeon, *Columba palumbus.* Unfortunately, the ornithological fauna is greatly impoverished due to the relentless hunting allowed under Italian law.

The most common reptiles are *Lacerta sicula*, present in enormous numbers, probably due to the lack of enemies, and the snake *Zamenis gemonensis* var. *carbonaria.*

The most frequent invertebrates are the Tenebrionid beetles *Tentyria italica, Stenosis intermedia, Asida bayardi* (an Italian endemic species), *Asida fiorii* (also an Italian endemic species), *Scaurus tristis*, an anthropophilous species limited to the so-called 'ruderetum', several species of *Blaps*, present in an enormous quantity due to the poor hygienic conditions of the inhabitants, the trans-Adriatic species *Dendarus dalmatinus*, the widely distributed *Pedinus meridianus*, the relatively thermophilous *Allophylax picipes*, living on the Tremiti in the same habitat as on the small Dalmatian island of Busi (or Biševo) under stones on calcareous sandy soils, and finally an apparently trans-Ionian species, *Raiboscelis azureus*, common to Greece, Asia Minor and Tremiti, whose presence on the latter is possibly due to passive transport. Very abundant also are Molluscs and the Diplopod *Glomeris pulchra.*

The Aeolian islands

Right in the middle of the Mediterranean Sea, the Aeolian Islands, an archipelago composed of small, ancient volcanoes, emerges from the sea. These are situated north of Milazzo (North Eastern Sicily) and west of Calabria. The archipelago has been known since ancient times: it was named in Greek Αἰόλου νῆσοι, Αἰολίδες or Λιπαραῖαι, in Latin Aeoliae, Hephaestiades or Volcaniae. The ancient peoples thought that it was the residence of the King of the Winds, Aeolus, and therefore it corresponded to the homeric Αἰολίη. The highest of the islands is Stromboli (924 m) one of the few active volcanoes in the Mediterranean area. The climate is a typically Mediterranean one (Fig. 50).

On these old volcanoes we find some very interesting soils, derived from pure lava, which were studied recently by Ferro & Furnari. More particularly, one form is associated with a vegetation type composed of *Cytisus aeolicum, Centaurea cineraria, Senecio cineraria* and broom, *Spartium junceum*: the sand comprises 94.4%, silt 1.9%, clay 3.7%. Humus is present to the amount of 0.44%, $CaCO_3$ is absent and the pH is 6.2. Associated with a vegetation association of *Quercus ilex, Rubus fruticosus, Pteridium aquilinum, Genista ephedroides, Spartium junceum, Asplenium obovatum* and *Erica arborea* there is a different soil

Fig. 50. The insular groups around Sicily.

Fig. 51a. Lipari. M. Pilato, the pumice-built mountain.

where we find sand = 88.9%, silt = 5.0%, clay = 6.1%, humus 1.7%, $CaCO_3$ absent, and a pH of 6.8 (Fig. 51).

The vegetation is very rich and varied. We will list here only *Calycotome spinosa, Pistacia, Cistus incanus* and *C. salvifolius, Artemisia arborescens, Inula viscosa, Smilax aspera* (subsp. *mauritanica*), *Erica arborea, Arbutus unedo, Phyllirea, Ruscus aculeatus, Asparagus* spp. plur., *Euphorbia dendroides, Coronilla emerus* and the cluster pine, *Pinus pinaster.* Extremely abundant is an imported perennial, *Opuntia ficus indica* or 'fico d'India'. Among the herbs we find *Matthiola sinuata, Cotyledon umbilicum-veneris, Lupinus perennis, Muscari comosum, Lupsia galactites* (= *Galactites tomentosa*), *Lamarckia aurea, Gnaphalium* sp., *Chrysanthemum coronarium* var. *concolor* and *C. myconis* var. *hybridum, Senecio ci-*

Fig. 51b. Stromboli. The solidified lava flow is perfectly distinguishable. On the left, a small village is visible.

Fig. 51c. Lipari. The mountain of pumice and the structures for its extraction.

neraria, Phagnalon saxatile, Rumex bucephalophorus, Silene gallica, Brassica fruticulosa, Anthirrhinum orontium, Satureja graeca var. *cosentina* (endemic to Calabria, Sicily and the Aeolian islands), *Euphorbia peplus, Convolvulus althaeoides, Catapodium loliaceum, Stipa retorta* (= *S. tortilis*), *Vulpia membranacea, Brachypodium ramosum, Trifolium arvense, Orchis* sp., etc.

Very interesting here is the presence of a parasite of *Cistus, Cytinus hypocistis*, a *Rafflesiacea* distributed throughout most of the Mediterranean, in Italy reaching some xerothermic oases of Northern Italy such as the Colli Euganei and the coast of Istria.

It is worthwhile mentioning the presence of some plants living on a hill of pure pumice (M. Pilato), and more precisely *Erica arborea, Cistus salvifolius, Spartium junceum, Inula viscosa*, together with an ornamental, introduced plant, *Mesembryanthemum crystallinum.*

On one of the smallest islands in the Archipelago, the so-called Strombolicchio, near Stromboli, several very rare species are to be found such as *Dianthus rupicola, Hyoseris radiata baetica, Senecio cineraria, Daucus carota* and the *Chenopodiacea Kochia saxicola*, an endemic species limited to the Island of Capri, Ischia and Strombolicchio (Fig. 52-54).

Among the birds of the Aeolian islands, we will just mention the Cory's shearwater, *Puffinus kuhli kuhli* and the manx shearwater, *Puffinus puffinus yelkouan*, both of them resident; some Falconiformes such as the peregrine, *Falco peregrinus brookei*, resident; the hobby, *Falco subbuteo*, the Eleonora's Falcon, *Falco eleonorae*, a summer and migratory bird; the kestrel, *Falco tinnunculus*, resident and migrant; the honey buzzard, *Pernis apivorus*, exclusively migratory, and finally the resident buzzard, *Buteo buteo.* The Galliform rock partridge, *Alectoris graeca*, has disappeared from this archipelago several years ago due to excessive hunting, whereas the quail, *Coturnix coturnix*, is still present, both as a migrant and as a

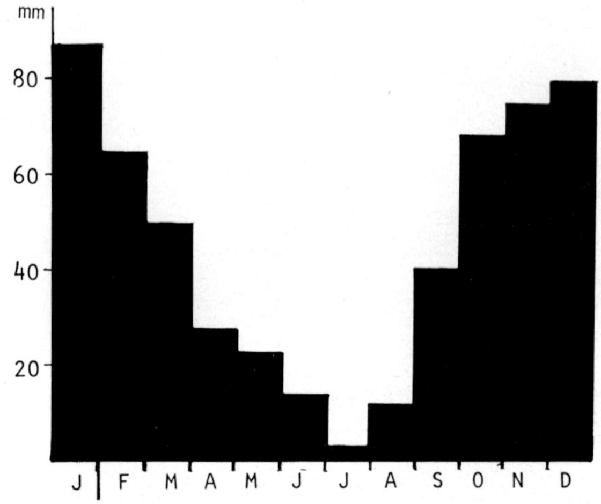

Fig. 52. Rainfall on the isle of Stromboli (Aeolian archipelago), based on the data from Ferro & Furnari.

resident bird. The migratory bird *Scolopax rusticola* (the common woodcock) is very abundant.

The Columbiforms are represented by the wild pigeon, *Columba livia*, and by the migratory *Streptopelia turtur*. Very abundant also now are the Passerine birds, among which there is the blue headed wagtail, *Motacilla flava* (= *Budytes flavus*) of which on the Aeolian Isles there are no less than four subspecies (*flava, thunbergi, cinereocapilla* and *feldegg*); each of them has a different winter and summer quarter. There are also *Monticola solitarius, Sylva atricapilla, S. melanocephala* and *Corvus corax*. Very interesting also is the presence of two forms of sparrow, *Passer hispaniolensis* and another form, considered a hybrid between the Italian

Fig. 53. Aeolian islands: Lipari. a. Pioneer vegetation on lava. b. *Cistus incanus* growing on lava (Photo: Prof. G.G. Lorenzoni).

Fig. 54. Aeolian islands: Vulcano. a. An Aeolian endemic, *Cytisus aeolicus*. b. idem: gramino-idetum with *Spartium junceum*, both arboreal and shrubby. (Photo: Prof. G.G. Lorenzoni).

sparrow, *Passer italiae*, and a subspecies of the Spanish sparrow, *P. hispaniolensis maltae*, living on Malta.

It is worthwhile pointing out the close relationship between the number of birds and the area of each island, as shown in Fig. 55.

Two species of lizards are present on these rocky islands, the common lizard *L. sicula* and an endemic species, *L. wagleriana*, restricted to Sicily and nearby islets. The latter is present on the island of Vulcano as a subspecies *antoninoi*.

Lacerta sicula has been split into a number of infraspecific forms, which we can call 'demes', such as *sicula alvearioi* (Faraglione Pollara, northwest of Salina), *s. lisca-biancae* (Lisca Bianca, East Panarea), *s. trischittai* (Scoglio Bottaro, near Lisca Bianca) and finally, *s. raffonei*, restricted to the basaltic, rocky, Strombolicchio. Perhaps the only snake of the archipelago is *Coluber longissimus*.

On Vulcano it was once possible to find the common rabbit, *Oryctolagus cuniculus* very easily, but it has probably completely disappeared due to unrestrained hunting.

Among insects, Tenebrionids are very likely the most interesting both from the ecological standpoint and from the biogeographical one. We can mention here *Pachychila dejeani*, a 'Rassenkreis', *Tentyria laevigata*, present in Sicily, Calabria and Malta, and maybe also in Campania; *Tentyria grossa*, a holo-Mediterranean relict, some species of *Stenosis*, the peculiar *Elenophorus collaris*, a beetle with a very large, broad posterior part of the body (elytrae) and a slender, long anterior part (thorax and head), present in Spain, the Balearic Islands, Southern France, Central and Southern Italy and Dalmatia, and closely related to a form in South America, with which it constitutes the tribe *Elenophorini; Alphasida grossa*, common under stones, related to a group of species spread all through the Western Mediterranean Basin; *Akis subterranea*, generally limited to houses, stables and pig-sties, restricted to Sicily and neighbouring islands (Ustica, Malta and Egadi); four species of *Scaurus*, one of which *S. aegyptiacus*, is a southern species found also in Africa, and restricted to rather human-influenced biotopes; the extremely common *Pimelia rugulosa*, walking during the hottest hours of the day along the lanes and roads of the Aeolian Islands, present only in Sardinia, Sicily and Southern Italy, but reaching Pelagruža Island, and therefore a trans-Adriatic element;

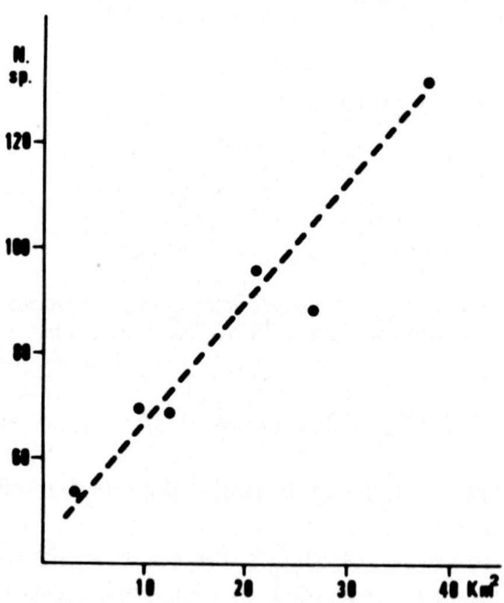

Fig. 55. Birds of the Aeolian islands.

three species of *Blaps*, widely distributed; the Italian endemic species *Dendarus lugens*; two species of *Pedinus*, one of which (*P. punctatostriatus*) is limited to Sicily, Tunisia and Southern France; the common *Allophylax picipes*; two species of *Opatrum*, one of which, *O. validum*, is limited to Sicily and Tunisia; the extremely rare, myrmecophilous, inconspicuous *Oochrotus unicolor*, limited to the Western Mediterranean Basin; a species unique for its ovoid, flattened, leaf-like shape, *Cossyphus moniliferus*, belonging to a genus widely spread in Africa, and finally the conspicuous, shining blue *Helops rossii*, limited to the Western Balearic Islands, Greece, Southern France and very few Italian localities, probably a trans-Adriatic species (Fig. 57).

The phenology of the Tenebrionid fauna of the Aeolian islands (and Ustica) shows a clear relationship with Spring, which in this region follows the wet Winter and precedes the dry, very hot Summer. This pattern is not dissimilar to that shown by the Tenebrionids of the Mediterranean region of Israel (Fig. 56).

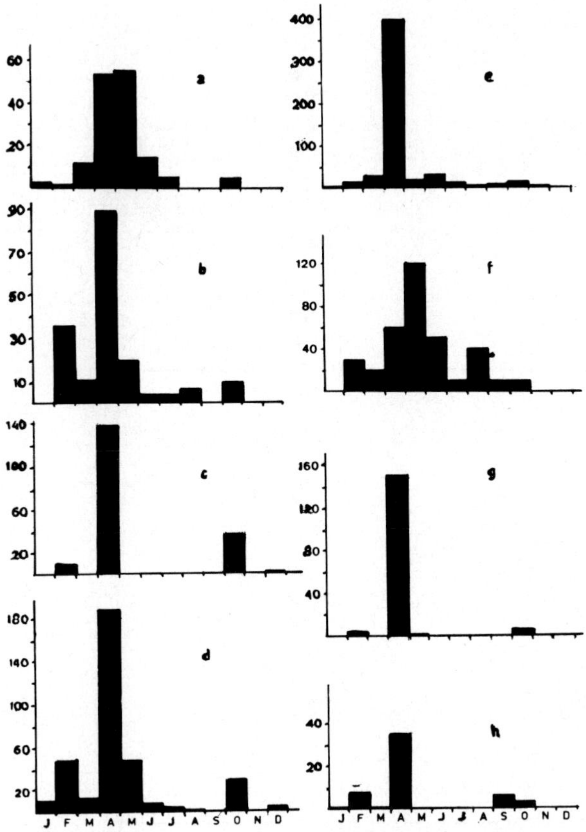

Fig. 56. Phenology of some Tenebrionid beetles of the Aeolian Islands and Ustica. a. *Tentyria laevigata*. b. *Tentyria grossa*. c. *Stenosis intermedia*. d. *Scaurus striatus*. e. *Pimelia rugulosa*. f. *Blaps gigas*. g. *Allophylax picipes*. h. *Opatrum verrucosum*.

Other invertebrates of the sclerophyllous forest – or its remains – of the Aeolian islands are various Coleopterous families, millipedes, centipedes, woodlice and spiders. Scorpions appear to be very rare if not absent. Very abundant are molluscs, which sometimes completely cover the soil or make clusters on the meagre xerophilous vegetation, particularly graminoid plants. There are also *Poiretia algira*, widely distributed in the Mediterranean area, in Italy limited to Sicily, Campania and Marche; *Rumina decollata*, a holo-Mediterranean element, in Italy extending as far as Venice (xerothermic oasis); *Papillifera bidens*, widely distributed in the Western Mediterranean area; the edible *Helix aperta*, very prized by

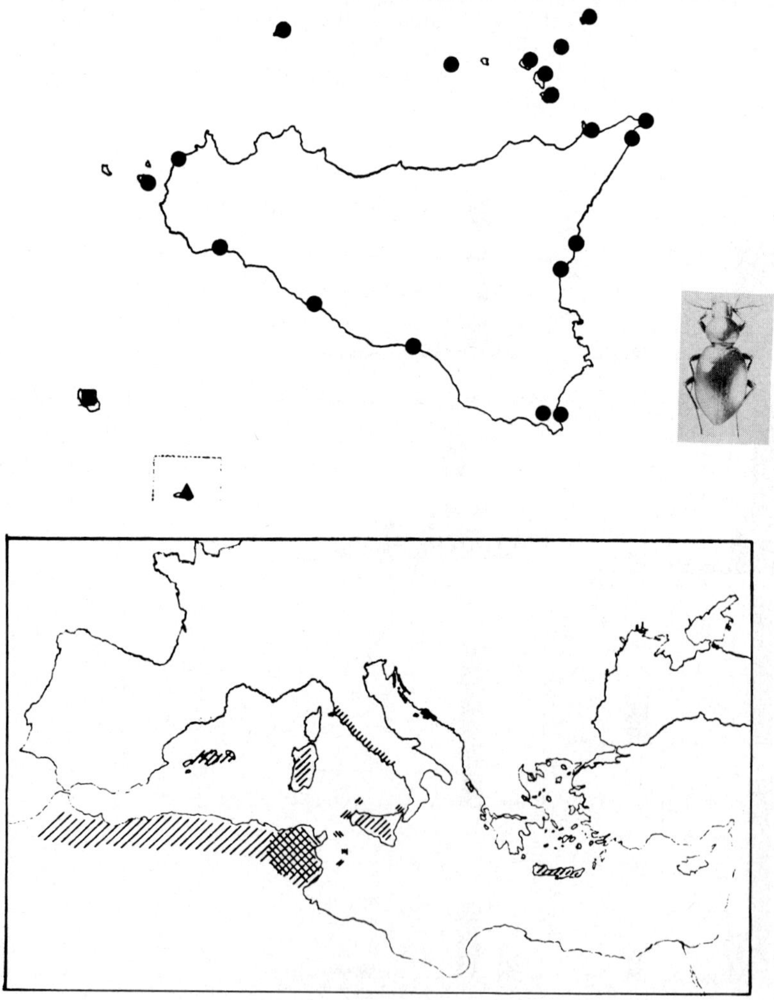

Fig. 57. Distribution of *Tentyria grossa*: top, the subspecies living in Sicily: ● *g.grossa;* ▲ *g.sommieri;* ■ *g.angustata;* bottom, //// *g.grossa,* |||||| *g.sommieri,* \\\\ *g.angustata;* ▦ *g.barbara.*

local inhabitants for its gastronomic properties, which constructs a thin, white, epiphragma in order to avoid dehydration in a very dry environment; *Cochlicopa lubrica*, a trivial species; and finally *Helicella conspurcata*, widespread in the Western Mediterranean basin, in Italy extending up to Piemont and Lombardy (Fig. 58).

The islands of the 'Sicilian Channel'

The island of Malta (Fig. 50) has a surface area of 246 km^2 and a maximum height of 239 m (Colli di Bengemma). The climate belongs to the Mediterranean type (rain factor = 28). It has been inhabited since Neolithic times (with megalithic constructions also more recently); Phoenicians, Greeks and Romans lived there. Today it is highly settled: a large part of the territory is cultivated (cereals,

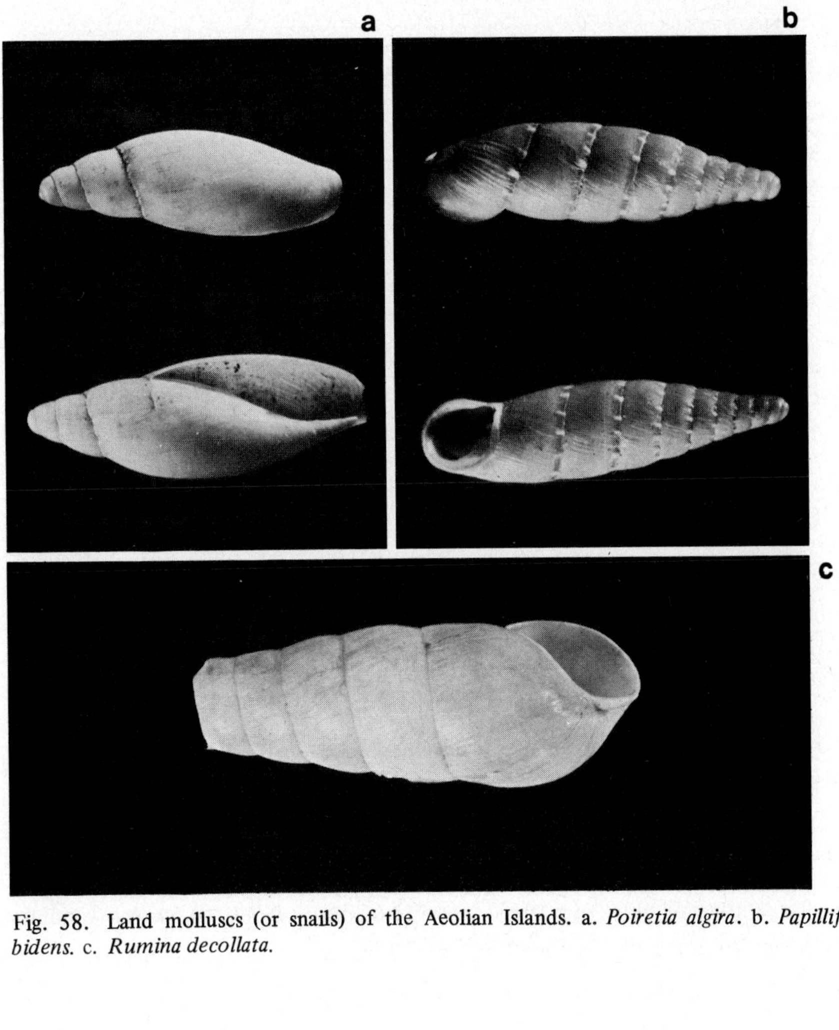

Fig. 58. Land molluscs (or snails) of the Aeolian Islands. a. *Poiretia algira*. b. *Papillifera bidens*. c. *Rumina decollata*.

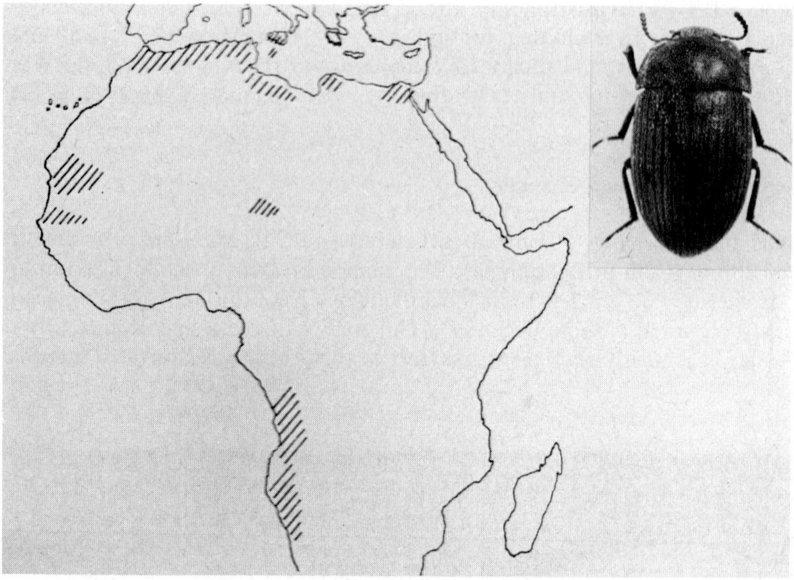

Fig. 59. Distribution of *Clitobius ovatus.*

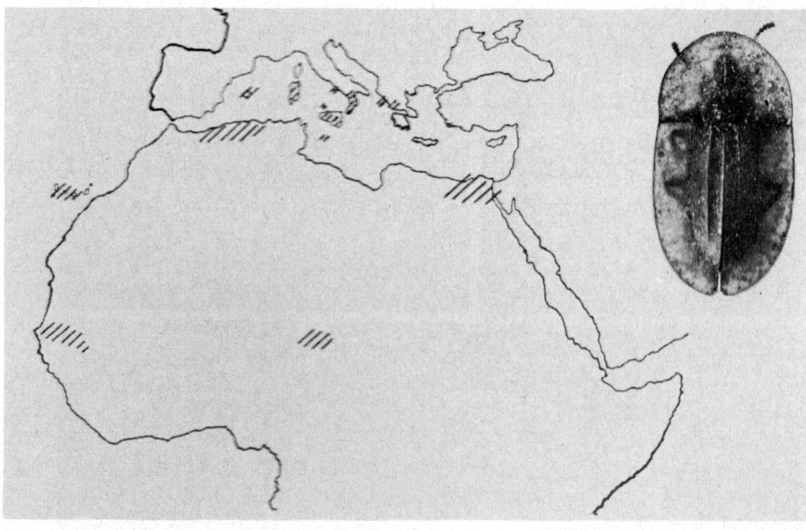

Fig. 60. Distribution of *Cossyphus moniliferus.*

potatoes, tomatoes, grapes and, of secondary importance, cotton and tobacco); also animal breeding is widespread (cattle, sheep, goats, pigs, etc.).

The reptiles of this arid island are the snake *Coluber leopardinus* and the lizards *Lacerta filfolensis* and *Chalcides ocellatus.*

The Tenebrionids are very abundant: sufficient to list *Tentyria laevigata, Alphasida grossa, Akis subterranea, Scaurus striatus, Allophylax picipes, Opatroides punctulatus, Clitobius ovatus*, a Southern European-African element, reaching as far south as Angola (Fig. 59), and *Cossyphus moniliferus* (Fig. 60).

Other beetles are *Lixus junci*, common throughout the Mediterranean, up to Central Europe, living on *Beta, Spinacia* and *Atriplex*, and the Carabid *Harpalus subquadratus*, extending from Spain to Greece and the Caucasus, already noted from Sicily.

Bugs common in Malta are *Dicranocephalus agilis moralesi*, a widely distributed species, the subspecies maghrebin-macaronesic; *Aphanus rolandri*, also widely distributed; *Reduvius pallipes*, a Mediterraneo-turanian species and *Nezara viridula*, an unimportant species.

The mollusc genus *Lampedusa*, present also in South-Eastern Sicily and on Lampedusa is interesting (see later), as well as an endemic species of the genus *Tudorella.*

In the so-called 'Channel of Sicily' there are several islands or groups of islands, of which we will consider only three, which are the best known from the ecological point of view: Pantelleria, Lampedusa and Linosa. The climate and vegetation are relatively similar on all these small islands.

Pantelleria is a volcanic island, with a surface area of 83 km^2 and a maximum height of 836 m (Montagna Grande); it has been inhabited since Neolithic times; has been called by the Greeks Cossyra (Κοσσύρα) and has also been occupied by the Phoenicians, so that its original vegetation has been greatly altered. World War II contributed to the major destruction of the island. In Pantelleria we find Reptiles such as the lizard *Lacerta s. sicula*, the snake *Zamenis hippocrepis* and the tortoise *Testudo hermanni.* The Tenebrionids are *Pachychila cossyrensis*, endemic to this island, *Tentyria grossa angustata*, also endemic; *Alphasida grossa cossyrensis; Akis spinosa barbara; Scaurus atratus*, a relict holo-Mediterranean species; *Blaps gigas; Heliopathes avarus, Opatrum validum schlicki*, a subspecies restricted to Tunisia and maybe Algeria; *Opatroides punctulatus, Probathicus anthrax*, limited to Sicily, Sardinia and Corsica, and therefore very interesting from the biogeographical standpoint.

Lampedusa has a surface area of 20 km^2 and a maximum height of 133 m (Albero del Sole). There are megalithic constructions, similar to those of Malta, and some huts dating from the Bronze Age.

In Lampedusa today the maquis has completely disappeared. The reptiles are represented by the lizard *Chalcides ocellatus*, the snakes *Macroprotodon cucullatus* and *Coelopeltis monspessulanus*, and the tortoise *Testudo hermanni.* On a small islet near Lampedusa a peculiar lizard is to be found, *Psammodromus algirus*, a species limited to South-Western Europe and North-Western Africa. The Tenebrionids are *Pachychila dejeani* subsp. *doderoi, Alphasida tirellii, Scaurus tristis* and *Blaps gigas*, both anthropophilous; *Heliopathes avarus* and *Allophylax picipes.* Probably in the past the Tenebrionid fauna was richer. Among *Hymenoptera* the *Mutillidae* are very frequent. Extremely interesting is the presence on this small island of two species of *Solifuga, Biton velox* and *B. ehrenbergi*, previously known from Tunisia and Cyrenaica (*B. velox*), northern Africa to South Somaliland and Cyprus and North Attica (*B. ehrenbergi*). Also interesting is the mollusc genus *Lampedusa*, present also in South-Eastern Sicily and on Malta, probably widespread during the Pleistocene.

Linosa is a volcanic island (basaltic, lapilli and lava), with a surface area of 5.3 km^2 and a maximum height of 195 m (M. Vulcano). It has been inhabited at least since Roman times (Fig. 62).

In Linosa the vegetation is relatively rich, composed of the introduced *Opuntia ficus-indica, Pancratium maritimus* (a psammophilous, circummediterranean species), etc. The remains of the original maquis today are *Periploca* (fam. *Asclepiadaceae*), *Pistacia lentiscus* and *Olea oleaster.* Very characteristic is *Mesembryanthemum crystallinum.*

The reptiles are probably represented exclusively by lizards, and more particularly by *Lacerta filfolensis* subsp. *laurentii-mülleri* and *Chalcides ocellatus zavattarii.* The Tenebrionids are *Tentyria grossa* sbsp. *sommieri, Curimosphena villosa* (Fig. 63), *Akis spinosa* sbsp. *barbara, Scaurus tristis, S. striatus* and *Blaps gigas,* three anthropophilous species, *Opatroides punctulatus*, a widely distributed species in all warm sectors of the Mediterranean area, extending as far east as Central Asia and as far south as Somaliland and India. Chilopods are very abundant, but there are apparently no millipedes, probably due to the severe environmental factors.

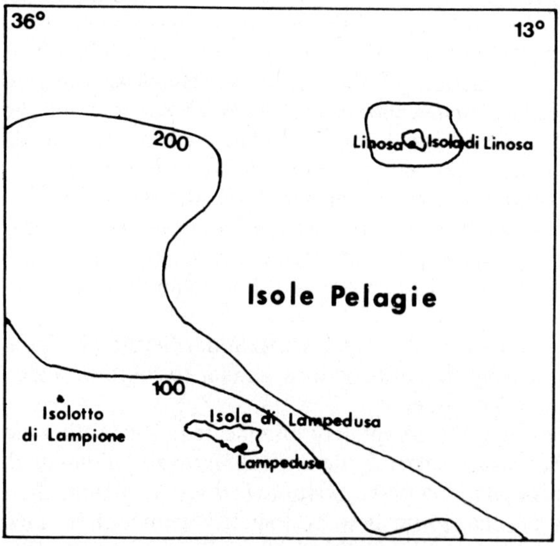

Fig. 61. Lampedusa, together with the two other Pelagian Islands, Lampione and Linosa. Lampedusa has a surface of 20.2 km^2 and a height of 133 m (Albero del Sole, literally the sun-tree). Its shores go down abrupty to the sea. There are very few springs and a very scanty vegetation. Interesting is the presence of remains of megalithic times (similar to those of Malta) and the presence of some huts of the Bronze Age, with an elliptical plant. Linosa, with a surface of 5.3 km^2, constituted of basaltic lapilli and lava, reaching a height of 195 m (Monte Vulcano). Though there are no springs, the original maqui with *Pistacia lentiscus* is quite well preserved. The smallest island is Lampione, uninhabited, with a surface of only 0.03 km^2 constituted of limestone. Highest point 36 m.

Fig. 62. The island of Linosa (photo Prof. A. Solazzi).

Fig. 63. Distribution of the beetle *Curimosphena villosa*, one of the few European species of the tribus *Epitragini*.

Western Sicily and the Egadi

On the extreme western part of Sicily, where once there was a flourishing sclerophyllous forest, but where today only a highly degenerate vegetation is present, alternating with poor cultivation, we find Trapani, situated at the foot of M. Erice, formerly called Drepanon. Trapani was then occupied by the Phoenicians.

Today the life in these localities must be considered as the remnants of what it once was: let us list some beetles for Trapani, such as the Tenebrionids *Tentyria grossa, Stenosis brenthoides, Akis spinosa, A. subterranea, Opatroides punctulatus.* The latter and *Stenosis brenthoides* have been found also on M. Erice (see later).

Among the Carabids there are *Amara aenea, Bembidion* (*Neja*) *ambiguum*, limited to the Mediterranean basin, *Nebria andalusiaca* a siculo-maghrebin element, present also in Calabria (Aspromonte), *Harpalus distinguendus*, a common species, *Drasterius bimaculatus* (fam. *Elateridae*) and *Pleurophorus caesus* (fam. *Scarabaeidae*). The most common bug is *Geotomus punctulatus*, a southern Mediterranean-Atlantic species (Servadei).

On M. Erice some beetles are still fairly abundant such as *Pachychila dejeani, Alphasida grossa* and *Opatrum verrucosum, Calathus mollis, Laemostenus algerinus*, a Western Mediterranean species, present from Morocco and Spain, to Dalmatia and as far north as Trieste; *Chrysomela banksi*, an Atlanto-Mediterranean species, together with some floricolous species belonging to the *Nitidulidae* (*Meligethes*) or *Chrysomelidae* (*Halticinae*). An interesting bug has been found on M. Erice, *Sciocoris ribauti*, a southern-european element not yet recorded for Sicily (Fig. 64).

The commonest lizard in the outskirts of Trapani is *Lacerta sicula sicula.*

Off the extreme western part of Sicily, i.e., off Trapani, there is an archipelago, the Egadi Islands, geologically joined to Sicily, composed of white calcareous 'tufo' in the lowlands and secondary limestone in the highest parts of the archipelago. The highest point is M. Falcone 684 m, on the Island of Marettimo.

The Egadi have been inhabited since Upper Paleolithic times, probably because they were a place where migratory birds nested. In a cave some important paintings have been found, of Franco-Cantabrian type, and a lithic industry which recalls the 'Gravettian'. Neolithic man also inhabited this archipelago.

The vegetation is that which has been preserved of the original sclerophyllous forest or maquis. The animal life is still fairly well preserved compared with that of the Trapani region. We will only mention some of the lizards, such as *Lacerta wagleriana marettimensis*, living along with *Lacerta s. sicula.* An endemic mollusc, *Siciliaria confinata*, is present on this archipelago. Among insects there are the weevils *Anysorhynchus barbarus*, south-European, *Brachycerus undatus*, euro-Mediterranean, present also in Algeria, the Chrysomelid *Timarcha pimelioides*, found on M.S. Caterina (Favignana) (27 March 1965), the Carabids *Calathus mollis* (Mediterranean area sensu lato), *Sphodrus leucophthalmus*, Europeo-Mediterranean, often living in caves, *Carabus morbillosus alternans*, a thyrrhenic-maghrebin species, although the subspecies is endemic to Sicily and Calabria (Aspromonte), the Tenebrionids *Pachychilina dejeani* (a 'Rassenkreis'), *Alphasida grossa, Akis subterranea, Scaurus tristis* and *S. striatus, Pimelia rugulosa* (subsp. *sublaevigata*, the form living in Western Sicily and Egadi), *Blaps gigas* and *B. mucronata, Allo-*

Fig. 64. M. Erice, 800 m, looking towards the south, with a very rich *Pinus* stand in the background and some *Ceratonia siliqua* (carob) trees in the foreground on the left. (Photo: Varvaro, Trapani).

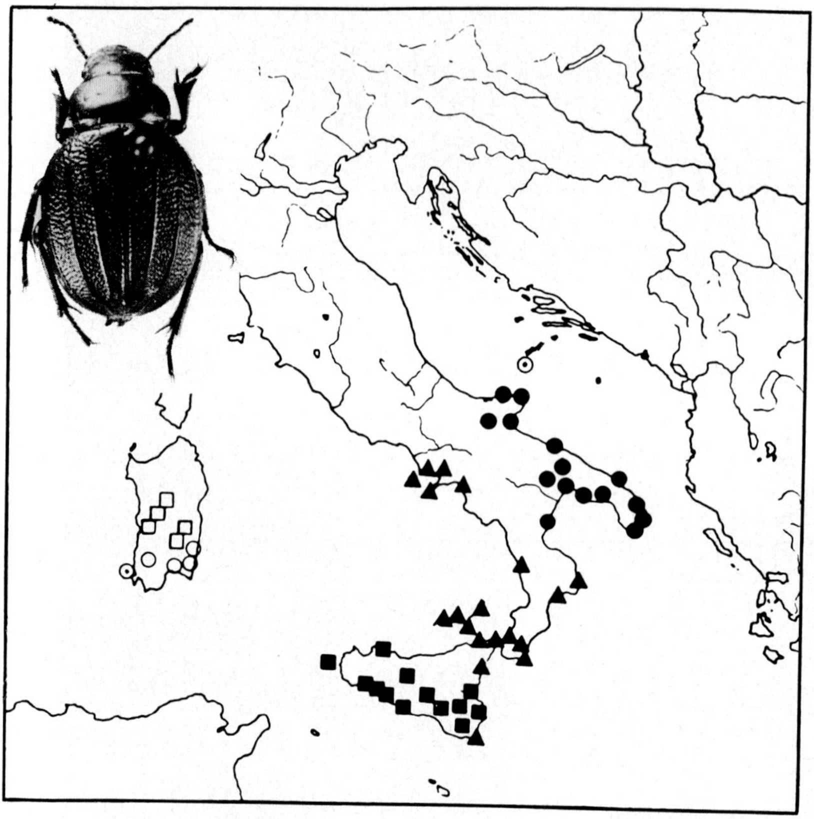

Fig. 65. Distribution of *Pimelia rugulosa:* □ *r.goryi;* ○ *r.sardoa;* ⊙ *r. pelagosana;* ▲ *r.rugulosa;* ▲ *r.sublaevigata;* ● *r.apula.*

phylax picipes, Heliopathes avarus (a species limited to Sicily, Sardinia and Calabria, and possibly also Lampedusa and Pantelleria) and *Pedinus punctatostriatus* (cf. Marcuzzi, 1970). The Heteropterous insect *Aëllopus atratus*, a euro-Mediterranean element, is also living in these islands (Fig. 65).

The Island of Elba

The fauna on Elba is a typical Mediterranean one, composed mainly of reptiles and arthropods, more especially insects.

Unfortunately many parts of the island have been heavily developed, especially through intense tourism, but some districts are still relatively well preserved. The dominant vegetation is the maquis with holm-oak (*Quercus ilex*), often very well developed; only on the North-western side of the island, constantly shaded and humid is a chestnut forest present. On the top of the mountain (M. Capanne) a pseudo-alpine, grassy, dwarf vegetation is present (Fig. 66).

The lizards are *Lacerta sicula campestris* and *L. muralis colosii.*

Among the beetles, which are very abundant, we will mention only some Tenebrionids such as *Akis bacarozzo* (named after the Roman dialect name to indicate beetles collectively), *Probathicus anthracinus, Blaps gibba, Opatrum sculpturatum* (maybe represented by a local subspecies, *O.s. ilvense*), *Pedinus meridianus, Stenosis* sp., *Pimelia bipunctata* and *Dendarus lugens.* Most of them are endemic to Italy or thyrrenian elements, sometimes relicts. The ecological valence is generally rather wide, so that they can be found in a variety of ecological niches, and this, together with their great euryphagy, is probably one of the reasons for their abundance.

Other beetles are the Chrysomelid *Galeruca interrupta*, a phytophagous species generally present under stones in dry, sandy places, restricted to *Artemisia campestris* and sometimes to *Lepidium graminifolium*, which can be completely defoliated by this insect; the Cerambycid *Parmena pubescens* (probably sbsp. *solieri*), a phytophagous species restricted to several plants (i.e. a euryphagous element), such as *Euphorbia characias* and *E. gerardiana, Crithmum maritimum, Centranthus ruber* and *Lavatera arborea*; the floricolous Scarabeid *Potosia morio*; the weevils *Otiorhynchus caudatus*, an Italian endemic species, extending from the Southern Alps to Puglia, *Thylacites canescens*, a euro-Mediterranean species, extending from the Alpes Maritimes, Var, Aude and Eastern Pyrenees as far east as Greece and Turkey; *Sitona gressarius* and *Donus crinitus*, holo-Mediterranean ele-

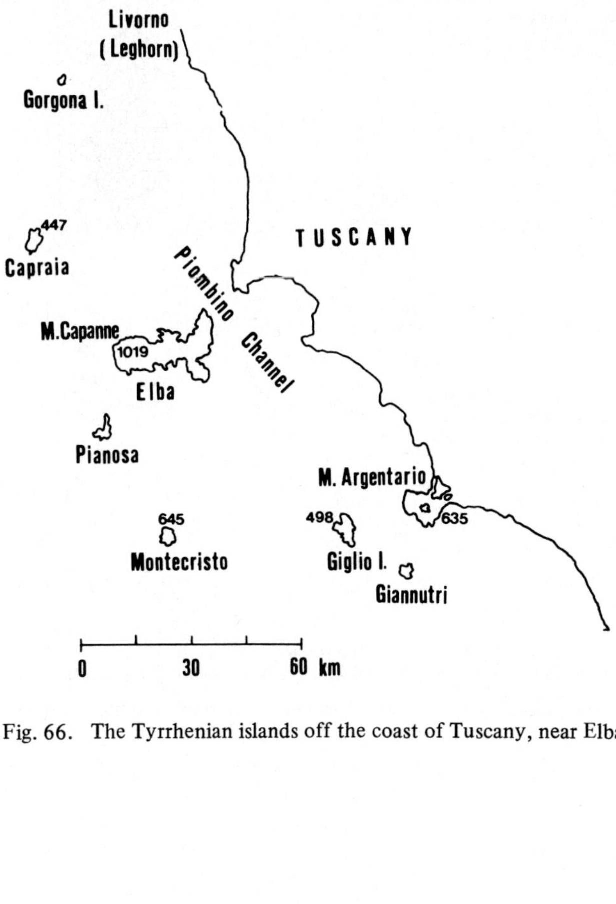

Fig. 66. The Tyrrhenian islands off the coast of Tuscany, near Elba I.

ments; the Carabids *Licinus silphoides*, typical of the northeastern Mediterranean basin and *Amara aenea*, a common species.

The bugs of the sclerophyllous forest (or of the remains of it) on the island of Elba are *Holcostethus albipes*, a Mediterranean element; *Aphanus rolandri* and *Lygaeus saxatilis*, widely distributed species; *Rhyparochromus saturnius* and *Macroscytus brunneus*, holo-Mediterranean elements, *Graphosoma lineatum italicum*, euro-Mediterranean; *Haploprocta s. sulcicornis*, holo-Mediterranean; *Emblethis griseus*, a common, holarctic species; *Sciocoris ribauti*, a southern-european species, new to Elba, where it has been found by the present author; *Eurydema ornatum*, a palearctic element and finally *Scantius a. aegyptius*, a holo-Mediterranean element.

Also some endemic Diplopods (or millipedes) live on this interesting island, as *Doderia elbana* and *Brachydesmus superus elbanus.*

The soil fauna includes an extremely interesting blind Carabid, *Reicheia ilvensis*, belonging to a genus with a holo-Mediterranean, relict distribution, present also in the Cape Region, witness of the existence of the old land called 'Thyrrenis' (Holdhaus). The same holds true for the Pselaphid genus *Trogaster*, an old inhabitant of Thyrrenis, closely related to some taxa living today in America and Australia. Geographical proof of these old connections is given by the presence on

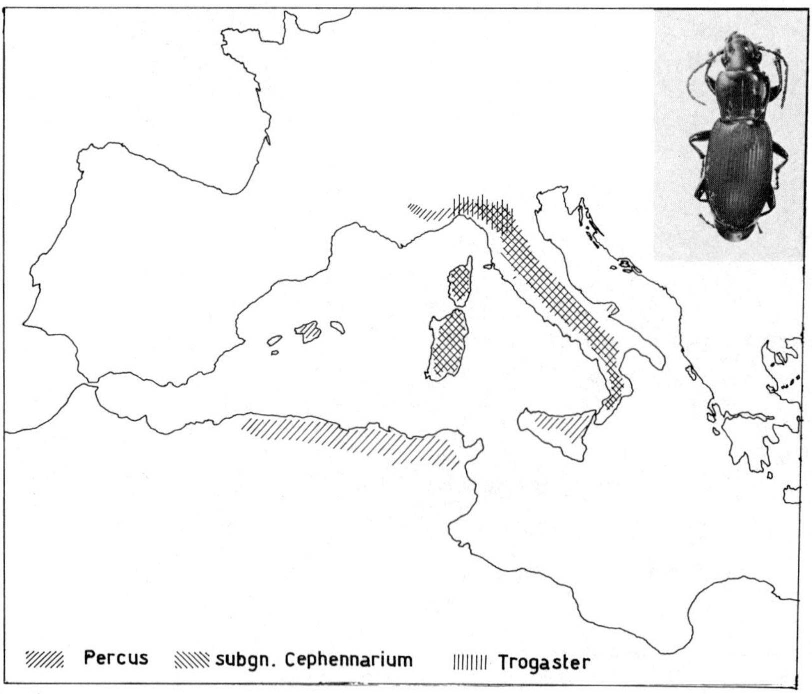

Fig. 67. Distribution of three 'paleothyrrhenic' taxa (genera or subgenera) living today on the island of Elba. They are all more or less bound to the sclerophyllous forest.

Elba as well as in Tuscany of faulted early Tertiary sediments, often intersected by diabase intrusions.

Among the insects there are several ancient genera present on Elba and in other parts of the Western Mediterranean area, once connected with it, such as *Percus* (Fig. 67) and subgen. *Cephennarium* (gen. *Cephennium*). All these elements point to a time in which all these countries were covered by forests: very scant indeed are the steppe elements, both on Elba and on the nearby large islands, Corsica and Sardinia.

Banyuls-sur-Mer (Côte Vermeille)

The peculiar position of Banyuls-sur-Mer, at the very border between Southern France and South-Eastern Spain, and at the same time on the extreme East of the Pyrenees, has allowed the co-existence in a small area of a number of species of great ecological and biogeographical interest. Not only that, but the environment is also quite different from that of most Mediterranean localities, since in a few square kilometers we find urbanized areas (the town of Banyuls-sur-Mer, the 'vignobles' or vineyards, famous all over France and all over Europe), a flourishing

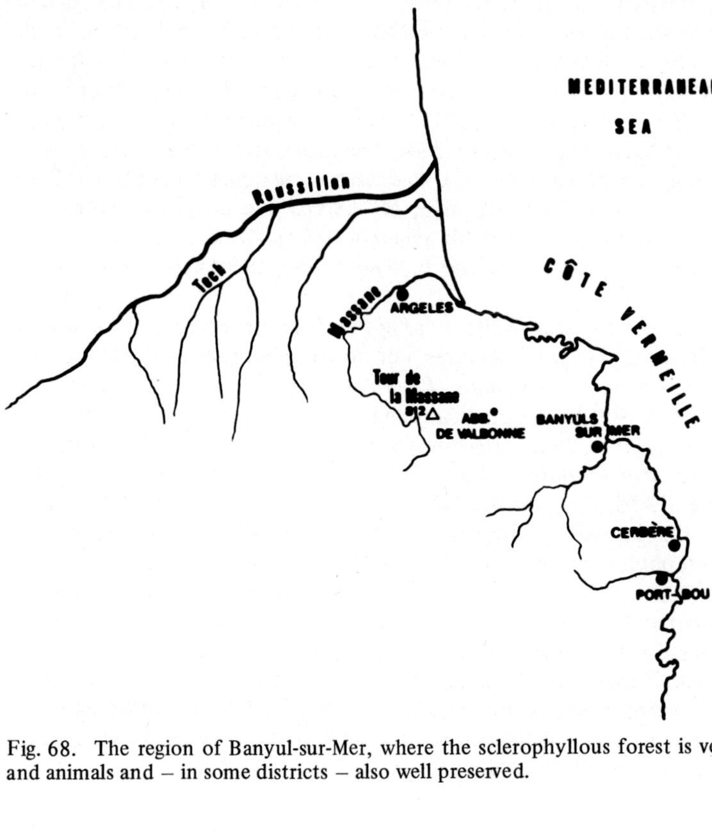

Fig. 68. The region of Banyul-sur-Mer, where the sclerophyllous forest is very rich in plants and animals and – in some districts – also well preserved.

sclerophyllous forest (as, for instance, at La Pouade where in the heart of the maquis, even in the hottest months of the year, a stream flows with a small waterfall and a large pond excavated by water in the hard rock, where a rich plant and animal life is present); some degenerate and almost denuded garrigues, just above the town on the road to Spain; and finally transitional forest, such as at La Valbonne, where it marks the boundary of maquis and heliophilous deciduous forest (or sub-Mediterranean forest) (Fig. 68).

For all these reasons, the animal life is extremely rich and varied, although nobody has so far catalogued it.

To begin with the mammals, we want to mention three small mammals, namely two small rodents, *Apodemus sylvaticus* (the common wood-mouse) and *Mus musculus spretus* (wild mouse), and one insectivorous species, *Crocidura russula* or white-toothed shrew.

Birds are extremely abundant, both resident and migratory: we will only list here *Sylvia borin* (garden warbler), *S. atricapilla* (blackcap), *S. undata* (Dartford warbler), *S. cantillans* or subalpine warbler, *Phylloscopus bonellii* (or Bonelli's warbler), *P. trochilus* (willow warbler) which probably reaches its western limit of distribution at Banyuls-sur-Mer, and *P. sibilatrix* (the wood warbler); *Hippolais polyglotta*, the melodious warbler, named for its ability to alter greatly its song; *Turdus merula* or blackbird; the robin *Eritacus rubecula*, the nightingale, *Luscinia megarhynchos*, the crested tit, *Parus cristatus*, the gold crest, *Regulus regulus*, often on holm oaks; the tree creeper, *Certhia familiaris*, also on holm oaks; the goldfinch *Carduelis carduelis*, and the predatory bird *Accipiter nisus* (or sparrow hawk). Furthermore there are many migratory birds, generally widely distributed, such as *Apus melba, A. apus* and *A. pallidus, Merops apiaster, Upupa epops, Riparia riparia, Hirundo rustica, Delichon urbica, Muscicapa hypoleuca, Acrocephalus scirpaceus, Phoenicurus phoenicurus* and *P. ochruros, Saxicola torquata* and *S. rubetra, Oenanthe hispanica, Motacilla alba, M. cinerea* and *M. flava* (represented by the subsp. *flava, iberiae* and possibly *cinereocapilla*), *Anthus pratensis, Lanius senator, Emberiza hortulana, E. cia* and *E. schoeniclus*, and finally *Fringilla coelebs.*

Reptiles are also fairly abundant, although in different ecological niches: we may mention *Vipera aspis* (in the Albères, not far from Banyuls-sur-Mer), *Coluber viridiflavus, Malpolon monspessulanus, Elaphe scalaris*, a species found only in Southern France and South-Eastern Spain and one of the most elegant snakes of the Mediterranean basin, *Coronella girondica, Anguis fragilis, Lacerta viridis* (not below 500 m), *Lacerta muralis; Lacerta lepida* (found in South-Western Europe and North-Western Africa), *Psammodromus algirus, P. hispanicus* (distributed like *Elaphe scalaris*, but limited to the sand dunes and therefore not a species characteristic of sclerophyllous forest) and finally the gecko *Tarentola mauritanica*.

Very common are molluscs, present under stones in the most humid situations, such as *Helix aspersa, Rumina decollata, Lauria cylindracea*, limited to France and Corsica, *Pomatias elegans* (in shady places), *Mastigophallus rangi*, limited to Côte Vermeille and Var, *Euparypha pisana* and *Eobania vermiculata*, two trivial species; *Azeca boissyi*, an endemic species confined to the Eastern Pyrenees and Var, found by the present author at Mas de la Serre, near Banyuls-sur-Mer (where there is the Laboratory for terrestrial ecology), and a species of *Aegopis.*

As far as the insect fauna is concerned, we will just mention some Orthopterous insects such as *Antaxius chopardi*, endemic to the Eastern Pyrenees, *Platycleis intermedia*, the Gryllidae *Pteronemobius lineolatus*, the hoppers *Calliptamus wattenwyllianus* (Western Mediterranean species), *Pezotettix giornai*, a common species, the earwig *Euborellia moesta* and the Mantoidea *Ameles decolor.* The hoppers are extremely abundant during the hottest months of the year in the garrigues near Banyuls-sur-Mer, where one can listen to their chirping.

Two species of termites are found in this region: *Reticulitermes lucifugus* and *Calotermes flavicollis*, living in the trunks of holm oaks, figs, vines and under stones, sometimes in dry, sunny places.

Among beetles, we can begin with an extremely interesting and rare species, the myrmecophilous *Paussus favieri* collected by Theodorides in a nest of *Pheidole pallidula*; the beetle is limited to the Iberian Peninsula and North Africa, and therefore its presence at Banyuls-sur-Mer is of paramount importance. The Tenebrionids are not very abundant: we can mention *Phylan abbreviatus, Bioplanes meridionalis* and some *Asida* (Fig. 69).

Staphylinids are represented by *Pseudocypus aethiops*, a Western Mediterranean species found for the first time in France by the present author, vicariant of the similar *P. mus; Xantholinus linearis*, a widely distributed species, and *X. translucidus*, maybe endemic to France. Scarabaeids are represented by a floricolous species, *Potosia oblonga*, very common in the maquis: its distribution is confined to Spain, the Eastern Pyrenees, Languedoc and from the Maritime Alps to Drôme; it is found on thistles; the larva is phytosaprophagous or lignivorous (Fig. 70-71).

The coprophagous Scarabaeids are *Scarabaeus pius, Gymnopleurus coriarius,*

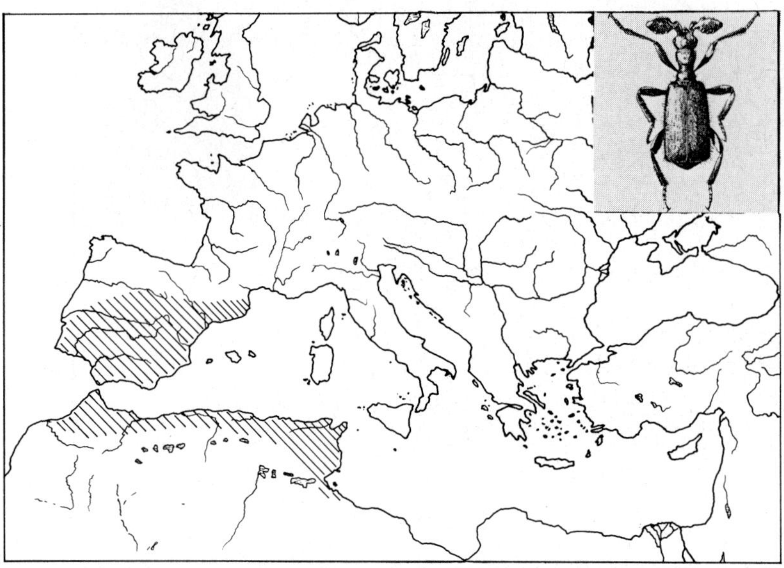

Fig. 69. Schematic distribution of the beetle *Paussus favieri*, limited to the dominion of the sclerophyllous forests of the Western mediterranean basin (photo of beetle from Jeannel).

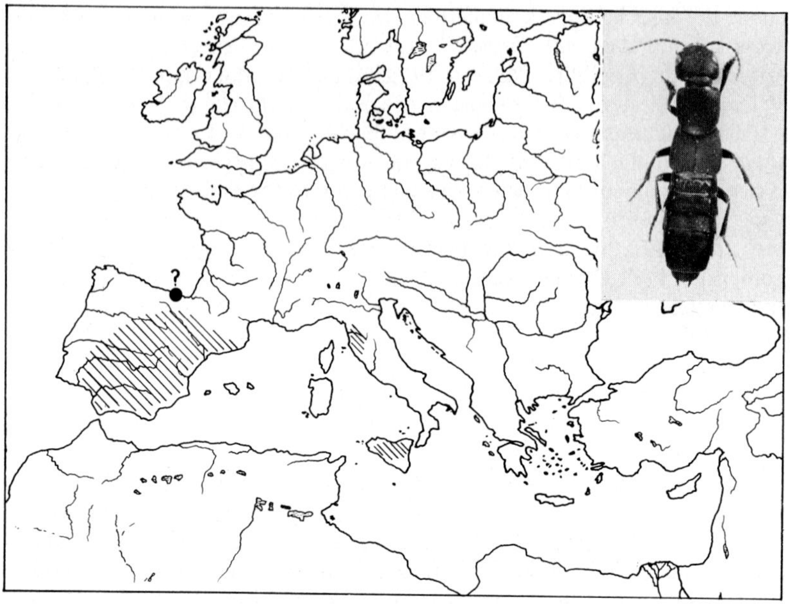

Fig. 70. Distribution of the Staphylinid beetle *Pseudocypus aethiops*, found by the author at Banyuls-sur-Mer. It is limited to the Western Mediterranean basin.

Fig. 71. Distribution of the Scarabeid beetle *Potosia oblonga*, limited to the Western Mediterranean Basin.

G. mopsus, Sysiphus schaefferi, Copris lunaris and *C. hispanus, Bubas bubalus, Oniticellus pallipes* and *O. fulvus* and lastly *Caccobius schreberi.*

The Chrysomelids are represented by one of the most beautiful beetles of the holm oak forests or maquis, *Lachnaea pubescens*, an orange-red shining insect, which catches the eye on the young, pale green leaves of holm oak, which strongly contrast with the old, darkgreen, shining leaves. Other species are *Timarcha* sp., *Tituboea biguttata* (var. *dispar*), a Mediterranean species, and *Sermyla halensis*, a widely distributed element.

Finally, we want to remember some Heteropterous insects or bugs, such as *Scantius aegyptius, Pirates hybridus* (a Mediterraneo-turanian element), *Piezoscelis staphylinus*, a Mediterranean element, *Dicranocephalus albipes*, an euro-Mediterranean species, *Cydnus aterrimus* and lastly *Eurydema ventrale*, a rather unimportant element.

BIBLIOGRAPHY

A. General part

Beuret E., Repartition geographique de quelques Arum des groupes maculatum L. et italicum Mill., *Bull. Soc. Neuchat. Sc. Nat.* 94, 1971, 29.

—, Presence d'un Arum diploïde en Italie, *Ibidem*, 95, 1972, 35.

Bodenheimer F. Animal life in Palestine, Jerusalem, 1935.

Braun-Blanquet J., Les groupements vegetaux de la France mediterranéenne, Montpellier, 1951.

Brink, van der F.H. & P. Barruel, Guide des mammiferes sauvages de l'Europe occidentale, Paris & Bruxelles, 1971.

Coiffait H., Les coleopteres du sol, Paris, 1958.

Colom G., Biogeografia de las Baleares. La formacion de las islas y el origen de su flora y de su fauna, Palma de Mallorca, 1957.

Comel A., La terra rossa italiana, *Ann. Staz. Chim. Agr. Sperim. Udine*, 2, 1933, 181.

—, Sul problema della terra bruna in Italia, *Boll. Soc. Geol. It.*, 53, 1934, 215.

—, Appunti pedologici sui dintorni di Corigliano d'Otranto, *Ibidem*, 65, 1946, 1.

—, I terreni climatici, *N. Ann. Ist. Chim. Agr. Sperim. Gorizia*, 1957, 211.

Dansereau P., Introduçao à Biogeografia, *Rev. Brasil. Geogr.*, 11, 1949, 3.

Del Villar H., Geobotánica, Barcelona, 1929.

De Philippis A., Classificazione ed indici del clima in rapporto alla vegetazione forestale italiana, *N.G. Bot. It.*, 9, 1937, 1.

—, Afforestation in arid zones. Italy, The Hague, 1970.

Eyre S.R., Vegetation and soils, London, 1963.

Favarger C., Relations entre la flore méditerranéenne et celle des enclaves à vegetation subméditerranéenne d'Europe centrale, Boissiera, 19, 1971, 149.

Gaussen H., Les étages méditerranées, Ass. fr. Avanc. Sci., Congres Rabat, Paris, 1934.

Kubiena W.L., Claves sistemáticas de suelos, Madrid, 1952.

Magistretti M., Fauna d'Italia, VIII, Cicindelidae, Carabidae, Bologna 1965.

Magistretti M. & S. Ruffo, Primo contributo alla conoscenza della fauna delle oasi xerotermiche prealpine, *Mem. Mus. St. Nat. Verona*, 7, 1959, 99.

— & —, Secondo contributo alla conoscenza della fauna delle oasi xerotermiche prealpine, *Ibidem*, 8, 1960, 223.

Marcuzzi G., Ecologia animale, Milano, 1968.

Martinoli G., La flora e la vegetazione del Capo S. Elia (Sardegna meridionale), *N. G. Bot. Ital.*, 42, 1956.

Mertens R. & H. Wermuth, Die Amphibien und Reptilien Europas, Frankfurt a. M., 1960.

Müller G., I Coleotteri della Venezia Giulia, II. Coleoptera Phytophaga, Centro Sper. Agr. Foest. Trieste, Pubbl. N. 4, 1949-53.
Odum E.P., Fundamentals of ecology, Philadelphia & London, 1959 (II ed.).
Opredeliteli nasekomich po povrezdeniyam kulturnich rastenii, Leningrad & Moskva, 1960.
Ozenda P., Biogeographie vegetale, Paris, 1964.
Peterson R. et al., Die Vögel Europas, Hamburg & Berlin, 1954 (transl. from English).
Rikli M., Das Pflanzenkleid der Mittelmeerländer, Bern, 1943-46 (2 vol.).
Polunin & Huxley, Mediterranean Flora, London, 1965.
Tischler W., Grundzüge der terrestrischen Tieroekologie, Braunschweig, 1949.
—, Synökologie der Landtiere, Stuttgart, 1955.
Toschi A., Mammalia, *Fauna d'Italia*, 7, Bologna, 1965.
— & B. Lanza, *Idem, Fauna d'Italia*, 4, Bologna, 1959.
UNESCO-FAO, Bioclimatic map of the Mediterranean Zone, *Arid Zone Res.*, Paris, 1963.
UNESCO-FAO, Vegetation map of the Mediterranean Zone, *Arid Zone Res., Paris*, 1970.
Walter H., Die Vegetation der Erde, Berlin & New York, 1973.
— & Lieth H., Klimadiagramm-Weltatlas, Jena 1960.
Zangheri P., Geobotanica, Encicl. Ital. Sci., Ist. Geogr. De Agostini, 1972, 375.
Zeltner L., Recherches de biosystématique sur les genres Blackstonia Huds. et Centaurium Hill. (Gentianacées), *Bull. Soc. Neuchatel. Sc. Nat.*, 93, 1970, 9.
Zohari M., Plant life of Palestine, Israel and Jordan, New York, 1962.

B. Special part

Benvenuti V. & M. Pavan, Riserve naturali italiane, Tip. Meroni, Albese (Como), 1972 (2nd edition).
Bigot L. & P. Bodot, Contribution a l'étude biocoenotique de la garrigue à Quercus coccifera. I. Étude descriptive de l'habitat et de la faune des invertébrés inventoriés, Vie Milieu, 23, 1972/73, 15.
— & —, Idem, II. Composition biotique du peuplement des invertébrés, Ibidem, 23, 1972/73, 229.
— & —, Idem, III. Dynamique de la zoocoenose d'invertébrés, Ibidem, 251.
Canzoneri S., Nuovi dati sui Tenebrionidae di 'Piccole Isole' italiane, etc. *Atti Soc. It. Sc. Nat.*, 113, 1972, 288.
Caron J. & D. Jarry, L'etude des endoparasites des petits mammiferes de Banyuls, *Vie et Milieu*, 7, 1956, 116.
Duguy R. & H. Saint Girons, Notes sur la faune herpetologique de la region de Banyuls, *Vie et Milieu*, 7, 1956, 413.
Ferro G. & F. Furnari, Flora e vegetatione di Stromboli (Isole Eolie), *Arch. Bot. Biogeogr. It.*, 44, 1968, 59.
Forcart L., Rezente Land und Süsswassermollusken der Süditalienischen Landschaften Apulien, Basilicata und Calabrien, *Verh. Naturf. Ges. Basel*, 78, 1965, 59.
Francini Corti E., Problemi di fitogeografia delle Puglie, *Arch. Bot. Biogeogr. It.*, 43, 1967, 25.
— & A. Messeri, L'isola di Marettimo nell'arcipelago delle Egadi e la sua vegetazione, *Webbia*, 11, 1956.
Gentile S., Memoria illustrativa della carta della vegetazione naturale potenziale della Sicilia, *Ist. Bot. Univ. Lab. Crittogamico Pavia, Quad. N.* 10, 1968.
La Greca M., Ricerche sulla fauna dele Tremiti e di Pianosa promosse dall'Istituto di Zoologia dell'Università di Bari. III. Il popolamento ortotterologico delle isole Tremiti e di Pianosa, *Boll. Zool.*, 22, 1955, 121.
—, L'ortotterofauna pugliese ed il suo significato biogeografico, *Mem. Biogeogr. Adriat.*, 4, 1959, 33.
Lomont H., Note sur la migration pre-nuptiale des oiseaux aux environs de Banyuls, *Vie et Milieu*, 7, 1956, 123.

Marcuzzi G., Studi ecologici e faunistici sui Tenebrionidi (Col. Het.) della Puglia, *Mem. Biogeogr. Adriat.*, 6, 1962, 1.
—, Osservazioni biocenologiche sulla Coleotterofauna pugliese, *Atti Accad. Naz. It. Entom.*, 11, 1963, 272.
—, Observations on the relationships between Tenebrionid fauna and soil, *Pedobiologia*, 4, 1964, 210.
—, Il Gargano, *L'Universo*, 44, 1964, 107.
—, Le Isole Tremiti: l'ambiente naturale, *Ibidem*, 807.
—, Aspetti naturalistici della Dalmazia, *Ibidem*, 48, 1968, 567.
—, Ricerche ecologico-faunistiche sulla coleotterofauna della Dalmazia. I. Descrizione dell' ambiente e stazioni di raccolta, *Atti Ist. Ven. S.L.A.*, 126, 1968, 281.
—, Osservazioni ecologiche sulla fauna del suolo di alcune regioni forestali italiane, *Ann. Centro Econ. Mont. Venezie*, 7, 1968, 209.
—, I Tenebrionidi (Col. Heteromera) delle Isole Eolie e di Ustica, *Atti Accad. Gioenia Sc. Nat. Catania*, 2, 1970, 1.
—, Contributo alla conoscenza dei Tenebrionidi delle Egadi, *Boll. Soc. Ent. It.*, 102, 1970, 87.
—, Contributo alla conoscenza dei Coleotteri Tenebrionidi di Sicilia, *Boll. Sed. Accad. Gioenia Sc. Nat. Catania*, 10, 1970, 405.
—, Contributo alla conoscenza dei Tenebrionidi delle Tremiti, *Boll. Soc. Ent. It.*, 102, 1970, 35.
—, Aspetti naturalistici delle Isole Eolie, *L'Universo*, 50, 1970, 915.
Marcuzzi G., Osservazioni ecologiche sulla tenebrionidofauna del Mediterraneo, Quad. Ecol. anim., Padova, La Ciclografica, 1976, N. 5.
—, Osservazioni zoogeografiche sulla tenebrionidofauna del Mediterraneo, Ibidem, N. 6.
—, & L. Dalla Venezia, Nuove indagini biocenologiche sulla Coleotterofauna pugliese, *Atti Ist. Ven. S.L.A.*, 124, 1966, 95.
— & —, Studi sui Tenebrionidi della Dalmazia. II. Prime osservazioni sui rapporti tra Tenebrionidi e suoli, *Atti Ist. Ven. S.L.A.*, 126, 1968, 305.
— & —, Osservazioni biocenologiche sulla malacofauna pugliese, *Arch. Bot. Biogeogr. It.*, 14, 1970, 1.
Marcuzzi G., L. Dalla Venezia, L. Messina & R. Beneventi, Nuove osservazioni biocenologiche sulla coleotterofauna pugliese, *Arch. Bot. Biogeogr. It.*, 43, 1967, 450.
Marcuzzi G. & E. Lo Casto, Osservazioni autoecologiche sulla Malacofauna pugliese, *Atti Rel. Accad. Pugliese Sci.*, 29, 1971, 3.
Marcuzzi G. & A. Vannozzi, Osservazioni sull'origine degli animali domestici, Quad. Ecol. anim., Padova, La Ciclografica, N.1, 1975.
Moltoni E., Escursione ornitologica alle Isole Tremiti, *Riv. It. Ornit.*, 27, 1957, 81.
— & S. Frugis, Gli Uccelli delle Isole Eolie (Messina, Sicilia), *Ibidem*, 37, 1967, 91.
Müller J., Tenebrionidae Dalmatiae, *Verh. Zool. Bot. Ges. Wien*, 71, 1921, 132.
Müller G., I Coleotteri della Venezia Giulia, Parte I. Adephaga, St. Entom., 1, Trieste, 1926, 3.
Novak P., Kornjaši jadranskog primorja (Coleoptera), *Jug. Akad. Znan. Umjetn.*, Zagreb, 1952.
—, I Coleotteri della Dalmazia, *Atti Mus. Civ. St. Nat. Trieste*, 24, 1964, 53.
Paget O., Beschreibung einer Molluskenausbeute aus Apulien (Süditalien), *Mem. Biogeogr. Adriat.*, 4, 1962, 169.
Paulian P. & G. de Liffiac, Un mois d'observations ornithologiques dans les Pyrenees orientales, *Vie et Milieu*, 3, 1952, 90.
Pavan M., Montecristo Riserva naturale, Tip. Meroni, Albese (Como) 1971.
—, Riserve naturali italiane: situazione e proposte di tutela dei poteri pubblici, Collana Verde, Min. Agric. For., N. 31, 1973.
Richard G., Les Termites à Banyuls, *Vie et Milieu*, 1, 1950, 95.
Servadei A., Rhynchota, *Fauna d'Italia*, 9, Bologna, 1967.
Susmel L., M. Cappelli, F. Viola & G. Bassato, Autoecologia del pino radiato al Grighini (Sardegna centro-occidentale), Ann. Centro Econ. Mont. Venezie, 9, 1968/69 (1975).

Susmel L., F. Viola & G. Bassato, Ecologia della lecceta del Supramonte di Orgosolo (Sardegna centro-orientale), Ann. Centro Econ. Mont. Venezie, 10, 1969/70 (1976).

Théodoridès J., Paussus favieri (Col. Paussidae) dans la region de Banyuls, *Vie et Milieu*, 1, 1950, 97.

Tomaselli R., Accenni alle successioni floristiche ed al dinamismo della vegetazione sicula, *Arch. Bot. Biogeogr. It*, 37, 1961, 3.

Zavattari E., Rinvenimento di Psammodromus algirus nell'isola dei Conigli di Lampedusa, *Boll. Zool.*, 21, 1954, 93.

—, Esplorazione biogeografica delle Isole Pelagie, *Ric. Scientif.*, 27, 1957, 458.

—, Rinvenimento delle Solpughe nell'isola di Lampedusa, *Boll. Soc. Ent.* It., 87, 1957.

THE DECIDUOUS FOREST

General characteristics

The deciduous forests occupy a very conspicuous portion of Europe, practically from latitude 40°N (Puglia, Albania) to 60°N (Scandinavia). They gradually grade through the intermediary of the mixed forests (Mischwald) into coniferous forests or taiga, so that their extension in the sense of longitude is from 10°W to 45°E. For these reasons, and especially because of the wide extension in latitude, we may expect to find – and actually do find – within the deciduous forest much greater variations in the biota than those in the sclerophyllous forest, whose limits in latitude are relatively modest (from 35°N – Crete – to 45°N – Istria, Crimea).

A description of the deciduous forest is therefore rather difficult. We can only say that it is a summer forest with a winter die-back in vegetation, with precipitation distributed rather regularly throughout the year, and with very variable thermal characteristics (Fig. 72, 73). Snow during winter is not a necessary feature, although it is fairly common. The mean rainfall varies from 750 to 1500 mm per year; the summer temperature may reach some 18–22°C, the winter temperature –5 to –10°C.

Broad-leaved trees such as oak, linden, beech, maple, ash, etc., are very characteristic of this forest type, which is widespread not only in Europe, but also in East Asia (China) and in the Eastern part of North America, where evidently the climate is relatively similar, i.e., temperate and ameliorated by the nearby sea.

The number of tree species, however, is not very high, especially if we compare this forest with some subtropical or tropical ones. Two main types of deciduous forests are recognizable: one developed essentially in the proximity of the Atlantic Ocean, and influenced by the oceanic climate (i.e. very small thermal variation during the year), characterized by beech; and another, developed towards the

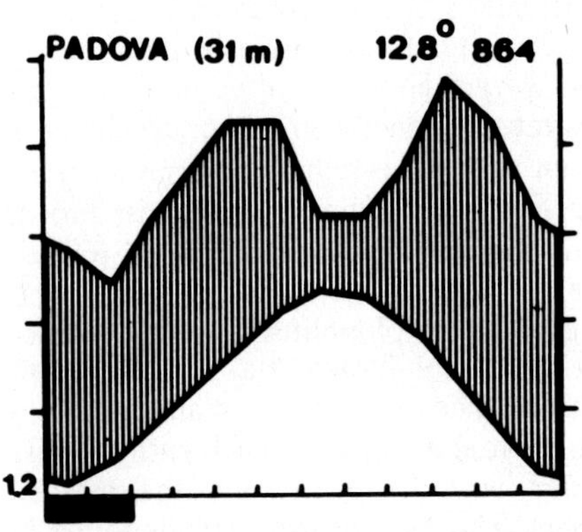

Fig. 72. The ombrothermic diagram of a region belonging to the deciduous forest, Padova (Northern Italy) (from Walther, modified).

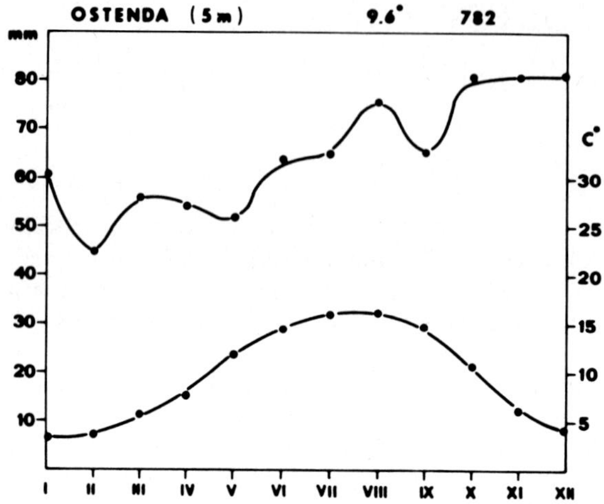

Fig. 73. The ombrothermic diagram of a locality of the dominion of deciduous forests, situated on the shores of the Dover street, Ostenda; the climate is similar to that of Padova, but with a milder winter due to the vicinance of sea.

interior of the Continent, or rather towards Western Asia, with very great annual temperature variation, characterized by oak (secondarily *Corylus, Castanea, Fraxinus, Ornus, Ostrya*, etc.). The former constitutes the so-called Fagetum (or Fagion of biocenology), the latter the so-called Quercetum (or Quercion of biocenology). Beech forests are generally defined as sciaphilous forests (from σκια, shadow), oak forests as heliophilous forests (from ἕλιος, the sun).

Very often, in place of the deciduous forest, in many parts of Europe (as well as in North America) we find fields, meadows, highways, or enormous towns: we shall see later, the dominion of the deciduous forest – second only to that of the sclerophyllous forests – has been inhabited by men for a long time and human influence is therefore very great. Where these modifications have been large one speaks today of 'Agricultural biocenosis' (Tischler's 'Agrarbiocenosen').

The soils present in the areas of the deciduous forests are essentially brown earth, in the West, and podzol in the East. Nevertheless very often they can exist side by side, according to the kind of substrate and microclimate (or ecoclimate).

The brown earth ('Braunerde' of German authors) is rich in humus, very thick and with a very deep cultivable layer, generally dark brown to almost brown black, with a large amount of nutrients and high biological activity. It is formed upon basic and easily weathered rocks, for instance, basic eruptive rocks and their tuffs and slates, mainly basalts, diabases, gabbros, amphibolites, etc. It is a loose soil, rich in pore spaces (eutrophic brown earth); the humus is a Mull. The plant cover consists generally of broad-leaved shrubs. The life, both plant and animal, is very intense, and very rich in species, with great abundance of earthworms. There is very good decomposition and humification, with the formation of the so-called humus-clay complex (Ton-Humus Komplex). The best brown earth is found in Bohemia, Vogelsberg (Hessen), Odenwald, Burgenwald, South-Eastern Styria and Carinthia (Fig. 74-75).

The oligotrophic brown earth, on the contrary, is very plastic, developed on rocks poor in calcium, light ochre coloured, poor in humus: it is widespread in Central Europe. The Central-European brown earth is widespread in the area of deciduous and mixed forests of Southern, Central and Northern Europe. Furthermore, Kubiena recognizes a calcareous brown earth, developed on calcareous loess and a ferrous brown earth, on non-calcareous rocks, not very plastic, dark brown ferrugineous coloured, developed on limonite, siderite and limestone rich in iron. It is found, for instance, in Eastern Styria where there are some metalliferous veins.

Fig. 74. Examples of brown earth.' Left: eutrophic centro-european brown earth, under a beech forest on basalt lava (Burgenland, Austria); right: oligotrophic centro-european brown earth under a mixed forest on sandstone slate (Admont, Styria) (from Kubiena).

Under oak and birch we very often find a podzol, named after a Russian term used to indicate a sandy soil, much leached, which occupies large areas in Northern Russia and Western Siberia. It is frequent in all countries with low temperature, heavy rains and a humid winter. The dense forest vegetation and the cold climate favour the conservation on the surface of a layer of plant debris, which can attain a thickness of 25-30 cm. The vegetation cover, because of the scant aeration, does not permit the development of a rich fauna and therefore the litter is not attacked by micro-arthropods, bacteria and other soil organisms which generally contribute to humus formation. Rain brings into the subsoil all soluble

Fig. 75. Examples of brown earth. Left: Eupodzolic brown earth in a montane meadow with *Crocus nudiflorus* on quarzitic slates (Cordillera Cantabrica, Asturias, 1400 m). Right: Podzolic brown earth under a beech forest with some whortleberries, on chloritoscists (Freyenthurn, Klagenfurt, Carinthia (from Kubiena).

mineral substances and part of the soluble organic matter, so that the upper horizon is deprived of its iron, aluminium and some alkaline and alkaline-terrous minerals, which are leached, while at a certain depth an impermeable brown-coffee coloured layer, rich in iron, is formed. It is the so-called hard pan ('Ortstein' of Northern Germany, 'alios' of the Landes).

The physical characteristics of podzols preclude cultivation because of the continuous leaching (which removes all mineral elements, and therefore all nutrients) and still more because of the existence at a certain depth of the impermeable hard pan (Fig. 75-76).

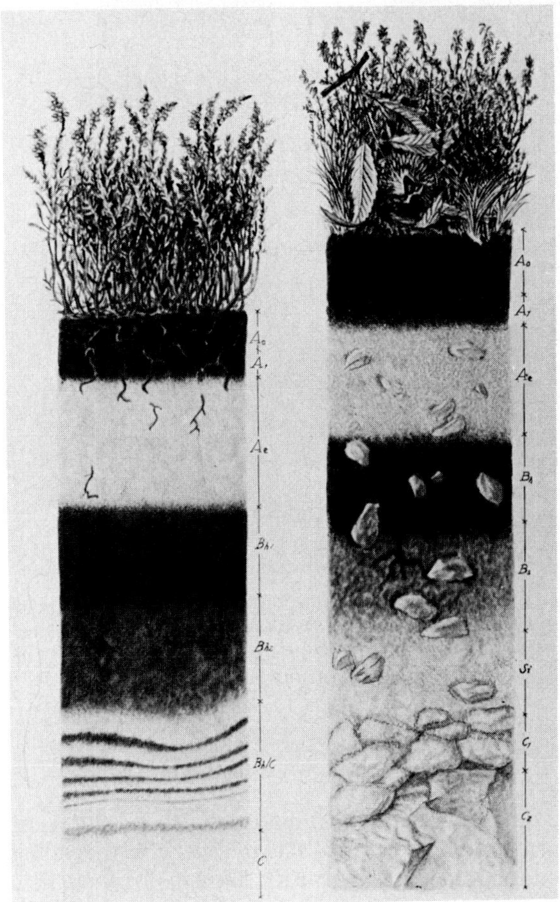

Fig. 76. Examples of podzols. Left: Humic podzol on sand dunes with *Calluna* with two layers of Ortstein (or hardpan) followed by several thin layers of Ortstein (Dorum, near Bremen on the shores of the North Sea). Right: Asturian ferro-humic podzol, under a heath with *Erica ciliaris*, with chestnut and oak, on sandy silicic rock (Cordillera Cantabrica, Asturia, 300 m) (from Kubiena).

Fig. 77. Examples of soils present in the European deciduous forests. Left: a mulliform rendzina, with *Pinus nigra austriaca* and *Sesleria nana*, on dolomia (near Vienna). Middle: Mull-rendzina with grassy vegetation, near Vienna. Right: Podzol ranker under a shrub with *Calluna*, on quarzite (Cordillera Cantabrica, Asturias, 1400 m) (from Kubiena).

Vegetation

In the heliophilous forests the dominant tree is oak: in the unmodified forests of *Quercus pedunculata*, many century-old trees grow side by side with the young trees grown from corns fallen during the preceding years. The dark brown bark is covered with mosses and lichens, and from the branches the mistletoe (*Viscum album*) hangs, the parasite plant sacred to the Celts (Fig. 78).

A very tall tree, practically ever-lasting, *Quercus pedunculata* shows a remarkable degree of gregariousness, often comprising almost pure, monospecific, forests on all clayey-sandy soils of Europe, reaching as far east as the Volga plains or the

Caucasus mountains. The continental climate is not unfavourable to it, and it can withstand very cold winters and heavy frosts. It is present also south of the Alps, such as in Northern Italy and the Appennines, where it grows from sea level and the Po plains up to the Picetum (or *Picea excelsa*) belt.

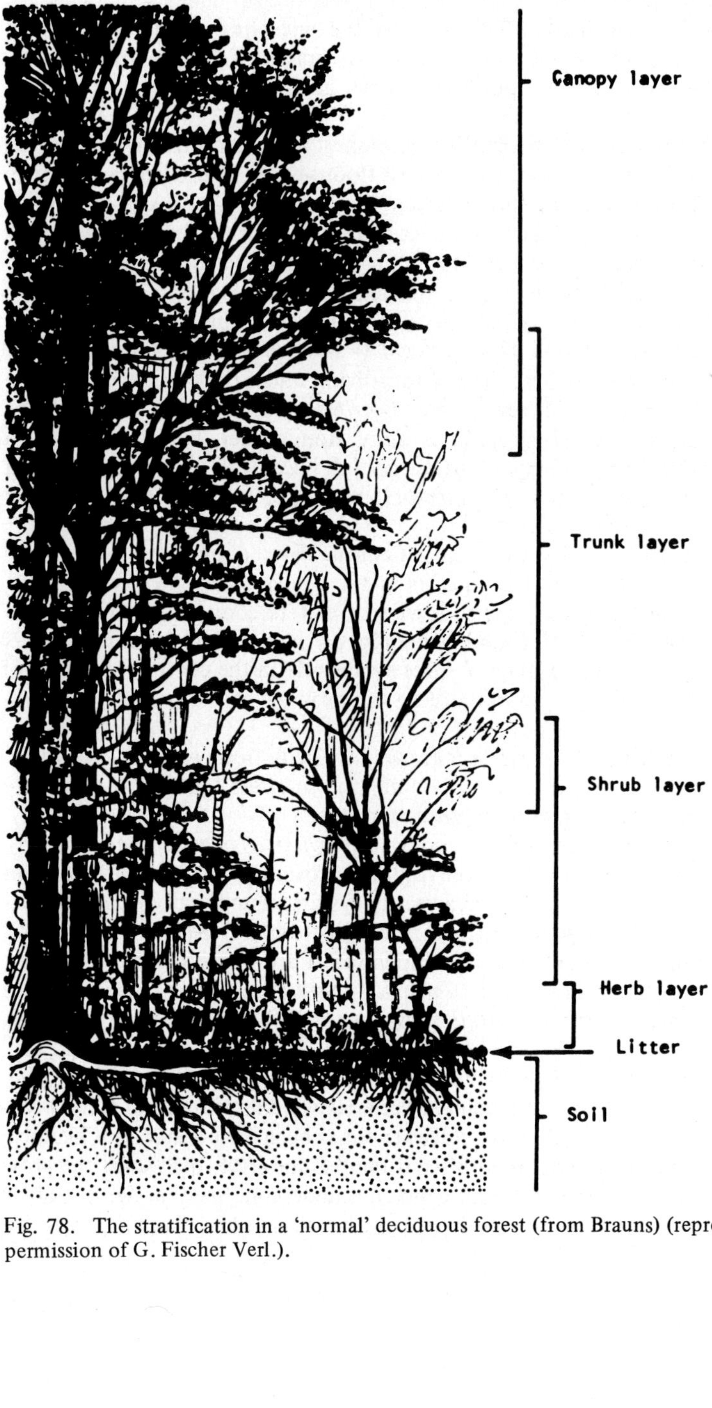

Fig. 78. The stratification in a 'normal' deciduous forest (from Brauns) (reproduced with the permission of G. Fischer Verl.).

Today, unfortunately enormous *Q. pedunculata* (or, as it is called in Italy, 'farnia') are very rare, as are all old plants; very few are the millenarian trees in the deciduous forests: let us remember the Aosta linden, the Macugnaga (Piemonte, east of M. Rosa) linden, the Mergozzo (West-Lake Maggiore) elm, but they are by far younger than the Platanus of Kos (Eastern Aegean Sea) under which they say Hippocrates used to teach (470-377 B.C.) or the olive tree of the Gethsemane garden (Jerusalem), which witnessed Jesus' last prayer. Today in Europe, and especially in the most densely populated areas, many broad leaved trees form coppices instead of forests.

Beech, birch and horn beam are the commonest trees associated with *Quercus pedunculata* in Eastern Europe. On the forest floor several small plants can thrive; some of them flower early in spring, when the leaves on the trees are not yet grown and do not intercept light: examples of these plants are *Anemone hepatica* or *Anemone nemorosa, Vinca* (the common periwinkle) and others, which in plant ecology (or better in phenology) are called 'short day plants'.

Particularly rich is the herbaceous vegetation living on the floor of forests of *Quercus petraea* (= *sessiliflora*), *Q. pubescens* (= *lanuginosa*), *Q. trojana* (in Italy known as 'fragno'), *Q. cerris*, etc. The shrubs living under these kinds of oaks are many: suffice to quote hazel, service tree (*Sorbus aucuparia*), several species of blackberry, some not very hygrophilous *Salix*, holly, alder, buckthorn, etc. In Northern or Central Europe only whortleberry (or *Vaccinium myrtillus*) grows on the floor of forests of *Quercus pedunculata*, birch and beech. Finally, the soil surface is often covered with moss.

South of the Alps, in Northern and Central Italy, a very common oak is *Quercus pubescens*, a south-eastern species of the genus, which thrives on sunny, dry, calcareous soils. *Quercus cerris*, however, prefers cooler and more humid situations and therefore it is found generally on hills or mountains. *Quercus trojana*, finally, in Italy called 'fragno', lives exclusively in the central, hilly part of Puglia (the so-called 'Murge') and in Greece (an example of transadriatic distribution).

A special race of *Fraxinus ornus* is cultivated in Sicily from which it is possible, by means of incisions in the bark, to obtain the so-called 'manna', a sugar like substance composed of the monosaccharide mannose.

Another interesting species of deciduous oak is *Quercus petraea* (= *sessiliflora*), the so-called 'rovere', dominant in the Illyrian provinces, i.e. in South Eastern Europe, from the Italian karst to Moravia, always associated with *Fraxinus ornus* and often with *Ostrya carpinifola* (Seslerio-Quercetum and Seslerio-ostryetum). Other elements of the heliophilous deciduous forest are maple (*Acer campestre*), hazel (*Corylus avellana*), hornbeam (*Carpinus betulus*), *Ostrya carpinifolia*, elm (*Ulmus campestris*), cornel or *Cornus sanguinea* with edible fruits, and *Cornus mas*.

An indigenous tree of Southern Europe and Asia Minor is the chestnut, *Castanea sativa*. Some particularly appreciated varieties were introduced in Italy, probably by Romans, and since then its cultivation has been widely spread. Chestnut, though preferring a mild climate with scant rainfall and fresh, acid soils, can grow also on calcareous or dolomitic terrain which has undergone weathering. In Italy is very widespread, going from sea level (e.g. on the Isle of Elba, where it thrives on the North-western side of the isle, in the heart of the sclerophyllous forest) up to mountain zones, from 400 to 1000 m. On Etna it reaches as high a level as 1600 m. Very often chestnut forms coppices today: from the enormous old stumps the young, slender stems rise, generally destined to furnish branches for agricultural practices or, in some countries near the sea, for mussel culture (Taranto, Naples, etc.). Its fruits are edible and are also utilized to make sweets. The wood, compact and resistant to humidity, is employed for the construction of dams, stockades, etc. The tannin obtained from its wood is used in tan-houses. Unfortunately, in recent years two chestnut diseases, the so-called cancer of chestnut and the ink-disease, have greatly damaged this beautiful tree in many parts of Southern Europe. The cancer is due to *Endothia parasitica*, an ascomycete *Diaporthales*; the ink disease, characterized by a blacking of the base of the stem and the roots, is due to a *Peronosporacea, Phytophthora cambivora.*

Very different is the dominion of the sciaphilous forest. The dominant tree is beech, a slender, straight tree, with a broad expanse of foliage. Generally, the forest is monospecific. Unfortunately the destruction of beech forests, from which people used to obtained wood and coal, has very much restricted the range of this arboreal species, and trees with trunks wider than 30 cm are very rare. Almost everywhere forests have been replaced by coppices. It must also be pointed out that beech litter, different from that of heliophilous trees, favours the formation of an acid soil, which in turn permits the development of a heath (*Calluna, Erica*, etc.).

Of course, the irrational and unrestrained deforestation has greatly influenced the life of the great mammals, which in the Middle Ages were very abundant in the dominion of the sciaphilous forest. Stags, roe-bucks and bears which found shelter and food in the old forests of Central Europe and Northern Italy, have

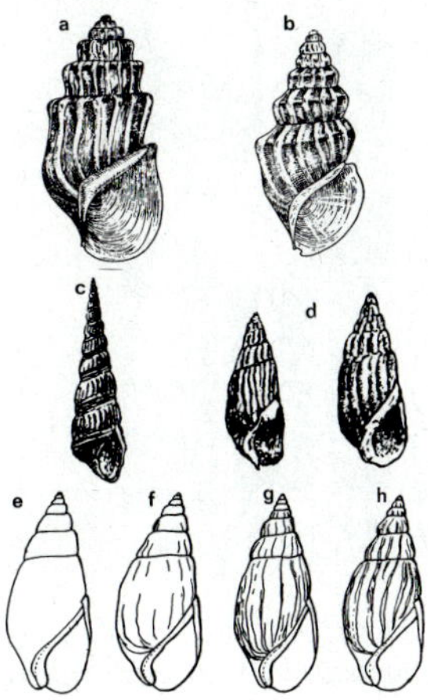

Fig. 79. a. and b. *Melanopsis parreyssi*, a Tertiary relict of thermal waters in Hungary and Romania (temperature 34-42°C (from Grossu); c. *Melania tuberculata*, Lake Kinereth (Tiberias), temp. 21.5-26.5° C (at –5 m); d. *Melanopsis costata*, from Lake Huleh; e-h. *Melanopsis praemorsa*, from Lake Huleh (from Bodenheimer).

disappeared and today only a few roe-bucks are still present, thanks to the strict laws protecting this elegant ungulate.

Today, where in Southern Europe there is heliophilous deciduous forest, in the Upper Miocene there was a vegetation and a fauna with typical tropical characteristics, with molluscs such as *Dreissenia* (of which one species still lives today, *D. polymorpha*, which, originating from the Caspian Sea, has invaded most of the rivers of Central Europe), *Melanopsis*, today partly relegated to thermal waters of South-Eastern Europe[1], partly in South-Eastern Asia[2], *Paludina* (still present in waters with a relatively constant temperature such as springs, caves, etc.) and mammals, such as tapirs (today relegated to the tropics of Eastern Asia and Southern America). Plants of that era were *Liquidambar europaeum*, closely related to *L. orientale* of Asia Minor (Fig. 81) belonging to the family *Hamamelidaceae*,

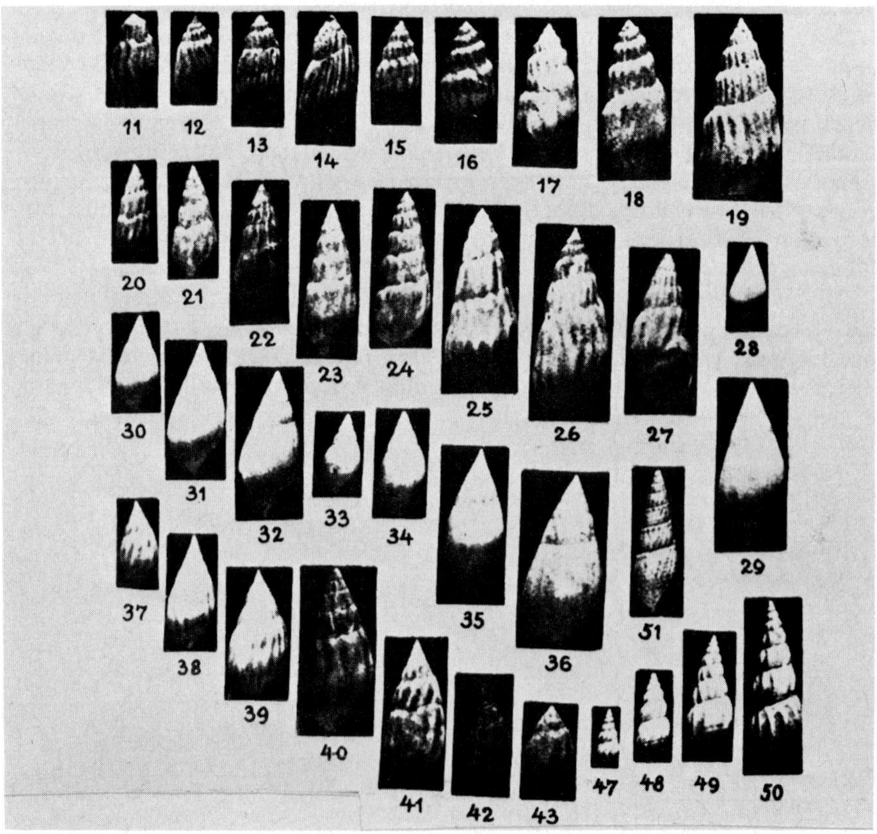

Fig. 80. Thermophilous freshwater snails: 11-14, *Melanopsis aaronsohni*; 15-27, *M. turriformis*; 28-42, *M. buccinoidea*; 43-46, M. spp.; 47-50, *Melania rhodiensis*; 51. *Melania tuberculata* (from Bodenheimer).

[1] *Melanopsis fareyssi*, Hungary and Romania, 34-42°C, *M. hungarica*, Hungary, 26-28°C, other species in thermal waters of Hungary, Croatia and Austria.

[2] Israel, Egypt, Algeria, Tanganyka, South-Eastern Asia, etc.

	Alleröd-interstadium (North Europe)	amelioration	*Betula* (north) *Pinus* (south)		9000-10.000	
	Finiglacial	Old Tundra-period	woods, heath, swamps & bogs dwarf birch, etc.		12.000	
	Bühl-Stadium (Central Europe)					
IV Glacial (Würm)	Zürich-Stadium . . . . (Central Europe)			*Upper Paleolithic* (60.000-10.000 BC) *Homo sapiens* or *H. sapiens fossilis* 3 main races: Cro Magnon (aurignacien), Grimaldi (aurignacien) and Chancelade (magdalenian)		9000 (10.000) 120.000 110
	Kittwangen-Stadium (Central Europe)					
III Interglacial	Riss-Würmian interglacial		mixed forest: *Abies nordmanniana, Zelkova, Castanea*	Mousterian (Neandertal) ex. Quinzano (Veneto), Swanscombe (Kent), Acheulean, and Fontechevade (Charente) Acheulean	120.000-180.000	60
II Interglacial	Mindel-Riss interglacial			Acheulean	235.000-425.000	190

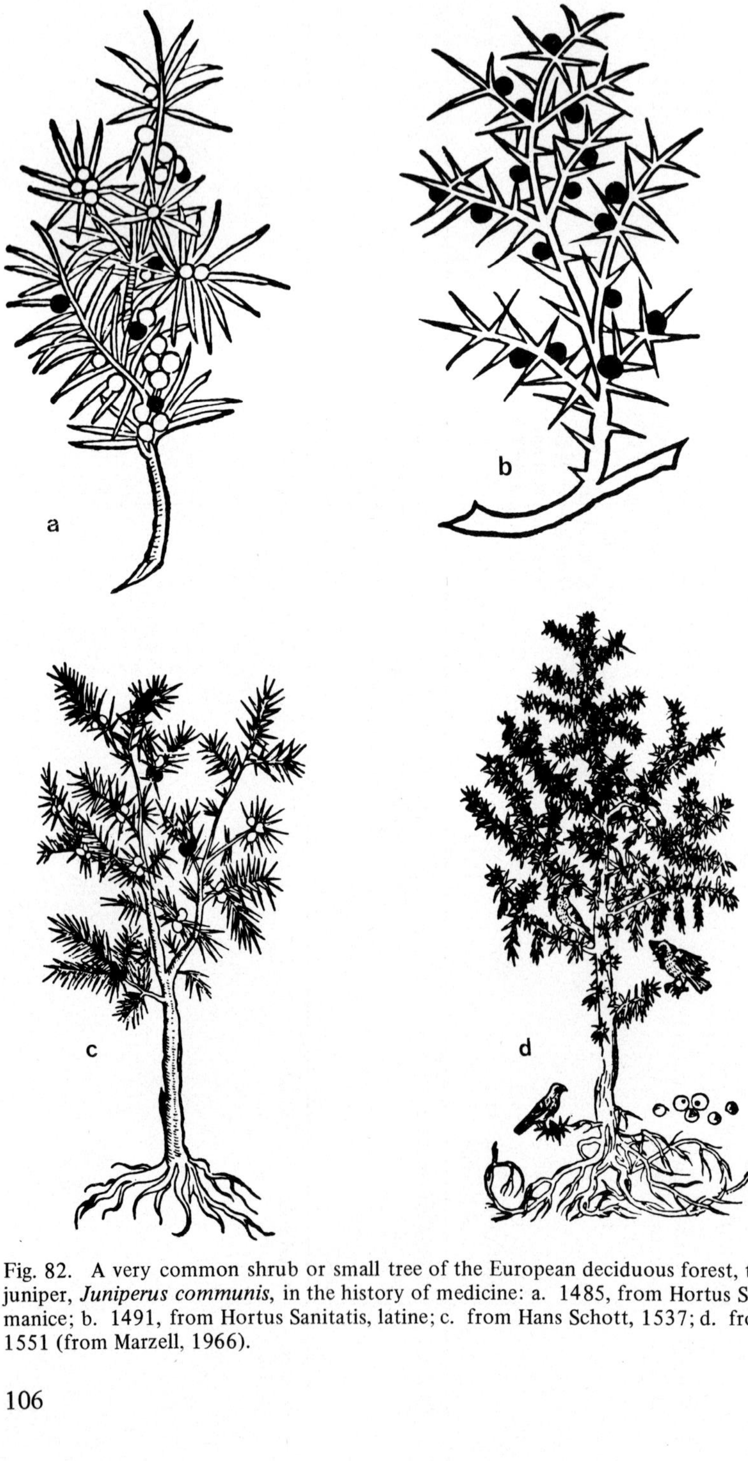

Fig. 82. A very common shrub or small tree of the European deciduous forest, the common juniper, *Juniperus communis*, in the history of medicine: a. 1485, from Hortus Sanitatis, germanice; b. 1491, from Hortus Sanitatis, latine; c. from Hans Schott, 1537; d. from H. Boch, 1551 (from Marzell, 1966).

The effects of man on European deciduous forests

During the 2nd interglacial period, the Acheulean culture, derived from Africa, had penetrated in Germany the region of Holberstadt, in Hannover and Saxony. The third interglacial corresponds to the Mousterian culture (180,000-120,000 years B.C.) in which the population, taking its name from Neanderthal (a small village near Düsseldorf) comprised the so-called Weimarer Kultur, named after the neighbourhood of Weimar.

During the last glaciation (Würm), in Germany there were many traces of Upper Paleolithic culture, attributed to *'Homo sapiens'* or *H. sapiens fossilis* (60,000-10,000 years B.C.). After that time, in the dominion of the deciduous forest there were many migratory currents, each corresponding to a given culture: so we find the Aurignacian in the caves of Rhineland and Bavaria, represented by Predmost man or Brünnrasse, a contemporary of Cro-Magnon (France) and Grimaldi (Italy). To this man is attributed the only artistic work of the Paleolithic in Germany, i.e. the antelope engraved in the cave of Schulerloch, near Kelheim. The Solutrean culture, coming perhaps from the Danube plain, is also found in Bavaria. The Magdalenian culture, very well represented in France (Lascaux) and Spain (Altamira), has left no remains in Germany: the evidence of this culture present in Thuringia, Brandenburg and Holstein suggests shepherds with reindeers who followed the retreat of ice-sheets towards the north (Fig. 83-84).

During the Mesolithic (8,000-3,000 years B.C.) most of Germany (and therefore most of the Central-European deciduous forest, which by now is an oak forest, succeeding birch due to climatic changes) was inhabited by man, who by then had diversified his activities: during the warm, pluvial period (5,500-2,500 B.C.) indeed the first agriculture appeared in Central Europe: proof is shown by the plough of Walle (Eastern Frisia), found in a mire and dated by means of pollen analysis at 3,500-4,000 B.C.

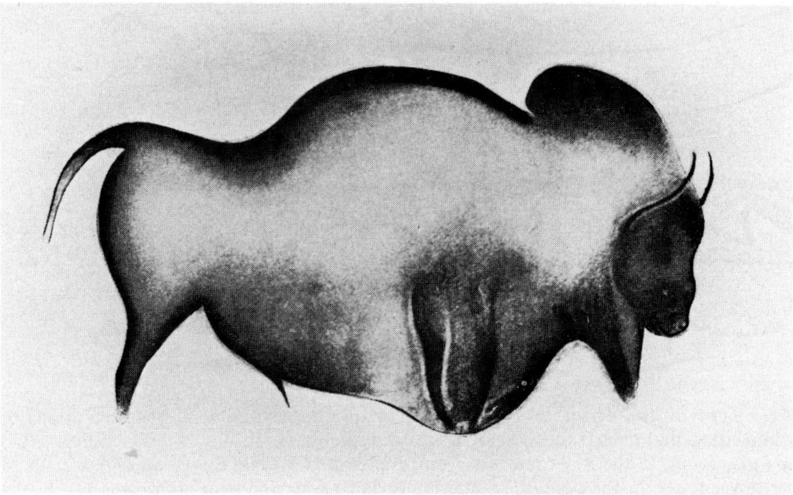

Fig. 83. Rock painting of *Bison* sp. in Font-de-Gaume cave (from Breuil).

During the Neolithic, agriculture showed tremendous progress due to the drier and warmer climate, with steppe characteristics, so that methods of agriculture existing for several millenia in the Near East were introduced into Central Europe.

Three main migrations reached Central Europe, the heart of the deciduous (generally sciaphilous) forest. One, coming along the Danube, characterized by 'ribboned' pottery, was over in Germany during 2000-1800 B.C. Another current, coming also from the East, but through Western Europe, i.e. Spain, France, Brittany, England and Scandinavia, is characterized by the megalithic culture, widespread throughout the sclerophyllous forest as we have already seen. A third migration, named after the campaniform (or bell) pottery, originally from North

Fig. 84a. Masks of men with the head of a chamois, possibly a totemic figure (engraving on bone, Teyjat, Dordogne) (from Breuil).

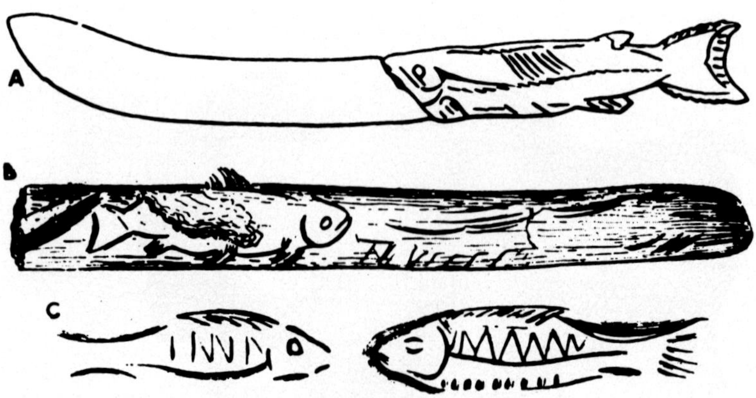

Fig. 84b. Signs of the effect of Paleolithic man on the environment: representations of fishes, indicating that at that time fishing was already known to man. a. salmon (*Salmo salar*) on a spatula, found in the Rey cave, near Combarelles; b. Cyprinid carved on a stick made of reindeer horn, found in the cave of La Mairie, Teyjat; c. beginning of stylization (La Madeleine) (from Breuil).

Africa and Spain, was responsible in Germany for the first palafitticolous settlement (1800-1400 B.C.) up to the 8th century.

From 1800 to c. 800 B.C. Germany was in the Bronze Age, but, whereas in the South-west the Celts discovered iron in 750 (Iron age – Hallstatt C), the true Germans stuck to bronze until 500 B.C.

The population density was always very low; even in the first century A.D. the total population of Germany was estimated around 2,000,000, so that we can imagine that no serious disturbance was caused to natural vegetation and local fauna, different from what happened in the dominion of the sclerophyllous forest.

The presence of man in France in prehistoric times is so well known, that many European cultures are named after the French deposits where the first remains were found (Chelleenne, Abbevillienne, etc.). The 'mousterian' is present in the Dordogne, Jura, Alsace, Lorraine, Côte d'Azure (Riviera di Ponente).

During the Upper Paleolithic there were populations living by hunting, and moving continuously during the year; they inhabited caves and other refuges under rock in the Pyrenean valleys, in Narbonnais, in Garronne, Ariege, Dordogne and Périgord: here we find remains of Mousterian (3rd interglacial), Aurignacian, Solutrean and Magdalenian cultures, and also sometimes of the Mesolithic time. The latter is represented by 'Azilian', frequent in the Ariege and Dordogne, whereas the microlithic Tardenoisian culture is present in the North of France. The Neolithic was very rich and wide-spread: the famous Campignan flint culture was very wide-spread in the North and the West.

In the period between the Neolithic and Bronze Age, almost the entire territory today corresponding to France was characterized by the presence of dolmen sepulchres. In this epoch, the influence of Iberian peoples is evident in the Midi, where we find for instance the so-called campaniform (bell-form) pottery (Fig. 85). Three groups were then present in France: a south-western, of Iberian pro-

Fig. 85. An example of bell-form pottery (from Mosso).

venance; a south-western, closely related to Swiss cultures, and a central-eastern one, in contact with Germany (see above). At the beginning of the first millenium B.C., the Iron Age appeared in France (Hallstatt culture). In this period there were again three main groups: one central or eastern-central, continuous with South Germany, Rhineland and Bohemia (Celtic peoples), a south-eastern, closely related to Switzerland, whereas in the south Greek influences were very clear (Μασσαλία, the present Marseille).

The Celtic culture and people were present, as well as in France and Germany, in the British Isles (insular celtic, as it is called by linguists as opposed to continental celtic). It is still debated whether the Celtic and Iberian are two different peoples or not. What is certain is that in Britain at that time we find both pure Celtic (i.e. French) elements and Iberian (i.e. Spanish) ones, due to the ease of sea passage from the British Isles both to the Iberian Peninsula and to France.

For a long period of time Celts have been the most abundant and widespread inhabitants of the deciduous forests of Western and Central Europe, that is France, Germany, Belgium, Northern Italy, etc. Thus we find Celtic remains in Engalnd, France, in the Iberian Peninsula and in the Danubian plain, where Celts were nevertheless mixed with peoples who lived there before them. For instance, in Ireland, where Celts have given their language to the present inhabitants of the island, they possibly represented only the dominant caste. (This phenomenon is rather frequent all over the world and within all races.) For this reason Celts today, rather than a racial entity, must be considered as a linguistic family, including Irish, Scottish and Welsh, the languages spoken in Cornwall and on the Isle

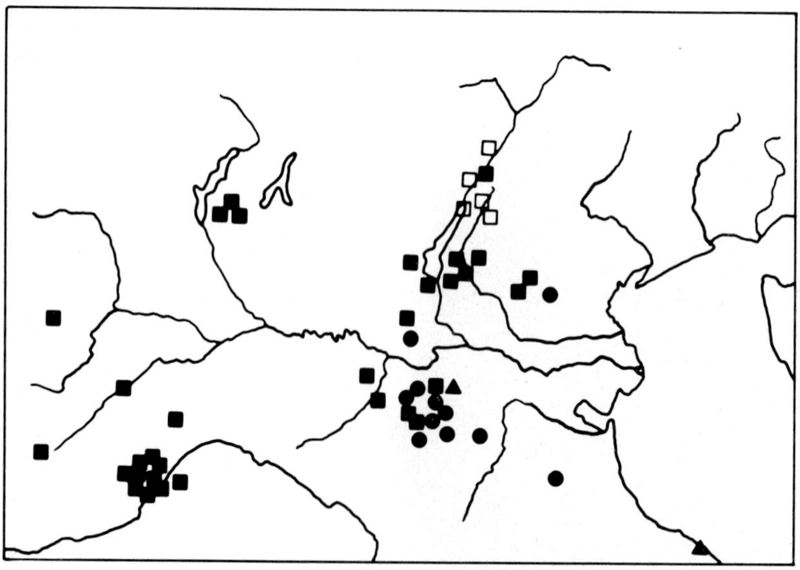

Fig. 86. Distribution of cultures during Middle Neolithic in Northern Italy (according to Barfield, 1967, from Broglio); empty squares, settlements of Trentino; black squares, Finale-Quinzano group; black circles Fiorano culture; black triangles, Ripoli culture.

of Man, Breton and the Celtic remains present today on the European continents (i.e. within the dominion of the deciduous forests), such as toponyms, hydronyms, some inscriptions and several surnames. The only Celtic language today recognized as 'official' is Irish.

Today we know that Celts must have been present also in Switzerland: at Zihl (parish of Cornaux, near Neuchatel), a wooden bridge of Celtic origin was discovered, built in 143 B.C. (Pilleri & Schwab). It means that Celts must have lived in this region at that time.

Many interesting remains of the Celtic language are found today in some dialect words in Northern Italy, for instance in the names of several plants. Let us remember only the whortleberry, *Vaccinium myrtillus*, which over a wide area has local names of Celtic origin, such as giásene, glásinis, etc., coming from the latinized glastum, which takes its origin from a Celtic word for blue, so that literally the whortleberry is called 'the plant with blue berries'. Other Celtic names conserved until today are lagena, for *Colchicum autumnale*, velme for *Alnus viridis* (from Celtic verna), larix for larch, brosorada for the heath *Erica carnea* (from 'brucus'), and several others.

It is certain that in the early Middle Ages in Germany there were still immense, impenetrable forests which were the source of myths and legends. Rivers in deep embankments and running through cuttings with steep shores, covered with forests; impervious hills; caves, everything was a reason for fear and superstition. Thus in the Rhineland, between the present towns of Worms and Mainz, in the 5th century A.D. the Nibelungs were born in the fantasy of people. These legendary inhabitants of the Rhine seem today to be referred to either as the descendants of the royal stock of the Burgundians, who at that time inhabited the Rhine shores, or as the demoniac masters of the fabulous Rhine treasure, which finally came into the possession of the Burgundian princes. The scenery of the Nibelungs' sage (from which Wagner took inspiration to compose his divine 'Tetralogy' (or 'The Ring of the Nibelung') is the wild shores of the Rhine, covered with impenetrable forests ('Urwälder'), with gorges and precipices, with which meanders and calm water alternated. From all this scenery – which today we are trying to reconstruct in vain by means of strict laws for nature conservancy or protection, and with the very 'American' conception of Natural Beauty – in the simple mind of ancient people could of course be born ideas of monsters and treasures.

In no way second to the Rhine for its fantasy-inspiring nature was the Brocken, an isolated mountain which rises in the very heart of the wild Harz, well known for its legends, where also today the phenomenon of the so-called 'Brocken spectrum' takes place. This phenomenon constists of an antisolar crown which we can observe if we turn our back towards the sun, when it forms a kind of halo around our head's shadow, when this shadow is seen projected upon a mist layer.

At Eichstätt (literally the town of oaks), a small town of Frankish Alb, on Altmühl River, a tributary of Danube, during the Middle Ages and since 871 was celebrated Saint Valpurgi's (or Walpurga) feast, called 'Walpurgisnacht', a true romantic Sabbath, the annual orgiastic meeting of devils, spirits and witches, on Blochsberg, a hill rising above the town. Saint Valpurga was born in Sussex in 710 and died in 778 in the monastery of Heidenheim, near Eichstätt. She was Villibald's and Vunnibald's sister. On the first of May of 871 her body was transported to St. Cross' church at Eichstätt, where, from October to February, an oily substance oozes from her grave, named Saint Valpurga's oil, which protects people against witches.

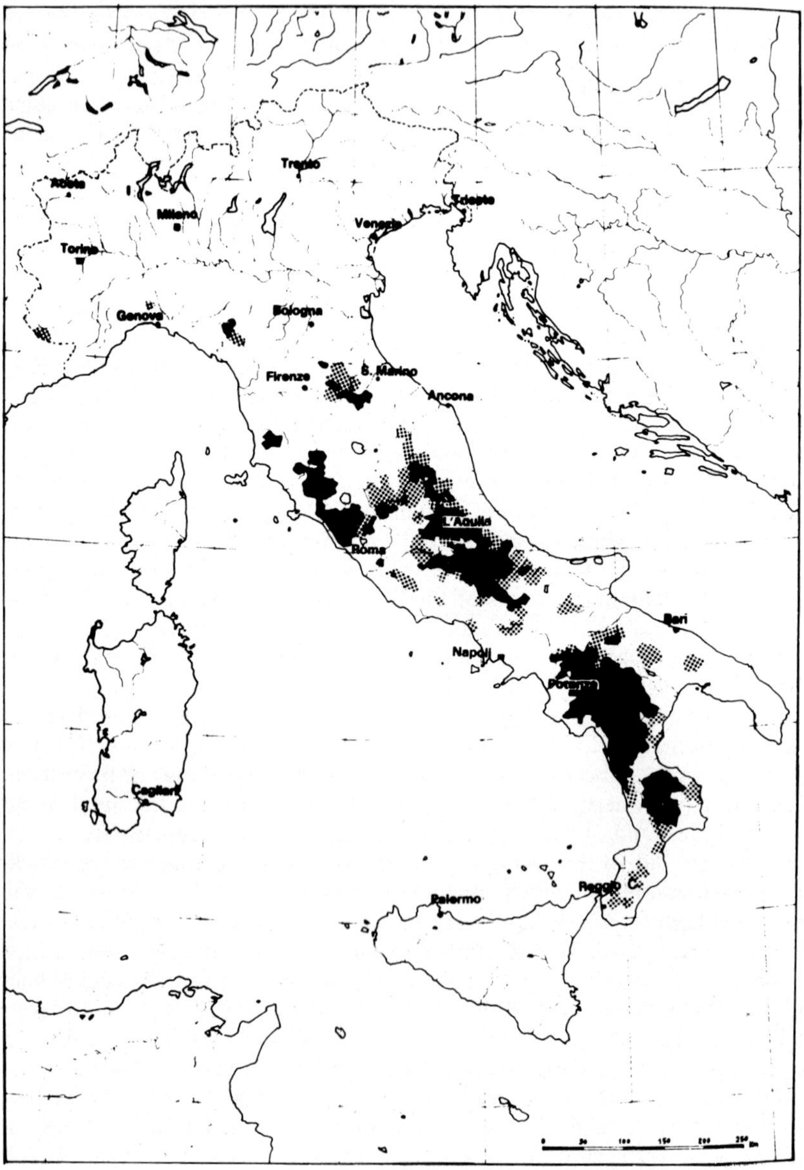

Fig. 87. The presence of the wolf (*Canis lupus* on the Appennines between 1900 and 1600 (hatched) and at present (black) from Cagnolaro et al.).

The fauna of deciduous forests

Mammals

The mammals of the deciduous forest are still relatively numerous, although in many places on the verge of extinction, as a consequence of human activities (especially in Southern and Western Europe).

To begin with the carnivores, once very widespread, we may recall the lynx, *Lynx lynx*, present, although very scattered, in the deciduous, mixed and coniferous forests of Central and Northern Europe, such as Sweden, Lapland, Poland, Romania and Russia. It disappeared from Germany in 1879; today some isolated specimen may come from the East to Zittau (Silesia) (1918). In Cadore (Dolomites) the lynx disappeared in historic times and on the Piemonte Alps and in the Langhe at the beginning of this century. They say the lynx is still present in the Western Pyrenees (Parque National des Pyrenées Occidentaux).

The wild cat *Felix silvestris* (= *F. catus*) is very rare in the mountains of Central Europe, where it should be protected. It is present, although very scattered, in the Iberian Peninsula and on the Caucasus. The otter, *Lutra lutra*, lives all over Europe, although it is very rare. It lives in passages dug either by itself or by foxes. It dives and swims with great agility but walks with difficulty. It has nocturnal habits and its food consists almost exclusively of fish, so that it can be very harmful to fish (more especially trout) farming.

The badger, *Meles meles*, digs a set with 4-8 openings on the southern slope of a hill, generally near fields. It is nocturnal, omnivorous, feeding on roots, fruits, insects, snails, earthworms and occasionally also young hares, birds and eggs. It has a hibernation which nevertheless may be interrupted (and therefore is not a true hibernation).

The polecat *Putorius putorius*, during summer lives in burrows, dug by foxes, rabbits, etc., and during winter in houses. It is nocturnal and its food consists exclusively of small vertebrates, such as hens, pigeons, and their eggs, snakes, lizards, frogs and fishes.

The so-called ferret, which some consider an albino form of polecat, once used in the hunting of rabbits, according to most recent students has originated from a Moroccan-race of polecat (*P. putorius furo*).

In the Baltic region lives another *Putorius, P. lutreola*, in German known as 'Nerz' or even 'Krebsotter', once present also in Germany, now only in the Baltic region and east of Germany, found in swamps and along water courses. In habit it is similar to otter: it feeds on crayfish and crabs, fish and fruits.

Other very common and well known carnivores, still present today in several deciduous forests, are the fox (*Vulpes vulpes*), the pine marten (*Martes martes*), beech marten (*Martes foina*) and the weasel (*Mustela nivalis*). Present here and there is the wolf, which of course is not exclusive to the deciduous forest. Today it is protected in several countries. It has an important rôle in the legends and superstitions of the old inhabitants of Central Europe, and, in Southern Europe, also of the present inhabitants. Particularly famous is the 'lycanthrope' from ἄνθροπος, man, and λύκος, wolf, i.e. half man and half wolf, which originated in the ancient popular mind, still bound to magic, since the wolf was considered an infernal or fatal being. The transformation of man into wolf happens on moonlit nights. The belief in lycanthrope still persists in some Southern Italian countries, such as Lucania (Fig. 87).

Rodents are also well represented, numerically more so than carnivores. We may remember here the interesting beaver (*Castor fiber*, from the Greek Kástor), today present in Norway, the lower Rhône valley, Poland, Russia and Germany. Here it lives only in the territory between Magdeburg and Wartenburg, near Wittenberg along the Elbe river. In 1919 there were c. 230 specimens, in 1955 c. 100, which means a diminution of the species; however, in 1962 210 specimens were present. It feeds on roots and bark; it constructs dams and dens in which it lives (the so-called beaver colonies or in German 'Burgen'). It has inspired some nice films such as 'The Beavers Valley', by Walt Disney. Once, the so-called 'castoreum' (from the Greek kastórion), a liquid secreted by the beaver from two glands in the subcutaneous sacks, opening near the urinogenital apertures, was employed in pharmacy to make perfume.

Fig. 88. The harvest mouse, *Micromys minutus*, and its distribution (from v.d. Brinck).

A Central European mouse is the harvest mouse, *Micromys minutus*, present in fields, gardens, meadows or on the edge of forests, as well as in mires. It builds up a sphaerical nest in shrubs, among grasses, sometimes among reeds. During autumn and winter it stays under hay-stacks, in hay lofts and stables. It feeds on grasses, roots and often also on small animals. In European deciduous forests (Fig. 88-89) the yellow-necked mouse *Apodemus flavicollis* can be found, reaching as far south as the Styrian Alps, above 1000 m altitude. Its habits are similar to those of *A. sylvaticus*, the common wild mouse. Furthermore, in this region another mouse can be found, *Apodemus agrarius*, or the striped field mouse, which, differing from the other two species, has more the habits of a vole, living in cultivated areas, edges of forests and during the winter under hay stacks, in hay lofts and stables. It reaches as far south as Veneto. Also in Central European forests, which have been much less affected by humans than those of Southern Europe, another interesting rodent can be found, the muskrat or Fiber (= *Ondatra zibethicus*), an animal introduced from North America and present in Bohemia

Fig. 89. The yellow necked mouse, *Apodemus flavicollis*, and its distribution (from v.d. Brinck).

in 1906, and whose habits are very like those of the beaver. It lives now in Saxony, Silesia, Oberpfalz, and also south of the Main and Danube, as well as in Northern Germany, east of the Weser. It has been found also in Italy. Sometimes it is a danger to dams because of the passages it digs under water. Its food consists mainly of plants and aquatic animals. In recent years another rodent has been introduced from South America to Central Europe, the coypu or *Myocastor coypus* (in German 'Biberratte', in Italy known as 'nutria'), because of the value of its fur. Today in Germany there are wild colonies of this rodent, in Schleswig Holstein, in Kreis Hameln, in the mountainous territory of Vogelsberg, etc. Like Ondatra, the coypu has also been found in Italy. This mammal swims like the beaver, lives in water and builds dens of earth on the banks of rivers. Because of its appearance, which is similar to that of a big water rat (*Arvicola terrestris*), but of larger size, this rodent has been exhibited in London, as a 'phenomenal product' of Londen sewers! (Fig. 90).

In the meadows and clearings in the deciduous forest, of couse, several voles can also be found: we will mentioned only one, *Microtus arvalis*, which has been

Fig. 90. The 'rat musqué' *Ondatra zibethica*, and its distribution in Europe (from v.d. Brinck).

the object of very interesting and stimulating research into ecogenetics.

Other mice belong to the genus *Pitymys,* of which we will remember two species, *P. subterraneus* (pine vole) and *P. multiplex* (Fatio's pine vole).

Two elegant, inconspicuous rodents, typical of deciduous forests, are finally the garden dormouse, *Eliomys quercinus* (literally 'bound to oaks'), widespread up to the Ural mountains, and the dormouse, *Muscardinus avellanarius* (literally 'bound to hazel') present in all Central and Eastern Europe.

Eliomys quercinus climbs trees very well; it is omnivorous but prefers small animals up to the size of a mouse; it has a long hibernation period, sometimes in treeholes, rock fissures or the dens of other mammals. *M. avellanarius* builds a spherical nest, the size of a fist, hidden in shrubs just above the ground, in which it hibernates. Another very peculiar species of rodent is the forest dormouse, *Dyromys nitedula,* which builds a round nest in trees, but hibernates in tree holes or small cavities in the soil. It is subdivided into several geographical races; let us mention here only two, *n. carpathicus*, present in Upper Silesia and Eastern Silesia, *n. intermedius,* living in Engadine, the Tyrol and Eastern Alps.

Clethrionomys glareolus is very interesting because of its geographical distribution, subdivided into many races, among them *g. glareolus* living in Central Europe, *g. nageri* in the Alps, and *g. italicus* in Alto Adige (Süd Tirol) and a couple of races in Jugoslavia. Finally, *Arvicola terrestris,* the so-called 'water rat', like moles builds long, flat, burrows in the soil and makes heaps of earth of coarse clods, often mixed with grass. It is harmful to meadows or clearings in the woods.

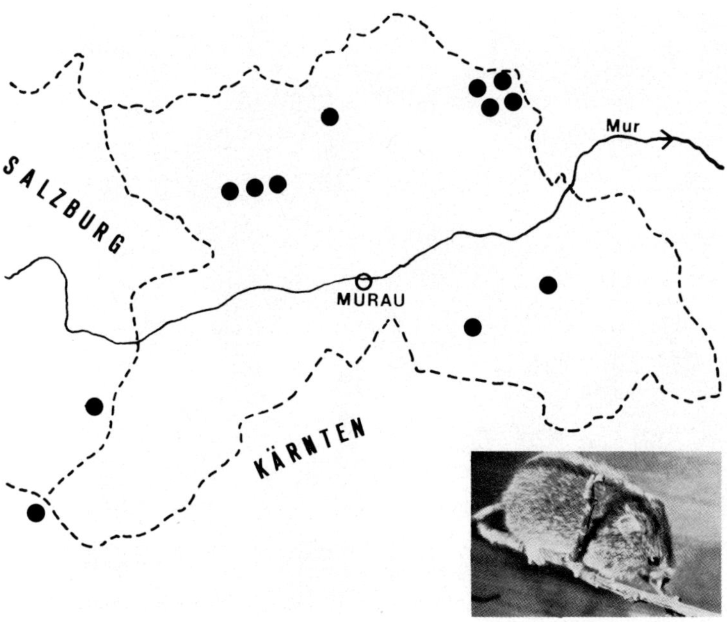

Fig. 91. New appearance of *Sicista betulina* in the Alps (Bezirk Mureau, Styria) (from Hable, 1974).

Three other small rodents present in the deciduous forests of Central Europe, though more typical of the steppes of Eastern Europe, are *Sicista betulina, Cricetus cricetus* and *Citellus citellus. Sicista betulina* or northern birch mouse is found in Eastern Prussia, Schleswig Holstein (Tolker Moor), the Bavarian Forest and Upper Silesia. It climbs, jumps and runs very fast; it lives in holes that it digs in the soil, in open places, not too dry, in fields of corn and oats, in mires, woods and meadows with scattered trees. *Cricetus cricetus* or 'hamster' is present in Central Germany, where it lives in cereal fields on silty soils or loess. It can carry cereals in its cheek pouches and also potato, turnip and cauliflower. *Citellus citellus*, or the European souslik, is present in Upper Silesia and Eastern Erzgebirge, where it lives in dry sandy, or silty, soils; it feeds on grass, fruits and insects; it has a long hibernation period (Fig. 91).

Two larger rodents in all kinds of deciduous forests (but not limited to this plant formation) are finally the *Lagomorpha Lepus capensis* (the common hare) and *Oryctolagus cuniculus* (the common rabbit). Both are the object of intense hunting in several countries and both are relatively harmful to cultivated areas, meadows, etc. The rabbit is also recorded in the Bible, where we read 'the coney because he cheweth the cud, but divideth not the hoof, he is unclean unto you' (Lev. XI, 5) and 'the conies are but a feeble folk, yet make they their houses in the rocks' (Prov. XXX, 26).

A magnificent animal still living in the ancient forests of Central Europe and the Alps is the red deer, or *Cervus elaphus*, present both in the plain and on

Fig. 92. A reconstruction of *Bos primigenius.*

mountains, where it lives up to the limit of the forest vegetation. In Virgil's Georgicae (L. III, v. 368-375) we read:

intereunt pecudes, stant circumfusa pruinis
corpora magna boum, confertoque agmine cervi
torpent mole nova et summis vix cornibus extant
hos non immissis canibus, non cassibus ullis
puniceaeve agitant pavidos foridine pinnae,
sed frustra oppositum trudentes pectore montem
comminus obtruncant ferro, graviterque rudentes
caedunt, et magno laeti clamore reportant.

An animal once very common in Central Europe, but now almost exterminated like its congeneric species in America, is the European bison, *Bison bonasus* (in German 'Wisent'). It used to live in the primitive forests (Urwälder) of Bialowicza (Lithuania) and in the preserve of Upper Silesia, protected, where today there are

Fig. 93. Data of extinction of *Bos primigenius* in the different parts of Europe. The continuous line indicates the northern limit of distribution of the species (from Lucaszewics).

maybe a few head. In Eastern Prussia the last bison disappeared in 1755, the last bison in Transylvania was shot in 1790 and the last in Saxony in 1793. At present, thanks to the 'International Association for the Protection of the Bison' there are in Europe some 160 head, of wich 25 are in Germany. It seems therefore that the species is increasing, which is very desirable since it is – like its congeneric species in North America – one of the most impressive and representative species of mammals of the mixed forest of the Northern Hemisphere (Fig. 92-93).

The wild boar is extremely widespread in Europe, not only in the dominion of the deciduous forests, although here it is maybe the most abundant. It is represented by several races whose systematics and origin are not completely clear. In South and Eastern Europe, according to some, there are indigenous forms originating from *Sus vittatus* of Southern and Eastern Asia, whereas in Central Europe there is *Sus scrofa ferus* and in Southern Europe a '*Sus mediterraneus*' or '*Sus palustris*'. Some think today that these forms are the product of a natural hybridization between European and Asiatic forms. The wild boar is often very useful since it eats an enormous quantity of insects harmful to forest species, particularly oaks, but if it enters a cultivated area it can cause very great damage. It is a gregarious animal.

The bats

Many are the bats of deciduous forests, where they can find food, shelter and where climatic conditions make their existence possible. Examples of bats are *Pipistrellus nathusii*, flying rather high up at sunset, hibernating in tree holes, rock fissures and houses; it is distributed throughout Central Europe, reaching as far south as Central Italy, *Nyctalus leisleri*, living in the forests of Central Europe; *Myotis bechsteini*, rare, flying late in the evening in woods and gardens; it hibernates in tree holes and rock fissures, spread out in Central Europa, together with the preceding species; *Myotis dasycneme*, or pond bat, rare, present in Central Europe near large water bodies; it flies only in darkness, very low near the water surface; it hibernates in rock fissures; it is particularly frequent in Silesia; the greater horse-shoe bat, *Rhinolophus ferrum equinum*, very widely distributed in Europe and Asia, flying only in darkness, hibernates in caves and old abandoned houses; the lesser horse-shoe bat, *Rhinolophus hipposiderus*, which literally means the same as ferrum equinum, from ἵππος horse and σίδερον iron. Other species are the whiskered bat *Myotis mystacinus*, extending from Europe to Indochina, found often near water, where it flies near the surface; it hibernates in rock fissures; *Myotis myotis*, or mouse eared bat, also present in Europe and Asia, flying along the roads and avenues of towns, slowly and low; it hibernates in buildings, tree holes and rock fissures; the pipistrelle, *Pipistrellus pipistrellus*, present in Europe, Asia, Morocco and maybe also on Madeira: it is the first species of bat to make its appearance in spring and the last to retire in autumn, and sometimes is active also on mild winter days. It flies from sunset to dawn, so that we must conclude that it is a very eurythermous animal; it hibernates in buildings and caves; the serotine, *Verpertilio serotinus*, from Europe to Kashmir, flies late in the evening (hence its name), very low, often in the streets of town, over gardens, etc. It has a long resting period in tree holes and buildings; the particoloured bat, *Vespertilio muri-*

nus, diffuse in Central and Eastern Europe, also in steppe areas; the common noctule, *Nyctalus noctula*, from Europe to Malaysia, flying very early in the evening before sunset; it hibernates in tree holes and buildings; the long-eared bat, *Plecotus auritus*, in Europe and Central Asia, often near houses, makes its appearance only in late spring and in the late evening, and hibernates in cellars and the like; and finally the barbastelle, *Barbastellus barbastellus*, a Euro-Caucasian-Moroccan species, rather rare, living particularly on hills, and hibernating in caves or cellars.

From all this we can conclude that all these bat species (though very similar in feeding habits) can also live in the same region since they have different 'ecological niches', being separated by phenology, or by circadian rhythm, or by micro-environment, both during active life and during dormancy, as we can appreciate from the following scheme:

Circadian rhythm				
Before sunset	sunset	twilight	darkness	dawn
N. noctula	*P. nathusii* *P. pipistrellus*	*P. pipistrellus*	*P. pipistrellus* *Myotis bechsteini* *M. dasycneme* *R. ferrum-equinum* *V. serotinus* *P. auritus*	*P. pipistrellus*
Environment during active life				
forests	forests on hills	towns (also gardens)	near water bodies	steppes
N. lesleri *M. bechsteini* *V. murinus* *P. auritus*	*B. barbast.*	*M. bechsteini* *M. myotis* *V. serotinus* *P. auritus*	*M. dasycneme* *M. mystacinus*	*V. murinus*
Micro-environment of dormancy				
houses (also aband.)	cellars	caves	tree holes	rock fissures
P. nathusii *R. ferrum eq.* *M. myotis* *P. pipistrellus* *V. serotinus* *N. noctula*	*P. auritus* *B. barbastel.*	*R. ferrum equin.* *P. pipistrellus* *B. barbastel.*	*P. nathusii* *M. bechsteini* *M. myotis* *V. serotinus* *N. noctula*	*P. nathusii* *M. bechsteini* *M. dasycneme* *M. mystacinus* *M. myotis*

All these animals are important members of the trophic cycles of deciduous forests as secondary or tertiary consumers, feeding on insects, most of which are harmful to agriculture and forestry, and they are therefore extremely useful to man and protected by law everywhere.

Fig. 94. *Plecotus auritus* (from v.d. Brinck).

Bats have two ways of reacting to unfavourable environmental conditions, either migration or dormancy. An interesting feature of migration is the fidelity with which these animals are bound to their refuges; out of 355 *Myotis myotis* marked in winter 1932-33 and re-captured the succeeding winter, only 3 had changed locality. Furthermore, a number of animals transferred at the end of the winter from their normal hibernating place to another, 150 km away, completed hibernation in the new place but the next year almost all of them went back to their customary winter quarter. This species of bat inhabits caves during winter which the individuals abandon as soon as spring arrives, when they start the migratory flight. They perform the return migration in only one or a few days. The summer quarters are situated in a semicircle, from North West towards South East, relative to the departure locality, and some 30-80 km distant from it. It is thought that bats are driven to perform these migrations in the quest for food and in the necessity of finding a shelter with a constant temperature to spend diurnal dormancy and the winter hibernation. These migrations, therefore, are quite regulated by trophic exigences, but it is not possible to exclude influences due to the sexual cycle or to climatic characteristics (Heim de Balsac, Bourlière & Grassé).

In some cases migrations are extremely short: according to Toschi and Lanza, for instance, the winter and summer quarters can be very close, for example the garret and the cellar of the same house, or, in other cases, they can be hundreds of km apart. Although there are great variations even within the same species, we may recognize quite residential species, and others which perform such regular and conspicuous movements that they assume all the characteristics of true migration.

Hibernation, which actually is more a dormancy than a true hibernation, is wide-spread within this animal group in our latitudes. Research carried out on European species, *Plecotus auritus*, has shown behaviour unique, not only in mammals, but even among all adult homoiothermous animals. Indeed, with the decrease in external temperature, the body temperature of *Plecotus* also decreases, so that it must be considered as a partially homoiothermous animal ('imperfect homoiothermous'), as is shown in Fig. 95.

Bats are also very remarkable because of their gregariousness, particularly during winter dormancy. In fact, it is easy to see in towers, belfries, cellars, caves, etc. large numbers of bats, which sometimes literally line the roof of a cave. Another interesting ecological feature of bats is their parasites; it is difficult to encounter another animal so rich in parasites as bats: – mites, fleas, unconspicuous blind apterous diptera, perfectly adapted to parasitic life on bats, etc., not to mention internal – or endoparasites (Fig. 94).

The flight of bats shows some interesting features, too, practically unique within the whole animal kingdom. One is the so-called echo-location, which consists of the perception of obstacles or flying objects in absolute darkness, which is possible due to the emission of ultra-sounds and the perception of the sonor waves reflected by surfaces still or moving nearby, similar to what happens when a boat employs radar to detect obstacles or other objects.

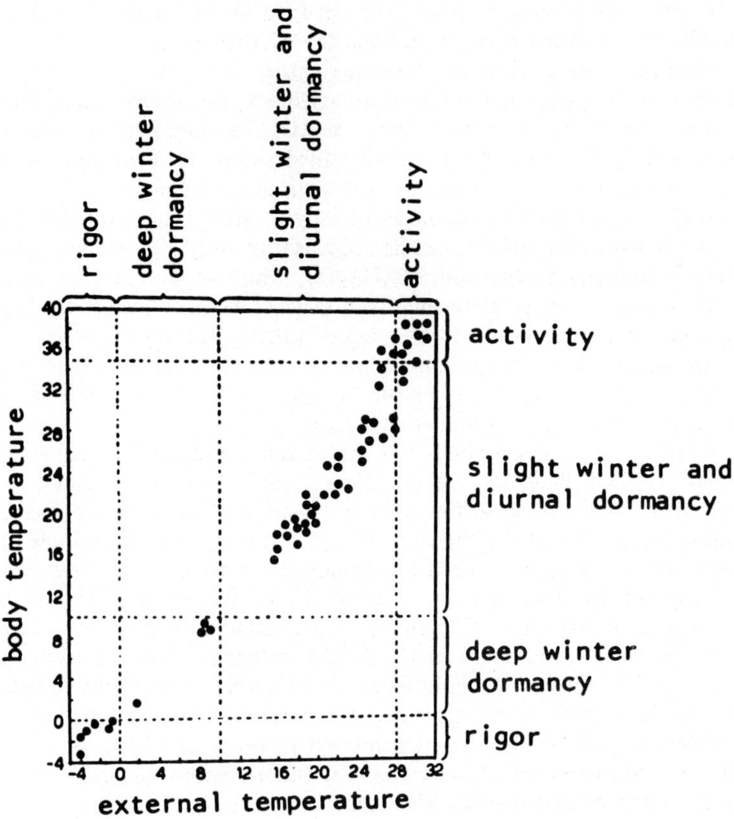

Fig. 95. Relationship between external temperature and body temperature in the bat *Plecotus auritus* (from Eisentraut).

The perfect adaption to flight, unique among all mammals and equalled only by birds, some rather peculiar habits such as gregariousnes, the ability to orientate even in complete darkness, the choice as abode of belfries, abandoned towers and castles, uninhabited houses, etc., and the crepuscular habits of bats inevitably strike the imagination not only of the most primitive men, but also of people nearer to our time or even – in some European countries – of contemporaries.

Like many other animals and plants, bats have nourished beliefs and magic practices according to which, at the same time, they served to prepare compounds with magic or medicinal properties, or even were raised to the rank of divinities (in China they are or were considered as propitious animals, and once they symbolized good fortune). More often, unfortunately, they produced in man a feeling of deep malevolence and hostility. In Italy, for instance, even today people think that if a bat gets entangled in the hair of a woman, it can be disentangled only through enormous efforts, and sometimes only by cutting the hair. In Central Italy, people still believe in vampires: in Abruzzo in 1920 people thought that bats were sent by witches and had the sad mission of tormenting them: they come to us to cause the most tedious diseases, they come down in the night through the chimney shaft, enter the rooms to attach themselves to sleeping children and to suck their blood, or to cut their faces, or to blind them (Altobello).

The insectivores of the deciduous forest are rather few: the hedgehog, or *Erinaceus europaeus*, a European-Central-Asiatic-element, the shrew, *Sorex araneus*, a eurasiatic geoelement, the white-toothed shrew, *Crocidura russula*, a European Central-Asiatic species, the lesser white-toothed shrew, *C. suaveolens*, and the common mole, *Talpa caeca*, a southern-European-anatolic element.

The hedgehog is know for its immunity not only to viper venom but also to insect venom, for instance that of the so-called Spanish fly, or blister beetle, *Lytta vesicatoria*, which contains cantharidin, $C_{10}H_{12}O_4$, which it can consume with no harm at all[1]. It is also able to withstand the attacks of enemies such as the fox or dog since it encloses itself in its strong, prickly armour. It does not resist, on the contrary, the attack of a large nocturnal bird of prey, such as the eagle owl, *Bubo bubo*. In some European countries the hedgehog is hunted because of its savoury meat, using tramp-dogs or other special dogs (bloodhounds).

Many superstitions have arisen about this useful and harmless little mammal. People say that the hedgehog, like other animals, sucks milk from cows. Its dried flesh has been employed against many diseases, such as incontinence (recipe due to Pedanius Dioscorides of the 1st century A.D.). It is a probable example of figurative magic, since it is known that the hedgehog, when attacked, urinates and gets all wet. Even in 1920 in Abruzzo (Central Italy), hog's bile dissolved in sugared water was administered to children affected by abdominal pain: evidently in this country people did not know that bile is a culture medium for bacteria localized in the gut, like those of typhoid fever and paratyphus, and that bile salts are toxic. We may not exclude the idea that this was also a case of figurative magic, in consideration of the close relationship between bile and belly. Amulets – made with hair and mandibles of hedgehog – infallible against general inflammation and particularly against mastitis, also come from Abruzzo.

[1] The use of this substance as an aphrodisiac can cause nephritis (Theodorides, 1950).

Strangely enough, shrews have also striken the popular imagination since earliest times; the ancient Egyptians embalmed and buried them with mummies. The bite and contact with these inconspicuous harmless creatures has always been considered with repulsion and horror, but at the same time the use of shrews does not cease in popular medicine. Its burnt flesh was employed in the treatment of ulcerations and sores, while the tail, reduced to ash, was considered a valid treatment against the bite of rabid dogs.

The mole has also been the object of myths and superstitions: according to an Abruzzo legend, the mole gave its eyes to the toad in exchange for its tail. In popular medicine the mole has always had great success: since earliest times its blood, fat and skin were employed in the care of women's beauty. It has also been applied in the therapy of conjunctivitis and of erysipelas, an infectious disease sometimes very serious! In Abruzzo even recently heart patients have been given mole flesh.

All these beliefs show that, in the past, man felt as a stranger in the face of nature, or superior to it, or an enemy. In many cases we find true magic practices and more especially applications or figurative and talisman magic. Nature offers us many 'similitudes' and the magician is able to take note of them to his benefit. Thus, ancient Egyptians prepared an amulet, which was applied to the feet of gout patients, with skin of red deer, an animal particularly fast at running; ointments made with the blood of a black calf or the horns of a black bull served to prevent one growing grey-haired. (We must not forget that all old Egyptians had black hair) (M. Bouisson). Ancient peoples thought that the forces of the external world must be turned to man's advantage: in a sense even the Arab conception of zoology – though very advanced from a scientific point of view – was not far from this position. Hāggi Khalīfa, who died in 1658, says, that 'the aim of zoology is to reach a medical treatment and to look for a benefit from animals, or to avoid the damage they can cause, as well as to learn how to know their admirable characteristics and their extraordinary actions'. But we must not forget – to understand and justify all these legends and superstitions about our animals, including the most beautiful and interesting – that not only during the Middle Age, but even during the first part of the Renaissance (up to the 17th century) the Church, which at that time was the only depository of philosophical and theological knowledge, prevented in every way the development of biology (and therefore of ecology), while it allowed physics, mathematics and astronomy to develop.

The totemism we can observe in some European peoples, both primitive and present, is also essentially an ecological phenomenon, since it is inherent in the relations man-animals or man-plants. Totemism also has religious or magic characteristics: thus we find the animal-god or the animal-man or animal-group. In this case man cannot eat the meat of the totemic animal, which practically represents his 'double' or, as we might say in modern terminology, his stand-in or alter ego.

Even today in Lucania (Southern Italy) there are some conceptions we could consider as totemistic, according to which persons identify themselves with an animal: as Levi tells us in his 'Cristo si è fermato ad Eboli', a woman was convinced she was the daughter of a cow; this belief did not prevent her from admitting she had also a woman mother.

Very few mammals are exclusive to sciaphilous forests, partly due to the fact that these animals are rather euryoikous and have a great facility of movement, partly because very few feed exclusively on beech, and partly because this tree is not limited necessarily to sciaphilous forests, but is present, although isolated, throughout most of Europe, both in heliophilous and in mixed forests.

We may point to some species whose area of distribution corresponds more or less to that of sciaphilous forests, for instance Daubenton's bat, *Myotis daubentoni*, one of the most frequent bats in Germany, flying immediately after sunset near the surface of water, which is also found sometimes in mountains; the dwarf weasel ('Zwergwiesel') *Mustela minuta*, according to some only a dwarf variety of the common weasel *Mustela nivalis*, limited to Western Germany and the Alps (also on the southern Italian slopes). Its size is 18-24 cm in the male, 16-21 in the female, except the tail, more or less 4 cm long. Two rodents with amphibious habits are also limited to the area of beech vegetation, namely Central Europe and the Alps: one is *Arvicola italicus*, exclusive to Italian Switzerland, Italy and Croatia; and *A. terrestris*, or ground vole (once called *A. scherman*), present in Central Europe and the Alps, except the Italian slope. The systematics of this rodent group are, however, not definite. The latter species lives both in water (where it exhibits amphibious habits) and in dry places. It is harmful to gardens because it digs in the earth; it is practically omnivorous. This is the 'water rat' of which Wilde speaks in his tale 'A devoted friend', where the rat represents an insensitive and selfish person.

Maybe the most typical mammal of the sciaphilous forests is the fat dormouse, *Glis glis*, living in forests, gardens and other similar environments and very rarely in coniferous forests. It thrives on trees or in shrubs, and is found from the plain up to 800 m or exceptionally 1500 m. It is a nocturnal creature, very restless, which can make remarkable leaps and can also climb on very smooth surfaces. During autumn it often enters barns of country houses. It has rather gregarious habits, though it cannot be considered as a social animal. The nest is made with mosses and plant fibres, generally in a hole in a tree, rock or well. It hibernates in soil up to 60 cm deep, but sometimes also in cellars. It has been introduced into England.

The birds

Although highly reduced compared with the original deciduous forests which covered most of Western and Central Europe, the present deciduous forest still constitutes a good environment for bird life. Practically all orders of European birds can be found within it, although sometimes as migrant or accidental birds.

Limiting ourselves to true terrestrial species, we may mention the hen harrier *Circus cyaneus*, a bird of prey found in most of Eurasia. In some places it is resident or even migrant (double passage). It can be harmful to game. The peregrine, *Falco peregrinus*, is also very common all through European deciduous forests; it is generally migrant and very harmful to game as a consequence of its dietary preferences (birds, including loft pigeons). Two other species of Falco are *F. vespertinus* (red-footed falcon) and *F. tinnunculus* (kestrel). The former is limited to Central and Eastern Russia, the latter extends all through Eurasia. The kestrel may be in some places resident or migrant. Two other diurnal birds of prey (or Falconiformes) are the goshawk *Accipiter gentilis* and the sparrow hawk, *Accipiter nisus.* They feed on animals as big as *Tetrao urogallus* or young hares. (*A. gentilis*) or, in the case of *A. nisus*, partridges, pheasants and several other vertebrates. Both of them inhabit most of Eurasia. Other birds of prey of the deciduous forest are the buzzard, *Buteo buteo*, the kite, or *Milvus milvus*, and the black kite, *Milvus migrans.*

The gallinaceous birds of the deciduous forests are few: let us mention here the red-legged partridge, *Alectoris rufa*, the rock partridge *A. graeca*, the common partridge *Perdix perdix*, the pheasant *Phasianus colchicus* and lastly the quail, *Coturnix coturnix*. The latter is well known because of the cryptic colour of the young. We want to call attention to the geographical distribution of two gallinaceous birds, *Alectoris graeca* and *Perdix perdix.*

Some Columbiform birds are also frequent: one is the stock dove, *Columba oenas*, present in the western part of Eurasia, in some places both resident or migrant; the rock dove, *Columba livia*, living in rocky places, such as caves and the like, whence its name; the wood pigeon, *Columba palumbus*, living and nesting on trees, and lastly the turtle dove, *Streptopelia turtur.* All of them are more or less harmful in fields and in general to cultivated areas.

A very peculiar bird, although not exclusive to deciduous forests, is the cuckoo, *Cuculus canorus*, which is named after its song, in practically every part of Europe (in German Kuckuck, in Italy cuculo, in French Coucou, in Dutch koekoek, etc.). In general, it is a migrant bird, nesting in Central-Northern Europe. Its most remarkable peculiarity is so-called nest-parasitism. The female, indeed, lays its eggs in the nest of another species and leaves them in the care of the female of this species. In the case of the cuckoo, there is furthermore an interesting case of colour polymorphism by which the eggs can be in various colours and each female lays in the nest of a species whose eggs are the same colour as its own. The means by which the cuckoo female, although laying eggs of different colours, can choose correctly the nest in which to lay, is the imprinting to which it has been subjected at birth (i.e. it knows the colours of its own eggs!). Not only that, but the young cuckoo, directly out of the egg, ejects from the nest the young nestlings of its adoptive parents, thus having at its disposal all the food otherwise destined for its step-brothers.

The nocturnal birds of pray or Strigiformes of the European deciduous forest are quite numerous: the tawny owl, *Strix aluco*, the long-eared owl, *Asio otus*, the short-eared owl, *Asio flammeus*, the barn owl, *Tyto alba*, the little owl, *Athene noctua* (source of myths and legends since the era of ancient Greece) and lastly the eagle owl, *Bubo bubo*. The latter is protected by law in several European countries.

Other birds present all through Europe, and therefore also in its deciduous forests, are the nightjar, *Caprimulgus europaeus*, a silvicolous nocturnal creature feeding mainly upon moths, and the swift, *Apus apus* (or *Micropus*, from *μικρός* and *πους* apus, from *α* and *πους*). Other non-passerine birds are the roller, *Coracias garrulus*, the king-fisher, *Alcedo atthis* (a piscivorous bird) and the hoopoe, *Upupa epops*, also named after its song in other European languages (huppe puput in French, hop in Dutch, etc.). This species – which is present all through Europe, Africa and part of Asia – was remembered by Foscolo in his 'Sepolcri', for the

Fig. 96. *Upupa epops.*

belief that it can live in cemeteries. More interesting, perhaps, are the woodpeckers, such as the green woodpecker, *Picus viridis*, found in Europe, Asia Minor and Iran, or the grey-headed woodpecker, *Picus canus*, or the great spotted woodpecker, *Dendrocopus major*. Similar to woodpeckers is the wryneck, *Iynx torquilla*, a migrant bird.

Of course, most birds belong to the Passeriformes. To begin with, let us mention the lark, *Alauda arvensis*, harmful to agriculture since it feeds on seeds (including wheat) and leaves of cultivated species. The wood lark, *Lullula arborea*, a migrant which estivates in the taiga (see relevant chapter) and hibernates in South Europe, North Africa and Palestine. The Hirundinidae are also present in most deciduous forests, although they sometimes prefer open places or even villages and towns: the swallow, *Hirundo rustica* (quoted by Oscar Wilde in his 'The Happy Prince'), the house martin, *Delichon urbica* (in German 'Hausschwalbe') and the sand martin, *Riparia riparia*. The Corvidae are also widely distributed within the European deciduous forest, although sometimes they are restricted to deforested

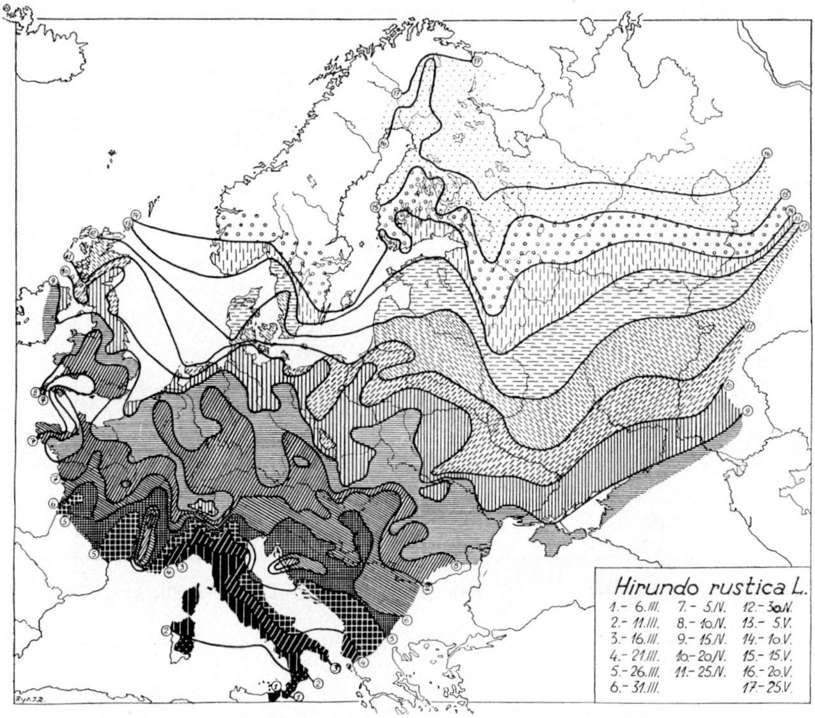

Fig. 97. Date of appearance of the swallow (*Hirundo rustica*) in the various regions of Europe, with particular emphasis for the zones once belonging to the biome of the deciduous forests (Central Europe, NW Europe and Southern Scandinavia) (from Sliwinsky).

areas or even cultivated ones: the raven, *Corvux corax* (which inspired Poe's famous poem 'The Raven'), generally a resident bird, nests in trees and on rocks, in inaccessible places. Its nest is a rough, solid construction, built with branches, roots and mud, inside lined with wool, down, feathers, etc. It is an omnivorous, or, more exactly, a carnivorous animal, sometimes very harmful to game. Two species which offer particular geographical interest are the crows, the carrion crow, *Corvus corone*, and the hooded crow, *Corvus cornix*. They are practically allopatric, but with an area of overlap where crosses between the two species are present. It is worthwhile noting that this area of crossing was demonstrated for the first time in 1887 and that by 1928 there had been no modification of the same. Other Corvidae are the rook, *Corvus frugilegus* and the jackdaw, *Coloeus monedula*, the jay *Garrulus glandarius*, strictly a forest bird, omnivorous and on the whole harmful, and lastly the magpie, *Pica pica*, a holoarctic element. It is extremely harmful to agriculture as well as to game, since it is very ravenous and eats everything it can catch, even young birds and hares.

Small passerine birds of the deciduous forests are the crested tit, *Parus cristatus*, spread all through Europe; the short-toed treecreeper *Certhia brachydactyla* (in German 'Gartenbaumläufer', which literally means running on the trees), a

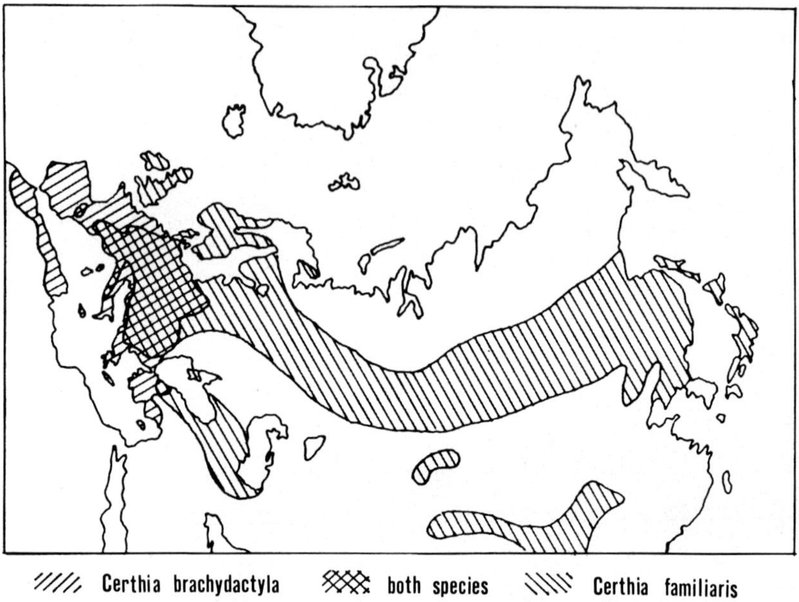

Fig. 98a. Distribution of *Certhia brachydactyla, C. familiaris* and area in which both species live (from Reinig, modified).

Fig. 98b. The retreat of forests in Eurasia during the Glacial Period (from Reinig, modified).

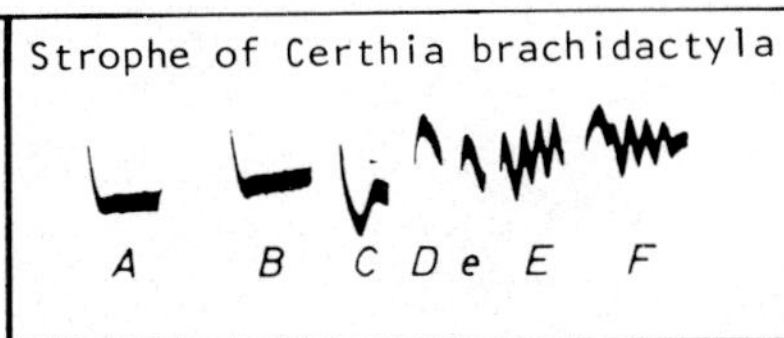

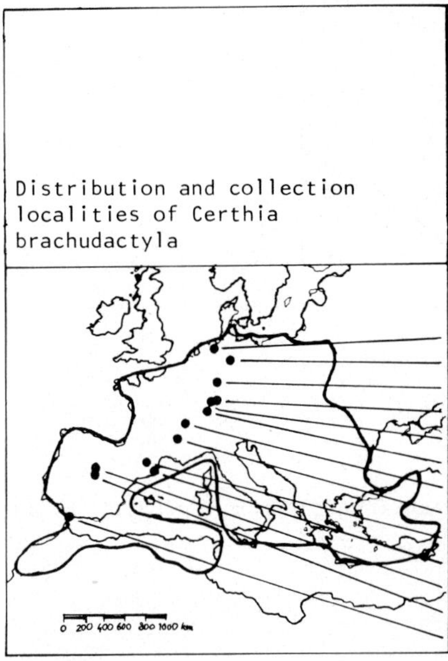

Locality	n	Types of F element 1	2	3	absent
Oldenburg	23	23			
Braunschweig	6	6			
Gießen	2	2			
Ludwigsburg	19	19			
Rastatt	7	7			
Freiburg	26	26			
Lyon	5	5			
Valence	10	9			1
Puigcerda	3	3			
Gerona	8	5			2
La Granja	25	1		23	1(1)
San Rafael	5				5
Jimena	14		14		

Fig. 99. Differences in the song of the Passeriform *Certhia brachydactyla*. Here again, as with *Aegithalos caudatus*, the Spanish population exhibits a different song type (especially in the F element) (from Thielcke).

resident, rarely migrant bird. It has evolved into two different physiological races, recognizable only by their song, namely a western, Iberian one, and a central-European race. Similar in habits is the nuthatch *Sitta europaea* (in German 'Spechtmeise', referring to its habits of climbing on trees like woodpeckers). Let us remember also the wren, *Troglodytes troglodytes*, inhabiting Europe, North Africa and Asia Minor, often associated with the robin of which it is a kind of miniature. It is entomophagous and therefore useful to agriculture. Very common and well known are the thrushes, and more particularly the field fare, *Turdus pilaris*, a migrant, *T. viscivorus* or mistlethrush (in German 'Misteldrossel') sometimes present in the albino form; the song thrush *T. philomelos* (= *ericetorum*), a migrant, the redwing *T. musicus* (= *iliacus*) in German 'Rotdrossel', referring to the colour of the wings, and lastly the common blackbird, *T. merula*, in some places migrant, in others resident. Since thrushes feed upon insects, spiders, centipedes, as well as small fruits from wild plants, they are on the whole useful to agriculture and forestry (Fig. 98-99).

One of the most beautiful European passerine birds is the blue throat, *Cyanecula wolfii* (= *Cyanosylvia svecica*) a migrant, nesting in Central Europe, generally in humid sites (ponds, swamps and the like). It is a true 'song bird' . Other species common but not exclusive to the deciduous forest biome are the stonechat *Saxicola torquata*, the orphean warbler, *Sylvia orphaea* (= *hortensis*), in German 'Sängergrasmücke', besides the congeneric species *atricapilla*, *communis, borin* and

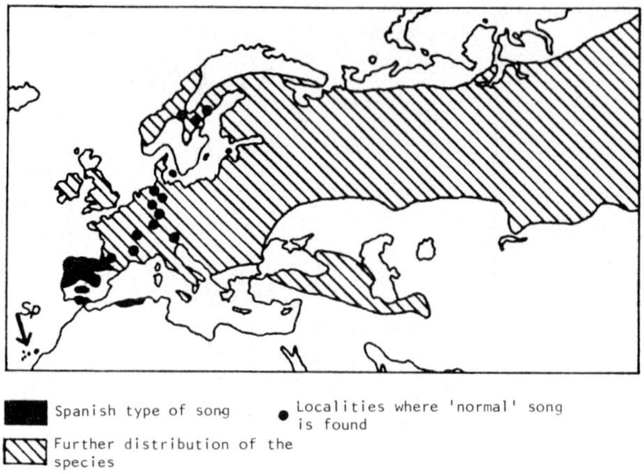

Fig. 100a. Distribution of the different types of song (true 'physiological' races) in the chiff chaff *Phylloscopus collybita* (from Thielcke).

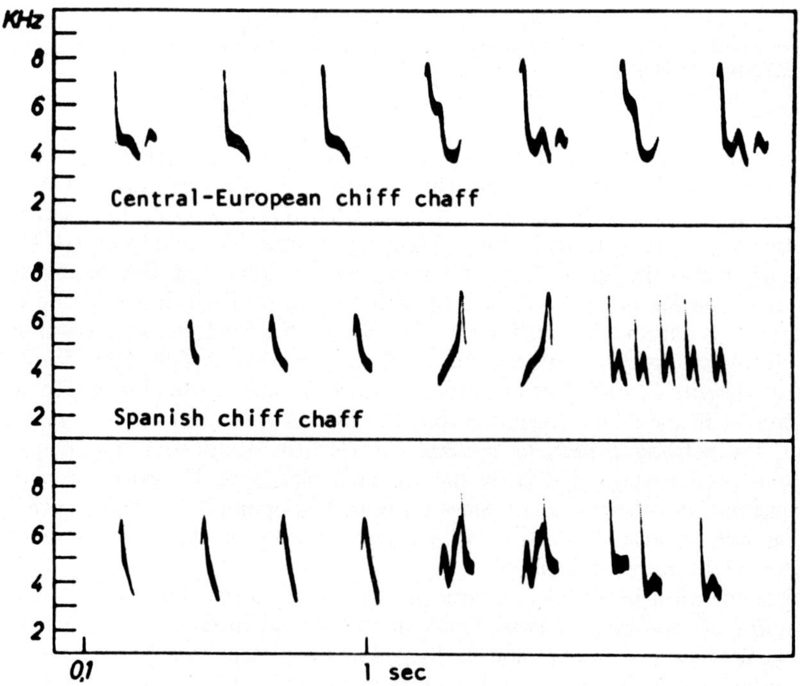

Fig. 100b. Differences in the song of the centro-European and Spanish populations of the chiff chaff *Phylloscopus collybita* (from Thielcke).

shape, size and feeding habits are the warblers, such as the willow warbler, *Phylloscopus trochilus*, the wood warbler, *P. sibilatrix*, the chiff-chaff *P. collybita* and Bonelli's warbler, *P. bonellii.* All these birds are useful to agriculture due to their feeding habits. A peculiar feature of *P. collybita* is the presence of a physiological race, shown by the song of the populations inhabiting the Iberian peninsula. This difference could theoretically lead to a species in statu nascendi, since the different song could prevent the reproduction between the Iberian populations and the Central European ones, regardless of the difficulty of their meeting due to mere geographical reasons (Fig. 100).

Other species are the hedge sparrow, *Prunella modularis* (in German 'Heckenbraunelle'), the pipits, meadow pipit, *Anthus pratensis*, tree pipit, *Anthus trivialis*, and rock pipit, *Anthus spinoletta*. They are also insectivorous birds.

The shrikes, the red-backed shrike, *Lanius collurio*, the woodchat shrike, *L. senator*, the lesser grey shrike, *L. minor*, and the great grey shrike, *L. excubitor* are among the most beautiful passerine birds in Europe. They are often limited to the deciduous forests of Central or Southern Europe and are all insectivorous. Also interesting are the species of the genus *Emberiza*, such as *E. citrinella* or yellow hammer, present in most Europe, *E. cia* and *E. cirlus* present in the south, and in a sense a vicariant species relative to the former. Well known and widely distributed are species belonging to the genus *Carduelis*, such as *C. carduelis*, resident or migrant, the green finch *Chloris chloris*, subdivided into several geographical races, such as *C. c. chloris*, a northern form, and *C. c. mühlei*, a southern element; the linnet *Carduelis cannabina*, resident or migrant and others. The serin, *Serinus s. serinus*, once known as *Serinus canaria*, typical of the deciduous forests of Western Europe, has shown in recent years a tendency to migrate eastwards (Fig. 101-103). To finish, let us mention a beautifully coloured creature, the hawfinch *Coccothraustes coccothraustes* (in German Kernbeisser), a migrant feeding on seeds with a hard shell (whence the German name), such as the seeds of beech, coniferous trees, *Taxus*, etc.

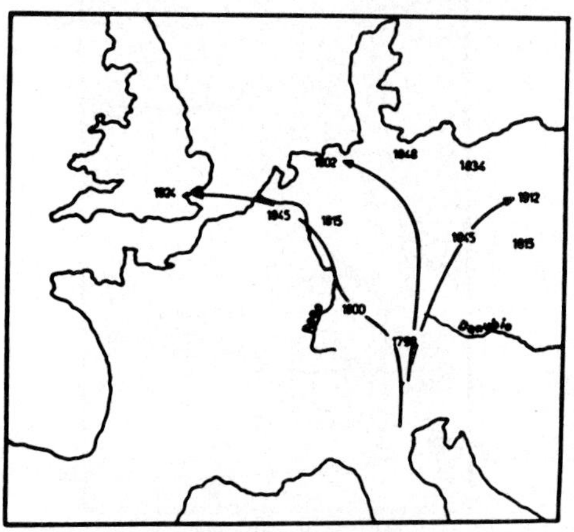

Fig. 101. Example of recent dispersion of an animal population in the deciduous forest, the Passerine bird *Phoenicurus ochrurus gibraltariensis* or *Ruticilla titys* (constructed on the basis of data by Schweiger).

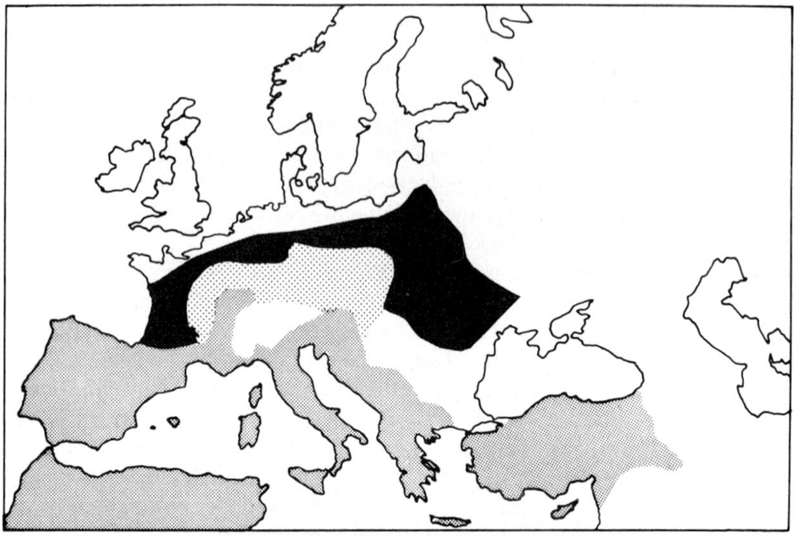

Fig. 102. Extension of the distribution area of *Serinus serinus* in the deciduous forests of Central Europe from 1800 (Southern Europe) to 1875 (all around the Alps) up to 1925 (centro-eastern Europe) (from Timofeef-Ressowsky, greatly modified).

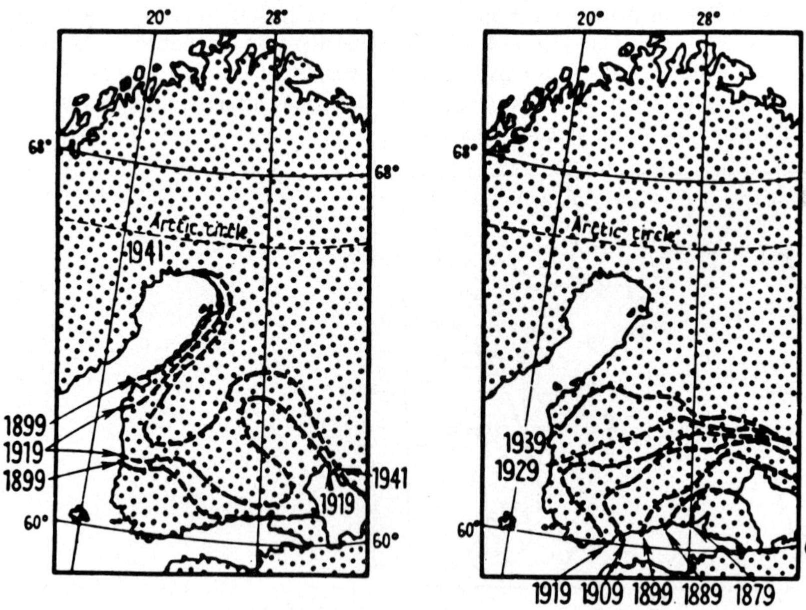

Fig. 103. Examples of invasions of Northern Europe by mammals and birds in present times. Left: the lapwing *Vanellus vanellus.* Right: the pole-cat *Putorius putorius* (from Kalela).

Reptiles and amphibians

The reptiles of deciduous forests are not very numerous, especially if compared with those of sclerophyllous forests. Let us recall *Lacerta agilis*, a species characteristic of Central Europe, thriving in dry, sunny places, from North and Central Europe to Central Asia. It is a 'Rassenkreis' subdivided into at least 6 subspecies.

The amphibians are, on the contrary, rather numerous. The most remarkable species are *Bufo calamita*, living in fields, dunes, and also on the small islands of the North Sea, from the Iberian Peninsula to the Western USSR; *Alytes obstetricans*, living in loose stones, in rock fissures, etc., present in the western part of Central and Southern Europe, and maybe also in Morocco; in Germany it is found in western and southern regions, up to Sauerland, Harz, Thuringia and Baden. *Pelobates fuscus* lives in sandy soils, generally buried during the day, and found from Central Europe to Western Asia. Finally, let us mention *Rana arvalis*, or the mire frog, living in humid, peaty meadows, a 'Rassenkreis' extending from North and Central Europe to Siberia, subdivided into three geographical races.

The invertebrates

Soil life in deciduous forests

In the deciduous forests of Central Europe (heliophilous, sciaphilous and mixed forests) soil life has been studied more than in all the other districts where deciduous forests are present, even more than in all other European biomes. Fig. 104 illustrates graphically the role of this life in soil.

The composition of soil flora and fauna, the number of specimens per volume or area, i.e. density, the biomass (and indirectly the production or productivity, both primary and secondary), the turnover of several organic and inorganic sub-

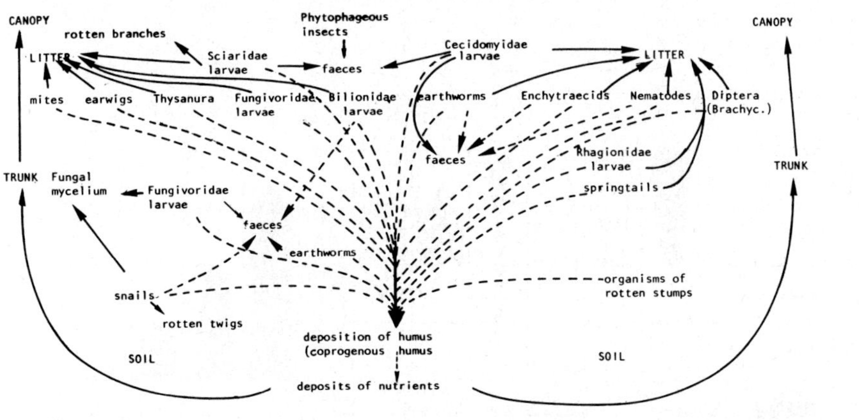

Fig. 104. Schematic representation of 'matter cycle' in a deciduous forest, mainly referring to primary degraders (inspired by Brauns, 1959). Continuous line: attacked by; broken line: elimination of excrements.

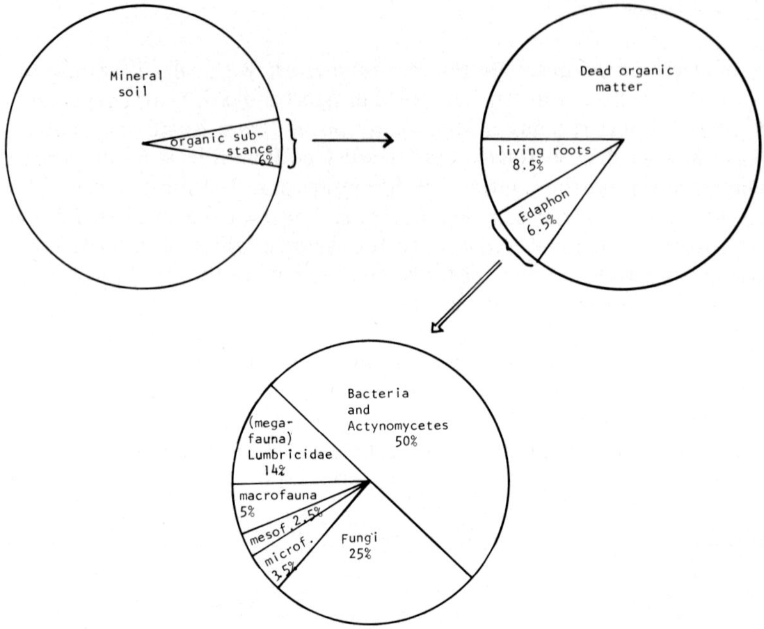

Fig. 105. Biomass of soil organisms in the upper layer of a deciduous-mixed forest with a humus of Mull type (from Dunger).

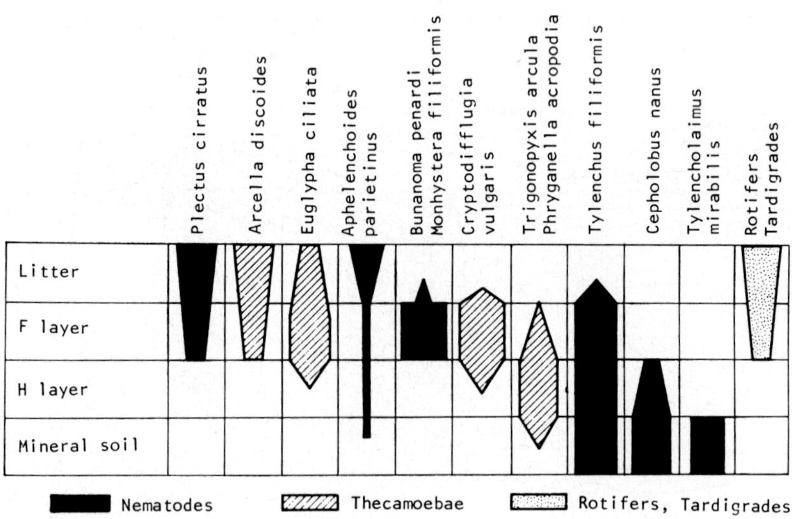

Fig. 106. Vertical distribution of some elements of the soil microfauna in a beech forest (from Brauns) (published with the permission of G. Fischer Verl.).

stances or elements in the soil, caused by soil organisms, and the relationships between soil life and soil fertility have all been thoroughly investigated by a number of authors (Brauns, Schaller, Dunger, Zachariae and others in Germany, Franz & Kühnelt in Austria, Van der Drift in the Netherlands, C.O. Nielsen in Denmark, Russel, Burges, Raw, Hale, Heal, Satchell, MacFadyen, Heath, Bocock, Edwards and many others in the British Isles, to quote only a few).

Among soil organisms let us remember bacteria, viruses, fungi and algae (Fig. 105). The fauna is often subdivided – although only conventionally – into micro-, macro- and megafauna. Examples of microfauna are shown in Fig. 106. The number of Protozoa is very difficult to estimate: Varga found 25000 Protozoa in 1 g of fresh forest soil, Nikoljuk 100,000 to 1,000,000. Franz considers as an average value for 1 g of soil 100,000 flagellates, 50,000 rhizopods and 1000 ciliates, but Stöckli speaks of 1,000,000 flagellates, 500,000 rhizopoda and 1000 ciliates.

The biomass of the *Thecamoebae*, a very interesting group of soil Rhizopods, is presented in the following table (in Pfalz):

Sciaphilous forest (*Fagus*) Raw humus		Heliophilous forest (Oak) Mull	
Litter	16 mg	Litter	51 mg
Raw humus, 2 cm depth	237 mg	Moder	21 mg
Mineral soil (up to 10 cm)	105 mg	Mull (up to 20 cm)	355 mg

The number of rotifers has been estimated in a very rich soil up to 600,000 specimens per square metre.

Number and weight of Nematodes are represented by the following figures (in Switzerland):

Ecosystem	Region	No. of Nematodes/m² (first 10 cm depth)	Weight of Nematodes (g/m²)
Deciduous forest	Mittelland	6,000,000	6.0
Mixed forest	Mittelland	4,100,000	4.1

The feeding habits of Nematodes are extremely varied, as one can appreciate from the following scheme (Dunger):

Soil type	% in weight of Nematodes classified according to their feeding habits				
	Bacteria	Roots	Algae	Small animals	unknown diet
constantly wet soil	14	6	71	8	2
normal soil	33	12	43	5	4
extrem. dry soil	7	2	88	1	1
raw humus	21	11	57	8	3
mosses	62	8	13	15	2

The following table gives an approximate idea of the number of specimens of each zoological group per dm³ of soil (Schaller):

Protozoa	1,000,000,000
Rotifers and Tardigrada	500
Nematodes	30,000
Collembolans	1,000
Acarina	2,000
Remaining arthropods	200
Enchytraeidae	50
Lumbricids	2

As far as the biomass of soil animals and therefore the secondary productivity (terrestrial productivity) is concerned, we may quote the following figures (g/m^2) valid for the forests of Pfalz (South Western Germany):

Sciaphilous forest (*Fagus*)	Lumbricids	Coleopterous larvae	Nematodes	Micro-arthropods	Thecamoebae
Total	2.015	8.605	4.052	1.817	0.873
Litter	0.075		0.048	0.935	0.010
raw humus (1–2 cm depth)	1.320		0.535	0.647	0.297
mineral soil (up to 25 cm)	0.620	8.605	3.469	0.235	0.566
Heliophilous forest (*Quercus*)					
Total	28.795		15.154	2.769	0.583
Litter	0.432		0.062	0.182	0.032
Loose soil	1.873		0.352	1.177	0.021
Mull (up to 25 cm)	26.490		14.740	1.350	0.530

The role of soil life in a deciduous forest is illustrated by Fig. 104. Here one can appreciate the richness of forms living in equilibrium with one another, exploiting all existing niches within the forest, each of them belonging to a given trophic level. Namely, we find 1) producers (i.e. autotrophic plants); 2) primary consumers or herbivorous animals, among which several are radicicolous (such as Nematodes, scale insects and Aphids, wireworms, Noctuidae larvae, Coleoptera larvae, Symphyla etc.). Sometimes they are very small, such as oribatid mites, springtails and Protura; 3) Secondary consumers, i.e. carnivores, feeding on herbivorous animals such as spiders, scorpions, centipedes, insects (both larval and adult) etc. Sometimes these animals can be very small such as, for instance, Protozoa, Nematodes, gamasid mites, pseudoscorpions and many micro-insects; 4) Tertiary consumers, i.e. carnivores feeding on carnivorous animals; and finally 5) saprophages or transformers. These can be subdivided in coprophagous, necrophagous, lignivorous, fungivorous, etc. Often saprophagous animals are divided – although only for convenience – into primary decomposers, or macrohumiphagous, and secondary decomposers, or microhumiphagous, according to the kind of substance they attack and the degree of decomposition of the same.

To the former belong insect larvae, millipedes, woodlice, earthworms and snails; to the latter oribatid mites, springtails, *Enchytraeidae* and many insect larvae (e.g. *Tipula*). The scheme in Fig. 104 represents in a synthetic way the main phases of the attack on litter by decomposers or transformers.

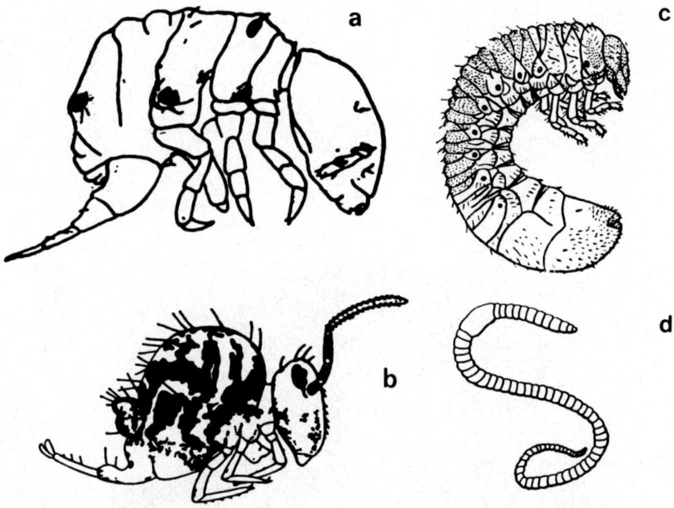

Fig. 107. Examples of soil meso- and macrofauna of deciduous forests. a. *Arthropleona* springtail; b. *Simphipleona* springtail; c. Scarab larva or grub; d. earthworm or *Lumbricidae.*

The attack on a beech stump by degraders is illustrated in Fig. 108.

Very well structured also are the faeces (or coprolithes) of many decomposers.

In the following table one can see the different rates at which leaves are decomposed, together with the C/N ratio of the same, which of course may be more or less the C/N values of humus or of its precursory substances.

kind of leaf	Duration of composition in years	C/N of leaf
Alnus nigra	1	15
Fraxinus excelsior	1	21
Ulmus campestris	1	28
Padus avium	1.5	22
Carpinus betulus	1.5	23
Tilia (linden)	2	37
Hazel	2	52
Quercus (oak)	2.5	47
Quercus pedunculata	–	36-39 (original data)
Castanea sativa (chestnut)	–	29-34 (original data)
Birch	2.5	50
Populus tremula	2.5	63
Beech	3	51
'Roteiche' (introduced)	3	53

An animal group which belongs to the soil and litter fauna and which generally takes part in the decomposition of litter is the molluscs or snails, a group which in the deciduous forest often finds optimal conditions ('Lebensbedingungen'). Particularly favourable are the sciaphilous forests of Central and Western Europe. Let us recall here only *Acicula fusca*, *A. polita* and *A. gracilis*, living in alluvial soils and

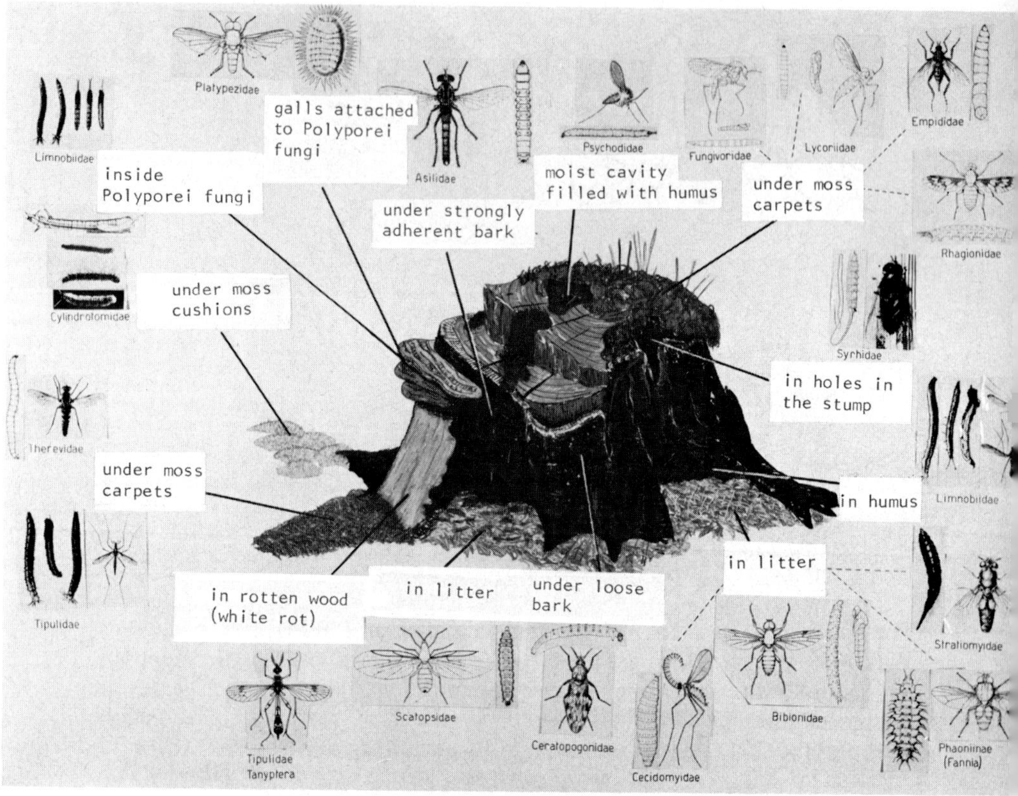

Fig. 108. The demolitive attack of a stump by *Diptera* larvae (inspired by Brauns; drawing by C. Friso; photos by U. Arezzini).

in small fissures in rocks, typical soil animals living in Germany since the Upper Miocene; *Azeka menkeana*, a small snail living in humid places, under stones, mosses, sometimes near the water; this genus was more abundant in the Tertiary, the Miocene species being similar to the recent ones. *Truncatellina costulata*, limited to Northern and Central Europe, south east up to the Caucasus, lives in relatively dry places, on rocks (generally calcareous), on the ground etc. It can be considered a soil animal; the genus has been present in Germany since the Upper Oligocene. *Vertigo alpestris*, in Central Europe, is a glacial relict, living generally in mountains or hills, e.g. in the Alps, Jura, Switzerland; in France it is found in the alluvium of the Rhône, north of Lyon. *Pagodulina sparsa* is found in shadowy places, under stones, mosses and litter, and therefore is a true soil animal; *Pupilla bigranata, P. sterri* and *P. triplicata* live on soil, under stones but also on plants, feeding probably on both fresh and decaying vegetation. *Vallonia alemannica* and *V. adela* live in the litter, under stones, on roots, sometimes near ants nests, and are both very important in the transformation of dead matter into humus; *Acanthinula aculeata* (Fig. 109a) is found in forests or more rarely on prairies, but not in wet places; it can live up to 1500 m in the Alps; it is found as a fossil since the Middle Pliocene, although the genus is known from the Oligocene; very similar to an *Acanthinula* is *Spermodea lamellata*, practically exclusive to beech forests of Sweden, Denmark, Baltic Germany, England and Ireland, known from the Holo-

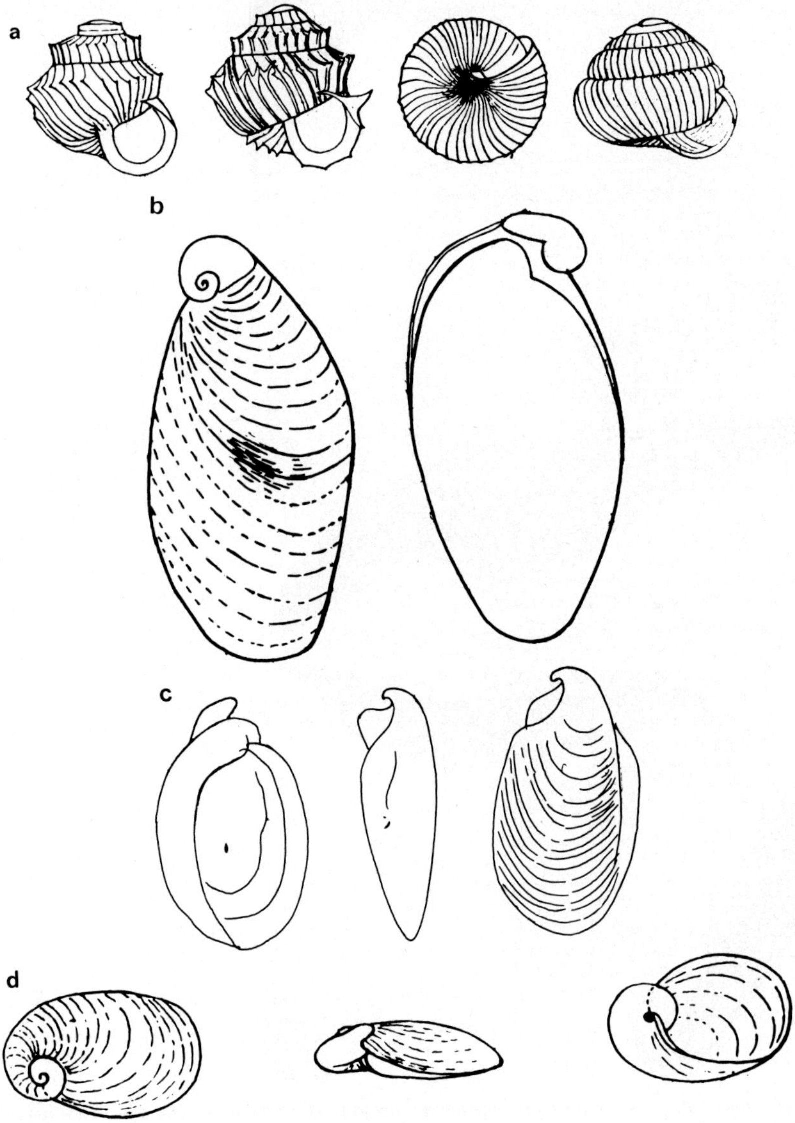

Fig. 109 a. A characteristic mollusc belonging to the soil fauna, *Acanthinula aculeata* (from Germain). b-d. Two typical families belonging to the soil fauna, *Parmacellidae* (b) and *Testacellidae* (c, d): b. *Parmacella moquini*; c. *Testacella pascali*; d. *Daudebardia brevipes* (from Germain).

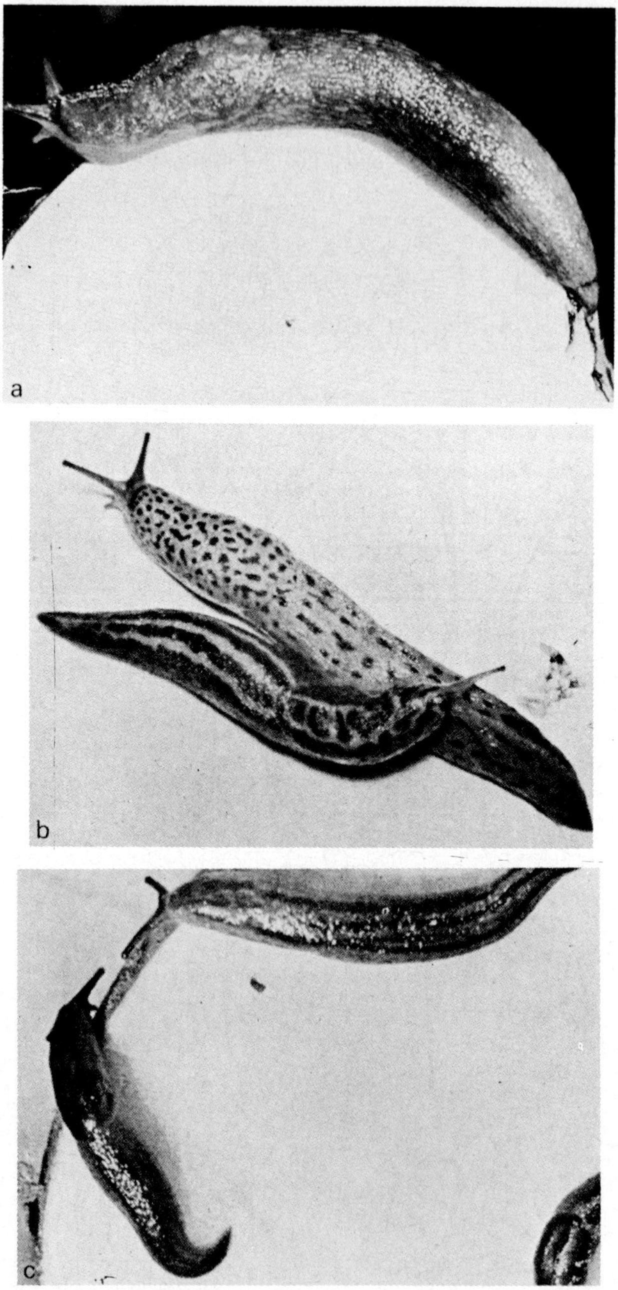

Fig. 110. Three slugs common to the temperate deciduous forest. a. *Arion subfuscus*, length up to 7.5 cm, widespread throughout Europe. It is essentially a polyphagous species, feeding on herbs, mushrooms and fruit, only secondarily on other animals. According to Frömming there are some ecological races, different by alimentary habits (herbivorous, microphagous and, perhaps, even carnivorous forms), by size, and life span. b. *Limax maximus*, reaching a length of 15 cm. It is also a polyphagous species. c. *Lehmannia marginata*, length up to 7.5 cm. Little is known of its alimentary habits, although it seems to be a phytophagous species (figures from Frömming) (with permission of Duncker & Humblot).

cene of Illertissen (Ulm) and probably originating from *S. plicatella* of the Upper Oligocene-Miocene of Bohemia.

Another species, *S. spermatia*, lives in the forests of Portugal (Bussaco). *Ena montana* and *Ena obscura* live under litter, wood, etc. in forests, often in mountain regions, particularly in beech forests. The former is probably a phytosaprophage. *Punctum pygmaeum*, a true soil animal, is also a phytosaprophage which helps the humification of litter. *Discus* (or *Goniodiscus*) *rotundatus*, one of the most common snails of deciduous forests, is restricted to wood, both of broadleaves and coniferous trees; although essentially a soil animal, it is practically indifferent to substrate (pH from 5 to 8), sometimes present also in caves or cellars. It feeds on humus, mushrooms, algae and pollen, so that it can be considered a secondary decomposer. *Vitrea diaphana, V. contracta* and *V. crystallina* are hygrophilous species living among mosses, under stones, in forests, sometimes near swamps or very humid places, with the exception of *V. contracta*, which is relatively xerophilous. They can be considered as soil animals. *Aegopis verticillus* is found in the soil of deciduous forests, and is probably a phytophagous animal, and therefore not important in humification processes. *Aegopinella* (= *Retinella*) *pura, A. nitidula* and *A. minor* live in humid, woody situations, under mosses, herbage and during dry periods, in the soil. It would be worth studying their feeding habits, which are so far completely unknown, since this genus sometimes represents an important component of forest malacofauna. *Oxychilus alliarius, O. cellarius* and *O. draparnaudi*, very easily recognizable thanks to their shining, almost transparent, reddish-yellow shell, highly flattened, live in humid places; they are nocturnal, with generally carnivorous feeding habits. *Daudebardia rufa* and *D. brevipes* are nocturnal animals, living during the day under stones, litter etc.; they are carnivorous, feeding mainly on earthworms. Let us remember finally the extremely interesting *Parmacellidae Boettgerilla vermiformis*, with a non-spiral, reduced, dorsal shell (Fig. 109d), present exclusively in Central Europe. It is taxonomically close to *Parmacella*, whose distribution is restricted to the sclerophyllous forests, and whose origin is probably very old (Fig. 109b, c).

Very abundant in the deciduous forest are the slugs, i.e. the species belonging to *Limacidae* and *Arionidae*: we will just mention *Arion ater, A. rufus, A. circumscriptus, A. subfuscus* and *A. intermedius*, often found on mushrooms, practically omnivorous, also *Lehmannia marginata, Limax tenellus, L. coerulans, L. maximus* and *L. cinereo-niger* (Fig. 110). All of them are polyphagous, as is also demonstrated by the large number of digestive enzymes detected by the present author and his co-workers in these forest animals.

As we have already said, the productivity and turnover of substances present in soil have been investigated by several authors. Let us recall here some observations made on earthworms, Glomeris and Diptera larvae in the region of the deciduous forest.

In a mixed forest it is possible to have more than 3 million earthworms per ha., with a total weight of 3000 kg. The faeces of earthworms (Fig. 112) can reach 80 t/ha in one year, a single lumbricid can eat daily a quantity of food corresponding to half its weight.

The millipedes of the genus *Glomeris* (Fig. 113) can reach the value of 200-300 specimens/m^2 (Dunger). In the Caucasus they seem to be the predo-

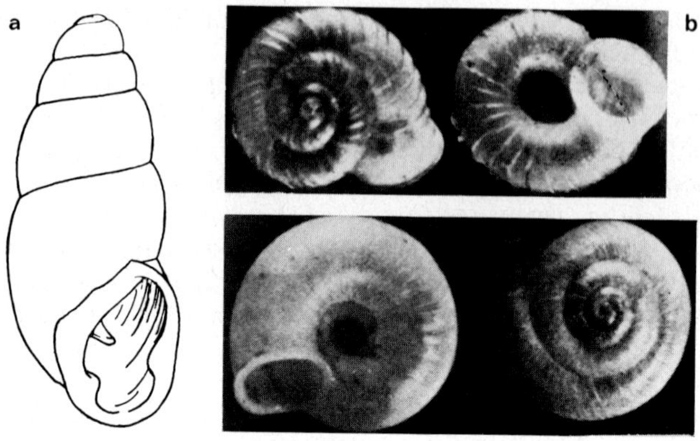

Fig. 111. Examples of molluscs belonging to the soil fauna: a. *Carychium minimum*; b. *Vallonia costata*; c. *Vallonia adela* (from Tierwelt Mitteleuropas).

Fig. 112. Faeces (or 'coprolithes') of an earthworm, attached to a dry leaf of hazel. It is an usual feature of heliophilous forests of Northern Italy, particularly during some months of the year (photo: U. Arezzini).

Fig. 113. Excrements of *Glomeris* together with the leaves eaten by these animals (left). Centre: the animal seen from above. Right: seen from the side. (Photo U. Arezzini)

minant invertebrates of litter (Striganova). Sometimes they can eat up to 3-4% of the litter produced in one year, and in the case of very humid forests, up to 2/3 of the litter. In the case of *Glomeris marginata*, 100 specimens can eat from 1.7 to 10% of the litter fallen on one square metre in one year (Bocock).

As far as the food preferences of this genus are concerned, we can report the data obtained by the present author in the study of some species of the deciduous forest of Euganean Hills:

Species	Tested leaves	Preference
G. euganeorum	holm oak-chestnut	holm oak
,, ,,	chestnut-*Rubus*	*Rubus*
,, ,,	oak-chestnut	none
,, ,,	oak-holm oak	none
,, ,,	chestnut beech	none
,, ,,	beech-holm oak	beech
,, ,,	beech-oak	none
G. undulata	holm oak-chestnut	none
,, ,,	beech-chestnut	none
,, ,,	oak-holm oak	oak

Very interesting is the ability of soil animals to change the C/N ratio of leaves during the passage of food through the alimentary canal, since the lower the ratio in the faeces, the nearer these are to humus or to substances from which humus originates. This ability of course varies according to zoological group and according to the kind of leaves which constitute the food, as can be seen from the two following schemes, one for *Glomeris* (based on original observations) and one for *Diptera* larvae.

Species	leaves utlized as food	C/N ratio of leaves	C/N ratio of faeces
G. euganeorum	chestnut	31.44	20.25
,, ,,	beech	28.00	25.56
,, ,,	holm oak	31.86	27.95
,, ,,	oak	36.32	27.30
G. undulata	chestnut	34.48	27.55
,, ,,	beech	30.50	26.80
,, ,,	holm oak	32.10	26.90
,, ,,	oak	39.10	29.50
G. conspersa	chestnut	29.82	26.31
,, ,,	beech	29.86	27.20
,, ,,	holm oak	38.79	25.61

Evidently, only a few species are able to reach during the digestion of leaves, C/N ratio values very near the values we find in humus (Moder).

Diptera larvae of the genus *Bibio* are able to transform the C/N ratio as follows:

Kind of leaves	C/N of leaves	C/N of faeces
chestnut	66.9	65.2
oak	56.0	33.5

As far as the turnover of chemicals of soil and higher plants is concerned, we will limit ourselves to mentioning the paramount importance of earthworms. Their coprolithes, compared with soil, are characterized by a higher cation-exchange ability, greater amount of exchangeable calcium, potassium and manganese, of phosphorous utilizable by plants, of total exchangeable bases of colloidal and organic substances. Since these substances are often transported to a great depth in the soil and since they contain humus in a form bound to inorganic materials (humus-clay complex) they are of enormous importance for soil structure and fertility.

Lastly, there are the enzymes. One can say that there is practically no constituent of litter which is not attacked by the digestive enzymes of soil animals (or of their intestinal bacteria), with the exception of resins, lignin and a few other substances. Indeed, the soil animals are equipped with a very rich and varied set of enzymes, even of entirely animal origin, by means of which they are able to digest oligosaccharides, polysaccharides[1] (with the possible exception of holocellulose), esters and lipids (including waxes, cork, cutine, etc.), most proteins (with the exception of casein, keratin and a few others) and practically all polypeptides. Recent observations (also by the present author) make us think of the great variety and high specificity of the digestive enzymes of soil animals.

[1] here also sodium alginate and lichenin are included.

Harmful invertebrates of deciduous forests

Many beetles attack the different parts of the trees of the deciduous forest. Beginning with *Cerambycidae*, we may recall *Aegosoma scabricorne*, whose larva lives in the trunks of many broad-leaved trees where it develops in three years; *Prionus coriarius*, living in dead trunks both of coniferous species and broad-leaved ones (euryphagous); *Rhagium mordax*, in dead trunks and stumps of both coniferous and broad-leaved trees; several species of the genus *Leptura*, among them *L. livida* (a 'Rassenkreis'), whose adults stay on the flowers of several plants and larvae in dry oak branches; *L. imitatrix*, an endemic, whose adult may be found on flowers; *L. aurulenta, L. maculata* and *L. quadrifasciata*, polyphagous but only on deciduous trees; *Cerambyx scopolii*, in living trunks and therefore often very harmful to beech, oak, horn beam, elm, maple and several fruit trees; its life cycle lasts two-three years; *Aromia moschata*, with a beautiful metallic greenish-blue colour, present in old willows (sometimes also healthy), characterized by odorous glands on the sides of the metasternum: the characteristic odour of rose and moss is due to an ether of salicylic acid (salol), which takes its origin from the leaves of willows. We find furthermore several species of the genus *Clytus* such as *C. rusticus* (on poplars), *C. arietis*, euryphagous on several broad-leaved trees, the adults on flowers; *C. sartor*, on broad-leaved trees but also on stems of *Robinia* and *Castanea*, the adults on flowers of *Achillea, Umbelliferae* and thistles; *C. figuratus*, on broad-leaved trees, the adult on flowers and *C. varius*, sometimes in stems of vines to which it can cause breakage; *Morimus asper*, a 'Rassenkreis' developing in dead branches of beech or oak; *Pogonochaerus hispidulus*, on several deciduous trees, secondarily on coniferous species and even on the mistletoe growing on oaks; and finally the elegant *Saperda scalaris*, on elm, birch and alder.

Some phytophagous species are more or less closely restricted to heliophilous forests, though in some instances they can also be found in the sciaphilous or mixed forests: let us remember here *Rhagium sycophanta*, on oaks, *R. inquisitor*, on coniferous species of Southern Europe, *Toxotus cursor*, on deciduous oaks of the Balkans; *Acmaeops collaris*, whose larva lives in dead oaks and chestnuts, *Leptura revestita*, on oaks, *L. melanura*, present also in mixed forests, *Cerambyx miles*, on oaks and fruit trees (also living ones, and therefore rather harmful). A congeneric species, *C. cerdo*, a 'Rassenkreis' is of particular forest importance as its larvae excavate galleries in trees such as oaks, and to a lesser extent walnut, ash, elm, willow and, on the isle of Cyprus, in carob. Its life cycle takes three years, the adult comes out of its hiding places during twilight, and flies around trees, mating during the day with the female laying on the bark of healthy trees.

Other Cerambycids of heliophilous forests are *Hesperophanes* (from ἕσπερα evening, and φαίνομαι to appear) on *Quercus robur*; *Stenopterus flavicornis* and *S. rufus*, in dead trunks or in vine-supporting poles; *Phymatodes kollari, P. testaceus* (in dead *Quercus*, both deciduous and evergreen); *P. lividus* in evergreen trees, which can also attack barrels made with chestnut timber; *P. glabratus* in mixed forests with *Juniperus*, in which its larva lives, and *P. alni*, in dead oak twigs; *Clytus antilope*, in oaks; *Plagionotus detritus*, in dead, felled trunks, and *P. arcuatus*, also in healthy trees; *Purpuricenus koehleri*, also in fruit trees; the genus *Dorcadion*, represented in Europe by several species among them *D. arenarium*, a

'Rassenkreis' whose larva lives in the soil where it eats the roots of several *Gramineae*, often in dry, karstic biotopes or in the clearings of heliophilous forests; *Oberea linearis*, a polyphagous species and therefore less dangerous than the others, and finally *Phytoecia cephalotes*, living in karstic grasslands, which in Southern Europe very often alternate with heliophilous forests.

Particularly important in the forest are all the stenophagous or monophagous species, among which generally we find the major pests; thus oaks can be attacked by the oak leaflouse *Phylloxera quercus*, by the lepidopterous insects *Thaumetopoea processionea, Lasiocampa quercus, Lymantria dispar* and *Tortrix viridana*, the weevils *Rhynchaenus quercus* and *Attelabus nitens*; the genus *Cerambyx* (several species), *Pyrrhidium sanguineum* (sometimes also on other species) and lastly, the scale insect *Kermococcus roburis*. The hazel is attacked by the Cerambycid *Oberea linearis*, the weevils *Byctiscus betulae* and *Apoderus coryli*, and the scale insect *Eulecanium coryli*. The elm is infested by the scale insect *Mytilococcus ulmi* (sometimes harmful to poplars), the Chrysomelid *Galerucella luteola*, which can completely defoliate adult plants. The ash is attacked by *Hylesinus* (= *Leperisinus*) *fraxini*, a polyphagous species; the chestnut by the weevil *Curculio* (= *Balaninus*) *elephas* and the birch by *Scolytus ratzeburgii*. Finally, the species of *Pinus* living in the heliophilous forest are victims of the caterpillar of *Thaumetopoea pityocampa*.

In the sciaphilous forests also there are many phytophagous insects which sometimes can be very dangerous to the vegetation, especially when the forest is monospecific, which is very often the case for beech forests.

Let us mention only the Dipteran, *Mikiola fagi*, a cecidogenous species, harmful especially to nurseries; the *Cerambycidae Pyrrhidium sanguineum* (polyphagous), *Rosalia alpina*, whose larva feeds inside old, withered beeches; this insect is one of the few species protected by law in some Central European countries, such as Bavaria and recently also Alto Adige (Süd-Tirol), *Evodinus clathratus*, widespread in Central Europe (Alps, Bohemian Forest, Sudetes, Carpathes and Bosnia), *Phytoecia rubropunctata, P. uncinata, P. nigriceps* (developing in *Umbelliferae*) and other congeneric species, generally linked to this plant family; the *Scolytidae Ernoporus fagi* and *Anisandrus dispar* and finally the weevil *Rhynchaenus fagi*, characteristic for its ability to make great leaps, resembling somewhat the *Halticinae* among Chrysomelids.

In England, where beech forests are relatively rare and limited to calcareous hills in Southern or South-Eastern parts of the territory, insects harmful to beech are *Rhynchota* such as the *Cicadellidae Typhlocyba douglasi*, the aphid *Phyllaphis fagi*, homoecious and hypophyllous, which causes wrinkling of beech leaves and sometimes appears in very great numbers; the scale insect *Cryptococcus fagi*, many *Lepidoptera*, some of which are rather stenophagous and therefore particularly dangerous, and finally some *Coleoptera* such as the Elaterid *Elater rufipennis*, the *Helodidae Prionocephalus serricornis*, whose larva thrives in the water collected in the big cavities of very old beeches (phytotelma), the *Erotylidae Diplocoetus fagi*, the beautiful *Scarabaeidae Sinodendron cylindricus*, and finally the *Scolytidae Ernoporus fagi*, living on beech but sometimes also on *Quercus cerris* or birch, which digs transverse maternal galleries, simple or with two arms.

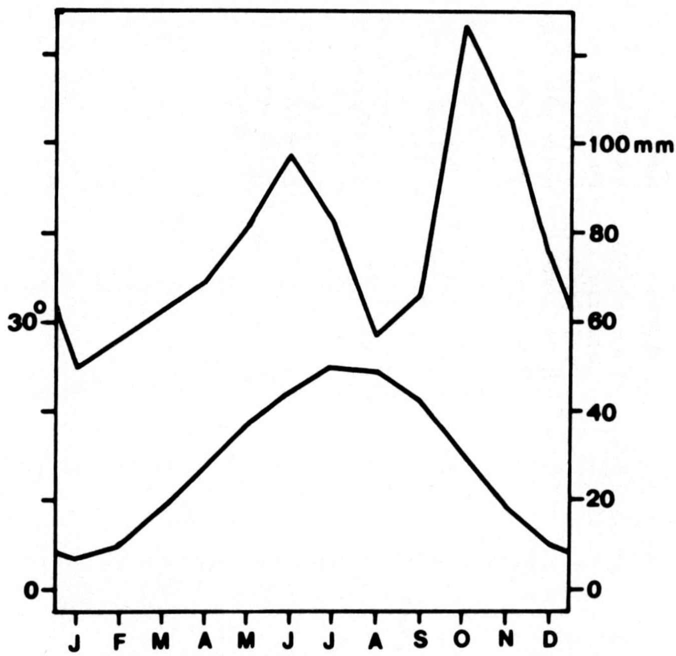

Fig. 114. Ombrothermic diagram for Venice.

Dynamism in ecology of deciduous forests

An example of a succession from sea shore to the Quercetum or deciduous forest is given by the region of Rosolina (or Rosapineda) not far from Venice, southwards, where the plant succession has already been studied by several students, and where the present author made some observations on animal succession (cf. the following scheme and Fig. 115).

Isopods	Pseudoscorpions	Opilionids	spiders	centipedes	millipedes	snails
Chaetophiloscia elongata	*Dactylochelifer latreillei*	*Opilio parietinus*	*Trochosa ruricola*	*Lithobius melanops?*	*Schyzophyllum sabulosum,*	*Helicella maritima*
C. hastata	*Neobisium carcinoides*	*O. saxatilis*	*Araneus ocellatus*		*Brachydesmus* sp.	*Cochlicella acuta*
Philoscia affinis	*Chthonius ischnocheles*	*Nelima doriae*	*Mangora acalypha*		*Pachiulus varius*	*Cepaea nemoralis* (gener. phytophagus)
			Tetragnatha extensa		*Ophiulus* sp.	*Trochoidea* sp.
			Pisaura mirabilis			*Oxychilus* sp. (omnivorous or carnivorous)
			Micrommata sp.			*Hygromia cinctella*
			Silla diodia			*Vallonia costata* (phytosaprophagous)
			Glyptogona sextuberculata			*Abida frumentum*
			Prolinyphus emphana			*Helix aspersa* (euryphagous!)
			Theridion sp.			*Lauria cylindrica*
			Episinus sp.			*Limax maximus* (omnivorous)
			Chiracanthium sp.			
			Heriaeus sp.			
			Philodromus sp.			
			Zora sp.			
			Dysdera sp.			
			Hyptiotes flavidus			

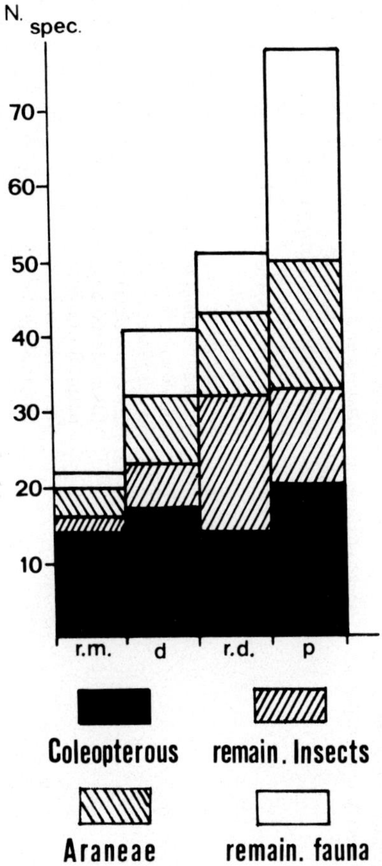

Fig. 115. Animal succession at Rosolina. r.m. = sea shores; d = sand dunes; r.d. = region immediately beyond the dunes; p = *Pinus* stand (from Marcuzzi et al., modified).

The climate of the region can be considered fairly similar to that of Venice, shown in Fig. 114. In these figures the climatic variations in the last 43 years are also taken into consideration. The soil shows the following characteristics:

— Insects

hoppers (*Orthoptera*)	roaches (*Blattaria*)	bugs (*Heteroptera*)	leaf hoppers (*Homoptera*)	beetles	*Neuroptera*
Acrida mediterranea, *Yersinella raymondi* *Chorthippus brunneus* *Phaneroptera falcata*, *Oecanthus pellucidus* (on *Mentha*, *Cichorium*, *Agrimonia*, *Verbascum*) *Paratettix meridionalis* *Arachnocephalus vestitus*, *Oedipoda coerulescens* (phytophagous), *Homocoryphus nitidulus* (hygroph., phytophagous),	*Loboptera decipiens* *Ectobius erythronotus*	*Eysarcoris inconspicuus* *Sciocoris macrocephalus* *Podos curvidens* *Aelia acuminata* (harmful to cereals) *Dolycoris baccarum* (sometimes harmful to cultivations) *Beosus maritimus* *Carpocoris mediterraneus* *Himacerus myrmecoides*	*Philaenus leucophthalmus*	*Calathus ambiguus* *C. melanocephalus* (saprophagous) *Dromius linearis* *Harpalus anxius* (granivorous) *H. serripes* (id.) *Carterus dama* (xerothermic relic) *Badister bipustulatus* *Amara aenea* (xero-heliophilous) *Melanophthalma distinguenda*, *Astenus pulchellus* *Xantholinus linearis* *Adonia variegata* *Thea 22-punctata* (eating the fungi of wheat mildew) *Cardiophorus equiseti* *Maladera holosericea* (psammophilous, sometimes harmful to crops) *Geotrupes vernalis* (coprophagous)	*Chrysopa carnea* (larva parasite of several insects!)

Tetrops praeusta
(larva on branches of several *Rosaceae*!)
Galeruca tanaceti
(on sev. *Compositae*!)
Crioceris paracenthesis
(xerothermic relic, on *Asparagus acutifolius*!)
Crioceris 12-punctata
(on *Asparagus officinalis*)
Spermophagus sericeus
(on sev. *Convolvulus* sp.)
Bruchidius marginalis
(on *Astragalus glyciphyllos*)
Otiorrhynchus ferrarii
(psammobiont)
Pissodes notatus (harmful to *Pinus* and *Picea*)
Lixus junci (linked to *Chenopodiaceae* and *Amaranthaceae*)
Apion pisi (on *Leguminosae*!)
Blastophagus piniperda (on dead *Pinus* and other Coniferous trees)
Stenosis intermedia
(psammobiont, probably thermophil.)

As far as the present vegetation – which can only approximately be considered as climax or near climax – is concerned, we may consider the holm oak as indigenous, or more precisely as a relict of a xerophilous woody formation, characteristic of the xerothermic period (more or less 4-5000 years ago) which successively has been replaced by a mesophilous formation, characterized by a deciduous species of *Quercus* (*Q. pubescentis*). According to some, this forest is a transition between the warm subzone of Lauretum and the cold subzone of Castanetum (Pavari, de Philippis).

Today we find what some (Marcello) call Pineto-lecceta, i.e. *Pinus pinea* – *Quercus ilex* formation: indeed, in 1928 5000 pines and 5000 holm oaks were planted.

An example of conservation in the deciduous forest

As far as the protection of the fauna of deciduous forests is concerned, we want to mention here measures taken in some countries for the protection of the local fauna.

Namely, in Bavaria (Germany) the following groups and species are protected by the laws of this 'Land': many birds and mammals which generally are also protected by law in other countries, for example Italy, including bats, hedgehogs, the small rodents *Eliomys quercinus, Dryomys nitedula* and *Muscardinus avellanarius*, the water tortoise *Emys palustris*, lizards (here including the blindworm or *Anguis fragilis*), snakes (with the exception of vipers), newts and salamanders, toads and *Bombina, Hyla arborea*, frogs (with the exception of *Rana esculenta* and *Rana temporaria*), and also the following invertebrates: *Papilio podalirius*, all the species of the genus *Parnassius*, the beetle *Lucanus cervus*, the ant *Formica rufa*, the longicorn beetle *Rosalia alpina* (one of the most beautiful beetles of the European fauna) and the carabid beetle *Calosoma sycophanta*. According to Bavarian law there are also some restrictive measures on the capture of other animals such as insects, which cannot be collected for commercial purposes, and the edible snail *Helix pomatia*, which can be collected as food only in some given circumstances and never sizes smaller than 3 cm, if it is found in an area which does not belong to the collector and without authorization of the owner. It is to be hoped that some laws of this kind will also be introduced by many other European countries to the dominion of deciduous forests.

Examples of European deciduous forests

Abruzzo: the region near Chieti

This region, like many other parts of Central and Southern Italy, is only what the peoples who have inhabited it since very remote times have left us, and this is very little, so that today agriculture and cattle breeding are extremely difficult and poor. Indeed, in this small district we find remains of the Lower Paleolithic (also very near Chieti, at Madonna del Freddo), Upper Paleolithic, Neolithic (caves,

Fig. 116. Example of the urgent need for protection of European temperate deciduous forests. Effects of a quarry in the Colli Euganei (Monte Ricco) near Padova (photo: E.P.T., Padova).

villages, flint workshops, also in the Chieti region), Eneolithic, Bronze Age and Iron Age. During the latter, Abruzzo was already inhabited by Italians, such as Sabinians, Picenians, etc.

The region belongs to Quercetum pubescentis, although highly degenerated. The climate is synthetically represented by an ombrothermic diagram (Fig. 117).

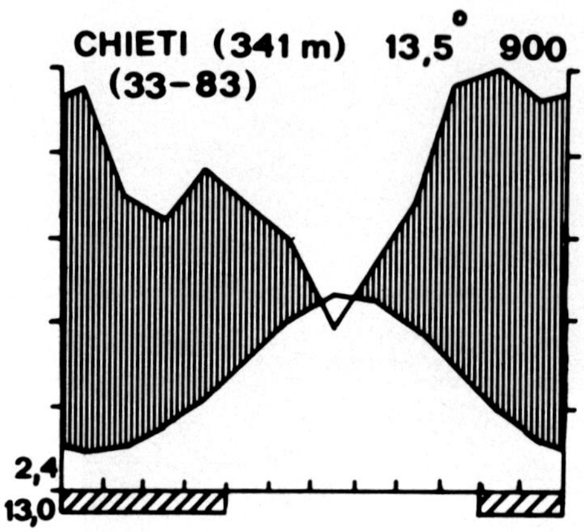

Fig. 117. The ombrothermic diagram for a locality of Central Italy, with an extremely reduced dry season (Chieti, Abruzzo) (redrawn from Walter's Klima Weltatlas).

The soil is of an intermediate type between brown earth and yellow earth; these are the main chemico-physical characteristics:

clay	silt	sand	$CaCo_3$	C	N	C/N	pH
24.2%	15.4%	60.3%	30.75%	1.9%	0.15%	12.8	8.27

The soil fauna, which has been thoroughly investigated by the present author, shows remarkable interest both from the ecological and the biogeographical standpoint. Within the mesofauna we find the whip-scorpions *Chthonius* sp., *Roncus lubricus*, the millipede *Lophoproctus lucidus*, the micro-coleopterous insects *Langelandia anophthalma*, living in the soil in rotting material, *Bythinopsis glabratus*, the protura *Eosentomon* sp., the thysanopteron *Pseudothrips meridionalis.*

Within the macro- and megafauna there are the woodlice *Chaetophiloscia elongata, Philoscia affinis, Metoponorthus pruinosus* and *Orthometopon dalmatinus*, the Opilionids *Trogulus coriziformis, Dicranolasma diomedeum*, the centipedes *Eupolybothrus fasciatus* and *Scolioplanes crassipes*, the roach *Loboptera decipiens*, the ant *Crematogaster scutellaris* (a Mediterranean species extending to the Sahara and East Africa), the neuropteran *Chrysopa prasina* (larval), the beetles *Harpalus pubescens* (carnivorous and granivorous), *Paederus litoralis, Stenus morio, Hister merdarius* (often nidicolous, but also on rotting matter), *Cylindronotus dermestoides*, under bark (phytosaprophagus, only sometimes phytophagous and therefore a potential pest), and finally the snails *Deroceras reticulatus*, a phytophagous species linked to crops ('Kulturfolger'!), during the day under litter, stones or wet wood, and *Helix aspersa*, the common 'Weinbergschnecke', edible and very euryphagous.

Here are data on number of specimens (per 500 cc of soil) and the biomass (mg/500 cc of soil) obtained in one year of research in this Quercetum pubescentis:

	Spring		Summer		Autumn		Winter	
	No.	Biomass	No.	Biomass	No.	Biomass	No.	Biomass
Nematoda					2	2	4.5	4.5
Enchytraeidae	0.5	0.1			8	1.6	5	1
Isopoda	0.5	5			1.5	15		
Pseudoscorpion.	1.5				1			
Acari	268.5	2.685	26	0.26	316.5	3.165	115.5	1.155
Pauropoda	5	2.5			3	1.5	1.5	0.75
Diplopoda	2	53.32	0.5	13.33			0.5	13.33
Symphyla	1.5	0.75					0.5	0.25
Chilopoda					0.5	4		
Protura	3						2.5	
Collembola	48.5	582 μg	7	84 μg	78.5	942 μg	19	228 μg
Thysanura	2.5	16.65						
Psocoptera					0.5	3.33		
Thysanoptera	0.5	3.33						
Homoptera	20.5	136.53			13	86.58	7	46.62
Diptera (l.)	2	20	1	10	2.5	25	4.5	45
Coleopt. (l)	1	15	2	30	2	30	0.5	7.5
Lepidopt. (l)			0.5	3.33				
Diptera	0.5	5			0.5	5		
Coleoptera	1.5	2.25	0.5	7.5	1	15		
Formicidae	2.5	16.85			1.5	99.9		
Hymenoptera (excl. Formic.)			0.5	3.33				
Gastropoda					0.5	1		

An interesting example of heliophilous deciduous forest is the Quercetum recently described by Pedrotti, also in Abruzzo, near the Sinello River (Vasto), 5 km from the sea and at a height of 30 m. The forest, which has the typical characteristics of a relict forest, is situated on an alluvial soil. The proximity of the river is responsible for a true succession.

The annual average temperature is 15.4°C and the rainfall 750 mm, so that the rain factor (or Lang's index) is 49. The latitude is responsible for the presence, within the deciduous forest, of some elements which are more typical of the European sclerophyllous forest (see first chapter) such as *Smilax aspera, Cotoneaster piracantha, Clematis flammula, Rosa sempervirens* and *Vitis vinifera*. Unfortunately, we know nothing about the animal life of this relict Abruzzan forest.

Here are the principal trees and shrubs of this Quercetum (cf. Pedrotti)

Trees	*Quercus cerris*-facies	*Equisetum maximum*-facies
Quercus robur	+	+
Carpinus betulus	+	+
Ulmus campestris	+	+
Fraxinus angustifolia	+	+
Populus alba	+	+
Acer campestre	+	+
Quercus cerris	+	+
Fraxinus ornus	+	+
Carpinus orientalis		+
Shrubs		
Cornus sanguinea	+	+
Carpinus orientalis	+	+
Lonicera caprifolium	+	+
Ulmus campestris	+	+
Evonymus europaeus	+	+
Crataegus oxyacantha	+	+
Carpinus betulus	+	+
Acer campestre	+	+
Prunus spinosa	+	+
Cotoneaster piracantha	+	+
Ligustrum vulgare		+
Quercus cerris	+	+
Pirus malus	+	+
Cornus mas	+	+
Fraxinus angustifolia	+	
Populus alba		+
Quercus pubescens	+	
Quercus robur		+
Vines		
Hedera helix	+	+
Smilax aspera	+	+
Clematis vitalba	+	+
Rubus ulmifolius	+	+
Rosa sempervirens	+	+
Clematis flammula	+	+
Vitis vinifera		+

Euganean hills (Colli Euganei)

> 'Licet aliqui eorum sint asperiusculi, plures nihilominus Coelo Terraque adeo benignis fruuntur, ut quod de Hesperidum, Alcinoique Viridariis fuit fabula, de ipsis sit historia'
>
> (Zannichelli, Venetiis, a. 1730, Opusc. Bot. posth., p. 79)

During the Neolithic, this region was inhabited by a people coming probably from the North East, the 'Liguri', followed by the Eneolithic, characterized by the introduction of copper. Indeed some remains of Neolithic huts and altars have been found in this region, and at Galzignano Eneolithic remains are present. Much more abundant were human settlements during the Bronze Age (c. 2000 years B.C.) when both palafitticolous and terramaricolous (i.e. people living in huts built on the land) were living together.

Later the so-called Euganean population is found, very probably belonging to the 'Liguri'. In the meantime the Celtic culture had reached the Euganean hills from the west, while the Euganean population still maintained its language. About 1000 B.C. some eastern populations arrived (the so-called 'Trojans') bringing with them their language and culture. Today they are called Paleovenetic or Venetic, to distinguish them from the present 'Veneti' who speak an Italian dialect with strong Roman influences. At the time of the Roman invasion, the Venetic allied with them, and they collaborated in order to eliminate the Celts (or, as the Romans said, the Gallic race) (around 390 B.C.).

The climate can be represented by Figures 72 and 118.

The vegetation as a whole can be attributed to the following types: Castanetum, growing frequently on fresh, deep, decalcified or silicified terrain, and therefore limited to the volcanic portions of the hills (Fig. 119). On north- or northwestern slopes, often mixed with beech, ash, linden, and accompanied by shrubs such as *Genista pilosa, G. germanica, G. tinctoria*, the juniper (*Juniperus communis*), the heath *Calluna vulgaris*, and by grasses such as *Epimedium alpinum, Spiraea aruncus, Oxalis acetosella, Teucrium siculum, Luzula pilosa, L. forsteri, Molinia coerulea, Stellaria holostea, Dentaria enneaphyllos, Asperula odorata, Senecio fuchsi, Solidago virga-aurea* and some ferns such as *Polypodium vulgare* or the common braker, *Pteridium aquilinum*. The Quercetum is less extended, localized mostly in the zone of Arquà Petrarca, the Teolo basin and the hill of Frassenelle. The dominant species are *Quercus pedunculata, Q. sessiliflora, Quercus pubscens*, associated with *Ostrya carpinifolia, Carpinus betulus, Corylus avellana* (hazel), and some elements more typical of the Mediterranean maquis, such as *Rhus cotinus, Dictamnus albus, Fraxinus ornus, Cercis siliquastrum, Viburnum tinus*, etc. The grasses are here represented by *Allium ursinum, Ruscus aculeatus, Asparagus acutifolius, Corydalis cava* (with beautiful little purple flowers), *Mercurialis perennis, Euphorbia amygdaloides, E. dulcis, Geranium sanquineum, Pulmonaria officinalis*, the orchid *Cephalanthera ensifolia*, with small white flowers, the vine *Tamus communis*, also *Helleborus viridis, Geum urbanum, Myosotis silvatica, Melittis melyssophyllum*, etc.

Widespread today – as a consequence of the deforestation which took place in the Euganean hills in pre- and historic times – are the prairies, generally dry, of the Brometum type, with *Festuca ovina, Koeleria cristata, Bromus erectus, Phleum boehmeri, Arabis hirsuta, Tunica saxifraga, Helianthemum chamaecistus, Coronilla varia* (with very conspicuous rose-pink flowers), *Salvia pratensis, Stachys*

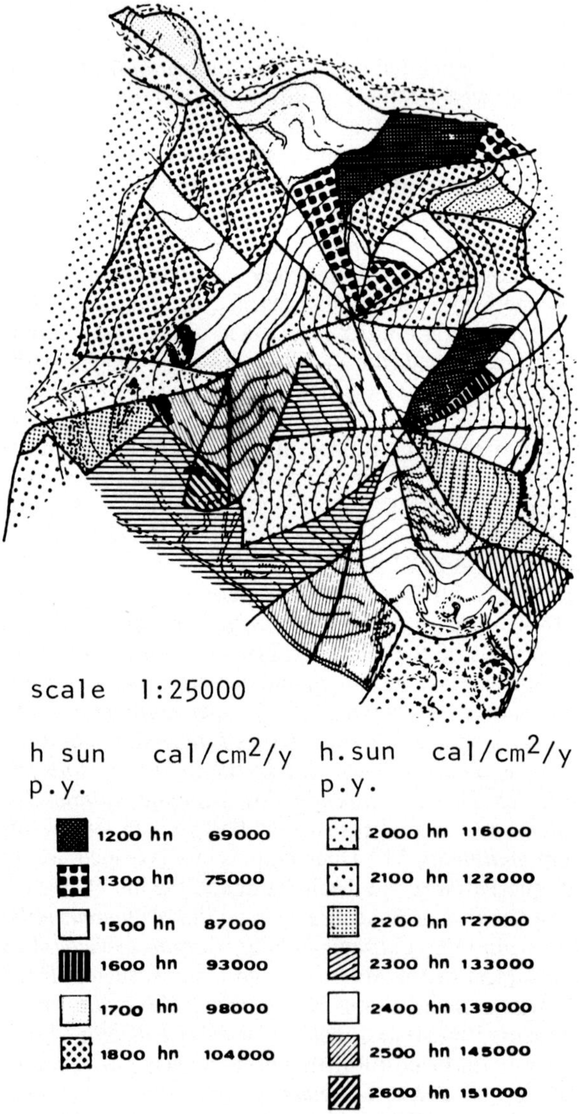

Fig. 118. Number of hours of 'normal sun' and annual radiation (with clear sky) on one of the southernmost Euganean hills, M. Ricco and Castello, near M. Calaone, where the author has carried out some ecological research (from Famiglietti).

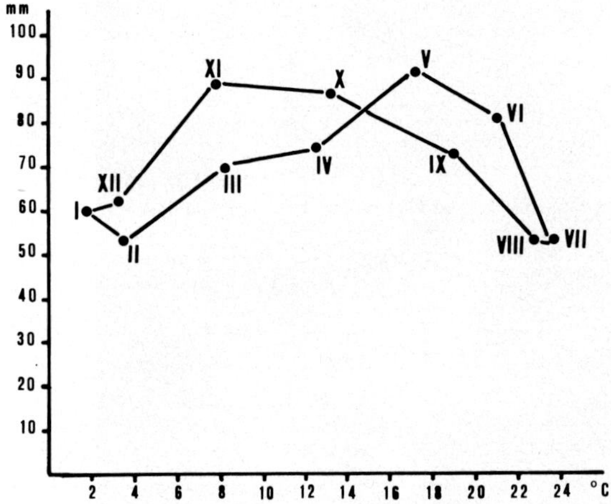

Fig. 119. A climogram for the Colli Euganei, near Padova.

recta, Teucrium chamaedrys, etc. The most interesting element of these 'prairies', localized to a few sites on calcareous soil, is an Eastern element, *Ruta patavina*, present in Italy exclusively on the Euganean Hills, and also present in several sites in Istria and the Balkans.

Very frequent are some examples of a rocky vegetation, which can be compared to Zangheri's Asplenietum and Adiantetum of deciduous forest of the Appennines. The main elements, more or less limited to rock, are *Sempervivum tectorum, S. arachnoideum, Sedum rupestre, S. album, S. maximum, S. villosum*, together with the ferns *Ceterach officinarum, Asplenium ruta-muraria, Adiantum capillus-veneris*, etc. In the southernmost, driest situations, it is possible to find the indigenous *Capparis rupestris* and the introduced *Opuntia vulgaris* (prickly pear).

The fauna has all the aspects of a relict one. The mammals still living in the Euganean Hills are the hare *Lepus capensis*, the fox *Vulpes vulpes*, the badger *Meles meles* (now on the verge of extinction), the squirrel *Sciurus vulgaris* (extremely rare if not disappeared), the dormouse *Glis glis*; more abundant, probably because of their nocturnal habits and because they are not the object of hunting, are the hedgehog *Erinaceus europaeus*, the mole *Talpa europaea*, several shrews (*Sorex*) and some field mice of the genus *Microtus*. Fenced in the Castello del Cataio, near Battaglia, lives the fallow deer *Dama dama*.

Birds are relatively numerous, partly due to recent measures taken by the Government and local agencies for their protection.

Reptiles, on the contrary, are rather rare: let us remember the green lizard *Lacerta viridis*, the common lizard *Lacerta sicula*, bound mainly to vegetation, and the wall lizard *Lacerta muralis*, often linked to dwellings, walls etc. Very frequent still is the blind worm *Anguis fragilis*, rarer the colubrid *Zamenis gemonensis* and extremely rare the vipers (*Vipera aspis* and maybe also *Vipera berus*). Two amphi-

Fig. 120. The region of Frassenelle. Top – the relict oak forest; left: general view of the forest; right: the litter at the foot of an oak. Bottom – the secondary prairie or meadow; left: general view of the meadow; right: a detail of the same.

bians still living in the Euganean hills are the common salamander, *Salamandra salamandra*, limited to cool, humid environments, and the toads, *Bufo bufo* and *Bufo viridis*, very often victims of the ignorance of people.

As far as the invertebrates are concerned, we will limit ourselves now to mentioning only a few of the most conspicuous, such as the scorpion *Euscorpius carpathicus*, common under stones; the big *Scolopendra cingulata*, the endemic millipede *Glomeris euganeorum*, extremely wide-spread in the foot-hill region, an endemic subspecies of *Abax ater*, *A.a. euganeensis*, and finally the extremely rare hopper *Micropodisma salamandra*.

First of all let us take into consideration the region of Frassenelle situated in the northernmost part of the Euganean hills, today occupied partly by cultivation of annuals, for instance wheat, partly by a relict Quercetum, with many clearings scattered here and there (Fig. 120-22).

The soils vary according to the biotope; namely, we may recognize a forest soil and a prairie soil:

	pH	$CaCO_3$	C	N	C/N	bed rock	sand	silt	clay
forest (Kubiena's terra fusca tipica)	7.2	3.0	13	1.04	12.5	11.4	–	–	– (*)
prairie (Kubiena's terra fusca or calcareous brown loam)	6.5	tr.	3.45	0.34	12	10.9	37.1	28.9	33.4

(*) humiferous layer only, with practically no mineral constituents, except skeletal

The vegetation today is rather impoverished, especially in the forest. Here we find *Quercus pedunculata, Acer campestre, Ulmus campestris, Fraxinus ornus, Taxus baccata, Crataegus monogyna, Ligustrum vulgare, Ruscus aculeatus, Hedera helix* (ivy), as well as the grasses *Galium aristatum, Iris pseudacorus, Galanthus nivalis, Anemone nemorosa, Lamium orvala, Pulmonaria officinalis, Ranunculus acer, Corydalis cava* and *Symphytum* sp.

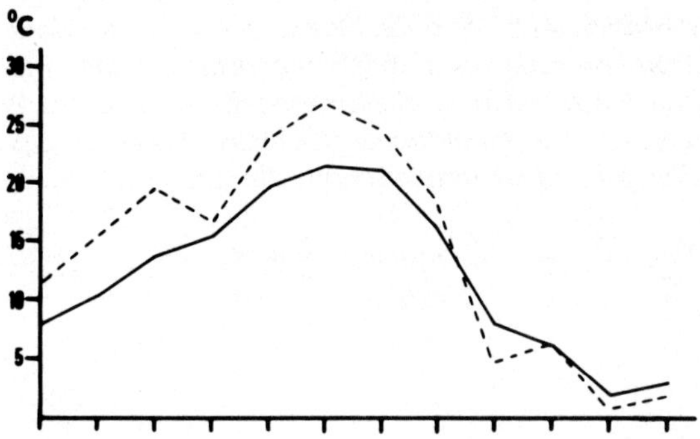

Fig. 121a. Frassenelle, forest: temperature of soil (continuous line) and air (broken line) at the moment of the collection.

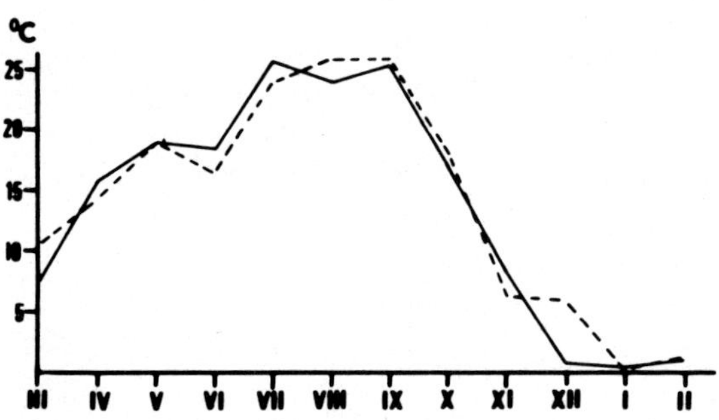

Fig. 121b. Frassenelle, meadow: see Figure 121a.

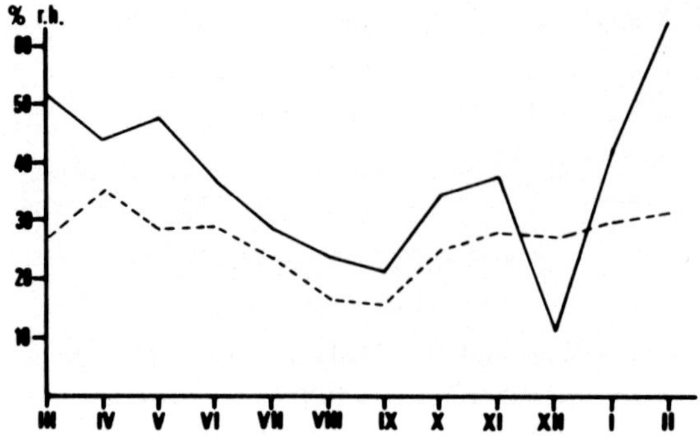

Fig. 122. Relative humidity of the air (broken line) and soil (continuous line) at the moment of the collection of soil fauna at Frassenelle (Colli Euganei).

In the prairie or secondary grassland we may find *Medicago minima, Potentilla tormentilla, Knautia sylvatica, Galium verum, G. mollugo, Poa pratensis, Galium aparine, Chrysanthemum leucanthemum, Lotus corniculatus, Dactylis glomerata, Rumex acetosa, Veronica chamaedrys, Achillea millefolium, Centaurea scabiosa, Trifolium campestre* and *T. repens, Fragaria vesca, Salvia pratensis, Plantago media, Colchicum autumnale* (the meadow saffron), *Bellis perennis, Carex verna* (a small delicate sedge), *Glecoma hederacea, Pulmonaria officinalis, Ajuga reptans, Viola canina, Viola hirta, Luzula forsteri, Anthoxanthum odoratum* and finally *Cerastium glomeratum*.

The quantitative aspects of the soil fauna are summarized here:

	March		April		May		June		July		August	
	No spec.	biomass	No spec.	biomass	No spec.	biomass	No spec.	biomass	No spec.	biomass	No spec.	biomass
Nematodes	58.5	58.5 μg	62	62	84.6	84.6	91	91	65	65	1.6	1.6
Enchytraeidae	82.7	16.54	102.6	20.52	96	19.2	130	26	26.6	53.2		
Isopoda	5.5	55	9.3	93	5	50	3.6	36	2	20		
Pseudoscorp.	4.2		2		0.6		2.6		3		2.3	
Araneae	2.2	8.8	3	13	2.3	9.2	3.6	14.4	2	8	2.3	9.2
Acari	1396.5	13.96	897.6	8.97	835	8.35	1231	12.31	1939.3	19.39	1544.6	15.44
Diplopoda	4	106.64	23.6	627.76	37.3	992.18	27.3	726.18	38	1010.8	21.6	574.56
Symphyla . .	1.5	0.75	6.3	3.15	1	0.5	2	1	7.3	3.65	8.3	4.15
Pauropoda . .	1.75	0.875	4.6	2.3	3	1.5	5.6	2.8	14.3	7.15	1.3	0.63
Chilopoda	8.25	66	10.3	82.4	4	32	8	64	3.3	26.4	36	288
Collembola	346.2	4154.4 μg	422.6	5071.2 μg	342	4104 μg	365.6	4387 μg	422.3	5067.6 μg	303.3	3639.6 μg
Campodeidae			0.3	1.99								
Japygidae					0.3	1.99			0.6	3.99	1.6	10.65
Protura	19.2	127.87	22	146.5	13.6	90.57	53	352.98	48	319.7	58.6	390.27
Psocoptera												
Thysanoptera					0.6	3.99	0.3	1.99				
Heteroptera			1	6.66			0.3	1.99	0.3	1.99		
Homoptera			0.3	1.99			0.6	3.99				
Hymenopt. (l)												
Hymenoptera	0.7	4.66	2	13.32	2	13.32	0.3	1.99	1.3	8.66	0.6	3.99
Formicidae	10	66.6	0.6	3.99	8.3	55.27	8.3	55.28	2.6	17.31	2.6	17.31
Coleopt. (l)	1	15	4.6	69	1.6	24	3	45	2.3	34.5	1.6	24
Coleoptera	1.2	18	2.3	34.5	6.6	99	2.3	34.5	3.6	54	3	45
Diptera (l.)	79.7	797	46.3	463	27	270	54	540	120	1200	36.3	363
Diptera							1	6.66	1	6.66	0.3	1.99
Lepidopt. (l.)	0.25	1.66			0.3	1.99						
Mecoptera (l.)					0.6	3.99	0.3	1.99	2.6	17.31	0.3	1.99
Mollusca G.							0.3	6				

	September		October		November		December		January		February	
	No spec.	biomass	No spec.	biomass	No spec.	biomass	No spec.	biomass	No spec.	biomass	No spec.	biomass
Nematodes	0.6	0.6	3.6	3.6	16	16	9.3	9.3	35	35	17.3	17.3 μg
Enchytraeidae			0.3	0.06	48.6	9.72	14.6	2.92	18.6	3.72	15.3	3.06
Isopoda			1.6	16	6	60			0.3	3	1.3	13
Pseudoscorp.			2		0.6				0.3		0.6	
Araneae	3	12	10.6	42.4	5	20	0.6	2.4			0.3	1.2
Acari	775.3	7.75	1923.6	19.23	656.3	6.56	238.6	2.38	478.6	4.78	662.6	6.62
Diplopoda	21	558.6	35	931	47.3	1258.18	24.6	654.36	25.6	680.9	44.3	1178.38
Symphyla . .	0.3	0.15	2.3	1.15	0.3	0.15	4	2	5	2.5	3	1.5
Pauropoda . .	0.3	0.15	6.6	3.3	1.6	0.8	1	0.5	2.3	1.15	3.6	1.8
Chilopoda	0.3	2.4	16	128	7	56	1.3	10.4	3.6	28.8		
Collembola	0.6	7.2	478	5736	217.3	2607.6	102.3	1227.6	214	2568	249.6	2995.2 μg
Campodeidae												
Japygidae												
Protura	0.6	3.99										
Psocoptera	8	53.28	17.3	115.22	3.6	23.97	2	13.32	4	26.64	1	6.66
Thysanoptera	1	6.66	4.3	28.64	1.3	8.66	0.6	3.99				
Heteroptera			0.3	1.99								
Homoptera	0.3	1.99									0.3	1.99
Hymenopt. (l)												
Hymenoptera	3.6	23.97										
Formicidae	3	19.98	4	26.64	3.6	23.97	1	6.66	0.3	1.99	5	33.30
Coleopt. (l)	2.3	34.5	4.6	30.63	26.6	177.15			2.6	17.31	0.3	1.99
Coleoptera	0.3	4.5	1.3	19.5	2	30	1	15	1	15	1	15
Diptera (l.)	2.6	26	4.0	60	4.3	64.5	0.3	4.5	0.3	4.5	1.3	19.5
Diptera			20.6	206	76.3	763	18.6	183	36	360	59.6	596
Lepidopt. (l.)	0.3	1.99	1	6.66								
Mecoptera (l.)												
Mollusca G.	0.6	12			0.3	6						

In the two different biotopes, although so close to one another, we find a rather different fauna, as one can see from the following list:
Forest: Nematodes such as *Aporcelaimellus obtusicaudatus, Aporcelaimus* sp., *Axonchium leptocephalum* (new to Italy), *Plectus cirratus, P. parietinus, Eudorylaimus* sp. (euryoikous genus), *Helicotylenchus* sp., *Mesodorylaimus* sp., *Anaplectus granulosus, Monhystera dispar* (possibly a vegetarian species), *Prodorylaimus* sp., *Rhabditis* sp. (generally feeding on bacteria, including living ones), *Suctonema* sp. (perhaps feeding on enchytraeids), *Nygolaimus brachyuris* (a carnivorous species), pseudoscorpions such as *Chthonius parmensis*; millipedes such as *Chordeuma silvestre* (petrophilous or 'montane' element), *Brachydesmus* sp., *Ophiulus* sp. and *Lophoproctus lucidus* (endemic to Italy); the earwig *Chelidurella acanthopygia* (feeding on fungal hyphae in the litter), the bug *Drymus brunneus*, the *Thysanoptera Haplothrips aculeatus, Thrips minutissimus, Odontothrips karnyi*, generally on *Leguminosae*; beetles such as *Trechus quadristriatus, Adalia 10-punctata*, a lady-bird very frequent in deciduous forests (Freude, 1966), *Cartodere separanda*, euryoikous species (Freude, 1967), *Micropeplus* sp. (*Staphylinidae*), *Orestia pandellei electra*, a very interesting Italian endemic, limited to the South-Eastern Alps and Prealps, *Langelandia anophthalma, Anommatus 12-striatus, Corticaria elongata*, and finally the weevils *Balanobius pyrrhocera* (the larva lives in the gallae of the Hymenopteran *Cynipidae Dryophanta folii*, on several species of oaks, particularly *Q. pedunculata*, which is parasitized by the Hymenopteran *Braconidae Bracon discoideus*), *Ceutorhynchus erysimi*, restricted to several *Cruciferae* whose roots are attacked by its larva, *Apion dichroum* (= *flavipes*) on several *Leguminosae, Apion elongatum*, on *Labiatae, Otiorrhynchus armadillo* (adult often on *Alnus glutinosa*, the common elder), *Barypithes gracilipes* new to Veneto. The number of parasite *Hymenoptera*, generally belonging to soil mesofauna, is very high: *Aphanogmus* sp., very euryphagous, *Ceraphron* sp., *Conostigmus, Dipara petiolata*, polyphagous but generally ectoparasitic, *Spalangiopelta brachyptera, Elachertus bisurmanus*, euryphagous, sometimes inside the larva of *Clysia ambiguella*, a *Tortricidae* harmful to the vine, *Stenopteromimar biciliatus*, generally oophagous and stenophagous (Grassé), *Leptacis* sp. (endoparasitic) and finally *Synopeas*, probably a parasite of *Diptera.*

Very different, as we have said, is the fauna of the clearing: particularly evident is the high proportion of planticolous, floricolous species compared with the fauna of the forest. Let us remember the earthworms *Octolasium, Eisenia rosea, Allolobophora rosea*, the nematodes *Aporcelaimellus obtusicaudatus* (feeding possibly on small Oligochaeta), *Aporcelaimus superbus* (idem), *Axonchium* sp., *Prionchulus* sp. (a genus feeding on Protozoa, nematodes, *Tardigrada*, rotifers and small *Oligochaeta*), *Plectus* (a genus feeding probably on bacteria), *Eudorylaimus, Helichotylenchus* sp. (some species of this genus are harmful to crops), the millipede *Brachydesmus* sp., the crickets *Pteronemobius heydeni* and *Acheta melas*; the bugs *Stygmocoris sabulosus, Catoplatus fabricii* (on *Chrysanthemum leucanthemum*) and *Lygus pratensis*, the *Aphidinae Baizongia pistaciae*, heterotopous on roots of *Gramineae* and on *Pistacia*, and *Coloradoa tanacetina*, on *Chrysanthemum*, the *Thysanoptera Aptinothrips rufus* (transmitter of the so-called depigmentation of several grasses), *Sericothrips abnormis*, generally in the flowers of *Leguminosae, Haplothrips aculeatus* (the larvae on several grasses), *H. tritici* (harmful to wheat in several South European countries), *H. kurdjumovi*, a carnivorous species, *Thrips tabaci*, living on grasses, tobacco, kapok, etc., *Frankliniella pallida*, on several *Leguminosae*, the beetles *Metopsia clypeata* (*Staphylinidae*), *Chaetocnema* sp. (*Chrysomelidae*), *Anommatus 12-striatus, Malachius australis* (present only in Italy and Southern France), *Sitona hispidulus* (the larva can be a real pest to several cultivated *Leguminosae*, more especially *Trifolium*, whose roots are severely injured), *S. sulcifrons*, often on *Leguminosae*, sometimes harmful to garden peas, *Apion millum* (on *Labiatae*), *Otiorrhynchus frescati*, an Italian endemic new to Veneto (the larvae of this genus are often harmful to the roots of cultivated plants) and finally the parasitic *Hymenoptera*, generally belonging to the soil mesofauna, *Aphanogmus* sp., *Trichomalus campestris* (possibly a parasite op *Diptera*), *Inostemma, Telenomus* (generally oophagous), *Trichopria* (endoparasitic) (Fig. 123).

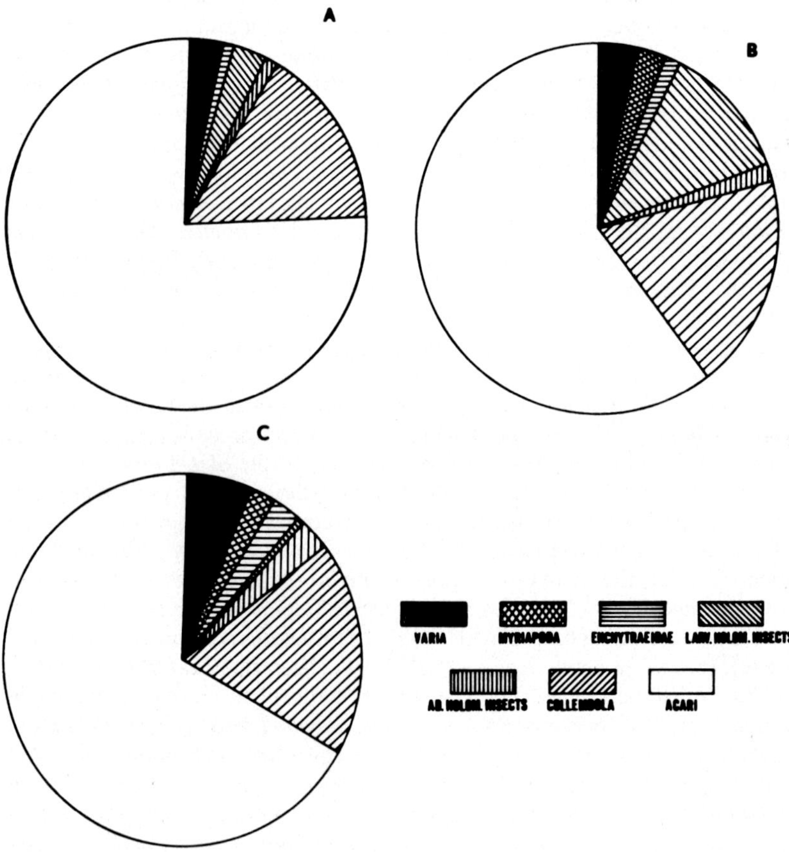

Fig. 123. Faunistic composition (%) of the soil fauna of Frassenelle: A, meadow, upper layer (0-4 cm); B, meadow, lower layer (4-10 cm depth); C, forest.

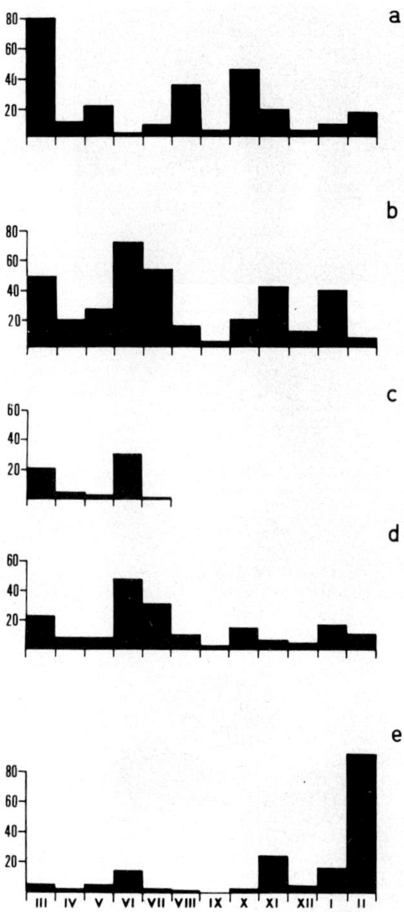

Fig. 124. Phenology of some Collembolans of Frassenelle (meadow, 0-4 cm depth). No. of specimens/500 cc of earth: a. *Tullbergia affinis*; b. *Onychiurus tetragramma*; c. *Isotoma minor*; d. *Isotoma bipunctata*; e. *Sminthurus elegans* (original, unpublished data).

Limited to the clearing, the *Collembola* have been studied by Dr Rusek, and we want to record them because in Italy, and more especially in Northern Italy, *Collembola* are rather poorly known: *Tullbergia affinis, T. quadrispina, T. krausbaueri, Onychiurus tetragrammatus, Isotomiella minor, Isotoma notabilis, Isotomurus palustris, Isotomina bipunctata, Pseudosinella sexoculata, Lepidocyrtus flexicollis, Anurophorus* sp., *Friesea mirabilis, Sminthurus viridis, Sminthurinus elegans, Folsomia quadrioculata, F. candida, Proisotoma* sp., *Lepidocyrtus* sp., *Heteromurus nitidus, Dicyrtomina* sp., *Onychiurus absoloni, Isotoma* cfr. *antennalis, Xenylla maritima, Entomobrya* sp., *Neanura* sp., *Cyphoderus albinus, Megalothorax minimus* and *Sphaeridia* sp. (Fig. 124).

If we compare now the collembolan fauna of this region with that of Recoaro

Fig. 125. Examples of soil fauna at Frassenelle: top left, *Trichopteron* larva; right, *Diptera* larvae; middle: two parasici *Hymenoptera*; Bottom: left, *Glomeris* sp.; right, juv. *Heteroptera*.

(see later) we see that only 11 species are common to the two localities, whereas 10 are present only at Frassenelle and 24 only at Recoaro. This shows the great difference existing between two localities, only 52 km apart, so that we must be very careful when we want to make generalizations in the description of a given European ecosystem or biome. Only very accurate research, if possible carried on with the same methods, will permit us in future to reach a more comprehensive understanding of the different ecosystems present in our Continent. Of course, in the case of Southern Europe the problem is much more complicated due to the presence of faunistic and floristic elements ever since preglacial times, differing from what we find in the other parts of Europe, where the Pleistocene Ice destroyed most of the pre-existent life.

Fig. 126. Examples of soil fauna at Frassenelle: top, left and middle, *Thysanoptera*; right, *Diplura Campodeidae*; Bottom: adult coleopterous insects.

Other localities whose soil fauna has been studied by the present author are M. Calaone near Este, Le Motte near Torreglia and Monte della Madonna near Teolo (Fig. 127).

The soils of these three localities are characterized as follows:

	Clay	silt	sand	$CaCo_3$	C	N	C/N	pH
M. Calaone (*)	13.8	36.2	49.9	–	3.84	0.25	16.8	6.7
Torreglia	31.9	26.5	42.5	–	1.32	0.09	14.6	6.5
M. d. Madonna	20.7	31.4	47.8	3.7	1.20	0.10	11.7	7.5

(*) Except the southern slope, where a Mediterranean forest is present

The edaphic microenvironment (i.e. soil temperature (at surface) and relative humidity (at surface) in the four seasons in the three biotopes) is represented by Fig. 129. Of course we may expect very great variations from one year to another. Fig. 118 shows the number of hours of insolation per year on the various slopes of the hills.

Fig. 127. M. Calaone: top, general view; middle left, *Quercus pubescens* and *Q. robur*; right, *Quercus ilex*; bottom: left, *Erica carnea*; right, *Castanea sativa* or chestnut.

The vegetation is rather different in the three biotopes. Namely, on M. Calaone we find a forest of *Quercus pubescens, Q. robur* (also hybrids between the two species), *Fraxinus ornus* and, on the north slope, *Castanea sativa* (chestnut). The most frequent shrubs are *Rosa gallica, Spartium junceum* (the common broom), *Coronilla emerus, Lonicera* sp., beside other perennials such as *Ruscus aculeatus, Asparagus acutifolius, Cistus salvifolius*, and some grasses such as *Poterium sanguisorba, Ranunculus* sp., *Achillea tomentosa, Potentilla erecta, Thymus serpyllum, Hieracium murorum, Melampyrum pratense, Serratula tinctoria* etc. Very frequent

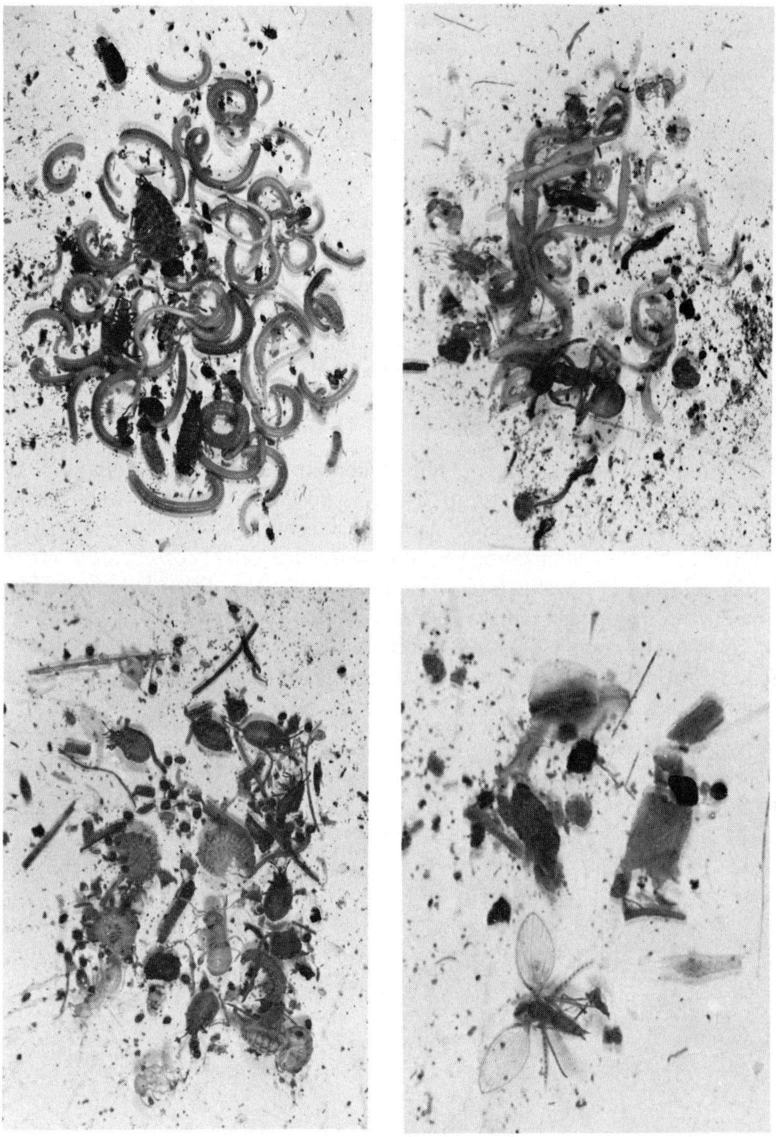

Fig. 128. Soil fauna of the Colli Euageni, just extracted by means of the Berlese' apparatus.

mosses are *Homalothecium sericans* and *Hypnum cupressiforme*. Furthermore, the fern *Polypodium vulgare* is common on the northern, cool, humid slope.

At Le Motte (Torreglia) on the northern slope we find a Castanetum, with *Fraxinus ornus, Calluna vulgaris*, the brake *Pteridium aquilinum*, and several mosses such as *Hypnum cupressiforme, Eurhynchium striatum* and *Dicranum majus*.

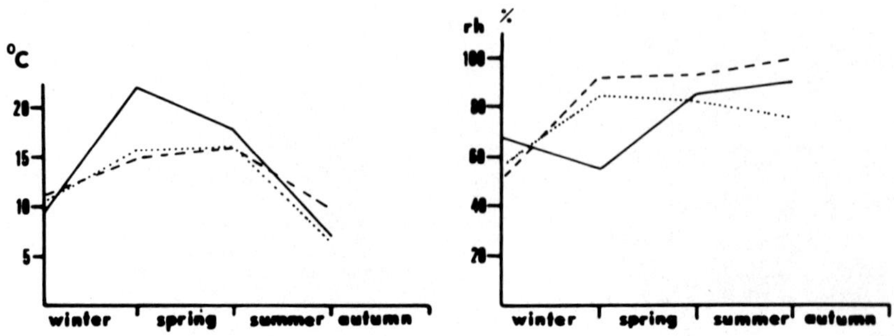

Fig. 129. Temperature (left) and relative humidity (right) at the soil surface at the moment of the collection of soils for animal research. Continuous line, M. Calaone (southern part of Colli); broken line, Torreglia (central part of the Colli); dotted line, Monte della Madonna, northern part of the colli.

On the Monte della Madonna, on limestone, the vegetation is composed of *Quercus robur, Cornus mas, Prunus* sp., *Corylus avellana, Castanea* (only on the northern slope), *Ostrya carpinifola* etc. At the surface we find (or rather we found before the recent heavy urbanization) *Ruscus aculeatus, Lotus corniculatus, Medicago minima, Onobrychis viciaefolia, Veronica officinalis, Scrophularia nodosa* and, on the northern slope, *Melampyrum pratense* and *Hedera helix*.

As far as the soil fauna of these three biotopes is concerned, we wish to represent here both the quantitative and the qualitative aspects. The quantitative data are summarized in Figure 130.

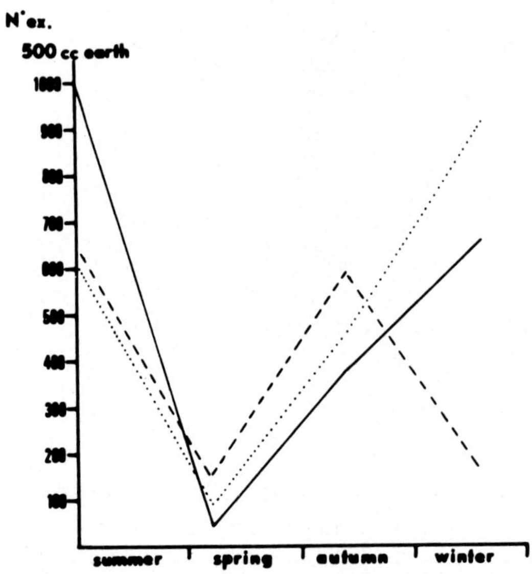

Fig. 130. Mean number of soil animals in four different seasons in the same localities (see Figure 129).

As far as the qualitative aspect of the soil fauna is concerned, we think it worth while to report the following data, also because they are practically the only information existing about the deciduous forest of Northern Italy, besides those we have already seen on Frassenelle.

1. M. Calaone: a) mesofauna: *Aporcelaimellus obtusicaudatus, Chthonius parmensis, C. tetrachelatus, Hysterochelifer tuberculatus, Eosentomon transitorium, E. delicatum, Acerentomon italicum, A. quercinum, Eosentomon romanum* (limited to Italy), *Acerentomon meridionale, Protentomon berlesei* (limited to Euganean hills and Cadore). *Acerentulus* sp., *Aptinothrips stylifer, Thrips physapus, Pselaphognathi* (larvae), *Polyxenus lagurus, Lophoproctus lucidus* (limited to Italy, Sicily, and islands off Naples), *Liposcelis* sp. and *Eretmocerus mundus* (a parasitic micro-hymenopteran).

b) macro- and megafauna: *Harpactes* sp., *Octolasium mima* (endemic of South Eastern Alps), *Trichoniscus* (*pusillus?*), *Philoscia affinis, Cylisticus convexus, Carinostoma argenteolunulatum, Chordeuma silvestre, Brachydesmus* sp., *Lithobius tricuspis, L. mutabilis, Monotarsobius microps, M. dubosqui, Chaetechelyne pharingealis, Scolioplanes crassipes, Himantarium gabrielis*, the earwig *Chelidurella acanthopygia*, the beetles *Quedius nemoralis* (?), *Tachyporus hypnorum, Luperus longicornis, Apion pisi, Helops coeruleus, Dromius linearis, Pedinus fallax, Barypithes gracilipes* and *Acalles lemur.*

2. Le Motte (Torreglia): a) mesofauna: *Aporcelaimellus obtusicaudatus, Neobisium carcinoides, Chthonius alpicola* (Eastern Alps), *Roncus alpinus* (Alps and Abrutian Appennines), *Eosentomon transitorium, Eosentomon delicatum, Acerentomon italicum, A. quercinum, Aptinothrips stylifer, Thrips physapus, Limothrips denticornis, Psyllipsocus ramburi* (Western-Mediterranean geoelement) and the parasitic microhymenopterous *Leptacis* sp.

b) macro- and megafauna: *Allolobophora* sp., *Philoscia affinis, Cylisticus convexus, Trochosa ruricola, Brachydesmus* sp., *Polydesmoidei* (indet.), *Schyzophyllum sabulosum, Ophiulus* sp., *Chilognathi* indet., *Lithobius nodulipes, Monotarsobius microps, Stigmatogaster gracilis* (Western-Mediterranean element) and the roach *Ectobius erythronotus.*

3. M. della Madonna: a) mesofauna: *Neobisium trentinum* (endemic to North Eastern Italy), *N. carcinoides, Chthonius orthodactylus, C. ischnocheles, Hysterochelifer tuberculatus, Roncus* sp., *Eosentomon transitorium, E. delicatum, Acerentomon meridionale, A. microrhinus, Acerentulus confinis, Acerella muscorum*, the *Thysanoptera Drepanothrips reuteri* (sometimes harmful to vineyards), *Megathrips lativentris* (possibly algivorous), *Taeniothrips picipes*, the micro-coleopteran *Raymondionymus fossor* and the micro-hymenopteran *Idris* sp.

b) Macro- and megafauna: *Allolobophora rosea, Octolasium* sp., *Cylisticus convexus, Trachelipus arcuatus, Lycosa radiata, Trochosa* sp., *Lacinius horridus, Glomeris* sp., *Chordeuma silvestre, Brachydesmus* sp., *Schyzophyllum sabulosum, Ophiulius* sp., *Chilognathi* indet., *Monotarsobius microps, M. aeruginosus* (a central European species), *Dicellophilus carniolensis*, the roach *Ectobius sylvestris*, the earwig *Chelidurella acanthopygia*, and finally the beetles *Xantholinus semirufus, Omophlus rugosicollis, Chrysomela haemoptera* (feeding on leaves of *Plantago*), *Athous* sp., *Otiorrhynchus armadillo, Calathus fuscipes* subsp. *latus, Platynus dorsalis, Pterostichys melas italicus, Harpalus azureus* (xerophilous), *H. dimidia-*

tus, Brachynus sclopeta, Asida sabulosa, Otiorrhynchus salicicola, O. fortis, O. globus and finally *Hypera variabilis.*

The Venetian Prealps

In the valleys which descend from the foothills at the Italian Alps, the deciduous forest – which we have already described for the Colli Euganei – grows up to the lower limit of the montane biome (i.e. 500 m). There it changes, with no clear boundaries, into the lower montane formations (chestnut, hazel, ash, etc.) still heliophilous, which only at higher levels (say 800-1000 m) are in their turn substituted by the beech forest, or sciaphilous deciduous forest. Needless to say the heliophilous forest of levels higher than 500 metres and, still more, the sciaphilous forests of these mountains, show a great similarity with those of Central and Northwestern Europe, and we will describe them in the chapter dedicated to montane biomes.

We will limit ourselves now to considering only the heliophilous forest of levels lower or equal to 500 m, and we will take as an example the Venetian Prealps north of Vicenza (not far from Colli Euganei), which include the locality of S. Quirico, in the Agno valley (Fig. 131).

Two facies are here recognizable, according to geology – represented by limestone (compact, massive limestone of Lower Ladinic) and volcanic tuffs and to soils (rendzina and brown earth, respectively).

The chemico-physic characteristics of the soils are the following

	pH	$CaCO_3$	clay	N	C	C/N	water
rendzina	6.8	13.1	19.9	0.63	7.05	11.2	27.43
brown earth	6.7	6.2	63.1	0.28	4.59	16.4	29.57

On limestone, the vegetation is a relict forest of *Quercus*, with *Ulmus scabra, Acer campestre, Fraxinus ornus, Fraxinus excelsior, Robinia pseudacacia* (introduced), *Carpinus betulus*, etc. and a herbaceous or shrubby vegetation with *Helleborus viridis, Epimedium alpinum, Vinca minor, Hedera helix, Euphorbia dulcis* and *E. amygdaloides, Lactuca muralis*, the vine *Clematis vitalba* (or traveler's joy), *Primula vulgaris* (or primrose), *Carex digitata* (a sedge), and the fern *Asplenium trichomanes*.

On vulcanic tuffs instead we find a vegetation with *Salix, Betula alba, Acer platanoides, Castanea sativa* and some grasses such as *Ranunculus repens, Fragaria vesca, Caltha palustris, Myosotis palustris, Platanthera bifolia, Urtica dioica* (a sign of human disturbance), *Oxalis acetosella, Vicia cracca*, some graminoid plants and the *Cryptogamae Pteridium aquilinum* and *Equisetum maximum*. In some places the forest has been eliminated and today we find meadows cropped for hay.

The soil fauna is illustrated in the following scheme (No. of specimens/500 cc of earth):

	February				April				June				August				October				December			
	300 m		500 m		300 m		500 m		300 m		500 m		300 m		500 m		300 m		500 m		300 m		500 m	
	limest.	tuffs	limest.	tuffs	limest.	tuffs	limest.	tuffs	limest.	tuffs	limest.	tuffs	limest.	tuffs	limest.	tuffs	limest.	tuffs	limest.	tuffs	limest.	tuffs	limest.	tuffs
Nematoda	2	2.5	36	3			7.3		4	1	15.2	1	43	7.5	52	19	25.5	7	0.5	3	29	5.5	52.5	1
Enchytraeidae	2.5	2	9	–	1.7	1.3	4.7	5.7	36.5	1	8.5	10.5	17	3.5	47		9.5	3		1	113	0.5	65	1
Lumbricidae														1				1.5			1			
Isopoda					0.3			4.7				0.5		0.5		1.5				2	2			
Pseudoscorpiones											1	0.5			2.5				0.5					
Acari	277	35	223	281	240	45.7	277	521.7	1120.5	22.5	427	163	782	96	958.5	492	1741	119.5	849	598	504	131.5	382	552
Aranaeidae	0.5						0.3												0.5					
Opilionida					0.3								1											
Diplopoda	0.5		0.5						2.5	0.5	0.5	0.5		0.5		0.5	2		0.5	0.5			1	
Symphyla	2				1		0.7		0.5	0.5	2.5		1		0.5	1.5	2	0.5	2			0.5		
Pauropoda	0.5	0.5			0.3		0.3		0.5		0.5				2.5				2					
Chilopoda					1				1.5		1	0.5			1		0.5						0.5	1
Collembola	73.5	21	19.5	17.5	127.7	14.3	54.3	21.7	162	11	153.5	17	78	17.5	160.5	19	308.5	7.5	76.5	25.5	170	9.5	80	83
Protura		0.5	1.5	1.5	1	0.7	4		7.5		2	1		4.5	38.5	4.5	6.5	13.5	11.5				2	1
Diplura							0.3																	
Psocoptera													1											
Thysanoptera															0.5	1								1
Heteroptera				0.5				1											13.5					
Homoptera			10	2			15	1.7		0.5	30.5	1.5			86	11			34.5	1.5			22.5	3
Coleoptera (l.)			0.5	1			1	2	0.5	0.5	1	2.5	3	1	1.5		2			1				
Diptera (l.)	7	2.5	53.3	4.5	9.7	3.7	37.7	5.7	10	2	62	6	9	2	60.5	4.5	8.5	9	112	5	5	6.5	69	6
Lepidoptera (l.)				0.3														0.5						
Coleoptera	1		2	5	0.3	2.6	0.3	1	1.5	1	1.5	2		0.5			1		1	1.5		1	0.5	
Diptera							0.3																	
Hymenoptera (no *Formicidae*)				0.5			0.3	0.3							0.5				1	2	2			
Formicidae					2	1		0.3		3.5		0.5		2.5				24						1
Mollusca																								

Fig. 131. Top. Anthropized aspects of the deciduous forest at almost 700 m. Bottom. Profile of soils of the same region, at almost 900 m: a brown earth and a reddish-brown earth, both podsolized, due to the desintegration of vulcanic tuffs.

The geophilous fauna of this woody region has been demonstrated to be very rich: earthworms, 6 species; isopods (or woodlice) 9 species; spiders 14; opilionids, 4; whip scorpions, 5; mites (limited to oribatids) 75 species; centipedes, 1 species; millipedes, 13; *Symphyla*, 4 species; pauropods, 5 species; Collembols (or springtails) 35 species; *Protura 5; Diplura Campodeidae,* 1 (*Campodea suensoni*); *Thysanura*, 1 (*Lepismachilis* sp.); *Psocoptera*, 2 species; *Thysanoptera*, 1 species (*Aptinothrips rufus*); bugs, 4 species; beetles, 41 species; ants 11; *Diptera* larvae 7; *Lepidoptera* larvae 2 and finally snails 28 species.

The composition of the fauna of this deciduous forest can be represented as follows (Fig. 132).

Among beetles, whose ecology and biogeography are generally better known, there are several species of great ecological interest, such as predators (*Pterostichus koyi, Pt. vulgaris*, several *Staphylinidae*), several phytophagous species such as *Sitona sulcifrons*, feeding on *Leguminosae, Otiorhynchus mastix, Apion seni-*

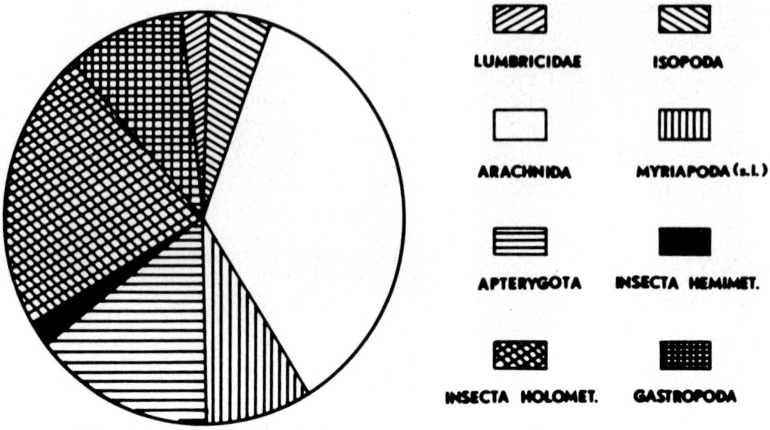

Fig. 132: Faunistic composition of total soil fauna of the Recoaro region.

culus (on *Leguminosae*), *Melolontha melolontha* (the common cockchafer), *Phyllopertha horticola* (harmful to fruit trees, *Rosa*, etc.), *Maladera holosericea* (radicicolous) etc. Other species are mycetophagous, for instance *Sphaerosoma* spp.; one species of weevil, *Smicronyx jungermanniae*, can be considered as useful since it attacks a parasite, *Cuscuta*, sometimes parasitic on flax. Some species finally are endogeous, like some *Staphylinidae* new to science or the weevil *Raymondionymus marqueti*.

Other beetles have a paramount importance to biogeography such as *Metrotyphlus castrii*, new to science, *Leptotyphlus* sp., also a new species; *Octavius castrii*, the same; *Meotica exigua*, known only from the Southeastern Alps and Prealps, *Arcopagus longulus*, from the South Eastern Alps, Istria and Croatia, a proof of the ancient origin of the Agno valley, belonging to an old land mass with no break in continuity to the North-western Balkans (the old 'North Aegeis').

The oak forests of South-Eastern England

Most of South-Eastern England in ancient times was covered by oak forests, as documented by several contemporary descriptions. The dominant species was *Quercus robur*, but *Quercus sessiliflora* (= *petraea*) and *Corylus avellana* were also present. Today, the hazel forms coppices. Remains of these forests are present today in New Forest (the term 'new' refers to the time of William the Conqueror, i.e. ca. 1080, which means almost 900 years ago) and Windsor Forest, part of which is accessible to the public, which regularly visits Windsor Castle, residence of the Royal Family. Often these magnificent forests have been reforested today, generally with the introduction of coniferous trees, and therefore their fauna is very probably impoverished, as always happens when man intervenes to alter the delicate natural equilibrium. Other arboreal species are the beech, which, as we know, is not heliophilous but can thrive if protected by oaks, the hornbeam

Carpinus betulus, the aspen *Populus tremula*, while the floor of the forest is covered by *Anemone nemorosa, Primula vulgaris* (the common primrose), *Mercurialis perennis, Ajuga reptans* and *Euphorbia amygdaloides*, common to all European oak forests, beside some rarer species, such as *Endymion nonscriptus*.

The climate of these forests is similar to that of the London area. The insects are still numerous as has been shown by the fine research carried out by Richards of the Imperial College of Sciences and Technology (London). Associated with oaks are the beetles *Calosoma inquisitor*, one of the most elegant European species, living up to its name (from *καλός* beautiful, and *σόμα* body), *Xylodrepa quadripunctata*, whose pronotum is vivid yellow with 4 black spots, from which its name (*4-punctata*) originates; *Agrilus pannonicus*, a small beetle with yellow-reddish legs, exclusive to Central Europe, and several weevils, such as *Attelabus nitens*, a small beetle looking as if varnished with Shellac, the species of the genus *Curculio venosus, glandium, villosum* and *pyrrhoceras*; the jumping *Rhynchaenus quercus* and *R. pilosus*, together with the dangerous *Scolytus intricatus*, a species particularly linked to oaks, in which it excavates a maternal gallery, transverse and short, 10-300 mm long, and *Anisandrus* (= *Xyloborus*) *dryographus* (literally 'which writes on oaks' from *δρύας* oak and *γράφο* to write, since its larva excavates passages under the bark which resemble strange writing or some odd and elegant hieroglyphics). This species feeds on oak, chestnut etc. and the passages have irregular and bifurcate lateral cuniculi. The Lepidoptera are also very abundant; some of them feed exclusively on oaks (stenophagous or even monophagous species). We find furthermore tenthredinid *Hymenoptera* (phytophagous and therefore more or less harmful), *Cynipid Hymenoptera*, causing the production of very peculiar galls on oak branches (Fig. 133) and lastly Cecidomyid Diptera (from *κέκις*, gall and *μυία* fly), also gallicolous.

The vegetation and fauna of these English oak forests are the pride of ecologists and of everybody who loves nature and who is worried about the future of

Fig. 133. Galls produced by the Hymenopteron *Cynips quercus tozae* (from Cecconi).

Fig. 134. The main streams of migration towards the British Islands at different times (from Scharff).

our landscape-beauty (what in America people call Wilderness and also Natural Beauty). Partly, they are autochtonous species, present in the same area even during the peak of Pleistocene glaciation (when the climate was decidedly colder than today), partly they are re-immigrants from the south or east after each glaciation, and more especially, after the last one (Würm). Indeed, as we can appreciate from Fig. 134, there have been periods in which only the southernmost part of England was free from ice, while all the rest of Britain and Ireland was covered by a sheet of ice, much the same as that which covered all the territories from the Arctic to Central Europe. Only in the final Würm advance was most of Britain free from ice, while in the penultimate advance more than half of it was covered by ice, so that any trace of plant and animal life was completely eliminated from this area.

Eighteen thousand years ago the glaciers retreated, and for the succeeding 10,000 years the climate became warmer, so that England could be re-populated from the south (both from the Continent and from the parts of England which were always free of ice), but 8,000 years ago – at the very moment at which the fauna was reaching its present aspect – Britain began to separate from the Continent, to be completely detached 6,000 years ago. Since that date, of course, no immigration of terrestrial animals and higher plants (i.e. vascular plants) from the

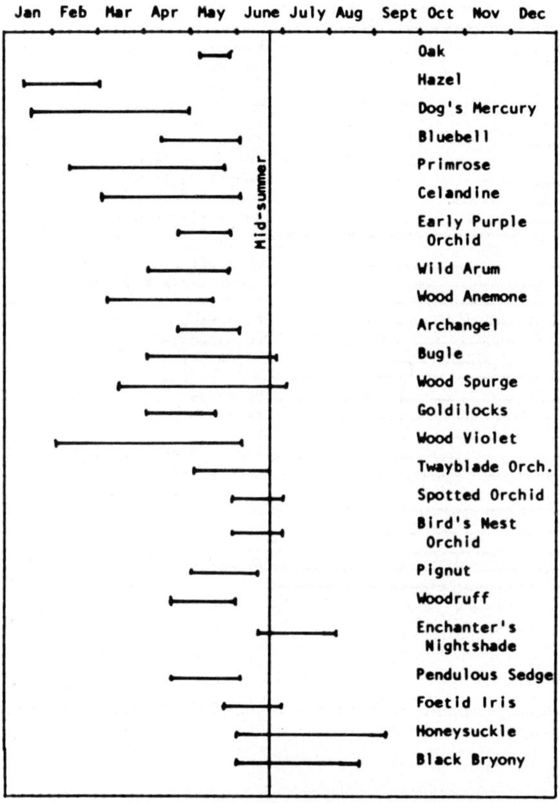

Fig. 135a. Time of anthesis of woodland plants in Southern England (from Neal) (with the permission of Heinemann Educ. Books).

continent has been possible. Some, however, say that small mammals and several higher plants were introduced by man, probably unwittingly, during the Neolithic.

Remains of beetles embedded in turf near Chelford, Cheshire, dating from a relatively warm period (interstadial) of the last glaciation (Würm), i.e., 57,000 years ago (the age has been estimated by the ^{14}C method), show that out of 100 species, all but 12 are also present today in the same locality. It is a fauna very similar to that present today in Finland at latitudes between 60 and 64°N. In deposits of mud and debris at St. Bees, in West Cumberland, dating from 10,500 years ago, out of 104 beetle species only 4 are absent today in Britain. Some of them, however, are found at higher latitudes, such as *Patrobus septentionis, Amara alpina* (present from Scotland and Norway to the Bering Strait) and *Carabus glabratus*.

Furthermore, in a stratum dated at nearly 13,560 years ago, situated near St. Albans north of London in Hertfordshire, there is a species, *Diachila arctica*, today living only in Northern Scandinavia, Siberia and North America but during the glacial era also present in Central Europe. 8% of the fauna of this stratum at

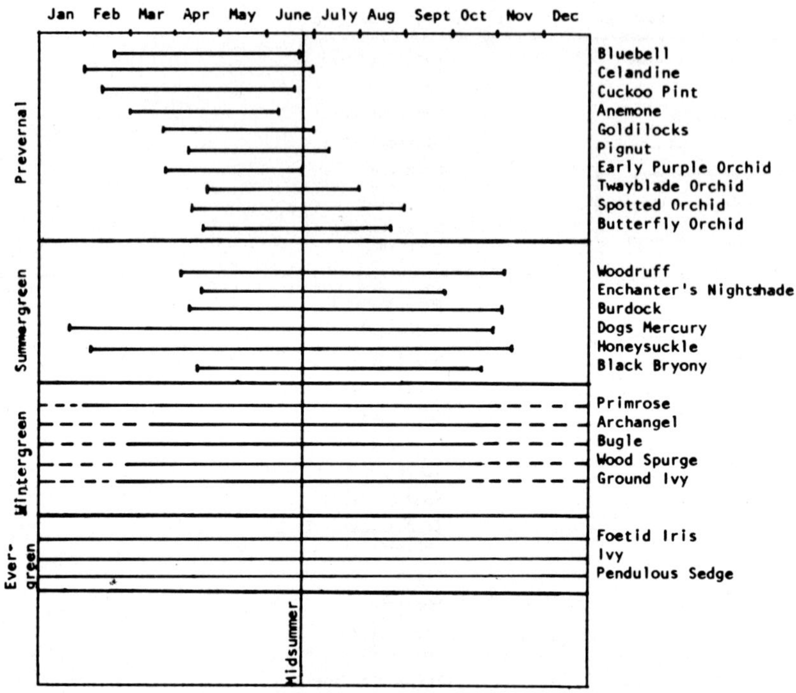

Fig. 135b. Leaf condition of woodland plants in Southern England (from Neal) (with the permission of Heinemann Educ. Books).

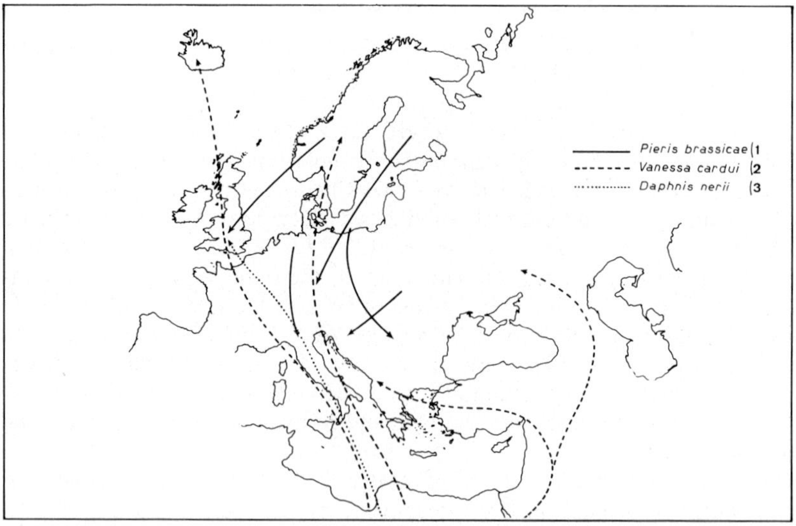

Fig. 136. The main routes of migration of three species of *Lepidoptera* common to Europe- and deciduous forests: the common cabbage butterfly, 1 *Pieris brassicae*; the common nettle butterfly, 2 *Vanessa urticae*, and the rose bay hawk moth, 3 *Daphnis nerii* (Photos of the insects from Grandi).

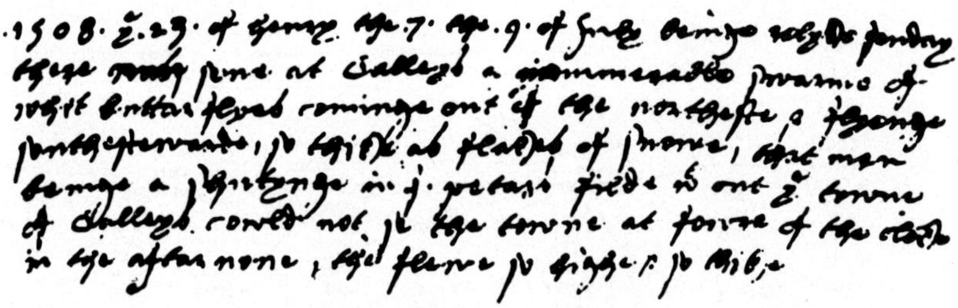

'1508, the 23rd year of Henry the 7, the 9 of July, beinge relyke Sonday, there was sene at Calleys an innumerable swarme of whit buttarflyes cominge out of the north este and flyinge south estewards, so thicke as flakes of snowe, that men beinge a shutynge in St. Petars fields without the towne of Calleys could not see the towne at foure of the clock in the aftarnone, they flow so highe and so thicke.'

Fig. 137. First record of migration of *Pieris brassicae* (the common cabbage butterfly) from the 'Chronicles of Calais in the Reigns of Henry VII and Henry VIII' (Turpyn).

St. Albans today has a more northerly distribution in Britain, a proof of the colder climate of that era.

We must point out that, in England, in the dominion of the deciduous forests, there are also some endemic species, among them 6 insects, of doubtful origin and provenance, beside some geographical subspecies which also seem exclusive to England. An example is *Papilio machaon britannicus*, which differs from the continental form in the colour and because its caterpillar feeds on a mire-*Umbellifera, Peucedanum palustre*, instead of several kinds of *Umbelliferae* as does the continental form.

Very interesting are the migrations of butterflies, which each year arrive regularly in England from the Continent (Fig. 136). Possibly the first reference to the migration of the common cabbage butterfly, *Pieris brassicae*, dating from 1508, appeared in Turpyn's 'Chronicles of Calais in the Reigns of Henry VII and Henry VIII' (Fig. 137). Still more interesting is the recent invasion of England by butterflies, a sign of a climatic change, which perhaps is also responsible for the extension northwards of some Central-European mammals and birds. Here are the names of the butterflies and the data of their first arrival in England:

Species	Family	year
Euphyia luctuata	Geometridae	1920
Eupithecia millefoliata	Geometridae	1939
Calophasia lunula	Noctuidae	1939
Hadena compta	Noctuidae	1948
Phalera fimbrialis	Geometridae	1950
Hydraecia hucherardi	Noctuidae	1952

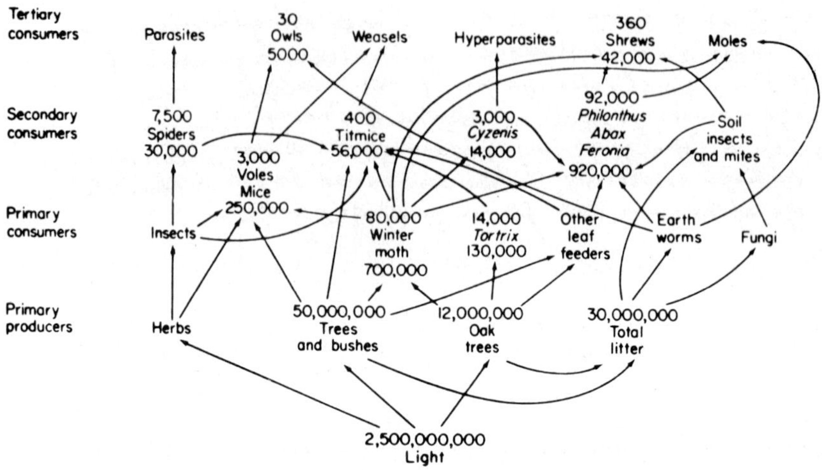

Fig. 138. Simplified energy-flow diagram for Wytham deciduous forest, near Oxford, England. Estimates of consumption and production in kcal/hectare-yr are inserted above and below each link for which figures are available (Varley, 1970).

Moravia

Another district where heliophilous forests are predominant is Moravia. In a region situated north of Brno the forest has developed on some karstic, calcareous substrate, with *Acer pseudoplatanus* and *Fraxinus excelsior*, which corresponds to the plant formation Acereto-Fraxinetum. Associated trees are *Sorbus aucuparia, Abies pectinata*, and others, while on the floor of the forest we find shrubs such as hazel, *Sambucus nigra, Sambucus racemosa, Lonicera xylosteum*, related to the well known honeysuckle, the cornel *Cornus sanguinea* and the spindle tree, *Evonymus europaeus*. Among herbaceous plants there are *Impatiens noli-tangere*, the *Crucifera Dentaria enneaphylla*, with nice small pink flowers and dark green, subdivided leaves, the *Aristolochiacea Asarum europaeum*, the monocotyledon *Polygonatum multiflorum*, the so-called Solomon's seal, some *Labiatae* such as *Lamium maculatum* and *L. galeobdolon*, the delicate and inconspicuous shamrock, *Oxalis acetosella*, containing calcium oxalate which gives the plant an acidic flavour, the lungwort, *Pulmonaria officinalis*, a *Borraginacea* with white-spotted leaves and blue purple flowers, a short day plant, and lastly several hygrophilous ferns, such as *Aspidium filix-mas, Asplenium trichomanes, A. viride* and *Cystopteris fragilis*. The soil where this forest has grown is a brown rendzina, with a pH = 7.1-7.5 (the high value is due to the presence of calcium carbonate in the substrate).

The soil fauna is represented by some collembolans, not very different from those present in North Eastern Italy, which points to the similarity of soil and climate and to the territorial connections, which have been possible since extremely ancient times (at least Middle Miocene).

We want to finish this small chapter with a brief description of the Querceta of

Belgium. As everybody knows (even from Caesar's description of 'Gallia' in his De Bello Gallico) most of Central and Central Western Europe, as well as Northern Italy (the Po plain) was covered with immense oak forests. Today some remains of these are still present in Belgium, for example in Haute Campine, an alluvial region with a temperate Atlantic climate. The vegetation is composed of *Quercus robur* and *Q. sessiliflora*, with *Betula pendula* and *B. pubescens*, i.e. a Querceto-Betuletum, representative of what phytosociologists call Quercetalia roboris-sessiliflorae, characteristic of acid, quartz soils. Other trees are *Pinus silvestris* (the common pine tree) and the aspen or *Populus tremula*. The forest floor is very rich in plants, both herbaceous and shrubs: let us remember the buckthorn *Rhamnus frangula*, the whortleberry *Vaccinium myrtillus*, the heath *Calluna vulgaris, Lonicera periclymenum*, the holly, *Sarothamnus scoparius*, and the broom *Genista anglica*. The herbaceous plants are *Molinia coerulea*, with linear, sharp and flat leaves and a violet inflorescence, after which the plant is named (*coerulea* means bluish), *Agrostis alba, Deschampsia flexuosa* (graminoid plants), the *Ciperacea Carex pilulifera*, the *Juncacea Luzula multiflora*, the beautiful *Majanthenum bifolium* and *Convallaria majalis*, the lily of the valley, which are rather similar in form, colour of leaves (shining, smooth and delicate) and flowers (whitish in the former, milk-white in the latter). The Dycotyledones are represented by *Teucrium scorodonia* and *Melampyrum pratense*, a semi-parasite as one can easily appreciate from the reduction in its chlorophyll containing parts. Ferns are represented only by the acidophilous *Pteridium aquilinum*, the common bracken. The soil is often covered with mosses.

The soil of this Central European forest is a brown podzol, with a very low pH (less than 5) and a raw humus (or Mor). At lower levels big lenses of clay have been observed. In the B-horizon there is ferric illuvation with the accumulation of abundant grains of quartz. The humus is represented at surface by a Mor and deeper by a Mull. The only mammals inhabiting this soil are the common mole and a wild mouse (*Apodemus* sp.). The invertebrate fauna consists mainly of earthworms (responsible for the formation of Mull), mites, springtails, dipteran larvae, centipedes and beetles.

Another Quercetum with *Galeobdolon* and *Oxalis*, near Mesnil-Eglise (Férage) has been studied very recently by Krizeli & Verstraeten. Some quantitative data on epigeous invertebrates of this Quercetum are reported here:

	Diptera	Hymenoptera	Coleoptera	(average No./trap)
Charme	230.2	136.6	47.6	
Coudrier	309.8	117	53.4	

Within the Diptera the following quantitative data are available:

	Sciaridae	Empididae	Tachinidae	(average No./trap)
Charme	42.	4.6	4	
Coudrier	23.16	31	0	

Of course the research should be extended to all insects or possibly to all arthropods.

Finally we want to report the number of specimens and the biomass (dry weight) of all the Diptera families studied by these authors, since they are practically the first data we have for European deciduous forests. It is needless to say that the biomass is an index – although only partial and indirect – of secondary productivity, and the knowledge of productivity – as everybody knows – is now one of the main aims of Ecology.

	Number of specimens/ha			Biomass (g/ha)		
	1967	1968	1969	1967	1968	1969
Psychodidae	2	2	16	0	0	2
Anisopodidae	30	0	2	42	0	3
Limnobiidae	674	100	240	1550	260	552
Tipulidae	10	10	16	130	130	208
Bibionidae	6	4	24	12	8	45
Mycetophilidae	414	374	254	248	225	152
Sciaridae	3654	300	434	658	54	78
Cecidomyidae	974	572	470	205	120	99
Ceratopogonidae	36	4	8	4	0	0
Chironomidae	548	26	52	219	10	20
Empididae	998	50	154	1796	90	277
Dolichopodidae	0	0	12	0	0	7
Lonchopteridae	170	14	20	68	6	8
Phoridae	260	298	372	65	74	93
Lauxaniidae	380	36	90	456	46	108
Chloropidae	2	0	0	0	0	0
Anthomyidae	32	10	30	64	20	60
Tachinidae	16	46	34	208	598	442
Total	8206	1848	2228	5725	1641	2154

BIBLIOGRAPHY

Albert R., Struktur und Dynamik der Spinnenpopulationen in Buchenwäldern des Solling, Verh. Gesell. Ökol., Göttingen, 1976, 83.

—, Biomasse von Chilopoden in einem Buchen-Altbestand des Solling, Verh. Gesell. Ökol., Göttingen, 1976, 93.

Altmüller R., Oekoenergetische Untersuchungen an Dipterenpopulationen im Buchenwald, Verh. Gesell. Ökol., Göttingen, 1976, 133.

Bertarelli L.V. & E. Boegan, Duemila Grotte, Touring Club Italiano, Milano, 1926.

Grimm R., Der Energieumsatz der Arthropodenpopulationen im Oekosystem Buchenwald, Verh. Gesell. Oekol., Göttingen, 1976, 125.

Haas H., Investigations on secondary productivity in a meadow in the Solling, FRG. Abundance, phenology and production of imagines, *Polish Ecol. Studies*, 2, 2, 1976, 153.

Hartmann P., Struktur und Dynamik der Staphylinidenpopulationen in Buchenwäldern des Solling, Verh. Gesell. Oekol., Göttingen, 1976, 75.

Hediger, Jagd Zoologie, F. Reinhardt Verlag, Basel, 1966.

Herlitzius R., Untersuchungen zum Streuabbau in Kalk- und Sauerhumusbuchenwäldern, Verh. Gesell. Oekol., Göttingen, 1976, 161.

Jeannel R., Coléoptères Carabiques, Faune de France, vol. 39, 40, 1941-2; vol. 51, 1949, Lechevalier, Paris.

—, Coléoptères Psélaphides, ibidem, vol. 53, 1950.

Koehler H., Nahrungsprektrum und Nahrungskonnex von Pterostichus oblongopunctatus und P. metallicus (Coleoptera, Carabidae), Verh. Gesell. Oekol., Göttingen, 1976, 103.
Magistretti M., Cicindelidae, Carabidae, in Fauna d'Italia, 8, Bologna, Calderini, 1965.
Marcuzzi G., Ecologia animale, Feltrinelli, Milano, 1968.
Marcuzzi G. & L. Dalle Molle, Considerazioni ecologiche sulla fauna del suolo di una regione piana dell' Italia settentrionale (Granze di Frassenelle, Padova), *Boll. Inform.* N. 35, Agrobiologia, Bologna, 1975.
Müller G., I Coleotteri della Venezia Giulia, I. Adephaga, Trieste, Tip. Mosettig, 1926.
Müller P., Voraussetzungen für die Integration faunistischer Daten in die Landesplanung der Bundesrepublik Deutschland, *Schrift. Vegetationsk.*, 10, 1976, 27.
Neal E.G., Woodland ecology, London, 1969.
Osella G. & A. Zanetti, La coleotterofauna dei nidi di Talpa europaea L. nell' Italia settentrionale a nord del fiume Po, *Boll. Zool. Agr. Bachic.* 12, 1974, 43.
Rusek J., Soil and caves as the environment of animal life, *Probl. Speleolog. Res.*, Prague, 1965, 163.
Schauermann J., Zur Abundanz und Biomassendynamik der Tiere in Buchenwäldern des Solling, Verh. Ges. Oekologie, Göttingen, 1976, 113.
Scherner E.R., Struktur und Dynamik der Avifauna des Solling, Verh. Gesell. Oekol., Göttingen, 1976, 145.
Thibaud J.-M., Biologie et ecologie des Collemboles Hypogastruridae édaphiques et cavernicoles, *Rev. Ecol. Biol. Sol*, 7, 1970, 283.
Vandel A., Biospéologie, La biologie des animaux cavernicoles, Gauthier-Villars, Paris, 1964.
Weidemann G., Struktur der Zoozönose im Buchenwald-Ökosystem des Solling, Verh. Gesell. Oekol., Göttingen, 1976, 59.

THE TAIGA

General features

The whole of Europe, from latitude N 58° to 70° and from longitude East 10° to 70° is covered by an immense forest, affected by the cold continental climate, called 'the taiga'[1]. The forest is also known as a coniferous forest, due to the dominance of coniferous plants, particularly *Abies* and *Pinus*. The broad-leaved trees are essentially birch (*Betula* sp. plur.), present sometimes in pure stands or more often mixed with coniferous species. The colour of the taiga changes from dark green, when the coniferous species form pure forests, to pale greenish in the growing season or bright yellow during autumn, if birch prevails. In winter, of course, only coniferous species (except larch) keep their leaves. The boreal forest belt constitutes the broadest forest cover of emerged lands (16.5%) present in Northern America, Greenland and Northern Eurasia. The Norway spruce *Picea excelsa*, and the birch reach the northern limit of tree vegetation, going from 59° in Norway (more or less at the latitude of isotherm 16° of the warmest month) to 72°N in Siberia. In these countries the coniferous species withstand temperatures as low as –30°C. Snow covers the soil with a thick blanket for several months (up to 7 or 8) and its progressive slow melting prevents an early growth of underwood, which would have very negative results. The very long winters, the persistence of snow, the thick fogs and the soil, which is often frozen at a certain depth even in summer, are the climatic conditions which coniferous trees – the dominant elements of taiga – must endure. The winter snow which accumulates upon the limited surfaces of evergreen leaves or needles, and the thick resinous barks preserve pines and spruce from frost. Their roots, although obtaining water only with difficulty from the soil, since water is immediately turned into ice as soon as it is deposited, are able to compensate for the loss of water due to the transpiration through the inconspicuous canopy.

In the dark and dense coniferous forest the plants of the undergrowth are very scarce and often only lichens and mosses cover the soil with a thick blanket. The larch forests, which during winter lose their leaves, are clearer and brighter and as a consequence juniper and whortleberry can grow vigorously. Here and there clearings appear, and herbaceous vegetation can grow undisturbed, so that, during summer, a magnificent flowering occurs. Very often these surfaces are wet and covered with mosses (see later).

The immense forest cover of the taiga is interrupted only by some cultivated areas (or here and there by human settlements) or more often by lakes, swamps and especially bogs. We must not forget that the great majority of the taiga is flat, often developed on alluvial soils, and that most of the area today belonging to this biome was covered with ice during the Pleistocene. The last glaciation – Würm – which of course in Northern Europe (as well as in Northern Asia and America)

[1] Russian word of altaic origin.

had a much longer duration than that in the Alps or other South European countries, ending only some 8000 years ago (recent Dryas period), has permitted or more often caused the formation of an enormous number of all kinds of water bodies. They vary from the biggest lakes in Europe (Onezhkoye Ozero, Ladozhkoye Ozero, Chudskoye Ozero, Rybinsk Lake in USSR, Vänern and Vättern in Fennoscandia, etc.) to small ponds or very often to bogs. It is useful to point out that the latter sometimes represent the final stage of a lake or a pond, which in the last 8000 years has been gradually filled with earth and then transformed into swamp and lastly into a mire. Here some small ponds can still exist (the so-called 'Mooraugen' of German authors).

The taiga is crossed by several rivers, interrupting its continuity, such as the Pechora, the Kama (one of Volga's tributaries), the Upper Volga, the Upper Dvina, the Ulla (in Finland). There are several rivers also in Sweden, all of them with a NW-SE direction, discharging into the Gulf of Bothnia.

From the climatological stand-point, the taiga belongs to the East European continental climate, with the exception of the Scandinavian mountains, where an Arctic or mountain climate prevails, and which of course are not covered with trees (see later). Norway and Sweden benefit from the Atlantic climate.

The climate is characterized by permanent cold or by mild summer and cold winter, and by rather sparse, seasonal, rainfall.

Of course, we may expect very great variations due to the different distance from the sea, but we must keep in mind that in the north the sea does not always have the effect that it has, say, on the Atlantic coasts of Europe, where it causes, in general, a milder, oceanic, climate, ideal for broad-leaved forests (Fagetum). If the sea is very cold, for instance in the case of the White Sea (and even the Baltic Sea), the continentality is not much affected.

As far as geology is concerned, the territory today covered by the taiga corresponds mainly to the Russian platform, and secondarily to the Baltic shield (peneplaned in the Precambrian, and successively modified by tectonic disturbances and uplifting in the Tertiary) and to the Russian granitic areas of Fennoscandia, with the exception of the Western portion of Norway, where the Caledonian plateaux (locally called 'Fjelde') are found, composed of gneiss and granite, and rising up to the highest mountains (2481 m). The easternmost portion of the territory, corresponding to the Urals, is represented by Hercynian massifs. We must keep in mind in this connection that the eastern border of Europe, covered by the taiga, is the eastern slope of the Urals (Južnyi Ural).

The soils, on the whole, belong to the podzol category, but of course other major soil types are also represented, due partly to climate, partly to the substrate characteristics (Fig. 139).

Most of the European taiga shows a ferric podzol, present in Northern Russia, in the regions of the North Sea, Baltic Sea and Fennoscandia and characterized by the presence of ferrous oxide (bivalent iron, and therefore acid). Near the coast of Western Norway and on the coast of Southern Finland ferric-humic podzol is present, on loose quartz sands, sandstone, quartzites, mica schists, gneiss and very acid granites, covered by *Calluna* and *Erica* (heath vegetation). Only at the extreme northern limit of the taiga has the so-called nordic nanopodzol developed, upon acid eruptive rocks and crystalline schists rich in silt.

Other soils well represented in the European taiga are eupodzolic brown earth, upon rocks poor in calcium, limited to Fennoscandia; centro-european brown earth, exclusive to the mixed forest ('Mischwald') of Southern Scandinavia, upon crystalline, non-calcareous rocks; oligotrophic brown earth, upon rocks poor in calcium in all Fennoscandia and some parts of Northern Russia. Upon non-calcareous rocks raw soils belonging to the category of rankers are also present, namely podzol-ranker, corresponding in practice to the upper layer of a normal podzol, without a B-horizon, present both in Northern Russia and in Scandinavia; it is covered by *Calluna* and *Empetrum nigrum*, besides coniferous trees. Dystrophic ranker (upon podzol) is very widely distributed all over the taiga, upon acid compact rocks, poor in nutrients and bases, such as sandstone, quartzites, siliceous

Fig. 139. Examples of podzols: left, Iron podzol under *Pinus silvestris* stand, above morainic soil in Finland; middle: Podzol-Ranker under *Calluna* and *Genista* above quarzit (Cordillera Cantabrica, Asturia); right, eupodzolic brown earth in a montane meadow with *Crocus nudiflorus*, above quarzitic slates (Cordillera Cantabrica, Asturia) (from Kubiena).

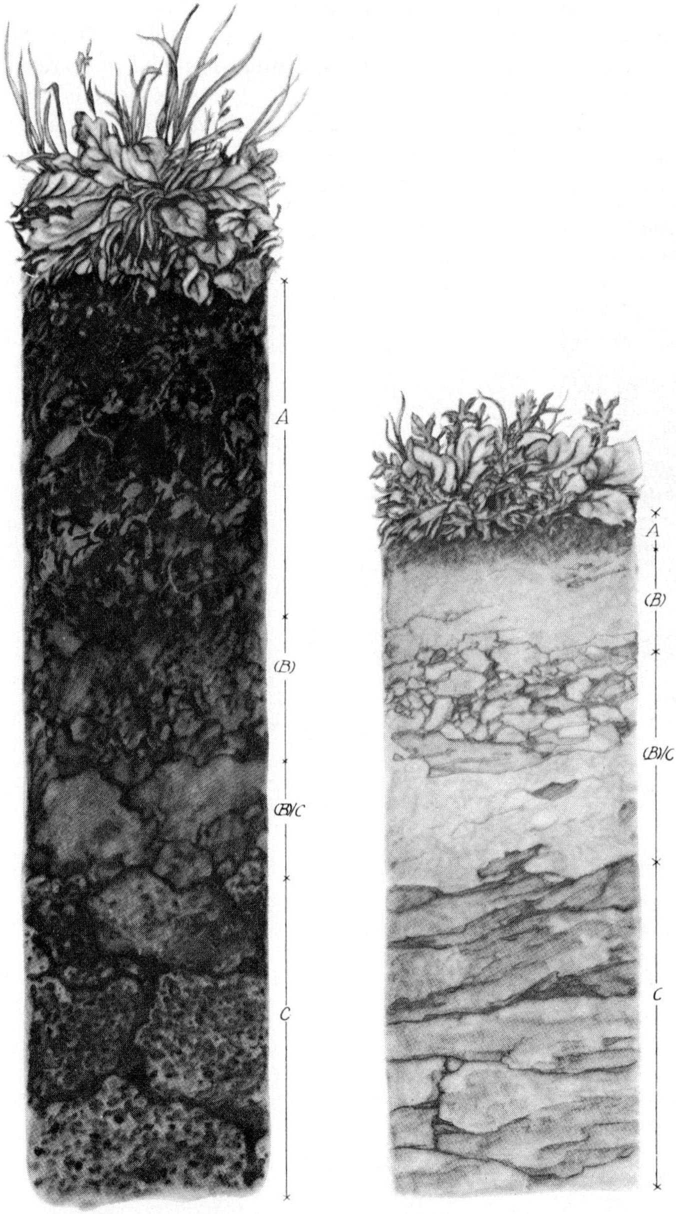

Fig. 140. Examples of brown earth: left, Centro-european eutrophic brown earth in a beech forest above basalt (Burgenland, Austria); right, Centro-european oligotrophic brown earth under mixed forest, above sandstone slates, in Styria (from Kubiena).

slate, acid eruptive rocks, crystalline slate, etc. This soil is covered mainly by mosses (*Hyloconium splendens, H. triquetrum, Hypnum crista castrensis*, etc.), besides *Calluna, Erica tetralix, Vaccinium vitis-idaea*, etc. (Fig. 140).

Finally, where limestone is present, and limited to Fennoscandia, soils of the rendzina category have developed, more precisely the mulliform rendzina, a rendzina rich in mineral substances but poor in clay, upon limestone poor in clay, siliceous limestone and dolomite, covered by heaths and *Pinus* stands or other coniferous trees, and the protorendzina (humid) (in German 'Urfleinserde') a raw,

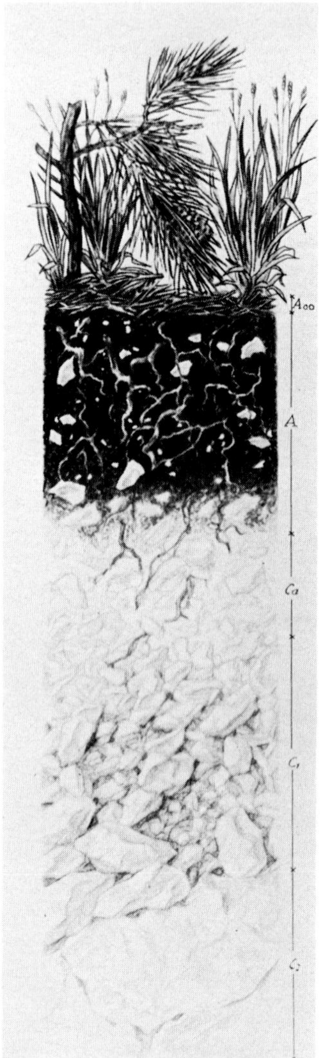

Fig. 141. Mulliform rendzina, under a *Pinus nigra austriaca* stand with *Sesleria varia*, upon dolomia (Mödling, Southern Wienerwald, Austria) (from Kubiena).

superficial, soil, composed of calcium humate and plant remains (Kubiena) (Fig. 141).

Needless to say the soil – as well as the vegetation – changes, with no clear limits, into the soil types of the tundra, in the North, and those of the mixed forest or even broad-leaved forests in the South (Fig. 142).

Most of the taiga belongs to the less densely inhabited parts of Europe (and in general of the world), the density being often 0-1 inhabitants per square kilometre. Nevertheless, the stability of the virgin forest landscape, as Björn Berglund says (1969), was already disturbed in Scandinavia 5000 years ago. The human landscape, characterized by open or half-open vegetation and by regenerated woods, has developed since then. Based on an interpretation of pollen diagrams from some areas of Southern Sweden, four general expansion stages of human influence have been described and dated within the period 3300 B.C. to 1100 A.D.

In Northern Russia there was a Neolithic culture characterized by nomad populations. The culture named after the comb pottery ('Kammerkeramikkultur') has been discovered in the Ural region. Here people used to extract copper in fairly recent pre-historic times. Nevertheless, we must keep in mind that prehistory is practically unknown as far as North Russia is concerned, where even today primitive populations exist, which show some affinities with the Madgalenian populations of the Upper paleolithic. There are also some Neolithic caves and stockades of poorly established ages. The wall engravings found near Nižnij-Tagil cannot be dated to a particular era.

About 2000 B.C. some ugro-finnic peoples inhabited the region between the Volga and the Dvina and the western slopes of the Urals. These peoples met the

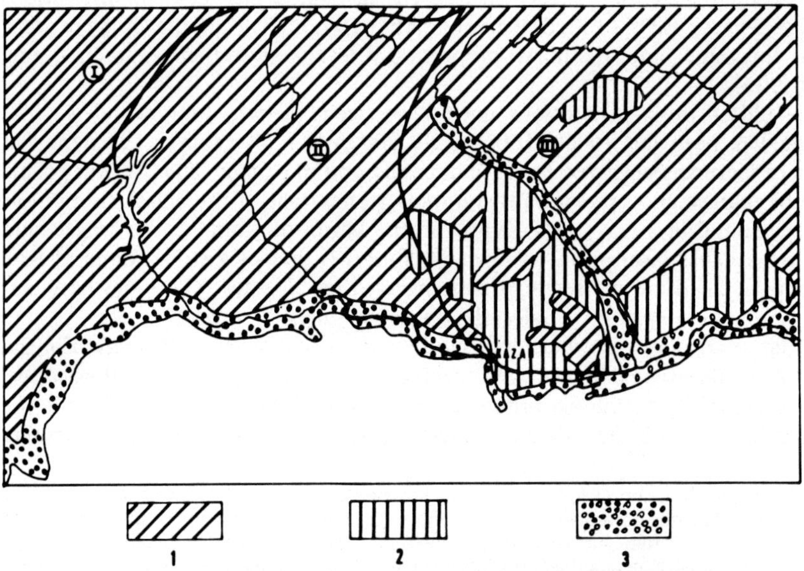

Fig. 142. Soils and vegetation in the southeastern part of the European taiga. I. Northern Uvalov. II. Southern taiga of the Transvolga plane. III. Southern taiga of the Vystako-Kama region. 1. Grassy podzol. 2. Grey soils (without forests). 3. Alluvium.

Indo-european peoples, who had reached the Baltic Sea from the south-east, in the first centuries B.C. Generally the forests were avoided, people preferring to dwell along rivers, shores of lakes and the dry parts of mires, which has undoubtedly contributed to preserving up to our times a great part of the taiga.

At the times of the great migrations (400-800 A.D.) some Germanic tribes reached Norway and partly incorporated the pre-existing peoples (not Aryan, maybe Finnish races), partly caused them to migrate towards the mountains and islands. Only at the beginning of the 8th century did the population show a remarkable increase, which corresponds to the beginning of the Viking expeditions (750-1000 A.D.). The Eddic poems and the so-called scalds (Bragi, Thorbjørn Hornklofi, Thiófold, etc.) belong to this period. The same happened in Sweden. Very interesting are the runic inscriptions, going from the 9th to 13th centuries A.D., which nevertheless, with the exception of Rök's inscription, have no literary qualities. Great stylized rock paintings date back to the Bronze Age (Tanum, near Bolsan and the neighbourhood of Norrköping).

A great number of figured stones, coming from the island of Gotland and others with runic inscriptions and interlaced decorations (Uppland) belong to the last Indo-european migrations and to the Viking epoch.

The runes are the letters of an alphabet with a meaning both phonetic and ideographic. Their origin might be due to magic practices, since people considered that, once cancelled, their effect ceased immediately. Runic magic seems indeed to constitute the most characteristic and highest feature of magic practices of the Northern Germans who inhabited the taiga of Fennoscandia. We may notice, in this connection, that the origin of the runic characters is extremely old, since some of them (20 out of 24) are present also in some very old inscriptions (4th-1st century B.C.) of the Italian populations who inhabited the alpine regions, and at the same time, the magic properties of letters are rather common among old peoples[1].

The oldest runic inscriptions must have been written upon wood, so that nothing has been left to us (100 B.C. – 200 A.D.). Successively, Eastern and Northern Germans engraved bones and metals, and in Norway and Sweden people began to engrave upon rocky walls and tomb stones. One of the oldest inscriptions has been found in a grave near Kylver on the Island of Gotland; the longest and most remarkable inscription, as to its antiquity, is that of a stone found at Rok (Östergötland), three metres high and engraved on all sides. The runes have survived among the peasants of Dalecarlia up to our times.

As far as Lapland is concerned, subdivided among Sweden, Norway, Finland and Russia, its original inhabitants are the Lapps (or as they called themselves, Sabme), belonging to the ugro-finnic populations. They keep several beliefs in the old shamanism and in the religion of their ancestors. Their sorcerer still uses today

[1] According to Bouisson (1964) sorcery magic appears to us in two forms of equal strength, the sung word and the formulae written on the walls of tombs and palaces, in the books of magic, etc. Not only that, but according to the Cabala each letter of the alphabet assumes its own life and is almost personified: the letters of the Hebraic alphabet reveal to us a teaching: alef is the symbol of the entire universe, since it has the form of two yods separated by a vau; beth is considered as representative of the opening of the Law (Zohar). For this reason the Bible begins with two words, each beginning with the beth.

the magic drum, similar to that used by Siberian tribes, and bear hunting is associated with some rites bound to the religion of this animal (maybe an example of zoolatry) common among all Arctic populations.

Very ancient peoples today inhabit the easternmost part of the European taiga, all belonging to non-Indoeuropean peoples. In the North, partially in the tundra, the Samoieds (or in Russian Nency) are found, belonging to a branch of the uraloaltaic family, and extending from the White Sea eastwards. Their religion is a form of shamanism. The Siriens (in Russian Zirjane), a north-eastern Finnic population, live along the rivers of the region, the Vyčegda, the Mezen′ and the Pečora. The Sirien language is – after the Hungarian – the ugro-finnic language with the oldest documents. Always in the taiga, the Permians (in Russian Permjaki) a Finnish population of the eastern group, who speak – like the Siriens – a language (the Permian) belonging to the ugro-finnic family, live immediately to the South of them.

The vegetation

The vegetation of the taiga is composed mainly of Scots pines (*Pinus silvestris*) and Norway spruce (*Picea excelsa*), larches (*Larix decidua*), junipers (*Juniperus communis*) among which, in the easternmost part, several Siberian elements appear (*Abies sibirica, Larix sibirica* and *Pinus cembra*). Several species of birch (*Betula*) are also present. From the vegetational point of view the southern limit of the taiga corresponds to the northern limit of oaks (*Quercus*).

Subarctic elements are *Angelica* sp., *Aconitus* sp., etc.; Atlantic elements, present only in Norway, are the holly *Ilex aquifolium, Digitalis* sp. etc.; subatlantic elements are *Gentiana pneumonanthe, Sanguisorba officinalis, Petasites albus*, etc. Widely occurring trees are *Alnus* sp. and *Sorbus aucuparia.* The beech (*Fagus silvatica*) is present in Norway up to latitude 61°N. In Sweden the upper limit of the coniferous forest is limited to 800-900 m, in southern and central parts, to 400 in the north. In North-Western Finland, near the border with Sweden and Norway (as at Kilpisjärvi), the tree line is at 650 m (Karppinen).

Higher than the coniferous forest, a birch forest is generally present. These two groups of trees constitute what is locally called 'regio subalpina'. Higher, a treeless region is present, locally called 'regio alpina', where several heath plants (*Ericaceae* and *Vacciniaceae*) are to be found ('Zwergstrauchheide' of German authors).

To give only a general picture of the variety of vegetation in the taiga, we report here some forest types taken from Cajander's classification.

Calluna-type. Moss vegetation (*Hyloconium parietinum* and *H. proliferum, Ptilium crista castrensis, Dicranum undulatum* and *D. scoparium*) and lichen vegetation (mainly *Cladonia rangiferina*) are present in varying degrees of abundance. Herbaceous and grassy vegetation is somewhat more abundant and richer in species than in the *Cladina* and *Myrtillus* type (see later). Dwarf shrub vegetation is abundant, *Calluna vulgaris* as a rule being the predominant species. Of the bushes *Juniperus communis* is fairly common. In most cases the forest-forming species is the Scots pine, but the birch (*Betula verrucosa* and *B. odorata*) and Norway spruce occur commonly as mixtures, and sometimes as the dominant tree species.

Vaccinium type. Moss vegetation, including feather mosses, is fairly continuous; lichens are common but as a rule of lesser importance. Herbs and grasses somewhat richer in species and more abundant than in the *Calluna* type. Dwarf shrub vegetation rather plentiful, consisting mainly of *Vaccinium vitis idaea*, accompanied in various measure by *Myrtillus nigra* and *Calluna vulgaris*; Juniper is very common. In most cases the forest is formed by the Scots pine, but the birch and the Norway spruce occur commonly together, often also as the main forest-forming trees. In some instances *Alnus incana* is the dominant species. This kind of forest occurs on moderately dry sandy ground and on glacial ridges and sometimes even on moraine (harju).

Myrtillus type. The moss vegetation as a rule is abundant and continuous, but it is not as luxuriant as in the *Hyloconium-Myrtillus* type. Its most important species are *Hyloconium proliferum, H. parietinum*, and various species of *Dicranum.* Herb vegetation is more abundant and richer in species than in the *Hyloconium-Myrtillus* type, consisting generally of very common species. Grasses are comparatively few. The dominant *Myrtillus nigra* is almost always accompanied by *Vaccinium vitis-idaea.* In natural conditions this forest type would always be covered by Norway spruce, but owing to burning-over for cropping purposes, fellings, etc. the Scots pine and birch occur very commonly as forest-forming trees. In some parts also *Alnus incana* (the common grey alder) and *Populus tremula* (the common aspen) can be dominant species.

Oxalis-Myrtillus type. This kind of forest represents a certain relatively self-contained intermediate form between the *Myrtillus* and the *Oxalis-Majanthemum* types. Moss vegetation is fairly abundant, but much scantier than in the *Myrtillus* type. Lichens are of no significance. Grass vegetation is somewhat luxuriant, and the herb vegetation much more abundant than in the *Myrtillus* type. Among the grasses and herbs, rather hygrophilous species such as *Oxalis acetosella, Melica nutans, Fragaria vesca*, etc. are also encountered. Some bushes are also present (*Rubus idaeus, Daphne*, etc.). Among the common species of trees a sprinkling of hardwood is often encountered and sometimes they may even be dominant (for instance *Quercus pedunculata* in Southern Finland).

Oxalis-Majanthemum type. Mosses rather scanty (*Hyloconium triquetrum*, various species of *Mnium* etc.). Among dwarf shrubs *Myrtillus nigra* and *Vaccinium vitis-idaea* are common, but generally in small numbers. Grasses and sedges are not very abundant (*Calamagrostis arundinacea, Aira caespitosa, Melica nutans, Carex digitata*) but the thin-leaved herbs such as ferns, *Oxalis, Majanthemum, Convallaria majalis, Rubus saxatilis, Fragaria vesca, Viola* sp., etc., are very abundant. If the conditions of the primeval forest had survived, Norway spruce would be the only dominant species, but in the conditions prevailing in some parts of the taiga – for instance in Finland – the birch (*Betula verrucosa* and *B. odorata*) is the most abundant forest-forming tree. Where burning over for cropping purposes has been practised, the grey alder and the aspen, besides some hardwood species of trees, also occur.

Cladina type. The soil is as a rule greyish white, owing to the abundance of *Cladina* vegetation (*Cladonia alpestris* in particular), the moss vegetation is rather scanty. Herbs and grasses are very few, nor are the dwarf shrubs very abundant, being interspersed among the lichens. The forest consists for the most part of Scots pine. This forest type occurs most commonly on the dry sandy heaths of

Lapland and the far North, but it is met with also on the poorest and driest soils of Southern Finland.

Empetrum-Myrtillus type. Moss vegetation dominant, including mostly the so-called feather mosses (*Hyloconium, Dicranum* etc.) and a certain proportion of *Polytrichum*; lichens are common, but less abundant than in the *Myrtillus-Calluna* types. Herbaceous and grassy vegetation rather scanty; dwarf shrub vegetation dominated by *Myrtillus*, on the contrary very common, and accompanied, although not in equal abundance, by *Empetrum nigrum.* In most cases the forest is formed by the Scots pine, almost constantly intermixed with the birch (*Betula verrucosa* and *B. odorata*), frequently with the Norway spruce and often with the aspen (*Populus tremula*). This forest type is widely distributed in dry heaths of Northern Finland.

Hyloconium-Myrtillus type. This forest type is characterized by a continuous and exceedingly luxuriant feather moss vegetation. Lichens, herbs and grasses are scanty, with dwarf shrub vegetation more or less abundant, including especially *Myrtillus nigra* and, apart from that, *Vaccinium vitis-idaea* and *Empetrum nigrum.* In most cases the forest consists of rather open Norway spruce forest, with an admixture of birch. Often the Scots pine is also met, mostly in the form of standard trees. This forest type takes up 9.9% of the economically productive forest land of the Province of Oulu, and it occurs especially on the upper slopes of the mountains in eastern parts of Northern Finland, where it is dominant. It is less common on the mountains of Western Lapland, occurring as a rule on their moist, northern, slopes.

Let us mention here – as a mere example – the composition of the forests of the Southern taiga in Russia, near Kazan (from Aleinikova, 1968):

Extension and foristic composition of a forest of the Russian taiga (Mariyskaya SSR) (from original data of the 'Dicaster' 1951-53):

% of forest	area in ha	Composition of the forests (%)						
		Picea excelsa	*Picea silvestris*	*Betula* sp.	*Populus tremula*	*Quercus* sp.	*Tilia* sp.	Remaining
52	1230	26	39	19	9	1	4	2

The pH of the soil in mixed forests varies from 4.9 to 5.3, in coniferous forests from 4.2 to 4.8.

When we come to examine the geographical distribution of the plants of the taiga, we find some very interesting species among the lower plants, such as Bryophytes and *Hepaticae*, not to speak of some higher plants, whose presence is dealt with in the section on nature conservation (p. 234).

Particularly interesting are perhaps the elements with a discontinuous distribution of boreo (or Arctic)-alpine types, and in some instance even of amphipolar-alpine type. Of 102 Bryales present in Switzerland nearly half are found north of the Arctic Polar Circle, beyond 80°N. Most lichens of the high mountains of South and Central Europe are not only actic alpine, but often amphiarctic or bipolar-alpine, since they are present both at the highest North and South latitudes, and on mountains. Examples are present among the crust-lichens of the families *Lecideaceae, Lecanoraceae, Buelliaceae*, leaf lichens of the families *Peltigeraceae, Umbilicariaceae, Parmeliaceae, Caloplacaceae, Physciaceae*, the so-called coral-lichens belonging to the genus *Thamnolia* and *Stereocaulon*, and several beard-lichens (*Alectoria, Usnea*, etc.).

Among liverworts we may mention *Blepharostoma trichophyllum*, present in the Northern hemisphere up to the Arctic region and on the Central-European mountains. Within the genus *Gymnotrimium* there are also some bipolar-alpine species. Among the *Andreaeales* there is the genus *Andraea* with its sectio *Chasmocalyx*, represented in Northern Europe and on the Alps by the species *A. nivalis.* Lastly, among the Bryales, and more precisely *Musci Arthrodontei*, we find the *Dicranales*, present both in Arctic regions and on the high mountains (arctic-alpine elements) and the *Grimmiales*, represented by *Rhacomitrium canescens*, an arctic-alpine element (Alps, Caucasus, Himalayas and Arctic regions) and *R. lanuginosum*, present on the central-western Alps and the costs of Northern Norway, besides several other parts of the world (evident proof of the great antiquity of the taxon).

Lastly, we want to mention the *Bryaceae*-genus *Mielichhoferia*, restricted to extremely acid soils, where some sulphuric acid is formed by the sulphides present in the substrate: in Europe we have two forms (species or subspecies of single species) one of which lives in the North (taiga), the other in central-southern Europe (cf. Gams, 1960).

An interesting tree of the taiga is *Betula nana*, which can be considered as a glacial relict, present also in Central Europe. Some insects have followed *Betula nana* in its spreading south; others on the contrary have adapted themselves to this plant only secondarily (i.e. after the last glaciation). The Lepidopterous *Tortricidae Ancylis tineana* belongs to this category, whereas the *Lithocollectidae Lithocolletis anderidae* should be a primary element, which would have followed *Betula nana* since the end of the glacial area. Other elements which look like glacial relicts since they are strictly monophagous species, are the mining Microlepidopterous *Stigmella nanivora*, and *Coleophora betulaenanae*, the *Cecydomyidae* midge *Massalongia bachmaieri* and the leaf louse *Calaphis arctica*, described from Greenland.

Changes of climate and vegetation of the European taiga, from Marchesoni & Paganelli (in Zangheri. p. 469)

duration (in years)	climatic-forest periods	climate characteristics	vegetation of the European taiga
+ 1000 0	Sub-atlantic 800 B.C. recent epoch	cooler and more humid	beech, spruce
– 1000 – 2000	Sub-boreal 2500-800 B.C.	decreasing temperature with some dry periods	oak; beech
– 3000 – 4000 – 5000	Atlantic 5500-2500 B.C.	warm with increasing humidity (climatic optimum)	mixed Quercetum
– 6000	Boreal 6800-5500 B.C.	dry warm	broad-leaved trees very wide-spread, maximum of the hazel
– 7000 – 8000	pre-boreal 8200-6800 B.C.	increasing temperature and dryness	Scots pine and birch
	recent Dryas period 8800-8200 B.C.	cooler and drier	subarctic tundra of park aspect
– 9000 –10000	Alleröd 10000-8800 B.B.	warm oscillation	limited spreading of forests
	middle Dryas period 10400-10000 B.C.	cold	tundra
–11000 –12000	Bölling 11300-10400 B.C.	warm oscillation	tundra with the park aspect
–13000 –14000 –15000	old Dryas period 15000-11300 B.C.	cold, with a slow ice retreat	treeless tundra (no forest pollen) with some shrubs of *Betula nana, B. humilis, Salix herbacea, S. reticulata, S. retusa*

The animals of the taiga

The mammals

Few mammals are exclusive to the taiga, but their form, physiology and still more behaviour make them very interesting creatures. All the other species – which as we are going to see are rather numerous – are common to other European biomes, such as the steppe in the South or the mixed or deciduous forest in the west.

The bear *Ursus arctos*, once frequent all through Europe, today is limited to the Pyrenees (where it is protected), the Eastern Alps, perhaps Graubunden (Grigioni) and the Balkan Peninsula. In Germany it has been extinct since 1700; in Silesia since 1770, in the Riesengebirge maybe only after 1804. In the Baltic region it is still present here and there, but only in inaccessible forests. It can eat everything: dry fruit and berries, acorns, insects; occasionally it attacks also cattle.

The European mink, *Putorius lutreola* (or *Lutreola lutreola*) (in German 'Nerz' or even 'Krebsotter') was once present also in Germany, but now only around and east of the Baltic Sea. It thrives in swamps and along rivers. Its habits are not very different from those of the common otter, feeding upon crayfish, crabs, fish and fruit. The otter *Lutra lutra* is present everywhere but lives exclusively along rivers. where it hunts fish.

The polecat *Putorius putorius* is limited to the southern part of the European taiga, often living near human settlements, more often in forests, particularly in dry lands but sometimes also in humid environments. It excavates its den in hidden places. When it is threatened, it expells a vapour secreted by the so-called moss glands, whence its scientific name (from the Latin word puteo). In the last century its area of distribution has been widely extended towards the North in Southern Finland.

The ermine *Mustela erminea* is also rather common in the European taiga. Its habits are well known and it is hunted for its valuable fur. The lynx, *Lynx lynx*, is limited to some districts of the taiga, but some individuals migrate in practically all directions.

The common fox *Vulpes vulpes* is present in all the area. Its habits are well known to everybody. An interesting carnivore of the European taiga is, on the other hand, the Arctic fox, *Alopex lagopus*, distributed above the Arctic Circle, during the winter moving southward. It is both a nocturnal and diurnal animal, living in small groups. It makes its den in crevices, rocks or in the ground, sometimes building a very wide complex of galleries, where it accumulates food. The wolf *Canis lupus* is present practically everywhere and its habits are well known. Today its behaviour is the object of very interesting research in North America, where radio tracking is very widely employed by researchers. Another trivial carnivore of the taiga is the marten *Martes martes*, also hunted because of its fur. An interesting carnivore, also, is the sable, *Martes zibellina*, three centuries ago living in the northernmost corner of Fennoscandia, today practically confined to Northern Asia, where it thrives in the coniferous forests (*Pinus* and *Abies*) but also in mixed forest, particularly near the upper courses of small rivers. According to Van der Brink, there is a case for reintroducing this species in Europe, but this would encounter very great difficulties. The animal is hunted because of the enormous value of its fur (which even 100 years ago was worth 300 Marks in Germany). Another carnivore of the taiga is the wolverine, *Gulo gulo*, present in all the northern part of the territory, as a fossil also in the Baltic Republics and in Poland. Some errant individuals wander from Norway towards Sweden and the Baltic Sea. The wolverine is both nocturnal and diurnal; the males live isolated except during the period of reproduction. It moves by meas of a short gallop, but can also jump and climb, although it is a bad swimmer and does not dive. The den is composed of a shallow hole in rocky slopes among shrubs (Fig. 143).

Fig. 143. The polar fox *Alopex lagopus* and its distribution and main lines of migration. Top: summer phase; middle: winter phase; c. particular constant dark phase (from v.d. Brinck).

A bat peculiar to the taiga is the *Vespertilio nilssoni*, or northern bat, present in most of the territory with the exception of Lappland. It comes out very early in the evening; it has a swift flight and it flies rather low (5-10 m). Probably it makes only one flight per night. It performs rather conspicuous migrations, but very little is known about these: it is certain that it can migrate, but probably there are no constant rules for this phenomenon, which furthermore seems to vary according to the populations and the localities.

A common insectivore of the taiga is the mole, *Talpa talpa*, whose habits are known to everybody.

Particularly interesting are the rodents of the European taiga, such as the blue hare or mountain hare, *Lepus timidus*, present all through the territory, in the southern parts only, living with the common hare, *Lepus capensis.* It is generally nocturnal, rarely living gregariously, more generally isolated. It moves by jumping, is a good swimmer, and rarely digs a den.

Much more interesting is the wood lemming *Myopus schisticolor*, limited to the northernmost sector of the taiga, a relative of the true 'lemming' living in the tundra. It dwells in the spruce forests of northern Russia, Southern Norway, Central Sweden and Finland, where it digs galleries in the moss. It shows – like the true lemming – population explosions, followed by migrations towards the south in search for food. In these cases it reaches the Gulf of Bothnia and Lake Väner. The explosions follow a ten year cycle, different from the lemming (which has a four year cycle) and similar to many North American homoiothermous animals of high latitudes. It migrates less than the lemming, but shows sometimes more local changes of territory (Fig. 144a, p. 200).

The flying squirrel *Sciuropterus russicus*, is a rodent similar to the common squirrel, *Sciurus europaeus*, provided like many other arboreal mammals (especially of the Tropics) with a sort of membrane or cutaneous fold extending from the anterior to the posterior limb. When the animal jumps, this fold is spread and acts as a parachute, thus allowing planed flight. It lives in birch forests or even in mixed forests (coniferous and birch forests) of Northern Russia, Siberia, going as far west as Northern Scandinavia and as far south as North Eastern Germany. It feeds on buds of birch and – if these are not available – of pine (Fig. 144b).

Another rodent typical of the European taiga is *Eutamias* (from the Greek *ευ* good and *ταμίας*, housekeeper) *asiaticus*, extended from the White Sea to Kamtchatka, the Isle of Sakkalin, Northern Japan and the Kurili Islands, replaced in the

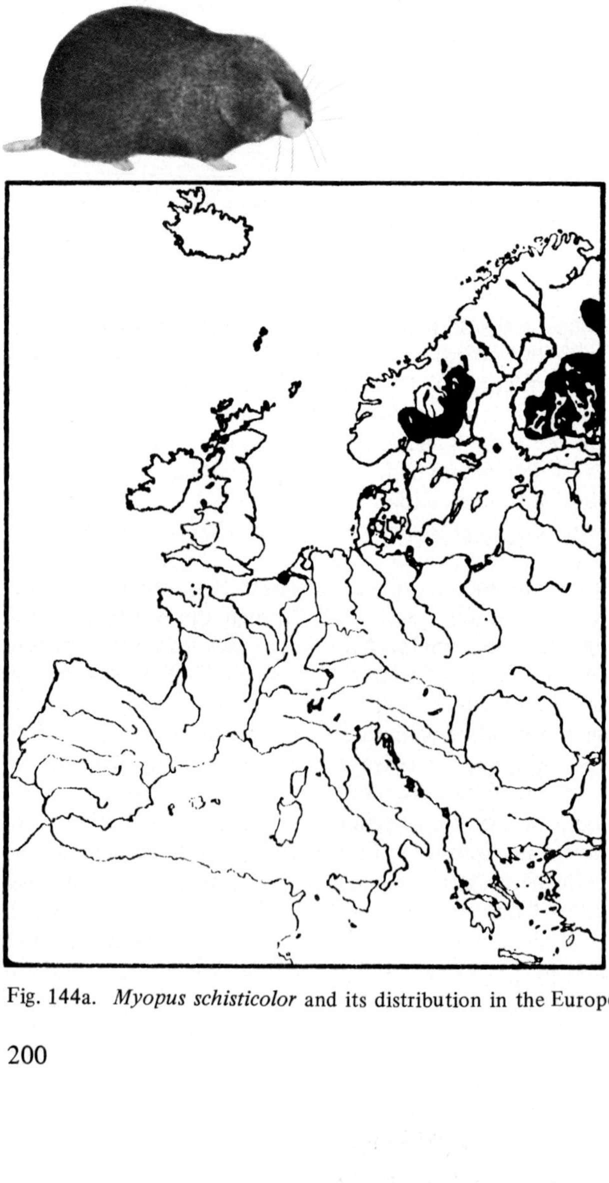

Fig. 144a. *Myopus schisticolor* and its distribution in the European taiga (from v.d. Brinck).

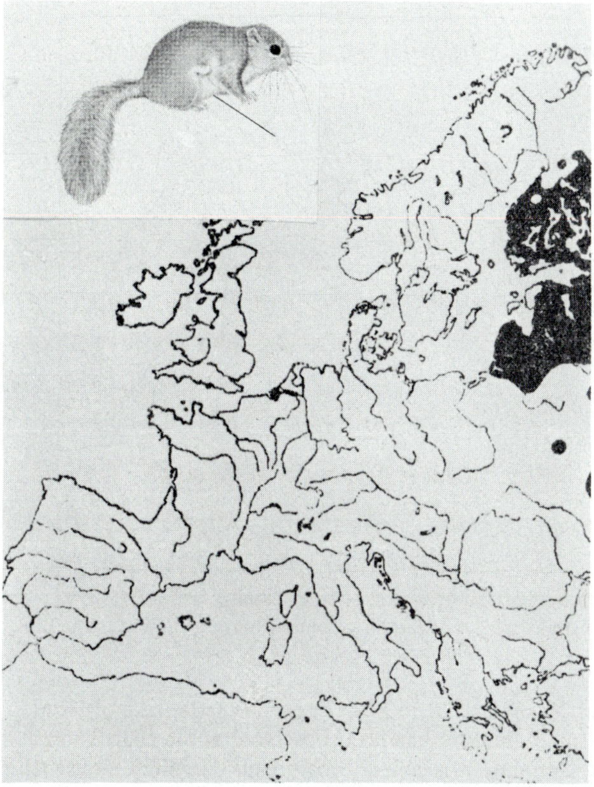

Fig. 144b. *Sciuropterus russicus* (= *Pteromys volans*) and its distribution in the European taiga (from v.d. Brinck).

Amur region by the oriental species *Eutamias orientalis.* It is locally known as 'burunduk'. It is a true ground squirrel, living exclusively on the ground, provided with two cheek pouches where it stores the food, whence its Latin name. It has a rather short, flat, tail and short ears, in harmony with the ecological rule known as Allen's rule. It climbs trees searching for seeds and dry fruits, but lives on the ground, where it digs tunnels. It winters in a chamber lined with moss and dry leaves, and filled with food reserves, since this small rodent occasionally awakens to eat (Fig. 145).

Another rodent very characteristic in its form and behaviour, though not exclusive to the taiga, is the northern birch mouse, *Sicista betulina*, a jumping mouse belonging to the family *Zapodidae*, present in Northern Asia, Eastern Europe, Scandinavia, Denmark and Germany. In this country it has made its appearance only recently, which permits us to think of a diffusion of the species towards the West still in progress. The first appearance was made along the lower Memel, at the border between Lithuania and Eastern Prussia in 1933; in 1936 it appeared in

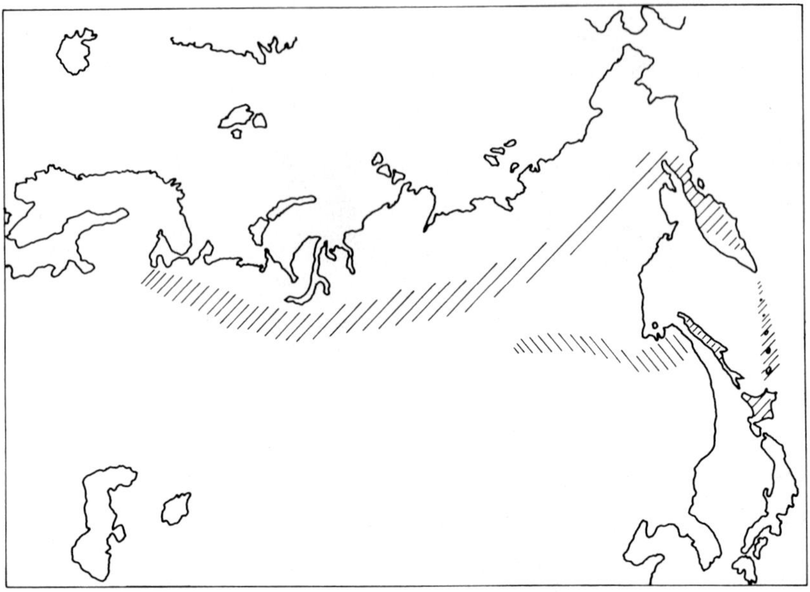

Fig. 145. Example of two vicariant species of *Eutamias*, of which one is present in the European taiga.

Schleswig Holstein (Tolker Moor) and in 1937 in Eastern Prussia (lake of Nordenburg). Recently this nice small creature has been observed on the High Tatra and in the Bavarian Forest. In Germany it is called 'Birkenmaus' i.e. birch mouse. It is the smallest indigenous European mouse: its body is 50-75 mm long, the tail 76-108 mm. It jumps and runs. It winters in a state of dormancy in holes dug by itself[1].

A beautiful animal of the holarctic taiga, i.e. of the coniferous forest which extends from Norway to Eastern Canada, is the elk (or moose) *Alces alces*, present in Eastern Prussia where it was introduced from Russia after World War II, since the individuals living there before had all been destroyed. In Silesia (Kreis Krebnitz) it was present until 1888. Before World War II the moose was living in Germany, introduced, more precisely in Eastern Prussia, on the eastern shore of Kur Haff and in the Frischings-Wald (South East of Königsberg), in the territory of lake Müritz and on the Darss (Pre-Pomerania or Rügen).

It dwells in great swampy forests, where it feeds upon bark, buds and leaves. Active particularly at sunrise and at sunset, but partially also during the daytime and during the night, it moves with a swift gallop (up to 65 km/h). It does not jump but dives and swims very well. It does not defend the territory and therefore is not a social animal: solitary in summer, it congregates during the winter.

The red deer, *Cervus elaphus*, is present in Scandinavia where some errant

[1] The nest building and the thermoregulation of this rodent have been studied by Pucek (Acta Theriol. 2, Bialowieza, pp. 1-154).

individuals can go as far north as beyond the Arctic Circle. The roe deer, *Capreolus capreolus*, is very spread out in Scandinavia, up beyond the Arctic Circle. The fallow deer, *Dama dama*, although imported to the taiga, according to van der Brink can be considered as belonging to the indigenous fauna of this biome. It reaches a latitude of ca. 62°N (Sweden and Norway). It is mostly nocturnal (in natural conditions) and rather gregarious.

The reindeer, *Rangifer tarandus*, is a typically migrant ungulate of the high latitudes, which moves during winter towards the South in search for food (trophic migration). The species has been subdivided by Lydekker into 6 varieties which probably are not all of the same systematic value (Scharff, p. 157). It performs not very extensive migrations, and aggregates in great herds composed principally of females, but also of young males, under the leadership of an old female. Some attempts have been made to introduce and acclimatize this beautiful ungulate in Scotland.

We may recall at this point that in Sweden some mammals have been re-introduced as game animals, such as the beaver, the hare, the wild rabbit and the fallow deer.

The birds

Very important is the bird life in the European taiga, especially in the fine season. In Sweden 240 species of birds nest, but only 60 are present in winter. Let us begin with the birds of prey, which constitute a very interesting part of food webs in the taiga. There is the golden eagle, *Aquila chrysaetus*, widely distributed up to very high latitudes; the spotted eagle, *Aquila clanga*, only in the southernmost part of the taiga, living on trees, generally near water bodies – rivers and swamps – nesting in the trees or on steppes with shrubs; the lesser spotted eagle, or *Aquila pomarina*, only in the southern part of the biome, generally near water, in forests where there are clearings, etc., it nests in trees; the buzzard *Buteo buteo*, very widespread except in the extreme north, living on rocky shores, in mires, plans, slopes of mountains, in cultivated areas and in forests, and making nests on trees and occasionally on the ground.

The rough-legged buzzard, *Buteo lagopus*, in Swedish fjällvråk, is present in all the northern part of Norway during the summer, going as far south as its southern limit. It has habits similar to those of the buzzard, living in open uncultivated areas, slopes of mountains, swamps and sand dunes. It nests particularly on rocks or on the ground (Fig. 146).

The sparrow hawk or *Accipiter nisus* is present all through the territory; the goshawk, *Accipiter gentilis*, also widely distributed, shows some 4 year cycles (see later). The black kite, *Mimus migrans*, only in the southern part of the territory, is limited to Finland and Russia; it lives near lakes and rivers, in forests or in places where there are some isolated trees; it nests on trees, and occasionally in old crow's nests. The honey buzzard, *Pernis apivorus*, is only found in the southern part, in clearings or at the edge of forests; it generally nests in old crow's nests. There are also the hen harrier, or *Circus cyaneus*, in Swedish Bla Kärrhök, a widely distributed species; *Circus aeruginosus*, the marsh harrier, in Swedish Brun Kärrhök, only in the southern part, living in mires, swamps with large areas covered with Phragmitetum, where it builds its nest surrounded by water; the short-toed eagle, or *Circaetus gallicus*, in Swedish Ormörn, exclusive to the southern part of the taiga, living on the slopes of mountains and in creeks, in forests, swampy plains, etc.; it feeds upon snakes, lizards and frogs, and nests on the trees; the osprey, *Pandion haliaethus*, in Swedish Fiskgjnse, present all through the territory, generally near lakes, rivers or on sea shores, it nests on small islets, rocks, trees, ruins, etc., occasionally also on sandy or rocky ground, it feeds on fish; the merlin, *Falco columbarius*, in Swedish Dwärgfalk, everywhere except the

Fig. 146. Distribution in Europe of *Buteo lagopus* (schematic).

southernmost part of Fennoscandia in open, hilly or turfy places, on the dunes at the coast and on rocks, it nests on the ground in grasslands and heaths, or on sand or even on trees in old crow's nests.

Also well represented are the nocturnal birds of prey or Strigiformes. Let us remember here the eagle owl, *Bubo bubo*, in Swedish Berguv, present everywhere with the exception of the extreme north, on rocky mountain slopes in forests, on rocks or even on open steppes. It nests among the rocks, in tree holes, or even in old abandoned nests of birds of prey. A bird characteristic of all the northern part

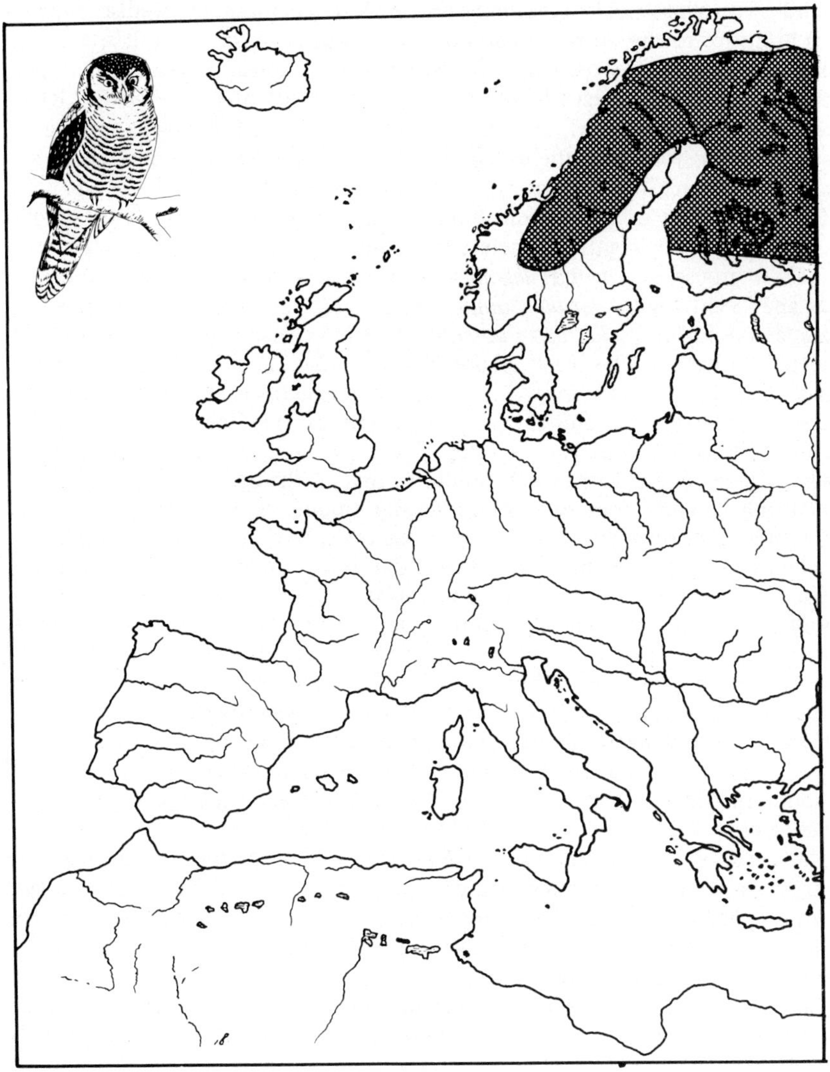

Fig. 147. Distribution in Europe of *Surnia ulula* (schematic).

of the taiga to the boundary with the tundra is the hawk owl, *Surnia ulula*, in Swedish known as Hökuggla, living in coniferous forests, in clearings, small birch stands, etc. It nests in tree holes or in old bird of prey nests. The pigmy owl, *Glaucidium passerinum*, in Swedish sparvuggla, is a widely distributed nocturnal creature, absent only in the south and north extremes of Scandinavia, living in coniferous forests, on mountains; it nests in tree holes and woodpecker nests. The Ural owl, *Strix uralensis*, in Swedish slaguggla, is spread over most of the territory with the exception of Norway and South Sweden, living in mixed forests and in forests where some clearings can be found. It nests in tree holes, occasionally in old bird of prey nests. The great grey owl, or *Strix nebulosa*, in Swedish Lappuggla, is exclusive to the northern parts of the European taiga, where it spends the whole year in dense forests. Only in some years it can spread to the southern part of the taiga. It lays its eggs in old nests of large birds of prey. Finally, let us remember Sengmalms's owl, *Aegolius funereus*, in Swedish Pärluggla, present in most of the territory with the exception of southern Sweden, in coniferous forests. It nests in three holes, sometimes in those produced by woodpeckers (Fig. 147).

The Galliformes or gallinaceous birds of the European taiga are fairly numerous and, in general, highly characteristic of this environment. Let us begin with the willow grouse or *Lagopus lagopus*, in Swedish known as dabripa; it lives in mires and healths with *Betula, Juniperus* and *Salix*, at lower sites like *Lagopus mutus*, at the higher as *Tetrastes bonasia.* It nests in small depressions in the soil, protected by bushes. The black grouse, *Lyrurus tetrix*, in Swedish orre, lives near trees at the edge of mires, in swampy sites with single trees, in peat on rocky hills covered with heath, in mountains and also above the tree line. It is also present, like the following species, in the Alps. The capercaille, or *Tetrao urogallus*, in Swedish Tjäder, has a distribution similar to that of the preceding species, but is present only in coniferous forests on hills and mountains. It nests on the ground in the undergrowth at the foot of coniferous trees, but sometimes also on heaths and among junipers. The hazel hen, *Tetrastes bonasia*, in Swedish Järpe, is distributed more or less like the two preceding species. It lives in mixed forests, particularly of aspen and birch, in hills and mountains, and locally also on the plain. It nests in small holes in the soil, well protected by the dense vegetation. Other gallinaceous birds of the taiga, although not very characteristic of this environment, are the partridge *Perdix perdix* and the pheasant, *Phasianus colchicus.*

Other birds belong to several orders, represented only by a single or a few species in the taiga: the wood pigeon, *Columba palumbus* (or ring dove) lives in the southern taiga, with the exception of treeless places; it nests in trees, hedges, old nests, etc.

The cuckoo, *Cuculus canorus*, in Swedish Gök, is present all through the territory and can be considered as a common species. The nightjar, *Caprimulgus europaeus*, in Swedish Nattskärra, is present only in the southern part, in mires, heaths, forests, particularly on the edge of clearings and in places rich in bracken. It lays its eggs on bare ground.

The swift, *Apus apus*, is present throughout the territory provided there are places suitable for nesting. It nests on buildings under the eaves, but occasionally also among rocks.

Of particular ecological interest is the lapwing, *Vanellus vanellus*, in Swedish tofsvipa, a species linked to water, present in swampy grasslands, meadows, swamps and moors. It nests in meadows. It has shown an extension of its area of distribution towards the north, which in Finland began in 1899 and continued until 1941 (Kalela).

The Piciformes or 'woodpeckers' are well represented in the taiga. Let us remember the green woodpecker, *Picus viridis*, limited to the southern part of Sweden and Norway, particularly linked to broad-leaved trees; the grey-headed woodpecker, *Picus canus*, also limited to the southern part, but present also in the mountains, although avoiding the coniferous trees; the black woodpecker *Dryocopus martius*, everywhere in old coniferous forests, particularly on medium or high mountains, locally also in mixed or beech forests; the great spotted woodpecker, *Dendrocopus major*, in most of the territory, in mixed forests; the white-backed woodpecker, *Dendrocopus leuconotus*, in broad-leaved forests, on some hills with rotting trees, locally also in dense coniferous forests, and in winter in the neighbourhood of towns; the lesser-spotted woodpecker, or *Dendrocopus minor*, in all the area, both in coniferous and broad-leaved forests; the three-toed woodpecker, *Picoides tridactylus*, present in woods of coniferous trees and birches, preferring old surfaces previously subjected to fire; it nests in tree holes and inside telegraph poles (Fig. 148).

All the remaining birds of the European taiga belong to the order Passeriformes, and, more particularly, to various families of this order.

Let us begin with the tree pipit, *Anthus pratensis*, in Swedish trädpiplärka, a widely distributed passerine bird living in heaths, clearings of forests, where it nests under bracken and high herbs; the red-backed shrike, *Lanius collurio*, in Swedish vaulig törnskata, present only in the southern part of the territory, in bushes, hedges, young forests, clearings, old stone bridges, etc., nesting in shrubs

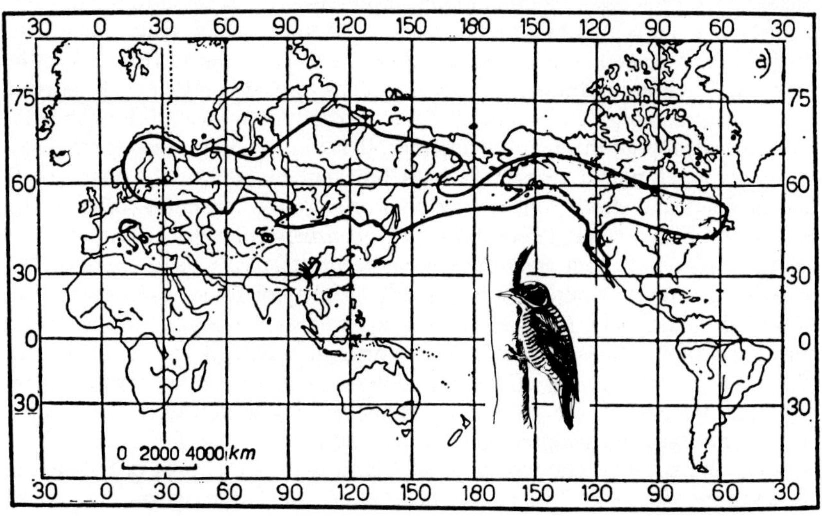

Fig. 148. Distribution of *Picoides tridactylus*, a typical bird of the holoarctic taiga (from Kühnelt) (with the permission of G. Fischer Verl.).

and small trees; the wren, *Troglodytes troglodytes*, present here and there in the southern part of the territory; the dunnock or hedge sparrow, *Prunella modularis*, present in most of the territory, in hedges, shrubs and small woods, where it nests in shrubs, young coniferous trees, brambles, etc. from near the ground up to the height of a person; *Locustella nevia*, or grasshopper warbler, is a passerine bird living in Southern Finland and Northern Russia, in swamps, humid grasslands, heaths, and dry wood clearings, where it nests near or upon the ground; the sedge warbler, *Acrocephalus schoenobaenus*, in Swedish sävsången, lives near water.

The icterine warbler, *Hippolais icterina*, present in all the southern districts and along the coast of Norway to high latitudes, lives in gardens, small woods and hedges; it builds a thick nest in shrubs, hedges and the like. Other passerine birds belong to the genus *Sylvia*: the garden warbler, or *Sylvia borin*, is present in most of the territory, in forests with a luxuriant undergrowth, in copses, shrubs, brambles, hedges, etc.; the flockcap, *Sylvia atricapilla*, present in Southern districts, lives in clearings rich in undergrowth, in very luxuriant hedges, and also in gardens, it nests in brambles, on evergreen plants, *Lonicera caprifolium*, etc.; the lesser white throat, *S. curruca*, lives in the Southern half of the territory, its habits are similar to those of the preceding species; the white throat, or *Sylvia communis*, distributed like the preceding species, is present in relatively open places, with shrubs, brambles, nettles, etc., it nests near the ground on low vegetation. The chiff-chaff, *Phylloscopus collybita*, is present in most of the area, generally on trees, it nests on the ground, in brambles, on evergreen plants, in open forests, etc.; the willow warbler, *Phylloscopus trochilus*, is present in all the territory up to the tundra included; differing from the preceding species, it prefers low vegetation, and nests on the ground in open places, at the base of hedges, etc. The wood warbler, *Phylloscopus sibilatrix*, more characteristic of deciduous than of coniferous forests, is present only in the southern part of the European taiga; the greenish warbler, *Phylloscopus trochiloides*, is present both in deciduous and coniferous forests, it nests on the ground, and occasionally in low stone walls. The pied flycatcher, *Ficedula hypoleuca*, is present in most of the territory, both in deciduous and coniferous forests, in parks, gardens, mixed forests etc. It nests in tree holes or upon walls. The whinchat *Saxicola rubetra*, in Swedish Buskskvätta, is present practically all through the territory, living in meadows, swamps, railway embankments, open places with single trees, bracken, etc.; during the migration it is found in fields and shrub zones; it nests on the ground, on herbs, often at the base of bushes. The redstart *Phoenicurus phoenicurus*, in Swedish rödstjärt, lives all over the territory as well as in the tundra, in forests, parks, heaths with shrubs and old trees, occasionally also in ruins. It nests in tree holes, dry walls, sheds.

The thrush nightingale, *Luscinia luscinia*, in Swedish Näktergal, is found in the southern parts of the taiga, particularly in Northern Russia; it lives in dense and humid forests, particularly of *Alnus* and *Betula*, or in the undergrowth and on swampy soils. It nests on the ground, on brambles, among nettles, etc. The bluethroat, *Luscinia svecica*, is represented by two subspecies, one, *L.s. cyanecula*, in Scandinavia, and another, *L.s. svecica*, in Russia. It is absent in most of Finland. It lives in swampy woods and heaths, in hedges, etc., and it nests on the ground beneath birches, willows and junipers, or in humid heaths often in the mountains. The wheatear, *Oenanthe oenanthe*, in Swedish stenskvätta, is present throughout the area in mires, on sandy hills, in meadows, on rocks and dunes. It nests in holes in walls, in rabbit burrows, stone piles, etc. The thrushes are also well represented in the taiga. The mistle thrush, *Turdus viscivorus*, in Swedish dubbeltrast, is present in most of the territory, with the exception of the western part of Scandinavia; it lives in large gardens, market gardens and forests. It nests in the forks of branches. The fieldfare, *Turdus pilaris* (in Swedish björktrast) is present throughout the territory; during winter it is found in open places, in search of food in fields and hedges. It nests generally in groups of several pairs near fields or on the edge of forests, generally of birch, but occasionally also in buildings and haylofts. The song thrush, *Turdus philomelos*, in Swedish taltrast, is present in the area but only as a summer bird, living near towns and villages, in parks, forests and hedges; it nests in shrubs, hedges, occasionally in buildings. The redwing, *Turdus iliacus*, in Swedish called rödvinge trast, is also present as a summer bird; it nests on tree stumps, wood piles, trees or shrubs, on the ground, in sparse forests and swampy places, often near families of *Turdus pilaris.* The long-tailed tit, *Aegithalos caudatus*, is present all over the southern part of the territory (*A. c. caudatus*). It lives in woods, heaths with some scattered shrubs, in the undergrowth and in hedges. It builds a nest made of mosses, oval shape, closed, generally hung on thorny bushes, brambles, etc.

The so-called 'tits' are also well represented. Let us remember here the crested tit, *Parus cristatus*, present in a southern part of the European taiga, generally in coniferous forests, but also in mixed forests. It nests in tree holes, often in rotten trees and in fences. The willow tit, *Parus atricapillus borealis*, in Swedish talltita, present all over the territory up to the tundra, prefers swampy, small woods where it can find rotten tree stumps. It uses for its nest rotten alders, birches and willows. The blue tit, or *Parus caeruleus*, lives in the southern part of the territory, and its habits are similar to those of the preceding species. The great tit, *Parus major*, lives in mixed forests all over the territory, in hedges, gardens, etc. It nests in tree holes, in walls or in cavities already occupied by birds. Lastly the coal tit, *Parus ater*, lives all over the territory with the exception of the extreme north of Fennoscandia; it prefers coniferous forests, whence its German name (Tannmeise). It nests in three holes, in earth banks and old stumps, generally near the ground (Fig. 149).

An interesting passerine bird is the tree creeper, *Certhia familiaris*, present in the southern part of the territory, in forests, parks, gardens with large trees; it nests under bark, which does not adhere very closely to the trunk, in tree holes and among ivy plants.

A passerine bird which has made its appearance in Europe only recently is the yellow breasted bunting, *Emberiza aureola*, as Timofeef-Ressowsky showed in 1940. The animal was confined to Western Siberia in 1825, and successively has moved westwards, so that after one hundred years it reached the Gulf of Finland. Today it thrives in open places, during summer near the water, in birch or willow forests but also on steppes. It nests in small shrubs and is now a nesting bird in Northern Russia and Finland (Fig. 150).

The yellow hammer *Emberiza citrinella* lives in open places, fields, at the edge of roads, on shrubs; it nests on, or near, the ground, at the base of hedges, etc. The ortolan bunting, *Emberiza hortulana* is present in most of the area, with the

Fig. 149. *Parus ater* (from Marcuzzi, 1975, with the permission of Manfrini ed.).

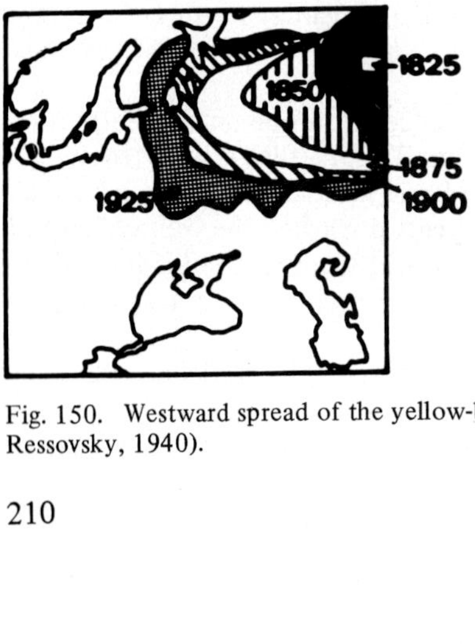

Fig. 150. Westward spread of the yellow-breasted bunting *Emberiza aureola* (from Timofeeff-Ressovsky, 1940).

exception of the northernmost part, in hilly, open places, often on the plains, in gardens, small woods or along the roads. It nests on or near the ground, in arable lands or hidden among the herbs. The red bunting, *Emberiza schoeniclus*, present throughout the territory but only during summer, lives in the Phragmitetum of ponds, lakes, swamps etc. and nests on or near the ground in the vegetation, more especially among vines. The chaffinch, or *Fringilla coelebs*, in Swedish bofink, is present almost everywhere during the summer, on hedges, in woods, on isolated trees in gardens, etc., practically everywhere there is also a single tree. It nests at a low level on shrubs or trees. The green finch, *Carduelis chloris* is present only in the southernmost parts of the taiga, as a summer bird, in gardens, shrubs, trees, in the cemeteries, etc. It nests in hedges, shrubs and small trees, particularly on evergreen species. The goldfinch, *Carduelis carduelis*, is present in the southern part, in gardens, market gardens, arable lands, searching for food upon thistles in uncultivated areas and along roads. It nests upon trees near the top of the branches. The siskin, *Carduelis spinus* lives in a large part of the territory, in coniferous forests, as well as in birch and alder stands; it nests very high in coniferous trees, on the topmost branches. The linnet, *Carduelis cannabina*, present during the summer in the southern part of the taiga, particularly in Russia, lives in open places with some scattered hedges; it nests generally gregariously in small forests on genista plants and hedges, occasionally also in fields and heaths. The scarlet grosbeak, *Carpodacus erythrinus*, is present only in the eastern part of the territory, being absent in Scandinavia; it lives in small woods, swamps and luxuriant undergrowth and near the water. It nests on the ground, or occasionally on swampy vegetation. The pine grosbeak, *Pinicola enucleator*, is distributed only in the northern part of the European taiga during the summer, and in the southern part during the winter. It lives in mixed or coniferous forests, and nests on coniferous trees. It feeds upon the seeds of the coniferous plants, dry fruits, young buds and perhaps also upon insects and their larvae. The crossbill, *Loxia curvirostra*, is distributed in most of the territory, and is a typical bird of the coniferous forests, particularly of *Picea* and *Larix*. Thanks to the form of its beak it can utilize the open, fallen, pine cones. It nests on coniferous trees during all

Fig. 151. *Carduelis chloris* (from Marcuzzi, 1975, with permission of Manfrini edit.); right, *Carduelis flammea* (idem).

the seasons of the year, even during winter. The parrot crossbill, *Loxia pytyopsittacus*, has habits very similar to those of the preceding species. It nests in Southern Finland and Scandinavia up to latitude 67°N; it is a summer bird; finally the two barred crossbill, *Loxia leucoptera*, is similar in habits to the two preceding species, but is linked particularly to larches. It nests in Northern Finland and Northern Sweden, whereas it winters in the Baltic region (Fig. 151).

The bullfinch, *Pyrrhula pyrrhula pyrrhula*, in Swedish domherre, lives in forests (also artificial), in small woods, hedges, gardens and market gardens; it nests on evergreen trees, often in gardens or hedges, or on small coniferous trees. The tree sparrow, *Passer montanus*, lives in the southern part of the territory, although it is present all along the coasts of Norway. Compared to the house sparrow it is less restricted to the artificial, human environment. It nests in tree holes, particularly in pollarded willows, haylofts, etc. The starling, *Sturnus vulgaris*, a common passerine bird, is also present in the taiga during summer, more particularly in towns and fields; it nests in tree holes, buildings, straw-roofs, but also in small cavities in the soil and in uncultivated areas. A beautiful bird, although not exclusive to the coniferous forests, is the golden oriole, *Oriolus oriolus*, present only in the eastern part of the territory, absent in all Scandinavia. It is linked to trees, and therefore present also in parks, old gardens, in the forests growing along the rivers and also in *Pinus* stands. It nests generally on horizontal forks of branches, and its nest has the shape of a hammock.

Numerous and very interesting are the *Corvidae* of the European taiga. Let us begin with the Siberian jay, *Perisoreus infaustus*, in Swedish known as lavskrikka, living in thick forests with coniferous trees and birches, in winter very near towns and villages. It nests particularly on pines, on branches near the ground (Fig. 152).

The jay *Garrulus glandarius*, in Swedish Nötskrika, is present all through the

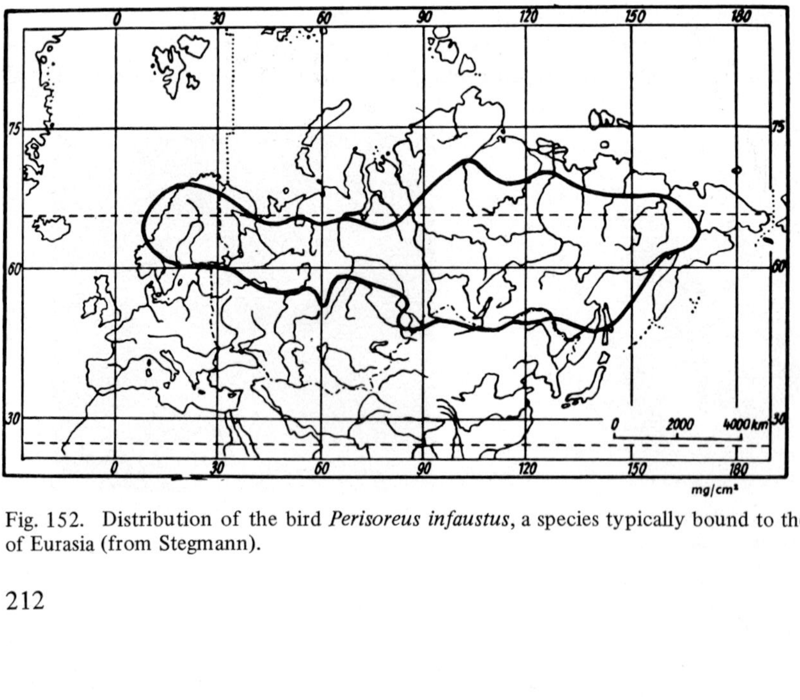

Fig. 152. Distribution of the bird *Perisoreus infaustus*, a species typically bound to the taiga of Eurasia (from Stegmann).

territory except the northernmost part, rarely far from trees, it nests generally in rather dense forests, both coniferous and deciduous. The magpie *Pica pica*, in Swedish skata, is a common *Corvidae* present throughout the territory, preferring cultivated areas and open places with shrubs and isolated trees. It builds a very characteristic nest in the form of a dome, situated on high branches or even – as in Scandinavia – under roofs or upon wood piles. The nutcracker, *Nucifraga caryocatactes*, in Swedish nötkråka, lives exclusively in the southernmost part of the taiga, in coniferous forests on mountains, in winter also in deciduous forests. It nests on coniferous trees.

The jackdaw *Corvus monedula* is also exclusive to southern districts, where it lives in parks, on rocks, old buildings (ruins), cultivated areas, etc. It nests gregariously in tree holes, on rocks, in buildings, inside chimneys, occasionally in earth holes and old nests of *Corvus frugilegus.*

To finish, we may mention the most common *Corvidae*, the rook, *Corvus frugilegus* (in Swedish råka) limited to the southernmost districts, the hooded crow, *Corvus corone cornix*, in Swedish grå kräka, presents everywhere, and lastly the raven *Corvus corax* (in Swedish korp) present all through the territory with the exception of most of Sweden.

Differing from other European biomes, both the taiga, and still more the tundra show explosive increases, in several homoiothermous animal populations, often with a cyclic aspect. The phenomenon is well known in Eurasia as well as in North America. As far as the European taiga is concerned, these cycles of population have been recognized in the following animals: *Lemmus lemmus, Buteo lagopus, Accipiter gentilis, Vulpes vulpes, Alopex lagopus* and *Lagopus lagopus* (Johnsen 1929, Elton 1942, Lack 1951). The data are summarized in the following table from Lack):

YEARS OF PEAK POPULATION

Lemming (*Lemmus lemmus*)	Rough-legged Buzzard (*Buteo lagopus*)	Goshawk (*Accipiter gentilis*)	Red and Arctic Fox (*Vulpes* and *Alopex*)	Willow Grouse (*Lagopus lagopus*)
1871-2	1872	1871-3	..	1872
1875-6	1875	1876	..	1876
1879-80	1879-80	1880	1880	1880
1883-4	1883-4	1883-4	1884	1883
1887-8	1887-8	1889	1887	1887
1890-1	1891	1891	1892	1891
1894-5	1894-5	1895	1895	1895
1897	1898-9	1899	1899	1897
1902-3	1902	1901-2	1903	1903
1906	1906	1905-6	1907	1906
1909-10	1909-10	1911-12	1910	1908-9
				1911-12

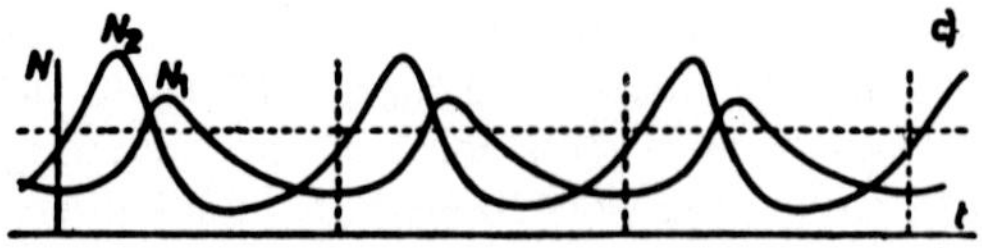

Fig. 153. Cycles of two homoiothermous animals of which one is the prey and the other the predator (from Volterra).

The explanation of the cycles is far from satisfying. Against a deterministic interpretation, which insists upon inner causes intrinsic to the populations, generally a stochastic interpretation prevails, in which the role of all external and internal variables is taken into consideration. One of these variables is the effect of predators upon the prey, studied from the mathematical point of view by Volterra. The fluctuations of two theoretical populations, one of a predator and another of a prey are represented by the graph of Fig. 153. The two curves correspond to the equation

$$\left(\frac{n_1}{e^{n_1}}\right)^{\epsilon_2} = C \left(\frac{n_2}{e^{n_2}}\right)^{-\epsilon_1}$$

where C is a positive constant, n_1 and n_2 values proportional to N_1 and N_2, which are the number of the individuals in the two populations; t is the time.

There has as yet been no detailed population study of the willow grouse *Lagopus lagopus* (Lack, 1968, p. 205). However, this bird shows a 4 year cycle when lemmings are present, and a 10 year cycle when the varying hare is present. According to Lack, the cycles of birds may depend on the cycles of predators – rodents, since when predators find no more rodents, they prey upon birds (p. 286). This would of course be sound proof of the probabilistic nature of the cycles.

A practical example of the effect of a predator upon a prey is represented in North America by the fluctuations in the populations of *Lynx canadensis* and *Lepus americanus*, studied by MacLulich.

Of course, a population may also fluctuate despite the effects of its predators. In this case a stochastic interpretation is also possible: namely, the fluctuation would correspond to what we can predict on the basis of a Markov chain, $g(z) = \lambda(z^2 - z) + \mu(1 - z)$, where λ is the natality coefficient and μ the mortality coefficient.

Amphibians and reptiles

Although the taiga is not the best environment for amphibians and reptiles, in the southern sectors of this biome there are several species which find suitable conditions for their life, i.e. temperature, shelter and food.

To begin with reptiles, we may mention first the lizard *Lacerta vivipara*, spread to the coasts of the Arctic Sea, together with the frog *Rana temporaria.* It is an extremely euryoikous species, very common also on our mountains. The blindworm *Anguis fragilis* is spread north almost like *Vipera berus*; it is also a very euryoikous species, living in broad-leaved and mixed forests, but also in orchards, fields and meadows near to forests. It hibernates in the soil where it follows the roots to a depth of 70 cm. The lizard *Lacerta agilis* is limited to the extreme southern limit of the taiga, in Swedish reaching latitude 61°N, it prefers dry and sunny places, living in steppes, forests, gardens, coppices, ravines, etc., often in *Pinus* forests (which of course are drier than the spruce forests). The water snake *Natrix natrix*, in Scandinavia extending as far north as 67°N, is more or less closely restricted to water, usually in rivers, lakes, ponds, etc. often in forests. It is particularly bound to water during summer. It hibernates in burrows of mammals, under roots of trees in fissures, etc. The small snake *Coronella austriaca*, also a southern element, in Sweden reaching the latitude of 64° and in Norway 63°N, lives in forests, places covered with shrub vegetation, rarely on steppes, with its main food consisting of lizards.

The most interesting is, without any doubt, the viper, *Vipera berus*, extending far into the Arctic Circle, living in broad-leaf forests, pine forests, mixed forests and marshy areas. It hibernates in the soil at a depth of 2 metres and perhaps more, and appears in the middle of spring, even when extensive areas are still covered with snow. There are also several amphibians although they are not very common.

One of the least known because of its behaviour, is *Pelobatus fuscus*, present exclusively at the southern limit of the taiga in Russia; it is an animal living in water only during oviposition: during the day it digs into the soil down to 1 metre. It prefers light soils of meadows, mossy swamps and pine forests.

The green toad, *Bufo viridis*, is also limited to the southern sections. It is better adapted to drought than all other amphibians of the taiga. It often lives in rodent burrows, in gardens, cellars, etc. and is active at night. The common toad, *Bufo bufo bufo* (historically the youngest subspecies of this 'Rassenkreis') extends to all the southern half of the European taiga. During the day it is found in barns, cellars, caves, under stones, roots, etc. It hibernates on land.

The water frogs, *Rana ridibunda* and *Rana esculenta*, are limited to the southernmost part of the taiga in Russia, Finland and, limited to the latter, Sweden. The first is always in water or near it, inhabiting ponds, lakes, rivers and hibernating on the bottom of water reservoirs; the second has very similar habits, and therefore is very probably competing with the former.

An interesting frog is *Rana terrestris* (= *arvalis*), present in Sweden, Finland, up to 68°40′N, and Northern Russia: it prefers swamps, wet meadows, edges of forests, forest clearings, sedge swamps, etc. It hibernates partly under water, partly on land. It seems to compete with *Rana temporaria.*

The fire-toad, *Bombina bombina*, is present only at the southern border of the taiga in Russia (57°N): it lives in ponds, pools, ditches, cut-off lakes in lowlands; it spends most of the time in water, but it hibernates on land.

Finally, we may mention the newts, and more precisely the common newt, *Triturus vulgaris*, a southern species although living also in Finland and in Sweden where it reaches almost 65°N; it lives in water only during the reproductive period and moves onto land where it hibernates; the crested newt, *Triturus cristatus*, is also a southern species although present also in Scandinavia up to 67°N. It thrives generally in water and hibernates in rotten stems and under leaves, more rarely under water slime.

Animals of forest interest

Among the very numerous animals which attack or can attack the coniferous trees, both of the taiga and of the mountains of Southern Europe, the most well-known are the beetles[1].

A typically phytophagous family is represented by the *Cerambycidae.* Let us mention *Rhagium bifasciatum*, found in dead coniferous trees; *Toxotus cursor*, distributed in North and Central Europe, Western Siberia, as far south as the Italian Alps; *Gaurotes virginea*, restricted to pines; the two species of *Evodinus, E. borealis* and *E. interrogationis*, both showing a discontinuous distribution of the boreo-alpine type, i.e. limited to the northernmost part of Eurasia and the mountains of Southern Europe; *Acmaeops pratensis*, whose larva feeds upon dead branches of spruce; *Spondylis buprestoides* (from the Greek word σπόνδυλος, dorsal vertebra) which lives in the larval stage in rotten stumps; *Asemum striatum*, living in dead trunks; *Criocephalus rusticus*, with habits similar to those of the preceding species; *Tetropium castaneum* and *T. fuscum*, both present in living trees and therefore sometimes dangerous to the forest; *Hylotrupes bajulus*, found in diseased, old trees as well as in logs, which can be damaged by this insect. Other interesting species are the beautifully coloured *Callidium violaceum*, whose larva lives in dry wood of both deciduous and coniferous trees; *Callidium coriaceum*, harmful to spruce, parasitized in Finland by the Hymenopteran *Pimplarinae Ephialtes manifestator* and *Xylonomus praecatorius*; and *Callidium aeneum*, feeding on dead wood. Another genus harmful to coniferous trees is *Clytus*: *C. pantherinus*, distributed in North Europe, Siberia, Bavaria and Northern Italy (Trentino); *C. capricornis* found in Russia, Siberia, Bosnia and Lower Austria; *C. ibex*, present in Russia, Siberia and although extremely rare, in Eastern Germany, and finally *C. herbsti*, present in Central Europe, the Caucasus and Western Siberia. All these species develop in living branches and trunks, so that they are relatively harmful to forests, notwithstanding their modest size. There are also *Monochamus sutor*, living in dead trunks; *Pogonochaerus decoratus*, found in living pines; *P. ovatus*, present also in deciduous trees and *P. fasciculatus*, present on pines and spruce, also living, and therefore often very harmful. Lastly let us remember the *Cerambycidae Acanthocinus edilis*, on dead pines, and *Oberea pupillata*, inside living plants of *Lonicera*, a very frequent genus in forests both of deciduous and coniferous trees, to which the common honey-suckle belongs.

Two other beetle families widely spread in the taiga are *Ipidae* and *Melandryidae.* The former is represented by *Ips typographus*, living on the Norway spruce, secondarily on pines and larches, sometimes even in healthy trees; *Pityogenes chalcographus*, bound to spruce, only rarely on pines and larches; this is one of the most harmful enemies of the Norway spruce; *Polygraphus subopacus,* sometimes attacked by some parasitic *Hymenoptera* such as *Dendrosoter* (from the Greek δένδρον, tree, and σότερ saviour) *middendorfi*, belonging to the family

[1] Just in Sweden, there would be about 15,000 species of insects of which 4,000 belong to the Coleoptera.

Doryctinae, and finally, *Hylurgops glabratus*, living on *Pinus cembra*, a boreal and oriental species, but present also in the mountains of the temperate part of Europe, where it can occur up to 2000 m and more.

To the *Melandryidae* belongs the genus *Pytho*. Let us mention only *P. kolvensis* and *P. abieticola.* The larva is extremely characteristic with its flattened shape, soft body wall and two large points or horns set on the posterior extremity.

Among the best investigated beetles of the Eurasiatic taiga we find the *Elateridae*, whose larva – known as wireworm – is often harmful to plants both arboreal and herbaceous. The most important ecological investigations are those of Cherepanov. In West Siberia – which means immediately east of the European taiga – this author has found 54 species of this beetle family, among which *Adelocera, Elater* and *Denticollis* are the most represented. *Harminius undulatus*

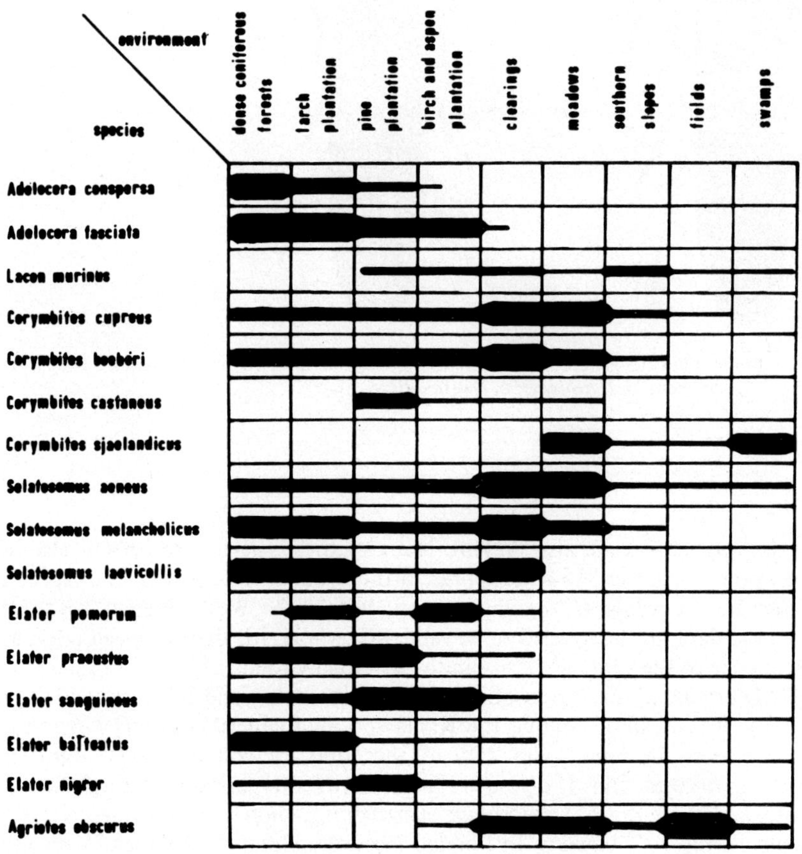

Fig. 154. Quantitative distribution of *Elateridae* in the western Siberian taiga according to different environments (from Tscherepanov, translated from the Russian).

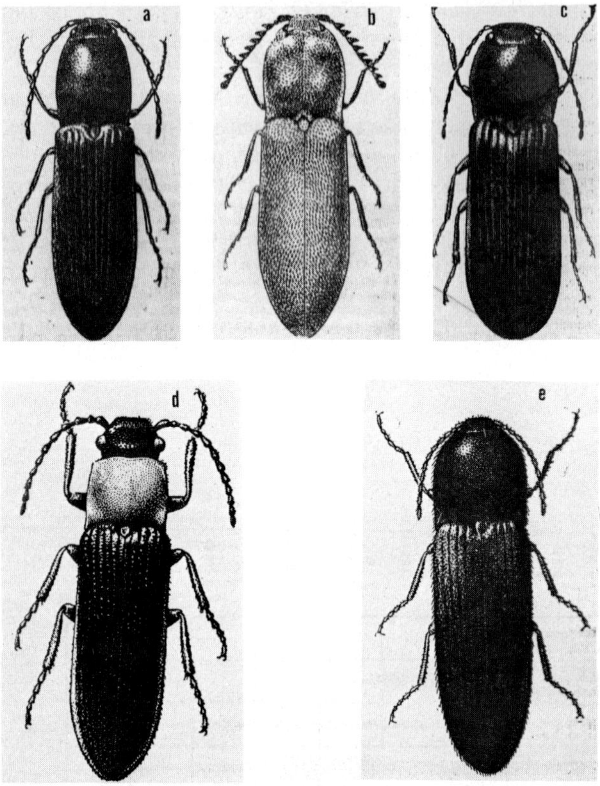

Fig. 155. Some Elaterid genera present in the European taiga (from Čerepanov, 1957): a. *Limonius*; b. *Adelocera*; c. *Harminius*; d. *Denticollis*; e. *Elater*.

and *Melanotus rufipes* are also present. Many of these species are present also in the European taiga. Figs. 154-157 represent the abundance of a number of species of *Elateridae* according to the environment within the West Siberian taiga. Of course these data are likely also to be valid, as a whole, for the European taiga, as a consequence of the great similarity in the macro-environment.

The *Elateridae* of the region of the rivers Kama, Čepz and Kil'mez have been studied by Roščinenko. Among the Kama species there is *Synaptus filiformis*, which can reach in some cases 25% of the total *Elateridae*, besides *Agriotes sputator, A. lineatus* and *A. obscurus*. In meadows alternating with small woods there are *Lacon murinus, Selatosomus nigricornis, Dolopius marginatus, Athous subfuscus, A. niger, S. incanus.* In flooded meadows along the Kil'mez 27 species of *Elateridae* live: *Limonius aeruginosus* can reach here up to 50-70% of the total fauna; other species are *Selatosomus aeneus, Dolopius marginatus, Prosternon*

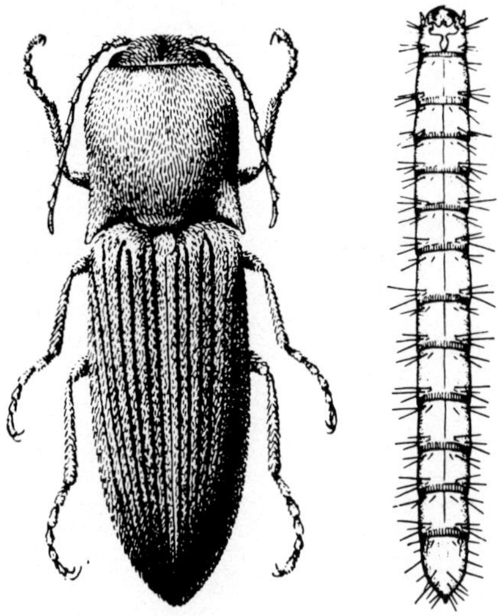

Fig. 156. *Agriotes lineatus*: adult (left) and larva (right) (from Čerepanov).

tesselatum, Elater sanguinolentus, E. pomorum, E. balteatus. Species present also along the Kama are *Agriotes sputator* and *A. obscurus.* Lastly, in the taiga through which the Čepz runs, 24 species of *Elateridae* have been found, among them *Dolopius marginatus, Selatosomus incanus, Corymbites pectinicornis, C. sjaelandicus, Lacon murinus, Selatosomus aeneus, Limonius minutus.*

It is worthwhile noticing, to show the enormous affinity between the environment of the taiga and that of the coniferous forest of the Alps, that all these species are present in a small district of the Austrian Alps, the Tyrol, mostly covered, at least in the past, by beautiful needle-leaved forests (*Picea, Pinus silvestris, Larix*, besides *Betula* spp.). Only one species, *Selatosomus nigricornis*, is present in other Central-European districts but not in the Tyrol.

The soil fauna

The soil fauna of the taiga is, after that of the deciduous forest, the best known in Europe, thanks to the voluminous work of Soviet scientists, more especially of the Moscow School headed by Ghilarov, and that of Kazan.

The huge amount of information these authors have gathered in many years of investigations has also permitted a very advanced level in the work these students have done within the plan of research for the International Biological Programme (IBP) so that now, as an addition to the many small articles issued in several

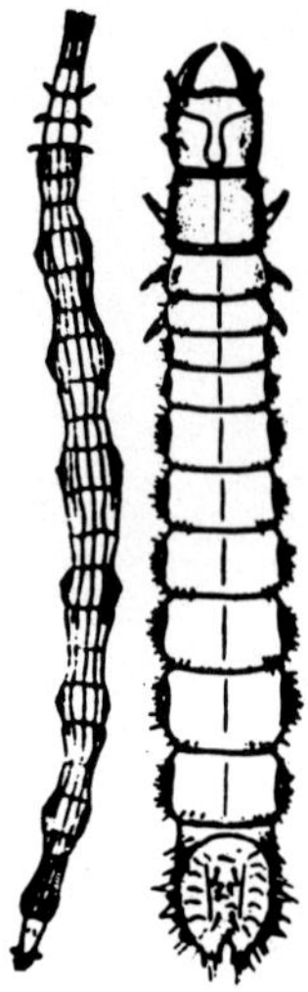

Fig. 157. The larvae of two species of *Elateridae* frequent in the European steppes. Left: *Cardiophorus ruficollis*, right: *Lacon murinus* (from Ghilarov et. al.).

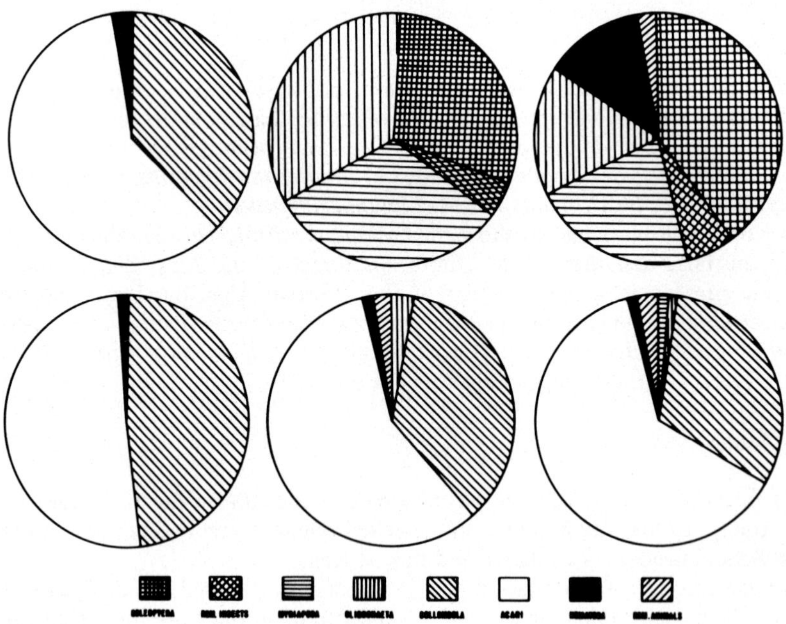

Fig. 158. Composition (in %) of the soil fauna of different forest types of the Vyatka-Kama region (constructed on the basis of Aleinikova's data, 1968, pp. 26-27).

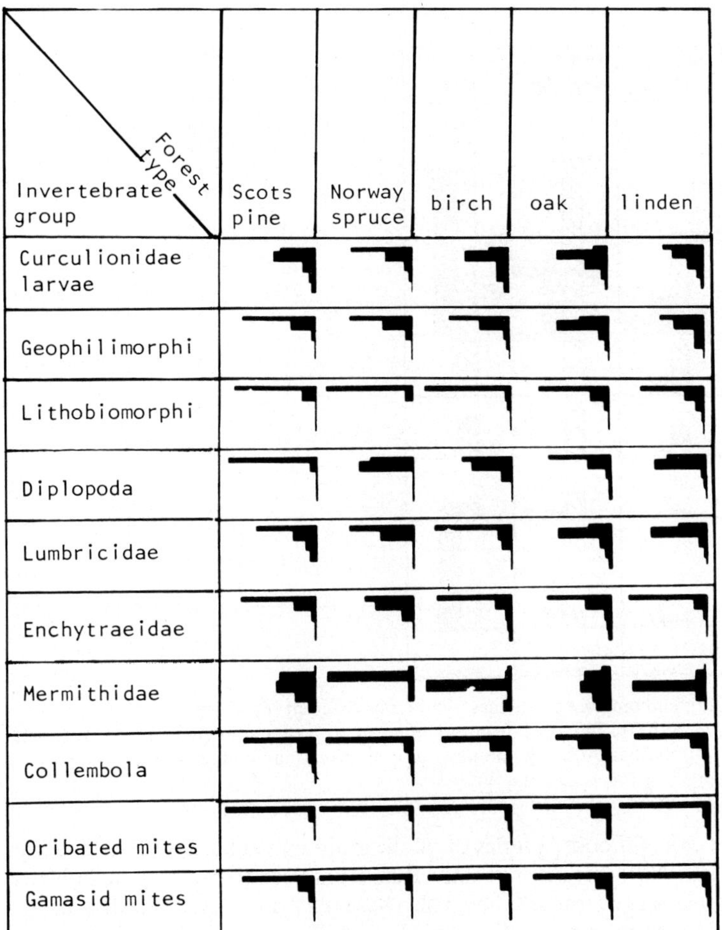

Fig. 159. Vertical distribution (in %) of the different soil invertebrates in different forest types of the Middle Povolz' (from Aleinikova).

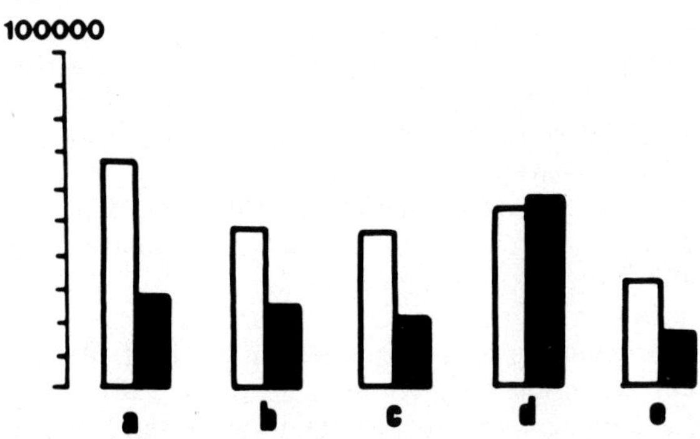

Fig. 160. Number of individuals per square meter of Oribatid mites and of Collembolans (white and black, respectively) in different forest types (from Aleinikova, redrawn). As can be clearly seen, the ratio mites/springtails shows quite abnormal behaviour in the Tilis forests which, according to the present author, would deserve further investigation. Forest types: a. spruce; b. Scots pine; c. birch; d. linden; e. oak.

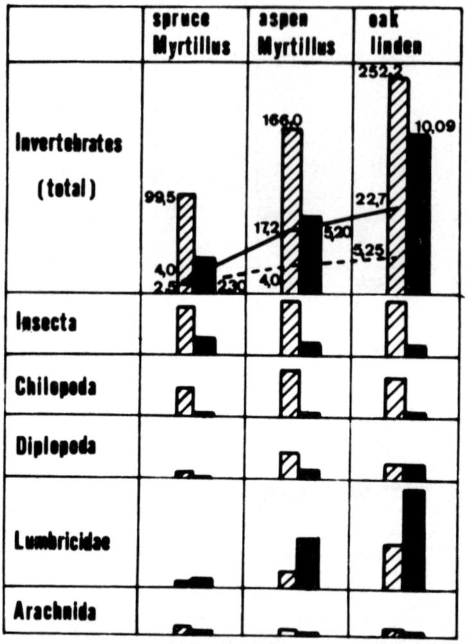

Fig. 161. Quantity and biomass of soil invertebrates in different forest types of the Trans-Volga planes (Mariiskaya SSR) (1963 data from Aleinikova, p. 51, translated from the Russian). //// quantity/m^2/40 cm depth; black: biomass in g/m^2; continuous line: soil humidity as %; broken line: humus as %.

annual reviews or handbooks, a series of publications edited by the IBP authorities is at our disposal. Nevertheless, not only Russia has been thoroughly investigated as far as soil fauna is concerned, Finland also has seen some very interesting papers dealing with its soil animals, more particularly with oribatid mites (Karppinen), not to mention the early contributions of Krogerus, Palmen & Platonoff and several others.

Since it is impossible to summarize even partially the enormous amount of data on soil fauna of the European taiga, we will quote only some results which will give an idea of the composition of the soil fauna, the relative abundance of each zoological group and of the biomass, determined for some biotopes and animal groups that we judge the most characteristic.

The quantity and composition of the invertebrate fauna of different kinds of forests of the taiga of the Mariyskaya SSR have been studied by Aleinikova. The data for 1965 are shown in Table 1, those for 1966 in Table 2.

The data concerning the region of Vyatka and Kama, belonging to the Tatar SSR are shown in Table 3, those belonging to the same region, but to its Mariyskaya part, with a rather different vegetation, are given in Table 4 (Fig. 158).

Fig. 161 shows the number and biomass of soil invertebrates in several kinds of forests of the Mariyskaya SSR. Fig. 160 shows the average number of oribatid mites and collembolans (or springtails) of different kinds of forests of the Middle

Table 1

	Norway spruce forests		Scot pine forests					Birch (*Betula* spp.)		*Populus tremula*	
	60-70 years old grassy strong podzolic soil upon a layer of clay	90-120 years old grassy middle podzolic soil upon a layer of sand	with lichen and moss	with *Vaccinium vitis idaea*		with *Myrtillus*		*Picea* and *Tilia* sp. grassy, strong podzolic soil	*Oxalis a.* and *Tilia* sp. grassy, middle podzolic	*Picea* and *Tilia* grassy strong podzolic soil	*Acer* and *Tilia* weak podzolic soil sometimes on alluvium
				weak podzolic soil on alluvium							
			100-120 y.	60 years	95-100 y.	60 years	75-100 y.	40-50 y.	50 y. old	50-60 years old	40-50 years old
Insecta Coleoptera (total)	16.5	14.5	13.0	22.0	13.0	21.0	29	16.5	17.0	16.5	36.0
Carabidae	1.5	3.0	1.5	4.0	2.0	3.0	2.5	2.5	1.5	0.5	–
Staphylinidae	1.0	3.0	0.5	2.5	0.5	0.5	1.0	1.0	2.5	2.5	3.0
Scarabaeidae	0.5	–	–	0.5	0.5	0.5	0.5	1.0	–	–	–
Elateridae	9.0	5.0	7.5	9.5	7.5	15.5	22.5	6.0	9.0	6.0	30.5
Curculionidae	4.5	2.5	1.0	5.5	2.0	1.5	2.5	4.5	2.5	5.0	1.0
Remaining families	–	1.0	2.5	–	0.5	–	–	15.0	1.5	2.5	1.5
Diptera	1.5	3.0	2.0	0.5	1.5	–	0.5	4.0	2.5	2.0	4.0
Hymenoptera	–	1.0	0.5	–	–	2.0	–	–	3.0	–	1.5
Lepidoptera	–	0.5	–	0.5	1.0	–	–	2.0	0.5	2.0	3.5
Remaining orders of Insects	–	–	3.0	3.0	0.5	2.0	–	3.0	2.5	–	–
Total	18.0	19.0	18.5	26.0	16.5	25.0	29.5	25.5	25.5	20.5	45.0
Myriapoda (s. lato) total	11.0	29.0	14.0	21.0	15.5	23.0	21.0	18.0	34.5	33.5	41.5
Geophilomorpha	9.5	10.0	0.5	6.5	2.5	7.5	6.0	10.0	13.5	18.0	19.5
Lithobiomorpha	1.0	15.0	13.0	14.5	11.5	12.5	14.0	8.0	20.0	14.0	18.5
Diplopoda	0.5	4.0	0.5	–	1.5	3.5	1.0	–	1.5	1.5	3.5
Ann. Oligochaeta											
Lumbricidae	5.0	4.0	3.5	1.5	1.5	5.5	5.0	9.0	5.5	7.0	15.0
Enchytraeidae	2.0	–	–	–	–	–	–	1.0	–	0.5	2.0
Nem. Mermithidae	–	7.5	–	–	–	–	–	17.5	1.5	7.0	1.0
Arachn. Araneino	0.5	8.5	3.5	6.0	4.0	4.5	1.0	5.5	8.5	4.5	–
Mollusca	–	0.5	0.5	–	0.5	0.5	2.0	–	–	1.5	0.5
Other invertebrates	–	1.0	–	–	–	1.0	–	0.5	–	–	–
TOTAL	36.5	69.5	40.0	54.5	37.5	60.0	58.5	77.0	76.0	74.5	90.0

Table 2

	Scots pine			*Tilia* sp.		*Quercus* sp.	Birch (with spruce)	
	lichens	*Vaccinium vitis idaea*	*Myrtillus*	110 years old	100–140	60–70 y.	60 y	65 y
	65–70 years old	75 years old	65–70 y. old	slightly sandy soil	slightly clayey podzol	granular sandy soil	slightly sandy podzol. s.	slightly sandy podzolic soil
	slightly sandy podzolic soil							
Ins. Coleoptera (total)	56.5	28.5	44.0	20.0	16.5	26.0	39.0	61.5
Carabidae	1.5	7.5	0.5	2.0	3.5	3.0	5.5	2.5
Staphylinidae	–	3.5	–	–	–	–	3.5	2.0
Scarabaeidae	8.5	6.5	–	–	–	–	2.0	16.0
Elateridae	24.5	6.0	29.5	7.5	6.5	11.5	8.0	15.0
Curculionidae	18.5	5.0	13.0	5.5	2.5	10.0	17.5	24.5
remain. *Coleoptera*	3.5	–	1.0	4.5	3.5	1.5	2.5	1.5
Diptera	0.5	–	–	2.0	1.0	6.0	6.5	–
Hymenoptera	1.0	–	0.5	–	–	–	1.0	–
Lepidoptera.	–	–	–	–	1.0	–	0.5	–
remain. *Insecta*	2.5	2.0	1.5	–	–	1.5	1.5	1.5
Total	60.5	31.0	46.5	22.0	18.5	33.5	46.5	63.0
Myriapoda (s. lato) total	9.0	14.5	12.5	13.5	22.0	10.0	12.5	17.0
Geophilomorpha	9.0	8.5	12.5	12.5	18.5	10.0	10.5	15.0
Lithobiomorpha	–	5.5	–	1.0	4.0	–	0.5	0.5
Diplopoda	–	0.5	–	–	–	–	1.5	1.5
Ann. Oligochaeta:								
Lumbricidae	–	1.0	2.0	10.5	11.5	31.0	15.0	13.5
Enchytraeidae	–	–	–	–	–	20.0	3.0	–
Nem. Mermithidae	–	–	–	–	–	2.5	–	–
Arachn. Araneina	3.0	6.0	2.0	–	6.0	3.0	8.0	5.0
Mollusca	–	–	–	–	–	–	0.5	4.0
remain. Invertebrates	2.0	0.5	2.0	–	2.5	3.0	2.0	1.0
TOTAL	74.5	53.0	65.0	46.5	61.5	103.0	89.5	103.5

Table 3: Number and composition of terrestrial invertebrates in different forest types of southern taiga in the Vyatka- Kama region (number of specimens/m²) (from Aleinikova)

	Pinus with *Vaccinium vitis idaea* 80-100 y.o. sandy podzolic	*Picea* with *Tilia* sp. 85 y.o. sandy soil	Birch with *Picea* 30-40 y.o. light clayed soil	*Pinus* with *Quercus* 20-40 y.o. light grey podzolic soil heavy clay	*Pinus* with Birch 60-80 y.o. sandy soil	Panovka *Quercus* sp. and *Tilia* sp. 40-60 y.o. grey sandy soil
MESOFAUNA						
Insecta Coleopt. total	8.5	12.5	103.0	24.0	12.4	43.5
Carabidae	0.5	4.0	21.5	2.4	1.2	7.2
Staphylinidae	2.0	1.0	3.0	2.0	1.2	5.1
Scarabaeidae	1.0	1.5	8.5	8.8	–	0.2
Elateridae	5.0	5.5	27.5	8.4	4.0	14.6
Curculionidae	–	–	42.5	2.1	4.8	14.1
Remaining *Coleoptera*	–	0.5	–	0.3	1.2	2.3
Diptera	–	1.5	4.0	0.8	1.6	130.0
Hymenoptera	–	–	6.0	0.4	1.2	3.5
Lepidoptera	–	0.5	2.0	0.4	0.8	0.7
Remaining Insects	–	–	3.0	–	–	0.5
Insects (total)	8.5	14.5	118.0	25.6	16.0	178.0
Myriapoda s. lato (total)	7.5	13.0	52.0	11.6	9.2	58.7
Geophilomorpha	2.0	13.0	44.5	7.6	3.2	31.7
Lithobiomorpha	5.0	–	6.0	0.8	4.4	23.7
Diplopoda	0.5	–	1.5	3.2	1.6	3.3
Ann. Oligochaeta:						
Lumbricidae .	6.5	13.5	36.0	27.2	12.4	140.5
Enchytraeidae .	–	0.5	1.0	1.0	3.2	2.5
Nem. Mermithidae	–	–	32.5	3.2	9.6	2.3
Arachn. Araneinae	1.0	1.0	6.0	2.5	2.4	7.5
Mollusca .	–	–	2.0	–	–	5.0
Remaining invertebrates	–	–	2.0	0.6	–	8.5
TOTAL.	23.5	42.5	249.5	71.2	52.8	403.0
MICROFAUNA:						
Collembola.	29850			18400	10050	15459
Acarina (total).	48675			19680	16660	31123
Gamasoidea	2912			2470	2240	1029
Oribatei	39263			12160	12740	27317
remain. *Acarina* .	6500			5050	1680	2777
Myriapoda .	112			90	130	594
Enchytraeidae .	75			70	790	347
Nematodes.	75			50	330	606
Remain. invertebrates	1988			610	820	1553
TOTAL.	80775			38900	28780	49482

Table 4

	spruce-*Vaccinium vitis idaea* 80–100 years old, light, clay, podzolic soil	birch spruce forest, 40–50 years old	spruce-*Oxalis* forest, 60 years old, sandy soil	spruce-bog 120–150 year old forest, swampy; peaty soil	pine-*Vaccinium vitis idaea* 75 year old forest sandy soil	*Pinetum hylocomiosum* 50–60 years old, sandy soil	birch-spruce forest, 40 years old, sandy soil
MESOFAUNA:							
Coleoptera (total)	8.5	15.5	44.4	22.5	22.0	31.0	65.0
Carabidae	0.5	1.0	2.5	5.5	0.5	–	4.5
Staphylinidae	2.0	0.5	5.0	1.0	7.5	2.5	1.0
Scarabaeidae	0.5	–	7.5	8.0	0.5	14.5	1.5
Elateridae	3.5	9.5	11.0	5.0	13.0	11.5	16.5
Curculionidae	0.5	1.5	18.5	2.0	–	2.0	40.5
remain. *Coleoptera*	1.5	3.0	–	1.0	0.5	0.5	1.0
Diptera	2.0	2.5	3.5	5.5	1.5	0.5	15.5
Hymenoptera	–	–	–	–	0.5	1.5	–
Lepidoptera	1.0	–	1.5	1.0	–	2.0	0.5
remain. *Insecta*	5.5	13.5	2.5	–	1.0	–	1.0
Total	17.0	31.5	52.0	29.0	25.0	35.0	82.0
Myriapoda s. lato (total)	8.5	9.0	26.5	18.0	15.0	19.0	24.5
Geophilomorpha	5.5	3.5	6.5	7.0	9.0	19.0	18.5
Lithobiomorpha	3.0	5.5	20.0	7.0	6.0	–	4.5
Diplopoda	–	–	–	4.0	–	–	1.5
Ann. Oligochaeta:							
Lumbricidae	3.0	1.5	2.5	6.5	1.0	7.0	15.0
Enchytraeidae	–	0.5	–	5.0	–	–	1.5
Nem; Mermithidae	20.5	3.5	–	2.0	–	–	12.5
Araneae	1.5	4.0	2.5	5.5	4.5	4.5	2.0
Mollusca	–	0.5	–	–	–	–	–
Remain. invertebrates	1.5	–	–	1.0	–	–	0.5
TOTAL	52.0	50.5	83.5	67.0	45.5	65.5	138.0
MICROFAUNA:							
Collembola	28925		21312		26057		21084
Acarina (total)	76512		79934		160285		45722
Gamasoidea	5187		11911		11314		4255
Oribatei	71063		63811		131314		29541
remain. *Acarina*	262		4212		17657		10813
Myriapoda s. lato	63		150		343		456
Enchytraeidae	–		12		14		571
Nematoda	50		100		300		571
remain. Invertebrates	2763		1833				1420
TOTAL	108312		103340		189185		69524

Table 5. Specific composition of the diplopod populations of forests of different landscapes in the Middle Volga region. In percentage (from **Aleinikova**)

Province	forest type	*Sarmatiulus kessleri*	*Strongylosoma pallipes*	*Schizophyllum sabulosum*	*Leptoiulus proximus*	*Chromatoiulus sjaelandicus*	*Chromatoiulus rossicus*	*Brachyulus jawlowskii*	*Polyzonium germanicum*	*Polydesmus complanatus*
Southern taiga	Spruce	42.1	10.5	42.1	–	5.3	–	–	–	–
Trans-Volga plain	Pine	22.8	–	66.3	2.9	7.0	–	–	–	1.0
	Birch	56.8	2.3	22.7	4.6	9.0	–	–	2.3	2.3
	Aspen	7.5	7.5	28.8	5.0	30.0	–	–	13.7	7.5
	Linden	1.2	74.3	5.5	2.0	16.6	–	–	0.4	–
Southern taiga	Spruce	12.5	–	75.0	–	12.5	–	–	–	–
Vyatka-Kama	Pine	32.1	11.9	17,2	–	30.6	–	–	8.2	–
region	Mixed forest	25.9	6.9	27.7	5.1	22.4	–	–	10.3	1.7
	Birch	31.0	–	38.0	–	31.0	–	–	–	–
	Oak	–	17.9	–	31.4	7.1	–	–	35.7	17.9

Table 6. Specific composition (in percentage) of the *Lumbricidae* population in different forest types and landscapes of the Middle Volga region (from Aleinikova)

Landscape, province and forest type	Trans-Volga plane			Vyatka-Kama taiga		
Species	coniferous for.	middle leaves forest	broad leaves forest	coniferous forest	middle leaves forest	broad leaves forest
Eiseniella tetraedra	–	–	–	–	–	2.7
Eisenia nordenskiöldi	34.6	48.3	52.5	39.3	75.0	4.3
Eisenia rosea	–	–	2.2	1.2	–	18.0
Eisenia foetida	–	–		–	–	
Eisenia uralensis	–	–	4.4	–	–	43.5
Allolobophora caliginosa typ.	–	–	6.8	14.2	1.6	0.1
Allol. caliginosa trapezoides	–	–	–	–	–	–
Allolobophora baschkirica	–	–	–	–	–	0.3
Dendrobaena octaedra	26.9	31.0	11.4	22.2	23.4	4.2
Dendrobaena subrubicunda	–	–	–	–	–	–
Octolasium lacteum	34.6	13.8	20.5	–	–	15.3
Lumbricus rubellus	–	–	–	3.1	–	–
Lumbricus terrestris	–	–	–	–	–	11.6
Bimastus tenuis	3.9	6.9	2.2	–	–	–
Number of species .	4	4	7	5	3	9
Average number of specimens per m² on the basis of excavation of soil pits	3.1	11.4	22.8	9.5	16.2	98.4

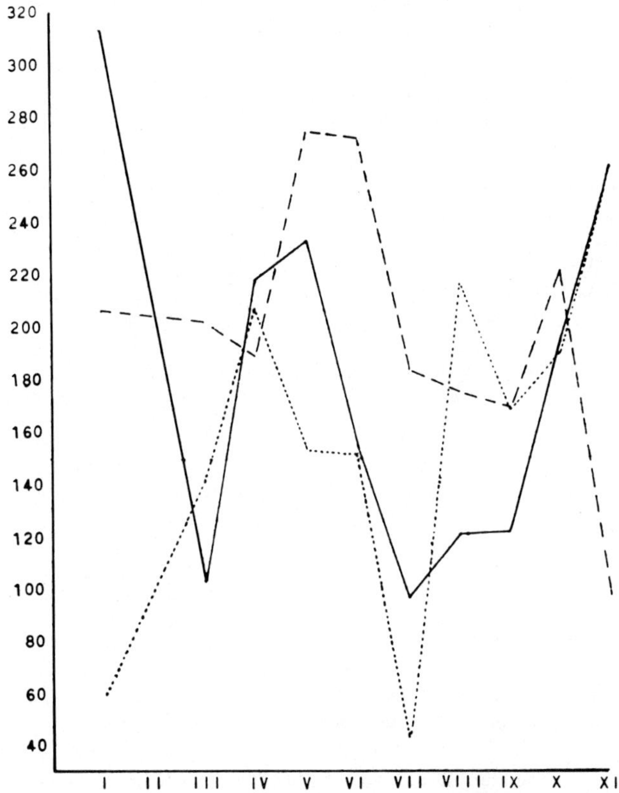

Fig. 162. Variation in the abundance of Oribatid mites in a *Myrtillus* type *Pinus* forest (——), a *Myrtillus* type Spruce forest (---) and in a *Vaccinium* type forest (. . . .) (from Karppinen, 1958).

course of the Povolz. The vertical distribution (in %) of the terrestrial invertebrates in several forest types of the Middle Povolz is represented in Fig. 159. Finally, the dependence of the millipedes on forest type is illustrated in Table 5, and that of earthworms in Table 6.

As far as the Finnish forest is concerned, we may mention some results obtained several years ago by Karppinen. The seasonal distribution of Oribatid mites in three different forest types (*Myrtillus* and *Pinus, Myrtillus* and *Picea, Vaccinium* forests) is represented in Fig. 162. We report here also the data furnished by this author for the variations in temperature and moisture of the soil during the year at Pasila, near Helsinki (Fig. 163). From a comparison of these figures it can be seen that in pine forests there are large numbers of Oribatid mites in winter, in the spruce forests there is a clear maximum in summer, whereas in a *Vaccinium* forest there are two peaks corresponding to spring and autumn. It is worth noticing the relatively large differences in temperature over the first 5 cm of soil, particularly in summer; this difference disappears almost completely from late

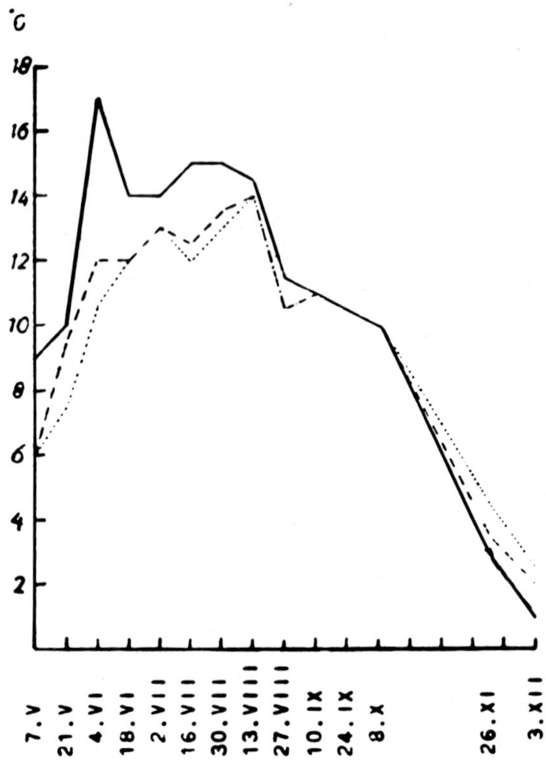

Fig. 163a. Seasonal course of temperature in a forest of *Myrtillus* type (from Karppinen, 1958).

August onwards. This differing behaviour of temperature even over so small a distance as 2 cm explains, of course, the various vertical distributions of some species of *Camisiidae* mites studied by Karppinen (Fig. 164), or the different abundance of Oribatid mites at various depths according to the different forests types (*Myrtillus-Pinus, Myrtillus-Picea, Vaccinium*).

Very remarkable results were obtained by Karppinen when he compared the abundance of Oribatid mites in the soil of many forest types of Southern Finland (which, however, do not belong completely to the taiga). From the figure he has given us (Fig. 165), we can notice the extreme poverty of *Oxalis-Myrtillus* forests when compared with all the other forest types. Among these, *Oxalis-Majanthemum*, associated with oak, provides the optimum conditions for Oribatid existence, from which we might infere that a mixed forest, at least in Southern Finland, is richer than a pure coniferous one, as far as soil fauna is concerned.

Still more interesting are the results obtained by Karpinnen when he compared the number of individuals of the dominant species (i.e. the degree of dominance among the biocenosis) with the number of species present in each kind of forest examined, which, in practice, means the 'specific diversity' (Fig. 166).

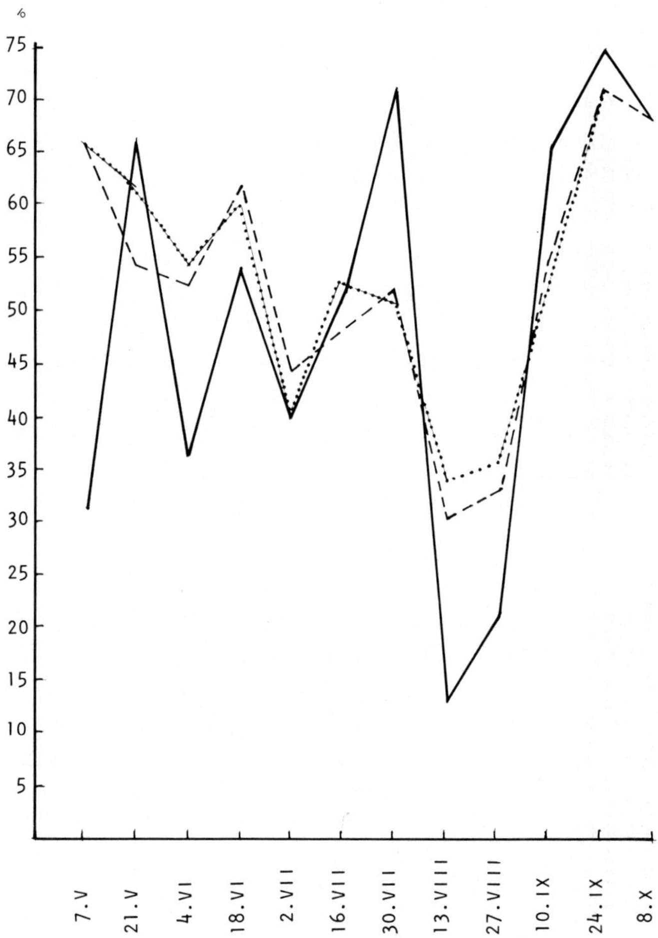

Fig. 163b. Changes of soil moisture of a forest of *Myrtillus* type (from Karppinen, 1958).

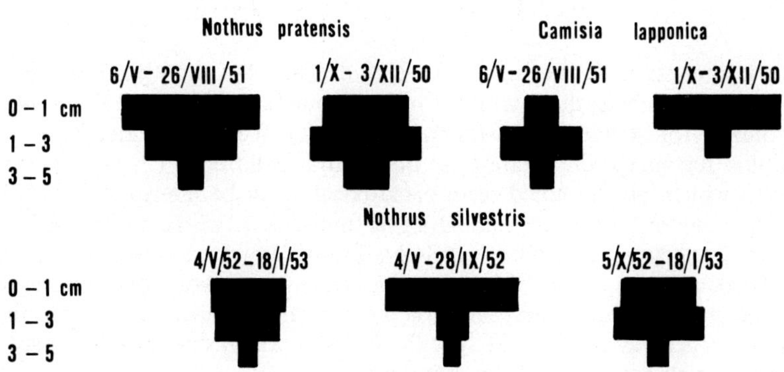

Fig. 164. Vertical distribution of the commonest Finnish Camisiid mites, in summer and in winter (from Karppinen, 1958).

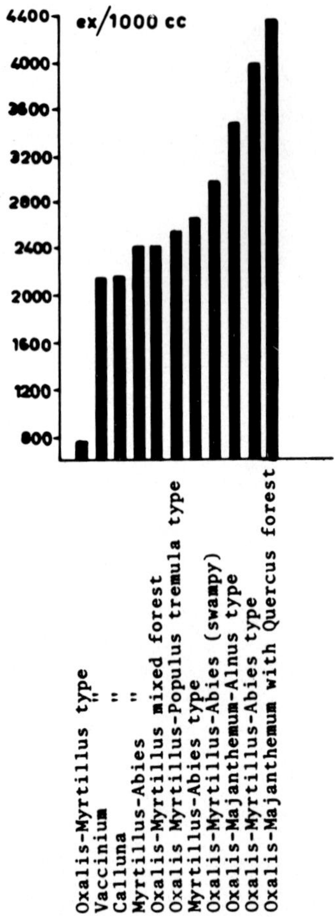

Fig. 165. Abundance of Oribatid mites in some forest types of Southern Finland (from Karppinen).

The environments with the highest specific diversity are the *Myrtillus*-spruce and the *Vaccinium* forests; those with the highest abundance values are the *Oxalis-Myrtillus* mixed forest with aspen (*Populus tremula*). We must notice that the latter, in this region, is the nearest to the natural conditions ('Urwald'): this phenomenon which 'per se' could seem paradoxical, must be interpreted according to Karppinen on the basis that Oribatid mites as a whole prefer slightly disturbed environments, and, more precisely, forests subjected to some forestry practices. Of course, we cannot think that this effect is a common one, neither in the taiga nor in the other biomes, since, on the contrary, conditions nearer to the climax show in general the highest specific diversity which does not necessarily coincide with a high population value or biomass.

The oribatid mites of the Russian taiga have been studied by Krivolutsky

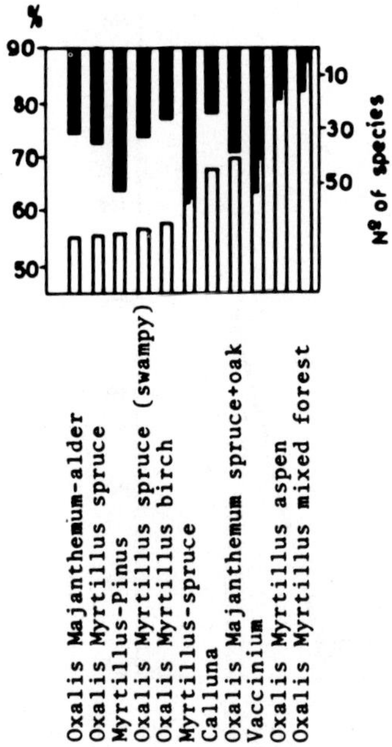

Fig. 166. Number of species and percentage in the composition of total fauna of Oribatid mites in several forest types of Southern Finland (from Karppinen).

(1967). The population density of this invertebrate in the taiga and mixed needle-leaved forests, where podzolic soils predominate, reaches 60,000-70,000 specimens per square metre (taking into account adult specimens only). The majority of the mites (about 95%) in the soil of this zone were concentrated in the litter and upper soil layer, but some of the Oribatids penetrated to 20-30 cm. The mites penetrated to the alluvial part of the soil profile in the western regions of this zone, such as Byelorussia, and a second maximum of Oribatid density was observed there at a depth of 30-40 cm.

According to this author, the number of specimens of Oribatid mites is dependent on a particular coefficient, named by Krivolutsky as 'coefficient of environmental suitability for Oribatid mites', in which both physical (temperature and rainfall) and biotic (litter production) factors are taken into consideration. Namely, the coefficient is represented by the formula

$$A = \frac{F+L}{L} - KR,$$

where $K = \frac{Q}{V}$ or the coefficient of moisture, Q the annual rainfall (in mm), V the annual evaporation (in mm), F the annual grass and leaf fall (dry weight in g/m^2), L the litter (dry weight in g/m^2) and R the solar radiation (K cal/m^2).

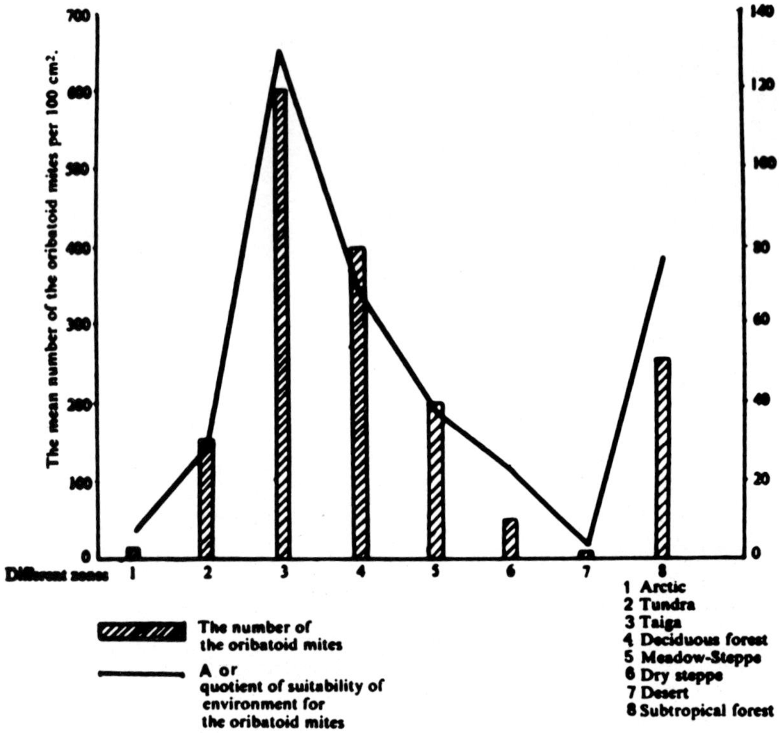

Fig. 167. The value of 'index A' (Krivolutsky, 1969) for different European biomes. Hatched: No. of Oribatid mites, continuous line: index of suitability of the environment for Oribatid mites.

It is evident from Fig. 167 that in the taiga A is exceptionally high, probably due to a large amount of litter and low solar radiation, so that Oribatid mites find here their optimum among all the biomes considered.

One group of animals of great ecological importance is represented by the numerous species living in old decaying stumps. Most of them are saprophagous or degraders (both primary and secondary) since they exclusively attack dead wood. With them live a great number of predators, belonging to the chilopods, spiders, beetles (more precisely *Carabidae* and *Staphylinidae*) and their larvae (the same families with *Cantharidae* and others).

The greatest part of the stump fauna is composed of *Ipidae, Cerambycidae, Buprestidae, Curculionidae* (*Pissodes*), also *Hymenoptera* (*Syricidae*), in the phloem, *Cerambicidae, Buprestidae, Ipidae* (*Xyloterus*) and *Syricidae*, in the xylem. In the rotten wood of an old stump we can find *Rhagium* and *Pyrochroa.* When the wood is invaded by fungi new groups make their appearance: the coleopterus families *Lymexilonidae, Elateridae* and *Cisidae*, the lepidopterous family *Tineidae.* In rotten, yellow or brown-red wood (when practically all of the cellulose has been attacked by animals, fungi and bacteria, and lignin is almost the

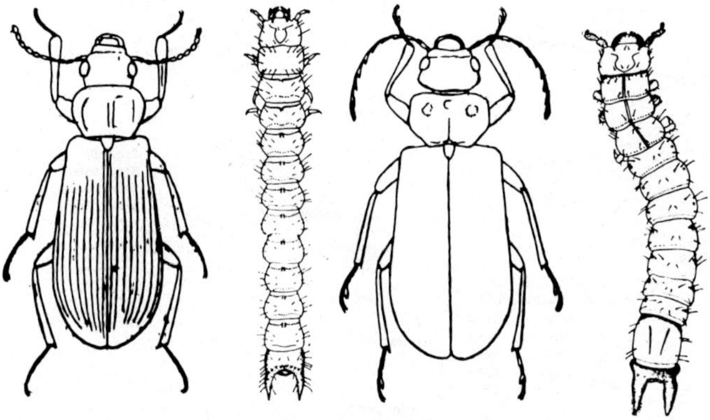

Fig. 168. Example of beetles which intervene in the demolition of stumps: *Pytho depressus* (adult and larva). *Pyrochroa coccinea* (adult); *Pyrochroa* sp. (larva) (adults from Portevin, larvae from Ghilarov et al., 1964).

only constituent) we can find the coleopterous families *Tenebrionidae* and *Lucanidae*, together with their predators such as ants (generally represented by the genus *Formica*), *Carabidae, Histeridae, Staphylinidae* and chilopods (or centipedes) (Fig. 168).

In 1939 Savely demonstrated the following feeding categories in the fauna of a decaying pine stump:

	species feeding on bark and phloem	species feeding on xylem	saprophages & mycetophages	predators	parasites
after 1 year	22	4	37	29	9
after 2 years	–	11	28	27	–

Another group of soil animals of extremely great ecological interest, and which so far have attracted very little attention from students, are the nidicolous species. As far as the European taiga is concerned, we have now a fair knowledge of this fauna thanks to the recent investigations of Sadekova, who studied the nidicolous Oribatid mites in the National Park of the region between the Volga and the Kama. The rodents are essentially *Microtus arvalis, Clethrionomys glareolus, Apodemus sylvaticus, A. flavicollis* (cf. p. 000). The number of the visited nests is as follows: 57 of *Apodemus sylvaticus*, 57 of *A. flavicollis*, 70 of *Clethrionomys glareolus*, 68 of *Microtus arvalis*, also 10 of *Micromys minutus* (maljutki) and 3 of *Glis glis* (lesnoi soni).

22,000 specimens were found (which means an average of 83 specimens per nest) belonging to 63 species and 30 genera. The maximum number of specimens found in one nest was 825; the maximum number of species 23. 53% of the total Oribatids was represented by *Scheloribates laevigatus, S. latipes, Oppia tu-*

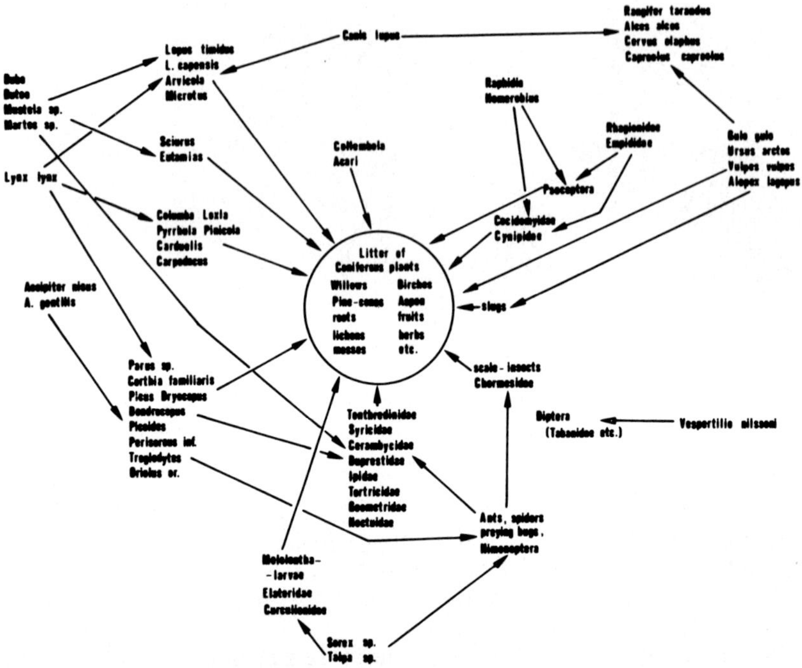

Fig. 169. Trophic relationships in the European taiga.

berculata. In most nests the following 11 species were found: *Scheloribates laevigatus, S. latipes, Oppia tuberculata, Xenillus tegeocranus, Autogneta longilamellata, Hermanniella granulata, Ceratozetes gracilis, C. mediocris, Belbodamaeus tuberculatus, Parachipteria punctata* and *Hypochthonius rufulus.* Probably most of them are euryoikous elements, generally present in humus and/or litter.

The problem of conservation

The problem of nature conservation in the taiga – although much less than in all the other European biomes – has become very serious in recent times as a consequence of the increasing human influence in several parts of it, both in Russia and Fennoscandia. The main factors of human disturbance are the presence of factories, mines, roads, new urban centres, harbours, etc. Also hunting, which during the last century was almost completely free and indiscriminate (particularly as far as fur animals were concerned), has contributed to an alteration of the delicate biological equilibrium existing in the taiga. This is more delicate than in other biomes because of the relatively recent origin of its biota (in general only postglacial), different from that in other parts of the continent and islands.

So, the necessity of protection and conservation of both the environment (or landscape) and single organisms has been felt more and more strongly, and measures have been taken and are being taken in order to maintain as undisturbed as possible the equilibrium still existing today within the taiga.

Three countries are particularly sensitive to this necessity: the USSR (or in our case the European part of Russia), Finland and Sweden. Of course the problems with which each country is confronted are different and therefore also the measures to be taken are different.

Particularly complete in every detail is the legislation on nature conservancy in the USSR, where many agencies are dealing with it, as legislative or consultative bodies. These bodies may be national, republican, provincial (oblast) or local. The relationships between such agencies are schematized in the following scheme from Borissoff (1970) translated from german:

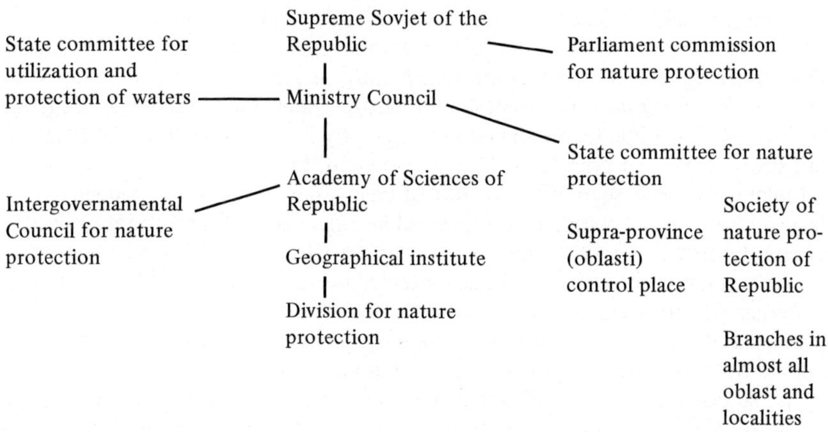

The law on the nature protection (October 27, 1960) has the following aims: protection of natural kingdoms, and of soil, subsoil, waters, forests, unforested vegetation, green spaces in urban centres, characteristic landscapes, and rare and interesting natural objects; Zapoviednik (integral natural reserves) and Zakaznik (partial natural reserves), defense of the landscape and of wild animals, nature protection from the health standpoint, quantitative and qualitative understanding of soil value; scientific research in the field of nature protection, teaching of nature protection in all schools, spreading of the nature protection idea; responsibility of the directors of Divisions and Operations; responsibility of the inhabitants for illegal utilization or deterioration of soil values.

The study of nature in the various reserves is very important for science, not only because it serves to investigate the relationships existing among all elements of the biogeocenosis, but also because these reserves are an obligatory fieldwork area for many university professors of Biology and Geography (different from many other European countries, where university teachers do not even know of the existence of natural reserves). Often the practical work of university students is done in the various Zapoviednik. The study of the elements of the various biocenosis includes, of course, also some animals of practical interest such as the zibellin or beaver, which can be fully followed only in a reserve. If for the study of a given species (say an ungulate or a water bird, or other species) the reserves are not sufficient, the law for nature protection foresees the creation of other suitable places for this aim.

As far as taiga biocenosis are concerned, we may mention here the reserve of the Ural Mountains and, in the Asiatic part of the taiga, the two enormous territories of Kronotzki (968,000 ha) and Altaiski (914,000 ha).

In the European taiga north of latitude 55°N there are 17 protected territories (Naturschutzgebiete). Particularly interesting is the reserve of Ilmenskji, situated 60 km west of Tacheljabinsk, with an area of 32000 ha. It is, as Borissoff says (1960, p. 68), a true mineralogical museum in the open, a 'true paradise for mineralogists' since some rare minerals or rocks can be found there, such as miaskite, graphite, syenite, granite and gneiss, nephelene, feldspar, 'Marienglas', ilmenite etc. The forests are composed mainly of pines and birches. The Asiatic ungulate *Cervus nippon* has become acclimatized here. The reserve was created on May 14, 1920, with a decree signed personally by Lenin. In 1960 there were more than 100 scientific or popular articles dealing with this reserve.

Finland has also seen the creation of many nature reserves (National Parks, etc.). Indeed, places completely unaltered by man can be found today exclusively in some limited districts of the far Lapland. In all parts of Finland the composition of forests, whose trees of forests interest are spruce, pine and birch, is altered as human disturbance, and the modern mechanical utilization of forests – with clearings and artificial replanting – is rapidly changing the preceding monotonous landscape into a planned semi-artificial or para-natural forest.

The largest reserve is the Pallas-Ounas-Fjeld territory in Western Lapland, with an area of 500 km^2. Besides coniferous and birch forests, there are large areas of heath (Zwergstrauchheide) with *Lycopodium alpinum, Betula nana, Arctostaphylos alpina, Empetrum hermaphroditum, Loiseleuria procumbens, Phyllodoce coerulea, Diapensia lapponica, Vaccinium myrtillus, V. uliginosum, Juncus trifidus* and *Luzula spicata*, most of which are also present in the Alps on acid soils.

Fig. 170 shows all the 23 State nature reserves, which cover an area of 1900 km^2, which corresponds roughly to 0.8% of the territory. Some of them are National Parks (white circles) other Natural Parks (black circles). Furthermore, there are some 75 private reserves, in general much smaller than the State ones.

As far as animals are concerned, the bear is present today in East Lapland and near the Soviet border; it is fairly common. Each year some 40 individuals are killed, most of them from the USSR. The wolf is also frequent in North Lapland and on the Russian border; the lynx on the contrary is extremely rare, and does not breed in Finland any more. All these species are not protected, even in the nature reserves. The golden eagle, *Aquila chrysaetus*, is protected, and is present in Southern Finland and in Lapland but is very rare. The sea eagle (Seeadler), protected, lives near the coast and its numbers are at present increasing. Unfortunately, the *Bubo bubo* or eagleowl is not protected and is becoming rarer. Only the Passerine birds are both loved by people and protected by law, so that they run no risk.

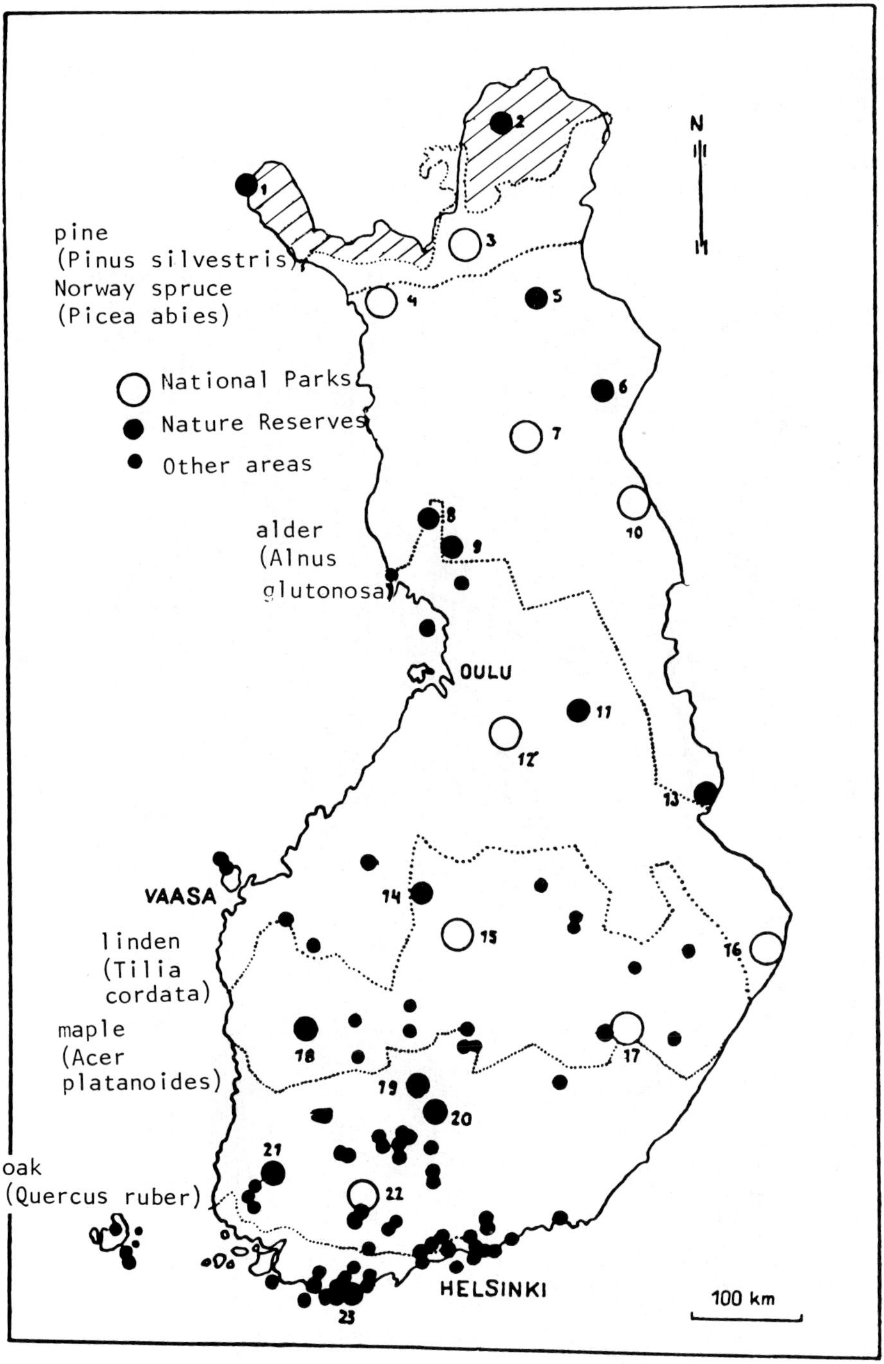

Fig. 170. The 'Naturschutzgebiete' in Finland, together with the northern limit of distribution of oak, maple, linden, alder, Norway spruce and pine (from Söyrinki).

Without any doubt the most interesting nature reserve in Finland is the National Park of Oulanka, in the territory of Kuusamo. This National Park is practically a true refuge for Arctic plants within the European taiga. The presence of these plants is due mainly to the nature of the substrate (dolomite, in general very rare in Fennoscandia). They have reached the territory of Kuusamo from the north or northeast, where they have survived the Pleistocene on small ice-free territories near the coast of the Arctic Sea and on the mountains of the Peninsula

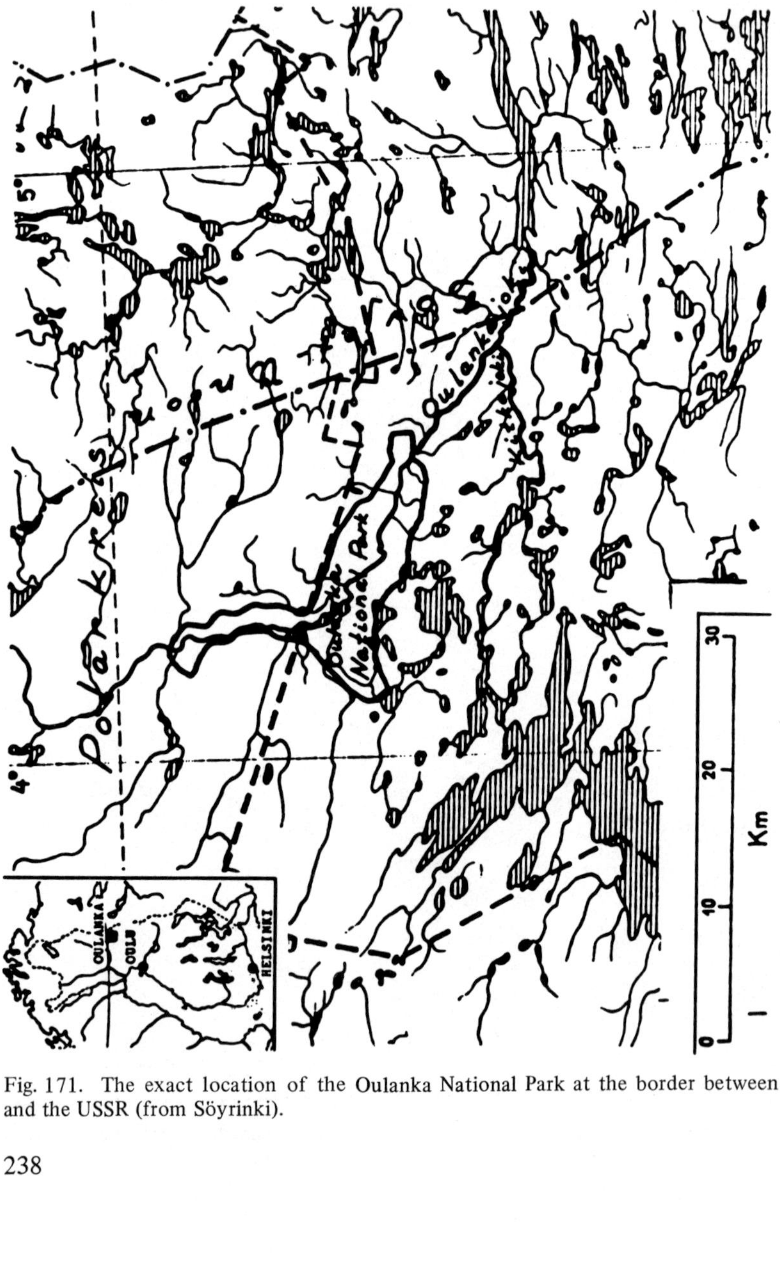

Fig. 171. The exact location of the Oulanka National Park at the border between Finland and the USSR (from Söyrinki).

of Kola. They came southwards, very probably immediately after the end of the last glaciation (Würm), and more precisely in the old Dryas period (10,700-10,000 years B.C.) (Fig. 171).

These Arctic relicts, which are entirely absent from the mountains of Central Europe, are as follows: *Roegneria borealis, Carex glacialis, Stellaria calycantha, Arenaria ciliata pseudofrigida, Silene furcata angustiflora, Draba hirta, Draba cinerea, Draba norvegica, Potentilla chamissonis, Astragalus alpinus arcticus, Oxy-*

Fig. 172. a: *Arnica alpina*; b. *Gypsophila fastigiata*; c. *Kobresia simpliciuscula*.

tropis campestris sordida, Epilobium davuricum, Epilobium nornemannii and *Arnica alpina.*

Arnica alpina is endemic to the Scandinavian mountains, akin to a species living in Novaja Zemlja and in the Northern Urals (Fig. 172a), *Arenaria ciliata pseudofrigida, Draba norvegica* and *Potentilla chamissonis* are present on both Western and eastern North Atlantic coasts; the other species partly show a north European-asiatic distribution, partly are circumpolar elements.

Other interesting species of the National Park of Kuusoma show an arctic-alpine discontinuous distribution (several plants and animals of the European taiga). They are: *Asplenium viride, Cystopteris montana, Woodsia glabella, W. alpina, Poa alpina, P. glauca, Eriophorum scheuchzeri, Kobresia simpliciuscula, Carex norvegica, Juncus triglumis, Salix reticulata, Cerastium alpinum, Arabis alpina, Thalictrum alpinum, Saxifraga aizoides, S. groenlandica, S. nivalis, Potentilla nivea, Dryas octopetala, Astragalus frigidus, Epilobium alsinifolium, Bartsia alpina, Pinguicula alpina, Gnaphalium norvegicum, Saussurea alpina* and *Mulgedium alpinum.*

The most interesting species from the phytogeographical standpoint is without any doubt *Kobresia simpliciuscula*, a small *Cyperacea*, a boreo-alpine element with an extremely fragmented area of distribution in the north (maybe of relict character), whereas in the south it is present on practically all the most important mountain systems (Fig. 172c).

As Söyrinki says (p. 224) many of the Arctic or Arctic-alpine species of Kuusamo have been represented for many thousands of years by small isolated populations, which possibly show some incipient signs of subspeciation or even speciation. Their study from a morphological, physiological and still more caryological standpoint would perhaps bring some very interesting results.

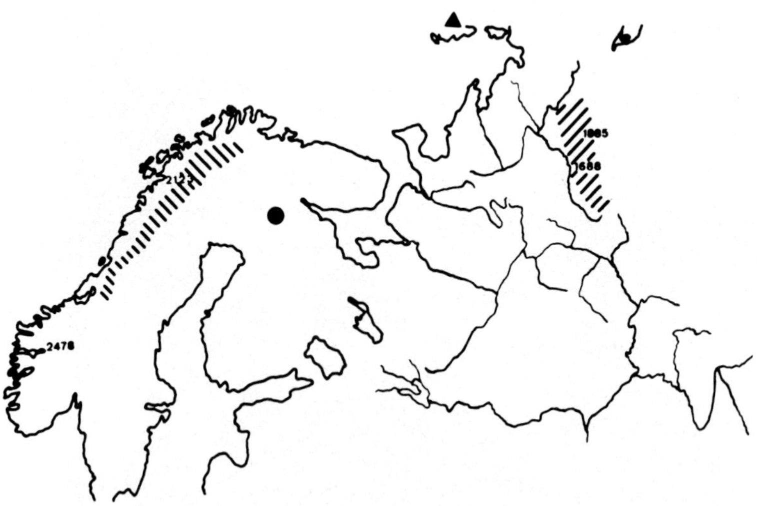

Fig. 173. Geographical distribution of *Arnica alpina* (West) and another species of *Arnica*, akin to *A.alpina* (Urals and Novaja Zemlja).

Lastly, among the plants of the Kuusamo territory, we find southern species, very rare in Finland, which have here their northern limit of distribution (*Asplenium ruta muraria, Schoenus ferrugineus, Gypsophila fastigiata, Cypripedium calceolus* (in German 'Frauenschuh'), one circumpolar species, *Calypso borealis,* a semiparasitic Orchid completely absent from Central Europe, and finally some Euro-Siberian species, characteristic of the taiga, which reached Kuusamo from the east, such as *Athyrium crenatum, Actaea erythrocarpa, Dianthus superbus* and *Lonicera coerulea*, besides the typically eastern *Salix pyrolifolia* (Fig. 173-174).

Another region where nature conservation is highly developed is Sweden, where, however, only some parts of the territory are covered by taiga, since the mountains belong mainly to the alpine zone (called locally 'regio alpina') and some warmer, southern parts are covered with broad-leaved forests.

We may notice in this connection that the concept of conservation or protection of nature had its origin in Sweden, at least in its normal, European, philosophy. Indeed, in 1877 the archaeologist and naturalist P.M.A. Säve proposed the creation of some 'freinseln' for birds (marine birds) and some 'Friedorte' for the conservation of terrestrial animals. He claimed that 'ein all- und zwischenvölkisches Gesetz zum Schutz des Tierlebens auszuarbeiten'. Three years afterwards, A.E. Nordenskjöld proposed the creation of Reichsparke, i.e. Royal parks 'wo Wald und Boden und See ganz unberührt bleiben sollten, wo Bäume nicht gefällt, Gebüsche nicht gerodet, Gras nicht gemäht werden dürfte, und wo alle Tiere, die nicht wirkliche Schädlinge waren, das ganze Jahr hindurch vor der Büchse des Jägers sicher sein könnten.' A park had already been created at this time in the USA, the famous Yellowstone Park, but based probably on a different philosophy of man and nature, so that very likely its creation cannot be compared with that of European parks.

Fig. 174. Geographical distribution of *Kobresia simpliciuscula*, a boreo-alpine species.

Today in Sweden there are 15 National Parks. The laws by which they are ruled and their control are not very different from those we have already seen for the USSR, with the difference that in Sweden, beside public land, there is also private land. Namely, we have:

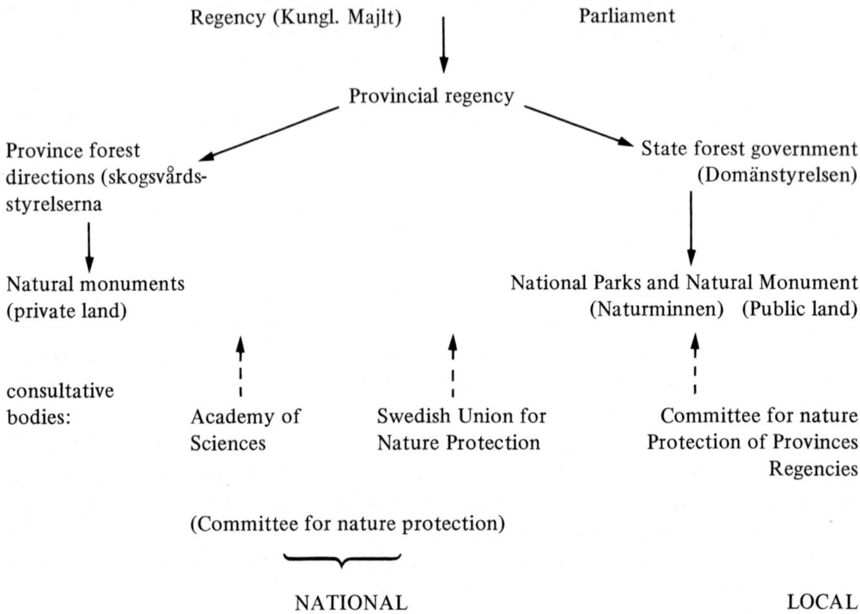

The largest Swedish parks are Sareks National Park and Stora Sjöfallets National Park, both in northern Lapland, whose area was originally (1909) 3400 km^2 but gradually has been reduced. These parks include all the biotic belts, i.e. pre-alpine coniferous belt, with a lower very dense wood with *Betula tortuosa* and an upper coniferous zone with *Picea* and *Pinus silvestris.*

Above that there are the subalpine belt with *Betula tortuosa* and then the alpine belt (subdivided into three zones).

Other national parks are: the Muddus territory, a coniferous forest in Northern Lapland, with an area of 492 km^2 – *Picea* and *Pinus* prevail and the bear lives here; National Park of Vadvetjåkko (24.5 km^2) and Abisko (50 km^2); National Park of Peljekaise in Middle Lapland (146 km^2) and the National Park of Sonfjällets (27 km^2) in Härjedalen; the bear lives here also.

In the southernmost part of the Swedish taiga there are two National Parks which should be two true reserves of the virgin forest ('Urwald'), the National Park of Töfsingdalens and that of Hamra.

Furthermore there are in Sweden natural reserves (Naturreservate), natural territories (Naturgebiete), natural monuments, etc.

Uppsala University includes several natural monuments among which are three beautiful broad-leaved forests and a 'Hochmoor' (Ryggmossen).

Single trees, groups of trees etc. on private soil can also be protected. Several plants are protected on a national or a local scale; there are some bird reserves and several animals are protected. Finally, the Natural Parks may be mentioned, whose aim is to promote an intimacy between nature and the people (cf. Du Rietz).

BIBLIOGRAPHY

Aleinikova M.M., Počvoobitaiuščie bespozvonočnie različnich tipov lesa v srednem Povolž'e, Materiali po faune i ekologii počvoobitaiuščich bespozvonočnich, Izdat. Kazansk. Univers., 1968, 3.

Bachmaier F., Die Zwergbirke (*Betula nana* L.) ein Glazialrelikt unserer Flora und Untersuchungen über ihre Insektenfauna, Jhrb. Ver. Schutze Alpenpfl. & Tiere, München, 51, 1966, 138.

Borissoff V.A. (Borissow W.A.), Vom Naturschutz und von einigen Berg-naturschutzgebieten in der UdSSR, Jhrb. Ver. Schutze Alpenpfl. & Tiere, München, 25, 1960, 64.

—, Naturschutz in der UdSSR. Probleme und Verwirklichungen, Jhrb. Ver. Schutze Alpenpfl. & Tiere, 35, 1970, 111.

Brink (van der) F.H. & P. Barruel, Guide des Mammiferes sauvages de l'Europe occidentale, Delachaux Niestlé, Paris & Bruxelles, 1971.

Cajander A.K., The theory of forest types, *Acta Forest. Fenn.* 31, 1926, 1-108.

Cecconi G., Manuale di entomologia forestale, Padova, 1924.

Du Rietz G.E., Naturschutz in Schweden, Jhrb. Ver. Schutze Alpenpfl. & Tiere, München, 25, 1960, 193.

Gams H., Die Herkunft der hochalpinen Moose und Flechten, Jhrb. Ver. Schutze Alpenpfl. & Tiere, München, 25, 1960, 85.

Karppinen E., Ecological and transect survey studies on Finnish Camisiids (Acar. Oribatei), *Ann. Zool. Soc. Zool. Bot. Fenn. Vanamo*, 17, N.2, 1955.

—, Ueber die Oribatiden (Acar.) der Finnischen Waldböden, *Ibidem*, 19, N. 1, 1958.

Krivolutsky D.A., Nekotorie zakonomernosti zonal'nogo raspredeleniia panzirnich kleščei (Some regularities in the zonal distribution of oribatid mites), Oikos, 19, 1968, 339 (in Russian with English summary).

—, Some observations on the quantitative distribution of Oribatid mites in the zonal soil types of the URSS, Proc. 2nd Int. Congr. Acarol., 1967, Publ. House Hung. Acad. Sci., Budapest, 1969, 39.

—, Rasprostranenie panzirnich kleščei v počvach SSSR, Problemi počvennoi zoologii, Izdatel. Nauka, 1969, 94.

Kubiena W.L., Claves sistemáticas de suelos, Cons. Sup. Inv. Cient. Madrid, 1952.

Lack D., The natural regulation of animal numbers, Oxford Clarendon Press, 1954.

Marcuzzi G., Ecologia animale, Milano, 1968.

Mertens R. & H. Wermuth, Die Amphibien und Reptilien Europas, Kramer, Frankfurt a. M., 1960.

Murr F., Der Tannenhäher, Jhrb. Ver. Schutze Alpenpfl. & Tiere, München, 23, 1958, 133.

Peterson R., G. Mountfort & P.A.D. Hollom, Die Vogel Europas, P. Parey, Hamburg & Berlin, 1954 (transl. from English).

Roščinenko V.I., Fauna ščelkunov v poimach rek Kam'i, Čepz'i Kil'mezi, Problemi počvennoi zoologii, Izdatel. Nauka, 1969, 139.

Sadekova L.Ch., Panzirnie klešči iz gnezd melkich mlekopitaiuščich saralovskogo učastka Volžsko- Kamsogo zapovednika, Problemi počvennoi zoologii, Izdatel. Nauka, 1969, 141.

Scharff R.F., The history of the European fauna, London & New York, 1899.

—, Distribution and origin of life in America, London, 1911.

Söyrinki N., Probleme und Leistungen des finnischen Naturschutzes, Jhrb. Ver. Schutze Alpenpfl. & Tiere, München, 25, 1960, 96.

—, Das Kuusamo-Gebiet, ein Refugium für arktische Pflanzen in der Nadelwaldstufe in Finnland, *Ibidem*, 35, 1970, 221.

Terentev P.V. & S.A. Chernov, Key to Amphibians and Reptiles, Jerusalem, 1965 (transl. from Russian).
Tscherepanov B.I., Die Schnellkäfer West-Sibiriens (Coleoptera Elateridae), V. Novosibirsk, 1957 (in Russian with German summary).
—, Provoločniki zapadnoi Sibiri (opredelitel') Izdat. Nauka, Moskva, 1965.
Zangheri P., Geobotanica, Encicl. Ital. Sci. Ist. geogr. De Agostini, 1972, II, p. 374-484.

THE TUNDRA

At the northernmost limit of Eurasia, in a semi-frozen landscape, a carpet of mosses and lichens, together with scattered and extremely stunted trees or shrubs, rests upon a soil permanently frozen at depth (permafrost, or locally tjaele).[1] This landscape is given the name 'tundra', a word of Lapp origin. This biome occupies in total 8 million square kilometres, of which only a very small part belongs to Europe.

The biome is limited to the extreme North of Fennoscandia, European Russia and to northernmost European islands of the Atlantic Ocean, the largest of which are Iceland, Jan Mayen, Svalbard and Novaja Zemlja.

Despite the poverty of the vegetation, the tundra environment varies considerably due to changes in climate, relief, geology and the morphology, which includes glaciers, mountains, volcanoes, bogs, grasslands, emergent cliffs and flat barren sediments. At one time musk-oxen lived on the sea-shore, and even today a polar bear may reach the north coast of Iceland on an ice-flow.

Many rivers cross the European tundra: in Russia we find the Dvina (only near its mouth), the Mezen and the Pečora; in Iceland we find a number of rivers running from the mountainous centre of the island (Hofsjökull) towards the coast in all directions (Eyjafiardara, Skialfandafljot, Jökulsa, Lagarfliot, Tjherma, Nordra etc.). The main lakes are present in Lappland (Enare and Imandra); in Iceland there are only some small lakes such as the Thingvallavatn (literally the lake of the Thing or Althing field, i.e. the site of the oldest Parliament in the world, the Althing), with a surface area of 120 km^2, Myvatn, in the north, or Hvitarvatn (literally the lake of the White River), formed by the melt-water of the Langjökull glacier.

During the very long winter season, the ice penetrates slowly into the soil, forming a rigid, impenetrable cover which often reaches a depth of 70 cm or more. The frozen soil prevents digging animals from breaking down and burying organic debris and only during the short mild summer (when the superficial thawing of the soil begins) is the detritus attacked by micro-organisms and animals. However, the presence of a layer of ice at depth prevents any mixing of humus (or, in general, organic substance) with the underlying mineral layer. The manifestation of this transformation of organic matter is a black soil, very acid, still rich in plant debris (see later).

Above latitude 65-70°N, the soil, covered by snow for 8-9 months per year and constantly frozen at depth (as we have seen), the humid and cold climate with temperatures very often as low as −35°C, and the very short summer, prevent the growth of any woodland vegetation. During the summer season, the weak sun stays constantly above the horizon, allowing a superficial thawing of the soil. The rain water, plus that coming from melting snow and ice, has no chance to infil-

[1] The Swedish tjaele has been translated in German by Pohle (1924) as gefrornis Schicht; in Russian it is called merslota.

trate and remains on the surface, and also does not evaporate because of the low temperature. The soil, already liable to movement because of thawing, becomes still more unstable in the upper layers because of these superficial waters, so that the phenomenon of solifluction is very common on slopes. For a period of 2-3 months the snow blanket of the tundra is transformed into an immense water meadow. The water accumulates in small depressions on the surface (locally called Rimpis) giving rise to the so-called 'humid tundra', where *Sphagnum* and large mosses (*Polytrichum, Webera nutans* etc.) grow. On the contrary, the elevated parts of the tundra constitute the so-called 'dry tundra', where lichens of the genera *Cladonia* and *Cetraria* develop. Sometimes noticeable hills may be formed, as high as 7 m, called in Finland palsat (singular: palsa), or even embankments or ramparts, called pounut, may be found.

The climate of the tundra belongs to the Arctic type, and is similar to that of the mountains of Scandinavia ('regio alpina'; see the chapter on taiga). As an example we may quote here data for Archangel (USSR):

Annual mean temp.	Temperature coldest month	warmest month	Rainfall Spring	Summer	Autumn	Winter	Total
0.3°C	-13.7°C	15.8°C	63	141	118	64	386 mm

The means of extreme temperatures in the Svalbards are +12°C (warmest months) and –41°C (coldest month), in Jan Mayen +10°C and –30°C. The annual temperature range in the Arctic varies from 30-40°C while diurnal variation is small everywhere.

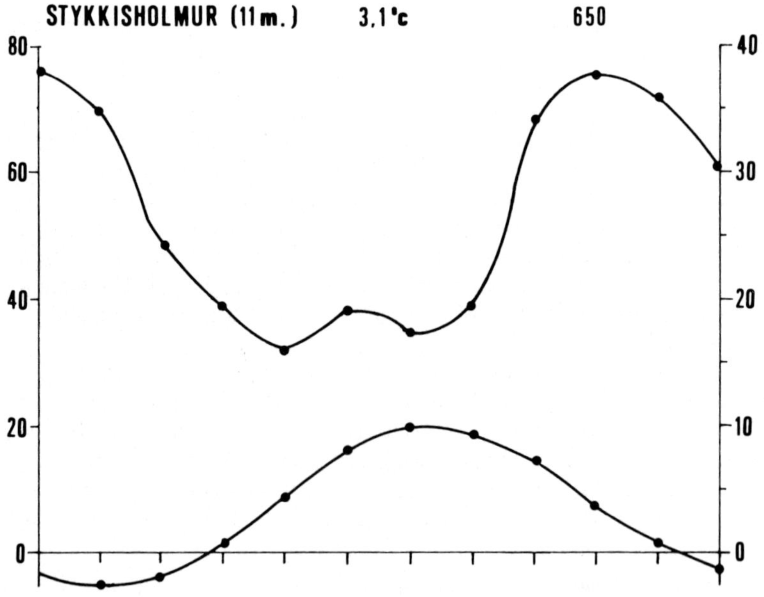

Fig. 175. Example of ombrothermic diagram for the tundra (Iceland).

On the Icelandic coast we have a winter mean temperature around −1 to −2°C, whereas in the north-east, away from the influence of the Gulf Stream, it is nearer −8°C or lower. The July mean temperature on the coast is 9°C (at Reykjavik 12°C). The annual average temperature in this locality, due to the presence of the Gulf Stream, is 6.5°C. In the southern part of the island the rainfall may reach 2000 mm per annum (Cape Portland) (Fig. 175).

As far as geology is concerned, we may recall for the tundra what we have mentioned for the taiga. Additionally, we must note the volcanic rocks of Iceland where there are many volcanic phenomena, for example active volcanoes, geysers (a kind of intermittent hot spring, which at regular intervals projects upwards columns of hot water – from the verb geysa – burst forth). Also present are warm or cold springs called reykjrs or hvers, plus thermal, sulphurous and radioactive waters. Other islands covered by a tundra-like vegetation are also volcanic (Jan Mayen and Faeroe – although the latter may only doubtfully be attributed to the tundra biome).

The pedology of European tundra is very complex and we will limit ourselves to describing only the major soil types, trying to conserve Kubiena's classification and terminology, although today this is not followed by all pedologists.

In this chapter we limit ourselves exclusively to the semi-terrestrial and terrestrial soils, leaving the saturated ones (bogs, swamps, ponds etc.) for another chapter.

The main semi-terrestrial types include the so-called Arctic rutmark, the tundra-anmoor and the tundra moss (or bog). The true terrestrial soils should be types of Arctic råmark (or Arctic raw soil), tundra-ranker or Nordic nanopodzol.

The Arctic rutmark (a Swedish word first used by Kjellman in 1879) is known also as Rautenboden, cellular soil (Högbom, 1914), polygonal soil (Nordenskjöld) or texture soil (H. Kaufmann, 1929). It is a raw Arctic fluvial soil, with a very characteristic surface aspect, not dissimilar from a kind of soil common to our highest mountains, where it is considered one of the many cryo-nival phenomena. The fissures extend only to the frozen layer or tjaele, and the plants grow generally only in these fissures, so that the centre of the polygons is bare.

The tundra-anmoor is a humid soil, with the presence of a muddy dystrophic anmoor making it a kind of alluvial gley. Generally it lies on the tjaele and the vegetation is represented by *Carex, Comarum palustre, Menyanthes trifoliata* and later on mosses. *Sphagnum* may be present or not. When the soil is not very humid there is growth of lichens, mosses, *Betula nana* and *Salix arctica.*

The tundra moss is very widely represented in the European tundra, composed of mosses which have recently invaded or colonized the layer of raw humus of the tundra ranker (see later). Generally the icelayer is very near the surface (30-40 cm). This soil type corresponds biologically to the most inactive soil in the world. To the normally unfavourable conditions of *Sphagnum* bog growth (poor in bases, nutrients and air-spaces, plus acidification and bad drainage) some other negative factors are added, for example an extremely cold and long winter, presence of a tjaele, lack of light during the long polar night, as well as the oblique summer solar radiation and high storm activity. Only in the older bogs do we find the hummocks already described (Palsmoor, Palsat). The vegetation is mainly composed of *Sphagnum* (more particularly *S. fuscum*), plus *Dicranum* and *Poly-*

trichum. On small hummocks flowering plants are present, such as *Empetrum nigrum, Rubus chamaemorus, Vaccinium uliginosum, Betula nana, B. alba*, etc. Lichens are very common.

The Arctic råmark (a Swedish word) is a raw soil of cold deserts, also a frozen Arctic soil, present as a climax in the region of thawing between the Arctic Circle and the tundra although it may be also found in the latter. It shows minimal chemical decomposition, soil organisms and weathering; on the contrary there is a remarkable mechanical desintegration due to repeated freezing and thawing. The vegetation is represented by mosses, lichens and a few flowering plants. The soil is present mainly on the islands of the Arctic Ocean (Novaja Zemlja and Svalbard). In some instances, when a locality is constantly exposed to winds, instead of an Arctic råmark a so-called Arctic hamada råmark is formed due to deflation of soil by wind and concomitant accumulation of a stone veneer (hamada is a name employed to indicate stone-deserts).

In the northernmost part of the tundra, and more especially on the islands of the Arctic Ocean a stoney polygonal soil may be present (Huxley & Oden, 1924) classified by Kubiena as 'structural råmark'. Also in this case the plants grow at the polygon edges. The presence of flowering plants has resulted in the German name 'Steingärtchen' for this soil type.

The so-called tundra-ranker is a soil typical of the dry tundra, with the formation of a terrestrial humus, but very inactive, and comprising scarcely decomposed plant remains. It is never soaked, nor transformed into turf, but generally is present on rocky hills (Fjälls). The humus is a raw or a coarse, dystrophic, moder. Only in some instances a mulliform moder is formed. The plants are represented by *Empetrum nigrum, Loiseleuria procumbens*, the graminoid plants *Stereocaulis paschale* and *Festuca ovina*, besides lichens such as *Cladonia rangiferina* and *Cetraria nivalis.* Sometimes this soil is completely covered with the faeces of lemmings (see later). As a consequence of the scarcity of micro-organisms, meso- and microfauna, the soil is very poor (Fig. 176a).

Also we must mention a peculiar form of podzol (word used by the old Russian peasants and applied to pedology for the first time by Dokutschajeff in 1879), the Nordic nanopodsol formed as an iron forest podzol with attenuated chemical decomposition and reduction of the profile. It is present at the limit between the taiga (see relative chapter) and the tundra and is known by German authors as 'Nordischer Zwergpodsol'. Different from most podzols, this type shows a profile of only 10 cm, so that each horizon is greatly reduced. It is formed generally above acid eruptive rocks and crystalline schists so that it is mostly sandy while the humus is raw. Only if there is a rich mesofauna present is there an acid moder (coprogenous moder). The vegetation is relatively luxuriant, compared with that on other tundra soil types (Fig. 176b).

Lastly we should mention two other soil types present only in Iceland, probably due to the climatic factors and still more to the geological conditions of the island: the iron humic podzol and the eupodzolic brown earth (Kubiena).

A quite peculiar feature of the tundra is the polar 'aurora' (better known as the 'Aurora Borealis') named after the night-time appearance of coloured lights in the sky. Sometimes the lower border of this light is situated 100 km from earth, and has a vertical extension of 20-100 km, but occasionally it occurs up to a height of 1000 km. Probably the Aurora Borealis is due to fast protons and electrons emitted by the sun during solar activity. Near Earth these electric particules are influenced by the geomagnetic field, and they bombard high atmosphere nitrogen and oxygen causing them to emit their characteristic spectra (green-yellowish by atomic oxygen (5577 Å), red by oxygen (6300 Å), violet or red by N_2^+ and O_2^+

Fig. 176. Typical soils of the tundra. Left: ranker under *Empetrum nigrum* on aplitic gneis (Lünahamari, on the Finnish coast of the Arctic Ocean). Right: Nordic nanopodzol (dwarf podzol) under a tundra wood with dwarf birch and *Cladonia rangiferina*, on sienite (Petsamo, Finnish Lapland) (from Kubiena).

The tundra in the past

The environment and the organisms (or biota) of the Arctic tundra of today assume a quite different significance if we imagine Europe 100-200,000 years ago, with two enormous ice centres, one an enlarged Polar ice cap, or Inlandsis, reaching at least as far south as Denmark, and the other covering the Alps like a blanket. The zone sandwiched between these two immense ice caps is exposed to icy winds blowing both from south and north, and the vegetation which can grow here is the same we find today in the tundra or – somewhat eastward – in the cold steppes of Central Asia.

The animals living in this special tundra of which we find today many remains in the aeolian deposits (loess), in the alluvial soils present along rivers, or in the peat which accumulated during thousands of years in bogs or on lake bottoms, were in part the same animals living today in the northernmost part of Europe,

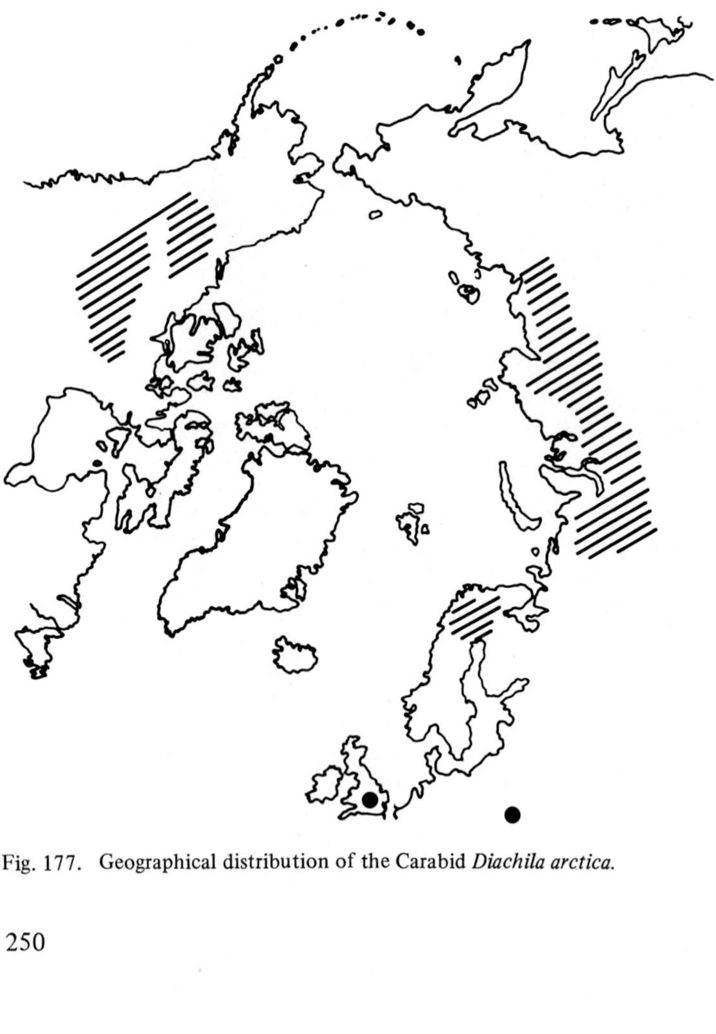

Fig. 177. Geographical distribution of the Carabid *Diachila arctica.*

although some are extinct not only in Europe but all over the world. Together they include the reindeer, the musk ox (today present as a wild animal only in North America), the polar fox, the variable hare, the lemming, plus two large species, the mammoth *Elephas primigenius*, whose remains have been found with those of the lemming, and the glacial rhinoceros, *Rhinoceros tichorhinus.* The musk ox was found as far as the Dordogne and Hermannstadt in Transsilvania; while the reindeer lived in the Alps and the Pyrenees. A very frequent bird of the tundra or the cold steppes of Central-Eastern Europe was then the rock ptarmigan *Lagopus mutus*, reaching as far south as the Apuane Alps in Italy and Santander province in Spain.

Very abundant were the frigidicolous invertebrates. Together with a number of zooplankters which have been preserved up to today, there were two freshwater snails, *Planorbis arcticus* and *P. strömi.* The beetles *Diachila arctica* (Fig. 177) *Hydroporus lapponum, Colymbetes dolobratus* and an extinct species of *Carabus*, *C. maeandroides*, have been found in the earth waks of Boryslaw, in Galicia[1].

The influence of human agency upon the tundra

Although very thinly inhabited, this biome has been influenced since early times by man. The density of population is less than 2 inhabitants/km^2 in most of the territory, and reaches values between 2 and 20 only in some narrow belts of Lappland and Northern Russia. In Iceland in 1956 there was a density of 1.5 people/km^2; more recently (1967) it rose to 1.9.

The indigenous population of the European tundra is made up mainly of Lapps, Samoieds and some other nomad tribes living in the northern parts of Russia (see the chapter on taiga). The Samoieds live also on Novaja Zemlja. Iceland was completely uninhabited until the arrival of Europeans (795 A.D.) and after the arrival of the Vikings (in the following century) the island has been colonized[2].

[1] At Boryslaw there was a swamp in which hydrocarbons (naphtha) had been formed, in which insects drowned in such a way that in some places the deposits are true insect conglomerates. Since the southern limit of the Inlandsis was not far away, the fauna had quite an Arctic character.

[2] We find today in Iceland interesting toponyms or hydronyms which have partly taken their origins from natural features of the island: Grimsey, a northern islet crossed by the Arctic Circle, is literally Grimr Isle, in remembrance of one of Nyal Thorgeirsson's sons, a hero of one of the most beautiful Icelandic sagas; Uxahver, a hot spring south of Husavik, means the spring of the ox, since once an ox fell in and was immediately cooked by the boiling water; Raudilaekur (red river) a small water course in the south of the island, named after the reddish colour of its waters, which transport ferrugineous particles, laxa, which means salmon (see German word Lachs) is the name of several rivers; hvità (meaning white) is given to several rivers because of the colour during the winter, when frozen over; nordrà, or river of the north (probably from an old Indo-european word for 'river' (ῥέο, rio, rjeka, etc.). Often the source of hydronyms has been independent of natural features: an example being the waterfalls (singular 'foss'), such as gulfoss (literally golden waterfall), godafoss (god-waterfall), geita foss (goat waterfall), ullarfoss (the waterfall of the wool), etc.

Fig. 178. Example of runic inscription (Museum of Reykjavik): it is written: Sator arepo tenet opera rotas, which can also be written as a cross in this way:

S	A	T	O	R
A	R	E	P	O
T	E	N	E	T
O	P	E	R	A
R	O	T	A	S

in which the sentence can be read four times (horizontally, vertically from the top left, and the right bottom, and horizontally from bottom right to left).

Nobody knows the meaning of this writing, which according to some should be a magic formula, according to others it would only be a means to draw a cross which, in some eras and in some places, was not allowed for religious reasons.

Once Iceland was considered the Thule (from the Celtic word thual, meaning 'septentrion') of which Pytheas (πύϑεας) has given us a description. Today Thule (Θουλη) – which according to Pytheas was to be found at 6 days of navigation from 'Britannia' (England?) – is interpreted as being Norway, or, according to other authorities, the Shetland Isles (maybe Mainland). Anyway, up to Roman times Thule was considered as the extreme northern limit of the Ancient World, whence the 'ultima Thule' (Virgilius, Georgicae, I, 30); from Virgilius Seneca transports the Thule to his Medea (II, 384). Pytheas discovered this far-northern land possibly between 310 and 306 B.C. (Marini, 1956). The indigenous populations (who possibly spoke Celtic) were according to Pytheas' description feeding upon a 'grain' (barley or oats?) and a root called 'kenkros' (maybe *Angelica archangelica*). They drank the so-called hydromel (from ὕδορ and μέλι) a kind of alcoholic drink obtained from the fermentation of water and honey.

The inhabitants of the tundra, of course, shared habits with those living on the taiga (see chapter Taiga) where we also find the runic alphabet and the scalds (Fig. 178).

If the influence of man on landscape has been rather limited in Northern Europe (except, of course, in the case of a big town e.g. Archangel, which today has some 1,400,000 inhabitants), it has probably been stronger in Iceland because of prevailing socio-economic conditions, at least in some centuries. In the XVIth century, for instance, Denmark permitted the formation of companies for the exploitation of the island and enormous quantities of dry fish, wool, skins and feathers of the duck *Somateria mollissima* (locally known as 'aethr') where gathered and sold. We must realize – as far as nature conservation on Iceland is concerned – that the presence of an enormous number of geysers allows an agriculture, horticulture and perhaps also cattle raising which otherwise would not be possible. This had an obvious effect on the original landscape of the island. To finish, let us mention that, according to Marini, the small groups of trees (*Salix* and *Betula*) today existing in the territory of Thorsmörk (literally Thor's forest) are the remains of more extensive forests or woodlands exploited in the first colonization period of Iceland.

The vegetation

Due to excess of humid acid in the soil, which is toxic to higher plants, only development of small birches (*Betula nana* and *B. glandulosa*), shrubs of *Pinus mugho* and dwarf willows (*Salix arctica* and *S. polaris*) is possible. Only in sheltered places where the extreme wind cannot reach, the silverish aerial inflorescence of *Eriophorum* together with the pale bluish *Myosotis*, sedges (*Carex*) and heaths (*Ericaceae*) bring a vivid note to the landscape. These perennials, however, have a very short growing season, limited to the extremely long days of the local summer.

In depressions in the soil, where water may stagnate *Sphagnum* species dominate; these are special mosses with a delicate, spongy body, forming big cushions soaked with water. The accumulation of the dead remains of *Sphagnum* gives rise to minuscule hills (the paltas we have already dealt with). If the extension of the *Sphagnum* vegetation is very large, a true bog is formed (see the chapter devoted to bogs).

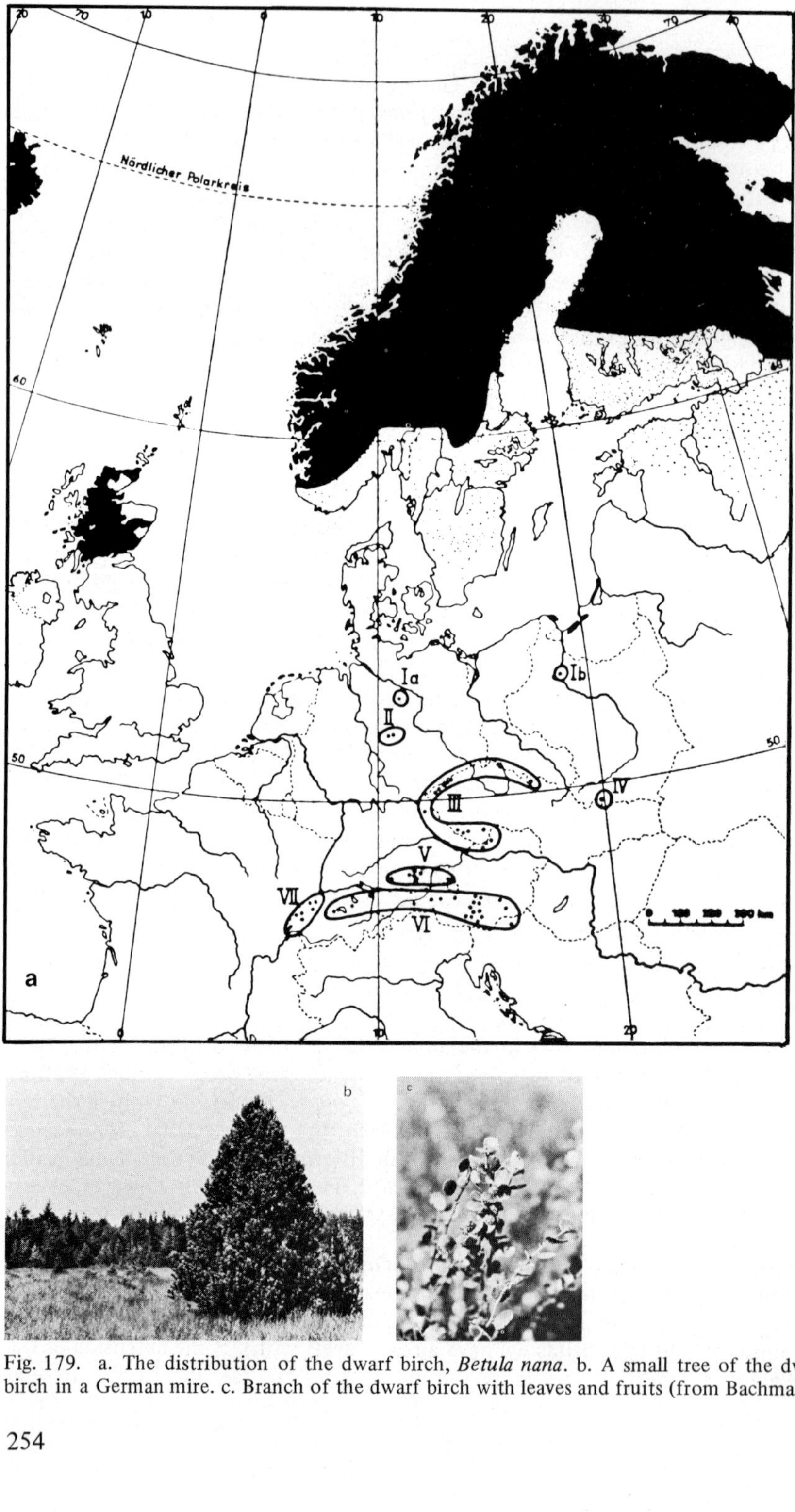

Fig. 179. a. The distribution of the dwarf birch, *Betula nana*. b. A small tree of the dwarf birch in a German mire. c. Branch of the dwarf birch with leaves and fruits (from Bachmaier).

On the sunny sheltered rocky slopes, true 'prairies' of lichens extend. These are among the most typical landscapes of the tundra, present along the coast of the Arctic Ocean, or in the 'fjelds' of Scandinavia. The same landscape, of course, is present in many non-European countries, for example Greenland.

All the tundra vegetation belongs to the Arctic floristic region (Kleopow) or, if we prefer, to the circumarctic region (Mensel, Jäger & Weinert). According to the latter, Iceland which we have described along with the northern European tundra (and that according to Polunin belongs with no restriction to this biome), must be placed in the boreo-European region. Mensel, Jäger & Weinert recognize another floristical limit, the circum-boreal one, south and parallel to the limit of the circumarctic zone. According to the present writer, this southern limit should correspond more closely to the true limit of the tundra. As a matter of fact, at the border between tundra and taiga there is a true 'echelon' of impoverishment of flora (in Gaussen's sense), since the border is not at all clear.

We have already dealt with *Betula nana* in the chapter on taiga. This species has a circumpolar distribution, since it grows both in Northern and Arctic Europe, in subarctic and Arctic territories of Asia and North America and on the coasts of Greenland, giving the tundra a character of 'Zwergstrauchheide' (literally 'heath with dwarf shrubs'). It is present also in Iceland. One of its enemies, the Aphid or blatt louse *Calaphis arctica*, is an Arctic-alpine species associated with the dwarf birch, so far recorded from south west Greenland (Grönnedal) and the Bavarian prealps (Bernrieder Filz near Starnberger See (cf. Bachmaier). Another shrub of the tundra is *Betula tortuosa* (Fig. 179).

In the tundra of Finland the following species are present, sometimes near the coast: *Woodsia glabella, Kobresia simpliciuscula, Carex glacialis, Carex halleri, Salix reticulata, Arenaria ciliata, Melandrium affine, Saxifraga aizoides, Dryas octopetala, Potentilla nivea, Lappula deflexa, Veronica fruticans, Pinguicula alpina* and *Arnica alpina.* According to Söyrinki (p. 99) these species very probably have survived either in the refuges of the Kola Peninsula or in those on the Norwegian coasts, at least from the last glacial period (Würm) (see also what we have said about conservation of the taiga, in connection with the flora of Kuusamo). This idea is quite in harmony with the hypothesis of the existence of pre-glacial Arctic endemics among insects (see later).

Let us remember here also the circumpolar species *Diapensia lapponica*, limited to ice-free rocky hills of the Arctic regions. Other interesting species are *Rubus chamaemorus, Cassiope tetragona* and *Phyllodoce coerulea* (Fig. 182a, b).

Arctic-alpine species that may be mentioned are *Dryas octopetala, Saxifraga oppositifolia, Silene acaulis, Linnea borealis, Arctostaphylos uva-ursi, Empetrum nigrum.* (Other Arctic-alpine species, however, are present – besides in the Alps – only in the North American tundra, as for instance *Potentilla nivea* and *Saxifraga hirculus.*) (Fig. 180-1).

An endemic species is perhaps *Cornus suecica*, closely related to the North American *Cornus canadensis* (Fig. 182c, d).

The presence of several circumpolar species (a category we find also among the animals) or of a pair of species of which one lives in the Old World, the other in the New, is interpreted as a remnant of the so-called Arcto-Tertiary flora, which, before the Pleistocene, covered part of the Northern Hemisphere, so that these

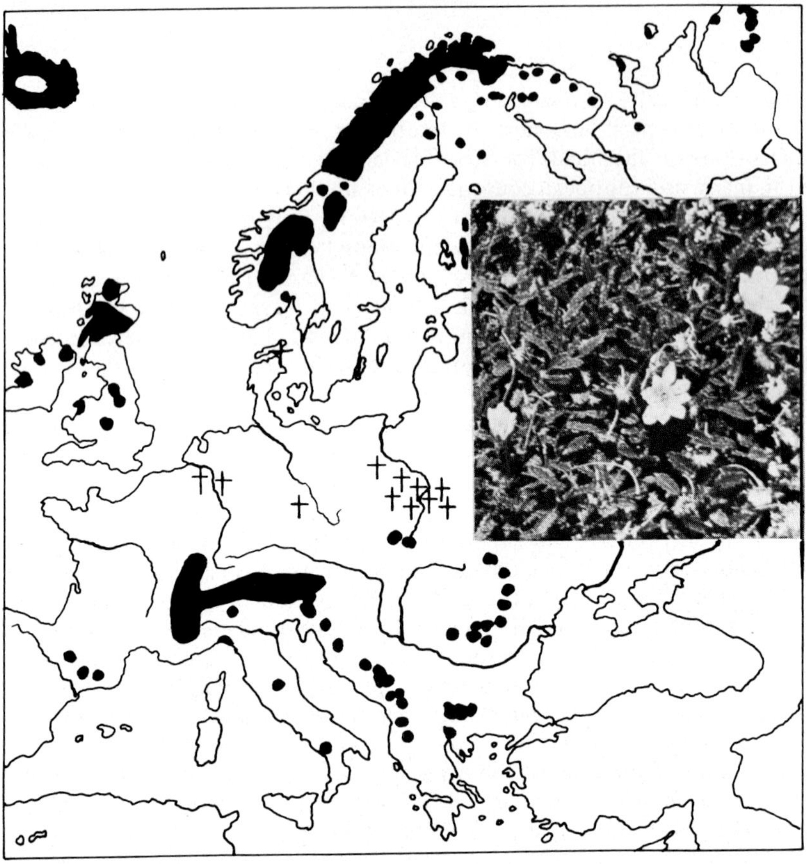

Fig. 180. Example of a plant with a distribution of the boreo-alpine type, *Dryas octopetala*. The crosses indicate fossil (Pleistocene) findings, which point to the presence of this species during the various glacial crises, in the lowlands of Central Europe (from Zangheri, greatly modified). Inset: *Dryas octopetala.*

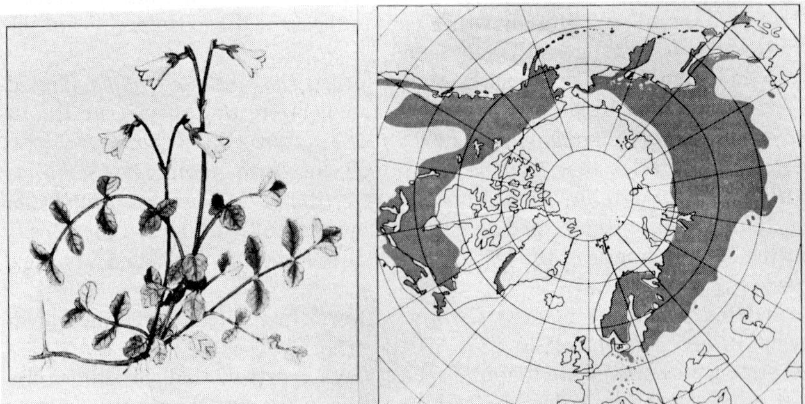

Fig. 181. Distribution and drawing of *Linnaea borealis* (drawing from Zangheri) (with the permission of Ist. Geografico De Agostini).

species may be considered a type of relict element. The problem, however, is still open to speculation.

The vegetation of Spitzbergen is particularly rich when compared with the high latitude. Besides some mosses (*Calliergon* and *Drepanocladus*), the following flowering plants have been found: *Festuca brachyphylla* and *Poa* sp. (on sand); *Dryas octopetala, Cassiope tetragona, Salix polaris* (which form a dense mat) and, finally, *Saxifraga oppositifolia, Phippsia* sp., *Salix reticulata, Papaver dahlianum,*

Fig. 182. a. *Cassiope tetragona*; b. *Phyllodoce coerulea*; c. *Cornus suecica*; d. *C. canadensis* (from Söyrinki, 1970).

Silene, Draba alpina, Oxyria, Polygonum, Cerastium, Saxifraga groenlandica, Carex saxatilis and *Poa alpina vivipara*, which practically correspond very closely to the vegetation of the morainic regions of alpine glaciers. In water *Eriophyllum* sp. have been observed.

Very many lower taxa exhibit an amphiarctic (or better amphipolar) – alpine distribution, as Gams demonstrated several years ago. This can be shown by the following groups (or species): the alga of the 'red snow' *Clamydomonas nivalis*, the green alga *Prasiola*, many families of lichens, such as *Lecideaceae, Lecanoraceae, Buelliaceae, Peltigeraceae, Umbilicariaceae, Parmeliaceae, Caloplacaceae, Physciaceae*, the so-called worm- or coral-lichens, *Thamnolia, Stereocaulon* and many beard-lichens, such as *Alectoria, Usnea* etc., the lichen species *Nephroma arcticum, Placopsis gelida* (Lecanoraceae), *Umbilicaria cylindrica* etc. *Usnea cylindrica* is present from Patagonia to Arctic regions, limited to the islands without reaching the mainland (Fig. 183). Other amphipolar-alpine taxa are the liverworts genus *Gymnomitrium*, the mosses *Andreaeales* (Klaffmoose) with the genus *An-*

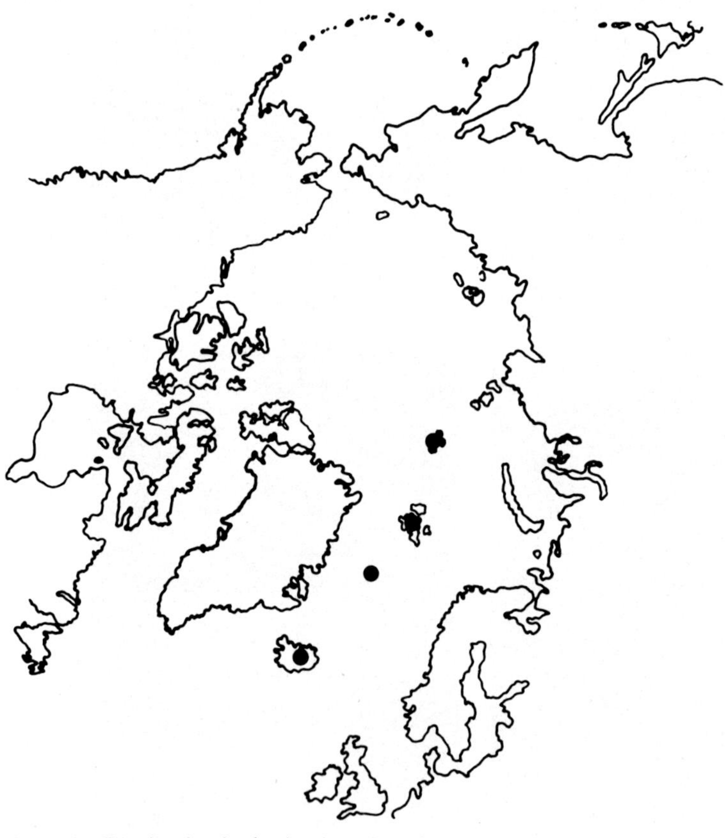

Fig. 183. Distribution in the Arctic region of the lichen *Usnea cylindrica.*

dreaea (one species, *A. nivalis*, is limited to the Arctic region and the Alps). Among the *Polytrichales* is the species *Oligotrichum hercynicum*; among the Musci *Arthrodontei* there is only one circumboreal-alpine species, *Ångstroemia longipes*, restricted to glaciers; among *Grimmiales* two genera are limited to the Arctic regions and the Alps: *Grimmia* and *Rhacomitrium*, both of them forming small cushion hummocks (Polstermoose).

Finally, let us mention here some Arctic-alpine species of mosses studied by Martinčič, more precisely: *Barbula icmadophila, Brachythecium collinum, Desmatodon laureri, Mniobrium sudeticum, Syntrichia mucronifolia, Tayloria acuminata, T. splachnoides*, etc. Many of them reach mountains relatively far south, such as the Karawanken, the Pohorje, the Velebit chain, Črna Gora, Macedonia (Šar Planina and Korab). Generally the more southern a locality, the higher the biotope at which the Arctic-alpine moss is present, as we can see from the following scheme:

	Slovenija	Dinaric chain	Macedonia
Amblyodon dealbatus	1000 m	–	1800 m
Catoscopium nigritum	1400 m	–	1900 m
Distichium inclinatum	500 m	700 m	2000 m
Hypnum fastigiatum	1000 m	1500 m	2000 m
Myurella julacea	800 m	1200 m	1800 m
Oligotrichum hercynicum	1200 m	1700 m	2200 m
Philonotis seriata	1200 m	1700 m	1900 m
Ptychodium plicatum	1000 m	1700 m	2300 m

It is very remarkable that, in the south, these species are bound to siliceous rocks (instead of limestone) which would point to the need for a cooler microhabitat. The same appears true for the lower pH these plants seem to prefer in the southernmost biotopes:

	Switzerland	Slovenija	Šar Planina
Desmatodon latifolius	7.0	7.0	5.0-5.9 pH
Timmia bavarica	7.0	–	5.0-6.0 pH
Polytrichum norvegicum	6.7	6.4	5.0-5.2 pH

A useful tool for describing any vegetational unit of the world is represented by Raunkjaer's spectrum. Here we can see the biological spectra of different parts of the European tundra:

Locality	Latitude °N	No. of species (flowering plants)	Ph	Ch	H	G	Th (%)
Franz Joseph Land	82°	25	–	32	60	8	–
Spitzbergen	79°	110	1	22	60	15	2
Vardö (Norway)	71°	134	2	15	61	13	9
Iceland	64-67°	329	2	13	54	20	11

It is evident that the tundra, as a whole, belongs to the climate (or the bioclimate) of Hemicryptophytes and Chamaephytes. The regular increase in Geophytes and Therophytes with the decrease in latitude is very interesting.

The ratio of flowering plants to lower Cryptogams is particularly low in the extreme North, as we can see from the following table:

	Flowering plants	Lichens	Mosses
Franz Joseph Land	36	94	40
Spitzbergen	128	450	–
Novaja Zemlja	182	450	–
Peary Island (Greenland)	94	–	150

To conclude, according to Zangheri the main physionomic divisions of the vegetation are: a) grasslands with *Cyperaceae* and *Juncaceae*; b) groups of dwarf willows (e.g. *Salix arctica*), *Empetraceae* and *Ericaceae* (*Empetrum nigrum* and *Vaccinium* sp.) or *Rubus chamaemorus*; c) steppe areas covered by lichens (particularly common *Cladonia rangiferina*, reindeer lichen), mosses and a few flowering plants such as *Luzula* and *Dryas.* At the southernmost limit of the tundra we find *Betula nana, Arctotaphylos uva-ursi* and *Loiseleuria* (= *Azalea*) *procumbens.* The Siberian tundra, or Kammenaia is much richer in species than the European one.

The primary productivity of the tundra is the lowest in the world, going from 0 (iced parts of Novaja Zemlja and Svalbard) to 4000 kg/ha/year (dry weight). Only on the shores of the White Sea can the productivity reach values as high as 4100-6000 kg/ha/year (cf. Bazilevič et al.). The annual energy fixation is $0.6 \cdot 10^6$ cal/m^2, representing the lowest value in the world, after desert scrub (Lieth)[1].

The fauna

Mammals

The mammals of the tundra are limited essentially to three groups, rodents, ungulates and carnivores. Secondarily, a few insectivores may be mentioned, but these have very little ecological significance. The two types of primary consumers (or herbivores) which are preyed upon by the secondary consumers (carnivores), together with birds, form the main food web, with the smaller vertebrates and invertebrates of lesser importance.

An exact knowledge of the productivity of these animals would be very useful, as well as knowledge of primary productivity, particularly to offer us a basis for understanding the characteristic population cycles of both birds and mammals of this biome.

Today the original mammal fauna has been greatly altered as a consequence of the destruction of the musk ox, domestication of the reindeer and hunting of carnivores for their valuable fur. Maybe the least disturbed mammals today are the rodents, because of their inconspicuousness and lack of value of their skin.

[1] It is strange that the values of primary production reported by Bazilevič et al. are 70% higher than those assumed by Lieth. For the tundra values calculated by this author (1-4 t/ha/year, averaging 1.4) correspond perfectly to those obtained by the Soviet authors.

The rodents are represented mainly by the lemming, *Lemmus lemmus*, one species or *Microtus* (*M. ratticeps*), two species of *Clethrionomys* (*C. rutilus* and *C. rufocanus*) and the variable hare, *Lepus timidus*. The lemming is a nocturnal creature, which runs close to the ground, does not jump, swims willingly and climbs only with difficulty. Its populations increase periodically (every 3 or 4 years); it migrates then toward Southern Scandinavia, looking for new territories. It crosses rivers and lakes without deviation and following definite routes, but when it encounters the sea it enters it and drowns. It digs a complex galleries burrow under stones and mosses, and during the long winter directly under the snow surface. It builds a nest with mosses and herbs, but it does not accumulate food reserves in it. The lemming exhibits no winter dormancy; during the cold season it remains in the elevated regions of Scandinavia (regio alpina).

The emigrations of the lemming in Scandinavian countries are spectacular; sometimes they pass through towns and villages and are so abundant that the dogs and the cats, after a short period, ignore them completely because they are simply ... bored! Often these emigrations lead to occupation of new areas elsewhere (Kalela & Kopenen). It is generally admitted that this system of emigration – practically unique in all the animal kingdom – is a very drastic way to reduce the standing crop of the population, so that the individuals which remain after this migration have sufficient food and space to live and reproduce. These population remnants will slowly bring about another population explosion, which will repeat the same phenomena (Fig. 184).

Another 'lemming' of the European tundra is *Dicrostonys torquatus*, different from the true lemming because of the presence of longitudinal stripes along the back. It is somewhat an intermediary between a vole and a true lemming. Its name is due to the fact that it has only a rudimentary nail on the thumb, whereas the nails of third and fourth toes of the anterior limb are very long. It thrives in the northern parts of the European and Asiatic continents from Lapland to Eastern Sibiria and Kamtchatka. It does not migrate.

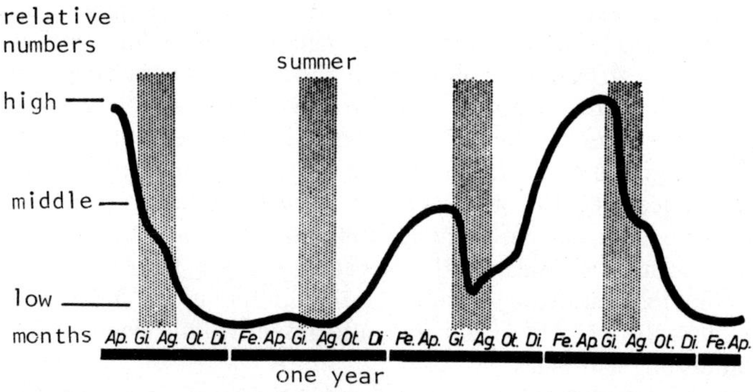

Fig. 184. Cyclic variations of population density in a North American species of the genus *Lemmus* (*L. trimucronatus*). General curve of an oscillation in a short term cycle, to show the important fluctuations or changes in trend and their approximate relationship with the summer seasons of successive years (from Pitelka).

The vole *Microtus ratticeps* lives in very humid places such as swamps and marshes. It remains active all winter, and is active both day and night; it swims and dives excavating its burrows with one chamber for the nest and others for provisions. The nest is lined with mosses, stems of dry graminoid plants, etc. The ruddy vole, *Clethrionomys rutilus*, lives in small woods of birches and shrubs; it climbs very well. It makes burrows and the nest is lined with herbs and mosses among the roots of small trees, stones, etc., but sometimes it enters houses. It is limited to the northernmost part of Europe. The grey-sided vole, *Clethrionomys rufocanus*, with a more extensive and southern distribution compared with the preceding species, lives in rocky, mountainous places among willows and birches, in the 'regio alpina' of Scandinavia, and therefore cannot be considered a true animal of the tundra. It digs its burrows among the grass in the shrubs and often enters houses. We should also mention *Microtus middendroffi*, feeding on *Rubus chamaemorus, Salix rotundifolia, Vaccinium vitis idaea, Pedicularis* and, secondarily, other plants. To this European environment belong also several *Anatidae*, more or less bound to humid environments.

Another rodent typical of the tundra is the variable hare *Lepus timidus* inhabiting only the northern parts of Europe (included Iceland, Faeroe, Shetland, Great Britain and Ireland) and the Alps although in the northern area it seems to be decreasing. Generally nocturnal, it is sometimes gregarious, swims very well; hides generally among rocks or big stones and rarely digs a den. During the summer it is often found in the mountains, during the winter on the plains.

Together with these species, more or less restricted to the Arctic or sub-Arctic environment, we find in the European tundra some common species, present almost all through Europe (and sometimes Eurasia). One of these species is the common squirrel, *Sciurus europaeus*, which does not rest in winter, making a nest with defoliated branches, lined with herbs, mosses and wool. Some mice present at these latitudes include the ground vole *Arvicola terrestris*, the short-tailed vole, *Microtus agrestis* and *Sylvaemus* (or *Apodemus*) *sylvaticus*, the latter limited in the north to Iceland.

The Carnivores are represented by some 6 species: the Arctic fox, *Alopex lagopus* (from λάγος rabbit and Πους foot), living both in the tundra and in the mountains (regio alpina); it is both diurnal and nocturnal as are many homoiothermous animals of the North. It is present also on Jan Mayen and Novaja Zelmja. It exhibits gregarious habits. Its den is a rocky crevice or a den previously occupied by another animal; in some cases it inhabits an extended complex of galleries. It stores provisions. The common fox, *Vulpes vulpes*, is also present in some parts of the tundra, but not in Iceland. The stoat, *Mustela erminea*, present all through temperate and cold Europe, is also absent in Iceland and the same is true for the weasel, *Mustela nivalis*, which is similar to the preceding species. The otter *Lutra lutra*, a common amphibious carnivore, is present in the tundra but not in Iceland. Much more interesting is the wolverine *Gulo gulo*, also absent from Iceland, which we have discussed in dealing with taiga.

The Ungulates are represented by the reindeer and the musk-ox: the former is found both as a wild animal and a domesticated one, while the latter had disappeared only to be recently re-introduced into Norway and the Svalbard islands and more temporarily into Sweden and Iceland. The musk-ox is indigenous to the

Fig. 185. The musk ox *Ovibos moschatus* (from Scharff).

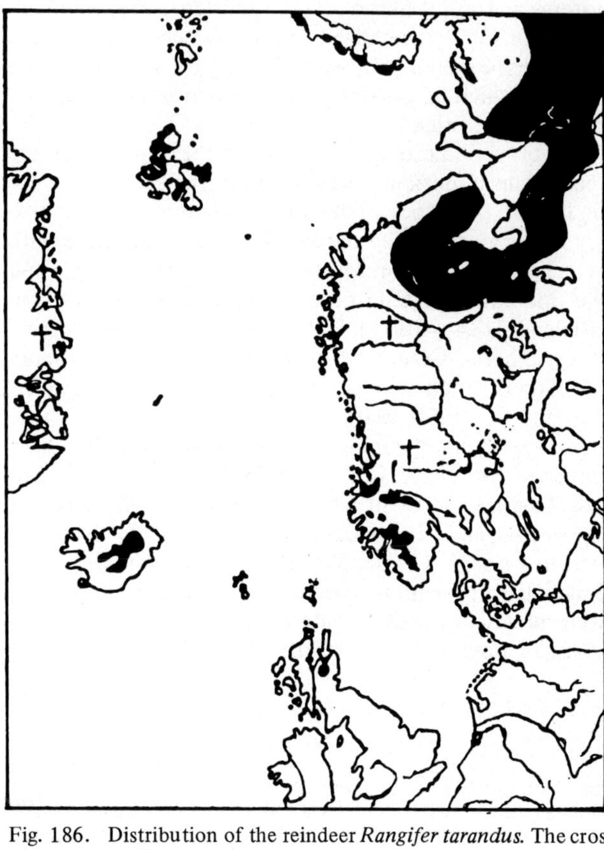

Fig. 186. Distribution of the reindeer *Rangifer tarandus.* The crosses indicate the past presence of the animal (from v.d. Brinck).

bare tundra, active during both day and night. In the summer it forms small herds, but during the winter it forms large closed groups, led by females ('cows') with a 'bull' assuring protection. When threatened, the members of the herd form a circle (Fig. 185).

The reindeer *Rangifer tarandus* is subdivided into several races, perhaps partly geographical, partly ecological. It can live on high mountains (regio alpina) of Scandinavia, in the tundra or in the taiga, provided that there is plenty of both running and standing water. Fig. 186 shows its area of distribution. It has been found as a fossil in Greenland, and at present an attempt is being made to introduce the animal into Scotland. The specimens living in Spitzbergen are smaller.

The reindeer moves like the horse, i.e. it can walk, trot or gallop; it swims moderately well; it does migrate, but its migrations are much reduced compared with those of the North American reindeer. It is gregarious, living in large herds made up mainly of females, and led by an old female. The males sometimes constitute smaller herds. In Lappland the reindeer is domesticated.

The importance of the reindeer in the tundra ecosystem has recently been the object of research both in Europe and North America. On the basis of previous knowledge of other ecosystems and other herbivores it has been possible to construct a schematic representation of a generalized grazing system valid for the reindeer (Fig. 187) in which all the major factors regulating the population are considered. The material flow (which is constantly accompanied by an energy flow) has also been recognized and described.

Since the reindeer is the only domesticated animal of the tundra and the adjacent ecosystems, its economic importance has also been the subject of very accurate analyses. Indeed, one of the types of land-use common to many high latitude and tundra ecosystems is grazing. The nature of man's relationship with the grazing herbivore may include exploitation for meat, wool, hides or active and passive recreation. Thus we find domesticated or semi-domesticated herds of reindeer in Norway, Sweden, Finland, USA (Alaska), USSR and Canada. Wild reindeer herds are present in Norway, Siberia, Greenland, South Georgia and, to a more limited extent, in Finland. Experimental reindeer research stations exist in Norway, Sweden, USA (Alaska) and the USSR. Mathematical simulation models have now been employed for about two years to compare and synthesize *Rangifer* data. The countries currently involved are Norway, Finland (in Europe), Canada and the United States in North America. A recent workshop in Trondheim was successful in updating existing models, and in providing comparisons of the dynamics of the grazing process between sites. It is reasonable to assume that modification of the currently existing models can facilitate synthesis of data. There is a clear need to supplement existing programmes to include the broader perspectives of the social sciences which are not now included in either the studies or the modelling efforts[1].

As a result of such studies a scheme has been constructed by the experts of the

[1] The first International Reindeer and Caribou Symposium was held at the University of Alaska in August, 1972. Communication between individual researchers was sustained largely through IBP, and the Trondheim Grazing Workshop was a logical outgrowth of the symposium.

UNESCO/MAB programme in which the information pertinent to understanding man's use of a large herbivore is illustrated (Fig. 188). Thanks to the research of a working group inside MAB we have today valuable guidelines to the study of this large herbivore. The parameters to be measured are many; without entering into details, we can mention the parameters, but not the functional relationships which would take us too far. The parameters are: food availability, indices of food intake, food intake, energetics, activity index, physiological blood parameters, condition, body composition and growth status, mortality, age of maturity, dominance index, breeding success, herd size and age/sex structure, productivity, har-

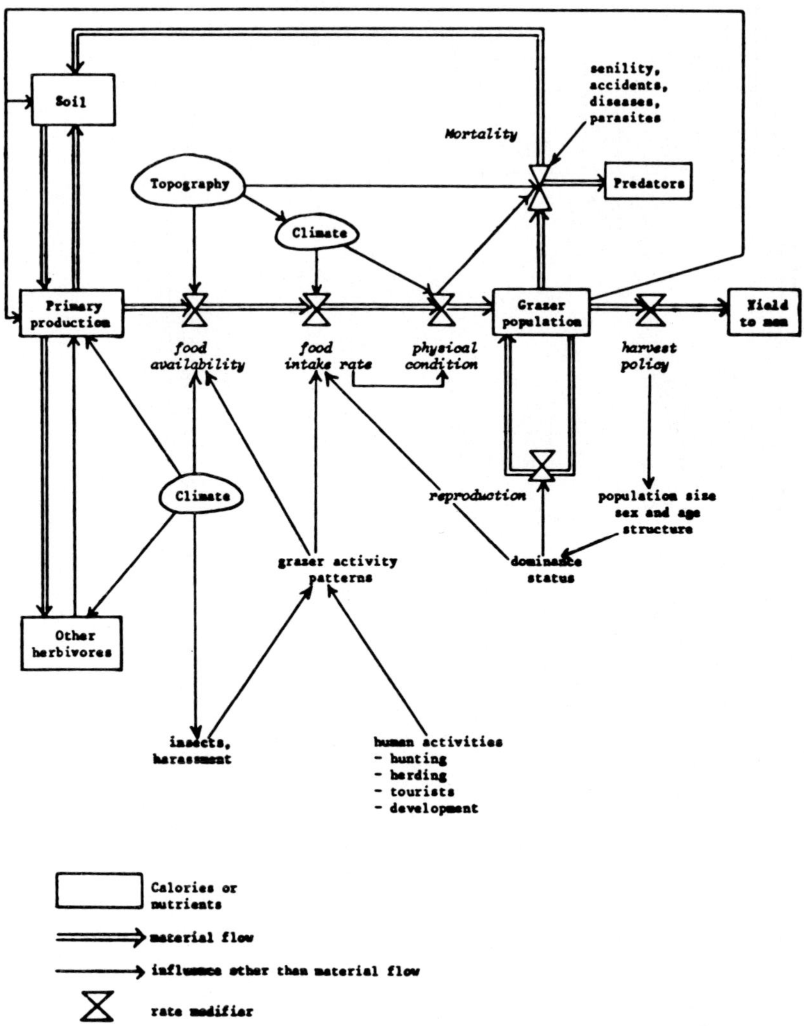

Fig. 187. Schematic representation of a generalized grazing system (from MAB 14, Nov. 1973).

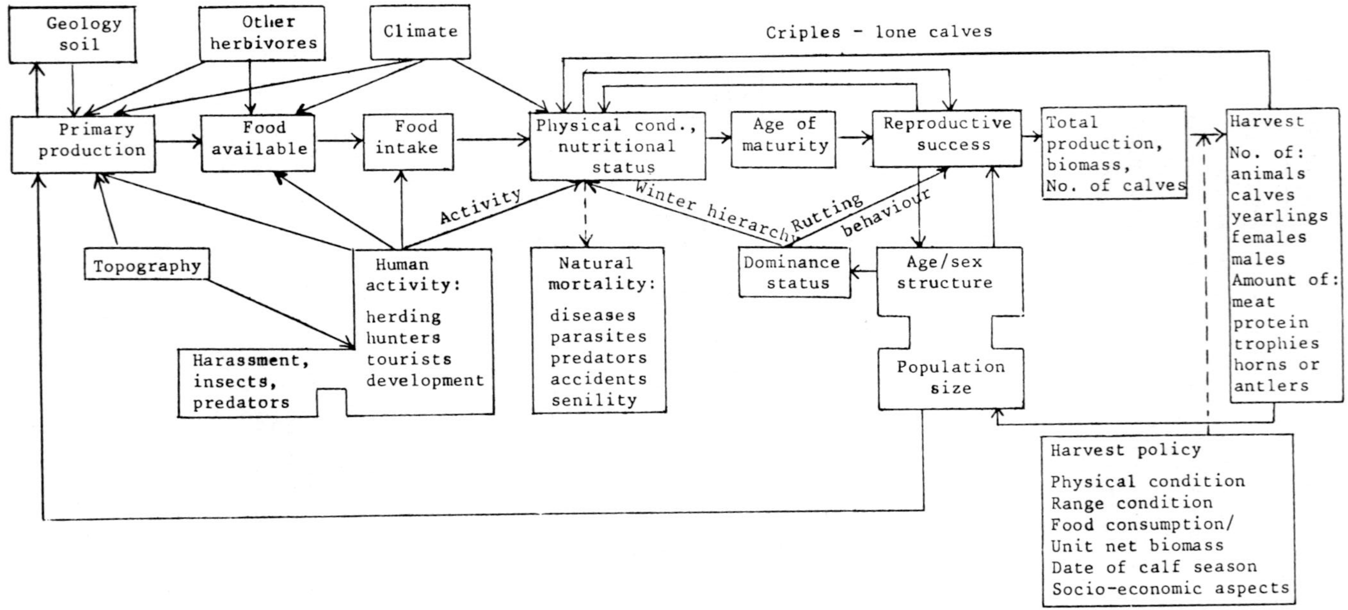

Fig. 188. Information pertinent to understanding man's use of large herbivores (Physical or 'natural ecosystem' considerations only; social side not easily modelled a priori, but will produce values relevant to understanding of problem) (MAB 14, Nov. 1973).

vested section or the net production of the herd and finally harvest policy (cf. MAB, 14, 20-23 Nov. 1973, pp. 104-105).

To finish this section we must mention another mammal order – Insectivores. In the tundra they are represented only by some common species such as *Sorex minutus* or the pygmy shrew, *Sorex araneus* or the common shrew (both of them without winter dormancy), *Sorex caecutiens* or masked shrew typical of bogs and woodlands of taiga and tundra, and finally the water shrew, *Neomys fodiens*, present in the northern tundra but not in Iceland.

The bird life of the tundra

Most birds of the tundra are migratory birds and present many problems for study. Unfortunately, the migrations of birds – though known for many years – are very poorly understood as far as the real causes of the migratory urge and the sense of orientation are concerned.

In the Bible we read of several migratory birds, some of which are also present during summer in the tundra: 'Lo! the winter is past, the rain is over and gone; the flowers appear on the earth; the time of the singing of bird is come, and the voice of the turtle is heard in our land' (Cant. II, 11-12) ... 'Yea, the stork in the heaven knoweth her appointed times; and the turtle and the crane and the swallow observe the time of their coming; but my people know not the judgement of the Lord' (Jer., VIII, 7).

Aristotle in his *περὶ τὰ ζωα ἱστορίαι* (Libris VIII, c. 14-15) also mentions several migrants such as the crane, the white pelican, the quail, the owl and the swan. Other birds, such as the kite, the stork and the lark, were thought to hide during the winter, while the swallow was thought partly to migrate, and partly to hide in holes in which it had been found without feathers during the winter.

There are many aquatic birds in the tundra, both marine (living at the coast) and freshwater (fowl etc.). The description we have given of the general features of this biome makes it quite understandable. Also there are many terrestrial birds, belonging to many orders. All of them are preyed upon by the birds of prey – both diurnal forms (or Falconiformes) and nocturnal (or Strigiformes) which abound in the tundra. Let us mention the golden eagle, the rough-legged buzzard *Buteo lagopus*, exclusive to these latitudes, and an essential element in the biological equilibrium of this ecosystem, as it kills the small rodents such as the lemmings, and therefore experiences, like its food source, oscillations in population density. Other predatory birds are the goshawk *Accipiter gentilis*, the white-tailed eagle *Haliaëthus albicilla* (in Swedish Havsörn) found exclusively at the coast; the osprey *Pandion haliaëthus* (in Swedish fisk-gjuse), the merlin *Falco columbarius*, found in the true tundra, and lastly the gyr falcon, *Falco rusticolus* (in Sw. jaktfalk) which nests on rocks and prefers rocky places (and is not therefore a typical tundra bird), but limited to Northern Europe, Iceland, Faeroe and, maybe, Greenland.

The nocturnal birds of prey are very common species present throughout Europe, such as the pygmy owl *Glaucidium passerinum*, the short-eared owl, *Asio flammeus* (in Iceland only during summer) and *Aegolius funereus.* A highly characteristic creature of the tundra is the snowy owl, *Nyctea scandiaca*, in Swedish fjälluggla, nesting on small bog hummocks (paltas) and present also in Iceland. Other interesting elements are the great grey owl, *Strix nebulosa*, a migrant or summer visitor and lastly the hokuggla (Sw.) *Surnia ulula.* Of particular interest in the biological equilibrium of the tundra is the snowy owl (Fig. 189).

The Gallinaceous birds of the European tundra are all resident and not very different from those which live on high mountains in Southern Europe, where the environment is similar, except possibly for the height (to which birds are generally not sensitive) and photoperiodism. Four species must be mentioned: *Tetrao urogallus* or capercaillie (Sw. tjäder), *Lyrurus tetrix, Tetrastes bonasia, Lagopus mutus* or rock ptarmigan (in Swedish fjallripa), a boreo-alpine species, and a congeneric species, *Lagopus l. lagopus*, or willow ptarmigan (or also willow grouse, in Swedish dalripa). The latter is exclusive to the tundra, and has a greater ecological importance than the former, though both of them are resident birds, with the same dietary habits and the same enemies. *Lagopus lagopus* (in Russian belo kuropatk) feeds on *Salix* (*S. rotundifolia, S. myrtilloides, S. polaris*), *Betula nana* (in Russian serežki), *Vaccinium uliginosum*, followed by *Equisetum, Empetrum nigrum, Dryas octopetala, Tussilago farfara, Carex, Geranium pratense*, insects and various plants. Other birds of the European tundra are the 'peregrine' *Falco peregrinus*, in Russian sapsan, the 'merlin' *Falco columbarius*, in Russian derbnik (Fig. 190).

All the other birds of the tundra are smaller, and are represented by a couple of woodpecker species and many passerine birds. They divide between themselves all the food reserves provided by this environment, which, as a matter of fact, is not as poor as it would seem at first glance: fruits, flowers, dry fruits and seeds,

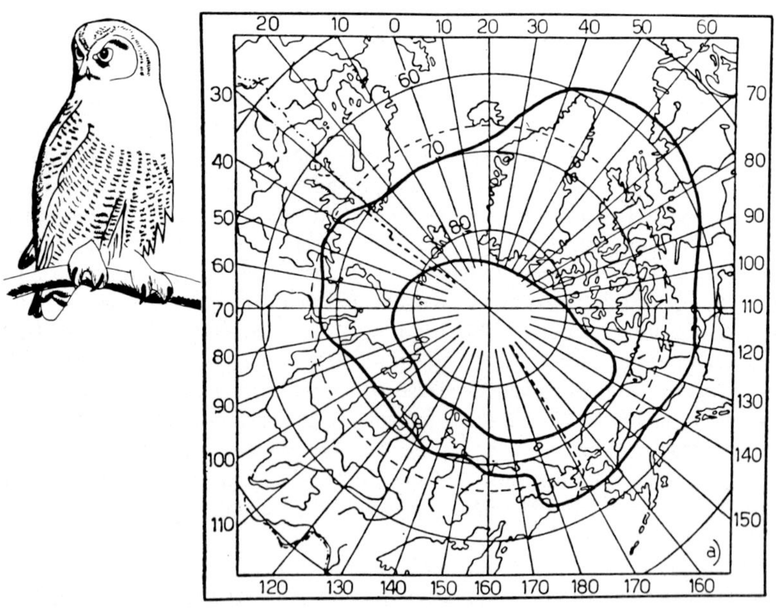

Fig. 189. Distribution of the snow-owl *Nyctea nivea*, a typical circumpolar bird (from Kühnelt) (with the permission of G. Fischer Verl.).

insects and other small invertebrates, buds and dead plant material. All these small creatures, which often compete with each other in the struggle for life, are hunted by the birds of prey already noted and the few carnivores living in the tundra. Of course we must not forget man who also hunts in the tundra – luckily for the birds they are not so extensively killed by men as in other European countries, particularly in Southern Europe and in Italy.

The woodpeckers are limited to the southernmost parts of the tundra where

Fig. 190. Distribution in Europe of a boreo-alpine bird, *Lagopus mutus*. Inset: the two forms of the bird, winter and summer phases. (Schematic).

there is an imperceptible transition into the taiga: let us remember the black woodpecker, *Dryocopus martius*, and the three-toed woodpecker (*Picoides tridactylus*) a boreo-alpine species represented in the North by a subspecies different from that living on mountains.

Most passerine birds visit the tundra only during summer, to nest and breed. *Eremophila alpestris* or shore lark (in Sw. berglärka) is a boreo-alpine species nesting both in the tundra and on the Balkan mountains. However, the Siberian jay, *Perisoreus infaustus*, is exclusively found in the tundra and the taiga, in small woodlands near the Barents Sea. The house martin, *Delichon urbica* (in Sw. hussvala) nests in the whole territory with the exception of the region of the North Cape. *Parus cinctus*, a tit (called in Swedish 'Lappmes') is a small creature nesting in the copses in the tundra, together with *P. atricapillus*, which is represented in the North by a peculiar subspecies (*P.a. borealis*) locally called talltita, living in swampy places or in small woodlands. The wren *Troglodytes troglodytes* lives only in Iceland where it may also winter. The thrushes are also well represented; *Turdus pilaris* or field fare (locally björk trast) nesting in birch, whence its name (björk = birch) present also in winter (and in Iceland only in winter); *T. philomelos*, locally tal trast, nesting in srubs, ivy etc. *T. iliacus* or redwing (Sw. rödwingtrast) present also in winter (and in Iceland only in winter); other passerine birds are the redstart *Phoenicurus phoenicurus* (Sw. rödstjärt), present during winter (also found on the Faeroes), the blue-throat *Luscinia s. svecica* (locally blåhake) in woody places, heaths, swamps etc. present also in winter. It nests on the ground, under birch and juniper, in humid heathlands. Other species are *Acrocephalus schoenobaenus*, the sedge warbler; *Sylvia borin* or garden warbler, found also during winter up to the North Cape; the willow warbler, *Phylloscopus trochilus* (locally lövsångara), present also in winter, and lastly the endemic species of warbler, *P. borealis*, or Arctic warbler (locally nordisk lövsångara) similar to *P. sibilatrix* of lower latitudes, nesting on the ground in small woodlands of birch and pine. The dunnock *Prunella modularis*, the pipit *Anthus trivialis* and the meadow pipit *A. pratensis* are widely distributed species. The red throated pipit, *Anthus cervinus*, locally known as rödstrupig piplärks, is characteristic of the humid tundra, often living near the coast of the Arctic Ocean. It nests on the ground and is easily recognizable because during the reproductive period it displays a vivid reddish colour on the throat. Also wagtails are present, including the white wagtail, *Motacilla a. alba*, present also in Iceland and the blue-headed wagtail, *Motacilla flava thunbergi* (in Swedish 'gulärla') the northern subspecies of this 'Rassenkreis'.

The waxwing *Bombycilla garrulus*, or in Swedish sidensvans, lives in small birch and pine woodlands, where it nests; the great grey shrike, *Lanius excubitor*, the largest European shrike, is an example of the validity of the so-called Bergmann's law; the twite, *Carduelis flavirostris*, extending to the North Cape, nesting on bogs and ödländ somewhat elevated. It is recognizable by the yellowish throat and in the males the red base of the tail. *Carduelis f. flammea* is also present (locally called gråsiska), representing proof of Bergmann's law, which states that in homoiothermous animals we find the largest forms (or subspecies) in colder areas, with the smaller in the warmer climates. It often nests in small temporary colonies. In summer it is found up to the North Cape and in winter it is also present in

Iceland. A typically northern species is *Carduelis hornemanni*, the Arctic redpole (or in Swedish sibirsk gråsiska) (in German 'Polarbirkenzeisig'), differing from the former because of the white base of the tail. It nests on the ground or in small bushes.

Let us conclude by mentioning three species which belong to the northernmost European elements, namely the Lapland bunting, *Calcarius lapponicus*, in Swedish lappsparv, practically an endemic species nesting in the tundra, in treeless Ödland and in mossy bogs; the snow bunting, *Plectrophenax nivalis*, locally snösparv, nesting in cracks in the rocks, often in mountainous places, in summer up to the North Cape and in winter also in Iceland. Finally, we should note the tree sparrow, *Passer montanus*, locally pilfink, also present up to the North Cape.

It would be extremely interesting to know the exact feeding habits of all these passerine birds (which are obviously different to those during winter in Central or Southern Europe) in order to calculate their role in the various trophic webs (or biocenotic connections) confronting us in various parts of the European tundra (cf. Fig. 203). Other Passerine birds are *Emberiza pusilla* (little bunting, in Swedish dvärgsparv) and *Emberiza schoeniclus*, the reed bunting, in Swedish sävsparv.

In the tundra, and more especially in Iceland, some foreign species are accidentally present, usually migrating along an abnormal route or more often transported by hurricanes or very strong winds. Some of them belong to marine or in general aquatic forms. Let us mention here *Diomedea chlororhynchos*, a kind of albatros, observed in Iceland; the ducks *Anas formosa* living in North Eastern Asia and observed in Iceland, and *Somateria fischeri*, a Siberian species found in the Arctic Sea up to Norway; the *Numenius borealis*, in German Eskimo-Brachvogel, a North-American element observed on Iceland and today possibly extinct; the gull *Rhodostethia rosea*, similar to the common *Larus ridibundus*, living in North East Siberia and found in Iceland, Faeroer, etc.; *Aethia cristatella*, a kind of *Podicipidae*, living in North Pacific but observed occasionally in Iceland; *Ceryle alcyon*, living in North America and observed in Iceland, and finally the *Sylviidae Parula americana*, exceptionally observed in Iceland.

All this comes about because Iceland is situated in an area of strong winds and storms so that birds may be transported to it, and find a temporary shelter.

Reptiles and amphibians

These are the vertebrate groups least represented in the tundra because they are heterothermous and – contrary to fish which are well adapted to cold water (as we shall see in the chapter devoted to freshwater environments), they require a high temperature for all their functions.

Two species are able to withstand the severe conditions of the tundra, both dormant in winter, the frog *Rana temporaria* and the lizard *Lacerta vivipara*, both of them reaching as far as the North Cape. A snake, *Vipera berus*, though not typical of the tundra, reaches the Arctic Circle in the Kola Peninsula and is found north of Archangel, up to the Mezen River.

Rana temporaria lives in forests, copses and among bushes, feeding on insects, molluscs, spiders, earthworms etc. It hibernates mostly under water. Egg laying begins almost immediately after hibernation, so that the time of laying is strictly dependent on latitude, as we can judge from the following table:

Britain	January
near London	February
Germany	February-March (in mountains: June)
near Kiev	end of February – mid March
near Moscow	March-April
near Leningrad	end of April – beginning of May
Lapland	between May 15 and June 17 (from Terentev & Chernov)

The metamorphosis occurs after 50-90 days.

Lacerta vivipara lives in marshes and shrubs; it swims, moves on the bottom and sometimes digs into the silt. It is a cold resistant and moisture-liking creature appearing immediately after snow-thaw in March or April but near the northern limit of its distribution as late as mid-June. It starts to hibernate towards the end of August or beginning of September and remains between roots, under tree bark, in rock cracks or soil fissures. It feeds on all kinds of terrestrial invertebrates.

The invertebrates

Insects

Insects have been fairly thoroughly studied in several parts of the European tundra, because of their zoogeographical importance (endemic species, boreo-alpine elements, etc.).

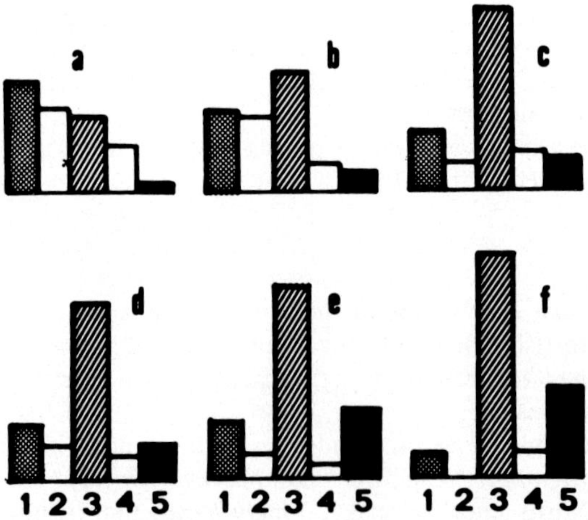

Fig. 191. Composition of the entomological fauna in different localities of the tundra biome, compared with that of Sweden. Greenland, although not belonging to Europe, has also been taken into consideration. According to Lindroth (from Tischler): a. Sweden, 17,000 species; b. Iceland, 700 species; c. Greenland, 530 species; d. Novaja Zemlja, 274 species; e. Spitzbergen, 108 species; f. Jan Mayen, 33 species. 1 = *Hymenoptera*; 2 = *Coleoptera*; 3 = *Diptera;* 4 = *Lepidoptera*; 5 = *Collembola*.

As far as the composition of the entomological fauna is concerned, Lindroth has given a synthetic representation (Fig. 191). It is evident that with increasing latitude or more generally with the increase of Arctic characteristics in the environment, *Diptera* and *Collembola* are favoured compared with the other orders. This can also be seen on the high mountains of the Alps or in general of Southern Europe, where we find an increased number of species and still more of individuals of both *Collembola* and *Diptera* (particularly *Chironomidae*).

Among the insects of the tundra, the category of the so-called boreo-alpine species exhibits a peculiar zoogeographical distribution. We will return to this point in the chapter devoted to the montane biome. For the moment it is sufficient to mention some beetles and butterflies whose distribution in Europe has been the particular subject of both investigations and speculation: *Simplocaria metallica*, present also in Greenland, lives in the tundra in sandy, generally humid places, and it has been found as a fossil in Saxony. *Geodromicus globulicollis*, a *Staphylinidae*, is limited to the banks of water courses or even water bodies, both sandy and stoney, generally without vegetation, but also in dry heaths. *Helophorus glacialis*, a *Hydrophilidae*, in the tundra living both near the coast and inland, is not closely linked to water (as it is in the Alps) since it may be found in the moist litter of coniferous trees, in birch woods or in *Trollius* grasslands. The *Elateridae Corymbites cupreus*, reaching as far east as Altai and Jenissei, has a polyphagous larva which nevertheless seems to prefer Graminoid buds; more than the vegetation, the soil is important for larva development, it must be silty-sandy, with some humus, while both peaty and sandy soils are avoided. The related species *C. rugosus* is distributed eastwards as far as Altai and Amur, and Japan and Alaska. *Hypnoidus hyperboreus* is an inconspicuous *Elateridae* found as far east as Kamchatka and Alaska. A very interesting species is the *Chrysomelidae Phytodecta affinis*, found as far east as Siberia and also in North America, from Alaska to the mountains of Southern Colorado (?). It lives on several willows, but particularly on *Salix lapponum* and *S. glauca.* According to Holdhaus, the specimens of this phytophagous beetle found in plains are possibly transported by rivers.

Another interesting species of phytophagous beetle is the weevil *Otiorrhynchus dubius*, which is also found in Greenland. It thrives at the coast, but it prefers morainic terrain to humid soils. The fact that this species has been found on the Eastern Alps on *Betula nana* makes it probable that it also feeds on this plant in the tundra. Whereas on the Alps *O. dubius* is generally amphigonic, in the North it is parthenogenetic. This phenomenon, which is fairly common in several boreo-alpine invertebrates, has not yet found a satisfactory interpretation. It has been found as a fossil near Dresden; *Otiorrhynchus arcticus*, present also in Greenland, lives near the coast (including those of the Arctic Ocean). It prefers sandy soils rich in quartz, often without vegetation. It seems to be restricted to the epilittoral zone (see the chapter on humid environments), but according to Holdhaus it has probably a double auto-ecology, since it is also found in the mountains of Scandinavia. Fossils of this species have been found near Pilzno in Western Galicia and perhaps also near Dresden. Another weevil is *Barynotus squamosus*, probably introduced into North-Eastern North America, while in Scandinavia generally present on the coast. It behaves as a grassland species, restricted to the sea-shores and almost certainly it is parthenogenetic in its northern area and amphigonic in the southern one. It is known as a subfossil from a *Sphagnum* bog on the Atlatic area of Jutland.

The most interesting boreo-alpine butterflies are *Pieris callidice*, an oriental species which reaches the European tundra only in the northern Urals; *Erebia lappona*, reaching in Asia as far north as the Altai. In the tundra of the Kanin Peninsula it is probably the most frequent diurnal butterfly; its larvae feed upon *Festuca ovina. Agrotis speciosa* exhibits a very wide, circumpolar distribution; the larva is polyphagous, but it is not a true tundra species. *Larentia munitata* also has a circumpolar distribution. In Iceland (where it is the only boreo-alpine lepidopterous insect) it is the widest diffused and most frequent *Geometridae.* It is a typical grassland species and in Sweden the caterpillar feeds upon *Alchemilla, Geranium, Galium* etc. Finally let us mention *Zygaena exulans*, reaching as far eastwards as Altai and Tarbagatai. The caterpillar feeds on lower plants and is very polyphagous (Fig. 192-195).

Also interesting is the insect fauna of the various islands belonging to the tundra biome.

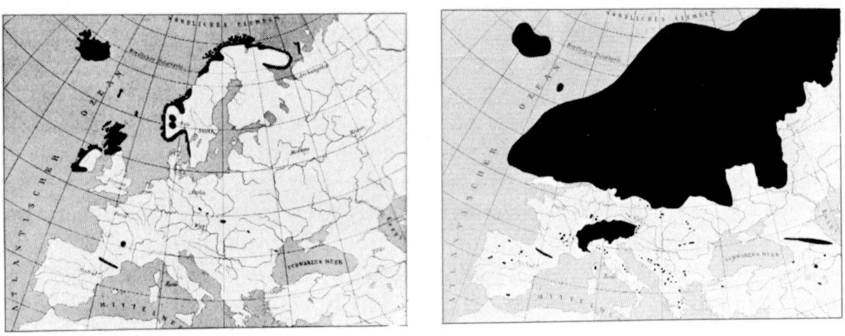

Fig. 192

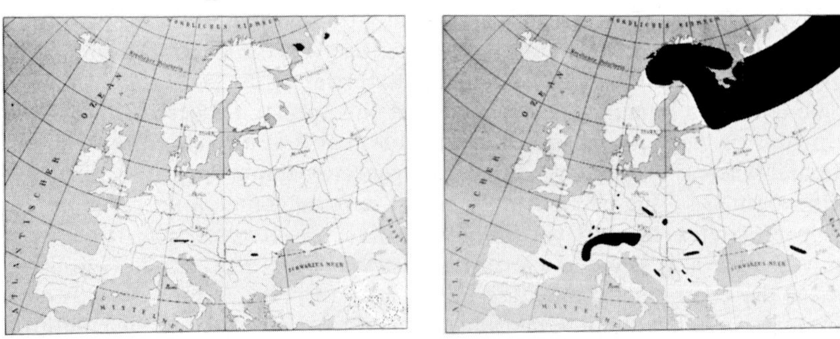

Fig. 193

Fig. 192-195. Examples of distribution of two insect orders, *Coleoptera* and *Lepidoptera*, of the boreo-alpine type. Fig. 192b shows the maximum extension of the glaciers in Europe during the Glacial Era (from Holdhaus).
Fig. 192. a. Distribution of the weevil *Otiorrhynchus arcticus* (from Holdhaus).
Fig. 193. a. Distribution of the Carabid *Pterostichus kokeili* (from Holdhaus). b. Idem of *Amara erratica* (from Holdhaus).

The insect fauna of Iceland is, according to Holdhaus, extremely impoverished, although the size of the island and its present climatic characteristics should permit the existence of a much richer insect life. Probably the insect fauna was largely destroyed during glacial episodes: proof of this should be the total absence of some freshwater groups such as *Gyrinidae, Hydrophilidae, Helodidae, Dryopidae* and *Ephemeridae.* On the whole, Iceland exhibits a northern European fauna with very few or no endemic species at all. Most of the entomofauna are beetles; the macrolepidopterous insects are represented exclusively by a few *Noctuidae* and *Geometridae,* probably due to the inclement summer climate of the island, with frequent rains and storms. There are additionally some microlepidoptera, *Diptera, Hymenoptera* (with the exception of ants) and a few other groups. Most of these animals have been introduced accidentally from the mainland since the first human immigrations. The local sagas say that these colonizers even brought with them the earth with which they built altars to their gods.

Only one species seems to be absent from Europe, *Agrotis quadrangula* (Fig. 196); another species on the contrary is present both in North America and Europe, *Hadena sommeri.*

Still more interesting from an ecological viewpoint is the entomological fauna

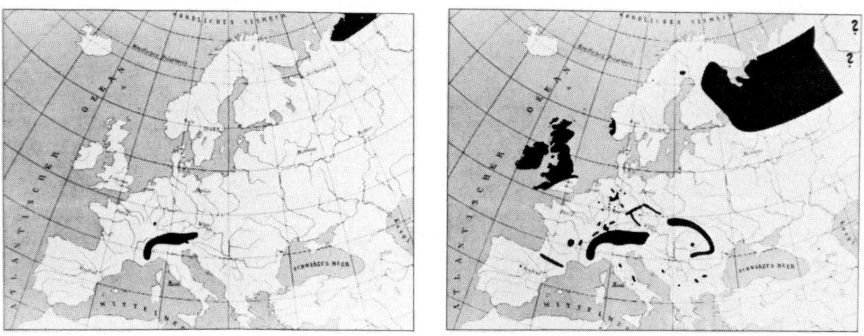

Fig. 194

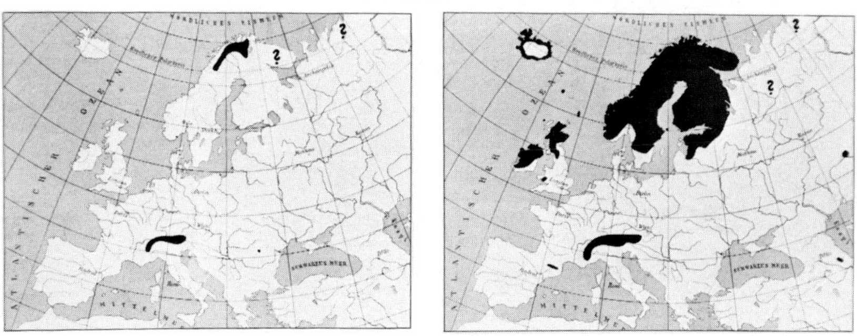

Fig. 195

Fig. 194. a. Distribution of the Elaterid *Corymbites rugosus* (from Holdhaus). b. idem of *Corymbites cupreus* (from Holdhaus).
Fig. 195. a. Distribution of the Lepidopteron *Arctia quenseli* (from Holdhaus). b. idem of *Larentia munitata* (from Holdhaus).

of the islands belonging to the true 'bare tundra'. Only three species of butterflies are reported from Novaja Zemlja: *Colias nastes* var. *werdaudi, Argynnis chariclea* var. *arctica* and *A. frigga* var. *improba.* Some insects are only recorded in the islands, but it would be unwise to speak of them as endemic (cf. Holdhaus, in Schröder, p. 993).

Our knowledge of Jan Mayen is too limited to permit any assumptions about the origin of the insect fauna which, on the whole, exhibits a Northern European character.

When we come to consider Spitzbergen, of interest is the small *Staphylinidae Atheta subplana*, which seems limited to the tundra ecosystem (Fig. 197). The beetles and butterflies are very poorly represented. The existence of endemics is improbable although one species of *Staphylinidae, Atheta subplana*, seems to be restricted to the eastern shores of the Arctic Ocean, and therefore it could theoretically be a pre-glacial endemic. The *Diptera* and the *Collembola* are the best represented orders; particular important without any doubt are the *Chironomidae,* of which 30 species are known from Spitzbergen.

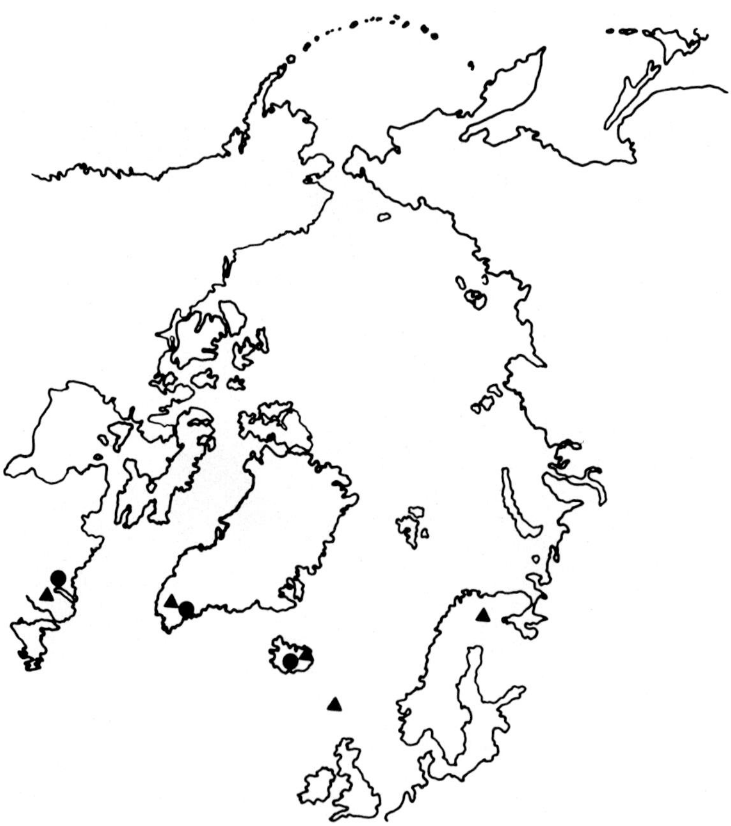

Fig. 196. Example of *Lepidoptera* with a circumpolar type of distribution: ● *Agrotis quadrangula*; ▲ *Hadena sommeri.*

From the Bear Islands, two species of *Coleoptera* have been reported, *Micralimma marinum* and *Arpedium brachypterum.* From Franz Joseph Land apparently only 7 species of Collembolans have been noted, which would prove the high tolerance of cold of these inconspicuous, almost microscopical, creatures, which on the high mountains occur higher than both other insects and multicellular animals, except Tardigrades. One species of Collembolan from Franz Joseph Land, *Isotoma brevicauda*, is limited to the Arctic European and Asiatic lands (Fig. 197) and it is not impossible to consider it endemic, as it survives glaciation on a 'nunatak', as did several arthropods in the Alps (see relative chapter).

One species of Collembolan of this land, *Hypogastrura tullbergi* (= *dubius*) is of peculiar ecological interest. The species exhibits an apparently arctic-alpine distribution (cf. Gisin, p. 43) and seems to be particularly adapted to low temperatures. Indeed, when the temperature to which the individuals are exposed during their development period is higher than 15°C, degenerative forms appear (var. *aberrans* Cass.), in which there is a reduction in eyes, mouth parts, furcae and

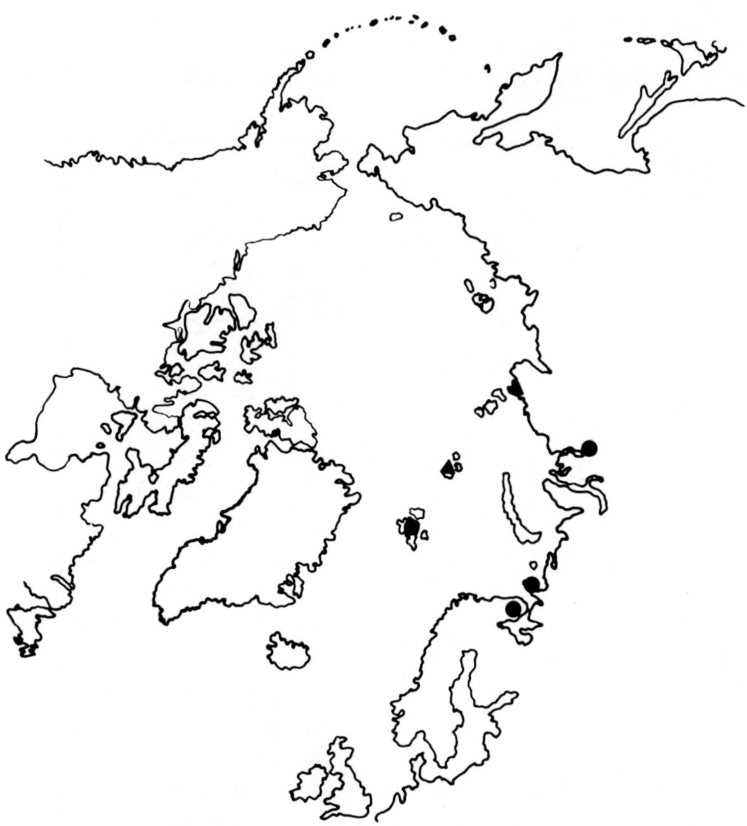

Fig. 197. Examples of insect with an arctic type of distribution: ● the beetle *Atheta subplana* (*Staphylinidae*); ▲ the springtail *Isotoma brevicauda*.

claws, and an enlargement of the tegument granules. This form is not able to reproduce as there is no genital opening.

One of the most typical *Diptera* families of the tundra is *Chironomidae* (or *Tendipedidae*), the common midges. In Lappland there are more than 35 lake *Chironomidae,* plus species whose larvae live in streams or terrestrial habitats. Proof that *Chironomidae* are well adapted to life in the extreme North is given by data on the Greenland Chironomid fauna (see table below):

Groups	Lappland (only lake-C.)	NE Greenland
Podonominae	2	1
Tanypodinae	4	2
Orthocladiinae	15	>40
Tanytarsini	4	6
Chironomini	4	3

According to Thienemann (p. 436), the Arctic lakes are inhabited by an impoverished subarctic fauna. We do not know whether there are some species restricted to these lakes. Also species living in ephemeral ponds (asthatic basins), are interesting: they are represented by 2 *Podonominae*, 4 *Tanypodinae*, 15 *Orthocladiinae*, 4 *Chironomariae* and 4 *Tanytarsariae*, together with a couple of spe-

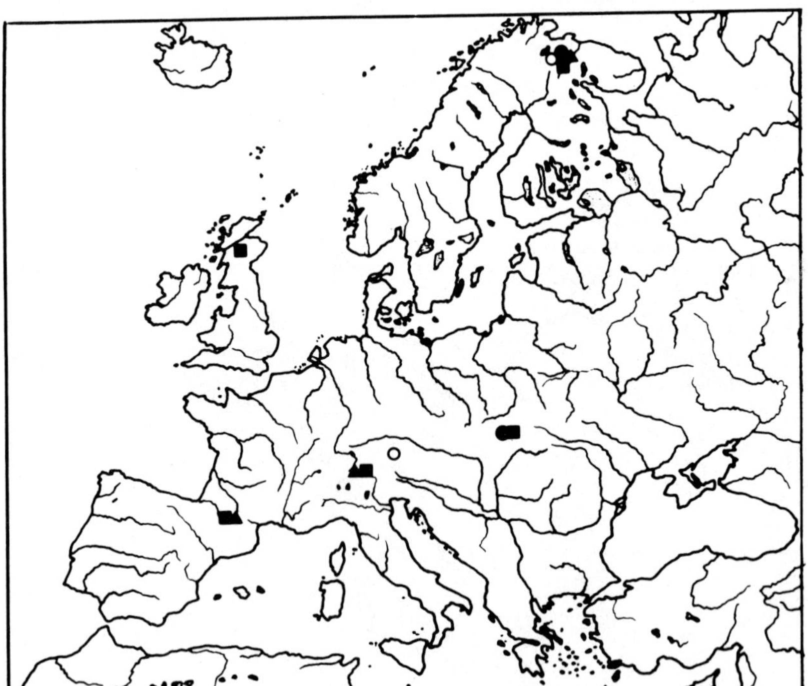

Fig. 198. Examples of some 'boreo-alpine' *Chironomidae* (or *Tendipedidae*): ○ *Eukiefferiella cyanea*; ▲ *Monotanytarsus boreoalpinus*; ● *Orthocladius tatricus*; ■ *Diplomesa lapponica.*

cies of *Ceratopogonidae* (see later). Wholly Arctic forms include two species of *Podonominae, Lasioderma gracilis* and *Trichotanytarsus posticalis*, as well as *Micropsectra groenlandica, Lapporthocladius abiskoensis, Abiskomyia paravirgo* and *Eukiefferiella suecica.* Several species are boreo-alpine: *Orthocladius tatricus, Monotanytarsus boreoalpinus, Diplomesa lapponica* and *Eukiefferiella cyanea* (Fig. 198-199). Probable additional boreo-alpine species (sensu lato) are *Euorthocladius rivicola, Rheorthocladius frigidus, Diamesa steinböcki, Eukiefferiella brevicalcar, Akiefferiella coerulescens, Pseudodiamesa branickii* and *Prodiamesa olivacea* (Thienemann, p. 357).

One species, *Podonomus kiefferi*, seems to be bipolar (Fig. 200). It is very remarkable that a species of this genus, *P. steineni*, is the winged insect which in the Antarctic reaches the highest latitudes (Janetschek) meaning that this genus, as a whole, regardless of its origin and age, is particularly adapted to life in cold deserts.

Another feature of the *Chironomidae* fauna of the tundra is the presence of terrestrial species: for instance, in the tundra of Alaska, larvae of *Chironomidae*

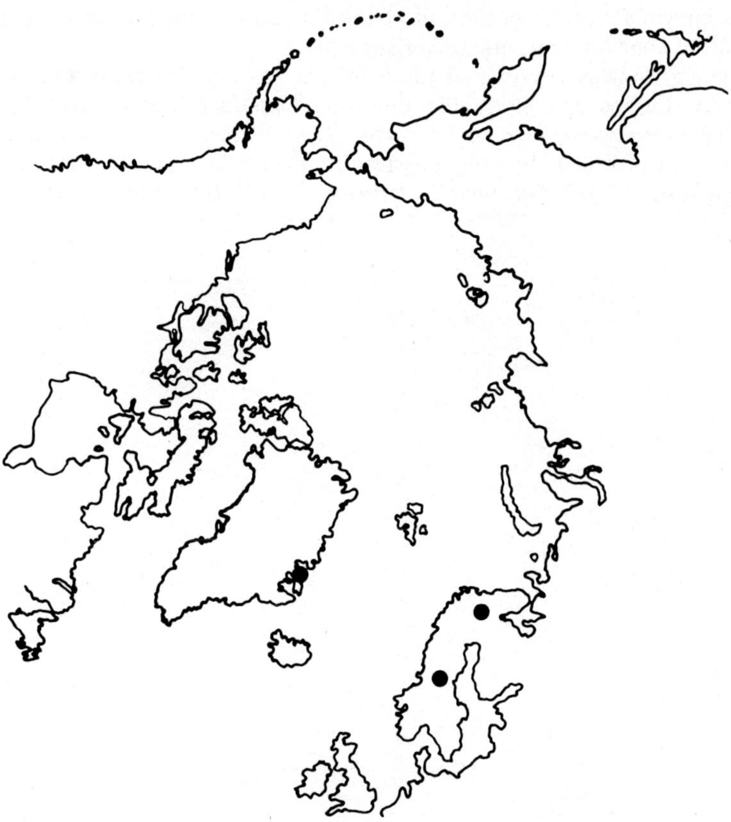

Fig. 199. Distribution of the Chironomid *Micropsectra groenlandica.*

have been found in lemmings' nests: from two nests Weber has found more than 2000 larvae of the genus *Spaniotoma* sensu lato. Also snow owl's nests are inhabited by terrestrial *Chironomidae* (Thienemann, p. 620). Unfortunately nothing is so far known about these terrestrial forms in the European tundra, but their presence is highly probable.

It is worth while recalling here the harmfulness of some small Chironomid-like midges, the *Ceratopogonidae*, whose larva lives generally with those of *Chironomidae*. The female stings to feed on the blood of higher animals, including man, so that their presence in the tundra is in some places very noxious. In 1713 Derham already knew these 'lästigen, gierigen Blutsauger' and Linné also described the 'Svidknott'-plague of Lappland. In 1919 Goetghebuer noted in writing about the sting of *Culicoides pulicaris* that 'the sting causes a sharp pain and a slight swelling of the skin, like a small papula produced by a nettle, surrounded by a reddish zone. The sting is generally associated with extremely disagreeable itching' (cf. Thienemann, p. 628).

Soil fauna

Very little is known about life in the soil of the European tundra, whereas more research has been done in the Asiatic (Siberian) zone.

In the tundra, immediately east of the Northern Urals (which constitute the border between Europe and Asia), the following *Elateridae* live: *Selatosomus rugosus, Corymbites cupreus, Sericus brunneus, Hypnoidus rivularius* etc. Quantitative research has provided the following results: a) in fields near Salehard: *Selatosomus rugosus,* 55.6%; *Corymbites cupreus,* 33.3%; *Hypnoidus rivularius*

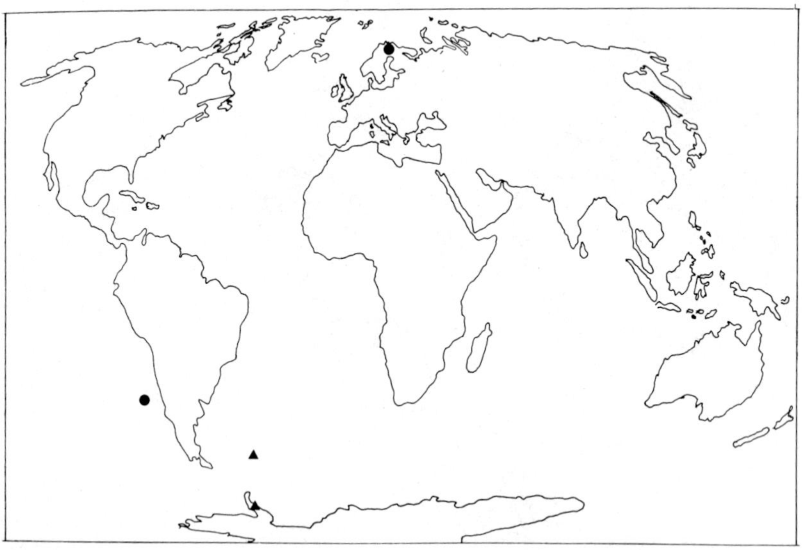

Fig. 200. Examples of circumpolar or amphipolar *Diptera*: ● *Podonomus kiefferi*; ▲ *P. steineni*.

11.1%; b) in the tundra: *Selatosomus rugosus* 83.9%; *Sericus brunneus*, 6.4%.

We may conclude that the most typical *Elateridae* of the tundra is *Selatomosus rugosus*, common on our mountains above the tree line. Other species of this family are *Hypnoidus hyperboreus* and *H. depressus*, which should be among the species most resistant to low temperatures (Cherepanov, 1965).

In the mossy tundra there is a special organic soil in which some particular animals live, such as *Onychiurus armatus, Eisenia nordenskjöldi, Sericus brunneus, Byrrhidae, Tipulidae,* etc. The biomass would be 1-17 g/m^2. Only if there is a podzol can the earthworm *Dendrobaena octaedra* thrive (Ghilarov, 1965).

As far as the mesofauna is concerned, some special collembolans can be found, such as *Sminthurides aquaticus* and *Desoria* sp. (Stebaev & Stebaeva) in moist alluvial tundra soils, with mosses. On the Spitzbergen, Loof recently has found a very remarkable number of Nematodes. Many of them are common to this archipelago and Novaja Zemlja, others to Spitzbergen and Northern Canada or to Greenland. One species, *Tylenchorynchus leptus*, has a very characteristic circumpolar, boreo-alpine distribution. Another species, *Tylenchorhynchus magnicauda*, has so far only been known from Spitzbergen and the Rocky Mountains (Fig. 201).

Thanks to the experimental work of Soviet authors we also know something about water and temperature relations of *Elateridae*, which abound in the soils of the tundra. The larva of *Selatosomus latus* (wireworm) can withstand temperatures as low as −6.6°C, whereas that of *Agriotes gurgistanus* (which lives in

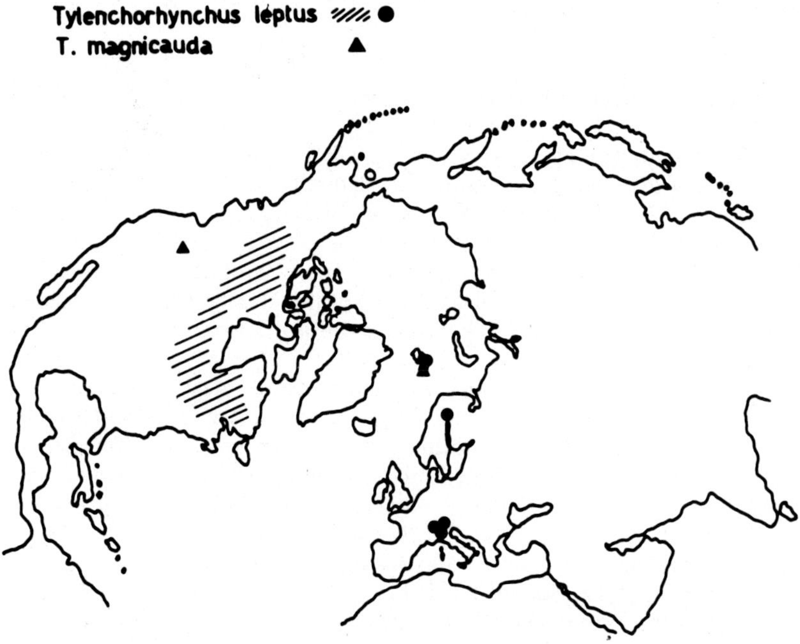

Fig. 201. Distribution of two Nematodes of the tundra, of which one (*T. leptus*) seems to be a true boreo-alpine element.

Southern Russia) dies at temperature c. 3.9°C. Some larvae of *Agriotes sputator* begin to die when temperature is lowered to 0.7°, but the critical death temperature is –3.9°C (Semenyak).

Other eco-physiological data are given by Cherepanov.

We may conclude from these figures that: a) before moulting *Selatosomus* and *Agriotes* larvae behave similarly by losing water rapidly (probably due to lack of a wax layer in the epicuticle) and b) after moulting *Agriotes* loses water only slowly and in small amount. It would be worth continuing this research, unique for tundra insects, and also to examine the relationship between loss of weight and the size of the larva, as Mellanby did many years ago for the meal worm *Tenebrio molitor* (cf. Marcuzzi, 1960).

To finish this small section on soil animals we cite a table from Cherepanov (1965) on the growth of larvae of two species of *Elateridae, Selatosomus aeneus* and *Agriotes obscurus* inhabiting the European tundra. The allometric development can easily be demonstrated by graphing these data.

Size of larva at different period of growth (in mm):

	Selatosomus aeneus		*Agriotes obscurus*	
Larval growth	head width	body length	head width	body length
Before metamorphosis				
Stage I	0.43-0.55	4.6- 6.0	0.23-0.24	3.6- 3.8
Stage II	0.80-1.2	8.8-12.5	0.37-0.52	6.0- 8.0
Stage III	1.30-1.58	13.5-16	0.75-0.92	11.0-13.8
Stage IV	1.85-2.25	19.5-22.0	1.0 -1.4	16.5-21.5
After metamorphosis	2.25-2.5	23.0-25.0	1.6 -1.7	25.5-26.5

The problem of conservation in tundra ecosystems

In the USSR the problem of conservation and wild-life protection in the tundra is under the aegis of the Central Laboratory for Nature Protection of the Ministry of Agriculture of the USSR, Moscow and more particularly of a special section of it, which deals with the theme 'Biogeocenosis in the polar territories', in collaboration with the Botanical Institute of the Russian Academy of Sciences, the University Lomonossow (Moscow), the Pedagogical Institute (Moscow) and other institutions. In 1969, for example, this Laboratory published a work on the polar bear and its protection in Russia.

Let us mention in this connection two small protected areas near the Finnish border, namely Heinässaret Island, near Petsamo, on the Arctic Ocean (once belonging to Finland) where there is a bird sanctuary important for puffins (*Fratercula arctica*) and other Arctic marine birds. On these islets *Rubus chamaemorus* grows. Another preserve is at Pummanginniemi, on the Fisher Peninsula, which includes the only European site of *Potentilla pulchella.*

In Finland there are two Nature Parks (Naturschutzgebiete) inside the tundra, namely the Park of Malla (30 km^2) and that of Kevo (346 km^2), both in Lappland, the first in the extreme north-west, the second in the north-eastern corner near the border with Norway.

Recently the problem of nature conservation in the tundra has received the attention of a programme of UNESCO, Man and Biosphere (MAB). According to this authority, the impact of modern industrial development employing large-scale technology on the relatively fragile ecosystem of the tundra is already important in relation to their effects on the quality of human environments. (We prefer to say simply 'environment'.) Since the movement of industrialization into the far north has only recently begun, the impact will spread and become even more noticeable in the years ahead. Coastal industrial developments, which especially impair aesthetic values and also involve water pollution hazards, are metallurgical industries using imported ores and local hydro-electric power, and pipeline terminal port developments which include tanks and docking facilities. Any water pollution that arises from these industries may have serious local consequences for fisheries and other marine life. In the future these are expected to increase and so too are the numbers of fish canneries sited on the coast.

MAB continues by noting 'Industrial developments in areas at some remove from present human settlements will involve intrusions into remote natural areas and so incur wilderness losses, wildlife habitat and migration disturbances and therefore perhaps wildlife reductions as well. These developments may also impair the subsistence economies of native peoples. Special environmental problems may result from pipeline oil spills in tundra areas or in northern streams. As yet there are no really effective recovery systems and the biodegradation of oil under Arctic

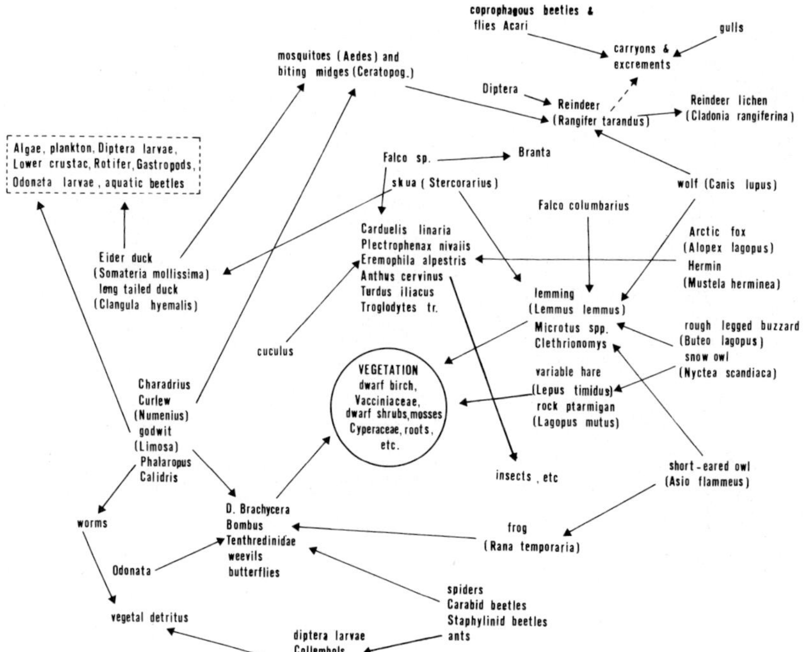

Fig. 202. Trophic relationship in the European tundra.

conditions is very slow. It would be especially unfortunate to have large-scale spills carried into the Arctic Ocean.'

As far as the recreational activities are concerned, the tundras have many attractions and problems common to those of the temperate mountains, but also some important differences. Accessibility is greatly reduced, and because of much more severe winter climates and limited daylight, winter sports are less likely to become a major part of human impact. These high latitude areas, however, have a very special value precisely because of their relative inaccessibility and degree of isolation, and they still constitute some of the least developed areas in the world. Thus very careful attention should be given to their utilization as foci for tourism and recreation. Some points of emphasis should include, according to MAB: the need to assess the tourist carrying capacity of these polar and subpolar lands; interaction between tourist and local native peoples; social and medical aspects; impacts upon wildlife and vegetation (cf. MAB, 14, 20-23 Nov. 1973).

Applied ecology in the tundra ecosystem

One applied aspect of ecology in the tundra biome is provided by the breeding of reindeer (cf. p. 187). Thanks to research by a working group of the MAB programme we have today a valuable insight into the major uses of the tundra ecosystem,

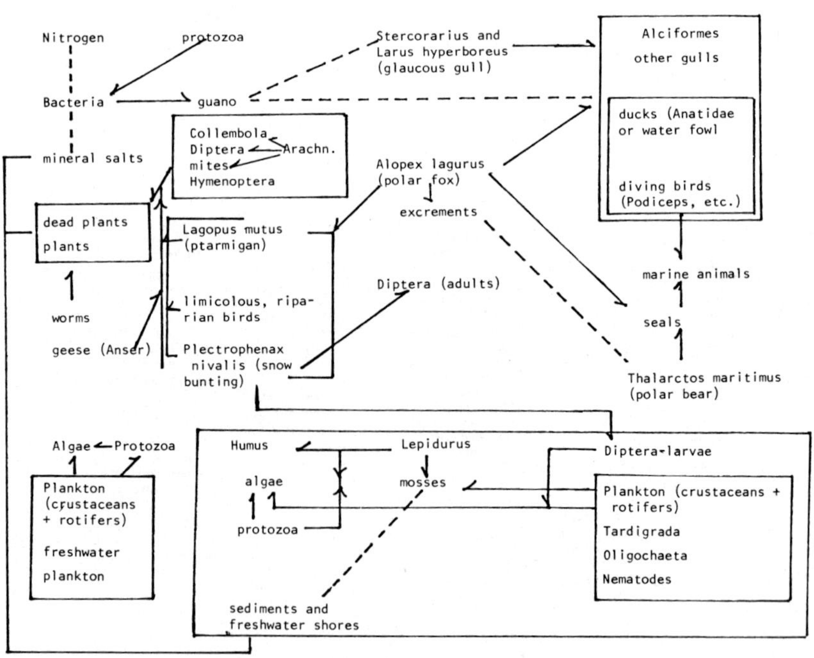

Fig. 203. Trophic relationship ('Biozoenotische Konnex') of the tundra on the Bear Island (south of Spitzbergen) (from Elton, modified).

namely grazing, industrial and other technological impact, as well as recreation and tourism.

In specifically discussing the grazing of reindeer, this working group suggests that the following variables must be considered and measured: 1) natural resource variables: land area, vegetation, animal, soil, climate; 2) economic variables: production variables, forms and amounts of taxation, land tenure, legal structure for land use, control and distribution of capital, organization of product sales/consumer purchase, legal structure for product sales, for worker mobility; 3) social system variables: demographic structure and population distribution, social structure, stratiphication system, degree of communication with others, education, mobility pattern (including migration), power structure (internal and external), inheritance structure, interest group formation and relevance to context, geographic density of personal networks; 4) cultural norms and factors: fixidity on land, alienation level, need for achievement, control over nature (vs. subjection), degree of universalism (vs. personalism), language and communication; finally, 5) health and human biology characteristics: mortality and fertility, causes of death, life span, incidence and prevalence of endemic and epidemic diseases, growth and physique (stature and weight), nutritional status, nutrition and diet, daily energy expenditures, gene characters and finally physiological fitness.

For each argument to be studied MAB outlines methods to be employed and the probable functional relationships.

Most work is being done by the Scandinavian countries, in particular Finland, Norway and Sweden.

As far as the reindeer are concerned, we wish to mention here the excellent research carried out in the Utsjoki area; here the number of reindeer is about 17,000, the area is about 4300 km^2 so that the average number of reindeer pasturing in the area is $4/km^2$. The three main problems the working group is determining are: a) grazing pressure in different areas; b) seasonal variation in pasturing; c) grading selection of different forage plants. Chemical analyses of reindeer forage plants are being made by the Polar Circle Agricultural Experiment Station of the Finnish Agricultural Centre. These analyses include dry matter, ash, organic matter, crude fibre, crude fat, crude protein, total nitrogen, sugar, P, Ca, K, Mg, Na, Cu, Mn etc. The material consists of reindeer lichen species, dwarf shrubs (*Vaccinium* sp., *Empetrum, Phyllodoce*), birch, grasses (*Deschampsia, Calamagrostis*) and also some mosses (*Pleurozium, Dicranum*) and many mushrooms. Material is studied repeatedly two or three times during summer. Furthermore, studies of the chlorophyll content of ground vegetation started in 1970. Some dwarf shrubs, mosses and lichens are included in the material. Phenological observations are made, including measurements of growth length of yearling pine, birch and dwarf shrub species. Thickness growth of tree roots was also measured. The growth length of mosses and lichens was measured as well. The aim is to discover factors controlling time and amount of growth.

At the Institute for Hunting, Game management and Freshwater fishing in Trondheim the plant food preferences of reindeer are studied measuring quality and quantity at all seasons. Samples from tame oesophagel fistulated animals, and from rumens of wild ones are utilized. Field observations show selection of certain plant communities, and studies of grazing time are used as a measure of dry matter intake. Correlation between grazing time and dry matter intake are examined by special experiments on tame reindeer. Extensive data on activity throughout the year are collected. On completion, data will be used to calculate the energy flow to the reindeer population from the different pastures they use through the year.

Also physiological conditions of the reindeer are the object of thorough analyses by Norwegians, in an IBP Grazing project. Namely, stomach samples were collected from shot reindeer. The lower jaws were also collected for determinations of age and condition, femurs for determination of condition by analysis of fat percentage, adrenal glands for possible investigation into stress/hormone production relationships, and the heart for study of possible parasites. (We know that adrenals are involved in the control of stress in general and in social stress in particular). For feeding studies by fistulation, experimental animals under constant supervision by veterinarians were used. Grazing determinations were made during the summer period where the feeding trials were carried out. Lastly behavioural studies were usually undertaken during and after hunting. The behaviour of the reindeer without human interference is of interest, and the researchers spent periods in the field at times and places other than those where the shooting and grazing trials took place.

We want to finish this chapter by mentioning the extremely interesting research the Scandinavians have done, and are doing, for IBP on native human populations, i.e. Lapps.

Some demographic research was carried out; for example, studies are being made on the history of Skolt Lapps. The scope of the investigations is to enable addition of demographic and genealogic data to biological studies on the Skolt Lapps. The following two aspects have been studied: a) examination of demography from 1925-1967 including absolute and relative population growth (natality and mortality); assimilation without migration into other ethnic groups and b) studies of genealogy, including ancestry, reproduction pattern, inbreeding patterns and detailed genealogic family trees. The Lapp enclaves are breaking up, especially the previously highly isolated Skolt communities. Intermarriages occur both between the different Lapp populations and also with Finns.

As far as genetics is concerned, blood group investigations have been performed with some 20 antisera on over 1300 Lapps in Inari. Specimens of erythrocytes, serum and plasma were frozen for future determination of red cell enzyme types, serum protein groups and for polymorphic traits, which still may be detected. A rare red cell enzyme variant was found among the Skolts. This indophenol oxidase or 'white patch' variant seems also to be relatively common among Finns. In the Lapps some other red cell variants seem to be entirely missing. The Skolts show gene frequencies which deviate considerably from those of other Lapps and other neighbouring populations. Studies of the haptoglobin subtypes have shown that the gene $Hp^{1\ F}$ has a low frequency among the Nellim Skolt Lapps. All Lapp populations so far investigated show a very low frequency of heterozygous Pi-phenotypes.

About 300 Lapps have been tested for some 20 HL-A antigens. Serum samples from child-bearing Lapp women were simultaneously tested for cytotoxic antibodies against lymphocytes. Both the frequencies of HL-A antigens and their combinations in the Lapps seem to differ from those of the surrounding populations. The Lapps have a higher frequency of rapid Isoniazid (INH) activators than the Finns. Analyses of the concentration of β-aminoisobutyric acid have shown that high excretion of this acid is frequent among Skolts. It has been stated that the methods available for quantitative investigations of this acid are not sufficiently reliable, and attempts are being made to improve the technique.

Finally, some interesting research on physiological conditions of Lapps are being made by these working groups to such an extent that is unimaginable among populations living in parts of Southern Europe.

Some observations on the diet inherent to Lapps were made when experts on nutrition visited Skolt households for five days in both summer and winter during which they studied qualitative and quantitative intake of food. In connection with the IBP/HA investigations in Inari, one-day calls were made, during which the Skolts were interviewed concerning the food they had eaten the previous day. These studies have shown that the diet of the Skolts is monotonous and heavy, and very poor in milk and vegetables. Owing to the low milk consumption, the intake of riboflavin and calcium, for instance, is insufficient. The Skolts mostly drink water with an extremely low mineral content. These findings and the results obtained by the experts on nutrition are correlated with the occurrence of caries and arterial sclerosis.

Finally, cold tolerance (which is obviously important in this biome) and work capacity have been studied by the researchers from the IBP. To study cold tolerance, the project is concerned with variation in age of maximal blood flow capacity in the skin of the hands and face, and also of some other circulatory parameters related to cold adaptation. Representative studies have been carried out on indoor and outdoor workers in Norway, and of Eskimos (North America) and Lapps. Work capacity and a number of cardio-vascular characteristics have been investigated randomly. Lapp men and women above the age of 40 years have been tested, and the variation in age, sex, and body size and composition are calculated. Work capacity is analyzed from measurements of maximal oxygen uptake and related respiratory and circulatory variables. Blood pressure and ECG at rest and in submaximal and maximal exercise have been measured, and the prevalence of hypertension and St-depressions in the exercise ECG is calculated. The fieldwork is now finished, and the results are now being evaluated and prepared for publication (Prof. J. Torgersen, University of Oslo).

BIBLIOGRAPHY

Bachmaier F., Die Zwergbirke (Betula nana L.) ein Glazialrelikt unserer Flora und Untersuchungen über ihre Insektenfauna, Jhrb. Ver. Schutz Alpenpfl. & Tiere, München, 31, 1966, 138.

Biasutti R., Il paesaggio terrestre, Torino, 1947.

Cherepanov A.I., Zhuki-ščelkuni zapadnoi Sibiri (Die Schnellkäfer West-Sibiriens) Novosib. Knizhnoe Izdateln. 1957.

—, Provoločniki zapadnoi Sibiri, Izdateln. Nauka, Moskva, 1965.

Dunaeva T. & V. Kutchernk, Etudes sur l'ecologie des vertebrés terrestres dans les tundras du Yamal méridional, Contrib. Conn. F. & Fl. URSS, Moscou, 1941 (in Russian).

Forcart L., Die nordischen Arten der Gattung Vitrina, *Arch. Moll.*, 84, 1955, 155.
Gams H., Die Herkunft der hochalpinen Moose und Flechten, Jhrb. Ver. Schutze Alpen Pfl. & Tiere, München, 25, 1960, 85.
Ghiljarov M.S., Zoologičeskii metodi diagnostiki počv, Izdat. Nauka, Moskva, 1965.
Huber B., Eine Fahrt in die Kanadische Tundra, Jhrb. Ver. Schutz Alpenpfl. & Tiere München, 27, 1962, 142.
Intern. Biological Programme, Progress report 1971, Scandinavian Countries. May 1971.
Idem, May 1972.
Kubiena W.L., Claves sistemáticas de suelos, Cons. Sup. Inv. Cientif., Madrid, 1952.
Lieth H., Modeling the primary productivity of the World, *Intecol Bull.*, 1974, 4, 11.
Loof P.A.A., Freeliving and plant parasitic Nematodes from Spitzbergen, collected by Mr. H. van Rossen, *Med. Landbouwh. Wageningen*, 71, 7, 1971.
Marcuzzi G., Il bilancio idrico nei Coleotteri Tenebrionidi, *Arch. Zool. It.*, 45, 1960, 281.
—, Rapporti tra equilibrio idrico e ambiente nei Coleotteri Tenebrionidi, Ibidem, 325.
Marini M., L'uomo e le isole. L'Islanda, isola europea, L'Universo, 36, 1956, 909.
Martinčič A., Zur Verbreitung der arktische oreophytischen Laubmoose in Jugoslawien, Jhrb. Ver. Schutze Alpenpfl. & Tiere, München, 31, 1966, 92.
Naess S.J., A.L. Steigen & T. Solhøy, Standing crop and calorific content in invertebrates from Hardangervidda Ecol. Studies. Analysis and Synthesis, Vol. 17, Fennosc. Tundra ecosystems, Part 2, edit. Wielgolaski, Springer Verlag, Berlin, Heidelberg & New York; 1975, 151.
Pöhlmann H., Oekologische Untersuchungen an Rentieren in Spitzbergen, Verh. Ges. Oekol., Wien, 1975, 89.
Schröder C., Handbuch der Entomologie, Jena, 1929 (II vol.).
Semenyak S.A., Zavisimost' migrazii nekotori'ch vidov provoločnikov ot ich choladostoikosti, Problemi počvennoi zoologii, Izdat. Nauka, Moskva, 1969, 146.
Söyrinki N., Probleme und Leistungen des finnischen Naturschutzes, Jhrb. Ver. Schutz Alpenpfl. & Tiere, München, 25, 1960, 96.
Stebaev I.V. & Stebaeva S.K., Zonal'nie spektri žiznenn'ich form nogochvostok i evoliuziia počv, Problemi počvennoi zoologii, Izdat. Nauka, Moskva, 1969.
Steigen A.L., Respiratory rates and respiratory energy loss in terrestrial invertebrates from Hardangervidda, Ecol. Studies. Analysis and Synthesis, vol. 17, Fennosc. Tundra ecosystems, Part 2, edit. Wielgolaski, Springer Verlag, Berlin, Heidelberg & New York, 1975, 122.
—, Energetics in a population of Pardosa palustris L. (Araneae, Lycosidae) on Hardangervidda, *Ibidem*, 129.
Stugren B., Grundlagen der allgemeinen Okologie, Fischer, Jena, 1972.
Terentev P.V. & Chernov S.A., Key to Amphibians and Reptiles, Jerusalem, 1965 (transl. from the Russian).
Thienemann A., Chironomus, Die Binnengewässer, XX, Stuttgart, 1954.
Zangheri P., Geobotanica, Encicl. Ital. Sci., Ist. Geogr. De Agostini, II, 1972, 374.

THE STEPPE

'... Und wie sie selbst diese Csárda so sichts auch um sie aus, Ringsherum endlose Sandbuzken[1] Im nackten Sande kommt kaum fort Ein Fliederbusch[2], Der seine schwarzen Früchte allsommerlich unlustig bringt ... bei der Csárda ist ein gelber Wald aus zwergigen Pappelbäumen[3] In mit Königsmelonen[4] bewachsenem Sand ... Dort blüht das melancholische Waisenmädchenhaar[5] und die blaue Blüte des Eselsbrodes[6]'.

A. v. Petöfi

The wetness of the soil surface, at least for a given period of the year, is the factor necessary for the development of a *Gramineae* formation, or better a grassland, in which generally only some shrubs or small scattered trees can thrive. These elements constitute immense prairies (or rather grasslands, since the former name is generally reserved for the North American formations) and steppes in places where – as we shall see later – neither a forest nor a desert can develop. The roots of the Graminoid plants constitute a compact superficial mat which continuously absorbs water, even where rainfall is scarce, water must not be absent during the development period, because the soil absorbs it very rapidly and evaporation (or in this case evapotranspiration) is generally high. In this period the temperature must be mild, but very cold winters, as well as prolonged periods of drought, do not effect the dormant phase of the vegetation.

The belt of herbaceous vegetation (i.e. grasslands and/or steppes) of the Old World extends from South Africa and Senegal to Mongolia, through Central Asia, and into Europe through South Eastern Russia. The true original prairies with a vegetation in which Graminoid plants dominate are present also in the Paris basin and in Western France (Landes), where during the warm season they cover the calcareous soil (partially with 'craie') with a thick green grassy carpet. In the forests of Western Europe also (see deciduous forest) the clearings on stony, dry, or even swampy soils, with *Gramineae, Centaurea pratensis* and *Succisa praemorsa* constitute fragments of grassland associations of natural origin.

In Southern Russia, the steppes, with a winter respite, cover immense areas. The tussocks of Graminoids do not form a continuous cover, so that here and there the soil is bare. The monotony of the landscape is interrupted only by some tall herbaceous plants, shrubs and sometimes by a group of trees. The vegetational composition varies according to the soil aridity and climate. Where the grassy cover is very thick and practically continuous, the soil is rich in organic matter and black: the two characteristic features of this major soil type are its black colour (so that even in the XVIIIth century the Russian peasants called it 'chernosem', from chornui, which means black) and the presence of a zone in which calcium carbonate forms concretions in the lower part of the soil profile. The chernosem (as well as burosem and castanosem; see later) is limited to zones

[1] dunes; [2] *Sambucus ebulus;* [3] *Populus nigra;* [4] *Tribulus terrestris;* [5] *Stipa pennata;* [6] *Echinops ruthenica.*

where the frequent rainfall keeps the soil constantly moist and where the temperature allows only very slow evaporation. The *Compositae* – also characteristic of steppes – flower only when the rainfall is abundant and the day very long (long day plants). Other elements on the contrary, for instance *Pancratium maritimum*, do grow in the most humid season, i.e. spring, and produce leaves until the period of flowering, which occurs generally in the late summer.

In Italy today, we find natural grasslands only in a few arid alpine valleys in the Val d'Aosta (Piemonte) and Val Venosta (Alto Adige or Süd Tirol). In the latter, near Lasa, there is a grassland area with *Stipa capillata, S. pennata* and *Festuca ovina*, a fragment of the original grassland present in this zone since the postglacial, with steppe, submediterranean conditions. Some insects, such as *Titubotea macropus*, correspond to these plants and according to some authors (Magistretti & Ruffo) these should also be considered as steppe elements. The karstic grasslands found today in Puglia (Southern Italy) in succession to woodland of holm oak or deciduous oak, and those found in the Italian Alps at intermediate heights (1200-1800 m) in succession to coniferous forest (mainly spruce), are very common but their origin is fairly recent and due exclusively to human activity (Fig. 204).

Ancient reclamation of areas – also in Italy – partly in freshwater, partly in brackish water, has given rise to grasslands or, very frequently to cereal fields. The latter have allowed the development of what is called by some authors 'cultural pseudo-steppes'. The hoppers which colonized these secondary 'steppes' (cf. La Greca) are very interesting.

The name pustz or pustza is reserved for all the Hungarian grassland communities, intermediate between steppe and prairie. This name probably comes from the slavic word 'pustoi' which means vacuum, or desert, or from poustina, solitude, because of the sensation that man feels in this environment. The Hungarian pustza extends across the Tibiscus basin, where it occupies very wide zones covered by a vegetation in which the Graminoid genera *Andropogon, Stipa, Bromus, Festuca* etc. dominate. It must be noticed, however, that in the pustza there is the possibility of reforestation and/or arboreal cultivation, whereas in the true steppe-prairie only herbaceous vegetation can grow as a consequence of the long winters, very short springs and very dry summers. This is quite understandable if we accept the idea of some authors, according to whom the vegetation of the pustza is a secondary one (Wendelberger, 1961) (Fig. 205).

Another example of secondary grasslands is given by meadows present on limestone in South-eastern England. These grasslands are due to very ancient deforestation and their presence has been continued because of grazing (by sheep) and the work of rabbits, which have prevented the regeneration of any form of shrubby and, in succession, arboreal vegetation. Characteristic of these secondary grasslands are the *Campanulaceae, Campanula glomerata* and *Phyteuma tenerum*, the *Gentianacea, Blackstonia perfoliata*, the *Compositae, Centaurea scabiosa* and *Hippocrepis comosa*, the *Labiatae, Oryganum vulgare* and finally the Graminoids, *Brachypodium pinnatum*, *Festuca ovina* (with violet or green flowers) and *Festuca rubra* (with violet flowers). Here and there it is possible to find a stunted shrub of box (*Buxus sempervirens*).

The climate of the steppe biome is due to the amount of rainfall, too scarce to permit the development of a forest but high relative to that which causes the formation of a desert. In Europe the values of annual rainfall go from 250 to 750 mm. Another fundamental feature of the climate is the wide annual thermal variation.

Of course, the macroclimate varies within the European steppes according to latitude, distance from the sea, form of the relief etc. The main climatic types,

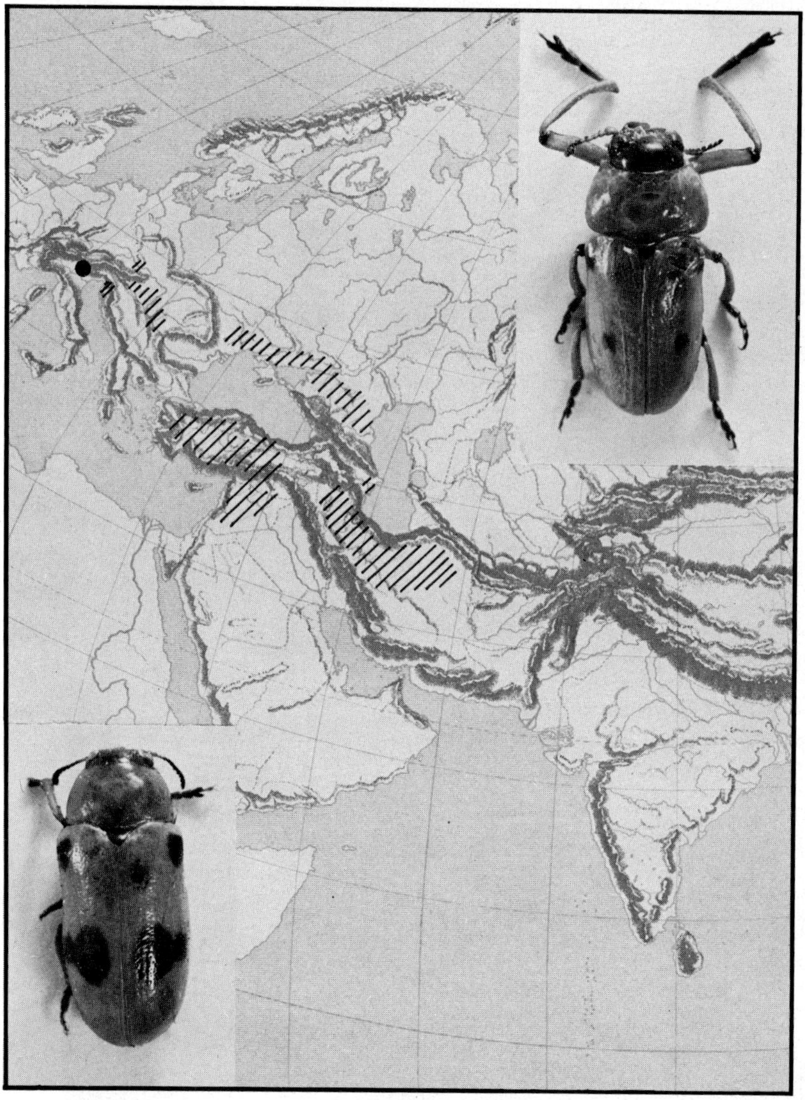

Fig. 204. Distribution of the Chrysomelid beetle *Tituboea macropus*. Top right: male, bottom left: female.

Fig. 205a. Detail of the puszta: a pit (or well) for the cattle (Photo: Dr. Kaszab).

Fig. 205b. General features of the puszta (photo Dr. Z. Kaszab).

based on the bioclimatic map of UNESCO/FAO, are the following: cold, steppe: the whole of Armenia and Azerbaidjan, the western shores of the Caspian Sea, in the south to Derbent; Southern Russia, from Odessa eastward to the Caspian Sea and the Caucasus and westwards to the mouth of the Danube and the lowest part of this river. Mediterranean, with several varieties, such as the xerothermomediterranean, along the shores of the Caspian Sea from Derbent to the mouth of the Kura; thermomediterranean, along the Kura valley; mediterranean, non-steppe, in central Crimea; mesomediterranean in Southern Bulgaria and European Turkey (to Istanbul), whereas the other parts of the Bulgarian lowlands have a submediterranean climate (deciduous forest).

Finally, the steppes of Hungary and Romania (middle lower Danube, near Bucarest) belong to the cold subaxeric climate, characterized by a period of frost and drought lasting from 2 to 4 months.

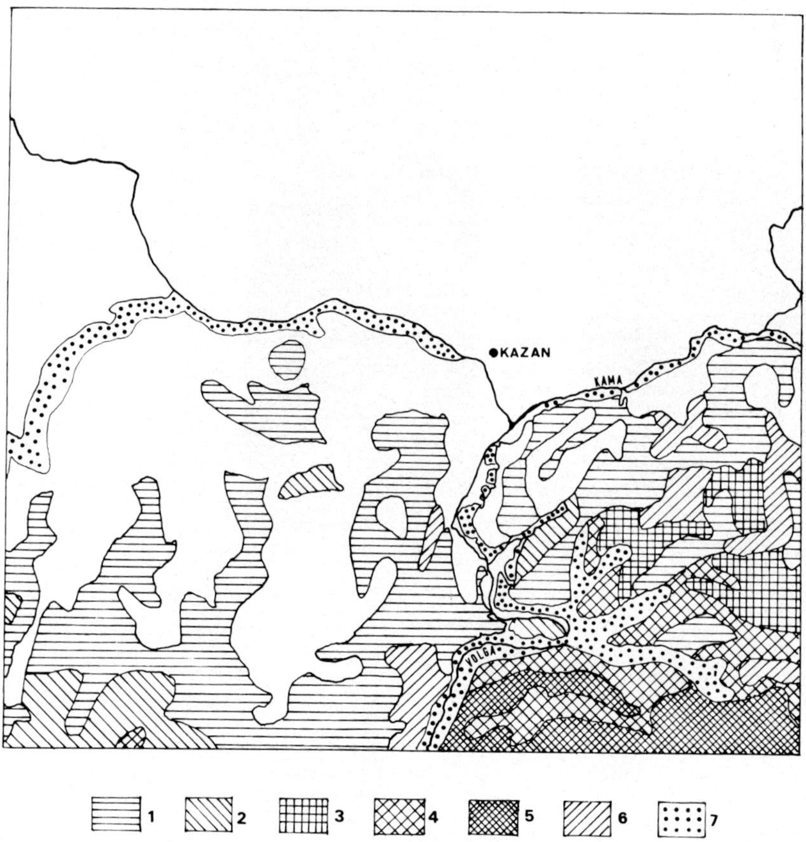

Fig. 206a. Soil types in the steppe region south of Kazan (from Aleinikova, modified). 1. leached, podzolized, chernozems; 2. typical medially humic, fertile and thick chernozem (of Middle Russia); 3. typical, fertile, medially thick chernozem (Oltre-Volga, Pre-Urals, Pre-Altai); 4. common chernozem; 5. southern chernozem; 6. chernozem with residual carbonates (ostatočno-carb. ch.); 7. alluvial soils.

The major soil types in the European steppes are chernosem, castanosem, burosem, serosem and solonchak. Only secondarily may we mention loess, syrosem, xerorendzina, brown loam and red loam (Fig. 206).

Chernosem (or chernosiom) is an AC soil, of grey, dark-grey to blackish colour, almost always rich in humus and with a deep humus layer. Superficially it is almost always decalcified.

Soil life is very intense: the earthworms contribute to the formation of the humic-clay complex; generally the faeces of the earthworms are conserved within the soil. The activity of subterranean rodents is also of paramount ecological

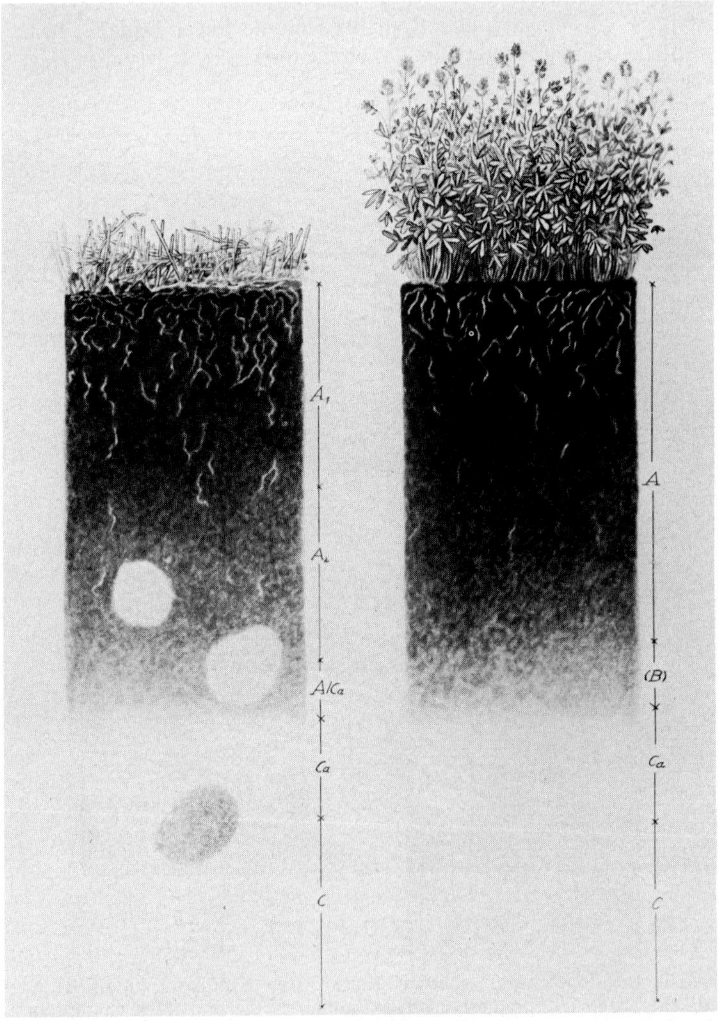

Fig. 206b. Chernozem, a typical soil of the European steppes. Left: chernozem on loess (Hangsdorf, Lower Austria). Right: degraded chernozem (Mistelbach, Lower Austria) (from Kubiena).

importance in the chernozem: we cannot forget the so-called 'krotovina', a kind of gallery made by *Spermophilus, Spalax, Marmota bobak* etc. filled with earth, where sometimes earthworms can be found. They can be easily recognized when their colour is different from that of the neighbouring soil surface. The chernozem is generally covered by a Graminoid steppe with *Festuca,* Stipa and *Koeleria,* more precisely *Festuca ovina, Stipa pennata, S. capillata, S. laessingiana, Koeleria gracilis, K. cristata* etc. In moist situations some small trees can develop (forested steppes).

Castanosem (or Kastannosem) is an AC soil of dark-grey brown colour, when humid or dark chestnut colour, whence the name ('castano' from chatain, castanea). The humus layer, generally of medium thickness, grades progressively into

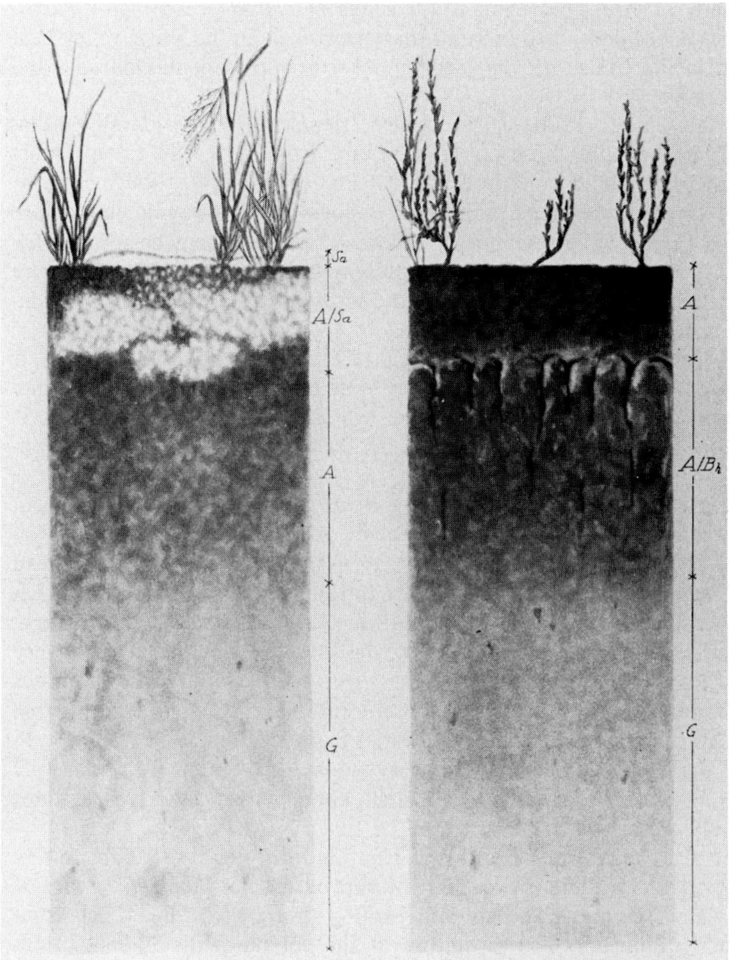

Fig. 206c. Typical soils of salty steppes. Left: Solonchak with *Atropis distans.* Right: Solonez with *Artemisia pauciflora* (from Kubiena).

the parent material (original mineral material) without any well-defined horizon in which limestone has accumulated (alluviated). It develops upon loess and loess-like sediments, rich in limestone; it is covered by a vegetation composed of *Artemisia, Festuca sulcata, Koeleria gracilis, Bromus inermis, Stipa* sp. plur., etc. Generally it shows a rich life, with the presence of earthworms. It is present mainly in South-Eastern Russia, south of the line Rostov-Saratov-Orsk, on the Crimea Peninsula, in Romania, and in part of Hungary.

Burosem or desert steppe brown soil, called by Dukutschajeff in 1889 'brown soil' (which of course has nothing to do with brown earth) is an AC soil, poor in humus, grey-brownish coloured, with a superficial humus horizon. In this, a more or less clearly formed limestone horizon always follows. It develops upon loess, loose marls and marly sands. It is typical of dry steppes, with a vegetation composed mainly of shrubs belonging to the genus *Artemisia* (*A. incana, A. austriaca, A. sieberi*). It has developed in South-Eastern Russia, in the Volga valley near Astrakhan and in the Uralsk province; furthermore it is present in Dobrudja and southern Bessarabia.

Serosem (Neustrujeff, 1900, called it also 'sierosem', which literally means grey earth) is an AC soil whose colour can vary from very light (almost pure white) to blackish, rich in limestone and with a recognizable, though very reduced, humus layer. It develops upon loess or loess-like loose sediments rich in limestone. It is typical of dry steppes; soil life is sparse. It can be utilized for agriculture only with irrigation. When it is present in an undulating region, there are all the forms of intermediate soil from true serosem to solonchak, with the intermediary of 'salty serosem'.

Solonchak – an edaphic soil type – was called after a popular Russian name; it is known also as szik or szek (Hungary), or even Zickboden, an Austrian name used in Burgenland (near the Neusiedler See, which is a brackish-water lake). It corresponds to Hilgard's white alkali soil (with the exception of the soda solonchak).

Solonchak is a soil belonging to the category of gley (alluvial, poorly drained, poorly oxygenated soils); it is alkaline or strongly alkaline, with a high concentration of soluble salts due to capillary rising and evaporation of the subterranean water, rich in salts. Some mobile iron in its ferrous (Fe^{++}) form exists, which sometimes separates as rusty spots. In spring time, when the water table is high, almost all of the soil is covered with water; the surfaces which emerge are very soon green, whereas during the summer the vegetation dries up and enormous areas are covered with salt (Sodium sulphate, chloride and carbonate) prevail, often accompanied by calcium sulphate. The pH varies from 7.2 to 8.5. We want also to mention here the soda solonchak (in Hungary 'sodaszik'), extremely alkalin (pH greater than 8.5), and the calcareous solonchak, characterized by a large amount of calcium carbonate, whose pH varies from 7.2 to 8.5.

Solonetz or black alkali soils are soils originating from slightly alkaline or neutral solonchak, in which there is a leaching of salts because of the deep descent of subterranean water, while the Sodium ion (Na^+) – strongly adsorbed by clay and humus – is not eliminated. The saturation of the colloids with Sodium brings about a strong dispersion of clay and humus, with easy mud formation, paucity of air and compact structure of the soil, which is therefore easily eroded. Life is represented almost exclusively by aquatic organisms during the period of flooding.

Finally, solod – or decoloured steppe earth – is the product of transformation of solonetz which loses its layer of clay bound to Sodium, substituted by colloidal solutions of silicic acid (as a 'sol'), together with the pre-existing sandy skeleton. In this way a soil is formed, similar to podzols, decoloured and superficially acid.

Loess syrosem (or loess raw soil) is an AC soil of light ochre, drab to light brown colour. Chemical decomposition is prevented and the soil is always sandy to silty-sandy, without any recognizable humus layer. It has developed upon loess deposits and is present in Moravia (Hanna), Czechoslovakia, Silesia, Hungary, Romania and Southern Russia. Xerorendzina is similar to serosem, but develops upon compact limestones, argillaceous schists, gypsum or marls rich in gypsum, on hills and mountains covered by an open xerophytic vegetation. It depends therefore on relief and geology.

Where brown loam is concerned, only some varieties of this major soil type are possible in the area of the steppes, verbigratia terrous brown loam, upon siliceous rocks, and earthy calcareous brown loam (= terra fusca terrosa; Kubiena) upon limestone. Secondary enrichment of the soil by limestone, gypsum or easily soluble salt brings about the formation of a superficial enrichment of the same substances.

We can also mention the decoloured brown loam present in the Paris basin (see p. 7) and in Southern Andalusia, as a relict soil, to which a relict vegetation corresponds. It would be worth while also examining the soil fauna to see whether there are also some relict animals. We must not forget that, according to Ghilarov, examination of the soil fauna sometimes permits a very accurate diagnosis of the soil.

To finish we must mention the red loam (Marbut, 1923) known also as Krasnosem (Affanasjeff, 1928) or Nyirok (popular Hungarian name). It is a very heavy, compact soil, with a complex of colloidal substances rich in siliceous acid and iron hydroxide (generally in colloidal form). The latter diffuses easily and only partially separates in an irreversible way, to give rise to red concretions, deposits and spots. The soil life is relatively poor, although sometimes some Mull is present.

The impact of man on the steppes

The steppes of the regions corresponding to Hungary, Romania and Bulgaria were inhabited by Danubians. The Danube has been a great European route of communication since very remote times, and its lowlands (which as we have seen are mainly occupied by steppes) were always inhabited, particularly during Neoeneolithic times. A Balkano-Danubian group, named after the locality Erösd – extending from Valakia and Transsylvania to Romania, Bulgaria and Tracia, as far as the Aegean Sea – is characterized by painted pottery and by small fictile feminine statues. Localities important from this point of view are Vinca (near Belgrade) and Lengyel, in Hungary. In the Upper Palaeolithic the dwellings, sunk in the soil, were elevated on perishable material (branches and skins of animals).

The steppes of Southern Russia, however, made communications between Central Europe, on one side, and Southern Siberia and Central Asia, on the other, very easy. In the Crimea some important prehistoric sites have been discovered, in which lithic industries of Clactonian and Acheulean type were represented. In the same region, and particularly in the caves of Kiik-Koba, a Mousterian industry is present, associated with skeleton remains of Neanderthalian type. In the Don lowland, important prehistoric deposits of the Upper Paleolithic have been discovered (Gagarino-Ukraine).[1] The Neolithic and Eneolithic remains are very abundant, for example the cultural aspect (or culture) of Tripolje (Kiev), or that of Fatjanovo in the Middle Volga.

The Tripolje culture spread all over the Ukraine and the Danube basin: its main characteristics are the rectangular dwellings dug in the soil (zemljanki) and the places destined for the offering and deposition of cinerary urns (ploščadki). The pottery of this era is characterized by a spiralo-meandric decoration.

Since the Eneolithic, on the steppes of the Black Sea (Kuban, etc.) the 'Kurgany' culture has occurred, named after some sepulchral tumuli in which skeletons have been found, crouching and painted red. This culture – continuing during the Bronze Age from 2000 to 600 B.C. – has been attributed to the so-called Cimmerians (or Kimmerians), a mythical people whose land was described by Homer as a region of perpetual mist and darkness. The Kurganys are present until the IIIrd century B.C., after which tombs appear with large slabs (associated with cremation) in which fibulae of Gallic (or Celtic) type and pottery with Greek influences are present.

The Celts certainly reached the Volga in the IVth century B.C., whereas the Sarmatians came in the IIIrd century.

A very advanced culture must have been present even in the V-IVth centuries, as it is documented by the style of some Kurganys (Kul-Oba, Kerč; Solocha, Dnepr, etc.).

Successively the migrations and colonizations of the area of the European steppes have been more and more frequent. The present territory of Romania, at that time inhabited by Thracians, was colonized by the Romans at the beginning of the IInd century A.D. In the IXth century the Magyars came from the East (Caucasus) and reached the steppes between the Don and the Dnieper, advancing further to occupy their present land (Hungary) in 896. The Slavs came from the East during the VI-VIIth centuries.

Thanks to its position between Europe and Asia, the Crimea has been for a long period of time a zone of transit and refuge for many people, such as the Scytes, Sarmati, Gotes and Unni. In the VIIth century, the Unni who inhabited the steppes situated between the Don and the Kuban moved westward and installed themselves in Bessarabia and Dobrudja i.e. the present Bulgaria.

[1] Very interesting are the 'Aurignacian' sites of Kamenec-Podolskij and Mezin-Černigov and the 'Solutreen' sites of Konev-Lubny. At Mezin (Ukraine) the Aurignacian hunters have left us some highly stylised fictile feminine statues, with a bird-shape. According to Maringer, these people were resident hunters. The Neolithic and Eneolithic remains are very abundant, for example the cultural aspect (or culture) of Tripolje (Kiev), or that of Fatjanova in the Middle Volga.

We may infer from all this that, as a consequence of human settlement and continuous migration, the steppe environment has been very much disturbed, as – and perhaps even more – have other South European environments (e.g. sclerophyllous forest).

Fig. 207a. Aspects of breeding in the puszta: grey oxen characteristic of the Hungarian puszta with very peculiar horns. Hortobágy National Park (Photo: Vitályos).

Fig. 207b. A rare 'race' of sheep at present living in the puszta, with highly characteristic horns (from Rakonczay).

Fig. 208. Attack of maize in Southern Russia by several insect larvae. a. Left, normal plant; right, plant attacked by *Oscinella frit* and *O. pusilla*. b. Plants attacked by three different species of insects. At right a plant attacked by *Oscinella frit* and *O. pusilla* (from Ščegoleva).

The vegetation

The floristic region extending from Hungary and Southern Russia through Asia minor, Iran and Transcaucasia, to Mongolia and Tibet, is called 'pontic', from the Latin Ponticus and the Greek ποντικός, i.e. of πόντος Εὔξινος.

It is represented by herbaceous and steppe communities, in which *Chenopodiaceae, Zygophyllaceae, Tamaricaceae* etc. prevail. The endemic species of *Astragalus* and *Salvia* are very abundant. At the beginning of the postglacial period some 400 pontic species reached Central Europe.

The European steppes belong to what plant ecologists call 'nordic prairies and steppes', typical of the internal areas of continents where a cold and dry continental climate prevails. The communities are mainly composed of xerophil *Gramineae,* characterized by a well developed root system; they must endure two principal adverse factors: the summer drought and the winter cold. They grow mainly as grassy carpets, not completely uniform nor continuous, in which several *Astragalus*, hemicryptophytes and geophytes are present. The growing season lasts some months, from the melting of snow until July; the rainfall varies from 250 to about 750 mm and is at a minimum in summer. The consequent drought is more serious because of the low atmospheric humidity and the winds which blow almost continuously, as during the summer the interior of the continent is a nearly permanent zone of cyclones or low pressure. During the summer most of the vegetation dies and the decomposition of the detritus brings about the deposition of humus, and therefore the formation of chernosems.

Analogous to the variety of climates within the European steppes there is also a remarkable variety of vegetation. Schematically only, we may recognize the following types or subtypes:

Lowland steppes of cold and temperate climates, with a dry summer, extending to the western shores of the Caspian Sea, with *Artemisia maritima, Festuca ovina, Stipa, Artemisia larscheana, Astragalus* etc. (Fig. 211).

The same is true for the Crimea, where we find *Stipa, Artemisia maritima* and

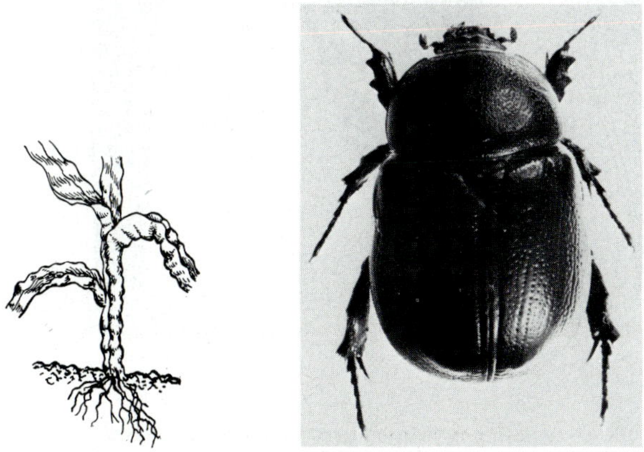

Fig. 209. Effects of the attack of the larva of the Scarab beetle *Pentodon punctatus* (right) From Servadei et al., with the permission of the Cedam editor.

A. taurica, Astragalus, Festuca ovina, and for Russia (and Kazakhistan) where *Gramineae* and *Artemisia* prevail.[1] The same vegetation is present also in the Kuma valley. As a whole the climate becomes warmer and warmer as one goes towards the Caspian Sea. In the South Eastern Caucasus we find a steppe characterized by *Artemisia meyeriana*, whereas in salty places specialized desert vegetation is present, with *Salsola dendroides, S. nodulosa* and *S. ericoides*.

High steppes, mainly arboreal, shrubby or not, in Mediterranean climate: at both the western and eastern extremities of the Crimea, characterized by *Festuca* and *Stipa*.

Shrubby or arboreal steppes and light forests, with a less dry, or Mediterranean climate: north of Kura, on the northern slope of the Caucasus.

Plateau steppes and mid-mountains, with cold climate and dry, temperate summer south of the Caucasus, such as in Armenia and Azerbaidjan; *Stipa*, a degradation of the stage of *Pinus* and *Quercus*, with relict stands of *Pinus kochiana, P. silvestris* and *Juniperus communis, Rubus idaeus, Sorbus aucuparia* and *Evonymus latifolia*.

Lowland steppes in cold climate with dry and temperate summer: all the

Fig. 210a. Attack of maize from the Lepidopteron *Pyrausta nubilalis* (from Ščegoleva).

[1] In the Eastern Crimea (Kerč) the vegetation is characterized by *Eryngium campestre, Gypsophila paniculata, Crambe tatarica, Centaurea* sp. plur., *Stipa pennata, Muscari, Artemisia absinthium* and *A. maritima*, Statice, *Carduus uncinatus, Heracleum* etc. In some salty steppes there are *Salicornia, Artemisia, Statice, Salsola, Crambe, Atropis* etc. instead (cf. Frank).

Fig. 210b. *Pyrausta nubilalis*: male-above; female-bottom. from Servadei et al., with the permission of Cedam editor.

Tibiscus basin (Hungary) as far as Belgrade, and then a wide belt along the Danube, east of Turnu Severin. The steppe is more or less continuous, with some stands of *Quercus pedunculata* on the hills (see heliophilous deciduous forest). The same vegetation is present in Romania, from the central part to Constantza: we find here agricultural land in a steppe region with *Stipa, Festuca vallesiaca* and *Artemisia austriaca*. In some zones of Hungary the steppe is replaced by edaphic formations, such as alkaline soils in a cold climate with hot and temperate summer (Szegedin) (Fig. 212).

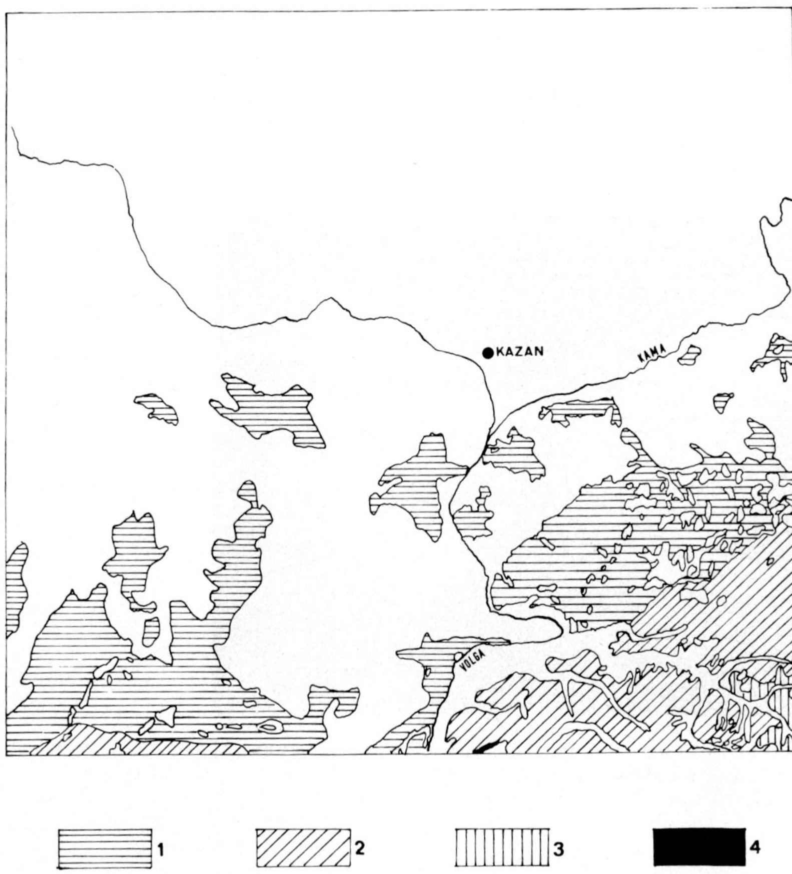

Fig. 211. Vegetation types in the steppe region south of Kazan (from Aleinikova, modified). 1. Meadow-steppes and slightly artificial meadows. 2. Steppes with characteristic and various herbs. 3. Steppes with characteristic herbs, on sandy soils. 4. Poliphytic steppe, with characteristic herbs on sand, or with weeds on mobile, bare, sands.

Fig. 212a. Typical vegetation of the puszta: *Camphorosma ovata* (Photo: Dr Z. Kaszab).

Fig. 212b. Typical vegetation of the Hungarian steppes (puszta): the Graminoid *Festuca pseudoovina* and the *Silenea Gypsophyla muralis*. The eggs of a bird are clearly visible (Photo: Dr Kaszab).

Fig. 212c. The puszta at Nagyiván: vegetation constituted by *Camphorosma ovata*. A pratincole *Glareola pratincola* is visible (Photo: Dr Kaszab).

The fauna

Mammals

One of the most interesting mammals of the European steppe is without any doubt the 'saiga', *Saiga tatarica* or steppe antelope, an animal with appearance and habits intermediate between those of the sheep and the reindeer. In the glacial episodes it lived in Central Europe and England; in the middle of the XIXth century it was present in Poland and the Carpathians and in 1865 was still living in the Kalmucks steppes between the Volga and the Don. Today its distribution in Europe is greatly restricted: some years ago it was living in the territory extending between the Manič and the Volga, whereas it had been exterminated between the Volga and the Ural. A reintroduction of this ruminant into Europe would be very auspicious. In Asia it is present south of latitude 55°N from Siberia and Turkestan to Djungaria, Mongolia and Altai, with the exception of the transcaspian region (Fig. 213).

During the winter the saiga goes south along the valleys of the Manič and the Sal, returning to its summer quarters at the end of May. It is gregarious, forming herds headed by a male or by several old individuals. It feeds mainly on halophilous plants. Its principal enemy is the eagle *Aquila chrysaetus*.

Beside ungulates, the rodents are without any doubt the mammals which best characterize animal life in the European steppes, as well as in other habitats. Let us mention first of all the genus *Spalax*, in Europe represented by species which

Fig. 213. The Saiga-antelope *Saiga tatarica*, a rare mammal of the European steppes (from Lydekker).

can be considered as descendants of some oriental species, such as *S. kirgisorum* (Kirgisi steppe). The asiatic element has furthermore sent descendants to North Africa (*Spalax aegyptiacus*) through species living in Syria (*S. beritensis*) and Palestine (*S. ehrenbergi*). As a matter of fact, this genus exhibits a typical Irano-turanian distribution. The shape of a *Spalax* is perfectly adapted to its habitat, so that we may speak of a true life-form (Lebensform): a cylindrical body without a neck, with a reduced tail and with the hairs bent forwards, different to all other mammals. They burrow with the head, which is broader than the back, differing from all the other geophilous rodents and the mole, which utilize anterior limbs. Where the vegetation is richer and thicker, their burrows are recognizable on the surface because of the mounds of earth which resemble those produced by the mole in other regions. Sometimes they can be very harmful to fields, gardens and kitchen gardens, but they can be easily controlled by means of traps.

In the European steppes live two species of *Spalax: S. leucodon* (= *dolbrogeae, hungaricus, graecus*) inhabiting all of Romania, part of Bulgaria, eastern Hungary (i.e. the area of the pustza) and Greece (exclusively Atthica), and *S. micropthalmus*, living in Southern Russia (Fig. 214). When we come to consider the distribution of the two species, we notice that they are completely discontinuous and very limited, which makes us think of a probable former geographical speciation and successive reduction of the area of distribution, perhaps as a consequence of some climatic change.

The genus *Spalax* exhibits another feature which may have ecological interest, for if we examine the genus as a whole, we notice that from North Africa to Hungary there is a progressive increase in the size of the various species or local races.

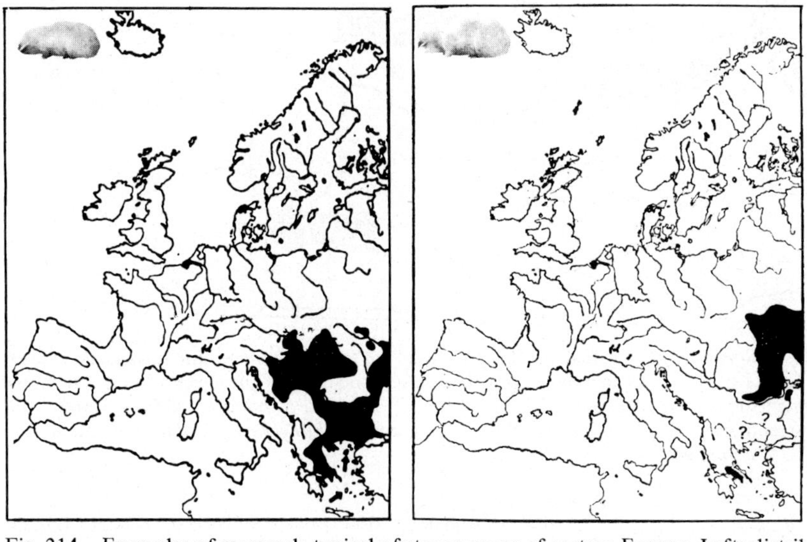

Fig. 214. Examples of mammals typical of steppe zones of eastern Europe. Left: distribution of *Spalax leucodon* (inset, top). Right: distribution of *Spalax microphthalmus* (inset, top).

This fact could be interpreted on the basis of Bergmann's ecological rule, according to which animals living in colder regions are bigger compared to those living in warmer regions (greater size means of course a reduced surface/volume ratio, which implies a reduction in heat dissipation) (Fig. 215).

The rodents belonging to the genus *Spalax* accumulate food in their burrows, consisting mainly of bulbs of hyacinths, *Asphodelus*, and other geophytes.

On the shores of the Volga, the so called 'Säkildäk turkan' or *Lagomys pusillus* lives, similar to a small rabbit, with short, rounded ears and a short tail, externally invisible. It accumulates huge quantities of hay for the winter, as high as 2 meters. Other rodents are the European souslik, *Citellus citellus*[1], digging dens in fields where it accumulates food and hibernates in a dormant state. It eats seeds, roots, fruits and rarely, animal substance. In some places such as in Tauria (Southern Russia) it is so abundant that it damages all kinds of vegetation: Brehm says that,

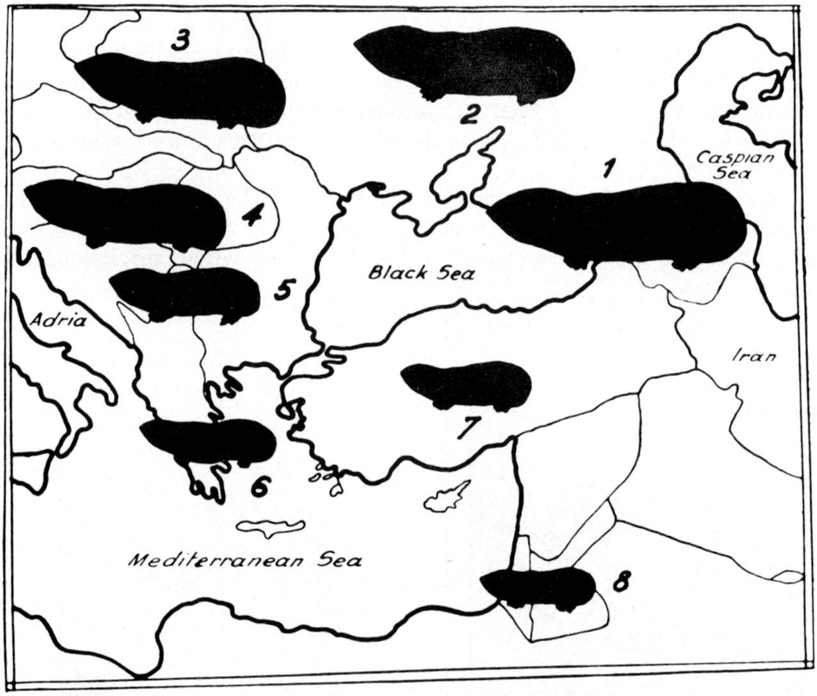

Fig. 215. *Spalax* grows gradually smaller towards the south. The Palestine form is the smallest known. The Irano-Turanian type of distribution is recognizable. The species can be attributed only with some doubts to the following elements: 1 and 2. *S. microphtalmus* (= *polonicus, typhlus* and *graecus* p.p.); 3,4,5, and 6. *S. leucodon* (= *hungaricus, dolbrogeae* and *graecus* p.p.); 8. *S. ehrenbergi.*

[1] In Russian suslik, in Polish susel, in Bohemian sisel; from this word the Eastern German Ziesel should derive, or, as Albertus Magnus wrote, Zizel, whence the latin name 'citellus (Von Martens in Brehm).

according to Fr. Falz-Fein, in Askania Nova (Southern Ukraine) each year a culling of 150,000 individuals of this small rodent was carried out.

In Southern Russia and the Caucasus two other steppe rodents are present, of Asian origin, *Citellus fulvus* and *Citellus rufescens* (Fig. 216).

The hamster or *Cricetus cricetus* is a rodent originating also in Asia, and reaching as far west as Belgium. Its spread depends on soil characteristics, being restricted to moderately solid, dry and fertile soils (loess, loam). It is present in Wurttemberg, Svevian Alb (where it reaches an altitude of 625 m), Central Germany, Brandenburg, Rhenania. A few individuals have been found in Bavaria and Necklenburg. At the end of the last century it was living also on Vosges; as a fossil it is known from Central France. Today it is kept in the home in several parts of Europe as a pet for children: it would perhaps be wise to call attention to its ability to transmit infectious diseases such as plague, since it is often infested with fleas. (We must remember in this connection that another species of *Cricetus*, *C. frumentarius*, is the reservoir of the agent of tularemia). The hamster feeds on all kinds of herbs, roots, and also on animal substances. It has a short dormancy period, and when it awakes it must eat immediately, so that it must store enormous amounts of grain: in a single den it is possible to find 50 kg of wheat! It is therefore extremely harmful; in 1817 in the Gotha lowland 111,817 individuals were captured, and in the period 1818-28 129,754. It is one of the few rodents (and maybe also mammals) which exhibits solitary habits, since the male and the

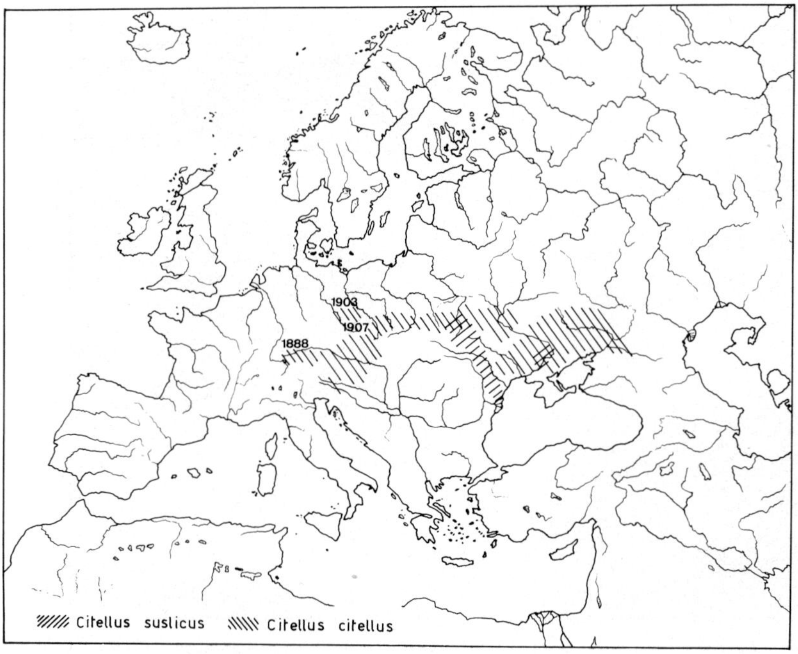

Fig. 216. The distribution in Europe of two steppe rodents of the genus *Citellus* or 'souslik'.

female unite exclusively for reproduction. As a consequence, in this animal aggression is much more developed than friendly behaviour.

An animal living perhaps more in an arboreal than in a steppe environment is the Bobak marmot, *Marmota bobak*, present from Asia to Poland, Galitia and Bukovina. In former times it probably also lived in France (Scharff). It can be considered as a true lowland marmot (its relative *M. marmota* will be dealt with in the chapter on mountains). Finally in Southern Russia we find two species of *Sicista, S. loriger* and *S. trigona*, beside the steppe squirrel *Sciurus anomalus*.

Very characteristic of the steppes is the 'Mull Lemming', as German authors call it, *Ellobius talpinus*, which spread out from the Crimea through the northern slopes of the Caucasus to Altai (Fig. 217). It feeds on peanuts, tubers of *Lathyrus tuberosus* and *Phlomis tuberosa*, bulbs of tulips etc. Occasionally it undertakes migrations; for example from Nečinka (Kazakhstan) to the Ural valley or from the Kazgalka valley to the hills of Obštii Syrt.

The Siberian jumping rabbit or alakdaga, *Alactaga saliens*, in contrast occupies the steppes between the Danube and the Don, Southern Russia (Cherson), the Crimea and the northern Caucasus. In Asia it is present to Mongolia, south of latitude 52°N and as a fossil it is known from West Germany. Its Russian name is semljanoi-salz (literally 'earth hare'), in Jaik they call it tuškančik (i.e. 'small hare'), Mongols and Buryäts call it alakdaga; the Kalmucks Morin-Ialma (or 'Horse jumper', whence the German 'Pferdespringer'); finally, the Tartars use the

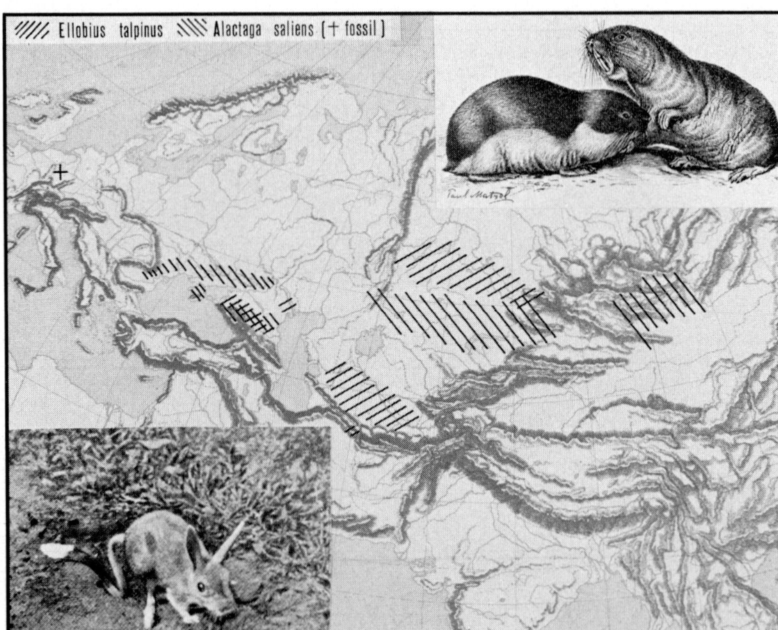

Fig. 217. Distribution of two rodent species living in the European steppes, *Ellobius talpinus* (top right) and *Alactaga saliens* (bottom left) (Animals from Brehm).

name tya-Ielman which means 'camel-hare'. It feeds on all kinds of herbs and all parts of them, but seems to prefer bulbs, like those of *Gagea lutea*.

All the other rodents are common species, present all over Central and Southern Europe, with the possible exception of *Apodemus agrarius*, widely distributed in the European steppes.

The horses are old inhabitants of the steppes of the south-eastern parts of Europe. *Equus caballus ferus* or *E. przewalski* is living in the wild in Central Asia (cf. p. 399); from it, according to some authors, the present Mongol horse and some of the present Russian breeds have originated. *E. caballus gmelini*, or 'tarpan', was wild in Southern Russia until 1850. It has given rise to the present Arabo-Persian race, the Landschläge race of the Balkans and some of the northern ponies (Highland Pony). Other races of *Equus caballus*, from which perhaps several present-day 'races' of horses have arisen were living in the past in European biomes other than the steppe.

An animal still frequent today, though not at all exclusive to the steppes, is the wolf *Canis lupus*, once spread all through Central Europe but today present only in Elsass-Lothringen, Rhenania and Eastern Prussia. In 1904 an individual entered Lausitz (Sudetenland). In Italy it is still present in the Appennines.

The jackal, *Canis aureus*, can be found on the steppes of Hungary and Bulgaria. All the other Carnivores are common species, present throughout Southern and Central Europe.

The insectivores and bats are also common species, with the exception of *Myogale moschata* (Fig. 218).

Myogale moschata or 'wychuchol' or also 'desman' (local names) is present in South-Eastern Europe, along the valleys of the Volga and the Don, and isolated in Central Asia (Buchara). This distribution points to an extremely ancient origin of the species as well as of the genus, which is present only in Eastern Europe (+

Fig. 218. The 'desman' or *Myogale moschata* (from Brehm).

Asia) and the Pyrenees (*Myogale pyrenaica*; cf. chapter on the montane biome). Both species and genus are an example of a ancient, relict, distribution. The genus is furthermore one of the most striking examples of discontinuous distribution.

The wychuchol is limited to water, like, although more so than, the insectivore genus *Neomys*. It resembles the otter in its habits.

We have seen that most of steppe mammals are much more widespread in Asia than in Europe. During the Pleistocene period many mammals came to Europe across the eastern plains, though not all of them necessarily had steppe habits. In Great Britain, more precisely in the deposits of Cromer Forest Bed, the following species are present, all of them according to Scharff coming from Siberia:

Alopex lagopus
Gulo gulo
**Mustela erminea*
**Mustela putorius*
**Mustela vulgaris* (= *nivalis*)
**Sorex vulgaris* (= *araneus*)
**Lagomys pusillus*
Castor fiber
Spermophilus eversmanni
Spermophilus erythrogenoides
Cricetus songarus
Lemmus lemmus
Cuniculus torquatus
**Mus minutus*
**Arvicola agrestis*
Arvicola amphibius
**Arvicola arvalis*
**Arvicola glareolus*
Arvicola gregalis
Arvicola ratticeps
Equus caballus
**Saiga tatarica*
Ovibos moschatus
Alces latifrons
Alces alces
Rangifer tarandus

The species indicated with a * are present today on the European steppes. *Equus caballus* was present until the last century.

We must not forget that in Siberia in that era there was a steppe with transitional communities to forested tundra in the northernmost parts (cf. Budel & Wolstedt or even Frenzel & Beug).

The migrations from Siberia towards Central Europe took place during the Pleistocene which, according to Scharff, were possibly due to lack of sufficient food after a few years of great increase in numbers of any particular species. 'It is not known – Scharff says – to what we owe the periodic visits of the Central Asiatic sand-grouse *Syrrhaptes paradoxus*, but it is certain that immense flocks of these birds invade Europe from time to time in the present day, just as those mammals may have done in past ages'.

Among mammals we find some examples of convergence in their evolution – sometimes completely independent one from another – which brought about an ecological equivalence, clearly visible in the following scheme (see page 313), in which the steppe mammals of the Old World have been considered (inspired by Tischler, 1955).

Birds

Not many birds are exclusive to the steppe environment: more often they are restricted to this environment as a consequence of their feeding habits, and so we can speak of elements which prefer the steppe habitat. For all the other species we can speak of common elements, often migrants, or even residents, widely distrib-

Life-form	Europe	Asia	Africa
zoophage subterranean digging anim.	*Talpa*	*Talpa*	*Chrysochloris* *Talpa*
Phytophage subterr. digging animals	*Citellus* *Cricetus*	*Citellus* *Cricetus*	*Georhynchus*
Phytophage cursorial (running) animals	*Saiga tatarica*	*Saiga tatarica* *Gazella subguttu-rosa* *Cervus* sp. *Equus* sp. plur.	*Gazella* sp. plur. antelopes *Connochaetes gnu* zebras
Zoophage cursorial (running) animals	*Canis lupus* *Canis aureus*	*Canis corsac* *Felis manul* *Acinonyx jubata*	*Lycaon* *Felis leo*
Phytophage jumping animals	*Alactaga*	*Alactaga*	*Pedetes* *Antidorcas*

uted all through the Southern Europe or more often throughout the Palearctic Region.

Beside the widely distributed *Aquila chrysaetus*, we may mention *Aquila nipalensis*, only occasionally coming from Asia, and very similar if not conspecific with *A. rapax* (= *naevioides*). The latter, known as the tawny eagle, is widespread in Asia, Africa and secondarily in Europe. In Europe it nests in Romania, and its large nests is built on the ground, generally on small earth mounds. The imperial eagle, *Aquila heliaca*, is widespread in south-eastern Europe, from Vienna to Orenburg, Cyprus, West Asia and Central Asia to Mongolia and China. During the winter it can be found in Egypt and Arabia. It builds its nest in trees.

The spotted eagle, *Aquila clanga*, can nest both in trees or, on the steppe, in small shrubs. It is present all through Central and south-eastern Europe, the Caucasus, and Asia minor; during the winter it can be found in East Africa. More typical of the steppe environment is the pallid harrier, *Circus macrourus*, present also in very dry environments (for instance in North-Eastern Iran, in a subdesertic habitat). It is a migratory bird, distributed in a wide belt going from Northern Europe and Asia to South Africa. The buzzard, *Buteo buteo* is very indifferent to environment (euryoikous species) and can also nest on the bare soil. Much more restricted to steppes is the longlegged buzzard, *Buteo rufinus*, also present in very dry places; it nests on the ground, only occasionally in trees. It can breed in Greece.

An interesting prey bird is the Saker Falcon or *Falco cherrug* living on open plains, semi-deserts and deserts. It nests in big trees but occasionally also on rocks. It breeds in Eastern Europe from Vienna to Russia and south-western Asia. In Eastern Europe it is used in falconry. It is partially migratory; in winter it may be found in the Mediterranean Basin, North-Eastern Africa and India.

Big birds with granivorous or at least phytophagous habits are rather common on the steppes. We can mention the red-breasted goose, *Branta ruficollis*, in winter generally on the steppes, near the sea shore. It nests in colonies on the tundra of Yalmal Peninsula and Western Siberia near the sea and over-winters in south-western Asia, southern Russia and sometimes in the Mediterranean Basin (Egypt). Exceptionally some individuals can over-winter on the steppes of Hungary.

It feeds on plant substances, marine algae, arctic herbaceous plants, small fishes and marine insects. It can also live in captivity. *Casarca ferruginea*, or the ruddy shelduck, is common during the winter on the sea shore (sandy shore) and along rivers, in fields and on steppes. It makes its nest in holes in dunes and rocks, in old trees or walls. It is a migratory bird, nesting in the Balkans, the Danube valley, Southern Russia, Asia as far east as China. It feeds on aquatic insects, mollusks and crustaceans, plant remains and occasionally small fishes. *Otis tarda tarda,* or the great bustard, nests on open plains, steppes, fields (for instance corn or turnip fields), and is often associated with *Tetrax tetrax* or the little bustard. It feeds on plants and insects, more rarely worms, toads and mice. Its area of distribution includes a great part of Eastern Europe and south-western Europe, west Asia and north-west Africa. It is generally a resident bird. Another interesting species of the steppes of south-eastern Europe (the lower Volga) is the *Otididae Chlamydotis undulata*, or Houbara bustard, typical of the desert regions of Africa and Asia, from the Canary Islands through the desert zone to Saissan Nor (north of Tarbagatai). It can be found in Europe in corn fields and is more or less polyphagous like the above mentioned species.

Much more characteristic of the steppe environment are species of the genus *Pterocles,* as well as *Syrrhaptes,* known as sand grouse. *Pterocles orientalis*, or the black bellied sand grouse, is a bird living in semi-deserts and stony regions; it is a resident bird, nesting on the ground and spread from the Canary Islands through North-West Africa and South-West Asia to Afghanistan. It has been seen in Spain, Portugal, Italy and Malta and just once in Germany. *P. alchata*, or the pin-tailed sand grouse, also a resident bird living in dry plains, high stony plateaux, sunny mud surfaces and also at the edge of swamps, has been observed in Southern Spain, Portugal, Italy, Sicily and Malta. Both species feed on plants (seeds, grain, buds) and more rarely on insects. They live also in captivity. The most interesting *Pteroclidae,* and maybe the most interesting bird of the European steppes, is without any doubt Pallas' sand grouse, *Syrrhaptes paradoxus*, characterized by very long, pointed, middle feathers of the tail, which make it a unique animal (Fig. 219). It thrives in sandy semi-deserts, where it makes its nest on the ground. It feeds on plants such as *Agriophyllum gobicum* (in the Gobi desert), seeds of *Salsola,* buds of *Salicornia,* etc. It is a resident bird in south-eastern Russia (Astrakhan and Ufa), the southern Transcaspian region and then in Central Asia to Mongolia, but it may migrate to the British Isles or Spain, as far north as Arkhangelsk, Norway or the Faeroes, or in Asia to China (Pe-chi-li). Its most remarkable feature is its habit – maybe unique among all birds – of performing invasions from time to time to Western and Central Europe. A great eruption took place in 1888 when many birds reached the extreme west of Ireland in May and June of that year following which some individuals bred in England. In 1863 a similar migration also occurred, though not perhaps on quite such a vast scale. More recently, 1908 saw the last great invasion; in this year the animal made its appearance in several Italian localities, such as Padova, Emilia, Lazio and Barletta, which would be the southernmost part of Italy in which the bird has ever been found. It is said that it bred in Italy in 1888 (Padova).

The family *Pteroclidae*, composed only of *Pterocles* and *Syrrhaptes*, and belonging to the Columbiformes or pigeon-like birds, is present today in Africa, Madagascar, southern Europe and Central and south-western Asia, as far east as India, mostly in deserts and plains. It is known from the European Oligocene.

On steppes in the Eastern Crimea (Kerč) the following species have been observed by Frank: *Otis tarda* (great bustard), *Melanocorypha leucoptera* (white winged lark), and *M. calandra* (calandra lark), *Calandrella brachydactyla* (short-toed lark), *Lanius minor* (lesser grey shrike), *Anthropoides virgo* (demoiselle crane), *Coturnix coturnix* (quail).

The following table gives an idea of the date of arrival in the Crimea of some migrants on their spring passage:

Species	Egypt	Cyprus	Smyrna	Crimea	East. Prussia
Emberiza hortulana	12-3	–	5-4	21-4	3-5
Calandrella brachydact. . . .	–	3-3	4-4	23-4	–
Anthus campestris	end of II	–	4-4	20-4	11-5
Motacilla flava	15-3	20-3	–	19-4	22-4
Motacilla feldegg	11-3	–	–	22-4	–
Muscicapa striata	13-4	11-4	–	3-5	8-5
Muscicapa hypoleuca	28-3	3-4	–	19-4	27-4
Muscicapa albicollis	30-3	end of III	31-3	19-4	–
Sylvia communis	21-3	26-3	22-3	19-4	6-5
Oenanthe oenanthe	3-3	middle of III	15-3	1-4	8-4
Saxicola rubetra	end of III	2-4	–	4-4	29-4
Phoenicurus p. phoenicurus	15-3	26-3	–	11-4	28-4
Luscinia megarhynchos	29-3	21-3	2-4	19-4	–
Hirundo rustica	4-3	21-2	2-3	12-4	27-4
Delichon urbica	middle of II	5-2	29-2	4-4	30-4
Micropus apus	–	3-3	–	7-5	10-5
Micropus melba	–	20-2	17-3	30-4	–
Upupa epops.	1-3	10-3	15-3	2-4	–
Merops apiaster	12-3	1-4	12-4	3-5	–
Coracias garrulus	–	2-4	14-4	7-5	–
Falco naumanni	1-3	–	19-3	19-4	–

A more complete picture of the phenology of the spring migrations in the eastern Crimea is shown in Fig. 220.

We also want to mention the species of birds which in Serbia are restricted to the so-called 'Kultursteppe', which – as we have already seen – has some similarities to the original steppe, including the micro-climate. The following species have been observed by Matvejev in Serbia: *Alauda arvensis, Galerida cristata, Emberiza calandra, Passer domesticus, Coturnix coturnix, Anthus campestris, Saxicola rubetra, Perdix perdix, Otis tarda, Corvus frugilegus, Falco vespertinus, Falco tinnunculus, Streptopelia turtur, Sylvia communis, Lanius collurio, Lanius minor* and *Emberiza hortulana.*

Beside all these species, more or less characteristic of the European steppes, we should finally mention some common birds whose presence in south-eastern Europe is more or less constant, since they are widely distributed throughout the rather arid districts of southern Europe and south-western Asia: *Sturnus vulgaris, Corvus corax corax, Corvus corone, Corvus cornix cornix, Corvus frugilegus, Pica pica, Circus c. cyaneus, Falco subbuteo, Falco naumanni, Accipiter gentilis, Falco biarmicus, Circaetus gallicus, Falco columbarius, Gypaetus barbatus, Neophron percnopterus, Gyps fulvus, Aegypius monachus, Asio o. otus, Bubo bubo, Athene noctua, Coturnix c. coturnix, Alectoris graeca, Alectoris barbara, Phasianus colchious, Caprimulgus europaeus, Merops apiaster, Alcedo atthis, Coracia garrulus, Carduelis carduelis, Lanius excubitor* and *Pastor roseus*.

Reptiles

Reptiles represent a very remarkable element in the biological equilibrium and in the trophic relationships (and the associated energy flow) of such a simple ecosystem as the steppe (simple if we compare it with the other ecosystem present in Europe, or more generally in Eurasia). In some extreme cases, for instance when the steppe borders the desert biome, the relationships between the steppe organisms (or biota) are as simple as those of the desert organisms, all the matter (and energy) flow is then contributed to by herbs (e.g. *Artemisia,* a typical genus of both the prairies of the New World and the steppes of the Old World) – insects (more particularly Tenebrionids or 'black beetles', easily recognizable by their short, plump body, dark – often black-coloured and slow movements) – lizards and

Fig. 219. The Central Asiatic sandgrouse, *Syrrhaptes paradoxus*, also living in the European steppes (from Scharff).

snakes. If we wish we may widen the chain a little to include also birds (granivorous, insectivorous and prey birds, which will feed on one or sometimes several links of the chain) and small burrowing rodents. Where the steppe flourishes in milder climate we find ungulates, carnivores (as big as the wolf) and, here and there, some human settlements with the associated domesticated animals (sheep, goats, donkeys and dromedaries).

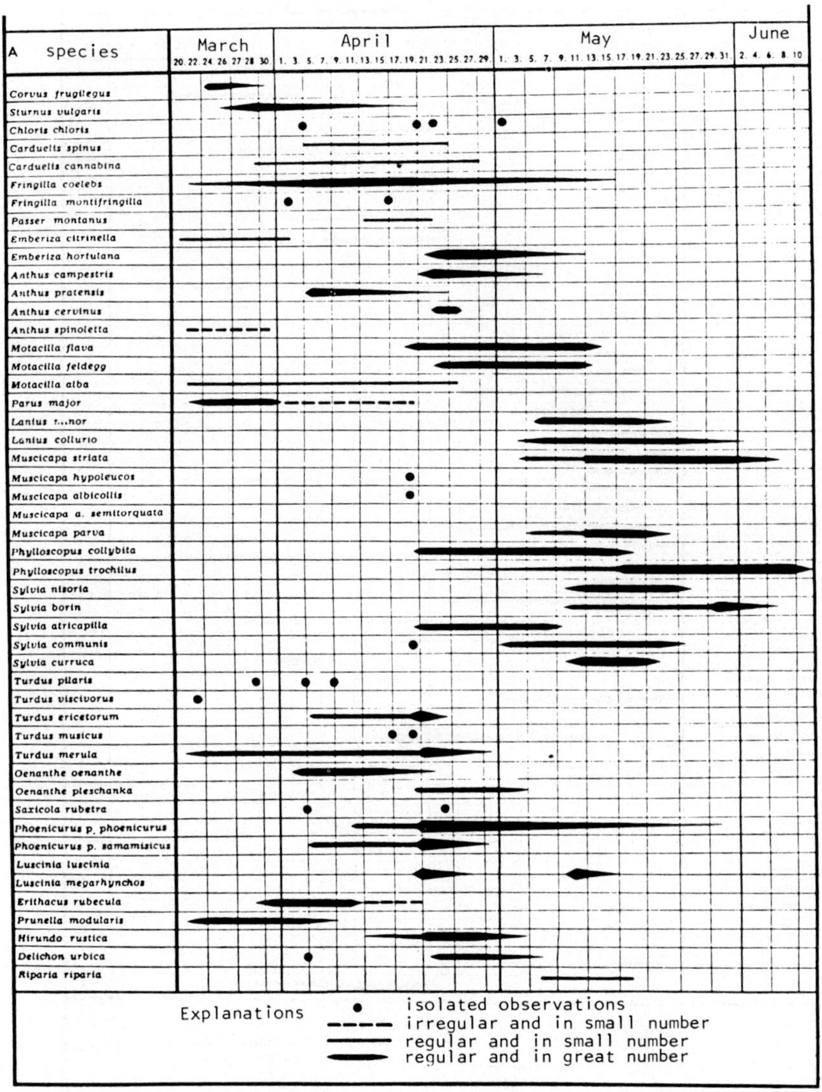

Fig. 220. a. Graphic representation of spring migration of Passeriforme birds in the Opuk-Kertsch region.

But let us go back to reptiles. Some snakes of the European steppes are quite harmless, for instance *Coluber jugularis* (= *gemonensis* according to some authors), with its vivid colours. It lives in the region from Albania, Jugoslavia, and Hungary to the Ural River and Iran and reaches an altitude of 1500 m in the mountains. It is a diurnal animal, feeding on small mammals, birds, lizards and snakes, but

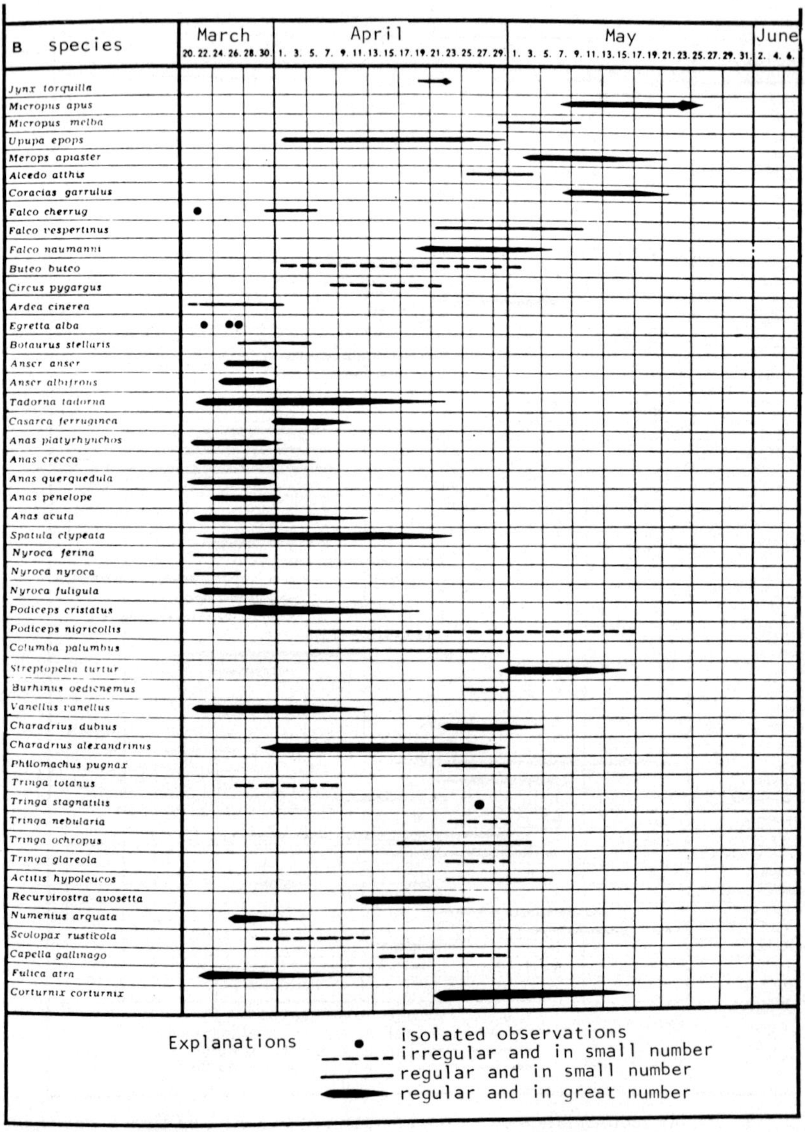

Fig. 220b. Graphic representation of non-Passeriforme birds (from Frank).

sometimes also on hoppers (generally *Acrididae* or 'locusts') and amphibians. It has fast movements and is aggressive, like its western relative, *Coluber viridiflavus*. In the spring it is the favoured food of storks. Not very different is *Coluber najadum* (= *Zamenis dahli*), grey-yellowish or grey-green, with some black and white spots. It prefers humid places as compared with its related species, living among herbs and near springs. It is found from the Adriatic Sea (Istria) through Asia minor, Syria and Palestine to SW Turkmenia and North Iran. It feeds on lizards (*Eremias, Ophisops, Lacerta*) and also, on small rodents and insects. *Coluber ravergieri* (= *nummifer*) present from Egypt to Turkestan, Afghanistan and India is limited in Europe to the extreme south-eastern sector, i.e. the shores of the Caspian Sea, such as Daghestan, near Georgia and Armenia. Its colour and the shape of the head resemble a little those of a poisonous snake, *Vipera xanthina*, with which it sometimes lives. It prefers semi-desert clayey or stoney localities, stoney canyons, hills composed of loess; it feeds on small vertebrates like the already described species. Sometimes it enters houses, hunting mice, but it is not aggressive. *Contia*, which includes small snakes no longer than 60 cm, with the head scarcely differentiated from the body, and with a rounded snout, is a genus characteristic of the steppes of the south-easternmost part of Europe. One example is *C. collaris*, present in Georgia, Armenia, Azerbaidjan, Daghestan, NE Turkey and northern Iran, where it lives in semi-desert, stoney places up to 1600 m lying under stones. It feeds on insects and spiders which are very common in these environments. Another example is *Contia punctatolineata* (Fig. 221).

Eryx and *Boidae* genus; it is relatively tropical, but is also present in Europe limited to the steppe environment. *E. jaculus* is the best known species, present in the south-eastern part of the Mediterranean Basin, more precisely Southern Jugoslavia, the islands of the Eastern Mediterranean, Transcaucasia and Daghestan, on the western shores of the Caspian Sea. It goes eastwards to Iran (Khorasan). It has a subcylindrical body, with the head not differentiated from the body, a rounded snout and a typical sand coloration, which can be considered cryptic. Its maximum length is 75 cm. It thrives in clayey or sandy places with a steppe or semi-desert vegetation. In sandy soils it moves very quickly, practically 'swimming' in the loose sand, so that it can escape its enemies very easily. If feeds on small rodents after having strangled them in the spirals of its body, which is a peculiar feature of the whole family *Boidae* (Fig. 222).

A very poisonous steppe snake which can scarcely be considered European is the *Crotalidae Ancistrodon halys*, which reaches its western limit of distribution in Europe, at the Volga mouth in the north, and in south-eastern Azerbaidjan in the south. It belongs to a genus widely distributed in Northern and Central America (the related species *A. piscivorus* is known as the 'mocassin snake') and in Asia. It belongs to the family which includes the rattlesnake, confined to tropical and subtropical America.[1] Our species hunts mainly after sunset and feeds on rodents

[1] The venom of *Ancistrodon* includes the so-called crotalotoxin or crotoxin, a crystalline proteic substance which acts as an enzyme on the phosphatides of the nervous system and the erythrocytes (or red blood cells). Specifically it acts on lecithine, from which – after the elimination of a fatty acid radical – lysolecithine is liberated, a substance with very great haemolytic activity.

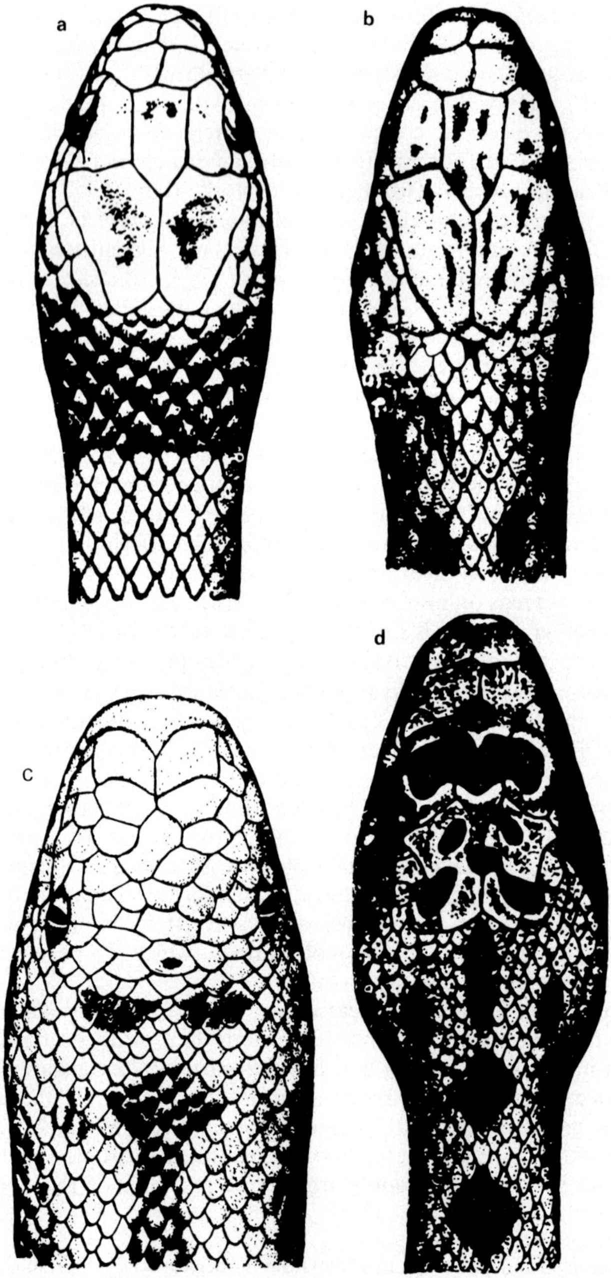

Fig. 221. Heads of some steppe snakes (non-poisonous). a. *Contia collaris*. b. *C. punctatolineata*. c. *Eryx jaculus*. d. *Coluber ravergieri* (from Terentev & Chernov).

as well as lizards, small birds, snakes and insects. People bitten by this reptile survive, but there have been several cases of mortality among domestic animals, particularly horses, so that in some parts of Kazakhstan this animal causes serious damage to horse breeders (Fig. 223).

Another poisonous snake common in the European steppes is *Vipera ursini*, more precisely the race *ursini renardi*, extending from Bulgaria, Romania and southern Russia as far east as Dzungaria and Mongolian Altai (Fig. 224).

This and other poisonous snakes are often a serious danger to man and domesticated animals living in the steppes. In Turkmenistan, according to Terentev & Chernov (1949), the number of snake bites is 0.1-0.3% of hospital patients; in some areas during some years this figure may rise to 0.7%. The majority of bites occur from May to August (92%) in people associated with in the fields (72%). In Turkmenistan the mortality of the victims is 12%.

Much more abundant – both in number of species and still more in number of individuals – are the lizards. This is easily understandable, if we reason in terms of energy flow, and think that we need a great number of lizards to feed a single snake. The *Eremias* belong to one of the most characteristic genera of the steppe environment: they are not very dissimilar from our common lizards, from which they can easily be distinguished by the presence of a fringe of small scales on the lower surface of the digits. This character is generally interpreted as a device by which the animal can run on sand without sinking, not dissimilar from what is assumed for some desert beetles of the genus *Pimelia*. *Eremias velox* is present in South-eastern Russia, Transcaucasia, the shores of the Caspian Sea etc. It reaches as far as Kansu. It lives in clayish steppes, where it hides in the fissures in the soil, sometimes also in grassy places, where the herbs stabilise the dunes with their roots. *Eremias arguta* is also widely distributed, from Southern Russia, Transcaucasia and the Transcaspian region to Dzungaria and Koshgaria. It lives in

Fig. 222. *Eryx jaculus* in its environment (from Brehm).

Fig. 223. The poisonous steppe snake *Ancistrodon halys* (from Brehm).

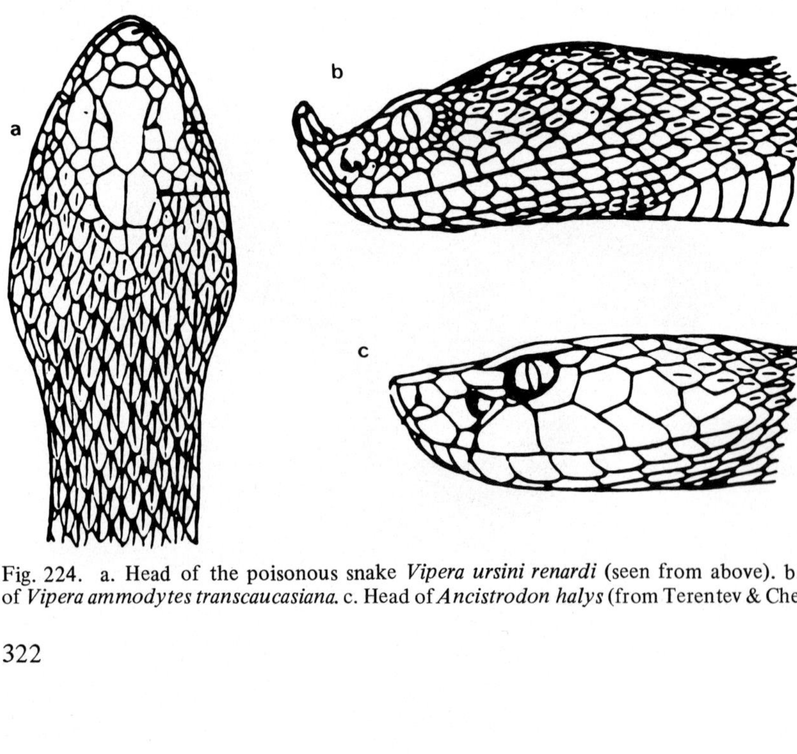

Fig. 224. a. Head of the poisonous snake *Vipera ursini renardi* (seen from above). b. Head of *Vipera ammodytes transcaucasiana.* c. Head of *Ancistrodon halys* (from Terentev & Chernov).

clayish deserts, steppes with *Artemisia*, etc. Both species feed mainly on Tenebrionid beetles and hoppers. *Mabuia* is one of the few tropical genera which have entered – though only just scarcely – Europe, since one species, *M. septemtaeniata,* can be found in the extreme south-eastern portion of Transcaucasia, at the very border with Asia, among steppe vegetation with *Tamarix* and *Haloxylon.* The genus *Ablepharus* exhibits a very interesting adaptation to fossorial habits and to an almost exclusively subterranean life, i.e. the lack of eyelids, replaced by a transparent, fixed circle which covers the eye – a sort of secondary or supplementary cornea – whence the name of the genus (from α and βλέφαρος). This genus is represented in the steppes of SE Europe by *A. bivittatus* (SE Transcaucasia) and *A. pannonicus*, from Hungary and Greece to Transcaucasia and from here to Arabia. In contrast to the sclerophyllous forest, which the steppe boundaries and into which it passes in many places, in the Eurasiatic steppes there are no geckos: one exception is *Alsophylax pipiens*, an elegant creature without dilated digits, found in the steppes from the left bank of the Volga to Turkestan, NE Iran to Ala Shan. It is frequent near Astrakhan. Maybe the habitat of this species of Palearctic gecko explains the form of its digits: it is worth while mentioning that in Western Asia (e.g. North-Eastern Iran) some species of this genus live exclusively in old, abandoned houses, where they find shadow and relatively high humidity associated with mild temperatures, whereas, in the open, only some species of *Agama* and *Phrynocephalus* can live (Fig. 225).

The genus *Agama* has a very particular ecological importance in all the steppe or desert environments of Asia, Africa and Europe. The various species live on the

Fig. 225. *Mabuya aurata* (from Terentev & Chernov).

ground; they are sometimes short and strongly built, with the limbs adapted to running and to digging in the sand. They feed mainly on insects. In Europe we find *A. sanguinolenta,* named after the colour of the back, reddish-grey with some dark spots, extending from the Caucasus to Irgiz in the north and Tarbagatai in the east. It lives in clayish or sandy deserts with some scattered shrubs, on which it climbs during the hottest hours of the day in order to escape the extremely high temperatures present at the soil surface. It feeds on big insects, principally belonging to the genus *Blaps,* characteristic of all the steppes of Eurasia. It can change its colour so that in Turkestan they call it 'chameleon'. *Agama stellio* is without doubt the best known European species of this genus, present in South-Eastern Europe (Saloniki, Cyclad Islands, Kérkira), Asia Minor, Syria, Palestine, Northern Arabia and Egypt. It reaches a length of 30 cm and has a tail characterized by spring scales which make several rings around the tail. It runs on rocks, walls and trees during the favourable season; during the winter it is found in cavities of the soil in a state of semi-dormancy, in groups of 6-20 individuals (Fig. 226).

Phrynocephalus is a genus including species similar to *Agama,* with a short

Fig. 226. *Agama stellio* in its environment (from Brehm).

body, a broad head (resembling a little that of a toad), with the snout extremely short, vertically cut, and with a strangulated neck. The toes are furnished with squamulae. The tail is short, sometimes prehensile, at the base broadened and flattened. We must mention *P. mystaceus,* characteristic of the Western Asiatic steppe and of those situated between the Don and the Urals, in the north, and between the Kuma and the Terek in the south (Fig. 227 a.b). *P. helioscopus* inhabits the steppes extending from South-Eastern Russia to Northern Mongolia. It thrives only on steppes with a clayish soil (which in Central Asia are called 'takyrs'), where it can hide in the fissures, and only rarely occurs on sandy steppes. Its colour is well adapted to that of the substrate (cryptic coloration) with the exception of the brilliant red colour of the tail in the male, which is often bent almost vertically, probably to be noticed by the females (Fig. 228).

The only species of tortoise present in the European steppes is *Testudo hornsfieldi*, in Europe limited to the mouth of the Volga, but in Asia extending to Western Pakistan. It thrives in arid, silty and sandy soils and feeds on the steppe plants. During the summer it often exhibits so-called 'aestivation', a kind of rest

Fig. 227. a. *Phrynocephalus mystaceus.* b. *Agama sanguinolenta.* (From Terentev & Chernov).

induced by heat. It hibernates in a state of dormancy, digging in the sand or the clay to bury itself. In spring it comes out, and directly it has fed on fresh plants it breeds. The female digs a shallow hole in the sand in a suitable sunny place and lays from 3 to 5 eggs, covers them with the sand and then abandons them. The baby tortoises come out probably at the end of June: they are preyed on by wolves, foxes and also by some birds, for example crows. *Testudo horsfieldi* can live without water and even avoids it, since – like many steppe animals – the water present in the food is sufficient for the physiological requirements of the animal.

The invertebrates

Phytophagous insects

The number of phytophagous insects living on the steppe is enormous and we will only mention two coleopterous families, *Cerambycidae* and *Chrysomelidae*, which

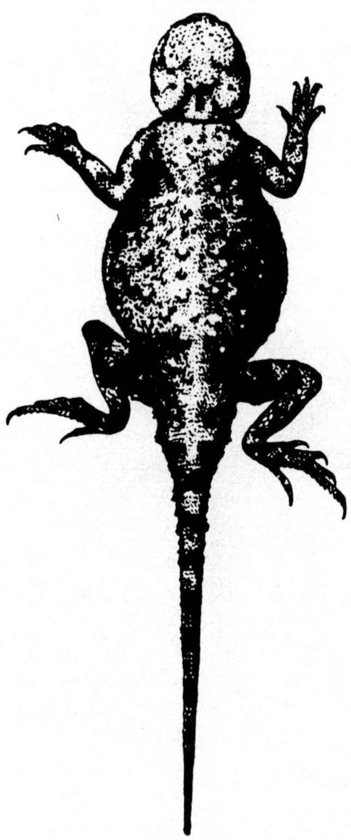

Fig. 228. *Phrynocephalus helioscopus* (from Terentev & Chernov).

have been very well investigated and which sometimes represent real pests to grasslands or to cultivated regions.

Among the *Cerambycidae* we find *Stenopterus flavicornis*, a ponto-mediterranean element very widespread in the Balkans, Asia minor, Syria and other Mediterranean countries. The adult lives on the flowers of *Umbelliferae* and *Achillea*, sometimes together with those of *Stenopterus rufus; Caenoptera marmottani* (present e.g. in Southern Russia) and *C. salicicola*, present in Hungary near Szeged. *Callimellum angulatum*, a ponto-mediterranean species whose adult lives on the flowers of *Crataegus, Sorbus* and several *Umbelliferae; Semanotus russicus*, a ponto-mediterranean element generally associated with junipers (closely allied to *S. persicum* of Northern Iran and the Transcaspian region) reaching as far north as Vienna (Hitzing) and Hungary (Pécs and Nemet-Bogson). *Rhopalopus hungaricus*, present also in Southern Russia and the Ukraine, and *R. spinicornis*, present also in South Eastern Europe and Bessarabia, i.e. in the area of the steppe; *Phymatodes köllari*, a ponto-mediterranean species; *Clytus rhamni* sbsp. *temesiensis*, limited to Transsylvania, Hungary, South Russia and Turkey, i.e. to the steppe districts of the European continent, and *Clytus speciosus*, with a pontic distribution; *Purpuricenus globulicollis*, a ponto-illyric species. Several species belong to the genus *Agapanthia*, such as *A. leucaspis*, present in Hungary and Southern Russia and Siberia; they develop in several *Compositae* and, secondarily in other herbs. An interesting species is *Oberea euphorbiae*, adapted to develop in herbs (*Euphorbia palustris*), a pontic species limited to Hungary, Russia (the Don basin), Transcaucasia, Slovakia, the Carpathian Ukraine, Moravia, going as far west as Venezia Giulia (Monfalcone), which is an area of encounter of several zoogeographical and ecological components, including the pontic one; in this region a particular sub-

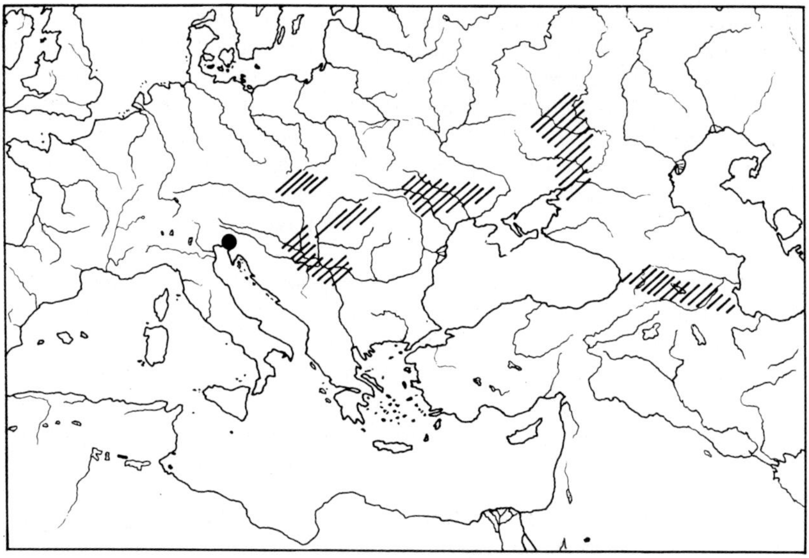

Fig. 229. Distribution of the Cerambycid beetle *Oberea euphorbiae*.

species (*imitans*) has envolved (Fig. 229). Finally we must mention the Cerambycid genus *Phytoecia,* which seems to be particularly well adapted to life on the steppes: it is enough to remember *P. scutellata*, widespread in Russia, Asia minor, Moravia, going as far north as Germany (Rheinbaden); *P. hirsutula*, found in Moravia, Hungary, the Balkans (as far west as the Quarnero islands), Southern Russia, Asia minor, Syria and Transcaucasia; *P. argus*, a faunistic element reaching as far west as the region of Rjeka (including Novi and Senj); *P. pustulata*, restricted to the *Compositae* genus *Achillea*, sometimes very harmful to *Chrysanthemum* cultivations, and present all through the steppes though not exclusive to this environment. A similar distribution is shown by *P. caerulea*, living on several *Cruciferae*, such as *Sinapis, Sysimbrium* and *Rapistrum.*

As far as *Chrysomelidae* are concerned, we could mention here *Lema,* a genus of polychromous and elegant small beetles, such as *L. cyanella*, limited to Central Europe, Russia and South-western Siberia, partcularly bound to *Gramineae* such as *Dactylis glomerata; L. tristis* extending from Mongolia and Siberia to France, living in dry, sandy places on several *Gramineae*, such as *Oplismenus crus-galli* (France), *Panicum italicum*, or *L. melanopa*, living on several *Gramineae* but particularly on wheat, oat and barley. The steppe zone of Eastern Europe is periodically subjected to great invasions of this beetle with consequential serious damage to cereal crops. The closely related genus *Crioceris* (characterized by a larva which covers itself with its own faeces, probably as a means of defence against enemies and against evaporation) is also found on the steppes. Let us mention *C. faldermanni*, found from Iran through Asia minor and the Caucasus to Greece (Parnasus); *C. quinquepunctata* (see later), *C. merdigera*, limited to several *Liliaceae* such as *Allium* and *Muscari, C.14-punctata* limited to Siberia and Eastern Europe, reaching as far west as the Karst near Trieste, living on *Asparagus* species, and *C. asparagi*, a 'Rassenkreis' represented by two races, a western one, limited to Central Europe and Northern Italy, and an eastern one (*maculipes*) extending from the Balkans (Zadar) to the Caucasus and Syria; the two races meet in Moravia (Trübau). This species can be very harmful to *Asparagus* cultivation. The already mentioned *C. quinquepunctata*, also living on *Asparagus,* is present in Bohemia, Moravia, Hungary, the Balkans (Croatia, Bosnia and Serbia) and Russia. One of the most remarkable Chrysomelids is without any doubt *Tituboea macropus*, in Northern Italy limited to the so-called xerothermic 'oases' (Magistretti & Ruffo) which it must have reached during some interglacial period or in the immediate postglacial.

To finish, let us mention some species of the genus *Chrysomela,* more or less restricted to steppe environments though not all necessarily exclusive to them. *C. chalcites*, known from Dalmatia (where the sclerophyllous forest can show similarities to steppe environments due to the dry, hot, summer), Hungary, Southern Russia and Syria; *C. cerealis*, an euryoikous species, limited to *Thymus serpyllum* (and maybe some other *Labiatae*), *C. salviae*, a ponto-mediterranean element, limited to *Salvia officinalis* and other *Labiatae, C. geminata*, linked to *Hypericum; C. marginata*, euryoikous, a 'Rassenkreis' limited to several *Compositae* such as *Achillea millefolium, Artemisia, Leucanthemum* etc., and finally *C. limbata* and *C. haemoptera*, both probably feeding on plantain.

We have already mentioned the so-called culture steppe or pseudosteppe ('Kultursteppe') and we have seen birds which, in Serbia, seem to be closely linked to this artificial ecosystem. Another group, still more abundant and ecologically interesting that the birds is the hoppers. We know a lot about these animals thanks to La Greca's research in Puglie (Southern Italy).

In this region this environment is the best developed, especially in the Foggia plain and on the 'murge' (a kind of karstic secondary prairie with some scattered groups of deciduous oaks) because of the great land areas devoted to wheat cultivation. Here we find plenty of orthopterous insects. According to La Greca, most of these species spread in Europe and Asia at the end of the Tertiary and later during the interglacial periods, when the steppes were widespread in the Palearctic Region. All these elements, coming from Central Asia, found in the regions cultivated with cereals (whose cultivation goes back to very remote ages, such as the Neolithic) a very favorable environment, not very different from their home environment, making allowance for some differences in climate (e.g. annual temperature range).

The most typical hoppers of culture land (or 'Kultursteppe') in Puglia are *Tessellana tessellata, Montana stricta, Platycleis intermedia, Sepiana sepium, Pholidoptera femorata, Decticus albifrons, Prionotropis appula, Tettigonia viridissima, Acrometopa italica, Tylopsis liliifolia, Calliptamus italicus, Calliptamus barbarus, Euchorthippus declives, Ephippiger apulus* and *Dociostaurus maroccanus*. After the harvest, on the dry stubble the community is impoverished in number of species and at the end of June or first half of July the most precocious species disappear (*Acrometopa italica, Ephippiger apulus, Dociostaurus maroccanus, Prionotropis appula*). The phytophilous or sciaphilous elements move towards the scarce shrubs growing near the fields, generally on their edge (*Platycleis intermedia, Sepiana sepium, Pholidoptera femorata, Tettigonia viridissima*), whereas the stubble represents the habitat of *Decticus albifrons, Tessellana tessellata, Montana stricta* and other more or less geophilous species.

We may observe at this point that, in Hungary, where rather similar conditions prevail, the most frequent element living on the stubble after the harvest is *Chorthippus brunneus*, followed by *Calliptamus italicus, Oedipoda coerulescens* and *Aiolopus thalassinus*. Evidently the picture is different from that described by La Greca for Puglie.

The soil fauna

The soil fauna of the steppe environment is fairly well known thanks to a number of studies by Soviet authors.

The mesofauna of Moldavia has been thoroughly investigated by Prochina. The number of individuals/m^2 is very variable, going from 38 during drought periods to 476 in the soil of a forested meadow during a rainy summer. The biomass is also very varied, depending on local factors: this author has found a value of 1.3-1.5 ts/ha in typical chernozem, 1.1-0.8 ts/ha in 'carbonates' and common chernozem, whereas the value can be as high as 5 in the steppes and 18 ts/ha in forested meadows. On the whole we can say that the true steppe soil fauna is rather impoverished as far as biomass is concerned. Prochina mentions

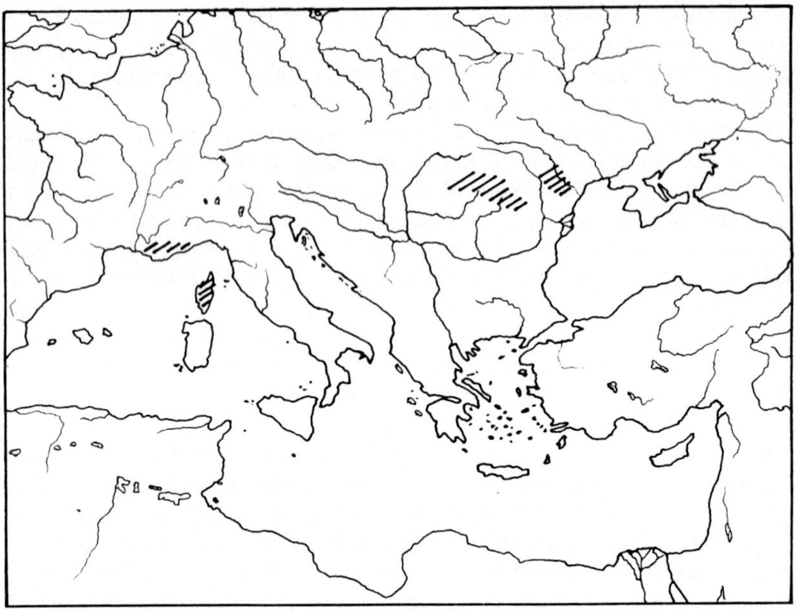

Fig. 230. Distribution of *Eophila dugesi*, a species restricted to steppe conditions in the eastern part of its distribution.

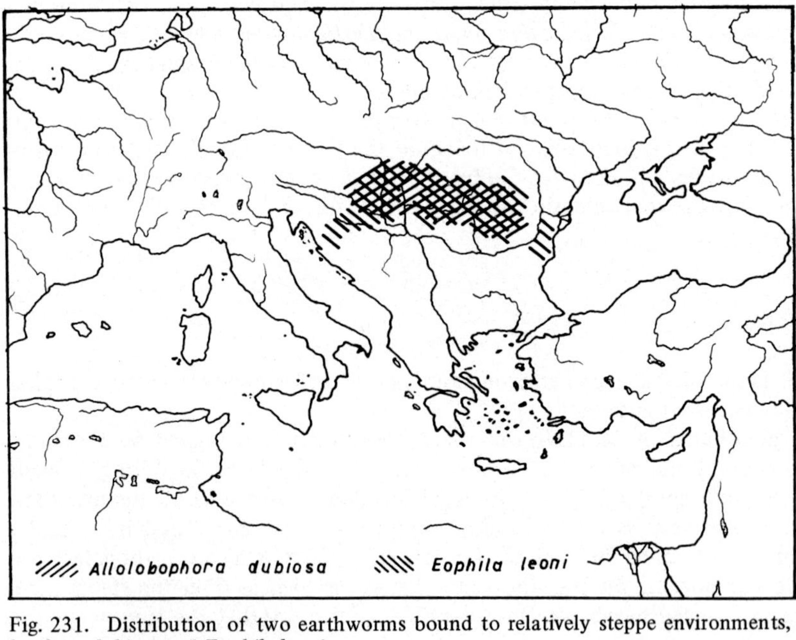

Fig. 231. Distribution of two earthworms bound to relatively steppe environments, *Allolobophora dubiosa* and *Eophila leoni*.

some interesting mediterranean elements belonging to the soil fauna of Moldavia such as *Eophila dugesi* var. *dacica* (a species with a discontinuous distribution including Nice, Provence and Corsica and then most of the Balcanian Peninsula) a very old relict species, today limited to the Alps Maritimes, Corsica and Transsylvania, Fig. 230), *Eophila antipai* (a species present in Central Europe, Italy and the Balcanian Peninsula), *Eophila leoni* (scattered in South Eastern Europe: Fig. 231), *Allolobophora dubiosa* (a species spread out in South Eastern Europe), *All. jassyensis* (present from South Eastern Europe to Afghanistan; Fig. 232), *Japyx confusus* (an Apterygote insect) and *Scolopendra* sp. (a centipede).

We also have some quantitative data for the Dniestr basin thanks to Striganova's research. The number of soil invertebrates is $370/m^2$. One must note that in forested areas this number may be as high as 529. The *Enchytraeidae* may reach $138/m^2$, which constitutes almost half of the total soil mesofauna. The mesofauna is composed of *Enchytraeidae* (40-60% of the total), earthworms (25%), myriapods (15-20%) and insects (5-10%). Of course in this classification Collembolans and Acari are not included. The *Oligochaeta* are *Dendrobaena octaedra, Octolasium complanatum, Eophila leoni* (widespread from Hungary to Croatia and Romania) *Octolasium lacteum, Eophila rosea.* The *Elateridae* found by this Soviet author are *Athous haemorrloidalis, A. incognitus, Synaptus filiformis, Agriotes sputator.* Furthermore there are *Curculionidae* and *Carabidae* (*Harpalini*).

The terrestrial *Oligochaeta* of Daghestan are also known thanks to Malevic's research. The following earthworms have been found in this extreme south-eastern region of the European steppes: *Allolobophora caliginosa typica, A. caliginosa trapezoides, Allolobophora jassyensis* (widespread from Central Asia, Southern Russia, Romania and Hungary to Palestine, the mouth of the Nilo and Italy) *Lumbricus terrestris, Eiseniella tetraedra typica, Eisenia foetida, E. rosea, Eiseniella tetraedra hercynia* (in humid situations), *Eisenia nordenskjöldi lagodechiensis* (also humid environments), *Dendrobaena subrubicunda, D. octaedra, Bimastus tenuis, B. eiseni* (widespread in the Holarctic Region), *Eophila montana* and *Octo-*

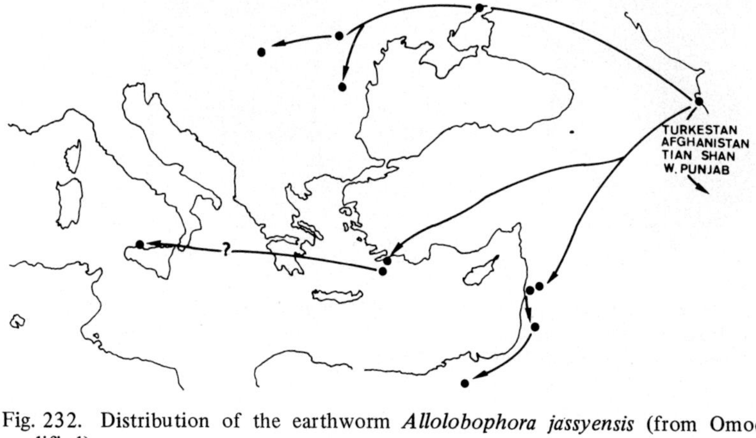

Fig. 232. Distribution of the earthworm *Allolobophora jassyensis* (from Omodeo, 1952, modified).

Fig. 233. a. *Opatrum sabulosum*, a 'Rassenkreis' of economic importance. b. *Gonocephalum pusillum*, in Russian 'mal'i medljak', of great economic importance. c. *Blaps halophila*, a typical steppe element. d. *Gonocephalum setulosum*.

lasium lacteum. The *Criodrilidae* are represented by *Criodrilus lacuum* (in humid situations, at the mouth of the River Samur) *Bimastus eiseni* (spread across Southern Europe, where it is fairly common).

The mites living in the soil of the Crimea are known through Gordeeva & Krivolutskii's research. The number of specimens per dm^3 varies from 18 to 70. There are 37 species; this paucity of mites in steppe environments, compared with forested ones in very interesting. Also *Elateridae* are rather scarce in the steppes, compared with forests: thus, in the Ukraine, Topčiev finds in the forest subzone 38 species but in the subzone of southern chernozem and dark castanozem only 15, which is less than half. The species characteristic of the first soil type are *Agriotes gurgistanus, Melanotus brunnipes, Selatosomus latus* etc.; the species characteristic of the second soil type are, however, *Synaptus filiformis, Melanotus fuscipes, Agriotes medvedevi* etc. They are all widely distributed species with the exception of *A. medvedevi*.

Very interesting results were obtained by Minoranskii in a study of the life history of a small Tenebrionid beetle, *Gonocephalum pusillum*, in Russian mal'i medljak (Fig. 233), a polyphagous species harmful to several annuals in the Rostov zone. It is controlled with DDT, G Ch Ts G and Vofatoksa.

The soil fauna of European salty soils, or, more precisely, the solonez of the steppes near Kuibičev, also fairly well known thanks to Buševa's research. The following Coleopterous species have been found: *Agriotes medvedevi, A. bifrons, A. binotata, Harpalus aeneus, H. distinguendus, H. caspius, Selatosomus latus, S. aeneus, Agriotes sputator*. Not all of them of course are closely restricted to salted places.

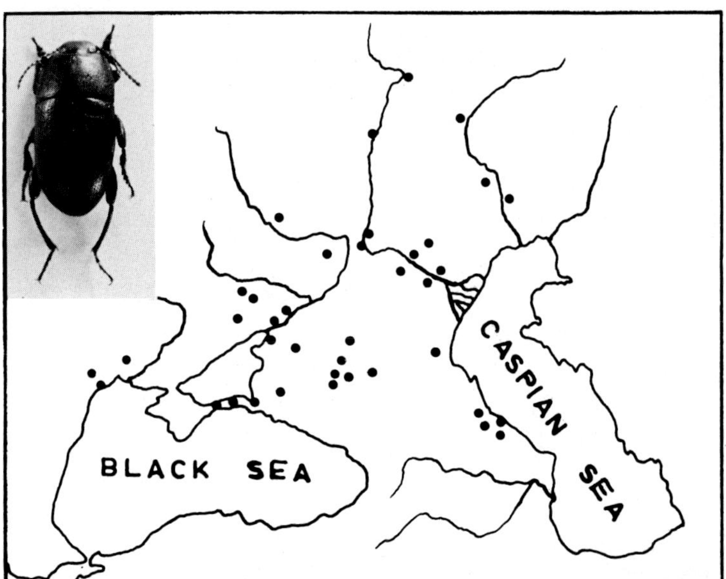

Fig. 234. Distribution of the Tenebrionid beetle *Pedinus volgensis* (localities taken from Medvedev).

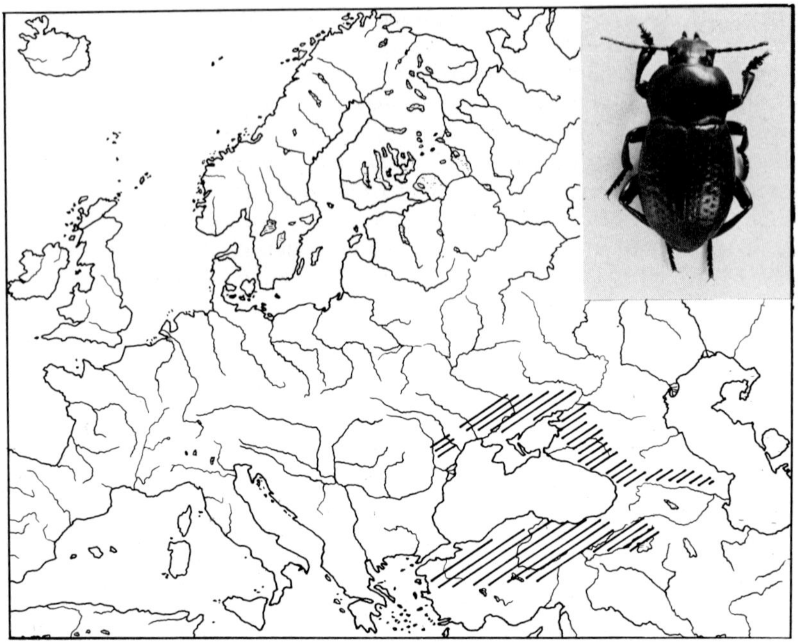

Fig. 235. Distribution of the Tenebrionid beetle *Dendarus punctatus.*

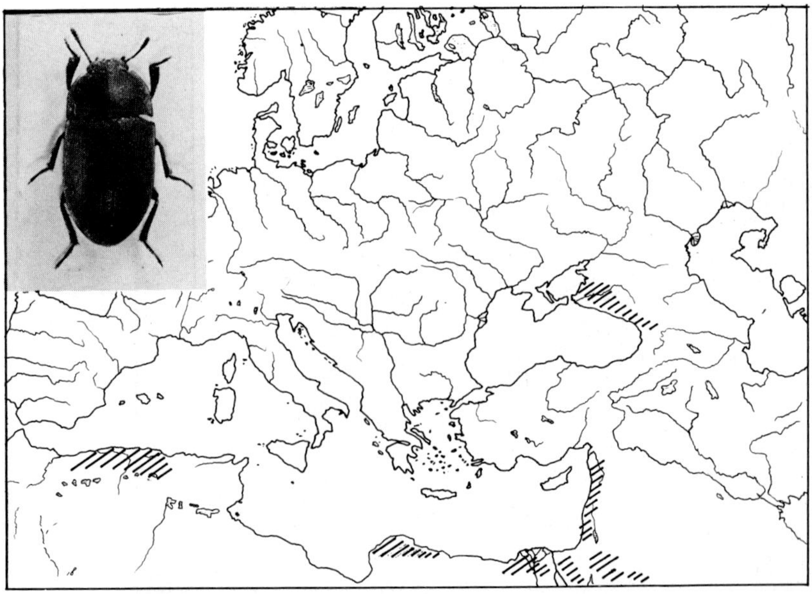

Fig. 236. Distribution of the Tenebrionid beetle *Clitobius oblongiusculus* (the species is widespread in Arabia and Egypt).

Finally, we should mention research on two insect groups living in the steppes immediately to the east of Europe, both of paramount ecological interest in the biological equilibrium of the steppe environment, namely Tenebrionid and Elaterid beetles (the former studied by Keleinikova, tha latter by Tscherepanov). It is a pity that such studies have not also been carried out on the European steppes (Fig. 234-239).

The best known part of the European steppes is very probably the Trans-Volga district south of Kazan, where Aleinikova has carried out very thorough inves-

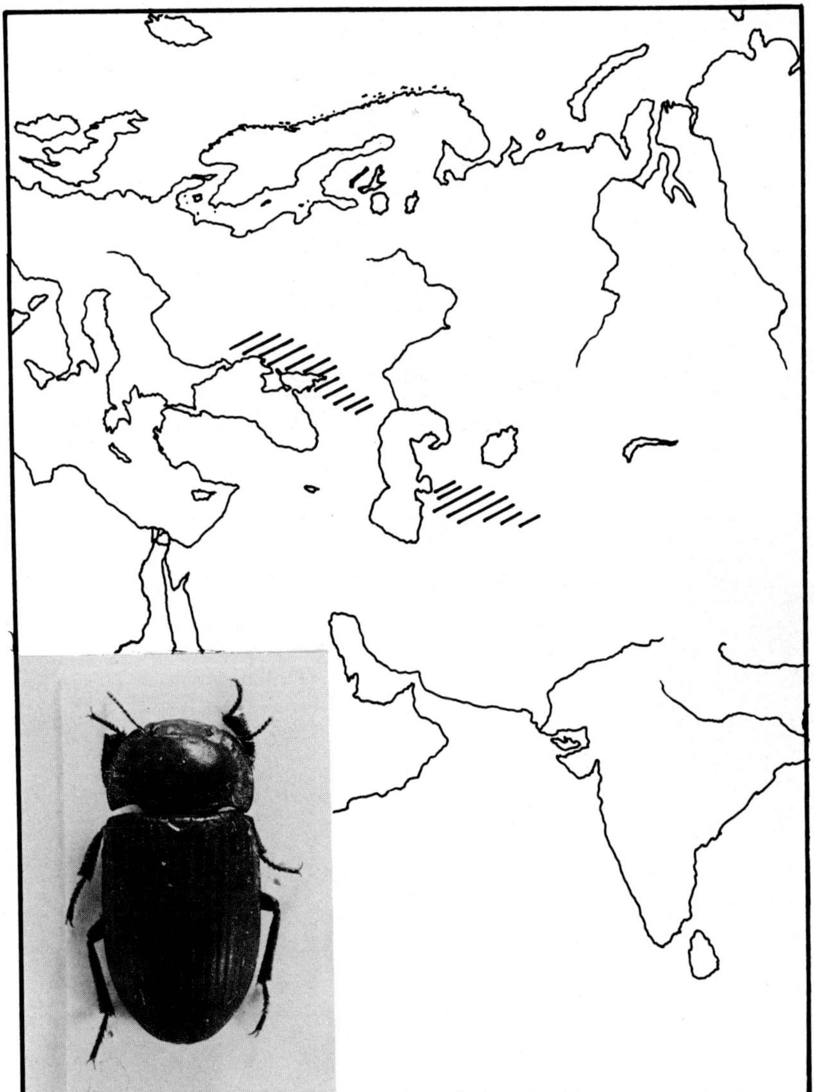

Fig. 237. Distribution of the Tenebrionid beetle *Lobodera dilectans*.

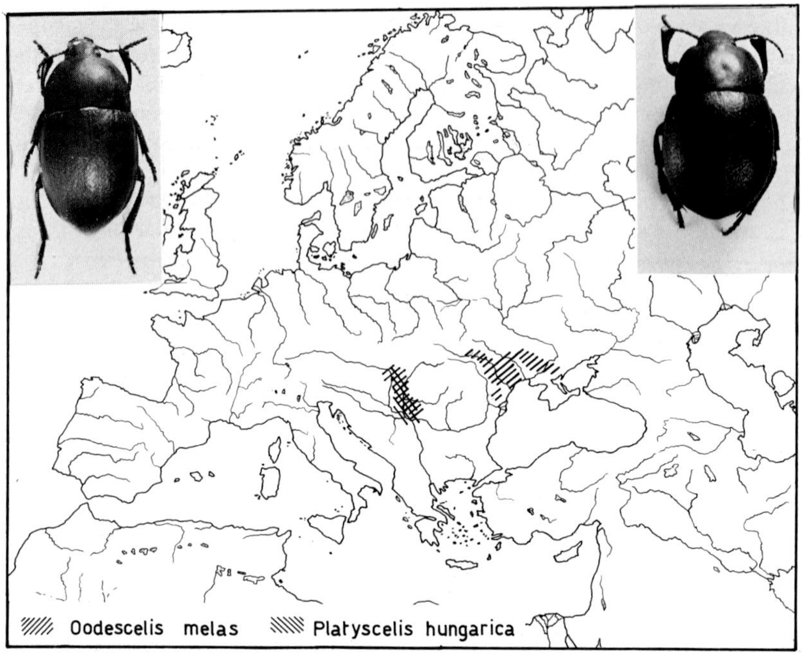

Fig. 238. Distribution of the Tenebrionid beetles *Oodescelis melas* (left) and *Platyscelis hungarica* (right).

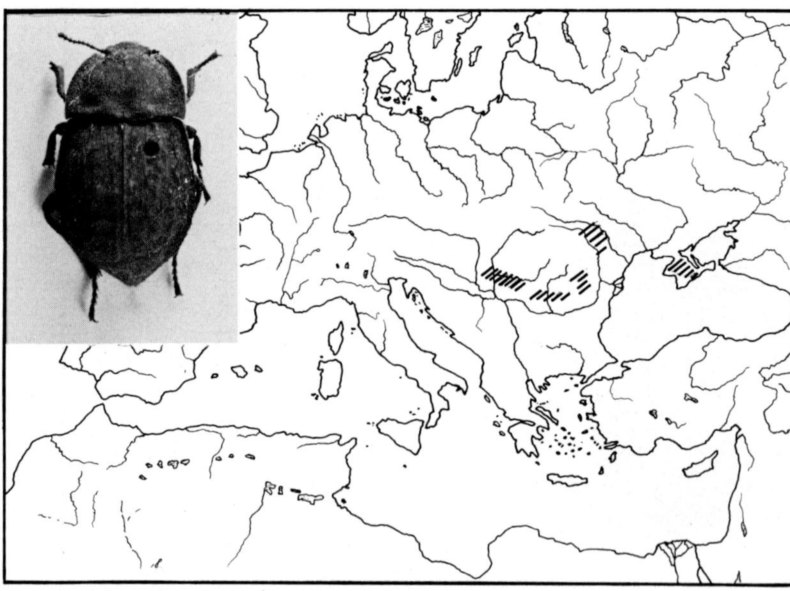

Fig. 239. Distribution of the Tenebrionid beetle *Asida banatica*.

tigations both on meso- and microfauna (or, as we might also say, macro- and mesofauna respectively, since the classification of soil fauna on the ground of animal size is very subjective). The main results obtained by Aleinikova are synthesized in the following schemes and tables. Specific composition (in %) and average number of *Elateridae*-larvae (wireworms) on the steppes of the lower Trans-Volga (based on the collection of soil samples:

Species	Presence (in%)
Cardiophorus cinereus	7.7
Athous haemorrhoidalis	38.5
Selatosomus aeneus	35.8
Selatosomus cruciatus	10.3
Dolopius marginatus	7.7
N° of species	5
Average N° of specimen/m² . . .	7.8

(from Aleinikova, 1968).

Specific composition (in %) and average number of earthworms (*Lumbricidae*) on the steppes of the lower Trans-Volga (based on collection of soil samples)

Eisenia nordenskjöldi	70.0
Eisenia uralensis	5.7
Dendrobaena octaedra	16.4
Bimastus tenuis	7.9
Number of species	4
Average N° of specimen/m² . . .	51.3

(from Aleinikova, 1968)

The following species of Carabid beetles are present in the steppe environments of the Middle Trans-Volga: *Carabus estreicheri, Ophonus rufipes, Harpalus distinguendus, Pterostichus anthracinus* and *Amara aenea.* All these species, with the exception of *Carabus* are present in different biomes in Southern Europe (Italy).

The following species of ants (*Formicidae*) have been found in steppe environments of the Lower Trans-Volga: *Tetramorium caespitum, Camponotus vagus, Lasius flavus, Cataglyphus aenescens, Formica fusca* and *Formica rufibarbis*. All these species are also present in non-steppe European environments.

Finally the number of specimens and zoological composition of the soil fauna of some forests of different type and some steppes of the lower Trans-Volga (number of specimens/m^2 and biomass in mg) are shown in the following table (see page 338).

Productivity

Not much is known so far about the productivity of European steppes since most of the results of the work done for the International Biological Programme (IBP), in which Soviet authors are particularly interested, have not yet been published. Concerning secondary productivity we have seen what we can now assume are average data for the soil fauna in the Trans-Volga. Of course the soil fauna represents only a fraction of secondary productivity.

Zoolog. group	Studenez specific oak forest		Mosti ravine forest with *Quercus*		Annino Aceretum with *Quercus*	
MESOFAUNA						
Coleopterous insects (total)	9.0		35.0		43.0	
Carabidae	1.0	100	–	–	4.0	400
Staphylinidae	–	–	3.0	240	4.0	320
Scarabaeidae	3.0	3.000	–		16.0	16.000
Elateridae	3.5	315	7.0	630	7.5	715
Curculionidae	1.0	50	16.0	800	10.0	500
remaining Coleopt.	0.5		8.0		1.5	
Diptera	2.5	25	27.0	270	5.0	50
Hymenoptera	1.5	9.9	–	–	1.0	6.6
Lepidoptera	2.5	16.5	8.0	52.8	4.5	29.7
Remaining Insects	19.0	129.36	2.0	13.2	–	–
Total insects	35.5		71.0		53.5	
Myriapoda (total)	12.0		10.0		23.5	
Chil. Geophilomorpha	0.5	8	10.0	160	19.5	312
Chil. Lithobiomorpha	0.5	8	–	–	2.5	40
Diplopoda	11.0	292.6	–	–	1.5	39.9
Lumbricidae	–	–	50.0	25.000	52.0	26.000
Enchytraeidae	–	–	113.0	22.6	9.5	1.9
Mermithidae	–	–	15.0		4.5	
Araneina	2.0	8	13.0	52	1.5	6
Mollusca	4.0	800	–	–	–	–
Remaining animals	0.5	65.0		2.0		
MICROFAUNA						
Collembola	5825	69.9				
Acrina (total)	58675	586.75				
Gamasoidea	2563	25.63				
Oribatei	38814	388.14				
remaining *Acarina*	17298	172.98				
Myriapoda	237	118.5				
Enchytraeidae	38	7.6				
Nematoda	–	–				
Remaining animals	4850					

For primary productivity, which from an economic standpoint is perhaps the most interesting, we have so far very scant information.

It is very likely that at least part of the results obtained by some North American authors in the study of the prairies of the USA will also be valid for our open vegetation systems.

According to these authors, primary productivity may in some cases be less than 100 $g/m^2/year$. The loss of water through evaporation (measured by means of a lysimeter) can reach 0.45 $mm/h/m^2$ in the soil and in plant 0.4 mm, which means up to 4 mm/h of water lost per gram of dry weight or plant biomass. In the desert prairie, a large amount of water is lost directly through evaporation from the bare soil, and plants are adapted more to enduring drought than to an efficient utilization of water. The dry prairies show a few changes in the specific composition as a consequence of grazing, but the distribution of fixed carbon is modified, since there is a diminution in the above-ground biomass and a proportional increase in the below-ground one. There is an interesting difference in the fixation of carbon between cold season plants and those which react to high temperatures: in the first group the fixation of carbon occurs through the C_3 chain (Calvin, principal cycle of photosynthetic carbon) in the second one through C_4 chain (Hatch-Slack).

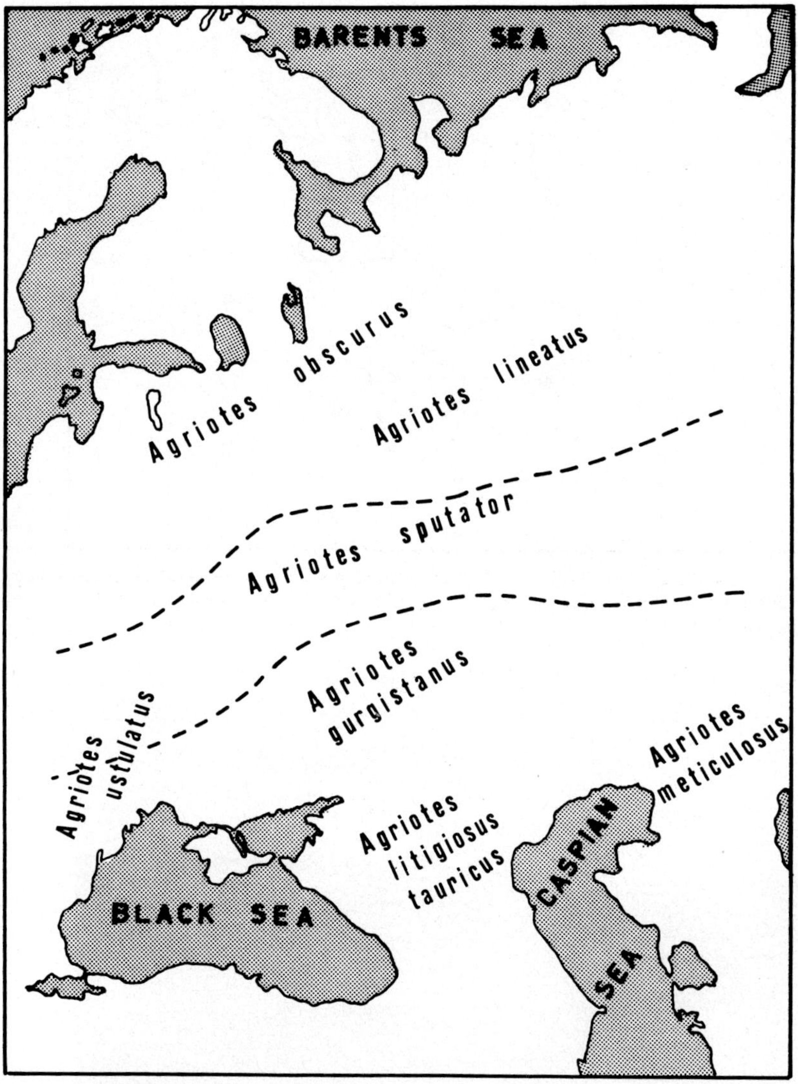

Fig. 240. Distribution of some beetles of the genus *Agriotes*, mainly dependent on soil and vegetation (from Čerepanov, redrawn and modified).

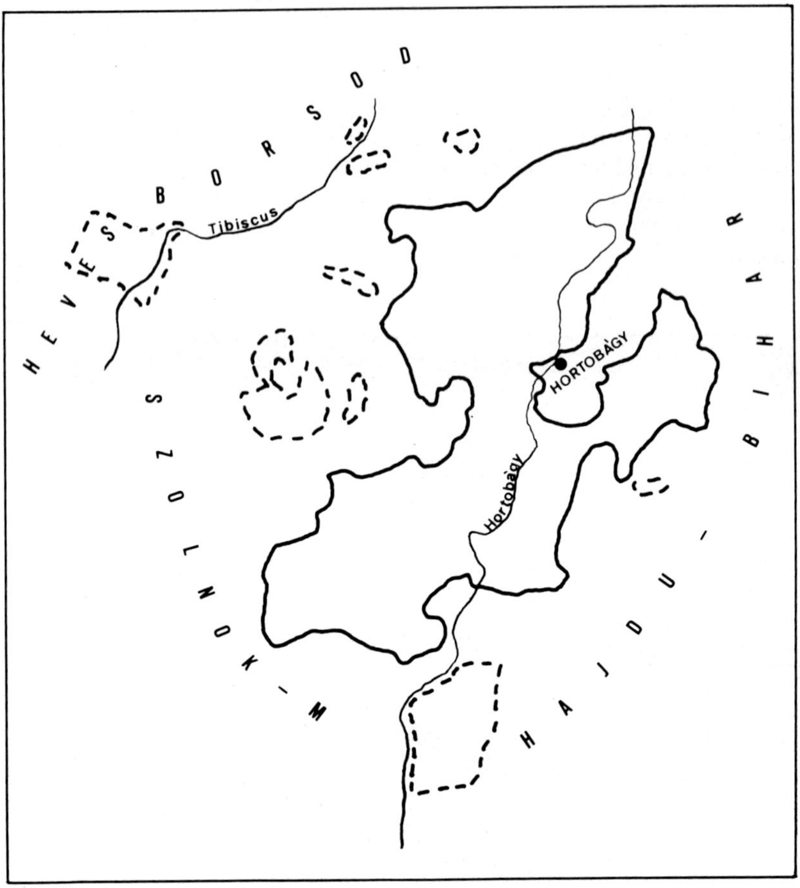

Fig. 241. a. Example of protection in the European steppes: the National Park of Hortobágy (from Rakonczay, modified). Continuous line: boundary of the Park; broken lines: boundaries of Nature Reserves.

Fig. 241. b. Hortobágy National Park: stud (Photo: Tamás).

The researchers of the Natural Resource Ecological Laboratory of Colorado State University have constructed a non-linear model to represent the dynamics of the grassland biomass, flexible enough to be adapted to different situations. The factors which control primary production are temperature, water, insolation and availability of nutrients. The model seems to be able to establish some interrelations between abiotic functions, productivity, consumers, decomposers and nutrients in the prairie ecosystem (Fig. 241).

BIBLIOGRAPHY

Aleinikova M.M., Počvoobitaiuščie bespozvoločnie različnich tipov lesa v srednem Povolž'e, Materiali po fauna i ekologii počvoobitaiuščich bespozvonočnich, Izdatl. Kazansk. Univ., 1968, 3.

Bodenheimer F.S., Animal life in Palestine, Jerusalem, 1935.

Buševa A.V., Nasekomie solonzovich počv kuibičevskoi oblasti, Problemi počvennoi zoologii, *Akad. Nauk, SSSR, Izd. Nauka, Moskva*, 1966, 25.

Corrain C., Il divenire biologico dell'uomo, Bologna, 1971.

Darskaya N.F., To comparative ecology of bird fleas of genus Ceratophyllus, Ectoparaziti, Izdat. Moskovski Univ., 1964, 31 (in Russian with English summary).

Formosov, Essays on ecology of rodents transmitters of tularemia, Contribut. to connaiss. F.Fl. URSS, Moscou, 1947.

Frank F., Die Vogel von Opuk (Schwarzmeer-Gebiet), *Bonn. Zool. Beitr.*, 1, 1950, 144.

Gordeeva E.V. & D.A. Krivolutskii, Materiali po pantsirnim kleščam, Problemi počvennoi zoologii, *Akad. Nauk, SSSR, Izd. Nauka, Moskva*, 1966, 41.

Hübl E., Die pflanzengeographische Stellung des pannonischen Raumes in Beziehung zu kontinentalen und mediterranen Klima-einflüssen, Verh. Ges. Oekol., Wien, 1975, 167.

Ioff I.G., V.E. Tiflov & O.A. Fedina, A list of flea species of the Stavropol region, Ectoparaziti, Izdat. Moskovski Univ., 1964, 24 (in Russian with English summary).

Jakucs P., G. Fekete & J. Gergely, Angaben zur Vegetation der Moldau und der Dobrudscha, *Ann. Hist. Nat. Mus. Nat. Hung.*, 51, 1959, 211.

Keleinikova S.I., O nekotorich ekologičeskich osobennostyach počvennich ličinok žukov-černotelok, Problemi počvennoi zoologii, *Akad. Nauk, SSSR, Izd. Nauka, Moskva*, 1966, 65.

La Greca M., L'ortotterofauna pugliese ed il suo significato biogeografico, *Mem. Biogeogr. Adriat.*, 4, 1962, 33.

Malevič I.I., K faune doždevich červei Dagestana, Problemi počvennoi zoologii, *Akad. Nauk, SSSR, Izd. Nauka, Moskva*, 1966, 83.

Maringer J., Le religioni dell'età della pietra in Europa, Torino, 1960 (transl. from German).

Matvejev S.D., Rasprostranjenje i život ptiza u Srbiji (Ornitogeographia serbica), Srpska Akad. Nauka, Pos. izd. Kn., 3, Beograd, 1950 (in Serbian with French summary).

Medvedev G.S., Zhestkokrilie, Zhuki-černotelki (Tenebrionidae), in 'Fauna SSSR', Izdateln. Nauka, Leningrad, 1968.

Mertens R. & H. Wermuth, Die Amphibien und Reptilien Europas, Frankfurt a.M. 1960.

Minoranskii V.A., K biologii malogo medlyaka, Problemi počvennoi zoologii, *Akad. Nauk. SSSR, Izd. Nauka, Moskva*, 1966, 87.

Nelzina E.N. & G.M. Danilova, Contributions to the biology of ticks Haemogamasidae etc., Ectoparaziti, Izdat. Moskovski Univ., 1964, 221 (in Russian with English summary).

Omodeo P., Particolarità della zoogeografia dei Lombrichi, *Boll. Zool.*, 19, 1952, 349.

—, Contributo alla revisione dei Lumbricidae, *Arch. Zool. It.*, 41, 1956, 129.

—, Oligochetes de l'Afghanistan II. Contribution a l'étude de la faune d'Afghanistan, *Atti Accad. Fisiocr., Siena*, 11, 1962, 3.

Prochina N.A., Ispol'zovanie počvenno-faunističeskich issledovanii dlya diagnostiki počv Moldavii, Problemi počvennoi zoologii, *Akad. Nauk, SSSR, Izd. Nauka, Moskva*, 1966, 105.

Rakonczay Z., A Hortobágy, elsö nemzeti parkunk, Búvár, 23, 1973, 67.

Rheitblat A.G., On the fauna of the Gamasoidea ticks of the Stavropol region, Ectoparaziti, Izdat. Moskovski Univ., 1964, 244 (in Russian with English summary).
Ščegoleva (ed.), Opredelitel' nasekomich po povreždeniyam kulturnich rastenii, Gosudarstvenno Izdatel. selskochozyastvennoi Literaturi, Leningrad & Moskva, 1960.
Striganova B.R., Počvennaya fauna poimi Dnestra, Problemi počvennoi zoologii, *Akad. Nauk, SSSR, Izd. Nauka, Moskva*, 1966, 130.
Terent'ev P.V. & S.A. Chernov, Key to Amphibians and Reptiles, Jerusalem, 1965 (transl. from Russian).
Tischler W., Agrarökologie, Jena, 1965.
Topčiev A.G., Sostav i raspredelenie provoločnikov v iskusstvennich lesach stepnoi zoni Ukraine, Problemi počvennoi zoologii, *Akad. Nauk, SSSR, Izd. Nauka, Moskva*, 1966, 138.
Tscherepanov A.I., Die Schnellkäfer West-Sibiriens (Coleoptera, Elateridae) Akad. Wiss. USSR, Novosibirsk, 1957 (in Russian with German summary).
—, Provoločniki zapadnoi Sibiri, *Akad. Nauk, SSR, Izdat. 'Nauka', Moskva*, 1965.

THE HEATHS

'... die Bahn, die dem Warthe-Netze-Bruch folgt und zwischen Cüstrin und Bromberg mit wenigen Unterbrechungen ein Bild traurigster Monotonie bietet' (P. Graebner)

An ecosystem which shows some relationships with grasslands (or steppes), i.e. open formations, is represented by heaths. It has mainly developed in north-western Europe, in areas where the true forest communities are not able to form as a consequence of persistent strong winds or perhaps the extreme poverty of the soil. Of the two factors the former is probably the more important, since some trees, if protected against the wind, can thrive in soils even poorer and more acid than those dominating in healths. It is possible to recognize acid heaths, on silicaceous rocks, and calcareous heaths on limestone. The former are characterized by *Calluna vulgaris, Erica cinerea, Agrostis* spp., *Deschampsia flexuosa* and *Nardus stricta*; the latter shrubs of hawthorn (*Crataegus oxyacantha*), juniper (*Juniperus communis*) and yew (*Taxus baccata*) besides graminoid plants of the genus *Avena, Bromus* and other.

On the whole, heaths are flat, sandy and have a small amount of humus, always acid. The vegetation is composed of lichens, the bracken *Pteridium aquilinum, Ericaceae,* gorse (*Ulex europaeus*) and other species able to thrive in such an environment. Often the term heath is employed for woodlands and small tree-covered areas, whence the German names Kiennheide, Föhrenheide and even Buchheide, as we find (or rather found in the past) near Templin, Berlinchen and Stettin. The term heath should be restricted, however, to an open, treeless territory, where the arboreal growth is limited to bush or low shrub vegetation and where a close grassy cover is lacking.

Heaths have a very limited agricultural value and their reclamation is difficult and costly because of the inherent low soil fertility. It is also more difficult if there is no water available for irrigation. One speaks sometimes of heath soil (in Italy 'Erica soil') for a very soft soil whose composition varies according to the regions, used in horticulture. It is composed of silicaceous sand and more or less decomposed organic matter; clay is absent. Sometimes the soil of areas with *Erica arborea* and *Arbutus unedo* is employed, but of course in this case we cannot speak of a true heath.

In France the heath (or 'lande')[1] would be but a transitional stage towards woodland if man did not intervene, voluntarily or not, to maintain, at a stage of

[1] In South West France the name 'tojal' is also used to indicate a heath.

quasi-stability, the vegetation which today characterizes the heath. The interventions can be fire (the deposition of ash on the surface of the soil is harmful to woodland and prevents the regeneration of a higher vegetation) or the periodic clearance of the gorse, which in some regions is utilized as litter. When these human activities cease, it is possible to see here and there some seedlings of oaks (*Quercus pedunculata*) or chestnuts, so that slowly a clear forest grows, which in turn permits the formation of a true forest, where only some scattered tussocks of *Ericaceae* persist.

In England according to some authorities the heaths are caused almost exclusively by repeated fires and grazing, which for centuries have prevented any return to the original woodland conditions. This is due also to the high population density (in Europe less only than that of Belgium, the Netherlands and – to a lesser extent – West Germany). Today the forested areas of Britain represent only 5% of the territory, a value which is probably the lowest in Europe. Not only that, but also the very heaths which were in a sense natural environments, have been replaced by cultivation, towns, mineral extraction, golf-courses etc. In the period from 1811 to 1960 alone, the heath in East Dorset and an adjacent part of Hampshire has been reduced from 30,000 to 10,000 ha and even what remains cannot be conserved for very long if one does not adopt conservational measures, such as those existing today in Germany (e.g. near Haichinger Heide) (Fig. 242a-c).

The heaths west of London are a constant attraction for recreation visitors who use them for picnics, and because their scant value from the standpoint of agriculture and cattle raising allows public access. This utilization of heaths – also making allowance for the respect Britons have for nature – compared with other European peoples – brings about a change in the precarious biological equilibrium generally present in heaths.

A particular kind of heath is present also in some parts of Southern Europe, for instance in Lombardia (Northern Italy) near Gallarate (Fig. 243). It occupies a terrace, remnant of a widespread alluvial terrace, which formed an uninterrupted border on the southern edge of the pre-Alps. It is a recent alluvium, composed of gravel and pebbles among which gneiss and mica schists prevail. There is also some fine, sandy detritus. The heath, locally called 'brughiera', is very ill-adapted for agriculture, despite all the attempts made since the beginning of the last century. Only afforestation would be possible, because very probably a forest represented originally the climax (*Pinus silvestris, Quercus* spp.). The landscape is monotonous as a consequence of the dominance of *Calluna* growth, in some places dense, in others less so, until it is finally replaced by *Molinia* growth. Some scattered trees (*Pinus, Betula pendula*, as well as the introduced *Robinia pseudacacia*) are also present. According to Giacomini, this Italian heath, probably secondary, is not comparable with those of north-western Europe. The most interesting species are the following: Trees – *Pinus silvestris, Quercus sessiliflora, Q. pedunculata, Pinus maritima, P. rigida, Betula verrucosa, Castanea sativa*; Shrubs – *Rhamnus frangula*; herbs: *Calluna vulgaris, Molinia altissima, Sarothamnus scoparius, Potentilla erecta, Jasione montana, Genista tinctoria, G. germanica, Hieracium pilosella, Festuca ovina, Agrostis alba, Luzula nemorosa* and *L. campestris*, etc. The bracken *Pteridium aquilinum* is not very abundant; the lichens are represented by a large number of *Cladonia*; also very abundant are mosses (*Polytrichum piliferum, P.*

juniperinum, Rhacomitrium canescens, Hypnum cupressiforme, etc.). The soil is a brown earth, in which only an early stage of podzolisation has taken place, so that this South-European heath is by no means comparable to analogous formations at higher latitudes.

Near Gallarate there is another heath with *Pinus*, at Besnate. This locality is famous because of the palethnological interest of two toponyms, Lagozzetta and Lagozza. The first is a small lake situated in the forest, the second has been con-

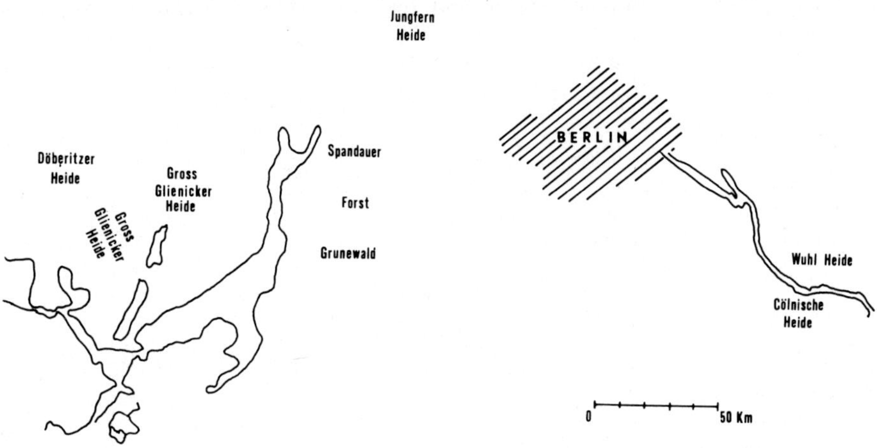

Fig. 242a. Extension of heathlands in Germany near Berlin in the last century (from Stiedler's Hand-Atlas, 1888).

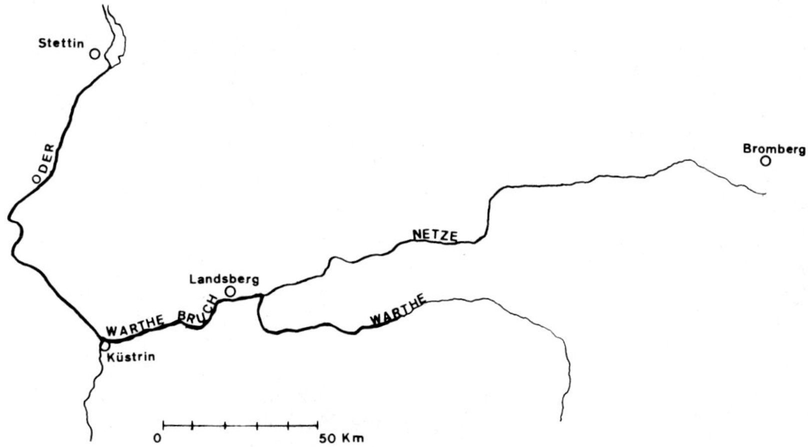

Fig. 242b. Extension of heathlands in Germany in the last century: the territory from Küstrin (or Cüstrin) to Bromberg, visited by Graebner at the end of the last century (from Stiedler's Hand-Atlas, 1888).

verted into a cultivated area. Very thorough palinological investigations have been carried out in these localities by Pasa and Durante Pasa. The culture of Lagozza is akin to that of Cortaillod (Switzerland) and Chassey (France); it is characterized by carinate cups, globose pots, with very simple ornamentation. The culture of Pescale (Isolino di Varese), not far from Gallarate and Besnate, also bound to an old heath, represents according to Corrain, a late phase of Neolithic, almost a continuation of the culture of Lagozza.

The geology of the territories today covered by heathland is varied, but on the whole, in Western-Europe (France and Britain), heath is found on Hercynian

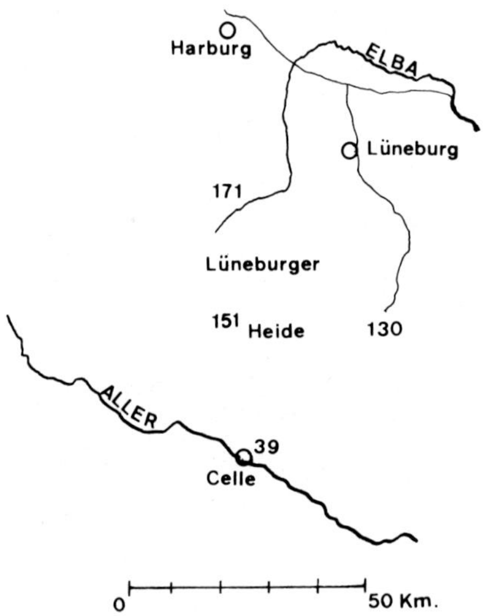

Fig. 242c. Extension of heaths in Germany in the last century: the Lüneburger Heide (from Stiedler's Hand-Atlas, 1888).

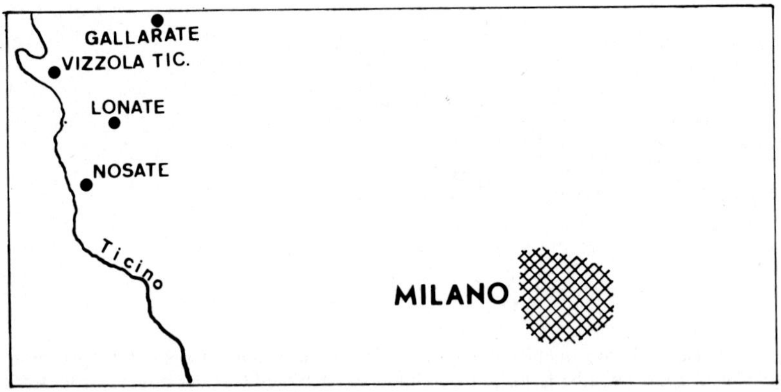

Fig. 243. The heath of Gallarate.

rocks, while in other lowland areas on more recent detrital sediments.

Climate also varies; in general we may distinguish three zones from West to East, namely a true oceanic, an attenuated oceanic and finally, in Northern Germany, a transitional one. The rainfall gradually decreases towards the East: so in Western England and Wales the annual precipitation is 1000 to 1500 mm; in Central Western England 800 to 1000 mm; in the French bruyères from 800 to 1500 mm and finally in Northern Germany 500 to 800 mm. The maximum is autumnal (November) but in some localities there can be two equinoctial maxima.

The temperature also shows areas with very great variations within the heathland: the mean annual temperature ranges from 7-8°C in the French Landes; the mean January temperature is from -2°C in Germany to 7°C in France and England; the mean July temperature finally reaches from 20°C (Landes) to 16°C (Britain).

Shown here are some climatological data for Brest (Brittany):

Temperature: annual average 17.7°C   Jan. 6.3 April 11.0 July 17.9 Oct. 12.3

Rainfall: annual total 824 mm	J	F	M	A	M	J	J	A	S	O	N	D
	84	75	57	54	49	51	53	54	78	91	96	82

The maximum corresponds to autumn (November); there is no hyperhumid period. A somewhat schematic precipitation diagram for Brest is shown in Fig. 244.

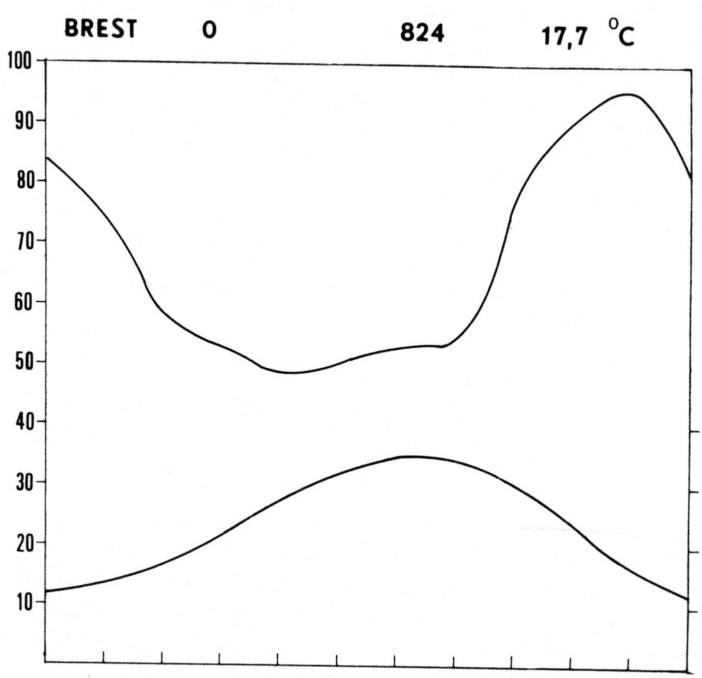

Fig. 244. Ombrothermic diagram for a locality belonging to the dominion of heaths.

Winds in many instances have paramount ecological importance: in the peninsula of La Hague (Cotentin) and in that of Crozon (Finisterre) the gorse, *Ulex gallii*, is a dwarf and forms balls offering the least resistance to the wind.

From a phytogeographical point of view the prevailling climate is that of the chamaephytes.

The major soil type is without any doubt the podzol (Fig. 245). Under conditions of high rainfall the surface of the soil tends to loose its soluble content, which is carried away by the rain water; this process is called leaching, and has the effect of lowering the base content of the superficial layers and also removes some of the humus content, since solid humus particles are carried down mechanically. Of course leaching will occur more readily where percolation is rapid than where it is slow, and consequently sandy soils are leached more readily than clay soils. Even on sandy soil, however, the process is greatly impeded if there is a close cover of grassy vegetation on the surface of the soil. The podzol as a whole consists of several horizons, which are of different colour and texture, produced from an originally more homogeneous material by the process of podzolisation.

We wish to reproduce here two analyses of heath podzols, one carried out by Friedlander in England, and another carried out in Northern Germany by Graebner.

In a heath at Pirbright, Surrey, we have

Constituents	Horizon				
	A1	A2	B1	B2	C
Loosely held water (% of weight soil)	41.5	9.6	16.7	15.7	15.9
Strongly held water (% of dry soil) .	5.4	0.8	2.1	1.6	1.1
Humus content (% of dry soil) . . .	23.5	3.4	33.1	1.9	2.2
Mineral content (% of dry soil). . .	71.1	95.8	64.8	96.5	96.7
pH	5.7	7.0	7.0	7.0	6.4

In the Lüneburger Heide we find

Components (%)	A1 humus containing sand, 45 cm	A2 pure quartz sand, 15-20 cm	B1 hardpan 5-10 cm	B2 red-brown sand, 50 cm	C white sand laying below
mono- and bivalent metals	0.066	0.045	0.066	0.068	0.038
Mn-monoxide	0.0074	0.0032	0.0056	0.005	–
Fe oxide	0.41	0.19	1.38	0.83	0.04
clay	0.42	0.07	0.71	0.71	0.17
phosphoric acid	0.03	0.01	0.07	0.03	0.01

Recently an interesting difference has been demonstrated between the B_h layer (i.e. the part of B horizon rich in humus) of forest podzols and that of secondary heathlands. Although the latter are more recent, they exhibit a better developed B_h layer: this is due to the fact that the organic substances of natural forest podzols are labile, those of *Calluna* heaths are much more stable, since they are composed by insoluble, not easily biologically degraded polymers (*Calluna* polymers). This should always be kept in mind when judging the age of a heath podzol.

Human impact upon heaths

Heaths were fairly densely inhabited even in Paleolithic times (Fig. 246). Also the Mesolithic of northwestern Europe has left us much evidence. The Azilian – named after the site of Mas d'Azil, a cave at the foot of the Pyrenees – represents a fishing culture which spread from Northern Spain through France, Belgium and the Netherlands, to England and southwards to Switzerland. The only artistic remains which Azilian fishermen and hunters have left us are some painted pebbles on which extremely schematized human figures are represented (Fig. 247a), not dissimilar from present day objects obtained from wood or stone in Australia and Tasmania, called 'churinga'. Probably these pebbles represent – as do their present Australian equivalent – the spirits of the ancestors (Maringer).

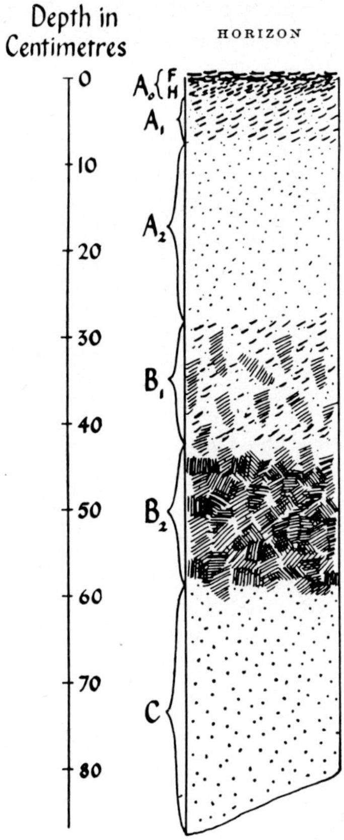

Fig. 245. Profile of a podzol from Surrey. F. *Calluna* remains and *Pinus* needles, almost unchanged. H. Very moist and dark humus. A_1. Light chocolate colour, sand mixed with humus. A_2. Very pale ashy grey, silica almost without humus. B_1. Blackish, due to deposition of humus, slightly ferrugineous. B_2. Dark brown due to presence of iron. C. Rich orange colour, very moist (from Friedländer).

The succeeding culture, or Tardenoisian, is also well represented all through the heaths of Western and North-Western Europe. In Jutland we find the Lingby industry, in Holstein the Ahrensburg industry (Fig. 248), in Lot-et-Garonne the Sauveterrian industry, in the Aisne Dept finally there is the true Tardenoisian industry. The Upper Tardenoisian and Campignan cultural periods are characterized all along the shores of the Atlantic Ocean and North Sea (for instance France, Belgium, the Netherlands and Germany) by the Kjökkermöddinger culture. Man in these regions and at that time dedicated himself to fishing: this is well documented by the remains of shells in places, even far from the sea, where primitive man is supposed to have lived. We must not forget in this connection that fishing is still today an activity of paramount importance in many towns and villages on the shores of the Atlantic Ocean, and that the use of marine molluscs and crustaceans in the diet of these people demonstrates the importance of sea resources in these regions. Very important Mesolithic sites belong to heathland: Téviec and Morbihan in Brittany, Bristol in England. In this connection we should also mention two mesolithic stations of Bretagne: Teviec, where microlites have been

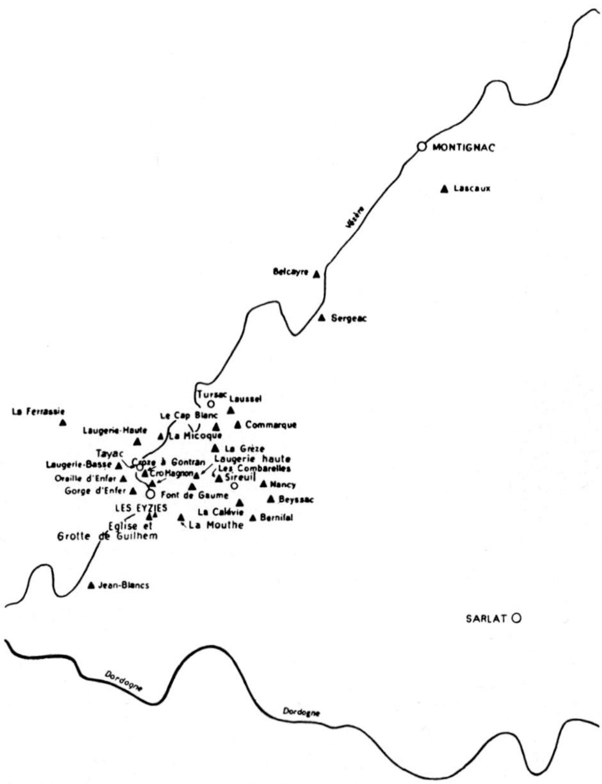

Fig. 246. The lower Vézère valley, were there are many sites of paleolithic hunters (from Maringer). The density of the human settlements in these early times of human history can give us an indirect idea of the human impact of some parts of the Continent.

Fig. 247a. Evidence of the presence of the bison (*Bison bonasus*) in heathlands of Western Europe (Dordogne): a carved bison with magic significance, belonging to the Chancelade culture (Upper Paleolithic).

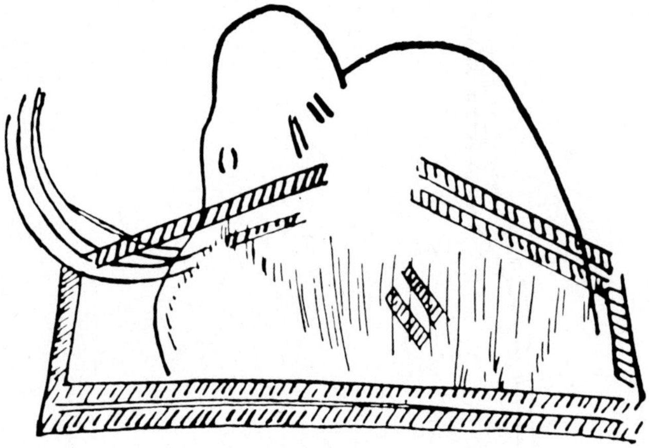

Fig. 247b. Evidence of the presence of the mammouth at Foret-de-Gaume, not far from a heathland territory during the Upper Paleolithic. This is also an example of magic (from Maringer).

Fig. 247c. Evidence of the presence of *Bos primigenius* (Ur) in French heathlands during the Upper Paleolithic: a bull following a cow; wall graffiti at Teyjat (Dordogne). Heaths were probably present immediately west of this territory.

found, and Morbihan, smaller than the former. Both of them were inhabited by shell-eaters, as the Kjökkermödinger testify. They economized the silex, which is a sign of a poor culture.

Even denser was the population on European heaths during the Neolithic time, when man showed signs of a further intellectual evolution, and therefore of the ability to alter nature to a greater degree.

The Neolithic is characterized in the area of heaths by the religion of the megalithic cycle, which began about 3000 years B.C., coming probably from the Mediterranean Basin (e.g. the monumental buildings of the shores of the Nile). The best known features of this cycle are the dolmen and the menhirs. The word

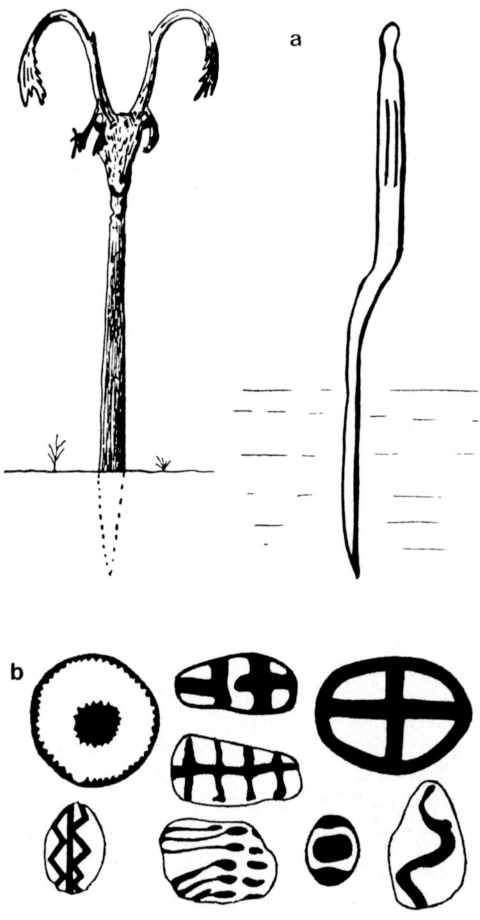

Fig. 248. a. Mesolithic remains in territories covered by heathland. Left, reconstruction of the so-called 'ritual pole' of Stellmoor; right, idol built with wood from Ahrensburg-Hopfenbach (Schleswig-Holstein). b. Painted Azilian pebbles, characteristic of Mesolithic of Western France (from Maringer).

Cultural periods in Western Europe	Northern Europe	Approximative absolute chronology
Upper Magdalenian	Hamburg-phase	15,000-10,000 years B.C.
Azilian, Lower Tardenoisian	Ahrensburg phase, Magdalenian culture	10,000-5,000 years B.C.
Upper Tardenoisian Campignan	Kjökken-möddinger culture	5000-3000 years B.C.
Beginning of Neolithic		
Full Neolithic	Neolithic	3500-2000 years B.C.

(from Maringer)

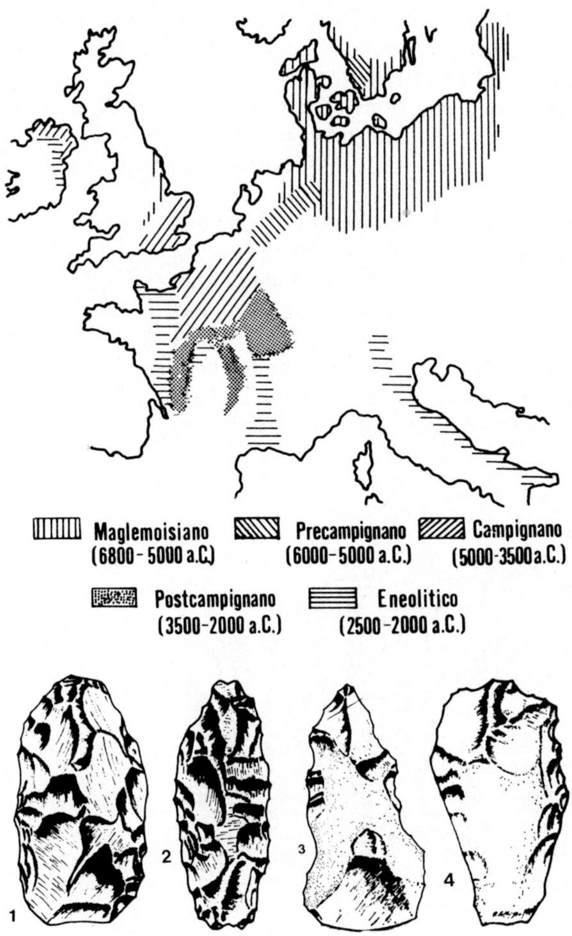

Fig. 249. Extension and subdivision of 'Campignan' sensu lato in Europe (from Corrain, modified): 1 and 2. tranchets; 3. axe; 4. scraper.

dolmen comes from the Breton words dol, for table and men, for stone; the word menhir from men and hir, for long. In Brittany there are nearly 3500 menhirs, in the region of Morbihan only. The Men-er-Hroech, near Locmariaquer, was 20.5 m high. When they are arranged in a circle, semicircle or oval they form a cromlech (from Breton words crom, circle, and lech, place), present both in Brittany and England. Famous is the cromlech of Stonehenge, near Salisbury, dating from 2123-1573 B.C. (^{14}C). Other similar monuments are present at Woodhenge, Avebury, Arminghaal and Arbor Low (Fig. 250). The pottery of this culture is characterized by the 'bell-shape cup'.

Sometimes menhirs are situated in several parallel rows, or 'Alignments': one example is that of Carnac (Fig. 251 and 252). The megalithic culture spread as far as Ireland.

In Brittany, more precisely in the standing stones of Kermarlo near Carnac, some signs exist which point to the existence of a celestial religion, besides that of the Earth: one is the axe, another the snake. The axe is the symbol of the thunderbolt hurled by the god of the heaven, the snake is the symbol of life. The presence in this period of some representations of red deer in Spain allows us to think that the Celtic religion – spread all through France and therefore the area of the heath, which considers this animal sacred – has its origin in an old religion of Neolithic times.

The megalithic culture also reached the Neolithic inhabitants of the heath regions of Northern Germany, such as Schleswig-Holstein (not far from Lüneburg) or Mecklenburg (east of Lüneburg). In this region a megalithic tomb has been found at Ziesendorf; in the former a tomb was found at Beldorf. In this tomb a stone was present with ornamentation which recalls the symbols of the idols of Western Europe. This means that at that time great migrations were performed all through the dominion of the heaths, which incidentally had more or less the same climatic and vegational characteristics as they have today.

Fig. 250. Megalithic construction in the region of old heathlands: Stonehenge near Salisbury (from Maringer).

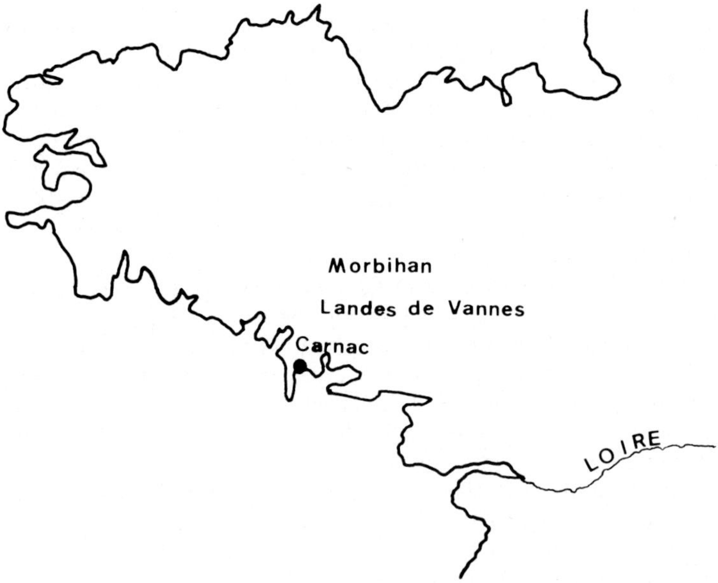

Fig. 251. Position of mesolithic (Morbihan) and Neolithic sites (Carnac) in Brittany, and presence of actual heaths (landes).

Fig. 252. The megalithic culture on the French heaths: 'alignments' at Carnac, in Brittany (from Maringer).

In the very heart of one of the greatest and most famous heaths, at Lüneburg, more precisely at Bavendorf, a demonstration of human sacrifices has clearly been revealed; other signs of this kind of sacrifice are present at Ziesendorf. According to Maringer, the origin of this ritual (sacrifical rites) must be looked for in the East, since similar phenomena are known for the regal tombs of Ur, 500 years older. According to this author, the religion would have come to Europe with the Indoeuropean (or Indogermanic) migrations, which reached Central Europe about 2000 B.C., thus having the chance of meeting the megalithic culture and all its religious manifestations.

The Nordic megalithic monuments also show symbolic signs relating to their religion of sun or heaven, such as the wheel and the hands. Finally, in the late Neolithic, a very remarkable feature is the axe made from amber or clay which, according to ethnologists must represent the celestial or solar god who hurls the thunderbolt. In this connection it is worth recalling the ancient religion of the Germans, with their god, Donar or Thor, called the Thunder God, who throws the thunderbolt hammer. We cannot say, however, that this religion had its origin in the religion of late Neolithic men, since in the Greek-Roman cycle also the divinity (Zeus or Jupiter) is the Thunder God, represented with a thunderbolt in his fist. Not only that, but from a linguistic point of view German is strictly related to Greek and Latin (comprising with these tongues the Centum- or Kmtom group of Indoeuropean languages) and not to the Celtic idiom which dominated in Western Europe, and still exists in some districts.

A sign that heaths are very old, and only in some cases due to human agency (felling) is given by: i) the ancient origin of the local names for heath, both in the countries where the Germanic languages are spoken (Germany, the Netherlands, England) and in the countries where today people speak French; ii) the abundance of toponyms or names of towns or villages which include the term or its equivalent in the different languages.

Whereas in the countries where a Germanic language is spoken only one name is used (heath, Heide), in France we find two quite different words employed for 'heath', namely bruyère and lande.

Bruyère comes in XII century from the Celtic name brucus, through the gallo-romanic 'brucaria' used also in late Latin for *Erica*. Brucus would correspond to Irish froech (Gamillscheg)[1]. The Italian name to indicate a heath (brughiera) also comes from brucus through the intermediary of Provençal 'bruguiera'. In Italy brugo is today the local name for *Calluna vulgaris*, but in several old dialects corresponding words are employed to indicate *Erica carnea* (Fig. 253). In France bruyère also means *Erica*; the names 'bruyère tetralix', for cross-leaved heath, and 'bruyère cendrée' for fine-leaved heath also exist. The flowers of *Erica tetralix* or *E. cinera* or heath bell in French is cloche de bruyère, which is the literal translation of heath bell. Heath cock or heath game (or even heath pout), i.e. *Lyrunus tetrix*, in French is called coq de bruyère or coq de bouleau; heath throstle or ring ouzel (i.e. *Turdus torquatus*) in French is grive de bruyère. As we

[1] The present brosse (shrub) is also derived from brucus; in the XII century we have broce for both heath and shrub; in Neo-Provencal we find brousso for *Erica* heath ('Besenheide') (Gamillscheg).

can see, the name bruyère is employed in France almost as heath or Heide is used in England and Germany respectively.

In France the term 'lande' is often used to indicate the territorries covered by heaths. The name comes from the Provençal landa which is a Celtic word (there is also the old Irish 'land' for open place) allied to some germanic words such as Land. In some places the lande passes gradually to a moor (see the chapter on humid environments). It is worth mentioning that in France a 'Department' has been named after lande (Landes, in South-Western France) (see later).

According to Devoto (1962) Lande is an extra-indoeuropean word penetrated into the indoeuropean area from the West. Examples are landa, in Catalan a low-land, lano in Bearnes countryside, landa, in the language of the Basques a field. From this territory the term would have reached Ireland (see above) and Germany (free space). According to Devoto, the case of heath is analogous: kaito appears in the local celtic names with the suffix -ceto, in the Welsh coit, adapted to designate an uncultivated area, whereas it has kept its original meaning in the Germanic languages (Heide, heath).

In German also there are several words derived from the name of the heath (Heide): Heidekraut means *Erica* as well as *Thymus, Myrice, Spartium scoparium, Empetrum nigrum, Ledum palustre, Pulsatilla patens, Daphne cneorum, Genista germanica* and even beech and birch. Besenheide means *Erica* and also groom (from Besen, the groom); Heidebrand is burning of stubble; Heideflachs is *Linaria*

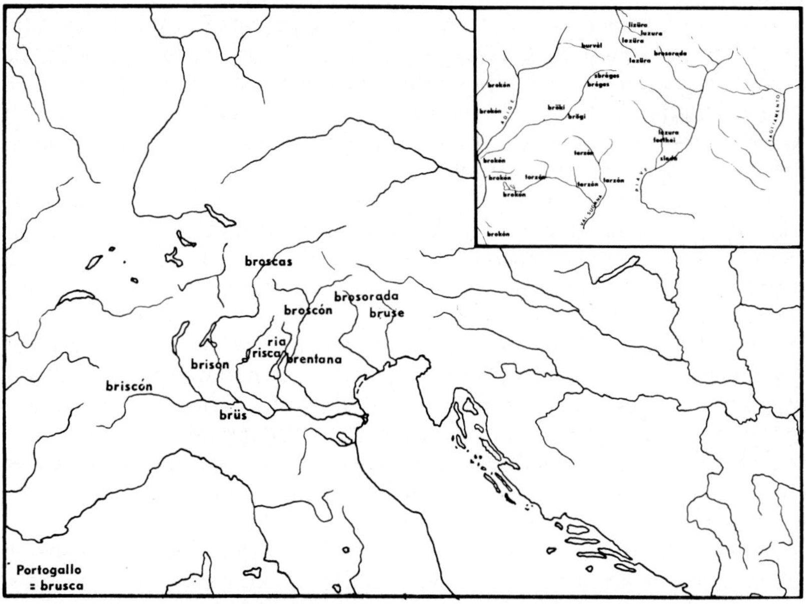

Fig. 253. Traces of the original Celtic 'brucus' for the Ericaceae *Erica carnea* in other parts of Europe. The name, of course, has been subjected to many alterations as always happens with old, pre-romanic, names (from Marcuzzi).

(a Scrophulariacea), Heideginster, *Ulex europaeus*; Heidegrütze, the same as Buchweizengrütze, pollard; Heidekorn, *Fagopyrum esculentum* (*Polygonum fagopyrum*); Hidelerche is *Alauda arborea*; Heiderose is the wild rose, *Rosa canina*; Heideschaf (or even Heidschnucke) is heath sheep; Heidegrab is dolmen; Heidegriess is semolina, made with grains of *Fagopyrum esculentum*; Heidelbeere is the fruit of the whortleberry, whereas Heidelbeerstrauch is the plant of *Vaccinium myrtillus*; Heidemehl (or even Heidenmehl) is the flour made with grains of *Fagopyrum esculentum*; Heiden is the plant *Fagopyrum e.*; Heidenröslein (or even Heideröslein) is the wild rose; Heidepfriem is *Ulex europaeus*, Heideschwamm is the mushroom *Psalliota campestris*, Heidehopfen means hop, Heidelilie is *Lilium martagon*, Heidemeise is a titmouse, Heidenelke is a *Dianthus* and finally Heideelster is the common magpie.

Not very dissimilar to the origin and meaning of Heide are those of heath, coming from the Middle English hethe or heeth and the Anglo-saxon heath, used both for heath and for the heath plant. The same name is used also for any of various shrubs and plants growing on heaths, and more especially for any plants of the genus *Erica*. We may observe at this point that, in Great Britain, both *Erica* and *Calluna vulgaris* are used for brooms, thatch and for heating ovens: it is very interesting from a semantic point of view that *Erica* in Italy is called 'scopa', which really means 'broom'.

In Britain we find – as in France and Germany – several flowers named after 'heath', such as heath bell (see above), heath berry, for any of various berries found on heath, e.g. bilberry or whortleberry (*Vaccinium myrtillus*) or crowberry (*Empetrum nigrum*); heath bird or heath grouse is the black grouse, *Lyrurus tetrix;* also heath cock or black cock is *L. tetrix*; heath corn or buck wheat is *Polygonum fagopyrum*; heath cypress is *Lycopodium alpinum*. Heather, from the Middle English haddyr, changed by association with heath (consonantic dissimilation, d ⟶ t) is *Calluna vulgaris.* The heath covered with heather is also called heather. Heather grass is *Triodia decumbens*; heath pea is *Lathyrus tuberosus*, a leguminous plant the tubers of which are eaten as vegetables.

We wish to mention at this point a very interesting semantic, and at the same time linguistic, problem which concerns a word derived from heath, even if the relation to ecology is barely apparent. There is an English word, heathen, which means 'pagan', coming from heath and more precisely from the Gothic haithi, due to a IV century mistranslation of the Armorican (the Celtic language spoken in north-western France) hethanos (heathen), by association with haithi (heath), and then misunderstood to mean 'wasteland dweller'. It is noticeable, again from a linguistic point of view, that the same happened with German language, where also we find two meanings for the word Heide (heath and pagan).

The origin of the terms heath, Heide employed in the countries where a Germanic language is spoken on one side, and that of the term tojal or lande, employed in SW-France on the other, is very ancient.

According to Kluge, in Middle Hochdeutsch and in Dutch we find Heide, in Old Saxon hetha, in Middle Dutch hede, in Anglo-Saxon heath, for uninhabited land, as well as for cultivated area (bebautet Land). The Old Nordic heidr means a desert high plateau (wüste Hochflach), the Swedish hed, Danish hede, Gothic haithi mean field and also cultivated field; koitia is an uninhabited land, or wilderness; in the ancient Celtic we find keito, from which comes the Gaulic root caeto – as in the toponym Kaitobrix, or ceto as in the toponyms Cetobriga, Etocetum. Finally, in old Cimbric we have coid, in Cimbric coed, in Old Cornish cuit, in Breton koad or coet (for forest).

According to Onion's Dictionary, heath is an open waste land, coming from the Gothic haithi[1], common Germanic chaithis[2], and Indo-European kait. In old Welsh coi and in recent Welsh coed the name is used to indicate a forest.

According to Pokorny, whereas kaito in Celtic and Germanic is forest, an area of land not built upon (whence the already mentioned terms existing in ancient Cimbric, Neo-Cimbric, Old Cornish, Middle Cornish and Breton employed to indicate a woodland instead of a heath), in Gothic and in more recent languages (i.e. German and English) the term is used to indicate a 'heath'.

A different position is assumed by Devoto. According to this author, the origin of words such as the Catalan landa (plain, lowland), Bearnais lano (country) the Basque landa (field) goes back to the era of the extra-indoeuropean culture characterized by the bell-shape pottery of which we have already spoken (par. 6). From Western Europe the term would have spread eastwards, so that we find it today in the Irish land, for 'fenced space', and in the German Land for 'free space'. According to Devoto the origin of kaito, heath, of which we have already spoken is parallel to this origin: whereas in the Gallic-ceto and in Welsh coit this term adapts itself to indicate 'an area not cultivated by man', and therefore a 'forest', in Germanic (Heide) this term maintains its original meaning. According to Devoto, both Heide and Land (as well as town, corresponding to the German 'Zaun', enclosure) have a western extra-indoeuropean origin, and would be characteristic of both Celtic and Germanic languages. (This theory is advanced also by Hubschmid 1954 for Land)[3].

The origin of 'tojal' is also very ancient and interesting, both from a linguistic and an ecological point of view. This term employed as we have seen (p. 343) in SW France comes from the Gascon touyo or toujo, or from some words existing in the Landes such as touje, touye, or at Quercy (Lot) toutcho, in Aragonese toya, toxa, Spanish tojo, Asturian toxo, Galician toxo, or Portuguese tojo, employed to indicate the gorse *Ulex europaeus*, a typical heathland shrub. These names are all of pre-Roman origin, and only after the Roman invasion 'tojal' originated (-al, frequent in Spanish for some plant formations). According to some, it comes from the Iberian toju or toja. In Aragon territory there is the toponym Toyagosa (near Teruel). According to some authors, the Basque ote or othe, employed to indicate the gorse, could have originated by a simple metathesis from toja (otea ⟶ otja ⟶ toja) (Devoto, p. 45).

It is needless to say that these words must be very old and therefore the presence of a heathland vegetation must also be very ancient, and not due to human disturbances (disclimax, as some say).

We may conclude at this point that Heide (or any equivalent term) in the past

[1] th more or less as ϑ

[2] ch in lieu of the Greek letter χ

[3] We wish point out in this context that even in 1901 Graebner advanced – though on very different grounds – a pre-Germanic origin for the word Heide.

had several meanings, but on the whole it always refers to an open formation or landscape, and not to a woodland. Thus in the ancient Dutch heide is an open, free, field, also of wheat, and heiden a peasant. The North Scandinavian word heidr is translated with saltus, montana, and in the Edda (early Icelandic work) it indicates a land destined for battles (battlefield) or indicates 'the appearance of the wolf'. In Anglo-Saxon there is something analogous to Old Dutch: Beowulf uses the term grau Heide for a wild, rocky territory with some crevaces near the sea, whereas haede is translated with campus incultus. Finally, according to Graebner, the term Magetheide would mean simply maget Heide, i.e. great heath, and more precisely the Lüneburger Heide.

The vegetation

The following heath types can be – though not always very easily – distinguished: pure heath; grassy heath (Grasheide) and woodland heath (Waldheide).

The pure heath is extremely varied and a general description is impossible. Let us mention here the *Calluna* heath, with absolute dominance of *Calluna*. This type comprises, according to Graebner, the heath κατ' εξοχην present e.g. in Northern Germany, Brandenburg, Western Prussia, etc. A raw humus, sometimes without a hardpan, is characteristic. The plants are lichens, *Carex, Genista, Gnaphalium, Hieracium, Boletus luteus*, small *Pinus silvestris* plants, etc. The *Erica tetralix* heath is also very important: it is more humid than the *Calluna* heath, sometimes composed of pure white sand (the German 'Bleisand'), with a vegetation formed by *Sphagnum* and, in some circumstances, *Juncus squarrosus, Scirpus caespitosus, Eriophorum vaginatum, Salix rosmarinifolia, Genista anglica* and *Pedicularis silvatica* (Fig. 254).

The *Empetrum* heath is generally mixed with other types: a particular case is that of an *Empetrum* heath on sand dunes, with *Ochroporus perennis, Polytrichum piliferum, Hypnum cupressiforme* var. *ericetorum* and var. *elatum, Carex* spp. etc. Much more developed is the so-called Moor-heath (Heidemoor or even Moosmoor) present exclusively in very wet places: the presence of *Sphagnum* (or, if the environment is drier, *Polytrichum juniperinum*) is characteristic together with *Eriphorum vaginatum* (forming some very strange tufts, called in German Bülten') and some plants typical of heaths, such as *Calluna, Myrica gale, Ledum palustre* and *Vaccinium uliginosum*. Examples of this kind of heath are present at Soltau and Munster in the Lüneburger Heide or in the Grunewald near Berlin. Analogous environments are present in the Alps also at about 1000-1500 m on siliceous rocks, for instance in the Dolomites at San Martino di Castrozza. In the most humid situations the vegetation approaches the dominion of the helophytes (see the chapter on humid environments and fresh water bodies), whereas at the edges of the heath are always some trees (*Pinus, Betula*) and shrubs (*Rhamnus frangula* and other). The number of plant species living in the heaths is very great. Many of them are 'characteristic species' ('Charakterpflanzen') such as *Myrica gale, Pulsatilla vulgaris, Ranunculus hederaceus, Genista pilosa, G. anglica, Ulex europaeus, Empetrum nigrum, Hypericum pulchrum, H. guttatum, Erica tetralix, Orobanche rapum genistae* (parasitic upon *Sarothamnus*) etc. Much greater, how-

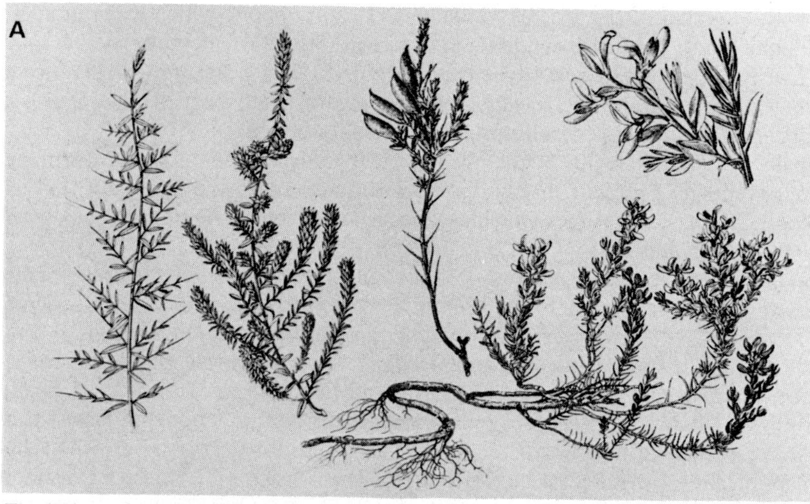

Fig. 254a. *Genista anglica*, a typical heath plant (see the map of its geographical distribution). a. Habitus of the plant; b. flower carrying branch; c. fruit carrying branch; d. plant grown in a dry environment; e. plant grown in water saturated air (from Hegi).

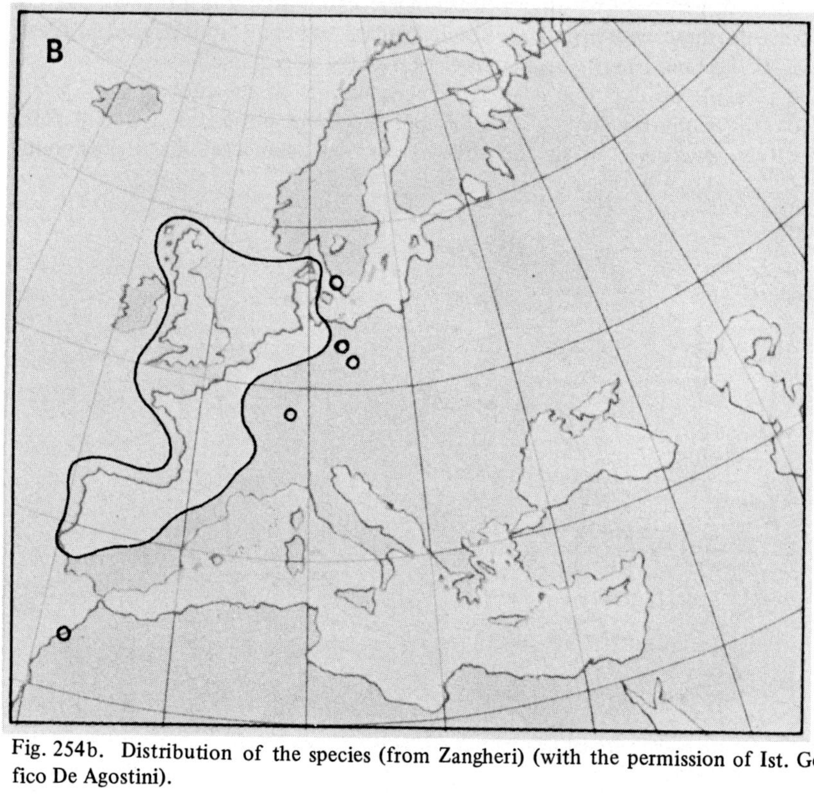

Fig. 254b. Distribution of the species (from Zangheri) (with the permission of Ist. Geografico De Agostini).

ever, is the number of the species which avoid heaths, among which is the juniper.

In many places the heath can grade into a Heath-moor (Heidemoor) which may have had its origin from a sandy barren area or from a grassland (in this case a Wiesenmoor): in Britain one speaks in some cases of heath peat, to distinguish a peat different from that generally found in mires (or bogs).

Finally we should mention the *Sarothamnus scoparius* heath (or in Germany 'Besenginsterheide') present in dry places, with loose sand, and a very scant vegetation of *Calluna* (sometimes missing), *Cladonia, Rhacomitrium canescens, Aira flexuosa, Potentilla cinerea, Viola tricolor*, etc.

The woodland-heath (Waldheide) shows all stages from pure heath to pine forest, or even broadleaved forest. We find scattered trees and the soil covered with moss (*Hypnum schreberi, Hyloconium* spp.) in which *Calluna* grows, as well as lichens, mushrooms (*Boletus luteus*), *Sphagnum,* graminoid plants (*Festuca, Agrostis, Calamagrostis*), *Luzula campestris* and *Salix aurita*. This type of heath can also be recognized in the Alps on siliceous rocks. Within woodland-heaths it is possible to recognize pine-heaths and broad-leaved forest heaths. The former are much more abundant at least in Northern Germany. The pine forest has a floor covered with *Calluna* (even on limestone), together with lichens, mosses, fungi and a great number of flowering plants (*Sieglingia decumbens, Molinia coerulea, Nardus, Eriophorum gracile, Carex* spp., *Pulsatilla* etc.). As far as broad-leaved forests are concerned, generally we find a heath in succession to a forest (beech, birch or even oak). In these circumstances, beside *Calluna,* we may find *Sieglingia, Carex arenaria, C. ligerica, Genista anglica*, etc. Also in the Alps some heaths of this type may be present: an example is the plant community at the foot of the Eastern Alps, on the Colli Euganei, where *Calluna* grows in close contact with *Erica arborea* (e.g. M. Rua). When the soil is very dry, subjected to a continental

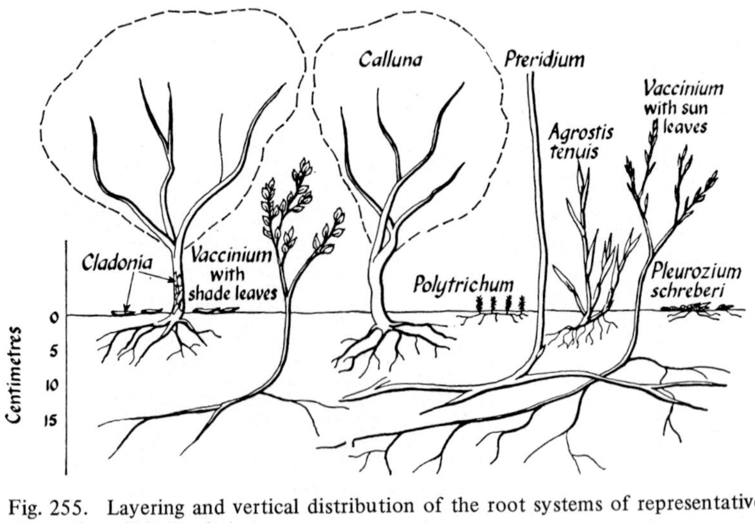

Fig. 255. Layering and vertical distribution of the root systems of representative heathland plants (from Friedlander).

climate, and is also poor in salts and nutrients, instead of a steppe, a particular heath may develop ('Heidekrautlose Sandfeld'). In these circumstances the only flowering plant will be *Weingaertneria canescens*, during the summer completely desiccated, together with *Spergula vernalis, Teesdalea nudicaulis* and *Erophila verna,* and the lichen *Cornicularia aculeata. Calluna* may be absent. Such arid formations can be seen near the Elbe at Pretzsch.

Finally, we must recall the environments which represent the transition from heath to halophytic vegetation, an Alnetum, a grassland, a grassy moor (Wiesenmoore), and lastly to a steppe formation, for instance in Eastern Germany or better in Eastern Central Europe, where a continental climate prevails.

Ecological implications of heath plants

As a consequence of the very severe environment prevailing in heaths, several adaptive mechanisms have evolved in the plants which have been successful in the colonization of this environment. The ling, *Calluna vulgaris*, is a small evergreen shrub with a shallow rooting system. It contributes a distinct layer giving shelter from wind and extreme temperatures and humidities to smaller plants (as well as animals) beneath it (Fig. 255).[1] The bilberry (or whortleberry), *Vaccinium myrtil-*

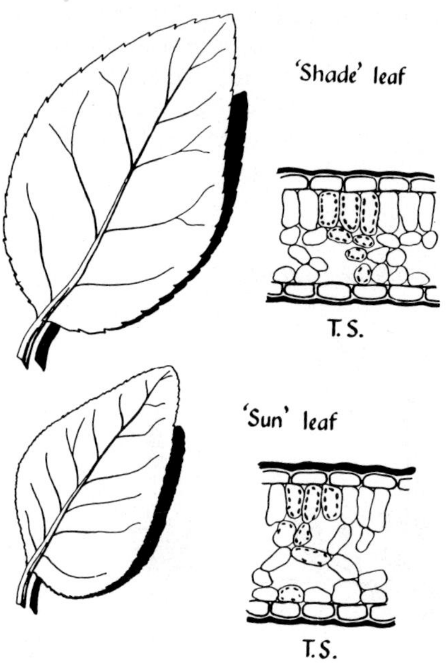

Fig. 256. Sun and shade leaves of *Vaccinium myrtillus*. Right: transverse sections of the leaves (from Friedlander).

[1] The species is widespread in Europe and Western Asia, as well as in the Moroccan Rif and on the Azores.

lus, often makes a subsidiary layer under the ling, but in some cases it is a dominant plant. It is able to live on the same soil as ling because it uses a different part of the soil profile, so reducing the competition. The bilberry is a variable plant, bearing 'sun' and 'shadow' leaves (Fig. 256). Sun leaves are a xeromorphic adaptation and are produced when the plant grows in exposed places, such as after burning of heathland, whereas shadow leaves are produced under sheltered conditions such as in woodlands ('Waldheide'). The shade leaves are larger, paler and less heavily cuticularized. The bracken, *Pteridium aquilinum*, is dominant in some heaths, being especially abundant near existing or felled woodlands. Its strong rhizome branches freely and leads to rapid colonization, and the deep shadow which it casts allows very little to grow beneath it, except for bryophytes. Bracken can in some cases shade out almost all other heathland vegetation. The graminoid plants are also very interesting (Fig. 257). *Deschampsia flexuosa,* the wavy hair-grass, is frequently dominant together with ling. It possesses slender rhizomes but usually does not occupy the whole of the fround, so that it allows other plants to grow with it. On the contrary, *Nardus stricta*, the mat-grass, grows so closely packed together that it excludes other vegetation. Finally the common bent, *Agrostis tenuis*, and the sheep's fescue, *Festuca ovina*, may replace ling on upland heaths; in the south they may become dominant as a result of excessive grazing.

The bell or Scots heather, *Erica cinerea*, may replace ling in many places, or may be co-dominant. Both plants may be found living close together, so *Erica* and *Calluna* must be ecologically similar. In moist places the cross-leaved heather, *Erica tetralix*, often replaces them, being of course dominant. Other plants more or less adapted to life in heathland are *Rumex acetosella, Plantago coronopus, Plantago lanceolata, Galium hercynicum* and the common tormentil, *Potentilla tormentilla* (= *erecta*).

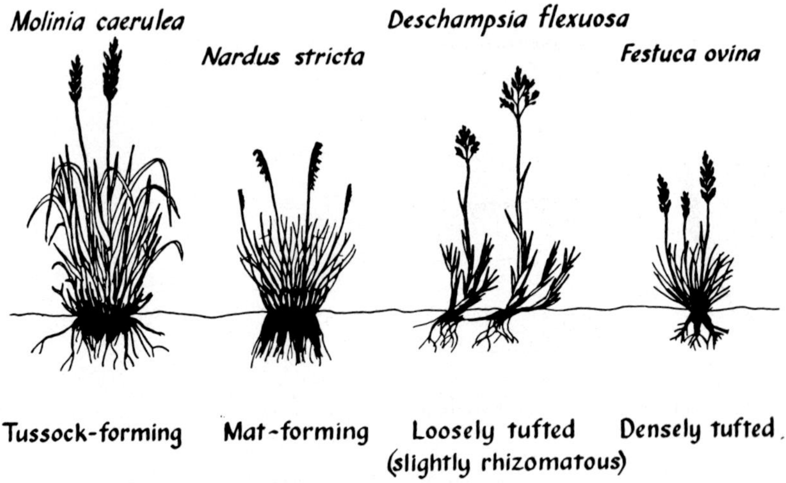

Fig. 257. The mode of growth of some heathland grasses (from Friedlander).

The gorse, *Ulex europaeus* is very common on heathlands, although by no means confined to them. In Europe it is limited to the western part, reaching as far south-east as Northern Italy. It prefers soils poor in bases. Sometimes it may form small communities but more frequently grows as isolated bushes. It is very well developed in Brittany. Another leguminous shrub, the broom *Sarothamnus scoparius*, is more confined to sandy habitats than the gorse. It is common but generally not as abundant as the gorse, even though it is one of the most rabbit-resistant plants in Europe. The gorse is a typical example of xeromorphic modification (Fig. 258) with both stems and leaves forming spines, thus reducing the area over which transpiration occurs.

Several Gramineae and Juncaceae prefer moist places: let us remember in this connection *Holcus lanatus*, the Yorkshire fog, the heath grass *Sieglingia decumbens*, the purple moor-grass *Molinia caerulea*, together with the reeds, *Juncus conglomeratus* and *Juncus squarrosus* (Fig. 259).

In the Atlantic section of heathland some 30 paleoendemic plant species can be localized. Two of them belong to two endemic genera, namely *Daboecia* (an Ericaea) and *Thorella* (an Umbellifera). Some of these endemic species grow only on the fixed dunes near the sea, though not all halophilous. The primitive origin of Ericaceae seems to be Africa, that of the genus *Wahlenbergia* West Africa; the source of *Ulex* and *Sarothamnus* on the contrary seems to be Iberian.

Origin of the heaths

The origin of heaths is varied and we are not always able to establish for a particular case what its origin was. In general one postulates a secondary origin for most heaths, due principally to felling and/or fires followed in some instances by land cultivation and/or cattle raising and grazing. This view, on the whole, is assumed by Hardy (1905) for the Scottish mires and heaths, De Martonne (1950) for heathland in general and more recently by Friedlander for English heaths.

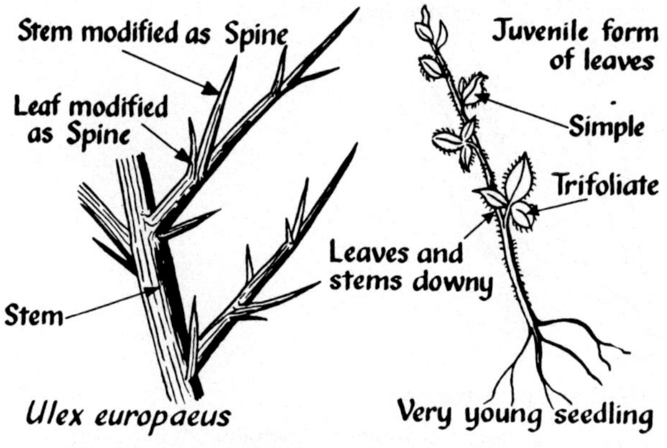

Fig. 258. Example of adaptation (xeromorphic) in a heath plant, the gorse *Ulex europaeus* (from Friedlander).

Palynological investigations have recently shown that originally, in Normandy, the post-glacial was characterized by the development of forests of different types. The beech was followed by *Alnus* and *Betula*. During the sub-boreal and the sub-Atlantic the effects of human agency on the plant landscape became stronger and stronger. During the sub-Atlantic the disappearance of the forest elements coincided with a 'boom' in *Gramineae, Cyperaceae* and *Ericaceae*. At the boundary between sub-boreal and sub-Atlantic, as a consequence of human presence, there is pollen of cereals, *Plantago, Rumex* etc. According to Frileux & Hualt, before the Atlantic there was already some felling which permitted the

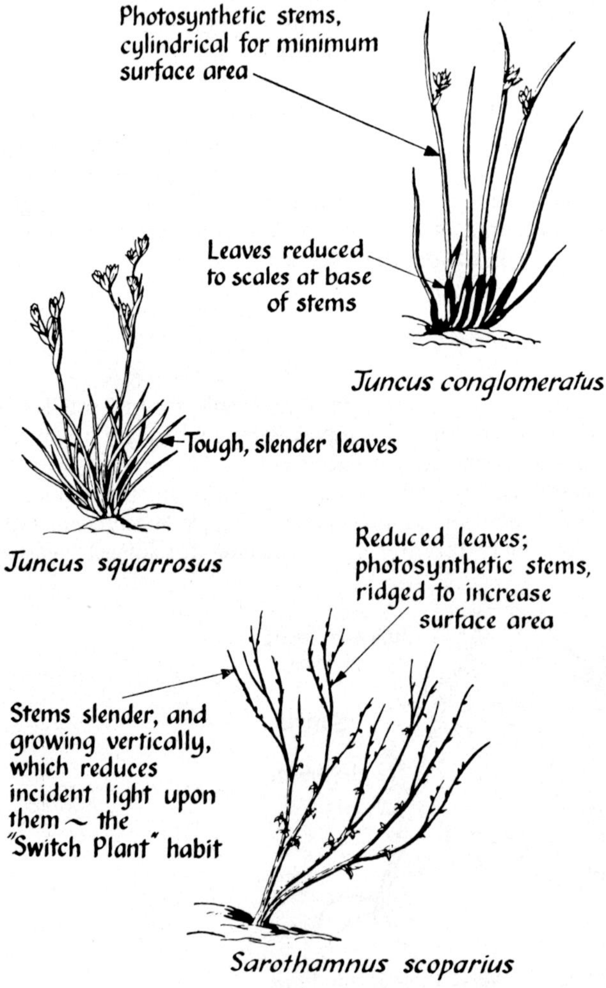

Fig. 259. Morphological adaptations to heath environment (xeromorphy) in three plant species: the soft rush *Juncus conglomeratus*; the heath rush *Juncus squarrosus* and the broom, *Sarothamnus scoparius* (from Friedlander).

extension of meadows. In the middle of the Atlantic, in the Bray, the cultivations make their appearance; in the sub-Atlantic the *Alnus* was very much utilized by man, which has also been observed in England during the Bronze and Iron Ages. The felling brought about development of birches. In the case of Normandy (Bray) the origin of heathland should therefore be a secondary one.

On the other hand we may imagine that many present-day heaths had a different origin. De Martonne, for instance, points to the origin of heathland from acid swamps and mires. More precisely (p. 1191) on the backshore of Fontainebleau sands, where water seepage occurs heath is formed with *Sphagnum, Carex* etc. This formation is called 'bruyère spongieuse' or literally spongy heath. If it is drained, a dry heath forms, with *Pteridium aquilinum* and *Calluna vulgaris* (or in some instances birch). According to De Martonne, 'les peuplements de resineux representent souvent le stade finale artificiel dont la destruction ramène la lande sèche'. Heath would therefore – at least in these situations – be a natural, let us say climactic, formation. The so-called spongy heath, with *Erica tetralix*, according to De Martonne is common in Western and Central France. It was studied by Allorge in 1922.

In other cases a heath can have its origin from a mire: in France, in the climate of Paris, the mire shows a series of facies in which the following plants appear: *Sphagnum, Carex, Eriophorum, Myrica, Erica tetralix* and *Erica ciliaris.* At the end of the 'sere' (or succession) some xerophilous species make their appearance, such as *Calluna vulgaris, Ulex nanus* and *Potentilla tormentilla*. 'La lande se trouve ainsi constituée (De Martonne, p. 1200).

Another argument for the natural origin of at least some heaths is furnished by Backman and other authorities, who have observed that a forest can acidify the soil on which it has developed, so preparing a suitable environment for a mire or heath. Backman has shown that the vast majority of *Bothnia* mires (Sphagnetum) has originated in the burying of forests. In order to have this transformation of a forest into a bog, the soil must be poor in soluble mineral substances. Of course, we may also imagine the successive transformation of a bog into a heath.

Not very different is the mechanism by which, in North Germany, the acid humus of woodlands brings about the formation of a heath. According to Graebner (1901), organic compounds from the humus of very old forests have caused the formation – in the deep layer of the plains – of grey soils which have sometimes been responsible for heath formation (cf. also De Martonne, p. 1150).

Finally, Friedlander admits the existence of a almost climactic heath in exposed places near the west coast of Cornwall, on soil derived from serpentine rocks, where very strong winds probably interfered with the growth of trees, so that dwarf shrubs are the natural dominants.[1] We think that probably wind, fog, the climate as a whole and perhaps soil have prevented, since the end of Pleistocene, the formation of forests in other parts of Western and North-Western Europe also, not only where the ice had destroyed any form of life, as in most of the British Isles, but also in a great part of Western Europe (The Netherlands, Belgium

[1] De Martonne (1950) thinks that there is a original heath only in places originally treeless, where soil is poor, as in fixed dunes covered with acid humus, alluvial soils or rocks covered with a scant amount of humus.

and North Western France) where during Pleistocene times only a tundra-like vegetation was present. In at least one area of these regions, the development of a forest has never been possible.

To conclude, if in some regions felling was an element which caused a deep modification in the environment, with the development of a heath and a consequently accentuated podzolisation – which, together with grazing and fire, made reforestation impossible – in other regions we must assume that heaths are very old and not secondary to a former deciduous forest.

One proof of that – even if not conclusive – is the existence of very old terms employed to indicate heaths. It is quite possible that in some regions, such as England, heaths are the consequence of felling which occurred only after the Roman invasion, as Friedlander points out, but the existence of pre-Germanic, extra-Indoeuropean names to indicate heaths both in England and in Northern Germany, where today germanic languages are spoken, as well as in France, is a confirmation that heaths did exist well before not only the Roman invasion (which would be mainly responsible for felling) but also before the first appearance of the Germanic peoples. (The same of course holds for France.)

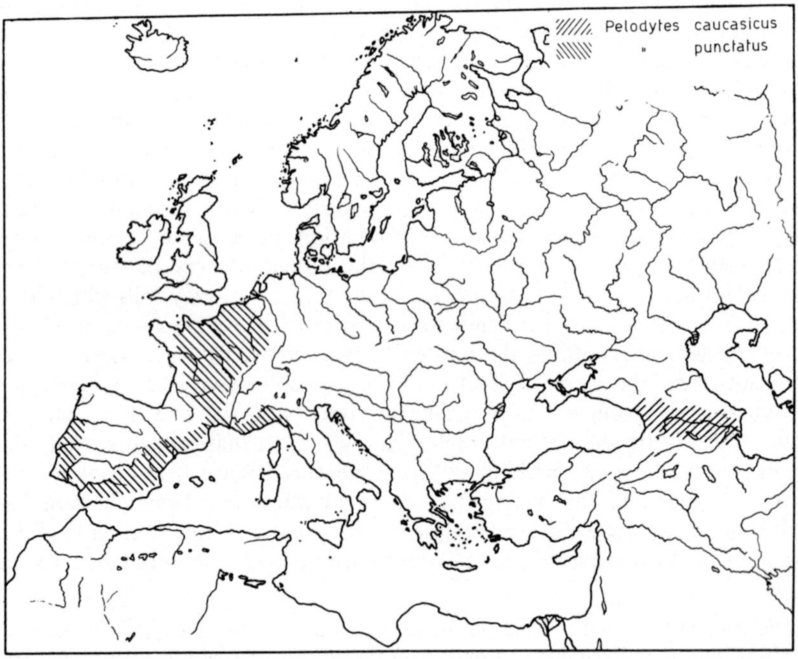

Fig. 260. Example of distribution limited to Western Europe, the Anuran *Pelodytes punctatus.*

The heath animals

The vertebrate fauna

On the whole, the heath vertebrate fauna is by no means typical of heaths, and may be considered an impoverished forest fauna. Indeed only common, rather euryoikous species are to be found.

The reptiles are represented by the adder *Vipera berus* and its prey, the common northern lizard, *Lacerta vivipara*. Both of them are characteristic heath species, and may be found there during much of the year. An interesting amphibian of heathland is the Anuran *Pelodytes punctatus* (Fig. 260).

Rather abundant are the birds, although few of them are characteristic of heaths: there is the linnet, *Carduelis (= Linota) cannabina*, so-called because it feeds on the seed of flax (linum), and secondarily *C. chloris, C. carduelis* and *C. flammea*. The latter is represented in the British Isles by the race *f. disruptis*. Very common is the lark (or skylark) *Alauda arvensis* together with several small insectivorous passerine birds. The main birds of prey are the kestrel, *Falco tinnunculus*, and the nocturnal bird *Asio flammeus* or the short-eared owl. One of the most interesting birds present in the British Isles, more precisely Cornwall, Wales and Western Scotland, is the black grouse, *Lyrurus tetrix*; less restricted to the heath environment is the red grouse, *Lagopus scoticus*, a species exclusive to the British Isles, living mainly on moors but breeding on heaths.

Mammals are represented mainly by mice, rabbits and shrews. The latter are essentially *Sorex minutus, Sorex araneus* and *Crocidura russula*; in the past *Crocidura suaveolens* was also rather common. The rabbit *Oryctolagus cuniculus* is still fairly well represented today. Very abundant are the mice, represented by the following species, each with a different ecological niche, circadian rhythm and behaviour: bank vole, *Clethrionomys glareolus*, water vole *Arvicola amphibius*, restricted to water and therefore to the most humid places within the heath; *Pitymys subterraneus*, or pine vole, gregarious but not social; *Microtus arvalis*, or common vole (known also as Orkney vole or Guernsey vole), gregarious but barely social, *Microtus agrestis*, gregarious and social, the harvest mouse, *Micromys minutus* and lastly the long-tailed field mouse *Apodemus silvaticus,* an ecological vicariant of *A. flavicollis*, limited to forests.

The invertebrates

The soil fauna

The soil fauna of heaths is composed of micro-arthropods, *Enchytraeidae,* beetles, centipedes (*Lithobiomorpha*). Nematodes, isopods (or woodlice), millipedes (*Diplopoda*) and gastropods are very rare. Earthworms may also be completely absent.

As a consequence of the large temperature and humidity variations, characteristic of heathland, and the presence of raw humus (which is sometimes the main humus type in heaths), the soil fauna of this environment is scarce compared to that of broad-leaved forests growing at the same latitudes,

On a heath in Surrey (Hadley Heath) Friedlander found the following data referring to air and soil humidity (May 1958):

Situation	day time temperature 16.5°C	night time temperature 7.2°C
1 m above ground level on open heath	40%	60%
at ground level among grasses	45%	69%
in the A_1 horizon under grass	74%	87%
at ground level under dead bracken	40%	62%
in the A_0 horizon under dead bracken	72%	85%

Friedlander has obtained the following data for the macrofauna of an English heath:

	at ground level under dead bracken (r.h. 40%)	in the litter under dead bracken (r.h. 72%)	in the A_1 soil horizon among grass roots (r.h. 74%)
annelids	0	0	2
snails	0	0	1
slugs	0	2	1
woodlice (*Philoscia*)	0	2	2
spiders	3	5	1
mites	0	4	0
centipedes	0	6	2
springtails	3	15	5
earwigs	1	2	1
carabid beetles	1	3	1
elaterid beetles	0	1	3
rove beetles (*)	3	9	4
Lepidoptera pupae	0	0	3
Total	11	49	26

(*) *Staphylinidae*

As a consequence of the low humidity woodlice are very scarce: in a heath studied by Friedlander, on a silicaceous sandy soil, practically only *Philoscia* was found, although this genus is much less resistant to dry air than the genera *Oniscus* and *Armadillidium*. The only possible explanation of this is the need of calcium shown by *Armadillidium* (at least in the British Isles). This would be one clear example of an interaction between two or more environmental factors. The lack of calcium carbonate is certainly the reason for the low snail count. Indeed, of the few species of molluscs which occur on heathland, the most abundant is the slug *Agriolimax agrestis*, a common, widely distributed species. Other molluscs may spread from neighbouring woodland to the heath in wet weather: an example is the brown-lipped snail, *Cepaea nemoralis*. On the other hand, the presence of sandy soil in heathland permits the existence of several soil (or geophilous) animals, which live in burrows or which lay their eggs in small pits in the ground. This is the case for the larva of the tiger beetle, *Cicindela campestris*, and for many solitary wasps and bees which use the sandy ground for their burrows: *Pompilidae* or spider-hunting wasps, *Philanthidae, Sphecidae* or sand-wasps and the solitary bee *Andrena*.

Orthoptera are not very common in heathland: two widely distributed species are *Metrioptera brachyptera* and *Chorthippus vagans*, present in most of Europe except for the southernmost part, and *Myrmeleotettix maculatus*, characteristic and very common on English heaths. Besides several bugs or *Heteroptera*, a great part of the epigeous insects are *Lepidoptera* (i.e. butterflies and moths) (Fig. 261). There are *Diacrisia sannio, Anarta myrtilli*, many *Geometridae, Tortricidae, Coleophoridae, Gelechiidae* (for instance *Gelechia betulae*), the small heath butterfly *Coenonympha pamphilus*, the common moth *Ematurga atomaria*, the cinnabar moth *Hypocrita jacobaeae*, furnished with aposematic coloration and – as is generally the case with warning colouration in most animals – with secretions which make it an unacceptable insect to many vertebrates, the grayling butterfly *Eumenis semele* and the grass emerald moth *Pseudoterpna pruinata*. Among beetles we find

Fig. 261. The small heath butterfly, *Coenonympha pamphilus*, on a branch of the cross-leaved heath, *Erica tetralix*. Note the form of the flowers (top left). They are somewhat bell-shaped, like those of the bell or Scots heather, *Erica cinerea* (from Friedlander).

Cicindela silvatica, many lapidicolous or sand-dwelling carabids, among them *Pterostichus angustatus* and *P. oblongopunctatus*, present all through Central-Northern Europe and Siberia, some weevils such as *Micrelus ericae*, 1.5 mm long, black, with a red rostrum, legs and antennae, present also in Southern Europe, the scarab beetle *Typhaeus typhaeus*, in which the male is characterized by two great horns situated on the sides of the prothorax. Also *Diptera* and *Hymenoptera* are fairly abundant.

A quite particular fauna is restricted to birches, which are present in heathlands either as scattered shrub trees, or constitute true small woods (woodland heaths). Let us remember some *Rhynchota,* many *Lepidoptera,* whose caterpillar is very often monophagous (i.e. feeding exclusively on birch), phytophagous *Hymenoptera* (*Tenthredinidae*) and *Coleoptera* such as the weevil *Bycticus betulae*, a small blue-green insect, with metallic elytrae, which causes a curling up of the leaf in which the larva develops. In Italy this species can develop also on vines. Perhaps richer is the fauna limited to pines. Very interesting are the parasitic *Rhynchota* belonging to the *Typhlocybidae, Aphidae, Adelgidae* and *Margarodidae,* several *Thysanoptera* (easily recognizable by a pair of elegant fringed wings), quite a few *Neuroptera* and *Lepidoptera.* The latter are mainly represented by some very harmful species such as the Noctuid *Panolis flammea*, the Geometrid *Bupalus piniarius* and the pine sphynx, *Hyloicus pinastri*. *Panolis* and *Hyloicus* are well known thanks to their population fluctuations which can have great importance in forestry, and can permit us to predict – at least within limits – the most heavy attacks on forests by these two species (Fig. 262). *Bupalus piniarius* has been studied for 13 years from the standpoint of population dynamics (density, reproduction and mortality rate of natural populations). It has been shown that the number of individuals is kept relatively constant by means of an auto-regulatory mechanism which acts through mutual contact among the larvae at a low population density for a given amount of food: an increase in number of contacts causes a dimunution in adult fecundity and young vitality. The beetles observed on pines on english heaths are partly useful since they destroy harmful insects (such as *Cleridae Thanasimus formicarius*, a reddish yellow insect with black legs and antennae, black head, part of pronotum and elytrae, the latter with red and

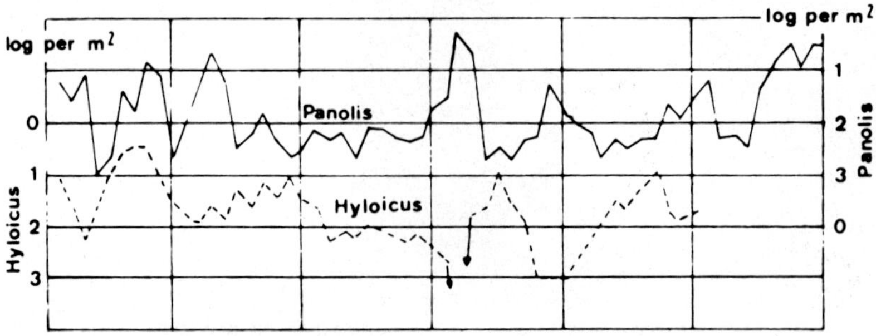

Fig. 262. Fluctuations in natural populations of *Panolis* and *Hyloicus* (larvae) at Letzingen (according to Verley, in Odum, modified).

white bands; some *Coccinellidae* preying upon leaf-lice and some *Carabidae*), partly harmful since they live at the expense of living trees or sometimes of dead wood (wood piles, timber). These animals are mainly *Anobiidae* and *Cerambycidae.* We may mention here only the *Cerambycidae Criocephalus ferus* (in felled trunks), *Rhagium bifasciatum* (in old stumps) and *Asemum striatum* (in dead trunks). The latter has recently extended its area of distribution to Southern England, from Northern Britain. Weevils and scolitids must also be included in this category.

We should mention also some chrysomelids present on English heaths: *Cryptocephalus 10-maculatus*, exclusive to Central-Northern Europe and Siberia; *Labidostomis tridentata*, widespread in Central Europe and Western Siberia, reaching the Alps and Prealps in the South; it is restricted to oaks, secondarily to hazel and willow; *Clytra quadripunctata*, showing aposematic or warning colouration (and therefore very probably furnished with some repellent substance) widespread in Central-Northern Europe and Siberia, southwards to the Alps, associated with willows, oaks and shrubs; the female lays its eggs on the ground. The two species of *Luperus* (*longicornis* and *flavipes*) exhibit a colouration we can also consider aposematic; *L. longicornis* is limited to Central Europe (including the Alps and Northern Italy) and Western Siberia; it lives on several shrubs, especially birch, alder and willow; *L. flavipes* is spread through all Central-Northern Europe, south to Tuscany, east to Central Asia, limited to the same plants as the preceding species, and lastly the elegant *Lochmaea capreae* (which literally means linked to *Salix caprea*, a montane willow) with aposematic colouration (black head and

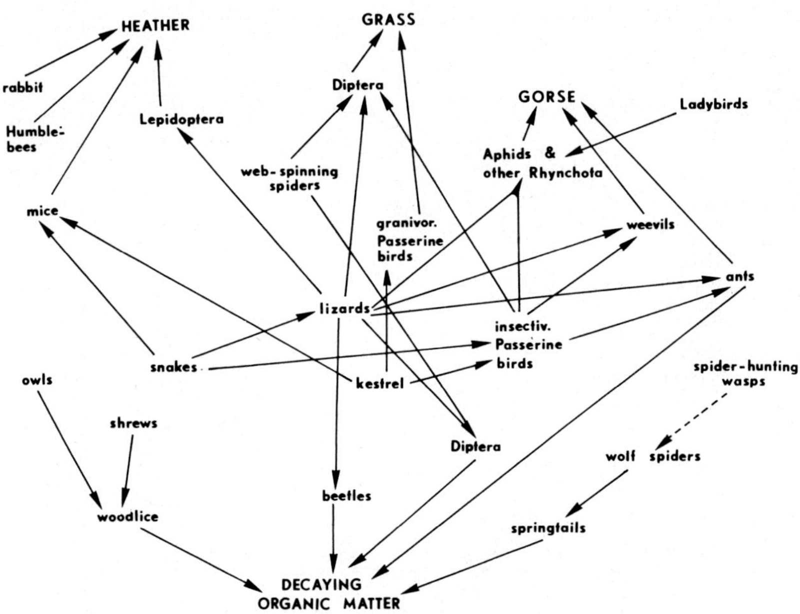

Fig. 263. Schematic trophic relationship (food-web) in an ideal European heath.

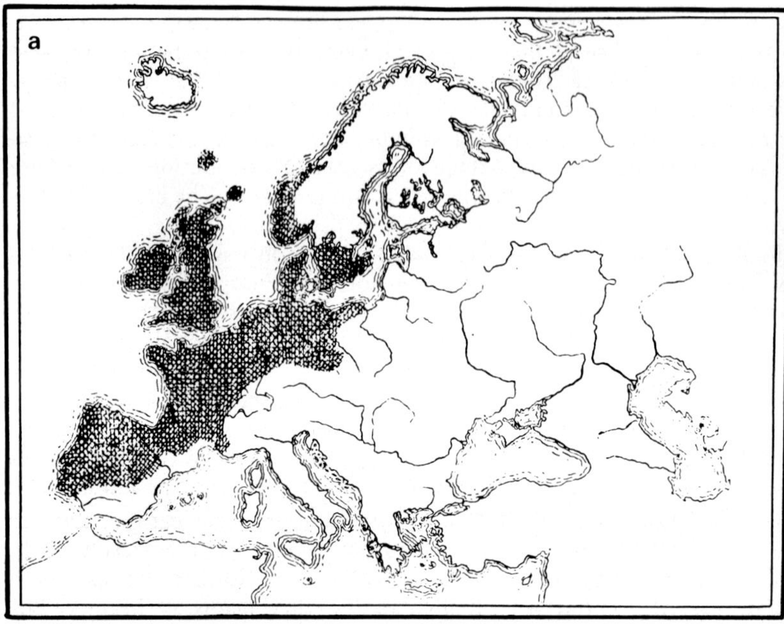

Fig. 264a. Distribution of the Carabid beetle *Nebria salina*, present in several territories covered with heath (from Jeannel).

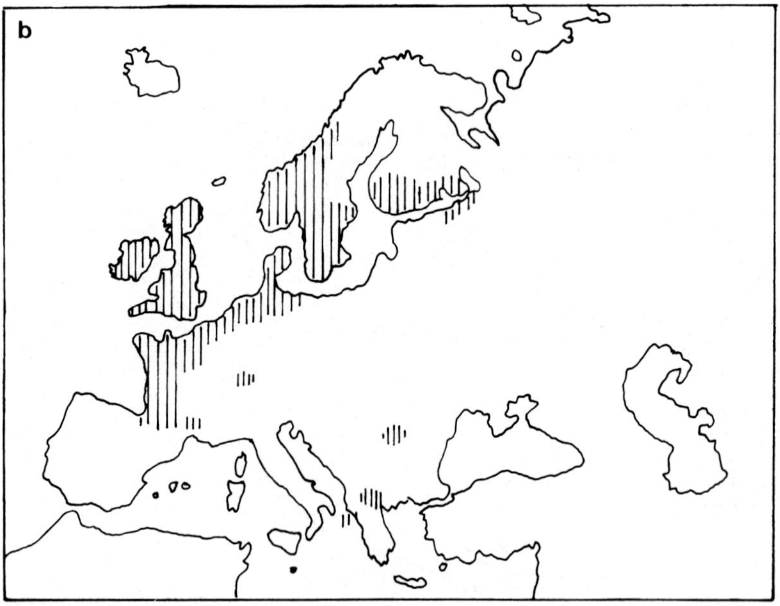

Fig. 264b. Distribution of *Choleva augustata.*

pronotum, yellowish brown elytrae) spread to all Europe and Asia, in the south limited to montane environments. It is limited to willow, poplar and birch, to which it can be harmful. This species lays its eggs upon the ground, which explains its presence also in heathlands.

Finally we should mention a few spiders, belonging to the secondary consumers or carnivores. Very typical are the jumping spiders or *Salticidae*, for instance *Evarcha arcuata*, and the wolf-spiders or *Lycosidae*, such as *Arctosa perita* and *Trochosa terricola*. The latter is the prey of the spider-hunting wasp *Anoplius fuscus*. Among the low vegetation of heathlands some spiders make their webs: this is the case with *Linyphiidae* (constructing the web among herbaceous plants like *Rumex acetosella*) or *Theridiidae*, which make a web among the short branches of gorse or flax.

From what is shown here we may infer that, in heathland, despite its monotony, all the ecological niches are occupied by animal groups, very similar as in other richer and more complex ecosystems (Fig. 263).

Elements of biogeographic interest within the heath fauna

Several European endemic species are limited to the north-western part of the Continent, and this kind of distribution has led Jeannel to speak of an Armorican distribution. These elements are frequent in West France, sometimes on the Cantabrian Mountains, and in Southern Britain. Generally they are spread west of

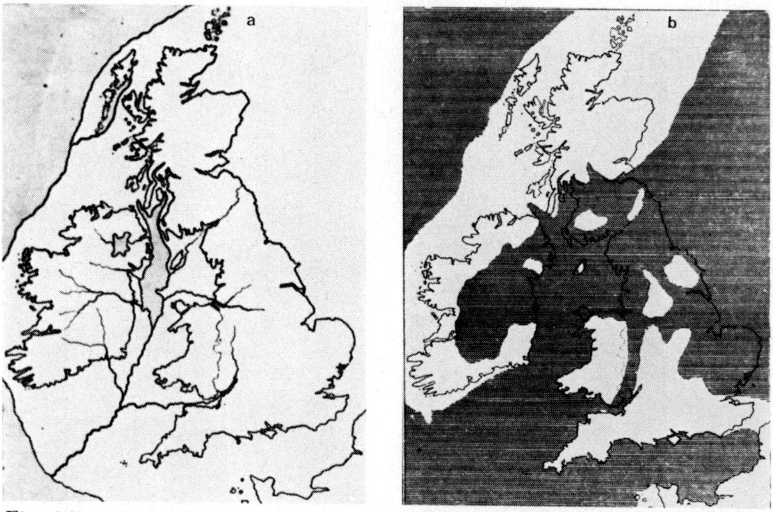

Fig. 265a. Map of the British Isles when the earlier members of the southern migration reached England. Shaded part: water; white part: land (from Scharff).

Fig. 265b. Map of the British Isles, showing approximately in what manner the sea may have invaded the country from the east during or shortly after the Forest Bed period. Darkly shaded parts indicate areas covered by water; white parts indicate land (from Scharff).

Paris. One example is the *Catopidae* beetle *Choleva angustata* (fig. 264b), but other species of this nidicolous genus show a similar distribution (*C. glauca, C. fagniezi, C. paskoviensis*). Probably all these elements are extremely old, coeval with some Hercynian species to which they are closely related, and therefore they represent the actual remnants of an ancient pre-Glacial European fauna. They are never present in Ireland. We must therefore deduce that they occupied Western Europe when England was united to France, but Ireland was already separated from Britain (Fig. 265). The carabid *Nebria salina* probably also belongs to the Armorican species. It has spread northwards and has been able to survive there (Southern Scandinavia, Scotland and the Faeroes); furthermore it was able to reach Ireland crossing during the Pliocene along the northern shores of the gulf of the Irish Sea (Fig. 264a). An interesting western European species is the millipede *Lepotoiulus belgicus* (Fig. 266).

South-western England and Southern Ireland particularly show some biogeographical peculiarities, such as the presence of old relict types of endemic species among the molluscs and other groups, which although they thrive today in heaths, originally inhabited an environment with tropical or subtropical characteristics. As Scharff says, the south-western flora of Britain migrated northward before the cold came, and survived in sheltered localities under the influence of the mild coastal climate. These elements are therefore pre-Glacial. Some elements, for instance the strawberry tree, *Arbutus unedo*, living today in Western Ireland, are of Miocene age according to Scharff, and were accompanied by other species

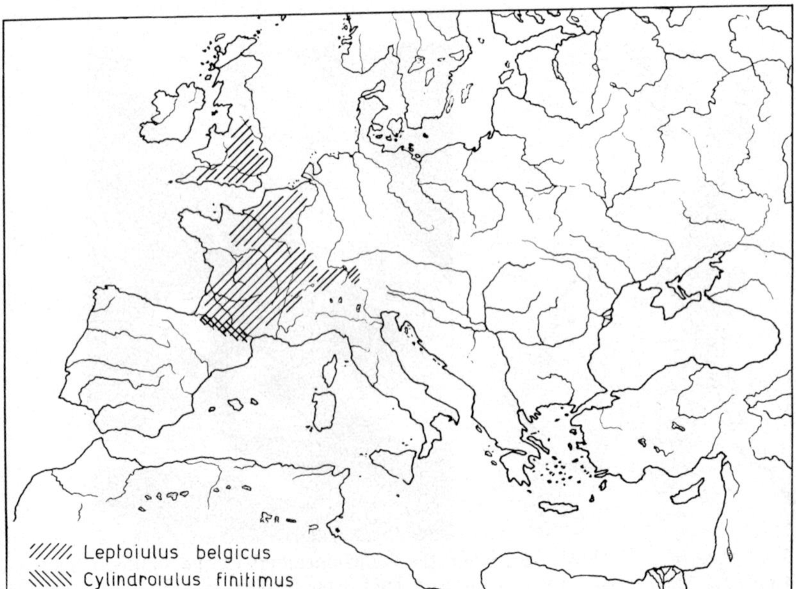

Fig. 266. Example of a distribution limited to Western Europe; two species of millipedes (*Diplopoda*).

which were completely destroyed during the glaciations. 'The survivors must be hardy species (saxifrages, heaths, etc.) which he (Forbes) considers to be the only relicts of this most ancient portion of our flora'.

Another example of ancient western European origin is the slug, *Geomalacus maculosus*, present only in Portugal and SW Ireland. The British species of *Testacella* (*maugei, haliotides* and *scutulum*), as well as the woodlice genus *Platyarthrus*, are also probably of south-western, pre-Glacial origin. The millipede *Polydesmus gallicus* may possibly belong to this so-called Lusitanian Fauna. *Lumbricus friendi* is an ancient element, typical of south-western Europe; *Dendrobaena veneta,* on the contrary, although present also in England, is an old element of eastern origin, as Omodeo has shown. Among the plants an example of south-western distribution is represented by *Euphorbia hiberna*, found in SW England, Ireland and Southern Europe, a little like the already mentioned *Arbutus unedo.*

Conservation problems in heathland

Heaths are today greatly threatened by many human activities, including afforestation, which of course would modify the physiognomy of pure heaths. The easy access to this environment, and often its proximity to settlements (towns, villages, etc.), different from moors, swamps and mountains, is a particular danger. Only very restrictive measures could therefore make certain the defense and conservation of this peculiar environment, regardless of its natural or secondary origin. The number of highly specialized plants and animals (invertebrates, and more precisely insects) deserves the most serious attention of all people concerned with land management, wildlife conservation, and the like such as geographers, ecologists, town planners etc.

In England recently some districts have been declared 'protected zones' (Hampstead Heath and Wimbledon Common). In Germany the Haichinger Heide has been also protected. It is necessary to pay attention also to the French 'bruyères' which are no less interesting than those of Germany or England.

BIBLIOGRAPHY

De Martonne E., Traité de Geographie physique, 1°, Paris, 1925; 3°, Paris, 1950.

Devoto G., Origini indoeuropee, Sansoni, Firenze, 1962.

Friedlander C.P., Heathland ecology, London, 1970.

Frileux P.N. & M.F. Hualt, Recherches preliminaires sur la vegetation actuelle et passée du Bray (Seine-Maritime), *Bull.Ecol.*, 2,1971,136.

Gamillscheg E., Etymologisches Wörterbuch der französischen Sprache, Univ. Verlag Heidelberg, 1969.

Giacomini V., Sulla vegetazione della brughiera di Gallarate, *Arch. Bot. Biogeogr. Ital.,* 34, 1958, 63.

Graebner P., Die Heide Norddeutschlands, etc., in die Vegetation der Erde (von Engler und Drude), V, Leipzig, 1901.

Guillet B., Relation entre l'histoire de la vegetation et podzolization dans les Vosges, Analyse de theses, *Bull. Ecol.*, 5, 1974, 79.

Jeannel R., La genese des faunes terrestres, Paris, 1942.

Kluge F., Etymologisches Wörterbuch der deutschen Sprache, Berlin, 1963.
Marcuzzi G., Un esempio di 'beni culturali': nomi dialettali di piante e relativi toponimi del Veneto, *Quad. Ecol. anim.*, N.2, Padova, La Ciclografica, 1976.
Maringer J., Le religioni dell'età della pietra in Europa, Torino, 1960 (transl. from German).
Omodeo P., Particolarità della zoogeografia dei Lombrichi, *Boll. Zool.*, 19, 1952, 349.
Onions C.T. (ed.) The Oxford dictionary of English etymology, Oxford, 1966.
Pokorny J., Indogermanisches etymologisches Wörterbuch, I, Bern & München, 1959.
Rohlfs G., Le Gascon. Études de philologie pyrénéenne, Niemeyer Verlag, Tübingen, 1970.
Scharff R.F., The history of European fauna, London and New York, 1899.
Winter, K., R. Altmüller, P. Hartman & J. Schauermann, s Forschungsprojekt Waldbrandfolgen: populationsdynamik der Invertebratenfauna in Kiefernforsten der Lüneburger Heide, Verh. Ges. Oekol., Göttingen, 1976, 225.

THE MONTANE ENVIRONMENT

εὕδουσιν δ' ὀρέων κορυφαί τε καὶ φάραγγες,
πρώονές τε καὶ χαράδραι

The tops of the mountain sleep, and all the valleys around,
and the slopes and the ravines

Alcman

It is very difficult to give an appropriate definition of mountain: practically all topography (or altitudes) higher than 500 m are 'mountains' and therefore their biota can be considered as montane biota. However, very often, rather than height, the morphology of the neighbouring landscape (or territory), the vegetation or the climate are essential or anyway important in defining mountains or montane environments. Therefore a small hill – though no higher than 500 m – can constitute an ideal environment for some 'montane' species – such as whortleberry or birch in Italy – when it offers some shaded, rather humid and cool micro-environments, generally present only at much higher altitudes. On the other hand, on the Sierras of Southern Spain and to some extent on the southern Spanish slopes of the Pyrenees, some lowland features (soil, climate, vegetation etc.) can reach very great altitudes, and we can hardly speak of a montane environment, unless we make allowance for the altitude and some physical factors such as lower pressure, higher evaporative power of the air, higher intensity of UV light etc. The montane environment is particularly less evident on the high plateau (Hochflach) where general lowland conditions prevail, especially when we take into consideration rainfall, soils, macro-climate, etc. Furthermore, from an ecological standpoint, we must always keep in mind whether the mountains are isolated or united to form chains, massifs, groups, high plateaux, etc., since here again physical factors act in a different way when compared to isolated mountains (winds, rainfall, insolation, presence of mist, etc.) (Fig. 267).

In the mountains the light is very different from that reaching the lowlands for several reasons: first of all the thickness of the air mass through which radiation passes, which diminishes with the increase in altitude. Thus, at a latitude 48°N in June we have the following conditions of total radiation (cal cm^{-2}):

altitude (m)	in sunshine	covered sky
200	691	155
3000	834	403

The increase in short wave-lengths (0.2-3μ) is similar:

	cal cm^{-2} day^{-1}	
altitude (m)	June	December
200	588	102
1000	659	124
2000	729	144
3000	773	152

The absorption of radiation due to the atmosphere also varies according to different wave-lengths:

Fig. 267. Dolomites, near Cortina d'Ampezzo (Cadore). a. Spruce forests (quite well preserved) with some dolomitic groups in the background. The area is without any endemic species because it was covered almost completely by ice during the Glacial Period (Photo: Dr Luisa Dalle Molle). b. Near preceding biotope, but looking westwards: the spruce have been felled and the effects of grazing are now clearly visible. The fauna is very poor. Tourism contributes to the greater human pressure (Photo: Dr Luisa Dalle Molle).

	visible light (550 mμ)		IR radiation (800 mμ)	
	% remained of original value	% absorbed by each stratum	% remained of original value	% absorbed by each stratum
higher atm.	100	7	100	3
c. 4000 m	93	18	97	9
Sea level	75		88	

The response of mountain animals to radiation (solar, calorific, visible and UV) is very complex and anything but uniform. Thus, we find insects with a shining surface (beetles of the genus *Chrysomela* or *Chrysochloa),* flies of the genus *Chilosia*) which reflects most of the solar radiation, whereas other insects exhibit a black, dull, surface, which can have a double effect, absorption of heat during the day time and during the short summer, and screening against UV rays, which, on the whole, are harmful to animals. This possibly could explain the great spread of black or dark colorations amoung mountain insects (*Pieris napi* f. *bryoniae, Argynnis pales, Endrosa irrorella* f. *andereggi* and f. *riffelensis, Endrosa aurita* var. *ramosa* among butterflies *Chrysochloa viridis* var. *lugubris*, *C. tristis* ab. *tenebrosa, C. gloriosa* ab. *atramentaria, Carabus auronitens* var. *atratus* among beetles, or the mountain melanism present in *Chilosia canicularis, Melanostoma mellinum* and *M. dubium*, among flies.

As we go to higher altitudes, the atmospheric pressure decreases according to regular physical laws. The mean air and soil temperatures decrease too, but the diminution is very irregular. The air, due to its rarefaction, retains to a lesser degree the solar radiation and therefore is cooler; the soil, on the contrary, when exposed to sun, receives a greater quantity of radiation, so that in mountains the air is fairly cool and the soil warm if exposed to the sun, cold if in the shade, with great thermal differences. The soil, however, is generally 2-3°C warmer than the air but the upper layers undergo very large diurnal temperature changes, more so than the air. Below the soil surface these changes gradually decrease, to almost nil at a depth of 20 cm. Of course this is very important for all soil animals able to live at that depth or still more to move during the day (Fig. 268-271).

A phenomenon common in mountain areas subject to a continental climate is the thermal inversion, during which on some winter days the temperature of the valley bottom is below that of the neighbouring valley sides; we must not forget that cold air is heavier than warm.

The differences between seasons decrease with height. Furthermore one of the most remarkable features of mountains is the influence of exposure on temperature and many other physical and biotic characteristics: due to different exposures we may have on one mountain (even a very small one) very different amounts of annual hours of insolation so that even on the same mountain live very different biota, according to the slope. It is pointless to say that this phenomenon is exclusive to isolated mountains.

On a mountain slope, generally the higher the altitude, the lower the temperature; for this reason the atmospheric humidity condenses into mist, fog or clouds with the increase in altitude, so that we very often have an increase in rainfall up to the upper limit of the forest (tree line), after which there is a large decrease both in humidity and rainfall. Of course, in this case the evaporative power of air

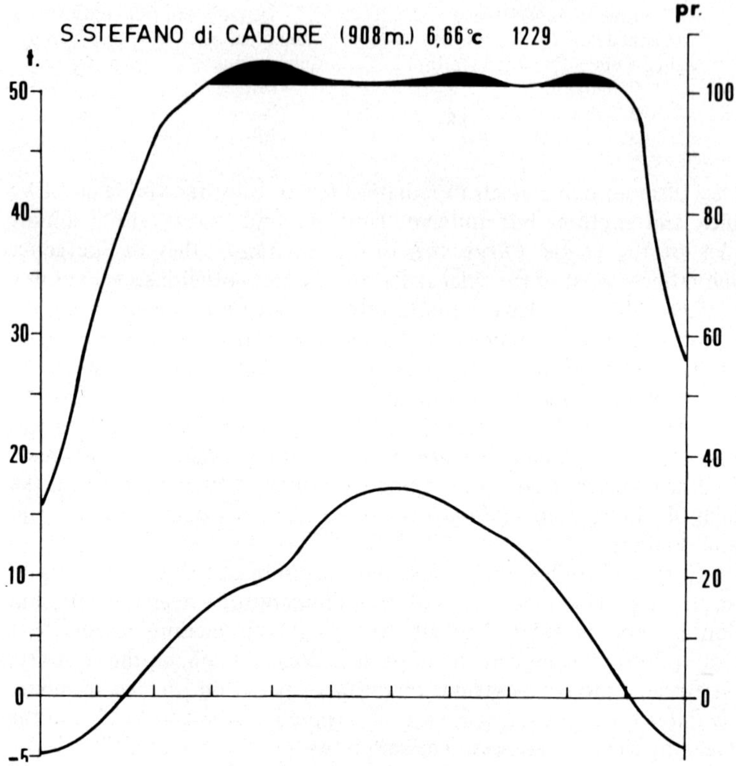

Fig. 268. Ombrothermic diagram for the Eastern Alps.

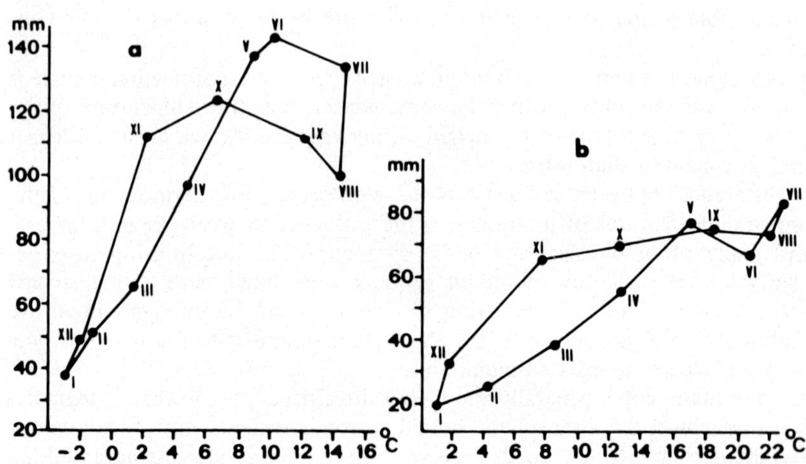

Fig. 269. Examples of climograms in the Eastern Alps. Left: Cortina d'Ampezzo (Cadore). Right: Bolzano (Bozen, Süd Tirol).

typical of high altitudes also contributes to actively reducing the atmospheric humidity. This explains the presence of xeric adaptations in high mountain plants (see later).

When winds (especially those from the sea) reach a mountain they may cause constant rainfall: rising air is subject to expansion and therefore to rapid super-cooling with altitude, so that all the water vapour transported by the wind is subject to very rapid condensation, whence the rainfall. In the lee of the mountain (or still more mountain chain) conditions of aridity very often prevail, so that one speaks of 'rain shadow'. In some cases, for instance in the Alps, which present their southern slope to the winds coming from two seas (on the east the Adriatic Sea, on the west the Tyrrhenian sea) as well as from the Po valley, on the northern slope we may very often observe the presence of a very peculiar, warm, dry wind, locally known as the Föhn (Fig. 272). This is due to an air mass which has undergone a loss of humidity on the southern slope, and which, when passing to the northern one, is rapidly compressed by the increase in pressure and therefore warmed. The great difference in heating of the valleys during the day compared with the higher altitudes and the cooling of the latter during the night causes constant winds, called breezes, blowing from the peaks toward the valley during the night and from the valley upwards during the day. Sometimes a strong lifting wind can be responsible for the transport of many flying insects (such as *Diptera Hymenoptera* and *Lepidoptera*) from the valleys up to the glaciers, where they meet a rapid death. The same currents are responsible for the transport of pollen to the glaciers, making possible the life of some pollenivorous creatures such as Collembolans (the so-called ice flea or *Isotoma saltans*).

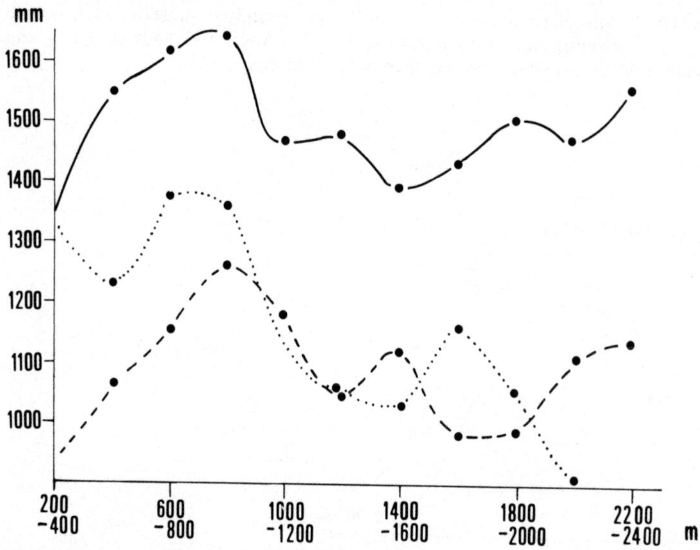

Fig. 270. Values of rainfall according to altitude in three different sectors of the Alps. ------ Western Alps, —— Central part, ····· Eastern Alps. In the Eastern Alps a clear increase is evident at 600-800 m. In the Central Section there are great differences at the different altitudes.

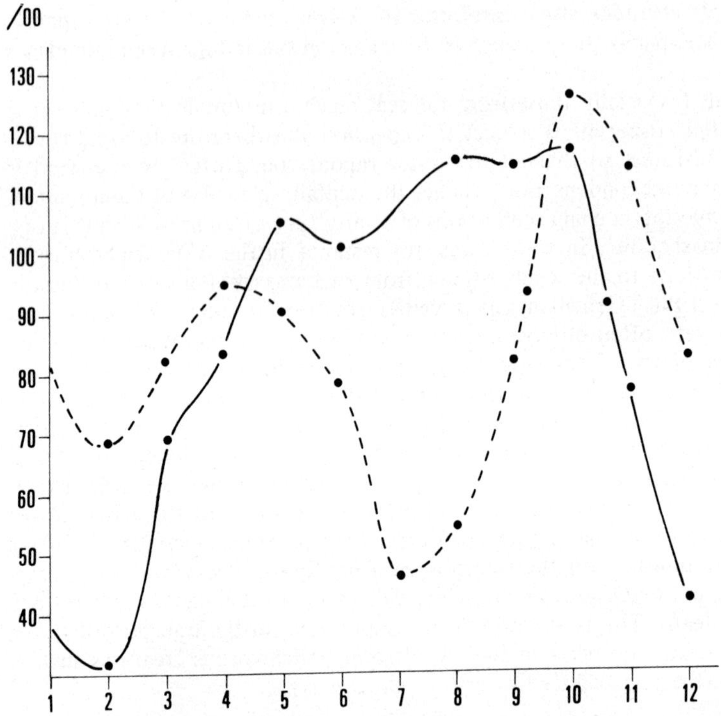

Fig. 271. Difference in rainfall (expressed as % of the mean annual value) in each month between Alps (______, Loveno) and Appennines (– – – – –, L'Aquila). A peak in spring and autumn is very evident on the Appennines, together with a summer drought.

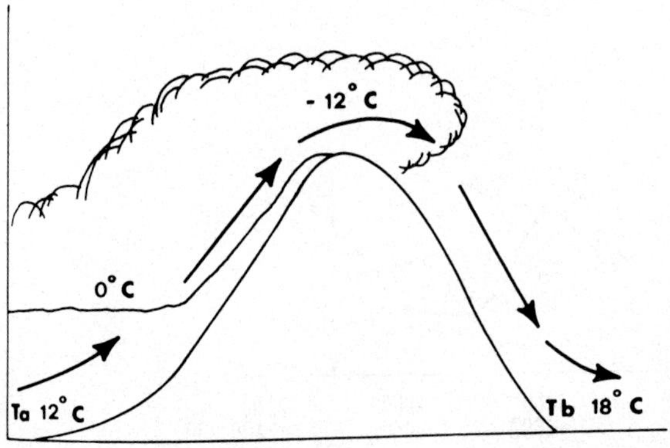

Fig. 272. Schematic illustration of the phenomenon of the föhn, a wind sometimes associated with electromagnetic oscillations and, inter alia, responsible for the so-called 'föhn disease' (from Barry & Chorley). Ta is the temperature on the windy side of the mountain, Tb the temperature on the lee side.

The soil

The soils differ greatly from one area to another as a consequence of the different geology, geomorphology, rainfall, exposure, etc., so that in the montane environment we find almost all the European types of soils. Only when there is a strong leaching, i.e. in humid situations, is the final product a relatively homogeneous soil type: in this case one can talk of 'climatic' soil, dependent more on the local climate than on the parent rock.

The most important soils present on European mountains are organic soils, podzols, forest brown soil, rendzina, and also several more or less raw soils such as råmark and rankers. The alluvial soils are also very widely represented in the valleys: gley, paternia, borovina, vega etc. We will emphasize here the importance of the soils of the high mountains, on the one hand, and that of alluvial soils, on the other (Fig. 273-274).

Let us begin with alluvial soils. Gley is an almost permanently sodden soil, sometimes even flooded, anyway with a great amount of interstitial water. The humus may belong to a moder or to a mull. The sub-soil is compact, and gleyed. Generally there is a variable amount of clay. Sometimes the gley horizon exhibits some bands (or deposits) of iron hydroxide. The vegetation is sometimes arboreal (*Alnus, Populus, Salix*), sometimes herbaceous. The soil fauna is always rich and almost universally includes earthworms. Whereas a moder-gley is found at higher altitudes, the mull gley is typical of the lower situations and more especially of the valleys of Eastern Europe, where the regime of the rivers is not very regular and some flooding is occasionally possible.

Another category of soils is composed of the so-called 'Auboden' (German) or warps, also warp soils (from a popular Lincolnshire name), similar to the former category, but without any gley formation. The soil is humid and sometimes also flooded. If there is never any flooding, the soil is more like true terrestrial soils: this difference is much more noticeable in Southern than in Central or Northern Europe. We should mention here also Kubiena's paternia (from Paterna, a small river in the Alpujarras south of Sierra Nevada) or in German 'grauer Auboden', immature, chemically little decomposed, generally light-grey AC soil with an evident humus layer. It is present upon sediments poor in limestone; generally it is sandy or silty-sandy. The soil fauna is well represented.

Kubiena's borovina (named after a popular Polish name) or 'Rendzina-ähnlicher Auboden' is similar to the last type, but is present upon sediments rich in limestone. In the South-European mountains, finally, another category of alluvial soils is present, the so-called 'vegas' (in Spanish literally shores of a river). They are chemically well decomposed, with an ochre-yellow, ochre, brown or even red colour (in this case due to the presence of iron hydroxide). Generally, the 'vegas' exhibit three horizons (A, B and C). According to the nature of the sediments upon which the vega has developed, we may recognize a vega parda ('brauner Auenwaldboden'), when it develops upon a brown earth eroded by rivers; a vega developed upon 'terra fusca', and finally a vega roja (in Spanish roja means red) developed upon a red loam (Rotlehm) or a red earth (or terra rossa), in German 'Roterde'. This category is widespread mainly in Greece and Spain. The soil life is particularly rich in vega parda where earthworms are generally present, thus contributing to mull formation. Of course this kind of soil is good for cultivations. The high mountain soils we will mention here are essentially råmark, ranker and rendzina.

Råmark in Sweden is used to indicate a raw soil of cold deserts. In the mountains we find the alpine råmark (or raw soil of the high Alps) corresponding to Neustrujeff's mountain desert soil, which is either a raw alpine, carbonate soil, or a raw alpine silicate soil, according to the geological substrate. It can be either a grey, loose AC soil, formed upon a compact parent rock, with little chemical decomposition and strong physical disintegration due to the frequent freezing and covered most of the year by snow and so frozen, or an AC soil, covered by a layer of stones or by some isolated rocks due to aeolian deflation. This latter category will permit the life of several kinds of plants, both herbaceous and shrubby. The herbaceous vegetation always has a cushion-like aspect and the roots can fix a

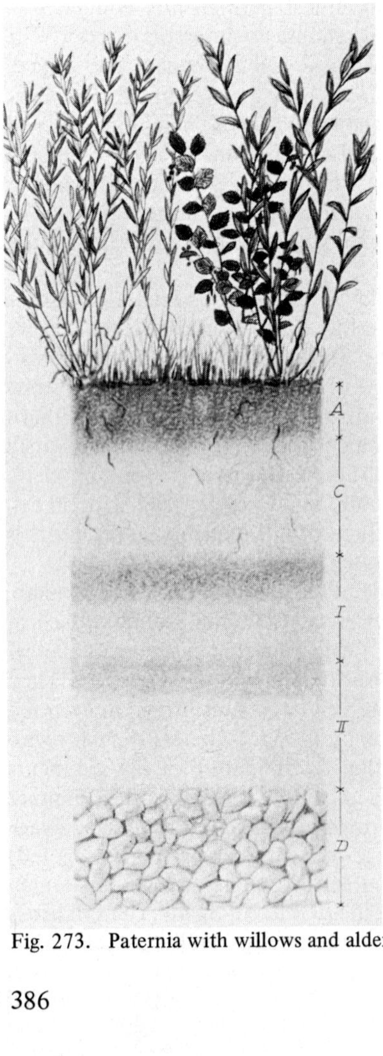

Fig. 273. Paternia with willows and alders (Valley of Enns, Styria) (from Kubiena).

certain amount of humus, where a thriving soil fauna is present. Kubiena has described a raw alpine soil beneath *Festuca duriuscula* upon sericitic phyllite-graphite rocks at the summit of Veleta Mt., Sierra Nevada, at 3740 m. Rankers are quite different.

Rankers – in Spain (Aragona) called 'tasca' since they can be separated from the parent rock like a carpet – are soils poor in limestone, whose humus layer is situated directly upon the parental rock, generally a siliceous rock or another kind

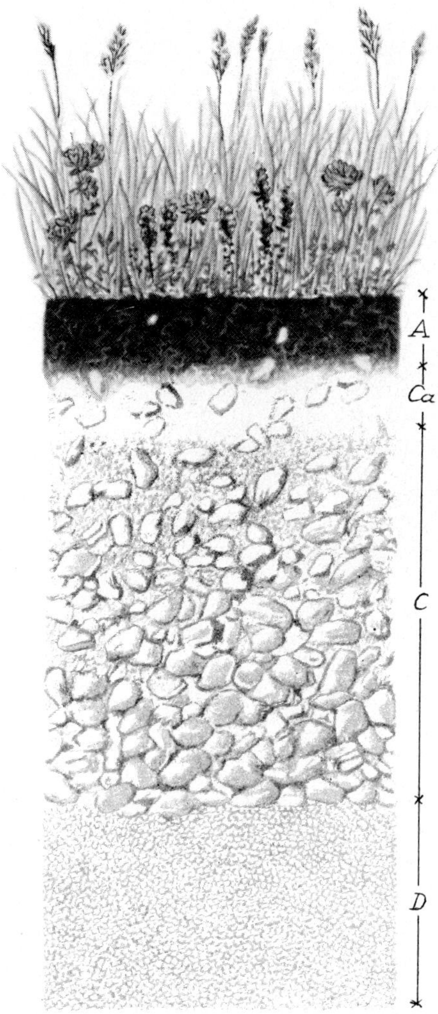

Fig. 274. Borovina under a dry meadow above limestone gravel and sand (Enns Valley, Styria) (from Kubiena).

poor in calcium (AC soils). Generally a ranker originates from a protoranker, i.e. a young ranker totally composed of loose elements, faeces of microfauna, small plant remains and poorly disintegrated mineral particles. The humus layer is very thin, only a few cm thick, disaggregated, generally blackish and situated directly upon the parent rock. It is covered by mosses, lichens, cushion-like plants, or meagre grasslands. Not dissimilar are 'eilag ranker' (Sacharov, 1905) or turfy soil of mountain summits (Glinka), and roots-ranker, formed by a mat of roots and the remains of cushion-like plants (mainly *Carex* and some high mountain

Fig. 275. Left: Ranker on chloritoschist with *Festuca varia* (Wöllaner Nock, Carinthia, 2000 m). Right: Ranker on granite, with *Poa bulbosa* (Cumbre de Guarramas, Sierra de Guadarrama, 2240 m (from Kubiena).

Gramineae), plus the faeces of micro-arthropods (generally Oribatid mites). This kind of ranker can also develop and exist on very steep slopes.

According to geology, presence or absence of iron, type of humus (moder vs. mull), microclimate, vegetation etc. several kinds of ranker can be recognized (Kubiena) (Fig. 275).

The rendzinas (Fig. 276, 277) which in practice represent the parallel of ranker on calcareous rocks, are also AC soils, represented generally by a humus layer (A horizon) resting upon the parent rock. Here again environmental factors (rainfall, temperature, steepness of the slope, exposure, type of vegetation, etc.) are responsible for the different types of rendzinas. The existence of many local, pop-

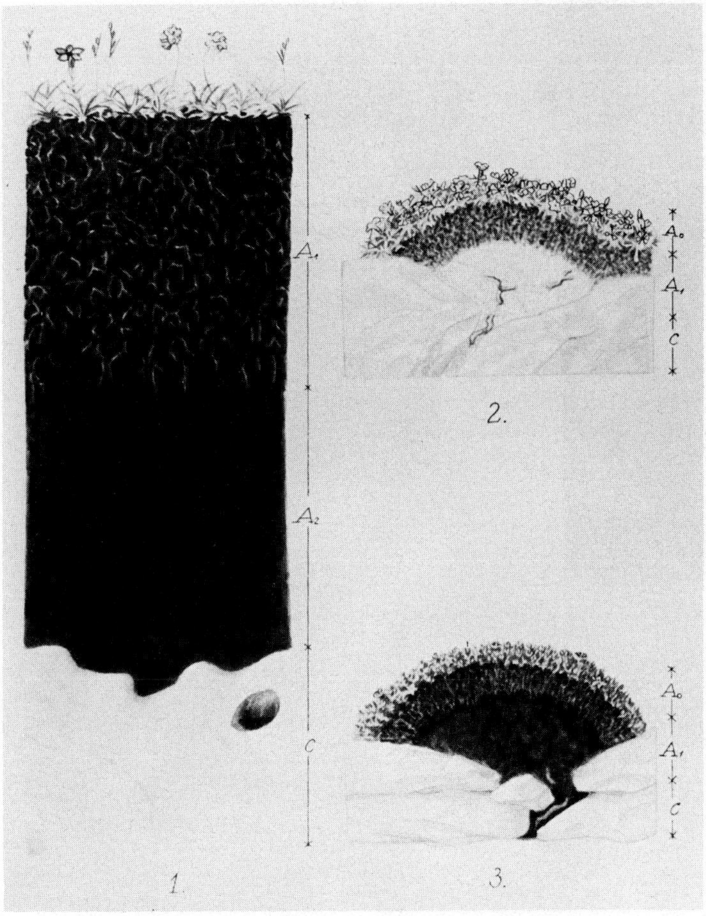

Fig. 276. Examples of rendzinas. 1. 'Pechrendzina', on limestone, with humid Firmetum (Hoch König, Salzburg, 2300 m). 2. Grey alpine rendzina on limestone, with cushion-like *Silene acaulis* (Dachstein, 2800 m). 3. Black alpine rendzina on limestone, with *Saxifraga rudolfiana* in its cushion-like form (Hoch König, Salzburg, 2700 m) (from Kubiena).

ular names in the different parts of Europe is proof of the universal presence of this major soil type: rendsina (successively changed to rendzina) is a Polish peasant name, employed in pedology by Sibirtzeff in 1896; the German authors have used the name 'Humuskarbonatboden'. In Russia a very stoney rendzina is called chrap; the rendzina present in the Caucasus, particularly upon gypsum, is called gadscha (gada) or gaza. Finally Akimzeff in 1931 employed the name

Fig. 277. Examples of rendzina of relatively low altitudes. Left: Rendzina on dolomia with *Pinus mugo* (= *P. montana*) and *Rhododendron hirsutum* (Dachstein, 1700 m). Right: Brown rendzina on dolomitic limestone, with *Quercus* and *Juniperus oxycedrus* (Guadalix, Madrid, 850 m) (from Kubiena).

'sulphate humic soil' for a rendzina developed upon gypsum. As a rule rendzinas form on limestone, marble, dolomite, magnesite, marl or gypsum. The soil fauna is generally very rich and varied.

In the periglacial zones, present at very high altitudes, in the Alps above 2400-2500 m, but in Northern Europe at much lower altitudes, there are some permanently frozen soils (true 'permafrost'; cf. chapter on tundra) which, because of the form they sometimes assume, are called polygonal soils. They are common in the present moraine territories (i.e. near the glaciers) and in the 'snow valleys' (Schneetälchen) (Fig. 278-279).

Another important phenomenon characteristic of high mountains is solifluction, i.e. the creeping of the upper layers of the soil, sodden with water, over the lower, frozen layers, in the case of sufficiently steep slopes. This phenomenon is one of the major causes responsible for the lack of tree growth above a certain height.

The following scheme will give us an idea of the evolution of soils in the montaine environment:

Zone of permanent snow
↓
high mountain humic soils
↓
brown earth, slightly podzolized soils ← podzols → degrated rendzina → rendzina

(brown earth, slightly podzolized soils ↑ Substrates poor in carbonates)

(degrated rendzina ↑ Substrates rich in carbonates limestone, dolomite, etc.)

Two physical features characteristic of mountains (though not exclusive to this environment) are glaciers and snow, both very important from an ecological point of view (Fig. 280-282).

A glacier is a natural ice accumulation due to the transformation of snow which falls during the winter time. For any glacier a so-called topographic limit of permanent snow exists, i.e. the altimetric level above which the fallen snow does not completely melt in summer. At the lower limit of a glacier the so-called tongue (or discharge channel) is present, from which a small stream generally originates. The surface of a glacier exhibits some peculiar features, such as crevasses, 'seracchi', 'funghi' (Italian), pits, streams, lakes, sand cones, etc. Generally glaciers show a descent or progression due to gravity. Only during summer is there a retreat. The retreat of many present-day glaciers causes the uncovering of wide areas of morainic debris, in which lakes can form. However, a lake can form also on the glacier, or between it and the debris accumulate a little way from its snout.

It is not the place here to talk about the consequences on relief or on the morphology of the alpine environment in general that the Pleistocene glaciation and the succeeding retreat of the ice has had.

As a rule, in the Alps we find the so-called 'alpine-valley' glaciers (or 1st order mountain glaciers) composed of a collector basin and a tongue descending to the valley. This kind of glacier is present also in the Caucasus. In the Pyrenees, on the contrary, today we have only cirque glaciers (glaciers de cirque) or 2nd order mountain glaciers, composed only of a mass of ice contained in an amphitheatre. Lastly, in Northern Europe, we have the Norwegian glaciers, composed of a very wide accumulation basin, covering a large high plateau (Hochplateau) with an ice field, from which one or more tongues (iceström) descend, often reaching the sea. The highest part of a glacier, where the snow transforms itself into ice, is called 'nevato' in Italy (maybe from the Latin nivatus), Firnfeld in the countries where German is spoken, névé in France. All these features, of course, have great importance for animal and plant life present on the edge of the glacier or beneath it.

'Nevaio' (Italian) is an area of soil permanently covered by snow which does not melt, but does not change into ice. Its size is very variable, and it is not possible to distinguish – as for glaciers – an accumulation zone and an ablation zone. Generally the 'nevai' are present on the northern slopes of mountains and more particularly in some nivation hollows. Their existence and extension have great importance for life developed at their edges or even upon them (see vegetation). The so-called 'Schneetälchen' (literally 'small snow valleys') are particularly important for animal life in the high mountains, since they keep all the surrounding soil very wet even in the warmest periods, so making possible the existence of a variety of insects, mites, millipedes and other hygrophilous invertebrates. In this case, the nature of the geological substrate is also very important (limestone vs. siliceous rocks): in the case of limestone the soil stays much more humid and animal life is much more abundant and varied compared with other rocks, such as, for instance, granite.

Fig. 278. Example of lithosoil: a detritic terrain on the Dolomites, partially covered with perennial snow.

The impact of man upon the mountain environment

The mountains of Central Europe were not generally inhabited until relatively recent times, probably because the environment was not very suitable to human settlements or, more generally, to human survival. Not only that, but in the past the lowlands and even more the foothills were rich in wildlife and offered sufficient food and suitable shelter to pre-historic men.

Among the most ancient human remains found at the foot of mountains or in foothills we must mention the bones of Fontechevade, found in a cave situated on the right bank of La Tardoire, a tributary of the Charente. These remains belong to the Riss-Würmian interglacial period, characterized by a temperate-warm climate, with some remains of *Rhinoceros mercki*, fallow deer, *Bovidae* and a *Testudo* very similar to *T. graeca.* The human remains can be attributed to a period between the Lower and Middle Paleolithic.

During the Middle Paleolithic, more precisely in the last interglacial and the beginning of the last glacial (Würmian), Neanderthal was living in the Vezère

Fig. 279. A typical crio-nival (or glacio-nival) phenomenon of the high mountain meadows (Girlandenboden, gradins de montagne) at Passo Rolle, Capanna Segantini, at nearly 2100 m (Photo: L. Dalle Molle, 1974).

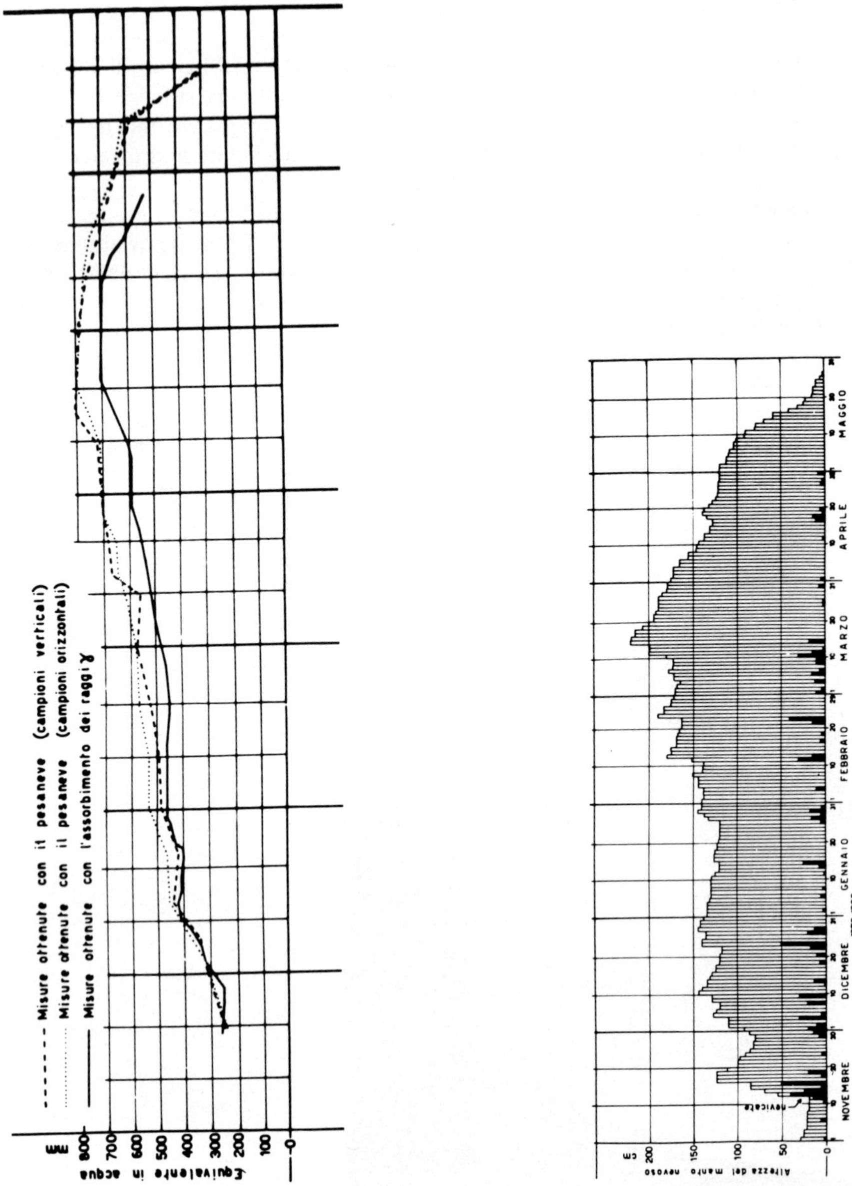

Fig. 280. Depth of snow and equivalent in water at the Fedaia (Marmolada, 2000 m). Continuous line: measures obtained by means of X-ray absorption; broken line and dotted line, measures obtained by means of direct weighting of the snow (from Castiglioni, 1959-60).

valley: a particularly famous site is Le Moustier, situated 10 km upstream of Eyzies (Dordogne valley), after which the Mousterian culture was named. This culture was present in South-Western France, for instance at Montmaurin, near Saint Gaudens, and in Croatia, at Krapina, situated 40 km north of Zagreb and west of the Ivančica mountains. Here man found shelter on the fluvial terraces on the shores of the Krapinica. In this region, remains of several vertebrates point to a warm period. Neanderthal man fed upon fruit and the catch from hunting and fishing; he probably practised ritual cannibalism; he used to sleep in shelters on the trees or in caves; his habits were nomadic. He knew about fire and buried his dead.

Much more interesting to us is the so-called 'alpine Mousterian', characterized by shelters in the open, found throughout the alpine chain, a poor lithic culture represented only by splinters of limestone, quartzite and flint, or by splinters of bones. The most famous sites are those at Drachenloch, at 2445 m on the Drachenberg near Vättis (St Gall), and at Wildkirchli, in the Säntis region. The former was inhabited by bear hunters (hunting the cave bear, *Ursus spelaeus*) now extinct species living in the Riss-Würmian interglacial. They have left us small heaps of stones with remains of bear bones, interpreted as offerings to some divinity. The latter site was also inhabited by bear hunters: some small boxes made with stone plates are interpreted as hunting magic, which is a very old and generalized kind of magic. The presence of bear hunters on one hand points to the abundance of *Ursus spelaeus* all over Central and Southern Europe, and on the other explains wide diffusion throughout Europe and Asia of the bear religion (see the chapter on taiga).

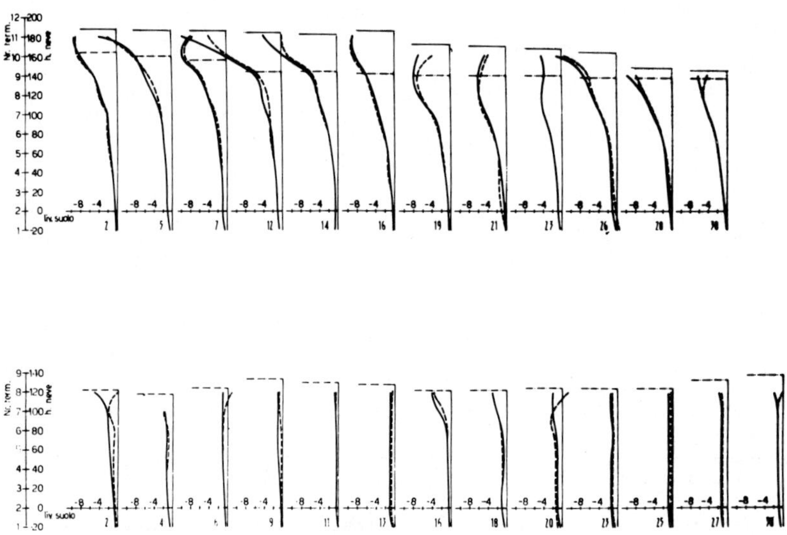

Fig. 281. Temperature of snow on the Fedaia (Marmolada, 2000 m) during the months of January (above) and March (below) in 1959. Continuous line, temperature at 08.00 h, broken line, at 14.00 h (from Castiglioni, Picozzi & Zunica, slightly modified).

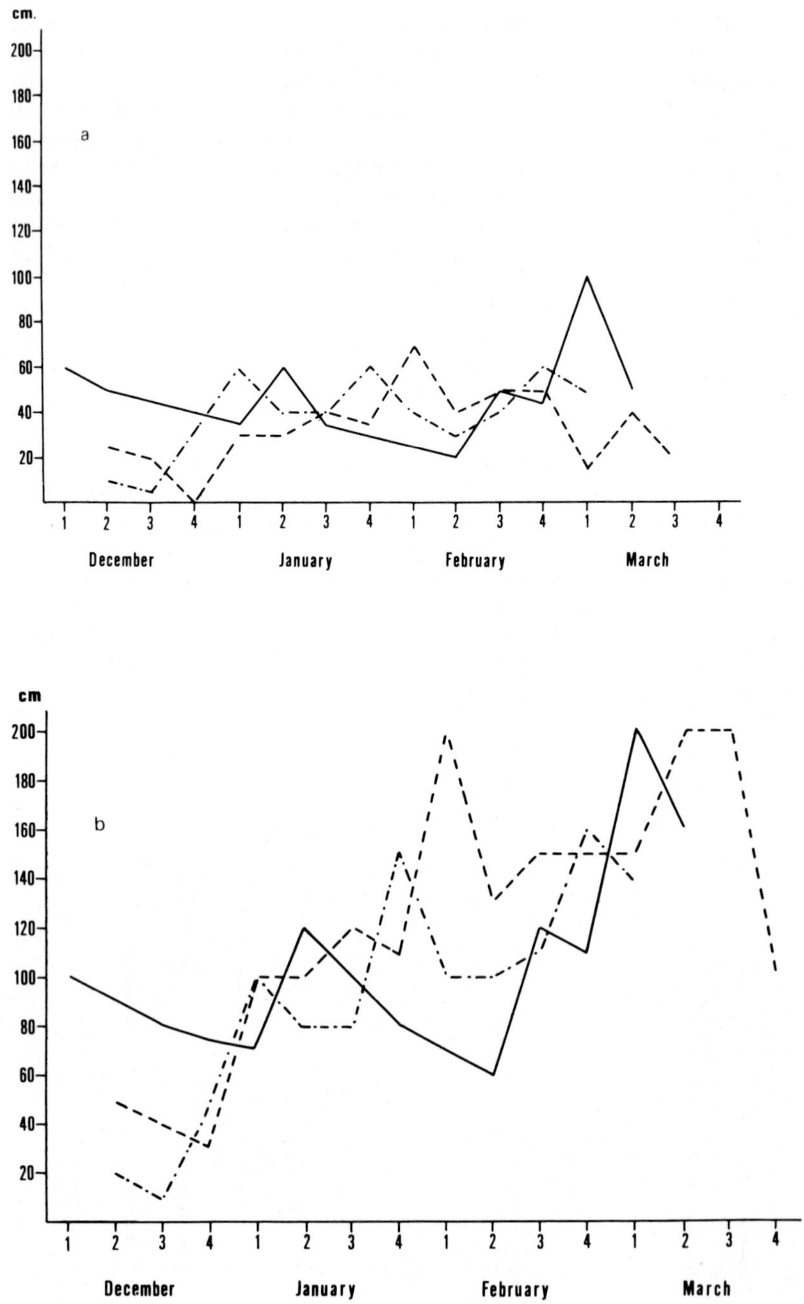

Fig. 282. Amount of snow from December to March in two Dolomite localities, a. Passo Croce d'Aune, 1010 m: b. Monte Avena, Casere dei bosch, 1253 m, in three successive years, 1969/70 (______), 1970/71 (–·–·–), and 1971/72 (– – – – –).

A cultural type extending into the heart of the Swiss Alps is named after Wildkirchli. The Mousterian is present also in the French Jura and in Poland (cave of Ciemna near Oicow).

It is worthwhile recalling that the presence of pre-historic man above 2000 m is an extremely unusual phenomenon, particularly when we realize that today the most elevated human settlement of the whole alpine chain is S. Martino in Passiria (St Martin Passeier) at 2325 m, and that only temporarily inhabited!

Fig. 283. Rock painting of *Rhinoceros tichorhinus* (above) and *Equus* sp. (below) (from Breuil).

In the Upper Paleolithic, a recent form of man sometimes called *Phaneranthropus* (or *Neoanthropus*) lived. We find – limited to the mountains of Central and Southern Europe – remains belonging principally to two cultures: the Aurignacian and the Magdalenian. These cultures do not necessarily correspond to different periods. The former is represented by remains found at Lespugne, in the

Fig. 284. Rock graffiti of *Equus* sp.: a. mediolineous horse from Combarelles; b. horse from Cave of Lascaux, with longilineous tendence; c. horse of celtic type from Cave of Font de Gaume; d. horse of nordic type from Combarelles (from Bourdelle).

Haute Garonne (famous for its steatopygian venus) and at Predmost (or Predmosti) in Moravia, on the right bank of the Bečva, a tributary of the Morava, near Přerov (according to some these remains should be attributed to the Solutrean instead of the Aurignacian culture). The famous graffiti of Lascaux (Dordogne) must also be attributed to Aurignacian people. Whereas in the latter locality essentially oxen (*Bos primigenius*), horses (possibly *Equus przewalski*), deer mixed with some bison and rhinoceros were present, Predmost is characterized by a large number of mammoths.

The Magdalenian (limited to the mountains) is represented at Raymonden, belonging to the municipality of Chancelade, near Périgueux: this culture is typical of a very cold period, witnessed by the presence of some arctic rodents (already present during the Mousterian age), reindeer, and bison, which were

Fig. 285a. A sign of the presence of a felid (*Leo leo*?) and of wild horses (*Equus* sp.) in Southern France: Font-de-Gaume (from Breuil).

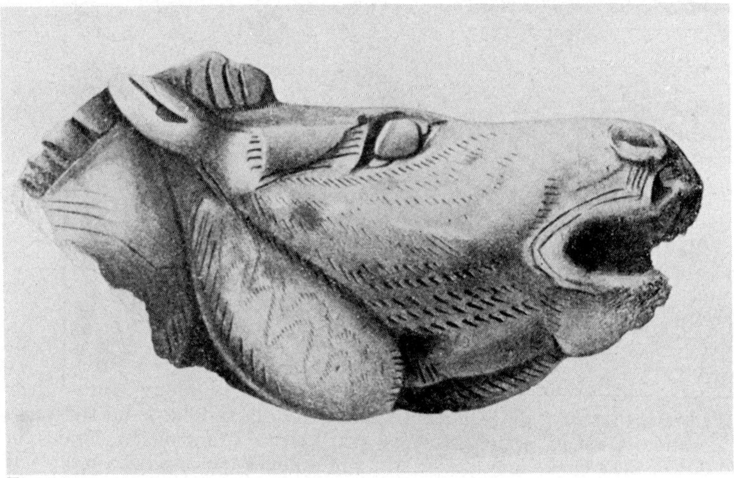

Fig. 285b. A sign of the presence of wild horses in Southern France during the Upper Paleolithic: Magdalenian of Mas-d'Azil (from Piette).

hunted by the men of this epoch. The remains found at Combe Capelle (Montfarrand, in the Puy de Dome) belong to a culture maybe intermediate between Magdalenian and Aurignacian. In Switzerland some Aurignacian and many Magdalenian remains are present in caves between Basel and Schaffhausen. Tardenoisian handicraft is also present. In montane Poland finally, we find both Magdalenian and Tardenoisian remains. In Poland also the Acheulean culture is represented (caves of Ciemna and Okiennik). This culture is present also in the Venetian Prealps (Lughizzano, Lessini Mts.).

The Mesolithic is not well represented in the mountains: there is the cave of Mas d'Azil, after which the Azilian culture has been named, and the station of Montardit, both in Ariège. According to some, the remains found at Ofnet (Bavaria, near Nördlingen, at 436 m) also belong to the Azilian culture, although some Magdalenian remains are also present. Here, brachycephaly makes its first appearance in the mountains of Central Europe. This morphological character will later spread slowly to the Alps and the circum-alpine regions. An Azilian culture is also present in a cave at Birstal (Switzerland) formed during the würmian epiglacial.

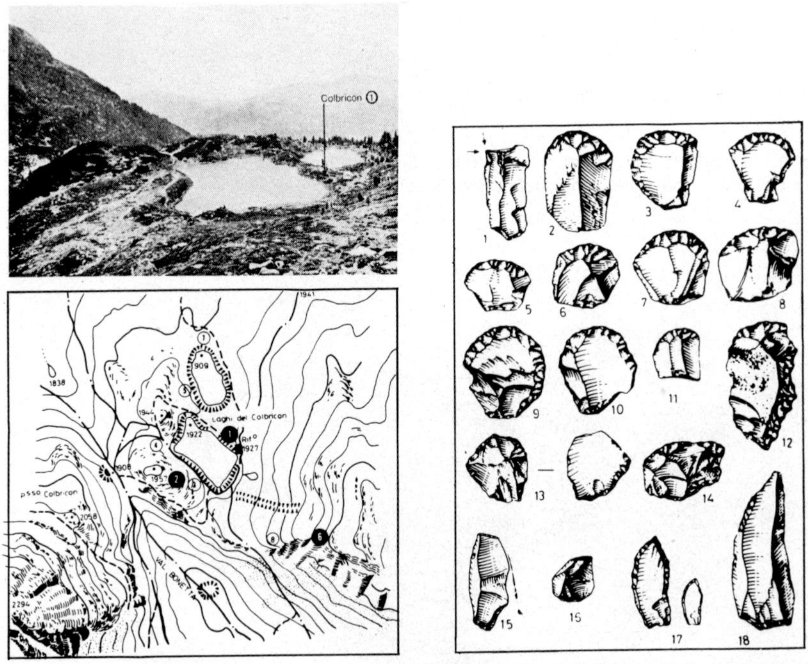

Fig. 286. a. The present aspect of the region of Colbricon, near Passo Rolle in the Dolomites (Trentino). b. Location of human, seasonal (summer) settlements during the Mesolithic near the lakes of Colbricon. The circles indicate the sites where human remains (silex manufacts and fireplaces) have been found. c. Examples of manufactured silex found near the Colbricon (from Bagolini et al.).

In the zone of Colbricon, near Passo Rolle, at a height of 1900-2200 m, eight pre-historic settlements have recently been found near two small lakes in the centre of this alpine zone. They belong to epipaleolithic (or Mesolithic) man, and more precisely to a complex of the sauveterroid type, whose presence in the Adige valley was already well documented. The culture is characterized by segments and triangles of flint and by the lack of trapezoid and rhomboid flint (Fig. 286). These settlements in the open are the highest known today for the European Mesolithic (or Epipaleolithic). The authors who have studied these remains think that some hunters passed the summer at this altitude (almost as high as Drachenloch!), probably associated with the seasonal movement of game. The settlements of Colbricon must at the same time be considered true lithic factories, documented by the great variety of artifacts, many in an intermediate stage of work. There are some similarities with settlements in the open Paris basin belonging to the Tardenoisian culture.

The Mesolithic is fairly well represented in North-Eastern Italy, sometimes at fairly high altitudes, and in Emilia (Passo della Comunella, at 1619 m, between M. Asinara and M. Sillano) where the settlements are situated only a little below the glacial cirques. The culture is similar in a sense to the so-called Castelnovian culture present in Provence at the Baume de Montclus and at the Abri of Font Pigeons. The site at Passo Comunella belongs to the Tardenoisian culture. In Abruzzo two sites have furnished mesolithic remains, one at the foot of the Maiella, and another near Fucino, at nearly 800 m. The man living at these sites belonged to a dolichocephalic race, attributed by Corrain to the Mediterranean variety.

Other mesolithic sites we should mention are those at Piancavallo (1300 m, near Pordenone, in the Venetian Prealps) and the Riparo Tagliente in Valpantena (Grezzana, Lessini Mountains) at more or less 200 m. The latter has preserved for us a complete skeleton belonging to the so-called epigravettian culture, characterized by segments and triangles. In the case of Riparo Tagliente a beautiful drawing on a pebble of a feline has recently been found (Fig. 287).

The Neolithic is characterized in the first phase by the so-called culture of square-mouth pottery, well developed throughout the whole of the alpine chain from Liguria to Friuli-Venezia Giulia. Both the variety of geographical position and topographic characteristics of these human settlements make us believe in the great environmental adaptability of the people belonging to this culture. The economy also shows a remarkable versatility.

Thus, in the Ligurian facies of this culture the settlements were in caves and in shelters under rocks; people were living by hunting and secondarily by fishing and the collecting of molluscs. In the inner parts of the Alps, more precisely in the Adige valley, the settlements belonging to the culture of the square-mouth pottery are localized in the detritus cones on the rocky slopes, where they found some shelter, but also in the open, on flat terraces at middle altitudes. The hunting of big game was predominant, but some fishing in the valleys – rich in lakes and swamps – was also undertaken. On the terraces agricultural activity was also possible so that human impact made its first appearance, though still limited.

The Neolithic is better and better developed in the European mountains: it is enough to mention the Massif Central (late Campignan culture), Cortaillod (on the shores of the Neuchatel lake), Horgen (on the shores of Zürich See) and Morges (on the shores of Geneve Lake). At Cortaillod there is evidence of agriculture and fishing. In the Appennine peninsula we have some Neolithic remains on the Lessini, at S. Anna d'Alfaedo, at nearly 1000 m, where Eneolithic remains have been found, and on Gargano, in Puglia, at nearly 900 m. The Neolithic present in the Alps is attributed by Corrain to the Danubian variety, characterized by mesocephaly.

On the mountains there are some interesting remains at Garniga on M. Bondone, at 750 m, belonging to the Middle Neolithic: a hut, probably oval in shape, has been found, belonging to the culture of the square-mouth pottery. Some think that in the Alps an early Neolithic characterized by impressed pottery, similar to a culture also present in Liguria and the Italian peninsula, developed upon a local mesolithic alpine substrate. Later in the Middle Neolithic square-mouth pottery developed: this was also a rather southern, more strictly 'italic' culture, which penetrated the Alps. Finally, in the late or Upper Neolithic, the culture of the southern Alps shows some affinities with the late-neolithic cultures of the Northern Alps, i.e. Aichbühl and Altheim (the latter a small town on the banks of the

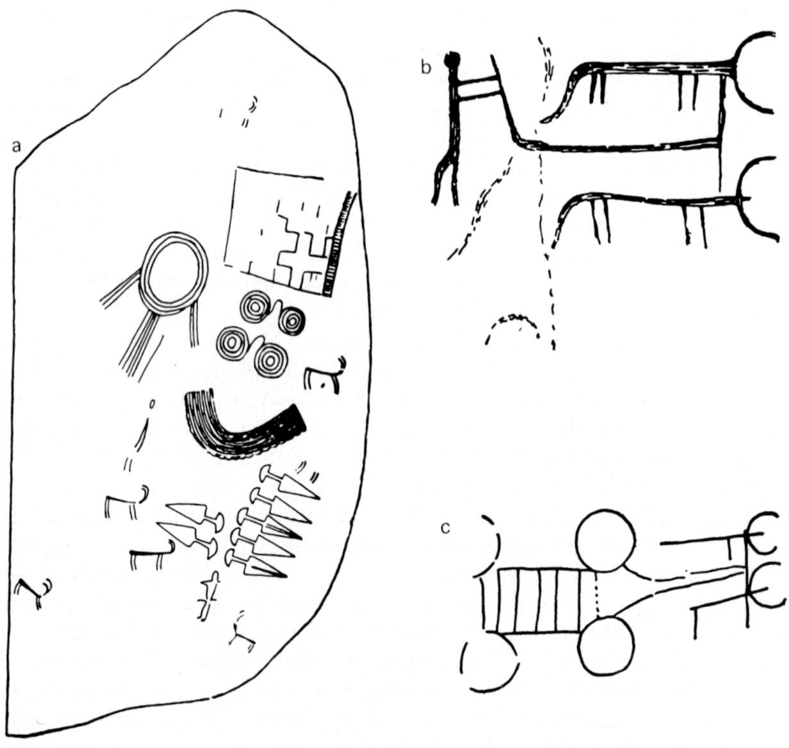

Fig. 287. Bronze Age in the Italian Alps. a. Drawing (schematic) on the so-called Masso di Borno (i.e. a rock found at Borno, Valcamonica). b. Oxen at the plough (Masso di Borno. c. A chart drawn by two oxen, from the Statue B of Lagundo (from Battaglia & Acanfora).

Inn on the border between Upper Austria and Salzburg). This points to the presence during that epoch of migrations responsible for the greater impact of man upon the alpine environment. Probably some old alpine names found today in the Eastern Alps are of very ancient origin, as a consequence of such migrations.

In Austria the Neolithic is represented by three different cultures, Danubian, Nordic and Western European. The former, characterized by ribbon pottery, is present at Lengyel and lasted until 1800 B.C. The second has been found in Baden (corresponding more or less to the Schwarz Wald) and at Mondsee (Salzburg). The Western European culture, characterized by bell-form pottery, was present at Michelsberg (near the present border between Western Czechoslovakia and Germany) from 1800 to 1400 B.C., where fortified villages and an agricultural system were present, which points to an impact upon the environment, at least since that epoch, though not at high altitudes. On the Monti Lessini (North of Verona) some signs of a migration from North-Eastern Europe are present on the northern slope, between 400 and 500 m. Some remains recently found on the Lessini, attributed to the post-Campignan and Chalcolithic, point to the presence in this elevated region of a culture based on pastoral work and agriculture; possibly some cannibalism and a crane religion were present. The animals found together with other remains of human presence are : goat, dog, ox, red deer, brown bear and wolf, together with two animals of cold climates, namely the ibex and the marmot, which constitutes a rather uncommon feature (A. & U. Sauro).

The Bronze Age is very widely represented in the territories today, corresponding to Austria and Germany from 1600 to 500 B.C. Let us remember that the remains found at Aunjetitz (today Unětice) near Roztoky, NW of Prague, were the first evidence of the Bronze Age in Central Europe, and where the practice of inhumation was present; Wieselberg; Adlerberg (today Hory Orlicke, at the border between Czechoslovakia and Germany) and Straubing, immediately south of the Bayrischer Wald (Bohemia). In Switzerland this epoch is represented at Morges.

As far as the Eastern Alps are concerned there is Doss Casteler, at 1150 m near Ossana, where a site has been found on a cone belonging to the early Middle Bronze Age; Doss de Lana, near the Lake of Ledro at 655 m, and Ledro Asat (the name of a small tributary of the lake). At Kohlgruben, near Barbiano, at 1375 m, a foundry of cupriferous ores has been found in a clearing of the coniferous forest, probably belonging to the latest Bronze Age. On Dosso delle Saette, at 1750 m in Val Vestino (Brescia) a settlement has been found not far from other sites of the same age, one on the western slope of M. Manos (1488 m) and another on Cima Igodello (1250 m). This is probably the highest site in the Eastern Alps belonging to the Bronze Age (Fig. 288).

In the Alps, the megalithic culture is well represented in Alto Adige (Süd Tirol) where some menhirs have been found, at Lagundo (near Merano, at 400 m), Termeno (in the Adige Valley, at some 276 m) at Santa Verena on Renon (c. 1000 m) and finally at Tötschling near Bressanone (or Brixen) at c. 1000 m.

In France, the megalithic culture reached some hilly regions, for instance the Petit Morin valley (nearly 200 m) where some signs of burial accompanied by funeral rites have been found in a cave dug in the soft rock. The presence of a dolmenic culture in the Iberian peninsula demonstrates the penetration from

France through passes in the Pyrenees, which means that, at that time, man was able to reach certain altitudes, at least in some circumstances: for instance megalithic remains have been found in the Ossay Valley near the Pic du Midi d'Ossau.

The Pyrenees seem to have been inhabited during the 1st millennium B.C. by tribes coming from the North or from the East, possibly belonging to the Ligurians, characterized by nomadic habits and patriarchal organization.

The Iron Age is represented at Hallstatt (Salzkammergut, at 500 m) from 800 to 400 B.C.: there are some fortified villages even at this altitude. In this epoch clear Celtic influence is shown by the Germanic tombs (inhumation). In Poland the first millennium B.C. is characterized by the arrival of the Western Slavs, whereas in Silesia (mostly mountainous) a Celtic population belonging to the La Téne period was present. In France in the 5th Century B.C. we find the Gallic or Celtic civilization named after La Téne, characterized by the burial (Celtic tombs, or 'Hügelgräber', whence the name 'Hügelgräberkultur'). In the 13th Century B.C. the Indo-European migration is responsible for the introduction in the alpine area of funerary urns (whence the name 'Urnenfelderkultur').

In Val Restela, near Vallarsa, at 730 m, NE of Rovereto, remains of a very ancient foundry for cupriferous ores have been found: this work seems to belong to the 2nd Late Iron Age, and therefore would be the most recent foundry of copper of the whole Alps.

The Doss dei Pingui near Mazzin of Fassa (the Avisio valley) at 1550 m, was the site of a settlement of the 2nd Iron Age (La Tène B), although some remains belonging to a former period were also present. It is the first pre-historic site discovered so far in the Fassa Valley, of which there is sufficient evidence. The remains show some similarities with those of the Castelir of Bellamonte (Travignolo valley, at about 1100 m), and belong to a culture common to Trentino, Alto Adige (Süd Tirol) and the Tirol, which could possibly explain the exis-

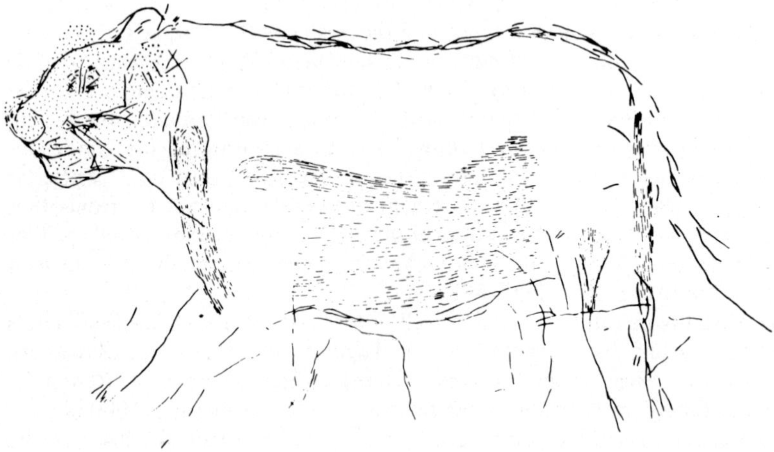

Fig. 288. Evidence of human presence and impact on the Alps during the Upper Paleolithic (Epigravettian). Raffiguration of a felid (*Leo leo*) on a pebble found at Riparo Tagliente, in the Lessini (Valpantena) from Bartolomei et al., 1974).

tence of some similar names for alpine plants in these three regions (cf. p. 000).

In pre-historic times, Illyrians, Celts and Danubians were living in the territory today corresponding to Austria, but only since the Iron Age (7-5th centuries B.C.) are archaeological remains fairly conspicuous in the Alps and is there some evidence of the passage of man through the alpine passes. The pre-Roman populations of the Alps were in the Western Section, Ingaunes, Taurines, Lepontii, Allobrogii; in the Eastern section Rhaetians, Tridentini, Norici and Vindelici.

The Val Venosta (Vintschgau) was inhabited by the Venostes (15 B.C.) closely related to the Vennonetes, living along the Rhine, and to the Venni, living along the Adda river. The Venostes extended from Lagundo (near Merano) up to Nauders (1365) beyond the Pass of Resia (1507 m), which was evidently used continuously. Human influence, though slight, must therefore have been present in this region even at this remote time. Gauls were also present in some southern portions of the Alps, as is very clearly shown by the present-day names of several alpine plants (see later). The passage of very ancient populations from one part of the Alps to another is also clearly revealed by the presence, in different sections, of the same name for several kinds of trees or herbs.

One of the most interesting and probably oldest names is that of *Pinus cembra*, although its etymology is very obscure. In Livinallongo it is called zirum, from the old stem gimru by means of a simple metathesis and change of consonant (g → z). It is worth mentioning that in Val Badia, *P. cembra* is tsier or zier, at Siusi (Seiseralpe) żirm or zirbm: according to some the former would be Latin, the latter German. In territories where the German language is spoken we find the names Zirbe and Zirm, both of unknown etymology and both recurring very often in the vernacular. Incidentally, we must note that in Lower Austria *P. mugho* is called Zerben, Zerm and Zürm, which could also come from gimru like the name of *P. cembra.* Certainly the Romans had no name for this tree: in the Rhaeto-Romanic of Engadina today we find names such as dschember, schember or gember. A *P. cembra* stand near Scarl (Lower Engadina) is called schambrina, and another in Val Trupchum (also in Lower Engadina) schembrina. Possibly from 'cembro' the geographic names Piz Cambrena in the Bernina group and Val di Cembra in Trentino also come.

Pinus mugho also has several pre-Roman, old, alpine names. In the Livinallongo it is called barancle (plur.), which is akin to barantl, a 'ladin' (Romansh) name used at Siusi (Seiseralpe) besides zuondra ('ladin') and mign (German). According to some, however, this term as well as migstauden used in the Sciliar (Schlern) near Siusi, could have its origin from the Italian mugo, used in Lombardia, Trentino and Veneto, and probably of Roman origin.

Another small tree or shrub common in the Alps, *Sorbus aria*, shows some interesting linguistic problems. In Livinallongo it is known as paremolè (plur.), in the Agordo region paremolar, paremoler, peromboler (similar to paromboler used in the Belluno region for *Prunus spinosa*). These terms would be related to parmuela of Engadina and parmuja of Val Bregaglia, a group of words to which the central 'ladin' paromole also belongs. According to Pedrotti & Bertoldi, 'reciprocal contacts and influences occured'. These contacts can be explained only on the ground of migrations of old alpine populations from one part of the Alps to another, as we have already postulated.

Much more interesting is the name given in the Eastern Alps to the green alder *Alnus viridis*: the gallic root amb or amp is widely spread (ampiadei in Comelico, ampedin at Cortina, àmbie at Laste on Rocca Pietore, ambi at S. Tomaso, etc.)[1]. According to Pellegrini & Zamboni, this root possibly comes from some Celtic languages, and can be connected to amblatium (existing since the 9th Century B.C.) or ambilatium ('Jochriemen', Meyer-Lübke's Romanisches etymologisches Wörterbuch), coming from the verb ambi-la (nerumlegen, Pedersen). On the contrary, the alpine names of *Alnus* (*glutinosa, nigra* or *viridis*) which include the root al-, would come from a Celtic or pre-Celtic word (Pellegrini & Zamboni). Still more interesting is the name *Alnus viridis* has in Livinallongo (velma) or in Piemonte and Liguria (verna, verno) which would derive from the gallic verna, used for any species of *Alnus* (in Irish fern). Finally another pre-Roman word is employed in the Alps to indicate *Alnus viridis* drause, used today in Piemonte (particularly the provençal and franco-provençal areas), Lombardia, western ladin areas etc. These names reached French-Switzerland and Savoie in the West, Val di Fassa (dróussa) in the East. In some localities names coming from both verna and drausa are present (Pinerolo, in Piemonte).

The small elegant *Viburnum lantana* also has very old alpine names: in Cortina it is called pagogna, which according to Pedrotti & Bertoldi could be related to mugne, mogne, mognine, a series of words of which there is a trace in mugnée, used in Carnia for *Rubus idaeus*. According to these researchers, it is an old alpine, pre-Latin name. Maybe it is worth mentioning that in Celtic muine means 'small woodland'.

Two alpine plants whose indigenous names have a definite Gallic origin are the whortleberry and the heath *Erica carnea.* The former is called in the south-eastern Alps and Prealps glésenes, glezenèi, giánesi, giánes, glásinis, etc., after a latinized gallic name, glastum, meaning blue, so that the whortleberry would be literally 'the plant bearing blue berries' which is quite true (therefore we find ourselves confronted by a beautiful case of semantic clarity). Many place names have taken their origin from the Gallic name for *Vaccinium myrtillus*. The heath *Erica carnea* is called brosorada in Cortina, from the Gallic brucus (see also the chapter on heaths). Pedrotti & Bertoldi, talking about these names, say in their text 'sediments have remained at the surface not as dispersed relics, but as compact areas, predominant from the Roman colonization up to the present day (Fig. 289).

And now a linguistic puzzle: a common *Liliacea, Veratrum album*, which grows in meadows from 1000 to 2000 m, especially in clearing in forests where cattle graze, and is easily recognizable in late summer because it is still green and flourishing when other plants have been eaten by the cows, has several common names in the Dolomites which seem to be extremely old. In Comelico (Padola valley) its name is rósči, at Cortina, rosco, similar to the rost of Val Gardena (Grödener Tal) and arosč da munt of Val Badia. In the neighbourhood of Belluno as well as in Emilia (northern Appennine) it is regoss; in the Agordo region aurosk, aorosk, arosk, orosk, oros, rosk. According to Pedrotti & Bertoldi these names could reflect, through metathesis and vowel dissimilation, a base gorost →

[1] In Austria some contamination occurs, such as Ampalterstauden and Ampferstauden (Pellegrini, 1973).

Fig. 289. The area in Northeastern Italy (the Dolomites included) where names of Celtic origin ('glastum') are employed for the whortleberry, *Vaccinium myrtillus*.

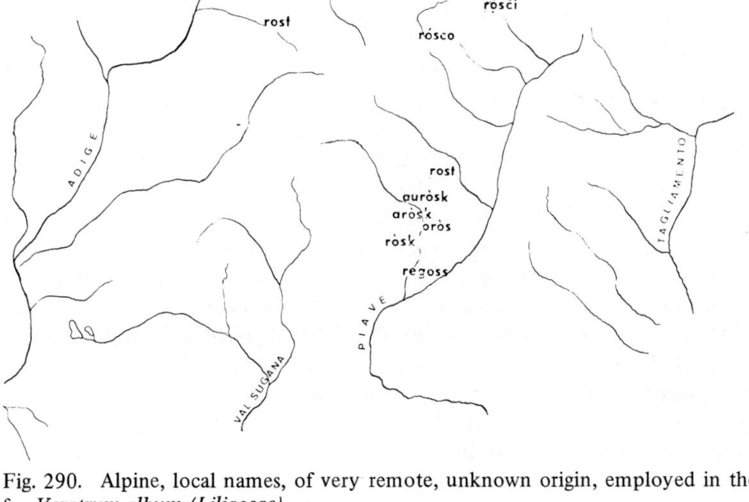

Fig. 290. Alpine, local names, of very remote, unknown origin, employed in the Dolomites for *Veratrum album (Liliaceae)*.

goross ⟶ regoss. Probably we are in the presence of an unknown, popular, archaic name. These students point also to the names golostiu, golostii and olostru, used in Sardinia for the holly, *Ilex aquifolium*, and gorosti used for this plant in Basque country. As a mere working hypothesis, the three nuclei of the Alps, Sardinia and the Pyrenees according to Pedrotti & Bertoldi would find their connection in a common root (Fig. 290).

As a whole in pre- and protohistoric times the mountains were not inhabited by man: he was restricted to the valleys or to foothill regions of the mountains, so that his impact on this environment began only recently and later than in other biomes. Also the passes were very rarely utilized: the first passages were probably those performed by merchants of metals and secondarily obsidian and amber. Later on the passes served in the migrations of people, generally from East to West and secondarily from West eastwards. We must not forget that some passes were very low: for example, the region of Dobbiaco (Toblach) which unites the Danube basin (Drau river) with the Adige basin (Rienza river), 1200 m high, or the region of Camporosso (800 m) which unites the Tagliamento basin with the Danube basin. The impact until Roman times was very light.

We may perhaps suggest that Mousterian man, and more precisely alpine Mousterian, was stronger and fitter to climb mountains and possibly more resistant to cold than *Phaneranthropus*, and of course present-day man: the presence of men at altitudes higher than 2000 m has never been demonstrated for other pre-historic populations.

Only in Roman times were the Alps permanently colonised and furthermore crossed by many routes. An old town, for example, is Aosta, at 583 m, a Roman 'colonia' in 25 B.C., a diocese in 451 A.D.

The main Roman roads were the Mons Matrona, going from Turin to Valence, through the Monginevro (1854 m), the Alpis Graia and Alpis Poenina, from Aosta to Lyon, the former through the Little San Bernardo (2188 m) and the latter through the Gran San Bernardo (2469 m); a road went from Milano to Coira (Chur), another from Verona to Augusta, passing through the Brennero (1372 m) (via Claudia Augusta).

In this connection we should mention Hannibal's journey through the Alps, when he went from France to Italy in 218 B.C., with 26,000 men and 317 elephants. It is uncertain whether he used the Monginevro (1854 m) or the Little San Bernardo (2188 m). In the Pyrenees, a Roman road passed through the Perthus, at 179 m, a depression within the Albères.

The presence of Romans in the South European mountains for a long period of time is well documented by the names of many trees and herbs: we will limit ourselves to mentioning here only some common species, such as pine, oak (*Quercus,* in Italian quercia), *Alnus* (which, however, seems to have a Celtic origin), poplar (populus, in Italian pioppo, in German Pappel), *Ulmus* (in Italian olmo), *Abies* (the fir, in Italian abete, whence many toponyms all through the Italian Alps), *Picea* (spruce, in Italian peccio, which has also given origin to many toponyms). *Salix* (willow, in Italian salice, in Spanish sauce, etc.), *Carpinus* (in Italian carpino), *Betula* (birch, in Italian betulla). Only doubtful is the Latin origin of the local names for *Sorbus aucuparia*, the common rowan tree, in which a root mal- or mel- is present: according to some these names could derive from the Latin mespilus (μέσπιλος).

In the Middle Ages human settlements were still at fairly low altitudes: the presence of old villages or towns on the Altopiano di Asiago (or Altopiano dei Sette comuni) in the Venetian Prealps, at a height of 800-1000 m, is due exclusively to the fact that the old Germanic populations were compelled to withdraw to the mountains by the populations of the lowland (Italic peoples). Indeed, today it is generally accepted that in the early Middle Ages some Germanic peoples (Goths, Eruli, Gapidi, Rugi, Longobardi – or Lombards –, Burgundians, Alamanni, Franks and Bavarians), who spoke the dialects of the 12th and 13th centuries (hochdeutsch), took refuge on this high plateau. They founded there several small villages still existing today, such as Gallio (from Gatium, or Gagium: gazo should mean a woodland), Rotzo, Roans (from robur or maybe from roan, a local word to indicate a steep slope, etc.). Today there are many place names and lake names of clear Germanic origin, and also the language spoken on this high plateau (a true linguistic 'isle') is a mixture of old Germanic languages. Let us recall only Spitz Bisa from Spitze (top) and Wiese (a meadow), Barenthal or 'bear valley', Parloch, from Bär, the bear and Loch, a cave (cf. Gorlato).

In relation to roads and passes many castles and monasteries have been built since very ancient times, each of them contributing in some manner to altering the original environment. We will mention here only two regions, one north of the Alps (Germany) and one south (Trentino).

According to modern historians, the genesis of alpine castles in Trentino must be looked for in the evolution of community 'castellieri'[1] (Volksburg) or in that of the fortresses built by the Goths, Longobardi, Carolingi, etc., as a residence for the feudatary or as a route control (Herrenburg). Let us mention here some of the oldest castles in Trentino (North-Eastern Italy, once belonging to Austria and before that to the Holy Roman Empire, or Germany).

The first castles in the sense of a fortified walled nucleus date from the 1st millennium. The castle of Castellano, built in the 12th century, is situated at 790 m in Vallagarina: to the same epoch the castle of Gresta (793 m) and that of Madruzzo in Val Cavedine, at 556 m, belong. Much older is Castel Campo, at 505 m in the Giudicarie (Val Duina, Lomaso) present in the 11th century and constructed in the site of a pre-historic castelliere. In Val di Chiese an old castle is Castel Romana, situated at 690 m, a 'tower-castle' with many old legends one of which is still alive (the countess Dina who in the moon nights rides a white

[1] Castelliere is the name employed to designate walled (fortified) villages built during the metal age in the two last millennia B.C. They were often transformed into Roman 'oppida' (178-177 B.C.) and later into Medieval fortresses and burgs. From 'castelliere' some toponyms present in Trentino, Puglia, Liguria and Provence are coming, such as Castellier, Caslir or Castellar (Battaglia). The 'castellieri' are generally present at rather low altitudes, so they cannot be attributed to the montane environment. In Venezia Giulia and present day Slovenia the highest castellieri are those of S. Michele di Senosecchia, at the foot of the Selva di Piro (Birnbaumer Wald) at 600 m (where the so-called Japids were living), at S. Michele di Storie (400 m, on the Carso or Kras, or in German 'Karst'), at Fontana del Conte (Knezak) at 580 m, at Obrovo S. Maria (Obrou) at 580 m, in Northern Istria. Here Proto-Illyrians and 'Histri' were living (Battaglia).

horse ...). At Stenico there is Castel Stenico, at 702 m, representing the evolution of a castelliere: here some very old alpine pre-roman population (the 'Stoni') was probably massacred by Romans in 118 B.C. In Val di Non we find Castel Belfort, situated at 746 m, built in the 14th century; Castel Bragher, built in the 13th century, situated at 640 m and the castle of Castelfondo, one of the most picturesque and wild castles of Trentino, situated at 864 m on a kind of rocky peninsula bound by the extremely deep ravines of the Rabiola and the Novella rivers. It was constructed near 1000 A.D. Other castles of this area are Castel Ceredo, 853 m, Castel Corona, 951 m, today reduced to some unaccessible remains, originally a castelliere, Castel Malgolo, at 780 m, Castel Malosco, at 1000 m, built in the 11th century; Castel Nanno, at 589 m, the most characteristic and 'Italian' feudal construction of Val di Non; there are some documents of the indictments of witches and sorcerers which occurred during the 17th century, i.e. after the so-called Counter Reformation which actually had its origin in Trento. Castel Thun is also an old Middle Ages castle in Val di Non, at 609 m, very well preserved (Fig. 291).

In Val di Sole we may mention Castel Caldes, at 682 m, from the 13th century; Castel Ossana, at 1019 m, of which today only some ruins and the tower are preserved, built in the 12th century; the Cevedale can be seen from this castle. In Val d'Avisio we may mention Castel Fornace, 747 m, and Castel Segonzano, at 566 m, both reduced to a few remains. Finally, in Val Sugana (i.e. Upper Brenta Valley) there are Castel Pietra at 1030 m, today represented by inaccessible ruins, and Castel Telvana, at 559 m partially destroyed.

The highest castles in Trentino are today reduced to ruins: Castel of Castellino, in the Low Sarca Valley, at 1268 m, and the castle of Altaguarda, in Val di Non, at 1280 m; this would be the highest existing castle in this southern alpine region.

Many castles are scattered all through foothills and mountains of Germany, although generally at very low altitudes. Let us mention for the Rhingau, S.W. Taunus, the castle of Johannisberg from the 12th century and the castle of Reichhartshausen also of the 12th century (near Hattenheim). Near Taunus there is the castle of Kronberg, of 1220, situated on a group of picturesque rocks; not far from the latter there is the castle of Falkenstein, of the 12th century, at 500 m, near Frankfurt: today only some ruins are left. In Baden we find Kloster Eberbach, of the 12th century, a cistercian building situated under the Katzenbuckel (m 626). Particularly old are the Altes Schloss of Baden Baden, at 474 m, near Battert, dating from the 3rd century, and the castle of Ebersteinburg, at 489 m, of the 10th century, built on some Roman foundations.

As a whole, in Trentino-Alto Adige the castles were built higher up than in Germany, and as a consequence the human impact upon the mountainous environment was greater. The same is true when we take into consideration the lowlands, both of Germany and Northern Italy (deciduous forests, banks of rivers, etc.), where there is a much more severe impact in Italy.

As a whole, during the Middle Ages the impact of man on the mountain environment has not been very great. The presence in the Italian eastern Alps of the so-called 'regole' (literally rules) prevented private holdings and represented a true 'common richness' (aquae, pasqua, nemora sint communia). With the arising of the Venetian Republic (9-10th centuries) a terrible exploitation of these moun-

Fig. 291. Examples of castles situated at relatively high altitudes in Trentino: a. Castel Ossana; b. Castel Thun; c. Castel Campo (with the permission of Provincia autonomia di Trento).

tains began; wood utilization, grazing and extractive industries had a true 'predatory' character, as Posocco says:

'Relative organization of the Belluno forests began only in the 14th century, when Venice instituted a wide forested public zone, with the high wood supply necessary for the dock-yards since 1501 under the direct control of the "Consiglio dei Dieci" (literally The Council of the 10).'

However, during the Middle Ages in all Southern and Central European mountains, besides this kind of exploitation (wood utilized for burning) which perhaps was not damaging the forests of that era, new forms of exploitation arose, namely deforestation in order to obtain a major extension of arable land and of areas for grazing, and the felling necessary to supply wood for burning in the small craft industries which slowly developed. Of course these major human activities increased with the demographic increase which occurred from the Middle Ages onwards. In some instances, deforestation caused serious damage to the mountains (soil erosion, etc.) for instance in the Venetian Alps (Cadore) and Prealps, where systematic wood exploitation for the ships of the Venice Republic lasted for many centuries; coniferous wood being utilized for ship construction and beach wood for the construction of paddles.

Today the height of permanent human settlements is very varied according to the different European mountains and particularly to the physical factors. On the Alps the highest village, as we have already said, is S. Martino in Val Passiria, associated with the presence of a mine.

In the south-eastern Alps there is Arabba di Livinallongo, at 1585 m, with some scattered houses up to 1688 m (Rallua); at greater altitudes we find only temporary abodes or buildings, partly used only for hay (tabià, see later), partly for human residence. In Alto Adige (Süd Tirol) some higher permanent settlements are present probably due to a different climate: there are Anterselva (Antholz), at 1422 m, Belprato (Schönau) in Passiria, Corvara (Rabenstein) at 1419 m, Melago (Melag) at 1918 m, in the Upper Val Lunga (Langtaufers) where there is only one house; Solda (Sulden) at 1905 m is one of the highest villages of this region. In Lombardia we have Livigno at 1800 m, in a sunny, south exposed valley, with some scattered houses as high as 1870 m (Tresenda). In Engadina finally some settlements can be found above 1700 m in V. Müstair (Tschierv, Lüsai, Lü, etc.) or in the Inn valley at Zuoz (1712), Madulain (1800), S. Moritz (1853 m). Also in the Ötztaler Alpen (Tirol) there are very high permanent human settlements such as Vent (1893 m) or Ober Gurgl (1927 m). These are among the highest alpine villages.

Among the temporary residences, we find in the Alps the so-called baite (sing. baita), of uncertain etymology, used as human residences only during the summer, sometimes only for very short periods. The 'baita' is a small building built with materials found in the area, generally a half-timber house, with the roof made of slabs (for instance schists, slates, etc.) or shingles (generally of larch or spruce). They are utilized as a shelter for men or cattle and also as a store.

A casera (from late Latin casearia, a feminine adjective meaning 'of the cheese', probably crossed with casa) a term used in Lombardia and Veneto, is a montane house where the milk produced by cattle during the summer is transformed into butter or cheese. Finally in the Alps, we should mention the so-called 'maso', a northern and 'ladin' word corresponding to the Middle Ages Latin mansum, which means abode, coming from the Latin mansus (inf. manere, to stay). It is a word used today in the Eastern Alps, comparable with the provencal or catalan 'mas', which is a colonial house. It is used to indicate a landed property, a 'masseria', a temporary abode associated with cattle raising in the mountains.

The height of these temporarily inhabited buildings is very varied. We find at 1000 m near Feltre Maso dei Marti, at 1350 Maso Vederna, Maso Simon, etc. In Alto Adige (Süd Tirol) there are some 'maso' at some 1400 m or higher (for instance North of Sesto or Moso in Pusteria). As far as casere (or 'malghe') are concerned, we have Malga di Lusia at 1899 m, at 1977 m Malga Cauvere, etc., near Predazzo; Casera Prademur (1900 m), Casera della Stia (1785 m) and Casera Focobon (1894 m) in the group of the Pale di S. Martino; but a malga can also be present at lower altitudes: for instance Malga Canali at 1302 m, or Malga Pradidali at 1428 m (south of the Pale group). Exceptionally low is Casera Rombaldi at 661 m near Feltre.

The hay-lofts are higher in the South-Eastern Alps: near the Marmolada, the so-called tabià (or hay-lofts) are present up to 2263 m (Prati Padon), 2122 m (Pian de Stanzon), 2182 m (Agnarezze); sometimes these hay-lofts constitute true settlements, as at Pian de Lobbia at 1949 m, or Tabià Palazza (1819 m) between the Marmolada and Malga Ciapela, on a southern slope, or Roschei (1898 m) and Ciampei (1865 m) on a northern slope, between Marmolada and the Livinallongo.

Other temporary abodes in the Alps are the so-called 'stavoli' (Cadore, from stabulum), châlets (in Swiss countries where French is spoken) and 'Almen', in countries where German is the spoken language.

In the Pyrenees the situation is fundamentally the same: there are permanent abodes reaching 1550 m in the Northern Pyrenees (Saussa en Lavedan) or 1688 m in the Southern Pyrenees (La Llagone, in Catalonia). Temporary villages are present between 1200 and 1800 m and are known as bordes, cortals or pardinas (Catalonia). Above 1800 m only refuges for shepherds are present, known as jasses, orrys or cayolars.

The human impact has been felt only in the most recent times, when heavy exploitation of natural resources – both renewable (felling, grazing) and non-renewable (mines, quarries, etc.), together with an increasing human presence, both in settlements associated with the above-mentioned primary and secondary activities, and as tourism (tertiary activity) has occurred.

Today most of the South and Central-European mountains – which as a whole are by far the most interesting from an ecological view-point – are more or less affected by man, and the menace against the environment and its biota increases every day. The necessity of protection has long been demonstrated and actually many areas are today protected against exploitation or in general human presence (see the subchapter 'Conservation in the montane environment').

The vegetation

Montane plants show some biological and systematic peculiarities by which we may easily recognize – with the changes of climatic and environmental conditions –

a subdivision of the vegetation into several altitudinal belts, characterized by fairly constant features. Furthermore, the evaporation potential, greater on mountains than in the lowlands, the greater intensity of light (particularly of UV) and other factors bring about some particular adaptations, which manifest themselves in the complex aspects of these features (Fig. 292).

In the mountains of Southern and Central Europe we may recognize essentially two vegetational belts: a lower, or montane one, mainly arboreal, and a higher, or alpine, generally herbaceous. The main altitudinal belts are therefore: a foothill belt (or belt of deciduous oaks, in southernmost Europe only also evergreen, sclerophyllous oaks, such as *Q. ilex*), a submontane belt (the so-called Castanetum, i.e. the belt of chestnut), a lower montane belt or Fagetum (i.e. the belt of beech), a higher mountain belt characterized by coniferous forests (mainly *Picea* and *Abies*), a subalpine belt (the belt of hysophilous, xeromorphous coniferous trees, i.e. *Pinus cembra* and even more *Larix decidua*), the belt of the stunted or dwarf shrub ('Zwergstrauch'), the belt of the alpine meadows and finally the snow belt (Fig. 293).

The oak forest is more or less xeromorphous (heliophilous forest); the beech forest is a sciaphilous formation, characterized by many shadowloving herbaceous plants; the Picetum is a mesophilous, xeromorphous forest, whereas the Laricetum is a hypsophilous xerophilous formation, also called by some authors 'cacuminal forested heath'. The stunted shrub belt is called by some 'treeless heath' due to the presence of *Vacciniaceae* and *Ericaceae*. The alpine belt corresponds to Brockmann & Rübel's class of Sempervirentiprata and the snow belt to the class of Frigorideserta. Some authors distinguish within the latter two subdivisions, namely

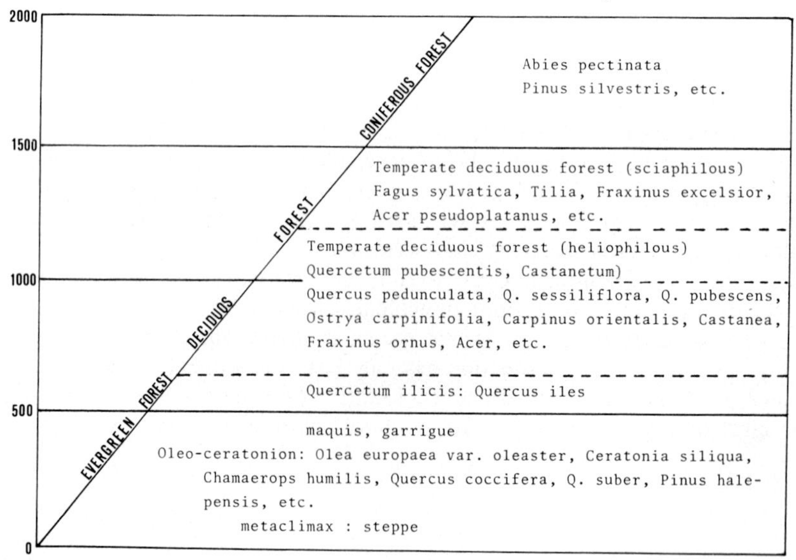

Fig. 292. Distribution of plant communities in the Mediterranean area according to altitude (from Polunin & Huxley, greatly modified).

a subnival one, above the lower limit of the isolated permanent snow spots according to Heer, and a nival belt proper, situated above the uninterrupted line of permanent snow. The ecological implications of this subdivision are very evident.

Some authors speak also, for a higher zone, of cryoplankton, which literally means the algal vegetation of the ice (see later).

The altitudinal limits of each belt vary enormously not only according to latitude as one might expect, but also according to rainfalls and its distribution during the year. We must keep in mind that the changes in rainfall from one place to another sometimes modify the altimetric or thermal upper limit of the vegeta-

Fig. 293. High mountain vegetation, at 2100 m in the Sesto region (Sextener Dolomiten) a. 'Zwergstrauchheide' with *Rhododendron hirsutum, Cyperaceae, Dryas octopetala,* etc. Dolomia with a rendzina at pH 7.0 and scarce organic substances. b. As above, at 2200 m, *Rhododendron, Horminum pyrenaicum* (broad-leaved plants), *Carex,* etc. Dolomia with a rendzina at pH 6.9, rich in organic substance. c. On the northern slope of Marmolada, at 2300 m, typical cushion-like plants (*Saxifraga, Salix, Carex firma,* etc.). Limestone with a rendzina at high pH. In the background a calcareous wall, partially covered with mosses, is well visible.

tion. Thus, in the Eastern Alps spruce and beech grow at a lower altitude, compared with the Western Alps, which is poorer in rainfall. Spruce is present as low as 500 m in Carnia and 1000-1100 m in Piemonte; beech is not present higher than 1000-1200 m in Carnia, whereas it reaches up to 1500-2000 m in the Central-Western Alps.

We will mention here as an example of such variations, the situation we find in Italy in the Alps and the Apennines:

PHYTOCLIMATIC ZONES	Thermal characteristics			
	mean ann. temperature	mean temp. of the coldest month	mean temp. of the warmest month	mean minimum temp.
LAURETUM II type (summer drought) middle and cold subzones	12-18	> 3	–	> –9°C
CASTANETUM a) warm subzone I type: without summer drought . . .	10-15	0-3°C	–	> –12°C
II type: with summer drought . . .	10-15	> – 1	–	> –15°C
b) cold subzone I type: ann. rainfall	> 700 mm			
II type: ann. rainfall	< 700 mm			
FAGETUM a) warm subzone	7-12°C	> –2	–	> –20°C
b) cold subzone	6-12	> –4	–	> –25°C
PICETUM a) warm subzone	< 6°C	> –6°	–	> –30°C
b) cold subzone	> 3°	≤ –6°	> 15	≤ –30°C
ALPINETUM	< 2	≤ –20	> 10	≤ –40°C

This is of course only one of the several schemes we may advance for mountain vegetation. In several localities, the prevailing conditions of climate, relief, exposure etc. also bring about very substantial modification, as e.g. the inversion of vegetation found on some mountains where the beech grows above the spruce, because the aspect of the lower parts is North and that of the upper parts of the mountain is South. This is the case in the valley situated west of Pic de Midi d'Ossau, near Bious-Artigues (Western Pyrenees). Not very dissimilar – though

Altitudinal limits	Forest species
southern Italy: 250-700 m in Calabria, Sicily and Sardinia up to 800 m	*Quercus ilex, Q. pubescens, Q. suber, Q. vallonea, Laurus, Arbutus unedo, Pistacia lentiscus, Phyllirea arborea, Myrthus communis, Thymus capitatus, Erica arborea, Olea, Pinus halepensis, Juniperus macrocarpa, J. oxycedrus*
Ligurian Alps and Northern Appennine: 300-600 m	*Castanea, Quercus pedunculata, Q. robur, Q. cerris, Q. ilex,* (upper limit) *Quercus vallonea, Acer* spp., *Fraxinus excelsior and F. ornus, Carpinus betulus*
Northern Italy: 200-600 m Central Italy: 300-800 m South. Italy: 600-900 m	
Alps: 300-900 m; North. Appennine 500-900 m; Centr. Appennine 700-900; Southern Appennine 800-1200 m	The same elements with the exclusion of the most thermophilous ones
In less rainy localities than those belonging to I type, mainly in North. Italy	Chestnuts and *Quercus pedunculata,* elms, poplars, willows, maples, linden, *Carpinus, Quercus robur, Q. cerris, Juniperus communis*
Alps: 700-1200; North. Appennine 700-1200; Centr. Appennine 800-1500, South. App. 1000-1700	beech, *Quercus pedunculata, Q. robur, Acer pseudoplatanus, Alnus nigra, Carpinus, Sorbus,* willows, poplars, *Abies pectinata, Pinus nigra, P. laricio, P. austriaca*
Alps: 800-1600; North. Appennine 1200-1600; Centr. Appennine 1500-1800; South. App. 1700-2000	beech, *Quercus pedunculata, Q. robur, Fraxinus excelsior, Sorbus aucuparia, Betula, Populus tremula, Abies, Pinus silvestris, Picea* and *Larix* (lower zone)
Alps from 1000 to 2200, in cold, shaded valleys as low as 700-800m, on the summits up to 2200 m	*Picea, Pinus silvestris, P. cembra, Larix decidua, Abies pectinata, Pinus mugho, Juniperus communis, J. sabina, Betula alba, Alnus viridis, Tilia, Fraxinus* (uppermost limit), poplars and willows
Alps up to 2200-2500 m	*Pinus mugo, P. cembra, Larix decidua, Pinus silvestris, Picea, Betula, Alnus viridis, Salix, Cytisus, Rhododendron* and herbs

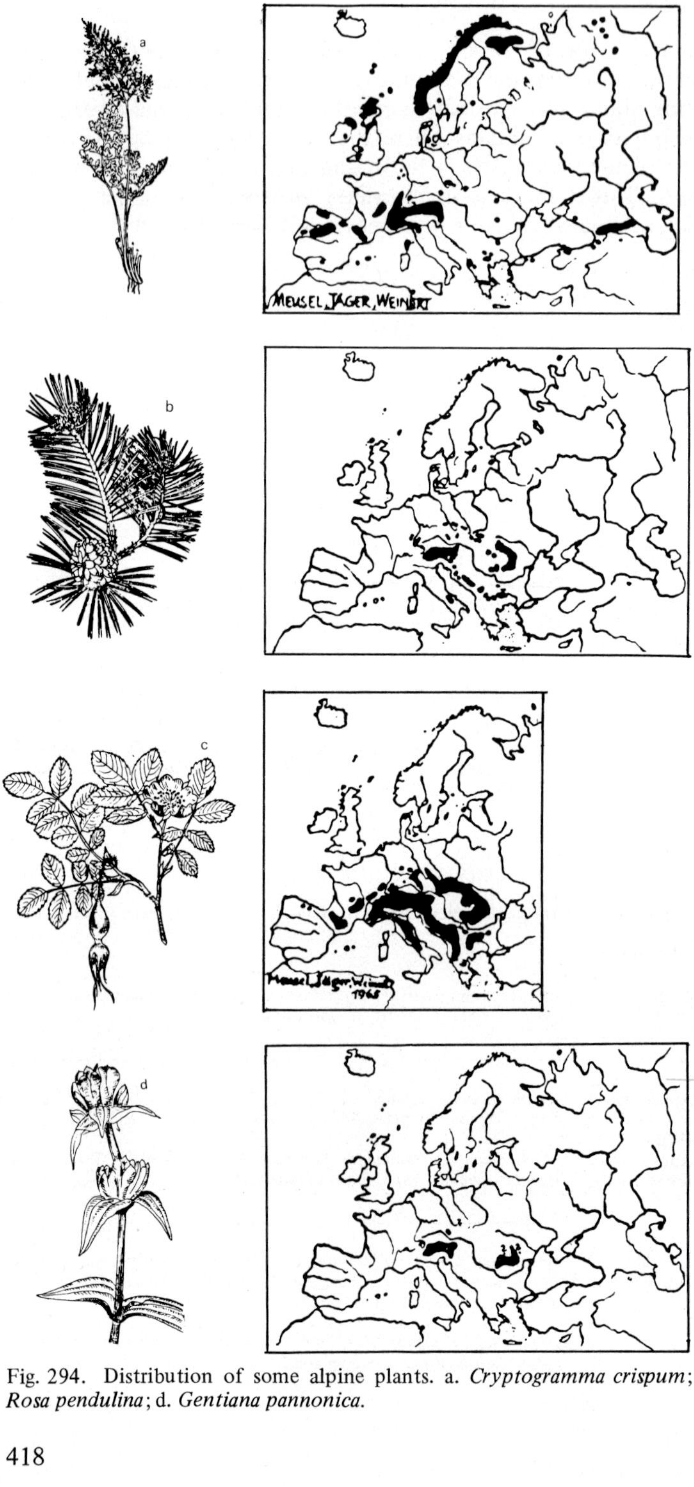

Fig. 294. Distribution of some alpine plants. a. *Cryptogramma crispum*; b. *Pinus mugo*; c. *Rosa pendulina*; d. *Gentiana pannonica.*

different as far as reasons are concerned – is the situation extant on the plateau of Cansiglio, in the Venetian Prealps. Here beech grows above the spruce and fir because of the presence of fog for a great part of the year and a great part of the day in the central, lower, part of the plateau which causes a particularly cool microclimate contrasting with that of the surrounding parts of the Cansiglio which is very often exposed to sun. Together with the fog, and partly responsible for it, we find a mass of cold air flowing from the higher marginal parts of the plateau, which cannot be warmed either in the cold winter or in the summer days because of the dense vegetation. As a consequence of this in the innermost part of the plateau there is a practically permanent layer of cold, heavy air. This microclimate is responsible not only for the presence of the Coniferetum (*Picea* and *Abies*) at a level inferior to that of the Fagetum, but also for some rather hypsophilous plants such as *Vaccinium vitis-idaea, Vaccinium myrtillus, Arnica montana, Hypericum quadrangulum, Homogyne alpina, Ajuga pyramidalis* and *Polytrichum commune.*

Another interesting feature is the lack of a Coniferetum on several mountains of Southern Europe, where a kind of alpine meadow follows the Fagetum. In this case in the upper part of the Fagetum we find some stunted, almost dwarf beeches, with the beech the common juniper is often associated. Such a condition is present on the three big European peninsulas, the Iberian, Italian and Balkan, as well as in some parts of the Northern, French slope of the Pyrenees, for instance in the Cirque de Troumouse, east of Gavarnie (Hautes Pyrenées), in the Appennines or even in the western Balkans (Velebit, Mosor, etc.).

A very peculiar feature is present in the easternmost part of the Pyrenees, for instance at Les Bouillouses, not far from Pic de Carlit: in this region, as a consequence of the relief and the rainfall regime, at a particular height – say at some 1800-1900 m – the Quercetum ilicis associated with *Genista purgans* is abruptly substituted by the Coniferetum (*Pinus uncinata* associated with *Rhododendron ferrugineum*). A little further within the plateau an Alpinetum (alpine meadows and bogs) follows, practically without any sort of transition.

In very rare instances the open vegetation above the tree line can belong to a steppe vegetation instead of a true 'alpine meadow'. This is the case on some summits of the North Western Caucasus, such as Mt Sheze, Mt Sober-Oash (725 m), Ostraya (650 m), Krestovaya (700 m), Tkhab (700 m). The high number of steppe elements within the soil fauna of these open environments – as revealed by Ghilarov & Arnoldi – points to a relict origin of this steppe mountain vegetation and the relative biota. Today the steppe plants are *Festuca sulcata, Koeleria gracilis, Phlomis tuberosa* and *P. pungens, Stipa pulcherrima* and several others.

As a result of adaptation to the different mountain environments we find very different patterns in the biological forms (represented synthetically by means of Raunkjaer's spectra) of montane stands. These spectra vary according to the different localities and environmental conditions. As an example we will refer here to the biological spectra of some montane localities of Italy and, for a comparison, of Switzerland:

Locality	Altitude (m)	Ph	Ch	H	G	Th
M. Argentario	635	12	6	29	11	42
Colli Euganei	30-600	7	3	43	14	33
Fagetum, Corletto M.	1100-1300	17	7	51	19	6
Degraded Picetum, Cadore	1100	17	7	56	19	1
Picetum with *Myrtillus*, Switzerland		20	19	51	2	8
xeroph. meadow, Colli Lessini	1200-1300	–	4	79	22	6
unfertilized meadow Asiago	1000-1300	–	6	77	15	–
fertilized meadow, Alpe di Siusi (Alto Adige)	1850-1950	–	–	87	7	6
degraded meadow, id.	1900-1950	2	13	76	4	7
Normal spectrum (world average)		46	9	26	6	13

The effect of increase in altitude on the biological spectra has been clearly demonstrated by Raunkjaer in Switzerland, near Poschiavo, as we can see from the following data:

	P	Ch	H	G	HH	T
> 2850 m	–	35	61	2	–	2
2550-2850	–	25	67	4	–	4
2250-2550	4	18	64	7	1	6
1900-2250	4	13	68	8	1	6
1550-1900	8	11	62	10	1	8
1200-1550	9	7	60	9	1	14
850-1200	10	5	55	9	2	19
< 850 m	10	5	55	8	1	21

As an example of alpine vegetation we will take the case of the South-Eastern Alps, and more precisely of the Dolomites, a montane group well-known from the ecological point of view because of much research carried out for more than a century.

The vegetation of the basal belt – limited to sunny situations – is represented by a thermophilous woodland, mainly composed of *Quercus pubescens* (locally known as 'roverella'), *Ostrya carpinifolia, Fraxinus ornus, Prunus mahaleb, Amelanchier ovalis, Cornus mas, Ligustrum vulgare, Rhus cotinus*, etc. *Quercus sessiliflora* can also be present. This belt runs all along the southern edge of Alps and penetrates deeply into the great valleys which converge on the Po valley (Tagliamento, Piave, Adige, Oglio, Ticino, etc.). In some instances it is easy to see that the penetration of these thermophilous elements occurred immediately after the retreat of the Pleistocene ice: thus, *Quercus pubescens* penetrates the Isarco valley to Fortezza (Franzenfeste); *Cornus mas* to Bressanone (Brixen); *Carpinus duinensis* up to Velturno (Feldthurns), *Rhus cotynus* to the lower part of Val Gardena (Grödener Tal).

The montane belt can be subdivided into two zones, a lower one, characterized by a mixed forest ('Mischwald') in which beech prevails; and an upper one, characterized by spruce and larch. The former has developed only where conditions of oceanic climate prevail, in this case all along the margin of the southern Alps: here we find pure stands of beech, sometimes mixed with coniferous species. The penetration of *Fagus silvatica* into the alpine chain is remarkable, since it reaches Prato all'Isarco (Blumau), Val di Tires (Tiers), Nova Levante (Welschnofen), Nova Ponente (Deutschnofen). It is present also at S. Martino di Castrozza (1500 m), Colle di S. Lucia in the Cordevole valley (1500 m) and in Comelico (i.e. the valley of the Padola) up to Selva Piana (1500 m). The fir *Abies pectinata* exhibits preferences similar to that of the beech, but its lesser dependence on an oceanic climate allows its penetration into more inland and higher localities.

The upper montane belt is mainly characterized by spruce (*Picea excelsa*) widespread everywhere from 1000 to 2000 m, in pure stands or mixed with the larch (*Larix decidua*), another coniferous tree highly characteristic of this environment. Larch can grow in particularly arid and sunny situations, higher than spruce, which after all prefers more fertile and humid environments. The undergrowth is homogeneous over very wide areas, and its species are widely distributed elements, common to the mountains of Southern and Central Europe and to the Northern European taiga (see chapter on the taiga). First of all we must mention the whortleberry *Vaccinium myrtillus* and the akin species *Vaccinium vitis-idaea,* forming broad patches of vegetation, pure or more often mixed with mosses such as *Ctenidium molluscum, Pleurozium schreberi, Hyloconium pyrenaicum, Eurhynchium striatum, Tortula norvegica, Gymnostomum rupestre, Cratoneurum decipiens, Plagiopus oederi*, etc. on limestone, *Hypnum cupressiforme, Dicranum majus, Mnium punctatum, Georgia pellucida* (a rare species), *Dicranodontium denudatum, Polytrichum juniperinum*, etc. on siliceous rocks (such as schists, phyllites, porphyrites, sandstones, etc.), or *Tortella tortuosa, Hyloconium splendens, Rhytidiadelphus triquetrus, Camptothecium lutescens*, common to both kinds of substrate. Other elements are *Pyrola uniflora, P. minor, P. secunda, P. chlorantha, Majanthemum bifolium, Oxalis acetosella, Homogyne alpina* and several ferns.

At the upper limit of the montane belt, that is above 2000 m, we find another coniferous tree associated with spruce and larch, and forming a more or less continuous belt, *Pinus cembra*, which has been defined by a student of alpine flora, Vittorio Marchesoni, as 'the most expressively alpine tree'. It is indeed found exclusively in the innermost and highest parts of the Alps, where the climate shows its greatest hygric continentality (Gams' index greater than 50°)[1]. The 'cembro' grows on small horizontal surfaces or slopes with a deep soil and a humus layer, in which case it can be an element of the so-called Rhodoreto-Vaccinietum cembretosum, composed of *Rhododendrum ferrugineum, Juniperus sibirica* (= *communis* var. *nana*), *Lonicera nigra, L. coerulea*, willows, etc., but it can grow also on almost bare rocks, isolated upon big blocks of limestone or dolomite.

Pinus cembra is present generally between 1900 and 2100 m, and in some localities it can reach, isolated, 2300 m, so that it determines the upper limit of forest (or arboreal) vegetation (tree-line).

The stunted, dwarf, shrubs (*Rhododendron, Juniperus, Pinus mugo*, etc) very rarely form a continuous belt, with a possible exception of some north-exposed, cool, humid slopes, where the snow cover lasts for a longer period. The main associations belonging to this belt are Rhodoreto-Vaccinietum (on acid soils),

[1] Gams' index is represented by the angle whose cotangent is equal to the ratio rainfall (in mm)/altitude (in metres).

Mugeto-Rhodoretum hirsuti, on basic soils, characterized by *Rhododendron hirsutum, Arctostaphylos alpina, Sorbus chamaemespilus* and *Pinus mugo*, sometimes growing also on unconsolidated gravels, and finally Alnetum viridis, with *Alnus viridis, Adenostyles alliariae, Mulgedium alpinum, Tozzia alpina*, etc. (on acid soils).

In the alpine chain, above the dwarf shrubs, two main herbaceous formations exist, meadows on silicic rocks (i.e. soils poor in calcium) and meadows on limestone (i.e. alkaline soils).

Within the latter category, when the gravel is not yet consolidated, for instance near glaciers or beneath some vertical walls, some peculiar species are present, furnished with a very well developed root system, such as *Thlaspi rotundifolium, Linaria alpina, Papaver rhaeticum, Poa minor, Trisetum distichophyllum, Doronicum glaciale* etc. These species are generally considered as 'pioneer species', since in many instances they prepare the soil and the environment for a more mature stable community. When the gravels and the isolated rocks are mechanically set, several species constitute the so-called Firmetum, where the grassy cover is still discontinuous and composed of small cushions and carpets of flowering plants: first of all *Carex firma*, associated with *Dryas octopetala* (see the chapter on tundra), *Saxifraga caesia, S. aizoon, Silene acaulis*, etc. In this environment the amount of organic material is still very low and the alkalinity (pH) very high. On northern slopes this formation can develop between 2300 and 2500 m.

When the soil is deeper and a definite humus layer has developed (generally of the moder type), we may find a Seslerieto-Semperviretum, highly characteristic of calcareous meadows, and named after the two more abundant species, *Sesleria coerulea* and *Carex sempervirens.* Several elements are generally associated with these two dominant species: we might mention here *Astragalus* spp., *Gentiana* spp. (e.g. *G. nivalis, G. verna, G. clusii*), 'Edelweiss' *Leontopodium alpinum*, several *Pedicularis* (such as *P. verticillata* and *P. rostrato-capitata*), *Anemone alpina, Aster alpinus*, etc. This association is more generally developed upon stoney, very arid slopes, with a southern exposure.

The meadows on siliceous rocks exhibit a relatively similar succession: the pioneer species in this case are *Ranunculus glacialis, Geum reptans, Oxyria digyna, Phyteuma globulariaefolium, Saussurea alpina, Eritrichium nanum,* etc. When the stones are consolidated and a fairly mature soil has had time to develop, we find the so-called Curvuletum, strictly corresponding to the above-mentioned Firmetum of the calcareous meadoes. Here *Carex curvula* prevails. Another association includes *Festuca halleri, Juncus jacquinii, Phyteuma hemisphaericum, Hypochoeris uniflora, Laserpitium halleri*. Finally we must mention, talking about siliceous rocks, the Nardetum, an association in which *Nardus stricta* prevails (sometimes alone), either present as a product of the degradation of the preceding association (*Festucetum halleri*), upon strongly decalcified soils where there is intense cattle raising, or as a product of deforestation and successive grazing. The Nardetum may develop also upon calcareous soils where leaching and acidification are very conspicuous. Associated plants are *Campanula barbata, Arnica montana, Geum montanum, Antennaria dioica, Nigritella nigra, Trifolium alpinum* and others.

As a whole, the associations upon siliceous rocks are poorer in species than those growing upon calcareous or dolomitic rocks.

The true nival zone is characterized by the exclusive presence of microscopical cryptogamae; only the so-called 'nunatakker' – small rocks or outcrops exposed within the glaciers or permanent snow – can be colonized by a few flowering plants and some macroscopic cryptogamae, such as mosses (only as protonema), lichens and liverworts. The most peculiar algae are blue-green algae or *Cyanophyceae*, diatoms, *Desmidiaceae*, and finally green algae or *Chlorophyceae*. Among these the best known is *Chlamydomonas nivalis*, which sometimes confers a pink or rose tinge to the whiteness of snow. Also some fungi are present: schyzomycetes and secondarily phycomycetes.

The plants of the high mountains exhibit a series of adaptations common generally to all the mountains of the world. One of these is the cushion-like form (Polsterpflanzen) present in many genera and families, which constitutes a true case of convergence in groups very distant from the taxonomic point of view, or even in species of the same genus or family living in very distant regions (for instance the Alps, Andes or East African mountains). Other adaptations are the presence of tomentum, down or felts, composed of recumbent hairs which practically cover the whole epigeous part of the plant. Examples of such structures can be found in *Leontopodium, Gnaphalium, Antennaria, Sempervivum arachnoideum, Potentilla nitida, Pedicularis* sp. plur. etc. Both these forms of adaptation can be interpreted as a defense against drought, and more especially against low humidity. In the case of *Crassulaceae* (*Sempervivum, Sedum*) and of some *Saxifraga,* the succulent character of the plant can be related to soil aridity (plants living upon granite outcrops or in general upon siliceous rocks instead of limestone). Also the kersophytes can be interpreted as a kind of adaptation to rocky, dry, environments, very common at high altitudes.

Another ecological feature particularly interesting within the alpine vegetation is the so-called 'succession'. At very high altitudes, not far from perennial snow or ice, often on the very edge of glaciers, some peculiar cold resistant, small flowering plants live, such as *Arabis, Thlaspi, Saxifraga, Ranunculus glacialis, Papaver alpinus*, generally cushion-like in form, sometimes with vividly coloured flowers. These species, which require very small amounts of soil and still less organic matter, and which withstand air dryness, strong insolation, very high UV doses etc. are very often known as pioneer species (see above). As we have already said, these species can be followed by others, more exigent, less euryoikous, as we proceed towards lower altitudes (*Poa, Festuca, Carex, Dryas octopetala, Salix* sp. plur., *Compositae, Pedicularis,* etc.), until a more 'mature' soil is formed, rich in organic matter (humus, generally of the moder type) which allows the growth of small shrubs such as *Rhododendron* and *Pinus mugo.*

The substitution of pioneer plants through an intermediate series up to Pinetum mugi (into which *Pinus cembra* and *Larix* can also penetrate) is visible or possible today only in a few places, where some peculiar morphological and climatological features prevail, but we must assume that a succession of this kind must have taken place everywhere after the glaciers of Pleistocene Age (Würmian, more precisely) retreated from the valleys and from the low altitudes, leaving behind them only bare morainic terrain and deposits of glacial silt. Indeed, in such situations today we find magnificent forests with Norway spruce, larch, fir, birch, etc., and only on their upper limit dwarf shrubs are present. It is only too obvious that this vegetation must have passed through several stages from the immediate post-Glacial to recent eras.

Since a succession can start either from bare rock or from a stoney humid place on the edge of a small brook or near water coming from melting snow, we may imagine two main succession types, the so-called xerosere and the hydrosere, respectively

However, if climate and soil do not vary appreciably we may safely assume that the vegetation does not change on its own and therefore that some types of vegetation we find today at very high sites or upon high mountain gravels, or even on rocky walls, are destined to preserve their physionomy without any evolution towards more 'mature' types. In such cases the name 'pioneer' would not be proper: this is actually the case with small cushion-like plants we find growing on limestone gravel or on the siliceous rock scree e.g. near the Passo dello Stelvio, at some 2700 m.

We may allow in a sense as many climaxes as there are stable conditions of montane vegetation at different heights: if we accept Lüdi's views about the Berner Alpen we may speak of a climax for the Picetum excelsae, for the Rhodo-

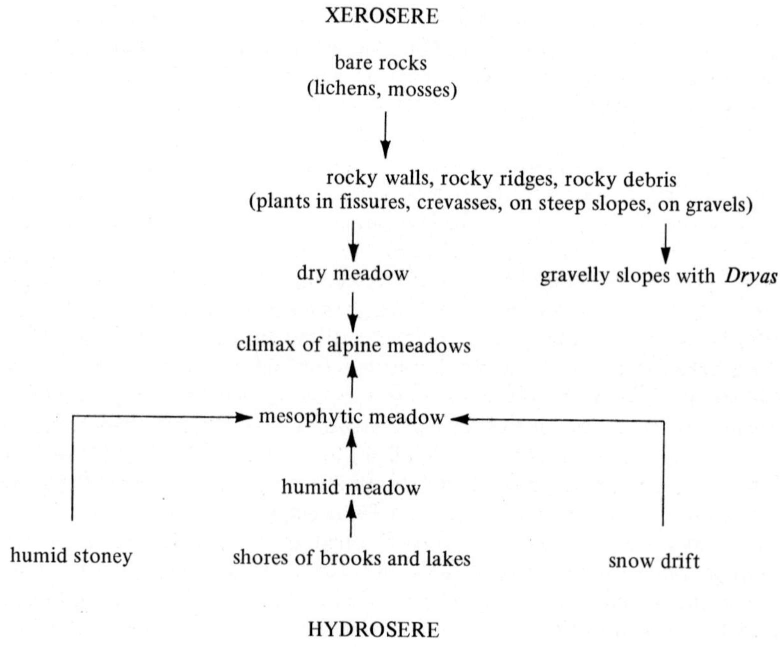

(from Cox, slightly modified)

retum ferruginei (1900-2100 m), for the Nardetum (2100-2300 m), for the Caricetum curvulae (2300-2700 m) and finally for the Gyrophoretum (higher than 2700 m). After all the concept of succession is a rather abused one.

As far as the phytogeography of the European mountains is concerned, we will just mention some cases of particularly interesting distributions among montane plants.

Boreoalpine species: *Dryas octopetala* and the fern *Cryptogramma crispa* (= *Allosaurus crispus*, in German 'Krauser Rollfarn'); endemic species of the Appennines (Abruzzo): *Trisetum villosum* (a Graminoid plant), *Brassica gravinae* (a Crucifera) and *Carduus chrysacanthus* (a Composita); species common to the Alps and Carpathians, the 'Ungarischer Enzian', *Gentiana pannonica*, whereas those common to the Alps, Carpathians and Balkan mountains are *Ligusticum mutellina* and *Senecio subalpinus*; a species spread throughout the Central European mountains is the dwarf pine, *Pinus mugo* (see Fig. 294). Finally we should mention some elements present on all the Southern European mountains with the exception of the Caucasus and the Appennines, such as *Soldanella montana* (in German 'Bergglöckchen'), or *Willemetia stipitata* (the latter with a very limited distribution); present on all South-European mountains with the exception of the Caucasus are *Homogyne alpina* ('Alpen-Brandlattich'), *Doronycum austriacum, Rosa pendulina* ('Gebirge-Rose'), and *Lonicera nigra* ('Schwarze Heckenkirsche').

In the Alps the number of endemic species is very high, due (1) to the relative antiquity of this mountain chain or at least of some parts of it; (2) to the effects of glaciation, which contributed to reducing the distribution areas of many species, generally eliminating them in the whole of the ice covered areas, so that today most endemic species are limited to the so-called 'massifs de refuge' existing along the Southern and to a lesser extent the Northern border of the alpine chain.

Together with this reduction in area, a speciation – or at least a subspeciation –

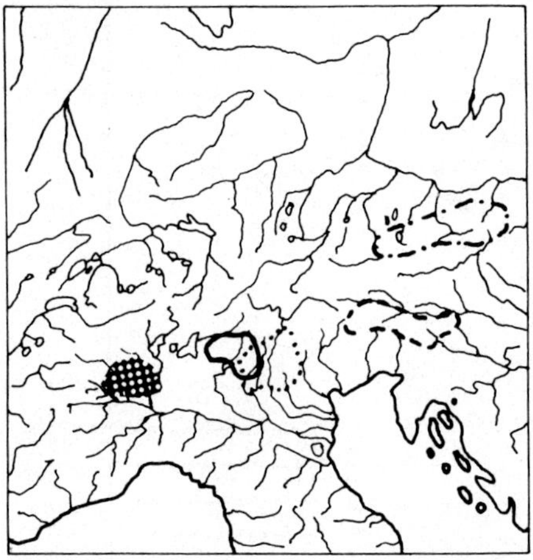

Fig. 295. Distribution of four endemic species of *Primula* and *Isoëtes malinverniana*: (lattice) *Isoëtes malinverniana*, —— *Primula glaucescens*; . . . *P. spectabilis*; - - - *P. wulfeniana*; -·-·- *P. clusiana*.

took place during the Pleistocene, when the different populations of a given species were isolated (geographic speciation). Examples of such endemic species are some primroses, such as *Primula glaucescens, P. spectabilis, P. clusiana* and *P. wulfeniana* (Fig. 295). Another interesting endemic is *Isoëtes malinverniana.*

In some instances within an alpine endemic species ecological races or subspecies have developed, as in the case of *Minuartia cherlerioides*, which shows a calcicolous race (nominal form, present in the Eastern Alps, such as Salzburg and the Cime di Lavaredo) and a calcifuge one, *rionii*, present on the Schwarzberg (Sass, Switzerland). According to Favarger (1973) the subspeciation took place in preglacial times.

Two closely related species, true geographically vicariant species, are *Euphorbia seguierana* (from Valais to Val Venosta) and *Euphorbia loiseleurii*, present in Southern France (Corbières, Vaucluse, Savoie) (cf. Favarger, 1973). Another very interesting case is represented by the two vicariant species, *Trisetum distichophyllum* and *T. argenteum*, both restricted to calcareous mountains, one more western, the other exclusively eastern alpine. Whereas *argenteum* is tetraploid, *distichophyllum* is also polyploid (Fig. 296). According to Beuret the origin of the two species is Pleistocene, but whereas *T. argenteum* has remained in its glacial quarters (endemic, relict species), *T. distichophyllum* has been able (possibly thanks to the presence of polyploidy) to recolonize the areas previously covered by ice (re-immigrant).

Other closely related species are present on the Iberian mountains, for instance in the Graminoid genus *Helictotrichon*, where three species have been demonstrated by Gervais on the ground of their caryotypes (Fig. 297).

As far as the Appennines are concerned, according to Favarger (1973) the

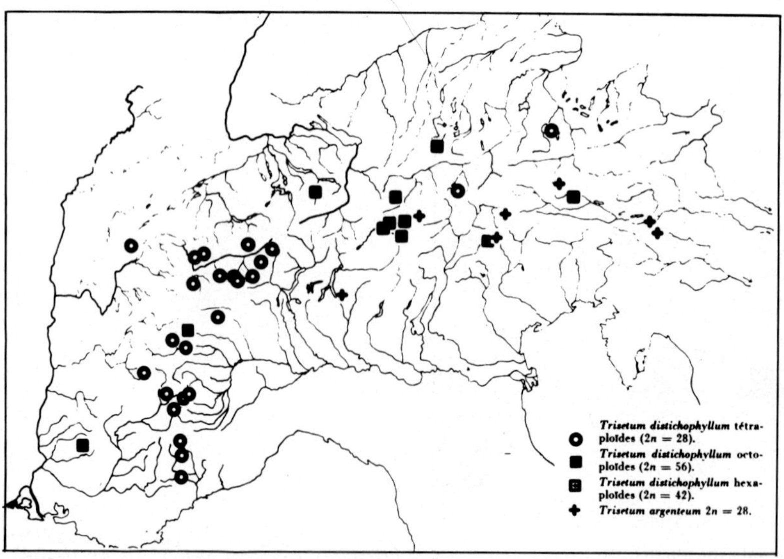

Fig. 296. Chromosome numbers of some species of the Graminoid genus *Trisetum*, present along the alpine arch (from Beuret, 1974).

most characteristic species of Abruzzo (endemics or Italo-Balkan species) developed during the Tertiary times through a gradual speciation starting from a flora common to the Centro-Mediterranean region, whence their affinities with some sub-alpine plants of the Southern Alps, of the Western Mediterranean or more frequently of the Balkan mountains.

This Appennine flora, impoverished by the local glaciations, has successively received a not inconspicuous contribution of alpine species, some of which show their allochthonous origin because of a polyploidy generally named migratory ('migration polyploidy').

Two main lines of migration have been followed by the present-day species: a first, western one, correlated with the Sardo-Corsican system, and a second, eastern one, correlated with the Balkan peninsula. To the first category the following species belong: *Globularia alypium, Anemone apennina, Stachelina dubia, Osmunda regalis, Sarothamnus scoparius, Polygala flavescens.* To the second category the following elements belong: *Podanthum limonifolium, Thapsia garganica, Asphodeline lutea, Cynoglossum creticum, Jurinea mollis, Colutea arborescens, Ballota acuta* etc. The following species seem to be truly endemic: *Carix levis, Potentilla dethomasii, Sideritis sicula, Alyssum bertolonii* (the latter a characteristic serpentine-limited species, or 'Serpentin-Endemit').

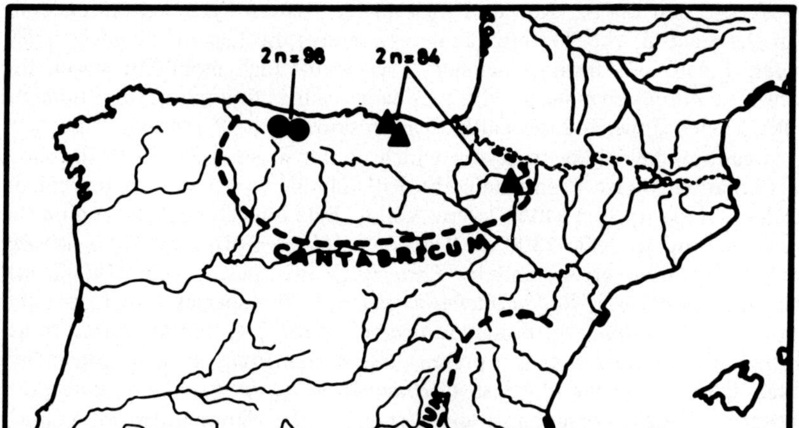

Fig. 297. Chromosome numbers of *Helictotrichon cantabricum, sarracenorum* and *filifolium* (Gramineae) (from Gervais, 1973).

We are also able today to recognize old elements which reached the Appennines in Eocene times, such as *Acer lobelii, Alnus cordifolia, Doronicum caucasicum, Muscari racemosum* subsp. *alpinum, Biarum tenuifolium, Saxifraga stabiana, Pinus leucodermis, Armeria macropoda, Onobrychis laconica* or *Ranunculus brutius* and another series, more recent, of elements which reached the Appennines only during the Pliocene, such as *Androsace mathildae, Anthemis barrelieri, Ranunculus illyricus, Drypis linnaeana, Paronychia kapela, Iberis pruitii* var. *tenoreana, Brassica gravinae, Cynoglossum majellense, Crocus imperati, Androsace australis, Armeria canescens, Campanula fragilis, Asyneuma trichocalycinum, Anthemis montana* sbsp. *columnae, Ajuga acaulis* and *Taraxacum apenninum.* The three last species reach as far south as Sicily (cf. Pignatti).

As far as the Pyrenees are concerned, the antiquity of this chain and the isolation to which it has been subject for very long periods of time have permitted – or better have been responsible for – the evolution of many endemic species, partly 'modern' (neo-endemics), partly also very ancient (paleo-endemics). We will remember only *Ramondia pyrenaica* belonging to the *Gesneriaceae,* an old element whose nearest relation is present in the Balkan peninsula. This species prefers valleys oriented N-S between 1000 and 1800 m, in the Central and Eastern Pyrenees. *Xatartia scabra*, belonging to the family *Umbelliferae* lives on siliceous rocks and schists at the Col de Nourré, in Val d'Eyne, Val d'Err, at the top of Mt Bac and on the Roc de l'Aigle in the Eastern Pyrenees. An endemic living in the Central, the Western (and, very rarely) the Eastern Pyrenees is *Iris xiphioides.* Finally we must remember a species of high mountain gravel, the Ranunculacea *Adonis pyrenaica*, with very large yellow flowers, present both on the north, French slope, and the south, Spanish slope of the Pyrenees.

The organisms belonging to seston which confer a green colour to the snow on the Carpathians have been studied by Kol. The perennial snow is present on the northern slope up to 1340 m (Kepy Valley, Bela Limestone Alps) and on the southern slope up to 1800-2300 m. The green colour is given by *Raphidonema* (Ulotrichales; a pure green) and by *Carteria* (Volvocales; a yellowish green). *Carteria* lives at pH 6,5, *Raphidonema* at pH 6,0. The species found on these mountains are *R. tatrae, R. transsylvanica, C. györffi, C. nivale* and *C. transsylvanica*. Furthermore a fungus is present, *Chionaster nivalis*, formely considered as an alga. On the Alps we find instead *Rhaphidonema vireti, chodati, sabaudum* and *bernium.* The phenomenon of 'green snow' in the National Park of Yellowstone (USA) is due to *Chlamydomonas yellowstonensis*, whereas in Antarctis the same phenomenon is due to a species of *Stichococcus.*

Animal life

The vertebrates

Mammals

Not many mammals are exclusive to the montane environment: many mammals are present all through the Centro-European forests, both deciduous, coniferous and even mixed, both in hilly or mountainous environments, and in the lowlands. This is quite consistent with their habits and still more with their great ability to spread themselves. Homeothermy also contributes to this 'vagility', present to a greater extent also in birds.

Within the insectivores there are Miller's water shrew, *Neomys anomalus* (= *milleri*), limited to the Alps, Pyrenees and Hungary, a shrew particularly common on mountains, so that in Germany it is called 'Alpen-Wasserspitzmaus' occurring up to 2000 m, but in Germany present also in the Eifel, Harz, Erzgebirge. The related species *Neomys fodiens* can reach localities as high as 1650 m in the Alps (Dolomites).

The alpine shrew, *Sorex alpinus*, a preglacial relict, is represented by at least two subspecies, *a. alpinus*, in the Alps and Svevian Jura, and *a. hyrcinicus*, in the Harz, Riesengebirge, Schwarzgebirge, Fichtelgebirge and the Bayrischen Wald. It is parasitized by the flea *Doratopsylla cuspis. Sorex araneus*, a widely distributed species is also represented in the European mountains by two races, namely *a. araneus*, in Germany, south as far as the Danube, and also in some places in the Alps which were covered by ice during the Pleistocene, and *a. tetragonurus*, in the Alps, Vosges, Rhineland, Schiefergebirge; between latitudes 40 and 50°N an intermediate form is present. It is parasitized by several fleas, such as *Ctenophthalmus congener, Palaeopsylla sorecis* and *Doratopsylla cuspis.*

A montane bat is *Pipistrellus savii*, an endemic species of Southern Europe, present for instance on the San Gottardo, in Oetztal, at Pasterze, in the Bohemian forest, etc. On the Alps it occurs up to 2600 m.

Much more restricted to the mountains is the variable hare, *Lepus variabilis* (= *Lepus timidus*), a boreo-alpine or better circum-alpine organism. It occurs generally higher than 1200 m, but in some places it co-exists with the common hare, *Lepus capensis*, with which it possibly competes for food. The variable hare is characterized by the phenomenon of moulting.

Some small rodents are also present in the montane environment, though not exclusive to it: there is the forest dormouse, *Dryomys* (= *Dyromys*) *nitedula*, in some places reaching 1200 m (Cadore) or in Italy even 1500 m. It is a 'Rassenkreis' (Fig. 298). The garden dormouse, *Eliomys quercinus*, also a 'Rassenkreis', can reach as high as 2000 m in Italy. The bank vole *Clethrionomys glareolus* reaches its southern limit in the Alps, where it can thrive up to 1800 m (Misurina, Dolomites) and even 2000 m. It is a very interesting 'Rassenkreis' (Fig. 299). It is the host to a number of fleas, such as *Ceratophyllus penicilliger, Ctenophthalmus agyrtes* and *C. congener, Doratopsylla cuspis, Hystrichopsylla talpae* and *Leptopsylla fallax*. The woodmouse *Apodemus* (= *Sylvaemus*) *sylvaticus* can reach much higher, in the Alps living up to 2300 m. A very characteristic rodent of the European mountains is the snow vole, *Microtus nivalis*, which in the Alps can occur as high as 3100 m: in the Dolomites it reaches 2400 m (Tre Cime di Lavaredo), near the upper limit of the alpine meadows or grassy vegetation. During the Pleistocene it retreated to the lowlands, for instance near Verona (Grotta della Fornace). It is parasitized by many fleas, such as *Ceratophyllus penicilliger, Amphipsylla sepifera, Ctenophthalmus agyrtes, C. congener, C. nivalis* and *Rhadinopsylla casta.*

Of course, the most 'montane' rodent is the alpine marmot *Marmota marmota*

(= *Arctomys marmota*), in German known as 'Murmeltier' or even 'Alpenmurmeltier', in the past very common in the Alps, Carpathians and Pyrenees. It was already known by Pliny (27-29 A.D.), but in the 17th century it was still considered by Athanasius Kircher a hybrid between the badger and the squirrel, and by Atmann a dwarf form of the former. During the Pleistocene it retreated to the foot of the Prealps (remains are found in several caves in Venezia Giulia, Veneto and Trentino), but it seems to have adapted to a cold climate only during the last glacial period (Würmian) since in the Middle Pleistocene it lived in a warmer climate, with *Palaeoloxodon antiquus, Bison schoetensacki, Dicerorhinus kirchbergensis* and *Ursus schertzi* (Achenheim, Elsass). Proof that the adaptation to cold is only a recent phenomenon is shown by its sensitivity to low temperatures, so that in the dens where it hibernates the temperature never goes below

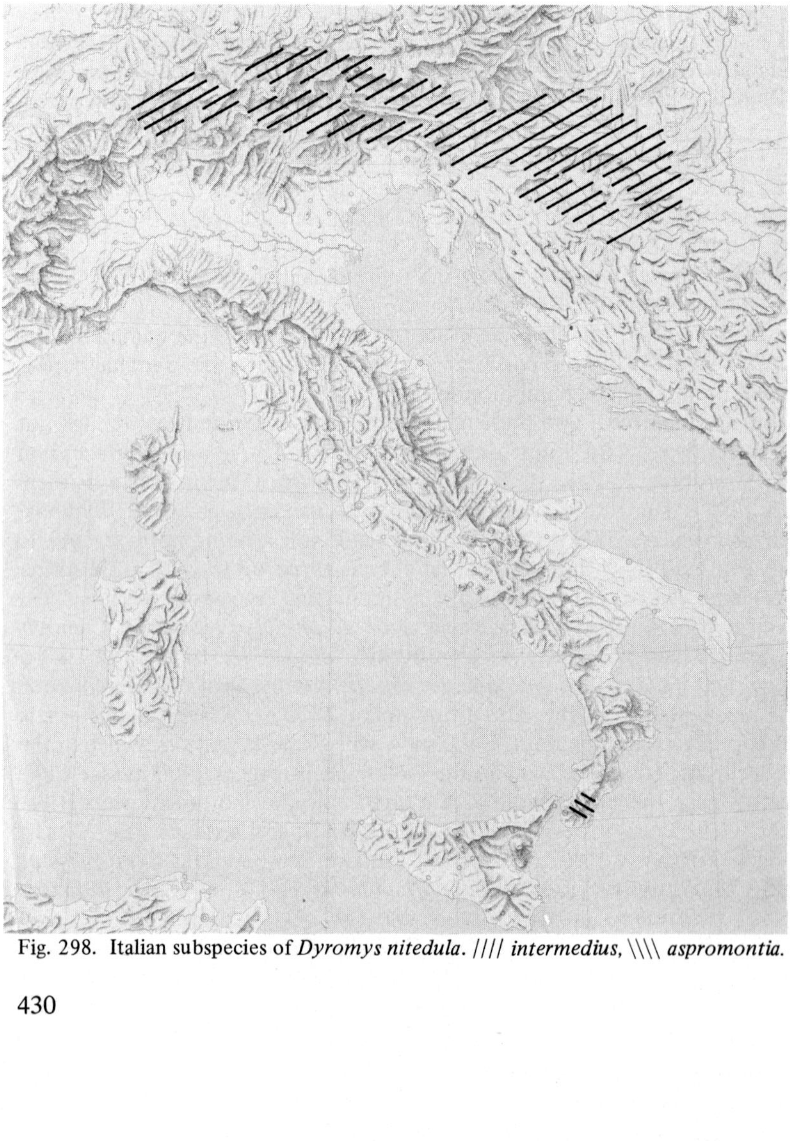

Fig. 298. Italian subspecies of *Dyromys nitedula*. //// *intermedius*, \\\\ *aspromontia*.

zero. In the meantime it also does not withstand relatively high temperatures.

Less than 100 years ago the marmot was widely distributed from the Western Alps to Berchtesgaden, the Eastern Tirol and Tatra. In 1888 it was extinct in the Carpathians and the Pyrenees, and in Germany it survived only in Allgau and near Berchtesgaden. Recently it has been reintroduced in several localities, so that today the number of individuals of this rodent is estimated to oscillate between 50 and 100,000.

The most typical montane mammals are without any doubt the Ungulates. The chamois, *Rupricapra rupicapra*, in German 'Gams', is spread throughout the Alps, the Appennines, the Carpathians and the Balkan mountains to the Caucasus. In the Alps it can thrive up to 3200 m. It corresponds to the 'caprae ferae' (Cato the

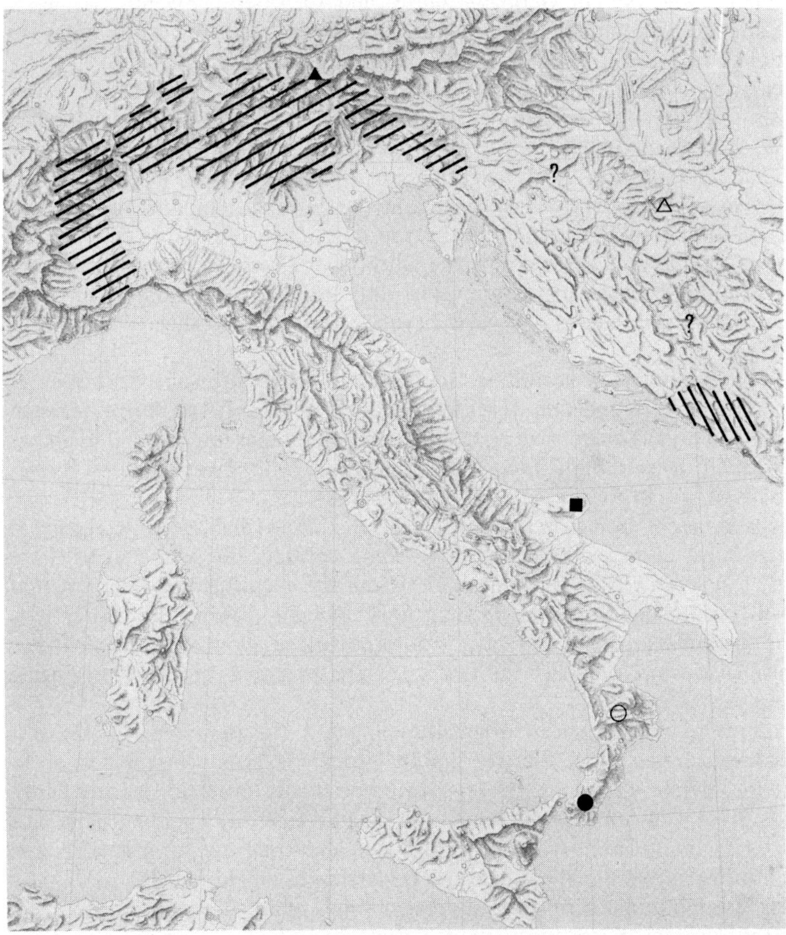

Fig. 299. *Clethrionomys glareolus* (Italian and Jugoslavian races only). //// *nageri*, ● *hallucelis;* ▲ *italicus*, ■ *garganicus*, ○ *curcio*, △ *sobrus*, \\\\ *gorka*. Example of a very old, relict distribution of a mountain mammal.

Censor) or to 'rupicapra' (Pliny, Naturalis Historia). In some ages it was named 'dama' after the fallow deer (*Dama dama*), for instance by Adam von Lebenwaldt (1693) in his 'Damographia oder Gemsenbeschreibung'. The chamois was earlier illustrated by Albertus Magnus, who also called it by the German name 'gemezen' and later by Konrad von Megenberg in his 'Buch der Natur' (1350). Konrad Gesner also deals extensively with this Ungulate in his Historia animalium, translated into German in 1583. Finally, the chamois is considered by the already mentioned A. von Lebenwaldt.

The name of this animal probably has a pre-romanic origin, witnessing the relationships existing even in very remote epochs between the peoples inhabiting the Alps and this Ungulate. In a Latin text of the 5th century, the word 'camox' appears, nearly corresponding to the althochdeutsch gamiza, to mittelhochdeutsch gemze (whence the present German Gemse) and to Albertus Magnus ' gemezen (1200). The althochdeutsch word gamiza has been preserved until today in Inneröztal (Tirol). In the Pyrenees it is known by the name 'isard'.

The chamois is parasitized by the Sporozoon *Eimeria rupicaprae*, which causes coccidiosis, and by the Nematodes *Haemonchus contortus, Trichostrongylus, Ostertagia, Dictyocaulus* and *Protostrongylus.* On the coat a Mallophage can be found, *Bovicola rupicaprae*; however, this is not a parasite.

Besides some Ungulates widely spread in the European mountains, but not especially confined to them, such as the roe deer, *Capreolus capreolus*, the red deer, *Cervus elaphus* and, more rarely, the fallow deer *Dama dama*, two species are worth mentioning because they are more strictly limited to the montane environment, namely the wild goat *Capra hircus* (= *ibex*) and the wild sheep *Ovis aries* (= *musimon*).

The ibex is subdivided according to some authorites into subspecies such as *C.h. ibex*, in the Alps and *C.h. pyrenaica* in the Pyrenees. It can thrive between 2100 and 3200 m, although exceptionally it can be found as low as 1000 m and as high as 4000 m. During the winter it is found generally at lower altitudes. It lives on the rocks, where it prefers the most abrupt slopes.

It has been nearly extinct in Portugal since 1892; in the Balkans (*C.h. aegragus*) it seems to have disappeared since 1891. Some consider the wild goat of Crete (2497 m), *C.h. cretensis* as indigenous. It is mainly a diurnal creature, but it is active well before dawn; it goes feeding until twilight. During the winter it is active for the greater part of the day. It does not move about very much. It has gregarious habits, but generally the two sexes are separated, uniting only during the mating season.

The mouflon or wild sheep originally inhabited the mountain forests, but lately has been driven above the tree line or into fields, although it is best acclimated at the middle elevations. It seems to be originally from Corsica and Sardinia, but is present also in Cyprus (maximum height 1951 m) where a special race lives (*O. aries ophion*). It is both diurnal and nocturnal, living in small herds composed of males and females, under the leadership of an old female.

Among the mountain carnivores, a great number of small or medium-size species, scattered more or less throughout Europe, is to be found in the European mountains. We have already dealt with them in the chapter devoted to deciduous and/or coniferous forests. As far as the mountains are particularly concerned, we will just mention two species, the brown bear and the pardel lynx. The bear lives in mixed forests mainly confined to mountains. It is a nocturnal creature but in places where it is not disturbed it is also diurnal. It has nomadic habits and moves about a lot. It lives a solitary and relatively sedentary life. It hibernates in winter, when it lives in a subterranean den. The pardel lynx lives in mountain forests but also in the lowlands. Its habits are very similar to those of the common lynx.

Some elements are indigenous to the Alps, and can be considered as preglacial relicts: an example is *Microtus nivalis*, belonging to a genus older than the Pliocene. The eastern species penetrated the Alps during an interglacial period, first of all the ibex, the chamois and the marmot (the latter maybe at the end of the Miocene). The arctic elements probably penetrated the Alps only after the amelioration of the last glaciation (Würmian): they are the bear, the variable hare, the ermine, the lynx and the marten. Some elements have an eastern, pontic origin: the marmot (whose origin is still very little known), the ibex (whose distribution centre is possibly the Himalayas) and the chamois, which seems to have originated in the mountains of Thrace and Northern Greece. The origin of the shrews (*Sorex* and *Neomys*) is completely obscure, though probably very remote (*Sorex alpinus* is present in all the European tertiary mountains with the exception of the Balkans). Only the Mediterranean element is absent from the mammalofauna of the European mountains, or at least of the Alps.

Mountain mammals exhibit several adaptations to life in the montane environment. One example is furnished by the altitudinal movements – kinds of migrations – shown regularly by some Ungulates. During the winter the chamois finds refuge in the forests, whereas in spring it descends to lower altitudes searching for fresh leaves to feed upon; afterwards it moves to the high altitudes.

Most mountain rodents – different from those inhabiting the lowlands – are diurnal, perhaps as a consequence of the low nocturnal temperatures. The small mammals, because of their limited powers of movement also remain in their quarters during winter. Some of them, such as the marmot, go into dormancy, during which the respiratory rate is slowed down and the body temperature is lowered. The snow vole *Microtus nivalis* is also active during the night, beneath the grassy carpet, covered by snow, which prevents the soil temperature from going below zero. The true hibernation, typical of the marmot in Europe, is characterized by changes in some endocrine glands such as the adrenal glands, hypophysis and the so-called hibernation gland, composed of the interscapular brown adipose tissue. The 'pars glomerularis' of the adrenal glands seems to be particularly significant for hibernation: animals without their adrenal glands cannot enter hibernation in a cold environment and remain normothermous (which practically means homoiothermous) until death. The hypophysis is involuted: it is characterized by the increase in the basophil cells in the spring, when reproduction is re-activated. Finally there is a maximum in the activity of the hibernation gland, demonstrated by an increase in the concentration of ribonucleic acid. This gland should be responsible for the hypothermy characteristic of hibernation, since a substance extracted from it causes hypothermia experimentally. After hibernation the mammals show some signs of imperfect homoiothermy, and only later it is perfectly re-established.

Among the diurnal birds of prey, a particular ecological role is filled by the griffon vulture, *Gyps fulvus*, in German known as 'Gänsegeiser', the largest bird of prey in Europe, characterized by the size of the beak. It is widespread in Central and Southern Europe, Western Asia and Northern Africa. It is scarce in the Eastern Alps and Maritime Alps, but common in Sardinia. It thrives in rocky, wild mountains, where it nests on bare rocks or in crevices, more rarely in trees. It feeds on carcasses, preferably of large animals such as oxen, horses or donkeys; it can eat living animals but only if driven by hunger. The bearded vulture, *Gypaetus barbatus*, is widespread in the high mountains of Northern Eurasia to the Himalayas. It nests in caves on precipices. In Europe it is limited to the southernmost parts, but in the past century it still lived in the Eastern Alps. The golden eagle, *Aquila chrysaetus,* is the mountain bird of prey par excellence, though not necessarily restricted to the montane environment. It is present in Eurasia and North America (holoarctic element). It is resident in the highest mountains but it can also migrate. It nests in high mountains, on rocks sometimes not accessible to man; the nest is made in the same locality for several years. It attacks any mammal and bird, but it cannot lift a prey exceeding its own weight (nearly 8 kg). Two species of *Hieraaëtus* live in the European mountains, *H. fasciatus* and *H. pennatus*; the former prefers rocky places, and sometimes also big caves; it nests upon very steep rocky walls, the latter prefers forests (both deciduous and coniferous) with some clearings where it can hunt. It nests in high trees. Their food is the same as that of eagles, but sometimes they also eat carrion.

The lanner falcon, *Falco biarmicus*, is limited in Europe to the south-eastern mountains (Italy, Sardinia, Sicily and the Balkan peninsula). It lives in a wide range of environments, generally in rocky places and nests upon rocks and only occasionally in trees. It feeds upon any kind of bird, hares, rabbits, mice and perhaps also amphibians and beetles.

The Galliformes include perhaps the most characteristic alpine birds, though they are not necessarily confined to mountains. They belong to different genera, each with a different size, habits and food preference, so that they can live together on the same mountain, separate as far as their ecological niche is concerned. Let us mention the ptarmigan *Lagopus mutus*, better known in the Alps as 'Alpenschneehuhn' (German) or 'pernice bianca' (Italian), a boreo-alpine element limited to Europe, subdivided into at least two geographical races, of which in the Alps we find *L.m. helveticus*. It lives generally above the tree-line, from 2000 to 3000-3500 m and nests in shelter among the rocks or even among plants up to 3200 m. It often hibernates in the high mountains, though it can descend to 1400 m. During the winter it forms big flocks of some 40-60 individuals. It nests in small holes or depressions in the soil, sheltered under rocks or among shrubs (generally *Pinus mugo* and *Rhododendron*). It feeds in the Alps upon buds, fruits and leaves of whortleberry and *Calluna* (heath), but in spring it feeds also upon insects. The black grouse *Lyrurus tetrix* often lives together with the following species (*Tetrao urogallus*). It is widespread in Europe up to latitude 68°N, going as far south as the Pyrenees, Alps, Carpathians, Caucasus and Central Asia. It lives in forests where spruce, larch, beech and birch are growing, but prefers coniferous ones. It feeds during the winter and spring upon buds, mainly of spruce and small roots (mainly of larch and pine) in summer and autumn upon insects, fruits (particularly whortleberry, strawberry, raspberry and juniper).

The caper-caillie, *Tetrao urogallus*, in the Alps known as 'Auerhuhn' or 'gallo cedrone' (or even 'urogallo') is the largest Tetraonid bird and general Galliform of our mountains. It prefers the coniferous forests of hilly or mountainous regions; it nests on the ground among the undergrowth at the foot of a big coniferous tree, but in some cases also among shrubs. It thrives generally between 1000 and 1500 m. It was already known – as *Tetrastes bonasia* and *Lagopus mutus* – by Pliny. Sometimes it will hybridize with other Galliforms such as *Lyrurus tetrix*: in this case the old females assume a masculine appearance, whence the German name 'Hahnefedrige Weibchen'. It lives essentially in coniferous forests where there is a rich undergrowth with raspberries (*Rubus idaeus*) and whortleberries (*Vaccinium myrtillus*) on which it generally feeds. It rarely goes higher than 1800 m. Its food, besides the above-mentioned berries, consists during the winter of plant remains (mainly of coniferous plants) in the other seasons of insects and fruits such as brambleberries, juniper, etc. It has always been widely hunted on account of its flesh, and only recently some control laws have been promulgated in order to avoid its total extermination. Its main enemies are the fox, the marten, the birds of prey (both diurnal and nocturnal) and also crows and magpies which can prey upon the eggs or the new-born individuals.

The hazel hen, *Tetrastes bonasia* (= *Bonasa betulina*), in German 'Haselhuhn', is an inhabitant of mixed and deciduous forests in hilly or mountainous regions. It prefers aspen and birch stands. It nests in depressions in the soil concentrated under the dense vegetation. Its food consists of buds, flowers and leaves of several plants, fruits of juniper, raspberry, whortleberry, strawberry, etc., but in spring also insects and earthworms enter the diet. In the past it has been very heavily hunted because of the flavour of its flesh. *Alectoris graeca* or the rock partridge is a Gallinaceous bird of the South-East European mountains living on rocky slopes, sometimes in the shrubs and grass between the treeline and snowline. It makes it nest on the ground among the rocks.

The eagle owl *Bubo bubo* is both nesting and migratory; in some parts of Central European mountains, such as Bavaria, this bird is protected by law. It lives in forests, on steep slopes or open land in hilly or mountainous regions, and makes its nest in depressions in the ground under rocks and debris, in hollow trees, or even in old nests of birds of prey. Its diet is composed of roe deer, wild cats, hares, squirrels, rabbits as well as many birds.

The pygmy owl *Glaucidium passerinum* lives in montane forests, which it leaves only in autumn. It hunts preferably at night. It lays its eggs in holes in trees (mainly deciduous – such as ash or beech), but does not build a true nest. Its diet consists of insects and mice, small birds, etc. so that it must be considered as a useful animal. Tengmalm's owl (*Aegolius funereus*), in German 'Rauhfusskauz', nests also in the Alps, though its true habitat is Northern Europe, Russia and Siberia. As a rule, it does not leave the high mountains where it lives in coniferous forests. Its food consists mainly of voles or rarely of small birds, such as trushes, finches and tits, so that it must be considered a useful animal.

Typical mountain loving birds are the woodpeckers as a whole; for example

Picus canus or the grey-headed woodpecker, whose distribution in Europe is very similar to that of *Aegolius funereus.* It lives in deciduous and coniferous forests, up to the tree-line, in all Central and Northern Eurasia. *Dendrocopus major*, the great spotted woodpecker, is very widely distributed, though in general rare, both in deciduous and in coniferous forests; *Picoides tridactylus*, the three-toed woodpecker (in German 'Dreizenspecht', which is a translation of *tridactylus*) exhibits a discontinuous distribution of the boreo-montane type. It lives in coniferous and birch forests, preferring the areas whose vegetation has long been destroyed by fire. It nests in holes it makes in big trees or in telegraph poles. Finally let us remember *Dryocopus martius* or the black woodpecker (in German 'Schwarzspecht') living at middle and high altitudes in old coniferous forests, but locally also in mixed or beech forests. It excavates its large nests in trees, often well above the ground.

Among the Passeriformes we should mention the shore lark *Eremophila alpestris* (in German 'Ohrenlerche') a small creature similar to the lark, extremely limited in distribution in Europe, nesting exclusively in the southern part of the Balkan peninsula above the tree line, and in the dry parts of the tundra. The nest sits on the ground. The male is easily recognizable thanks to three black spots, one behind the eye, another on the throat and a third on the forehead.

A martin, present sometimes at very high altitudes (up to 2300 m), though not exclusive to mountains, is the crag martin, *Ptyonoprogne rupestris.* It nests in fissures in rocks or in holes in rocky walls and is limited to southern Europe. In the Alps it is a migratory bird.

The *Corvidae* are represented mainly by three species, the nutcracker *Nucifraga caryocatactes* (in German 'Tannenhäher') and two species of *Pyrrhocorax.* The former is limited to montane coniferous forests, which it leaves during the winter to go to live in deciduous forests. It nests in coniferous trees, whence its German name ('Tannen' is in fact *Abies pectinata*). Two species of *Pyrrhocorax* live in the European mountains, *P. pyrrhocorax* or the chough, and *P. graculus*, or the alpine chough. The former also lives on the sea shore, the latter never reaches this environment being a much more montane animal. Both nest in rocky environments generally not accessible to man, in caves, occasionally in ruins. *P. graculus* can be seen flying near glaciers or the permanent snow. Both species are gregarious, forming great flocks; they are omnivorous, feeding upon hoppers (very numerous on mountains) or grain. In some instances they feed also on small carcasses.

Among the true passerine birds we will mention only a few, limited more or less strictly to montane environments, at least during the reproductive period. The *Certhiidae* are represented by *Certhia familiaris*, the tree creeper, living in forests, parks and gardens, where large trees are present. It prefers the mountain forests of Central and Southern Europe, with the exception of the Iberian peninsula. The wall creeper, *Tichodroma muraria*, is easily recognizable because of the vivid red coloured wings, a not very usual feature among European birds. It inhabits rocks on which it climbs just as Certhias climb on trees. It nests in fissures in rocks covered with mosses and lichens. A very interesting disjuncted distribution is that of the ring ouzel, *Turdus torquatus*, in German 'Ringdrossel', which can be considered as a boreo-montane element. It is a migratory bird, nesting in Northern

Europe and in the Centro-European mountains. In the Alps it is resident only in the west. It feeds on insects, spiders, millipedes, as well as the fruits of many plants (*Taxus, Lonicera,* hawthorn, ash, ivy, juniper, mistletoe, etc.). On the whole – like its congeneric species – it is a useful bird for agriculture and forests.

The grey wagtail *Motacilla cinerea*, in German 'Gebirgstelze', is mainly confined to rivers and brooks in hilly or mountainous regions. It nests on high mountains. It is, on the whole, an insectivore feeding upon gnats and flies, but also beetles and other insects. It may, therefore, be considered useful. The rock thrush, *Monticola saxatilis*, limited to the southern part of Europe, nests in open, rocky territories from 900 to 2400 m, only occasionally among trees or in ruins. The black wheatear, *Oenanthe leucura*, is exclusive to the mountains of Western Europe, namely the Iberian Peninsula, Alpes Maritimes, Sardinia and Sicily, so that it is sympatric with the congeneric species *O. hispanica.* It nests in caves among the rocks and often the entrance to the hole is protected by a small wall of stones. Like its congeneric *O. oenanthe* (which in the Alps can go up to 2600 m) it feeds on insects. A very characteristic montane bird is the alpine accentor *Prunella collaris*, in German 'Alpenbraunelle', in Europe limited to the mountains of the Alps and the main peninsulas; during the winter it lives in the valleys. It nests in holes in rocks protected by vegetation.

Three species of *Carduelis* are worth mentioning, *C. spinus* (the siskin), *C. flammea* (redpoll) and *C. citrinella* (citril finch). The first species lives in forests with coniferous species, birch and alders, and nests in coniferous trees, generally far from the trunk. The redpoll is represented in the Alps by a peculiar subspecies, *f. cabaret*, which is smaller than the other, nordic, subspecies, so that we can see in this 'Rassenkreis' an example of Bergmann's rule. It has a disjunct distribution, which in a sense could be considered as boreo-montane or better circumboreal. In the Alps (at least in the eastern part) both subspecies *f. flammea* and *f. cabaret* are present, the former as migratory birds, the latter as residents. *C.f. cabaret* lives on rocky sites, near and above the tree line; it nests often gregariously on birch, alders, willows and juniper. It feeds on the seeds of birch, elders etc. (granivorous birds). Finally the citril finch (in German 'Zitronenzeisig', calles as in English, after the yellow colour of the ventral parts of the body) prefers mountains with sparse coniferous trees and open rocky ground: during summer it reaches 700 m in Schwarzwald, in the Alps 1500 m, but in the Dolomites it has been found on the Alpe di Faloria at nearly 1900-2000 m.

We want to finish with this animal group – so important in the economy of the montane ecosystems – by mentioning one of the most typical high mountain birds, *Montifringilla nivalis*, or the snow finch (in German 'Schneefink'), characterized by a grey head and black throat. It thrives on bare mountain tops above 1800 m, only during the winter coming a little lower. *M. nivalis* nests in fissures in rocks, on walls and beneath the gutters on alpine refuges (Hütte) or baite. It has a very reduced distribution area which is quite an unusual feature in this vertebrate group (Fig. 300).

Amphibians and reptiles

Terrestrial amphibians and reptiles are not characteristic of mountains on the whole, probably because of their heterothermy (or poikilothermy) which does not enable them to be active for most of the year, which on the contrary is possible for birds and mammals.

Among the best adapted species to the montane environment are the black salamander *Salamandra atra*, limited to the Alps and Jura, present in the Alps up to 2200 m, and never present below 1000 m; the viviparous lizard *Lacerta vivipara*, widespread in Central and Northern Eurasia, and living in the Alps up to 3000 m; the viper *Vipera berus*, distributed more or less like *Lacerta vivipara*, and

Fig. 300. Distribution of a typical mountain bird, *Montifringilla nivalis* (schematic).

occurring in the Alps up to 2750 m; *Vipera aspis*, a Centro-South Eastern European species, in the Alps present up to 2950 m, and finally *Vipera ammodytes*, limited to South-Eastern Europe in the Eastern Alps reaching more or less 1000 m, and therefore not to be considered as a true montane element.

Other amphibians and reptiles present on mountains thanks to their euryoikous habits are *Salamandra salamandra*, in the Eastern Alps present up to 1400 m, found active in the snow in December; the common toad *Bufo bufo*, in the Alps reaching 2200 m; the smarag toad *Bufo viridis*, in the Alps reaching as high as 2000 m, but found in the Himalayas up to 4560 m; *Rana temporaria*, present in all central and northern parts of the Palearctic Region, in the Alps as high as 3000 m; *Lacerta viridis*, in the Eastern Alps up to 1000 m (Alto Adige or Süd Tirol); *Lacerta muralis*, the wall lizards, in the Alps present up to 2000 m; the common blindworm, *Anguis fragilis*, reaching in the Eastern Alps at least 2000 m (Alpe di Fedaia); the Aesculap-snake, *Elaphe longissima*, a widely distributed snake, in the Alps present up to 1600 m; *Coronella austriaca*, also widely distributed, known from some localities at 2000 m (Paganella, near Trento) and finally *Coluber viridiflavus*, a rather stenotopic species, found by the present writer in the South-Eastern Alps at 1700 m (Comelico).

Some of these species, for instance, *Lacerta vivipara* and *Vipera berus* and secondarily *Rana temporaria* and *Salamandra atra*, enter actively into the trophic webs at high altitudes, and represent in some places a considerable part of the biomass. Unfortunately, so far no research has been carried out on the energy budget of these interesting heterothermous creatures.

The invertebrates

The altitudinal distribution

As for plants, we could also demonstrate for animals in some sectors of the Alps, e.g. the Eastern Alps, a subdivision of the fauna into altitudinal belts, each characterized by specific elements. Unfortunately the number of animals is so much greater than that of flowering plants that we cannot even deal with a small zoological group like the Coleoptera.[1] We can only say that the number of beetles is greatest within the coniferous forest belt (Picetum), where many small meadows are present, followed by that in the alpinetum; many species are common to the two belts. On the other hand, very few species are limited to Fagetum (lower montane belt) or to the collinar belt (Castanetum). Few species extend from the alpine belt to the Fagetum, fewer still from the alpinetum to the collinar zone.

It is worth mentioning the species of two zoological groups, beetles and snails (and slugs), which reach the highest altitudes in the Dolomites (Eastern Alps), though are not exclusive to them. The beetles are as follows: Carabids *Cychrus caraboides*, 2500 m; *Carabus bertolinii*, 2300 m; *Carabus creutzeri*, 2900-3000 m

[1] The reader interested in this subject can find some lists of species in Marcuzzi, 1956 (p. 342) and 1961 (p. 76).

(Tofane glacier); *Notiophilus biguttatus,* 2500 m; *Bembidion lampros*, 2300-2500 m (Cima Plose); *Bembidion nitidulum*, 2300-2500 m; *Nebria jockischi*, 2400 m (north slope of Marmolada); *Calathus melanocephalus*, 2400 m (idem); *Amara alpestris*, 2300 m; *Amara erratica*, 2500 m; *Harpalus luteicornis*, 2300 m (Marmolada); *H. quadripunctatus*, 2500 m (Pfannspitze); *Asmerynx laevicollis*, 2300-2500 m; the *Staphylinidae Staphylinus ophthalmicus,* Croda del Becco 2500 m; *Philonthus nimbicola*, 2500 m (Marmolada); the *Silphidae Silpha tyrolensis,* 2300-2500 m (Padon, north of Marmolada); the *Byrrhidae Byrrhus fasciatus*, 2300 m; *Crepidodera melanostoma*, a *Chrysomelidae*, 2300 m; the weevils *Otiorrhynchus prolixus*, 2300-2500 m; *O. montivagus*, 2300-2500 m; *O. gemmatus*, 2300-2500 m; *O. nodosus*, 2300-2500 m; *O. chalceus*, 2600 m (northern slope of Marmolada); *O. schmidti*, 2600 m (Rif. Mulaz on the Pale di S. Martino); *O. hadrocerus*, 2500 m (northern slope of Marmolada); *O. cadoricus*, 2600 m (Rif. Mulaz); *O. subquadratus*, 2500 m (northern slope of Marmolada) and finally the Scarab beetle *Aphodius fossor*, 2300-2500 m on the Plose (Alto Adige).

Someone claimed that the following species are confined to the alpine belt in the Dolomites ('eualpine species', Holdhaus) *Carabus alpestris, Nebria castanea, N. diaphana, N. germari, N. hellwegi, Trechus dolomitanus, T. pallidulus, Amara alpestris, Pterostichus panzeri, Chrysochloa frigida, Phytodecta nivosus, Otiorrhynchus cadoricus, O. chalceus, O. hadrocerus, O. schmidti, O. auricapillus, O. alpicola, Dichotrachelus stierlini* and finally *Aphodius schlumbergeri.* However, some of these species, in particular circumstances, can reach lower levels, as in the case of a very cold, north-exposed valley, Val Fonda (north of M. Cristallo, and south of Carbonin), where some 'eualpine' species can be found or at least were found when the glacier had not retreated very far (e.g. *Trechus pallidulus*, found at only some 1450 m).

The following molluscs have been found at very high altitudes in the Dolomites (Eastern Alps): *Cochlicopa lubrica*, 2400 m; *Vertigo alpestris*, higher than 2000 m (Passo Gardena), *Vertigo pusilla*, 2200 m and *V. moulinsiana* at 2300 m (Monzoni), *Truncatellina strobeli*, 2200 m (Marmolada); *Columella edentula*, 2500 m, not far from the Marmolada glacier; *Pupilla cupa*, 2400 m (the same mountain); *Pupilla muscorum*, 2200 m or higher; *Pyramidula rupestris* and *Delima stentzi*, 2740 m (Cima Lausa, probably the highest point in the Dolomites where mollusks have been found); *Clausilia dubia*, 2200 m (Pordoi); *Punctum pygmaeum*, 2200 m (Marmolada); *Retinella petronella*, a rare species, 2172 m (Passo Limo in Fanes Group); *Vitrea subrimata*, 2200 m; *Euconulus fulvus*, 2400 m; *Semilimax nivalis*, 2300 m (Marmolada); *S. pegorarii*, a very rare species, the Lago di Antermoja (2496 m) and *S. kotulae*, also a very rare species, 2500 m (Col Loggia); *Helicolimax annularis*, 2000 m; *H. pellucidus*, 2200 m and *H. glacialis*, 2516 m (Seejoch, Lago di Antermoja); the slugs *Arion subfuscus, Deroceras agrestis, Limax cinereoniger* and *Lehmannia marginata* also go higher than 2000 m; finally *Arianta arbustorum* (see also p. 449) reaches 2500 m on the northern slope of Marmolada.

It is very difficult, indeed almost impossible, to establish now which species are 'eualpine'. Only very thorough research (hardly possible in Italy today) would allow us to reach firm conclusions about this.

Among the molluscs, due to the more limited number of species compared with beetles, it is possible for the mountains of Veneto (Prealps and Alps) to recognize some species restricted to the lower montane belt (Fagetum) and others restricted to the upper montane belt (Picetum), though only in a general way. The former are *Acme gracilis, Abida illyrica, Orcula gularis, Agardhia truncatella, Vallonia excentrica, Jaminia bidens, Laciniaria vetusta, Cochlodina grossa, Arion circumscriptus, Deroceras planarioides, Cepaea vindobonensis.* The latter are: *Chondrina clienta, Vertigo substriata, Iphigena mucida, Fusulus varians, Retinella pura, Semilimax semilimax* and *Fruticicola hispida.* No species have been found exclusively in the alpine belt. Of course in other sectors of the Eastern Alps the situation is likely to be very different.

The number of species present in each altitudinal belt is as follows:

500-1200 m (Fagetum)	ca. 120 species
1200-2000 m (Picetum)	ca. 80 species
higher than 2000 m (Alpinetum)	ca. 45 species

This regular decrease is probably a general feature (Fig. 301-302).

In Puglia – as an autoecological investigation carried out by the present author has shown – there are not true montane molluscs: Fig. 303 illustrates the quantitative altitudinal distribution of several species living in Puglie. None of them is exclusive to high altitudes; some are fairly independent of altitude; only one, *Monacha rizzae* an endemic species of the southernmost part of Italy and of eastern Sicily (Fig. 304) seems to prefer montane situations, though thriving also in the lowlands.

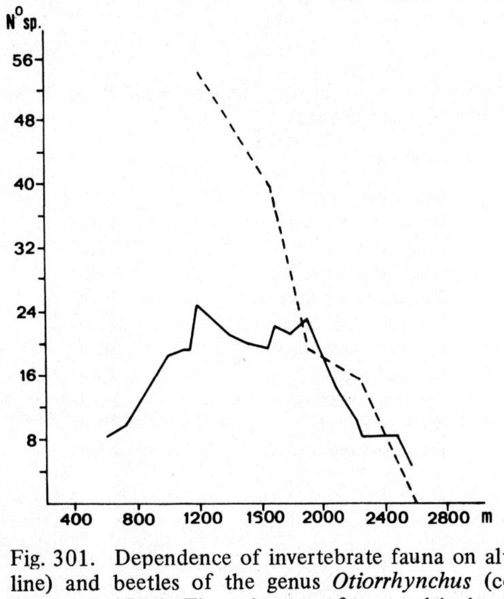

Fig. 301. Dependence of invertebrate fauna on altitude: No. of species of molluscs (broken line) and beetles of the genus *Otiorrhynchus* (continuous line) on the Dolomites (from Marcuzzi, 1951). The existence of many alticolous species of *Otiorrhynchus* is very evident, especially if compared with molluscs.

The following scheme shows the altitudinal distribution on the Dolomites of the some species of *Otiorrhynchus* (*Curculionidae*), some of which behave as stenohypsous, some as euryhypsous:

Alpine belt Subalpine belt Montane belt Hilly belt

dubius
varius
schmidti
chalceus
hadrocerus
cadoricus
globulus

  dolomitae
  fortis
  foraminosus
  montivagus
  azaleae
  subdentatus
  pauxillus
  sensitivus

    mastix
    inflatus
    geniculatus
    caudatus
    scaber
    vestitus
    chrysocomus
    ligustici

← armadillo →
 ← ovatus →
← bisulcatus →

The maximum height at which the invertebrates can live in the Alps (National Park of the Gran Paradiso) is illustrated by the following data (from Erhard):

Lumbricidae		Araneae	
Allolobophora caliginosa	2500 m	*Drassodes heeri*	3000 m
Dendrobaena octaedra	2800	*Drassodes microps*	3000
Bimastus constrictus	2800	*Gnaphosa badia*	3000
Lumbricus terrestris	2800	*Gnaphosa petrobia*	3000
Octolasium lacteum	2800	*Xysticus erraticus*	3000
Mollusca		*Coelotes fabulator*	3000
Limax tenellus	2300	*Pardosa torrentium*	3000
Arion subfuscus	2900	*Pardosa nigra*	3000
Ariunculus mortilleti	2900	*Xysticus cristatus*	3000
Vitrina pellucida	2900	*Pardosa frigida*	3000
Hyalina pura	2500	*Thanatus arenarius*	2800
Lepidoptera			
Parnassius apollo	2600		
Pieris rapae.	2800		
Colias phicomone	3000		

Bäbler has found the following upper limits of distribution of the different zoological groups in the nival zone (or belt) in Switzerland:

Molluscs	2950 m	Acari	4275 m
Lepidoptera larvae	3237	Pseudoscorpions	2880 ca.
Coleoptera	3237	Diplopods	2900
Diptera	4275	Chilopods	3237
Collembola	4275	Lumbricidae	2950
Araneae (exuviae)	3550	Enchytraeidae	3237
Opilionids	3000	Platyhelmintes	2700

Finally, we want to quote Burla's data on the upper limit for the *Drosophila* species living on the Alps:

D. repleta	660 m	*D. funebris*	1700
D. simulans	1010	*D. obscuroides*	1800
D. immigrans	1240	*D. transversa*	1800
D. busckii	1240	*D. alpina*	1800
D. testacea	1700	*D. subobscura*	2000
D. melanogaster	1700		

The decrease in the number of species belonging to the different animal group is almost universal. For instance, for the Rhopalocerous butterflies of Switzerland and Germany, Speyer & Speyer have obtained the following data:

Lower region	165 species
Montane region	122 ,,
Subalpine region	85 ,,
Alpine region	54 ,,
Subnival region	20 ,,

Fig. 302 shows the diminution with the increase in altitude of the number of species in the soil fauna on the Marmolada massif (from Marcuzzi) and of the number of Rhopalocerous butterflies on M. Baldo (from Milano). Fig. 301 shows the same relationship for the number of Molluscs and species of *Otiorrhynchus* (Col. Curculionidae) in the Dolomites. Finally, Fig. 305 represents the altitudinal quantitative distribution of species of *Glomeris* (Diplopoda) in the Dolomites.

Alimentary or 'functional' categories

As far as the feeding habits of montane invertebrates (limited to macrofauna) are concerned, we can distinguish herbivores, predators (or carnivores) and saprophages. Herbivores are mainly represented by insects (and more particularly beetles) and molluscs, predators by beetles, chilopods (or centipedes) and arachnida, saprophages by some beetles, diplodops (or millipedes) and some molluscs. We will list only a few species in each category. Herbivores. Beetles: *Elateridae* are represented by *Agriotes lineatus* living on several plants including wheat; it can be harmful both to cereal fields and young seedlings. The larva of *Corymbites cupreus* can also be harmful; among the genus *Athous* (larva) on the other hand, some species can be harmful, some useful because they have predatory habits. The damage done by *Agriotes* larvae is less severe in soils rich in humus, since these animals behave as saprophages and therefore can utilize the humus. A *Buprestidae* living on mountains is *Buprestis rustica* (up to 1800 m) generally confined to coniferous forests. The larva lives inside felled firs (*Abies pectinata*) without causing serious damage. Several *Scarabaeidae* can be more or less harmful to mountain plants, e.g. the larva of the common cockchafer *Melolontha melolontha*, which

Altitudinal distribution of spiders (*Arachnidae, Araneae*) in Tyrol (Austria, Nordtyrol). The diminution of No of species with increasing altitude is clearly shown (according to Thaler, from Janetschek 1973)

	Web-spinning species					Hunting species					16 more families
vegetation belt / Total in Tyrol No (%)	Theridiidae	Erigonidae	Linyphiidae	Araneidae	Agelenidae	Lycosidae	Gnaphosidae	Clubionidae	Thomisidae	Salticidae	
Cryptogamae 14 (2,6)		6 (5)	6 (6)			2 (5)					
Cushion-like plants . . . 40 (7,4)	3 (7,5)	15 (13)	10 (10)			4 (10)	1 (3)	1 (3)	2 (5)	4 (11)	
Grassy heath 100 (19)	6 (15)	28 (24	25 (25)	3 (8)	1 (5)	8 (20)	7 (23)	5 (14)	6 (15)	9 (26)	2 (4)
Tree line and dwarf shrubs 205 (38)	12 (30)	45 (39)	44 (44)	12 (34)	8 (40)	19 (47)	13 (43)	10 (28)	13 (32)	15 (43)	14 (28)
Total No 540 (100)	40 (100)	115 (100)	100 (100)	35 (100)	20 (100)	40 (100)	30 (100)	35 (100)	40 (100)	35 (100)	50 (100)

reaches in the Alps about 1300 m. It lives on several plants in nursery gardens where it attacks the young roots. *Polyphylla fullo* has very similar habits. *Amphimallon solstitialis* and *A. assimilis* are also fairly common: the larva of the former feeds upon the roots of herbaceous plants, the adult on the leaves and young stems of *Pinus* and, secondarily, of other coniferous and deciduous trees. *Serica brunnea*, one of the commonest Scarab beetles at high altitudes (in Eastern Alps up to 2000 m), when adult eats the leaves of deciduous trees, when a larva the roots of several plants including young plants in nursery gardens. Phytophaga (*Cerambycidae* and *Chrysomelidae*) of course include the most dangerous beetles, some true pests, particularly those belonging to the former. For instance, there is *Hesperophanes* (literally 'it which appears in the evening', from Ἑσπέρα and φαίνομαι) *cinereus*, occuring up to 1800 m which lives on deciduous species. The female lays its eggs in dead wood, thus representing a hazard to timber or worked wood. *Tetropium castaneum* is one of the most dangerous beetles as it attacks healthy trees, the preferred species is the Norway spruce. *Pogonochaerus fasciculatus*, a small insect no longer than 1 cm, has been found in stumps of spruce but probably also attacks pines and other species. *Callidium violaceum* can be harmful to conifers and other species; *Hylotrupes bajulus* lives only on dead plants, poles (including telegraph poles) and beams in houses, the damage it causes can be compared with that of termites in other environments. The beautiful *Rhagium*

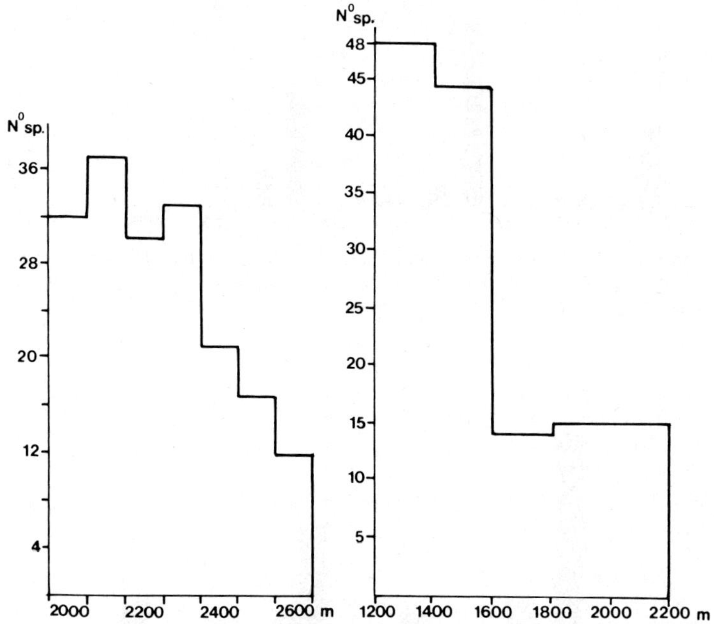

Fig. 302. Decrease of the number of species with the increase in altitude on some Southern European mountains. Left, soil fauna of the Marmolada (Dolomites). Right, Rhopalocerous butterflies on Monte Baldo (near Verona, from Milani).

and *Oxymirus* live under the bark of dead and decaying tree trunks: *R. bifasciatum* under the bark of decaying trees, *R. inquisitor* under that of dead, not decaying, larches.

Ipidae are also economically very important since they attack living trees, and sometimes decaying ones. *Ips cembrae* can reach 2200 m; it is restricted to *Pinus cembra* (p. 421) of which it eats the seeds. It is one of the most harmful Ipidae to forests of *Pinus cembra* and *Larix. Ips curvidens* attacks mature, generally decaying, firs (*Abies pectinata*). It is monophagous. *Orthotomicus laricis*, occurring up to 1900 m, lives on spruce and larch, in trunks abandoned in the winter without previous decortication. *Xyloterus lineatus*, found in the Dolomites up to 1500 m, lives in coniferous forests, where there are felled and not decorticated trunks, but it attacks only the superficial layers. It is a polyphagous species.

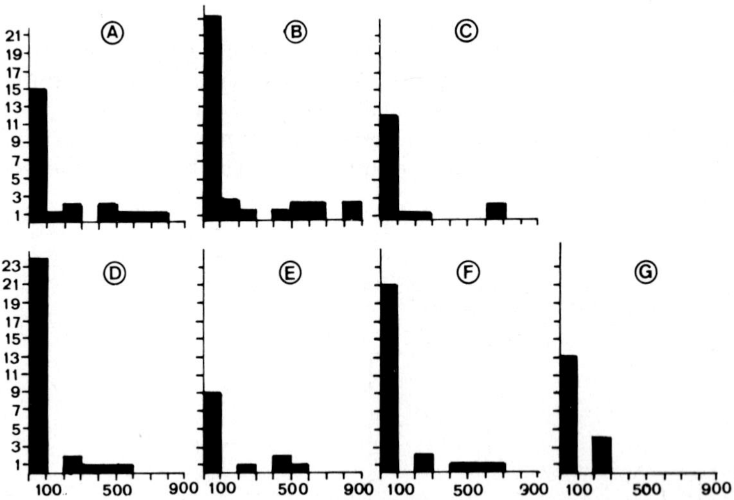

Fig. 303a. Altitudinal distribution of some molluscs in Puglie (Southern Italy) (From Marcuzzi & Lo Casto). a. *Pomatias elegans,* b. *Rumina decollata,* c. *Candidula jonica*, d. *Cernuella virgata*, e. *Helicella conspurcata*, f. *Theba pisana*, g. *Helix aperta*.

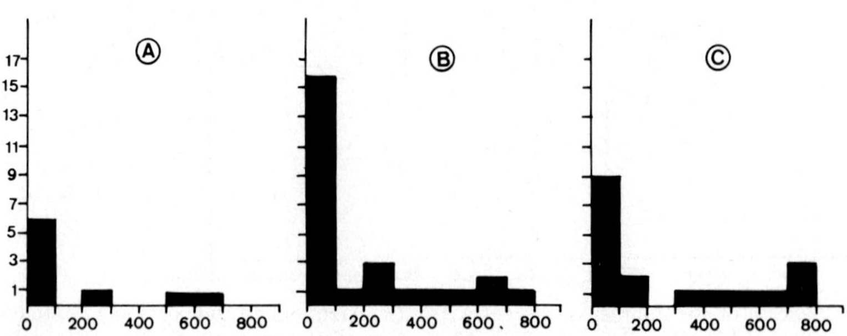

Fig. 303b. a. *Trochoides pyramidata*, b. *Monacha parumcinta*, c. *Monacha rizzae*.

Hylastes cunicularius is well known all over the Alps because of the galleries it digs, both at the larval and the adult stage (maternal, longitudinal simple galleries, larval irregular galleries and wide, subcortical maturation galleries). It is found up to 1800 m but it is not very harmful. *Hylastes ater* lives in all species of *Pinus* generally on young plants and in young afforested areas; it can be considered as the precursor of *Pissodes pini* (see later). The most harmful species are perhaps *Ips typographus, I. sexdentatus*, with those belonging to the genus *Pityogenes. Ips typographus* is present up to 1700 m and is harmful to *Picea* and *Abies*, particularly if the tree is decaying as often happens when these species live out of their normal environment. *Ips sexdentatus* prefers forests of *Pinus silvestris*, as well as the two above mentioned species. Finally there are *Pityogenes chalcographus*, living up to 1700 m, a terrible enemy of the Norway spruce, *P. bidentatus*, preferring the young parts of *Pinus silvestris*, and finally *P. bistridentatus*, living up to 2100 m, one of the most harmful insects to *Pinus mugo.* Recently some very interesting work has been done on the chemical substances isolated from pine phloem which atract these *Ipidae* (α-terpineol, and secondarily cis-carveol and trans-carveol) as well as on the reactions to temperature and light (Perttunen and collaborators, 1957-1968).

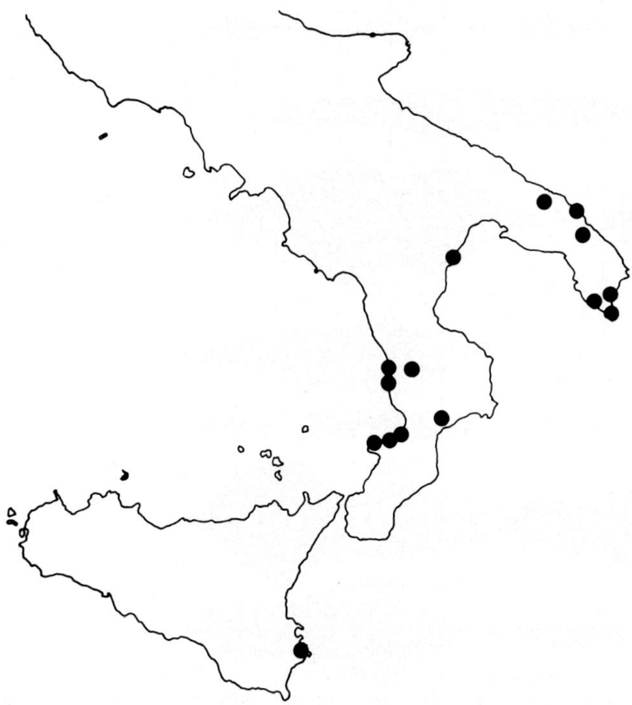

Fig. 304. Distribution of the Southern Italian endemic mollusc *Monacha rizzae* (from Marcuzzi & Lo Casto).

We will mention only a few of the *Chrysomelidae. Melasoma aenea*, which feeds mainly on alder, secondarily on hazel, exhibits aposematic colouration, characteristic of dangerous animals; in this case the insect is protected by the secretion of salicylic aldehyde. *Phytodecta intermedius* (= *quinquepunctatus*), lives on *Sorbus aucuparia* (the common rowan tree), to which it causes defoliation. *Phyllodecta vitellinae* eats the leaves and buds of willow and poplar; *Zeugophora flavicollis*, found up to 2000 m, is harmful to poplar, willow and secondarily hazel; and finally *Luperus pinicola*, which also exhibits an aposematic colouration (black and red) is harmful to pines.

Among weevils there is *Otiorrhynchus ovatus*, which in the larval stage lives on spruce, also in nursery gardens; probably other species of this genus, so common in all European mountains, have practical importance for forestry. *Hylobius abietis* when adult is very harmful to spruce, since it eats the bark of young trees (6-10 years old) and the young branches of mature trees, which it can kill as it attacks the alburnum. Lastly the genus *Pissodes: P. notatus* is one of the most harmful species to pines: its larva digs a gallery in the cortical layers and even in the alburnum of the trunks of young pines (and other coniferous species) 3-15 years old, and in the branches of older trees. *Pissodes pini* is harmful only to forests in

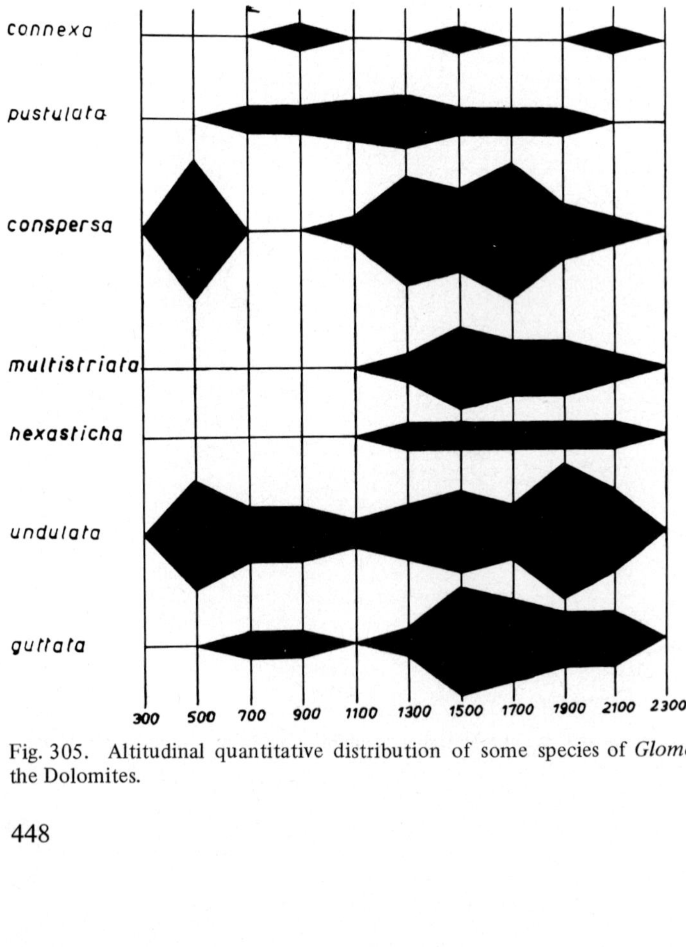

Fig. 305. Altitudinal quantitative distribution of some species of *Glomeris* (Diplopods) on the Dolomites.

which people do not remove the decaying trees and do not cut the trees when they have reached their economic maturity.

Among other phytophagous invertebrates are molluscs (snails and slugs), which are very common in humid environment, such as forests, northern slopes of mountains, cool, wet places, in proximity to water or snow etc. Unfortunately we do not know much about their feeding habits or, as a whole, about their ecology, but we may list the following species as herbivorous (of course algivorous, fungivorous, mycophagous species are also included in this category). *Vertigo moulinsiana* lives on several hygrophytes such as *Glyceria aquatica, Carex paniculata, Iris pseudacorus, Sparganium* etc. feeding on the fungi which attack these plants (*Ustilago, Puccinia* etc.); *Columella edentula*, an euryphagous species lives on *Dryas octopetala, Petasites alba, Phyteuma spicatum, Spiraea filipendula, Adenostyles, Doronicum, Cirsium* etc.; *Vallonia pulchella* and *V. costata*, very common on forest ground, seem to prefer slightly decaying leaves; *Pyramidula rupestris* living in the Alps up to 3000 m, restricted to the fissures of calcareous rocks, is a mycophage eating lichens, microscopical algae and particularly fungal hyphae. The xerophil *Zebrina detrita* lives upon a number of plants included *Echium* (a *Borraginacea* genus) and *Urtica* (the common nettle). Among the *Vitrinidae*, which are generally omnivorous or even carnivorous, some species can be also herbivorous: an example is *Helicolimax pellucidus*, which is also active during the coldest months of the year and very euryionic (pH from *4.5 to 8.0).* It is the prey of harvestmen (Opilionids).

The slug genus *Arion* is represented in the mountains by many species, which – although polyphagous – can eat a number of herbs, mushrooms and fruits. The genus *Deroceras,* on the contrary, is truly herbivorous, and some of its species can be considered true pests when they invade gardens or kitchen-gardens. *Deroceras agreste*, for instance, eats *Leguminosae, Gramineae,* cabbages, beets, carrots, potatoes and several other crops usually cultivated in the montane environment. Its predators are *Carabus,* toads, newts, the blind-worm and several invertebrates.

Fruticicola sericea is an example of an euryphagous snail, feeding on *Heracleum, Cirsium oleraceum*, elder, nettle etc. *Monacha incarnatta* is also an euryphagous species, restricted in Central Europe to *Petasites, Urtica, Heracleum spondylium, Phragmites* etc. The striking *Isognomostoma holosericeum* lives on the ground under litter and feeds mainly upon the higher fungi (Basidiomycetes). Finally the largest snail living on high mountains, *Arianta arbustorum*, which reaches 2600 m, in the Alps is a herbivorous species found often on *Cirsium oleraceum, Petasites* and *Urtica.* The dependence of the size of this animal upon the geological substrate has been clearly demonstrated by Goldfuss, as the following data (taken from Frömming) show:

Substrate	average weight of the shell
graywacke . . .	0.22 g
alluvial soils . . .	0.28 g
granite	0.32 g
porphyry	0.40 g
humid meadow . .	0.64 g
limestone ("Muschelkalk")	1.28 g

Predators. We will begin with beetles, represented mainly by *Carabidae* and *Staphylinidae.* In the Eastern Alps the most common genera and/or species are *Cychrus* (3 species), *Carabus* (*creutzeri, alpestris, violaceus*), *Leistus nitidus, Nebria* (4 species), *Notiophilus* (2 species), *Bembidion* (many species), *Trechus* (several species), *Calathus* (2 species), *Pterostichus unctulatus, P. jurinei, P. metallicus, Abax* (*ater* and *exaratus*) and *Cymindis vaporariorum.* The most common *Staphylinidae* in the Dolomites are two species of *Anthophagus, Staphylinus megacephalus, S. alpestris, S. ophthalmicus, Philonthus nimbicola, P. montivagus, Atheta tibialis, Quedius ochropterus, Q. alpestris* (exclusively alpine), *Tachinus rufipes.*

Most Arachnids are predators, with the exception of some groups of mites (e.g. Oribatids). They are generally widely distributed species, very 'vagil', euryoecious, though there are also some species particularly restricted to the high mountain environment. The principal groups are spiders (*Araneae*), scorpions (gen. *Euscorpius* in the mountains of Central Europe, but also other genera in the Caucasus), whip-scorpions (*Pseudoscorpiones*) and harvestmen or Opilionids. Several mites are also predators, but they belong to the soil fauna (see later).

We will just mention here some spiders which in the Dolomites seem to be specific to high altitudes. They are *Gnaphosa badia, Lycosa nigra, Hilaira montigena, Oreonetides glacialis, Entelecara media, Erigone tirolensis* and *Xysticus glacialis*

All Chilopods (or centipedes) are carnivores: let us mention here only *Dicellophilus carniolensis, Lithobius lucifugus, L. tricuspis, L. forficatus, L. rupivagus, Scolioplanes* (2 species) and *Bothropolis* (2 species).

Saprophages are extremely abundant in all the montane environments, from the collinar to the alpine belt. We may mention among the macrofauna some beetles, such as *Silphidae, Hydrophilidae, Bostrichidae* (lignivorous; see later), *Ipinae, Cerambycidae, Elateridae, Nitidulidae, Scarabaeidae, Tenebrionidae* etc. and some dipteran larvae. Another saprophagous group is the *Diplopoda* or millipedes: in the Dolomites (Eastern Alps) the most common species are *Polyzonium eburneum, Glomeris guttata, G. conspersa, G. undulata, Haploglomeris multistriata, Chordeuma silvestris, Polydesmys edentulus, Cylindroiulus grödensis, Unciger foetidus, Schizophyllum sabulosum* (a very trivial species), *Leptoiulus alemannicus, L. montivagus, Dactylophorosoma nivisatelles* (particularly at high altitudes), and *Orotrechosoma cornuigerum* (only at high altitudes).

The molluscs also contain some saprophagous species which take part in the decomposition of litter and therefore in humus formation. In the Dolomites for instance the commonest species are *Cochlicopa lubrica*, feeding upon humus, living plants and fungal spores; *Pupilla muscorum*, feeding upon humus, secondarily dead or living plants and the spores of lower fungi; *Ena montana*, essentially saprophagous; *Goniodiscus ruderatus,* present in rotten wood and litter, probably responsible for the transformation of wood into humus; *Retinella nitidula*, an essentially saprophagous *Zonitidae*, is characteristic because of its red-amber colour; *Zonitoides nitidus*, euryphagous but also feeding on humus and spores of fungi.

Within the saprophagous species we can recognize some special categories of animals with highly specialized feeding habits, such as necrophagous, coprophagous, fungivorous and lignivorous.

As far as necrophagous insects of montane environments are concerned, it is sufficient to mention *Necrophorous, Catops, Sciodrepoides watsoni, Thanatophilus rugosus* and *T. sinuatus, Necrobia violacea* etc. All these animals are preyed upon by a number of beetles, particularly *Staphylinidae* (*Quedius mesomelinus, Tachinus, Aleochara* etc.). The presence of one species or another depends, more than on the type of carrion, on its degree of decomposition.

Coprophagous beetles are – in alpine mountains – the Scarab beetles belonging to the genus *Aphodius, Onthophagus* (e.g. *fracticornis*), *Geotrupes* (*stercorosus* and *alpinus*) and the *Hydrophilidae* belonging to *Sphaeridium, Cercyon* etc. Their predators are mainly represented by the genus *Hister* (*unicolor, quadrimaculatus* etc.).

The category of lignivorous insects is very well represented in the mountains. Let us mention here only the *Bostrychidae Sinoxylon* (larva in many species of trees), the *Ipinae* already mentioned when talking about herbivores; the *Cerambycidae,* such as the larvae of *Prionus, Ergates faber, Tetropium* (already mentioned when talking about herbivorous species), *Elateridae*, e.g. the larva of *Melanotus rufipes,* living on rotten stumps and under bark of oaks, hazel beech, spruce and pine; larva of *Athous* living in rotten stumps. *Nitidulidae* (larval stage) as *Ipidia*, under rotten bark of old trunks of coniferous trees, *Amphotis*, under the bark of old stumps inhabited by the ant *Lasius fuliginosus, Soronia,* under bark of several deciduous trees, *Scarabaeidae*, such as *Oryctes*, living on wood debris ('Mulm') of oak or even of rotten trunks; *Dorcus*, whose larva feeds on the rotten wood of several deciduous trees, such as beech, oak, linden, willows etc. Finally *Tenebrionidae*, represented in European mountains by *Platydema violaceum*, on beeches, oaks etc. *Arrhenoplita bituberculata*, up to 1000 m, under barks; *Hypophloeus*, under bark in the galleries excavated by *Ipidae; Uloma* in the debris of the galleries excavated by large *Cerambycidae,* under the bark of coniferous trees; *Neatus picipes*, in rotten stumps of oaks and in the debris of pines; and finally *Helopini,* e.g. *Enoplopus* living in wood debris. In some instances unfortunately it is very difficult to say – without very painstaking observations (also experimental) – whether these animals feed on dead wood or on the fungi which develop in this micro-environment.

Finally we can mention some fungivorous beetles, although some of them also feed upon living mushrooms, and therefore cannot be considered totally as decomposers: among the *Staphylinidae, Mycetoporus, Bryoporus, Bolithobius, Bolitochara lunulata* and several species of *Atheta* (*sodalis, amicula*, etc.); among the Endomychidae *Endomychus coccineus*; among the *Mycetophagidae Typhaea fumata.* They are all very common species throughout Europe, not particularly restricted to the montane environment, though in it they often find very suitable micro-environments.

Comments on Pyrenean invertebrates

The variety of invertebrates in the Pyrenees is almost as great as that in the Alps, as a consequence of the antiquity of this chain and of the enormous differentation of habitats, no less than that in the Alps. It is enough to point out the difference in the macroclimate and the vegetation between the Spanish, southern slope of the chain, and the French, northern one. Furthermore, some Pyrenean biota can also live in the Cantabrian mountains, without any real evidence of discontinuity: this is the case with some *Carabus, Trechus, Zabrus* etc.

We will confine ourselves to mentioning only a few localities chosen from those in which the author has made personal collections and observations.

In the easternmost part of the chain there is an isolated mountain, La Massane, only 812 m high, rich in endemic species, partly because of its vegetation (a beach forest completely isolated within an area of sclerophyllous forest, or, as Dajoz says, 'une île au milieu de l'ocean'), partly because of its antiquity. Among the beetles, we find the following endemic species: 4 soil-dwelling species, the *Staphylinidae Phloeocharis subclavata*, the *Scydmaenidae Cephennium albericum* and *Scydmaenidae dobyi* and the *Pselaphidae Mayetia sphaerifera*; also the *Ptiliidae Ptilium schuleri massanae*, living in dead wood, the *Mycetophagidae Mycetophagus fulvicollis massanae*, fairly common in *Pleurotes* and the dead wood of beeches deeply penetrated by the mycelium of these fungi; the *Staphylinidae Bolitochara dajozi* seems also to be endemic.

La Valbonne, at the foot of La Massane, is a deciduous forest with some *Quercus ilex, Q. suber, Acer monspessulanus, Rhamnus alaternus, Ulmus*, etc. The ground-living fauna consists of a great number of beetles (*Calosoma, Carabus, Staphylinus*, etc.), *Embioptera, Isoptera, Diplura* (*Campodeidae*), earthworms etc. The following species can be mentioned as belonging to the zoocenosis present in this transitional forest: the snails *Pomatias elegans* and *Euomphalia strigella* (rare in SW France), the millipedes *Glomeris marginata, Ommatoiulus rutilans* (extending almost from sea level up to 1500 m). *Omm. sabulosum*; the centipedes *Geophilus carpophagus* and *Lithobius inermis pyrenaicus*, the Opilionid *Odiellus brevispina*, limited to France and Spain; the beetles *Pseudocypus aethiops* (a Western Mediterranean element), *Xantholinus translucidus, Timarcha* sp., *Sermyla halensis*, the orthopteron *Antaxius chopardi* (endemic to the Eastern Pyrenees), *Gryllulus*

Fig. 306. Pyrenees, Gavarnie. A general view of the Cirque de la Gavarnie from the north (Gave de Pau).

sp., the bugs *Dicranocephalus albipes* and *Cydnus aterrimus*; the slug *Arion lusitanicus*, the earthworm *Dendrobaena rubida* and the Pseudoscorpion *Histerochelifer tuberculatus.*

In the Ariege, near Moulis (where the Laboratoire souterraine du CNRS has been constructed inside a small limestone hill) a submontane zoocenosis is present

Fig. 307a. Vallée d'Aure. Right, Pic Méchant, 2944 m; left, Grand Pic d'Estaragne, 3186 m.

Fig. 307b. Vallée d'Aure, N.W. of Estiber.

in which the following species appear to be fairly common: the snails *Cochlostoma nouleti, Abida pyrenaearia* (St. Liziere d'Ustou), *Abida* cf. *polyodon, Vitrea subrimata, Clausilia rugosa, Cl.* cf. *abietina, Cl.* cf. *dubia, Hygromia cinctella, Helicella ericetorum, Cepaea nemoralis, Goniodiscus rotundatus, Pomatias elegans*; the millipedes *Glomeris marginata* (in a beech forest), *Gl. intermedia* (on the Pyrenees extending from 450 to 1300 m or more), *Polydesmus inconstans* (from 450 to 1250 m), *Leptoiulus garumnicus* (scarcely a montane species, being present from 550 to 750 m), *Cylindroiulus finitimus* (St. Liziere d'Ustou), in the Pyrenees living from 750 to 2350 m and therefore a rather height tolerant species, *Cyl. spinosus* (extending between 350 and 1250 m); the centipedes *Strigamia crassipes, Lithobius piceus, L. doriae, L. melanops, Stigmatogaster gracilis*; the Opilionids *Scotolemon lespesi* and *Odiellus brevispina*, the beetles *Otiorrhynchus singularis* (under bark?), *O. morio, O. ovatus, O. auropunctatus, Cionus schönherri* (a Western Mediterranean species), *Quedius obscuripennis* sbsp. *pyrenaeicola, Onthophagus nutans, O. fracticornis* (two coprophagous species), *Timarcha* sp., also the roach *Ectobius livens*, the hopper *Tettigonia viridissima* and the *Heteroptera* (or bugs) *Peritrechus gracilicornis* and *Capsodes flavomarginatus.* Furthermore we may mention from this interesting submontane region *Arion lusitanicus, Lehmannia marginata, Deroceras agreste*, the Pseudoscorpion *Neobisium doderoi*, and the earthworms *Lumbricus friendi, Dendrobaena octaedra, D. rubida* and *D. attemsi.*

Near Le Carlit, in the Eastern Pyrenees, a high plateau, granitic in nature,

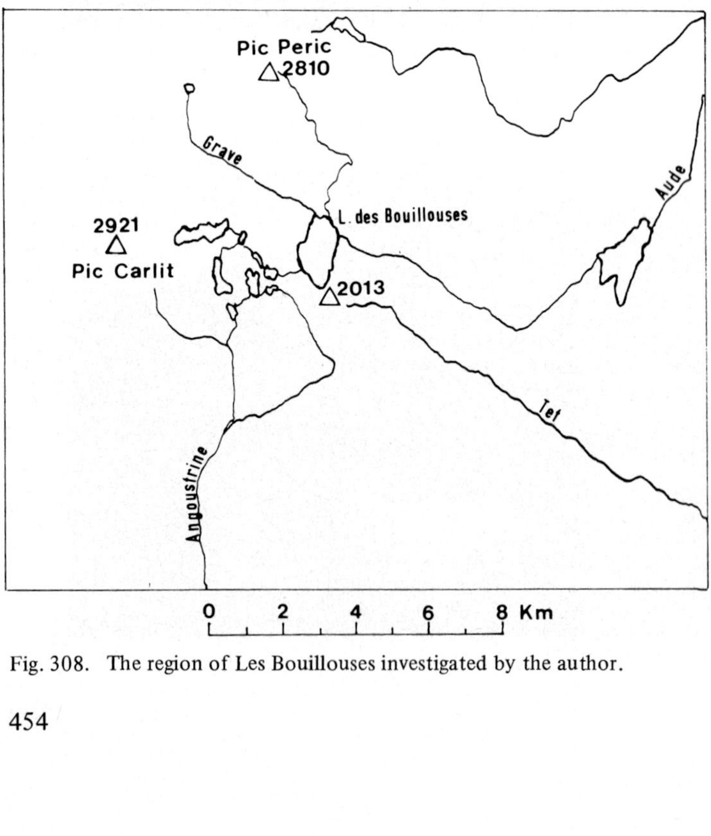

Fig. 308. The region of Les Bouillouses investigated by the author.

known as Les Bouillouses (Fig. 308), extends at about 2000 m, covered partly by a *Pinus uncinata* stand and partly by very peculiar bog in which there is an enormous number of small ponds, irregularly scattered over the vast high plateau. In the forest growing on granite, *Calluna vulgaris, Genista purgans, Rhododendron ferrugineum, Juniperus* sp., *Gentiana* sp., *Pedicularis* sp. are present. The fauna is very poor because of the geological nature of the substrate: we may mention the beetle *Pseudocypus picipennis nevadensis* (new to the Pyrenees), the millipedes *Glomeris marginata, Cylindroiulus silvarum* (on the Pyrenees between 1100 and 2100 m), *Cyl. chalandei*, the centipedes *Geophilus pyrenaicus, Lithobius doriae, Monotarsobius* sp., the Opilionid *Mitopus morio* (a common species), the beetles *Otiorrhynchus monticola, O. singularis, Diodyrrhynchus austriacus* (fam. *Nemonychidae*), *Philonthus montivagus, P. pseudovarians, Othius melanocephalus, Tachinus pallipes, Tachyporus nitidulus* (some mycetophagous species), and the coprophagous *Aphodius fimetarius, A. depressus, A. alpinus, A. rufipes, A. fossor, A. haemorrhoidalis* and *Geotrupes niger*; also the slugs *Arion subfuscus, A. lusitanicus, Lehmannia marginata* and the earthworms *Lumbricus friendi, Dendrobaena octaedra* and *D. attemsi.* Extremely interesting is the paucity of snails as a consequence of the acid substrate.

Not far from Moulis is a very extended beech forest, known as Bethemale, extending from the valley (1000 m) up to the Col de la Core (1409 m), where the beech is replaced by *Pinus uncinata* and *Abies pectinata.* Whereas the fauna of

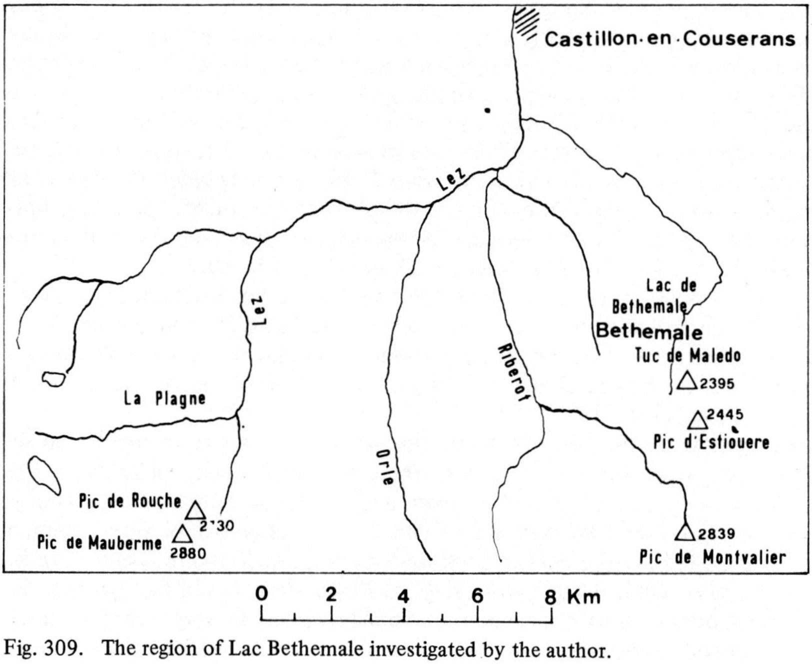

Fig. 309. The region of Lac Bethemale investigated by the author.

the latter formation is very poor, the fauna of the floor on the beech forest is very rich: we can mention the snails *Cochlostoma* cf. *obscurum, Clausilia rugosa, Hygromia limbata, Cochlodina laminata, Goniodiscus rotundatus*; the beetles *Aphodius vaccinarius, Geotrupes stercorosus*, the millipedes *Glomeris marginata, Loboglomeris pyrenaica* (in the Pyrenees extending from 1100 to 2100 m), *Glomeridella kervillei, Polydesmus inconstans, Cylindroiulus spinosus*; the Opilionids *Nemastoma scabriculum, Paranemastoma quadripunctatum* and *Mitopus morio*; the beetles *Philonthus splendens, Tachinus humeralis, Ontholestes murinus, Callicerus rigidicornis* (a rare element, spread across Central and Southern Europe), *Atheta aquatica, Aphodius rufipes* and *A. fossor*; the slugs *Arion circumscriptus, A. hortensis, A. subfuscus, A. lusitanicus, Lehmannia marginata, Deroceras reticulatum* and *D. agreste*, and finally the earthworms *Eisenia eiseni, Dendrobaena rubida, D. octaedra, Lumbricus friendi* and *Allolobophora* sp.

Not far from Moulis there is a very interesting locality at La Plagne, 1200 m high, from which one goes up to the Col de la Catanere, through a beech forest called Bois de Ruerech (1550 m); the area is just north of the Pic de Maubermé, close to the Spanish border. The region is relatively undisturbed, and the fauna is very rich and varied. We can list the molluscs *Discus rotundatus, Vitrinobrachium* cf. *pyrenaicum, Retinella* cf. *incerta, Aegopinella* cf. *nitidula, Clausilia rugosa, Hygromia limbata*, the beetles *Geotrupes stercorosus*, the millipedes *Glomeris marginata, G. intermedia, Loboglomeris pyrenaica, Glomeridella kervillei, Cylindroiulus finitimus, C. spinosus*, the centipedes *Chaetechelyne corsica, Stigmatogaster gracilis, Strigamia crassipes, Cryptops hortensis, Lithobius piceus, L. tricuspis, L. pellicensis, L. doriae* and *Monotarsobius duboscqui* (one of the richest places in the Pyrenees as far as centipedes are concerned), the Opilionids *Mitopus morio, Dicranopalpus martini* (endemic to the Pyrenees and Portugal, i.e. to the Iberian Peninsula); the beetles *Otiorrhynchus pyrenaeus* (reaching the Pyrenees and SW France) *Plinthus caliginosus* (limited to France, Central Europe and England, maybe an 'armorican' species, probably very old), *Homapterus subnudus, Orchestes fagi* (on beech leaves), *Philonthus pseudovarians, Ocypus pedemontanus* sbsp. *pyrenaeus, Tachyporus nitidulus, Emus hirtus, Leptusa pulchella, Aphodius rufipes, A. fossor, A. haemorrhoidale* and the planticolous *Trichius fasciatus.* Furthermore the cenosis contains the slugs *Lehmannia marginata* and the earthworms *Lumbricus friendi, Dendrobaena octaedra, D. rubida* and *D. attemsi.*

The animal population of the Hautes Pyrenees is rather different. In the very heart of this part of the chain, on the shores of the Lac d'Oredon, a magnificent spruce forest extends, rising on all sides towards the massifs which surround this beautiful lake, on whose shores the Station d'Hydrobiologie of the Toulouse University is sited (Fig. 310).

In the valley of the Neste d'Aure, the stream coming from the Massif de Néouvielle and from the Lac d'Orédon, there is a small town, Vielle Aure. The environment here is much altered by man, and the fauna of the valley, which is exposed NS and therefore very sunny and dry, is composed of very common elements: included are the beetles *Ontholestes murinus, Aphodius haemorrhoidalis, A. fimetarius, Onthophagus ovatus, Homaloplia ruricola* (an herbivorous element, here represented by a melanic form which seems to be very rare in France), *Dorcus parallelopipedus* and *Larinus* sp., the slug *Arion lusitanicus*, the millipede

Ommatoiulus sabulosus, the earthworm *Allolobophora* sp. and the *Diplura Campodea kervillei.*

South of Vielle Aure, at about 1100 m, near Ens, there is a fairly well preserved beech forest (today mainly represented by a coppice), whose invertebrate fauna is fairly rich and varied; the underlying rock is schists. Most of the fauna is concentrated in old decaying stumps and under fallen trunks: there are the slugs *Arion hortensis, A. lusitanicus* and *Lehmannia marginata*, the millipedes *Loboglomeris pyrenaica* (strictly endemic), *Cylindroiulus finitimus* (common to most of Pyrenees), *C. silvarum, C. spinosus, Haplopodoiulus spathifer* and *Polydesmus coriaceus*, and the earthworms *Eisenia eiseni, E. foetida, Dendrobaena rubida* and *Lumbricus friendi.*

The zoocenosis near the shores of Lac d'Orédon contains the following elements (only the commonest of course are named): the molluscs *Vitrinobrachium pyrenaicum*, an old, lusitanic element, limited to the Pyrenees and Ireland, witness of very different geographical and climatic conditions compared with those existing today; it is new for the region of Lac d'Orédon; *Clausilia* cf. *dubia, Goniodiscus rotundatus*; the millipedes *Loboglomeris pyrenaica, Ommatoiulus sabulosum*, (extending from 0 to 1850 m), *Tachypodoiulus albipes*, extending from 1850 to 2480 m, the centipedes *Lithobius doriae* and *Stigmatogaster gracilis*; the beetles

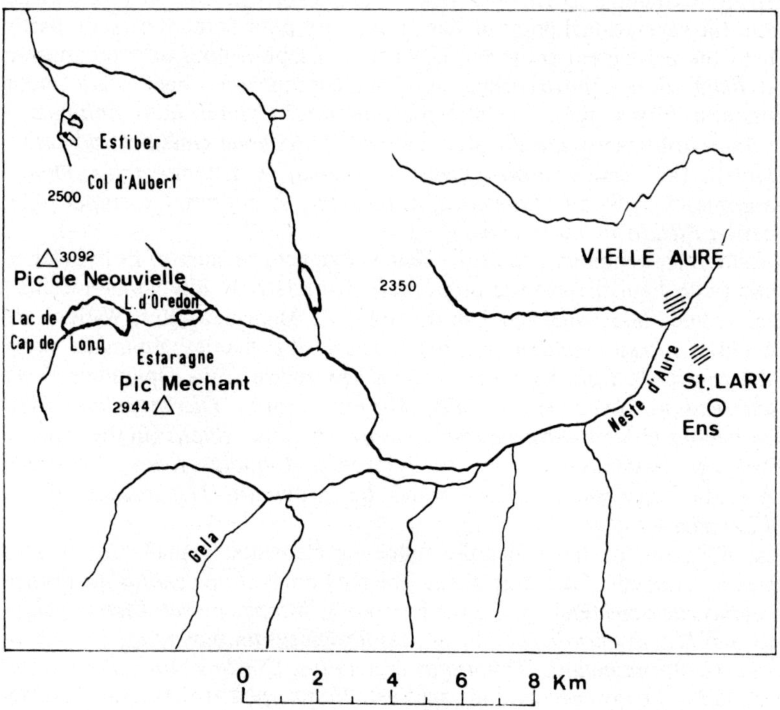

Fig. 310. The region of Lac d'Oredon investigated by the author.

Otiorrhynchus monticola, O. singularis, Barynotus squamosus (a boreo-alpine species), *Plinthus imbricatus* (limited to the Pyrenees and Central Spain), *Zyras humeralis*, a myrmecophilous *Staphylinidae,* associated with the ant genera *Lasius* and *Formica,* new for the Pyrenees; *Aphodius obscurus, Chrysomela marginata*; the hopper *Podisma pedestris* (on small shrubs) and finally the slugs *Arion lusitanicus* (present in an enormous number near the lake) and *Deroceras agreste* and the earthworms *Lumbricus friendi, Octolasium lacteum, Dendrobaena rubida* and *D. octaedra.*

Much more interesting is the fauna of the Esteragne, a massif rising directly south of the Lac d'Oredon, where a very rich alpine fauna is present between 2300 and 2500 m, in some alpine meadows and near the snow line: here we may find the molluscs *Abida* cf. *pyrenaearia, Vitrina* cf. *pellucida, Phenacolimax pellucidus, Euconulus fulvus* (which reaches here its highest altitude in the Pyrenees), the millipedes *Proteroiulus brolemanni,* Pyrenean endemic in the zone of permanent snow between 2000 and 3000 m, the beetles *Otiorrhynchus monticola, O. morio, Aphodius obscurus, A. alpinus*, together with a great number of Carabids not yet determined.

Finally the zoocenosis is completed by slugs such as *Arion lusitanicus, Lehmannia rupicola* (new to the Pyrenees) and the earthworm *Allolobophora caliginosa* (near the snow).

North of Lac d'Oredon is the Col d'Aubert (2500 m), formed from granite: in Val d'Estiber, extending north and east of Col d'Aubert, is a very interesting region from the vegetational point of view, but a very poor fauna is present, partly restricted to the permanent snow and consisting of lapidicolous or coprophagous elements: *Bembidion, Otiorrhynchus,* ants and coprophagous beetles are among the commonest forms; there are also the millipedes *Cylindroiulus finitimus*, a common and eurihypsous species, the centipedes *Strigamia crassipes* and *Lithobius castaneus,* the beetles *Otiorrhynchus monticola, O. auropunctatus, Philonthus pseudovarians, Aphodius fimetarius, A. obscurus, A. rufipes, Geotrupes niger,* and the earwig *Pseudochelidura sinuata.*

Near Cauterets, a famous area of the Hautes Pyrenees because of its hydrothermal springs (with high therapeutic properties) two relatively low alpine localities have been visited, Marcadau and Lac d'Estom. At Marcadau (Ref. Wallon) the fauna includes *Vitrinobrachium* cf. *pyrenaicum*, the beetles *Aphodius vaccinarius*, the centipede *Lithobius marcuzzii* (a new species), the Opilionids *Nemastoma scabriculum, Lacinius ruentalis, Mitopus morio, Dicranopalpus pyrenaeus*, the beetles *Otiorrhynchus auropunctatus, Hylobius abietis* (in the forested zone), *Philonthus alpinus* (new to the Pyrenees), *Aphodius fossor, Geotrupes stercorarius*, the bug *Eremocoris abietis* and the earthworms *Dendrobaena octaedra* and *Eiseniella tetraedra* (Fig. 311).

At Lac d'Estom the fauna has the following elements: a snail, *Goniodiscus rotundatus*, a centipede, *Lithobius doriae*, the Opilionids *Homalenotus quadridentatus, Lacinius ruentalis* (endemic to the Pyrenees), *Mitopus morio, Dicranopalpus pyrenaeus* and *Megabunus dianae*; the beetles *Barynotus unipunctatus, Otiorrhynchus morio, O. auropunctatus, Philonthus cruentatus, Quedius bonvouloiri, Astilbus canaliculatus* (a myrmecophilous species), *Homaloplia ruricola* (an herbivorous element) and the *Chrysomelidae Timarcha strangulata.*

On the border between France and Spain is the magnificent Cirque de la Gavarnie, one of the few regions of the Pyrenees (Hautes Pyrenees) whose geological nature is sedimentary: it represents a typical glacial geomorphological phenomenon, now with no glacier, but with some spots of perennial snow which provide a suitable environment for many calcicolous invertebrates such as *Trechus* and others. The fauna is extremely varied and interesting, including several strictly endemic species. Let us mention here only the molluscs *Abida* cf. *pyrenaearia, Vitrina* cf. *pellucida, Nesovitrea hammonis, Pyrenaearia carascalensis* (endemic to the Pyrenees), *Phenacolimax pellucidus*, the millipedes *Loboglomeris pyrenaica* (Troumouse), *Ceratosphys nivium* (extending between 1900 and 2250 m, and therefore a truly hypsicolous or mountain-dwelling element), *Ommatoiulus sabulosum* (Troumouse only), *Leptoiulus juvenilis* (Tromouse), *Cylindroiulus iluronensis*, a high mountain species present on Col de la Gavarnie or Col de Boucharo[1], at

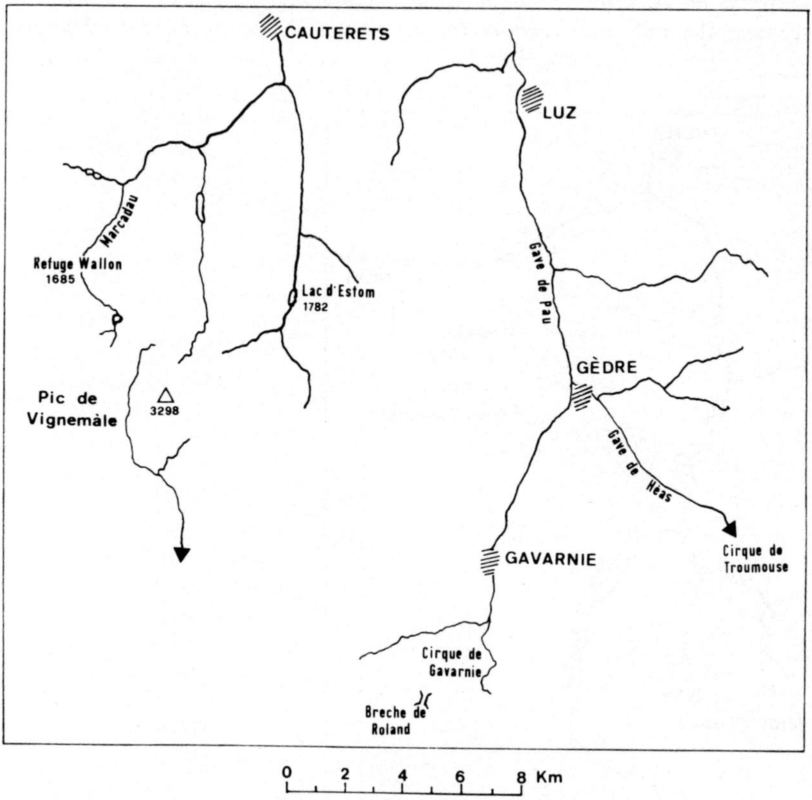

Fig. 311. The region of Cauterets and Gavarnie investigated by the author.

[1] Known also as 'Breche de Roland' after the name of the legendary hero of the Chanson de Roland, famous for his strength, courage and chivalry, killed while fighting with the Saracens at Roncesvalles in 778 A.D.

2250 m, *Tachypodoiulus albipes*, the centipeds *Stigmatogaster gracilis* (2250 m), *Lithobius tricuspis, L. lapidicola, L. doriae*, the Opilionids *Mitopus morio, Dicranopalpus pyrenaeus* (limited to Southern France and Spain), the beetles *Otiorrhynchus monticola, Barynotus unipunctatus, Ot. morio, Homapterus subnudus* (endemic to the Lower and High Pyrenees, Cantal, Puy de Dome), *Quedius anceps* (near the snow), *Q. attenuatus* (a northern element), *Aphodius ater, A. alpinus, A. rufipes, Timarcha strangulata, Chrysomela pyrenaica, Protorina ludovicae* (endemic); the earwig *Pseudochelidura sinuata*, the slugs *Arion lusitanicus* and *A. intermedius, Lehmannia marginata* and finally the earthworms *Lumbricus rubellus* and *Dendrobaena rubida.*

In the Western Pyrenees there is more limestone than in the Hautes Pyrenees and Eastern Pyrenees, and its presence is clearly revealed by fauna and flora. Penes Blanques (Fig. 312), 2483 m, is SE of Pic de Ger (2612 m) and exhibits a rich hypsicolous and nivicolous fauna, consisting mainly of Carabids. The snails are *Cochlostoma* cf. *partioti, Vitrea subrimata, Candidula* cf. *gigaxii, Cernuella* cf. *augustiniana*; the millipedes *Tachypodoiulus albipes*; the centipedes *Lithobius acu-*

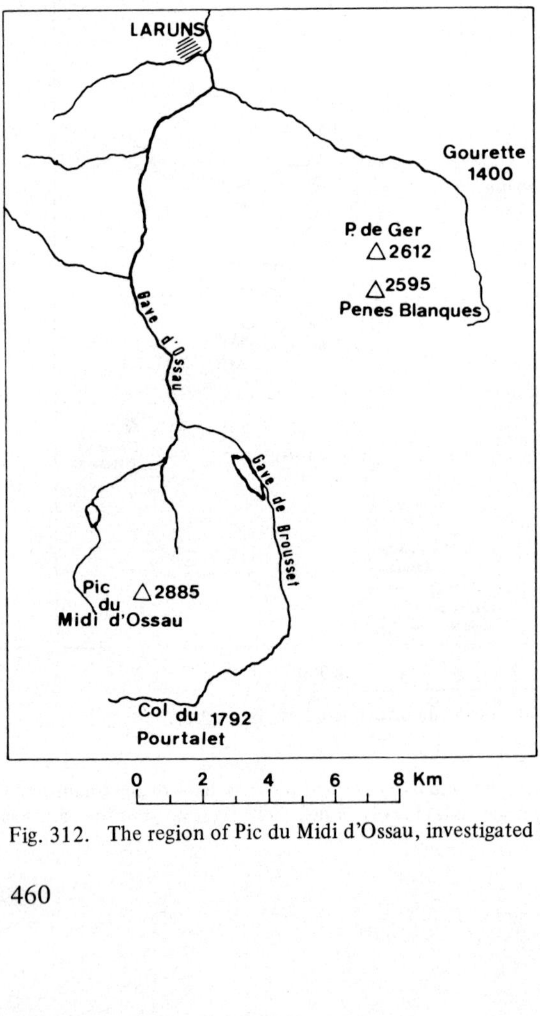

Fig. 312. The region of Pic du Midi d'Ossau, investigated by the author.

leatus (a new species discovered by the author), *L. tricuspis*, the Opilionid *Mitopus morio*, the beetles *Otiorrhynchus monticola, O. malefidus, O. morio, Quedius punctatellus, Timarcha strangulata*, the earwig *Pseudochelidura sinuata*, and the *Diplura Campodea ilixonis* (endemic).

At the foot of this beautiful mountain, in the beech forest (which in this region reaches as high as 1500 m) a zoocenosis is present in which the following species appear: *Cochlostoma* cf. *obscurum, Abida secale, Clausilia rugosa, C.* cf. *dubia, Helicella* cf. *stiparum, Chondrina avenacea, Goniodiscus rotundatus*, the millipede *Ommatoiulus rutilans*; the centipedes *Cryptops hortensis, Lithobius pilicornis* and *L. doriae*; the beetles *Otiorrhynchus* sp.; the *Diplura Campodea pieltaini* (a Pyrenean endemic species) the slugs *Arion subfuscus, A. lusitanicus, Lehmannia marginata, Deroceras reticulatus* and *D. agreste* and the earthworms *Dendrobaena octaedra* and *D. rubida.*

Near the summit of the Pic du Midi d'Ossau, in an alpine meadow on limestone at about 1700 m, an interesting zoocenosis is composed of *Cochlostoma* cf. *obscurum, Abida secale, Goniodiscus rotundatus*, the millipedes *Glomeris marginata* (in the Pyrenees extending from some 500 to 2000 m), *Ommatoiulus lienhardti* (in the alpine belt only), the centipedes *Strigamia crassipes* and *Lithobius doriae*, the Opilionids *Homalenotus quadridentatus, Mitopus morio, Ischyropsalis hellwigi*, present in the forest, under big stones, a species limited to South France and Spain; the beetles *Otiorrhynchus morio, O. auropunctatus, Geotrupes niger* (a coprophagous species), the earwig *Anechura bipunctata*, etc.

The environment, as well as the biota, is completely different in the westernmost part of the Pyrenees, where the Atlantic climate prevails and causes a high relative humidity and an almost permanent fog. Some species absent in other parts of the chain or restricted to water can thrive here on the soil, often clayey and damp. Among the vertebrates the common salamander *Salamandra salamandra* can be found on the floor of the beech forests; among invertebrates a terrestrial Turbellarian, *Microplana pyrenaica*, exclusive of the French slope of Western Pyrenees is present (found by the author in the Forêt d'Issaux at nearly 1200 m). In a cave in this enormous forest – maybe one of the biggest of the Pyrenees – the Opilionid *Ischyropsalis pyrenaea* has been found.

On La Gouffre de la Pierre St. Martin, there is practically an alpine meadow at about 1700 m because of the peripheral position of the mountain (cf. p. 416) and its exposure to wind: at this height, in fact, the last dwarf beeches can be found together with the bracken *Pteridium aquilinum* and the juniper. The substrate is limestone. On the clayey soil (a kind of mull very rich in clay), because of the high relative humidity of the air, some aquatic beetles, such as *Helophorus* sp. and *Dryops* sp., can thrive, quite an unusual feature on European mountains.

The zoocenosis of La Pierre St. Martin is composed of the snails *Cochlostoma* cf. *obscurum, Abida secale, Chondrina avenacea, Clausilia pyrenaica, Trissexodon constricta* (exclusive to the Basses Pyrénées), *Helicella ericetorum, Phenacolimax pellucidus*, the beetles *Geotrupes stercorosus* (restricted to the faeces of large mammals), the millipedes *Loboglomeris rugifera, Protoglomeris vasconica*, exclusive to the western part of Pyrenees, reaching as far east as Troumouse and Vielle Aure, *Polydesmus angustus, Cylindroiulus londinensis* (extending from 0 to 1650 m), *Haplopodoiulus spathifer*, a mid-montane species (1100-1650 m, at least judging from the personal collections of the author), the centipedes *Strigamia crassipes, Cryptops hortensis, Lithobius pilicornis, L. doriae, Monotarsobius curtipes*; the Opilionids *Homalenotus quadridentatus, Mitopus morio, Dicranopalpus pyrenaeus, Megabunus dianae* (a Western European species) and the *Diplura Campodea pieltaini.* Furthermore the following species of beetles appear in the community: *Otiorrynchus morio, Ocypus pedemontanus* sbsp. *cantabricus, Aphodius obscurus, A. alpinus, Timarcha strangulata* and *Chrysomela pyrenaica.* The slugs are *Arion hortensis, A. lusitanicus, Deroceras agreste* and *Semilimax pyrenaicus*, endemic to the Pyrenees and Ireland, i.e. a 'lusitanian' or atlantic element of very old age and great biogeographical interest.

Comments on the Carpathian Invertebrates

The glacial period has not had the same effect in the Carpathians as it had in the Alps. The 'massifs de refuge' are practically absent here. In these mountains, in contrast, there is a gradual impoverishment of the fauna from the south northwards, which cannot be explained on the basis of present-day climatological and pedological conditions. The endemic species are well known, particularly among the carabid beetles, thanks to the work of Holdhaus, published about 20 years ago. Here are the endemic species of this coleopterous family: *Carabus planicollis, C. obsoletus, Leistus gracilis, L. banningeri, Nebria heegeri, N. reichi, N. reitteri, N. fussi, N. transsylvanica, N. tatrica* and *N. carpathica*; many species of the genus Trechus (cf. Holdhaus, p. 274), *Duvaliopsis bielzi, D. pauperculus, D. meliki, D. pilosellus*, many species of the genus *Duvalius* (l. cit. p. 276), *Dicropterus brevipennis, Deltomerus tatricus, D. kaszabi, D. carpathicus, Patrobus roubati* and *P. quadricollis, Pterostichus negligens, P. rufitarsis, P. foveolatus, P. findeli* and *P. fossulatus, Platynus glacialis, P. banaticus* and finally *Amara misella.* The blind endogeous species belong to the genus *Dicropterus, Duvalius* and *Duvaliopsis.* Other endemic species belong to the *Silphidae, Scydmaenidae, Pselaphidae* and to the genus *Otiorrhynchus.*

The situation is somewhat different in the western Carpathians, and more particularly in the Lower Tatras, where the soil fauna has been studied by Stach, and more recently by Nosek.

The Apterygota fauna occupied the Tatras very early and its composition during the Middle and Upper Pliocene was perhaps somewhat similar to the fauna inhabiting the territories of present-day south-east Asia. Indeed, in that era, forests of a similar type covered the slopes of Tatras. It may be supposed that some representatives of this ancient Carpathian fauna have survived in these mountains up to now, e.g. *Tetrodontophora bielanensis, Neanura carolii, Podura aquatica* and *Sminthurides aquaticus.* A great devastation of this ancient fauna of the warm Pliocene climate and many changes in its composition were caused by the Pleistocene glaciations. The devastation of the fauna of the Tatras was probably greater than that of the Alps because of their very small surface and because the northern ice-masse ('Inlandsis') were more distant from the Alps than from the Tatras. Very likely a great proportion of the primitive forms of the families *Poduridae* and *Isotomidae* was able to survive during the glacial period in the natural conditions of Central Europe. Some of the species have a boreo-montane distribution, e.g. *Agrenia bidenticulata, Anurophorus binoculatus, Tetracanthella arctica* etc. (Fig. 313). On the other hand a number of Collembolans have been introduced in the post-glacial period (e.g. many *Sminthuridae* and *Entomobryidae*). *Hypogastrura tatrica* can be considered an endemic Carpathian species. *Morulina verrucosa* (Fig. 313c) is another species with a restricted area of distribution: it seems to be a Carpathian element, a relict, somewhat modified, from the glacial period; its near relatives show a circumpolar distribution. Another endemic species is *Hypogastrura breviempodialis.*

Comments on the Caucasus invertebrates

With reference to the Caucasus, we want to mention the soil arthropod fauna of the N-W mountains (M. Sheze, Mt. Sober-Oash, Kotsegur, Tkhab, etc.) recently studied by Ghilarov & Arnoldi. The vegetation – which has a steppe, relict character – is composed of *Festuca sulcata, Koeleria gracilis, Phlomis tuberosa, P. pungens, Allium rotundatum*, with several Graminaceous plants. Its origin, as the two previously mentioned authors have shown on the basis of the study of the soil fauna, is mediterranean. We can quote the *Elateridae Selatosomus latus, Melanotus brunnipes, Agriotes ponticus, A. gurgistanus* and *A. circumductus* (a Caucasian endemic species), the Tenebrionids or 'black beetles' *Opatrum sabulosum, Pedinus femoralis, Oodescelis polita, Crypticus quisquilius, Cylindronotus brevicollis* and *Laena starki*; the *Scarabaeidae Amphimallon solstitialis, Rhizotrogus aestivus, R. aequinoctialis* and *Pentodon idiota*, the Carabids *Zabrus spinipes, Harpalus flavicornis, Ophonus azureus, Harpalus caspius, Licinus cassideus, Amara communis*, etc.; several of these species are granivorous; the weevils *Otiorrhynchus schoenherri, O. edithae* and *O. erinaceus* (endemic to the Caucasus), also *O. caucasicus, O. fullo, O. ligustici, Brachycerus junix, Stomodes tolutarius* and *Eusomus*

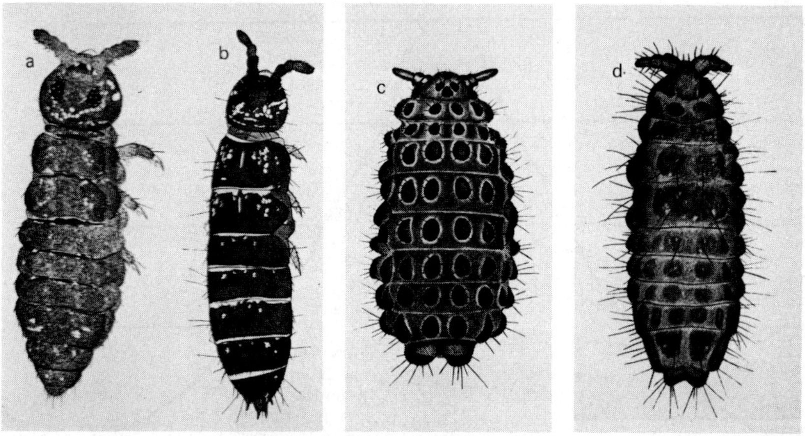

Fig. 313. Some Collembols characteristic of the soil fauna of the Carpathians. a. *Hypogastrura crassaegranulata* (a 'Rassenkreis'); b. *Tetracanthella arctica*, a boreo-alpine element; c. *Morulina verrucosa*; d. *Neanura caroli* (from Nosek).

mirabilis, and finally the Chrysomelids *Timarcha tenebricosa, Chrysomela limbata* and *C. lurida* sbsp. *reitteri.*

The ants living in this region of the Caucasus are *Lasius alienus, Solenopsis fugax, Leptothorax unifasciatus, Camponotus aethiops, Messor rufitarsis*, also *Leptothorax nylanderi* (a mesophilous species) and *Tapinoma tauridis* (a Crimean-Caucasian species).

The Myriapods are very interesting, i.e. Diplopods and Chilopods, namely *Chromatoiulus rossicus*, a predominating species in the steppes of the Eastern Ucraina, *Escaryus retusidens, Clinopodes flavidus, Brachygeophilus sukacevi* (endemic to the NW Caucasus and Southern Crimea), also *Monotarsobius baloghi* and *M. turkestanicus.* The dominant Isopod here is *Protracheoniscus giliarovi.* All these animals are relict species which in the past were able to colonize these mountains in a period in which steppes had a much wider distribution than today (cf. chapter on steppes).

Zoogeography of mountain invertebrates

The distribution of mountain animals (as well as that of vegetation) is so varied that we can hardly attempt a classification of zoogeographical categories. Indeed, no two species have quite the same distributional pattern, and this is both for historical reasons (place and time in which the species originated and reached its present distribution) and for ecological reasons (autoecology of the species and in some cases ecology of the community to which it belongs). In general, the categories we can establish – very often conventionally and arbitrarily – can be based either on the present type of distribution, or on the presumed origin and past distribution. In this chapter we will try to make a classification of some montane European invertebrates based only on the first criterium. Remarks about the possible history of the species will be made only in a few cases.

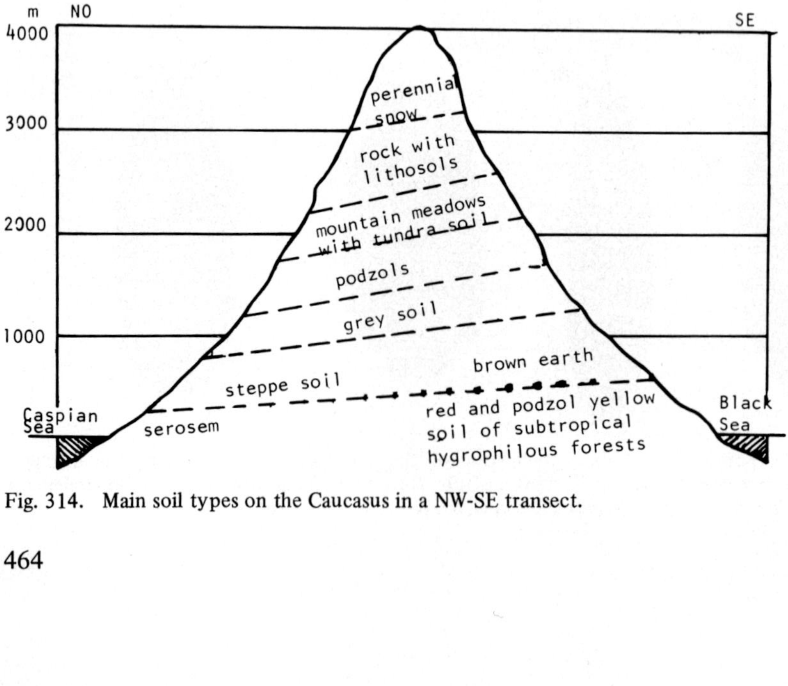

Fig. 314. Main soil types on the Caucasus in a NW-SE transect.

We will intentionally limit ourselves to three zoological groups, among the most interesting and best known from the zoogeographical point of view, namely mollusks, beetles and millipedes. From what we are going to show we will see how different are the zoogeographical categories according to the animal groups, for the reasons we have seen above, so that we cannot always expect to apply the same chorological subdivisions to *all* animal groups.

Let us begin with molluscs. Species present in the greater part of the Alpine chain, or at least both in Eastern and Western Alps: *Chondrina megacheilos, Truncatellina striata, Cochlodina fimbriata, Semilimax nivalis, Limax albipes, Fruticola ciliata, Helicigona cingulata* (mainly south), *H. planospira* (south), *Carychium mariae* (south, partially vicarious species with *C. minimum*).

Species limited to Western Alps: *Arion alpinus, Ariunculus mortilleti, Limax lachensis, L. millepunctatus, Milax ochraceus, M. guebhardi* (subfossil), *Vitrea zanclea, Chilostoma glacialis* (up to 2500 m), *C. alpina* (between 1700 and 2800 m), *C. fontenillei* (ecological vicariant of preceding species, at lower altitudes), *C. millieri* (up to 2500 m), *C. nicatis*, etc.

Species limited to Eastern Alps: *Chondrina circumplicata, Truncatellina salurnensis, Orcula gularis, Delima stentzi, Cochlodina commutata* (a southern element with relict stations at Schneeberg), *Iphigena tettelbachiana, Caecilioides aciculoides* (only south), *Zonites gemonensis* (idem), *Retinella hiulca* (idem), *Fruticicola leucozona* (idem), *F. lurida* (idem), *Helicigona aemula* (idem), *H. intermedia* (idem), *Cochlostoma henricae* (only North and South), *C. philippianum* (only

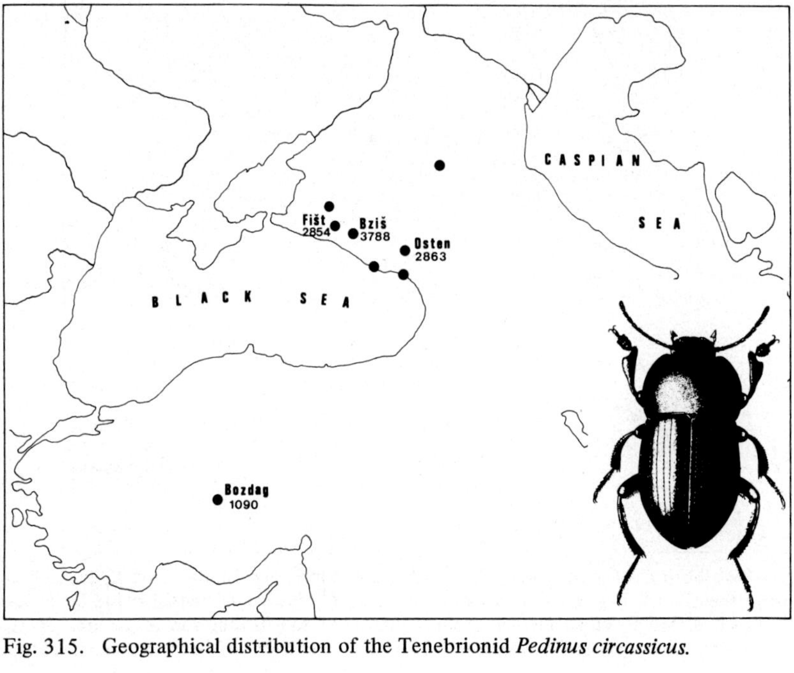

Fig. 315. Geographical distribution of the Tenebrionid *Pedinus circassicus.*

south) and *Pleuraeme spectabilis* (North and South, eliminated in the formerly ice-covered central Alps) (Fig. 316).

Species present in the Eastern Alps and Balkans (generally NW part): *Agardhia truncatella, Zonites verticillus* (in the last interglacial extended up to central and western Germany), *Acme gracilis* (up to Croatia).

Alps and Carpathians: *Orcula dolium, Fruticicola unidentata, Isognomostoma isognomostoma, I. holosericeum.*

Circum-polar, present also on the high mountains of Southern Europe: *Vertigo arctica.* True boreo-alpine species seem to be absent.

Germany (montane), Alps, Carpathians and Balkans: *Laciniaria vetusta.*

Central Europe and Alps: *Helicolimax diaphanus, H. kochi* and *Acme polita.*

Pyrenees – *Trissexodon constricta, Cryptazeca monodonta, Elona pyrenaica, Laminifera subarcuata* and *L. pauli, Clausilia abietina* and *C. nana, Helicella orcina, Hygromia tassyi, Mastigophallus rangi, Abida bigerrensis, A. pyrenaearia, A. portioti, Testacella companyoi, T. bourguignati, T. servaini, Iberellus companyoi, Vertigo baudoni, Sandahlia cylindrica, Azeca cylindrica* and *A. dupuyi* and finally the species of the genus *Cochlostoma mabillei, berilloni, bearnica* and *alloglypha.* Molluscs seem to be more localized than beetles but less than millipedes.

Diplopods (millipedes).

Species limited to Eastern Alps: *Bericostenus fagi* (endemic), *Glomeris longaronensis, Onychoglomeris tyrolensis, Polydesmus monticola, P. edentulus, Craspedo-*

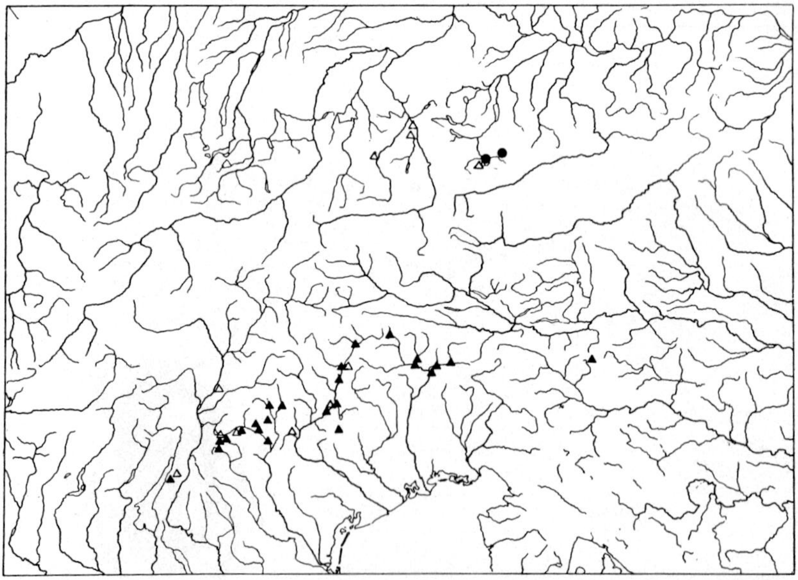

Fig. 316: Distribution of two species of land snails eliminated in the central, crystalline Alps during the Glacial Era: *Pleuracme veneta* (△) and *Cochlostoma henricae;* the latter has evolved to two different subspecies as a consequence of the disjunction of its area of distribution since the Pleistocene, *C.h. henricae* (▲) and *C.h. huettneri* (●) (from Marcuzzi, 1956).

soma slavum, Dactylophorosoma nivisatelles, Orotrechosoma cornuigerum (endemic), *Asandalum larii, Leptoiulus frigidarius, L. braueri, L. brentanum* (endemic), *L. riparius, L. tosanus, L. catascaphicus* (endemic), *L. oribates, Microiulus muscivagus, Ophiulus collaris* (endemic), *Cylindroulus meinerti, C. fulviceps, Ypsiloniulus grödensis, Leptophyllum karawankianum.*

Species limited to Western Alps are *Craspedosoma doriae jucundum, Oxydactylon tirolense cisalpinum, Microchordeuma voigti, Parachordeuma brolemanni, Crossosoma peyerimhoffi, Semiosoma devillei, S. bordei, Ceratosoma gallitorum, Brolemanneuma pectinifer, B. palmatum* and *B. furcatum, Floria tuberculata, Callipus foetidissimus gallicus, Polimicrodon uncinatum* and lastly *P. digitatum*. One species present on Western Alps and Appennines is *Glomeris romana.* An Appennine element in contrast is *Hirudisoma aprutianum* (Fig. 317).
Limited to Eastern Alps and Balkans (mainly the extreme north western portion): *Gervaisia noduligera, Asandalum pusillum, Leptoiulus marmoratus, Ypsiloniulus molybdinus.*

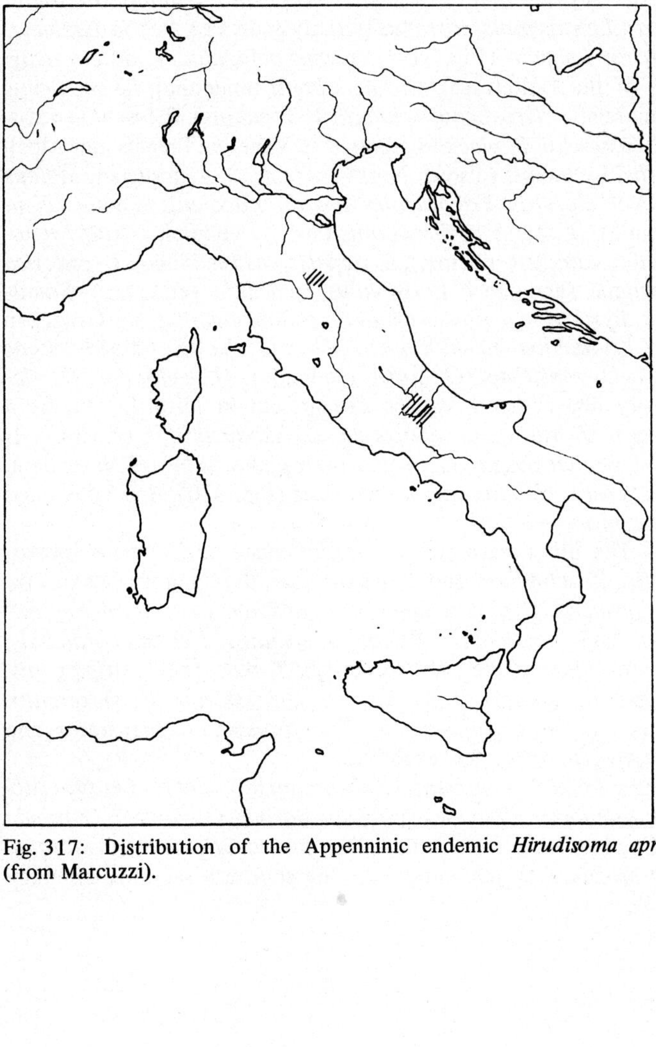

Fig. 317: Distribution of the Appenninic endemic *Hirudisoma aprutianum* (Myriapoda) (from Marcuzzi).

Pyrenees. – *Hirudisoma pyrenaeum* and *H. latum, Ommatoiulus haacheri, Vascoblaniulus cabidochei, Pireneosoma ribauti, P. digitatum, Chamaeosoma brolemanni, Vascosoma coiffati, Hypnosoma pallidum, H. exornatum* and *H. juberthieorum* and finally *Adenomeris.*

The beetles are the best known montane animals, not only in Europe, but all over the world. It is possible therefore to draw some very detailed distributional patterns which, as we have already said, represent only the present-day situation, with no reference or implication as to the origin of the species or their paleogeography.

In a very schematic, in a sense conventional way, we may recognize the following zoogeographical categories:

a) *Iberian Peninsula.* – The Carabids *Trechus bordei, T. distigma, Hygrocarabus melancholicus*; the Tenebrionids *Phylan foveicollis, P. indiscretus* (on the Sierra Nevada reaching 2000-3000 m), *Phylan foveolatus, P. collaris* and *P. mulsanti.*

b) *Pyrenees.* – The Carabids *Trechus bordei, T. distigma, Carabus punctatoauratus; Cechenius punctatulus, Byrrhus auromicans, B. nicolai, B. españoli,* etc.

c) *Alps.* 1: Eastern Alps: *Carabus linnei folgariacus, Carabus bertolinii* (Fig. 318), *C. alpestris, Leistus punctatissimus* (strictly endemic), *Nebria diaphana, Nebria hellwigi, Trechus longulus* (Fig. 319), *Trechus pallidulus, T. dolomitanus* (an endemic species of the Dolomites), *Duvalius breiti* (endemic), *Laemostenus schreibersi* (a 'Rassenkreis'), *Pterostichus schmidti* (= *cognatus*), *P. josephi* (= *pascuorum*, endemic), *P. schaschli, P. placidus* (limited to Venetian Prealps and Alps), *Abax eccheli, Abax beckenhaupti* (also in North Jugoslavia), *A. springeri, Molops stolzi, M. edurus, Amara alpestris, Tricotichnus knauthi, Euconnus longulus, Neuraphes coecus* (endemic), *Bythinus bothrophorus* (strictly endemic), *Ancyrophorus carnicus, Leptusa monticola* (endemic), *L. trapezicollis* (endemic), *L. cadorica* (endemic), *L. dolomitana* (endemic), *Boreaphilus melicharii* (endemic), *Stenus holdhausi* (endemic), *Byrrhus signatus, Aphodius schlumbergeri* (from Grigna to Venetian Prealps), *Chrysochloa melanocephala, Orestia electra, Otiorrhynchus dolomitae* (endemic), *O. sensitivus, O. fortis* (endemic), *O. cadoricus, O. globulus* (one of the very few endemic species also present in Tirol, i.e. far from the massifs de refuge), *O. varius, O. eremicola, O. subquadratus* (reaching as far west as Engadina), *O. montivagus, O. prolixus, Tropiphorus bertolinii* (endemic), *Dichotrachelus stierlini, Neoplinthus mucronatus* (Fig. 320), *Brachyodontus kraussi* and *Stomis rostratus.*

2: Western Alps. – The blind, cavernicolous *Bathysciinae* of the genus *Isereus, Royerella, Speodiaetus, Troglophyes* and *Troglodromus*; the Carabids *Carabus vagans, C. cenisius, C. fairmairei, C. putzeysianus, C. latreillanus* etc.; *Stomis elegans, Cychrus angulicollis* and *cordicollis, Leistus ovipennis, Nebria cordicollis, N. kochi, N. morula, N. laticollis*, etc.; *Trechus aubei, T. putzeysi, T. strigipennis*, etc.; *Duvalius ghidinii, D. canevai, D. gentilei, D. clairi, D. roberti, D. carantii, D. ochsi, D. cailloti*, etc.; *Laemosthenus alpinus, L. obtusus* and *L. angustatus*, and finally *Molops senilis*, to give only a few examples.

3. Entire alpine range: *Carabus depressus, Cephennium montanum, Leptusa subalpina* (south. Alps), *Adaliopsis alpina, Chrysochloa bifrons, C. cacaliae, C. frigida* (only south. Alps), *Crepidodera peirolerii, C. melanostoma, Otiorrhynchus foraminosus, O. salicis, O. frigidus, O. gemmatus* (on the southern sections generally

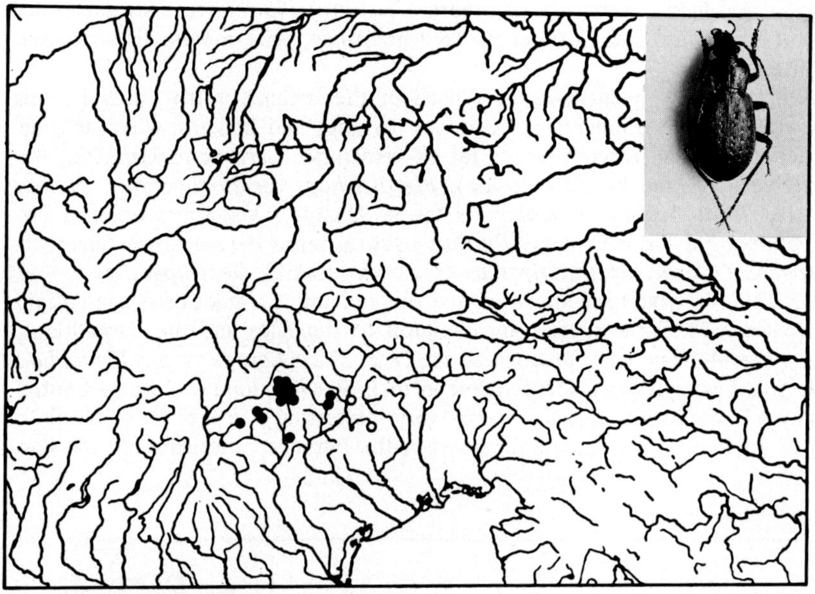

Fig. 318. Distribution of the endemic beetle *Carabus bertolinii*: the presence of the Piave River has brought about – after the formation of the species – its subdivision into two races, one east of Piave, one west.

Fig. 319. *Trechus longulus.*

amphigonic), *Liparus grappensis* (southern Alps only), *Plinthus tigratus, Euconnus claviger* (from Nice and l'Esterel up to Alto Adige, with a very discontinuous distribution).

d) *Balkans.* A very peculiar genus typical of the Balkans is the Carabid genus *Tapinopterus.* Some elements are common to Alps and Balkans, which in practice present no real discontinuity: let us remember *Carabus carinthiacus, Nebria germari* (as far as Herzegovina), *Pterostichus fasciatopunctatus, Molops austriacus, Trichotichnus laevicollis* (as far as Albania), *Neuraphes ornatus, Trimium emonae* (as far as Croatia), *Byrrhus gigas* (as far as Herzegovina), *Chrysochloa vittigera, Otiorrhynchus auricomus* (as far as Croatia), *Geotrupes alpinus.*

e) *Carpathians.* – We have already listed the endemic Carabids of this mountain range (page 462). Some species are common to the Alps and the Carpathians, which in the Miocene were practically one (Fig. 322): *Leistus piceus, Bembidion glaciale* (to Transsylvania), *Pterostichus jurinei, Byrrhus luniger* (also in Central Germany), *Phytodecta kaufmanni.* Some very interesting elements are common to the three great European mountain systems, the Pyrenees, the Alps and the Car-

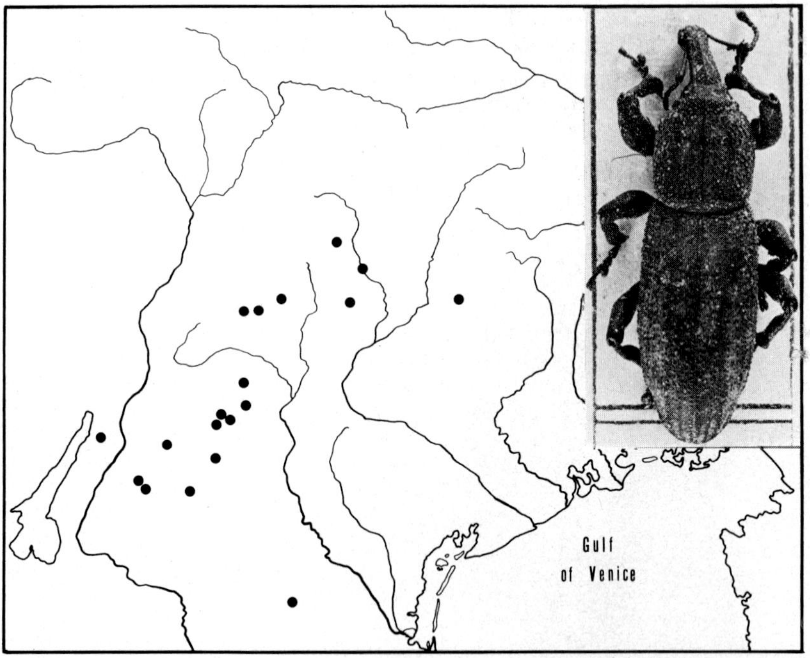

Fig. 320. Distribution of an endemic weevil, *Neoplinthus mucronatus*, limited to the 'massifs de réfuge' of a small part of the Southeastern Alps. The habits of these species – as those of several other weevils – have prevented a wider distribution of the species both before and after the Glacial Period. Its origin must be very old, since the species is present west of the river Adige and east of the river Piave, so that it must have preceded the formation of the present-day courses of these rivers. (Most of the localities shown were communicated to the author by Dr G. Osella of the Natural History Museum, Verona.)

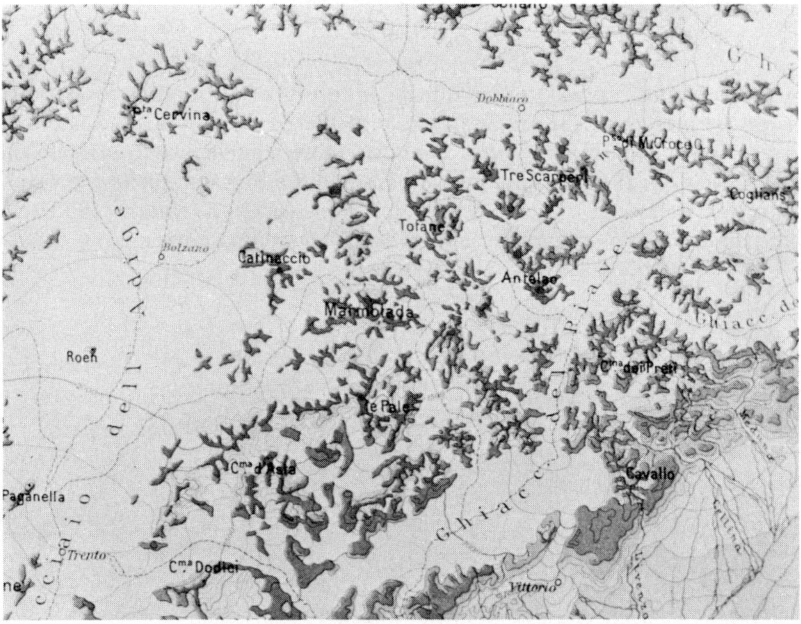

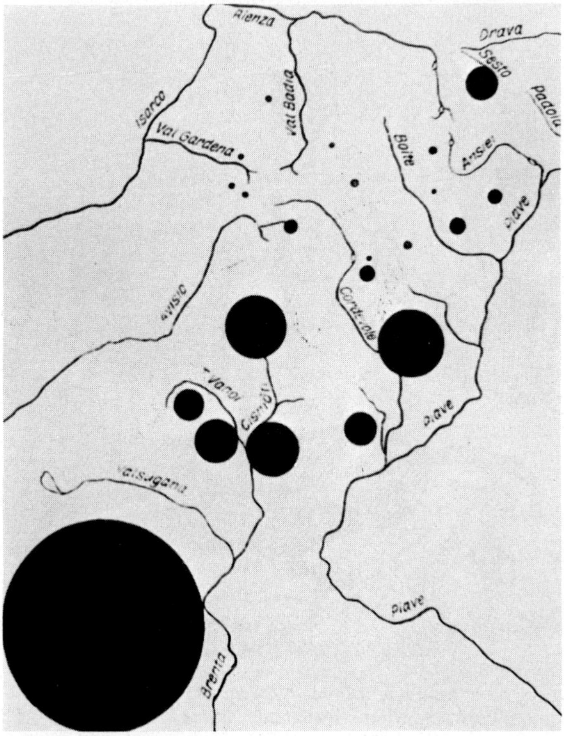

Fig. 321 bottom, Number of endemic coleoptera of the various localities of the Dolomites and Lessini (South of Valsugana) represented by means of circles of different diametres. A close relationship is recognizable with the presence of areas free from ice during the Pleistocene. Top, Extension of the glaciers in the Dolomites and nearby territories during the Pleistocene glaciation. Plant and animal life could only survive in mountains free from ice, the so-called massifs de refuge (from Atlante Fisico-Economico d'Italia, Touring Club Italiano, published with the permission of the latter.

pathians: e.g. *Leistus nitidus*, a Carabid living in forests of the montane belt. Some species can be present in the same systems plus the Balkans: this is the case with the Chrysomelid beetle *Chrysochloa virgulata.* More abundant of course are the elements present on the Alps, the Carpathians and the Balkans, due to the virtual continuity (as today) of these three systems: *Pterostichus unctulatus* (Sudetes), *Euconnus motschulskii, Pedilophorus auratus, Chrysochloa intricata* and *C. variabilis.*

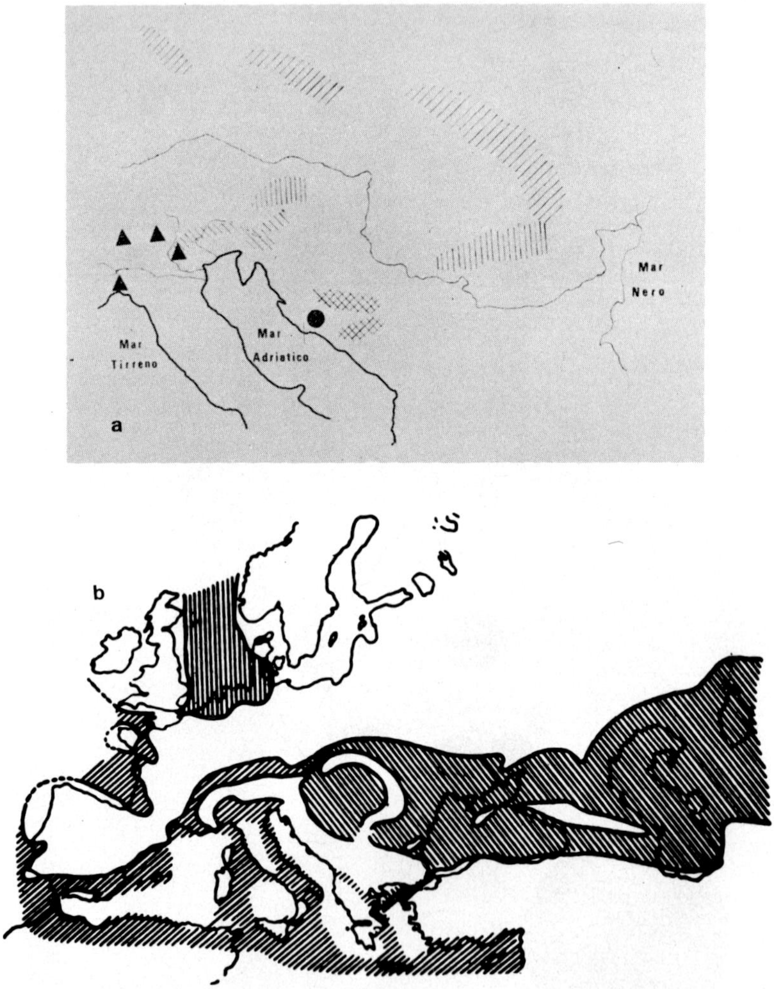

Fig. 322. Distribution of a montane beetle, *Chrysomela haemisphaerica*, a 'Rassenkreis', subdivided into many subspecies, whose systematic status is still obscure. For comparison a geological reconstruction of Miocenic Europe (from Gignoux) is added. (Some data on the geographical distribution of *C. haemisphaerica* are due thanks to a personal communication by Prof. S. Ruffo.)

f) *Appennines.* – The Tenebrionids *Colpotus strigosus* (Fig. 323), *Asida pirazzolii.* etc.
Some species are common to the Pyrenees, the Alps, the Balkans and Appennines: *Otiorrhynchus anthracinus* (= *helveticus* Boh.).
More abundant of course are the elements common to the Alps and the Appennines: *Stomis roccai, Pterostichus cristatus*, a 'Rassenkreis'; one species (*Pterostichus morio*) is present furthermore on Sudetes and the Carpathians.
Lastly some elements can be common to the Alps, the Pyrenees, the Carpathians and the Appennines: this is the case with a coprophagous species, with a strong means of dispersal whose wide distribution can only be of recent origin and of scarce biogeographical significance, *Aphodius mixtus.*
g) *Corsica* and *Sardinia.*
Corsica. – Some endemic beetles live in the lowlands, reaching only 500-600 m: *Aphiliops aubei, Pselaphus kiesenwetteri, Dasytes grenieri, Otiorrhynchus grenieri;* four elements are exclusive to the alpine belt: they are *Trechus varendorffi, Amara bickhardti, Atheta varendorffi* and *Otiorrhynchus corsicus.* Some Corsican ele-

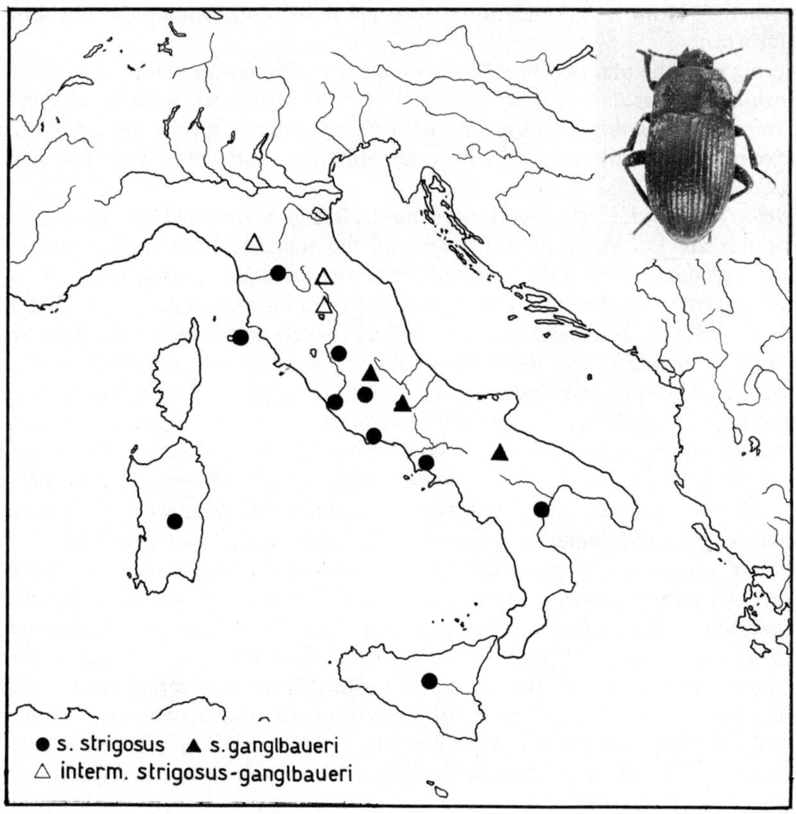

Fig. 323. Distribution of *Colpotus strigosus.*

ments exhibit some affinities with Pyrenean taxa: the *Pselaphidae Mirus permirus* is a relative of *M. lavagnei* living in Hérault; the myrmecophilous *Silphidae Attumbra josephinae* is common to Spain, Southern France (Banyuls sur Mer) and Corsica. All these elements date from a period in which all these territories were joined. In Corsica, as in Sardinia, there are some recent endemics (neo-endemics) among the Tenebrionid genus *Asida*, whose systematics are not quite clear.
Taking only the planticolous, often corticicolous, Tenebrionids, we will mention here *Helops coeruleus, Probaticus ebeninus, P. anthrax, P. superbus* (endemic), *Cylindronotus genei, C. abeillei* (endemic), *C. dryadophilus, C. assimilis, Nephodinus corsicus* (endemic) and *N. metallescens.* The latter could have originated when Corsica was still joined to Sardinia.
Sardinia. – We may mention here the Carabid *Duvalius sardous*, a relative of some species living in Catalonia, Cevennes and Provence; the *Scydmaenidae Eudesis aglena*, a relative of a species living in the Eastern Pyrenees, the *Pselaphidae* genus Trogasteropsis, common to Sardinia and Catalonia; the genus *Speonomus* (*Silphidae*) common to the Pyrenees and Sardinia, where it is represented by three species, analogous to the Diplopod genus *Typhloblaniulus,* a Pyrenean taxon with one Sardinian species. They are all very old elements, representative of the so-called 'Tyrrhenis'. More recent endemic species are to be found among the Tenebrionid genus *Asida.*
Considering only the planticolous Tenebrionids, we will mention *Probaticus ebeninus, P. anthrax, P. gibbithorax, P. sphaericollis, Cylindronotus aemulus, C. genei, Nephodinus metallescens, Allardius sardiniensis*, an endemic species belonging to a genus exclusive to Sardinia and Sicily, originating therefore when these two territories were united.
h. Caucasus. – Besides the beetles we have already mentioned (p. 463), some Tenebrionids are known from this chain and the adjacent mountainous regions: *Dendarus extensus*, from Azherbaidjan, *Dendarus crenulatus*, from the same region and Armenia, *Pedinus circassicus* (Fig. 315) from Caucasus (M. Fišt, M. Ošten, M. Bziš, M. Ačišcho, M. Abago etc.), as well as Georgia and Turkey (Mt. Bozdag).

We want to mention also some planticolous, generally corticicolous, Tenebrionids, such as *Hypophloeus longulus, H. basalis* (endemic), *Neatus subaequalis, Belopus tibialis, B. sulcatus, B. araxides* (endemic), *Cryphaeus cornutus, Catomus tagenioides* (endemic), *Hedyphanes dejeani* (endemic), *Helops coeruleus steveni* (a localized race), *Probaticus dorsalis, P. viduus* (endemic), *P. arboreus, P. pullus* (endemic), *Stenohelops carinatus* (endemic), *Cylindronotus femoratus, C. bellator* (endemic), *C. gibbosus* (endemic), *C. gibbicollis* (endemic), *C. funestus* (endemic), *C. glorificus* (endemic), *C. gloriosus, C. bosphoranus, C. faldermanni, C. diteras* (endemic), *C. adriani* (endemic), *C. brevicollis, C. vexator* (endemic), *C. sareptanus, Adelphinus ordubadensis* (endemic), and finally *Laena starcki, L. justinae, L. lederi, L. quadricollis, L. baudii* and *L. piligera*; all of them, with the exception of *L. starcki*, are endemic to the Caucasus. From this we may safely assume that this mountainous chain is a centre of speciation for both *Cylindronotus* and, secondarily, *Laena.* One species, *Aphodius montanus*, winged and therefore capable of active dispersal, is present in the Caucasus, as well as in the Alps, Carpathians and Balkans.
i. Crete. – Several Tenebrionids are present in the montane centre of the island,

whose maximum height is 2497 m (Idhi Mt.). We want to mention here only some planticolous, often corticicolous, Tenebrionids, namely *Helops glabriventris, Probaticus granipennis, P. euboeicus, Cylindronotus cretensis* (endemic), *Euboeus mimonti* (belonging to a genus limited to Euboea and Crete, dating from the era in which these two territories were joined), *Laena oertzeni* (endemic) and *L. ferruginea.*

j. Central European mountains (Ercynian species *sensu lato*): *Cychrus angustatus, C. attenuatus, Carabus auronitens, C. coriaceus, Abax carinatus, Amara montivaga, A. schimperi, A. nitida, A. cursitans, Euconnus carinthiacus, E. styriacus, E. rutilipennis, Bythinus glabricollis* and *B. femoratus, Sphaerosoma seidlitzi, Otiorrhynchus inflatus, O. vestitus* (very limited distribution), *O. uncinatus, O. subdentatus, O. fraxini, O. chrysocomus, Tropiphorus tomentosus, Liparus dirus, L. glabrirostris.* All these species probably find the same environmental conditions in the Alps and in the lowlands or hilly regions of Central Europe. They have in general been able to occupy, after the Pleistocene, the areas formerly occupied by ice (re-immigrants).

Some elements are common to the Pyrenees, Alps and Central Europe, for instance the weevil *Otiorrhynchus morio*, and the Carabid *Nebria castanea* present also on the Vosges and Jura.

Some elements, however, are present only in Central Europe, in the Alps, Carpathians and Balkans, i.e. in a relatively continuous and homogeneous region: this is the case with the *Chrysomelidae Timarcha metallica* (in Central Europe only in the mountains).

A few species can be found in the Alps, Central Europe and Balkans, up to Banate: *Otiorrhynchus geniculatus* and *O. auricapillus.*

Other species, probably of different origin, are present today in the Alps, Centro-European mountains, Balkans and Lesser Asia: this is the case with the weevil *Otiorrhynchus pauxillus.* This distribution probably dates back to the era in which Lesser Asia was united with the Balkan Peninsula.

Finally, a few species may be present in Central Europe, the Alps, Appennines, Balkans and Caucasus, as for instance *Otiorrhynchus alpicola.*

A category of scant zoogeographical interest is the European one, comprised of animals common to most of Europe, though generally restricted to the montane environment: there are *Cychrus caraboides, Pterostichus strenuus, Euconnus fimetarius, Euplectus karstenii, Pselaphus heisei, Otiorrhynchus rugosostriatus, O. singularis* and *Tropiphorus obtusus.* Some of these species can also reach very high altitudes, for instance *Otiorrhynchus singularis* on the Pyrenees (p. 458). Let us list some more examples of montane beetles also present in the Vosges: one of them, *Chrysochloa viridis*, is present in the Vosges, Carpathians, Balkans and Alps; another, *Plinthus megerlei*, is present in the Vosges, Carpathians, Alps and Pyrenees. This distribution is probably very ancient and it deserves a proper study.

Another interesting lapidicolous beetle is the Tenebrionid *Cylindronotus convexus*, probably a very old preglacial element, found from Switzerland to Hungary, present both in northern and southern Alps, Bavaria, Vorarlberg, and also in the central, crystalline, portion of Alps as in the Tyrol, west of Silltal. In some parts of the Alps it is present on the mountains previously covered by ice rather than on the massifs de refuge, as is the rule for many beetles (Giogo di S. Vigilio,

Campo Magno near Madonna di Campiglio, Alpe di Siusi, Pizzo Coca) so that it must be considered as a post-glacial re-immigrant. Of course no explanation is today available for this unusual type of distribution.

Several European montane animals (both vertebrates and invertebrates) exhibit a peculiar geographical distribution, which – differing from that which we have seen so far – is not limited to mountains. They belong to the so-called boreo-alpine category, of which we have already seen some examples when discussing vertebrates (*Lepus timidus, Lagopus mutus, Turdus torquatus* and a few more). We have also seen that these animals are not necessarily montane, since in the north they often live in the lowlands; furthermore many of them are not limited to Europe, but are also present in Northern Asia and, in some instances, in North America (circum-polar alpine species). We will also find this kind of distribution among freshwater animals. We should point out in this connection that the boreo-alpine category only has a biogeographical value, and not an ecological one, since the autecology of the different species is extremely varied. Possibly differences in ecology could be detected if we could consider separately the geographical types we have already mentioned, such as boreo-alpine, boreo-altaic, boreo-mongolic etc. which to the best of my knowledge has never been done before.

We will confine ourselves now to mentioning some examples of boreo-alpine invertebrates, arbitrarily chosen from several groups, since it would be impossible to enumerate them all.

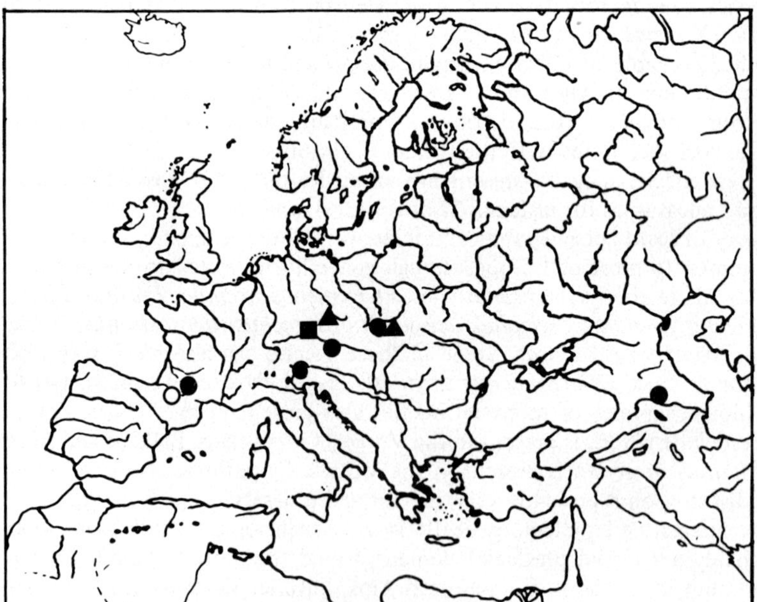

Fig. 324. *Hypogastrura crassegranulata*, a 'Rassenkreis' present also on the Caucasus and the Pyrenees, both emerged during the Miocene.

Among the spiders there is *Erigone remota* (Novaja Zemlja, Siberia, Alps and Pirin, Bulgaria); among mites a boreo-alpine distribution can be ascribed to *Lealaps clethrionomydes*, associated with the bank vole (cf. p. 429) and *Histriomyssus gudaricus*, associated with the snow vole and bank vole. A boreo-alpine or circumboreal distribution is possibly that of the Oribatid mite, *Tectocepheus sarekensis* and of some Collembolan, such as *Anurophorus binoculatus, Folsomia sexoculata, F. diplophthalma* and *Onychiurus procampatus.*

Among dragon-flies there is *Somatochlora alpestris*, present in Fennoscandia, Northern USSR, Hokaido and the mountains of Central Europe, as far south as Bulgaria. Among hoppers we should mention here at least *Aeropedellus variopedatus*, present in Northern Eurasia, Central Asia up to Kamchatka and Alaska, the Caucasus, Alps, Bulgaria (Rila Mt.) and the Appennines. The origin of this element (a typical Angarian element) must be very old, certainly pre-glacial. *Melanoplus frigidus* (= *Podisma frigida*) lives in Northern Eurasia (eastwards to Sakalin) the Alps and Bulgaria. Very interesting is the riparian bug *Salda littoralis*, which lives in all northern Holarctis, as well as in the high mountains of Central and Southern Europe. In Northern Europe it is a halophil species.

Among the many Lepidopterous insects we want to list for their geographical peculiarity there are *Erebia epiphron*, in the north limited to the British Isles; *Arctia flava*, in the south limited to the Alps and Rila (which proves to be a very old and conservative territory), *Hadena maillardi*, in the north limited to Fennoscandia; *Gnophos dilucidaria*, in the north only in Fennoscandia, in the south only in the Alps, Carpathians and Balkan mountains, and finally *Orenaja alpestralis*, with a rather discontinuous distribution (Scandinavia, Ural Mts., Pyrenees, Alps, Carpathians, Korab and Rila), possibly because of some ecological reasons (cold-loving species?).

Among Hymenoptera, *Bombus lapponicus* exhibits a circum-polar distribution. In the mountains of the Balkan peninsula this insect is represented by a peculiar relict form, *B.l. scandinavicus m. hypsophilus*, limited to the mountainous massifs (Rila, 2000 m, Ossogovska Planina, 1600-1800 m) with no possibility of interchange. The genetics of this geographical morpha deserves particular study.

The best known boreo-alpine insects are, without any doubt, beetles. We will confine ourselves here again to listing only some interesting cases, such as *Nebria gyllenhali*, a petrophilous element, extending to coniferous and alpine belts; *Patrobus assimilis*, soil-dwelling, alpine or present in the coniferous belt confined to moorlands; *Amara erratica* and *A. quenseli*, a soil living, alpine and subalpine species; *Arpedium brachypterum*, a soil living, alpine species; *Geodromicus globulicollis*, a soil living and riparian, alpine species; *Atheta islandica*, riparian and soil living, alpine or even subalpine but on moorlands; *Silpha tyrolensis*, soil living species, both alpine and subalpine; *Neuraphes coronatus*, a soil living, subalpine element; *Corymbites rugosus*, soil living, alpine, in the Alps up to 3000 m (cf. chapter about taiga and tundra); *Hypnoidus rivularius*, soil living, subalpine and alpine, and *H. hyperboreus*, in the Alps limited to the Dolomites (cf. chapter on tundra). The *Cerambycidae Evodimus interrogationis* a planticolous, subalpine species, present in Northern Europe, Alps, Tatra, Caucasus and Siberia; *E. borealis*, present in Northern Europe, Siberia and the Alps, and *E. variabilis*, a siberian species present in the Alps exclusively on Col de Vars in the Western Alps, must also be considered as boreo-alpine elements. The history of the latter would prove to be extremely interesting. Other species are the *Cerambycidae Acmaeops septentrionis*, planticolous, restricted to pines and larches, limited to the coniferous belt, *Phytodecta affinis*, planticolous (on *Salix retusa* and *S. herbacea*) a typically stenophagous species, alpine, *Otiorrhynchus dubius*, soil living, both alpine and subalpine, parthenogenetic in the northern area, amphigonic in the southern: this is consistent with our knowledge about the much greater ability of parthenogenetic races to invade areas previously covered by the Pleistocene ice sheet, although the reason for this ability is not yet known. *Otiorrhynchus salicis*, soil living, subalpine, parthenogenetic in the northern area, in Switzerland, the Sudetes and the Carpathians, amphigonic in the Southern Alps and in Würtenberg. The two forms can be considered true biological races, since in the females there are no appreciable morphological differences.

The origin of the boreo-alpine type of distribution is extremely complex and differs according to the different taxa or even to the different species, since both the time and the place in which they originated are varied. For many of them the retreat of the Pleistocene ice sheet, both towards the north and towards the Alps, will have been sufficient to determine the present distribution, but for many species a preglacial, often asiatic, origin must be postulated. For this reason one speaks now of boreo-altaic, boreo-mongolic, circum-polar distributions, etc.

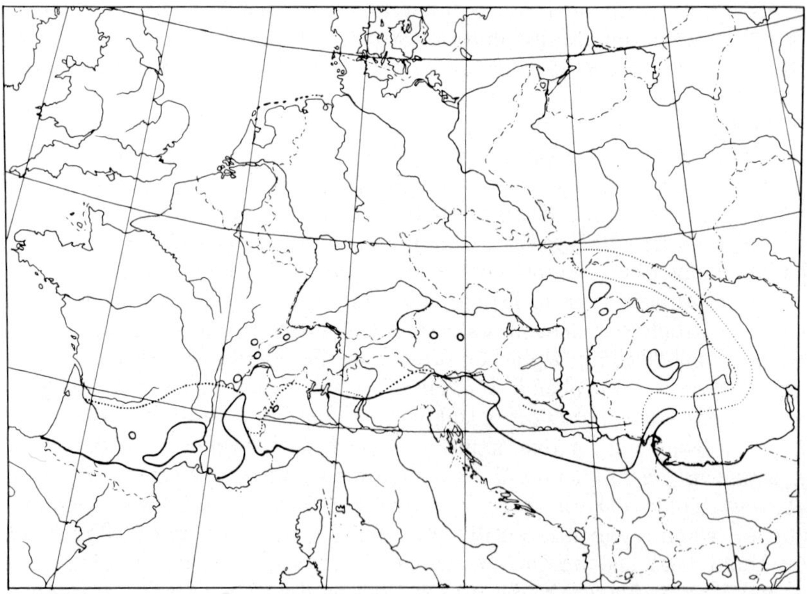

Fig. 325a. The effect of the glaciation on the animal life in the hypogeous dominion. Continuous line, northern limit of the distribution of the true cavernicolous insects; dotted line, northern limit of distribution of the blind coleopterous insects of the montane forests. o = small areas of distribution of some cavernicolous animals north of the above-mentioned limit (from Holdhaus, 1954).

Soil fauna

The soil fauna is extremely varied according to substrate, soil, vegetation and history of the region involved. As a consequence of the poor means of locomotion of true soil animals, they are more confined to their micro-habitat than the epigeal

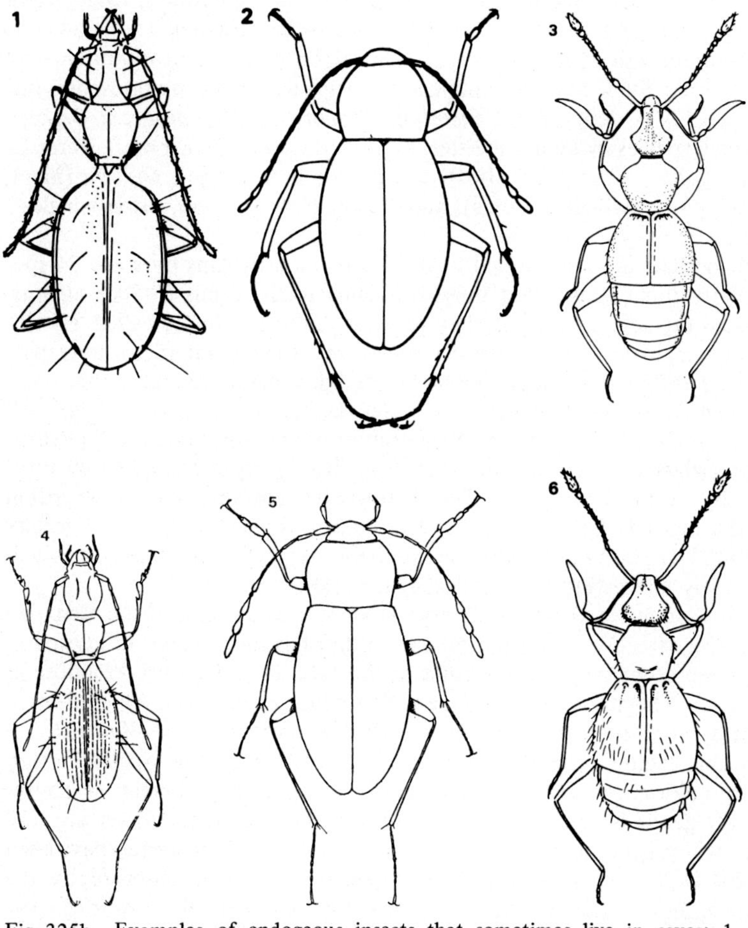

Fig. 325b. Examples of endogeous insects that sometimes live in caves: 1. the Carabid *Orotrechus müllerianus*, living in the soil, under stones, as well as in caves of the karst region near Trieste; 2. the Silphid *Orostygia* sp., a genus present both in the soil of beach forests under deeply earthed stones, and in caves of the Southeastern Pre-Alps; 3. the Pselaphid *Lophobythus cristatus*, a microphthalmus, endogeous species living in the opening in caves in Ariège (Southwestern France). True cavernicolous insects: 4. the Carabid *Anophthalmus hirtus*, present in caves in Slovenija and in the karst region near Trieste; 5. the Silphid *Aphaobius latitarsis*, from a cave near Montenero d'Idria (Idrija, Jugoslavia); 6. the Pselaphid *Amaurobythus remyi*, from a cave in Northern Corsica, near Pietrocorbara (some species of this genus are completely eyeless). (1,2,4 and 5 from Müller, 3 and 6 from Jeannel; 1, 3 and 6 redrawn by C. Friso).

ones, so that in the soil fauna a very clear difference between the formerly ice-covered areas and those permanently free from ice is recognizable. In this soil animals are similar to cavernicolous ones (Fig. 325). For all these reasons many endemic species can be found among soil animals; some relict, disjunctive types of distribution are also present (*Koenenia, Campodeidae, Teneriffiidae,* etc.).

The animal groups which constitute the bulk of the soil fauna of European mountains are *Protozoa,* Turbellarians or flat worms, Nematodes, *Oligochaeta* (both *Enchytraeidae* and *Lumbricidae*), *Hirudinea* or leeches, and Tardigrades, as well as Arthropods and Molluscs (snails). The Arthropods in our mountains are mainly represented by woodlice, millipedes, centipedes, mites, pseudoscorpions, *Protura, Campodeidae, Japygidae, Collembola, Psocoptera, Thysanoptera, Orthoptera* (only small species living in the litter, such as *Gryllulus, Tetrix* etc.), *Heteroptera* and *Homoptera, Coleoptera* (both as adults and as larvae), *Diptera* (idem, often represented by wingless forms), and finally *Hymenoptera,* often wingless, generally parasites of other insects.

The Turbellarians are represented in the North Eastern Alps by some 35 species, particularly interesting being those belonging to the genus *Rhynchodemus.* The Nematodes are very numerous; they generally belong to the so-called 'Hydrobios' (microscopical species) and can live therefore either in aquatic or terrestrial environments. *Hirudinea* are represented in the Alps mainly by the rather common land leech *Xerobdella lecomtei*, restricted to beech forests, very hygrophilous, a relict, preglacial element. The Nematomorphs are represented in the North Eastern Alps (whose terrestrial fauna has been thoroughly investigated by Prof. von Franz) by some 19 species. The Tardigrades are very numerous, often present at the highest altitudes: on the Marmolada, for instance, several species have been found at 3000 m. Woodlice are probably very ancient organisms, which exhibit extremely interesting geographical distributions.

The number of mites is extraordinary: in only a small area of the Venetian Prealps, near Recoaro (M. Spitz) the present author has found some 130 Oribatid mites, which represent only a proportion of the total of this group. Pseudoscorpions are represented mainly by the genera *Chthonius, Roncus* and *Neobisium*; in the Dolomites more than a dozen of species are present, some of which are endemic to this region. The Palpigrada are represented in the North Eastern Alps only by *Koenenia austriaca*, a cavernicolous species, whereas in Southern Europe the genus is equally endogeous or cavernicolous. The *Protura* are very little known: whereas in the whole of North Eastern Alps only 9 species have been found, in the Dolomites (Cadore) some 9 species have been observed by the present author, and in the Appenines at only one locally in Abruzzo 4 species.

Campodeidae are represented by several species, probably very old and often with a relict, disjunctive area of distribution: there are *Campodea fragilis*, *C. suensoni, Plusiocampa exsulans, P. caprai*, the latter cavernicolous in the lowlands, endogeous in the mountains, very much the same as some other soil invertebrates. *Japygidae* are very rare in European mountains: on the North Eastern Alps only two species have so far been reported, both living rather at very low altitudes; in the Pyrenees, however, the present author has found some Japygids at La Plagne at some 1500 m.

Collembolans are almost as numerous as Acari: to mention again the region of

Recoaro-M.Spitz, some 50 species have been found there by the present author. Some of them are of very great zoogeographical interest. *Psocoptera* are rather rare in the soil fauna of our mountains: in the Dolomites the species *Liposcelis* sp. prope *simulans* can be found; in the Apennines (Abruzzo, at 1500 m) no *Psocoptera* have been observed. Characteristic soil *Heteroptera* are *Tingidae*, small delicate creatures, almost transparent, living in mosses and in the upper layer of the soil. *Homoptera* are represented both by purely soil organisms, such as the *Coccidae* or scale insects *Orthezia cataphracta*, and by some elements which pass part of their life cycle in the soil, part on plants (the *Aphidoidea Phylloxerini*).

Coleoptera, Diptera and *Hymenoptera* account for practically all the soil holometabolic insects (Fig. 326).

Soil gastropods belong to very different families, both of *Basommatophora* and *Stylommatophora,* and to very different alimentary and ecological categories. *Carychium minimum* belongs to an aquatic group (*Basommatophora*) but it can be found in very humid environments, both in meadows and in forests. It feeds on algae and decaying plants. A typical soil gastropod is *Caecilioides acicula*, a blind, unpigmented species living generally only in the deeper layers of the soil, up to 40 cm; it feeds mainly on fungal mycelia. Other species closely associated with the soil are *Vallonia pulchella* and *Vallonia costata,* feeding on decaying vegetable matter. Some carnivorous molluscs, more or less restricted to the soil, at least as far as their feeding is concerned, are *Vitrina* and *Daudebardia*, which prey on *Enchytraeidae*, whereas *Testacella haliotidea* and *Glandina algira* feed on earthworms. Finally, some small molluscs, which always stay in contact with the soil and therefore belong to the soil fauna sensu lato, are *Pyramidula rupestris, Punctum pygmaeum* (p. 143) and the very peculiar *Acanthinula aculeata* (Fig. 109a).

Obviously, the abundance of each zoological group in the soil fauna varies greatly from one locality to another, as a result of environmental factors which favour one group rather than another, and perhaps also because of the history of the biotope, which has allowed the arrival of, and successful colonization by,

Fig. 326. Example of a weevel living at high altitudes on the Dolomites, *Otiorrhynchus montivagus.* Left, larva; middle, pupa; right, adult.

SPRUCE FOREST

No of specimens and biomass of soil fauna in a locality in the Dolomites (Borca di Cadore). The biomass of Nematodes, mites and springtails is expressed in μg, all the others in mg

	spring				summer				autumn				winter			
	litter		soil		litter		soil		litter		soil		litter		soil	
	No	b.	No	b.	No	b.	No	b.	No	b.	No	b.	No	b.	No	b.
Nematodes	–	–	1.5	1.5	0.5	0.5	–	–	0.5	0.5						
Enchytraeidae	3	0.6	3	0.6	5	1			0.5	0.1			0.6	0.12	1	0.2
Lumbricidae																
Isopods																
Araneae	0.5	2	0.5	2					0.5	2						
Pseudoscorp.	0.5															
Acari	1951	19510	218	2180	1066	10660	11	110	1652	16520	70	700	1014	10140	437	4370
Pauropods	2	1	0.5	0.25	1	0.5					0.5	0.25				
Diplopods					0.5	13.3										
Symphyla	3.5	1.75	0.5	0.25	2	1					2.5	1.25			0.3	0.15
Chilopods	1	16	0.5	8	2	32					1	16				
Protura	14	90.4			4	26.4							0.3	1.98	1	6.6
Collembola	588.5	7062	167	2004	452	5424	5	60	202	2430	21.5	258	99	1191	232	2792
Psocoptera																
Thysanoptera					0.5	3.3										
Homoptera									1	6.6			0.6	3.96	0.3	1.98
L. Diptera	16	160	1.5	15	9	90			1.5	15	1	10	11.3	113	1.7	17
L. Coleoptera	4.5	67.5	2.5	37.5	5	75			1.5	22.5	0.5	7.5	1	15	1.7	25.5
L. Lepidoptera																
Diptera . . .	0.5	3.3							0.5	3.3			0.6	3.96		
Coleoptera . . .					0.5	3.3									0.3	1.88
Hymenoptera . .			0.5	3.3												

MIXED FOREST

No of specimens and biomass of soil fauna in a locality in the Dolomites (Borca di Cadore). The biomass of Nematodes, mites and springtails is expressed in μg, that of the other groups in mg

	spring				summer				autumn				winter			
	litter		soil		litter		soil		litter		soil		litter		soil	
	No	b.	No	b.	No	b.	No	b.	No	b.	No	b.	No	b.	No	b.
Nematodes	0.5	0.5	7	7	2.5	2.5	1.5	1.5	0.5	0.5	1.5	1.5			6.3	6.3
Enchytraeidae	11.5	2.3	17	3.4	7	1.4	0.5	1.0	5	1	5	1			2	0.4
Lumbricidae					1.5	750	0.5	250	2.5	1250						
Isopods	2	20	0.5	5	2	20										
Araneae	0.5	2	1	4									0.3	1.2		
Pseudoscorp.	1.5				1.5				1		0.5		0.3			
Acari . . .	1234	12340	1100	11000	980.5	9805	487.5	4875	1105	11050	350.5	3505	600.3	6003	209	2090
Pauropods	3	1.5	18	9	3	1.5	2	1	1	0.5	–	–	3	1.5	5	2.5
Diplopods	0.5	13.3	1	26.6	1.5	39.9	1	26.6							2.6	69.1
Symphyla	6	3	13.5	6.75	2.5	1.25	2	1	4	2	1	0.5	1.3	0.65	2.3	1.15
Chilopods			2,5	40	4.5	72	0.5	8							0.3	48
Protura . . .	2	13.2	16.5	108.9	53.5	353	22	145	8	52.8	7.5	49.5	1	6.6	10.3	68
Collembols . . .	396	4752	943	1306.8	468	5616	127	1524	275	3300	130.5	1566	123.3	1479.6	98.6	1183.2
Psocoptera											1	6.6				
Thysanoptera											1.5	9.9				
Homoptera			3	19.8												
L. Diptera . .	13	130	14	140	3	30	2.5	25	6.5	65	4	40	5.6	56	4	40
L. Coleoptera	1	15	8	120	3.5	28	0.5	7.5	5	75	2	30	2	30	1	15
L. Lepidopt.							0.5									
Diptera	0.5	3.3	1	6.6	1	6.6	0.5	3.3								
Coleoptera			1,5	9.9							0.5	3.3				
Hymenoptera					1	6.6	0.5	3.3								
Mollusca . . .					1	200										

some particular taxa rather than others. A thorough study of the biocenological aspect of montane soil fauna and, more particularly, of its origin is, however, still lacking. Fig. 327 shows the variation in the faunistic composition of two fairly close stations on the Venetian Prealps (Cansiglio and M. Spitz).

Still more varied is the number of individuals (and relative biomass) of the soil fauna in the different parts of European mountains. Just in the Eastern Alps, the present author has found some values (N° of specimens per 500 cc of soil) as low as 144.8 (on the Marmolada at 2600 m) and as high as 7426 (at Geralba near S. Vito di Cadore, in the litter of a spruce forest at 1100 m). Of course the difference can be still greater.

The following tables represent the composition of the soil fauna in two different but very close biotopes (a spruce forest and a mixed one, with spruce plus beech) in Cadore (Dolomites). A richer and more varied fauna is evident in the mixed forest, where a mull is present, differing from the spruce forest, where the humus is a moder (Pag. 482-3).

Figs. 329a-c represents synthetically the main quantitative features of the soil fauna of three regions of the Eastern Alps and Prealps, differing in altitude as well

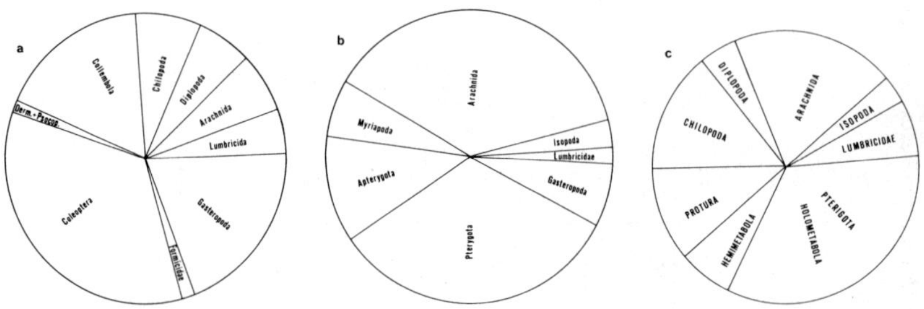

Fig. 327. a. No. of species of the different groups (*Acari* not included) on the Cansiglio. b. No. of species of the different groups in the zone Recoaro M. Spitz. c. Composition of the soil fauna of Cadore (Dolomites, Southeastern Alps).

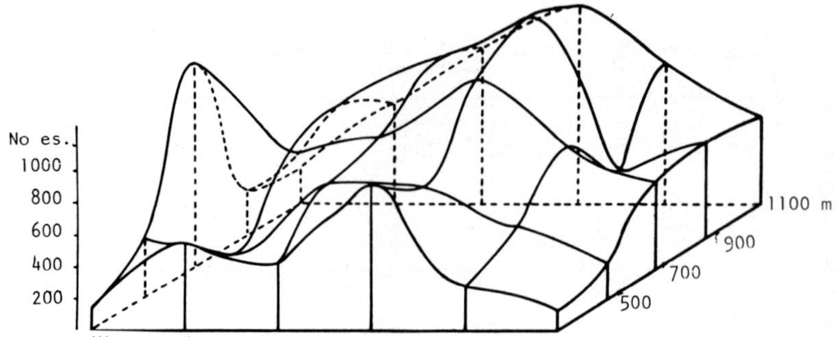

Fig. 328. Soil fauna of the Recoaro M. Spitz region (Venetian Pre-Alps). No. of specimens of soil mites according to season and altitude (the collinar belt is also included).

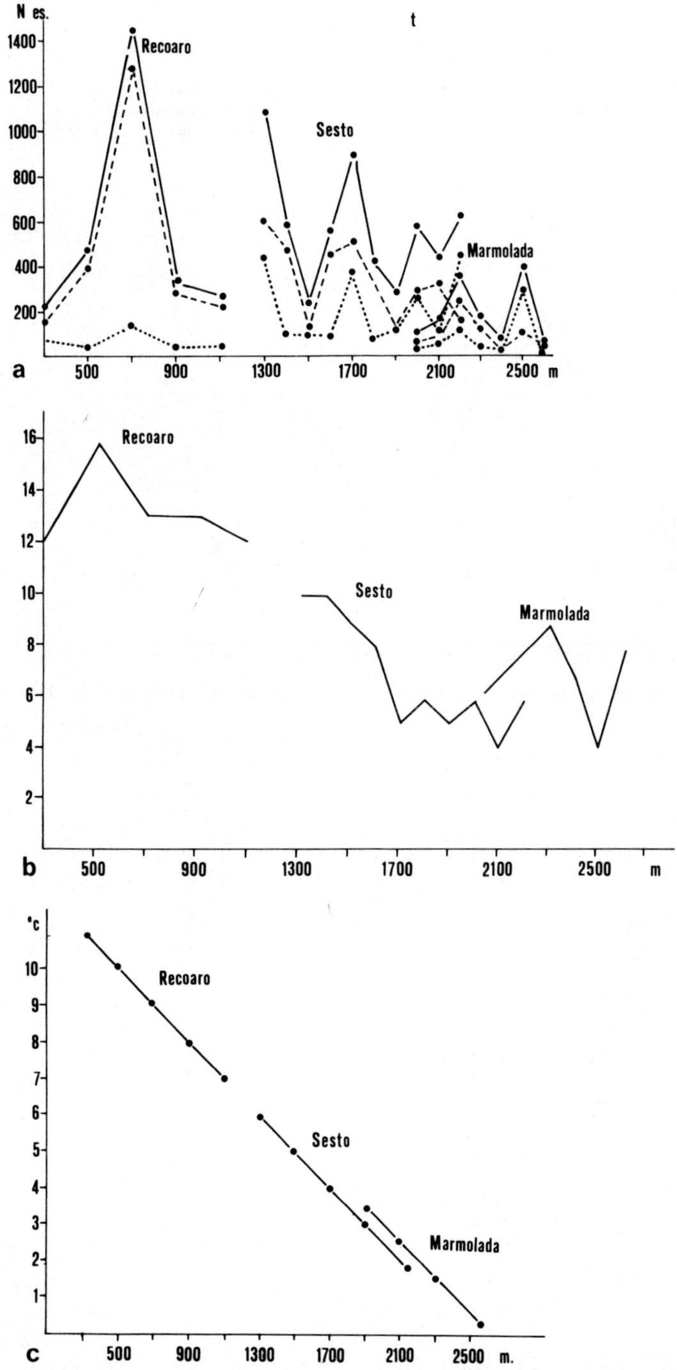

Fig. 329. Comparison (qualitative as well as quantitative) of the soil fauna of three different localities, which differ in mean temperature (dependent on latitude and altitude): Recoaro (Venetian Pre-Alps), Marmolada (Central Dolomites) and Sesto (or Sexten) in the northern Dolomites). a. density of soil fauna (No. of specimens/500 cc of soil); continuous line, total fauna; broken line, Acari; dotted line, Collembols; b. No. of zoological groups in the three biotopes; c. mean annual temperature of the three biotopes.

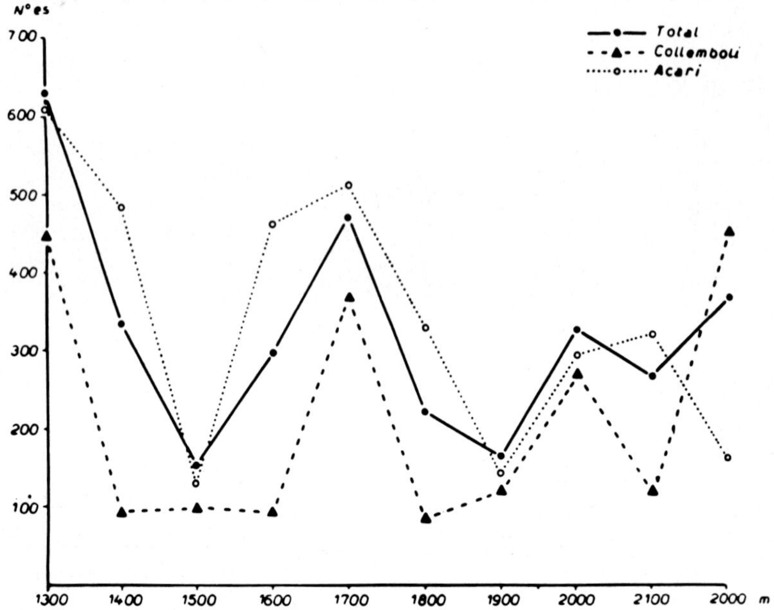

Fig. 330a. No. of ex/500 cc of soil in the Sesto Dolomites at different altitudes (from Marcuzzi, 1964).

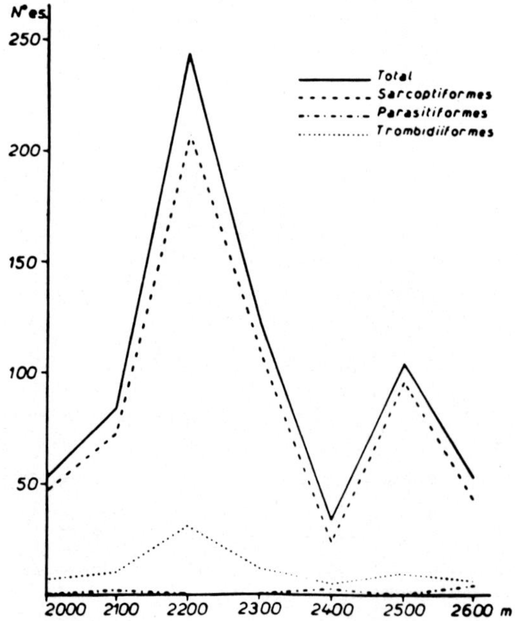

Fig. 330b. Number of mites according to altitude (No. of ex/500 cc of soil) on the Marmolada (from Marcuzzi, 1959).

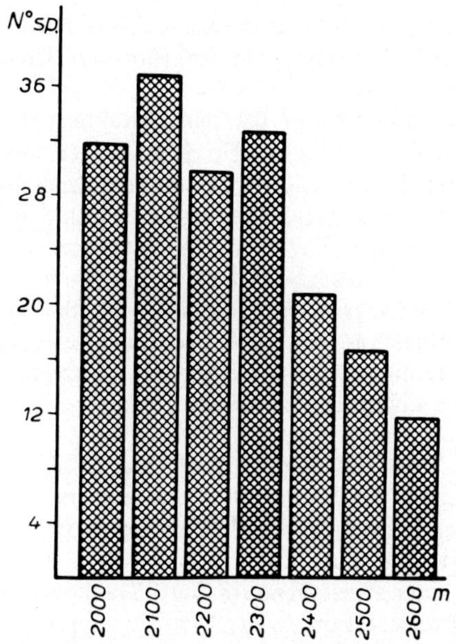

Fig. 330c. Number of species of soil animals found at different levels on the Marmolada (from Marcuzzi, 1959).

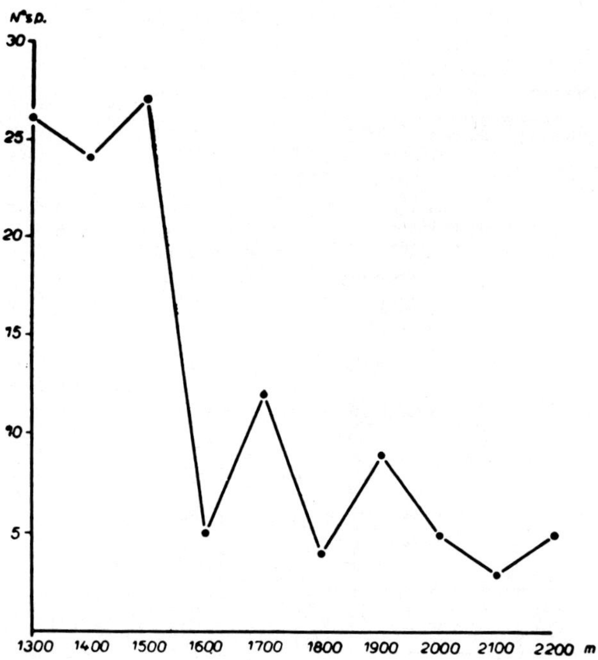

Fig. 330 d. No. of species of the soil fauna at different altitudes in the Sesto Dolomites (from Marcuzzi, 1964).

as in latitude, Recoaro M. Spitz (Venetian Prealps), Marmolada (central Dolomites, belonging to Trentino and Veneto) and Sesto or Sexten (northern Dolomites, belonging to Alto Adige or Süd-Tirol).

Fig. 330 shows a comparison of the soil fauna of the same localities; it is evident that the main factor controlling density and faunistic richness is temperature, associated both with the latitude and the altitude. We can say that generally low temperatures are not favorable to soil life, maybe because most true soil animals are very ancient and originated when the climate was much warmer than today and mountains had not yet reached the present-day heights. Of course we must not forget that temperature is also associated with the history of the biotope, which in high mountains is much shorter, as they were generally covered by ice during the Pleistocene, and that on the mountains as we proceed northwards we have conditions gradually farther from a climax, which could explain both the less dense and less rich soil fauna.

The distance from the climax has been thoroughly illustrated by Janetschek for the Tyrolian Alps where, during the last century, all glaciers have exhibited a retreat, so that an area formerly covered by ice has undergone colonization by pioneer species, not very unlike what must have happened when, in the whole alpine region, glaciers retreated after the Pleistocene era. Fig. 331a shows the altitudinal distribution of several taxa (true succession) in the Tyrolian Alps which were freed from ice during the last century; Fig. 331b shows the altitudinal distribution of some Collembolans in a part of Marmolada (northern slope) similarly

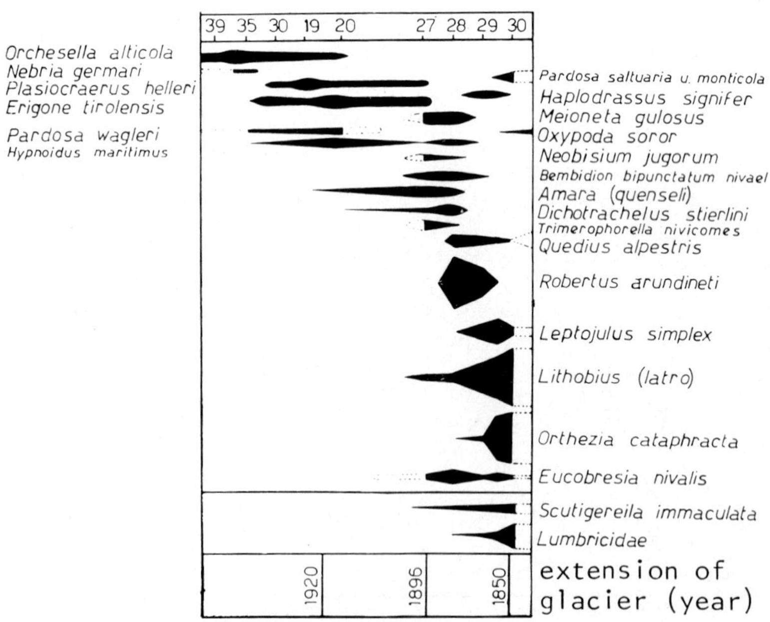

Fig. 331a. Example of a succession in a bare area of the Alps, the Niederjoch-Vorfeld (Tirol) (from Janetschek, slightly modified).

freed from ice during the last century. The effect of the distance from climax on specific diversity of the soil fauna has been studied by the present author applying the information theory, and the results are shown in Fig. 332.

Finally, we want to mention the results Rusek has recently obtained from the study of the Collembolan fauna of two adjacent forest formations on the high-plateau of Cansiglio (Venetian Prealps). From a comparison between a beech forest (present in the southern part of the plateau) and a spruce forest (present in

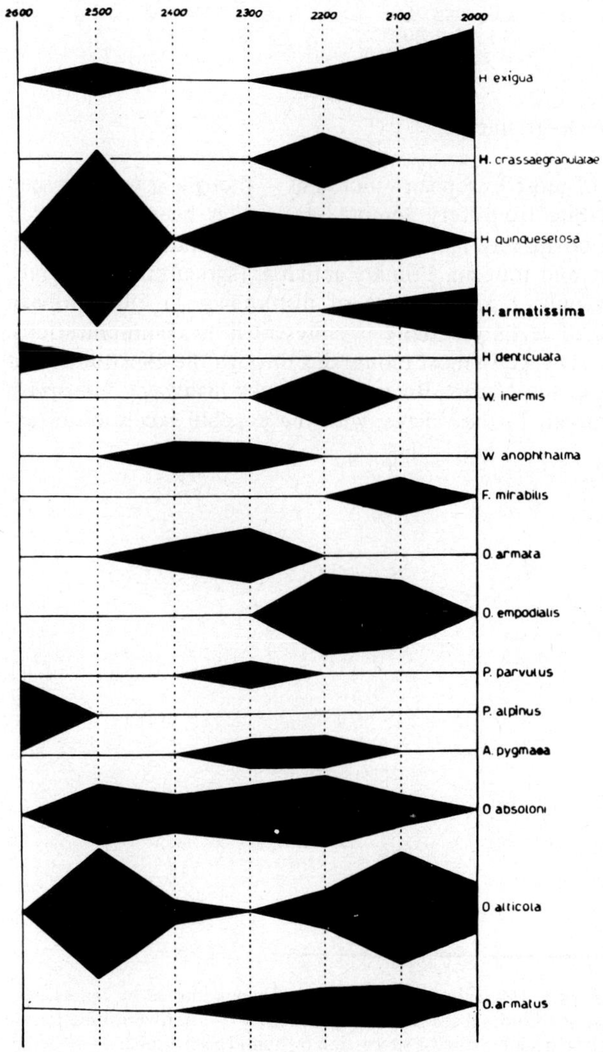

Fig. 331b. Abundance of the different species of springtails on the Marmolada at different levels; some true succession can be recognized easily (from Marcuzzi, 1959).

the northern part), separated by some meadows, a very remarkable difference is evident in the zoogeographical composition of the soil fauna of the two forests, i.e. percentage of endemic, alpine, boreo-alpine and montane species as one can see from the following scheme:

	Picetum		Fagetum	
	N° species	ex.	N° species	ex.
Endemic sp.	1	1	3	72
Eualpine sp.	4	62	0	0
Boreo-alpine sp.	4	79	1	2
Montane species	6	60	2	3
Total	15	202	6	77

Protection in the montane environment

The human colonization of most European mountains – though, as we have seen in foregoing chapters, dating from very remote eras – has been accentuated recently as a consequence of the very rapid development of secondary and tertiary activities, namely industry and tourism. Primary activities (agriculture, cattle raising) have generally been only a minor cause of disturbance in the European mountains. Another kind of tertiary activity is represented by communications, which have carried serious damage to most mountains through the construction of roads, motorways etc. (e.g. the Verona-Bolzano-Innsbruck highway). Industries, however, are generally limited to the valleys, with the possible exception of ex-

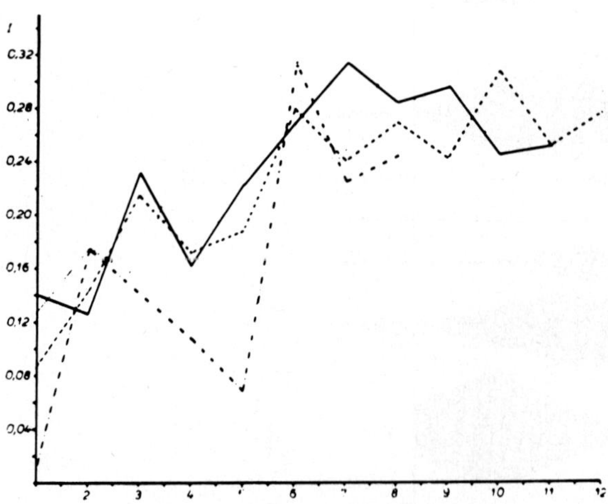

Fig. 332. Index of biotic diversity (Margalef's) in several successions studied by Janetschek: .... Hintereis, main succession, longitudinal section on glacial soil; ---, transversal section on sunny slope; –·–·–, Hintereis, particular case of succession on humid glacial soils; —— Niederjoch, main succession, longitudinal section on glacial soils. The numbers along the abscissa indicate the seral stages (from Marcuzzi, 1964).

tractive industries (mines) and the damage they cause does not affect the true 'alpine' environment very much. Tourism is quite different, as it reaches – both in winter and summer – very high densities, especially when a mechanical means of transport is present (chair lift, cable cars, funicular railway, ski-lift etc.). Also alpinism as such represents in some instances a menace to mountains, particularly when the latter are easily accessible and the climbing degree of difficulty is not excessive, so that great numbers of people are able to reach the montane environment, with all the consequences on its biota and its physical components (rock, gravel, rock-faces, glaciers, permanent snow etc.).

Last but not least, an ill-managed forestry and/or cattle raising can do serious damage to the montane environment, in some instance causing irreversible phenomena (e.g. erosion).

In order to lessen all these dangers for the montane environment the idea of protection for the territory has long been put forward. It is not the intention here to give a history of conservation in the mountains: we will instead limit ourselves to illustrating some of the major achievements which some nations have been able to accomplish. We should point out in this connection that the field of conserva-

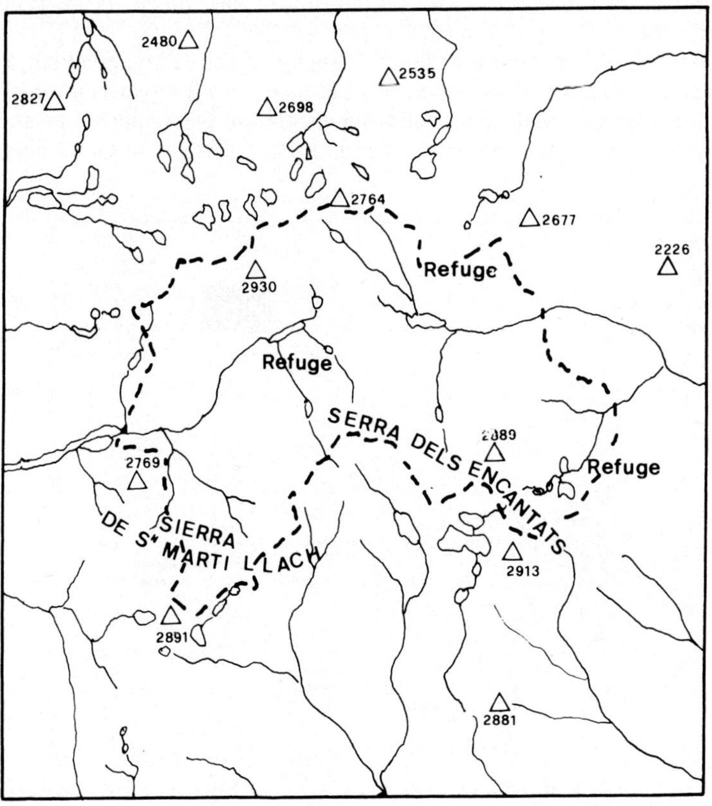

Fig. 333. National Park of Aiguës Tortes (Spain) (scale 1:200,000).

tion is one in which the work of ecologists is often supported by that of other students (geologists, geographers, pedologists, economists, sociologists) as well as by technicians, administrators, politicians etc. This is also one of the reasons why the ecologist's work is often very difficult.

Before mentioning some concrete examples of effective protection of montane environments in Europe, it is opportune to recall that in the last decades some people, motivated by the best intentions, have made some very pessimistic forecasts, which luckily for us have not come true yet and probably will not in the near future. Without going into the whole philosophy of conservation – which is not the aim of this work – we want only to remember a few 'catastrophic' descriptions which, in a world not sufficiently grounded in ecology and in the study of ecosystems, have perhaps had an excessive effect. Rachel Carson's 'Silent spring' is one of these descriptions: maybe an alert against the indiscriminate use of pesticides was – and still is – useful, but in our mountains, and in most parts of Europe, there is no real menace of a 'silent spring'. Another work, although not so famous, is Gretter's L'ultimo verde (The last green). In Italy, where the book was written, and more particularly in the Alps, we are not confronted with the last green: the mountains of the entire Alps, the Apennines, Sicily and Sardinia are still covered with magnificent forests, which are far from being the 'last' ones.

We do not want to delve here into the philosophy of nature conservation: as far as European mountains are concerned, we can mention a few papers in which the interested reader can find some solid arguments on the opportunity and necessity of protection of the montane environments. Sufficient to quote Sepp

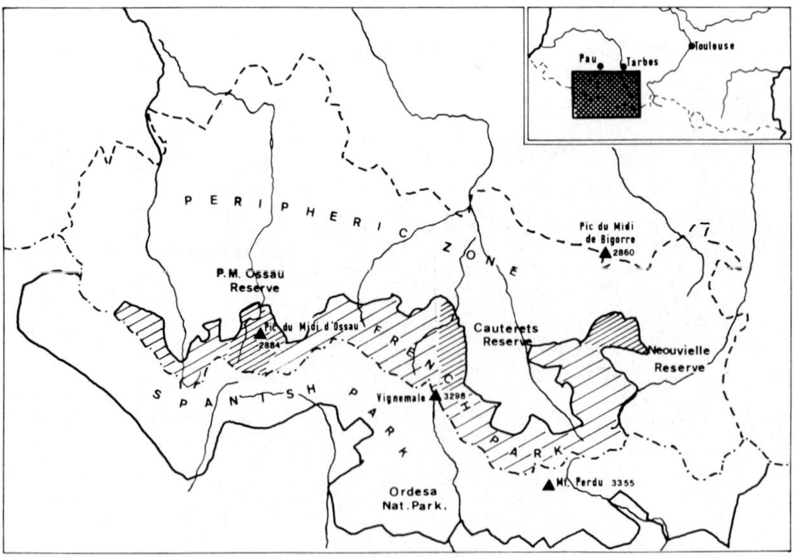

Fig. 334. The National Park of the Western Pyrenees. A central, more protected, and a peripherical part are visible. In the central part there are three 'reserves'.

Fig. 335a. Punta della Sforzellina from Passo di Gavia (Photo: Saibene).

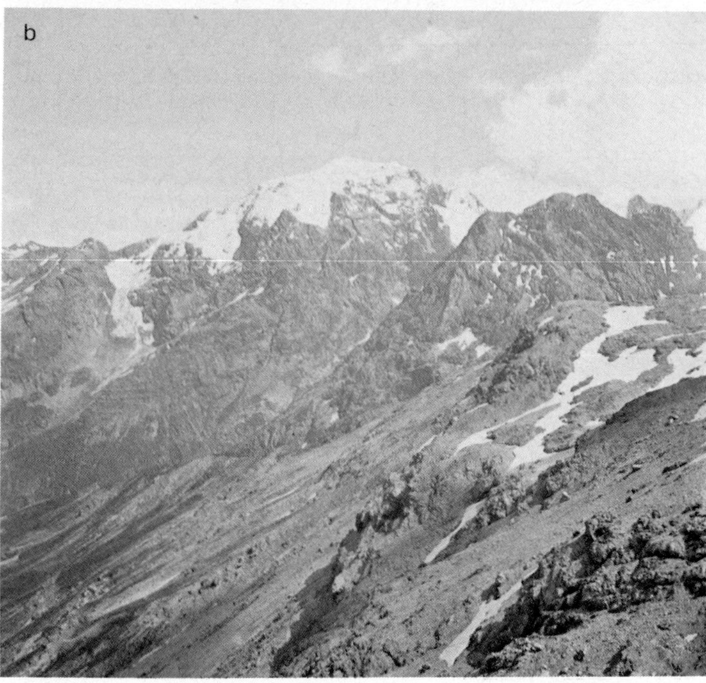

Fig. 335b. The Ortles from Passo dello Stelvio (Photo: Saibene).

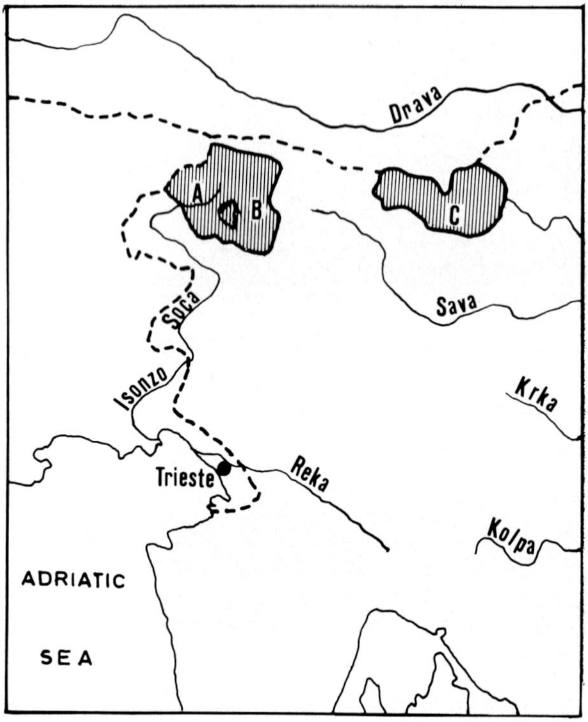

Fig. 336. National Parks. A. Triglav, B. Julian Alps, C. Slovenija (Jugoslavia)

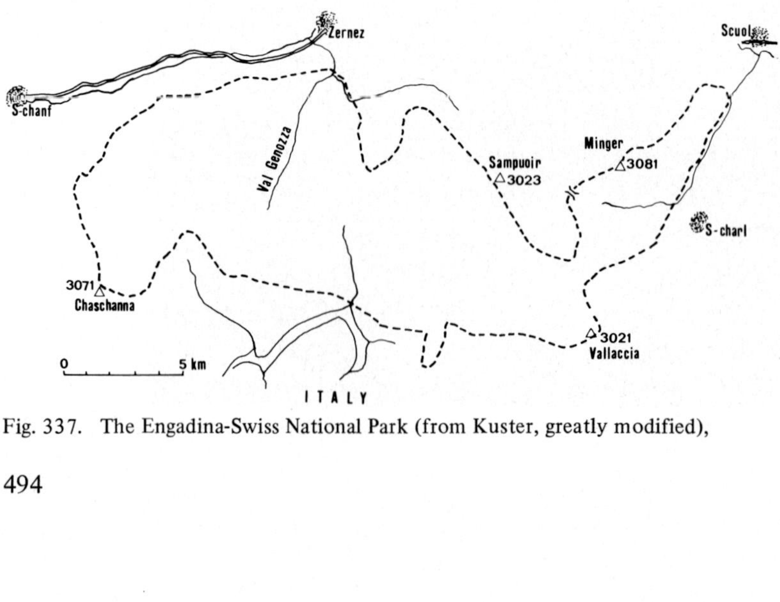

Fig. 337. The Engadina-Swiss National Park (from Kuster, greatly modified),

(1959), Wendelberger (1965), Oechslin (1965), Hoegner (1970), Engelhardt (1970).

We prefer to mention here some concrete accomplishments, existing already in Europe, such as the National Parks, the Naturschutzgebiete (literally territory for nature protection) the Integral Natural Preserves and similar institutions. Thus in

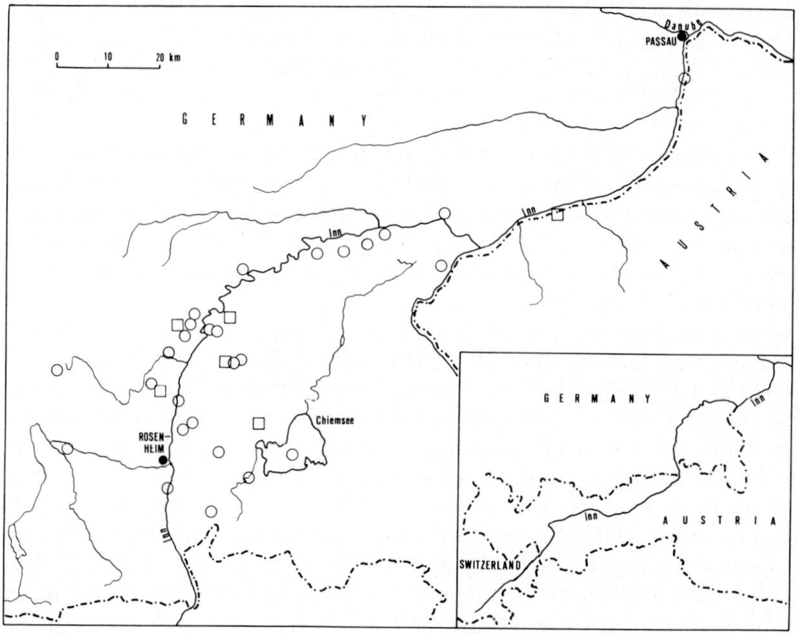

Fig. 338. Chiemgauer Alpen (Germany): localities of naturalistic interest proposed for protection.

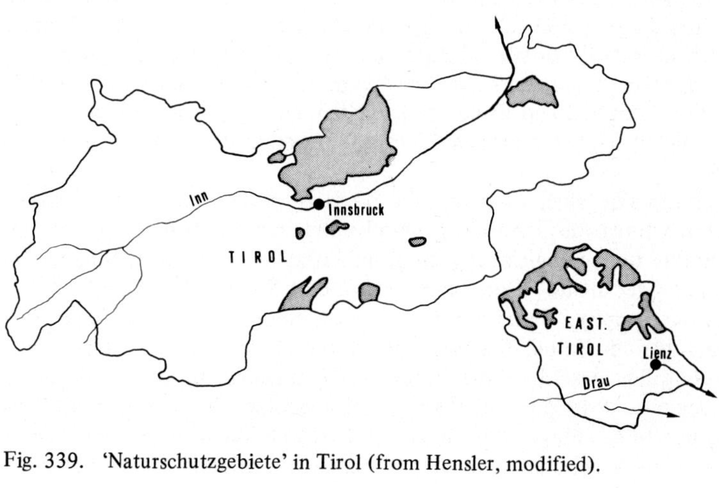

Fig. 339. 'Naturschutzgebiete' in Tirol (from Hensler, modified).

Spain there are the National Parks of Ordesa and of Aigües Tortes near the French border (Fig. 333). In France there is the Western Pyrenees Park (Fig. 334). In Italy we find the following National Parks: Gran Paradiso, Stelvio, Abruzzo, Circeo, Gennargebtu (cf. Fig. 335) besides several Natural Parks (Fusine near the border with Austria and Jugoslavia, Adamello-Brenta in Western Trentino, and Paneveggio- Passo Rolle in Eastern Trentino, etc.) and some Integral (or partial) Natural Preserves created mainly in the Alps and in the Appennines.

As far as Jugoslavija is concerned, there are the National Parks recently created in Slovenija (Fig. 336). In Switzerland we have the Swiss National park (Fig. 337) and the Preserve of Obwalden. As far as Austria is concerned we may mention only as an example the several Naturschutzgebiete existing in the Tyrol (Fig. 339), the reserves of Salzburg (including all the lakes of this region) (cf. Conrad), the preserves existing in Carinthia (cf. Hansely), or that of Hoher Ifen in Allgäu. In Austria we find also the so-called 'Landschaftsschutzgebiete'. In Germany alpine conservation is very well developed: sufficient to list the Chiemgauer Alpen (Fig. 338), the Ammergebirge in the region of Garmisch-Partenkirchen, or the beautiful region of the Königssee.

In USSR protected montane environments existing in the Crimea, Ural Mts. and Caucasus. In these territories there are hundreds of preserves, whose protection is entrusted to Republic or Province administrations (cf. Borissoff). In Romania there are now very interesting reserves in the Carpathians: one is Retezat National Parks, another is the protected territory of Rodnei Mts.; in both the chamois *Rupicapra rupicapra carpatica* is living (cf. Puscariu).

There are many proposals for the creation of National Parks or at least protected territories, but generally they meet with very great resistance from the economic interest (real or illusory) of the local population or rather of business-men. We can remember in the Appennines the proposal for a National Park of the Pollino, at the border between Basilicata and Calabria (cf. Bettini et al.); in the Alps the proposal for a protected territory for the forest and and heath near Augsburg (cf. Bresinsky) and another for Zillergrund near the Italian border. In Italy a list of all biotopes which deserve protection for their vegetational characteristics has been published by ASFD (a state agency for state-forests). In this book a general description is made of several montane biotopes in both the Alps and the Appennines. Recently, some work has been done in this country by town planning people, who described and listed the so-called 'cultural goods' which include also montane wild regions or biotopes (Piemonte, Lombardia and the Appennines near Bologna).

Much researches has been done on the biota of several European National Parks or Protected territories, so that, generally, the creation of these institutions is often a stimulus for ecological and environmental research. A good environmental study has been made of the National Park of Stelvio (Northern Italy) in which geology, geography, flora and fauna (particularly Vertebrate fauna) are described. Dietz has illustrated the avifauna of the National Park of Gran Paradiso; Karl (1969) has described the biota of Chiemgauer Alpen; Mayer et al. have done an accurate description of the vegetation ecology of the Naturschutzgebiet 'Ammergauer Berg'; Mayer has studied the vegetation ecology and paleoecology of the forests which surround the Königssee, included in the Naturschutzgebiet. The flora and fauna of a forest and a heath near Augsburg were the subject of a description by Bresinsky, etc.

We must not finish this chapter without mentioning the paramount importance of the 'Verein für Schutze der Alpenpflanzen & Tiere' Munich, to all the problems inherent in the protection and conservation of montane environments not only in Europe but in the whole world.

BIBLIOGRAPHY

Acanfora M.O., Le statue antropomorfe dell'Alto Adige, *Cult. Atesina*, 6, 1952, 5.

ASFD, Studi per la valorizzazione naturalistica del Parco Nazionale dello Stelvio, Bormio, 1969.

ASFD, Censimento dei biotopi di rilevante interesse vegetazionale meritevoli di conservazione in Italia, Camerino, 1971.

Baccetti B., Considerazioni sulla costituzione e l'origine della fauna di Sardegna, *Arch. Bot. Biogeogr. It.*, 40, 1964, 37.

Bagolini B., Risultati delle ricerche del 1968-69 nella stazione preistorica sul Doss dei Pigui presso Mazzin in Val di Fassa, Dolomiti, *St. Trent. Sc. Nat.*, 47, 1970, 18.

—, Primi risultati delle ricerche sugli insediamenti epipaleolitici del Colbricon (Dolomiti), *Preist. Alpina, Rend.*, 8, 1972, 107.

— & P. Biagi, L'insediamento di Garniga (Trento) etc., *Preist. Alpina*, 11, 1975, 7.

Battaglia R., Nuove statue antropomorfe scoperte nell'Alto Adige, *Atti Mem. Accad. Patav. S.L.A.*, 65, 1953, 3.

—, Ricerche e scoperte paletnologische nel Gargano (1955-1956) *Riv. Sc. Preistor*, 11, 1956, 1.

—, I Castellieri della Venezia Giulia, in Le Meraviglie del passato, Mondadori, Milano, 1958, 419.

Battaglia R. & M.O. Acanfora, II masso inciso di Borno in Valcamonica, *Bull. Paletn. Ital.*, 64, 1954, 225.

Beron P., Sur les éléments boréo-alpins de la faune bulgare, *Bull. Inst. Zool. Mus., Acad. Bulg. Sci.*, 30, 1969, 115.

Bettini V., P. Frenna & A. Castiglioni, Ipotesi di parco nazionale nel Monte Pollino, *Acqua Aria & Ecologia*, 39, lugl. ag. 1974, 368.

Beuret E., Cytologie et distribution de *Trisetum distichophyllum* (Vill.) P.B. et *Trisetum argenteum* (Willd.) Roem. et S., *Bull. Soc. Neuchatel. Sc. Nat.*, 97, 1974, 235.

Blaser P., Standortsfaktoren und Bodenbildung in einem subapinen Lawinenanrissgebiet (Stillberg bei Davos), *Verh. Ges. Oekol., Wien*, 1975, 49.

Borissoff V.A., Naturschutz in der UdSSR. Probleme und Verwirklichungen, *Jhrb. Ver. Schutze Alpenpfl. & Tiere, München*, 35, 1970, 111.

Bresinsky A., Wald und Heide vor den Toren Augsburgs, *Jhrb. Ver. Schutze Alpenpfl. & Tiere, München*, 27, 1962, 125.

Cagnolaro L., D. Rosso, M. Spagnesi & B. Venturi, Inchiesta sulla distribuzione del lupo (Canis lupus L.) in Italia e nei Cantoni Ticino e Grigioni (Svizzera), *Ric. Biol. Selvaggina*, Bologna, N. 59, 1974.

Cassagnau P., Ecologie du sol dans les Pyrénées centrales, *Actual. Sci. Industr.*, 1283, 1961.

Cernusca A., Standörtliche Variabilität in Mikroklima und Energiehaushalt alpiner Zwergstrauchbestände, *Verh. Ges. Oekol., Wien*, 1975, 9.

Conrad K., Natur- und Landschaftsschutz im Lande Salzburg, *Jhrb. Ver. Schutze Alpenpfl. & Tiere, München*, 33, 1968, 75.

Corrain C., Ricerche paleo-antropogeografiche nel Gargano, *Ric. Scientif.* 29, 1959, 2142.

Dietz J., Vogelkundige Streifzuge im Gran Paradiso-Nationalpark, *Jhrb. Ver. Schutze Alpenpfl. & Tiere, München*, 23, 1958, 179.

Engelhardt W., Der Mensch des technischen Zeitalters und seine Umwelt, *Jhrb. Ver. Schutze Alpenpfl. & Tiere, München*, 35, 1970, 15.

—, Rund um den Königssee, *Ibidem*. 37, 1972, 176.
Erlacher K., Das Naturschutzgebiet 'Kaisergebirge', *Jhrb. Ver. Schutze Alpenpfl. & Tiere, München,* 30, 1965, 66.
Favarger C., Recherches cytologiques sur quelques phanerogames de la flore valaisanne, *Bull. Murithienne*, 90, 1973, 3.
—, Cytotaxonomie de quelques orophytes des Abruzzes, *Acta Bot. Academ. Sc. Hungaricae*, 19, 1973, 81.
Fischer K., Naturschutzgebiet 'Kaisergebirge'. Ja oder nein?, *Jhrb. Ver. Schutz Alpenpfl. & Tiere, München*, 27, 1962, 52.
Franz H., Die Nordost-Alpen im Spiegel ihrer Landtierwelt, etc., I, Innsbruck, 1954; II, 1965.
—, Die Rolle der Böden in den hochalpinen Oekosystemen, *Verh. Ges. Oekol., Wien*, 1975, 41.
Frey G., Hoher Ifen – des Allgäus grösstes Naturschutzgebiet, *Jhrb. Ver. Schutze Alpenpfl. & Tiere, München*, 30, 1965, 138.
Gauckler K., Einstrahlungen der Alpenflora im Bayerischen Wald und Oberpfälzer Wald, *Jhrb. Ver. Schutze Alpenpfl. & Tiere, München,* 37, 1972, 1.
Gervais C., Nouvelles determinations de nombres chromosomiques chez des avoines vivaces, *II. Bull. Soc. Neuchatel. Sc. Nat.*, 96, 1973, 81.
Ghilarov M.S. & K.V. Arnoldi, Steppe elements in the soil arthropod fauna of North-West Caucasus mountains, *Mem. Soc. Entom. It.*, 48, 1969, 103.
Gorfer A., Castelli del Trentino, E.P.T., Trento, 1966.
Hansely H., Natur- und Landschaftsschutz in Kärnten, *Jhrb. Ver. Schutze Alpenpfl. & Tiere, München*, 36, 1971, 181.
Hensler E., Natur- und Landschaftsschutz in Tirol, *Jhrb. Ver. Schutze Alpenpfl. & Tiere, München*, 32, 1967, 66.
Hoegner W., Freunde der Natur am Ende?, *Jhrb. Ver Schutze Alpenpfl. & Tiere, München*, 35, 1970, 13.
Holdhaus K., Die Spuren der Eiszeit in der Tierwelt Europas, *Abh. Zool. Bot. Ges. Wien*, 18, 1954.
Huber F., Respiratorischer Kohlenstoffverbrauch alpiner Zwergstrauchbestände, *Verh. Ges. Oekol., Wien*, 1975, 31.
Janetschek H., Tierische Successionen auf hochalpinen Neuland, *Ber. Naturw. Med. Ver. Innsbruck*, 48/49, 1949.
—, Das Problem der inneralpinen Eiszeitüberdauerung durch Tiere, *Oesterr. Zool. Z.*, 6, 1956, 421.
—, Kältezonen und Hochgebirge, in Grzimek's Tierleben, 1973, 184-200.
Jeannel R., La genese des faunes terrestres, Paris, 1942.
Jochimsen M., The development of pioneer-communities on raw soil above alpine timberline, *Verh. Ges. Oekol., Wien*, 1975, 61.
Karl H., Das Ammergebirge – endlich Naturschutzgebiet!, *Jhrb. Ver. Schutze Alpenpfl. & Tiere, München*, 29, 1964, 157.
—, Das Naturschutzgebiet in den Chiemgauer Alpen, *Jhrb. Ver. Schutze Alpenpfl. & Tiere, München*, 34, 1969, 115.
Kofler A. & J. Kollmann, Zur Molluskenfauna Südtirols, *Mitt. dtsch. malak. Ges.*, 3/27, 1974, 101.
Kol E., On the green snow of the Southern Carpathians, *Ann. Hist. Nat. Mus. Nat. Hungar.*, 51, 1959, 161.
Körner Ch., Wasserhaushalt und Spaltenverhalten alpiner Zwergsträucher, *Verh. Ges. Oekol., Wien*, 1975, 23.
Kuster A., Der Schweizerische Nationalpark heute, *Jhrb. Ver. Schutze Alpenpfl. & Tiere, München*, 28, 1963, 48.
Larcher W., Produktionsoekologie alpiner Zwergstrauchbestände auf dem Patscherkofel bei Innsbruck, *Verh. Ges. Oekol., Wien*, 1975, 3.
Lienert L., 'Erholungsraum Obwalden' (Schweiz), Kantonaler Natur- und Landschaftsschutz, *Jhrb. Ver. Schutze Alpenpfl. & Tiere, München*, 37, 1972, 115.
Marcuzzi G., Fauna delle Dolomiti, *Mem. Ist. Ven. S.L.A.*, 31, 1956.
—, Die Fauna der Dolomiten, *Verh. Zool. Bot. Ges. Wien*, 96, 1956, 95.
—, Supplemento alla 'Fauna delle Dolomiti', *Mem. Ist. Ven. S.L.A.*, 32, 1961

—, L'endemismo nella fauna delle Alpi Orientali, *Arch. Bot. Biogeogr. It.*, 38, 1962, 85.
—, Observations on the biogeography of the soil fauna of Dolomites, *Verh. D. Zool. Ges., Wien*, 1962, 689.
—, Ecological considerations on the Collembolan fauna of South-Eastern Alps and Prealps. *Rev. Ecol. Biol. Sol*, 10, 1963, 327.
—, Osservazioni sul popolamento animale delle Alpi, *Arch. Bot. Biogeogr. It.*, 44, 1968, 173.
—, Osservazioni ecologico-biogeografiche sulta fauna del suolo dell'Appennino Abruzzese, *Lav. Soc. It. Biogeogr.*, 2, 1971, 583.
—, Considerazioni zoogeografiche sulla Collembolofauna del Cansiglio (Prealpi Venete), *Boll. Mus. Civ. St. Nat. Venezia*, 24 (Suppl.) 1973, 71.
—, Fauna delle Dolomiti ed. Manfrini, Calliano, 1975.
Marzell H., Der Wacholder in der Namengebung und im Brauchtum der Alpenländer, *Jhrb. Ver. Schutze Alpenpflanzen & Tiere, München*, 31, 1966, 126.
Mayer H., Zur Waldgeschichte des Steinernen Meeres (Naturschutzgebiet Königssee), Jhrb. *Ver. Schutze Alpenpfl. & Tiere, München*, 30, 1965, 100.
Mayer H., Feldner R. & W. Gröbl, Montane Fichtenwalder auf Hauptdolomit im Naturschutzgebiet 'Ammergauer Berg', *Jhrb. Ver. Schutze Alpenpfl. & Tiere, München*, 31, 1967, 20.
Moser W., Neues von der botanischen Forschungstation 'Hoher Nebelkogel', *Jhrb. Ver. Schutze Alpenpfl. & Tiere*, 33, 1968, 125.
—, Einblickein das Leben von Nivalpflanzen. *Jhrb. Ver. Schutze Alpenpfl. & Tiere, München*, 32, 1967, 101.
—, Neues von der botanischen Forschungsstation 'Hoher Nebelkogel', Tirol, *Ibidem*, 33, 1968, 125.
Myczkowski S., Restoration of forest in the Tatra National Park, *Verh. Ges. Oekol., Wien*, 1975, 107.
Nosek J., The investigation on the Apterygotan fauna of the Low Tatras, *Acta Univ. Carol., Biol.*, 1967, 349.
Notiziario-Mitteilungen, *Preist. Alpina, Rend., Trento*, 9, 1973, 223 (several articles).
Oechslin M., Probleme und Sorgen auch in der Schweiz, *Jhrb. Ver. Schutze Alpenpfl. & Tiere, München*, 30, 1965, 31.
Pellegrini G.B. & A. Zamboni, Commento a nomi friulani di piante raccolti nell'ASLEF, *Studi Ling. Friuli*, IV, 1974.
Pignatti S., Arealtypen und die Entstehung der apenninischen Gebirgsflora, *Mitt. ostalp. din. pflanzensoz. Arbeitsgem.*, Camerino, 9, 1969, 107.
Piskernik A., Naturschutz in Jugoslavien, *Jhrb. Ver. Schutze Alpenpfl. & Tiere, München*, 32, 1967, 118.
Puscariu V., Die naturgeschützte Fauna der rumänischen Karpaten, *Jhrb. Ver. Schutze Alpenpfl. & Tiere, München*, 35, 1970, 239.
Rehder H., Phytomasse- und Nährstoffverhältnisse einer alpinen Rasengesellschaft, *Verh. Ges. Oekol., Wien*, 1975, 93.
Rusek J., Zur Collembolenfauna (Apterygota) der Prealpi Venete, *Boll. Mus. Civ. St. Nat. Venezia*, 24, 1973, 71.
Sauro A. & U. Sauro, Boscochiesanuova, ed. Vita Veronese, E.P.T., Verona, 1968.
Schönenberger W., Ueber den Erfolg einer Hochlagenaufforstung auf verschiedenen Standortstypen (bei Davos), *Verh. Ges. Oekol., Wien*, 1975, 55.
Schrott F., Kofler A. & J. Kollmann, Zur Kenntnis der Schneckenfauna von Sappada in der Provinz Belluno, *Mitt. dtsch. malak. Ges.*, 3/26, 1973, 51.
Sepp K., Naturschutz und Landschaftspflege von heute, *Jhrb. Ver. Schutze Alpenpfl. & Tiere, München*, 24, 1959, 144.
Wagner H., Botanisch-oekologische Unterschutzungen im Glocknergebiet, *Verh. Ges. Oekol., Wien*, 1975, 37.
Wendelberger G., Naturschutz als kulturelle Verpflichtung unserer Zeit, *Jhrb. Ver. Schutze Alpenpfl. & Tiere, München*, 30. 1965, 19.

HUMID OR ALLUVIAL ENVIRONMENTS

General characteristics

The humid environments or 'alluvial environments' have been the subject of much research all over Europe. The variety of these environments, going from simple marshy meadows to swamps, from ponds, littoral brackish lagoons, periodic waters of very different nature (the so-called 'Klein- and Kleinst-gewässer' of German authors), to the 'fontanili' (Northern Italy), to the 'marcite' (Northern

Fig. 340. Humid environments near the Lagoon of Venice: 1, region of Altino: a canal going towards the Lagoon; 2, the same ending in the same, brackish water (11,87‰ of salts), with no reptiles or amphibians; 3, inland biotope, slightly brackish water (1,29‰). Here is present the frog *Rana esculenta*; 4, the Sile River with freshwater; here also is present *Rana esculenta*.

Italy), the rice-fields, the various kinds of karst waters, to springs, to finish with the 'barene' of the Lagoon of Venice or the 'sansouires' of the French Midi, causes an enormous range of problems involving structures, environmental adaptations, physiology and behaviour. Suffice to quote as an example of environmental adaptation respiration of the larvae of the genus *Donacia*, which utilize the air present in the pore spaces of aquatic plants; of the physiology, respiration in poorly oxygenated waters, favoured or made possible by the development of special haemoglobins which exhibit a great affinity towards oxygen.

Someone has rightly said (Paludi: risorse sconosciute, published by the WWF) that 'thanks to the great variety of the fauna and the flora, the complexity of the relations among communities, the high rate of productivity, the quick succession of the natural chains, the indirect effects which involve the least changes in the physical factors, swamps are the best among natural laboratories' (free translation from the Italian).

These are only the purely ecological features of humid environments, but there are also the aesthetic and economic aspects. Pratesi has said, in 'Italia Nostra' (Vol. 54, 1967), that 'because of their intact beauty, swamps have a strong touristic appeal'. Unfortunately, not in all parts of Europe have the humid environments conserved their 'intact beauty'.

The humid environments may be roughly subdivided into swampy places, such as swamps and bogs; the shores of inland waters (i.e. lakes, ponds and rivers) and those of the sea. Lakes, ponds and swamps may be composed of freshwater or brackish water; although in nature there are any number of possible transitional conditions from one to the other, so that in some areas we may find several – if not all – of them within a few square kilometres.

Corresponding with the great variety of humid environments, the soils are very various. We will limit ourselves to mention here only some of them, particularly common in Europe.

The raw semiterrestrial soils, sometimes known as rambla (in German 'Rohauboden') from the Arab word ramla, which means coarse sand, are young, gravelly-sandy or sandy-silty, little desintegrated or transformed fluvial sediments. They may be partially colonised by higher plants. Although it is not possible to recognize in them a humus layer (Fig. 341a) since the remains of plants are decomposed but not structurally modified, these soils exhibit furthermore the excreta of soil animals or insects boring in the leaves (miners). When the level of the river increases, the soil may be flooded. In some parts of Europe this kind of soil is extensive, and confers a peculiar note on the landscape: this is the case on the banks of some large rivers, e.g. the Loire in France, the Tiber or the Arno in Italy, or the edges of small islets present in some large rivers (e.g. the Po in Italy).

Along the rivers flowing into the Arctic Ocean, the banks are covered with a polygonal soil known as Rutmark, which can be translated as cellular or polygonal soil (see the chapter on the Tundra). It is a slightly gleyed, raw soil, practically without macroscopic vegetation, frozen for most of the year, even on the surface, and covered with snow. It thaws only at the surface during summer ('permafrost').

In the alluvial regions of Central Europe and Northern Italy there is an interesting and widely diffused type of semi-terrestrial or alluvial soil, called by German authors 'Wiesenboden' (literally meaning 'meadow-soils') and by Kubiena valley-Anmoor. It is a gley, almost always wet, with a humus not dissimilar from gyttja, blackish muddy, and smelling like ink. It is spread in the valleys and on the flat shores of the lakes (Fig. 341b). When some earthworms are present (as is very often the case), there is a transition from Anmoor towards a Mull (Mull-gley). The iron compounds present in this soil, because of the scarcity of oxygen, change to the ferrous form (Fe″) and therefore some water soluble compounds are formed,

Fig. 341. Left: Rambla with *Myricaria germanica.* In the soil there is an earthed-up A-horizon (Enna Valley, Styria). Right: Anmoor under a meadow with *Iris germanica* and an undersoil strongly gleyized (Enna Valley, Styria) (from Kubiena).

which in peculiar conditions (as for instance along the burrows dug by earthworms) may be oxidized and change irreversibly into iron hydroxide (Fe″), which is brown-rust coloured and not soluble in water. (See C. Jeanson's research on the technique of demontsrating the presence of Fe″ compounds in earthworm burrows). Valley-Anmoor has a soil fauna in which aquatic animals prevail, as well as specialized earthworms.

The so-called raw gley (syrogley) is a permanently wet soil, lacking a perceptible humus layer, and often crossed by furrows and spots of rust, with a generally bluish or greenish colour. The productivity of organic matter is very poor as there are neither burrowing animals nor excreta of terrestrial animals.

Pseudo-gley is a very compact soil, scarcely permeable, sandy-silty, with strong

Fig. 342. Left: Paternia under shrubs of willows and alders, with two unearthed profiles (I and II) (Enna Valley, Styria). Right: Borovina under a dry meadow upon limestone gravel (C) and limestone sand (D) (Enna Valley, Styria) (from Kubiena).

changes in wetness at the surface. It is more-or-less brown with light spots. It is present in Central Europe, for instance in Saxonia, in oak forests where the floor is covered with *Molinia coerulea*.

Borovina is a soil poor in humus, rich in calcium carbonate (which develops bubbles when treated with diluted HCl), with a perceptible light coloured humus layer, either because of the low content of humus, or because of mixing with the lower inorganic layer as a consequence of cultivation. Its permeability is good; its name is a popular Polish name, corresponding to the German 'Rendsina-ähnlicher Auboden'. Inside the humus layer during the rainy season there is an intense soil faunal activity, with good decomposition and humification.

Finally Paternia is a soil poor in humus and calcium carbonate (so that it does not develop bubbles if treated with diluted HCl), but in all other things similar to borovina. The name comes from the river Paterna, in the Alpujarras (South of Sierra Nevada, Spain), and corresponds to the German 'Grauer Auboden'. The soil fauna is richer and more active than that of the ramblas; in some cases a colonization by earthworms is possible (Fig. 342).

The organic soils of Central and Northern Europe will be described when talking about moors.

In Southern Europe an important category of alluvial soils is represented by the so-called 'vegas' (literally banks of a river), and more precisely by brown loam-vega. This is developed in valleys, with an extraordinarily intense colouration, with a strong ability to expand and easily transformed into mud, from a hard, compact stone-like soil. Due to heavy rains there is considerable erosion, with the formation of channels, pot-holes and deposition of sediments of very variable size. The soil life is concentrated in the uppermost layer, as the compact undersoil prevents root growth, so that presence of trees is limited.

The brown vega is similar to the preceding type, but is an intense brown colour, lumpy, with no tendency to expand and therefore minimal erosion. This soil is present mainly in mountain valleys covered with forests, where siliceous rocks outcrop. It exhibits a normal soil fauna, with decomposition and humification of the organic matter, and the presence of earthworms, so that the presence of a Mull is a normal feature (Fig. 343a).

The red vega is red to brown-red, present in the river valleys, almost saturated with water so that it easily expands into mud, with a tendency to form a dense structure. This soil is distributed mainly in Greece and Spain and also probably in Southern Italy: generally it is very good for cultivation.

Smonitsa is a soil rich in calcium carbonate, fertile, with a good humus, dark grey to blackish, with a thick humus layer. Below the humus there is generally a well-developed white horizon of limestone enrichment. It occurs along rivers and brooks. The soil life is very active, with good decomposition and humification, and consequent formation of the humus-clay complex. Also calcium humate may be formed (Fig. 343b).

Swamps

Because of the enormous variety of swamps, the very definition of this kind of humid environment is difficult: swamp is an environment characterized by a range of sizes, variable climate (especially micro-climate) going from brackish swamps along the coastlines to completely freshwater swamps situated in hilly or even montane places (where the swamp may become a mire – if the soil is acid and the climate cool), to marshy meadows completely dry during the summer.

The size may vary from several square kilometres (for instance in Italy the so-called 'padule' of Fucecchio, in Tuscany), to a few hundreds of square metres; but even in this case a great number of plants may colonize the environment, utilizing all the existing ecological niches.

Very often, within a swamp, there are some small ponds, sometimes deep

Fig. 343. Left: Brown loam vega, with a bizonal profile (Gallego Valley, Zaragoza). Right: Smonitsa under a field of buckwheat above limestone gravel (Vienna Basin) (from Kubiena).

enough to contain permanent water. In some cases the swamp is crossed by a channel, a stream etc. with running water, which makes possible the establishment of a peculiar plant and animal life, typical of running waters (see later).

For these reasons, sufficiently extensive swamps (over 0.2 km^2) may contain plants from a wide range of terrestrial and aquatic environments, grassland, marsh, swamp, still and running water, e.g. *Veronica anagallis, Mentha aquatica, Potamogeton* spp. Obviously there are corresponding differences in animal species, some living in the mud, others under stones, on plants or swimming on the surface, like the beetle *Gyrinus*. Fish, amphibians and reptiles are also always present in swamps and birds and mammals live in the Phragmitetum (reed-bed) and Caricetum of European swamps at least for part of the year.

The swamps are exploited by man almost everywhere: cutting canes, willows or sedges, collecting water lilies and other flowers for local markets, and grazing and watering of cattle. Also fishing may be productive in some European swamps, at least where overfishing has not occurred. For these reasons there is today some measure of human influence in most swamps of our Continent. Also, the fact that this environment is, or was, particularly common in the southern countries, characterized by high densities of population (higher than in the North) and where there are ancient human settlements, has contributed to a disturbance dating over a long time (see later). Again in Southern Europe swamps have been often, if not always, synonymous with malaria (the word indeed signified that the disease, whose causative agent was not known until the last decade of the past century, is linked to the air of the swamps, considered – rightly or wrongly – unhealthy) explaining the considerable swamp-land reclamation in many countries, such as Central-Southern Italy or the Po-Delta region.

Therefore, today the environment of swamps is, as we will see, one of the richest and most typical of our landscapes (provided one is able to look at it in a spirit of 'natural beauty') but very often different from the original conditions, of which we have some idea from old descriptions and pictures.

We will concern ourselves here exclusively with a Southern European swamp, with a relatively mild climate, permanent water, fed by small springs and not yet disturbed too much by man (such an example is not easy to find).

Winter is the season of silence and rest. At most, one will be able to hear the call of a resident bird looking for food in the meadows around the swamp. The vegetation is completely quiescent: only some withered *Phragmites*, some *Scirpus* without flowers or with the remains of the fruits of the preceding year are visible, or the shrubs (*Salix, Rhamnus frangula*, etc.) rising from the middle of the swamp.

The surface is often covered with snow or ice where the ponds are very deep. Only the streams are free from ice, and flow quietly and, at first glance, free of any form of life. Under the ice larger animals winter: frogs, *Dytiscus, Hydrous piceus*, besides a number of larvae and eggs of all kinds of invertebrates.

When the last snow has melted on the shores of the swamp beetles begin to move; one is an inconspicuous Carabid, *Bembidion rivulare*.

In March no plants have started to grow with the exception of the flowering of the willow (called catkins) and some sedges which, however, do not flower until the rest of the vegetation does. At the shore a larger number of Carabids is present: *Bembidion articulatum, Drypta dentata* (under stones), *Odacantha me-*

lanura (both on the shores and on the herbs), as well as *Pterostichus vernalis*, faithful to its name. On the plants still without flowers, the first phytophagous insects are present, such as the Chrysomelid *Prasocuris phellandri*, living on the Kingcup or Marsh marigold, *Caltha palustris* and on some *Umbelliferae* in which the larvae develops (*Oenanthe, Berula*).

The plants of the middle latitudes are extremely sensitive to photoperiodism: despite the temperature of the air, or, in the case of the swamp plants, of the water, if the day does not include a certain number of daylight hours, the plants will have no buds, no leaves, and no flowers. If this is evident in forests, it is still more perceptible in an environment completely exposed to the light from dawn to dusk, as is a swamp or a bog. Whereas in March only some species show leaves, in April a great number of plants are flowering, particularly among the true swamp elements, such as *Carex stricta* (sometimes a dominant species, after which an association – the strictetum – is named) and *Carex distans*, *Caltha palustris*, a very showy species because of its large, rounded, shining dark green leaves and its large, vivid yellow flowers, besides some trees such as the willow *Salix cinerea* and the white poplar, *Populus alba*. Immediately, flowering plants are invaded by a number of insects, some of whom are phytophagous, some predators or parasites. On the flowering willows, the Megalopteron *Sialis flavilatera* is extremely abundant, whereas the delicate Ephemerid *Cloeon dipterum* flies here and there. On

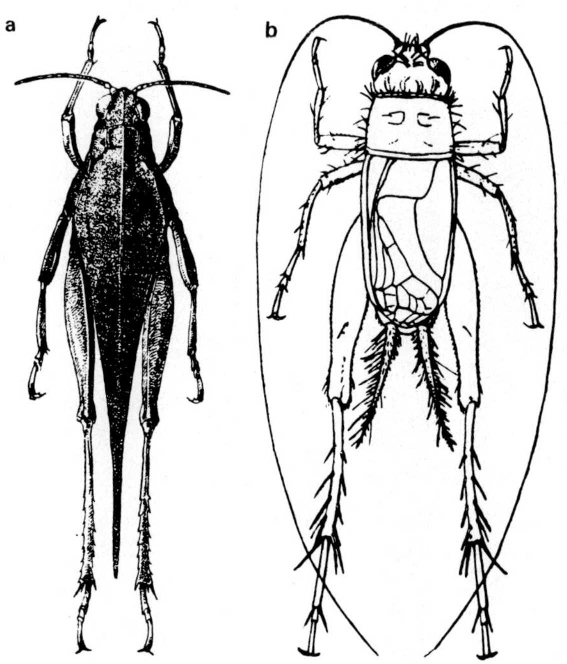

Fig. 344. Orthopterous insects of swamps: a. *Tetrix subulata*, b. *Pteronemobius heydeni* (from Chopard).

the bottom of the swamp we may find in April the first hoppers of the year: two small *Acrididae* or *Chelifera, Tetrix ceperoi* and *T. subulata*. The former is present all over South Europe and North Western Africa; it is greyish or brownish coloured, with or without spots; the latter, present in Europe and Asia, exhibits an extremely variable colouration, and often a characteristic median longitudinal light stripe over all its length (Fig. 344). On the plants the first phytophagous species may be found now: *Hydrotassa marginella*, bound to several species of *Ranunculus* and to *Caltha palustris*, also *Phaedon cochleariae*, which winters below earth clods and debris as an adult, coming out in April, so that in June the first new generation of the year makes its appearance. It lives on *Nasturtium officinale* (the common water cress) and *N. amphibium*, on *Cochlearia armoracia* and *Veronica beccabunga*. The *Chrysomelidae Donacia sericea* lives on *Carex* as well as on the flag *Iris pseudacorus* and some *Juncus; D. consimilis* upon *Carex* and perhaps *Caltha palustris; D. rustica,* restricted to several sedges, *Cladium mariscus* and some *Juncus* (limited to Central Europe and Italy); and lastly *D. aquatica*, frequent on the elegant, slender, *Sparganium simplex* (bur-reed), *Ranunculus lingua* (a strictly swamp species) and some sedges. Willows and poplars are inhabited by two beautiful Chrysomelid species, whose colouration may be considered as aposematic, since they are extremely showy. Like all species exhibiting these colourations, they are provided with substances keeping their enemies away: *Melasoma populi*, with red elytra and metallic green head and pronotum, present on poplars (*Populus pyramidalis, P. alba* and *P. nigra*), and willows (*Salix purpurea, S. alba*, etc.) and *Plagiodera versicolor*, with a metallic bluish-green colour, present

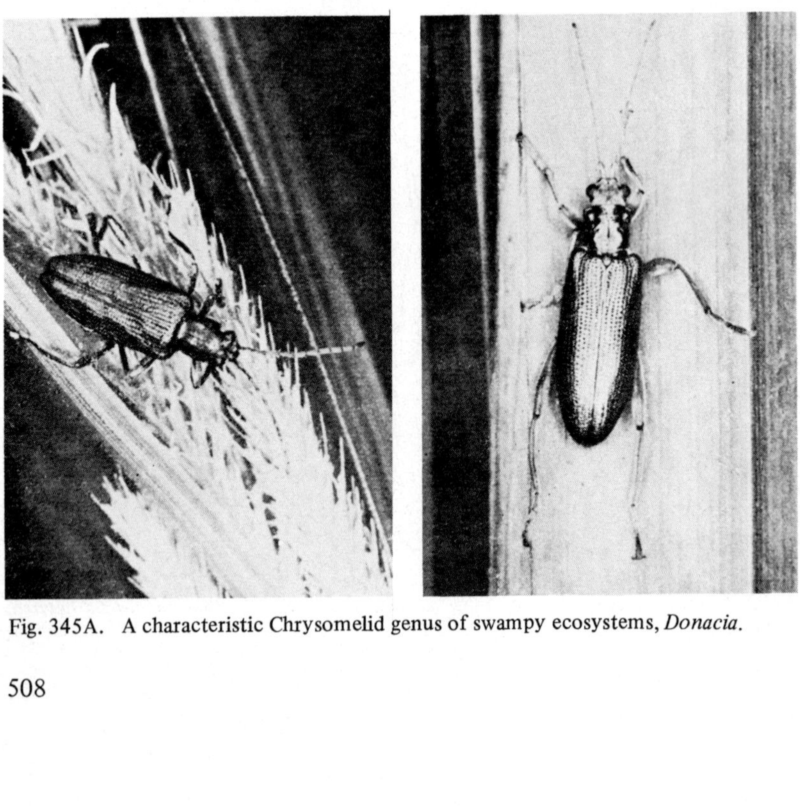

Fig. 345A. A characteristic Chrysomelid genus of swampy ecosystems, *Donacia*.

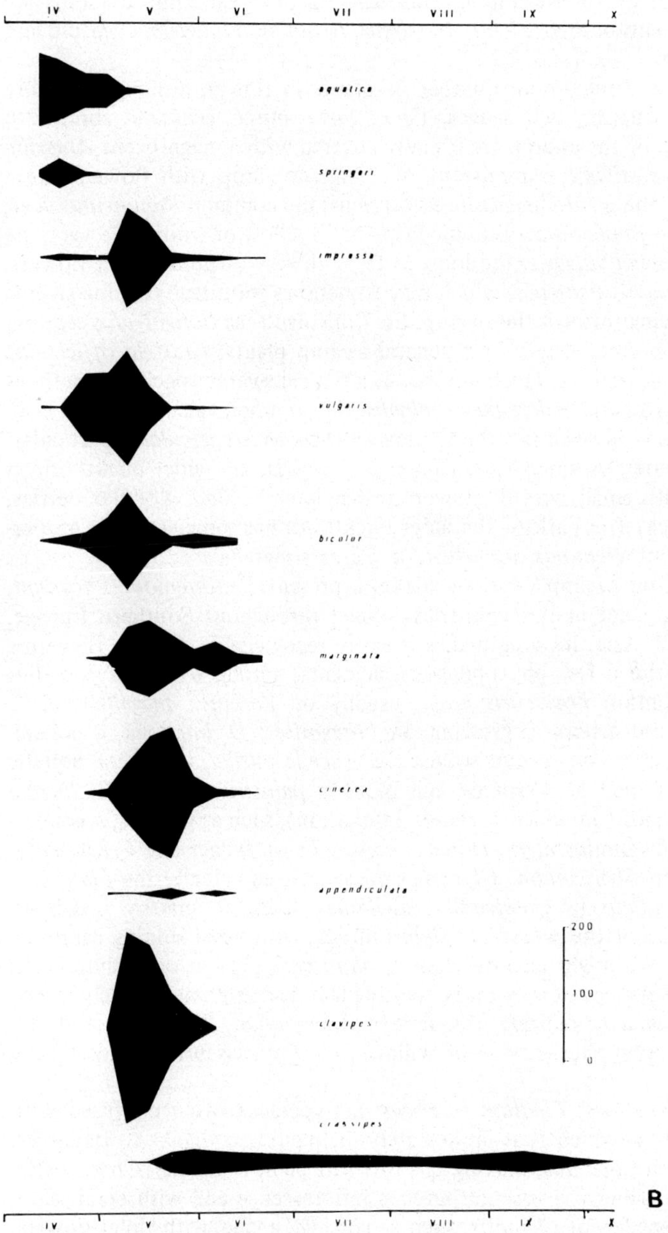

Fig. 345B: The phenology and abundance of some Italian species of this genus (from Ravizza).

on willows (*Salix alba* and *S. viminalis*) and *Populus nigra. Melasoma populi* contains salicilic aldehyde (Fig. 345).

On the bottom of the swamp in this part of the year some Carabids are frequent, let us mention *Pterostichus aterrimus, Bembidion quadrimaculatum* and *B. octomaculatum*.

May brings us the maximum number of plants in flower, although possibly exceeded in July-August: two sedges, *Carex hirta* and *C. echinata*, come into flower; some parts of the swamp are literally covered with a magnificent *Amarillidacea*, *Leucojum aestivum*, reminiscent of a big snowdrop with flowers like a short, plump bell; the gladiolus *Gladiolus illyricus*; the common *Ranunculus acer*; the Rosacea *Spirea filipendula*, with unconspicuous yellow or whitish flowers; the Primulacea *Lysimachia vulgaris*, the loosestrife, with very obvious yellow flowers, the myosotis, *Myosotis palustris*, which may sometimes form true floating swords along streams flowing through the swamp; the Convolvulacea *Convolvulus sepium*, climbing on *Phragmites*, sedges and general swamp plants; *Gratiola officinalis*, with small lilac-pink flowers; *Valeriana dioica*, a typical swamp species, sometimes living with the congeneric *Valeriana officinalis*, the common valerian (which is by no means found only in swamps), the Cyperacea *Heleocharis acicularis*, particularly restricted to water. A small tree, *Rhamnus frangula*, the alder buckthorn, is now flowering with small whitish flowers which later become globular berries, first red, then black. The bark of the alder buckthorn has some cathartic properties similar to that of *Rhamnus cathartica*, or 'Cascara sagrada'.

On the floor of the swamp a curious cricket is present, *Pteronemobius heydeni*, the only swamp cricket in our countries, spread throughout Southern Europe, NW Africa and SW Asia. Its song makes it easily recognizable. On the flowering plants we find quite a few phytophagous elements, among which Crysomelids prevail: let us mention *Prasocuris junci*, usually on *Veronica anagallis* and *V. beccabunga; Donacia braccata,* growing on *Phragmites; D. impressa*, a pollenivorous element present on several sedges, *Galerucella pusilla*, on several aquatic and swamp plants such as *Veronica* and *Stachys palustris; Galerucella lineola* found on several trees (on which it causes defoliation), such as willows, secondarily black alder and *Populus nigra, Donacia consimilis* on sedges and *Caltha palustris, D. simplex* on *Sparganium, Glyceria* and sedges, and finally the *Elateridae Corymbites sjaelandicus* (= *Ctenicerus tessellatus*), living in marshy meadows, swamps and bogs, and the weevil *Attelabus nitens*, with a red shining carapace, living on willows. We might also mention *Cyphon padi* present on swamp herbs (see p. 524), three species of lady-birds, hunting leaf-lice on plants, namely *Coccidula rufa, Hippodamia 13-puntata*, frequent on *Sparganium, Phragmites* and sedges, and feeding upon the leaf-lice of willows, *Hyalopterus pruni*, and *Adonia variegata*.

June sees other plants: *Cladium mariscus*, a Cyperacea often confused with *Phragmites*, but from which it is immediately distinguished thanks to its leaves, with minutely serrate margins, making it a difficult plant to handle; *Orchis incarnata* with a dense cylindrical rose-purpureous inflorescence and with erect, elongate leaves; some species of *Clematis*, such as *viticella*, a vine, with violet flowers, *recta* with erect stem and white flowers; and lastly *vitalba*, the common traveller's joy, a vine with whitish flowers. Furthermore we should mention two species of

Ranunculus strictly limited to water, *flammula* and *lingua*, with linear lanceolate leaves (different from all other European species), *Lycopus europaeus*, or gipsy-wort, flowering until September, *Galium palustre, Eupatorium cannabinum,* the hemp agrimony, closely associated with water; *Senecio paludosus*, a great fen ragwort, with very vivid yellow flowers, *Cirsium palustre*, or marsh thistle, very characteristic of humid situations, together with two Pterydophyte^s^, the fern *Nephrodium thelypteris*, present also in Central European mires, and the horse tail *Equisetum palustre*, also common in bogs.

Some plants grow directly in the water: they are *Nasturtium vulgare, Scirpus lacustris* (up to 1-1.5 m high, with a characteristic dark brown terminal inflorescence) known as bulrush and the cattail *Typha*, often collected and brought to the markets. On the bottom we may find a Carabid beetle, *Oodes helopioides*, an Euro-Sibiric element typical of swampy soils, mainly of the lowlands. On the plants interesting species may be found, for example, the aphidivorous lady-bird *Propylaea 14-punctata*, the Chrysomelids *Clytra laeviuscula*, generally on willows, *Labidostomis pallidipennis, Cryptocephalus janthinus*, characteristic of marshy meadows, where the larva eats the epidermis of the leaves of *Phragmites* and other plants.

July sees a new maximum for anthesis: it is the period when at noon one hears no sound, no voice in the calm of the swamp immersed in the heat where practically only the plants seem at their ease. Not even the slightest trace of air movement makes the leaves move. Only hoppers may inform us of their presence by chirping. *Thalictrum flavum*, the common meadow rue, *Lythrum salicaria* the purple loosestrife and *Epilobium hirsutum* or codlin now flower and the same is true for some semi-aquatic plants (which in general flower after terrestrial species) such as *Phragmites communis*, several species of *Cyperus, Scirpus holoschoenus, Alisma plantago*, or water-plantain from *αλισμα*, plantain), *Butomus umbellatus*, of *Butomaceae, Hippuris vulgaris* or mare's tail, *Mentha aquatica*, etc.

This is the time of flying species: butterflies, flies and gnats, *Hymenoptera* and dragonflies, besides hoppers. Among dragonflies we must mention *Sympecma fusca*, sometimes occurring far from water; *Pyrrhosoma nymphula, Agrion pulchellum* and *A. puella*, two *Zygoptera* with very elegant slender bodies and almost transparent wings; *Calopteryx splendens*, to which the name is perfectly adapted (literally 'with beautiful wings', from *καλός* and *πτερός* and 'shining', from the Latin splendens), found also in running waters and rice fields; and lastly *Somatochlora flavomaculata*, flying over swampy soils, and in small woodlands, at the edge of forests, but never upon the water surface. It is mainly a Nordic species, but extends as far south as Tuscany.

The hoppers are very common: *Homocoryphus nitidulus*, belonging to *Tettigoniidae* or *Ensifera* (i.e. the kind of hoppers with long antennae and a long, often recurved, ovipositor); it is a nocturnal species; the song of the male is prolonged and croaking, constituting a continuous, rather acute vibration; it feeds on seeds of *Gramineae;* the small *Chorthippus dorsatus*, with a colouration varying from green to brown, and with two lateral dark stripes. Its song is composed by 4-6 notes, followed by a pause (several years ago the BBC recorded the songs of most British grass-hoppers, and today there are many interesting records available for research in ecology and/or ethology). Other insects common in the summer are

Rhynchota and *Diptera* – the latter are represented by several species of *Dolichopodidae*, continuously flying over the water, hunting small mud-dwelling animals, for example the *Oligochaeta*; *Eristalis sepulchralis; Dictya umbrarum* (a *Sciomyzidae*), observed on alder and *Angelica* (an *Umbellifera*); some species of the genus *Chlorops* restricted to swamps and bogs; *Ephydra riparia*, present all over Europe, with an aquatic larva living in all types of water (even brackish water an mineral springs); and finally, the *Tabanidae Chrysops caecutiens*, flying during the hottest hours of the day in search of some mammal from which to suck blood: unfortunately in some cases man is the mammal.

A Lepidopteran particularly found in humid environments is *Eustrotia uncula*, sometimes present on sedges.

In August some additional species of hoppers make their appearance: *Conocephalus fuscus*, a *Tettigoniidae*, with long wings and long slender filiform antennae; the female is recognizable because of its ovipositor, as long as the body. The colour is delicate green, with a brown stripe extending from the head to the elytrae. It feeds upon small insects and the song of the male is deep, vibrating and discontinuous, with 4-6 calls per minute, each of 260 vibrations. It is a diurnal animal, different from *Homocoryphus nitidulus*, to which it is similar in appearance. Two *Chelifera* are also present: *Mecostethus grossus* is very big (the length may reach 4 cm), with an olive green colouration interrupted by a very marked yellow stripe near the anterior margin of the elytron (i.e. the first pair of wings) and the hind femurs a vivid red colour. It has a very weak stridulation. *Aiolopus thalassinus* is a multi-coloured hopper, often green, found in Southern Europe and North Africa as far south as the Niger.

In autumn, life is still rather luxuriant in Southern European swamps: on those plants still flowering, or fruiting, several beetles may be found; one is an aphidivorous species, the lady-bird *Semiadalia 11-notata*, living at the margin of forests, in clearings, grasslands, meadows, etc. upon *Compositae* and *Umbelliferae*; other species are phytophagous, such as the Chrysomelids *Laema melanopa*, living and developing upon *Gramineae,* both cultivated (oat, barley, wheat) and wild (*Agropyrum, Lolium, Dactylis, Hordeum murinum* etc.); and two species of *Chaetocnema, concinna,* upon *Polygonum, Rumex* and sometimes *Rheum* (the common rhubarb), and *tibialis*, a Mediterranean species, reaching as far north as Poland and as far east as Turkestan, occurring on several *Chenopodiaceae*, such as *Atriplex, Beta, Chenopodium, Salsola* and *Salicornia*; it can also cause serious damage to sugar beet (Fig. 346).

In Italy swamps are widespread and have been the subject of several investigations. In a brackish Phragmitetum (reed-bed) near Fregene (Lazio) with an area of not more than 400 square metres, the following numbers of Coleopterous insects have been found: *Carabidae*, 15 species, of which *Bembidion dahli* is the dominant species; *Staphylinidae*, 2 sp., *Pselaphidae*, 2 species, *Endomychidae*, 1 species, *Coccinellidae*, 1 species, *Malachiidae*, 1 species, *Anthicidae*, 4 species, *Chrysomelidae*, 2 species. The most characteristic species is the Carabid *Bembidion dahli*.

The *Heteroptera* (or bugs) have been studied in the swampy places all around the Venice Lagoon. The following associations and related *Heteroptera* fauna can be recognized (Giordani Soika, 1949):

1. Association of humid swampy soils, characterized by *Carex gracilis: Eusarcoris aeneus, Berytus montivagus, Pamera lurida, P. fracticollis, Microplax interrupta, Halticus apterus, Heterocordylus genistae, H. farinosus*.
2. Phragmitetum (reed-bed): *Eusarcoris inconspicuus, Carpocoris pudicus, Dolichoris baccarum, Terapha hyosciami, Lygaeosoma reticulatum, Nysius ericae, N. senecionis, N. graminicola, Dimorphopterus spinolae, Serenthia atricapilla,*

A.

B.

Fig. 346. The karst swamp of Pietra Rossa, near Monfalcone (Friuli-Venezia Giula). A. General view of the environment (today almost completely drained and transformed to cultivated fields). B. Detail of an emerged part of the swamp, where the marsh fern *Nephrodium thelypteris* and the buck bean *Menyanthes trifoliata* are easily recognizable.

Orius minutus, Adelphocoris lineolatus, Notostira erratica and *Trigonotylus ruficornis*.

3. Humid, almost freshwater (oligohaline) marshes; characterized by *Carex oederii: Aelia acuminata, Eusarcoris inconspicuus, Carpocoris pudicus, Dolichoris baccarum, Eurydema festivum, E. oleraceum, Terapha hyosciami, Stictopleurus abutilon, Nysius ericae, N. graminicola, Cymus melanocephalus, Serenthia confusa, Nabis ferus, Poecyloscitus cognatus* and *Notostira erratica*.

A very peculiar swampy or humid environment extensive in Northern Italy is provided by rice fields. The most remarkable invertebrates are the spider *Argyroneta*, the *Diptera Stratiomys, Culex, Anopheles, Tabanus, Tipula*, the *Odonata Calopteryx* and *Libellula*, the genus *Ephemera*, the *Trichoptera Limnophilus, Phryganea, Triaenodes*, the Megalopteron genus *Sialis*, and the Lepidopteron genus *Hydrocampa*. According to Moretti, the rice field exhibits no peculiar species, since its elements are present also in swamps, rivers and brooks. Several insects cause serious damage to rice (Fig. 347).

The vertebrates of the rice fields are some species of *Rana, Tropidonotus* (the water snakes), the birds *Ardea* and *Alcedo*, and the mammals *Neomys fodiens, Lutra lutra* and *Arvicola amphibius*.

Malaria is a disease which has greatly afflicted people in Southern Europe and to some extent in Central Europe (Fig. 348).

The disease is necessarily linked to the presence of still or slowly flowing

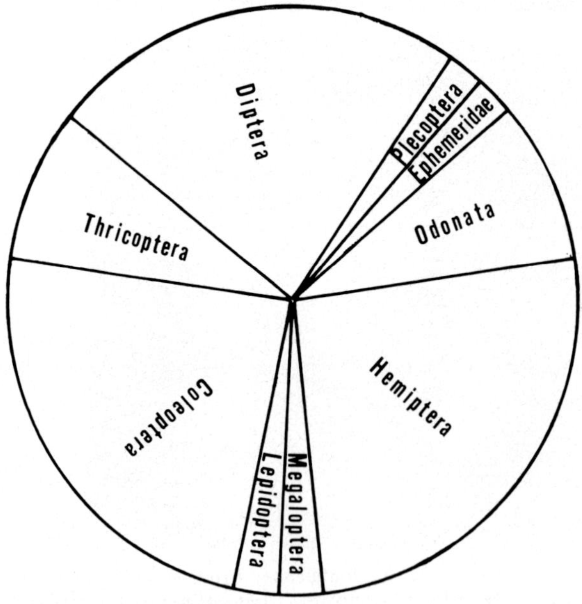

Fig. 347. Quantitative composition (number of species) of the insect fauna of the Northern Italian rice-fields (constructed on the basis of Moretti's data).

waters, regardless of whether they are fresh or brackish in which live the *Anopheles* larvae; malaria is caused by a blood parasite, a Protozoon belonging to the genus *Plasmodium*, transmitted by the adult *Anopheles*.

Malaria has been known for a long time, particularly in Italy. They say that this disease has caused the disappearance or the decline of several Italian cultures, such as the Nurhagic and the Etruscan, of some flourishing towns in Sicily (Selinunte, Σελινοῦς) in Magna Grecia (Sibari, or *Σύβαρις*, Locri Epizefiri, or *Λοκροι οι* Επιζεφύροι), in Campania (Poseidonia, *ποσειδωνία*, or Paestum), as well as in the Po valley, for instance Adria and Spina, both near the Po-Delta.

Horatius, when describing his journey from Rome to Naples (which at that time was a true adventure) speaks of 'mali culices', i.e. bad mosquitos, but of course we do not know whether he was referring to insects of the genus *Culex* (today by far more numerous) or of the genus *Anopheles*, to which the transmitters of malaria belong, and which certainly in that period abounded in these regions.

The idea that malaria was transmitted by *Anopheles* is old (Lancini speaks of it at the beginning of the XVIIIth Century), but the real agent was only discovered in 1888 by Laveran and then described by several Italian authors and also by Ross. The demonstration that only mosquitos of the genus *Anopheles* can transmit living and active *Plasmodium* was furnished in 1898 by Grassi. As a consequence of this discovery, the control of malarial mosquitos and the land reclamation – actually the draining of all swampy areas – began. However, an almost complete eradication of malaria succeeded only after World War II thanks to the introduction of DDT by Allied troops in Sicily and the Italian mainland.

The presence of mosquitos in Europe is documented by an illustration of Ortus Sanitatis (Augsburg, 1536).

Another genus of mosquitos, *Aëdes*, represented by *A. aegypti* or *Stegomyia fasciata*, is responsible for the transmission of yellow fever, a human disease due to a virus, the so-called amarillic virus, so named after the Spanish word amarillo (yellow). Yellow fever is endemic in several tropical countries, but in the past it

Fig. 348. The status of malaria in Europe. White: originally not malarial areas; //// originally malarial areas; black: phase of strengthening of the control; dotted: complete eradication of the disease (from the Atlas of Reader's Digest Selection).

reached Spain and from there also Italy (epidemics of Leghorn of 1804). The disease is highly lethal.

Land reclamation is performed by means of: 1) natural drainage, possible when the territory is higher than that of the surrounding channels into which the waters must be transferred; 2) by means of an alluvial fill – when the territory to be reclaimed is situated below the level of the highest surrounding waters, all the territory must be filled with fluvial deposits up to, at least, the level of the surrounding waters; 3) by means of mechanical drying: in this case the water must be raised: generally a special system of pumping is utilized; 4) mixed systems.

Littoral ponds are linked with the sea, so changing them into saline areas or into the so-called 'valli' utilized for fishing (see p. 686), while the water balance between the sea and the lagoon is controlled (see chapter on Continental waters).

Let us remember the land reclamation that occurred in the Paludi Pontine (from Pomptinus or Pontinus, possibly derived from the old town Suessa Pometia); it began, though not in a very rational way, with Pope Pius VI (1717-1799) and was completed only recently. Other famous Italian swamps subjected to land reclamation are those of the Maremma, where malaria was a serious difficulty in the establishment of human settlements (malaria was indeed known in Tuscany as 'Maremman fever'). The land reclamation began with Pietro Leopoldo or Leopold the Second (1747-1792) but was finished only recently.

Within swamps, vertebrates, amphibians, reptiles and birds have a very important role. The reptiles are mainly represented by snakes and tortoises. Two species of the latter inhabit European swamps, *Clemmys caspica*, characteristic of Southern Europe (including the South-Western shores of the Caspian Sea, after which it was named), North-Western Africa and the Near East. It is sometimes abundant, with hundreds of individuals, near the shores of ponds, streams and lakes. It is a 'Rassenkreis'. The great naturalist, the Rev. Tristram, says that this tortoise captures and transports any dead or wounded bird found under water. It feeds mainly on birds, secondarily upon frogs, tadpoles and worms, and it is often caught by fishermen. In some instances it has been seen to feed on water melons growing near the water. It winters under the water in a state of hibernation. Its main enemies are the storks (which feed upon the young tortoises) and the Egyptian vulture *Neophron percnopterus*, which attacks the adults. In 1832 this reptile species was found by Ménétriés in a sulphurous spring at 40°C: this is not surprising when we consider that during the summer this species lives in very warm waters and that it is almost indifferent to the chemical composition of the water in which it thrives.

The other European tortoise is *Emys orbicularis*, present in Central and Southern Europe, West Asia, and NW Africa. In the past it was very abundant all through Northern and Central Europe, up to Scandinavia, Denmark, Switzerland, Württemberg and South Bohemia. It feeds mainly on small fish, snails, insects, worms etc. but also on some plants. It winters, like the Caspian tortoise, in the mud under water. It exhibits very interesting behaviour in egg deposition.

The snakes are essentially two or three species of the genus *Natrix*. *Natrix natrix* (= *Tropidonotus natrix*), a 'Rassenkreis', is spread out in Europe, West Asia, and North Africa. On the Alps it reaches the level of 2300 m. It lives along rivers, lakes and ponds, in swamps and rice-fields. Only occasionally can it be found far from the water; in some instances it has been observed in the sea, 4 miles from the coast. It swims very well and it can remain under water for several hours. Generally it undergoes a very short hibernation. It feeds mainly on frogs and small fish which it swallows alive.

The related *Natrix tessellata* (= *Tropidonotus tessellatus*) seems to be still more closely bound to water than the preceding species; in the Alps it has been found up to 1200 m. It is found in Southern Europe, Western Asia and Egypt feeding on fish, frogs, tree frogs, as well as newts and toads, so that probably it does not compete much for food with the other water snake, with which it very often lives.

The Amphibians of the swamps are essentially frogs (*Rana esculenta, Rana ridibunda*) and tree frogs (*Hyla arborea*); the other species are more or less aquatic, and will be taken into account when describing the aquatic environments.

We could not have a complete picture or idea of the life in swamps if we did not take into consideration bird life. The limits of this book impose on us an extremely brief description of this ecological aspects of swamps and ponds, interesting both from a qualitative and a quantitative standpoint. We can only say that all extant ecological niches are in one way or another utilized – better exploited – by birds. These in their turn exhibit many adaptations (structural, physiological and behavioural) to life in swamps.

Let us mention the pigmy cormorant, *Phalacrocorax pygmaeus*, nesting in swamps; the white pelican *Pelecanus onocrotalus*, nesting in swamps in Bulgaria, Romania and in winter in Northern Greece; the heron *Ardea cinerea* (in German 'Fischreiher') common in humid meadows, river banks and sea shores, and the little egret, *Egretta garzetta* (in German 'Seidenreiher') living in swamps and lagoons; the night heron *Nycticorax nycticorax* (in German 'Nachtreiher') looking for food in swampy territories; the bittern (from the Middle English bitter, bitoure, and the Old French butor) *Botaurus stellaris* (in German 'Grosse Rohrdommel') nesting in swamps; the well known white stork, *Ciconia ciconia* (in German 'Weisstorch') from an old term of uncertain origin nesting on houses and barns; the beautiful spoonbill, so-called also in German (Löffler) because of the shape of its beak, *Platalea leucorodia*, limited to very few European countries nesting often in swamps; the teal, *Anas crecca* (in German 'Krickente', from kriechen, to crouch) living in reed ponds and nesting in mires, swamps, etc.; the very elegant King elder, *Somateria spectabilis* (in German 'Prachteiderente', from aethr (Iceland) eider (Swedish), ederfugl in Danish, in Dutch Koningseidereend, in Swedish Praktejder) nesting in ponds on the tundra, both of Eurasia and North America, so that it may be considered a circumpolar bird. It differs from the related species eider, or *Somateria mollissima* (in German 'Eiderente') because the latter is strictly a seabird.

Two species of geese are common in our swamps, the grey-lag goose *Anser anser* (in German 'Graugans') and the white-fronted goose, *Anser albifrons* (in German 'Blässgans') both nesting in swamps, the latter generally in the Arctic tundra.

The *Rallidae* are represented by several very characteristic species highly adapted to the life in this humid environment: one is the water rail (from the old French rasle, raale) *Rallus aquaticus* (in German 'Wasserralle'), nesting in sedges or reeds on the water; another, the spotted crake, *Porzana porzana*, in German 'Tüpfelsumpfhuhne') common in moors and swamps where it also nests; the purple gallinule, *Porphyrio porphyrio*, in German Purpurhuhn, so called because of its dark purple-blue colour on the dorsal surface, in French Poule sultane and in Italian 'pollo sultano', present exclusively in southernmost Europe, where it nests in the swamps of Spain, Sicily and Sardinia and possibly Puglia. The moorhen *Gallinula chloropus* (in German 'Teichhuhn') nesting in reed beds and upon shrubs near the water, and finally, the coot, *Fulica atra*, in German 'Blässhuhn', having the same habits as the moorhen. Other aquatic birds are the snipe (from the Middle English snype and Old Norse snifa) *Capella gallinago*, in German 'Bekassine' (from the Latin beccus, probably of Celtic origin) nesting in sedges or reeds, only occasionally in heaths (which in many parts of Europe are adjacent to moors or swamps); and the jack snipe, *Lymnocryptes minimus* (in German 'Zwergschnepfe', in Dutch 'bokje' and in Swedish 'Halvenkel beckasin') nesting in swamps and mires; the black-tailed godwit (probably from the Anglo-Saxon god, good and wiht, a creature), *Limosa limosa*, (in German 'Uferschnepfe') during winter in swamps or at the river mouths, during summer in humid meadows and swamps, occasionally also in mires and sand dunes; the dunlin (from dunling, coming from dun = dark-brown and diminuitive) *Calidris alpina* (in German 'Alpenstrandlaufer') nesting in swamps or mires near the sea; the elegant inconspicuous broad-billed sandpiper, *Limicola falcinellus* (in German 'Sumpflaufer', in Dutch 'breedbekstrandloper', in Swedish 'myrsnäppa'), whose length reaches only 16.5 cm, nesting in swamps where it also lives during migrations.

The red-necked phalarope (from φαλαρις meaning coot and πους, foot) *Phalaropus lobatus*, nests in small groups in swamps or near the tundra coasts, often together with its relative *P. fulicarius*; the pratincole, *Glareola pratincola*, in German 'Brachschwalbe', in swamps often near the sea, where it also breeds.

Quite different animals in shape and still more in habit are the pomarine skua (from the Faroese skugver (ca. 1604) coming from Old Norse skufr) *Stercorarius pomarinus*, in German known as 'Mittlere Raubmöve', which literally means predator gull, in Dutch middelste jager, in Swedish Bredstjärtad labb, whereas in North America it is known as pomarine jaeger, all names referring to its predatory habits (in German Jäger means, in fact, hunter), a sea bird nesting in the swamps of the tundra of Northern Russia, Northern Siberia, North America, as well as in Greenland, but during the winter spreading as far as the Mediterranean Sea. The black tern, *Chlidonias niger* (in German Trauerschwalbe, in Dutch zwarte stern and in Swedish Svarttärna), extremely elegant, with a very variable colour during the year makes a floating nest in swamps and lagoons. One true bird of prey common in swampy places, albeit not exclusive to these, is the nocturnal short-eared owl, *Asio flammeus* (in German 'Sumpfohreule') living in humid environments and nesting on sedges or other herbaceous plants. Finally we should mention in this connection some Passerine birds, sometimes very characteristic of swamps: let us note the marsh warbler, *Acrocephalus palustris* (in German 'Sumpfrohrsänger') living in all kinds of humid environments. Its nest is an unorderly construction hanging from the sterns of plants above the dry soil. Very similar are *A. scirpaceus*, reed warbler (in German Teichrohrsänger), *A. schoenobaenus*, sedge warbler (in German Schilfrohrsänger) and, lastly, *A. paludicola*, aquatic warbler (or Seggenrohrsänger) generally found upon sedges. They are all easily recognisable from their songs.

Mountain swamps

If swamps, marshes and ponds are far more frequent in the lowlands, due to the nature of the substrate and soil, in some montane localities the morphology, associated with a given type of substrate and soil, has also allowed the formation of swamps and ponds. On the whole, we may safely assume that on carbonate soils a swamp is generally formed, on siliceous soils a moor.

We would like to bring an example of an alpine swamp, developed in a wide alluvial region (an 'Alm') at 2000 m, between the massif of the Marmolada (at South) and the volcanic groups of Mesola and Padon (at North), separating the Piave valley (East) from the Adige valley (West). The substrate is mainly calcareous (Marmolada limestone), only in the northern part mixed with or substituted by a brown earth derived from the volcanic tuffs. Today, unfortunately, the swamp has disappeared as a consequence of the creation of a man-made lake, the Fedaia Dam, constructed for hydro-electric purposes, with a volume of 16 millions cubic metres. What we will describe here is therefore only a recollection of the past.

The plant and animal life of this swamp was extremely rich. The vegetation was dominated by *Cardamine, Epilobium, Umbelliferae, Polygonum* sp., *Ranunculus acer*, many Graminoid plants and mosses. The fauna included beetles (*Quedius, Atheta, Cytilus sericeus, Sericus brunneus, Anthobium alpinum, Helophorus nivalis* and *H. glacialis*), *Lepidoptera* (*Gelechia, Pyralidae*), *Diptera* (both adults and larvae), Collembols, mites, Enchytraeids, earthworms (*Eiseniella tetraedra, Lumbricus rubellus, Octolasium lacteum, Dendrobaena rubida*, and *D. octaedra*) and finally Nematodes.

Other alkaline swamps extant in the Eastern Alps ought to be protected; there are very few and they are constantly under the threat of construction of hydro-electric schemes, if not simply of drainage for agricultural purposes or – still worse – for tourist recreational aims. Indeed, not many present-day tourists are able to understand the beauty of a mire or a swamp preserved in its original entirety.

Moors

> ... they stood upon a bleak and desert moor,
> where monstrous masses of rude stone were cast about,
> as though it were the burial-place of giants;
> and water spread itself wheresoever it listed,
> or would have done so, but for the frost that held it prisoner;
> and nothing grew but moss and furze, and coarse, rank grass.
>
> Charles Dickens

Whereas in Southern Europe the higher temperature of the water and of the soil causes a complete mineralization of all organic substances produced annually in the swamp, so that new plants can germinate and the perennials can put up their new stems, leaves and flowers each spring, in Central, and even more in Northern, Europe the organic material produced each year cannot be completely mineralized and an anaerobic transformation of subterranean and/or subaquatic parts may occur (fermentation), causing the formation of a peculiar organic substance, rich in carbon, generally called peat or turf.

Whereas in Southern Europe we find marshes, lagoons and swamps in which the prevailing vegetation is made up of flowering plants, and only to a lesser extent by Pteridophytes (ferns and horse tails), in Central, and even more in Northern Europe, instead of swamps we find in practically all humid, freshwater environments an unusual habitat, in which Bryophytes prevail, and more especially some species of the genus *Sphagnum*, whose remains, after anaerobic fermentation, form peat. This humid environment is known as a moor (also mire or bog).

Pines, birches, alders and very occasionally oaks are trees which may live on a moor, replacing willows, poplars and buckthorn which grow in swamps. *Carex*, on the contrary, is present on both moors and swamps, although in general the species differ between the environments.

Whereas the water and soil of swamps are often alkaline or, at least, neutral, in the enormous number of mires we find all over the Central and Northern parts of our Continent as well as in the British Isles and Iceland, the pH is very often more acid than neutral. This is partly due to the presence of humic acids, which are constantly present on the moors but not in the swamps, partly due to the substrate, which in Northern Europe and the British Isles is so often formed by acid or iperacid rocks (for instance granite). (We must not forget that whereas Southern Europe has had a relatively recent geological history, the mountains having originated during the Tertiary, most of them being of sedimentary origin and more or less calcareous, Northern Europe generally has an older geological history, including the oldest 'shields' existing on Earth surface (these shields are very often granitic).)

What are the most remarkable differences between swamps and moors? High productivity in the former, low in the latter; presence of an enormous number of plants and animals, euryionic or in some instances calcicolous in swamps, exclusive presence of titanophobe (or calcifuge) species or at most very euryionic elements on moors, where the pH may reach values as low as 3.2. Finally in the swamps there is often a much higher temperature than on moors, so that this must be considered a limiting factor, strongly reducing the number of species able to live in moors, with acid cold water, and poor nutrient status, but rich in humic acids. For these reasons the physiological or ecological category of plants belongs to oxyphytia (Del Villar), characterized by a strong acidity of soil (and water); the limiting factors are the pH and the paucity of oxygen in the soil. As far as aquatic plants of mires are concerned (see the chapter on quatic environments) one can speak of oxhydrophytia (Del Villar).

Of course we must not think of moors in terms of a general paucity of life: the only constant difference is a lesser specific diversity and a poorer flora and fauna, i.e. a greater monotony when we compare moors with swamps. To all the abovementioned negative factors which intervene, and have intervened since the end of the glacial periods, in the formation of moor biota, we may add shortness of the growing season, which still further reduces the number of species thriving in this peculiar and interesting environment.

Proof of the universal presence of moors all over Central and Northern Europe (and of course also on the mountains of Southern Europe, where, as we have already seen, the environmental legacy of the glaciations is similar to that of northern countries) is shown by the wealth of names used for this environment, their antiquity and the many plants and animals and toponyms and hydronyms named after moorlands. This of course shows us that man has been acquainted with moors since his first appearance in the northern regions.

Initially we should note that moor is the name used both by Germans and by Britons (or at least Englishmen) to refer to the same environment.

Moor comes from the Anglo-Saxon 'mor', used for a wasteland, a moor, a fen; in Iceland we still find 'mor', in the Netherlands 'moer', in Denmark 'mor', and, as we said, in German speaking countries 'Moor'. In the British Isles moor is indeed a tract of relatively open country usually covered with heather (see also the chapter on Heaths) and often marshy or peaty. Sometimes however, moor is used as a synonym of heath.

Many common names are formed with 'Moor': among plants we have moorball for the Alga *Cladophora holsatica* found in sponge-like balls at the bottom of freshwater lakes; moor besom for *Calluna vulgaris*, moorflower for buck bean (in Dutch boksboon) or even goat's bean, the *Menyanthes trifoliata*; moor grass for *Sesleria coerulea*, and finally moorwort for any of a group of low evergreen shrubs growing on a moor, but particularly for *Vaccinium oxycoccus* (from the Middle English wort and the Anglo Saxon wyrt). Among animals we have moor buzzard, the marsh or moor harrier, also called moor hawk, *Circus aeruginosus*; moor cock or moor game for the red grouse *Lagopus scoticus* (moor cock is also black cock, or *Lyrurus tetrix*), moorfowl in Britain is also the red grouse; moor frog is the edible frog *Rana esculenta*; moor coot is any gallinule; moor hen is both the common gallinule *Gallinula chloropus* and the female of the red grouse. Other common names referring to some peculiarities of mires are: moorband (or moorpan) for the hard clay layer underlying many moors, moor coal is a kind of easily broken friable lignite coal; moor evil (liver fluke) is a disease of sheep and cattle (British dialects), or even a quagmire, i.e. soft, wet mirey ground that shakes or yields under foot (corresponding to the German 'Schwingenderasen' or 'Schwingmoore'); finally in British dialect moorstone is a granite, evidently because, as we have said, some moors in the British Isles lie upon old, granite rocks of the Caledonian massif.

Mire, a name also employed to designate moors, comes from the Iceland myrr, or myri, used to distinguish a bog or swamp.

Very common is the term bog (coming from the Irish 'bogach' and the Gaelic 'bog', which means soft, moist).[1] From 'bog' we have today bog bean (or also bog nut) for *Menyanthes trifoliata*, the goat's bean, called also marsh trefoil (in Italian trifoglio fibrino), bogberry for *Vaccinium oxycoccus*, called also bogwort; bog moss is *Sphagnum*; bog oak (or also bogwood) serves to indicate trunks, and large

[1] Here the semantic clarity is evident.

branches of oak, pine, etc. found well-preserved in bogs, of a shining black or ebony colour; bog rush is a sedge growing in moors.

Fen is also very common in England: the name comes from the Middle English fen or fenne, the Anglo-Saxon fen or fenn, for a marsh, a bog, or a fen. From it we find the words fenberry for *Vaccinium oxycoccus* (the small cranberry); fen cricket, for the mole cricket, *Gryllotalpa gryllotalpa*; fen duck for the shoveler, *Spatula clypeata*, a species of wild duck frequently found in fens and marshes; fen goose for the common grey goose or grey-lag, *Anser anser*. Finally fenland is a marshy land: as toponyms we find Wicken Fen, The Fens, etc., both near Cambridge.

Moss, from the Anglo-Saxon Mos, for a swamp, means today a bog, especially in Scottish and Northern English dialects. The same word is of course employed to define the Bryophytes. Moss also appears very frequently in British toponyms.

As far as the product of the slow fermentation of organic substances deposited on the bottom and in the underlying layers of moors is concerned, two names are most usually employed, peat and turf. Peat comes from the Middle English pete, whence the Medieval Latin peta (a piece of turf, probably a specialized variation of petia, which in Celtic means a piece). Peat is used to indicate the true turf, used as fuel in many parts of Europe in the past. Peat bog is actually a bog or marshy place in which peat occurs (we must realize that peat is produced only in acid places, rich in *Sphagnum*).

Turf is a much more interesting term, if we consider it from the linguistic standpoint. The word 'turf' is used in several ways: it comes from the Middle English turf, torf, and the Anglo-Saxon turf and means a surface layer of earth containing grass plants with their matted roots; it is practically the same as sod, from the Middle English, probably in turn from the Middle Dutch sode, akin to English seethe, sodden, referring to the wet ground from which the turf was cut. Practically turf corresponds to sward, or even greensward (from Anglo-Saxon sweard, akin to Dutch zoord, Danish svoer, Iceland svördhr and German Schwarte), all signifying the skin of bacon, and at the same time the surface of the earth.

From turf, the turf ant, *Lasius flavus* has been named; turfworm (or even sodworm) is probably an earthworm; turf moss is a tract of turfy mossy or boggy land.

Turf has its equivalent in other languages: tourbe in French, torba in Italian and Torf in German. Tourbe comes from the French turba (in turn coming from the German Torf) whence tourbière for turf-pit, turf moss, peat moss or peat bog. In Italy we find exclusively torba (peat) and torbiera (moor) which evidently has been named after the former: this is easily understandable when we think of the extreme rarity of this environment in Italy (apart from the Italian Alps).

In Germany Torf is a common word: from it several common names derive, such as Torfboden, a swampy, peaty soil, a moor; Torfbruch, a moor; Torferde, a swampy earth, the peat; Torfgrube, a moor and also torflayer, torfland; Torfmoor, with the same meaning. Torfmoos is the same as moor moss, i.e. *Sphagnum*. Fairly analogous is the German Moor, whence Moorboden, Moorgrund, Moorland, Moorerde, Moorgegend etc. Moor-schnepfe is the common woodcock *Scolopax rusticola*; Moorteich is a pond inside a moor.

Many toponyms, both in the U.K. and in Germany, have had their origin from the term (and corresponding environment) moor (or Moor): in the U.K. we find Moor Park, North Liverpool, as well as NW London, near Rickmansworth; Moor Row in Cumberland near the Solway Firth, not far from Cleator Moor, a town on the coast; The Moors is an extensive region situated in the extreme SW of Scotland (Galloway); Moortown is a small town in Lindsey (E. England). In Belgium and more precisely in the Flemish speaking portion of it (Flanders), we find Moorseele and Moorslede (Western Flanders). Finally in Germany we have Moorlake SW Berlin and NE Potsdam on the shores of the Havel; Moorburg SW of Hamburg on the left bank of the Elbe, Moorende, in the same region, in Das Alte Land, an evidently reclaimed area, Moorfleth, SE Hamburg, on the right bank of the Elbe, Moorrege, near Ütersen, at the Elbe mouth (on the right bank), Moorwärder, South of Hamburg, between Northern and Southern Elbe, and finally Moorwettern, a brook or canal going from Moorburg (East) towards the Este, a tributary of the Elbe (left bank) in Das Alte Land.

This concentration of toponyms and hydronyms named after Moor near Hamburg seems very interesting and maybe would deserve particular research both in palynology and history.

According to morphology, climate, geology and substrate an enormous variety of mires is possible, and classification is very difficult since it is not always possible to recognize the main, decisive, external factor which has determined the present day aspect and development of the mire. Indeed many intermediate, and still more many mixed types, are recognized today.

A modern classification of mires (IBP, CN) gives the following categories. Blanket mire, or also blanket bog ('Terrainbedeckendenmoore'): peat formation is initiated in basins, stream axes and on watersheds where the drainage slope is not too great, the peat forming a blanket over all gently undulating terrain; ridge-raised mires or ridge-raised bogs; peat formation initiated in basin complexes, forming domes which overflow by seepage across very flat water partings between the basins; raised mires, or raised bogs ('Hochmoore'); peat development initiated in basins, peat growth producing a dome or cupola rising above the mineral ground watertable; the classic case is the Baltic raised bog with ring lagg; continental raised mires, or continental raised bogs ('Waldhochmoore') peat formation initiated in small closed basins and terminating in a flat dry surface; they are often dominated by *Pinus sylvestris*; 'Schwingmoore': peat formation initiated as a floating crust or mat, which may cover the entire water surface; Flachhochmoore: peat formation initiated in basins and drainage tracts spreading by seepage over flat terrain, between the basins and drainage tracts to form extensive mires with a flat surface; Talmoore: peat formation is confined to the floors of valleys usually of small dimension; Hangmoore: peat formation occurred in relation to springs, seepage lines and perched watertables, usually along the sides of small valleys or in the valley heads; Rankin/Rose valley mire are complexes of the above two types, formed in small valleys and valley head positions; *Schoenus nigricans* and *Eriophorum gracile* are characteristic species; Flachmoore, or fen: peat formation in the valleys and on flood terraces of large watercourses subjected to the ground water flowing through them; Beckenmoore, or fen: peat formation occurs in open basin with permanent inflow and outflow streams, and subject to the ground water flowing through the system; they are characterized by the fern *Dryopteris cristata*;

Quellenmoore, peat formation occurs around a spring; finally, upland mixed mire complexes, a heterogeneous group of mires usually consisting of complexes or blanket mire, where ground water enriched with calcium bicarbonate comes to the surface in spring or seepage areas. Although, in reality, a mixture of the preceding types, they appear distinct enough to deserve separate treatment, and include the mixed mires of Du Rietz (1949).

Besides these main categories, so-called mire complexes have been recognized and described; they are a series of adjacent mires or different mire parts that are not completely separate but together orm a more complex unit (Cajander, 1913). They can be divided into bog mires complexes and fen mire complexes. In the former, the larger part of the mire or mire complex consists of areas directly, or nearly exclusively, fed by incident precipitation although smaller areas may receive other water. In the second category the whole mire or most parts of it receives water inputs that have been in earlier contact with mineral soil. However, the 'minerotrophy' of some of them is open to question.

Animals characteristic of mires are the dragon-flies *Leucorrhinia pectoralis* and *L. dubia* (the latter only in Hochmoore), *Sympetrum flaveolum, Aeschna juncea,* the beetles *Cyphon, Plateumaris, Pterostichus diligens*, *Diptera* of the genus *Dolichopus*, *Orthoptera* and molluscs. In the Hochmoore several groups such as Amphibians are almost completely absent. Other terrestrial animals common in mires are Diplopods, Chilopods, Apterygota insects, predatory beetles, ants and spiders. Some species seem to be associated with *Vaccinium*, some with *Calluna*, others with forested mires.

We will limit ourselves, in view of the aims of this book, to describing only the two major mire types, i.e. Flachmoore and Hochmoore.

'Flachmoore'

We will consider two cases as examples of the enormous variety of this environment: the environment of very low altitudes in the Eastern Alps, in a calcareous region where the water is not very acid and no *Sphagnum* occurs, and the environment of high altitudes in non-calcareous regions, where species of *Sphagnum* are present. In the former very remarkable differences due to the seasonal cycle are present; in the latter the biota are present and mostly active only during the summer and early autumn (Fig. 349).

Let us consider first a lowland mire. The vegetation is varied due to both the fluctuating water level of the mire and the distance from lakes or streams within the mire, which determine differences in soil composition, presence or absence of peat, different humidity of the soil, etc. The main vegetational associations are the *Phragmitetum* and the *Caricetum flavae*. Within the former we find the purple loosestrife *Lythrum salicaria, Convolvulus sepium, Mentha aquatica, Carex pendula*, the horse tail *Equisetum palustre*, the blackberry *Rubus* sp. ex gr. *fruticosus*, the bittersweet *Solanum dulcamara* (a vine), etc. This vegetational zone changes gradually into that of the Magno-caricetum, represented by *Carex elata, Alisma plantago, Lythrum salicaria, Menyanthes trifoliata, Myosotis palustris, Convolvulus sepium, Lysimachia vulgaris, Parnassia palustris* (the so-called grass of Parnassus), *Salix incana* and *Triglochin palustris*.

Fig. 349. A 'Flachmoor' of a lowland locality, Lipoi, near Feltre (Veneto). General view of the mire and two small ponds where some ecological investigations have been carried out by the author (Photo: Dr G. Dalla Corte).

Fig. 350. Profile of turf in the mire of Fiavè (South-Western Trentino) (photo Dr. G. Dal Doss).

The Caricetum flavae is characterized, instead, by the presence of *Equisetum palustre, Eriophorum polystachium, Tofieldia calyculata, Lotus corniculatus, Potentilla tormentilla* (the common tormentil) and *Potentilla* (or *Comarum*) *palustris* or marsh cinquefoil, as well as *Schoenus nigricans* or bog rush and *Molinia coerulea* or purple moor-grass. In the late spring, *Eriophorum* or cotton-grass gives to this formation the aspect of a true Eriophoretum.

Another, although less well represented, association is a kind of Schoenetum, possibly a Schoeneto-Serratuletum, characterized by *Schoenus nigricans, Serratula tinctoria*, several sedges, *Tofieldia calyculata, Pinguicola vulgaris* or common butterwort (an insectivorous plant), *Molinia coerulea, Parnassia palustris* and *Lysimachia vulgaris* or 'loosestrife'.

Mixed into the preceding formations are small areas of meadow, in which it is possible to recognize a drier facies, characterized by *Dactylis glomerata, Lychnis flos-cuculi, Centaurea jacea, Chrysanthemum leucanthemum, Trifolium pratense, Anthoxanthum odoratum, Brunella vulgaris, Lotus corniculatus, Festuca ovina, Luzula forsteri, Avena pratensis, Ranunculus acer* etc., and another facies, more humid and luxuriant, though represented by a lesser number of species such as *Trifolium pratense, Myosotis palustris, Lamium album* (white dead-nettle), *Holcus lanatum, Artemisia vulgaris* (mugwort), *Onobrychis viciaefolia, Lychnis rubra, Rumes acetosa* (the sorrel), *Rumex obtusifolius* (the broad-leaved dock), *Ranunculus bulbosus, Knautia arvensis*, etc.

Whereas in the Flachmoore of montane zones *Sphagnum* is always present, in the lowlands, where often there are some alluvial areas in which limestone is present if not dominant, the mosses are represented by many species and genera, but not by *Sphagnum*. Here are some representative species of mire mosses in the lowlands: *Acrocladium cuspidatum, Campylium stellatum, Polytrichum glaciale, Cratoneuron commutatum, Mnium affine, M. rostratum, Thamnium alopecurum, Bryum caespiticium, Auracomnium palustre* and *Philonotis marchica*.

The animal population is also completely different in lowland mires from those in the mountains.

In the soil earthworms are common, such as *Dendrobaena octaedra* and *Eiseniella tetraedra*, the latter with amphibious habits, generally living in wet soil at the water's edge. In drier places a millipede, *Schizophyllum sabulosum*, may be present, plus some *Julinae* and *Cylindroiulinae*. Insects are abundant and omnipresent, although not as interesting as those of the montane mires. Let us mention the dragon-flies *Enallagma cyathigerum*, a circum-boreal species reaching as far south as the Appenines where it thrives only above 1000 m; in the Alps it is present up to 2270 m; *Ischnura elegans* and *Agrion puella*, two common species; *Agrion pulchellum*, in Italy rather sporadic; *Cordulia aenea*, a rather northern species, in Italy limited to the Alps, and finally an as yet undetermined species of *Libellula*.

The hoppers of the mire are *Tetrix vittata*, a form limited to Central Europe and Western Asia and *Omocestus viridulus* and *Chorthippus dorsatus*, two rather unimportant species, not at all typical of mires. The entomofauna of this kind of mires also includes *Heteroptera* or bugs, such as *Stenodema calcaratum*, a phytophagous species, *Leptopterna dolobrata*, feeding on the lymph of *Gramineae* e.g. *Phleum* and *Alopecurus; Pachybrachius fracticollis*, restricted to damp meadows, generally on *Symphytum*, and *Eurygaster testudinaria*, a common species. The *Homoptera* are represented by some *Cicadinae* or leaf hoppers such as *Tettigella viridis,* often found on *Scirpus* and *Juncus, Philaenus leucophthalmus* and *Lepyronia coleoptrata* (often in humid situations), as well as by the leaf louse *Rhopalosiphum nymphaeae*, a dioecious species, whose primary host is a prune (*Prunus*) and secondary hosts are several swamp plants such as *Nymphaea, Typha, Alisma*, etc.

The ants of the mire are rather common elements: they are *Tapinoma erraticum, Lasius brunneus, Myrmica laevinodis*, the latter often restricted to wet moss.

Of course, among the most interesting and best known insects are beetles. Staphylinids are represented by *Anthophagus caraboides*, a floricolous species, *Stenus boops* and *S. juno*, ripicolous species bound to rather humid environments such as decaying leaves, vegetation debris, the foot of trees, etc. *Cantharidae* are represented by several generally carnivorous forms such as *Cantharis livida*, *Cantharis rufa, Rhagonycha femoralis* and *Charopus concolor*, none of which are particularly restricted to mires. Much more interesting is the inconspicuous *Telmatophilus* (from τηλμα a small pond) *typhae*, limited to Northern and Central Europe, where it lives gregariously upon the flowers of *Typha* and *Sparganium*, in whose seeds its larva develops. The *Chrysomelidae* are almost as numerous as those of the swamps (p. 504): let us mention *Cassida prasina*, generally present on *Achillea* and *Tanacetum, Cryptocephalus hypochoeridis*, often on yellow *Compositae, Lochmaea capreae*, on willows, poplars and birches; *Chaetocnema arida*, found in humid meadows with sedges and *Juncus*, and finally *Mantura obtusata*, generally related to *Rumex* and therefore not specifically characteristic of humid environments. A very common Megalopteron is *Sialis*, whose larva lives on the muddy bottom of the water. The only snail is *Succinea pfeifferi*, a phytophagous, secondarily phyto-saprophagous species, feeding on dead plants, particularly *Alisma plantago*.

Leaving aside the numerous birds living in mires, either as resident or migrant species, the only vertebrates are Amphibians and Reptiles. Common species, although not particularly restricted to mires, are the common frog, *Rana temporaria*, the edible frog, *Rana esculenta*, the toad *Bufo bufo*, the tree frog *Hyla arborea*, plus the salamander *Salamandra salamandra* (the newt *Triturus alpestris* is a strictly aquatic element, and therefore we do not mention it among the elements of humid environments). The snakes are the water snake, *Natrix natrix*, and the racer *Zamenis* (= *Coluber*) *viridiflavus* var. *carbonarius*.

Mountain 'Flachmoore'

We will consider an environment developed between 1000 and 1500 m, upon siliceous rocks, with *Sphagnum* or more precisely *S. laricinum, S. subsecundum* and *S. magellanicum*, three widely distributed species. The higher vegetation is represented by the horse tail *Equisetum palustre*, *Eriophorum polystachyum* (= *angustifolium*) typical of Central and Northern Europe, the sedges *Carex davalliana, C. stricta* (= *elata*), *C. limosa*, already known from moors, both 'Flachmoore' and 'Hochmoore' (Hegi), *C. rostrata*, present in 'Flachmoore' of Central Europe, *C. lasiocarpa* (constituting, as in the Central European moors, a floating carpet or 'schwingende Decke', together with *Carex rostrata* and *Scorpidium scorpioides*), *Sparganium natans*, the alder *Alnus glutinosa*, an element which in Central Europe causes the formation of the so-called 'Alnetumtorf', *Viola palustris*, according to Hegi characteristic of both 'Flachmoore' and 'Hochmoore'. *Drosera rotundifolia*, the so-called sundew (from δρόσος the dew), has a fairly circum-polar distribution in the Apennine going as far south as Sibolla, characteristic of

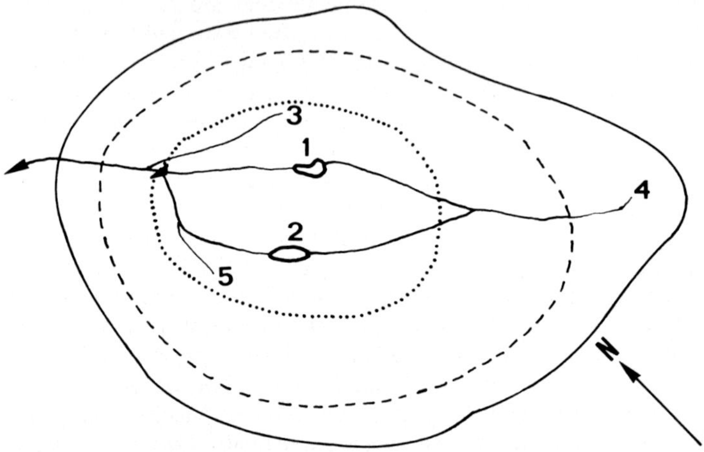

Fig. 351a. The 'Flachmoor' of Lagabrun, Trentino (Northeastern Italy). 1 and 2, the two ponds ('Moorteich'); 3. a small pond with Zygnemales; 4. perennial source; 5. temporary source. Continuous line: limit of the depression of Lagabrun; broken line, limit of the marshy meadow; dotted line, limit of the canebrake.

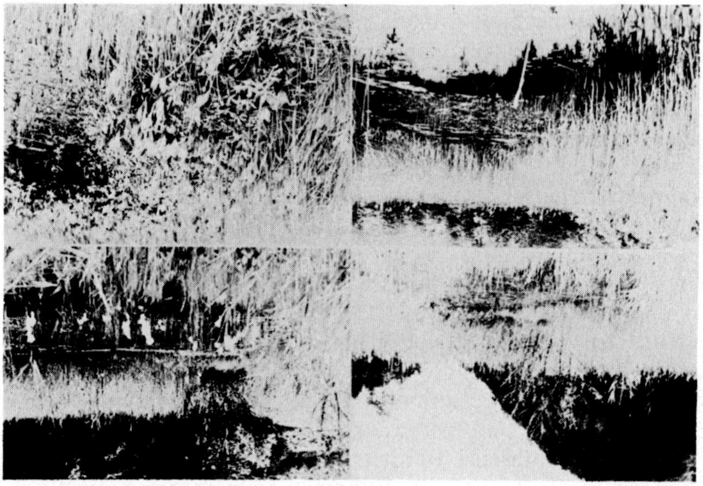

Fig. 351b. details of 351a: 1, a small pond with some trees of *Alnus glutinosa*; 2, the same with *Eriophorum*; 3, another pond with *Potamogeton*; 4, vegetation with *Menyanthes trifoliata* and *Carex* sp.plur.

mires and more precisely of what Rübel calls Sphagniherbosa; it is a well known carnivorous plant; *Potentilla* (= *Comarum*) *palustris*, *Parnassia palustris* or grass of Parnassus, a common hygrophilous species, and lastly the marsh trefoil *Menyanthes trifoliata,* very common in mires ('Flachmoore'), and many other herbs (Fig. 351).

The animal life, limited to invertebrates, is varied. Let us mention the beetles *Pterostichus minor*, and *P. diligens*, both present in wet mosses of the reed bed (Phragmitetum), *Stenus* sp., a Staphylinid, *Cryptobium fracticorne*, present in wet mosses, *Cyphon variabilis*, closely bound to humid situations; the Chrysomelids *Donacia rustica*, a species living upon several sedges, limited to Central Europe, extending as far south as Lombardia (Northern Italy), the inconspicuous *Halticinae Haltica oleracea*, feeding upon *Epilobium, Polygonum aviculare*, etc. and *Aphtona lutescens*, feeding upon the *Lythrum salicaria* and the lady-bird *Coccidula rufa*, a typical swamp species.

Diptera are represented by two species of *Chironomidae* or non-biting midges, *Cricotopus* sp. and *Smittia pratorum*, both characteristic of highly oxygenated waters, by two craneflies, *Tipula autumnalis* and *T. scripta*; by two mosquitos, *Aëdes vexans*, extremely abundant and very annoying, and *Culex* sp.; and finally by a number of *Dolichopus* flying over the water bodies within the mire. Also the scorpion-fly *Panorpa communis* and the *Planipennia Chrysopa vulgaris* are present within the mire vegetation.

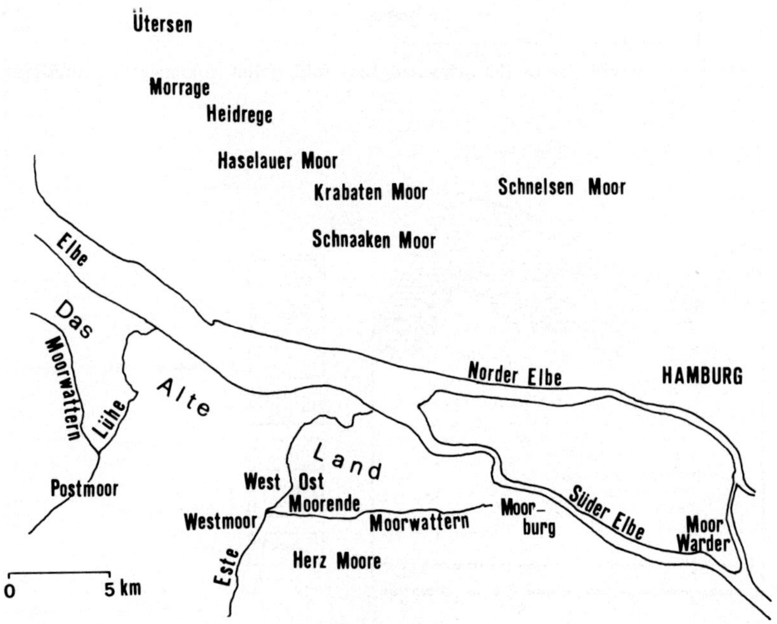

Fig. 352. The old aspect of the Elba River west of Hamburg. Note the abundance of toponyms or hydronyms in which the word 'Moor' (mire) is present (from Stiedler's Atlas, greatly simplified).

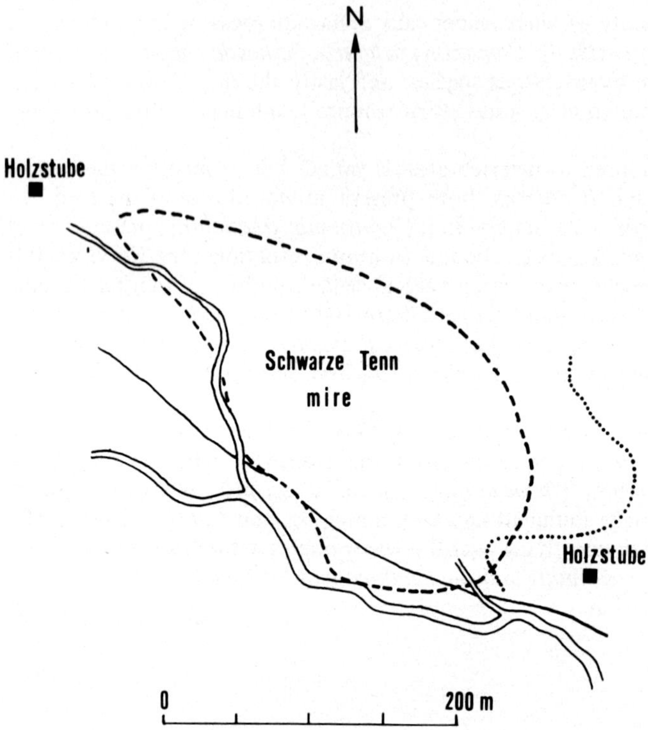

Fig. 353a. The position and size of the Schwarze Tenn mire (from Hohenstatter, modified).

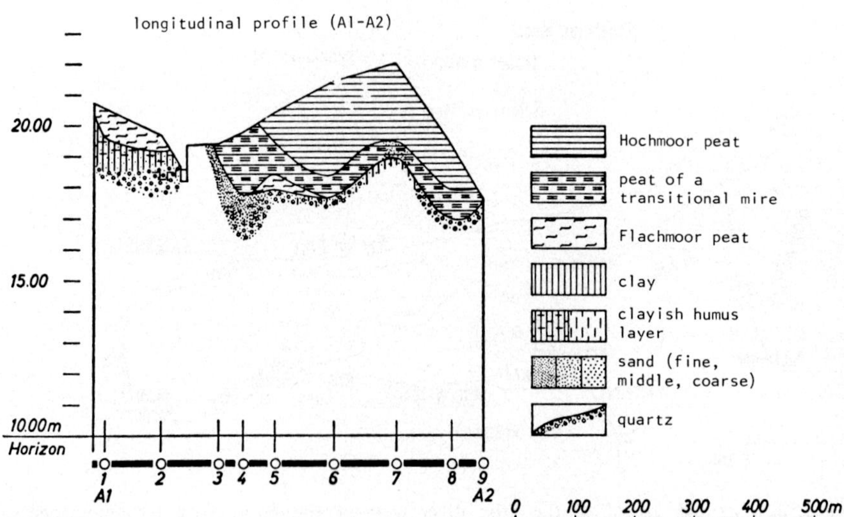

Fig. 353b. The profile of a mire (Schwarze Tenn-Moor) (from Hohenstatter).

Very interesting are the caddis-flies, or *Trichoptera*, for example *Chaetopteryx gessneri*, a species typical of relatively high mountains, whose larva can live only in cold, running waters. *Halesus auricollis* is also limited to montane environments and to cold waters (we must not forget in this connection that the microclimate of mires is usually cooler than that of the adjacent zones); *Limnophilus borealis*, a species limited to Central and Northern Europe, and finally *Limnophilus lunatus* and *L. stigma*, two rather trivial species. Also hoppers are characteristic of montane Flachmoore: let us mention *Chorthippus dorsatus*, also present near water, *Chrysochraon brachypterum*, generally restricted to montane environments, *Mecostethus grossus* commun in moist grassy places, *Conocephalus fuscus*, typical of reed bed and in general of humid environments.

The dragon-flies of this environment are *Sympetrum flaveolum*, a form belonging to a characteristic genus of mires, *Leucorrhinia dubia*, a boreo-alpine element present only in bogs in the Alps and the Pyrenees, reported in Italy exclusively from Cadore, Trentino and Alto Adige; *Somatochlora metallica*, generally limited to Central and Northern Europe, but noted in Southern Europe as far south as Lazio (Italy) and Greece. It is also present in Asia Minor, according to Tümpel almost exclusively in mires, living at the water edge together with common elements such as *Aeschna juncea, Lestes sponsa* and *Ischnura elegans*.

A hygrophilous, but not aquatic, snail is *Succinea putris*, common upon many hygrophilous herbs, feeding upon living plants and only occasionally on dead and decaying material.

'Hochmoore'

For this category of mires we will discuss examples from the Northern Alps, or more exactly from Bavaria (Fig. 353).

One example is offered by the Barenmoos, at 1380 m, near Pfronten Kappel, characterized by *Trichophorum caespitosum, Eriophorum vaginatum* (a cotton grass), *Andromeda polifolia* or marsh andromeda, *Carex limosa* and *Carex rostrata*, the two moss species *Sphagnum acutifolium* and *S. rubellum* and the sun-dew, *Drosera rotundifolia*. At the margin of the mire a belt of marshy meadows is present, with *Scirpus silvaticus, Ranunculus acutifolius, Adoxa moschatellina, Veratrum album, Viola lutea, Atropa belladonna* (the dwale) and *Carex acutiformis*. Beyond this meadow a spruce forest occurs, with *Vaccinium myrtillus*, very luxuriant, and the mosses *Entodon schreberi, Dicranum scoparium* and *D. undulatum, Dicranodontium denudatum, Polytrichum strictum* and *P. juniperinum*, besides several *Hepaticae* (liverwort).

The soil profile of the turf is very interesting and is illustrated by the following table:

Depth (in cm)	Botanical composition	Degree of decomposition (1)	Water (%)	Ashes (%)
0- 50	Hochmoor-*Carex* turf with well-decomposed *Sphagnum*, some *Eriophorum* and many *Rhizopoda*-remains	5-6	80.57	27.56
50-100	Idem	6	80.70	12.62
100-150	Hochmoor turf with *Sphagnum* and *Pinus* wood remains	4-5	84.02	4.55
150-160	*Pinus* forest turf	7	81.21	8.63

[1] According to von Post's scale, ranging from 1 (no decomposition) to 10 (complete decomposition).

Underneath there is a quartz fine sand.

Another example is provided by a mire on Galgenbühl at 840 m; here vegetation is present in which the beautiful *Gentianacea Sweertia perennis* dominates, with *Tofieldia calyculata, Gentiana pneumonanthe* (marsh gentian), and *G. asclepiadea, Parnassia palustris* (the grass of Parnassus), *Succisa pratensis, Galium boreale, Briza media, Molinia coerulea, Schoenus ferrugineus, Comarum* (= *Potentilla*) *palustre* and *Astrantia major*. Where the soil is wet, *Carex rostrata, C. fusca, C. echinata* and *Menyanthes trifoliata* grow.

However, the most typical Hochmoor plants are *Sphagnum acutifolium, S. medium, S. rubellum, Vaccinium oxycoccus, Andromeda polifolia, Potentilla erecta, Aulacomnium palustre, Trichophorum caespitosum, Calluna vulgaris, Carex panicea, Sweertia perennis, Gentiana pneumonanthe* etc. Also the extremely rare sedge *Carex microglochin* is present in this mire. Outside the mire a marshy meadow has developed, with *Symphytum officinale, Geum rivale, Ranunculus sceleratus, Epilobium palustre, Equisetum palustre,* the moss *Entodon schreberi* and the marsh trefoil *Menyanthes trifoliata*. The soil profile of the turf is as follows:

Depth (cm)	Botanical composition	Degree of decomp.	Water (%)	Ashes (%)	pH
0-50	Transition-Hochmoor-turf with remains of *Sphagnum acutifolium* and *Trichophorum c.*, single giant remains of *Calluna* somewhat earthy	6	85.00	12.68	4.0
50-100	Hochmoor-transitional mire turf with *Sphagnum, Eriophorum v.*, giant remains of *Vaccinium, Carex* roots, pollen of spruce, *Rhizopoda* etc.	5	90.70	3.68	4.19
100-150	Hochmoor turf with *Sphagnum, Eriophorum vag.*, remains of birch bark, *Rhizopoda* etc.	5	91.09	3.38	4.30
150-200	Idem, with remains of *Pinus* bark	5	90.53	5.04	4.80
200-250	Hochmoor-forestal turf, with many remains of wood of spruce, pine, *Sphagnum, Eriophorum v.*, etc.	4-5	90.66	4.02	5.19
250-300	Transitional-Forestal turf	4	91.81	4.28	5.82
300-350	*Hypnaceae-Sphagnum* turf very rich in plant and animal remains	4	91.60	5.75	5.88

Depth (cm)	Botanical composition	Degree of decomp.	Water (%)	Ashes (%)	pH
350-380	Muddy turf with many remains of wood of birch, spruce, and of small plants and animals	7-8	91.09	12.50	6.39
380-400	Mineral mud, with single remains of *Hypnaceae*	9	83.44	50.23	6.41
400-440	Muddy turf with many remains of wood of spruce, many plant and animal remains	8	87.70	25.66	6.89
440-470	Humus containing lake-chalk with shells of *Pisidum, Valvata piscinalis*, remains of mosses, etc.	–	77.36	60.33	7.6

The third Hochmoore worth mentioning is a mire situated on the so-called Schwarze Tenn at some 1000 m (Fig. 353), with *Sphagnum rubellum, S. recurvum, S. papillosum, Drosera rotundifolia, Vaccinium oxycoccus, Scheuchzeria palustris, Eriophorum vaginatum, Rhynchospora alba, Carex limosa, C. canescens, Tofieldia calyculata,* and the moss *Hypnum* sp. Near the moor there is a *Pinus montana* grove with *Vaccinium myrtillus* (the common whortleberry), *Vaccinium-vitis-idaea*, the heath *Calluna vulgaris, Carex dioica* as well as the mosses *Aulacomnium palustre* and *Polytrichum strictum*. Also *Molinia coerulea, Eriophorum latifolium, Menyanthes trifoliata* and *Carex flava* are present.

In the turf of this mire, with the remains of *Scheuchzeria*, many animal remains are also present: the *Rhizopoda Arcella vulgaris, Centropyxis aculeata* (Fig. 354), *Hyalosphenia* sp. and *Euglypha* sp., the rotifer *Callidina angusticollis*, shells of *Phyllopoda*, remains of beetles and larvae of *Hydracari*. According to Erika Hohenstatter, who has undertaken excellent research in this mire, it represents almost the ideal type of an alpine Hochmoor, and it has been preserved in its original conditions. The *Scheuchzeria palustris* is here in such a condition that is very difficult to find elsewhere in the Alps today.

Below is the soil profile of the turf:

Depth (cm)	Botanical composition	Degree of decomp.	Water (%)	Ashes (%)
0-50	Hochmoor-turf with *Spagnum, Eriophorum v., Vaccinium ox.*	2	91.53	1.82
50-100	Turf with *Spagnum* and *Eriophorum, Trichophorum c.* and wood	4	86.06	2.03
100-150	*Sphagnum* and *Eriophorum*	3-4	91.85	1.78
150-200	Idem plus *Trichophorum* and *Scheuchzeria*	4	90.55	3.19
200-250	Idem plus some wood	4	90.26	2.76
250-300	'Halbtorf' somewhat slimey	8	82.63	62.92
300-320	Humus containing clay, with some quartz	–	44.50	77.78

As the life on moors is poorer than that in swamps, few birds are also present, both in terms of number of species and individuals. We should mention the bittern, *Botaurus stellaris*, the teal, *Anas crecca*, the beautifully coloured *Aythya nyroca*, ferrugineous duck (also white-eyed pochard), in German 'Moorente' (literally meaning moor-duck) whose habits, however, are the same as those of *A. ferina* which is not characteristic of mires. A very interesting duck of this wet environment is the common scoter, *Melanitta nigra*, in German 'Trauerente', nesting exclusively in small lakes of the tundra Hochmoore. Together with the preceding species, albeit not so closely linked to mires, is the grey-lag goose. Also rather characteristic of the Arctic mires is the whooper swan, *Cygnus cygnus*, in German 'Singschwan', the only swan also present in Northern Italy during summer.

Two other species encountered, but not so characteristic of mires, are the spotted crake and the lapwing. A species particularly restricted to Hochmoore is the golden plover, *Pluvialis apricaria* (in German 'Goldregenpfeiler), nesting on herbs in heaths. Other elements present in European mires are the snipe, the jack snipe, the bar-tailed godwit, *Limosa lapponica*, or 'Pfühlschnepfe', restricted as a nesting bird to the mire of Lappland, and the dunlin, *Calidris alpina*, already mentioned when discussing swamp ecosystems.

An interesting predatory bird from the extreme Northern European mires is the Arctic skua, *Stercorarius parasiticus* (in German 'Schmarotzerraubmöve', in Dutch 'kleine jager' and in Swedish 'Vaulig labb', whereas in North America it is known as 'Arctic jaeger') flying over the sea far from the coast, nesting gregariously on the tundra moors. It is an enemy of the lemming *Lemmus lemmus* (see the chapter on the tundra).

A gull, the lesser black-backed gull, *Larus fuscus* is also present in summer in the mires of Northern Europe, where it nests, but it is not closely associated with this environment. The only bird of prey frequently found is the short-eared owl. Lastly, a passerine bird worth mentioning is the meadow pipit, *Anthus pratensis*, generally present in mires, sand dunes and scrub-land.

Alluvial environments such as moors or swamps are sometimes subjected to slow changes during which they assume more and more attributed of the nearby environment, becoming stable, well drained, mature both in the biocenological and the cybernetic sense of the word.

Thus, if an alluvial, eutrophic lake is subjected to infilling, it will first become a pond, then a swamp and finally a meadow, which initially will be wet, but later on dry. The latter will be colonized in time by shrubs and young trees of the species

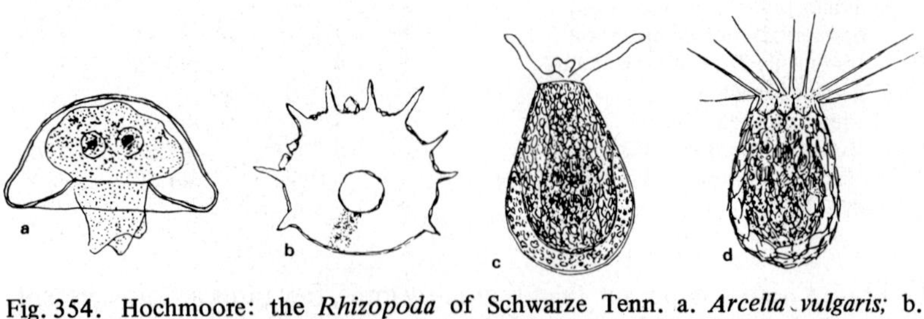

Fig. 354. Hochmoore: the *Rhizopoda* of Schwarze Tenn. a. *Arcella vulgaris;* b. *Centropyxis aculeata;* c. *Hyalosphaenia* sp.; d. *Euglypha* sp. (from Harnisch, Tierwelt Mitteleuropas).

present in the nearby forests so that, at the end of the succession, the lake will be completely replaced by a forest.

In the case of a swamp we may envisage a slow change due to infilling in which sedges (particularly *Carex stricta*) and reeds (*Phragmites communis*) have a paramount importance, so that from the ponds existing today a swampy environment will develop, and the present-day swamp will change to meadow.

Considering the swamp we have described on p. 506, and assuming that it will be subjected to infilling (which is highly probable) we could imagine a transformation like this:

↓

Water several meters deep: *Ranunculus aquatilis, Potamogeton natans, P. lucens, Hottonia palustris, Elodea canadensis, Nymphaea alba, Nuphar luteum, Utricularia minor, Vallisneria spiralis, Ceratophyllum demersum* and *Chara*

↓

Water 20-30 cm deep: *Cladium mariscus, Phragmites communis, Carex stricta, Scirpus lacuster, S. holoschaenus, Juncus, Nephrodium thelypteris, Leucojum aestivale, Gratiola officinalis, Alisma plantago, Caltha palustris, Hippuris vulgaris, Lysimachia vulgaris* and *Typha*

↓

Humid places: *Valeriana officinalis, Menyanthes trifoliata, Valeriana dioica, Cirsium palustre, Thalictrum flavum, Convolvulus sepium, Ranunculus acer, Equisetum palustre, Rhamnus frangula, Senecio paludosus, Plantago lanceolata, Alisma plantago, Potentilla tormentilla, P. reptans, Carex echinata, Hypnum stellatum* and *Scorpidium scorpioides*

↓

Dry places: *Sonchus asper, Hypericum quadrangulum, Peucedanum venetum, Epilobium hirsutum, Solanum dulcamara* (a vine), *Clematis recta, Clematis vitalba* (a vine), *Salix cinerea, Populus* sp. plur, and *Ulmus campestris.*

We must not forget that a similar succession was described by some British students several years ago for an alkaline mire, a fen, which is not fundamentally very different from a swamp – Wicken Fen in Cambridgeshire (The Fens):

aquatic plants

↓

Phragmites (Phragmitetum)

↓

Cladium mariscus (Cladietum)
colonization by shrubs

↓

Rhamnetum

As a demonstration that the swampy environment is similar almost everywhere, not being conditioned by the regional climate or macroclimate, but by microclimate, soil (humus, turf, gyttja, etc.), soil pH and water, etc., we may point to the presence in two such distant environments as an Italian swamp (the one we have described on p. 506) and an English fen (Wicken Fen) of the same plant elements, namely *Phragmites communis, Scirpus lacuster, Typha angustifolia, Potamogeton natans, Hottonia, Utricularia, Cladium, Ranunculus lingua, Nymphaea, Lysimachia, Eupatorium, Thalictrum, Cirsium palustris, Lythrum, Valeriana dioica, Succisa pratensis, Iris pseudacorus, Senecio, Rhamnus* and *Nephrodium*. Some genera are present in both environments, although represented by different species: *Myriophyllum, Convolvulus, Galium, Spirea, Carex, Scabiosa, Juncus, Peucedanum, Lathyrus* and *Hypnum*. In both environments *Characeae* are present; in both the common alder *Alnus glutinosa* is missing.

Proof that the microclimate in the case of damp freshwater environments is of paramount importance is furnished by the presence in Italy (i.e. in Southern Europe) of some northern elements or even true glacial relicts. The microthermous element is represented by *Galium palustre, Potentilla tormentilla, Cirsium palustre* and *Menyanhtes trifoliata*, present in Northern Italy, in Central Italy (Sibolla) and on the Massif Central. A glacial relic must finally be considered *Serratula tinctoria* or saw-wort, present in the northern Italian swamp we have described, in the humid environment at Sibolla (Tuscany), and in Languedoc.

The shores of lakes

The shores of eutrophic, alluvial lakes share their plants and animal life with those of swamps and moors, accordingly the characteristics of soil, substrate, pH etc. are also similar. Only the oligotrophic, often montane or high alpine lakes exhibit a rather bare shore, rocky or stoney, with only very scanty macroscopic vegetation and animal life.

On the shores of one of the largest Italian and Southern European lakes, Lake Maggiore, three main cenoses have been recognized, namely an association of sandy shores with *Stenus bipunctatus*, a very widespread Staphylinid beetle and *Omophron limbatum*, a Carabid present along the sandy shores of most European rivers, together with three secondary synusiae (or sub-associations), one characteristic of the transition towards shore woodland ('Auen'), characterized by *Phyllodecta tibialis*, and *Chaetocnema semicoerulea*, two Chrysomelids, the former a Central European species common on willows, and particularly on *Salix purpurea*, the latter a widely distributed species common on willows, whose larva feeds on some tall Graminoid plants; another synusia is characteristic of the transition towards meadow associations, exemplified by a bug of the genus *Cymus* and by the beetle *Corticarina gibbosa*, a very common element, and finally a third, characteristic of the transition towards the reed bed, characterized by the above-mentioned *Omophron limbatum* and the weevil *Lepyrus palustris*, a holoarctic element living on *Rumex* as a larva and on several species of *Salix* as an adult.

A second cenosis is typical of shore-meadows: it is characterized by the hopper *Homocoryphys nitidulus* and the bug *Adelphocoris seticornis*, a Euro-sibiric element. The third cenosis is restricted to the reed bed, and is characterized by the Dipteran *Ceresa bubalus* and the weevil *Apion viciae*, widely distributed and living on several species of *Vicia*.

Along the shores of lowland lakes, near the water, we find earthworms which can withstand prolonged submersion, such as *Eiseniella tetraedra*, particularly hygrophilous snails, such as several species of the genus *Succinea*; Collembolans or springtails particularly adapted to life in humid environments, such as *Podura aquatica* and *Sminthurus aquaticus*, besides several adult insects whose larvae live and develop in water, such as *Ephemeridae*, dragon-flies, caddis-flies, *Diptera*, plus a number of birds and only to a lesser extent other vertebrates (mammals, reptiles and amphibians) not so typical of this environment.

Maybe the best known European lake as far as ecology is concerned is the Neusiedler See at the border between Austria and Hungary. The enormous variety of soils along its shores is responsible for the variety of vegetation and animal life. On solonetz soil (see the chapter on steppes) a vegetation typified by *Camphorosma annua* (the so-called 'Kampferkraut'), *Puccinellia limosa* ('Sumpf-Zickgras') *Pholiurus pannonicus* ('Schuppenschwanz'), etc. is present. On solontchak vegetation is represented by the following species: *Lepidium cartilagineum* ('Salzkresse'), *Puccinellia salinaria* (or 'true Zickgras'), *Suaeda maritima* (the so-called 'Salzmelde'), and *Crypsis aculeata* ('Dorngras'), *Cyperus pannonicus* ('Ungarisches Zypergras'), *Bolbochoenus maritimus* or 'Knollenbinse', etc.

Where flooding occurs during summer, a salty meadow rich in species develops, with *Juncus gerardi, Scorzonera parviflora* etc. *Carex distans* lives only in the driest places (Fig. 355).

Many of these elements are exclusive to salty steppes of Central and Western Asia and their presence along a European lake is due largely to the geology and pedology of this interesting region.

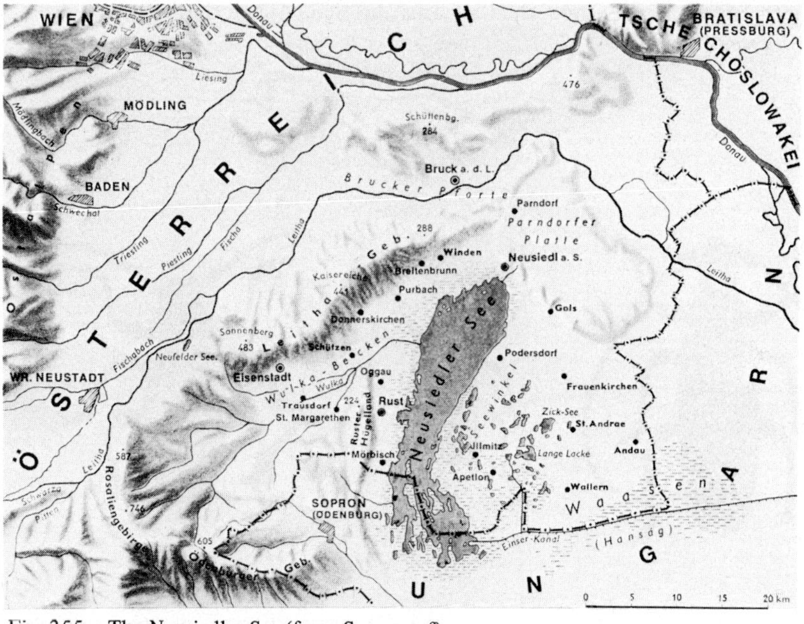

Fig. 355. The Neusiedler See (from Sauerzopf).

The snails of the shores of Neusiedler See are very rare: in the Caricetum or sedge belt *Bithynia tentaculata* is present, a very common species.

In the marshy meadows the following species can be found: *Arianta arbustorum*, a common species, *Helicella obvia, Helicella striata, Trochoidea geyeri*, an element limited to Western Europe, with Neusiedlersee as its eastern limit. Where trees are present *Vitrina pellucida* and *Succinea oblonga* can thrive.

Extremely varied and rich is the arthropod – or more especially the entomofauna – of the lake shores, recently described by Kühnelt.

Where woodland is present, with poplars (*Populus alba* and *P. nigra*) and the introduced black locust tree *Robinia pseudoacacia*, we may find the scale insect *Lecanium robiniarum*, often parasitized by *Anthribus nebulosus*, an *Anthribidae*. Upon poplars *Anomala aenea*, a Scarabeid, the *Cerambycidae Saperda carcharias* and the Chrysomelid *Cryptocephalus fulvus* (whose systematic status is still obscure) are present. In the soil mites are very abundant (especially Oribatei) and also a pseudoscorpion, *Atemnus politus*, may be found. In the small woodlands on moist soils along the shores a Cariceto-remotae-Fraxinetum is present (i.e. a woodland with alder and ash), where the Chrysomelid *Agelastica alni*, typical of alders, and the weevil *Lignyodes enucleator* restricted to the litter are present, whereas in the soil some large beautiful Carabids live, such as *Carabus glabratus* and *Carabus variolosus.*

In more humid environments, and more exactly where *Carex elongata* is associated with *Alnus nigra*, live the *Buprestidae Dicerca alni*, restricted to alder, and some elegant *Cerambycidae* such as *Rhamnusium bicolor, Aegosoma scabricorne* and *Lamia textor*, found on willows and poplars. Other elements of this association are the *Chrysomelidae Gastroidea viridula* restricted to *Rumex* (exclusive to Northern and Central Europe, the Caucasus and Siberia) and the *Helodidae Scirtes hemisphericus* (Fig. 356).

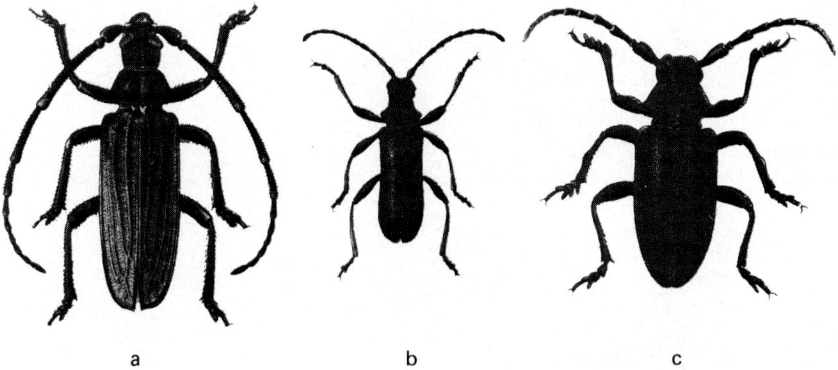

a b c

Fig. 356. a, *Aegosoma scabricorne*; b, *Rhamnusium bicolor*; c, *Lamia textor* (from Schaufuss).

In the transition association between meadows and Caricetum (constituted by *Carex elata* and *C. gracilis*) some peculiar insects thrive, such as the crickets *Gryllus campestris* and *Acheta deserta*, and mosquitos belonging to the genera *Culex* and *Aëdes*.

The most interesting associations, however, are, as one would expect, the Caricetum and the Phragmitetum (or reed bed). The former, which on the western shore of the lake reaches some 4 km of width, is inhabited by the Aphid or leaf-louse *Hyalopterus pruni*, the scale insect *Antonina phragmites*, the Dipteron *Lipara lucens*, a gallicolous species, besides some *Cecydomyidae* (also galligenous elements) for example *Giraudiella inclusa* and *Thomasiella flexuosa*. In the stems of the sedges some *Lepidoptera* develop, such as *Leucania* and *Nonagria*, whereas the leaves are eaten by the caterpillar of *Laelia coenosa*. In the soil the Carabid *Odacantha melanura* is very abundant and in the driest places within the Caricetum the spider *Trochosa singoriensis* (the so-called Russian tarantula) may be found. Finally, within the reed bed some elements such as *Malachius vulneratus* (probably restricted to salty places), *Bombus muscorum*, often nesting in the soil, the two hoppers *Tetrix bolivari* and *T. subulata*, and the predator *Pirata piraticus*, a spider, are present.

The vertebrates of the shores of the Neusiedler See are as interesting as the invertebrates from the ecological standpoint.

The Amphibians are *Pelobates fuscus, Bufo bufo, Bufo viridis*, the tree frog *Hyla arborea*, the moor frog *Rana arvalis* (in swampy meadows) and the edible frog *Rana esculenta*. The Reptiles are represented by the pond tortoise *Emys orbicularis*, the moor lizard *Lacerta vivipara*, the two water snakes, *Natrix natrix* and *Natrix tessellata*, and finally, though very rare, the smooth snake *Coronella austriaca* and the viper *Vipera ursinii* (locally known as 'Wiesenotter'). There are an enormous number of birds[1].

Mammals are represented by the mole *Talpa europaea*, the common shrew *Sorex araneus* (the commonest mammal of this region), *Neomys anomalus, Neomys fodiens* or watershrew (often found in abandoned nests of birds such as *Porzana* and *Panurus*), the fox *Vulpes vulpes*, the ermine *Mustela erminea*, the weasel *Mustela nivalis*, the polecat *Mustela putorius*, the badger *Meles meles*, the otter *Lutra lutra* (today very rare), the wild boar *Sus scrofa*, the deer *Cervus elaphus* and the red deer *Capreolus capreolus*, the hare *Lepus europaeus*, the dwarf mouse *Micromys minutus*, very abundant especially in the Caricetum, the water rat *Arvicola terrestris*, closely associated with water, the introduced muskrat ('Bisamratte') *Ondatra zibethica*, which came to the Neusiedler See from Bohemia in 1923-26 and today is widely distributed in all Caricetum and swampy biotopes, and finally the swamp mouse ('Sumpfwuhlmaus') *Microtus oeconomus*, very abundant everywhere around the lake.

Mountain lakes

In the montane regions lakes are very often oligotrophic and the shores are stony or rocky, so that life in this marginal environment is extremely scarce. In contrast,

[1] We recommend the publication 'Das Buch vom Neusiedler See' if the reader is interested in this topic.

if the water body is not too deep and is situated on alluvial ground so that the shores are not steep, some vegetation may grow near the water, and the basin will be eutrophic. Sometimes this kind of water body can undergo eutrophication through the presence of cattle grazing near the water and watering in it. Of course, in this case life is more luxuriant.

We will cite only two examples, both taken from the Dolomites, one small lake of the alpine zone and a pond of the coniferous region. The lake is situated at 2000 m in the Alnetum viridis-belt, on volcanic tuffs (Lago di Fedaia); the pond is situated upon dolomite and is not far from the Passo M. Croce di Comelico, at 1550 m. Let us outline briefly the main characteristics of life on the shores of these two alpine water bodies; we will return to them in discussing life in European freshwaters.

On the shores of Lago di Fedaia the vegetation mainly consists of an Alnetum, i.e. a formation with the alpine alder, *Alnus viridis*, which succeeds altitudinally forest regions on siliceous soils. The lake edges are barren, muddy, and only in some parts represented by a coarse calcareous sand mixed with clay. The only herbs are a species of *Arabis* and an *Equisetum*. The fauna is exclusively earthworms (*Dendrobaena rubida, Lumbricus rubellus, Octolasium lacteum, Eiseniella tetraedra*), mites, collembolans, Coleopteran and Dipteran-larvae, Enchytraeids, a few snails of the species *Euconulus trochiformis*, a few beetles (*Anthophagus fallax* and *A. melanocephalus* near the alders, *Atheta tibialis* and *Simplocaria semistriata* near the water) and finally one species of bug, *Saldula orthochila*, a typical hygrophilous animal, eurytope but stenoikous.

The small pond, situated in the spruce forest, has an extremely luxuriant vegetation on its shores: *Carex goodenowii, Juncus articulatus, Caltha palustris*, and many mosses, such as *Cratoneurum commutatum* and *Acrocladium cuspidatum*. A little farther away from the water, although still within the wet environment, we find *Parnassia palustris, Potentilla tormentilla, Orchis* sp., *Eriophorum polystachium, Agrostis alba, Carex flava* and *C. vescicaria, Potentilla anserina, Rhinanthus minor*, the beautiful red-cattle *Pedicularis palustris* and *Valeriana dioica*.

The faunal constituents are earthworms (*Dendrobaena octaedra, D. rubida, Octolasium lacteum, Allolobophora* sp.), snails (*Cochlicopa lubrica, C. lubricella, Aegopinella pura, Euconulus fulvus, Trichia hispida, Arianta arbustorum, Vitrea subrimata* and *Succinea oblonga*, besides the slug *Limax maximus*), *Diptera*-larvae (*Dolichopodidae, Tipulidae* and *Fannia* sp.), *Coleoptera*-larvae, adult *Coleoptera* and more particularly the *Staphylinidae Quedius unicolor, Tachinus corticinus, Stenus tarsalis* and *Actobius cinerascens*, and the Carabid *Pterostichus diligens* (very characteristic of mossy wet places), spiders (such as *Centromerita bicolor* and *Oedothorax retusus*), centipedes, millipedes (*Cylindroiulus grödensis* and *Unciger foetidus*), mites and springtails. Needless to say the pond is eutrophic.

On the shores of the lakes are sites – permanent or not – of a great number of birds. Just as an example we will mention the most typical ripicolous species (albeit not necessarily limited to lakes) such as the great crested grebe (*Podiceps cristatus* (in German 'Haubentaucher') with a nest floating upon aquatic plants, the horned grebe or *Podiceps auritus* (Ohrentaucher), and the blacknecked grebe *P. nigricollis* (in German 'Schwarzhalstaucher'), nesting gregariously; the cormorant *Phalacrocorax carbo*, an euryoikous species nesting sometimes gregariously upon trees growing on the small islands in the lakes; the heron *Ardea cinera*; the great white heron, *Casmerodius albus*, limited to South-Eastern Europe, the little bittern, *Ixobrychus minutus*, nesting upon the water, occasionally in small colonies; the scaup, *Aythya marila*, in German 'Bergente' generally a sea bird, but nesting gregariously in small islands in the lakes.

More typical for lakes is the tufted duck, *Aythya fuligula* (in German 'Reiherente'); restricted during the breeding period to small islands in the lakes of the tundra is the longtailed duck, *Clangula hyemalis* (in German 'Eisente', in Dutch 'IJseend', in Swedish 'Aifågel', and in North America 'old-squaw'), characterized by a very peculiar summer distribution of boreo-alpine type. To the same humid environment belongs the common scoter *Melanitta nigra* with the same summer quarters, but particularly found on the Hochmoore. Also, two geese may be found along the lake shores: the grey-lag goose, *Anser anser*, and the bean goose, *Anser fabalis* (in German 'Saatgans'), in Dutch 'rietgans' and in Swedish 'sädgås') nesting in the Arctic. Still more northern is the snow goose, *Anser caerulescens*, in German known as 'Schneegans' (which means exactly snow goose), in Dutch 'sneeuwgans' and in Swedish 'snögås', limited as a nesting bird to circumpolar regions, where it has given origin to several geographical races (Fig. 357). A lake bird in its wild state is the mute swan, *Cygnus olor*, in German 'Höckerschwan', present for example in some lakes of Eastern Prussia, a partially migratory bird, reaching as far south as Spain and Italy. Not dissimilar in habits, but more northern, is the whooper swan, *Cygnus cygnus*. Finally let us mention two repicolous species, the common sandpiper *Actitis hypoleucos*, belonging to the *Scolopacidae*, and the red-necked phalarope, *Phalaropus lobatus*, belonging to the *Phalaropodidae*

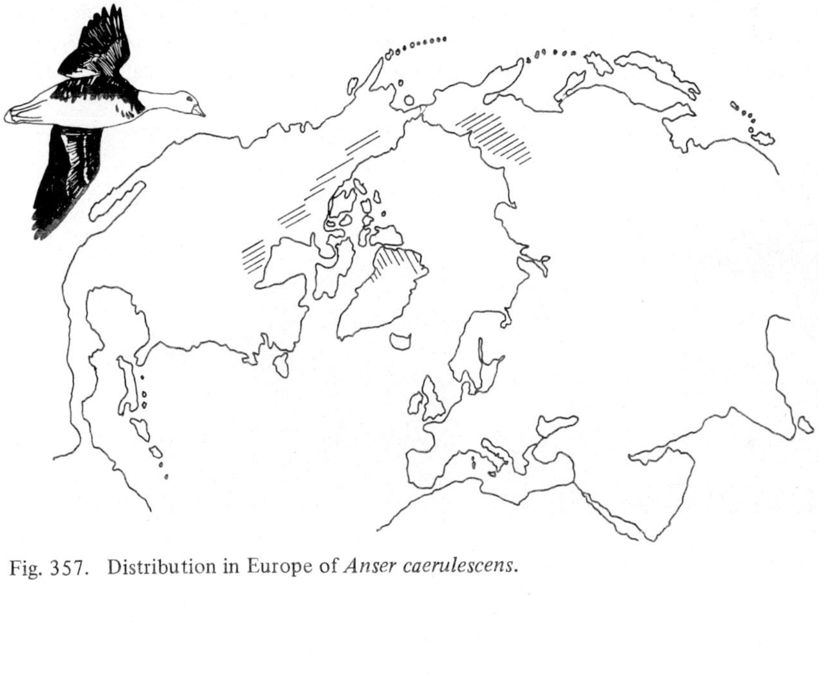

Fig. 357. Distribution in Europe of *Anser caerulescens*.

τοῦ μὲν πετάλοισιν ἐπ' ἀκροτάτοισ'
ἰζάνοισι ποικίλαυ
πανέλοπες καὶολόδειροι λαθιπορφυρίδες
καί ἀλκυόνες τανυσίπτεροι

(Upon its highest leaves magnificant ducks slight,
with shining necks and with the colour of porphyry,
and Kingfishers with very, very long wings)

Ibycus

We will take into consideration as an example of a river of Southern Europe the Adige River (Etsch), recognising that the biota is similar to that of most rivers, provided that, as often occurs, the geomorphology, substrate, soil, climate, etc., are the same. We will limit ourselves to the invertebrates, since vertebrates are in general widely distributed and are able to spread out not only along a river but also from one catchment to another.

An earthworm, *Dendrobaena rubida*, almost cosmopolitan, inhabits the soil, present also in very humid situations such as lake shores. The woodlice are *Philoscia affinis*, characteristic of forests in which there are some streams, although present also in fields and meadows; it may reach 1400 m in mountains. It was originally found in Italy from where it spread to several parts of Europe; *Trachelipus rathkei*, a eurytopic species, apparently indifferent to soil type, may reach in Switzerland an elevation of 1800 m. It is widely diffused along the streams, migrating both actively and passively, since it withstands submersions of several days and can be transported by floods for great distances. An Opilionid, present on the banks of the Adige, is *Nelima doriae*, a form limited to the Western and Central Mediterranean basin. Very many spiders occur, partly living in the soil, and partly on flowers, leaves, etc. Let us mention here *Araneus ocellatus; Mangora acalypha*, in Central Europe living in the forests; *Singa hamata*, found in Central Europe on

Fig. 358. Two aspects of the River Sile towards the mouth. On the shores several poplars are well visible; in the foreground at left *Phragmites communis.*

shrubs and herbs, *Singa heri*, generally on swamp plants, more particularly reeds (*Phragmites*), *Pachygnatha listeri*, generally under mosses, stones and on shrubs; *Tetragnatha extensa*, common on herbs and shrubs growing on the river-banks; *Pisaura mirabilis, Clubiona* sp., *Micrommata* sp., *Misumenops tricuspidatus,* generally on white and yellow flowers[1], *Xysticus* sp., *Myrmarachne formicaria*, in Central Europe generally at the base of trees; *Dictyna pusilla*, a very euryoikous species, in Central Europe found on sandy heaths and in spruce forests. The centipedes are *Lithobius forficatus*, a widely distributed, common species and *Monotarsobius microps*, typical of SW Europe. Millipedes are rare, represented only by some *Cylindroiulidae*. The hoppers are interesting: *Zeuneriana marmorata* is a very relict species, present in a few parts of Southern Europe, *Conocephalus fuscus* and *Homocoryphus nitidulus* are two rather trivial, very hygrophilous species, *Tetrix ceperoi* is also common, but more thermophilous than the preceding species; the bugs living here are *Eysarcoris aeneus*, a eurosibiric mediterranean element frequent in marshy meadows, particularly upon *Stachys* and in some cases on sedges; *Orius minutus* is a palearctic element, in Italy present in meadows, sedges, reed beds and also on coastal dunes; *Adelphocoris lineolatus* is a widely distributed species living on *Leguminosae* and *Compositae*, but also in Caricetum, Scirpetum and Phragmitetum; *Adelphocoris ticinensis* is particularly restricted to wet places, where it lives on *Lythrum salicaria, Achillea* etc., *Nabis ferus*, a common European species; *Acompus rufipes*, common in humid meadows, *Coreus marginatus*, also frequent in wet conditions, particularly upon *Rumex* or shrubs. The *Cicadinae* or leaf hoppers are represented by some widely distributed species such as *Empoasca pteridis* (= *Typhlocyba rosae*), *Tettigella viridis*, in marshy meadows on *Scirpus* and *Juncus* (sometimes harmful to cultivation of *Sarothamnus*) plus *Eupteryx vittata* and *Philaenus leucophthalmus*. Beetles are much more interesting. Let us mention the inconspicuous *Bembidion pygmaeum*, mainly ripicolous occurring both in lowlands and on mountains; *Demetrias monostigma*, in swampy areas, characteristic of reed beds; in the Netherlands this species lives in the yellow dunes, up to 75 m from the sea, among the leaves of *Ammophila arenaria*, the Marram grass, where there is a constant temperature and a relatively high level of humidity; *Oodes helopioides*, on swampy soils, generally in lowlands, a very hygrophilous element, sometimes completely submerged by water; *Asaphidion stierlini*, ripicolous, hygrophilous element, generally on wet soils; in Northern Italy it exhibits ecological requirements similar to those of the similar *A. flavipes*, i.e. shady banks of streams, in the Populetum-Alnetum in lowlands, in the Alnetum in upland localities; it is relatively stenothermous; *Agonum moestum* and *Europhilus micans*, ripicolous, present also on swampy soils; *Anisodactylus binotatus*, in humid places, often ripicolous; *Drypta marginata*, with similar habits; *Platynus obscurus*, near water courses in very humid woodlands; hygrophilous; *Platynus dorsalis*, a species reaching altitudes as high as 2000 m, on humid soils, although in laboratory studies it proved to be rather thermophilous and photophilous; *Melanophthalma distinguenda*, a very small cosmopolitan beetle, living both in the lowlands and in mountains; *Catops nigricans*,

[1] A case of colour resemblance (possibly also adjusted resemblance) belonging to what Cott calls, 'adventitious allurement'.

a nidicolous species, generally associated with the burrows of rabbits, but sometimes in abandoned wasps' nests; *Scaphosoma agaricinum*, very common in tree-mushrooms, under bark and in rotten stumps; *Coccidula rufa*, typical of swampy plants, particulary reeds; both larva and adult feed upon leaf-lice; *Coccinella septempunctata*, the common lady-bird, an important destructor of aphids, and therefore very useful to cultivated plants; *Propylaea 14-punctata*, another kind of lady-bird, also aphidivorous although much less common than the preceding species; *Cantharis rustica* and *C. livida*, not particularly limited to humid environments; *Elater pomonae*, generally upon trees such as oaks, beeches, birches, pines, etc., *Cyphon* sp., a characteristic element of reed-beds; *Drilus flavescens* whose larva is a parasite of snails; although the male lives on flowers, the female is found under litter or among herbs; *Tropinota hirta*, a floricolous Scarabeid beetle living on the flowers of *Viburnum, Genista, Taraxacum, Ranunculus, Plantago*, etc., whereas the larva lives in plant detritus, among roots, etc. The Chrysomelids are very characteristic: *Plagiodera versicolor*, on several species of willows and poplars (*Salix alba, S. viminalis, Populus nigra* etc.) to which it can be harmful; *Aphtona lutescens*, whose adult eats buds and leaves, whereas the larva develops inside the same plants attacked by the adult, often *Lythrum salicaria; Longitarsus suturalis,* on *Lithospermum officinale, Calluna vulgaris* etc.; *Chrysomela graminis,* a Central-European species generally living in humid places on *Mentha aquatica; Epithrix pubescens*, a very inconspicuous insect living on *Solanaceae* such as *Solanum nigrum, S. dulcamara, Lycium barbarum, Hyosciamus niger*, etc., *Haltica oleracea*, a widely distributed species feeding on *Polygonaceae* and *Oenotheraceae*; the beautiful *Cassida viridis*, bound to *Labiatae*, such as *Mentha, Lycopus, Galeopsis, Stachys, Salvia* and *Melissa*; the larva feeds upon the leaves; two species of *Chaetocnema, concinna*, on *Polygonum* and *Rumex*, sometimes on *Rheum* (the common rhubarb), where the adult lives on the leaves and the larva on the roots, sometimes harmful to sugar beet; and *tibialis*, bound to *Chenopodiaceae* such as *Atriplex, Beta, Chenopodium, Salsola, Salicornia* etc., harmful like the preceding species to sugar beet; *Lema melanopa*, both on cultivated (wheat, oat, barley, not on rice) and wild *Gramineae* (*Agropyrum, Lolium, Dactylis, Hordeum*, etc.); it prefers rather cool, shady places, and finally *Emarhopa rufa*, a Ponto-Mediterranean species bound to *Convolvulus* (*arvensis, sepium*) generally not far from the sea. Other elements common on the shores of this river are *Spermophagus sericeus*, an elegant small beetle living in the fruits of several *Convolvulus* such as *C. soldanella, C. arvensis, C. sepium*, etc. and finally many weevils. Let us mention only *Ceutorhynchys symphyti*, living in humid environments, swamps, along the breaks of forests etc. generally on *Symphytum* (both *tuberosum* and *officinale*); *Cidnorrhinus quadrimaculatus*, all over Europe on the nettle (*Urtica urens* and *U. dioica*), *Lepyrus armatus*, rather large, more or less restricted to humid places, where it lives on shrubs and herbs; *Nanophyes marmoratus*, a very inconspicuous elegant animal, generally in humid places on the purple loosestrife *Lythrum salicaria*, in whose ovaries the larva develops; *Phytobius waltoni*, upon the water-pepper *Polygonum hydropiper*; the genus is bound to humid meadows or to the edge of waters, where its members live on the plants; the larva feeds freely on the stem of these plants and is generally stenophagous; *Lixus iridis*, generally on *Umbelliferae* such as *Chaerophyllum, Sium latifolium, Conium maculatum*, etc., the larva in the stem of *Cicuta* (cow bane) and *Heracleum spondyloides* (cow parsnip), but also on reeds and iris; the beautiful *Polydrosus coruscus*, limited to Central Europe, living on several species of willows, on the shores of waters both in the lowlands and in mountains; two species of *Sitona, humeralis*, whose larva lives on the roots of several species of *Medicago* while the adult occurs on the same plant and *Ononis repens, Lathyrus aphaca, Trifolium repens* and *Pisum sativum; suturalis*, whose larva feeds upon the roots of several plants and adult lives on *Lathyrus pratensis, Vicia villosa, Vicia cracca, Ononis* etc. Also *Mononychus punctum-album* is very typical of humid environments: it is generally found on several species of *Iris*, both wild and cultivated; finally we must mention *Rhamphus pulicarius*, whose larva is a miner living on *Betulaceae* and *Salicaceae* such as *Betula alba, Salix viminalis, S. alba, Populus nigra*, while the adult is found on alders and hazels.

Another insect very common on the banks of these rivers – though not at all characteristic of this environment – is *Panorpa communis*, the scorpion-fly, a very common species.

Finally we must not forget molluscs, which however are not typical of this environment, such as *Cepaea nemoralis, Eulota fruticum* and *Limax maximus*. The former inhabits gardens, fields, small woodlands, and cultivated areas as well as human settlements. In the Alps it reaches 1300 m; it feeds on many higher fresh plants, occasionally also on the litter of deciduous species, as well as on many fruits. It exhibits both aestivation and hibernation. *Eulota fruticum* is a European Northern-Asiatic element, on the Alps up to 1700 m, common in gardens, fields, at the edges of woodlands, on herbs, shrubs, etc. It feeds upon nettle, hop, rasberry and blackberry, sometimes also on *Sambucus ebulus* and other plants, both arboreal and herbaceous. It exhibits hibernation and sometimes also a short aestivation. Finally, *Limax maximus*, reaching on the Alps 1900 m; it is found near human settlements, in cellars, gardens, parks, yards, but also in woodlands, under stones or on old stumps; it feeds on the leaves of many plant species, as well as on several species of higher mushrooms. It is also partially carnivorous.

Mountain rivers and brooks

Quite different, of course, is the life along a small mountain stream. We want to introduce here, as an example, a small stream in the Dolomites, near the Passo M. Croce di Comelico, and more precisely the Rio Bianco (literally 'white brook'), so called because it flows over dolomite and limestone and therefore sometimes its water is milk-white from the presence of extremely fine calcareous particles. The vegetation of the banks includes small willows (*Salix caprea* and *Salix purpurea*) and larches (*Larix decidua*), associated with herbs such as *Equisetum variegatum, Glyceria* sp., *Poa* sp., *Carex paniculata, Juncus compressus, Trifolium repens, Ranunculus montanus, Carum carvi, Polygonum bistorta, Tussilago farfara* (the common coltsfoot), *Caltha palustris* (or king cup), the tormentil, *Potentilla tormentilla*; some mosses are mixed into the flowering plants, partially covering the bank gravel. The fauna is very poor, probably because of frequent flooding; some earthworms (*Lumbricus rubellus*), Enchytraeids, snails (*Cochlicopa lubrica* and *Nesovitrea hammonis*), bugs (*Pachycoleus* (= *Cryptostemma*) *waltlii* (an European element, in Italy present as far south as Emilia), and *Acalypta musci*, an inconspicuous, extremely elegant, *Tingidae*, exclusive to Europe, in Italy restricted to northern and central areas; beetles (the carabic *Asaphidion pallipes*, the *Elateridae Hypnoidus dermestoides*, the *Staphylinidae Cryptobium fracticorne, Quedius attenuatus, Atheta aterrima, A. debilis* and *Oxytelus 4-carinatus*) plus *Diptera-* and *Coleoptera*-larvae, spiders and mites.

The banks of European rivers, large and small, are constantly visited or inhabited by a great number of birds, equal to, if not greater than, the numbers present on the shores of European lakes and ponds.

Let us begin with *Podicipidae*, represented by the horned grebe, *Podiceps auritus*, the little grebe, *P. ruficollis*, during winter present only at the mouth of rivers; also the common cormorant, *Phalacrocorax carbo* is present at river mouths, whereas the pygmy cormorant *P. pygmaeus*, lives along rivers and swamps. Interspecific competition between cormorants, namely *P. carbo* and *P. aristotelis*, was studied experimentally by Lack several years ago, and it has been shown that – due to different feeding habits – there is no particular intense competition between the two species, although occupying the same ecological niche.

Ardeidae are also common inhabitants of river-banks; they are the heron, the great white heron, the night heron and the little bittern, none of which is exclusive to this type of humid environment. The *Anatidae* or ducks sensu lato are represented by the red breasted *Mergus serrator* (in German 'Mittelsäger'), living on the sea when not breeding; the goosander *Mergus merganser* (in German 'Gänsesäger'), found along rivers only in winter, the bean goose and the whooper swan, already mentioned when talking about lake shores.

The beautiful and characteristic crane, *Grus grus* (in German 'Kranich') is also present along rivers, where it may nest; the *Charadriidae* are also represented: let us mention only the little ringed plover, *Charadrius dubius*, characteristic of the gravel bars in rivers, where it sometimes nests; a *Scolopacidae* frequent along both rivers and brooks of montane regions is the common sandpiper *Actitis hypoleucos* (in German 'Flussuferläufer'), which makes its nest in this environment, where it feeds upon insects, crustaceans, and also amphibians and fish. The *Phalaropodidae* are represented by the red-necked phalarope, *Phalaropus lobatus*, not usually typical of this environment. Two families which are practically exclusive to, or at least very characteristic of, river banks are the *Alcedinidae* king fisher, *Alcedo atthis*, and the *Meropidae*, bee-eater, *Merops apiaster* (in German 'Bienenfresser'). The former is closely restricted to water courses nesting along the, often rocky, vertical or very steep banks. It feeds by hunting fish whence its common name (in Italian 'martin pescatore'); the latter feeds upon insects, and particularly *Hymenoptera* (whence its common name) and is generally found near rivers, particularly because of its habit of making its nest in holes present in vertical, sandy, gravelly or earthy banks. *Merops apiaster* is a southern bird which sometimes also breeds in the British Isles, in Denmark and Southern Germany, and is present sporadically as a visitor in Northern Europe up to Lapland.

Finally several passerine birds live permanently or as migratory birds along the banks of rivers and brooks: the sand martin, *Riparia riparia*, in German 'Uferschwalbe' (or in North America bank-swallow); it nests in holes dug by itself in sandy, gravelly banks, and more rarely in holes in rocks. Other characteristic birds of montane rivers and brooks are the so-called wagtails, namely the white wagtail, *Motacilla alba alba* ('Bachstelze'), the grey wagtail, *Motacilla cinerea* ('Gebirgstelze') and finally the Rassenkreis *Motacilla flava*, in the British Isles represented by the race *M.f. flavissima*, or yellow wagtail, in Germany by the race *M.f. flava* or blue-headed wagtail (the true 'Schafstelze'), in other parts of the Continent by other races.

The most common shore passerine bird, probably in the world, is the *Cinclidae* dipper (from the Anglo-Saxon dyppan or dippan, to dip) *Cinclus aquaticus*, in German 'Wasseramsel', literally meaning 'water blackbird' (Fig. 359), present all through Europe near clear, rapidly flowing waters, particularly in montane regions, continuously searching for water animals (mainly insect larvae), for which it has to dive and sometimes to 'walk' on the bottom of the river searching for its food, a habit unique among European birds.

Mammals are also frequent along rivers, and some of them show very interesting adaptations to this environment, both structural and behavioral.

The beaver *Castor fiber* lives along rivers, old river beds and lakes, in forested areas (with oaks, ashes, alders, willows, poplars, birches, etc.). Today it is limited to very few rivers; recently it made its appearance near Lyon, in the Saône valley and in the gorges of the Tarn. It has been re-introduced in Finland and Sweden, and partially also into Eastern Europe. Proof of reintroduction is still being undertaken near Geneva. In Finland the Canada beaver (*Castor fiber canadensis*) has been introduced and it seems to have acclimatized itself[1].

The beaver has mainly nocturnal habits, spending most of its time in the water. It is a solitary animal, but may live in small family groups. It makes a subterranean den but in the North it constructs special huts; it also builds up dams and it digs canals. It does not hibernate in winter.

Two carnivores live along European rivers: the European mink and the otter. The former, *Lutreola lutreola*, in German 'Nerz' or 'Sumpfotter', lives in the vegetation of slowly running rivers and in ponds in the proximity of forests; it is a nocturnal, solitary, amphibious animal. It feeds mainly on crabs, plus frogs, fish and other aquatic animals. In some European countries it has disappeared in the last

Fig. 359. *Cinclus aquaticus.*

[1] It was present in Northern Italy in Neo-Eneolithic times at Postojna and in the Bronze Age in Lombardia (Solferino). At Adria it was present in the 5th Century A.D.

few years (Germany, Austria, Hungary). The otter *Lutra lutra* (in Spanish nutria, in Italian lontra, from the Latin lutra, whose origin is rather obscure) is present all through Europe wherever it can find some shelter along rivers, small brooks, lakes and canals, also in open swamps, near brackish waters and at rivermouths. Sometimes it can stray far from water; it is mainly nocturnal and migrates diurnally (pendular migration) along a river; it is a solitary animal but may also favour family groups. The den is located in a natural cavity, or a hollow tree, or among the roots of big trees, and sometimes it digs its own. It feeds on fish, crustaceans (crayfish and crab), molluscs, water rats, frogs and reptiles.

Other mammals present on the shores of rivers are water shrews, *Neomys fodiens* and *N. anomalus*, the latter a little less restricted to water than the former, and some small water rodents of the genus *Arvicola*. We should mention the water vole, *Arvicola amphibius* (in German 'Wasserratte') present in small rivers, streams and standing waters (including brackish); and the ground vole, *Arvicola terrestris*, more or less limited to water, in streams, swamps and lakes. Both swim very efficiently.

Extremely interesting, on the other hand, is an insectivorous genus living along the rivers of the Pyrenees, in Western Europe, and in Southern Russia, in the eastern zone. Formerly it was considered a single genus (*Myogale*) although today people prefer to think of it as belonging to two separate genera: *Desmana* (= *Galemys*) *pyrenaica* or desman (in Spanish 'almizclera', in Portuguese 'rato almiscareiro', from almizele, moss) is an extremely peculiar creature (Fig. 360). It lives at altitudes between 300 and 1200 m along streams near water mills, and in swampy meadows where water can freely circulate. In Eastern Europe (Southern Russia) the genus (or taxon) is represented by *Desmana* (= *Myogale*) *moschata*. The two elements originated from only one ancestor, endemic in Europe (possibly in Western Europe) since there is a fossil evidence of this genus in Jugoslavia and in the Pleistocene of England, Belgium and Germany.

Life on the sea-shore

Before describing plant and animal life on the sea-shore, it is necessary to have some concept of what constitutes the sea-shore. This is particularly necessary when we take into consideration the sea-shores of the Atlantic coasts of Europe where there are enormous differences in tide-levels.

The terrestrial life on the sea-shore (with which we are actually concerned) is essentially limited to the supralittoral fringe, plus (as we will see later) the epilittoral one. The supralittoral fringe, as its name says, is present immediately above the maximum high tide level (or Spring tide, the name having nothing to do with the season of the year, but deriving from the Old English 'springan', meaning rise). We will deliberately not consider the mesolittoral zone, since it may be invaded by terrestrial animals only during low tide, restricting ourselves to the 'true' supralittoral zone. We may define this, for the purpose of this book, as that part of the sea-shore above the extreme lowest high-water Neaps (from Middle English neap and Anglo-Saxon nep, scanty, or lacking), i.e. the lowest of the recorded high tides or Neap tide (ELHWN). Below this line the shore is inundated by every tide.

Three more levels have an ecological interest for our purposes: the mean high water Neaps (MHWN), i.e. the average of the lowest high-waters reached by the tide during each fortnightly set of tides (the fortnight is, of course, related to the moon cycle), the mean high-water springs, or MHWS, i.e. the average of the higher levels reached by the sea during each fortnightly set of tides; and finally the extreme high-water springs (EHWS), i.e. the highest level reached by the sea at the highest tide or tides of the year, sometimes called 'high-water equinoctial Spring tide' because it corresponds more or less to equinoxes. The latter of course will have only a very limited, if any, ecological significance.

Fig. 360. A characteristic mammal bound to the shores of rivers, limited to the Pyrenees and the mountains of Northern Spain, *Desmana* (= *Galemys*) *pyrenaica*. (from v.d. Brinck).

*Rocky shores**

'Built upon a dismal reef of sunken rock, some league or so from shore,
on which the waters chafed and dashed, the wild year through,
there stood a solitary lighthouse.
Great heaps of sea-weed clung to its base,
and storm-birds-born of the wind one might suppose, as sea-weed of the waterrose and fell about it, like the waves they skimmed.'

Charles Dickens

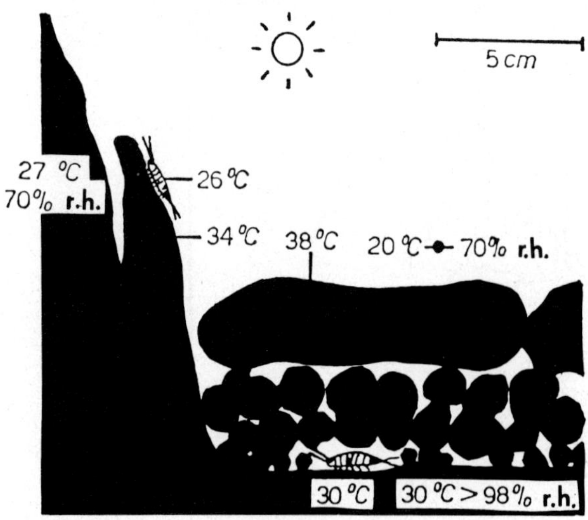

Fig. 361. Example of a micro-climate: vertical section of a reef of red sandstone and of the niche inhabited by the woodlouse *Ligia*. The difference between the conditions prevailing under the stones and those existing above them, both in temperature and in relative humidity, are clearly shown (from Edney).

* The engraving, taken from a Victorian work on Natural History, shows the rocky and sandy shore at Kynance Cove, Lizard, Cornwall (from Southward).

On rocky shores, true terrestrial life is present only in the supralittoral fringe and, of course, in the so-called epilittoral zone (see later).

The microclimate of this zone is very distinct due to the conjunction of air and water, two extremely different environments. The following data gives us an impression of the temperature in this humid environment:

	sunny day in summer	dull day in summer	sunny day during frost	dull day in winter
Sea temp. inshore	14.6	14.3	6.3	7.6°C
Temp. of tide pools	17.4	14.5	4.9	4.9
Shade air temp. on land	16.5	14.5	0	4.5
Shade air temp. close to rocks	17.8	15	3.9	5.8

Fig. 361 gives us an idea of the actual microclimate on some rocky shores where animals live.

In Western Europe, for instance on the coast of England, we find the following plant and animal species (limited to the supralittoral fringe): the periwinkles *Littorina neritoides* and *L. saxatilis*, the lichen *Verrucaria maura* and the Isopod *Ligia oceanica* (Fig. 362).

Periwinkles exhibit perhaps the most remarkable features of adaptation of marine animals to terrestrial life, concerning nitrogen catabolism and reproduction. On the whole we find – as a result of long evolution – two mean features, both characteristic of land animals, uricotely (i.e. the transformation of all nitrogen catabolic products into uric acid, instead of ammonium salts) and viviparity. The following data shows us this extremely peculiar evolution (maybe unparallel-

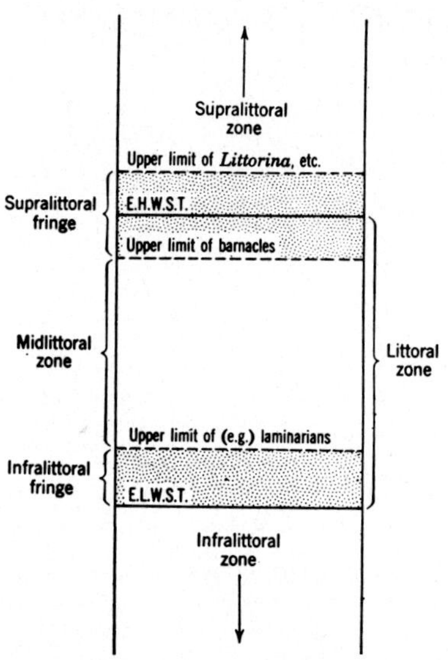

Fig. 362. The general zonation of European rocky shores. EHWST = extreme high water spring tide; ELWST = extreme low water spring tide.

ed by other animals with the possible exception of the transformations of respiratory organs in tropical semi-terrestrial crustaceans):

Freshwater species of Gastropoda	0.1 mg uric acid/g
Intertidal and sublittoral species	0.5-5.0 mg uric acid/g
Littorina neritoides	25
Terrestrial species (*Helix, Limax*)	31 – 700

On the whole, species of *Littorina* show a correlation of larval type with the tide level at which the adult lives on the shore, starting with planktonic eggs and larvae in *Littorina littorea*, living near mean low water Neaps, through egg laying and direct development in *Littorina littoralis* living at mean tide level, to incubation in a brood pouch and viviparity in *Littorina saxatilis* (living at mean high water Neaps). However, *Littorina neritoides*, the highest of them at all (mean high

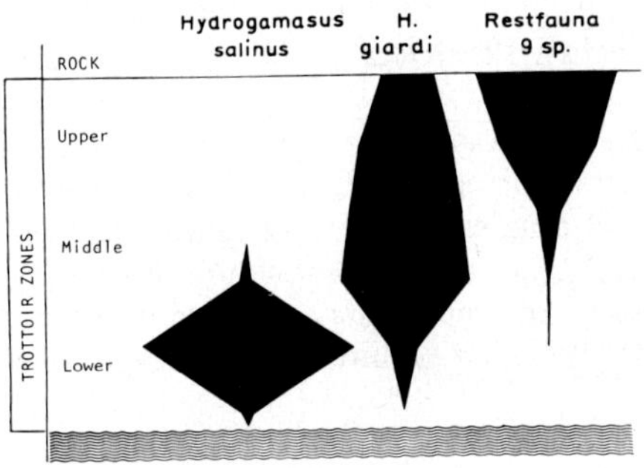

Fig. 363. Vertical quantitative distribution of the terrestrial arthropods found on a Mediterranean rocky shore (trottoir) (From Schuster, 1965).

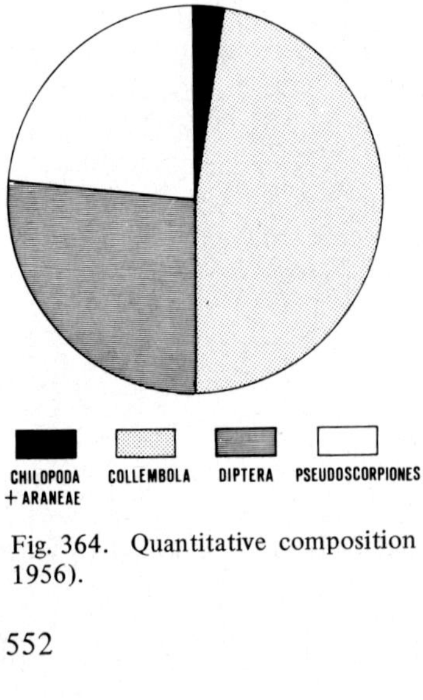

Fig. 364. Quantitative composition of the terrestrial fauna of the trottoir (from Schuster, 1956).

water Springs and above) has eggs that are shed into the sea, with a larval phase in the plankton.

On the Mediterranean coasts of Europe, life on the rocky shores is much poorer than that of the sandy ones; the arthropods here are the best represented animals. An example is provided by the so-called trottoir built up by calcareous algae in several parts of Southern Europe. Here the following terrestrial species can be found: a spider, *Desidiopsis racovitzai*, an inconspicuous *Agelenidae*, the acari (mites) *Hydrogamasus salinus* and *H. giardi*, thalassobionts belonging to *Gamasinae, Nanorchestes collinus*, belonging to *Trombidiformes*, not strictly bound to sea shores, and one species of *Protereunetes*, a *Trombidiformes; Hydrogamasus salinus* and *H. giardi* behave as ecologically vicarious species. Other elements are the Pseudoscorpions *Pselaphochernes liroralis* and *Paraliochthonius singularis*, known only from Southern Italy (Portici near Naples) and Southern France (Cap l'Albeille), both probably stenoikous species: several thalassic centipedes such as *Henia bicarinata*, the springtail *Axelsonia tittoralis*, a thalassobiont with a rather wide distribution (eurytope) but stenoikous; *Diptera* larvae, mainly belonging to *Dolichopodidae*, and *Chironomidae* (*Clunio* and *Trichocladius*, the former however with aquatic larvae). The quantitative aspect of this particular fauna is represented by Fig. 363-364.

The ecology of the arthropod fauna of rocky shores has not been much investigated, especially if compared with that of sandy or muddy shores. We know, however, something about the ability of its elements to withstand submersion, as the following tables show:

1. *Geophilidae* centipedes (at 17° C)

Species	10% sea water	60% sea water	100% sea water
Haplophilus subterraneus	30-46 h.	12-21 h.	4- 8 h.
Strigamia maritima	36-72	12-36	12-24
Hydroschendyla submarina	48-84	36-48	12-36

2. Mites (A, thalassobious; B, terrestrial euryoikous sp.)

Species	type of water				
	sea water	distilled water	running water (brook)	tap water	
Hydrogamasus giardi	80	3	3	4 days	A
H. salinus	80	4	–	–	A
Xenillus tegeocranus	19	60	60	–	B
Damaeus auritus	31	80	80	80	B
Platynothrus peltifer	24	60	70	–	B
Euzetes seminulum	18	50	–	–	B
Steganacarus magnus	12	50	–	–	B
Pergamasus runciger	1	8	15	16	B
Pergamasus sp.	1	12	–	14	B

Very often rocky shore arthropods are very widely distributed as we can see from the following data:

Taxon	Sofar known localities
Dendrouropoda sp. (*Acari Uropodina*)	European Mediterranean coasts, Red Sea, Maldives, Atlantic coast of South America
Rhodacaropsis inexpectatus-gr. (*Acari Gamasides*)	North Sea, Baltic Sea, European Mediterranean coasts, Red Sea, Atlantic coasts of S. America and Pacific coasts of Central America
Nanorchestes amphibius gr. (*Acari*)	Atlantic and Mediterranean coasts of Europe, North Sea, Spitzbergen, Norway, Jan Mayen, Atlantic coast of South America, and maybe Red Sea
Selenoribates-Thalassozetes-group (*Acari*)	European Medit. coasts, Red Sea, Maldives, Atlantic coasts of South America and Pacific coast of Central America
Anurida maritima (*Collembola*)	All European coasts, Atlantic coast of North Africa, Black Sea, Atlantic coasts of North and South America, South Africa, Galapagos
Archisotoma sp. (*Collembola*)	Atlantic coast of Europe, North Sea, Baltic Sea, Spitzbergen, Jan Mayen, Greenland, Atlantic coast of North America, Bahamas, Mediterranean coasts, Madagascar, Falkland, South Orkneys
Pseudanurida sp. (Coll.)	Red Sea, Persian Gulf, Sunda-Archip., Australia, Tokara-islands, Atlantic coast of S. America
Cafius (Staphylinid genus)	The coasts of all continents
Hydroschendyla submarina (Chil.)	All European coasts, Bermuda

As far as the epilittoral zone is concerned, we have here one of the most active and luxuriant habitats, at least during summer, not only of humid environments, but also of most European ecosystems. This richness of life is mainly, if not exclusively, due to bird life.

As far as the vegetation of the epilittoral zone is concerned, it varies according to the latitude. Suffices to say that in Europe the vegetation of the rocky shores

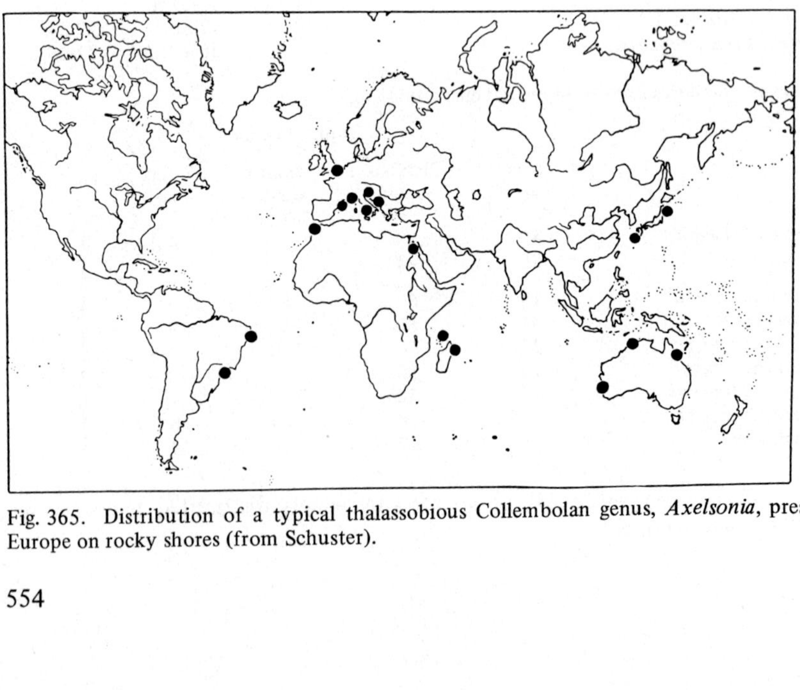

Fig. 365. Distribution of a typical thalassobious Collembolan genus, *Axelsonia*, present in Europe on rocky shores (from Schuster).

may go from a steppe xerophytic vegetation, present on the sunny hot shores of Southern Spain, the Balearics, Sicily, Greece or Crete, to that covering the borders of the Arctic tundra, reaching the Arctic Ocean, where only a stunted vegetation is present, mainly mosses and lichens. In some instances the shore is not rocky but gravelly or simply stoney. The number of plant species, however, is not great: we may mention *Crithmum maritimum* (the rock samphire), *Capparis spinosa* (the so-called caper, from French capre, derived from the Greek καππαρις) whose flower buds are eaten in several European countries, *Limonium cancellatum* (and many other species of this genus), *Atriplex halimus*, several Graminoid plants, *Glaucium luteum, Ecballion elaterion, Picridium vulgare, Apocynum venetum, Beta vulgaris* and *B. maritima, Euphorbia peplis, Polypogon monspeliensis* and several others (Fig. 366).

We will of necessity confine ourselves to some of the most unusual birds: we

Fig. 366. The Umbellifera *Crithmum maritimum*: a and b. plants growing on a vertical cliff near Trieste; c. plant growing on stoney soil in the Tremiti Islands. d. General view of the sclerophyllous forest in Dalmatia, near Dubrovnik, today highly degraded.

may note that this life is not very different in various European countries, with the obvious exceptions. This is partly due to the euryoikous character of these sea creatures, partly to the fact that most of them migrate, and therefore are not bound to any particular territory, climate or ecosystem.

Rocky shores have often stimulated the imagination of poets and writers, or simply of people, as have few other water environments, and their creatures, beginning with birds, have been the subject of tales and fables.

Thus the Tremiti islands, a small archipelago off the coast of Puglie in the Southern Adriatic Sea, were known as 'Diomedeae Insulae', since it was believed that on one of them, S. Nicola, there was Diomedes' sepulchre. These islands are indeed associated with the myth of Diomedes' death and with the metamorphoses of his companions into birds, the 'Diomedeae Aves'. Actually, Aldovrandi's 'Avis diomedea' described in 1603 as 'Diomedea avis cum unco acuto vulgari' is

Fig. 367a) Rocky shores: Färoer. General view: some flying birds are clearly visible in the foreground. (Photo Prof. G. Pilleri, Bern).

Puffinus kuhli kuhli, in Italy called 'berta maggiore'. We must not forget that its related *P. kuhli yelkouan* emits noises closely recalling children's screams. Of course all this is more than sufficient to create a legend.

Thus *Procellariidae* have often been named after 'Storms': Sturmschwalbe, stormvogeltje (Dutch), stormsvala (Swedish), stormfågel (Swedish), 'uccello delle tempeste' (Italian, etc.). The albatros, *Diomedea exulans*, the largest sea-bird, sometimes seen as an occasional visitor in Europe, has inspired one of Baudelaire's poems. The storm petrel *Hydrobates pelagicus* (= *Procellaria* or *Thalassidroma pelagica*) has been named 'St. Peter's bird' because of its habit in suspending itself over the waves so that it appears to be walking upon them, but actually hovering with its wings (as everybody knows, in the Bible (Matt., 14: 28-31) we read of St. Peter walking upon the waves of the lake of Bahr Tobarie). *Puffinus p. yelkouan*, because of its habit of flying touching the waves, chasing one another in

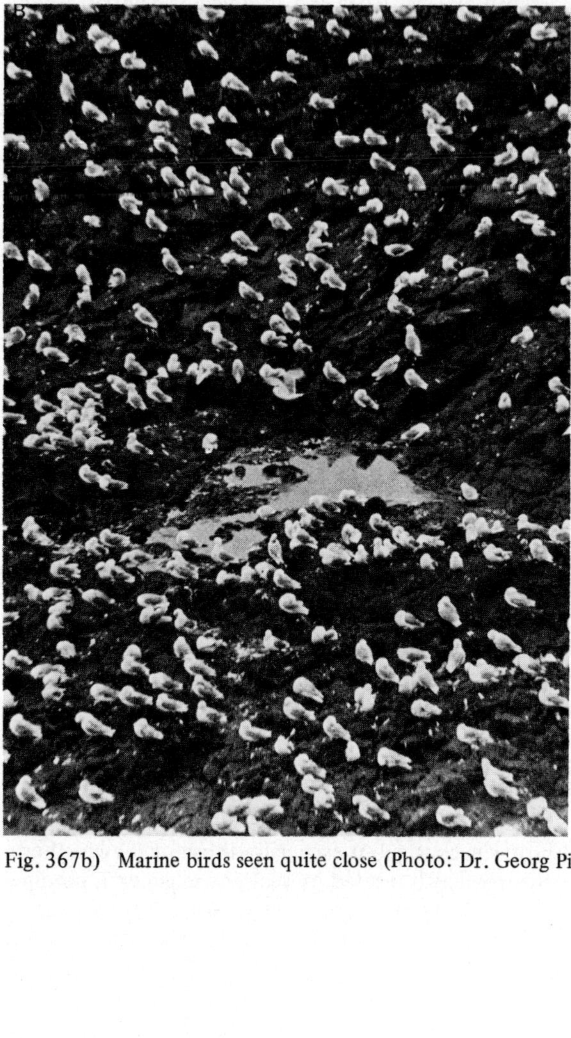

Fig. 367b) Marine birds seen quite close (Photo: Dr. Georg Pilleri, Bern).

single file as if they were being pursued, has been called 'damned souls' in Eastern Europe.

Also the generic scientific names of rocky shore birds are often inspired by these legends, as for instance *Pelagodroma, Oceaonodroma, Pterodroma, Oceanites, Fregata* (from the Greek ἄφρακτα), *Phaethon* (in Greek literally 'shiner').

Let us now more objectively consider which species lives permanently, or at least temporarily, on rocky shores, often in the epilittoral zone, if not in the supralittoral one. The *Procellariidae* are represented by petrels and shearwaters (from the Anglo-Saxon scear, for shear) such as Wilson's petrel *Oceanites oceanites,* breeding exclusively on the coast, Leach's petrel *Oceanodroma leucorrhoa,* in German 'Wellenläufer' (meaning 'running on the waves), Manx shearwater, *Puffinus puffinus,* a Rassenkreis; Cory's shearwater *Puffinus kuhli,* a Southern European species, the fulmar (coming probably from the Old Norse full, foul and mar, sea gull) *Fulmarus glacialis* (in German 'Eissturmvogel', literally meaning the bird of the ice-storms), sometimes found in enormous numbers. The *Phalacrocoracidae* are represented by the shag, *Phalacrocorax aristotelis*, present upon rocky shores and islands with vertical cliffs and in holes or cavities eroded by the sea. It nests gregariously among rocks.

Quite different as to shape, habits and more particularly feeding habits are the *Charadriidae*, among it we might mention the turnstone *Arenaria interpres* ('Steinwälzer') nesting upon bare rocky soil on islands near the shore and feeding upon insects, small snails, crustaceans, remains of fish, etc.

The *Scolopacidae* are represented mainly by the purple sandpiper *Calidris maritima* (in German 'Meerstrandläufer') present during the winter on rocky shores. More abundant and characteristic are the *Laridae*, such as the greater black-backed gull, *Larus marinus*, the lesser black-backed gull, *Larus fuscus* (less restricted to rocky environments), the herring gull *Larus argentatus*, and the glaucous gull *Larus hyperboreus*, exclusive to the Arctic.

A well-adapted family is the *Alcidae*, to which several European species belong, for example the razorbill *Alca torda* (in Dutch alk, in Swedish tordmule, in North America razor-billed Auk), nesting on rocks, gregariously, often together with *Uria aalge*. The latter guillemot (from the French Guillaume, in German Trottellumme, in Dutch Zeekoet, in Swedish Sillgrissla) has similar habits and like the *Procellariidae* is a true sea bird.

However, the most peculiar and perhaps best known sea bird is the puffin (from puff, because of the shape of the beak), *Fratercula arctica*, in German Papageitaucher) characterized by its triangular, red blue-yellow beak, whence the German name. It nests gregariously in holes dug by rabbits, by shearwaters or by itself. It is one of the most remarkable European birds.

Among rocky shore, or truly marine, birds we must also include a diurnal bird of prey, the white-tailed eagle, *Haliaeetus albicilla* (in German Seeadler, and in North America grey sea-eagle) which, however, is not as closely restricted to this environment as the afore-mentioned species. If feeds on fish, which it catches by diving, on mammals as big as a small roe buck and birds as big as a duck, although it may also eat carrion.

The rocky shores of all European coasts, but especially those of the Northernmost part of the Continent, are frequently visited by seals, or in general, pinnipedes. One of them is the walrus, or sea horse, *Odobaenus rosmarus*, in German Walross (from Wal = whale and Ross = horse), in Dutch walrus, in Norway known as Rosmar, in Lapland as Morsk. It lives in the Arctic Ocean from Jenissei, through Novaja Zemlja, Spitzbergen, Greenland up to Hudson Bay, Canada. It was observed near the coast of Scotland in 1817, 1825 and 1902, and near Ireland in 1897. In 1857 it was seen near both the Hebrides and the Orkneys, and sometimes it may reach the coasts of Norway. It lives upon floating sea ice and icebergs but sometimes resides on land (Fig. 368).

The ringed seal, *Pusa hispida* (= *Phoca hispida*) is present not only in the Arctic Sea but also in the Baltic Sea and in some lakes and inland seas (Caspian Sea, Aral sea, etc., Fig. 369). The presence of this species (represented probably by several races) in waters completely unconnected today indicates that at no very distant date (at least since Pliocene times) a closer connection with the Arctic Ocean

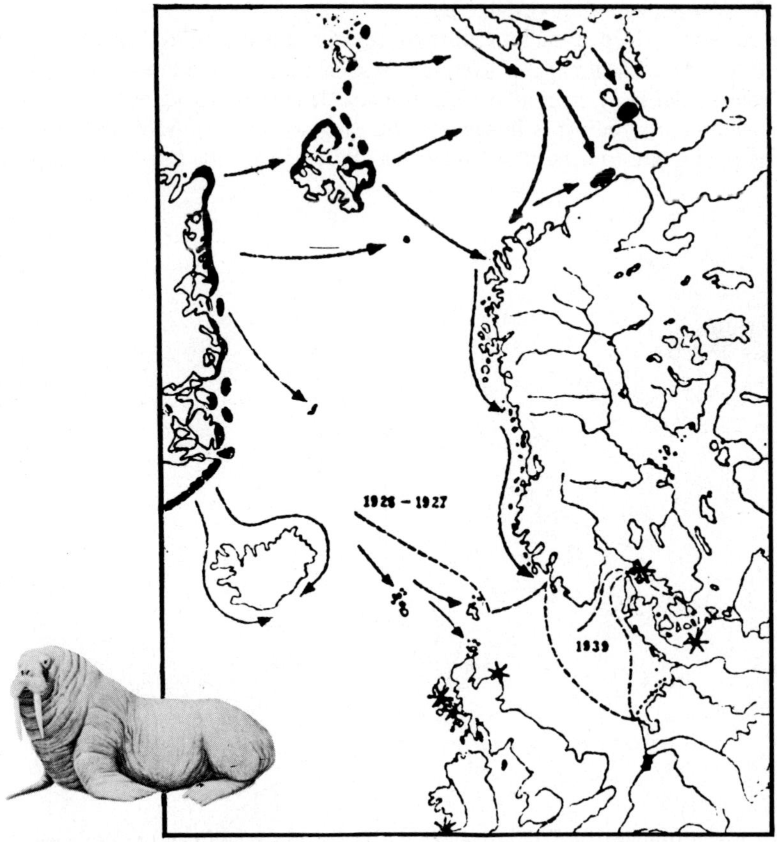

Fig. 368. The morse, *Odobaenus rosmarus* (male) and the distribution of the species in Europe; the arrows indicate the migrations which occurred in recent times (from v.d. Brinck).

existed than at present. According to Scharff, the connection must have occurred at a time when the Caspian Sea extended far to the north – when indeed the so-called post-Tertiary Caspian deposits were laid down (Fig. 370). Since the boulder-clay which covers the plain of Northern Russia is assumed to be a ground-moraine of an extensive northern ice-sheet, we would expect that the Caspian deposits were non-contemporaneous with it. Probably, while the boulder-clay was being laid down in Northern Europe the Aralo-Caspian Sea had some communication with the White Sea (see also the life of the Caspian Sea in the chapter on aquatic environment).

The grey seal or Atlantic seal, *Halichoerus grypus*, in Spanish foca parda (literally brown seal) is a gregarious animal. It dives up to 145 m down and can stay under water up to 22 minutes.

The hooded seal (Bladdernose, or simply bladder) *Cystophora cristata* (in German Klappmütze, in Finnish kiknebb, known in Lapland as avjor, or fattenuorgo or even dado), is found throughout the Arctic Ocean, from Greenland to Spitzbergen, North America and Northern Europe, occasionally reaching England and France. It is a pelagic seal, often migratory, easily moving on ice and land. It dives up to 300 m deep. Finally a more southern seal is the monk seal, *Monachus monachus* (= *M. albiventer*), in German Moncherobbe, in Spanish foca fraile (which means friar-seal), in Italian foca monaca. It lives on rocky shores where it can find shelter. In Italy it is limited to the eastern shores of Sardinia, those of Montecristo and the Southern Adriatic Sea. It is gregarious, but often appears

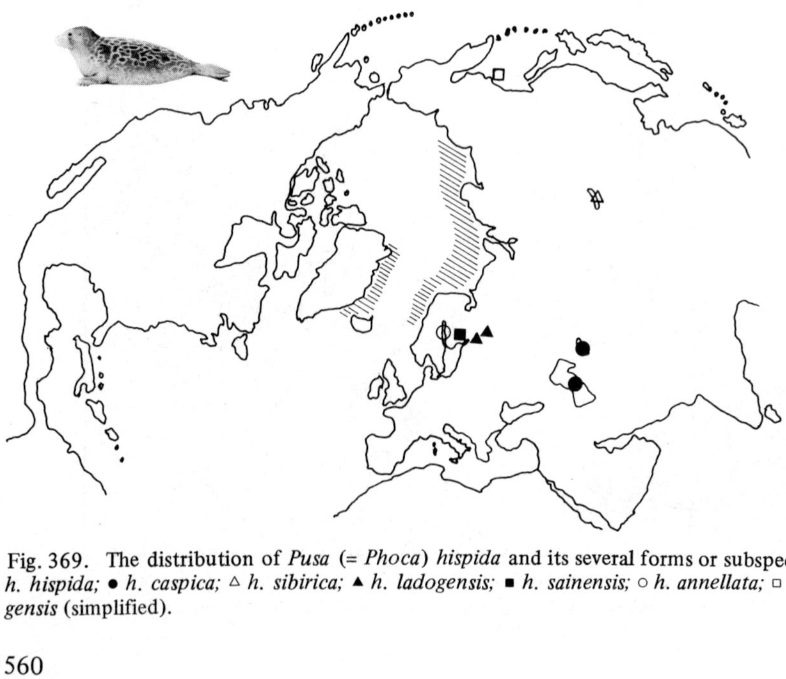

Fig. 369. The distribution of *Pusa* (= *Phoca*) *hispida* and its several forms or subspecies: \\\\ *h. hispida;* • *h. caspica;* △ *h. sibirica;* ▲ *h. ladogensis;* ■ *h. sainensis;* ○ *h. annellata;* □ *h. gichigensis* (simplified).

solitary as a consequence of the reduction of its population. According to Niethammer & Kremer, it is a mammal threatened by extinction, and it should therefore be protected by law (Fig. 371).

When talking of the supra- and epilittoral zones of the rocky shores, we must not forget the marine caves (grottos) whose openings are situated in this zone and in which there are usually some terrestrial organisms. Another very peculiar feature of this environment are the so-called 'rock pools', frequently found on all kinds of rocky shore, regardless of geological composition, although maybe slightly more frequent on calcareous rocks. The rock pools are inhabited exclusively by aquatic animals, characterized by exceptional euryoiky, withstanding enormous

Fig. 370. Map of European Russia. Light-dotted parts indicate the areas covered by boulder clay. Heavily dotted parts those exhibiting Aralo-Caspian and other Post-Pilocene deposits (from Scharff).

changes in salinity and temperature of the water, as well as drought (Fig. 372). One of these organisms, the beetle genus *Ochtebius*, is derived from freshwater insects. The same is true for mosquitos.

Sandy shores

The vegetation of the sandy coasts is similar throughout Europe, and around the Mediterranean, with the exception of the northernmost parts of Europe. The cause of this relative monotony, or at least uniformity, is substrate (raw mineral soil), microclimate, characterized by the exposure to sunlight (making allowance for the wetter parts of the Continent), and therefore strong irradiation (also of UV light) relatively high temperatures (at least on the sun-exposed dune sides), and finally, exposure to winds (sea-breeze, land-breeze, or even strong winds coming in summer from the sea and in winter, generally, from the land (Fig. 374).

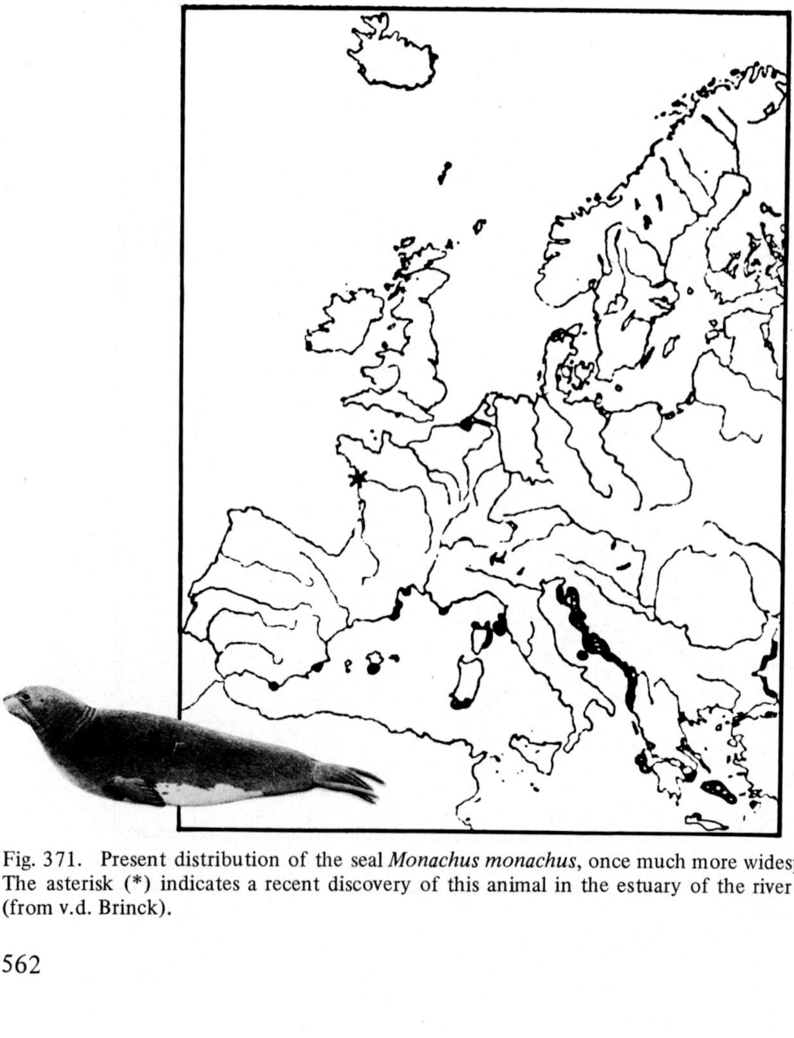

Fig. 371. Present distribution of the seal *Monachus monachus*, once much more widespread. The asterisk (*) indicates a recent discovery of this animal in the estuary of the river Loire (from v.d. Brinck).

We will mention here just a few species of flowering plants of Southern European dunes, but we must keep in mind that the situation is not very different from the Atlantic coasts of Europe. The Umbellifera *Eryngium maritimum*, or sea holly, greenish blue coloured and spiney; the Crucifera *Cakile maritima*, with small rose-pink flowers; *Matthiola* sp., with pink or violet flowers; the *Euphorbia paralias*, looking like an ornamental plant; the Graminea, *Ammophila arenaria* or Marram grass, and the Cyperacea *Cyperus mucronatus*, the Silenea *Silene sericea*; the Graminea *Sporobolus pungens*, plus *Agropyrum junceum* (sand crouch-grass), *Lepturus incurvatus* and *Lagurus ovatus* (hare's tail), the Liliacea *Ornithogalum narbonense* (a star of Bethlehem), the beautiful *Mesembryanthemum* or Hottentot, an introduced succulent from South Africa with large red or yellow flowers; the humble Dipsacacea *Scabiosa maritima*, the thorney Umbellifera *Echinophora spinosa*, the small and elegant *Medicago marina*, with yellow flowers; the beautiful conspicuous Liliacea *Scilla maritima*, flowering in late summer when all its leaves have withered. Away from the sea, the common Tamarisk (from the Arab Tamir, a date), *Tamarix gallica* and some species of *Juniperus*, such as

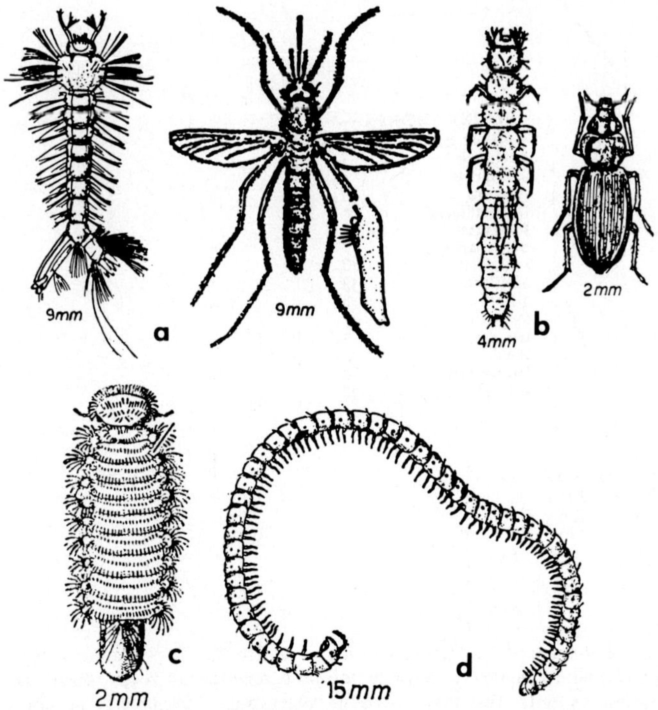

Fig. 372. Animals characteristic of the so-called rock pools, typical of rocky shores: a. mosquito *Aedes mariae* (larva and adult); b. the beetle *Ochthebius* sp., larva and adult. Animals of the Mediterranean rocky shores: c. the Diplopod *Polyxenus lapidicola*, and d. the Diplopod *Thalassisobates adriaticus* (from Riedl).

Juniperus macrocarpa may be found (in northern localities substituted by the common juniper, *Juniperus communis*).

The physical characteristics of the ground are mainly responsible for the type of vegetation (and subsequently of animal life): there is a low water content and it is difficult for hypogeous organs to penetrate the substrate. In the case of salty sandy shores one can speak of halopezophytia (Del Villar); in the case of the sand dunes of Northern Europe with *Empetrum* and *Calluna* one can speak of oxyphytia. This kind of vegetation association exhibits the characteristics of a Graminetum (*Elymus arenarius* or lyme-grass, *Psamma* (= *Ammophila*) *arenaria*), a fruticetum (*Hippophaë rhamnoides* or sea buckthorn of Northern Europe) or an arboretum (pine stands on the Baltic coasts, or those on the SW coast of Spain).

On the shores of the Northern Adriatic Sea, not far from Venice, a rather poor cenosis lives near the sea, in the organic debris zone, called 'wrack' around the North Sea (from the Middle English and Middle Dutch 'wrak').

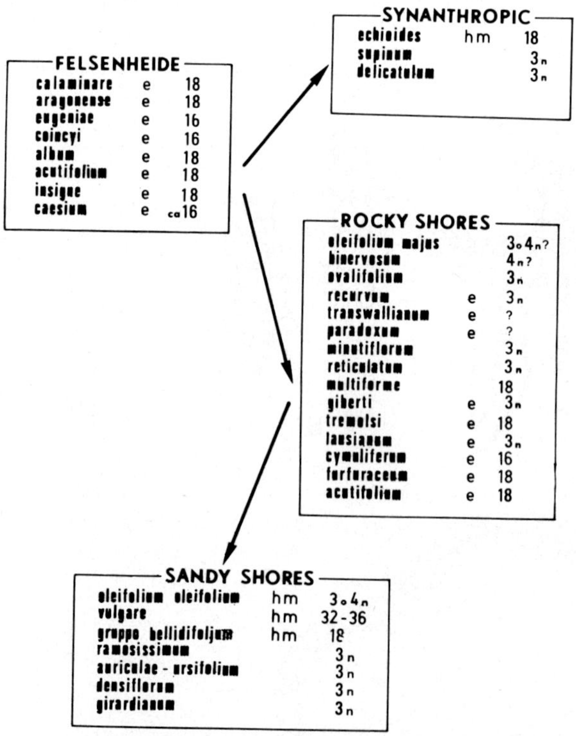

Fig. 373. The species of *Limonium* of the Mediterranean, for which the chromosome number is known, classified according to ecology. 'Felsenheide', and secondarily rocky shores are contiguous and/or continuous with the sclerophyllous vegetation. The species of sandy shores and the synanthropic ones belong to different ecosystems. For each species the somatic chromosome number is shown; e = endemic; Hm = holomediterranean species. All other species have a relatively wide distribution within the Mediterranean area (from Dolcher & Pignatti, 1971, modified).

Fig. 374a. The remains of a beautiful *Pinus maritima* stand at Migliarino (Tuscany) near the sea shore, now practically destroyed by soil and air pollution due to substances coming partly from the sea, partly from the land. In the foreground some plants of *Euphorbia paralias* still in good condition. (Photo Prof. G. Pilleri, Bern).

Fig. 374b. Plants of *Eryngium maritimum* with specimen of *Scolia* probably *flavifrons*, the largest European species of the genus. (Photo Prof. G. Pilleri, Bern).

Fig. 374c. A typical beetle of sandy shores, *Scarabaeus semipunctatus*, the well known insect which constructs pellets with faeces in which it lays eggs. (Photo Prof. G. Pilleri, Bern).

We find here some Amphipods such as *Talitrus saltator*, present along the coasts of Europe from Southern Norway to the Mediterranean (Sicily, Tunisia and Algeria), as well as in the Azores. *Talorchestia deshayesii*, on most European coasts from Denmark to the Mediterranean and the Black Sea; and *Orchestia mediterranea* spread on the Atlantic coast of Europe from the British Isles southwards to the Mediterranean Basin. Some spiders thrive in this humid environment, hunting especially beetles and flies: they are *Erigone dentipalpis, Misumena* sp., *Heliophanus tribulosus* and *Zelotes* sp., widely distributed species. A millipede, *Schizophyllum sabulosum*, can be present, although not at all typical of this cenosis. Very interesting is the earwig *Labidura riparia*, a conspicuous, cosmopolitan species very common under wrack and pebbles on sea shores and along rivers. It frequently digs burrows which may extend for 2 metres characterized by a small mound of sand at the entrance. It is an essentially carnivorous form. Very characteristic are beetles: *Cicindela lunulata*, a Central-Asiatic, Euro-Mediterranean form, also found in some interior regions of South-Eastern Europe as far as the Caspian Sea; it is a halobious species, although sometimes found far from the sea and on non-saline soils; *Harpalus anxius* can be found on sandy beaches although not a typical species of them; *Bembidion ustulatum*, holoarctic, ripicolous and euryphipsous, and therefore not at all characteristic of this environment; *Tachys micros*, Euro-South-Mediterranean/Macaronesian species, ripicolous and hygrophilous, but not necessarily halophilous species; *Trechus quadristriatus*, a very euryoikous, common species; *Asaphidion stierlini*, ripicolous and on humid soils but not halophilous. Two *Histeridae* deserve maybe particular attention: *Hypocaccus rugifrons* (from ὑπό and κακκή) in sandy soils on dead fish or on corpses of small animals; and *Pachylopus dimidiatus*, on decaying substances, probably halobious. Sometimes near the sea a scarab, *Scarabaeus semipunctatus*, is very common, though commonly more restricted to the dune environment (see later); it is a psammophilic species. In comparison, the Tenebrionids or 'black beetles' are highly characteristic: one of them is *Phaleria bimaculata*, extensively found along the shores of Southern Europe, Morocco and Egypt; it is a halopsammophile feeding upon any kind of decaying organic substances; the larva has the same habits as the adult, with which it lives. *Trachyscelis aphodioides* is also a halopsammophile, which, however, is more restricted to dunes than to wrack; it is very convex, black and shining, with very characteristic fossorial limbs. *Xanthomus pallidus* is a halopsammophile widely distributed on the shores of Atlantic Europe from the Netherlands to Gibraltar, Spain, France, Italy and Northern Africa.

The animal life of a dune is rich and varied. Two species of woodlice are present: *Chaetophiloscia elongata* and *Acaeroplastes melanurus*. The former is a more euryoikous species, an oriental element present in most of the Mediterranean, in the North reaching the January isotherm 5°C, equivalent to an average summer temperature of 18-19°C. It is not necessarily associated with sandy environments. *Acaeroplastes melanurus* is present in the littoral zone of the Mediterranean basin, with psammophilous and perhaps slightly halophilous habits, although in some places (Languedoc) it has been found up to 150 km from sea. It is generally found in association with *Astragalus tragacantha-Plantago subulata*. It requires, as do Isopods in general, a high level of humidity.

Two species of Pseudoscorpions thrive on the dunes of the Northern Adriatic Sea: *Dactylochelifer latreillei* and *Rhacochelifer disjunctus*. The former is rather widely distributed, the latter is exclusive to Italy, Austria, Hungary, France and Morocco (a relict disjuncted distribution). *D. latreillei* has been found in England on rocky coasts.

Only one species of Opilionid, a very trivial species, *Phalangium opilio* has been found. Instead, spiders are very common, both on the soil and on plants. We must mention *Mangora acalypha*, a common species, *Tetragnatha extensa*, *Dietyna arundinacea*, very euryoikous forms found on heaths in Central Europe. *Theridion aulicum*, extending to Centro-South Europe and the Mediterranean Basin, *Chiracanthium* sp., *Hyptiotes flavidus*, a fairly interesting species limited to Southern Europe and Madera, *Erigone dentipalpis*, a common, widely distributed species, *Tibellus macellus, Misumena* sp., *Heliophanus tribulosus* and *H. cupreus*, two widely distributed elements, *Dendryphantes* sp., *Phlegra bresnieri* (holo-Mediterranean species), *Pellenes geniculatus*, an endemic to Southern Europe, and *Marpissa* sp.

No centipede has been observed in this particular humid environment while the only millipede is the extremely common *Schizophyllium sabulosum.*

Both numbers and species of insects are very abundant. Let us mention the earwig *Labidura riparia*. The hoppers include *Acrida mediterranea*, present in the European Mediterranean region (Spain, France, Italy, Dalmatia and Greece), *Yersinella raymondi*, limited to Southern Europe, but very rare in France, and *Chorthippus brunneus* a very euryoikous species. *Arachnocephalus vestitus* is exclusive to Southern Europe and Algeria, its presence on dunes (or more exactly on mature dunes of the Northern Adriatic Sea) must be considered as a thermophilous relict, since in Northern Italy it is considered as a typical element of the so-called xerothermous oasis. The microclimate should be responsible for its presence in this otherwise rather sub-continental cold region. Two more grass-hoppers must be mentioned: *Oedipoda coerulescens* and *Locusta migratoria*. The latter is present in its 'solitary' form or phase, living generally on sandy, swampy or heathy soils. As is well known, the gregarious phase of this insect has migratory habits.

The bugs are *Ischnodemus quadratus*, an Atlantic-Mediterranean species, common on *Cyperaceae*, infrequent but widely distributed: near Venice it has been found in mesophilous meadows, Caricetum gracilis, Scirpetum maritimi as well as in the Salicornietum. In the dunes it is one of the most abundant insects. *Lygaeosoma sardea; Zicrona coerulea; Adelphocoris lineolatus*, extremely euryoikous; *Rhopalus parumpunctatus*, near Venice in mesophilous meadows and in the Salicornietum; *Lygus basalis*, on *Umbelliferae; Aelia acuminata*, a characteristic, conspicuous species, very frequent on cereals to which it may cause serious damage; *Nabis feroides*, a rather rare Southern European form, and, finally, *Nysius thymi*, common on dry soils.

The *Cicadinae* or leaf-hoppers are *Tettigella viridis* (only in mature dunes), *Philaenus leucophthalmus*, *Delphacodes* sp., *Idiocerus ustulatus*, in Central Europe common on *Populus alba*, and *Neophilaenus infumatus*, an Euro-anatolic form, in Central Europe present in steppe territories maybe as a glacial relict.

The beetles are, as usual, the most interesting and physiognomical elements. Let us mention *Cicindela lunulata, Harpalus serripes*, generally restricted to dry areas; *Tachys scutellaris*, a Euro-Turanic element, psammo- and luto-halobiont, i.e. a halophilous species living both on sand and mud; *Trechus quadristriatus*, a common element; *Arthrolips aequalis* (= *Sacium dorsatum*), *Hypocacculus metallescens*, a Mediterranean form sometimes present in *Aroideae* flowers to which it is attracted by the smell; *Micraspis 16-punctata*, often considered as halophilous, never above 400-500 m, sometimes on the flowers of *Ranunculus acer*, *R. repens*, or *R. bulbosus*; *Coccinella 7-punctata*, the common lady-bird; *Chilocorus bipustulatus*, aphidivorous as is the preceding species; *Hyperaspis concolor*, *Thea 22-punctata*, a rather thermophilous species, helpful to agriculture because it feeds on the fungi responsible for wheat rust; *Anthicus fenestratus*, an inconspicuous psammophilous species limited to Southern Europe and Syria; *Colotes maculatus*, a more southern element; *Psilothrix cyaneus*, a beautiful floricolous species extending as far north as England and Ireland; *Malachius spinipennis,* a rather rare species; *Scarabaeus semipunctatus*, a psammobiont generally found on excreta of men and donkeys, active up to 15.00; the inconspicuous, very characteristic, Scarabeid *Psammobius sulcicollis*, a psammobiont present under stones and dry faeces, coprophagous or phytosaprophagous. The *Chrysomelidae* are represented by *Galeruca tanaceti*, generally on *Compositae* such as the milfoil *Achillea millefolium*, the tansy *Tanacetum vulgare, Chrysanthemum, Taraxacum* and *Cirsium* (a thistle), in some instances harmful to cultivated plants; *Galeruca interrupta*, generally restricted to field southernwood. *Artemisia campestris*; *Haltica oleracea*, on *Polygonum* and *Oenothera* (immediately behind the active dunes, in the mature dunes, an introduced *Oenothera* is actually plentiful), *Psylliodes* sp., probably limited to some *Cruciferae*, and finally *Cryptocephalus fulvus signaticollis*, common on dunes near the Northern Adriatic Sea. No weevil has been found in this environment. The Tenebrionid *Phaleria bimaculata* (occasional, since it is linked to the presence of wrack), *Trachyscelis aphodioides* (very rare at this latitude) an halopsammobiont, and finally *Ammobius rufus*, a psammohalobiont probably very rare for the same reasons. The *Neuroptera* are well represented by *Chrysopa formosa* (on the mature dune) an aphidovorous insect both as an adult and as a larva, and the two ant lions *Creoleon lugdunense* and *Synelisis baetica*, both rather southern thermophilous elements. The larva of the latter feeds on small insects which it captures in a pit dug in the sand by the enormous mandible, the so-called forceps.

Dunes are a very suitable environment for snail life; *Euparypha pisana* and *Helicella maritima* are the most important species in this part of Europe, particularly the former. *Euparypha pisana* is present all along the Mediterranean coasts and along the Atlantic littoral to the English Channel; it is a xeroresistant species, able to escape the excessive heating of soil by climbing on plants, where it often forms very characteristic clusters whose formation has been studied in detail by Bigot (Fig. 375).

It is only relatively thermophilous, thalassophilous if not halophilous. *Helicella maritima* is a Mediterranean species very xeroresistant, in France common on low littoral species such as *Suaeda* and *Salicornia*. In the Var it reaches 900 m altitude. Both seem to be phytophagous elements.

Another small although rather rare snail present in the rear dunes, rather than in the foredunes, is *Helix aspersa*, a very euryoikous species, drought resistant thanks to its ability to bury itself in the sand or earth or to hide under mosses and

Fig. 375. Sandy shore formation of 'cluster' of *Euparypha pisana* (from Bigot).

litter. It is extremely euryphagous, feeding upon the leaves and flowers of many plants.

The dunes of the Northern Adriatic Sea represent, as a whole, a dynamic component of a succession, which we may postulate for most of the sea coast of this region, where there is a continuous contribution by rivers (Isonzo, Tagliamento, Livenza, Bacchiglione, Brenta, Adige and Po) and therefore a continuous advance of land. In a general way we may outline a succession for this part of Southern Europe, following this route

Cakiletum

↓

Agropyretum

↓

Ammophiletum
xanthietosum

↓

Ammophiletum
medicaginetosum

↓

Scabiosetum var. a

↓

Tunica and *Silene angustifolia*

↓

Scabiosetum var. b

Koeleria and *Silene otites*

↓

Scabiosetum clematidetosum

↓

Juniperus stage

↓

Orneto-Quercetum ilicis

↓

(Quercion pubescentis?)

↓

CLIMAX
Querceto-Carpinetum boreoitalicum

↓

(mesophilous facies)

building up of dunes ↓ (Ammophiletum xanthietosum to Tunica and Silene angustifolia)

leaching of Calcium carbonate and humus formation ↓ (Scabiosetum var. b to Querceto-Carpinetum boreoitalicum)

(from Pignatti, 1959)

To each vegetation stage some particular chemical and physical characteristic of soil corresponds, as is evident from the following scheme:

	Barren soil	Agropyretum	Ammophiletum xanth.	Ammophiletum medicag.	Scabiosetum	Orneto-Quercetum ilicis	Quercetum pedunculatae	Schoeneto-Erianthetum	Schoeneto-Plantaginetum	Schoeneto-Molinietum
pH	6.8	6.8	6.8	6.9	6.7	5.8	5.4	6.9	7.5	7.6
$CaCO_3$ (%)	55.6	73.3	71.1	72	60.5	16	11.9	70.0	63.6	–
NaCl (%)	0.4	0	0	0	0	0	0	0	0.29	–
Humus	0	0	0	0	0	1.2	2.3	0.8	1.7	–
Water	12.1	4.4	2.9	2.8	4.9	–	–	12.9	13.2	–
Grain size analysis:										
2000-1000 μ	1.6	0	0	0	0	–	–	0	–	–
1000-200 μ	83.6	81.7	65.3	73.9	36.9	–	–	27.0	–	–
200-50 μ	14.8	18.2	34.5	24.6	59.1	–	–	65.1	–	–
50-20 μ	0	0	0	1.1	4.0	–	–	5.2	–	–
< 20 μ	0	0	0	0	tr.	–	–	2.8	–	–

Strictly associated with this plant succession, as one might expect, there is an animal succession.

We think that it might be useful to compare, in this context, the animal population of sandy shores in different Italian regions, namely Veneto in the Southern Adriatic Sea and the Ionian Sea. The following table allows us to compare (rather euryoikous and eurythermous species) and contrast (probably stenothermous elements) these regions. Of course, the presence or absence of a given species may be due to other than mere ecological reasons, for example, a paleogeographical one, since Veneto and Southern Italy have experienced a very different recent geological history.

ADRIATIC SEA					IONIAN SEA		
Veneto	Gargano North coast	Gargano East coast (Vieste)	Torre Sabina	Torre S. Foca	Porto Cesareo	Taranto	Metaponto
Amara aenea	*A. aenea*						
Harpalus serripes	*H. serripes*		*H. serripes*				
Trechus 4-striatus						*T. 4-striatus*	
Carterus dama	*C. calydonius*			*C. calydonius*			
Tachys scutellaris	*T. scutellaris*						
Tachys micros							
Bembidion ustulatum	*B. subfasciatum*						
Calathus melanoceph.		*C. mollis*					
Dyschirius nitidus	*D. rugicollis*						
	D. numidicus						
Cicindela lunulata		*C. lunulata*					
Colotes maculatus	*C. maculatus*						
Malachius spinipennis				*M. spinipennis*			
Astenus pulchellus	*A. pulchellus*						
Stenosis intermedia	*S. intermedia*		*S. intermedia*	*S. intermedia*		*S. intermedia*	
Ammobius rufus	*A. rufus*	*A. rufus*	*A. rufus*				*A. rufus*
Trachysc. aphodioides	*T. aphodioides*	*T. aphodioides*	*T. aphodioides*	*T. aphodioides*		*T. aphodioides*	*T. aphodioides*
Phaleria bimaculata	*P. acuminata*	*P. acuminata*				*P. bimaculata*	
Xanthomus pallidus				*X. pallidus*		*X. pallidus*	
Anthicus fenestratus	*A. fenestratus*			*A. fenestratus*	*A. 4-guttatus*		
	A. transversalis						
Otiorrh. ferrarii	*O. juvencus*	*O. ferrarii*	*O. ferrarii*	*O. ferrarii*		*O. ferrarii*	
				O. juvencus		*O. juvencus*	
Scarabaeus semipunct.	*S. semipunctatus*		*S. semipunctatus*	*S. semipunctatus*	*S. semipunctatus*	*S. semipunctatus*	
Psammobius sulcicollis	*P. porcicollis*					*P. sulcicollis*	*P. porcicollis*
						P. porcicollis	
Galeruca tanaceti			*G. pomonae*			*G. pomonae*	
Galeruca interrupta							
Hypocaccus rugifrons		*H. metallicus*	*H. apricarius*			*H. rufipes*	*H. apricarius*
Cardiophorus equiseti			*C. collaris*			*C. collaris*	
			C. exaratus				
Chilocorus bipustulatus			*C. bipustulatus*		*C. bipustulatus*		
Adonia variegata						*A. variegata*	*A. variegata*
Coccin. septempunctata						*C. 7-punctata*	*C. 7-punctata*

The arthropod fauna of the sandy coasts of the Baltic Sea is well known thanks to Krogerus' thorough research. In the zone between the high tide line and the rear dunes an enormous number of arthropods has been found, as we may see from the following data:

Taxon	Number of species	Number of specimens	(per 177 square meters)
Myriapoda s. lato	2	–	
Apterygota Insects	13	181	
Orthoptera	11	52	
Odonata	9	6	
Corrodentia	2	5	
Thysanoptera	5	3	
Hemiptera	147	1019	
Neuroptera	8	39	
Trichoptera	23	51	
Coleoptera	216	1287	
Lepidoptera	118	257	
Diptera	418	1466	
Hymenoptera	388	538	
Arachnoidea	83	295	
Total	1443	5199	

Very few environments in the world – not only in Europe – have been studied so accurately.

The sandy shores of Finland exhibit the following microenvironments, which in practice almost constitute a true succession, quite similar to what we have already noted around the Mediterranean Sea (p. 571): salty surfaces near the Sea only with algae; suprasaline dunes or foredunes with succulent vegetation (*Honkenya* or sea sandwort, *Salsola* etc.); Suprasaline foredunes with low grasses such as *Agrostis stolonifera, Puccinellia, Juncus balticus*, etc.; deflation surface of Supralittoral zone with a very scant vegetation; Supralittoral dunes and foredunes with *Elymus arenarius*; supralittoral dunes with *Salix repens*; supralittoral foredunes with a vegetation dominated by *Festuca*; 'Mature' dunes with varied, open, grassey vegetation; and finally mature dunes or links with heathery closed vegetation, with the cranberry plant, the bearberry *Arctostaphylos*, etc.

Fig. 376 shows the numerical abundance of species and individuals of each micro-environment (true seral components) and stratum.

Here are the most characteristic species of each microenvironment or seral stage:

Salty surfaces: *Hesperophilus arenarius* and *Dyschirius arenosus*; Suprasaline foredunes and dunes: the Dipteran *Hylemyia candens*; Suprasaline foredune: *Chironomidae* undetermined, *Cicindela maritima* (a beetle), the *Diptera Tetanops myopina, Fucellina griseola* and *Hylemyia candens*; deflation surface of the supralittoral zone: the small beetle *Anthicus flavipes*; supralittoral foredunes or dunes: *Tetanops myopina, Liburnia excisa, Atheroides elymi, Chrysopa abbreviata, Hylemyia candens, Erythraeus regalis, Hylemyia coarctata, Artanes fallax, Cneorrhinus plagiatus, Arenostola elymi, Philonicus albiceps, Thereva annulata, Oscinella frit*. Supralittoral dunes with *Salix repens: Microsynamma bohemani, Melasoma col-*

laris (a beetle) and the dipteran *Sapromyza setiventris*. Supralittoral foredunes with *Festuca*: the hopper *Gomphocerus maculatus, Ochtiphila aridella, Hylemyia candens, Phimodera humeralis, Thereva annulata*, and the inconspicuous *Oscinella frit*, the wheat fly.

Mature dunes with open vegetation: *Episyron rufipes, Hylemyia candens, Gomphocerus maculatus* and the elegant *Thereva annulata*. Finally in the mature dunes with closed vegetation (kind of heath) the most characteristic species is the Microlepidopteron *Scythris siccella*.

The arthropods of this sandy coast of Finland may be subdivided into ecological categories, such as littoral species (halophil), xerophilous species, euryoikous or 'ubiquitous' species and finally hygrophilous, not halophilous species. The figure shows the percentage of the total artropod fauna represented by each ecological category. Practically all the arthropods on the sandy Baltic coast are xerophilous or euryoikous forms, and the same is true for the sandy shores of the Mediterranean (Fig. 337).

As it has been shown for the few organisms on rocky shores (p. 553), Krogerus has noted the ability of the arthropods of the Baltic Sea to withstand immersion by seawater. The data obtained are listed in the following table:

Species Group	Approximate immersion times	
Hesperophilus arenarius – Staphylinid	9-10	days
Dyschirius obscurus – Carabid	7- 8	
Cneorrhinus plagiatus – a weevil	6- 7	
Aegialia arenaria – a Scarabeid	6	
Dyschirius arenosus – Carabid	5	
Amara silvicola – Carabid	3- 4	
Ischnodemus sabuleti – a bug	3	
Anthicus sellatus – a beetle, Anthicid	2	
Phimodera humeralis – a bug	1.5- 2	
Adrena argentata – an *Apidae*	1.5- 2	
Paratinus femoralis – a *Cantharidae* beetle	1.2- 1.5	
Eonius bimaculatus – an *Anthicidae* beetle	1.5	
Cicindela maritima – a beetle	1.5	
Psammochares plumbeus – a Hymenopteron	1.5	
Philonicus albiceps – an Asilid fly	1.2	
Sphingonotus coerulans – imago (hopper)	1- 1.2	
,, ,, – nymph	6-18	hours
Artanes fallax – a spider	8-15	,,

At one time sandy shores, in many European countries by far more extensive than the rocky ones, were frequented by an enormous number of birds, found both on sea and land, including the inter-tidal zone which twice a day is free of water, attracted by plentiful food. Differences, sometimes only slight, in size and feeding habits made possible the co-habitation of closely related species sometimes in the same ecological niche. Today the constantly and rapidly increasing human population density in some areas (especially on Mediterranean coasts) and still more the increasing industrialization of all European countries, have brought about a steady decrease in the number of coastal birds, on sandy shores more than on others, because of their easy accessibility to man.

Finally, in recent years sea pollution has caused changes on sandy shores (and, of course, also on rocky and muddy shores) especially when the polluting agent is oil: we will come back later to this argument.

For these reasons bird life is made more difficult each day on this type of sea shore, and only protective measures will be able to arrest the declines in population.

Several families are present in this humid environment: *Gaviidae* (the so-called divers) such as *Gavia arctica* (black-throated diver), *G. immer* (great Northern diver), *G. adamsii* (white billed diver) and *G. stellata* (red-throated diver). The distribution of the great northern diver (in German called 'Eistaucher') both as a breeding and a wintering bird, is very interesting.

The family *Anatidae* is represented on sand dunes by the ruddy shellduck, *Casarca ferruginea*, in German 'Rostgans', limited to the southernmost parts

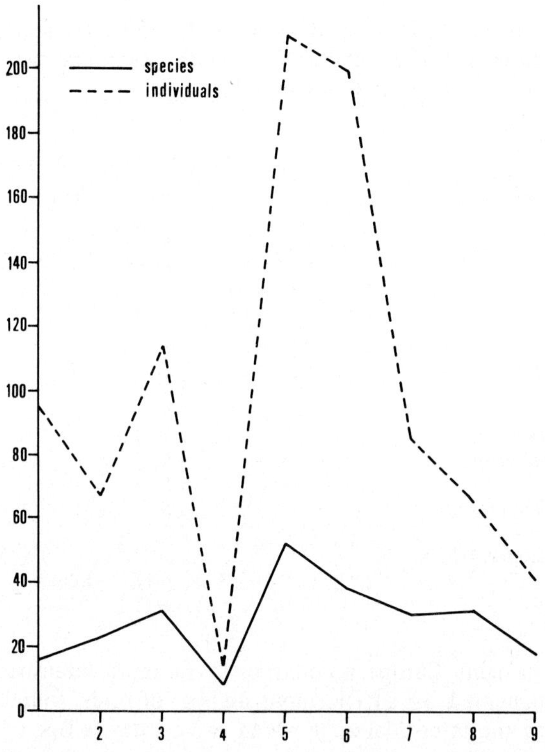

Fig. 376. Number of species and individuals present in the different microenvironments (parts of a succession) of the sandy shores of Finland. 1, salty surface with algae; 2, supra-saline zone with some succulent plants such as *Honkenya* and *Salsola*; 3. supra-saline grassy dunes with *Agrostis stolonifera, Pucciniella, Juncus balticus*, etc.; 4. supra-litoral deflation surface; 5, supralitoral dunes and pre-dunes, mainly with *Elymus arenarius*; 6. dunes with *Salix repens;* 7, dunes with *Festuca;* 8, terminal dunes with a variable grassy open vegetation; 9. terminal dunes or dune-land with heathery, closed vegetation (*Empetrum, Arctostaphylos*, etc.). (From Krogerus).

of the continent but accidental in Sicily (where it is known as 'fiscuni imperiali' or even 'fiscuni furastieri') and as a partial migrant up to Iceland, Finland, etc. Another *Anatidae* is the Brent goose, *Branta bernicla* (in German 'Ringelgans', in Dutch 'Rotgans', in Swedish 'Prutgås) in winter on the shores of Atlantic Europe, in summer in the Arctic tundra.

The family *Charadriidae* is well represented: let us remember *Charadrius hiaticula*, ringed plover (in German 'Sandregenpfeiler'), often nesting on sand; *Charadrius alexandrinus*, Kentish plover (in German 'Seeregenpfeiler') nesting on sand, present also on the shores of brackish water (Neusiedlersee, cf. p. 537); the grey plover *Pluvialis squatarola* (of uncertain etymology) in German 'Kiebitzregenpfeiler', in winter on sandy shores, nesting in the tundra, and finally, the turnstone, already mentioned when discussing rocky shores.

The family *Glareolidae* is well represented in Europe by the cream coloured courser, *Cursorius cursor*, an extremely elegant bird. One of the most widespread and omnipresent families is of course *Laridae*, of which we will mention only the herring gull *L. argentatus*, the little gull *L. minutus* (in German 'Zwergmöwe', in Dutch 'Dwergmeeuw', in Swedish 'Dvarmås', and in French 'mouette pygmée') reaching a total length of 28 cm; the black-headed gull *L. ridibundus*, in German 'Lachmöwe', present like the former species on sandy gravelly shores; and finally the extremely elegant and beautifully coloured common tern, *Sterna hirundo.* In Italy, where it is named 'rondine di mare', it is present, and breeds in the Po valley, in some central provinces and in Sardinia.

The passerine birds are also common on sandy shores, but none of them is sufficiently restricted to this environment.

Finally a predatory bird, present sometimes on sand dunes, albeit not typical of them, is the short-eared owl.

The sandy shores of Northern Europe are frequently visited by a mammal, the harbour seal, *Phoca vitulina*, in German 'Seehund', in Spanish 'foca comun' or

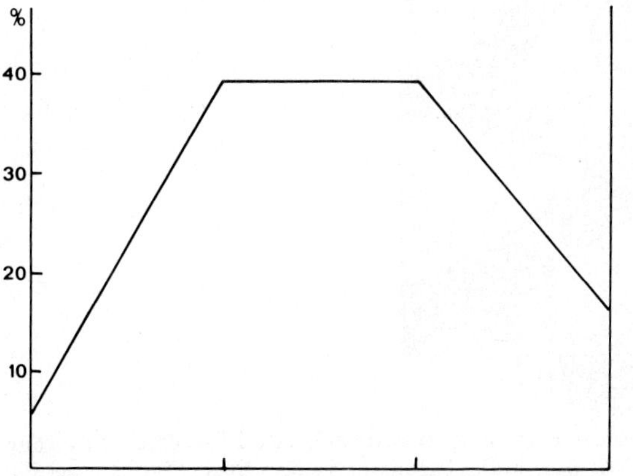

Fig. 377. Percentage of the various ecological categories in the total arthropod fauna of sandy shores of Finland (Krogerus).

'lobo marino' (meaning sea-wolf). It is found on sandy shores, in estuaries and, more rarely, on rocky shores. It exhibits gregarious habits.

Sometimes also *Cetacea* (whales and dolphins) are found on European sandy shores, where they are stranded by the sea, because they are tired, or more probable through illness or wounds. Let us remember *Mesoplodon mirus*, stranded once on the French coast, once in the Outer Hebrides and several times on the coast of Ireland. In the Adriatic Sea the sperm whale, *Physeter catodon* has been stranded on several occasions; the narwhal *Monodon monoceros* has been found on the coasts of Great Britain, the Netherlands and Norway. In the Adriatic Sea the blue whale, *Balaenoptera musculus* has also been stranded once, whereas in the Tyrhhenian, near Salerno a finwhale or common rorqual *Balaenoptera physalus* has been stranded. On the Tyrrhenian coast the Risso's dolphin *Grampus griseus* has also once been stranded near Ladispoli (Rome), a *Ziphidae* has been stranded recently at Finali Ligure.

Finally, very rarely, large tortoises may be found stranded on our shores: in Italy for instance the *Dermochelis coriacea* has been found several times.

Fig. 378. A most peculiar muddy shore environment: extremely small barena called locally 'velma', emerged generally only at low tide, in the Lagoon of Venice. During the summer the velma can be as hard as stone and crack, with the appearance of some salty efflorescences, recalling somewhat the true solonchaks. In this microhabitat *Salicornia stricta* finds an environment similar to that of deserts (salty deserts) (from Marcello & Pignatti).

Muddy shores

Muddy shores are present only where water is still, little influenced by currents, in bays and gulfs, and where there is a great amount of clay and silt transported by rivers. More especially, muddy shores are present near, or along, estuaries, which show all these physical characteristics. Furthermore, muddy or silty shores may develop in litteral ponds or lagoons, more or less isolated from the sea. These littoral lagoons are present both along the North Sea ('Haffen') and in the Mediterranean Basin ('barene' of Northern Adriatic Sea, 'Sansouires' of Southern France, 'limans' of the Black Sea).

The vegetation of the 'barene' is very rich and varied, depending on the distance from the water (and therefore the probability of flooding) and on the degree of soil moisture (which can go from dry to extremely wet, almost saturated). We shall mention here only the most characteristic species growing in this wet environment (Fig. 378 and 380).

Phragmites communis (= *Arundo phragmites*), a cosmopolitan element; *Spartina stricta* or broom sedge, sometimes partly submerged, a Mediterranean-Atlantic element; *Aelurops litoralis*, only in dry places; the reflected poa *Glyceria distans*, typical of humid 'barene', subjected to flooding, one of the halophilous-hygrophilous species most resistant to salt; *Schoenus nigricans* or bog rush, only in dry places, together with the sedge *Cladium mariscus; Carex extensa* and *C. flava*, in dry 'barene' together with *Sonchus maritimus*; several species of the genus *Juncus, Atriplex hastatus*, var. *triangulare; A. laciniatus* and *A. portulacoides*, extremely frequent; *Salicornia fruticosa*, the marsh samphire, very characteristic of this environment; *Arthrocnemum glaucum*; two species of *Salsola* or salt wort, *Salsola kali* and *S. soda*, the latter generally in flooded places; *Suaeda maritima* or herbaceous seablite, extremely abundant, often associated with *Salicornia herbacea, Statice limonium* (= *Limonium vulgare* or sea-lavender), characteristic of 'barene', a halophilous, Mediterraneo-Atlantic element conferring on this environment its peculiar bluish-violet colour in late summer and autumn, visible at great distance, together with two less frequent species, *S. minuta* and *S. bellidifolia*; several halophilous species of *Plantago* such as *cornuti*, south-European, *coronopus*, euro-asiatic, and *maritima* var. *crassifolia*, circum-Mediterranean form; the extremely peculiar *Aster tripolium*, the sea aster, a halophilous euro-asiatic Composita of which G. von Martens in his 'Reise nach Venedig' (Ulm, 1824) said 'an allen Ufern der Lagunen und Valle, die es im Herbst mit seinen freundlichen hellblauen Blumen verschönert', *Artemisia caerulescens*, *Inula crithmoides* (golden samphire) and *Sonchus arvensis* var. *maritimus* (field milk-thistle) are the remaining *Compositae* of 'barene'. The former is a Southern European species; it was mentioned by von Martens in his travels through the Lagoon of Venice; the second is named after an Umbellifera, *Crithmum maritimum*, because of the succulent aspects of its long, cylindrical, perfectly glabrous leaves, it flowers from July to September; it is a halophilous Mediterranean-Atlantic form; the yellow hue of the barene in summer is due to this plant. The third species, *Sonchus arvensis* var. *maritima* (= *Sonchus maritimus*), is also present in environments subjected to frequent flooding.

A chorological analysis of this florula shows that part of the association has a western Atlantic or sometimes American affinity, while part of it exhibits a very clear Asiatic, recent, origin (steppes, halophilous elements). Whereas the former are generally halophilous-hygrophilous elements but rarely true halophytes, the latter are constantly true halophytes, sometimes also nitrophilous, characteristic of more or less unstable edaphic conditions.

It is necessary to remark at this point that the true halophilous plants (halophytia; Del Villar) are associated with excess alkalinity in the soil: the limiting factors are pH, the excessive osmotic pressure and the toxic effect of some salts. Only succulent *Chaenopodiaceae* are able to survive (*Salicornia, Chenopodium, Suaeda* etc.) and *Plumbagineae* (*Statice*), together with shrubs such as *Tamarix* in Europe.

It is perhaps worthwhile mentioning that the halophilous vegetation has attracted the attention of botanists since the very beginning of Botany: Pietro Andrea Mattioli (1501-1577) from Siena (Tuscany) in his Commentarii in sex libris Pedacii Dioscoridis Anazarbei de Medica materia, Venetiis, 1554, in Epistolarum medicinalium libri quinque, Pragae, 1561 and lastly in Compendium de plantis omnibus una cum eorum iconibus de quibus scripsit sui in Commentariis in Dioscoride editis, Venetiis, 1571, records *Atriplex portulacoides* ('Halimus vulgaris'), *Atriplex hastata* ('Atriplex marina'), *Salsola kali* ('Atriplex kali'), *Crithmum maritimum* ('Crithmum'), *Artemisia caerulescens* ('Absinthium marinum'), *Statice limonium* ('Limonium'), and *Salsola kali* var. *tragus* ('Tragum'). These elements, all present in the 'barene' of Venice, have been collected by Mattioli

Fig. 379. Andrea Mattioli (1501-1577).

in the saltern near Trieste, where climate and soil are very similar to those near Venice.

The animal population of a barena is mainly composed of winged insects, on the whole widely distributed, psammophilous or more often lutobiont (i.e. bound to a muddy soil), practically constantly halophilous elements.

Taking into consideration the 'barene' of the Lagoon of Venice, we may mention the following invertebrates.

The woodlice are *Armadilloniscus littoralis*, present on sea shores in France, Italy, Jugoslavia and Greece, probably characterised by a discontinuous distribution due to the discontinuity of its habitat (i.e. muddy shores). *Halophiloscia fucorum* is a typical species, present on shores in the supralittoral zone, found under stones or in wrack; it has a very wide distribution. *Armadillidium assimile*, an euryoikous species present both along sea shore and rivers, in the mountains up to 1200 m.

Fig. 380. a. A 'barena' of the Lagoon of Venice, b. barena with *Statice limonium* in flower, c. barena with a rich vegetation of *Aster tripolium*, and d. Canebrakes (Phragmitetum) in the Lagoon of Venice.

The hoppers are *Conocephalus fuscus* and *Homocoryphus nitidulus*, two common hygrophilous species, besides *Epacromius coerulipes* (Fig. 381), a Rassenkreis, extended from Central Asia to Northern Italy, probably a steppe element of oriental origin, like its congeneric *E. tergestinus*, and many plants of this peculiar humid environment. Very interesting is *Zeuneriana brunneri*. The common earwig *Labidura riparia* and the praying mantis *Mantis religiosa* are also plentiful: the latter is an active predator on hoppers, and is generally found in open, sunny, places, on herbs and shrubs. Another earwig is *Anisolabis maritima*.

Bugs and *Cicadinae* (or leaf hoppers) seem to be rare in this habitat. The most interesting bugs are the *Saldidae* such as *Saldula pilosella*, present exclusively in the Salicornietum, a European species, in Italy found exclusively in Veneto, Friuli-Venezia Giulia (i.e. the north-easternmost part of this country), besides Puglie and Basilicata; *Salda adriatica*, present both in the Salicornietum and Scirpetum maritimi, an Eastern Mediterranean species, in Italy limited to its north-eastern borders (Friuli-Venezia Giulia and Veneto) and therefore both stenotopic and stenoikous; and finally *Salda* (= *Halosalda*) *lateralis*, also exclusive to the Salicornietum, a Euromediterranean form, in Italy limited to Veneto (‚barene'), Tuscany, Puglia, plus Sicily and Sardinia.

The commonest beetles are the Carabid *Dichirotrichus obsoletus*, an Atlanto-Mediterranean form, halobious, and *Cicindela lunulata*, present also on sandy shores. *Dischyrius nitidus*, a Euro-Asiatic element, psammo-ripicolous, in some instances present together with some strictly halophilous species such as *D. salinus*, *D. chalybaeus* and *D. apicalis*; *Pogonus riparius*, a Ponto-Mediterranean luto-

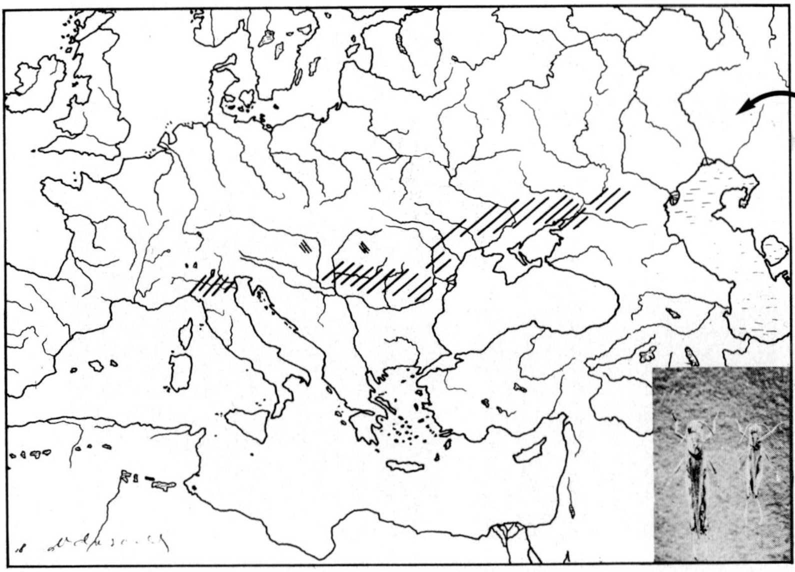

Fig. 381. Distribution in Europe of the grasshoppers *Epacromius coerulipes* (inset) //// and *E. pannonicus* \\\\.

halobiont form, very characteristic of this environment; *Philhydrus agrigentinus*, a small, extremely rare hydrophilid, so far known only from Sardinia, Sicily, Spain and Italy, where the Northern Adriatic Sea represents the northernmost point of its distribution; *Ochthebius dilatatus*, halophilous, extending from the European Atlantic coasts to the Caspian Sea; the Staphylinid *Remus sericeus*, a widely distributed halophilous species, present on sea shores under wrack; three lady-beetles, *Adonia variegata* (a Palearctic species) living on flowers of *Eryngium campestris, Beta vulgaris, Cirsium, Chenopodium* and *Atriplex; Coccinella 11-punctata*, a widely distributed species, particularly frequent on sea shores, on dune herbs etc., or often on *Umbelliferae* and *Achillea*, maybe a halophilous species; and lastly *Bulaea likatschovi*, a widely distributed form, feeding probably on the pollen of *Chenopodiaceae*, but limited to a few sites in the Northern Adriatic Sea (Fig. 382-3). One of the most interesting beetles of the 'barene' is *Heterocerus flexuosus*, a true life-form because of the form of the body and still more of the fossorial limbs (Fig. 384). It is a halobious species, present on the sea shore and in salty inland places, in Europe and North Africa; the adult seems to feed on organic remains present in the mud; eco-physiological research on the feeding habits of this animal might prove interesting. The most interesting species is probably the endemic *Chrysomela schatzmayri*, present exclusively on the sea shore of Veneto and Istria (Fig. 385) where it lives on several species of *Artemisia* and on *Inula viscosa*, a Mediterranean form. Other Chrysomelids are *Cassidula nobilis*, a European form living upon several *Cariophyllaceae* (*Silene, Spergula*) and *Chenopodiaceae* (*Chenopodium, Atriplex, Beta*); it can be harmful to sugar

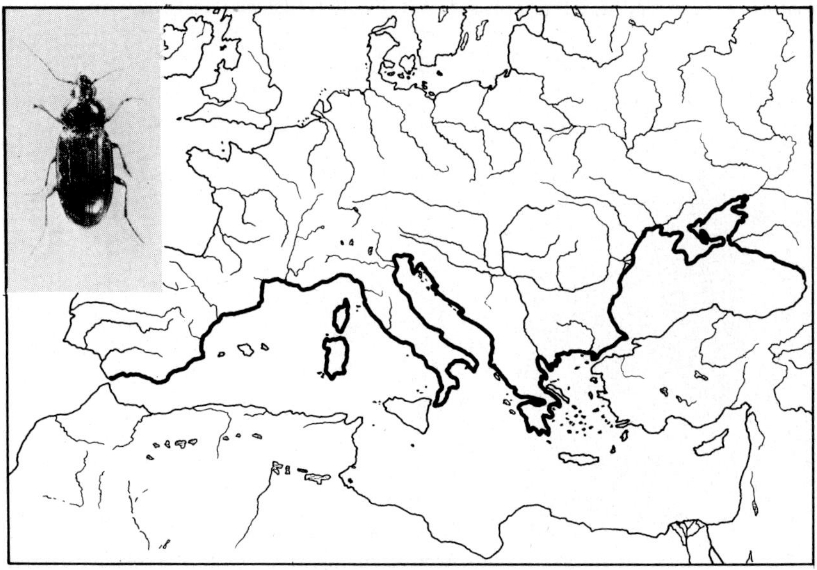

Fig. 382. Distribution of the Carabid beetle *Pogonus riparius*.

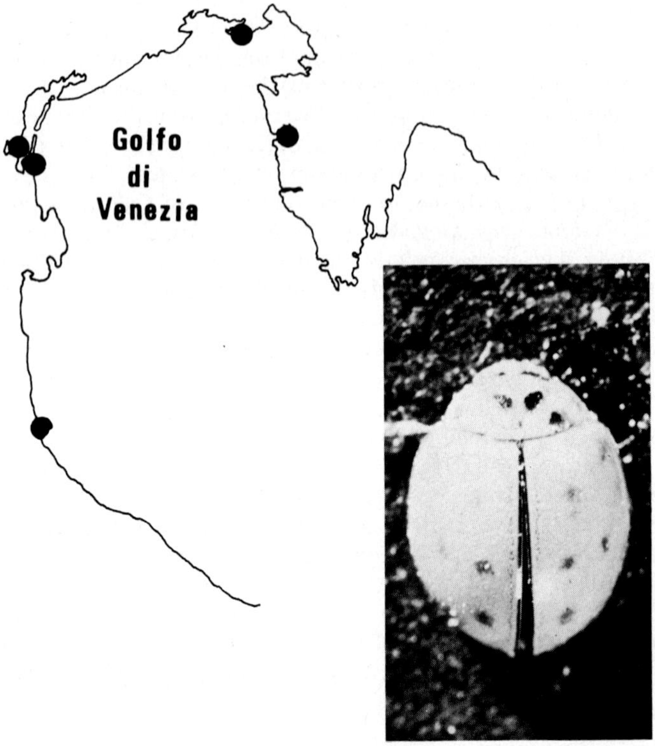

Fig. 383. Localities of the northern Adriatic Sea from which the Coccinellid beetle *Bulaea lichatschovi* has been reported. The species is limited to Italy and Western Istria.

Fig. 384. A life form of muddy shores, *Heterocerus flexuosus* (note the dilated anterior tibiae).

beet, where it lives together with *Cassida vittata*; *Chaetocnema tibialis*, a southern species living on several *Chenopodiaceae* such as *Atriplex, Beta, Chenopodium, Salsola, Salicornia*, typical forms of the 'barene' or of the 'sansouires'. It can be a real pest on sugar beet. A Tenebrionid is *Phaleria bimaculata*, which, however, is not at all typical of muddy environments; it is a halopsammobiont.

Other beetles worth mentioning because of their rarity or their ecology are the Staphylinid *Achenium depressum*, the weevil *Baris scolopacea*, restricted to *Atriplex halimus*, and the Cerambycid *Calamobius filum*, present along the coast of the North Adriatic Sea, as well as in Dalmatia; it is a Southern European-Mediterranean form, present also in some xerothermic oasis of Central Europe (Slovakia, Germany); it is limited to several Graminoid plants both natural and cultivated, and sometimes it is harmful to wheat.

Finally, one species of Collembolan deserves our attention, *Anurida maritima*, so plentiful on the 'barene' that it forms broad bluish patches on the clay soil; when the soil is submerged (during high water, e.g.) the animal lies on the water

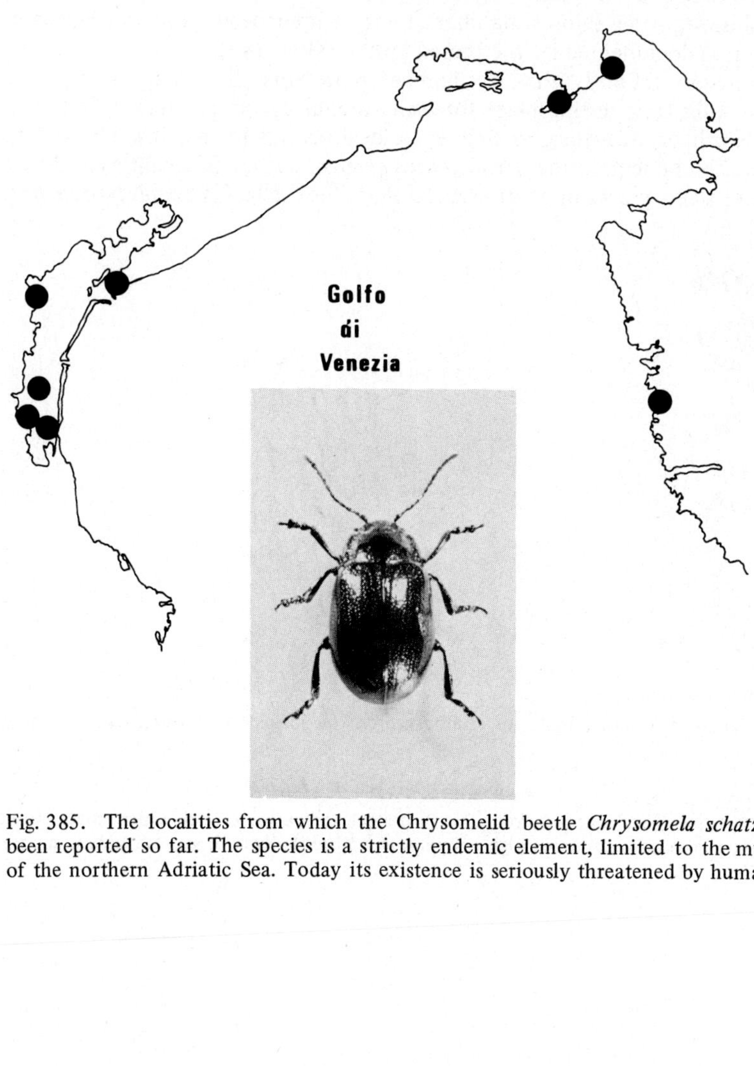

Fig. 385. The localities from which the Chrysomelid beetle *Chrysomela schatzmayeri* has been reported so far. The species is a strictly endemic element, limited to the muddy shores of the northern Adriatic Sea. Today its existence is seriously threatened by human presence.

surface without either being submerged or wetted, but it can also withstand submersion under stones covered by the water (Fig. 386).

The only species of snails commonly living in this humid environment are *Truncatella subcylindrica*, living on sea shores of Atlantic Europe, the Mediterranean Sea and the Black Sea under stones and algae, on the muddy sand at the mean tidal level. The other species is *Phytia myosotis*, extensive on the shores of the Mediterranean, at Madera, in the North Sea to Ireland and Scotland. It lives in the littoral zone near sea or brackish waters and is a phytophagous species (Fig. 387-388).

A very similar environment present in Southern Europe is given by the so-called 'sansouires'[1] of the Camargue and other parts of Southern France. Pader in 1890 said of the sansoires 'during winter it is a swamp, during summer a desert. The swamp is inhabited by innumerable aquatic birds, the desert by clouds of mosquitos'. Plant and animal life of 'sansouires' are very much the same (at least at the community level) as those of the 'barene' of the Northern Adriatic Sea (Fig. 390).

The reptiles include two lizards, *Lacerta sicula* and *Lacerta muralis*, often living together, showing some interesting phenomena of interspecific (and intrageneric) competition, as documented by the present author (Fig. 389).

The Venetian estuary with its 'barene' represents, like the Camargue in France, an obligatory resting place for many migratory birds, which travel from South to North or vice-versa, as well as an ideal habitat for nesting. We wish to mention in this connection the heron, *Ardea cinerea*, nesting in localities in the Po valley and in a few places in North-Central Italy; the mallard *Anas platyrhynchos*,

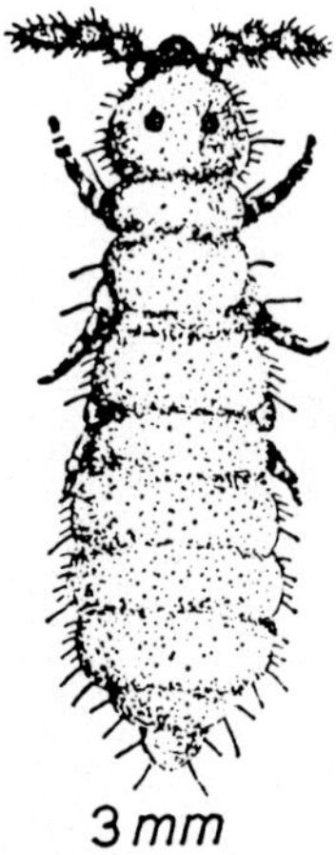

Fig. 386. An insect, characteristic of muddy shores, the springtail *Anurida maritima* (from Riedl).

[1] sansouire seems to be derived from the Provençal sansouiro, a land which covers itself with salt and efflorescences during high temperatures, a sterile land: 'li sansouiro de Camargo' (Mistral, 1932).

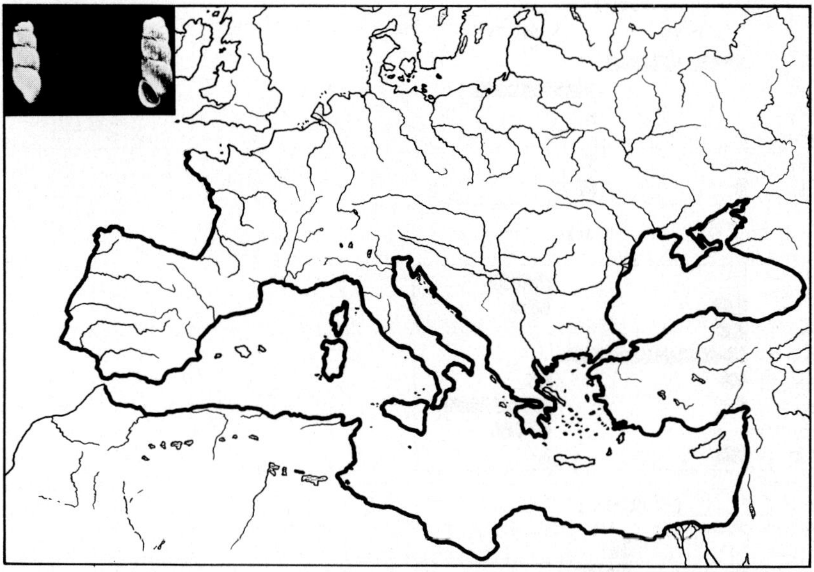

Fig. 387. Distribution of the Mesogastropoda *Truncatella subcylindrica*.

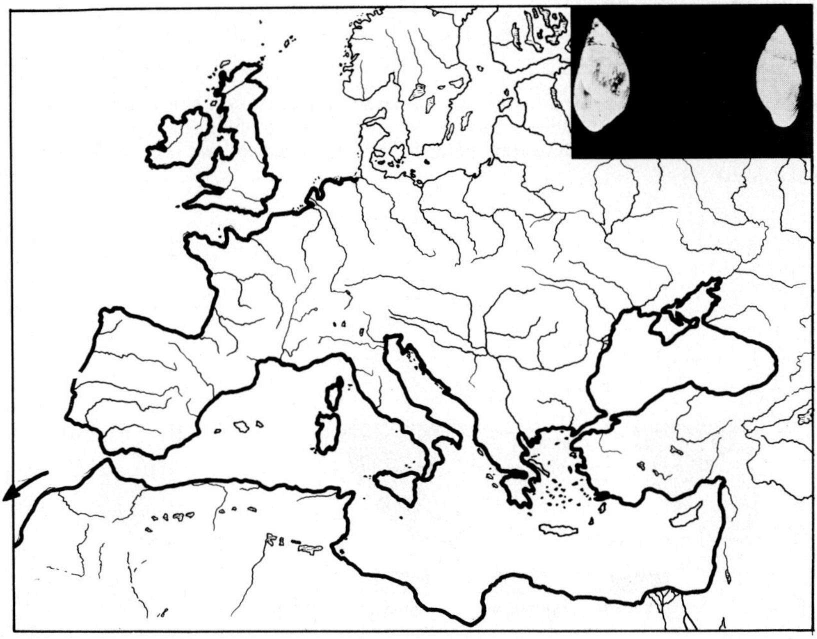

Fig. 388. Distribution in Europe of the Pulmonata *Basommatophora Phytia myosotis* (also found on Madeira).

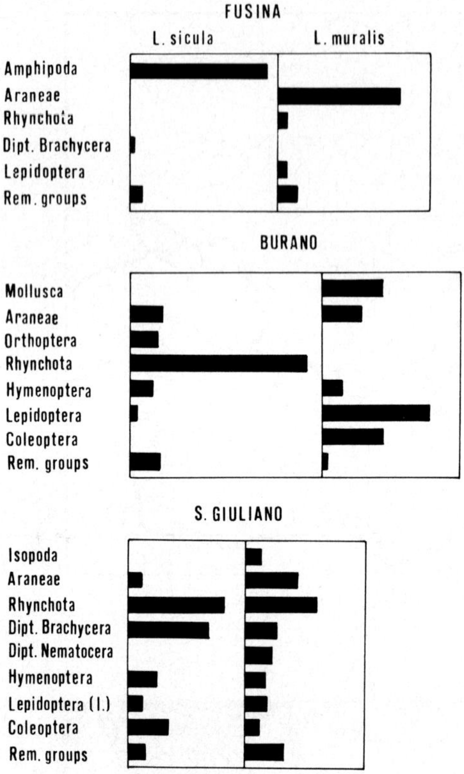

Fig. 389. Differences in the food of the two species of lizards, *Lacerta sicula* and *L. muralis*, that live together in some halophilous habitats (muddy shores) of the Lagoon of Venice. The slight difference in the food composition permits the co-existence of two very similar species (from Marcuzzi).

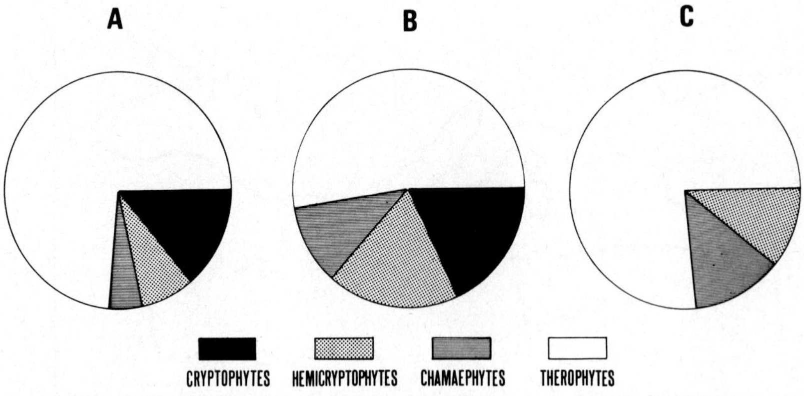

Fig. 390. Raunkjaer's spectra of three different aspects of the sansouires (from Bigot, 1971): A. *Arthrocnemetum glauci;* B. *Salicornietum fruticosae*; C. *Thero-brachypodion.*

the wigeon *Anas penelope*; the garganey *Anas querquedula*; the shoveler *Spatula clypeata*, in declining numbers as a consequence of land reclamation; the red-crested pochard, or *Netta rufina*, characteristic of this kind of environment although today decreasing because of land reclamation and spring hunting; the moor hen, *Gallinula chloropus*, in lower numbers each year; the avocet *Recurvirostra avosetta*, typical of estuaries and brackish swamps, but decreasing; the spotted redshank *Tringa erythropus*; the great crested grebe, *Podiceps cristatus*, of which there are today a few surviving couples in the Po valley and in Sardinia; the black winged stilt, *Himantopus himantopus* (from ἱμάς, lace or string and πους, foot) a partially migratory bird, but declining; the oyster catcher, *Haematopus ostralegus,*

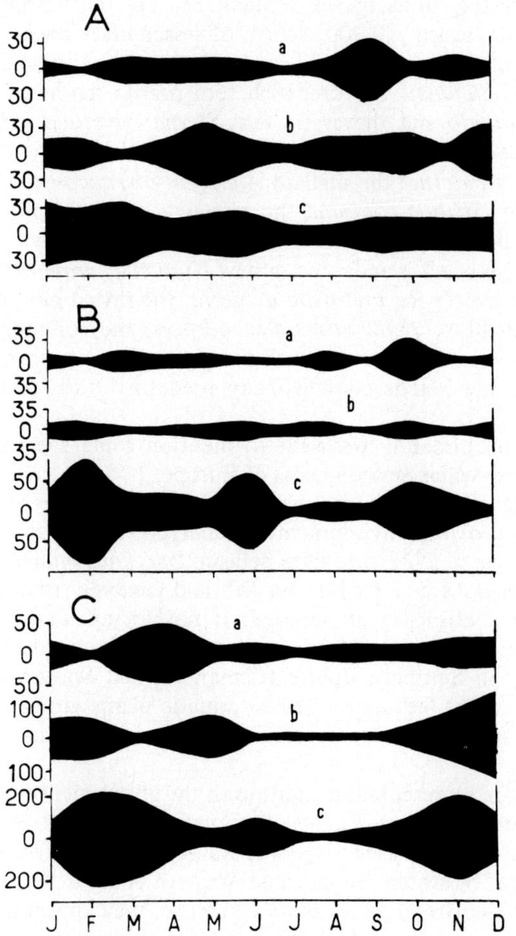

Fig. 391. Annual variations in the abundance of three groups of soil animals in three different plant associations. a. No Oribatid mites; b. Oribatid mites; c. Arthropleones springtails. A = *Arthrocnemetum glauci*; B = *Salicornietum fruticosae;* C = *Thero-brachypodion* (from Bigot).

a ripicolous bird, whose rare pairs in the Po delta are decreasing through excessive hunting; the common tern *Sterna hirundo*, a nesting bird here; the little egret *Egretta garzetta*, a partially migratory bird, nesting in small colonies in the Po valley and in the Tuscan 'maremma' and at present notably increasing; the coot, *Fulica atra*, typical of brackish lagoons linked to the sea, so that often its food consists partially of eel grass (*Zostera*) and sea lettuce (*Ulva lactuca*); it nests in our lagoons but it is now decreasing; and finally, the lapwing *Vanellus vanellus*, of double passage, also found in fewer numbers.

The large number of species (or genera) common to this Mediterranean region and similar habitats on the islands of the Wadden Sea on the North Sea cost, is proof of the relative uniformity of this kind of humid environment between relatively distant regions. The nesting birds of the Wadden Sea are the herring gull, *Larus argentatus*, numbering nearly 10,000; scores of lesser black-backed gulls (*Larus fuscus*); hundreds of common gulls (*Larus canus*), and thousands of black-headed gulls (*Larus ridibundus*), the sandwich tern *Sterna sandvicensis*, the common tern *Sterna hirundo* and the arctic tern *Sterna paradisea* and lots of little terns *Sterna albifrons*; over 5000 pairs of the eider duck *Somateria mollissima*, the sheldrake *Tadorna tadorna*; the mallard *Anas platyrhynchos*, plus the teal *Anas crecca*, the shoveler, *Spatula clypeata*, the garganey, *Anas querquedula* and the pintail *Anas acuta*. In addition to these nesting species, the following birds have been observed on these small islands; the curlew *Numenius arquatus*, the redshank *Tringa totanus*, the avocet *Recurvirostra avosetta*, the ringed plover *Charadrius hiaticula*, the Kentish plover, *Charadrius alexandrinus*, the common snipe *Capella gallinago*, the ruff *Philomachus pugnax* and the corncrake *Crex crex*. Due probably to their isolation, the islands contain many predatory birds that breed on the ground.

To finish this small section on bird life, we want to mention some species present, more-or-less, in all brackish water environments of Europe.

We will limit ourselves to mentioning a *Podicipidae*, the black-necked grebe, *Podiceps nigricollis*, all but typical of this environment; a *Phalacrocoracidae, Phalacrocorax carbo*, or cormorant (see p. 558); the white pelican, *Pelecanus onocrotalus*, nesting in reeds beds in lagoons and feeding on fish and generally found with gulls, terns and cormorants, affording an unusual, if not to say unique, subdivision of work; the elegant flamingo, *Phoenicopterus ruber* (from φὸινιξ purple red) a bird resident only in Southern France (Camargue) and Southern Spain ('marismas' of the Guadalquivir) feeding on plants (aquatic plants growing beneath the surface of water) small crustaceans (*Artemia salina*, the common brine shrimp), snails and clams.

Anatidae or ducks sensu lato are represented in addition to the above-mentioned species by the white-headed duck, *Oxyura leucocephala*, in German 'Ruderente', nesting amond the reeds of brackish water lagoons. Its geographical distribution embraces Northern Africa, Southern Europe and Western Asia as far as Turkestan and Beluchistan. Also the Brent goose, *Branta bernicla*, may appear in this environment outside the breeding season. The crane, *Grus grus*, may be considered typical of lagoons, where it makes its nest in reed beds (in the past it was also present in Italy: in the 1920's it was still living near Venice and Caorle). More frequent although not very characteristic, are the *Charadriidae*, such as *Charadrius hiaticula*, or ringed plover, *C. alexandrinus* or Kentish plover and the grey plover, *Pluvialis squatarola*.

It is needless to say that muddy shores like all the other sea shores are anything but self-sufficient from the standpoint of energy-flow, and therefore they can scarcely be considered as an ecosystem, their life depending on both marine and terrestrial environments.

As we have described a succession for the sandy shores (p. 571) we may, though with far greater difficulty, describe some pathways followed by plant and animal successions in brackish waters. If we compare in this connection the lagoon of Venice with the muddy shores of Schleswig-Holstein (studied by Tischler), we may reach a synthesis represented by the following scheme:

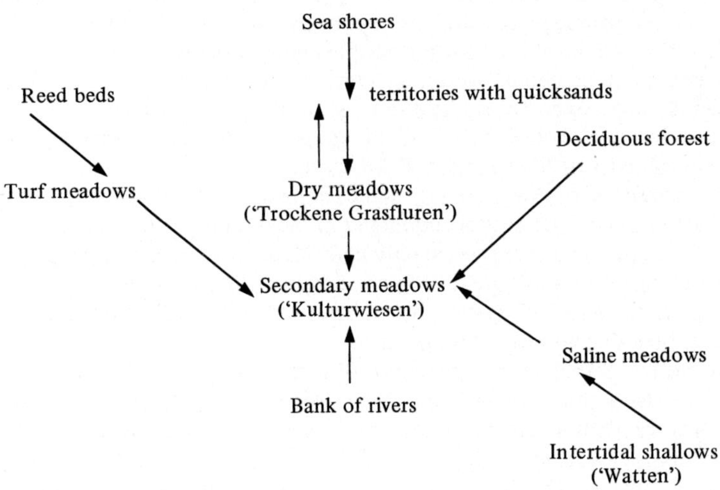

The human impact upon humid environments

Freshwater environments

When discussing the impact of man on the shores of freshwaters, we must think of past environments in Europe. In many countries lakes were more numerous and larger than today: many of them formed during the Glacial Period, and are surrounded by glacial moraines. The smallest have been completely colonised by aquatic plants, while the biggest become smaller: reeds and other plants invading the shores and superimposing each year a new layer of vegetation, formed the mires. Actually, many human remains, from Neolithic to early Bronze Age, have been found in mires. In the deep layers of some mires the humus and the surface of the fields cultivated during Neolithic have been brought to light. Also abodes constructed on pallases, similar to those present today in several parts of the world, have been discovered[1]. Among the poles implanted deeply in the clay, flint weapons, remains of pottery, and wagon wheels have been found in some mires, for instance that of Mercurago near Arona. Furthermore, from the bottom of the old lake, invaded successively by the mire, a pirogue, paddles and a wooden anchor have been discovered.

Many palafittes were present in the Neolithic of Switzerland, where the dense forests left room for human settlements only around water bodies. Palafittes were still in use during the transalpine aeneous culture (Eneolithic).

To the Neolithic belong the following palafittes: Lagozza near Besnate, Palude di Brabbia, Monate and Bardello near Varese. To the Eneolithic belong the palafitte of Arquá Petrarca (in the region of the Colli Eugani near Padova) and that of Lagozza near Vho di Piadena.

At the end of the Eneolithic several palafittes had been transformed into 'terramare' (i.e. palisades constructed upon soil instead of water) whence the name of 'terramaricolous' used for the people living in them.

The palafitticolous people were busy with breeding and agriculture. They had domesticated dogs, oxen, pigs, sheep and goats; they grew wheat, barley, grapes and they wove lint. Today this kind of human settlement is very common in several tropical parts of the so-called Third World.

In the palafittes of Switzerland, some remains of Neolithic vegetation are present, and in the excrements accumulated in the sediment the seeds of the fruits man ate at that period are visible (mainly cherries and prunes). The people belonging to this culture had made great progress in agriculture, since the seeds of wheat, barley, rye, millet, lentil, chick-pea, broad bean and other vegetables were present. Also animal breeding was well developed.

The mires (or perhaps turfy meadows) of Fimon (in Eastern Veneto) were also inhabited by Neolithic people living on palafittes. Human skeletons have been found, together with a great number of bones, teeth and horns belonging to various animals, cereals (probably barley) and carbonized seeds. Furthermore several pirogues were excavated in tree trunks, fishing implements (probably a bow-net) a number of remains of pottery (generally attributed to Eneolithic), worked flint of all kinds and finally bronze tools (axes) (Trevisiol). The prevailing culture belonged to Neolithic and more precisely to the culture characterized by 'square mouth' pottery.

The wood utilized for these palafittes – or better for the land reclamation of the zone in which it was erected – dates from 4500 B.C. In the huts near the fireplace several shells of water bivalves (*Unio*) and land snails have been found, together with mammal bones, vertebrae of fish and seeds. The mammals belonged to *Cervidae, Bovidae* and *Ovidae*. The plant remains belong to wild, edible plants (*Trapa natans* or water caltrop, hazel) and a few seeds of wheat.

Let us remember now a few localities of the Bronze Age, famous for their palafittes. The palafitte of the mire of Chizzoline belongs to the culture of Polada, i.e. of

[1] A palafitte (from the Latin palus) is a construction made on poles (palus) or a settlement formed of huts placed on a kind of horizontal platform supported by poles driven vertically into the bottom of a lake, a swamp, or into the shores of a lake.

swampy habitats within which were erected a complex wooden raft at the beginning of the Bronze Age. It is not dissimilar from the culture of Barche di Solferino and Carriana (south of the morainic amphitheatre of the lake of Garda) and of some swampy stations of Brianza (Bosisio), of Isolino di Varese, or even of the palafitte site of Ponale, near the Lake of Ledro. At Chizzoline, however, successive cultures are represented of the late Bronze Age and of Iron Age II.

Finally, at Polada, near Lonato (Brescia) according to the most recent investigations, there were no true palafittes, since there are neither poles implanted in the bottom, nor horizontal poles usually employed for land reclamation, but a village whose huts were built directly upon the ground. The culture is attributed to Bronze Age sensu lato.

In some cases (Le Morandine, Valli Veronesi) an old palafitte formerly belonging to the early Bronze Age was re-utilized by new people (Romans) to build their constructions using the same methods employed by the previous inhabitants. The animals represented here are partly domesticated (ox, sheep, goat or *Capra hircus palustris*, pig and dog), partly in a wild state (wild boar, stag, roe-buck). The ox corresponds to *Bos taurus brachyceros*, the same as found at Fimon, another palafitte of this region.

According to Zorzi (1955), the region where the palafitte was found (Le Morandine) was covered by arboreal vegetation, with some thickets interrupted by swamps and water courses. The natural conditions allowed the cultivation of plants in the clearings created by felling, raising of animals, hunting and fishing. According to Zorzi the inhabitants of this palafitte adapted themselves to their chosen environment.

In general, the spread of palafittes both in humid environments (palafitticolous people) and in rather dry environments (terramare) occurred in Northern Italy from the Neolithic to Iron Age or even Roman times. The palafitticolous people lived in a wide area from Piemonte (Varese) and Lombardia (Barche di Solferino, Peschiera etc.) to Trentino in the North (Lago di Ledro) and Veneto in the South (Lago di Fimon). The remains found on the bottom of the lake of Bor di Pacengo (near Peschiera) point to relationships between the Danubian area and the Po valley area, making us think of the possibility of exchanges between these two relatively distant areas[1].

The most interesting terramare are those of the lake of Garda, with two different cultural aspects belonging to the Middle and Recent Bronze Ages (Tombola di Cerea, Franzine, Bovolone, Povegliano and Monte Lonato). On the contrary, the terramare of Bellanda near Mantua belongs to the culture of Polada (eneo-padan environment).

Terramare of the Bronze Age are also present at Montata di Reggio Emilia, S. Polo di Enza, Castione, Casaroldo, Gottolengo and other villages of the provinces of Modena, Reggio Emilia, Parma and Piacenza (Säflund, 1939).

According to Aspes (1970), in the Po valley there was during the Bronze Age a slow evolution from the culture characterized by interment to that characterized by cremation. However, the mere presence of one type of burial or another is not

[1] The study of the wood employed in the construction of palafittes showed that in the Bronze Age both the chestnut and the horse chestnut were widespread in Italy.

sufficient to establish precisely the age and the type of a given culture. The remains found both in palafittes and in terramare makes us think of active exchanges between distant geographical groups in Mycenae, Sicily, Sardinia, Danube basin and Switzerland. All this points to a great mobility of man of the Bronze Age, representing on one hand a certain impact of man upon nature, on the other explaining the presence of ethnic, cultural, etc., elements in different parts of Central-Southern Europe. The names of many plants (particularly alpine) reflect very often these old migrations.

In Central Italy swamps have been inhabited since very remote times; particularly well known in this connection is the Paludi Pontine, in Lazio, named after Pomptinus or Pontinus, a name probably derived from the old town Suessa Pometia, in a territory occupied by Romans in 358 B.C. Ager Pomptinus has been called a territory covered by swamps. These were inhabited by a transient population of shepherds and charcoal burners living generally in groups of huts called 'lestre', situated in the maquis clearings or sclerophyllous forest growing near the swamps.

At the beginning of the Würmian glaciation, the region was inhabited by several types of animals, typical of a warm climate such as elephants, rhinoceros, hippopotamus etc., all hunted by Neanderthal man who lived both in the open and in caves. These people were engaged in an extensive stone industry called 'pontinian'.

In the Upper Paleolithic Neanderthal man was replaced by the present day *Homo sapiens*, who in that period hunted steppe animals. Man was present also during the Neo-Eneolithic (remains of 'greenmarble', flint and obsidian). The Iron Age is documented by the remains of a blackish pottery; finally in Paludi Pontine there are numerous Roman remains.

Another Central Italian swampy area is in the Maremma. The name 'maremma' comes from the Latin maritima (neuter plural of the adjective maritimus) and is employed to indicate a littoral pond or swamp separated from the sea by a littoral sand bar, similar to the 'lagoons' of the littoral region of the Northern Adriatic Sea. We find this name in Foscolo ('Nè la maremma Elea ricca di pesce') and in Giusti ('Come mandre a svernar nelle maremme ...').

The life of man in freshwater and marine or brackish water humid environments, and its impact upon nature can be imagined also through the presence in many European countries of wooden boats, the so-called 'pirogues' not unlike those used today in several parts of the Third World. For instance, the double pirogue of Polada (Brescia) could be compared with those used today in the Society Islands (Micronesia) (Fig. 392). The simple pirogue or monoxylon has been found at a great number of sites, both in Italy and in Central Europe (e.g. Pesse, in the Netherlands). In Italy monoxylons cannot be dated earlier than the early Bronze Age. On the contrary, in the Netherlands they belong to the Mesolithic (Pesse, in Drenthe) and more exactly from 6315 B.C. (^{14}C method). In Neolithic there are pirogues in Denmark (e.g. Sjaelland or Seeland), Federsee (South Germany) and Switzerland (Saint-Aubin, Egolzwill, Bouldry and Neuchatel). Very likely the pirogue came to Northern Italy from the Alps. Today the number of definite localities in which pirogues have been found in Italy are: Piemonte, 2; Lombardia, 11; Trentino, 1; Veneto, 2; Emilia, 1; Toscana, 1; Umbria, 1 and

Latium, 1. None of these data allow us to think that pirogues were used at sea. Today some assume that maritime navigation (which dates from ancient times) was possible with a double pirogue, but not with a monoxylon. In this connection we might add that in the Aegean Sea, boats were made with cloth or leather. In Ancient Egypt, during the Neolithic, boats were made with reeds and canes (Fig. 393). Finally, during Minoan times the boats were made of wood, generally coming from Italy.

Sea shores

> Cretans say that the honours rendered to Gods, the sacrifices or the mysteries of religion are of Cretan origin, and that the other peoples learned them from Crete. Demeter passed from Crete to Atthica, and then to Sicily, and then to Egypt, bringing with her the cultivation of wheat.
>
> Diodorus Siculus, or
> *Διοδορος ο Σικελιώτης*
> Βιβλιοθήκη v. 77

The impact of man on sea-shore environments is very old.

The collecting of marine animals in prehistoric times must have been high judging by Quarternary remains. Both remains of Echinoids (sea-urchins) and edible molluscs (snails and clams) have been strongly smoothed by marine agencies. These shells were collected for uses other than a dietary one, and chosen on the ground of 'les coloris les plus brillants ou les formes les mieux appropriées à leur destination' (Barral, 1954).

The observations of the habits of some present-day primitive cultures, where we find collecting associated with fishing, makes it rather probable that a similar kind of activity was undertaken by primitive European peoples. Let us remember the so-called 'pêche à l'assommoir', consisting of shoals of fish driven inland to the beach and stranded. In the Orkney Islands, for instance, the local inhabitants know that many cetacean are stranded on some beaches, and they institute a special rota of watching for the cetaceans, so that they may go immediately to sea and direct the animals towards a trap, where they are stranded. Some say that prehistoric man, and more particularly paleolithic man, fed himself first with sea food and only later with animal flesh, although this does not seem likely.

Recently near Nice (at 'Terra Amata') some remains of huts erected by hunters who periodically stopped on the shore have been discovered. These remains are dated at 300,000 years B.C. and palynological investigations show that these people reached the coast at the end of each spring or beginning of summer and later returned to the interior. The temporary inhabitants of the sea shore collected mollusc (remains of oysters, *Mytilus* and limpets) and the presence of vertebrae and other bones of teleosteans points, in the meantime, to fishing or the collecting of fish as weel. Algae were furthermore used as litter or as food. Together with the bones of birds and mammals, there are also remains of marine turtles.

Also we must realise that, together with human artifacts, the presence of molluscs (snails and clams) remains is a constant feature of all the bone caves

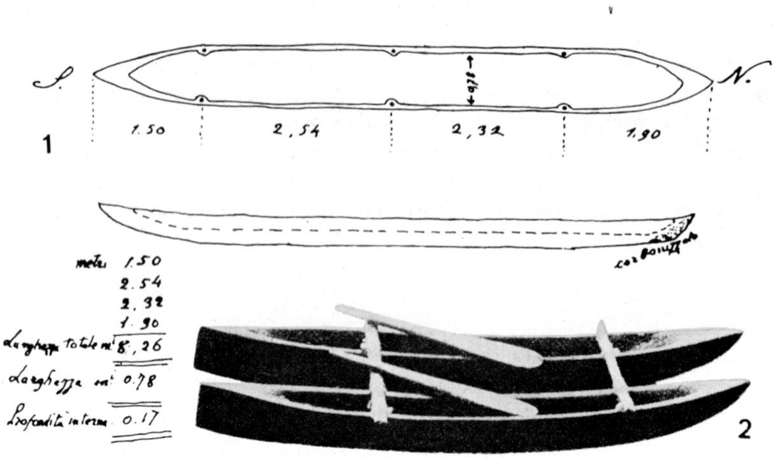

Fig. 392a. Italian pre-historic double pirogues (top). For comparison, a double pirogue as used today in the Society Islands (bottom) (from Cornaggia Castiglioni, 1967).

Fig. 392b. Italian prehistoric wood monoxylons (from Cornaggia Castiglioni, 1967).

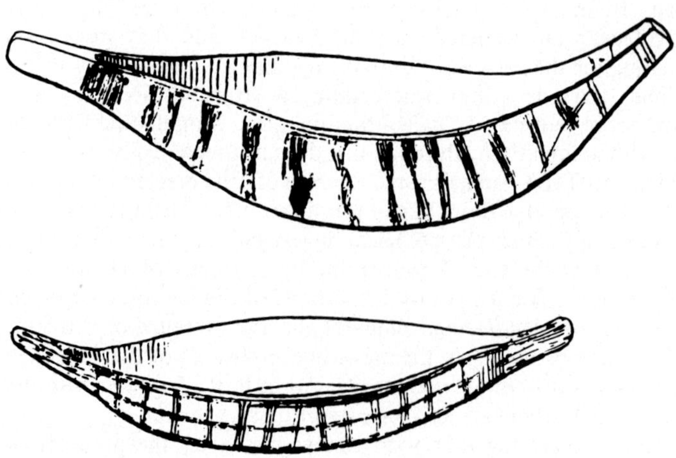

Fig. 393. Neolithic Egyptian boats built with reeds and canes (from Mosso).

utilized as an abode or as a sepulchre, by stone-age man, which shows that the gathering of these animals was very frequent along all kinds of sea shores. Thus, in some localities along the coast of Southern Italy (Moscerini, near Gaeta, a cave at Capo S. Maria di Leuca in Puglia, the Horse-cave near Lecce, also in Puglia, a microlithic Mousterian facies is present, characterized by the presence of tools (points and scrapers) made from the shells of the clam *Meretrix chione* (= *Calliste chione*).

In a deposit, referred to the Epigravettian of Arene Candide (Liguria), a kind of cap has been found in a sepulchre, made up of hundreds of shells of the Gastropod *Nassa neritea* (= *Cyclope nerites*) together with other shells of Gastropods (including *Cypraea* sp.) and exoskeletons of small starfish. In the same cave lower valves of *Spondylus* have been found, which were presumably used as spoons (Issel, 1884) (Fig. 394).

By the Upper Paleolithic, fish had acquired an increasing importance in the economy; in the 'kiökkenmoeddlings' of Mesolithic age in addition to nets, fish-

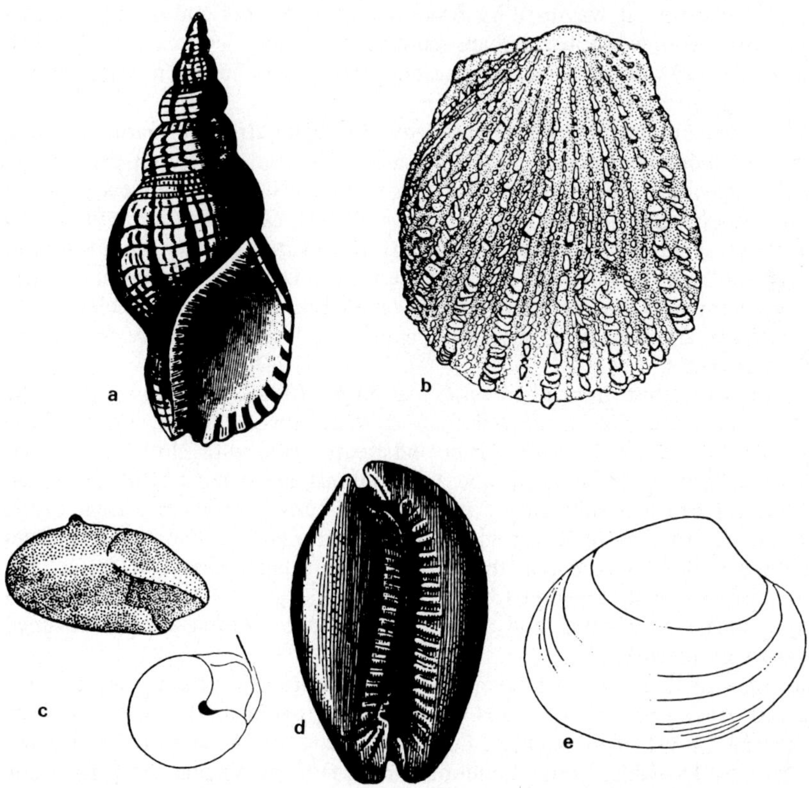

Fig. 394. a. *Tritonium tritonis* (from Leunis); b. *Spondylus gaederopus* (from Riedl); c. *Nassa neritea* (from Riedl); d. *Cypraea* sp. (from Leunis); e. *Pitaria chione* (or *Meretrix chione*) (from Riedl).

hooks and various fishing tools, skeletons of teleosts have been found.

The Solutrean and Magdalenian artists, who, as Graziosi says, used to reproduce particularly those animals which were their basis for life, very frequently drew fish, sometimes with very remarkable care. Fish are indeed reproduced much more frequently than bear, fox, amphibians and snakes, and only a little less than stags, *Bos primigenius* and *Capra ibex*. We find today some maginificent illustrations of fish, seals and an aquatic animal, possibly a fish or a cetacean, in the graffiti on rocks at Niaux (Ariege) and Pindal (Oviedo) and on horn found at Montgaudier (Charente).

In a Minoan seal, a woman is playing the shell of a *Tritonium*. These shells, used for sacred purposes, have been found at Palaikastron (Crete). Probably the shell of *Tritonium nodiferum* (the common sea trumpet or triton shell or even conch shell) has been utilized as a bugle throughout Mediterranean countries. Many of them have been found in the cave of Arene Candide, as well as in those of Balzi Rossi, Galluzzo and Pollera. The shells of several snails were used by the Minoan religion. The Romans employed *Tritonium nodiferum* as a war bugle, with the name of 'buccina' ('buccina jam priscos cogebat ad arma Quirites'), and also, in the past century, it was used by fishermen as a trumpet and by hunters as a hunting horn. Also, we should perhaps remember that very few years ago the shell of *Tritonium nodiferum* was played in several Italian churches (Piemonte, Liguria) (Mosso).

An inflated importance was acquired by the many stranded marine animals, such as cephalopods (squids, cuttle-fish) and teleostean fish (even pelagic and eurybathic species). Indeed today, during summer, it is easy to collect stranded *Atherina* (smelt), *Diplodus* (= *Sargus*) *vulgaris* (base), *Diplodus* (= *Sargus*) *sargus* (base), *Boops* (= *Box*) *boops* (bogue), *Sarga* (= *Box*) *salpa* (goldlin), *Mullus surmuletus* (red mullet), *Trachurus* sp. (hors mackerel) etc. During winter, after a storm, it is very common to find many crustaceans on Mediterranean shores such as *Leander, Pasiphaea, Euphausiacea, Anomura*, crabs etc. We have already treated the stranded cetaceans.

At some early, but unspecified time, man began to sail, and many parts of the Mediterranean coast saw a succession of cultures, towns, trades, etc. We have already mentioned the double-pirogues and boats made with cloth and leather. The more solid wooden boats allowed a continuous and dense traffic all through the Mediterranean Sea and outside it, as some natural substances present in the South and not in the North (obsidian) or vice-versa (Baltic amber) dating from Minoan times, demonstrate that there was already a fairly heavy impact of man upon sea shores in that period (3000-1100 B.C.). Also in this period mythical marine animals were created, and the belief in strange sea monsters has remained until only very recently.

The edges of rivers and seas have always been locations for human settlements, as the numerous present day towns, small and big, demonstrate. As far as rivers are concerned, most of the greatest European towns (often capitals) are situated along their banks (Paris, Rome, London, Torino, Orleans, Vienna, etc.). The same holds, of course, for sea-shores. In some instances estuaries have been chosen as a location for human settlements (Hamburg, Nantes, etc.).

Furthermore, some people have caused great changes in the morphology of

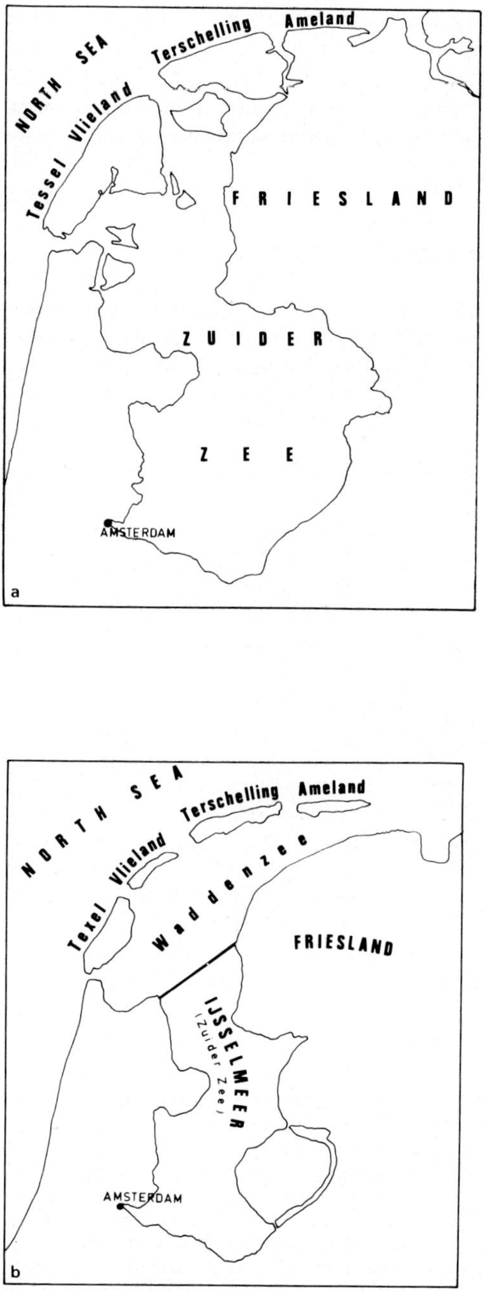

Fig. 395. Land reclamation in estuarine waters (muddy or muddy-sandy shores): the Zuiderzee in the past century (Stiedler's Atlas) (a) and at approx. 1960 (b) (Reader's Digest Atlas, 1961) (schematic).

coasts in order to reclaim land for urban and agricultural use. Suffice it to quote only two examples which are perhaps the most significant in Europe; the land reclamation of the Zuider Zee (Fig. 395) and the works the inhabitants of Venice have done since the Middle Ages in order to conserve Venice in its original condition. To this aim they have controlled sedimentation in the Venice Lagoon by diverting the rivers (Fig. 396).

Conservation problems in humid environments

We have seen how most humid environments are separated one from another: one might say that their major characteristic is their discontinuity, differing from most other European environments, with the exception maybe of mountains and also freshwater environments.

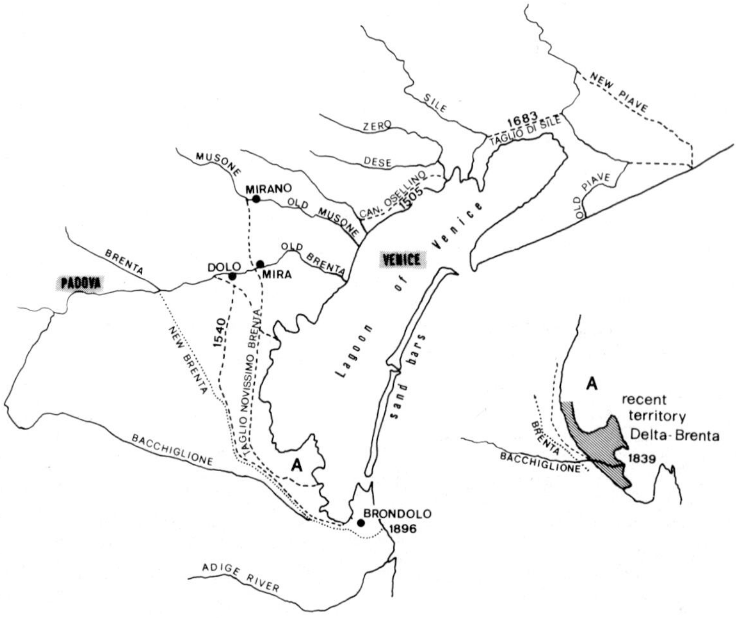

Fig. 396. Demonstration of the main alterations to the hydrography of the region of Venice in order to prevent the deposit of alluvial materials in the Lagoon of Venice, and subsequent filling. In 1693 the Taglio di Sile was made in order to make the waters of the Sile flow towards the Piave; successively a new Piave was made, in order to let most of the water flow eastwards, and so far from the Lagoon. In 1505 the Canale dell'Osellino was dug, in order to divert the course of the Marzenego eastwards, thus protecting Venice from alluvial material. In 1457 the Brenta was diverted from Dolo to Lugo and in 1540 from Dolo to Brondolo (near Chioggia). In 1839 the Bacchiglione was made to flow eastwards, together with the New Brenta in the region of Conche where in a very few years the so-called 'Delta Brenta' was formed thanks to the deposits transported by the two rivers. For this reason, in 1896, the Brenta (plus Bacchiglione) was made to flow southwards, towards the present-day mouth, south of Chioggia and north of the Adige mouth (slightly schematic).

To this discontinuity is due the patchy character of several biota of humid environments. Not only that, but the greater the stenoiky of a species (regardless of whether it is plant or animal,), and the lower its ability to spread (active or passive), the greater its rarity and the danger of extinction at least in Europe. One example is given by the beaver, in Europe limited to very few sites (allowing for those into which the beaver has been reintroduced), whereas in North America it is still widely distributed (albeit represented by a different geographical race). The same holds for several plants living in mires (see later).

Besides the character of discontinuity of humid environments (suffice it to quote high mountain lakes, high mountain 'Hochmoore', alkaline mires, estuaries, the highly characteristic environments of 'barene' and 'sansouires', brackishwater lagoons, isolated rocky islands in the sea, the 'faraglioni' and other similar features), relatively easy accessibility to man of most humid environments, the greater and greater facility we have to drain and reclaim land for any use, increasing population density all over Europe, constitute a major threat to humid environments and a possibility of destruction. This destruction of humid environments automatically implies the destruction of many of their biota (Fig. 397).

Waterfowl are particularly threatened by industrialization, land reclamation, increased human pressure, not even taking hunting into consideration.

In all European countries pleas have been put forward in favour of waterfowl (though ecologists believe that all organisms living in humid environments deserve attention, and not only birds) and something has been achieved although with great difficulty and obstacles of all kinds. (Needless to say that obstacles vary from one country to another as a consequence of environmental education, educational policy, and general politics (Fig. 398-400).

Let us recall two examples of how waterfowl may be protected. One is the resolution at the 2nd European Meeting on Wildfowl Conservation, which in part, says '... The 2nd European Meeting on Wildfowl Conservation draws the attention

Fig. 397. Specific problems in conservation of humid environments: the Lagoon of Venice where industrialization (Porto Marghera) causes heavy air, soil and water pollution. The life of the Lagoon is seriously affected.

of the Netherlands' Government to the exceptional biological value of the Waddenzee, especially as a breeding and resting ground for the *Anatidae*, Waders and Spoonbill (*Platalea leucorodia*); emphasises that this extensive shallow maritime area is one of the most important breeding and resting grounds of these birds and consequently the Netherlands have a great responsibility to the countries of Northern and Eastern Europe; recommends that no project for the modification of

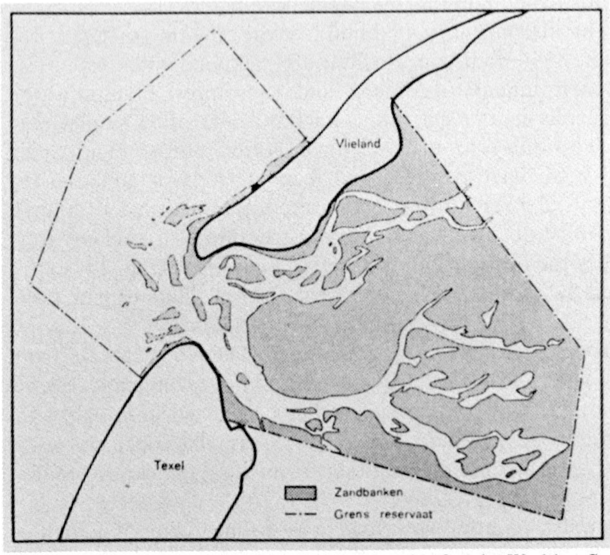

Fig. 398. Texel's seal sanctuary (from 'A plea for the Wadden Sea').

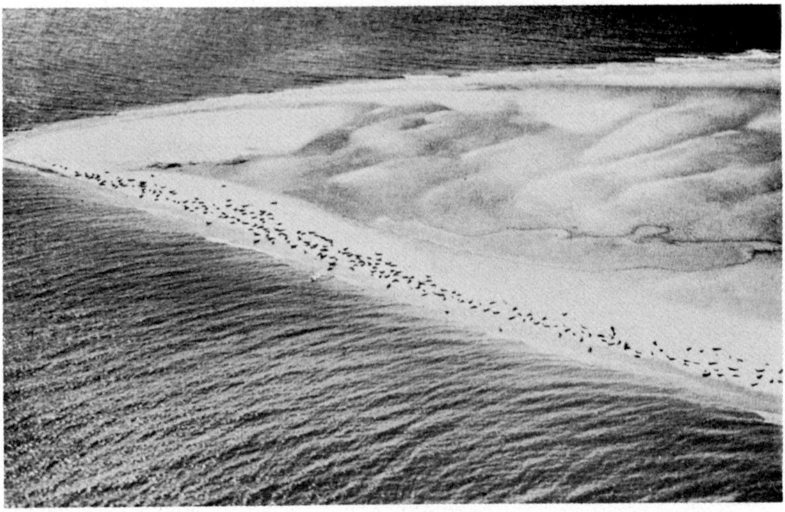
Fig. 399. Seals on the shores of the Wadden Sea (from 'A plea for the Wadden Sea').

Fig. 400a. Flying *Sterna sandvicensis* (locally 'grote stern') in the territory of the Wadden Sea (from RIVON-Verhand. 4, 1967).

Fig. 400b. The common stern in the territory of the Waddensee (from RIVON-Verhand., 4, 1967).

the Waddenzee be undertaken without previous consultation with ecologists, as to the effect such works might have on the biological resources of the Waddenzee, and that no modification be carried out which might threaten the existence of the *Anatidae*, Waders or Spoonbills'.

A second example is the creation in Italy of a private animal refuge at Bolgheri, near Leghorn. The refuge is of paramount importance for Italian avifauna; however, its judicial status does not permit too much optimism as to its future. Only if it were public and would become a National Park or a Natural Reserve, would the protection of wildfowl be assured forever.

Today the animal refuge of Bolgheri is an oasis for the protection of the fauna and includes c. 400 ha in which an area of 70-150 ha, according to season, is swampy. Fifty-nine species of birds are present, of which a part represents aquatic species, namely 2 *Podicipidae,* 5 *Ardeidae*, 9 *Anatidae*, 2 *Accipitridae* (diurnal birds of prey), 3 *Rallidae*, 2 *Charadriidae*, 6 *Scolopacidae*, 1 *Recurvirostridae*, 3 *Laridae* and 1 *Alcedinidae*. The percentage variation of *Ardeidae, Rallidae, Anatidae* and *Charadriiformes* during the year has also been studied.

Marine, or aquatic, birds and mammals living on our coasts are also in great danger. Suffice it to quote one species that has become extinct in Europe only in recent times, the *Alcidae Pinguinus impennis* present in Iceland up to 1844. Of course, the number of tropical species in danger is comparatively greater.

Mammals seriously threatened in Europe are the desman *Galemys pyrenaica*, the maritime bear *Thalarctos maritimus*, the walrus *Odobaenus rosmarus rosmarus* and the Mediterranean seal, *Monachus monachus*. Here again the endangered tropical species are far more numerous.

Sea shores in particular are under the constant threat of human change, through changes in their physionomy and destruction of their biota. Sea birds die in millions particularly because of hydrocarbons discharged by industries and tankers, and floating on the sea and in brackish water. Very often these animals come to die on sea shores.

On the sandy shores of Italy a destruction of pines has occurred over the last years. The *Pinus* species are mainly *Pinus pinea* (stone pine or domestic pine), *Pinus halepensis* (Aleppo pine, Jerusalem pine), and *Pinus pinaster* (sea pine, wild pine, the latter more isolated and utilized to fix sand). The causes of this destruction, according to Fanfani, are the pollution of the sea and inland waterways, wind, the levelling of the protecting dunes, man's influence on the soil and the vegetation (particularly the destruction of the indigenous Mediterranean sclerophyllous forest), damage caused by cattle, uncontrolled townplanning, all kinds of developments and facilities (both temporary or for residential purposes), road construction, in some cases lowering of the water table, or in others its elevation and mixing with saline water. Needless to say mires have an enormous importance in conservation problems, since they may as a whole be considered very fragile ecosystems, whose existence is permanently menaced by human agencies such as drainage, fertilization, burning, grazing, agriculture, peat extraction, afforestation, etc. Also proximity to cities or industrial areas (that today are expanding even in the Far North) and the educational or other visitor-uses (not least the research in palynology and pre-history) constitute a continuous risk for the preservation of this particular form of humid environment. For these reasons, in several countries measures have been taken, and also the IBP(CN) is planning conservation schemes. Also a study project of mires throughout the Northern Hemisphere (Project TELMA) exists.

In this connection we might consider it useful to list the plant species which it should be necessary to protect. Unfortunately no list of animals has been made so far, to the best of my knowledge.

Fig. 401. Human impact today on sandy shores. La Versiglia, once a beautiful sandy shore of Tyrrhenian Sea (Photo: Prof. G. Pilleri, Bern).

The species in Europe as a whole are, *Carex buxbaumii, Eriocaulon septangulare, Hammarbya paludosa, Liparis loeselii, Rancunculus ophioglossifolius, Saxifraga hirculus, Spiranthes aestivalis* and *S. romanzoffiana*. Species becoming rarer in the British Isles are *Andromeda polifolia, Betula nana, Calamagrostis neglecta, Carex appropinquata, C. aquatilis, C. atrofusca, C. chordhorriza, C. flava, C. microglochin, C. papercula, C. rariflora, C. stenolepis, C. vaginata, Cicuta virosa, Corallhoriza trifida, Deschampsia setacea, Dryopteris cristata, Eriophorum gracile, E. latifolium, Gentiana pneumonanthe, Isoëtes echinospora, Lathyrus palustre, Lycopodium innundatum, Naumburgia thyrsiflora, Nuphar pumila, Osmunda regalis, Pinguicula grandiflora, Rhynchospora fusca, Scheuchzeria palustris, Stellaria palustris, Stratiotes aloides* and *Viola stagnina*. Several other species are rapidly becoming rare due to drainage, eutrophication, pollution and very soon will be included in a list of extinct species.

The protection of humid environments has demonstrated, in recent years particularly, to be necessary, and several national or international agencies have been created to this aim, unfortunately not all of them equally effective. The international Biological Programme (IBP, CN) has dealt, and is still dealing, in European countries with such problems but unfortunately IBP is not organized on the same basis in all the various contributing nations and many are not ready to cooperate equally.

Probably the most effective means to achieving protection of humid environments are through National Parks, the Nature Reserves, the Naturschutzgebiete, the Landschaftsschutzgebiete and similar organisations. In Italy, for instance, among the proposed nature reserves there are marine parks which would serve to protect (besides the marine species) terrestrial or amphibian elements living on sea shores. Some territories are put under bond according to Article 67bis of the Law on Hunting (Art. 28 of the Law 2.VIII. 1967, n. 799). These areas form 'oases' for the protection and refuge of resident and migratory birds (also waterfowl) within which hunting and decoying are forbidden. At the end of 1972 there were c. 200 oases, totalling 200,000 ha.

The humid environment is particularly well protected in the Lake of Sabaudia, within the National Park of Circeo (now a nature reserve). Two more brackish water lagoons, Lago di Caprolece and Fogliano are excluded from the Park, so that the migratory birds find no protection there. We most note, however, that the Circeo is the most irrational 'Park' in Italy.

Other reserves created recently in Italy, and which may serve as a refuge for waterfowl, or in general for the organisms living in humid environments are the integrated Nature reserve at Piscina della Gattuccia and Lestra della Coscia, in the province of Latina; the Forest reserve 'Duna Feniglia' near Grosseto; the Oriented Natural Reserve 'Falasconi' near Foggia; the Natural Reserve Valle dell' Ofanto in the province of Pescara; the reserve for animal population of Vincheto di Celarda, near Belluno (too small for its purpose); and the Zoological reserve 'Sacca di Bellocchio' near Ravenna.

BIBLIOGRAPHY

A guide to Wicken Fen, The National Trust, London, 1950.

Aguesse P & L. Bigot, Observations floristiques et faunistiques sur un étang de moyenne Camargue: la baisse salée de la Tour du Valat, *Vie et Milieu*, 11, 1960, 284.

Alessio M., F. Bella, C. Cortesi & B. Turi, University of Roma Carbon-14 dates VII, Radiocarbon, 2, 1969, 482.

Backlund H.O., Wrack fauna of Sweden and Finland, Opusc. Ent. Suppl. V, Lund, 1945.

Battaglia R., Scoperte paleontologiche nelle torbiere del Veneto e del Trentino e nelle grotte della Venezia Giulia, *Hist. Nat.*, 1, 1946, N.2.

Bigot L., Essai d'ecologie quantitative sur les Invertebres de la sansouire camarguaise, Impr. M. Declume, Lons-Le-Saunier, 1965.

—, Recherches sur les groupements de Gasteropodes terrestres: la constitution des 'grappes', *Vie et Milieu*, 18, 1967, 1.

—, Ecologie des milieux terrestres salés, *Boll. Soc. Ecol.*, 1971, 2, 99.

Böck F., Die Bedeutung der Donau als Rastplatz überwinternder Entenvögel, *Verh. Ges. Oekol., Wien*, 1975, 241.

Broglio A., Ricerche 1971 nell'insediamento neolitico del Molino Casarotto (Valli di Fimon, Colli Berici), not published data.

Cagnolaro L., D. Rosso, M. Spagnesi & B. Venturi, Inchiesta sulla distribuzione della Lontra (Lutra lutra L.) in Italia e nei Cantoni Ticino e Grigioni (Svizzera) 1971-1973, *Ric. Biol. Selvaggina*, Bologna, N. 63, 1975.

Contact Committee for Nature and Landscape Preservation (editor), A plea for the Wadden Sea, Amsterdam.

Cornaggia Castiglioni O., Una nuova stazione 'palustre' dell'anfiteatro morenico gardesano: la Torbiera di Chizzoline in Comune di Sojano del Lago (Brescia), *Natura*, 47, 1956, 131.

—, Le piroghe preistoriche italiane. Problematica ed inventario dei reperti, *Ibidem*, 58, 1967, 5.

—, & F. Toffoletto, Il bacino lacustre della Polada ed il suo insediamento preistorico, *Natura*, 49, 1958, 54.

Corrain C. & G. Malgeri, Le stazioni neo-eneolitiche dell'Italia nordappenninica. Le sepolture ed i resti scheletrici umani, Borghero, Padova, 1975.

Corti U., Störche ziehen durch die Alpen, *Jhrb. Ver. Schutze Alpenpfl. & Tiere München*, 25, 1960, 22.

D'Ancona U., Ricerche idrobiologiche sul Padule di Fucecchio. I. Nota introdutiva, *Boll. Pesc. Idrobiol.*, 14, 1938, 3.

Delamare Deboutteville C., Biologie des eaux souterraines littorales et continentales, Hermann, Paris, 1960.

Dietl W., Die Landschaftsoekologische Bedeutung der Flachmoore. Beispiel Davallseggenrieder, *Jhrb. Ver. Schutze Alpenpfl. & Tiere, München*, 40, 1975, 47.

Dobesch H., Das Bestandsklima im Schilfgürtel des Neusiedlersees, *Verh. Ges. Oekol., Wien*, 1975, 173.

Dürkop H., Die Tierwelt der Anwurfzone der Kieler Förde, *Schr. Naturw. Ver. Schl. Holst.*, 20, 1934, 480.

Fanfani A., Condizioni dei pini del litorale italiano in relazione ai fattori ecologici, *Min. Agric. For., Collana Verde*, 32, 1973.

Festetics A. & B. Leisler, Oekologie der Schwimmvögel der Donau, besonders in Niederoesterreich, *Arch. Hydrob.* Suppl. XXXVI, 4, 1971, 306.

Frömmhold E., Vom heimlichen Leben in Moor, Wittenberg Lutherstadt, 1956.

Gams H., Die Alpenmoore, *Jhrb. Ver. Schutze Alpenpfl. & Tiere, München*, 23, 1958, 15.

Giordani Soika A., Studi sulle olocenosi. III. Gli Emitteri Eterotteri nelle olocenosi della Laguna di Venezia. *Boll. Soc. Venez. St. Nat., Venezia*, 4, 1949, 62.

Halaskova V., K. Hurka, M. Kunst & P. Stys, Sooser-Moor und Salzgebiet in Westböhmen, *Acta Univ. Carol. Biol.*, Suppl., 1960, 1.

Harnisch O., Die Biologie der Moore, Bie Binnengewässer, VII, Stuttgart, Schweizerbart, 1929.

Hohenstatter E., Untersuchungen einiger unbekannter Moore im bayerischen Alpenraum, *Jhrb. Ver. Schutze Alpenpfl. & Tiere, München*, 38, 1973, 30.

Hurka K., Die Carabidenfauna des Sooser Moores in Westböhmen (Col. Carabidae), *Acta Univ. Carol. Biol., Suppl.* 1960, 59.
Karl H., Des Naturschutzgebiet in den Chiemgauer Alpen, Jhrb. Ver. Schutze Alpenpfl. & Tiere, München, 34, 1969, 115.
Kraus O. Vom bayerischen Moorschutz: Das Hochmoor am Kesselsee bei Wasserburg/Inn, Jhrb. Ver. Schutze Alpenpfl. & Tiere, München, 22, 1957, 20.
Krogerus R., Ueber die Oekologie und Verbreitung der Arthropoden der Triebsandgebiete an den Küsten Finnlands, *Acta Zool. Fenn.*, 12, 1932.
Leentvaar P. (edited by), De Zuidelijke Vechtplassen, Flora en Fauna, RIVON, Verh. N.7, 1969.
Marcello A. & S. Pignatti, Fenoantesi caratteristica sulle barene nella Laguna di Venezia, Mem. Biogeogr. Adriat., 5, 1963, 189.
Marcuzzi G., Considerazioni ecologiche sulla palude carsica di Pietra Rossa (Monfalcone), *Atti Ist. Ven. S.L.A.*, 115, 1957, 127.
—— La laguna di Venezia, Aspetti fisici e biologici, *L'Universo*, 51, 1961, 78.
—— Un ambiente carsico da conservare: il Lago di Doberdò, *Italia Nostra*, 58, 1968, 46.
—— Osservazioni ecologiche qualitative sull'erpetofauna della Laguna Veneta, *Mem. Accad. Patav. S.L.A., Padova*, 80, 1968, 333.
—— & L. Dalla Venezia, L'ambiente di barena considerato alla luce dei problemi conservazionistici, Atti XI Congr. Ass. Naz. Laur. Sc. Biol., Venezia, 1970, 79.
Marcuzzi G. & L. Dalle Molle, Secondo contributo alla conoscenza dell'ecologia della torbiera piana di Fiavè (Trentino occidentale), *Studi Tr. Sc. Nat.*, 52, 1975, 187.
—— & ——, Contributo alla conoscenza del popolamento animale di una torbiera piana del Veneto (Lipoi, Feltre), *Ibidem*, 53, 1976, 123.
——, —— & L. Nigro, Contributo alla conoscenza del popolamento animale della torbiera piana di S. Martino di Castrozza, *Ibidem*, 54, 1977, 93.
Marcuzzi G., G.G. Lorenzoni & G. Dal Doss, Primo contributo alla conoscenza di una torbiera piana del Trentino occidentale (Fiavè), *St. Trent. Sc. Nat.*, 48, 1971, 124.
Micheler A., Der ausseralpine Inn: Naturleben einer Flusslandschaft, Jhrb. Ver. Schutze Alpenpfl. & Tiere, München, 35, 1970, 167.
Palmen E. & S. Platonoff, Zur Autökologie und Verbreitung der ost-fennoskandischen Flussuferkäfer, *Ann. Ent. Fenn.*, 9, 1943, 74.
Pavan M., L'uomo nell'equilibrio della natura, *Min. Agric. For., Collana Verde*, 17, 1967.
——, Riserve naturali italiane: situazione e proposte di tutela dei poteri pubblici, *Ibidem*, 31, 1973.
Reichholf J., Die quantitative Bedeutung der Wasser-vögel für das Oekosystem eines Innstausees, *Verh. Ges. Oekol., Wien,* 1975, 247.
Sauerzopf F., Grossraum Neusiedlersee – Land zwischen der Ebene und den Bergen, *Jhrb. Ver. Schutze Alpenpfl. & Tiere, München,* 31, 1966, 107.
Schaefer M., Trophische Beziehungen in einem Küstendünen-Oekosystem, einer 'natürlichen Monokultur' von Ammophila arenaria, *Verh. Gesell. Oekol., Göttingen,* 1976, 241.
Schuster R., Das Kalkalgen-Trottoir an der Cote des Albères als Lebensraum terricoler Kleintiere, *Vie et Milieu*, 7, 1956, 242.
Schuster R., Die Oekologie der terrestrischen Kleinfauna des Meeresstrandes, *Zool. Anz.* 28 *Suppl.*, 1965, 492.
Southward A.J., Life on the sea-shore, Heinemann Educ. Books Ltd. London 1967.
Spanò S., Rifugio faunistico di Bolgheri (Livorno). Ardeidi etc. Atti I Simp. Naz. conserv. natura, Bari, 1971, 163.
Stys P., Die Wanzenfauna des Moorgebietes Soos in Böhmen (Heteroptera), *Acta Univ. Carol. Biol., Suppl.* 1960, 83.
Tamino G., Ricupero di un Grampus griseus (Cuv.) arenato sul Lido di Ladispoli (Roma) il 5 aprile del 1953, *Boll. Zool.*, 20, 1953, 45.
——, Ricupero di una balenottera arenata sul Lido di Salerno il 10 febbraio 1953, *Ibidem*, 51.
Tobisch S. & W. Dunger, Carabiden des Neissetales bei Ostritz (Oberlausitz) und ihre Reaktion auf Industrie- Emissionen, *Abh. Ber. Naturk. Mus. Görlitz*, 48, 1974, II/1.
Trevisiol G., Rinvenimenti preistorici nelle torbiere delle valli di Fimon nel Vicentino, *Atti Ist. Ven. S.L.A.*, 104, 1946, 745.
Zorzi F., Stazione palafitticola preistorica e romana alle Morandine di Cerea (Verona), *Atti Accad. Agric. Sc. Lett. Verona*, 5, 1955, 1.

FRESHWATER ENVIRONMENTS

Chiare, fresche, dolci acque ...
Petrarca

General characteristics

The freshwater environments are extremely varied and a classification can only be arbitrary and somewhat artificial. The very distinction between running and standing water is sometimes very difficult; furthermore many inland waters are as salty as, and sometimes more so than, sea water, so that the distinction between freshwater biota and marine biota is also in a sense arbitrary.

For instance, a river does not exhibit a uniform character from source to mouth, since in each section of its course the physical and chemical characteristics vary greatly; a lake or even a pond also is not a uniform structure or ecosystem.

Even more than in the terrestrial environment, where all animals are mobile and where there is never a scarcity of oxygen or of light, in the aquatic environments animals show many adaptations to cope with several negative or limiting environmental factors, such as temperature, drying up of a water basin during summer, light, oxygen, salinity and, recently, toxic substances present in polluted waters. Very interesting adaptations can be observed in respiratory mechanisms: it is enough to mention two, the device through which the Chrysomelid larvae, living permanently in water, can breath atmospheric air (*Donacia*, e.g.) and that by which adult beetles can breath atmospheric air under water (*Dryops, Hydraena* etc.). Interesting mechanisms have also evolved for feeding in animals which do not move very much and utilize special filters to catch minute organic particles or organisms (filter feeding) as in the Ephemeridae (*Oligoneuria rhenana*) or Trichoptera (*Wormaldia occipitalis*) or beetles (*Scirtes hemisphericus*) (Fig. 402, 403).

We could add the alternation of generations in Cladocerans and Rotifers of astatic basins (bisexual and parthenogenetic reproduction), the varying number of generations per year in many zooplankters as a direct response to temperature, the cyclomorphosis present in Cladocerans and Rotifers, the tolerance of very high values of salinity in brackish water animals (see brackish water, p. 579) and that developed to pollution both organic or urban (saprobic organisms) and industrial (Pb, e.g.).

The sediments

Subaquatic sediments or lacustrine sediments are very varied, both in genesis and chemical nature, according to the various environmental factors we find in Europe, such as geography (latitude, altitude, climate etc.), geology, geomorphology, vegetation (both of the lake and of its shore) and according to the history of the region to which the water basin belongs, including here the impact that the glacial period had on many European lakes.
The main subaquatic sediments we may recognize are as follows:

Non calcareous protopedon, or non calcareous subaquatic raw sediment, an inorganic lacustrine sediment, formed upon clastic sediments poor in calcium, colonized by organisms but without a true humus layer. This type is present in acid waters; often there is irreversible separation of iron hydroxide (non soluble iron compounds) caused by oxidation.

Ferrugineous dystrophic raw sediment: it is composed of sediments extremely rich in limonite originating from slow flocculation in acid, ferrugineous, waters. This type is typically present in Fennoscandia upon acid eruptive rocks and cristalline slates. Generally humic acids are so abundant as to almost completely prevent any form of life.

Lacustrine marl raw sediments: a frequent type of sediment rich in calcium

a

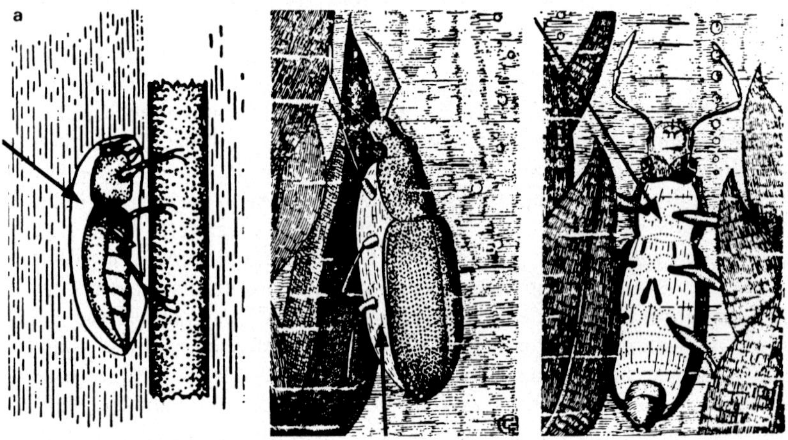

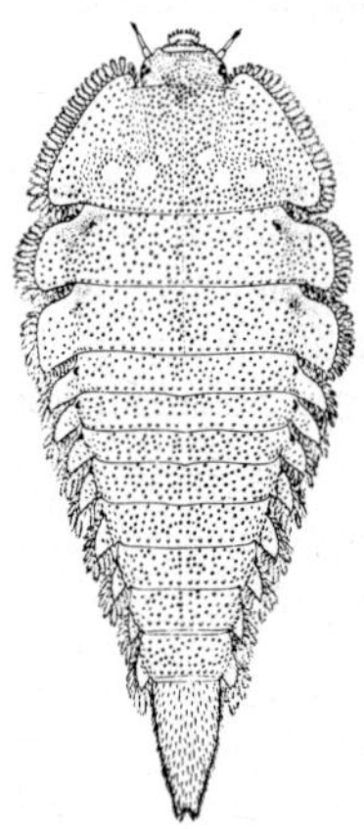

Fig. 402. Adaptations for breathing in aquatic invertebrates. a. Three beetles which utilize for breathing a bubble of air; left – *Dryops* (from Thorpe), middle – *Hydraena testacea* seen from the side (from Binaghi), right – *H. heterogyna*, seen from below (from Binaghi). The arrows indicate the airbubbles.
Bottom. larva of *Helmis latreillei* from Wesenberg-Lund modified.

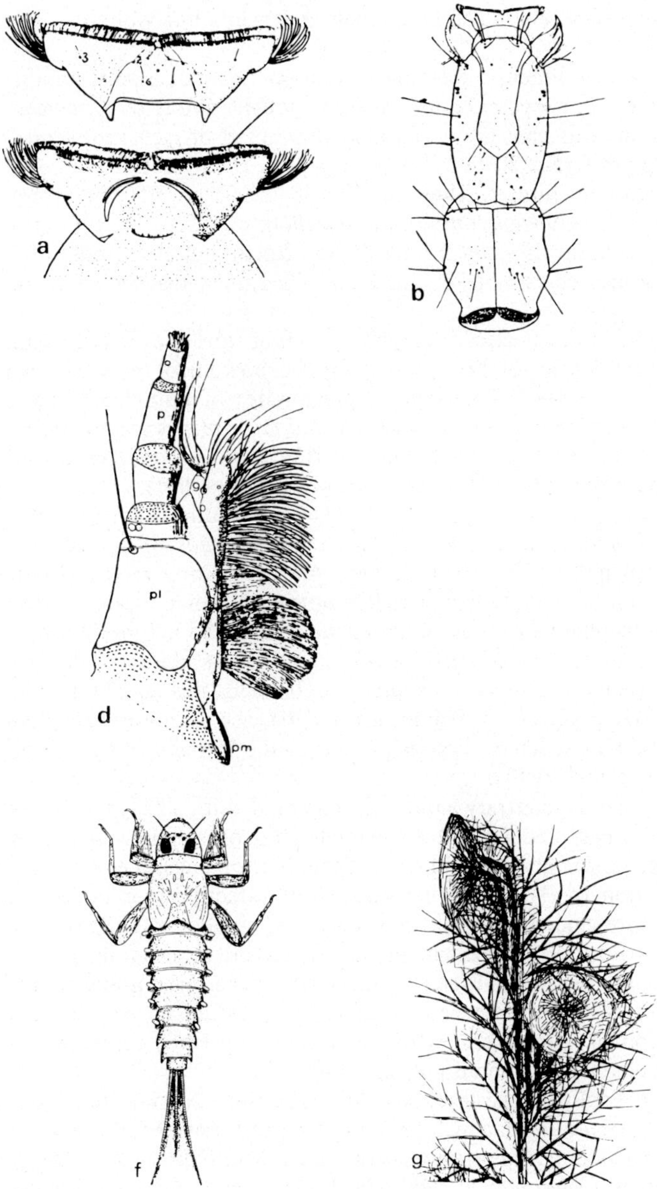

Fig. 403. Adaptations for feeding in aquatic invertebrates. a, b, c, and d. Larvae of the caddish-fly *Wormaldia occipitalis*: a. dorsal lip (or labrum), b. head and prothorax, c. case of the larva, d. outer part of the right maxilla (from Wesenberg-Lund); f. pre-imaginal stages of the Ephemerid *Oligoneuria rhenana*, in which the front limb shows a fringe of setae adapted to filtering the food (from Wesenberg-Lund); g. case of the larva of the caddish-fly *Holocentropus* with a net for filtering the food particles (from Wesenberg-Lund).

(from 20 to 70%) and clay, colonized by animals and plants but without a true humus layer.

Lacustrine 'craie' raw sediments: they are composed either of a great number of shell remains or by the separation of calcium brought about by *Characeae* (macrophytic algae): in this case they are called chara marl or lacustrine 'craie'. They are rich in calcium (from 70 to 90%), poor in clay, at the surface colonized by organisms, but with no humus. They are almost always covered by macrophytes such as *Characeae, Myriophyllum, Ceratophyllum* etc. They are typical of waters very rich in calcium; they reach latitude 55° North and an altitude of c. 1100 m. Along the shores of some lakes this kind of sediment reaches a remarkable thickness.

'Dy' or brown mud, known also as tyrphopel, limnic turf etc. It is an acid, swamp sediment, not suitable for life, present on the bottom of peaty, brown waters. It is mainly composed of amorphous precipitates of humic gels (i.e. a semifluid or even solid colloid) and of animal remains, since plants are practically absent. This type of sediment is only present in the region of podzols and of 'Hochmoore', and therefore only in Northern Europe (Fig. 404, left).

Gyttja (a Swedish popular name) or grey mud is a subaquatic sediment rich in organisms, composed almost exclusively of faeces and remains of animals and secondarily of phytoplankters. The colour is grey, grey-brown or even blackish. It is formed in waters rich in oxygen, rather rich in nutrients (which means a eutrophic or at least mesotrophic lake) and rich in organic matter. Sometimes it can be several metres thick. The surface of gyttja is often covered by plants, both algae and flowering plants such as *Nymphaea, Nuphar*, etc. It smells slightly of ink; only if there are putrefactive processes (which means the presence of sapropel) is there a production of H_2S. This sediment type is present in the lowland (if there is no formation of 'Dy') (Fig. 404, right).

Limnic sapropel is on the contrary a putrid limnic mud, with a bad, sometimes unbearable odour, strongly reduced. It is characteristic of standing waters in which putrefactive processes abound and where there is a total absence of oxygen and abundant formation of H_2S and other gases of putrefaction (e.g. methane). Sometimes red spots are evident, due to the presence of sulphur bacteria (purple sulphur bacteria, cf. page 622). Plants and animal remains are attacked by anaerobic bacteria and secondarily by sapropelic animals (Protozoa, Nematodes, Copepods and *Corethra*-larvae). The presence of H_2S and the absence of oxygen make possible the formation of iron sulphides in chemically very labile forms, which confer on sapropel a characteristic bluish-black colour.

Fen (or turf) is an English name used only in recent times to indicate all soils rich in calcium and nutrients, with a high content of organic matter formed under water or due to strong sodding. It corresponds to the German Fehn and the Dutch veen. Practically, fen is a synonym of 'submerged mire', valley mire, hard water mire (which means high content of calcium) or even 'Flachmoor' or 'Verlandungsmoor'. It is formed in still waters with little biological activity, and humus poor in mineral matter. There is an excess in supply and a strong accumulation of slightly decomposed plant remains, due to the lack of oxygen, acid reaction and other conditions not favourable for quick humification. According to prevailing vegetation, several types of fen are recognizable. We will mention here only reedbeds

fen, accompanied by horse tails *Equisetum palustre* and *E. limosum*, the bulrush *Scirpus lacustris*, the bur reed *Sparganium ramosum*, etc. It is well developed in Central and Northern Europe (e.g. at Wicken, Cambridgeshire).
Carex-fen, often accompanied by marsh trefoil, is frequent in Central and Northern Europe and in Northern Italy at the foot of the Alps or in some alpine valleys.
Hypnum-fen has several mosses of the genus *Hypnum*, often accompanied by *Carex*: it is frequent in the Alps, in Scandinavia and in some parts of Northern Germany.

Finally, let us mention the forest fen, or swampy forest mire, or even alder-mire because it is often formed under thickets of alders in standing, hard, water or more frequently in slowly flowing waters. In these conditions a turf rich in remains of leaves, fruits (strobiles) and alder wood is formed. With this plant, we find willows and secondarily swampy plants which grow very luxuriantly in the shadow of alders. The animal life of this soil is fairly rich. This kind of fen is frequent in Central Europe (particularly Northern Germany) and Northern Europe, and possibly also in Northern Italy at the foot of the Alps.

Vegetation

Aquatic vegetation (present mostly in lentic environments) is represented by the ecological category of *Hydrophytia*. In fresh waters *Limnophytia* are present, subdivided into *eulimnophytia*, in standing waters, *potamophytia*, in running waters, *nerolimnophytia* on the shores of a water body and *pelagolimnophytia* in open water. Phytoplankton is represented mainly by *Schyzophyceae* (*Cyanophyceae* or blue-green algae), *Schyzomycetes, Peridineae,* Diatoms, Flagellates, *Conjugatae (Desmidiaceae)* and *Protococcales Chlorophyceae*. The Bryophytes are represented by liverworts (*Riccia*) and mosses such as *Hypnum, Amblystegium* etc. Pterydophytes are represented in European waters by *Azolla* (of North American origin), *Salvinia natans*. Finally flowering plants are represented by *Hydrocharis* (whence Hydrocharitetea for the ecological unit), *Lemna minor* (whence *Lemnea minoris*), *Wolffia arrhiza, Pistia stratiotes* (belonging to the epipleon) and *Ceratophyllum, Myriophyllum, Utricularia, Ranunculus* sp. plur., *Aldrovandia* etc., belonging to the hypopleon. Some species (*Lemna trisulca*) can be attributed to the heteropleon, since it can belong both to the epipleon and the hypopleon.

Some species are restricted to rocky, others to sandy or muddy bottoms. Within the former we find *Schizothrix, Tolypothrix, Rivularia, Spirogyra, Cladophora, Oedogonium, Ulothrix*, the *Phaeophycea Pleurocladis lacustris*, some *Florideae* such as *Lemanea, Batrachospermum* and *Chantransia*, together with mosses (*Fontinalis, Cinclidotus* etc.).

On sandy bottoms we may find, on the contrary, *Hypnum, Marsilia, Pilularia, Isoetes*, and many flowering plants such as *Potamogeton pectinatus, P. natans, Najas, Alisma natans, Elodea canadensis* (introduced from America), *Vallisneria, Sparganium, Hippuris, Nymphaea, Nuphar, Ranunculus* sp. plur., *Callitriche palustris, Elatine paludosa, Trapa natans, Myriophyllum* and many others. Of course, we will not take into consideration here the swampy plants we have already discussed when talking of semi-aquatic environments.

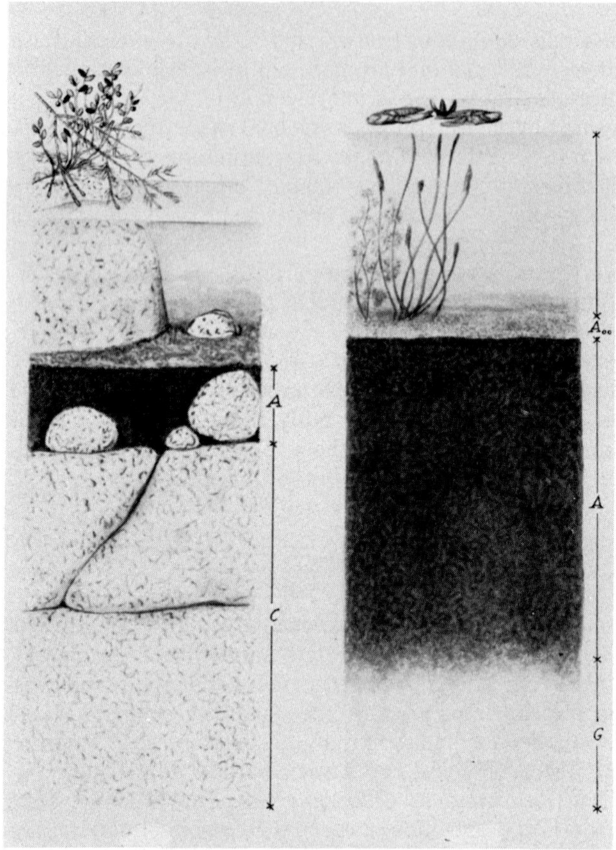

Fig. 404. Left: a 'dy' in a swamp region with podzols and Hochmoore, on granite and under a pine stand (Böhmerwald). Right: a eutrophic gyttja with *Nymphaea alpha* and an 'afja' of Chlorophyceae (Wies, Styria) (from Kubiena).

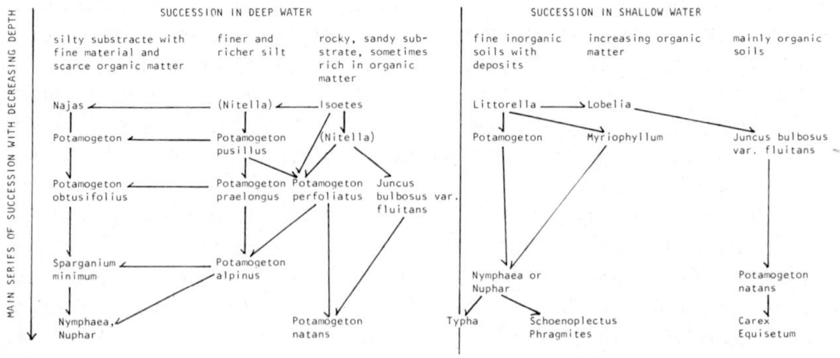

Fig. 405. Succession of aquatic flowering plants in English Lake District (Cumbria) (only the dominant species have been indicated) (from Margalef).

In the brackish waters of lagoons, shore pools, 'marismas' etc (cf. page 579) the ecological category of the Halo-hydrophytia is present. On rocky substrates we may find some true marine groups such as *Fucus, Laminaria* within the *Phaeophyceae, Lithothamnion, Litophyllum, Corallina* etc. within the *Rhodophyceae*. On sandy or muddy substrates (which prevail by far in brackish waters) the most common genera or groups are *Enteromorpha, Monostroma, Ulothrix, Vaucheria, Caulerpa* etc. within the *Chlorophyceae, Lyngbya, Rivularia* and others within the blue-green algae. The flowering plants living in this peculiar environment are *Potamogetonaceae* such as *Cymodocea, Zostera, Potamogeton, Ruppia, Posidonia, Zannichellia* (characteristic of lagoons), and the *Hydrocharitaceae*. In this case we also will not discuss swampy vegetation.

Generally one speaks of a succession in the slow changes in aquatic vegetation that have occurred since the first colonization of a water body.

A very interesting case of succession we can observe today is that occurring in some man-made lakes: a very remarkable example is given by the succession of diatoms which occurred over 9 years in an artificial lake in NE Spain, at Sau.

The impact of man on freshwater environments

Since the most remote times fresh water has attracted men living exclusively on the products of hunting and fishing.

In the Lower and Middle Palaeolithic, fishing was performed with the hands: fish were captured directly or were made to swim into a narrow place where it was easier to catch them. During the Mesolithic the so-called fish-collecting was done, without any instrument at all, performed both in freshwater and the sea, although, in the present-day territory of Northern Poland, implements for fishing made of bone or horn and attributed to the Mesolithic have been found on the bottom of lakes and rivers. Neolithic man of Poland has left us some harpoons and floats belonging to nest. During the Upper Eneolithic or in the early Bronze Age, the first metallic hooks make their appearance among the palisades of Biskupin, near Poznan.

In historic times we may mention Aristotles' observations about a poisonous white juice extracted from a plant called Plomos or Phlomos which the Phoenicians used also for sea fisheries. Aristotle describes the use of bow nets, still today very much used for fishing, and also a method the Phoenicians used to catch mullet. They released male mullet to gather and capture females, and then they released some females to go and capture males. This method was still employed in the past century on some Mediterranean shores.

In the High Middle Ages fishery and fish culture were highly specialized, although fish culture existed in China even in remote times, for in 475 B.C. Fan Li was able to write a treatise on fish culture. In the High Middle Ages a Lex Visogothorum was issued, which forbade people to stop salmon from running up rivers. In the High Middle Ages in Poland the following fishes were caught: the sturgeon *Acipenser sturio*, the catfish *Silurus glanis*, the pike *Esox lucius* and the salmon (*Salmo salar*). In 1100 the Cyprinids *Vimba vimba* and *Aspius aspius* were also fished. In the Low Middle Ages the following species were fished: *Perca fluviatilis, Lucioperca lucioperca, Carassius carassius, Tinca tinca, Abramis brama, Scardinius erythrophthalmus, Rutilus rutilus, Coregonus albula, Osmerus eperlanus, Misgurnus fossilis* and *Anguilla anguilla* (Chmielewski).

The oldest hooks go back to the Neolithic. Nets are known from the Swiss Neolithic; harpoons were also employed in pre-historic times.

Boats or ships are extremely old: possibly the oldest so far known are the monoxylons of the Swiss Neolithic.

Besides fisheries in fresh waters, since very remote times there has been some fishing and fish culture in brackish waters, such as those near the sea in several parts of the Mediterranean coasts of Europe, separated from the sea by sand bars, so that several species of fish and other marine animals can enter to feed for a given period, going back to the sea for reproduction.

The temporary presence of marine fish in the lagoons was known to Romans: Marcus Terentius Varro indeed writes Sic nostra aetas luxuria propagavit leporaria ac piscinas protulit ad mare, et in eas pelagios greges piscium revocavit.

Fish culture in shore ponds has been performed since Roman times: Columella, C. Plinius Secundus and Varro speak separately of the 'peschiere' of Campania. Some claim that Sergius Orata, was been named after the *Chrysophrys auratus*, and Lucius Licinius Murena would have been so named after '*Murena*' or *Muraena helena*.

Today the remains of Lucullus' peschiere in the brackish waters of the Lake of Paola are clearly visible.

The so-called peschiere (fishing basket made with canes) go back to very ancient times: Romans probably learned their use from the Etrurians, and they from the Phoenicians. Practically the same system is present in different parts of the Mediterranean coast of Europe: Bordigues or 'chile' of the Great Lake of Berre (Strabo's Insigne Stagnum) and of the shore pools of Corsica, the Encanisadas of Mar Menor and the Albufera in Spain; the Bramade of the Greek archipelago, the 'lavorieri' of the north Adriatic lagoons, the Bondanoni and the cannare of Orbetello and the Giostre of the brackish ponds of Sardinia.

The so-called 'valli', particular sections of the lagoons utilized for fish culture have existed for a long time: formerly they were called Piscariae aquae. A legend claims that St Mark the Evangelist 'fu costretto dal mare burrascoso a fermarsi in laguna, legando la barca ad uno dei pali che sostenevano le grisiole della Valle di Zuane Bono ed ivi pernottò' (Braccolani, 1664). Cassiodorus in 514 A.D. speaks of 'piscinae neptuniae' in which here and there oysters develop. Actually, in some excavations people make today for agricultural purposes near the northern Adriatic Sea, they find, with Roman and Pre-Roman remains, some traces of poles fixed in the same way as the present day 'lavorieri'. In 537 A.D. Cassiodorus sayd that the inhabitants of the lagoon of Venice feed exclusively on fish and that in the lagoon there are many piscinae neptuniae and many 'garismazie' (from the Greek) i.e. beds in which oysters, shrimps and scallops grow. However, only after 1000 A.D. is there sure documentation of fish culture in the 'valli'. As far as the term 'valle' is concerned, it appears officially only in 1425 ('clausura vallium'): today people think that this term has nothing to do with the Celtic val, the Greek avlon or the Latin vallis (a valley) but might derive from the Latin valleum and the Celt-gallic balla or ballah, which means parapet, fence, shelter made in an open space, with crossed branches and poles, so as to prevent any entrance (G. Bullo, 1940).

The fantasy of the primitive inhabitants of Europe has created living creatures in brooks, rivers and lakes, always feminine, strictly confined to waters, such as naiads and undines.

Undine, a name coined by Paracelsus (1493-1541) from unda, a wave, is a legendary protagonist of European folklore especially Germanic, half woman, half fish. With their songs and their charms they seduced fishermen and sailors, bringing them to ruin. In Germanic mythology they were called nixe (sing. nix or nis) and they had more or less the same characteristics. Some claimed that the undine was without a soul until married to a mortal and the mother of a child. According to some, on the contrary, the nix is a frolicsome water spirit.

Naiads (from Ηα ίας from *ναειν*) were generically the nymphs of waters (sources, rivers and lakes), Zeus' daughters; according to Homer they brought fecundity to earth and protected marriages.

Nymph, from the Latin nympha and the Greek *νύμφη* literally a maiden or a young married woman, is a minor feminine divinity of classical mythology, associated with springs, rivers and lakes (naiads, see above), to forests (dryads) and to mountains (oreads). They were friendly to mortals whose love they did not disdain. Romans saw in them tutelary goddesses of water and springs (Camene, Carmenta, Juturna, etc.). Leopardi says 'già di candide ninfe i rivi albergo, placido albergo a specchio furo i liquidi fonti' (which literally is more or less 'already hostel of candid nymphs the rivers, the liquid sources have been a tranquil hostel and a mirror...')

Famous for the nymph Cyane (*κυάνη*) living in the river of the same name (*κυανῆ* in Greek means bluish, which actually is the colour of its water), is a tributary of the Anapo, on whose shores a magnificent plant, the papyrus (*Cyperus papyrus*), grows. It is the place where Hades' charriot – after Persephon's rape – opened its way to Avernus. The nymph Cyane had her cult and feasts at Syracuse for a long time.

Countless are the legends and myths concerning European waters.

In Germanic mythology we find the so-called Rheingold (or Rhine gold), the hoard of gold guarded by the Rhine maidens and afterwards owned by the Nibelungs and Siegfried. The story is told in varying forms in the Volsunga saga, Nibelungenlied and – lately – in Wagner's Ring of the Nibelung. The ring was made from the Rheingold by Alberich, leader of a race of dwarfs called the Nibelungs.

Some particularly wild, rocky lakes, set in the middle of mountains, often the seat of clouds, winds and snow-storms, have given rise in the fantasy of primitive inhabitants of mountains (who in some regions have resided at high altitudes for many centuries) to fantastic legends and stories. Some tranquil lakes, on the other hand, have given origin to more joyful, or even religious, legends.

In the heart of the Dolomites, in the group of Catinaccio (or Rosengarten, which literally means a garden of roses) is the Antermoia Lake, one of the most picturesque lakes of this region because of the wildness of its rocks and the clearness of its waters. The origin of this lake (which is partially karstic) was once the nucleus for a terrific legend of witches and demons. 'In the Fassa valley once a man lived, who one day, a little for the love of adventure, a little to follow his wife, found himself amid a diabolic sabbath celebrated in the amphitheatre of Vaiolon, whose walls now bare and rocky, were then a fantastic and agreeable garden. Here indeed on some particular nights, the fairies and elfs living in the Fassa Valley joined to dance and to sing, during which some satanic orgies occurred. The night on which the man was in the Vaiolon, was one of these nights, and sounds and songs were heard in the fragrant air ... Unwillingly, the man pronounced a word which was a magic conjuration, and instantly all the spell disappeared. Garden, flowers, woods, tranquil waters and murmuring brooks, everything changed into a wild and horrid interlacing of pinnacles, small valleys, fissures and precipices, which do have a beauty, but due only to a wild and majestic nature. Witches and sorcerers are compelled to escape and take refuge in the Antermoia Lake, where they will be busy in the future giving rise to winds and storms. Even today, when the storms rise and burst forth upon the waters of the lake and amid the gorges of Vaiolon, some believe that they see the witches and the sorcerers confined within it making an uproar, raving stirring the waves of the lake, shaking the points of the rocks, and darting furiously amid the cold fantastic fog (Bolognini, Leggende del Trentino).

In South-Eastern Trentino there is a quite different lake, amidst beautiful forests of spruce and beech, in a sunny basin, the Lago Santo (literally the saint lake) near Cembra, whose calm waters are moved now and then by a light breeze. At Cembra people still tell that once, where today is the surface of the lake, there was a dense forest. The owner of this forest, near to death, bequeathed it to his three sons. The agreement on the partition caused such a furious quarrel, that one of the sons came out with the following curse 'I ask God to convert our forest into a lake'! The next day the sons went to the mointain, and instead of their property they found a lake that was getting bigger and bigger. Then the people of Cembra went there in procession and the priest of the village – while everyone stood on the shores of the lake – threw into the water the ring of the Virgin venerated in the parish church. Instantly, by enchantment, the lake got smaller until it reached the present-day level (Tomasi, 1963).

Much older legends exist in the region of Cortina d'Ampezzo, in the Dolomites. The lake Ghedina was once called Lee de Lunédes, which literally means the lake of the spirits. In fact, according to the Ladin tradition, anyone was allowed to know his destiny by beating the waters of the lake with a stick: out came the Mjanines, small magic beings who kindly took trouble to disclose the secrets of the future (Traversi, 1956).

In the Lake of Costalares, at the feet of Faloria, the good witch Svalazza once lived, who took care to stir the waters of the lake before the eyes of a person threatened by some disgrace.

Finally, near Cortina, a local term has been preserved until now, 'anguanes', the virgins of the waters, or nymphs, who lived around the numerous lakes and on the banks of the brooks of this magnificent region: the so-called Rio delle Vergini (literally the brook of the virgins) a tributary of Rio Costeana, having its origin near the Malghe Fedarola (m 1750), has preserved, in the present-day hydronym, the memory of the old pagan cult (Traversi).

The impact of man upon the freshwater environment has been very strong only in recent times – say in the last one hundred years – on the one hand because of fisheries and fish culture (see later when discussing standing and running waters), on the other, because of heavier and heavier pollution, so that today only a very few European water bodies are in natural or semi-natural condition.

Furthermore, besides the impact on extant waters, man has created many new aquatic systems, such as the canals (for irrigation, navigation, etc.) and the so-called 'man-made lakes', whose number increases each day. Some erroneous lines of hydrogeological policy have caused, in some instances, not only an alteration in the pre-existing fluvio-lacustrine system, but even alterations in the terrestrial environment, and in some cases (Vajont valley, in Veneto), destruction of human life and dwellings.

Aquatic ecosystems today therefore are seriously menaced by man and all his activities (industry, agriculture, communications and tertiary activities in general).

Standing waters

Standing waters, constituting the so-called lentic or lenitic environment, from the Latin lenis, calm are mainly represented by lakes, ponds, swamps, mires and periodic waters. The main characteristics of their biota depend on their size, only secondarily on the geological nature of the substrate and geomorphology. Sometimes the life of a water body is relatively short (especially in term of geological time or chronology): this is particularly true for alluvial, lowland, postglacial lakes. In some instances, as in astatic basins, physical and biological conditions vary also during the year if not during the season (summer, e.g.)

Lakes

The difference between lakes and ponds does not depend on the area or surface of the water body, but on its depth: lakes are generally so deep that there is a difference between the lower, constantly cold layer (the so-called hypolimnion) and the upper light layer (the so-called epilimnion). The temperature of the former is nearly 4° C, that of the latter is higher during the summer, lower during the winter and limited to some cold regions or to mountains (Alps, Pyrenees etc.). In some instances (sub-arctic regions) the temperature of the epilimnion reaches only 4° C during the summer, so that we have homeothermy. The temperature of the two layers does not regularly decrease, but drops quickly within the so-called thermocline, so that from a thermal point of view a lake behaves as two distinct bodies, epi- and hypolimnion.

Generally lakes with a large epilimnion are rich in organisms (eutrophic), those with a large hypolimnion are very poor as far as life is concerned (oligotrophic). Only in very acid bogs can the presence of humic acids and other toxic substances be so high as to prevent practically any form of life, so that one speaks of dystrophic lakes or waters. Some exceptions are shown by high alpine lakes which – although with a large hypolimnion and a great depth – have a very rich planktonic life: these lakes are called crio-eutrophic (ex. Antermoia Lake, in the Dolomites).

Eutrophic lakes (generally lowland, alluvial lakes) are characterized by some particular genera of the midge family *Chironomidae* (e.g. *Chironomus*, characterized by the presence in its blood of a haemoglobin with a very great affinity for oxygen, which in this category of lakes is very scarce), whereas oligotrophic lakes, always rich in oxygen even on the bottom, are characterized by other genera and subfamilies of this Nematoceran family (e.g. *Tanytarsus*, without haemoglobin).

The macrophytic vegetation of a lake and more particularly of a eutrophic lake is stratified as follows: a) a belt of shore vegetation generally in part aerial (swampy vegetation); b) a belt of floating plants or *Laminetum* (*Nymphaea, Nuphar, Potamogeton natans* etc.); c) a belt of submerged plants (*Myriophyllum, Cheratophyllum, Vallisneria, Chara* etc.); d) a transitional zone, and, finally, e) a very deep zone, completely without any form of photosynthetic plants.

Of course in Europe there is an enormous variety of lakes, and not all their organisms are equally well known. Particularly well known are zooplankton and fishes.

If we confine ourselves to this latter animal group, we may distinguish species exclusive to the pelagic layer, and others – by far the majority – more or less restricted to the bottom. To the former some planktophagous species belong, such as a small *Coregonus* species, *C. albula*, besides the big *Coregonus wartmanni* of lake Constance and a form of char, *Salvelinus alpinus salvelinus.*

Some species live exclusively in the hypolimnion of very large lakes, such as the smelt *Osmerus eperlanus f. spirinchus*, and some Cyprinid fishes such as *Pararutilus frisii meidingeri* and *Chalcalburnus chalcoides mento*. Medium depth lakes with a limited hypolimnion and a rather flat shore (generally eutrophic) are inhabited by some peculiar fishes present also in not very rapid rivers. Let us remember the bream *Abramis brama*, essentially on muddy bottoms, and *Perca fluviatilis*, living in clear waters over rocky bottoms. The eutrophic, lowland, alluvial lakes, finally, are inhabited by *Acerina cernua* (upon sandy marly bottoms), the *Gadidae Lota lota* (or burbot) which eats the eggs of other fishes and is therefore often very dangerous to a fishery, also *Thymallus thymallus* or grayling, *Padogobius nigricans*, the dace *Leuciscus leuciscus* and the chub *Leuciscus cephalus cabeda* (in Italian 'cavedano')(Fig. 406b).

Finally, in shallower lakes, poorer in oxygen, some euryoxibiont and eyrythermous species live: among others the carp *Cyprinus carpio*, the roach *Rutilus rutilus*, the rudd *Scardinius erythrophthalmus*, the green tench *Tinca tinca*, the crucian carp *Carassius carassius* the bleak *Alburnus alburnus* and lastly the pike *Esox lucius* (Fig. 406a).

The phytoplankton of lakes is very well known and its importance in the feeding of zooplankton has been well documented. It varies greatly both in density and in specific composition according to the different type of lake: the highest densities, of course, are present in eutrophic lakes. Very interesting is the problem which has arisen only in recent times of the colonization by phytoplankton of man-made lakes. Another very recent problem – whose importance is increasing each day – is that of productivity, since lakes are a real or at least a potential reservoir of food for man: so, fertilization of lakes has been studied and performed in several countries in order to increase the productivity of phytoplankton and as a consequence the productivity of zooplankton and of animal life in general, here included the fishes utilized by man as food.

As examples of lacustrine phytoplankters we may mention *Asterionella, Synedra, Oscillatoria rubescens, Gloetrichia echinulata* (Fig. 407) *Anabaena, Microcystis* (Fig. 408), *Mallomonas caudata* (Fig. 409), *Lyngbya compressa*, besides the

Fig. 406a. Fishes of shallow, alluvial lakes, poor in oxygen: top – the rudd *Scardinius erythrophthalmus*; bottom – the green tench, *Tinca tinca* (from Ladiges & Vogt). (Reproduced with the permission of Paul Parey).

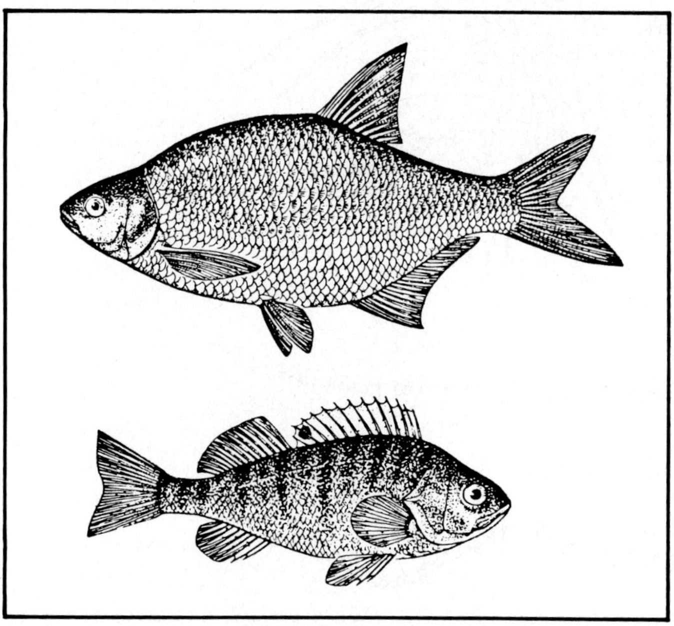

Fig. 406b. Fishes of middle deep lakes, with a limited hypolimnion. Top – *Abramis brama*; bottom – *Perca fluviatilis* (from Sterba).

phytoflagellates *Chrysococcus porifer, Ceratium hirundinella* and some other species, *Gymnodium mirabile* (Fig. 409), *Volvox aureus*, etc. An almost unique case of red colour in an entire lake due to a blooming of phytoflagellates is that of the lake of Tovel (in Trentino) in which the flagellate *Glenodinium purpureum* lives.

Mallomonas caudata is typical of the metalimnion, i.e. the layer dividing epi- from hypolimnion.

In the hypolimnion true phytoplankters are, of course, very rare because of the scarcity of light. Let us remember the *Cyanophyceae* (or *Schyzophyceae*), the common blue-green algae *Oscillatoria rosea* and *Lyngbya compressa* and some colourless Flagellates such as *Astasia*. In compensation there is an enormous number of bacteria, such as Eubacteriales, the colourless sulphur bacteria (Fig. 410) the purple bacteria (Fig. 411), the *Phaeobacteria, Chlorobacteria* (Fig. 412) which

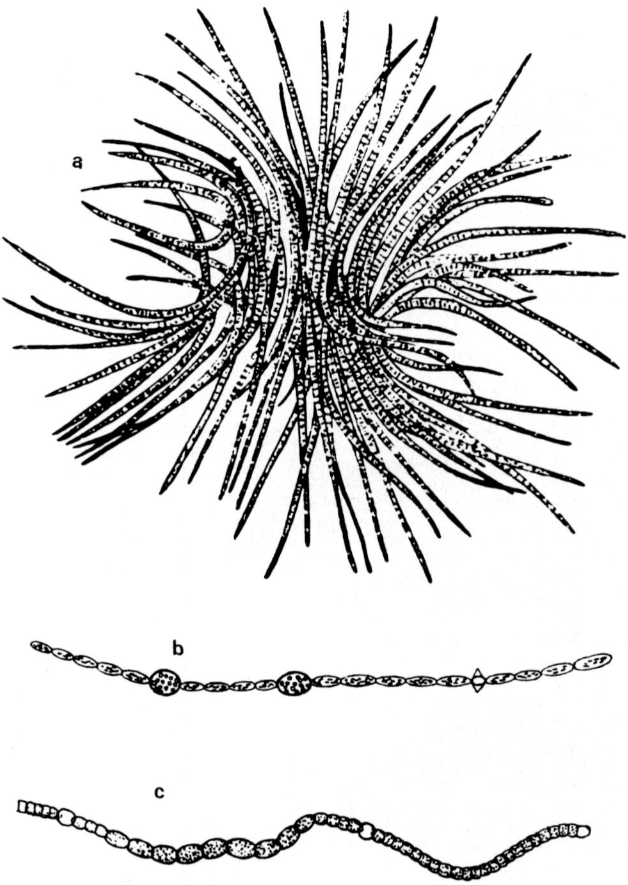

Fig. 407. Blue-green algae or *Cyanophyceae:* a. *Gloeotrichia echinulata;* b. *Anabaena elliptica* (Northern Germany, Sweden); c. *A. variabilis* (from Huber-Pestalozzi, with the permission of Schweizerbart).

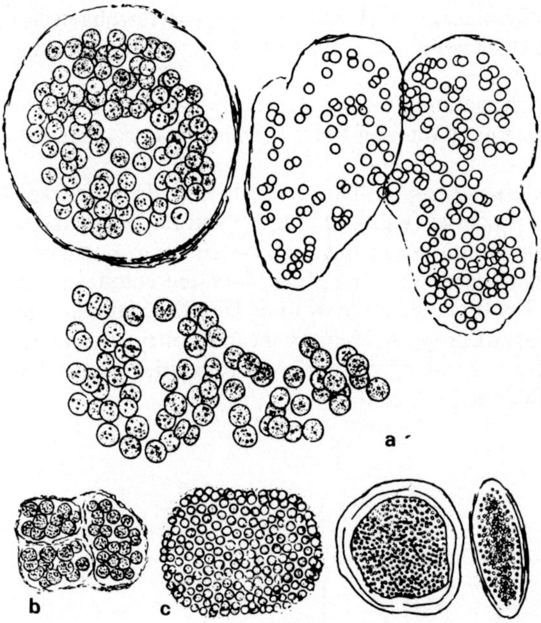

Fig. 408. Examples of *Microcystis* from European lakes: a. *M. robusta*; b. *M. viridis*; c. *M. marginata* (from Huber-Pestalozzi, with the permission of Schweizerbart).

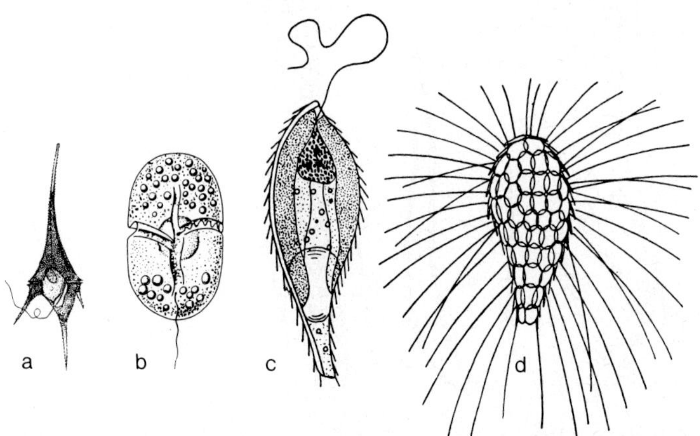

Fig. 409. Examples of phytoplankton: a. *Ceratium hirundinella* (from Kudo); b. *Gimnodinium rhomboides* (from Oltmanns); c. and d. *Mallomonas caudata* (from Huber-Pestalozzi).

show a weak ability to photosynthesize, the iron bacteria (or ferrobacteria) (Fig. 412) and the so-called 'giant bacteria' such as *Steineckea*, of uncertain systematic status.

Some plankters thrive in the hypolimnion only during the warmest months, whereas in winter they belong to the epilimnion flora: *Oscillatoria rubescens* and *O. prolifica*.

Very often some phytoplankters are responsible for a 'bloom': the blue-green algae are most concerned in such processes (e.g. *Microcystis*,), but Phytoflagellates can also undergo such sudden population explosions (or blooms): an example of a bloom is that of *Glenodinium sanguineum* to which the blood-red colour of the Lake of Tovel (Trentino) is due. The first indication of toxic Dinoflagellates and of a red colour of the water in an extra-European, Mediterranean country is accredited to Moses (ca. 1491 B.C.), who – speaking of the Egypt calamities – says 'He changed their waters in blood' (Psalms, CIV, 29-31).

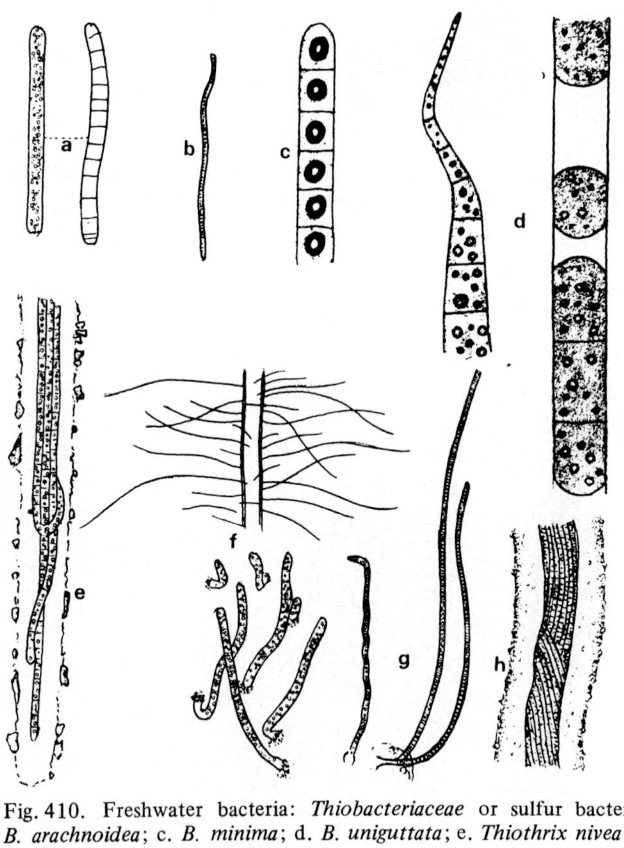

Fig. 410. Freshwater bacteria: *Thiobacteriaceae* or sulfur bacteria: a. *Beggiatoa alba*; b. *B. arachnoidea*; c. *B. minima*; d. *B. uniguttata*; e. *Thiothrix nivea*; f. *T. tenuis*; g. *annulata*; h. *Thioploca schmidlei*; i. *T. ingrica* (from Huber Pestalozzi) (with the permission of Schweizerbart).

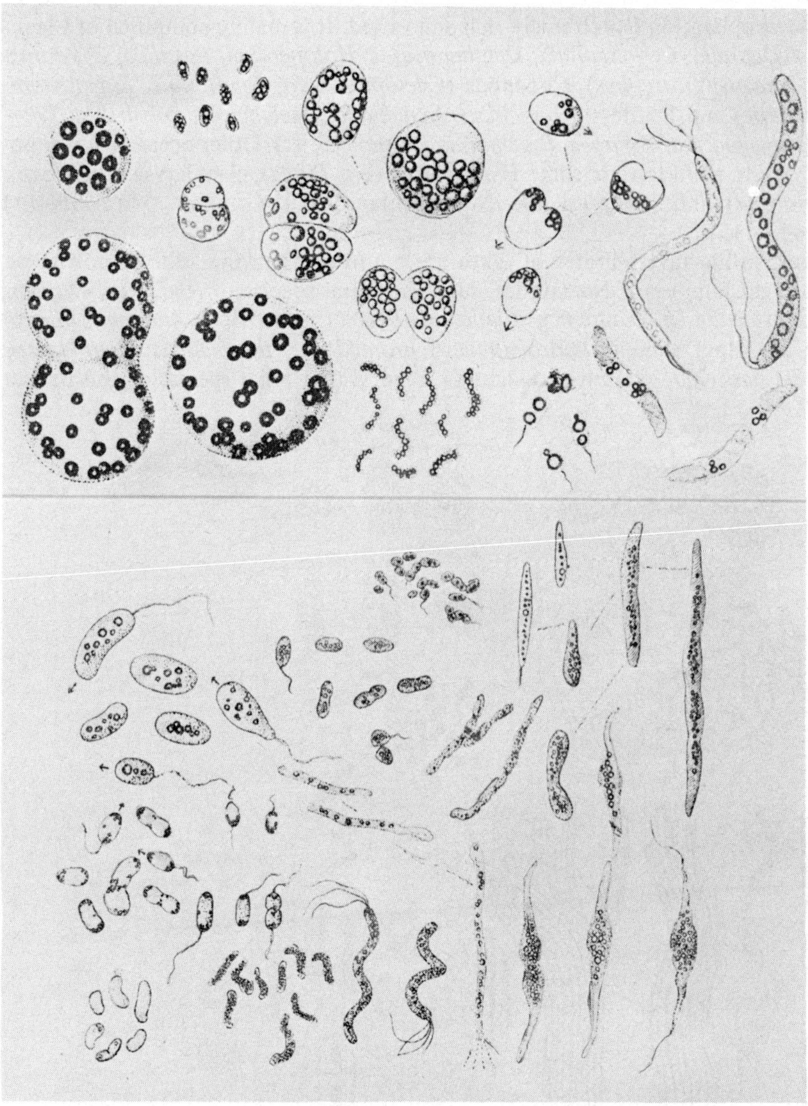

Fig. 411. *Rhodo-Thiobacteriacea* or purpur bacteria: Top: a. *Spirillum kolkwitzii*; b. *Pseudomonas hyalina*; c. *Thiovulum majus*; d. *T. minus*; e. *T. mülleri*; f. *Thiophysa macrophysa*; g. *T. volutans*; h. *Chromatium fallax*; i. *Thiospira winogradskii*; j. *Thiospira bipunctata*; k. *T. agilissima*; bottom: l-o. *Chromatium*; p-t. *Rhabdochromatium*; u-w. *Thiospirillum* (from Huber-Pestalozzi, with the permission of Schweizerbart).

The zooplankton is extremely rich and varied. It is mainly composed of Cladocerans (*Daphnia, Ceriodaphnia, Diaphanosoma, Holopedium, Bosmina, Chydorus, Leptodora, Bythotrephes*), Copepods (*Cyclops s. lato, Diaptomus, Eurytemora, Heterocope*) and Rotifers (*Conochilus, Synchaeta, Asplanchna, Polyarthra, Triarthra, Mastigocerca, Anuraea, Brachionus, Notholca* etc.). Other occasional components of the plankton are some Hydracarina (e.g. *Piona*), clam larvae (*Dreissensia polymorpha*) and larvae of the Nematoceran genus *Corethra* (= *Chaoborus*) (Fig. 413-415).

The benthic invertebrates of lakes are countless. Leaving aside Protozoa, we find a great number of Nematodes, such as several species of *Rhabditis, Ditylenchus brevicauda* (in Denmark), *Anaplectus blanci* (in the Alps), *Anonchus mirabilis* (in the Alps), some *Cylindrolaimus, Achromadora, Ethmolaimus, Microlaimus, Ochridia bathybia*, exclusive to Ochrida Lake, a true relict species of this Balkan

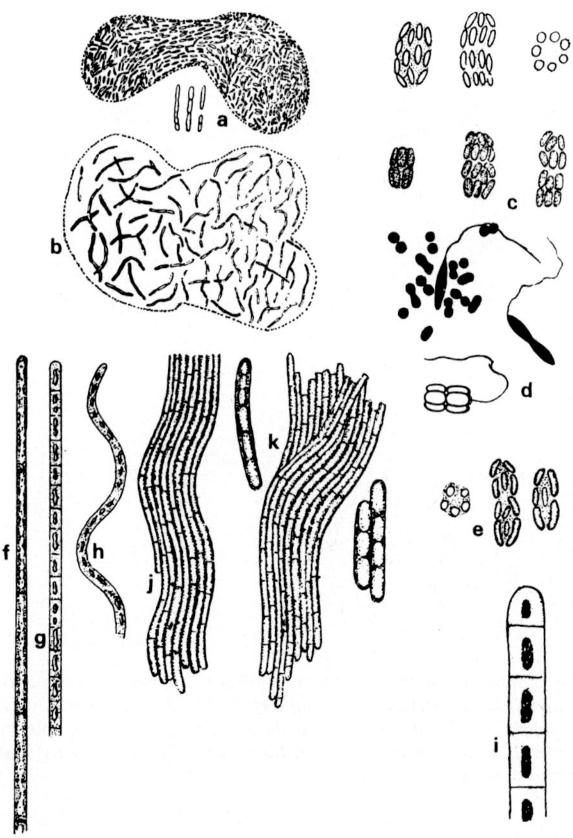

Fig. 412a. Freshwater bacteria: Chlorobacteriaceae: a. *Pelogloea bacillifera*; b. *P. chlorina*; c. *Chlorochromatium aggregatum*; d. *C. aggregatum*; e. *Pelochromatium roseum*; f. *Pelonema tenue*; g. *P. pseudovacuolatum*; h. *P.* (?) *spirale*; i. *P. hyalina*; j. *Peloploca undulata*; k. *P. taeniata* (from Huber-Pestalozzi; with the permission of Schweizerbart).

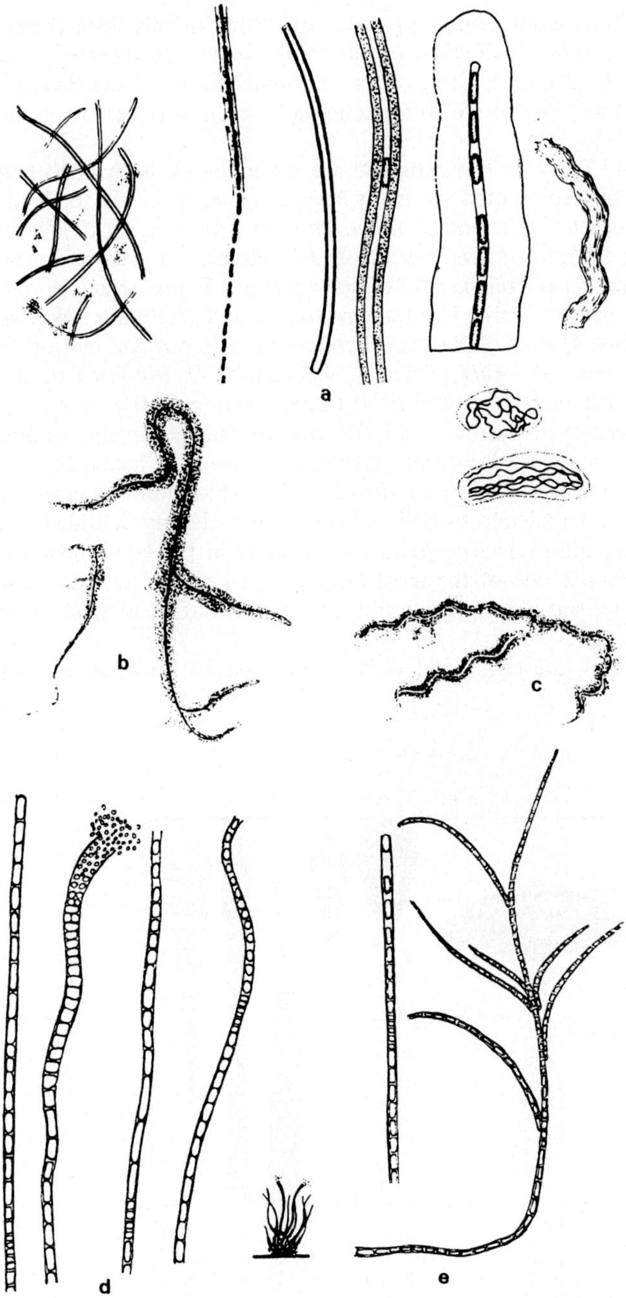

Fig. 412b. Freshwater bacteria: iron bacteria. a. *Leptothrix ochracea*; b. *L. discophora*; c. *L. pseudovacuolata*; d. *Crenothrix polyspora*; e. *C. fusca* (from Huber Pestalozzi, with the permission of Schweizerbart).

lake (Thienemann), *Punctodora ohridiensis*, also exclusive to this water basin, *Ironus colourus* and *I. rotundicaudatus*, of the Alps, *Tripyla gigantea* of Lake Balaton, some species of *Tobrilus, Campydora balatonica*, also of Lake Balaton, three species of *Iotonchus*, one of which is exclusive to the Alps (*I. simmenensis*) and a few others.

Very interesting and fairly well investigated are the molluscs, both snails and bivalves. The former are represented by many species endemic to the Balkans, such as *Costovalvata hirsuticostata* and *C. rhabdota* of Ochrida lake, or *C. klemni* of middle NE Greece; *Ohridotropidina relicta* of Ochrida lake, present in several ecological forms; *Tropidina ochridana* also of this lake and *T. pisana* and *porna*, from the Tyrrhenian part of Italy; *Valvata stenotrema* and *Hydrobia ochridae* from Ochrida lake, some species of *Pseudamnicola* such as *P. carinata* and other species from Ochrida lake, *P. exilis* of South Macedonia, *O. insubrica* of the sub-alpine lakes of Northern Italy ('Insubria'), *Gocea ochridana*, five species of *Ohrigocea* and four species of *Lyhnidia*, all endemic to this extremely old and interesting lake; a great number of endemic species from the same lake belong to genera *Pyrgula, Chilopyrgula* and *Pyrgohydrobia* (Fig. 416); several species of *Gyraulus (Planorbidae)* are endemic to Balkan lakes, here including Ochrida lake; finally, some *Acroloxus* and some *Ancylus* live exclusively in this lake whose age (at least Miocene) makes it one of the most interesting lakes – if not the most interesting – in Europe, and also in the world, as far as evolution of freshwater invertebrates is concerned.

Bivalves are much less interesting and only rarely exclusive to lakes: we can

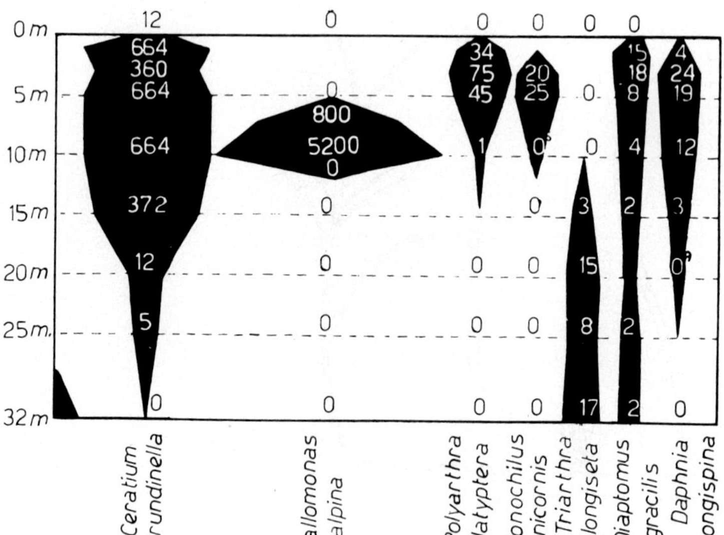

Fig. 413. Vertical distribution of plankton in the Lunzer See (Austria) (from Ruttner).

mention *Sphaerium parenzani* of Ochrida lake and *Pisidium lilieborgi*, a holoarctic species whose presence in the Iberian Peninsula, however, seems due to passive transport by birds, besides some non-endemic species of this genus.

Oligochaeta are represented by many species some of which are of great zoogeographical interest, for instance those belonging to *Peloscolex*, one of which is present in Skadar Lake (Albania), three in Ochrida lake and one in the lakes of Ochrida and Prespa (Fig. 416). *Euclyodrilus* is also represented by some endemic lacustrine elements; some other endemic species of Ochrida furthermore belong to the genera *Lamprodrilus* and *Criodrilus*.

Isopods are present in European lakes, with some species of *Asellus* esclusive to Ochrida lake. Amphipods are much more interesting: *Gammaracanthus lacustris* is known only from lakes in Fennoscandia and the central European lowlands: it is a glacial marine relict (Illies). Practically the same is true for *Pallasea quadrispinosa*. Other elements are *Pontogammarus* (a genus of Ponto-Caspian affinities) *boeoti-*

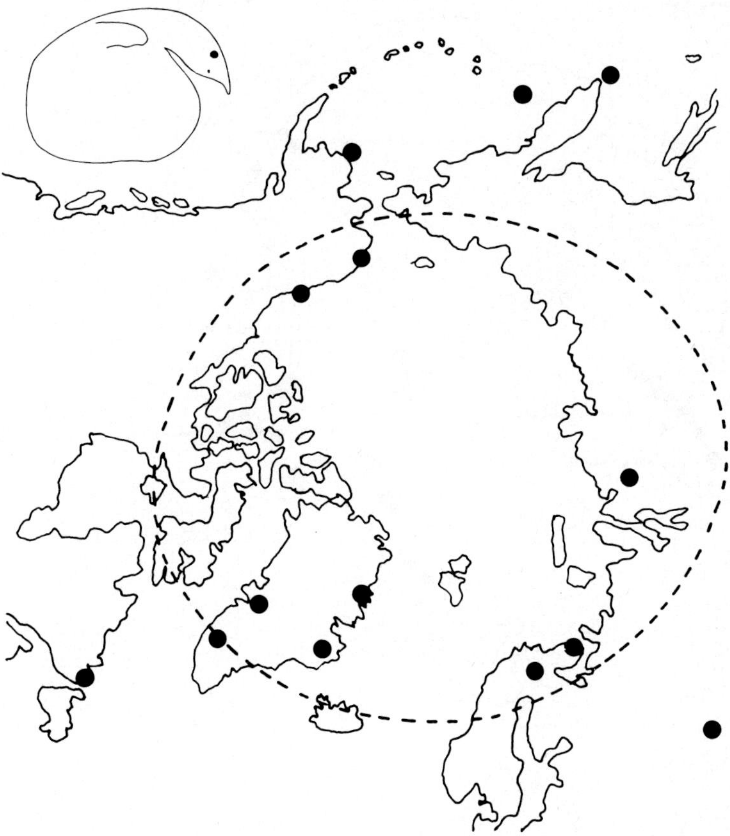

Fig. 414. Circumpolar distribution of the Cladoceran *Eurycercus glacialis* (figure from Thienemann).

cus from Paralimni lake (Greece, Boeotia), *Rivulogammarus ochridensis* from Ochrida and Prespa lakes.

Insects are not very typical of lakes: there are some Ephemerids such as *Ephemera paulae* from Garda lake, *Baetis saliens* of Fennoscandia, *Arthroplea congener* and *Caenis lactea*, two very euryoikous species, and finally a rare centro-northern-European species, *Caenis undosa*, with a highly disjuncted distribution.

Stone flies are completely absent: one exception would be an arctic species, *Capnia zaicevi*, present in Novaja Zemlja as well as in Siberia: its origin deserves particular study since it seems to be pre-glacial. Dragon flies on the other hand, as a consequence of their ecological requirements, are fairly abundant in the limnic environment. Several genera are represented, such as *Sympecma, Erythromma,*

Fig. 415. Circumpolar distribution of the Phyllopod *Branchinecta paludosa* (figure from Zhadin & Gerd).

Coenagrion (with some endemic species such as *hylas* in the Alps and *syriacum* in the Caucasus), *Aeschna, Lindenia tetraphylla, Somatochlora sahlbergi*, endemic to Fennoscandia, a pre-glacial relict which survived in that country during the Pleistocene or perhaps followed the retreat of ice after the glaciation, as did many animals, and finally *Brachythemis leucosticta* and *Leucorrhinia circassica.*

Bugs are also rather abundant: *Arctocorisa, Callicorisa, Corixa, Hesperocorixa, Sigara* (*S. semistriata* represented in Corsica by a local subspecies), *Glaenocorisa, Cymatia*, etc.

Beetles are not typical of lakes: let us mention *Potamonectes canariensis*,

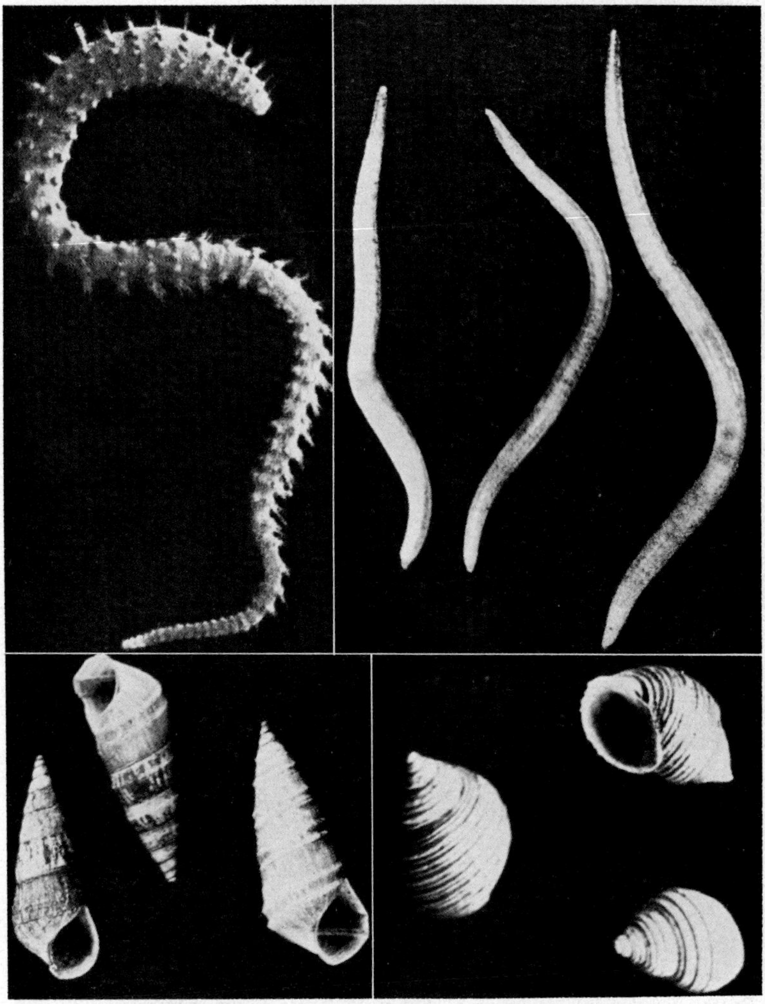

Fig. 416. Endemics of Lake Ochrida: a and b, *Oligochaeta*; c-d. snails: a, *Peloscolex stankovici*; b. *Lamprodilus pygmaeus*; c. *P. sturanyi*; d. *Sinaia munda* (from Thienemann).

Fig. 417. Benthonic elements endemic of European lakes; a. the Amphipod *Gammarus (Rivulogammarus) ochridensis*; b. the Amphipod *Synurella ambulans longidactylus*; c. the trout *Trutta ochridana*; d. the trout *Trutta balcanica* (= *letnica*) (from Thienemann).

known only from the Canary Islands and the Outer Hebrides, where it possibly represents a pre-glacial relict; *P. multilineatus*, from Fennoscandia, also probably pre-glacial; *Copelatus ruficollis*, extremely euryoikous, probably furnished with an active means of diffusion, *Colymbetes koenigi*, from the Caucasian lakes, and finally the conspicuous and interesting *Dytiscus latissimus*, wide-spread and extending into Sibiria.

Caddis flies are very numerous and show many adaptations, both morphological and physiological-behavioral, to life in still waters. Many families are present in European lakes: *Hydroptilidae, Polycentropidae* (including also some endemics species), *Phryganeidae* (particularly well represented although generally by rather common species), *Limnophilidae* (e.g. *Apatania hispida* of Fennoscandia; the genus *Limnophilus*, represented by widely distributed species but also by some endemic, especially in the Caucasus, which probably deserves particular zoogeographical study because of its isolation, relatively old age and the existing connections both with the East (e.g. Iran) and the West; the genus *Asynarchus* including some nordic, apparently circumboreal species, and probably also some boreo-alpine species (for instance *A. lapponicus*); *Leptoceridae*, including some endemic species of *Athripsodes* in the Caucasus and some endemic species in the Alps (for instance *Mystacides leucoptera* from Trentino or Süd Tirol). The family *Calamoceratidae* is also very typical of lakes, represented by widely distributed species, which is possibly due to the ability of these animals for active dispersion. Lepidoptera larvae are also present in our lakes, though none exclusive to them.

Chironomids are very well represented in the limnic fauna and have a paramount importance in the determination of the degree of eutrophy of a lake because of their ability to thrive, even in waters very poor in oxygen, because of the presence of haemoglobin in their blood (tribus *Chironomini*). The subfamilies present in our lakes are *Tanypodinae* (generally carnivorous) very often with a very restricted distribution, including also endemic species in the Alps or in the mountains and lowlands of Central Europe, such as the genera *Procladius, Psilotanypus, Tanypus*, etc. Also *Orthocladiinae* can be very strictly limited to lakes: one species, *Chaetocladius festivus* is present only in Fennoscandia and on Spitzbergen: here again we are confronted with the problem of the origin of many species today confined to the extreme North, which probably has not yet been satisfactorily solved. Still more interesting is the distribution of *Cricotopus glacialis*, which includes Iceland, the tundra of the extreme north, Greenland and Spitzbergen. Several species seem to be boreo-alpine (e.g. *Diamesa culicoides*) but the majority are characteristic of the lakes of Central Europe. Many species belong to the *Chironomini: Cryptochironomus, Einfeldia, Harnischia*, etc.; *Tanytarsini* and more especially the genus *Tanytarsus* are also very abundant and – as we have already said – characteristic of oligotrophic lakes.

Ponds

In one sense the opposite of lakes, where water is permanent and where there is a constant division between cold, deep water, and surface water, with changes in temperature and oxygen content (and therefore pH), there is a category of eutrophic, shallow waters, with no difference in temperature between surface and bottom, rich in vegetation over their entire area, very often poor in oxygen, with a temperature which in some regions can vary during the year from very high (even 35-40°C) to freezing point (0° C), and subjected to complete drying up during some particularly long, warm, dry summers. This category in Europe includes ponds, swamps (see also page 504) mires, ephemeral waters (or astatic basins, or 'Kleingewässer'), phytotelmas and lithotelmas. We may say that nearer the tropics these kinds of waters are much more frequent.

Fish life in ponds is very rich because of the high degree of eutrophication of these small shallow aquatic environments, some of which are occasionally brackish.

In contrast to lakes, in ponds there is rarely a major difference in temperature between the surface and the bottom; the latter is covered with flowering plants or

Fig. 418. The rôle of the Crustacean *Mysis relicta* in the nutrition of fish in Lake Onega. The shaded area indicates the percentage of fish feeding on *Mysis*. 1. Smelt (*Osmerus eperlanus*); 2. Grayling (*Thymallus thymallus*); 3. 'sig yamnyi' (*Coregonus* sp.); 4. Ludoga whitefish (*Coregonus lavaretus ludoga*); 5. freshwater white fish (*Coregonus lavaretus*); 6. Onega whitefish (*Coregonus albula kiletz*); 7. burbot (*Lota lota*). 8. sculpin (*Myxocephalus quadricornis onegensis*); 9. perch (*Perca fluviatilis*); 10. ruff (*Acerina cernua*) (from Gerd).

with macrophytic cryptogamae (e.g. *Characeae*). Ponds, of course, are fit only for the most euryoikous fishes, since they may dry almost completely during summer and in some regions freeze almost completely during winter.

Where the bottom is rather muddy and completely covered with macrophytic vegetation the tench and *Misgurnus fossilis* find their optimum. If, on the contrary, the bottom is at least partially sandy and without vegetation, other species appear: *Alburnus alburnus, Leucaspius delineatus, Scardinius erythrophthalmus, Rodeus sericeus amarus Gasterosteus aculeatus, Pungitius pungitius* and *Phoxinus phoxinus*. Some species, such as the green tench, *Misgurnus fossilis, Pungitius pungitius* and *Carassius carassius* can also live in very small ponds, swamps and mires. In Southern Europe in latter years a North American fish, the sun fish, *Eupomotis sol* has been found in swamps, whereas in ponds and small slowly flowing rivers and canals another North American fish, the cat-fish *Ameiurus nebulosus* has spread to the detriment of indigenous species. Some species which are among the most euryoikous can thrive in ponds subjected to dessiccation during the summer, since they can survive for some time in wet mud (*Tinca, Misgurnus* and *Carassius*) (Fig. 420).

Finally some species such as *Umbra krameri*, typical of the Danube and Dnepr basins, prefer flood-waters to the rivers themselves. Several fishes live in this peculiar environment (Fig. 421).

All ponds, with the exception of some very dystrophic waters of Northern

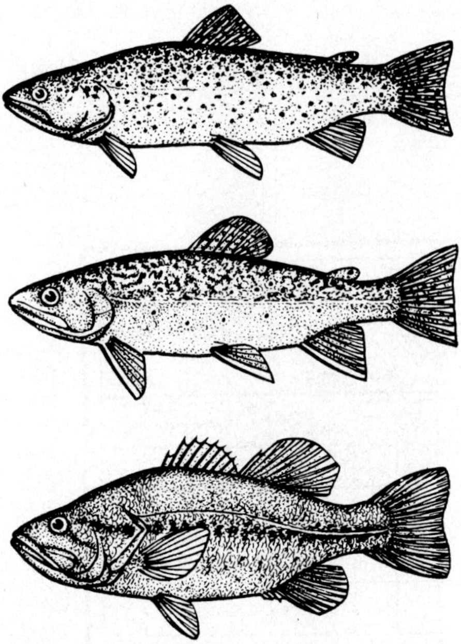

Fig. 419. Examples of lake fishes: the rainbow trout *Salmo gairdneri*, an imported species; the charr, *Salvelinus fontinalis*, living in all kinds of cold lakes; and the black bass, *Micropterus salmoides*, an imported species (from Ladiges & Vogt). (Reproduced with the permission of Paul Parey).

Fig. 420a: A pond in a swamp with water-lilies (top). 420b: Examples of pond fish (below): left – *Gasterosteus aculeatus*, the common stickleback; right – *Pungitius pungitius* (from Ladiges & Vogt). (Reproduced with the permission of Paul Parey).

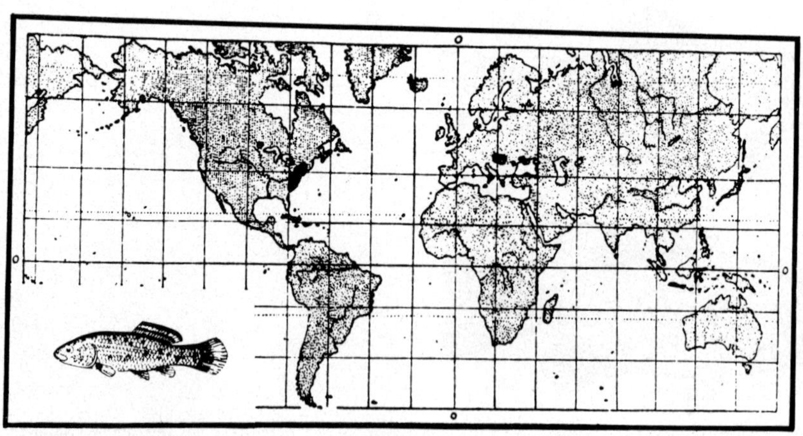

Fig. 421. *Umbra krameri* and geographical distribution of the genus *Umbra*, of a highly disjuncted type (figure of the fish from Ladiges & Vogt, map from Sterba). (Reproduced with the permission of Paul Parey).

Europe, are highly eutrophic and offer a suitable environment for an enormous number of plants (both lower and higher, both micro- and macroscopical) and animals. The productivity of ponds is sometimes very high: in Europe from a pond of 10,000 square metres it is possible to have a yield of carp of 4-500 kg per year, and if the technique is refined even 8-1000 kg. Of course the biota of the ponds are extremely varied in the different regions; not only do ponds offer a practically infinite range of conditions, size, depth etc. from the example of the

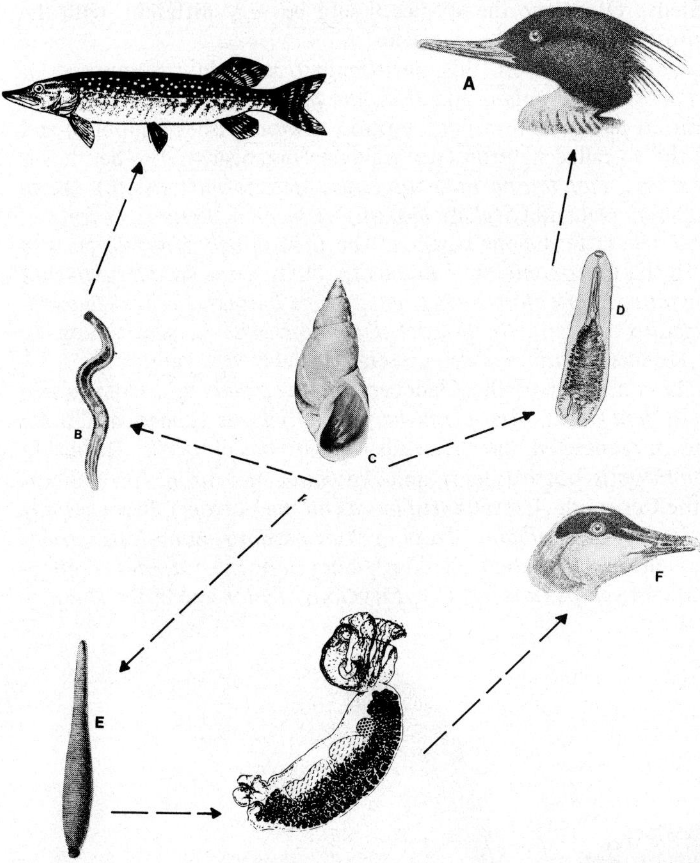

Fig. 422. The main *Trematodes* transmitted to European ponds by the snail *Lymnaea palustris*: B, *Azygia lucii*, as an adult parasite of a freshwater fish, the pike or *Esox lucius*, where it lives in the stomach and the intestine; D, *Notocotylus attenuatus*, parasite of some birds such as the goosander *Mergus serrator*, where it lives in the caecum, more rarely in the rectum, and only exeptionally in the duodenum, and *Cotylurus cornutus*, passing through the intermediary of some leeches, to some birds as, for instance, the beautiful *Somateria mollissima* or eider duck. In the latter the worm lives fixed to the intestinal villi; the host is not seriously damaged, however, because the worm feeds on the intestinal content and not on blood (the fish is from Ladiges & Vogt; the leech from Wesenberg-Lund, *Lymnaea palustris* from Frömming, the birds from Arrigoni degli Oddi. All these editors have given permission for the reproduction of the figures).

Trasimeno Lake (which is actually a 'pond-lake' because of its limited depth) to some water bodies of a few square metres only 30-40 cm deep. For this reason, a treatment of pond biota is nearly impossible or at least useless: practically most of their organisms live also on the edges of lowland, eutrophic lakes, as well as in swamps, provided these offer an almost permanent water body.

Just as an example, we will consider here the case of some small ponds in a large karstic swamp in Venezia Giulia (Pietra Rossa; cf. also on page 506) but for the aims of this book we could examine, say, a pond in Fennoscandia, or the British Isles, or Sicily. Of course the species would be very different, with the exception of the most eurytopic and euryoikous.

The macrophytic vegetation of this environment is mainly composed of *Nymphaea alba, Potamogeton natans* and *Hottonia palustris*. In our 'ideal' pond we may distinguish, at least theoretically, benthos, nekton (fishes, tadpoles) and plankton, besides the so-called neuston (surface-living organisms). The benthos is Nematodes (*Plectus* sp., *Punctodora ratzeburgensis, Mesodorylaimus* sp.), Oligochaeta (*Slavina appendiculata* and *Stylaria lacustris*), *Cloeon dipterum, Agrion* sp., and *Diptera* larvae. The latter belong partly to the *Ceratopogonidae* (*Bezzia* sp.) but the majority to the *Chironomidae*: *Tanypus* sp., *Tanytarsus* sp., *Corynoneura* sp., *Cricotopus algarum, Eucricotopus* sp., *Crenopelopia binotata, Trichocladius s. lato, Pseudorthocladius* sp., and *Synorthocladius semivirens* (a species new to Italy). An Isopod, *Asellus aquaticus* is also present, although very rare.

The plankton is composed of the Cladocera *Simocephalus* sp., *Drepanotrix dentata, Eurycercus lamellatus, Acroperus harpae, Chydorus sphaericus, Alona quadrangularis* and *A. rectangula*, the Ostracods *Notodromas* cf. *persica* (probably new to Italy), *Cypridopsis obesa* (idem), and *Darwinula stevensoni* (on the bottom) and finally the Copepods *Atteyella trispinosa* (on the bottom), *Bryocamptus minutus* (idem), *Macrocyclops albidus*, *Tropocyclops prasinus*, *Eucyclops serrulatus* (the most frequent and abundant plankter), *Eucyclops macruroides, Ectocyclops phaleratus, Microcyclops varicans, Cryptocyclops bicolor* and lastly *Thermo-*

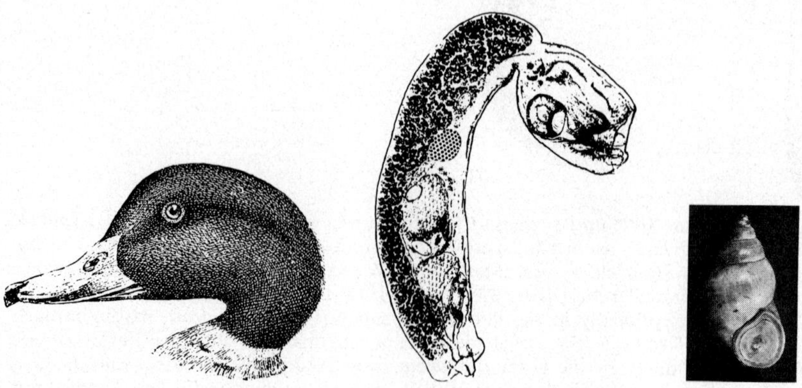

Fig. 423. Transmission of the trematode *Apatemon gracilis* from the snail *Bithynia tentaculata* to a duck. In this bird the worm lives in the intestine, in some instances causing serious trouble.

cyclops hyalinus. The hyponeuston is represented by mosquito larvae (*Anopheles maculipennis*), the epineuston by two springtails, *Podura aquatica* and *Sminthurides aquaticus.*

Swamps

Life in a swamp is quite various and we will not even try to make a census, which would make very little sense from an ecological standpoint, of the principal vegetal and animal components. It is enough to say that in a relatively large swamp (say some 0.2 km^2) we find plants of still waters, together with species of running waters such as *Veronica anagallis*, (the speedwell) *Mentha aquatica, Potamogeton* (the water spikes) etc. We find invertebrates on the bottom, in the mudd or under stones, on the submerged parts of plants, inside the latter (endophytic animals, as for instance the larvae of the Chironomid genus *Endochironomus*), animals typical of running waters, animals which pass all their life, or most of it, skating or swimming on the surface of the water, such as some bugs (*Gerris, Velia, Hydrometra*) or some beetles (*Gyrinus*). The same holds for fishes, amphibians and reptiles, which exploit all ecological niches and all the trophic levels extant in the swamp.

If there is a marked difference in the terrestrial bios of a swamp during the different seasons (page 506) a difference, though less pronounced, is present also among aquatic life, as we will see now.

Let us consider the same swamp we have discussed in the chapter about humid environment, so that a comparison is possible. During winter we may find a large water beetle, *Dytiscus marginalis* or the rarer *D. dimidiatus* swimming also under the ice, or the very little, shining, dark blue *Gyrinus substriatus* swimming very quickly on the surface of the water in some whirlpools.

In spring a smaller Dytiscidae, *Colymbetes fuscus* makes its appearance under the water surface covered by the water lentils, *Lemna trisulca* and *L. minor*, whereas on the bottom the larva of *Sialis flavilatera* (which lives two years as a larva) may be found (cf. Fig. 402). The latter is a very euryoikous insect, which can withstand a high degree of organic pollution together with the water isopod *Asellus* and Oligochaeta such as *Nais* and *Tubifex*. A very remarkable, rather conspicuous, water insect is the nympha of *Cloeon dipterum*, an Ephemerid or may fly, which swims very quickly in an elegant manner in the water. It has a low requirement of oxygen and sometimes it is found together with the small Cladoceran *Chydorus sphaericus*, one of the most euryoikous Cladocerans. Both these animals can tolerate organic pollution.

Other beetles making their appearance in spring are *Ilybius fuliginosus* and *Peltodytes caesus*. The former has a larva which in the first stage crawls on the bottom and successively swims freely in the small ponds of a swamp. In some instances it can thrive also in brackish waters Peltodytes on the contrary prefer small waters rich in vegetation: *P. caesus* indeed spawns upon the single filaments of *Spirogyra*.

In May the vegetation is richer: some species are already in flowers: *Ranunculus aquatilis*, the water crowfoot, (in Italian called 'finocchio d'acqua' because of the form of its submerged leaves similar to chose of fennel), which can literally cover a wide extension of water as a prairie with white flowers, or *Veronica anagallis* the speedwell, whose pale blue flowers project from the water. In the water the nympha of a very elegant and delicate insect, *Ephemerella ignita*, can be present, crawling on the plants and the bottom: it is not resistant towards organic pollution and can therefore be considered as an indicator of the purity of water. Another interesting inhabitant of swamps is the larva of *Cyphon*, frequent in small ponds covered with *Lemna*: the larva is carnivorous and swims with the ventral surface upward. It is fairly tolerant towards pollution. *C. simile* for instance, can withstand a quantity of lead equal to 3-6 ppm. Some large water beetles make now their appearance: *Hydrous piceus* (the biggest European water beetle whose larva is known as 'water devil'), *Hydrophilous caraboides* (similar to the former but much smaller), both of them living also in rice fields (page 514) and the *Dytiscidae Ilybius obscurus*. On the bottom, in the mud or among the intricate mass of filamentous algae, a large number of larvae of not-biting midges or Chironomids live, which according to the presence or not of haemoglobin can tell us whether the water body is eutrophic or oligotrophic. Also larvae of Simuliidae can be found, both on stones or on large leaves, as for instance those of *Caltha palustris*. They are filter feeding animals, utilizing their large fan-like antennae as a filter. They prefer waters almost saturated with oxygen, but some species can also live in slightly polluted waters (β- mesosaprobious) and tolerate a lead pollution up to 0.2-0.5 ppm. The adult female of Simuliidae has haematophagous habits and bites men and house animals; when the bites are very numerous they can bring about intoxication and even death. Exceptional mortalities of cattle have been observed in Hungary, due to *Simulium columbatzense*. In the mud of small ponds, poor in oxygen, very often larvae of *Tabanids* (or horse flies) are found: they are among the insects the most resistant water towards pollution, since they assume directly atmospheric oxygen. The same is true for the pre-maginal stages of mosquitoes such as *Culex pipiens* (the common mosquite, frequent also today during summer in many Southern European countries, as for instance Italy) and *Anopheles maculipennis* (with its numerous varieties) (cf. Fig. 426). Whereas the latter is a surface dwelling form, generally living in ponds rich of vegetation, thy larva of *Culex* can thrive also in the septic zone of a polluted water course, and therefore its presence is not of help as far as water qualities are concerned.

The diptera family more represented are, however, *Chironomidae*, or non-biting midges, sometimes present both as larvae and as adults in tremendous numbers (the adults often exhibit the habit of swarming). Some of them are endophytic elements, for instance *Endochironomus dispar*, mining the leaves of aquatic plants, particularly *Sparganium* and *Sagittaria*. It is furnished with haemoglobin, but it can thrive also in the litoral fringe of lakes, together with the genera *Cricotopus, Parachironomus, Lenzia* etc. which do not possess haemoglobin. In the slowly running waters very often present in swamps, two more genera are present: *Tanytarsus* and *Tanypus*. Some species of the latter can withstand a lead pollution of 0.2-0.5 ppm, together with *Simulium, Cloeon, Velia* and *Sigara*.

May sees another animal group very abundantly represented in swamps, the leeches or *Hirudinea*. The best known is the common leech *Hirudo medicinalis*, so named because once is was very much employed in pharmacy for blood-letting. It attaches itself very quickly to the legs of a person who enters the water of a swamp barefoot, for ecological purposes or not, and it is practically impossible to detach it until is well fed. During the last century it was intensively bred because of its medical utilization: only at Hildesheim (near Hannover) in 1860, 3.5 millions individuals have been obtained! It can transmit some bacterial infections. Another leech, *Glossiphonia complanata*, characteristic because of its greenish colour with small, elegant, black spots, is present in swamps: it is characteristic of waters in which a domestic pollution is present, and in a sense it can be considered as a biological indicator.

Finally, in swamps another group of 'worms' is present, the flatworms or Turbellarians, such as *Planaria torva, Polycelis tenuis* and *Dugesia lugubris*. These species are found on plants, in mud or on stones where they slide as small slugs, thanks to the presence of cilia all over the body surface. Among flatworms, *Polycelis tenuis* is a true swamp organism which, however, avoids vegetation, present often in eligotrophic waters with an amount of Ca equal to 5 mg/l. It also withstands humic substances and therefore a certain degree of acidity. *Dugesia lugubris* is also characteristic of still waters and avoids both vegetation and too low temperatures. It seems to be exclusive of eutrophic waters, with an amount of Ca at least equal to 10 mg/l. It can withstand even a certain salinity, rather high temperatures and also some toxic substances. It is therefore an euryoikous species. *Planaria torva*, on the other hand, is rarer: it withstands some salinity and it is extremely euryoikous.

Two interesting groups are rather frequent in spring: Amphipods and Isopods, the latter generally represented by *Asellus* (*A. aquaticus*). Amphipods are represented in Southern Europe by some species such as *Synurella ambulans, S. philareti, Rivulogammarus balcanicus, R. fossarum* etc., *Niphargus valachicus* and *N. hrabei* (limited to the Danube plain) and a few others. Both groups are detritivorous or transformers and intervene in the demolitive attack of dead organic substances.

In May several vertebrates appear: tadpoles of the frog *Rana esculenta* (which had laid eggs in April), young fry of some fishes such as the pike, tench and carp, associated to a species introduced from North America at the beginning of this century, the Pumpkin seed sunfish *Eupomotis* (or *Lepomis*) *gibbosus*. Due to the great prolificity and the great power of adaptation, this element had proved very harmful for other – more valuable – species, which in many water bodies have been supplanted as a consequence of the struggle for life. From May to August it lies in a kind of nest dug in the sand. It is rather lazy and slow in its movements. Aquatic plants are still scarce in May: let us mention two beautiful flowering plants, *Sparganium erectum* with leaves like long and pointed knifes, and a terminal floral cluster; and a Nymphaea-like plant, the spatterdock *Nuphar luteum*, projecting from the surface of the water like buttons.

In June we may find some larvae of Chironomids such as *Cricotopus biformis* or *Thienemaniella vittata*, together with some very interesting small beetles such as *Hydrochus impressus*, limited to the Centro-Western Mediterranean, characteristic of still waters, and *Scirtes hemisphaericum*, with a wider distribution area. Its larvae live in relatively dinty waters where they stay attached to the surface film with their abdomen upwards, in a sense constituting a kind of hyponeuston. The larva metamorphoses in hollow stems of aquatic vegetals. Also the adult stays with the ventral part upwards.

The summer has arrived: many aquatic plants are now visible: *Potamogeton natans*, with long, elegant, extremely shining leaves, violet-greenish or brown-greenish coloured, floating on the water, with stems of very various length according to the depth of the water body; *P. pusillus, P. crispus* and a few other species; *Ceratophyllum demersum* and *Myriophyllum verticillatum*, light-green coloured with very elegant verticillate filiform leaves, and finally the beautiful albeit modest *Utricularia minor*, or bladderwort, a carnivorous plant belonging to the Lentibulariaceae. It has a number of utricles with which it captures small aquatic animals, often Copepods or other plankters.

Odonata are now metamorphosing: let us remember *Sympecma fusca*, whose eggs are attached to the Graminoid plant *Glyceria fluitans, Pyrrhosoma nymphula*, whose larva lives among the aquatic plants but before metamorphosis passes to the soil; *Agrion puella*, whose eggs have been observed upon the inferior page of leaves of *Potamogeton natans*, and finally *Calopteryx splendens*, whose larva walks upon objects laying on the bottom, such as small pieces of wood or branches.

Other insects making their appearance in summer are Diptera larvae. Among the most interesting, from an ecological standpoint, are Syrphidae like *Eristalis sepulchralis*, living in the mud of waters sometimes highly polluted, since it utilizes for respiration only the atmospheric oxygen; *Ephydra riparia*, (Fig. 440b) living in all kinds of waters, even in brackish water and in mineral sources, developing upon submerged parts of aquatic plants, and finally Tabanids such as *Chrysops caecutiens*, living as a larva on the bottom of swamps. Finally, during summer another insect order makes its appearance, sometimes reaching a remarkable density, caddis flies. Let us mention here only three genera, though their number in swamps is much higher: *Oxyethira, Grammotaulius* and *Silo. Oxyethira* larvae live indifferently in standing waters or in small brooks: they thrive in the submerged prairies constituted by mosses as, for instance, *Fontinalis antipyretica*. The case is constituted by a very thin, transparent kind of capsule oval-trapezoid in which the larva is clearly visible. *Grammotaulius atomarius* builds up a case with small pieces of vegetal debris (parts of *Carex, Phragmites*, etc.) as well as with pieces of shells or even with entire empty snails. Finally *Silo pallipes*, is present only in brooks flowing slowly through swamps, or feeding the same: it has a very characteristic case so that it is very easy to determine this caddis-fly genus.

Several organisms, however, are present in a swamp all through the year, both plants and animal: among the latter we should mention some unconspicuous arthropods, namely the collembols or springtails *Podura aquatica* and *Sminthurus aquaticus. Podura* is common all over the water surface (also of brackish waters) and has been observed in Switzerland up to 2500 m. It does not wet at all because the tegument is covered with a hydrofuge substance (wax). Sometimes it sticks to some aquatic plant or object and then it dives into the water.

Another order of insects with uniform alimentary habits (predators) and widely diffused in swamps – though not as a rule exclusive to them – is given by *Heteroptera* or bugs. The bugs we may find and actually find in European swamps have very different sizes, shapes and habits, so that they occupy different ecological niches and can live side by side without any kind of competition among them. When we make such an assumption we think of *Nepa cinerea* and *Ranatra linearis*, crawling upon the bottom, one short and very much depressed (almost like a dead dry leaf), the other long, slender and cylindrical, very much alike to a stick. They both feed on small aquatic invertebrates. *Ranatra* is present also in rice-fields; it has been described, probably for the first time, by Aldrovandi in 1602. Swimming here and there, often also in highly polluted water, we may see a *Notonecta*, named after the habit to swim with the ventral side upwards (from the Greek *νῶτος* the back and *νηκτός*, swimming). It feeds on dead or dying insects present on the surface of the water, and when it is picked up by hand it often stings with very painful effects because it injects a highly injurious substance, containing some inflammatory and some anti-clotting substance, beside some digestive enzymes such as proteases and lipases. The effects are partly primary and, after some time, secundary, given by an inflammation of the skin, which can be 12 cm long and 5 cm wide, with a local temperature rise up to 37.8° C.*

Finally, two surface genera: *Gerris* and *Velia*, both skating on the surface of the water looking for small living or dead insects to prey upon; the former is not able to submerge because of its small weight compared with surface, which prevents the insect from disrupting the surface film (where there is a strong surface tension) and entering the water. The latter, much shorter and therefore heavier, if menaced by something can swim freely in the water and escape the enemy. Some species of *Velia*, e.g. *V. currens*, can withstand lead pollution up to 0.2-0.5 ppp. We could not abandon insect life without mentioning the adaptations to aquatic life present in the Chrysomelid genus *Donacia* and other allied genera, a group almost exclusive of swamps (see page 508). The larvae of these beautiful beetles, as an adult characterized by a metallic lustre and greenish cupreous colours, are constantly monophagous or stenophagous, feeding on one species or a few species belonging to the same genus or family. It eats, attaching itself to the roots of a plant: some claim that the animal is assisted in digestion by some symbiont microorganisms living in the Malpighian tubes, where they constitute true 'mycetomes'. The respiratory function, however, exhibits the greatest adaptation to aquatic life, unique not only among insects but perhaps among all animals. The respiratory tubes of the larva (or tracheae), filled with air (as is the rule in all terrestrial insects) merge into the last pair of abdominal spiracles communicating with two hawks penetrating into the vegetal tissues, and more precisely in the aeriferous spaces of the plant. In this way the animal, though living permanently in the water, breaths the air contained in a plant, which in turn communicates with the atmosphere through its stomata. Also egg-laying is the fruit of adaptation of *Donacia* to an amphibious life: the female lays eggs upon the submerged surface of a vegetal. Before metamorphosis, the larva builds up a cocoon of a paper-like aspect filled with air, attached to some aquatic plant. Some holes in the surface of the plant permit the passage of air (and oxygen) from the latter to the cocoon.

* In the case of another water bug, *Belostoma cordofanum*, present in the eastern part of the Mediterranean basin up to Southern Dalmatia (Neretva) and Greece, the stinging has a paralizing effect on the prey: the venom injected by this enormous insect (length up to 7 cm) is a very active, neurotropic salivary venom, similar to that of the snake genus *Crotalus*.

And, as a last group, molluscs, i.e. gastropods or snails and bivalves. On the bottom of the ponds, on plants, towards the shores, water snails abound: one is *Bithynia tentaculata*, short, conical, greenish-yellow coloured with small dark spots, generally living in waters rich of organic debris; it feeds on plankton (filter-feeding animal). It can also transmit some helminthiasis to fishes and also to ducks, in which a Trematode, *Apatemon gracilis*, is frequent (Fig. 423). *Planorbis* are also very common: the spires are developed practically only in one plane so that the animal looks a little like a coin. One species, *P. carinatus*, present in all kinds of freshwaters, another, *P. spirorbis*, living in all kinds of preferably pure waters, on mountain up to 1200 m. A third species is *P. planorbis*, a very euryoikous element, though preferring waters in which there is a rich vegetation and a muddy bottom, on which it lives in enormous quantities. It seems to feed upon algae and purpureous bacteria. It buries itself deeply in the mud.

In slowly running waters and in the springs which feed swamps, as well as in completely still waters, a very elegant, small snail, in shaped like a Phrygian cap, dark olive-green coloured, with small pale spots is *Theodoxus fluviatilis*. Sometimes it is present in enormous quantities together with flatworms, leeches and other snails (generally *Bithynella*). It is mainly a predator, feeding upon other molluscs as, e.g., *Sphaerium*, but it utilizes for its diet also diatoms and algae. A more trivial species is the snail *Galba* (= *Lymnaea*) *palustris*, widely distributed, present in all kinds of still waters as well as in very slow water courses, rich in plants. It cannot withstand high amounts of iron or humic acids. It can withstand a partial dessiccation, being one of the few European *Pulmonata* with amphibious habits, which recalls the genus *Succinea*. It feeds on aquatic or semi-terrestrial dead, rotten, plants such as *Alisma plantago, Lysimachia vulgaris, L. nummularia, Stachys palustris, Mentha aquatica, Lycopus europaeus, Veronica beccabunga* and *Symphytum officinale*, but it also exhibits necrophagous alimentary habits. It transmits several helminthiasis (given by *Trematoda*) to ducks, geese, swans and goosanders so that when a swamp is frequently visited by these birds the molluscs can be harmful (Fig. 422).

Another very abundant species is *Viviparus viparus* (once known as *Paludina vivipara*) limited to Centro-Northern Europe, reaching as far south as the feet of the Alps and probably Central Italy. It prefers ponds rich in vegetation (which during day time develop oxygen) but still more running waters, since, different from most *Pulmonata*, it has a great requirement of oxygen (it is indeed a *Prosobranchia*). It feeds on debris and algae and only secondarily on animal substances and living or dead plants. The digestive enzymes of this snail have been investigated by the present author. It can transmit a fluke *Fasciola hepatica*, a parasite of sheeps.

Finally, on the muddy bottom of swamps, though also of lakes (up to a depth of 30 m) and brooks, a small elegant *Bivalvia* (the group which includes the common clams), *Sphaerium corneum*, is present. It exhibits a perfectly symmetrical shell, circular or slight elliptical shaped. It is able to thrive also in some environments very poor in oxygen (as it is the case of a muddy bottom, where some putrefactive phenomena occur as well) since it is furnished with a peculiar haemoglobin very able to keep also extremely small amounts of oxygen. It is a microphagous animal.

Among the smallest invertebrates we find Protozoa: suffice to quote Ciliates or 'Infusoria', zooflagellates (very rare) and Rhizopoda. They play a particular role in the ecology of a swamp, for instance, *Amoeba, Difflugia, Arcella, Lecquereusia*, etc. The latter is a very characteristic small creature, with two processes of the protoplasma (ectoplasma), true pseudopodia, which resemble somewhat the horns of a snail, hyaline cylindrical, with rounded tip. These pseudopodia are long, nearly twice the length of the body (Fig. 425). This is protected by a lorica or cuirasse subglobose, little transparent, furnished with a very peculiar vermicular microsculpture. Among Phytoflagellates we should mention here *Ceratium* (Fig. 424 and 427).

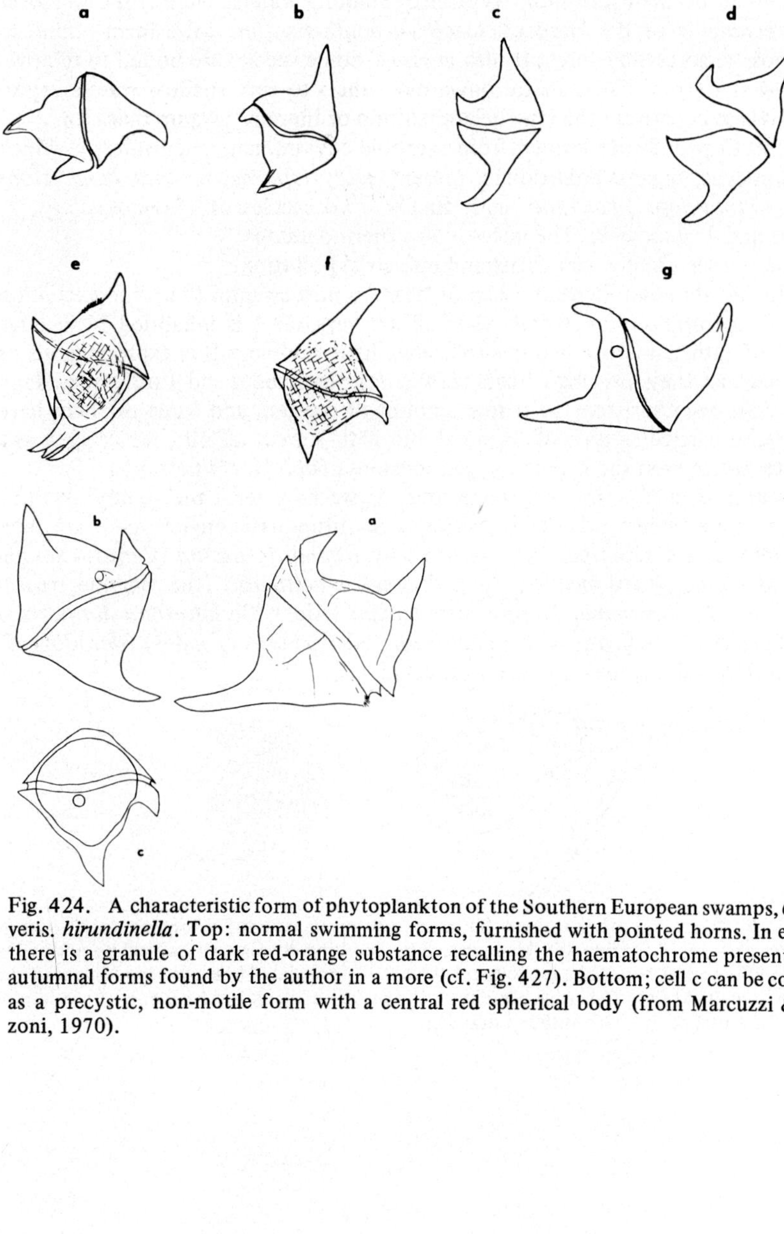

Fig. 424. A characteristic form of phytoplankton of the Southern European swamps, *Ceratium* veris. *hirundinella*. Top: normal swimming forms, furnished with pointed horns. In element g there is a granule of dark red-orange substance recalling the haematochrome present in some autumnal forms found by the author in a more (cf. Fig. 427). Bottom; cell c can be considered as a precystic, non-motile form with a central red spherical body (from Marcuzzi & Lorenzoni, 1970).

Zooplankters are very common in swamps: let us remember Rotifers like *Brachionus*, recognizable by a cuirasse furnished with some highly characteristic points; *Hydracari*, with a globular body and long, slender legs which permit the animal to swim very fast, and finally crustaceans, represented by Copepods, Cladocerans and Ostracoda. Their number is enormous: many are true plankters, and can be found also in the litoral fringe of lakes (see page 626), some of them crawl, swim or walk on the bottom or on the submerged vegetation (*Ostracoda, Harpacticoida*, some Cladocerans). Many species exhibit in our latitudes an alternance of generation, in summer being represented only by parthenogenetic females, in autumn by males and females which provide the formation of fecundated eggs, destined to survive all winter ('Dauereiern' as the first German limnologists called them). Some of them are more typical of running waters, such as the Cladocer *Alona rectangula* or the Harpacticidae*Bryocamptus pygmaeus*, a form bound to some mosses, sometimes present also in caves. Some species are bound to relatively deep water (from half a metre onwards), others to very shallow waters (up to 50 cm): these constitute the true heleoplankton or literally 'swamp plankton'.

Several Copepods are known from saprobic environments, i.e. waters in which some domestic, organic pollution is present: *Eucyclops macruroides, Paracyclops affinis, Ectocyclops phaleratus* and, finally, two species of *Thermocyclops, T. hyalinus* and *T. dybowski*. The latter is also thermophilous.

Some species, finally, can withstand industrial pollution.

To finish, the microbenthos. The bottom of most swamps (with the exception of the oligotrophic ones, which after all are very rare) is inhabited by a huge amount of rather microscopic invertebrates, little mobile, often saprophagous or microphagous: they are Oligochaeta (*Limicola*), Nematodes and Tardigrada. Many of them are very resistant to a low amount of oxygen, and some of them have developed a particular type of haemoglobin with a great affinity for oxygen as a device to utilize even the lowest oxygen tensions (*Tubifex, Nais* etc.)

If swamp elements are very numerous, as we have seen, the stenoykous animals, i.e. the elements practically exclusive of this aquatic environment are rare.

Nematodes are represented in swamps by *Rhabditis insolita* (Centro-European), *Craspedonema styriacum* (Alps), *Psilenchus aestuarius* (the Danube basin), one species of *Criconema, Monhystera demani* (Alps), *Diplohystera dintheriana* (also terrestrial as happens with many Nematodes present in water), *Mesodorylaimus luci* (Central Europe) and a few others.

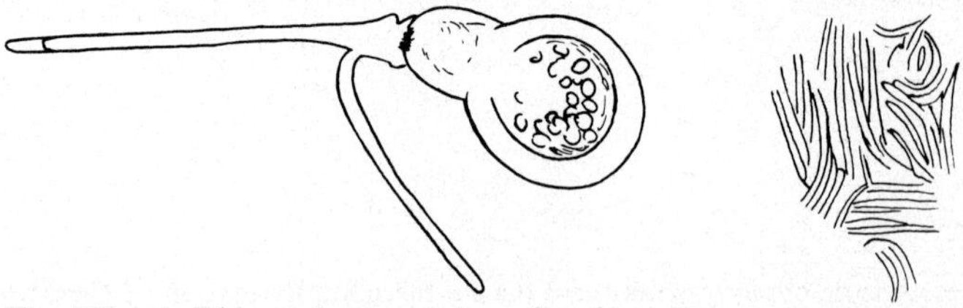

Fig. 425. A Rhizopod present in a Southern European swamp, belonging to the genus *Lecquereusia* (veris. *L. extranea*), characterized by two long, hyaline, cylindrical pseudopodia, of which alternatively one is extended and the other retracted. These pseudopodia recall the horns of a snail. At right the striae present on the surface of the lorica have been drawn (from Marcuzzi & Lorenzoni, 1970; author's drawing).

No Gastropod is typical of swamps, although they abound sometimes in this environment (euryoikous species like *Lymnaea, Physa, Galba, Radix, Planorbis* etc.). The only known examples (several species of *Pseudomelampus*) are present on the Atlantic islands off the European coast (Madera, Azores etc.).

Collembolans are found in swamps but of course are limited to the epineuston and are not true aquatic animals: *Isotoma viridis* (a very common species), *Isotomurus alticolus* (widely distributed on mountains), *I. ciliatus* and *I. palustris* (also a common species).

Only one beetle, the *Hydrophilidae Chaetarthia seminulum*, seems so far to be exclusive to swamp.

Chironomids or non-biting midges, although so common in all kinds of waters, do not prefer swamps: exclusive species are perhaps *Camptocladius stercorarius* (terrestrial widely distributed species), several species of *Gymnometriocnemus*, from the Alps to Northern Europe, *Limnophytes virgo* and *Metriocnemus corticalis*, both in the lowlands of Central Europe, *Orthocladius latescens* (idem), *Paraphaenocladius camptoneurus* (idem) and a few more. On the whole, this environment has not favoured their evolution.

Mires

We have already discussed mires when talking about semi-terrestrial environments (page 519). What we can say here is that mires are characterized by acid resistant plants and animals, able to live in oligotrophic waters, poor in nutrients, so that

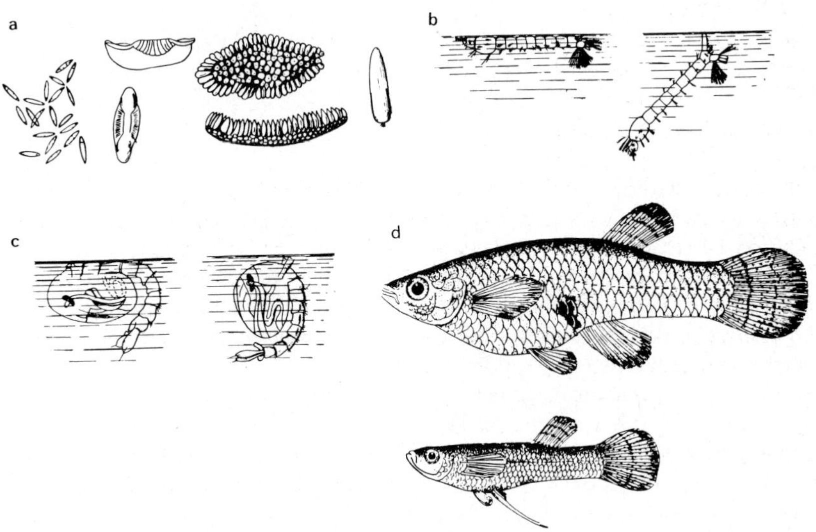

Fig. 426. a. eggs, b. larvae, c. pupae of mosquitos belonging to the genus *Anopheles* (left) and *Culex* (right); d. the guppy *Gambusia holbrooki* (top – female, bottom – male) used in the past in the struggle against *Anopheles* and malaria (insects from Grassi, fish from Neveu Lemaire).

rather than true tyrphophilous organisms, we have in general euryoikous, euryionic elements which can thrive in mires, partly because they do not find here the competition present in other, more 'normal', kinds of waters.

Since the water of mires is often colder than that of similar environments situated at the same latitude and altitude, mire organisms in general are either eurythermous or even stenothermous frigidicolous.

Although not exclusive to mires, we find essentially pre-imaginal forms of insects such as dragon flies (*Leucorrhinia pectoralis, L. dubia, Sympetrum flaveolum, Aeschna juncea*), beetles (*Cyphon, Plateumaris*) and some molluscs. On 'Hochmoore' some zoological groups are practically completely missing: Ciliates, *Brachionidae* (Rotifers), Nematodes, *Oligochaeta, Hirudinea*, Molluscs, *Culicidae*-larvae, Ostracods, as well as fish and amphibians.

We may present here some quantitative data obtained by the present author for a mire in Trentino, the Lagabrun mire (in %):

Animal group	Pond with *Sphagnum* and *Carex limosa*			Pond with *Potamogeton*
	from bottom to surface	only at surface	near the shore	
Rotifers	0.62	49.7	82.2	20.87
Nematodes	8.69	4.8	3.2	57.7
Oligochaeta	19.87	1.9	–	1.09
Tardigrades	0.62	0.48	–	0.54
Copepods	24.84	6.76	4.0	7.14
Nauplii	4.34	1.92	1.6	1.09
Cladocerans	32.3	14.0	2.4	7.69
Ostracods	–	1.44	–	–
Chironomid-larvae	8.07	18.84	5.6	3.84

'Kleingewässer'

More interesting without doubt are the 'small waters' (or 'Kleingewässer'). Nematodes are very rare here, or perhaps they have been little investigated: *Mesorhabditis spiculigera* is a terrestrial species able to live also in small waters. *Andrassya vivipara*, present in the lowlands of Central Europe, is much more typical; *Dorylaimus vixamictus* is present only in the Danube basin, *Enchodelus arcuatus* seems to be a terrestrial species, so that it seems that nematodes have not developed any particular system for living in periodic waters.

The same, and even more, holds for Oligochaeta, among which we find only one species, *Lamprodrilus mrazeki* of the mountains of Central Europe.

A couple of amphipods are known only from small waters: *Niphargus potamophilus*, living near the mouth of the Don (Caspian Sea), and *Synurella philareti*, living along the lower course of the Dnjepr. A study of their ecology might prove very interesting.

Heteroptera are fairly well represented: let us mention as telmatobious (i.e. living in small ponds) *Callicorixa wollastoni* of North and Central Europe, *Sigara limitata* limited to Southern and Central Europe (in Italy present in some northern localities as well as in Campania), *Gerris argentatus*, also limited to South and Central Europe, in Italy present all over the country, *Gerris lacustris*, extremely euryoikous, although restricted only to ponds. Many species of course live in either lakes and ponds, which is easily undertandable when we consider that they have no respiratory problems, breathing atmospheric air. Some species live also in brackish waters.

There is a very great number of chironomids, which have chosen as their habitat the small waters, eutrophic, rich in algae and other organisms which make up their diet; the presence of haemoglobin in their blood has allowed their adaptation to life in such a peculiar environment. Among the *Tanypodinae* (carnivorous animals) we find *Telmatopelopia nemorum* (= *Ablabesmyia n.*) frequent in Holstein and on the Lüneburg heath, *Xenopelopia nigricans, Zavrelimyia nubila* (Central Europe), and also several *Orthocladiinae* such as *Chaetocladius acuminatus* (Fennoscandia), *Limnophyes constrictus* (Central Europe), *L. italicus* (so far known only from Italy), *Orthocladius bidenticulatus* (Alps), *Paralimnophyes arcticus* (Fennoscandia), *Pseudosmittia lacunarum* and *P. longicrus*, of Central Europe, one species of *Trissocladius* limited to Central and Northern Europe as far as Spitzbergen and the Bäreninseln. Finally, *Chironominae* are represented by several *Chironomus* spp., *Lenzia* (*L. connectens*), *Microtendipes distalis, Paratendipes nigrimanus. Tanytarsini* also are sometimes exclusive to small waters.

On the whole, we may say that, rather than stenoikous, adapted species, the pond species are euryoikous, able to withstand conditions not very favourable to life. We cannot exclude the possibility that, at least in some cases and some zoological groups, the lack of competition has permitted flourishing development in this peculiar micro-environments.

Phytotelma are very rare in Europe, or at least have been very little investigated. Exclusive forms seem to be rare, one is the larva of the chironomid *Metriocnemus cavicola*, living in hollow trees, which is considered by some as a subspecies of *M. atratulus*; it is present from Sweden to the Alps and can withstand freezing for several months.

Lithotelma have also been very little investigated: let us remember the biota observed by the present author on some karst rocks (limestone) near Trieste, at the border with Jugoslavia (Opicina or Opčina): Rotifers such as *Philodina erithrophthalma* and *Adineta (= Callidina) vaga*, two euryoikous species; the Cladocerans *Moina rectirostris* and *Alona rectangula*, both euryoikous species, and two Ostracods, *Heterocypris incongruens* and *Darwinula stevensoni*; some Nematodes, numerous colonies of the blue-green alga *Nostoc* and finally a great number of larvae of *Ceratopogonidae* (biting midges), *Dasyhelea* sp. (Fig. 428, 429).

In a lithotelma of the Pyrenees (Ariège) a biocoenosis composed of chironomid larvae (*Ablabesmyia* sp.), *Ceratopogonidae* (*Dasyhelea* sp.) *Chydoridae, Cyclopidae*, the rotifer *Philodina roseola*, and *Nostoc* is also present. Other organisms observed in such environment – though more rare – are larvae of Diptera (*Chironomus, Polypedilum, Theobaldia, Culex*) and Ostracods.

Finally, in a lithotelma of Hungary the only organisms present were larvae of the *Ceratopogonidae Dasyhelea geleiana* (probably endemic), Protozoa (especially ciliates), rotifers (*Philodina*), turbellarians and crustaceans. The similarity between these three regions (Pyrenees, Italian Karst and Hungary) is striking, which allows us to think that only a few genera or groups are able to withstand conditions as severe as those prevailing in the lithotelma.

Karst lakes

A particular category of water bodies must be included among astatic basins, namely the karst lakes and ponds, very common in some calcareous regions of Southern Europe, and more particularly of the North-Western Balkans, where the karst phenomena reach their acme in Europe. We will limit ourselves here to mentioning only a few examples of karst waters present in Slovenia (Northern Jugoslavia) near the Italian border. (Fig. 430, 431).

The temporary lake of Cirknic, known to the Romans as Lugea palus, and even mentioned by Strabo, is a perfectly flat area at c. 550 m, situated at the foot of a mountain (Javornik, 1270 m). It is partly a grassland, partly cultivated (potato fields) but once or twice a year it is covered by water, whose depth oscillates between zero and 5-6 m, when it can be crossed by boats. After a period of some months, the water sinks back into the same cavities from which it originated. During these periods many fish can develop or be introduced from the subterranean water, among them pike and tench and the inhabitants of the region can fish for them. It is possible to catch 40 tons in one day. There are more than 80 cavities through which the water can flow and which contribute to the filling and

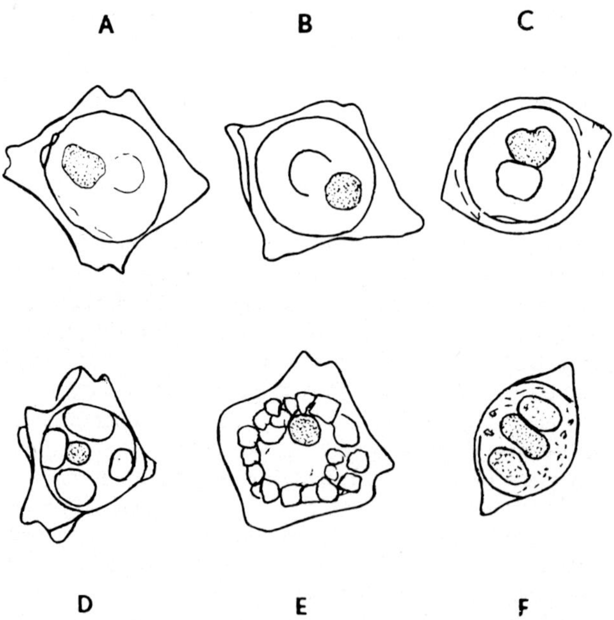

Fig. 427. Autumnal aspects of a phytoflagellate belonging to the genus *Ceratium*, found by the author in a mire (Lagabrun 'Flachmoor'). The forms must be considered as hypnozygotes: in each a granule of haematochrome is clearly visible; cell f must be considered a true wintering form (from Marcuzzi & Lorenzoni, 1968; author's drawing).

successive emptying of the lake, but only two (Mala and Velika Karlouca) are really efficient. When completely filled, this lake reaches a volume of c. 100 million cubic meters and a surface of c. 38 square km. The average depth is 2.75 m, so that the term 'lake' is not correct for this water body, which is actually a pond.

Torquato Tasso described Cirknic lake in his 'Terzo Giorno del Mondo Creato' with verses which could not be more exact from an ecological stand point. 'A la palude Lagia, onde si vanta la nobil Carnia, lunga età vetusta non ha scemato ancor l'onore e 'l grido; quivi si pesca prima, e poi ch'è fatta secca ed asciutta, in lei si sparge il seme e si raccoglie, e tra le verdi piante prende l'abitator gl'incauti

Fig. 428. *Lithotelma* in limestone near Trieste on the Jugoslavian border.

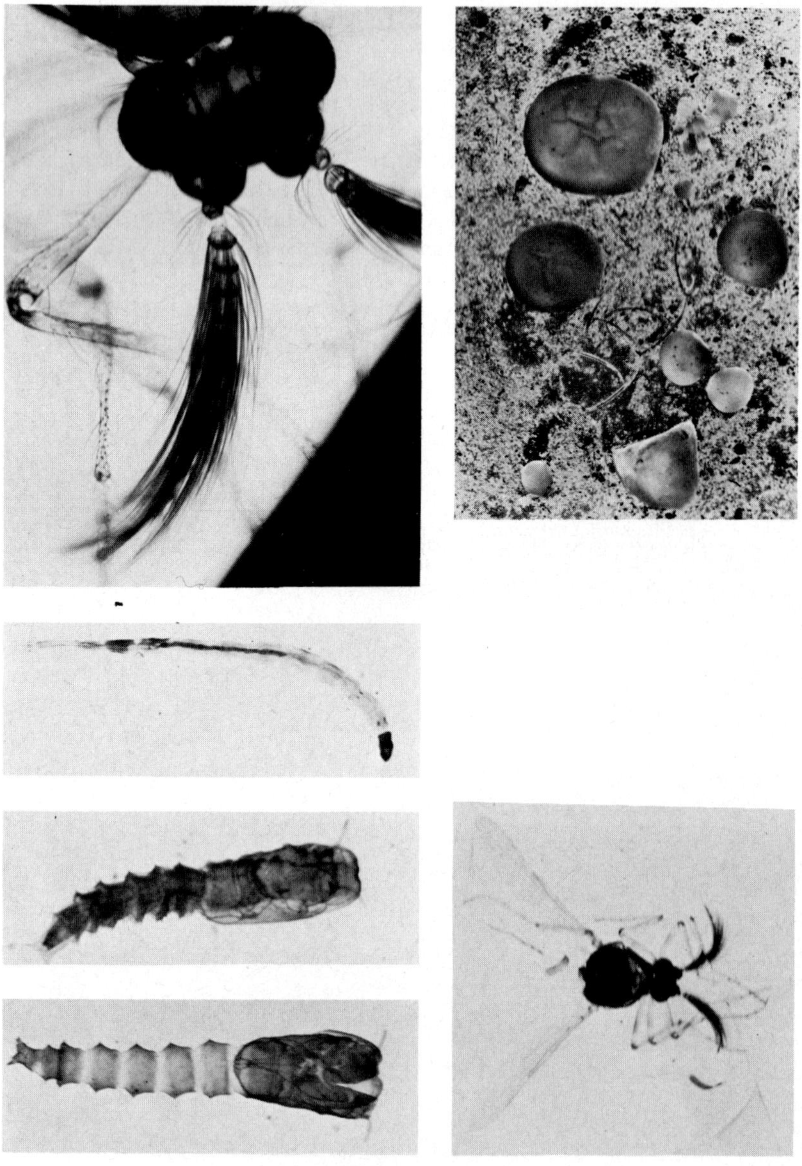

Fig. 429. Organisms living in a limestone lithothelma (rock pool) of the karst region near Trieste, on the border with Jugoslavia. a. *Nostoc* spheres of various sizes (*Cyanophyceae*) and larvae of *Ceratopogonidae diptera*; b. adult of *Dasyhelea* (*Ceratopogonidae*) (male), the abdomen has been asported for determination; c. head of the same, much enlarged; the antennae characteristic of the male are clearly visible; d. larva of of *Dasyhelea*; e. pupa of the same; f. exuvia of the pupa after the emerging of the adult.

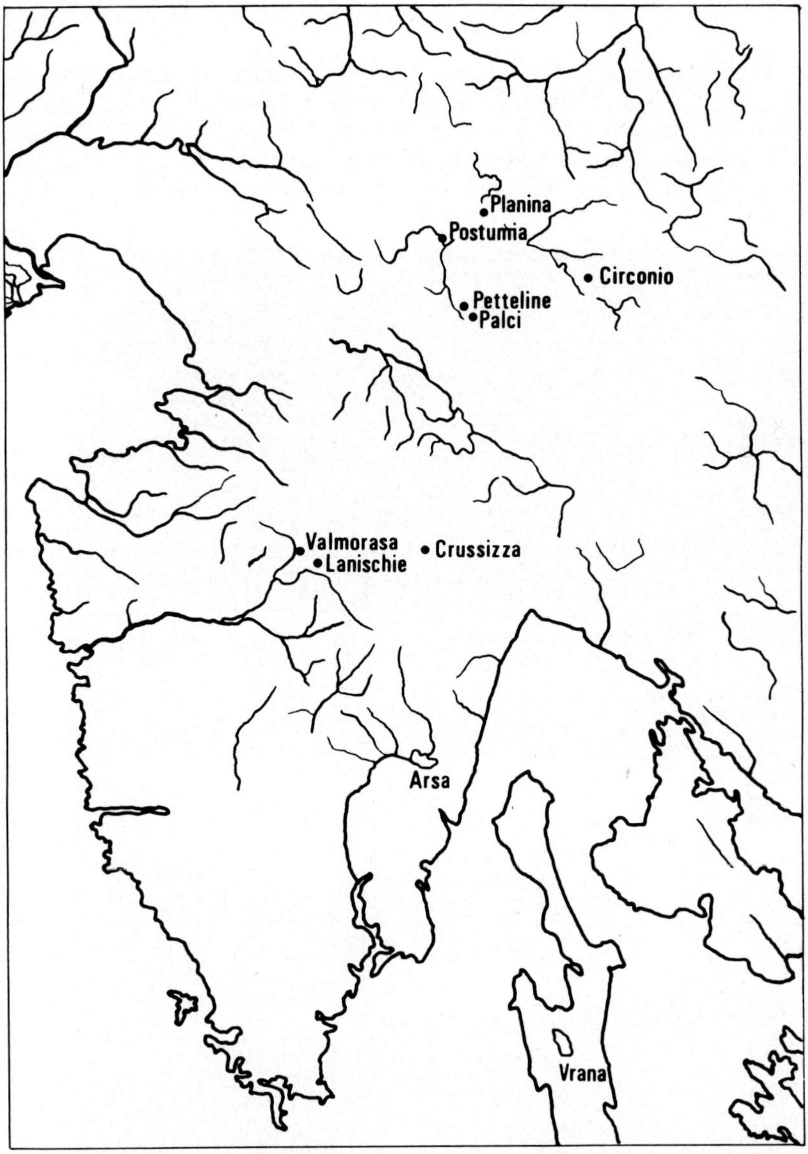

Fig. 430. Location of karstic lakes in Northwestern Slovenija. Note the presence of rivers generally without an ending, which means that they first have a subaerean or superficial course, and then a subterranean or hypogeous one. Several hypogeous animals live here, among which the blind Urodel Amphibian *Proteus anguinus*. The lakes are identified by their Italian, pre-bellic names.

augelli. E 'n tal guisa addivien, che 'n vari tempi l'istessa sia palude, e campo, e selva'*.

It is worth mentioning a few other temporary karst ponds situated in Slovenia, Petteline, NE- St. Petar, a small 'polje' of 0.36 square kilometer, 1.2 km long and 0.5wide. Water generally reaches a depth of 15 m; Palči, or Paličje, E of St. Petar, at ca. 560 m with a surface of 0.36 square km, 12 km long and 0.5 wide. The water comes both from a subterranean karst system and from the impervious

Fig. 431. A temporary water body (or asthatic basin) in the Trieste carso (or 'Karst') near the Jugoslavian border. This water body, thoroughly investigated by the author, has recently disappeared because of lack of consideration for nature conservation in this region.

basin in which the lake is situated. The depth is 10-15 m, but on November 11 1859 it reached 25 m.

Other temporary karst lakes are those of Hrušica, near Obrou in Northern Istria, Valmorasa, NW of Pinguente, in Istria, Lanischie, E-Pinguente, and Postojna (or Adelsberg). The latter forms only when the homonymous cave (one the biggest in Europe) is not sufficient to drain all the water of the Pjuka river (a karst water course). Also the polje of Planina can be sometimes transformed into a lake, whose area reaches 10.5 square km.

'Marcite'

A very peculiar artificial environment, half way between standing and running waters, is that of the so-called 'marcite' (Italian) present in Lombardia (Northern Italy), a kind of permanent irrigated meadow constituting a hydrological system rationally controlled. The water supply is irregular from spring until the end of summer, whereas in autumn and winter it is continuous. The water present in the 'marcita' is generally warmer than the air during winter and cooler during summer. Of peculiar interest are the caddis flies, of which the following species have been recorded: *Odontocerum albicorne, Limnophilus rhombicus, L. flavicornis, L. lunatus, Anabolia lombarda, Micropterna nycterobia, M. fissa* and *Halesus digitatus*. These elements – and probably also other animals – come to the marcite from ditches and springs which bring water to the marcita. One species of caddis fly, *Micropterna fissa*, as a larva can cause some damage to cultivated Graminoid plants.

Rice fields

Rice fields are characterized by shallow water whose depth changes strongly over very short periods of time, by the intensity of light changing according to the height of the rice, and by an extremely variable temperature, in northern Italy going from 9.5° (April-May) to 40° C (June). The oxygen content varies from 4.5 to 6.5 cc/l; the pH varies from 7.7 to 7.0, but in some places (Piemonte) it can also be lower. The degree of eutrophy is very high. The fauna as a whole is allochthonous, representing practically all animal groups. Among the Protozoa *Euglena, Amoeba, Pelomyxa, Difflugia, Arcella, Gromia* and several other *Rhizopoda* and *Ciliata* are found; among the flatworms several planarians are present; among Nematodes the genus *Dorylaimus*, among leeches the species *Piscicola geometra*. The plankton is represented by rotifers (*Monostyla, Rotaria, Scaridium, Cyrtonia, Testudinella*), copepods of the genus *Cyclops* s. lato, *Cladocera* (*Daphnia, Ceriodaphnia, Chydorus, Acroperus, Pleuroxus* etc.) and ostracods (*Cypris*). A very characteristic crustacean is *Apus cancrifomis.* Hydracari are represented by *Hydrachna* and *Hydryphantes*. The commonest insects of rice fields are Diptera

* 'The great age of the Cirknic swamp, a boast for the noble Carnia, has not yet diminished its honour and fame. Here before one fishes, and when it is all dry one sows and then one harvests. And among the green plants the inhabitant catches the incautious birds. So that it happens that in different times it is a swamp, a field, a forest' (free transl. by the author).

larvae (*Stratiomyia, Culex, Anopheles, Tabanus, Tipula*), beetles (*Cybister, Dytiscus, Hydrophilus, Gyrinus*); bugs (*Hydrometra, Notonecta* etc.), nymphs of dragonflies (*Libellula* and *Calopteryx*), ephemerids (*Ephemera*) and caddis flies (*Limnophilus, Phryganea, Triaenodes*). *Planipennia* (*Sialis* larvae) and *Lepidoptera* (*Hydrocampa* larvae) are also present in this peculiar artificial ecosystem. The most frequent snails belong to the genus *Planorbis* s. lato, *Lymnaea* and *Viviparus*. Among rice field vertebrates *Tinca tinca, Lepomis auritus* (sun fish), *Cobitis taenia, Exos lucius, Rana esculenta* and *Triturus* sp. plur. must be mentioned. The rice field is potentially very suitable for fish culture: the species which can thrive best here are *Cyprinus carpio, Tinca tinca, Grystes (= Micropterus) salmoides, Eupomotis, aureus, Ameiurus nebulosus* and *Carassius auratus*.

Brackish waters

When freshwater mixes with sea water, as in the case of littoral ponds (lagoons, limans of the Black Sea, Haffen of the North Sea, 'lòn' of Iceland etc.) and estuaries, we find a brackish water environment, whose salinity varies from levels scarcely higher than that of freshwater, to levels typical of sea water. Sometimes freshwater, which is lighter, stratifies upon salt water, which lies on the bottom and does not mix with the former. In this case, we may find at the bottom a great paucity of oxygen and even putrefactive phenomena, for instance in the Black Sea or in some littoral ponds of Tyrrhenian Italy (Lago di Fondi, Lago di Caprolece). Brackish waters are in general characterized by the variability of salinity and temperature, so that their inhabitants are euryhaline eurythermous.

Some have made a distinction between oligo-haline waters (salinity from 0.5 to 2-3%) and mesohaline waters (salinity from 2-3 to 15-20%). In the former, freshwater animals and plants prevail, which in this environment can develop much better than in freshwater habitats because of lack of competition. Thus, together with some species of marine origin, we find here pre-imaginal forms of insects (*Diptera*, beetles), molluscs (*Anodonta*), fish and amphibians (*Rana, Discoglossus*).

In meso-haline waters, on the other hand, we find species either absent in other kinds of waters, or present only at both limits of salinity, which find here their optimal conditions: these are the autochthonous brackish water elements. We may mention two Polychaeta, *Nereis diversicolor* and *Mercierella enigmatica*, the coelenterate *Cordylophora lacustris*, the copepods *Eurytemora hirundo* and *Temora longicornis* and some fishes belonging to the family *Syngnathidae*. The phytocenosis is characterized by *Ruppia maritima*.

Species typical of brackish waters are in several zoological groups, though of course they vary according to region and, still more to type of water (coastal, internal, lacustrine, sea water etc.).

Among the Nematodes several species are typical of brackish environments, such as *Panagrolaimus salinus* of the British Isles, *Leptolaimus papilliger* of Central Europe, *Aphanolaimus pulcher* of Northern Europe, *Monhystera gerlachi* (Italy) and *Paracyatholaimus intermedius* (Central Europe). A few more species are known from brackish waters of Central and Northern Europe. Of course, it is possible that further investigations will prove that some of these elements are not exclusive to this environment.

One gasteropod, *Smaragdia viridis*, is an extremely interesting, relict of the Tethys Sea which extended, up to the Mesozoic, between Gondwana and Angara. Its presence is limited to the margins of South Western Europe (Spain, France, Italy). *Theodoxus pallasi* is a Caspian- Sarmatian species, relict of former geographical conditions, today present only in the Caspian Sea, the Danube basin and Varna lake (Bulgaria). The genus *Hydrobia* is particularly adapted to life in brackish waters, both in the Black Sea and the Mediterranean Sea, for instance the étang de Berre (*H. acuta*), Western Mediterranean (*H. procerula*, in Cataluna, Pyrenees Orientales, Aude, Herault and Bouche du Rhõne), in North Sea and Baltic Sea (*H. neglecta*, able to withstand a salt content of $10^{0}/_{00}$) or on the Atlantic coasts of Europe (*H. ulvae*, named after the sea lettuce *Ulva lactuca*).

Another genus limited in Europe to brackish water is *Truncatella*, represented by one species limited to the Bidassoa mouth (Basses Pyrenees) and the coast of Provence (*T. juliae*, Fig. 432), another exclusive to Madera and therefore not European proper, a third exclusive to the Adour mouth (Basses Pyrenees) (*T. minuscula*) and finally one species, *T. truncatula*, known from Madera and Southern Spain. These are probably very old species, witnesses of an ancient land bridge from the Iberian Peninsula to the Atlantic islands, well documented by several organisms, for instance earthworms.

Some Gastropods are exclusive to the Caspian Sea (see later), whose marine origin is well documented. They are *Micromelania*, with 7 Caspian endemics and one species, *M. lincta*, present in the Don, Dnjepr and other rivers of Bulgaria, probably an old relict species; genus *Caspia*, with 5 Caspian endemics, *Clathrocaspia*, with 4, *Caspiella* with 3 and finally *Clessiniola* with three species. A few of these elements are present also on the shores of the Black Sea, witness to

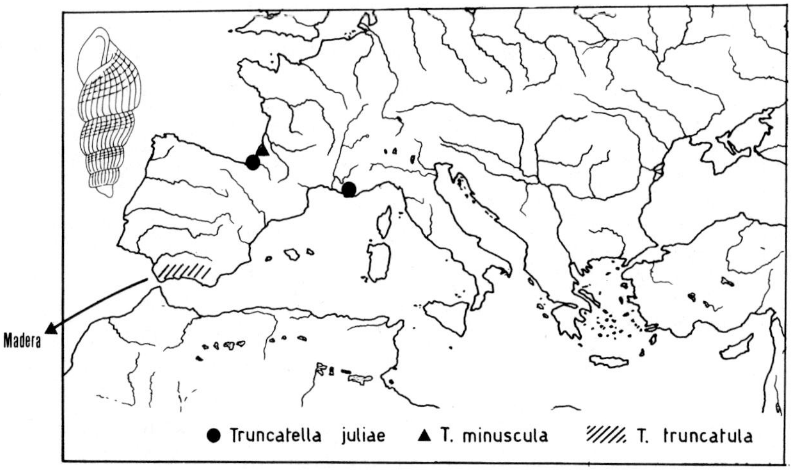

Fig. 432. Distribution of three species of molluscs of the genus *Truncatella* living on muddy shores (figure of *T. juliae* from Germain).

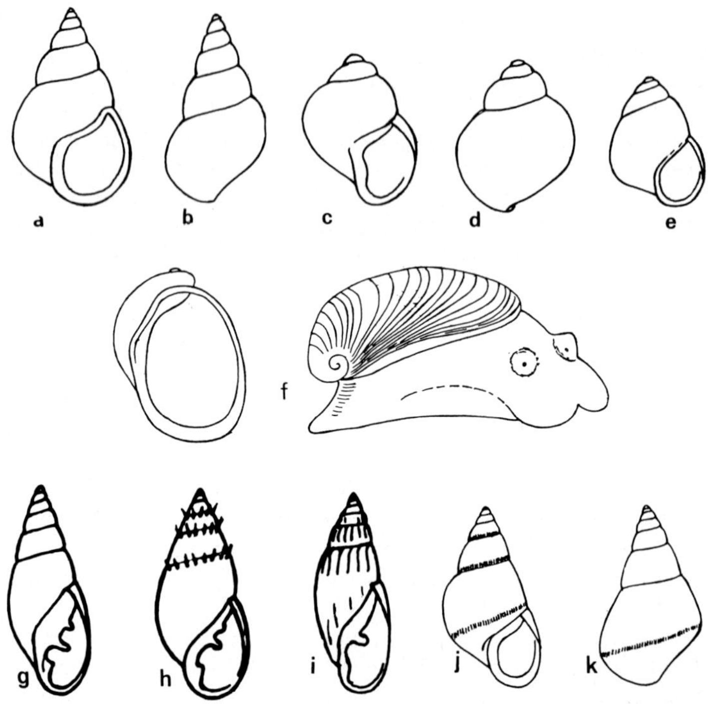

Fig. 433. Examples of snails highly characteristic of European brackish waters: a,b. *Assiminea eliae*; c,d. *A. littorina*; e. *A. ostiorum*; f. *Otina otis*; g,h. *Phytia myosotis*; i.*Leuconia micheli*; j,k. *Assiminea grayi* (from Germain).

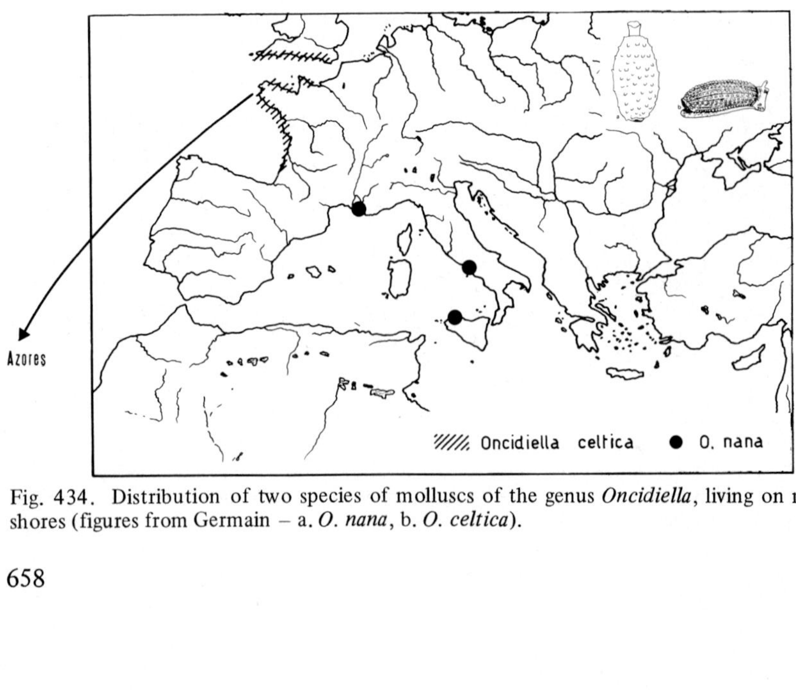

Fig. 434. Distribution of two species of molluscs of the genus *Oncidiella*, living on muddy shores (figures from Germain – a. *O. nana*, b. *O. celtica*).

ancient communications between these seas (most of the Tertiary to the Lower Pliocene). Among the family *Assimineidae* (Fig. 433) there are some very interesting forms, present on the shores of the Mediterranean and its islands, as a witness to old land connections: the Balkans (*Assiminea cardonae*), Atlantic France and Portugal (*A. eliae*), Arcachon (*ostiorum*), Bulgaria and Romania (*rufostrigata*), Sardinia, Elba and Calabria (*Paludinella caroti*), as well as the North Sea (the Channel and the Baltic Sea, where *Assiminea grayana* or *A. grayi* lives).

The family *Nassariidae* is also present as a stenoikous halophilous group with the relict species *Proneritula westerlundi*, present near Tejo in Portugal and in some swamps near Astros (Balkans, Arcadia). The same holds for the family *Onciidae*, represented by *Oncidiella celtica* in the Azores, on the Atlantic French coast and in the British Isles, maybe as a witness to an old land connection (analogous to the case of some slugs) living among the tufts of *Lichina* together with *Otina otis* and *Lasaea rubra*, and *O. nana* of the Western Mediterranean Basin (Palermo, Naples, Marseille). *Oncidiella* lives on the shores of the sea or estuaries, where it may be submerged twice a day (Fig. 434). The family is a typical tropical one, and its present-day distribution in Europe can be considered as the witness to old geological periods. *Ellobiidae*, a family to which some famous hypogeous freshwater forms belong (*Zospeum*), is represented in brackish water by *Ovatella* and *Auriculinella*, the former all around Europe up to the Baltic Sea, the latter exclusive to the Western Mediterranean, from Sicily to Madera.

Pelecypods, or bivalves, are also very important components of brackish water communities, although they are limited to Eastern Europe, as a witness to the enormous extension of sea and brackish lagoons during Tertiary times to Lower Pliocene in this part of the Continent. In fact, *Dreissena andrusowi, D. caspia, D. distincta* and *D. grimmi* are exclusive to the Caspian Sea. *Adacna luciae* is exclusive to the Danube mouth, *A. grimmi* is limited to the Caspian Sea, *A. laeviuscula* is present in several forms (probably ecological) in the 'limans' of the Black Sea from the Danube to the Kuban mouth, and then along the shore of the Caspian Sea, which means that speciation in this genus occurred when the Black and Caspian Seas were interconnected.

Finally, some species are today limited to the Caspian Sea, such as *Monodacna colorata, M. edentula* and *Didacna crassa*. Also the Oligochaeta *Limicola* may be limited to brackish waters: let us mention *Tubifex costatus* and *T. nerctus, Psammoryctes deserticolous, Peloscolex benedeni, P. heterochaetus, Thalassodrilus* (literally 'sea-worm', from the Greek) *prostatus, Phallodrilus parthenopeus, Spiridon insigne* and *S. scrobiculariae*, besides some marine species which in Europe thrive in brackish water, belonging to the genus *Monopylephorus* and *Clitellio*.

An Isopod, *Saera nordmanni*, living in the Azores, Baleares (Mallorca) and the British Isles is perhaps a relict species (cf. some gastropods with an analogous distribution). Several other Isopods live in European brackish waters although they are not exclusive to them.

Amphipods are represented by some Ponto-Caspian *Corophium*, several Ponto-Caspian *Chaetogammarus, Dikerogammarus*, apparently limited to the Danube and the Volga mouths, *Gammarus aquicauda* typical of the Mediterranean coasts, besides several Ponto-Caspian *Gmelina, Niphargoides, Pontogammarus, Stenogammarus* and a few others.

Brackish water bugs are extremely few: *Aepophilus bonnairei*, present, although rare, in South- Central Europe, and *Gerris odontogaster*, very eurytopic although in Italy limited only to the northern part of the peninsula.

Beetles are also very few: among the *Adephaga* we find *Haliplus apicalis,* limited to Central and Northern Europe, a Western Mediterranean element, *Metaporus meridionalis*, limited to Portugal, Corsica, Sardinia and Northern Africa, *Agabus conspersus*, present also in Africa, Asia and the Canary Islands; and finally *Acilius duvergeri*, exclusive to Portugal and North Africa.

Hydrophilidae are also few: *Ochthebius* are particularly well adapted to this peculiar environment, represented by some 8 species, some of which (*O. lejolisi*) are limited to rockpools (see page 561), *Helophorus elongatus*, a common species, *H. fulgidicollis*, more restricted in distribution, *Hydrobius rottenbergi*, and finally *Enochrus politus*.

Chironomids are very interesting and common inhabitants of brackish waters, both littoral, interion and brackish seas such as the Baltic. Among *Tanypodinae* we find *Procladius breviatus*, limited to Central Europe; *Orthocladiinae* are represented by many elements, such as *Cricotopus braunsi*, found only near Amrum (Schleswig Holstein); other species of this genus are present on the sea shore or in interior waters, such as *Cricotopus halobius* and *C. halophilus*, though not exclusive to this environment; *C. stagnorum* lives in Central Europe, *C. vitripennis* is present also in Greenland, *Microcricotopus balticus* is limited to Northern Europe, and finally three species of *Smittia*, described by the present author in the lagoon of Venice, namely *S. malarodai, S. gridellii* and *S. d'anconai*; the former lives together with *S. gridellii* and *Cricotopus vitripennis, S. d'anconai* lives with *Chironomus cingulatus.*

Chironominae are represented by *Chironomus triseta, Fleuria lacustris* and *F. nigra* and by the *Tanytarsini Paratanytarsus confusus* (Northern European).

Some Chironomids are halophilous although not exclusive to brackish waters: in littoral ponds in the Netherlands, for instance, there are at least 16 species (Kruseman, 1933), one of them, *C. plumosus*, present also in the 'limans' of the Black Sea; in the interior waters of Westphalia live at least 16 species (Thienemann, 1915); in the Baltic Sea (a brackish water sea) some 11 species are present; finally in the lagoon of Venice, the present author (1948) found 10 species, belonging to three genera, *Chironomus, Cricotopus* and *Smittia*. Three species of the latter were new to science.

The Caspian Sea

O Volga... my cradle,
none has ever loved you as I do.
At daybreak, alone
while the universe sleeps
and the scarlet light creeps
o'er the dark blue waves,
I went to my native river

N. Nekrasov

Waterless land of sun – a yellow shawl
without roads, fields or villages;
only hilly verges, hot sand
and dry bunches of saksaul

Vs. Rozhdestvenskii

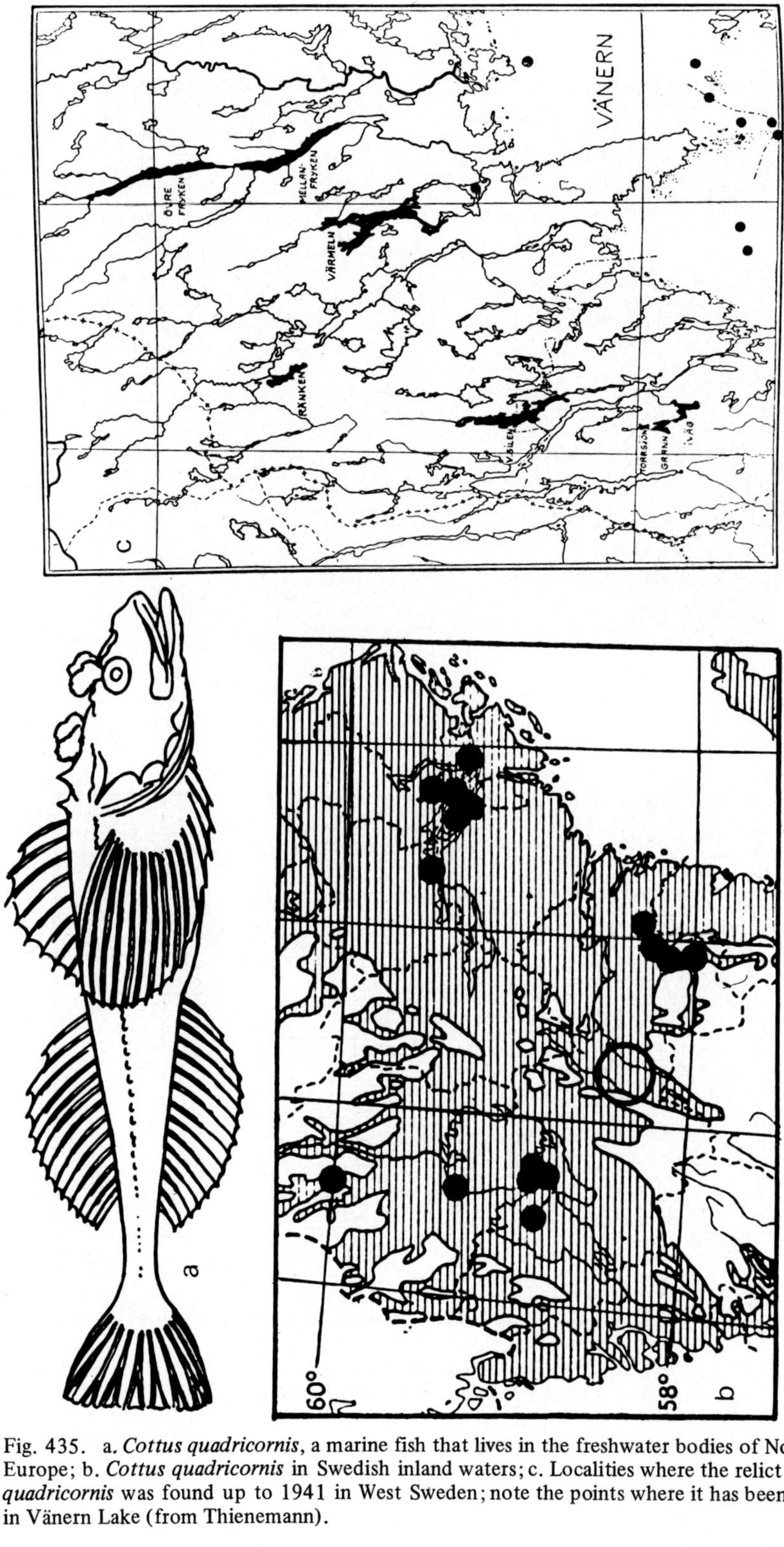

Fig. 435. a. *Cottus quadricornis*, a marine fish that lives in the freshwater bodies of Northern Europe; b. *Cottus quadricornis* in Swedish inland waters; c. Localities where the relict *Cottus quadricornis* was found up to 1941 in West Sweden; note the points where it has been found in Vänern Lake (from Thienemann).

a

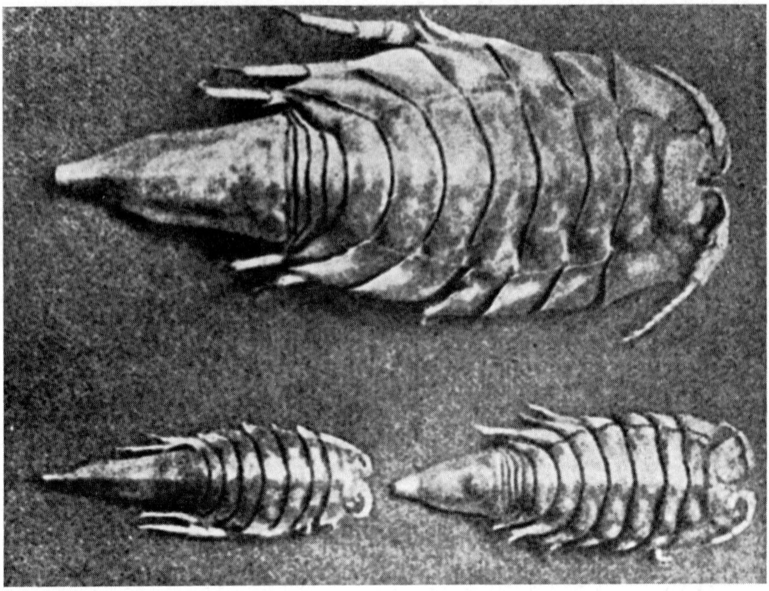

b

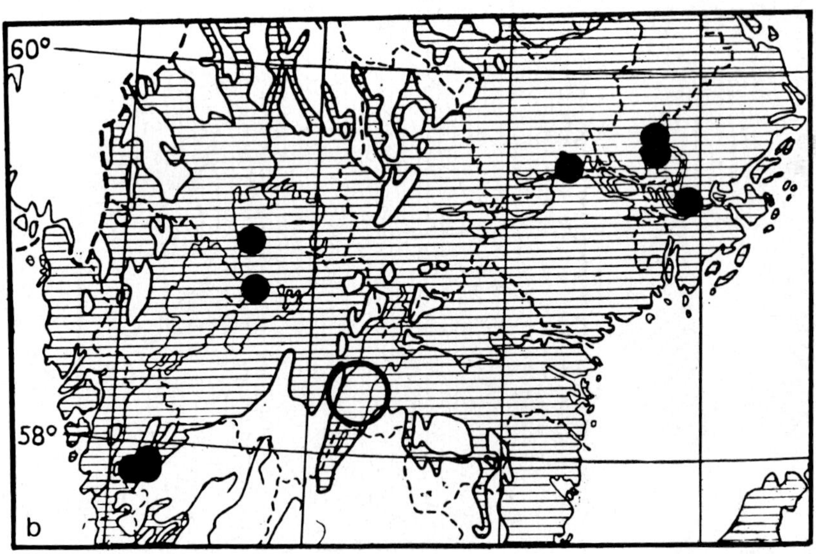

Fig. 436. a. *Mesidothea sibirica* and *M. entomon vetterensis*; b. presence of *Mesidothea entomon* in Swedish inland waters. The large empty circle corresponds to Vättern Lake (from Thienemann).

A peculiar brackish water body is the Caspian Sea, whose Western part (and shores) belongs to Europe. An enormous number of fish are adapted to life in this inland water body: we will just mention *Caspialosa kessleri volgensis, C. caspia caspia, C. saposhnikovi, C. brashnikovi agrachanica, C. brashnikovi brashnikovi, C. sphaerocephala* and *C. caspia salina*. For each species larval development in the laboratory and in nature has been studied by Pertseva-Ostroumova. *Gobiidae* are also very abundant: in the northern part of the Sea alone there are 15 species; The eggs, larvae and larval development of each of them has been studied by Kazanova (Fig. 437).

The relation between the presence of the most important fish (many of which have economic importance) and environmental factors have also been thoroughly investigated, as the following table shows:

Species	temperature (°C)	salinity (o/oo)
Caspialosa b. brashnikovi	19.6-27.6	24 -25
C. caspia salina	17 -24	17.6 -34.9
Clupeonella delicatula caspia	19.8-23.8	20.37-34.39
Atherina mochon pontica	17 -27	17 -54.7
Potamoschistus caucasicus	17 -27	17 -54
Neogobius fluviatilis pallasi	17 -28.5	20 -36.9
Proterhinus marmoratus	18 -25	24 -31.4
Syngnathus nigrolineatus caspius	17 -30.8	23 -54

The following invertebrates have been collected, by a quantitative method, from June to December: among the molluscs *Monodacna, Adacna, Didacna, Cardium, Dreissena*, and several unidentified Gastropods; among the crustaceans *Corophiidae, Gammaridae, Cumacea, Mysidacea* and Ostracods; among Annelida *Oligochaeta, Ampharetidae* and *Nereis* sp. (a polychaete); finally among the insects *Chironomidae*.

The benthos has been studied quantitatively, both biomass and percentage composition as the following table shows:

species	biomass (g/m²)		% composition	
	1935	1937	1935	1937
Dreissena caspia	0.08	–	0,6	
D. polymorpha	0.51	–	4.1	
Monodacna	1.77	4.04	14	61.3
Adacna laeviuscula +				
A. plicata	5.33	0.14	42	2.1
A. minima	0.23	1.67	1.8	25.3
Cardium edule	–	0.14	–	2.1
Mytilaster lineatus	0.2	–	1.5	–
Cumacea	0.13	0.01	1.0	0.1
Corophiidae	0.6	0.02	4.7	0.3
Gammaridae	1.2	0.02	9.5	0.3
Mysidacea	0.27	0.03	2.2	0.5
Chironomidae	0.98	0.09	7.7	1.5
Oligochaeta	1.26	0.39	10	5.9
Polychaeta	0.04	0.02	0.3	0.3
Archaeobdella esmonti	0.08	0.02	0.6	0.3

The following species have a particular ecological (and sometimes also economic) importance in the Northern Caspian Sea: the fish *Abramis brama*, the coelenterate *Cordylophora caspica*, the Gammaridae *Pontogammarus robustoides Niphargoides caspius* and *N. corpulentus, Pandorites* etc. and finally the two crayfish *Astacus leptodactylus* and *A. pachypus* (Fig. 438, 439).

The plankton, both phytoplankton and zooplankton, is also very interesting. The zooplankton is mainly composed of the rotifers *Asplanchna priodonta, Anuraea aculeata, Brachionus pala* and *B. bakeri, Euchlanis dilatata, Ploeosoma truncatum, Rattulus caspicus* etc., the cladocerans *Evadna trigone, E. camptonyx* and *E. hircusi, Cercopages gracillima, Alona rectangula, A. quadrangularis, Moina micrura, M. rectirostris, Chydorus sphaericus, Bosmina coregoni*; the copepods *Calanipeda aquae-dulcis, Heterocope saliens, Eurytemora grimmi* and *E. affinis, Limnocalanus grimaldii, Cyclops strenuus, C. bicuspidatus* and *Nannopus palustris*. Furthermore there are the Mysidacea *Mesomysis intermedia, M. kowalevskii, Paramysis bacuensis* and *Metamysis grimmi*. The phytoplankton is *Coscinodiscus, Compilodiscus, Aphanizomenon, Chroococcus, Botriococcus* and *Spirogyra*.

The Azov Sea

Not a cloud-the sun shines
on the scorched meadows
not an oak in sight-only the bare steppe
stretching endlessly above the sea

Pushkin

It is worth mentioning the great similarity between the biota of the Caspian Sea and that of the Azov Sea, as a witness to the period in which the two great basins, the Caspian Sea (plus Aral Sea) and the Black Sea (including Azov Sea) were united.

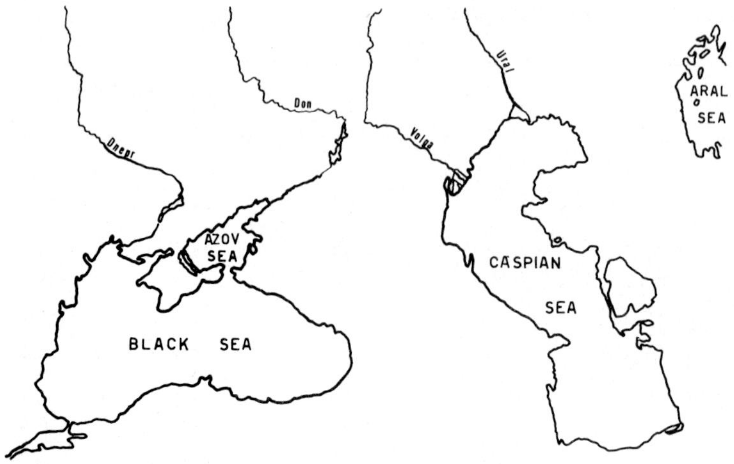

Fig. 437. The position of two brackish water bodies, the Azov Sea and the Caspian Sea.

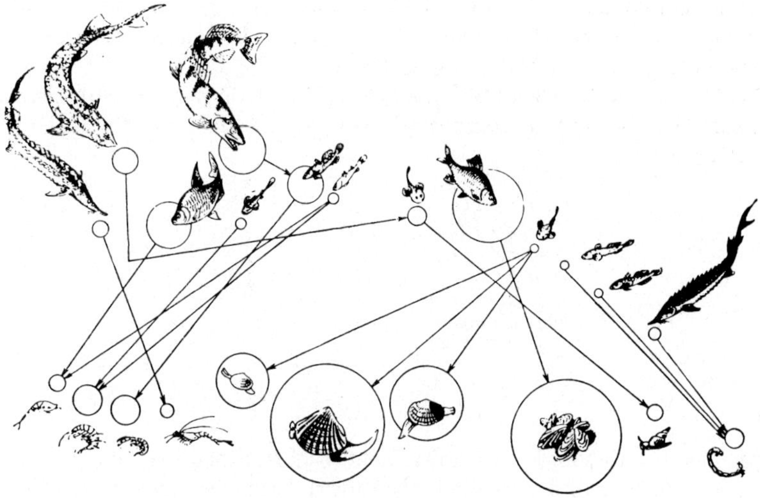

Fig. 438. Main food of fish in the North Caspian (from Shorygin).

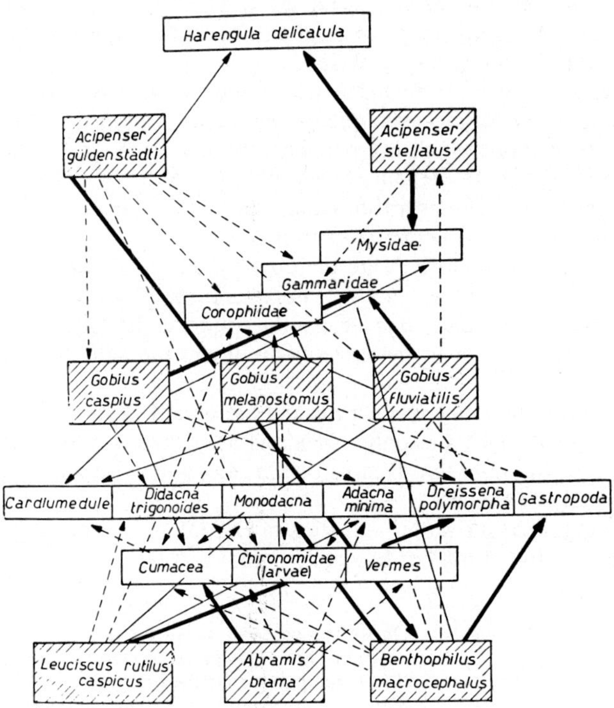

Fig. 439. Trophic relationships of benthophagous fishes in the northern part of the Caspian Sea (from Sheltenkova).

The following groups have been reported (Komarova) for the Azov Sea: copepods, amphipods, *Cumacea, Pterocuma, Gammaridae, Corophiidae, Harpacticoidea, Brachynotus*, Ostracods, *Chironomidae* (including *Tanypus*), *Polychaeta* (*Nereis* and *Hispaniola*) and molluscs (*Cardium, Syndesmya, Mytilaster, Monodaena* and *Hydrobia*). The percentage composition (by weight) of the benthos of this brackish water body is:

Gammaridae	2.7
Chironomidae	18.5
Dreissena	48.0
Adacna	27.0
Cardium	1.2
Phryganidae	2.2

Saltern

A rather peculiar salt water whose salinity is much higher than that, not only of brackish water, but also of sea water, is the saltern (from the Anglo-Saxon 'seal taern'). They are – as are all supersalt or hyperhaline waters – characterized by a lack of stability: the environment changes continuously and the organisms must be sufficiently euryhaline to include in their tolerance range both freshwater (after heavy and prolonged rain, for instance) and sea water. These organisms, very few in number of species, reach enormous numbers of individuals, but so far no study has been done on the productivity of the saltern.

Mammals and reptiles are rare in comparison with piscivorous birds. Fish of supersalt waters are generally euryhaline, estuarine species, apparently more easily adapted to these waters than those which live permanently in the sea. When the salinity reaches 100°/oo all the biota die; but at some 40°/oo, for instance in the Caspian Sea, some fish can live and their introduction can be of some economic importance, since these waters cannot be otherwise utilized.

The commonest organisms of the saltern are the phyllopod *Artemia salina* or brine-shrimp (Fig. 440a), *Ephydridae*-larvae (Fig. 440b) and the phytoflagellate *Dunaliella viridis* (Fig. 440c). *Artemia salina* drinks the water of the saltern and dilutes it by absorbing the salts through the intestinal wall and then eliminating them through the gills.

To finish this small section we want to mention the life extant in salt mines, for instance at Wieliczka in Poland: the only organisms are *Amoeba salina* and some zooflagellates such as *Amphimonas, Pleurostomum*, etc. whose biology and systematics is still completely unknown.

We may say at this point that no water – at least in Europe – is completely sterile. If it is so, it is due not to nature, but to man.

Running waters

Ebrus-the most beautiful of the rivers-
which in Thrace flowest with a strong sound,
along lands famous for their horses-
thou goest down silent to the purpurous sea

near Ainos ...

(Alcaeus)

Generally running waters originate from springs or wells. Only in some alluvial regions a very peculiar type of source exists, the so called 'risorgiva', or resulting water in which the water does not come out directly from the rock, but arises suddenly from underground, after having run for many kilometres sub-surface, originating from rain water which permeates the calcareous rocks of the adjacent mountains and the alluvial deposits. These 'risorgive' are very common at the foot of the South calcareous Alps and Prealps, where one speaks also of 'line of risorgive.*

In other cases the origin of the water course can be melting snow or ice. In the case of normal sources, the temperature of the water corresponds roughly to the

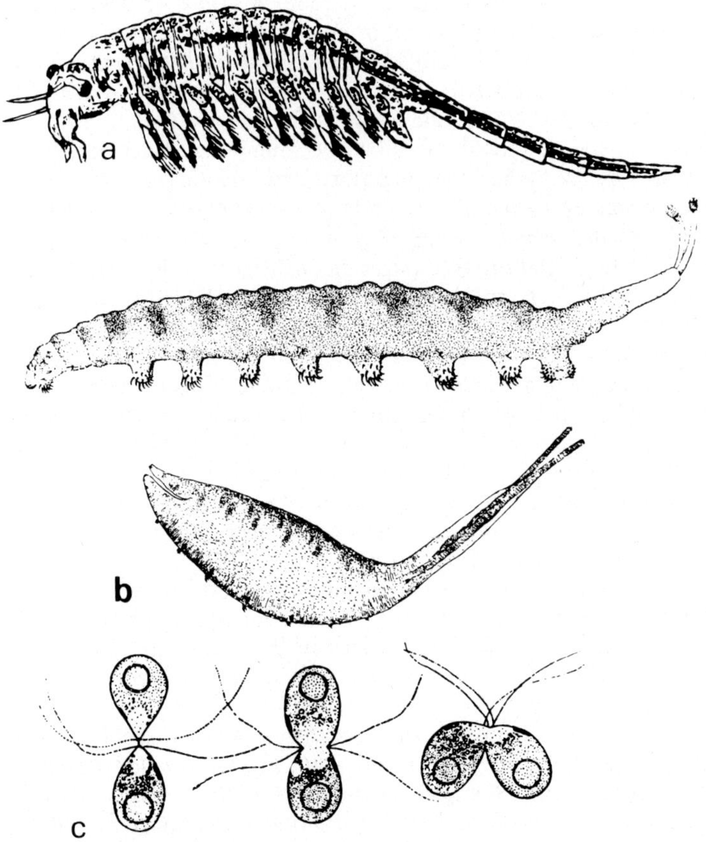

Fig. 440. Typical organisms of salterns: a. *Artemia salina* (from Thienemann); b. larva and puparium of *Ephydra riparia* (from Thienemann); c. the phytoflagellate *Dunaliella viridis* (from Lerche).

* In some places the term resultive (from the Latin resultare, resilire, to spring out, to burst) is used for resulting water.

average annual temperature of the region; in the case of melting snow or ice of course the temperature of the brook is near zero.

Brooks

Generally brooks which take their origin from a spring are very rich in oxygen as a consequence of the low temperature and the turbulence of the water, which causes active mixing of air and water. Brooks generally exhibit a rocky, stoney or gravel bottom, partially covered by a moss vegetation (often composed of *Fontinalis*). Only the less rapid brooks of hilly regions or of lowlands have a vegetation composed of flowering plants, both aquatic and swampy. To this category the 'risorgive' also belong (Fig. 441).

Brooks, regardless of whether they belong to mountains, hills or lowland, are characterized by trout (*Salmo trutta* f. *fario*, or *Salmo fario*), generally accompanied by the millers thumb, *Cottus gobio*, the loach, *Noemacheilus barbatulus*, some gobbid fishes such as *Padogobius panizzai* (in Veneto known as 'gò'), an endemic species of the Southern Alps, and often by the minnow *Phoxinus phoxinus*. Often in brooks other Salmonid fishes can be found: the 'Huchen' *Hucho hucho*, limited to the Danube basin, and a migratory, amphibiotic fish, the salmon *Salmo salar*, which swims up to the highest parts of rivers only for reproduction.

In the lower parts of the brooks – which may overlap the upper portion of the middle course of the river – the brook changes gradually into a river. Width and depth increase, the current decreases, temperature increases (17-18° C) accompanied by a lower oxygen content. The bottom is sandy or gravel-sandy, and some small bends and meanders can form in which the water is almost still and some peculiar animals, not adapted to rapid water can be found. This is the so-called *Thymallus* zone, in which this fish is accompanied by several cyprinids such as *Chondrostoma, Gobio* and *Alburnoides*.

The invertebrates of European brooks which share space, and partly food with fish and often constitute their prey are countless. We will limit ourselves to mentioning only the most important and best known among them, following a systematic criterion instead of a zoogeographical one.

Turbellarians or flat worms are very abundant and sometimes very obvious: *Dugesia (= Planaria), Dendrocoelum* etc. They are very interesting primitive animals, gliding upon the stones and feeding on dead or living animals. Nematodes are not typical of this environment: we might mention *Mesodorylaimus subtiliformis* (Central European mountains).

Molluscs, on the contrary, are very numerous both in number of species and in number of individuals (high density): let us mention *Theodoxus peloponesianus* (Southern Greece) and *T. gighi* (Stampalia Island), *Paladilhia cylindracea* (near Siena), *P. lusitanica* and *P. heussi*, in Portugal, *P. targioniana* (near Florence), *Bythinella armoricana* (near Nantes), *B. dacica* (the Carpathians), *Pseudamnicola reatina*, in the Peschiera River (near Rieti), *Bithynia boissieri* (Greece, Southern Italy and Sicily), *B. hellenica* (Ionian Islands), *B. socialis* and *B. amnicolina* (Sicily), *Emmericia patula* (North Eastern Adriatic Sea, for instance the swamp of Pietra Rossa), *Melanopsis loreana*, in South Europe, *M. luciae*, in Rhodes island and *M. tricarinata*, near Valencia in Spain, and finally two species of a very specially shaped snail genus, *Ancylus hispanicus* and *A. moreleti* (Western Europe).

Very few Oligochaeta Limicola are exclusive of this lotic (from Latin lotor I wash myself) environment: *Rhynchelmis granuensis* (Centro-European mountains), *R. veydovskyi* (idem), *Bythonomus subcarpathicus* (idem).

Crustaceans are not particularly rare; examples are two species of crayfish, *Austropotamobius torrentium* and *A. macedonicus*, the latter limited to the Southern Balkan Peninsula; the Isopod *Saera ortizi*, exclusive to the Western Iberian Peninsula. Much more common are amphipods: *Dikerogammarus spandli* (Thessaloniki), *Echinogammarus klaptotzi* (Ibiza, Baleares), *Homoeogammarus lo-*

Fig. 441. Lotic environments – a brook.

chites and *H. simoni*, endemic to the Western Iberian Peninsula, the latter present also in North Africa, which represents a very common distributional pattern; *Jugogammarus kuščri*, of the Krka River (Central Dalmatia), *Parhomoeogammarus anisocheirus* of Hauts Pyrenees and of the lowland of Central Europe and *P. lusitanicus* (of Portugal), and finally a great number of *Rivulogammarus*.

Insects constitute the bulk of the brook invertebrate fauna. An important role is played by ephemerids, among them *Parameletus chelifer*, with a circumpolar distribution, *Metreletus hessei*, of Central Europe (both in the lowlands and on mountains), a great number of species of the genus *Baetis*, some species of *Procloeon, Pseudocloeon, Centroptilum, Epeorus, Ecdyonurus*, some of them endemics, such as *E. peterseni* of Fennoscandia and the arctic part of the Urals, *E. zebratus* of Corsica, genus *Rhithrogena* (literally 'borne from the brooks') some of them endemics (*R. brennerana* of Brennero, *R. eatoni* of Corsica, *R. fiorii* near Bologna (Northern Appennine) *R. nivata* from the Alps; *Cinygma caucasica*, living in the brooks coming from the Caucasus, *Paraleptophlebia ruffoi*, endemic to Umbria (M. Sibillini) and several other genera and species so far known only from this type of water.

Ephemerids generally feed on microscopic algae, but some, such as *Oligoneuria rhenana* seem to utilize their anterior limbs to filter the plankton which serves as food (cf. Fig. 403).

Another very interesting group in the food chains of our mountain or hilly brooks is the *Plecoptera* or stone-flies. The number of species is too great to be enumerated: it is enough to quote some genera exclusive to brooks, such as *Brachyptera* (some of them endemic to Sardinia, or the Balkan mountains), *Amphinemura, Nemoura*, one of which has a circumpolar distribution (*N. arctica*), several species being endemics (*carpathica, ceciliae*, proper of Spain, *monticola*, exclusive to the Tatras, *peristeri*, of Peristeri mountain, *pirinensis*, of Pirin Mt., *silana*, of Sila Mt., in Calabria, etc.), *Protonemura* and *Leuctra*, also represented by many very strict endemic species, including the most characteristic stone-flies of European brooks, reaching altitudes of 2300 m, *Isogenus, Perlodes, Isoperla* and *Perla*, among them the biggest European species. They feed mainly on algae and plant detritus, but the last mentioned genera are all carnivorous, feeding on insect larvae. A study of the dietary requirements and food preferences of these elegant aquatic creatures would prove very interesting, because stone flies are one of the main ingredients of the natural diet of trouts.

Dragon flies are not so characteristic of brooks, albeit some species seem to be exclusive to them: *Agrion virgo*, three species of *Cordulegaster, Macromia picteti* (with a disjunct distribution) and some others. They are all carnivorous.

An extremely interesting group of insects – though not very well represented in brooks – are the bugs, and more precisely the genera *Hesperocorixa, Micronecta, Gerris* (*G. lateralis*, widely distributed), *Hebrus transversalis*, exclusive to brooks of the Caucasus, a very inconspicuous creature, and some species of *Velia*, closely linked to running water, such as *V. affinis, filippii, peloponensis* and *rhadamantha*. Together with *Gerris* and the beetle genus *Gyrinus* they constitute a very peculiar 'Lebensform', swimming all day long on the surface of water, looking for food which consists of small living animals which the bugs catch and generally bring to the shore to suck.

Among beetles – which are practically omnipresent in all terrestrial, semi-terrestrial and aquatic environments – we find some very interesting forms, highly specialized for this lotic environment, besides others not so well adapted to it; of course the former will be present in the most rapid water, the latter in the meanders and in relatively calm water. Practically every ecological niche of a brook is inhabited by some family or genus of beetles. Let us remember the *Adephaga* (in general carnivorous) *Brychius, Graptodytes* (particularly well represented in Western Europe), *Stictonetes*, typical of brooks of the westernmost part of Iberian Peninsula, *Deronectes*, with several endemic species (*depressicollis* in Southern Spain, *doriai* in the Caucasus, *lareyniei* in Corsica) which indicates a relatively ancient origin for these taxa, *Potamonectes* (also with some endemic species) and a few species of *Agabus* (which in general prefers standing or slowly flowing water).

The *Hydrophilidae* are also well represented: let us mention here *Hydraena*, well known because of its respiratory mechanisms involving the presence of a big bubble of air on the ventral side of the animal which supplies it with oxygen until it must go back to surface to change the old bubble for a new one (cf. Fig. 402). It is very characteristic of brooks all over Europe and it is also represented by many endemic species, which is easily understandable considering the inconspicuous nature of these animals and the difficulty they have in moving from one brook to another, which causes their geographical isolation.

The genera *Ochthebius* and *Limnebius* are also well represented in brooks, although not as well as *Hydraena*. Hydrophilids have a very varied diet, but on the whole the larva is carnivorous, the adult phytophagous.

Besides *Adephaga* and *Hydrophilidae*, several other families of beetles inhabit brooks and exhibit behavioural patterns practically unique among all the insect fauna. The genus *Helodes* has an aquatic larva, living in springs and brooks, whereas the adult lives near the banks. The *Psephenidae* have an aquatic larva, often restricted to springs, and a terrestrial, planticolous adult. Among *Dryopidae*, highly specialized for aquatic life, although not as strictly an aquatic family as the *Dytiscidae* or *Hydrophilidae*, the apomorphous species and genera spend the water at least in the larval stage. *Elminthidae* spend all their life in water. In Central Europe, and probably also in the remaining parts of Europe, each zone of a brook or rivulet has its own species; so that we may think of a speciation linked to ecology – possibly true ecological speciation – different from most other water beetles, which evidently have undergone geographical speciation as have probably most animals. Indeed, among this family the following categories are clearly recognizable: epirhithron, mesorhithron and metarhithron species, besides epipotamon, mesopotamon and metapotamon species. According to Steffan, in general, the most primitive representatives of this family live in the lower course of a river, the most evolved ones in the upper course or even in springs. Very few are the epigeous environments in which so close a relationship between habitat and evolution is visible (Illies, pp. 273-275).

Another small group of insects whose larva thrives in brooks is *Megaloptera*, represented by *Sialis fuliginosa*. Among the *Neuroptera* a couple of species live in this aquatic environment: *Osmylus elegantissimus* (Caucasus) and *O. fulvicephalus*, present all over Southern and Central Europe.

Epirhithron	Mesorhithron	Metarhithron
Dupophilus brevis	*Dupophilus brevis*	*Potamophilus acuminatus*
Elmis coiffaiti	*Elmis coiffaiti*	*Normandia nitens*
E. latreillei	*E. aenea*	*Elmis maugetii*
E. aenea	*E. rietscheli*	*E. obscura*
Esolus angustatus	etc.	etc.

The other two insect groups sharing the 'Lebensraum' with beetles, *Ephemeridae* and *Plectopera*, living mainly on stones (above and beneath them) are *Diptera* and *Trichoptera*, both represented only by the larval stage and both very important in the feeding of trout and other fish which love rapidly running waters.

The number of these types is enormous, so that we may even try to enumerate them. Perhaps an indication of the main families present in this lotic environment will be sufficient for the aims of this book. Trichoptera or caddis flies are mainly represented by *Rhyacophilidae* (from ῥύαξ a brook) *Glossosomatidae*, *Philopotamidae, Psychomyidae, Brachycentridae, Limnephilidae, Drusus, Cryptothrix nebulicola, Potamophylax, Stenophylax, Micropterna, Chaetopteryx, Silo* etc.), *Sericostomatidae*, etc. Many are endemic. Their feeding habits are mainly planktophagous, the larva constructing nets for catching plankters (Fig. 403e) or being furnished with some apparatus for catching plankton. In the case of *Wormaldia* the peculiarly built maxilla is utilized as a net for the collection of small algae (Diatoms and unicellular algae).

Diptera larvae belong mainly to two Nematocerous families, *Simuliidae* (or biting gnats) and *Chironomidae* (non-biting midges). Of course several other families are also present, such as *Blepharoceridae, Ceratopogonidae* (or biting midges) etc. *Simuliidae* are mainly brook-loving animals, so that it is impossible to enumerate them. The larva and the pupa exhibit a semi-sessile habit, adhering the hind part of the abdomen to a stone, or a leaf (e.g. a leaf of *Menyanthes trifoliata*), or even a tree stump. They are never present on sandy or muddy bottoms. Their food is composed of detritus which the larva filters by means of its enormous fan-like antennae, which are a very peculiar feature and can be considered as a true 'life form' (filter-feeding animals). Some species may prove to be indicators of pollution, although not very much work has been done in this field.

Chironomidae (= Tendipedidae) are represented in the majority by some subfamilies such as *Orthocladiinae* and the tribus *Tanytarsini* of the subfam. *Chironominae*. All of them, of course, do not possess haemoglobin. There are many endemic species and an evolutionary study would prove to be extremely interesting. The feeding habits of this family are mainly algivorous, although one subfamily (*Tanipodinae*) includes only predatory species.

Rivers

Another section of the river is represented by what is generally known as the lower part of the middle course. Here the bottom is sandy and at some points also muddy, with rich macrophytic vegetation. Some gravel areas or banks can appear in the course of the river, where some arboreal or shrubby species can also thrive (poplars, willows, alders). The shores are often covered with small thickets composed of these species (the well known 'Auen' of German-speaking countries). Oxygen is still present in good quantities; the temperature is a little higher (more or less 19° C) and turbulence may be present here and there. This is the so-called *Barbus*-zone, since in Central Europe the most characteristic fish is *Barbus barbus,* in other parts of the Continent replaced by vicarious species such as Barbus *meridionalis* (Fig. 442). Some fish live on the bottom, hunting insects, snails, bivalves, worms (Oligochaeta Limicola and leeches) and crustaceans (Isopods, Amphipods, crayfishes): the bream *Abramis brama*, the white bream *Blicca björkna*, the perch *Perca fluviatilis*; restricted to the Danube basin, we may also mention *Acerina schraetzer* ('Schratz'), *Aspro zingel* ('Zingel') and *Aspro streber* ('Streber'). Many predatory fishes are also present such as *Aspius aspius* (from the Elba to the Urals), *Leuciscus, Esox lucius* and *Lucioperca lucioperca*.

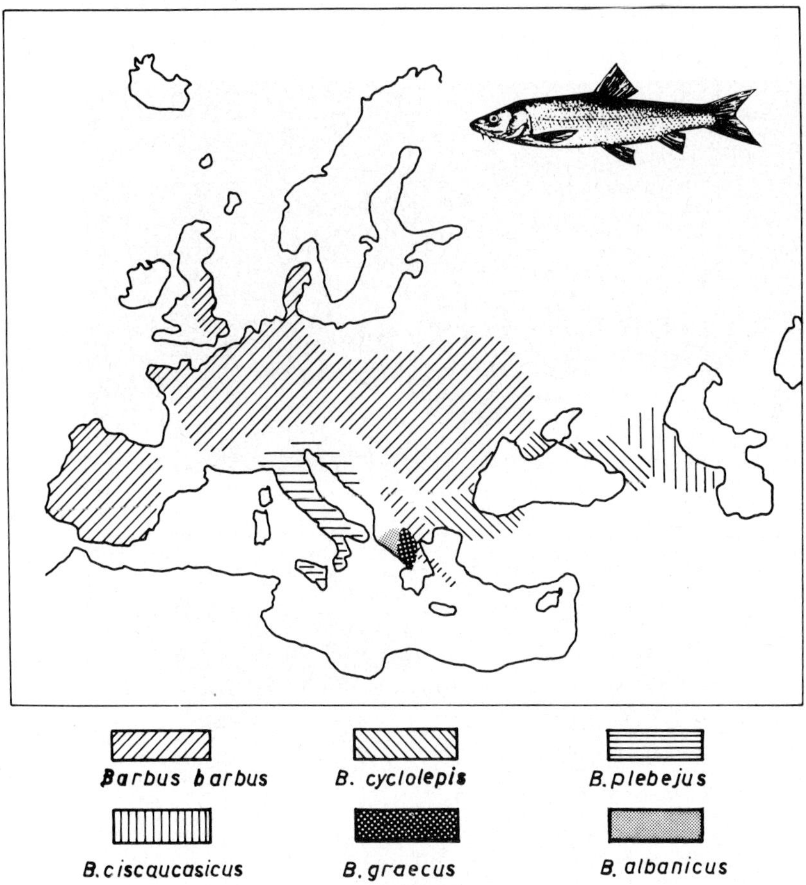

Fig. 442. Example of an 'Artenkreis' in which geographic speciation probably occurred after segregation of each population (inset: *Barbus graecus*). (Figure of fish from Ladiges & Vogt, with the permission of Paul Parey).

Finally we reach the lower course-wide and slow and sometimes murky as a consequence of the silt and clay in suspension in the water. Oxygen is relatively scarce, the temperature shows some differences between bottom and surface, but on the whole is rather high, the bottom is frequently muddy, without higher plants in the middle, whereas it is covered by many flowering plants and *Cryptogamae* (mainly *Characeae*) near the shores. The latter do not greatly differ from those in a swamp or a pond of the nearby lowland (*Caricetum, Phragmitetum*, etc.). Sometimes in spring a beautiful flowering of *Ranunculus aquatilis* and *R. trichophyllus* is visible all along the shores, mixed with some *Potamogeton*, such as *P. fluitans, P. crispus* etc.

Fish abound, thanks to the conditions of general eutrophy of the environment: let us mention *Abramis brama, Blicca björkna, Esox lucius, Lucioperca lucioperca, Tinca tinca, Cyprinus carpio*, several species of *Leuciscus* (for instance *L. cephalus cabeda* or 'cavedano'), the 'Wels' *Silurus glanis* (in all Central and North

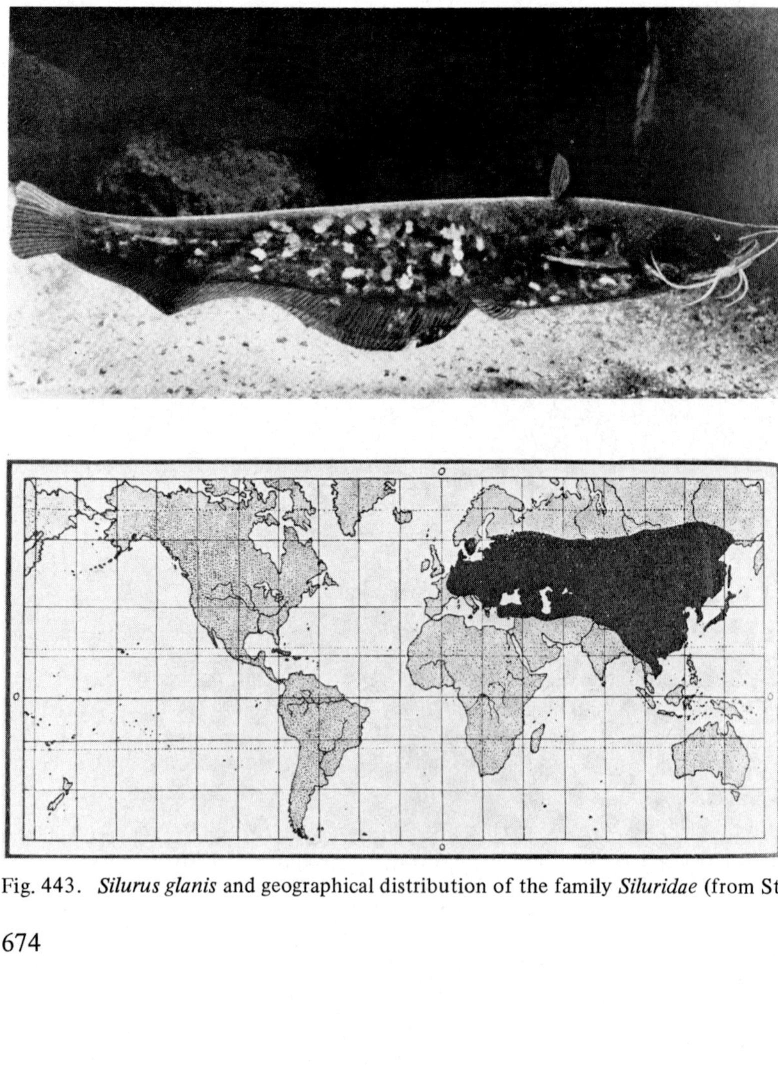

Fig. 443. *Silurus glanis* and geographical distribution of the family *Siluridae* (from Sterba).

Eastern Europe) which can reach a length of 3 metres, and finally the eel *Anguilla anguilla* (Fig. 443).

Several migratory marine fsihes are present in this part of the river: *Acipenser* or the sturgeon, *Alosa* (allice shad, twaite shad), *Caspialosa* (in Russian 'Sel'd'), *Osmerus eperlanus* or smelt, and some species of the genus *Coregonus (C. oxyrhynchus* or 'Schnepel', typical of the Elba and the Rhine).

At the mouth of the river life and physical conditions change: the water is practically still; it is mixed, sometimes twice a day (in relation to high water), with sea water; there is a remarkable amount of suspended organic material, besides clay and silt, so that life is possible only for a limited number of highly euryhaline species.

The fish are *Acerina cernua* (the 'pope' from pope futher, or 'ruffe', probably from rough) the stickle back *Gasterosteus aculeatus*, the flounder *Platychthys flesus* and at the mouth of the Danube also some Gobiid fishes. The invertebrates – mostly saprophagous – are sometimes very abundant, although limited to a small number of species of *Oligochaeta Limicola* (e.g. *Tubifex*), insect larvae (Chironomids), molluscs, crustaceans (Amphipods, Isopods) and the bivalve *Dreissenia polymorpha*. Near the banks of some rivers an enormous quantity of plankton develops, some of which can also reach the centre of the river.

Although necessarily somewhat conventional, as are all classifications, Illies' classification of the different zones of a river from its source to the mouth can be of some use:

Rithron (from ῥειθρον a brook)	Epirithron	Upper course of the Salmonid region, or upper trout region
	Metarithron	Central course of the Salmonid zone, or central trout region
	Iporithron	Lower course of the Salmonid zone, *Thymallus*-zone
Potamon from ποταμός a river	Epipotamon	Upper course of the Cyprinid region, *Barbus*-zone
	Metapotamon	Lower course of the Cyprinid region, *Abramis brama*-zone
	Ipopotamon	Region of *Acerina cernua* and *Platichthys flesus*

Epirithron and Metarithron correspond more or less to the Upper course of a river; Iporithron to Upper-middle course; Epipotamon to Lower-middle course; Metapotamon finally corresponds to the lower course of the river. Ipopotamon is practically the estuarine zone.

Together with fish, in some rivers cetaceans may be present at least in some periods of the year. One of them is the 'beluga' (from the Russian bel, white) *Delphinapterus leucas*, living in arctic and sub-arctic seas, circumpolar, in shallow waters, often ascending some several hundred miles up rivers. Limited to Europe, it is found in the Northern Dvina, Mezen and Pečora; at the mouths of these rivers it is intensively fished. It has also been found in the estuary of the Loire. It feeds on Arctic cod, herrings, smelt and Pleuronectids. The narwhal, *Monodon monoceros*, also sometimes ascends rivers (Fig. 444).

The invertebrates of the true rivers often belong to the same groups we have found in brooks, but generally to different genera or at least species. Of course some taxa are present in both types of lotic environments, as well as in springs and other types of water.

Planarians or flat worms are represented by several species: exclusive to rivers seem to be *Macrostomum mosquense* (Moskva river), *Otoplana antipa* (Danube near Lom, in Bulgaria), *Gieysztoria capreola* (in the British Isles). However, most species live also in the upper course of rivers and in brooks. Nematodes are much more abundant than in brooks: some genera worth mentioning are *Cephalobus, Caenorhabditis, Metadiplogaster, Nothotylenchus (N. danubialis)* and *Aphelenchoides (A. fluviatilis)* both in the Danube, *Tobrilus vistula* (in the Vistula), *Mononchus tenuicaudatus* (in the Rhône) etc.

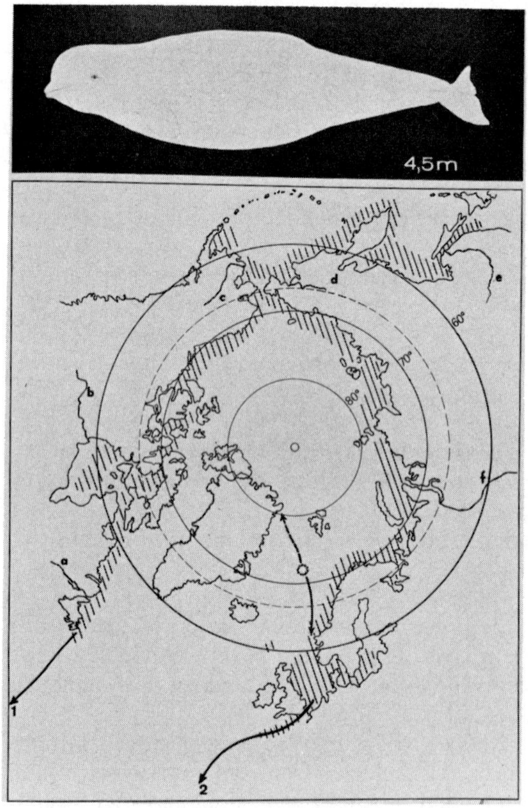

Fig. 444. Distribution of *Delphinapterus leucas*: a. St. Lawrence, b. Churchill, c. Yukon; d. Anadir; e. Amur; f. Ob. ····· summer migrations; —— winter migrations; → (N.1) prolonged route extending southwards to Massachusetts; → (N.2) prolonged route extending southwards to the Bay of Biscay (from Marcuzzi & Pilleri).

The molluscs, of course, are much more characteristic and vary according the region and the river zone or river conditions. Among Gastropods or snails we should mention here the genus *Theodoxus*, some species of *Viviparus* (such as *V. amblyus* living near Braila in the Balkan Peninsula, Romania). The southern genus *Tropidina* is represented in Sicily by two species, *anapensis* (i.e. of the Anapo River, East-Sicily) and *monterosatoi*; some *Pseudamnicola* such as *P. penchinati* live in the lower Danube; the interesting *Melanopsis cariosa* (= *costellata*, represented by several forms) lives in the Guadalquivir, etc. Bilvalves, however, are represented by some conspicuous species such as *Margaritifera auricularia* and *M. margaritifera*, *Potomida littoralis* (characterized by a very discontinuous distribution), *Unio baeticus* exclusive to Andalusia, and by some small species such as *Sphaerium solidum*, extending from Northern France to the Ural River.

Oligochaeta exclusive to rivers are extremely rare: we could remember one species of *Pelodrilus, P. cuenoti*, found so far only in the Moselle. Of course, here again, we have several euryoikous species, occasionally very resistant to pollution.

Insects abound: the Ephemerids represent a conspicuous part of the fluviatile entomofauna, much more so than in brooks. Many of them are endemic: *Siphlonurus croaticus* to Plitvice, *S. vernalis* to the Oka river, *Baetis pavidus* to Italy (Bologna and Calabria) *B. subatrebatinus* to Emilia Romagna, *Cloeon languidum* to the same region (Cesena, not far from the sea) *Procloeon concinnum* to Portugal, *Centroptilum pulchrum*, characteristic of the Garonne, *Heptagenia trimaculata* of the Vardar River, *Caenis pygmaea* endemic to some rivers of Sardinia (of doubtful systematics) and finally *Caenis valentinae* and *Caenodes felsinae* of the environs of Bologna (the former along the river Reno). Of course not all European rivers are sufficiently and uniformly investigated from this point of view.

Stone-flies are also very abundant – though less so than Ephemerids – because of their great rheophily (which practically means being stenoxybiont and stenothermous, which makes this animal group a typical constituent of the fauna of brooks). Of course for this group, as for fish, a distinction based on the different parts of rhitron and potamon would theoretically be possible, and research in this aspect highly desirable.

We can mention some species of the genera *Brachyptera, Rhabdiopteryx, Taeniopteryx*, etc. One of the most euryoikous species seems to be *Capnia bifrons* present all over Europe to the Ural River; some species, on the contrary, are limited to Central and Northern Europe (*Isogenus nubecula*, extending up to Sibiria, *Perlodes dispar, Isoperla difformis, I. obscura*, also up to Sibiria, *Isoptena serricornis*).

Dragon flies are also more abundant in rivers than in brooks: examples are *Gomphus flavipes* (a widely distributed species) *G. graslini, G. simillimus, Onychogomphus flexuosus*, limited to the rivers of the Caucacus and *O. uncatus*, of Western and Central Europe.

Bugs are few: *Micronecta distans* of Western and Central Europe, very rare, *Gerris najas*, a very euryoikous and eurytope species.

Beetles do not abound in rivers, which means that the evolutionary mechanisms in these animals have been directed to making life possible in brooks but not in rivers: an explanation of this phenomenon would be very difficult.

Practically the only beetle found so far only in rivers is the *Dytiscidae Grapto-*

dytes pseudopictus, of the Caspian Region; some species, however, are present both in rivers and in brooks, or also in rivers and lakes (very euryoikous, not very specialized, elements).

Among *Megaloptera* we should mention *Sialis nigripes*, limited to Southern and Central Europe.

Caddis flies, of course, are very abundant: they can be often easily distinguished from those living in the upper course of rivers on in brooks because of the materials utilized in the construction of the case, which can be of organic nature (small plant particles, small shells of snails and bivalves, etc.), whereas in brooks the case is almost exclusively inorganic. Exclusive fluviatile species are *Hydropsyche bulbifera, H. jordanensis* (Bulgaria and Banate), *H. nevae* (Centro-North-European), *H. subguttata*, the almost ubiquitous species *Cheumatopsyche lepida*, present also in Africa and Asia, *Athripsodes leucophaeus* (Southern and Central Europe), *Triaenodes kawraiskii* extending eastwards to Afghanistan, and a few others. Here again we must admit the development of evolutionary mechanisms for life in brooks but not in the rivers, where less differentiated elements thrive. All these insect groups have a paramount importance in the feeding of many fish, and their economic significance must always be kept in mind when studying the effects of pollution on the life of a river.

Diptera larvae are not so abundant: among *Simuliidae* we find some species within the genera *Cnephia, Eusimulium, Schönbaueria, Wilhelmia, Gnus* (exclusive to Northern Central Europe). *Simulium* (some species limited to the Caucasus, such as *S. tarnogrodskii* or *S. kurense*), some limited to Fennoscandia, such as *S. reptans* etc. Among chironomids, the subfamily *Chironominae* tribus *Chironomini*, i.e. forms with haemoglobin appear: let us mention *Cryptochironomus burganadzcae, C. forficula, C. hirticeps, C. imberbipes, Parachironomus fridmanae* (the Danube basin), and finally a couple of *Tanytarsini*. The ratio *Chironomus/Tanytarsus* or still more *Orthocladiinae/Chironominae* seems to be very important in the distinction of brook from river chironomid fauna.

The springs

> An der sonngewohnten Strasse, in dem
> hohlen halben Baumstamm der seit lange
> Trog ward, eine oberfläche Wasser
> in sich leis erneuernd, still' ich meinen
> Durst: des Wassers Heiterkeit und Herkunft
> in mich nehmend durch die Handgelenke
>
> R.M. Rilke
>
> (On the sunny road-in a hollow, old, trunk,
> which since long is used as a source and slowly
> renews in itself a water mirror-
> I calm my thirst: the limpid water and its flow
> I take in the hollow of my hand...)

Springs or wells deserve a special treatment due to their uniqueness in nature and to the differences in the rivers which emerge from them. According to whether the water comes directly from underground and gives rise to a small brook, or it makes a small pond or lastly it stagnates to form a sort of swamp, from which the brook takes its origin, one speaks of rheocrenous, limnocrenous and heleocrenous sources. In the first category the bottom is rocky, stony or gravelly, very rarely sandy, poor in vegetation, which – when present – is mainly composed of mosses. In the second category the water flows from the bottom of a small basin; the bottom is muddy or sandy and often very rich in higher plants. Finally, in the case of a heleocrenous source, the water filters from the soil, so creating a swamp. Of course all transitions from one type to the other exist.

The fauna is very varied according to the height at which the spring is situated, the geological nature of the substrate and still more the geography. As a general rule, it seems that, in Europe, the true 'crenobious' animals, i.e. those living exclusively in springs, are missing in the parts of the Continent covered by ice during the Pleistocene (e.g. *Bythinella, Hydracari* etc.).

Among spring animals we find many zoological groups. Turbellarians or flatworms are represented by *Microdalyellia minima* (near Ohrid lake), *Castrada ochridensis* (idem), *Castrada papii lapponica*, a subspecies restricted to Kittilä (Fennoscandia), *Typhloplanella sexmaculata* (near Leningrad); Nematodes are also present: let us mention *Criconemoides hygrophilum*, of the Danube basin, *Tobrilus lomnickii*, of the Central European lowlands, *Prionchulus thiocrenobius* (idem). Gastropods ary much more adapted to life in springs: in Spain (Estremadura) a *Hydrobia (H. rivularis)* lives; in Northern Jugoslavia a *Valvata (V. schmidti)* has been found near Töplitz (Krajn); in the same region a *Paladilhia (P. robiciana)* lives together with an enormous number of hypogeous endemic species.

A very great number of snails belong to the genera *Paladilhia* (subgen. *Belgrandia, Belgrandiella* and *Moitessieria), Bythinella* (several subgenera scattered all through Southern and secondarily Central Europe, probably eliminated by the glacial period in the North, most of them endemic species, which is consistent with the hypothesis of the persistence of old, preglacial, species in territories never covered by Pleistocenic ice), *Pseudamnicola* etc. A genus with a very interesting geographical distribution is *Horatia*, limited to the Balkans, extending as far NW as the Natisone River, on the border between Jugoslavia and Italy; several species live exclusively in springs (i.e. they are crenobious): one is present in Tempe Valley (Thessalia, *H. exigua*), one near Dubrovnik (*H. klecaciana*), another in Dalmatia (*praeclara*) and one in Hercegovina and Črna Gora (*jadertina*). Possibly the disjunct character of surface hydrography of the Western Balkans, mostly of a Karst character, has caused and favoured the speciation in these interesting snail genera. Probably hypogeous freshwater species (as well as terrestrial ones) have a more continuous area of distribution and have not undergone geographical speciation as have the epigeous ones.

Oligochaeta are represented by *Peloscolex zavreli* (in Central Europe) and *Epirodrilus michaelseni* (near Janina in Greece). Isopods are also present in wells: *Asellus infirmus* in the Caucasus, *Saera schellenbergi* in Istria, Amphipods typical of springs belong to the genus *Sinurella: S. dershavini* near Saratov, *S. donensis* in Caucasus, *S. meschtcherica* near Mestchera.

Beetles are rather abundant: let us remember *Hydroporus alpestris* of Portugal, *H. dobrogeanus* of Dobrogea (strictly endemic) *H. jakobsoni* of the Caucasus and *H. longulus*, a very euryoikous species. A *Hydraena, H. vedrasi*, is known so far only from a spring in Greece (Euboea). Extremely interesting is a weevil, *Bagous perparvulus*, living in several parts of Southern and Central Europe to the Caucasus; it lives in water in its larval stage and maybe also as an adult.

Of course most spring insects belong to the *Trichoptera*, and more precisely to the genera *Microptila, Plectrocnemia* but more particularly *Apatania* (including *Apatidea*), represented by several endemic species in many parts of Europe. *Drusus mixtus* is known only from a spring in Savoy. Other genera of caddis flies are also represented.

Diptera larvae belong, as in brooks, to the families *Simuliidae* and *Chironomidae* (non-biting midges). Among *Simuliidae* we find many species of *Eusimulium*, all strictly endemic to a small region; the genus *Boophthora* seems to be restricted in Europe exclusively to springs; several crenobious *Simuliidae* belong to the *Odagmia, Tetisimulium* and many more to *Simulium*.

Chironomidae share with *Simuliidae* the 'Lebensraum' of springs: the subfamilies *Podonominae* and *Orthocladiinae* are the only representatives of this important family. Particularly well represented is the genus *Eukiefferiella*.

Thermo-mineral springs are fairly frequent in most European countries, from Iceland to Italy, but their biota are very little known. Often the chemico-physical characteristics are well investigated because of their utilization in therapy (hydrology, hydrotherapy, etc.): thus for instance very accurate analysis exist for the thermal springs of Cauteret (Hautes Pyrenees), for the Colli Euganei (near Padova: Abano, Montegrotto, Monteortone, etc.), for the waters of Valdieri (Piemonte). The water of the geysers, much utilized for heating (houses, glasshouses etc.) has also been the subject of research.

In the springs of Valdieri the water belongs to the sulphate type: the salts are the following (μg/l): sodium chloride, 40; sodium sulphate 87.4; sodium silicate 33; potassium silicate 41.9; calcium carbonate 9, magnesium carbonate 0.8; Fe and Mn oxides, 1.3; aluminium oxide 2.0; phosphoric acid 0.8; silicic acid 25.3; H_2S 1.4. The numerous sources come from the layers of gneis and quartz-micaceous schists.

The biota are extremely interesting: first of all the number of plant species increase regularly with the decrease in temperature (Fig. 445a). A true succession is evident for several plants, going from 55° to 30° C as it is shown by Fig. 445b. *Phormidium valderianum* is a typical thermophilous species.

The most thermophilous group is represented by blue-green algae or *Cyanophyceae.* These elements belong to Del Villar's Hydrothermophyta.

In Centro-European thermal springs there are many animals including 40 species of Nematodes, but only two of the latter live at temperatures higher than 40° C (*Dorylaimus carteri* and *Chromadorita leuckarti*). Only 8 species live at a temperature higher than 45° C: 5 Protozoa, 2 rotifers and a beetle. One Protozoon, *Cothurnia* sp. thrives in the thermal springs of Central Europe at a temperature of 63° C.

In the thermal springs near Abano (Padova) at 46° C the following animals have been found: the snail *Hydrobia aponensis* (so named after Abano = Aponum) the Nematode *Dorylaimus atratus*, the Ostracod *Cytheridea (= Cyprideis) torosa*, the beetle *Bidessus thermalis* and the larva of a *Stratiomyia*.

Many kinds of mineral springs exist in Europe: 'acidulae', gypsum, alkaline,

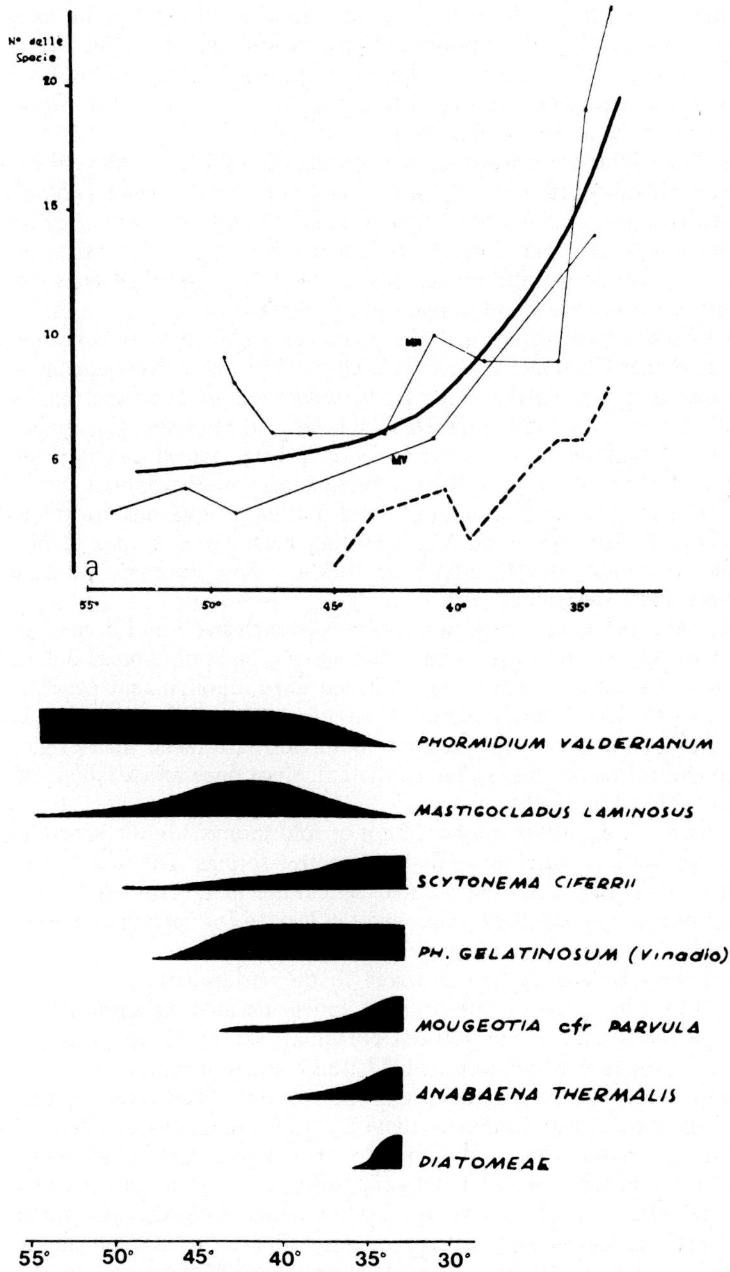

Fig. 445. Quantitative aspects of the distribution of algae in some thermal springs (Valdieri, Piemonte): a. dependence of floristic richness on temperature; b. distribution of the most typical organisms according to temperature. *Phormidium gelatinosum* comes from the Vinadio thermal springs (from Pignatti & Ciferri).

calcium chloride, 'Bitterwässer', ferrugineous, soda (Sodium chloride), Iodic, arsenical, magnesic ('Bittersalz') and sulphurous (H_2S). Very frequent are the Nematodes in all these kinds of springs: in sulphurous springs 33 species have been found; in acid springs 8 species, in radioactive springs 5, in ferrugineous springs 2. Many of these species are common inhabitants of springs.

Finally, in some sulphidric springs of the Marche (Central Italy) Moretti has found some animals furnished with respiratory pigments: *Naididae* and *Tubificidae* (among *Oligochaeta*), Cyclopids and Cladocerans among Crustaceans, besides some species breathing atmospheric air (*Eristalis* and *Stratiomyia* larvae). In salsoiodic- ferrugineous springs Moretti found only a few ciliates and flagellates, besides *Diptera*-larvae (*Stratiomyia* and *Dasyhelea*) and *Hydracarina*.

A few water beetles seem to be particularly adapted to life in thermal waters: at Fischau (near Wiener Neustadt) the beetle *Riolus wichmanni*, a *Dryopidae* lives; in Western Mediterranean basin the beetle *Hydroscapha crotchi* is present; in the Western Mediterranean, the Balkans the Danube basin and the Caucasus, *Hydroscapha granula* is found, present also in Western Turkey, which points to an extremely disjunct distribution just as the fractionation of the habitat would suggest. In the Western Mediterranean and Central European lowlands, as well as in Western Turkey, *Hydroscapha gyrinoides* lives; they belong to a peculiar family, *Hydroscaphidae*. It would be very interesting to know how these species have been able to reach their present day sites.

Very peculiar hot springs are the geysers, practically exclusive – in Europe – to Iceland. The term geyser comes from the verb 'geysa', burst-out, break-out or spring-out, but locally they are called reyk, whence some toponyms such as Reykianes, Reykiavik etc. The Great Geysir emits an intermittent gush up to 40 m in height. From the hot springs of Iceland some toponyms have taken their origin: for instance Uxahver, literally the 'spring of the ox', since once an ox fall in and was immediately cooked by the hot water.

In Iceland there is a countless number of hot or cold springs, known as reykirs and hvers, besides thermal, sulphurous and radioactive springs. The presence of many thermal springs makes possible limited agriculture in the so-called 'tun', grassey surfaces near the houses. Hot water serves in Iceland for heating in houses, glasshouses, stables, open-air swimming pools etc.

As the geysers in Iceland bring hot water to the surface, other analogous phenomena exist in other parts of Europe. Well known are those of Larderello in Tuscany, such as the 'fumarole', or soffioni boraciferi, spurts of vapour hissing from the earth, or others, the 'bulicami', small basins containing a mass of muddy water in continuous motion as a consequence of the pressure of gases coming from the bottom of the basin. Near Larderello there are also emanations of sulphuric acid coming from crevasses, whereas the 'mofete' are clouds of CO_2 which sometimes burst out from masses of water. Finally the 'solfatare' are volcanic activities, not eruptive, and limited to the emission of sulphureous gas mixed with water vapour. These solfatare are present also in the Campi Flegrei, at Pozzuoli and on the Eolian islands (Vulcano). All these thermal activities are of volcanic origin, not dissimilar from that which occur in the geysers; however the water table also takes part in these processes.

The most interesting, and important from an economic standpoint, are the so-called 'soffioni boraciferi', hot water fumaroles containing borax, today utilized in the production of boric acid. The average chemical composition of these waters is the following:

water vapour	55.50%
CO_2	42.60
fuel gas	0.19
H_2S	0.88
N	0.16
Boric acid	0.30
NH_3	0.30

These peculiar hydro-geological phenomena were already known by the Etrurians; a description of them is due to Titus Lucretius Carus (99-55 B.C.) in his De Rerum Natura; passing comments are found in the works by C. Plinius Secundus and Strabo (στράβων). Other works in which notes about these hot springs exist appeared in the VIIth century A.D. and near 1000 A.D. In a codex of 1320 we read 'Prope lacum Veliene agente terremotu corruit magna terrae quantitas moxque apparuit ingens calidus, deinde aquae lacus, quae plus quam brachiis quadraginta iaculabantur in altum...'.

Fishing and fish culture in continental waters

There are five ways to produce richness, one of which is fish culture, i.e. the breeding of fish in ponds. A pond of 8 acres is constructed and subdivided in 9 compartments. Here and there some aquatic plants are deposited. At the second moon of March 20 lovely female carp full of eggs and 4 males are introduced. If you let the water stand, the carp will lay their eggs (Fan Li, 475 B.C.)

Fishing in European freshwaters depend on many factors, which are only partially ecological, since they are an economic activity (we talk of course only of commercial fisheries, not of the sport, and like all economic activities they depend on the general politics of the different countries).

As an example of a country in which fishing is very well developed we may quote Poland, where in 1963 the production of fish was so distributed:

lakes	7300 t
ponds	12300 t
rivers	1000 t with a total of 20,660 t.

Here are the most frequent freshwater fish of Poland, with and indication of their ecological characteristics (from Backiel):

Species	Migrations	Main food	Maximum weight (kg)
Lampetra fluviatilis	river-brackish water	benthic. invert.	2
Salmo trutta	river-sea	fishes	1.5
Esox lucius	–	fishes	15
Tinca tinca	–	benthic. invert.	3
Cyrpinus carpio	–	benthic. invert.	15
Barbus barbus	short distances	benthic. invert.	6
Carassius carassius	–	benthic. invert. + plankton	2
Abramis brama	short distances	benthic. invert.	5
Leuciscus cephalus	short distances	plants (summ.) fishes (wint.)	5
Leuciscus idus	short distances	invertebrates	8
Vimba vimba	river-brackish water	benthic. invert.	2
Chondrostoma nasus	short distances	algae + benthic. invert.	3
Aspius aspius	short distances	fishes	6
Silurus glanis	–	fishes	40
Anguilla anguilla	sea-fresh water	invert. + fishes	6
Lota lota	–	invert. + fishes	3
Lucioperca lucioperca	–	fishes	10
Perca fluviatilis	–	invert. + fishes	2
Blicca björkna	–	benthic. invert.	1
Rutilus rutilus	–	invertebrates	1

The data on fishing in Polish lakes are synthesized in Fig. 446. As far as river fishing in Poland is concerned we can get an idea from Fig. 447, which shows the relationship between the spread of some fish and the level of the water course in which they live.

Finally, the productivity of Polish rivers is illustrated by the two following tables:

Production of exploited fish

Region	kg/ha
Vistula estuary	60
Odra estuary	100
Vistula (middle course)	10-30
Middle course of the Odra and lower course of the Warta	10-30
Middle and lower course of the San	10

Estimated quantity of fish in several Polish rivers (kg/ha)

River	part of the course		
	upper	middle	lower
Upper Vistula	124	180	375
Sola	32.8	29.2	31.5
Upper San	83.3	17.8	7.7
Rogoznik	16.9	134.2	165.8

As we have already had the opportunity to say, productivity – both primary and secondary, both terrestrial and aquatic – is today of extremely acute importance as a consequence of the increase in human population and the consequent demand for food.

As far as Southern Europe is concerned, fish are not so intensively exploited. Not only that, but, in recent years, pollution (see later) has increased so much that today exploitation would be rather difficult and in some cases even dangerous to human life or at least human welfare.

We wish also to report here the quantity of each fish species (in kg) caught in the principal Italian lakes from 1925 to 1929. Today the values would be very different due both to thy policy of Italian fishery (and agriculture) and to the pollution of many lakes among those considered. For comparison the data obtained for the Trasimeno Lake during 1956-59 are also reported.

Should we compare the productivity of inland waters in Poland with those in Italy, we would find that, whereas in Poland (1960) there is an annual production of still water fish = 0,62 kg/ pro capita, in Italy the production is = 0,08 kg. This, of course, comes mainly from the government policies of the two countries.

As far as fish culture is concerned, in Europe the most important species from an economic stand point are trout (salmoni-culture), carp (cypriniculture), both formed in freshwater, *Mytilus* sp. and *Ostrea* sp. both formed in brackish water (besides of course sea water) and finally astaci-culture, or breeding of crayfish, limited to a few rivers.

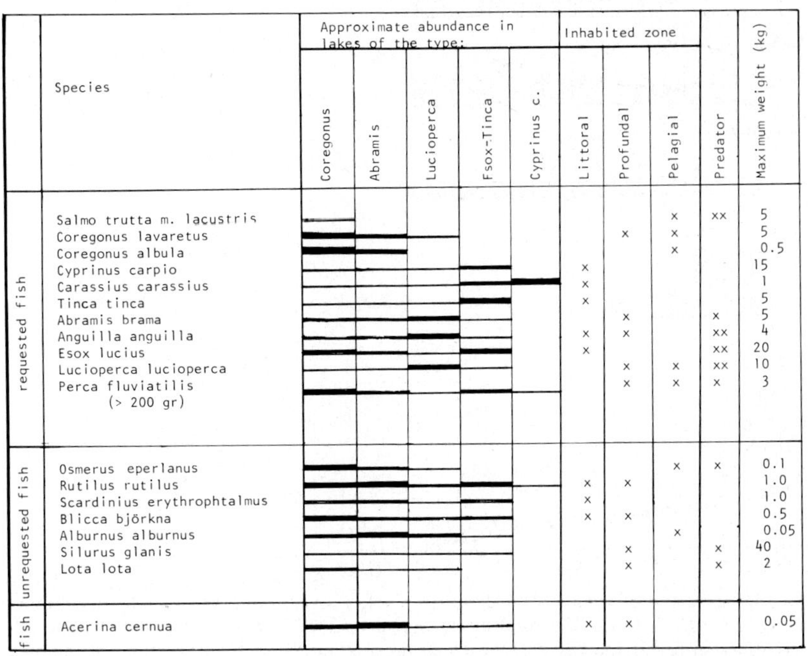

Fig. 446. The main ecological characteristics of lake fishes exploited in Poland (from Zawisza).

Other fish which on our Continent are the subject of culture are the green tench *Tinca tinca*, the roach *Rutilus rutilus* and *R. rubilio* (the latter in Italy), the pike *Esox lucius*, the pike-perch *Lucioperca lucioperca* (particularly in Jugoslavia) and the trout (mainly *Salmo fario, S. lacustris, Salmo marmoratus, S. irideus* (an introduced species), as well as the char, *Salvelinus alpinus*.

Practical problems connected with fish or pond culture are the fertilization of the water, the 'lake management' (every day of increasing importance) and culture in man-made lakes.

A peculiar form of exploitation of water, with the aim of human food supply, is the so-called valliculture (from 'valle', a term used in the lagoons of the Northern Adriatic Sea to indicate an area of water completely or only partially closed,

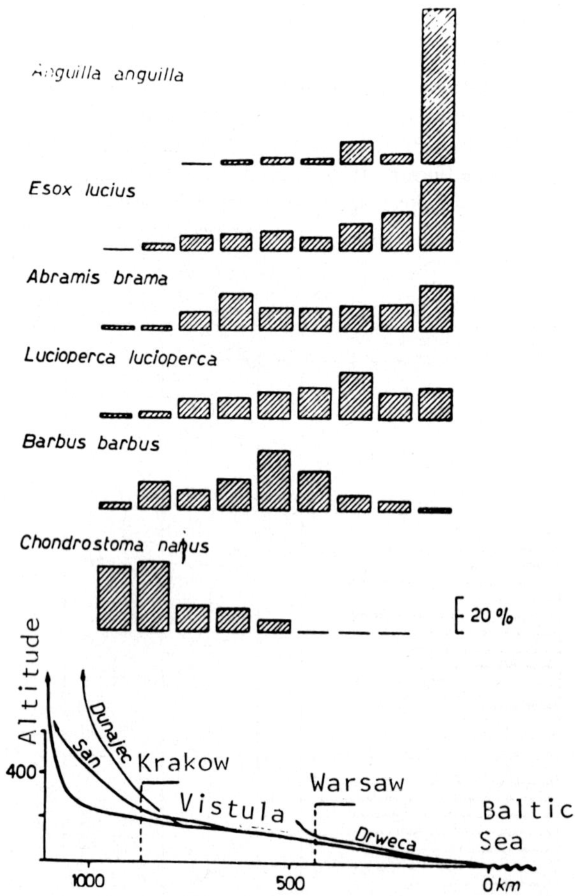

Fig. 447. Distribution of some fishes along the Vistula River, based on the data on commercial fishing from 1952-1963. The total mean annual fishing has been taken as 100. The different columns represent the percentage of fish taken from the different sections of the river (from Backiel).

in which fish culture is performed). The high degree of eutrophy of these brackish-water type of littoral ponds attracts fish which migrate from the sea to the lagoons or 'valli', where they stay until sexual maturity. Today there is also artificial introduction of fry.

A particular system of closing of the valli (by means of the so-called 'lavorieri', kind of sluices through which water can pass but not fish) prevent fish from returning to sea for reproduction, and in the mean-time permits their capture in a special 'death chamber', where they come attracted by a relatively high concentration of salts in the water. Some particularly euryhaline species live permanently in the lagoons: *Gobius* sp. plur., *Atherina mochon, Blennius, Syngnathus, Aphanius (= Cyprinodon)* or *Gasterosteus*, (the common stickleback) but the species of economic importance are marine species which stay in the valli exclusively to feed. Some species stay only for a short period: the turbot *Bothus maximus*, the sole *Solea impar* and *S. solea*, the toothed gilthead *Dentex dentex*, the red mullet *Mullus barbatus* and *M. surmuletus*, the umbrine or bearded umbrina *Umbrina cirrhosa*, and several others. Other species, on the contrary, stay in the lagoons until complete sexual maturity: the mullets (*Mugil* sp. plur.), the eel *Anguilla anguilla*, the gilthead *Chrysophrys auratus*, the bass *Labrax lupus* (sea dace or sea wolf) and the flounder *Pleuronectes flesus*. These are the most important fish from the economic standpoint.

In this connection we must mention that, in the great littoral lakes of the Baltic Sea, there are several marine fishes, such as the salmon *Salmo salar*, the sea trout *Salmo trutta*, the plaice *Pleuronectes platessa*, eels, herrings, shads etc. side by side with freshwater fish such as the smelt *Osmerus eperlanus* and the bream *Abramis brama*.

For fishing in inland waters the usually employed tools are hooks, harpoons, snares, bow nets and nets (simple, vertical, monthle vertical nets and other types specialized for the catching of a particular kind of animal). Many kinds of boats (from the Middle English boot, bote, bot, and the Anglo Saxon bat) are used for fishing in inland waters (in Italian boat is barca, from the late Latin). Let us remember the hoop boat, used in the alpine lakes; the flat for shallow waters, the punts (in Italian sandalo) and the 'sandolino', a kind of small canoe; the lighters and barges in Italian 'burchi' the wherries, in Italian 'burchielli' (today existing only as a memory near Venice in the old course of the Brenta) the so-called 'gondole' used exclusively in the lagoon of Venice, of which Shakespeare talks in his Merchant of Venice, and many other types varying from one country to another.

The boats used for fishing in inland waters are very old; the tools employed for fishing also date from very ancient times the hooks go back to Neolithic; nets were known from the Swiss Neolithic; harpoons were present in even older times (Paleolithic).

In Trentino (Northern Italy) a very peculiar system is utilized to catch trout: the so-called 'pescaia' (from Medieval Latin piscaria) a barrage along the course of the river made with wood and stones and sometimes furnished with bow nets. In Tuscany 'pescaiolo' is a barrage used for fishery in rivers, made with many materials such as branches, neshes etc. The pescaiolo must let water flow through without changing its course.

Even today many superstitions are attached to fishing boats: in North Europe for instance it is forbidden to name salmon; amulets and charms are used for boats and nets against witchcraft and sorcery. Venetian fishermen say some argot formulae when cleaning tubs or when pitching boats...

Pollution of inland waters

'Fish are a very useful barometer of the real state of purity of a water. No river should be considered as in a satisfactory condition unless fish will live and thrive in it'

H.D. Turin, 1947

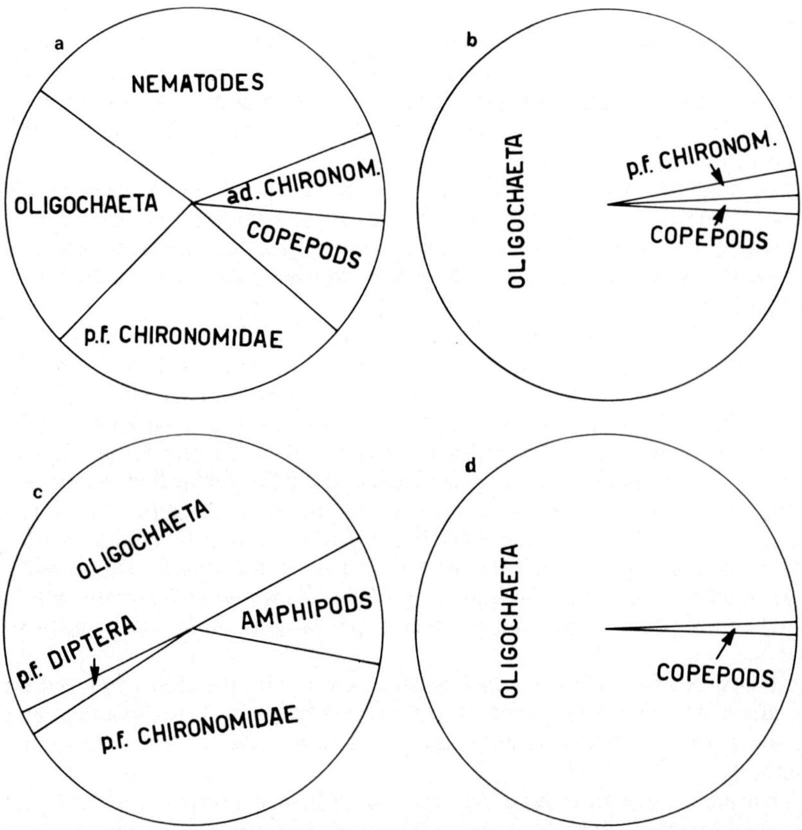

Fig. 448. Composition of animal populations of the Pescara River at different levels; a, upstream l'Aquila; b. immediately downstream l'Aquila – the abundance of *Oligochaeta* is a clear sign of pollution; c. near Bussi sul Tirino, where there is the least pollution of the entire Pescara Basin; d. near the mouth of the Pescara River, where *Oligochaeta* are the only animals able to withstand pollution.

When industries and human settlements discharge their by – products into a river the dissolved substances – both mineral and organic-which are sometimes present in very small amounts, increase greatly, so that all river life, beginning with fishes, may be destroyed. The effects can be manifold: from one part we may have toxic substances such as sulphuric acid discharged by industry, a decrease in oxygen consequent on the fermentation of organic substances present in the discharges of several industries (distilleries, sugar refineries, breweries etc.) and in the waters coming from a urban settlement (sewage). Finally, the effect can be due to the deposition of solid materials suspended in the water: this deposit can prevent the development of all biota and in some cases it can sterilize a water-course.

In Europe each year the discharge into rivers (and lakes) of several chemicals employed in agriculture, such as insecticides, herbicides, fungicides etc. becomes more dangerous. Some herbicides are also used against the natural vegetation of a river (*Elodea*) or to control the development of mosquitoes or *Simulium*-larvae. Although these substances are not dangerous to fish, they bring about a disturbance of the natural equilibrium in a river because they destroy the plants necessary to release oxygen to the water and to provide some shelter to young fishes. So, for instance, copper sulphate employed as a herbicide in the Lake of Orta (Piemonte) caused a diminution of fish because of the impoverishment of the phytoplankton. Fertilizers in general cause eutrophication but are not dangerous. On the contrary, insecticides such as DDT, HCH or E 605 are dangerous to insect larvae and crustaceans, which constitute the main natural food of fish.

Probably the best known pollution is urban, so far organic bio-degradable substances, but in the few last years represented also by several non-degradable substances such as synthetic detergents (e.g. Manoxol O.T.).

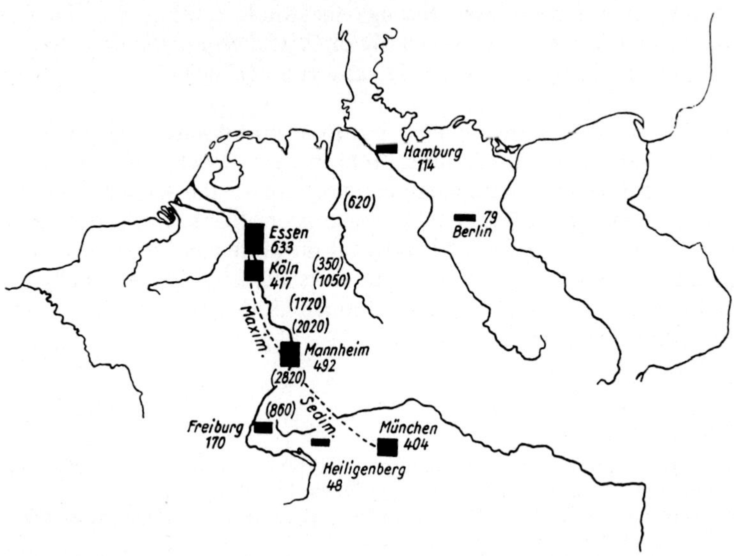

Fig. 449. The radioactivity of the waters of West Germany in 1957, measured in 10^{-4} $\mu C/m^2$. The values in parentheses refer to terrestrial environment (from Herbst et al.).

Generally urban pollution undergoes a natural auto-decomposition through the intermediary of several anaerobic and aerobic organisms both plant (microorganisms) and animal. At the end of the process only inorganic, soluble, non-toxic substances are formed. The anaerobic part of the decomposition implies the presence of elements such as bacteria, fungi, algae, rhizopods, flagellates, ciliates, rotifers and Gastrotrichia. To the aerobic part chlorophyll-containing plants, Oligochaeta. Insect larvae and many other saprophagous animals, besides filamentous bacteria such as *Sphaerotilus* and *Leptomitus*, contribute. Once the decomposition is almost complete, a state of eutrophication is reached: the rich chironomid fauna permits an abundant development of fish. Thus, in Munich there are some ponds destined for the culture of rainbow trout (*Salmo gairdnerii*) into which enriched water is allowed to enter, which causes eutrophication with an annual yield of 500 kg/ha.

In other towns near the discharge point of sewage *Rutilus rutilus* and other Cyprinid fishes abound.

In some instances a true succession can be be demonstrated along the cenosis which participate to autodecomposition.

Along the Pescara River (Abruzzo, Central Italy) where several industries discharge untreated by-products containing, inter alia, Hg, Pb, Cd, Cr Fe, Bo and where some towns such as L'Aquila (capital of the region Abruzzo-Molise) discharge all the sewage directly into the river, heavy pollution is present, demonstrated by the overhelming predominance of Oligochaeta and secondarily of leeches of the genus *Clepsine*. Fig. 448 a and b show the change in the fauna before and after the passage of the river through the town of L'Aquila; Fig. 448 c shows a relatively 'normal' faunistic composition in a not very heavily polluted biotope; whereas Fig. 448 d shows the extreme monotony of the plankton at the mouth of the Pescara River, after it has flowed through the whole town of Pescara. The benthos of this biotope is composed exclusively of *Oligochaeta Limicola*, extremely euryoxibiont and probably also euryhaline as a consequence of the vicinity of the sea.

Two other kinds of changes are present in European inland waters: the discharge of hot water from industries and still more from nuclear power stations (nuclear reactors) and the radioactive contamination due both to the fall-out (existing since 1945) and to some wastes from nuclear reactors or other sources of radioisotopes. It is needless to say that the latter is by far the more dangerous since there are no external, recognisable, signs of it, whatever its intensity. Unfortunately the destiny of radionuclides has been investigated much more in the sea than in the freshwater environment (Fig. 449).

BIBLIOGRAPHY

Aguesse P. & L. Bigot, Observations floristiques et faunistiques sur un etang de Moyenne Camargue: la baisse salée de la Tour du Valat, *Vie Milieu*, 11, 1960, 284.

A plea for the Wadden Sea, Contact Committee for Nature and Landscape Preservation, Amsterdam.

Berg L.S., Freshwater fishes of the USSR and adjacent countries, Jerusalem, 1965 (transl. from Russian).

Bertarelli L.V. & E. Boegan, Duemila Grotte, Touring Club Italiano, Milano, 1926.
Bigot L. & F. Marazanof, Notes sur l'ecologie des Coleopteres aquatiques des marismas du Guadalquivir etc. *Ann. Limnol.*, 2, 1966, 491.
Calvino F. & G. Barrocu, Notizie sulle prime esplorazioni degli 'Angurtidorgius' con osservazioni geologiche sull'idrografia carsica ipogea del Salto di Quirra (Sardegna sud-orientale), *L'Universo*, 44, 1964, 865.
Das Buch vom Neusiedlersee, Wien, Wollzeilenverlag, 1961.
Delamare-Deboutteville C., Biologie des eaux souterraines littorales et continentales, Hermann, Paris, 1960.
Dogiel V.A., G.K. Petrushevski & Y.I. Polyanski, Parasitology of fishes, Oliver & Boyd, Edinburgh & London, 1961.
Dokulil M., Horizontal- and Vertikalgradienten in einem Flachsee (Neusiedlersee, Oesterreich), *Verh. Ges. Oekol., Wien*, 1975, 177.
Donner J., Randbiotope von Fliessgewässern als Orte der Anpassung von Wasserorganismen an Bodenbedingungen, *Verh. Ges. Oekol., Wien*, 1975, 231.
Ghetti P.F. & G. Gandolfi, Progetto per il censimento e l'analisi della dinamica della fauna ittica nelle acque superficiali della Provincia di Parma, *Atti III Simp. Naz. Conserv. Nat.*, Bari, 1973, 184.
Ghetti P.F., R. Antonietti, G.C. Artusi, G.M. Curti & A. De Marchi, Definizione della qualità delle acque correnti della Val Parma mediante l'analisi comparata di alcuni parametri chimici e biologici, *Atti III Simp. Naz. Conserv. Nat.*, Bari, 1973, 263.
Giannotti F., Il carico biologico convogliato dalle acque delle fonti del Clitunno, *Boll. Pesca Pisc. Idrob.*, 12, 1957, 265.
Hadl G., Oekologische Auswirkungen von Umweltveränderungen auf Fliessgewässer in Grossstadtnähe, *Verh. Gesell. Oekol., Göttingen*, 1976, 463.
Herzig A., Der Neusiedlersee, charakteristiche Eigenschaften und deren Auswirkungen auf des Zooplankton, *Verh. Ges. Oekol., Wien*, 1975, 189.
Hinz W., Ueber die Siedlungsdichte von Pisidien in subarktischen Gewässern Skandinaviens und die Bedeutung dieser Muscheln im Oekosystem, *Verh. Ges. Oekol., Wien*, 1975, 139.
Huber-Pestalozzi G., Das Phytoplankton des Süsswassers, Die Binnengewässer, XVI, Stuttgart, Scheizerbart, 1938 (I), 1941 (II, 1).
Illies J., Limnofauna europaea, Fischer, Stuttgart, 1967.
Janković M., Die Besiedlung der Wasservegetation im Apatiner Überschwemmungsgebiet, *Arch. Hydrob., Suppl.* XXXVI, 1971, 384.
Jungk E., K. Kilian, L. Steubing & R. Uebers, Auswirkungen der Abwasserbelastung auf die Flora und Fauna im Raum Siegen, *Verh. Gesell. Oekol., Göttingen*, 1976, 315.
Kinzelbach R. & U. Schmidt, Zur Oekologie abwasserbelasteter Altrheine, *Verh. Gesell. Oekol., Göttingen*, 1976, 455.
Klein L., River pollution. II Causes and effects, Butterworth, London 1962; III. Control, London, 1966.
Kohler A., Makrophytische Wasserpflanzen als Bioindikatoren für Belastungen von Fliessgewässer-Oekosystemen, *Verh. Ges. Oekol., Wien*, 1975, 255.
Labus B., Nobel W., R. Smetana & A. Kohler, Der Einfluss der Abwassersubstanzen Marlon A (Anionenaktives Tensid) und Bor auf die Photosyntheserate einiger submerser Makrophyten, *Verh. Gesell. Oekol., Göttingen*, 1976, 325.
Ladiges W. & D. Vogt, Die Süsswasserfische Europas, Hamburg & Stuttgart 1965.
Lenz F., Einführung in die Biologie der Süsswasserseen, Berlin, J. Springer, 1928.
Marcuzzi G., Ecologia animale, Feltrinelli, Milano, 1968.
Marcuzzi G., Considerazioni sulla protezione di alcuni ambienti acquatici carsici, *Boll. Zool.*, 39, 1972, 521.
Marcuzzi G. & L. Dalle Molle, Secondo contributo alla conoscenza dell'ecologia della torbiera piana di Fiavè (Trentino occidentale), *Studi Trent. Sc. Nat.*, 52, 1975, 187.
– & –, Contributo alla conoscenza del popolamento animale di una torbiera piana del Veneto (Lipoi, Feltre), *Ibidem*, 53, 1976, 123.
Marcuzzi G. & A. Lorenzoni, Osservazioni biologiche sul popolamento animale del Lago Santo di Cembra (I nota), *Studi Trent. Sc. Nat.* 45, 1968, 54.
– & –, Osservazioni ecologiche sul popolamento animale della torbiera piana di Lagabrun (Cembra) (II nota), *Ibidem*, 165.

– & –, Considerazioni ecologiche sul popolamento animale della torbiera di Lagabrun, *Ibidem*, 46, 1969, 247.
– & –, Considerazioni ecologiche sul popolamento animale del Lago Santo di Cembra, *Ibidem*, 47, 1970, 81.
– & –, Der Lago Santo bei Cembra (Trentino, Italien), *Ver. Schutze Alpenpfl. & Tiere*, 35, 1970, 1.
– & –, Osservazioni ecologico-faunistiche sul popolamento animale della palude carsica di Pietra Rossa (Monfalcone) (II nota), *Vie et Milieu*, 21, 1970, 1.
– & –, Osservazioni ecologiche sul bentos di alcune acque carsiche, *Atti Mem. Accad. Patav. S.L.A.*, Padova, 82, 1970, 425.
Marcuzzi G., G.G. Lorenzoni & G. Dal Doss, Primo contributo alla conoseenza di una torbiera piana del Trentino occidentale (Fiavè), *Studi Trent. Sc. Nat.*, 48, 1971, 124.
Melzer A., Einflüsse einer Gewässerbelastung auf die Makrophytenvegetation der Osterseen, einer Seenkette in Oberbayern, *Verh. Gesell. Oekol., Göttingen*, 1976, 473.
Negrea Ş. & A. Negrea, Ecologia populatiilor de Cladocera şi Gasteropode din zona inundabilă a Dunării, Edit. Academ. Rep. Soc. Rom., Bucureşti, 1975.
Nemenz H., Physiologische Anpassungen an Salzstandorte bei Land-und Wassertieren des Binnenlandes, *Verh. Ges. Oekol., Wien*, 1975, 221.
Parisi V., Studio sulle biocenosi lotiche del Po a Trino Vercellese negli anni 1967-68, *Rend. Ist. Lomb.*, B, 105, 1971, 5.
Penso G., I prodotti della pesca, U. Hoepli, Milano, 1950.
Pignatti S. & R. Ciferri, Osservazioni microbiologiche effettuate nelle terme di Valdieri, *Atti Ist. Bot. Lab. Critt. Univ. Pavia*, 18, 1960, 234.
Popescu-Marinescu V., Die Benthos-Biocönosen der unter dem Einfluss der Donau stehenden Gebiete des Voedeltas, *Arch. Hydrob., Suppl.* XXXVI, 191, 377.
Reichenbach-Klinke H. Die wichtigsten Nutzfischarten der Donau und ihre mutmassliche Verteilung, *Arch. Hydrob., Suppl.* XXXVI, 1970, 263.
Reichholf J., Biotopstruktur und oekologische Funktionen der Staustufen am unteren Inn, *Verh. Gesell. Oekol., Göttingen*, 1976, 447.
Ribi Kaspiiskogo Morya, Mosk. Obščestvo Isl. Moskva, 1951.
Ruffo S., Lo stato attuale delle conoscenze sulla distribuzione geografica degli Anfipodi delle acque sotterranee e dei paesi mediterranei, Comm. I Congr. Intern. Spéléol., Paris, III, 1953 (1956).
– –, Le attuali conoscenze sulla fauna cavernicola della regione pugliese, *Mem. Biogeogr. Adriat*, 3, 1955, 1.
Sampl H., Das Limnische Oekosystem des Millstätter See in Kärnten in den letzten 40 Jahren, *Verh. Ges. Oekol., Wien*, 1975, 129.
Schäfer A. & P. Müller, Auswirkungen der Saar-belastung auf die Speziesdiversität von Benthosbiozönosen und die Verweildauer exponierter Organismen, *Verh. Ges. Oekol., Wien*, 1975, 277.
Schaeperclaus W., Fischkrankheiten, Wenzel & Sohn, Braunschweig, 1941.
Schindler O., Die Fische unserer Hochgebirgsseen, *Jhrb. Ver. Schutz Alpenpflanzen & Tiere, München*, 22, 1957, 98.
Sterba G., Freshwater fishes of the world (transl. from German), Vista Books London, 1962.
Streit B. & J. Schwoerbel, Experimentelle Untersuchungen über die Akkumulation und Wirkung van Herbiziden bei benthischen Süsswassertieren, *Verh. Gesell. Oekol., Göttingen*, 1976, 371.
Tasovac T. & R. Radosavljević, Ergebnisse systematischer Messungen der α und β-Radioaktivität in der Donau in den Jahren 1958-67 im Gebiet von Vinča, *Arch. Hydrob. Suppl.* XXXVI, 1969, 85.
– – & – –, Die Verteilung der Gesamt-β-Radioaktivität auf die flüssige Phase und die suspendierten Teilchen im Donauwasser, *Ibidem*, 92.
Thienemann A., Die Binnengewässer Mitteleuropas, I, Die Binnengewässer, Stuttgart, Schweizerbart, 1926.
– –, Verbreitungsgeschichte der Süsswassertierwelt Europas, Die Binnengewässer, XVIII, Stuttgart, Schweizerbart, 1950.

Vandel A., Biospéologie, La biologie des animaux cavernicoles, Gauthier-Villars, Paris, 1964.
Vendegna V., Studi sull'inquinamento delle acque italiane. 1. Carico zoologico ed evoluzione dell'inquinamento del Fiume Lambro, *Riv. Idrob.*, 7, 1969, 135.
Vogel S., Der Einfluss von Phenol und von phenolischen Verbindungen auf den Gaswechsel autotropher und heterotropher Planktonten, *Verh. Gesell. Oekol., Göttingen*, 1976, 361.
Wesenberg-Lund C., Biologie der Süsswassertiere, Wien, 1939.
– –, Biologie der Süsswasserinsekten, Berlin & Wien, 1943.
Witt U., Effects of artificial fertilization on a high-mountain lake (Vorderer Finstertaler See, Tirol), *Verh. Ges. Oekol., Wien*, 1975, 161.
Zhadin V.I., Mollusks of fresh and brackish waters of the USSR, Acad. Sci. USSR, Jerusalem, 1965.
Zhadin V.I. & S.V. Gerd, Fauna and flora of the rivers, lakes and reservoirs of the USSR, Jerusalem, 1963 (transl. from Russian).

THE ENVIRONMENT OF HUMAN SETTLEMENT

The environment of human settlements is composed of the dwellings, stores and many other buildings within which an enormous number of animals and, secondarily, plants live. The former are often anthropophilous species, the latter belong to several categories, including the ruderal plants. The animals living in human settlements are particularly ecto- and endoparasites of man and house animals

Fig. 450a. A large town (Milan) not very well planned, with an historic centre easily visible in the middle, and a not harmonious growth around the historic centre, occurring over a long period of time, with very different conceptions of town planning or with no conception at all (from Guida Turistica d'Italia, Casa ed. De Agostini) (reproduced with the permission of Ed. De Agostini).

(dogs, cats, mice, rats, swallows, pigeons, sparrows etc.) plus many saprophages or detritivorous species (Fig. 450-53).

The house may, in general, be subdivided – at least in most European countries – into three ecological units: the roof space (or attic), the living area and the cellar. The former is characterized by a dry-hot climate during the summer and dry-cold climate during the winter; the second exhibits a dry, rather warm, climate all through the year; the latter is constantly humid and cool. In the roof we find pigeons (*Columba livia*), bats (*Rhinolophus hipposideros, R. ferrum-equinum, Plecotus auritus, Vespertilio discolor*, etc.), mice (*Mus musculus*), cerambycid beetles developing in the wood present in the roof (particularly in the past), some zoosaprophages such as *Dermestes* and *Anthrenus*, some moths etc. In the living premises two microenvironments can be recognized, i) the kitchen, where mainly insects are found (cockroaches, beetles belonging to the genus *Tenebrio, Tribolium, Latheticus, Calandra, Stegobium paniceum, Ptinus* and *Bruchus*) and ii) the living and sleeping rooms where lice, fleas and bugs, moths, *Dermestidae* (*Anthrenus museorum* and *Attagenus piceus*), *Lictus* (coming from USA), the psocoptera *Liposcelis divinatorius* and the pseudoscorpion *Chelifer cancroides* can be found. Other insects very common in our houses are flies (*Musca domestica*) and mosquitos (*Culex, Anopheles*) and, only secondarily, gnats of the genus *Phlebotomus*, a vector of several leishmaniosis. In the cellar we may find, expecially in old houses, rats, mice, beetles (belonging to *Mycetophagidae* and to the Tenebrionid genus *Blaps*) isopods or woodlice, *Gryllidae* or crickets (*Gryllulus domesticus* and *Gryllomorpha dalmatina*) and slugs (*Limax maximus* and *Limax flavus*) (Fig. 454-457).

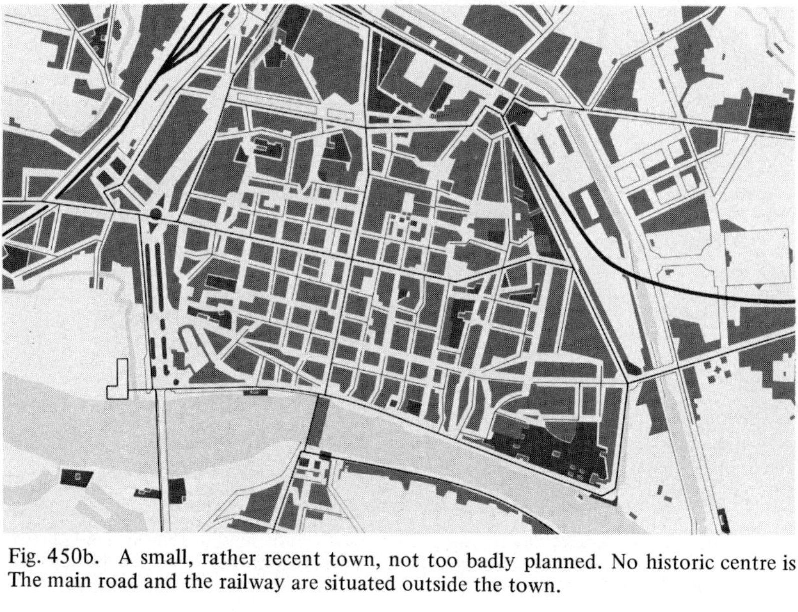

Fig. 450b. A small, rather recent town, not too badly planned. No historic centre is visible. The main road and the railway are situated outside the town.

The rats belong to two species, *Rattus rattus* and *Rattus norvegicus*. The latter has probably come from Northern Europe and not, as some formerly claimed, from the Volga. The common mouse is represented in Europe by several subspecies or 'races': *M. musculus musculus*, present in North Eastern Europe, living both in houses and in the open, and possibly originating from the wild race *musculus spicilegus*, today present in Brandenburg, on the Baltic Sea coasts, etc. *M. musculus domesticus*, present in Western and Central Europe, probably evolved from the wild race *m. bactrianus*, of Asia minor. This form lives exclusively in houses. In Southern Europe a third domestic race exists, *M. musculus brevirostris*, spread throughout the Mediterranean region; and finally, *M. musculus praetextus* present in North Africa and some Mediterranean islands, such as Sicily, Lampedusa, Crete etc.

The two typical house animals, dog and cat, have a very different history and origin. The domestication of the former is extremely ancient, and the dog seems to have been the first domestic animal; it must have originated from the wolf (*Canis lupus*). The domestic cat probably comes from *Felis ochreata* of Egypt, so that today in some Mediterranean regions, e.g. Palestine, it is not always easy to

Fig. 451. A characteristic village on the shores of the Eastern Mediterranean Sea (Aegean Sea), Myconos. In the foreground a windmill. The ancient town was inhabited by Ionians. (from Enciclopedia 'Il mare', Ed. De Agostini).

distinguish between these two forms, particularly in the more arid regions. It should be mentioned that the Bible does not speak of domestic cats, and this could suggest that the domestication of this animal is relatively recent, probably after the beginning of the Christian Era. Incidentally the Palestine domestic cat exhibits habits different from those of its European relatives, for example it is less clean, it does not hunt mice but prefers to steal food from houses, and damages the soil of gardens. Furthermore it is a carrier of the rabies virus.

The genus *Passer* to which the house sparrow belongs is one of the most studied passerine genera in the world, since it is the object of research by the working Group on granivorous birds, belonging to the PT section of the IBP (International Biological Programme), headed by the Institute of Ecology of the Polish Academy of Sciences (Warsaw). Due to this research, our knowledge about practically all aspects of the ecology of house sparrows is today enormous, when

Fig. 452. Aspects of European abodes: a. modern house inhabited by high income people, Naples; b. modern houses of Padova; c. and d. rural houses in English countryside.

compared with other European and North American birds. The knowledge covers anatomy, behaviour, breeding biology, distribution, food, growth, morphology, movements, pathology, physiology, population dynamics, psychology and voice, relationships with man and other animals, taxonomy (research is, of course, carried out on all *Passer*-species), evolution and finally ecology (i.e. the study of habitat and phenology). Most work has been done on *Passer domesticus*, followed

Fig. 453. Human settlement in the north Färöer (photo Prof. Georg Pilleri, Bern).

by *P. montanus* and *P. hispaniolensis*, and to a lesser degree on *P. griseus, P. moabiticus* etc.

It would be impossible to summarize all but a few of the data gathered about sparrows since this Working Group began functioning.

We could, however, mention some results obtained in Czechoslovakia by Turček on ecology and population dynamics of the house sparrow *Passer domesticus*. The observations were made within a section of roughly 150 m of street be-

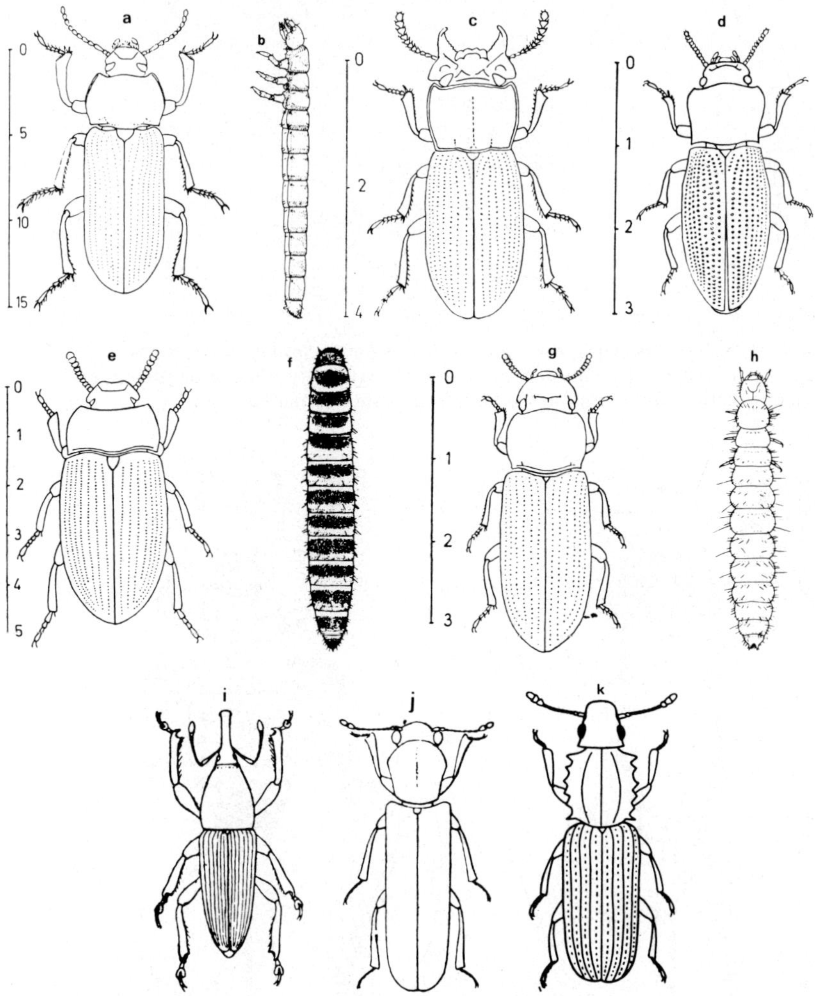

Fig. 454. Synanthropic beetles harmful to stored food: a. *Tenebrio molitor;* b, *T. molitor* larva (the yellow mealworm); c. *Gnathocerus cornutus*; d. *Palorus depressus*; e. *Alphitobius laevigatus*; f. *Alphitobius* sp. larva; g. *Tribolium confusum*; h. *T. confusum* larva; i. *Calandra granaria*; j. *Lyctus brunneus*; k. *Oryzaephilus surinamensis* (a, c, d, e, g, k from Freude & Lohse; f. from Hayashi; b, h redrawn from Peterson, i, j from Portevin).

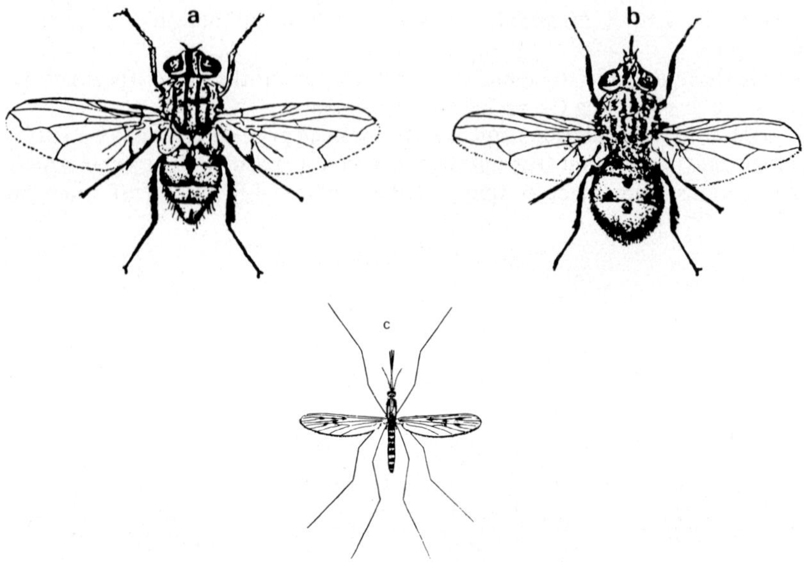

Fig. 455. a. *Musca domestica*, the common house fly; b. *Stomoxys calcitrans*, the common stable fly; c. *Anopheles maculipennis*, the mosquito that is very often responsible for the transmission of malaria in Southern Europe (a, b from Brain; c, from La Face).

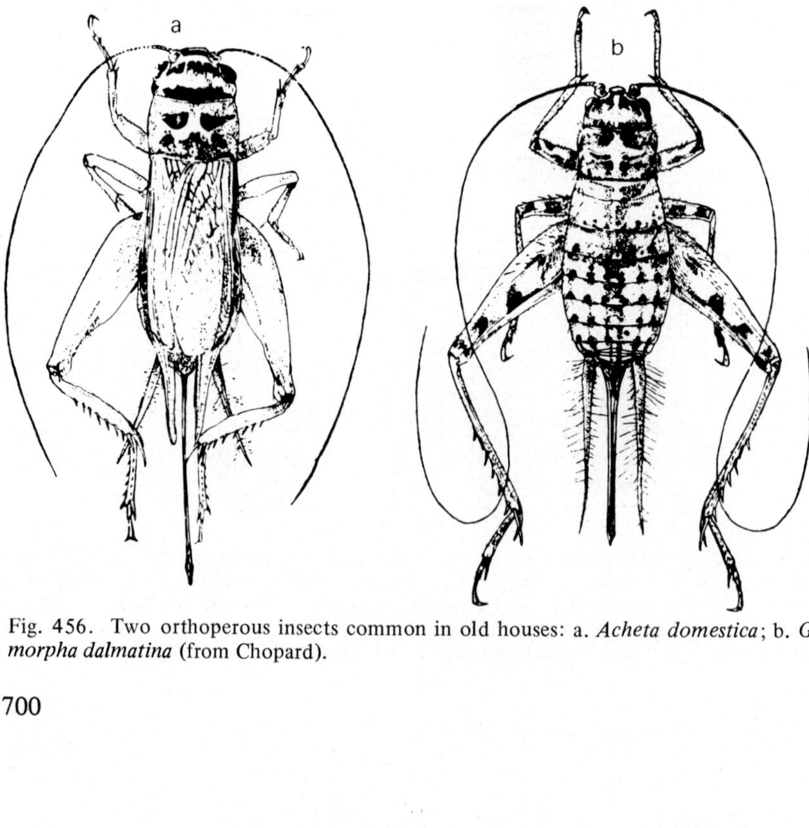

Fig. 456. Two orthoperous insects common in old houses: a. *Acheta domestica*; b. *Gryllomorpha dalmatina* (from Chopard).

a

b

c

Fig. 457. a: the black or house rat, *Rattus rattus*; b. the house mouse *Mus musculus*; c. the brown or Norway rat, *Rattus norvegicus*.

tween two rows of buildings in a small Slovakian town, Banská Stiavnica, about 600 m above S.L. and with a population of slightly over 10,000. Within the town, the entire sparrow population is divided into separate micro-populations throughout almost all the year. The sparrows congregate in various places for roosting, or in extreme weather conditions for foraging. Forty-six percent of the birds have been observed on the ground, mostly on paving, while 54% were noted above ground, mostly on buildings, but with some on fences. The difference between 'on the ground' and 'above ground' is statistically not significant. Fourteen percent of the birds were present on wires (power and telephone) and TV antennae. With regard to street paving, it is worth noting the relative abundance of foraging sparrows on andesite covered streets, compared to a lack of sparrows on the asphalted street surfaces. The reason might be the accumulation of food between the andesite blocks and almost no food on the smooth asphalt surface. The maximum number of birds was observed under clear sky at 21-25°C and 16-20°C under cloudy sky; minimum numbers were noticed under clear sky at 11-15°C and under cloudy conditions with an air temperature of 21-25°C.

During July and August occasional observations were made on a nest located in a cavity of a gambrel-type roof. The cavity was utilized for many years. The same pair of sparrows bred there twice between June and August. During the feeding of fledglings there were 7 exchanges of male-female per hour, with a minimum of 1 and a maximum of 20. At nearly two thirds of the nest feeding periods the male did not participate. The maximum of exchanges and thus the maximum feeding rate, derived from bringing the young birds un-natural food, e.g. kitchen-made food such as rice, bread etc. Thus synanthropy of the house sparrow might be an advantage even in the rearing and success of young. During the incubation period and later in the early period of feeding young, the parents carried almost daily to the nest green parts of plants, mostly buds of *Artemisia absynthium*. The adults fed on the aphids living on this plant species.

The segregation of urban sparrows into sub-populations is apparently advantageous and, presumably, adaptive. Any sub-population retains its own territory for foraging (except during mass flocking from late summer to early spring) and own roosting places for which, even in urban habitats, heavy competion takes place. The sub-population is stable and possibly self-regulating. In urban habitats the distribution of sub-populations of sparrows indicates that old, profiled, ornamental, or even neglected, buildings are by far preferred; they may also indicate food reserves or sources (mostly waste organic material) available, not only for sparrows, but also for other synanthropic animals such as house mice, brown rats, collared turtle-doves and others.

Very interesting from the bio-energetic point of view, is the relatively high expense in finding and consuming food. On a road or street with a high traffic density a vast amount of energy must be spent as to consume a given amount of food the sparrows must repeatedly fly out and return as they are continually disturbed by pedestrians or vehicles. There are no details available about this extra energy requirement nor about ways in which it is compensated for. It can be surmised that either this movement is a regulating factor in urban populations in a restrictive sense, or the urban populations are adapted to these conditions. This latter may have an evolutionary meaning.

It is worth mentioning a rather unusual insect present in human settlements because of its ecology. The habit people once had of burying dead in church cellars was responsible for some rather disgusting phenomena, for example, the invasion of churches or other buildings by flies whose larvae fed upon decaying organic substances, including corpses. An example, although a little disgusting, is the description of one of these cases reported by the 'Hamburger Bericht' of June 16th, 1795, which we translate literally as:

'In one of the local churches a fact has occurred that again constitutes the intricate problem of how dangerous it is to bury dead in churchcrypts. The practice that the wine to be offered during the Holy Communion, is placed in the chalice on the altar immediately before the service was followed in this church. A little after the Communion had started, the wine, which as usual, had been consecrated by the preacher was found, to his great astonishment, to be filled with small flies. He emptied the chalice and took fresh wine, but this also was full of small flies, so that he again threw it away and for the third time consecrated some fresh wine. However, this too, was not completely free of insects, but it had to be utilized because there was no more wine available. After the service a thorough search took place, and they found a small opening in a fresh tomb; the tomb was opened and immediately a whole swarm of small flies flew out, having hatched on a recently buried body'.

The fly is the Phoridae *Conicera tibialis*, in which there can be several generations inside a tomb or a coffin, so that the adults never see daylight.

If it is true that human activity causes a change in the environment and on the whole, a reduction in the number of plant and animal species, which means a reduction in specific diversity, the number of animals attracted by human settlements in recent times is continuously increasing. In Poland, for instance, the raven *Corvus corax* nests in the immediate neighbourhood of houses; in Czechoslovakia the crow *Corvus frugilegus* is more and more accustomed to towns. Also several passerine birds nest increasingly upon and inside human dwellings (Ringleben, 1958). According to some authors, this is due to the fact that birds are becoming less afraid of man. Incidentally we would mention that the black bird, which in many European countries is more and more restricted to towns does not nest in the towns of Northern Czechoslovakia: due to the fact that in this region there are large populations of the song thrush *Turdus musicus*, with which the black bird enters in competition.

One of the most abundant animal groups – both qualitatively and quantitatively – in towns is the birds. At Milan for instance, Moltoni finds the following numbers of species, sub-divided according to their ecology:

Resident species	16
Summer visitors	8
Winter visitors	10
Regular migrating species .	7
Irregular migr. species . .	65

giving an approximate total of 106 species. In Central Europe, according to Strawinski, there are some 28 species more or less restricted to towns.

Obviously the specific composition of town faunas varies accordingly to latitude and other geographical conditions, as the following table shows:

Central Europe	Vienna	Forlì (Northern Italy)
	Birds	
–	*Falco tinnunculus*	*Falco tinnunculus*
	Corvus frugilegus	–
Ciconia ciconia	–	–
Tyto alba	*Tyto alba*	*Tyto alba*
Athene noctua	*Athene noctua*	*Athene noctua*
Iynx torquilla	–	–
Apus apus	*Apus apus*	*Apus apus*
Coloeus monedula	*Coloeus monedula*	–
–	*Turdus merula*	–
–	*Parus major*	–
–	*Carduelis chloris*	–
–	*Columba livia*	–
Chelidon rustica	–	*Chelidon rustica*
Hirundo urbica	–	*Hirundo urbica*
Phoenicursus ochruros	*Phoenicursus ochruros*	–
Motacilla alba	–	–
Hippolais icterina	–	–
Galerida cristata	–	–
Serinus canaria	–	–
Passer domesticus	*Passer domesticus*	*Passer italiae*
	Mammals	
Crocidura russula		
Rhinolophus hipposideros	–	*Rhin. ferrum-equinum*
Plecotus auritus	–	*Plecotus auritus*
–	*Vespertilio discolor*	–
Mustela nivalis	–	–
Mustela foina	*Mustela foina*	–
Mus musculus	*M. musculus spicilegus*	*M. musculus*
Rattus rattus	–	*Rattus rattus*
Rattus norvegicus	*Rattus norvegicus*	*Rattus norvegicus*

Very interesting from the ecological standpoint is the quantitative composition of the different zoological groups constituting the fauna in town centres, in allotments and kitchen gardens of the suburbs, and in the outskirts, so that some true gradients are recognizable. We will just report here Schweiger's data obtained for Vienna:

Zoological group	No of species		
	marginal zone outskirts	gardens suburbs	central, zone
Carabus (beetles)	17	7	0
Bombus (Hymenoptera)	13	7	1
meadow Orthoptera	32	6	1
forest Orthoptera	20	4	1
nesting Fringillidae	12	7	1
Reptiles	7	4	0
Rodents	17	12	4
Dermestidae (beetles)	24	15	27

A quite peculiar aspect of European synanthropous species is the presence of ectoparasites or haematophagous insects, which attack man. In the past all our dwellings, except those situated at the high latitudes, were inhabited by several parasites, such as lice (both of the head and of the body), bugs and fleas. Even today mosquitos are abundant. This partly explains the frequent occurrence, in the past, of several diseases due to microscopic agents transmitted by these insects, the occurrence in some parts of Europe (particularly Southern Europe) of epidemics and the presence even today of primitive malaria in several countries. Plague (in Italian peste, from the Latin pestis) and exanthematic typhus, known also as dermotyphus, or by English-speaking people simply as typhus (from the Greek tyfos, which means fancy, or even a fever with torpor) were by far the most lethal diseases. The first, due to a flea-transmitted bacterium, is still endemic today in many tropical countries but it has appeared in Europe at various times: in Athens (430-429 B.C.), described by Thucidides, where it caused Pericle's death; in Italy in 1348 (Boccaccio's description) and again in 1656, from where we have an admirable description in Manzoni's 'Promessi sposi'. Fleas are living – besides on rats – upon mice, marmots and occasionally dogs and cats, to which they pass from dead rats. As a prophylactic measure, the so-called de-ratting must

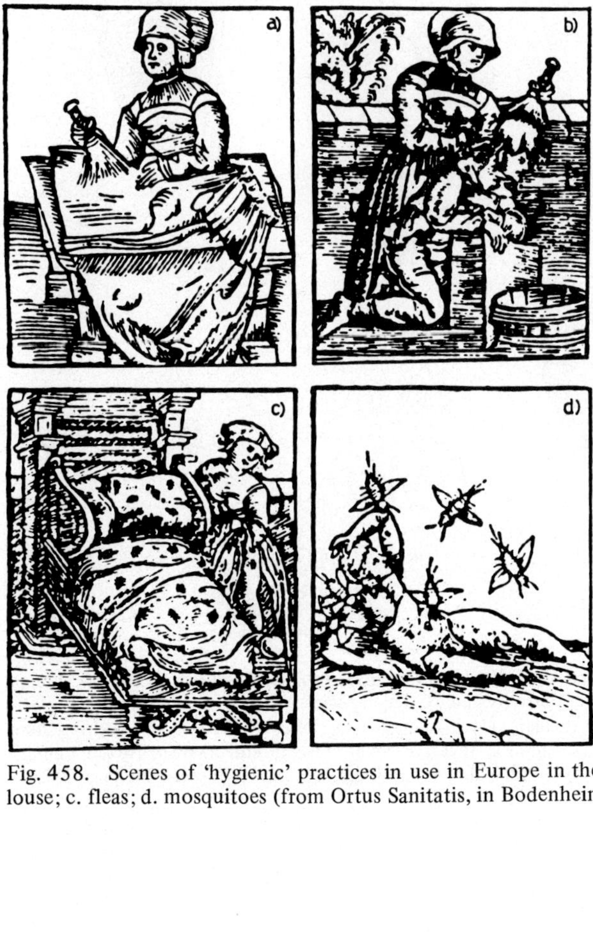

Fig. 458. Scenes of 'hygienic' practices in use in Europe in the past: a. cloth moths; b. head louse; c. fleas; d. mosquitoes (from Ortus Sanitatis, in Bodenheimer, 1928).

be mentioned (the last one was performed in Europe after World War II in Trieste by the Allies). The plague is also mentioned in the Bible (Samuel, II,3).

The first quarantine was been performed in 1377 by the Republic of Ragusa (today Dubrovnik, Jugoslavia). Generally plague is due to the infected ship rats arriving from countries where the disease is endemic. In the past plague was able to spread easily in Europe because of the poor hygiene conditions. The presence of fleas is documented in Augsburg's 'Ortus Sanitatis' (1536) (Fig. 458). Rats, beside plague may in Europe transmit another infectious disease, the so-called sodoku, a Japanese word coming from so, mouse, and doku, poison, due to a spirochaete, *Spirillum morsus-muris* (or *Spirillum minus*). This disease is transmitted by the bite of rats which can come in contact with man in cellars, roofs, warehouses etc., but also through bathing or swimming in water where rats are present, may be sufficient for this disease to be transmitted to man.

Typhus is due to a *Rickettsia* transmitted by the body louse, *Pediculus humanus corporis*. It is typical of subtropical countries, but it has often made its appearence in Europe, particularly Russia, Galicia and Silesia. Known since early times,

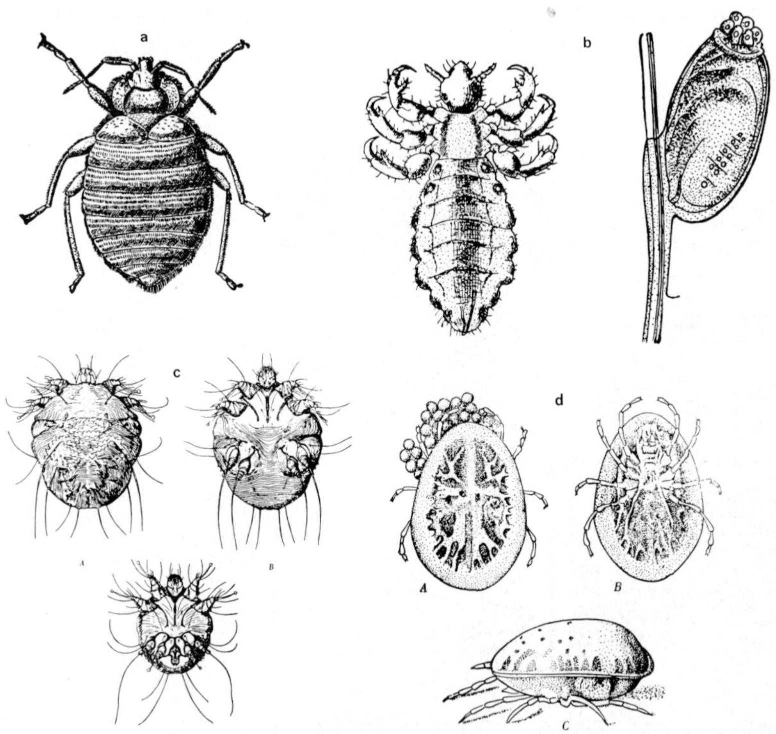

Fig. 459. a. *Cimex lectularius*; b. *Pediculus humanus capitis*; c. *Acarus siro* (a and b: female; c: male); d. the tick *Argas reflexus* (fermale). This species is common on the pidgeons at S. Marco Square, Venice, and can also attack man. It can cause a local oedema and erythema but it can also transmit the so-called Q fever, a rickettsiosis (a and b, from Neveu Lemaire; c, from Berlese; d, from Brumpt, Précis de Parasitologie, VI. ed. 1949, with the permission of Masson et Cie).

it was described first in 1546 by Fracastoro. Epidemics of typhus among troops during World War I are notorious, the trench war, associated with over-crowding and to very poor hygienic conditions, was mainly responsible. During World War II there were practically no cases of typhus thanks to the vaccination of both German and Allied troops and their prisoners of war. The louse (*Pediculus humanus capitis*) is already mentioned in Augsburg's 'Ortus Sanitatis'. (Fig. 459).

Perhaps it is convenient to mention, en passant, the presence in human settlements of many parasites found on man, or in some instances restriced to two or more hosts, man and other animals, both vertebrates or invertebrates (heter-

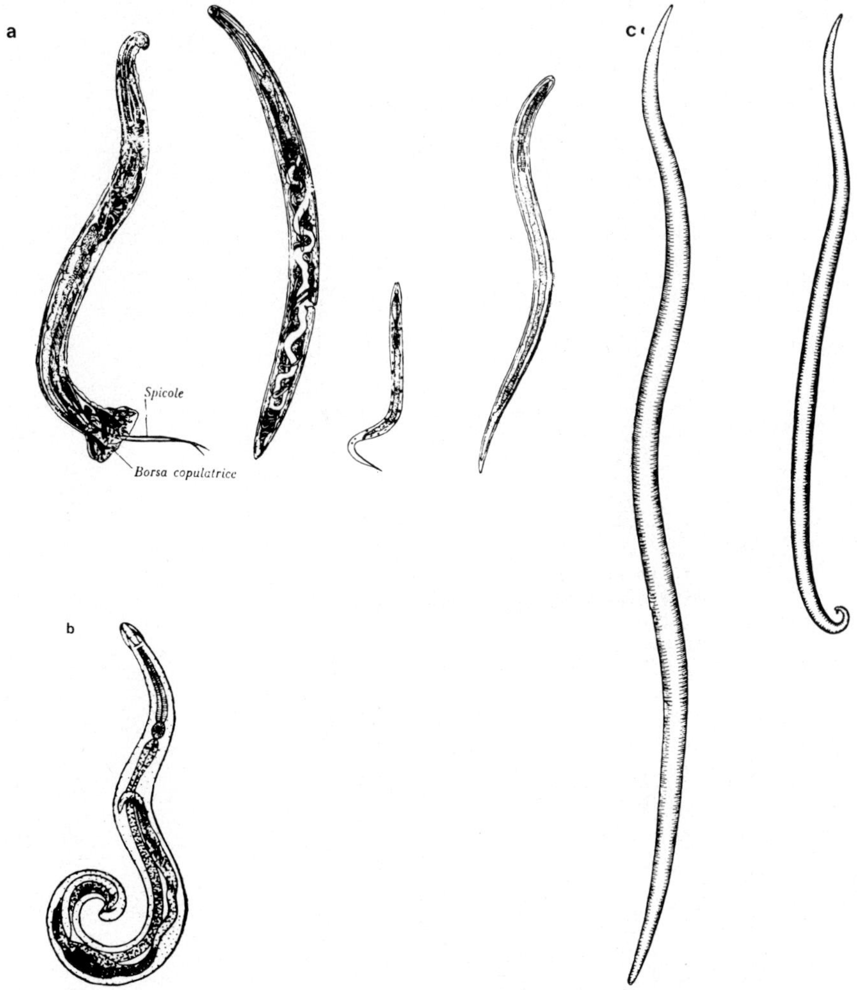

Fig. 460. a. *Ancylostoma duodenale* (male, female and two larvae); b. *Enterobius vermicularis*, male; c. *Ascaris lumbricoides*, female at left, male at right (from Leuckart, except Ascaris).

oxenic parasites). Among the former we may mention Protozoa, nematodes and ticks (generally haematophagous species). The heteroxenic parasites are represented by *Trichinella spiralis* (Fig. 460-463), *Moniliformis moniliformis*, an Acanthocephalan bound as a larva to Tenebrionid beetles (*Blaps* sp. plur.) and as an adult to rodents (rats) or to some omnivorous species such as dogs, cats and man. Also several Platyhelminthes or flat worms are heteroxenic parasites: let us remember *Hymenolepis diminuta*, bound to Tenebrionid beetles and man (or rat), or *Rodentolepis straminea*, present in *Tenebrio molitor* (the common meal worm) and man (or rat). Also the trematode *Opistorchis felineus* is a heteroxenic parasite living at different stages in its life cycle in a small freshwater snail (*Bithynia leachi*), in freshwater Cyprinids such as *Tinca vulgaris* and then in some mammals, among them man.

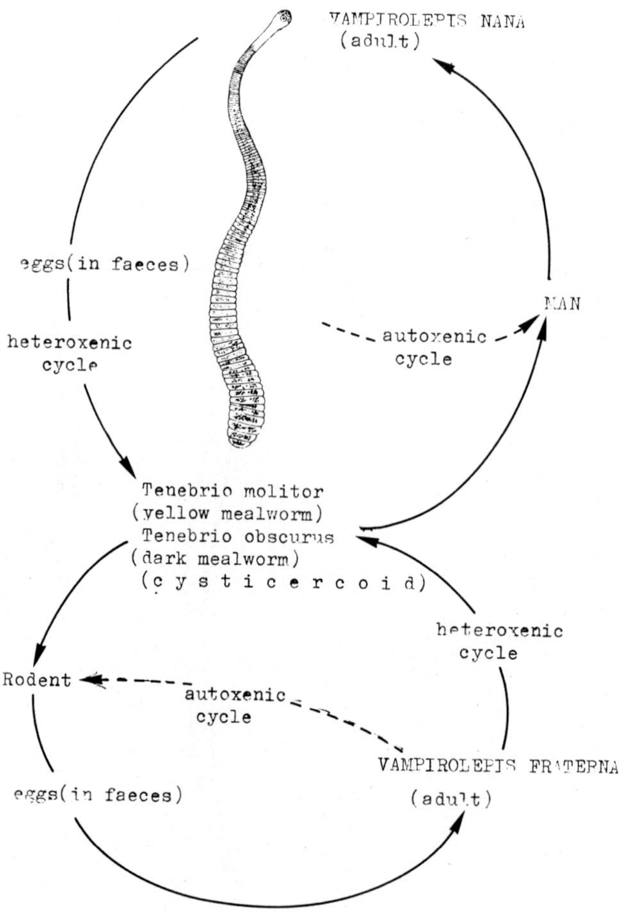

Fig. 461a. Transmission of the Cestodes (tapeworms) *Vampirolepis nana* and *V. fraterna* (figure from Leuckart).

The domestic cricket, *Gryllus* (or *Acheta*) *domesticus*, already the object of tales and poetry of our childhood, seems to play a rôle in the transmission of some parasites of domestic rodents, such as *Hymenolepis nana* var. *fraterna*, a parasite also found in man and more particularly in children of many countries, particularly in the South (e.g., Sicily). It is worthwhile mentioning Vallisneri's contribution to human helminthology (De vermium humanorum generatione, Patavii, 1710) and to the knowledge of some microscopic parasites of cattle, living in both blood and viscera, anguilliform, also found in the aorta of horse, where 'an enormous aneurism' may be present.

The climate of towns and their outskirts offers many interesting peculiarities, which however are not much studied, at least in Europe. Temperature in urban areas during summer may be much higher than in the adjacent countryside. This fact can cause heavy rains on some days. Of course, particulate matter emitted by various factories, houses and in general urban buildings is particularly concentrated in the central areas of the town, especially when we compare them with the countryside (Fig. 464). Of course this has an undoubted relevance

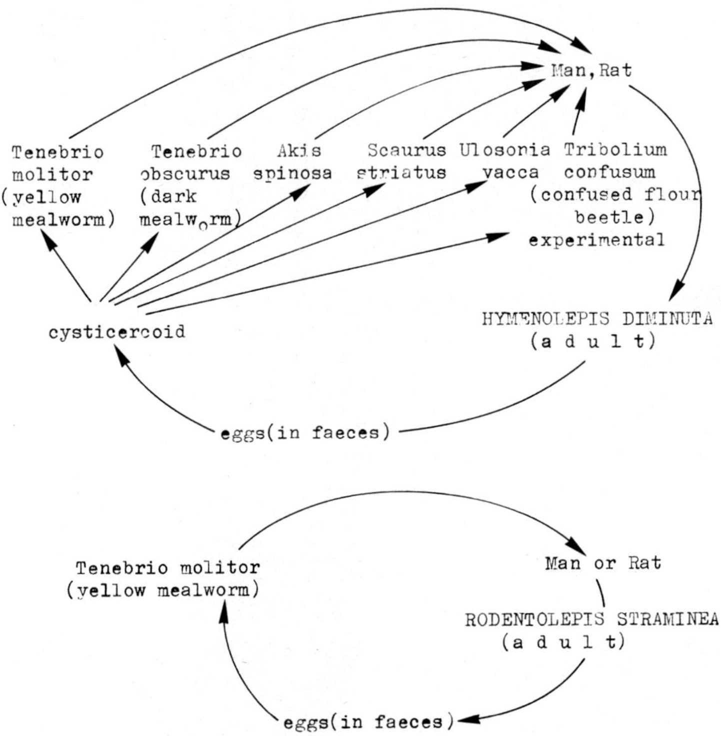

Fig. 461b. Top: Scheme of transmission of the tape-worm *Hymenolepis diminuta* to man. The intermediary hosts are several members of the Coleopterous family *Tenebrionidae*. Bottom: Scheme of transmission of the tape-worm *Rodentolepis straminaea* to man. The intermediary host is the mealworm, *Tenebrio molitor*.

when we consider human health and welfare. Still more serious is the presence of several chemical substances in the air near towns, although this of course mainly depends on the kind of industries. Also the use of the conventional fuels (coal, petrol, naphta, etc.) contributes greatly to air pollution (Fig. 465-467), which in some towns today reaches dangerous levels. A whole branch of health care deals today with these problems. For instance, in some factories (as at Porto Marghera, near Venice) several chemicals surpass the maximum allowed concentration (or MAC): sulfurous anhydride, sulfuric acid, hydrogen sulfide, ammonia, beside dust and noise. The continuous development of industry and the increasing use in agriculture of many chemicals (herbicides, insecticides, fungicides, fertilizers etc.) have determined the diffusion of several diseases formely very rare or even absent, called 'work diseases'. Unfortunately there is today, in some coun-

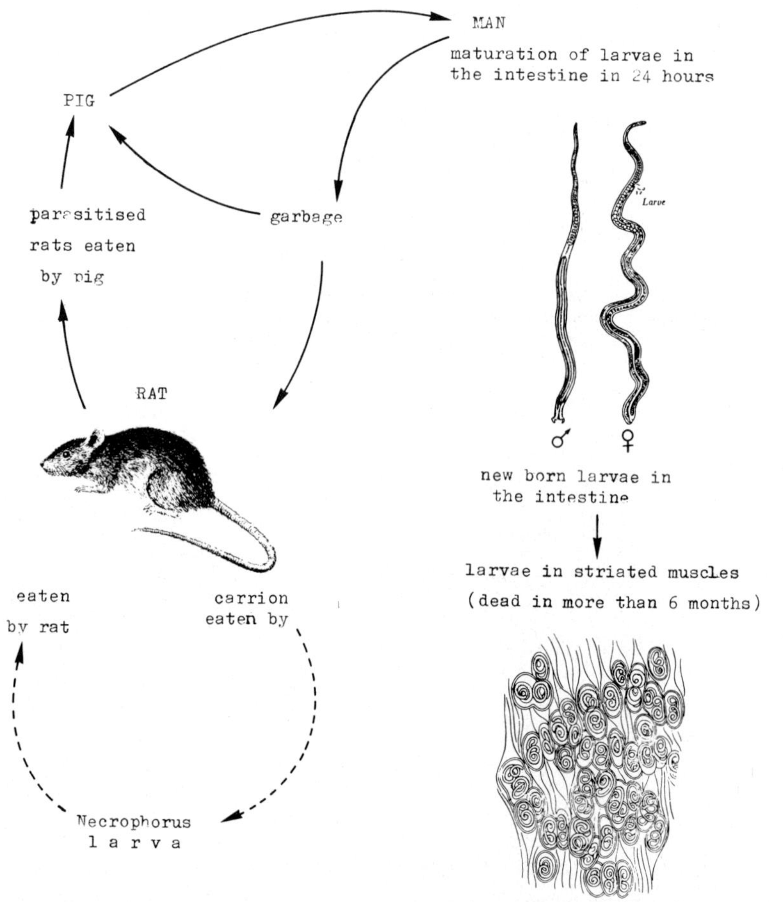

Fig. 462. Schematic cycle of *Trichinella spiralis*, a parasite very frequent in countries where there is a large use of swine products. (*Trichinella spiralis* from Brumpt, Précis de Parasitologie, VI. ed. 1949, with the permission of Masson et Cie.).

tries (even in technically well developed countries), a tendency to confuse ecology tout court and the study of pollution.

Finally, beside the climate of the towns, we are confronted with many micro-climates in the various kinds of dwellings, the sitting, height, presence of window etc. Also the study of micro-climates can have some sanitary implications. It is enough to mention the different systems today existing to acclimatize the premises in which man lives (air conditioning etc.).

The biological productivity of urban areas is very little understood: we wish to mention in this connection some data obtained about secondary productivity, and more precisely about bird populations in some areas of Central Europe:

Biotope	Number of pairs/km²
Gardens and houses in the agricultural region of Rhenania	2700-2000
Berlin cemetery	1630
Frankfurt zoological garden	1459
Kitchen gardens of Rhenania	384-334
Industrial region of a large town of the Ruhr	338

It is evident that the urban environment is extremely suitable to bird life. Of course it is not possible to extend these results to other animal groups.

The birth of the town was the immediate consequence of agriculture, since before the appearance of this type of human activity man was characterized by nomadic habits. Therefore it is only during the Neolithic in the Mediterranean basin that the first towns appear. Originally they were very small: in classical Greece we find the so-called 'town-state' including an urban centre (Athens) and

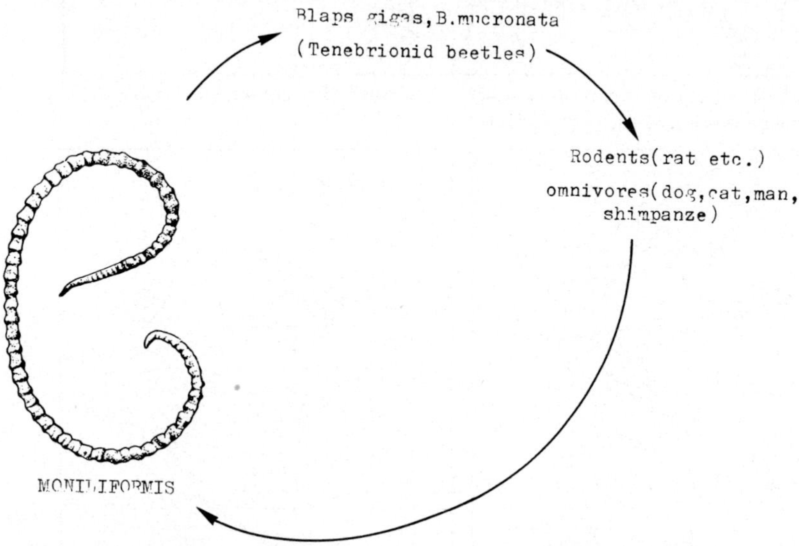

Fig. 463. Transmission of the Acanthocephal *Moniliformis moniliformis* (figure from Brumpt). (Précis de Parasitologie, VI ed. 1949. With the permission of Masson et Cie.).

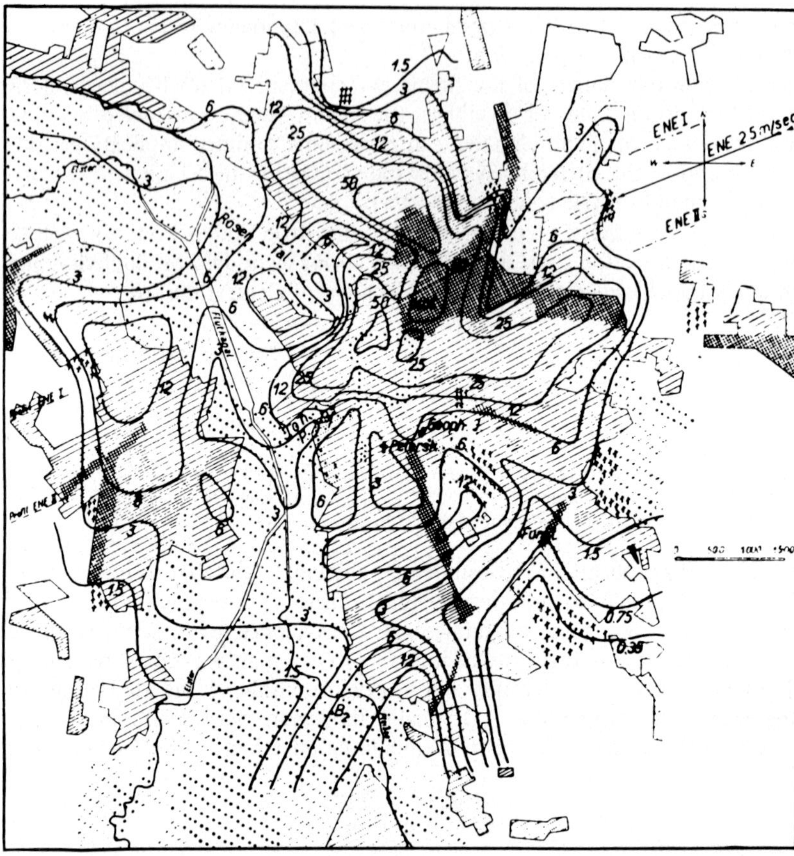

Fig. 464. Distribution of dust in Leipzig (E. Germany) with an ENE wind (from Geiger). (With the permission of Viehweg Verlag).

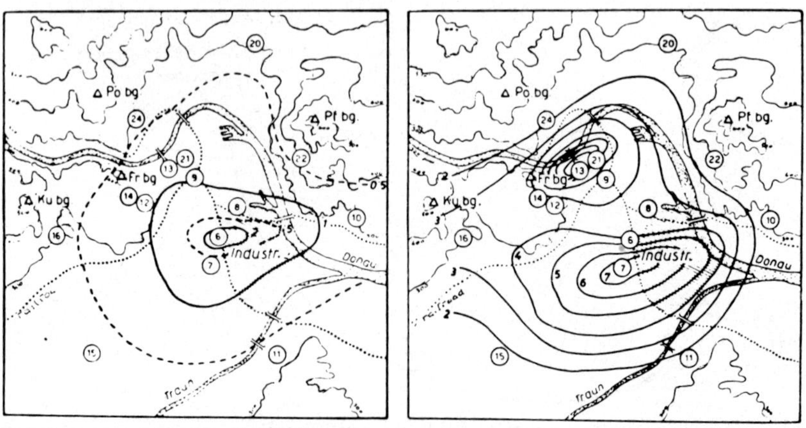

Fig. 465. Distribution of SO_2 at Linz (Austria) during summer (left) and during winter (right) (from Geiger). (With the permission of Viehweg Verlag).

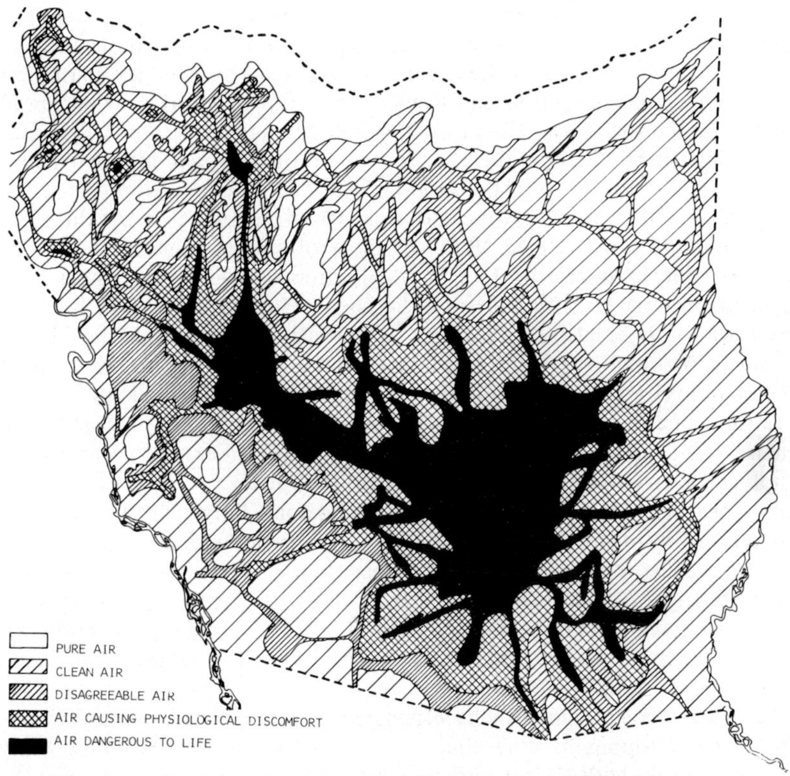

Fig. 466. Example of air pollution in a large European town (Milano).

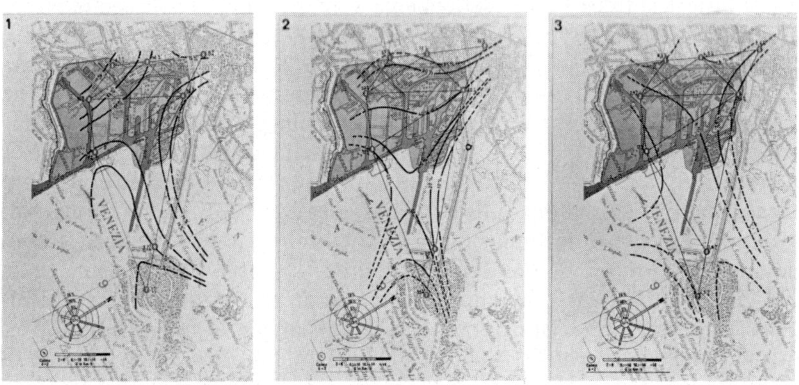

Fig. 467. 1. Isoprobability to surpass 0,01 mg/m³ of fluorides (1971); 2. Isoprobability to surpass 0,075 ppm of SO_2 (1971/72); 3. Isoprobability to surpass 0,05 ppm of N-oxides (1971) at Venice (from MONTEDISON, Ricerche sull'inquinamento atmosferico della zona di Porto Marghera-Mestre-Venezia).

its rural environment; its harbour (Piraeus) was also included in this 'town'. The total population amounted to nearly 60-70,000 persons, as estimated on the basis of the grain stores which, when the town was blockaded by enemies, permitted the occupation of Athen. During the Roman times also 'territorial organization' was based upon the town: the 'civitas' corresponded to a town with its neighbouring area. During the Middle Ages the market square is often closely associated with the church, whence the modern definition of 'cathedral-town'. More often the town coincided with a fortified nucleus: this is well documented by the great number of present towns called 'Burg' or more often whose name finishes in 'chester' (from the Latin castrum), or in 'burgh'. Burg, or even burgh comes from the Anglo-Saxon burh or burg and from the Middle English burgh, which means fortified town. It is practically the same as borough, coming from the Middle English borwe or borgh, or from Anglo-Saxon burh, which also means fortified town. These terms have their origin in the word beorgan, to protect. Thus, we find today, as the name of town, Burg in France in the Hautes Pyrénées, near Bourg; Burg in the Netherlands on the island of Texel in the North Sea; Burg near Magdeburg, another one in Dithmarschen and another in the Spreewald; Burg auf Fehmarn on an island near Kiel,Hohe Burg, a small hill near Rostock, Burgau in Schwaben etc. As a suffix, we find it in Vicksburg, Pittsburgh, Edinburgh etc. or even in Peterborough. Chester as a suffix appears in Manchester. In Eastern Europe also the town takes its origin from a fortified nucleus (gorods). During the Middle Ages the population of a town was rarely greater than 10,000: in England only London and York reached such dimensions. In 1400 London had reached 50,000 inhabitants and in XIV century Paris 250,000.

In the Middle Ages many towns originated from markets ('market-town'): many present day toponyms show this origin e.g. Neumarkt, near Salzburg. Many German towns begin with 'Markt', others are formed of two words, of which the first is Markt: Lustenau in Franconia, Oberdorf in Western Bavaria, Rettenbach in Schwaben. In England we find Market Drayton, Market Harborough, Market Weighton etc.

In early historical times human settlements were generally established in places well defined topographically, such as the top of a hill (acropolis, Castellieri etc.), on the shores of a river or a lake (Stratford-on-Avon, Kingston upon Hull, Frankfurt a. Main, etc.) or in harbours well protected against winds and waves. The establishment of towns along rivers, was often determined by crossing points. This is the reason why today there are many toponyms in which we find Bruck or bridge respectively, or, in Great Britain, the term ford. There is Bruges in Belgium (from Brugge, in Flemmish bridge), Bruck a.d. Leitha in Lower Austria; Bruck a.d. Mur, in Styria, Brückenau in Franconia, Saarbrücken, the capital of Saar, etc. In England we find 'bridge' in Cambridge, Stourbridge (near Birmingham) or in Bridgend, a small town in Wales and another one on an island of Western Scotland. Furthermore, we have Bridge of Allan, B. of Cally, B. of Dee and B. of Weir in Scotland. The name 'ford' (Anglo-Saxon and Middle English) is present in Guilford (Surrey), Hereford (West of Birmingham), Kingswinford and Stafford near Birmingham, or Stratford-on-Avon (Shakespeare's town).

Very often towns developed as resting places along routes particularly at crossroads. Towns according to Smailes are 'nodes of road systems and their importance is faithfully reflected by the measure of 'nodality' (Fig. 468) they possess. After the XVIII century, as a consequence of the industrial revolution in Britain, and the enormous progress in agriculture including animal breeding and mechanization, there was a diminution of the number of persons employed in agriculture and a parallel increase of those employed in secondary and tertiary activities. As an immediate consequence 'urbanization' occurred increasing the size of pre-existing towns. Today the town may be considered as a part of the earth surface differentiated through the presence of buildings and other characteristic structures. Generally a gradient between town and country is easily recognizable. To speak of a town today it is sufficient to include a shopping centre, three or four banks, a large department store, plus cultural, recreational and sanitary services.

Often two or more urban centres, originally separate, unite: this phenomen is called a 'conurbation'. Examples are Greater London, with 8-9 million inhabitants, composed of the city (4210 inhabitants in 1968) and 32 boroughs (Fig. 469). English towns are 'urban agglomerations': Birmingham, Leeds, Liverpool, Manchester, Newcastle-upon-Tyne (besides the already mentioned London). Paris has 2,590,000 inhabitants in the town, 8,196,000 inhabitants in the urban agglomeration. Greater Vienna in 1966 had 1,638,000 inhabitants; the urban agglomeration of Moscow has 6,507,000 inhabitants (1967).

Today a town is subdivided at least in some parts of Europe into three zones: the central business district (City), the residential zone and the industrial zone.

Finally, the 'urban catchment' is composed of a town and all the rural areas which look to the town and are influenced by it. These areas utilize institutions and services located in the town. Closely related to and dependent upon the 'urban catchment' there is the phenomenon known as 'commuter movements' (Fig. 470).

Closely associated with human settlements, with the possible exception of the northernmost ones, situated in the Arctic, and some isolated dwellings situated on

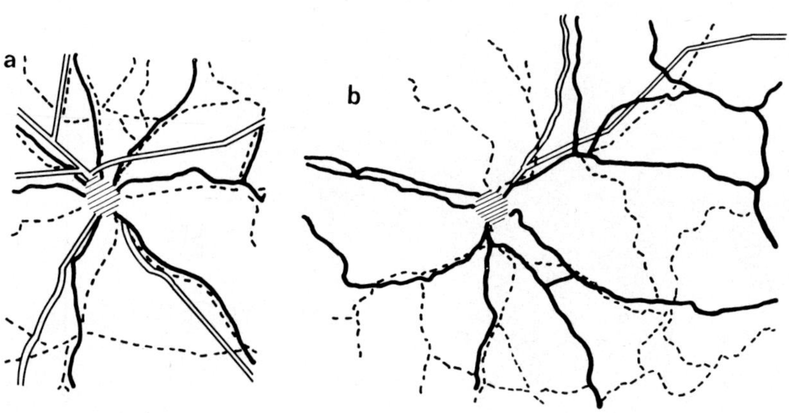

Fig. 468. Examples of modality. Left: Milano, right: Torino. ——— roads, ═══ highways, – – – – – railways.

rocky sea shores or in some hot/dry regions of Southern Europe, we find gardens, parks and often kitchen gardens. Indeed, in town planning today one speaks of 'public green'.

The variety of plants growing in these environments – both spontaneous and cultivated – is enormous. As a matter of fact this environment shows a range from natural conditions (ancient parks with spontaneous trees, such as beech, birch,

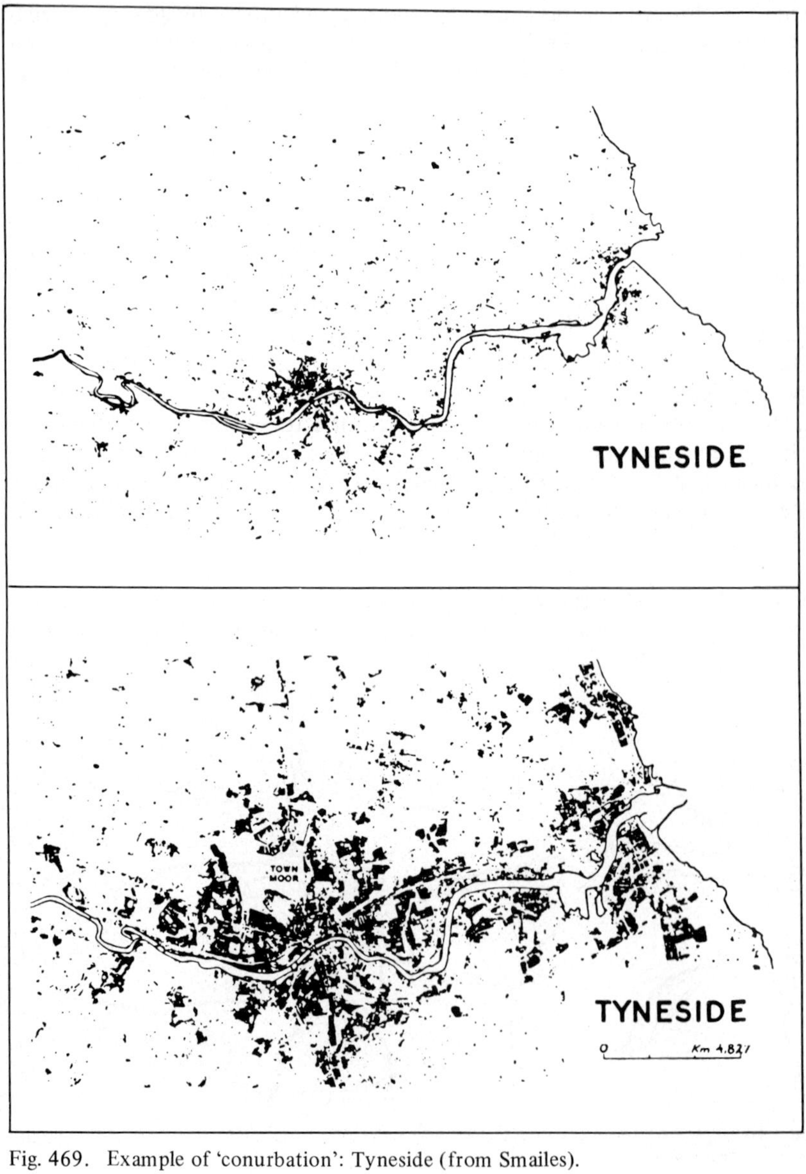

Fig. 469. Example of 'conurbation': Tyneside (from Smailes).

spruce, fir, pine etc.), to rather artificial gardens, in which there are a variety of plants coming from all parts of the world, associated with some spontaneous plants, to the modest kitchen gardens, in which only a few edible plants are cultivated.

Let us begin with kitchen gardens. Although most species are the same as those cultivated in fields on an large scale for both local and regional markets, plants cultivated in kitchen gardens are restricted to the local market and very often exclusively to the grower's family (subsistence agriculture).

The same holds for flower gardens, in which, together with plants cultivated on an agricultural basis (clover, roses, chrysanthemums, etc.) people cultivate ornamental plants which grow exclusively in gardens or, in some instances, only on the window balconies of houses.

Most of the plants cultivated in our kitchen gardens are food plants, very few having industrial uses, for instance the poppy, *Papaver somniferum* from which opium is extracted, a beautiful plant of oriental origin, with rose-coloured flowers, generally represented by a 'double' variety. The vegetables include *Raphanus sativus*, or radish, with two varieties, *radicula*, with a red or white root, and *niger*, with a root which is employed as a relish to give savour to meat; the chick-pea *Cicer arietinum*; the broad bean *Vicia faba*, or the lentil *Lens esculenta* or *Vicia lens*, the sweet pea, *Lathyrus odoratus*, the pea *Pisum sativum*, the bean *Phaseolus vulgaris* and the runner bean, *Phaseolus coccineus*; the bottle gourd, *Lagenaria vulgaris*, whose fruits are used to make containers, and the common marrow, *Cucurbita*

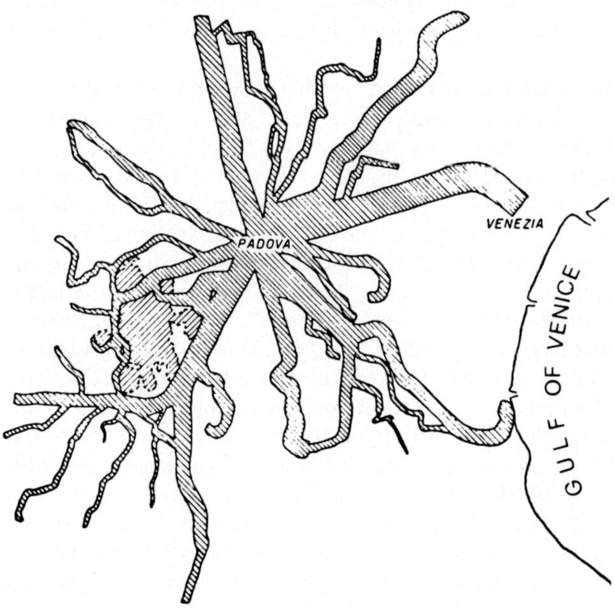

Fig. 470. Example of commuter pattern to and from Padua: the width of the belts which unite the various towns and villages is proportional to the number of hours necessary to make the journey (from Marcuzzi & Camuffo).

pepo; the cucumber *Cucumis sativa*; the melon *Cecumis melo*, with its varieties *cantalupo* (cantaloupe) and *maltense*; the water melon *Cucumis citrullus*, the garlic *Allium sativum*, the leek, *Allium porrum* and the onion, *Allium cepa*. The latter have been used since Roman times, particularly in Southern Europe, when a poet said 'dura sunt vobis ilia, messores, quia estis allia nocentia cicutis', which is completely false!

Other plants cultivated in some parts of Europe are the red currant, *Ribes rubrum*, the sun flower *Helianthus annuus*, originally from Peru and used for its oil, employed for cooking as well as to make soap, varnish etc. The seeds of the sun flower are utilized to feed chickens. *Helianthus tuberosus* is originally from Brasil and now grows in a semi-wild state along roads, canals etc.; it has an edible root; *Carthamus tinctorius*; the safflower (from the Arab qurtum, whence the Medieval Latin Carthamus), is grown for its oil to make soap. The plant is originally from the Eastern Mediterranean or Africa. The endive *Cichorium endivia* is much used in Italy; the benne plant (in Malay bijen) *Sesamus orientalis* (of oriental origin and etymology) is employed to make an oil (sesame oil) used both as a food and to make margarine.

The snapdragon *Antirrhinum majus* with variously coloured flowers is present in some parts of Europe as a spontaneous plant; marjoram, *Origanum majorana*, which together with sage, *Salvia officinalis*, is typical of the European sclerophyllous forest as is also rosemary, *Rosmarinus officinalis*. The sweet basil *Ocymum basilicum* is an ingredient of the savoured South European cooking. In Italy, these herbs are called 'aromi' – which literally means 'odours' – which is quite understandable when we think that all these species belong to the family Labiatae, characterized by the presence in their green parts of aromatic substances (essential oils).

In most vegetable gardens fruit trees and in Southern Europe, vines grow: let us remember the cherry tree, *Prunus avium*, and a related species, *Prunus cerasus*, whose fruit (for example, the so-called 'marasca') are used to make liquors.

Flower gardens are characterized by an enormous variety of cultivated plants often grown because of their beautiful flowers: we will mention here only the larkspur, *Delphinium ajacis* , a Ranunculacea, the wallflower *Cheiranthus cheiri*, a Crucifera, *Alcea* (or *Althaea*, from the Greek althaia, the wild mallow) *rosea*, or garden hollyhock, together with the Malvaceae *Lavatera trimestris, L. punctata* and *Hibiscus syriacus* (from the Greek ἴβισκος, marsh mallow), or rose of Sharon. The Compositae *Santolina chamae-cyparissus*, the *Pyrethrum indicum*, together with other species are used to produce the well known insect powder. *Cineraria maritima* indigenous to Dalmatia and *Calendula officinalis*, or pot marigold are other Compositae. Some species have a bushy habit, for instance the white jasmine *Jasminum officinale* (from the Arab yasmin) and *Jasminum fruticans*, utilized to make hedges; the oleander *Nerium oleander* (of uncertain etymology) native to the Mediterranean region; *Lycium barbarum*, native to Africa, a shrub, and some herbs such as marigold, *Tagetes* (from Tages, an Etruscan deity) in the Netherlands cultivated on an industrial scale (very interesting research has been done on the nematodes parasite in the roots of this plant); the Indian cress, *Tropaeolum majus*, growing often on the balconies of windows (in Greek tropaios, from trepein, to turn); *Amaranthus caudatus*, or amaranth, with a vivid magenta red colour; the poke weed *Phytolacca decandra*, characterized by its conspicuous, red fruits, *Plumbago europaea*, native to Southern Europe, with vivid blue flowers, often visited by the moth *Macroglossa stellutarum*, which, when it is suspended in the air during flight, recalls somewhat a small honey-creeper; *Salvia splendens* with dark red flowers; *Canna indica* (known as canna) a monocotyledon plant whose flowers exhibit a tremendous variety of colours, as one may appreciate for example at the Jardin des Plantes in Paris; the iris (or flags) *Iris florentina*, with perfumed, white conspicuous flowers, and *Iris germanica*, with dark violet flowers; the many species of crocuses (*Crocus*) some of which are indigenous to South Eastern Europe, flowering in early Spring (short-day plants), and the Amaryllidaceae of the genus *Narcissus* such as the daffodil *N. tazzetta*, the jonquil *N. jonquilla*, besides *N. incomparabilis* and *N. pseudonarcissus*; several of them grow wild in South-Eastern Europe or Western Asia. Finally we should mention the lily, *Lilium candidum*, and the tulips.

The variety of trees cultivated in the parks and gardens is also great: among Gymnospermae let us remember a juniper, *Juniperus virginiana*, some cultivated firs such as *Abies cephalonica, Abies canadensis, Abies cilicica, A. pinsapo* etc.; *Araucaria embricata*, native to South America; the cedars, *Cedrus deodara* and *C. libani* from the Near East and *Cedrus atlantica* from North-West Africa; the elegant *Gingko biloba*, a true 'living fossil' since the family dates from Paleozoic and the genus from Jurassic; *Sequoia gigantea; Taxodium distichum*, not dissimilar from the preceding species, cultivated in wet areas such as the edges of ponds and streams in our gardens and parks; it is known as 'Virginia cypress', and is characteristic because of its aerial roots or pneumatophores.

Most ornamental trees belong of course to the Angiospermae. In Southern Europe several palms grow, such as *Phoenix dactylifera, Chamaerops humilis* (in some parts of Europe still today spontaneous), together with the palm-like Gymnosperm, *Cycas revoluta* etc. Widely distributed species are *Negundo fraxinicola* (or *Acer negundo*), native to America, *Aesculus hippocastanus,* with white flowers, and *A. rubicundus*, with dark pink flowers, both known as horse chestnut because of the close similarity of their fruits with those of the common chestnut, *Melia azadarach* (from the Greek melia, ash) called Persian lilac, pride-of-India or even bead tree, originally from the Tropical areas of the Old World, *Evonymus* (or *Euonymus*) *japonicus, Pittosporus tobira*, native to Australia and New Zealand, with some whitish flowers giving off a strong, sweetish, perfume; the tree of heaven *Ailanthus glandulosa* (from ailanto, Malacca) originally from China and the Moluccas, well know because of the so-called ailanthus moth, *Platysamia cynthia*, native to China and cultivated in the Eastern USA because of its silk producing larva. This is highly characteristic because before metamorphosis it weaves a cocoon composed partly of silk and partly of small pieces of ailanthus leaves, a habit rather unique. Some ornamental trees belong to Leguminosae sensu lato: let us mention *Amorpha fruticosa*, or bastard indigo; a North American tree *Gleditchia triacanthos*, or honey locust, with fruits resembling carobs and a thorny trunk, originally from North America; and *Albizzia iulibrissim*, an oriental species, allied to *Mespilus japonica*; *Tamarix africana*, not dissimilar from the common tamarisk *Tamarix gallica*, a European species; *Bignonia catalpa* and *Paulownia imperialis*, both belonging to *Bignoniaceae*, with very beautiful pale lilac flowers, an evergreen shrub, *Arbutus andrachne*, native to Greece, the holly (from Anglo-Saxon holen or holegn) *Ilex aquifolium*, wild in several parts of Europe with a mild rather humid climate, *Ligustrum japonicum*, an evergreen shrub used to make hedges, the lilac (from the Arab lūlăk and the Persian lilaj or even lilianj) *Syringa vulgaris*, in Italy known as 'fior di maggio', which literally means 'may-flower', with strongly perfumed flowers, white or lilac coloured; the paper mulberry *Broussonetia papyrifera*, belonging to Moraceae, in some parts of Southern Europe turning to a semi-wild state, native to China; the sweeping willow (or Babylonian willow) *Salix babylonica*, and many others. Together with trees, some peculiar monocotyledon plants are cultivated, for example *Agave* sp.plur. and a small ornamental shrub, *Ruscus hypoglossum* belonging to Asparaginaceae, similar to its relative *R. aculeatus*, the common butchers broom.

Hedges are generally formed by *Robinia pseudacacia*, the false acacia or black locust tree, originally from North America and now growing in an almost wild state in many parts of Europe, the box tree, *Buxus sempervirens*, wild in several parts of Southern Europe, the already mentioned *Ligustrum japonicum*, and others. Vines are also very widespread: let us mention *Ampelopsis quinquefolia* and several species of *Cissus*' native to America, characterized by a simple, non subdivided leaf. Sometimes they literally cover walls of old houses, as in Paris, Rome, Venice and other towns.

Finally in towns trees have been planted along main roads or avenues: examples are the plane tree *Platanus orientalis* (from the Greek πλατανος) native to the Far East; the cypress or *Cupressus sempervirens*, of Mediterranean etymology, coming from the East and introduced into Italy by the Romans, *Thuja occidentalis*, originally from North America, and several others.

The European parks have an aura reminiscent of a bygone age: Schönbrunn near Vienna; Versailles and Les Tuilleries in Paris, which entertained the last King of France before the Revolution; the magnificent park of Windsor Castle, kept in the English style merging with the natural vegetation, which,because of the mild climate, is extremely luxuriant, composed both of coniferous and broad-leaved trees; or even the park of Villa Reale at Stra, near Padua, along the Riviera del Brenta, modelled on the French Settecento; the 2400 ha. estate of La Mandria, near Turin, dating from 1500, when Emanuele Filiberto was alive. It is said that Tasso, exiled from Turin, took inspiration here to describe the 'Giardini di Armida'. Also there is the park of Villa Reale of Portici near Naples, with practically subtropical vegetation comprising enormous holm oaks, Mimosas, giant *Agave* etc., spread out at the foot of Vesuvius, whose winter snow-covered cap contrasts with the luxuriant evergreen of the surrounding vegetation.

The parks of London (Hyde Park), Cambridge (Park Parade) and many other English towns are also famous. Victor Hugo has given us a detailed and long description of the park of Luxembourg in Paris; a beautiful, although a little sophisticated, park is that of Miramare, near Trieste, created by the archduke Maximilian Ferdinand Joseph of Habsburg, who lived there before his departure for Mexico. Famous for its beauty is the garden of Boboli, at Florence.

Often in Italy many parks are exploited, as are many other things of this country, for recreation and tourism, and therefore they are too commercialised, with clashes of colours (both of flowers and trees) not always in good taste: this maybe is the case at Stresa, Chianciano, Montecatini, Sirmione etc.

Some parks on the contrary are characterized by their almost natural conditions and vegetation: Fontainebleau, Vincennes etc.

Often the European parks have been subjects for description by writers and poets: 'Il Giardino dei Finzi Contini', 'I Sepolcri' etc. Some concealed, abandoned corners of old, great parks, neglected by time, invaded by solitude and overgrown with weeds have inspired some 'crepusculari' writers of 'fin de siecle': Victor Hugo offers us a vivid description of a garden of Rue Plumet in his 'Les Miserables', which, as a matter of fact, demonstrates his very keen naturalistic knowledge.

However, the most delicate descriptions of writers and poets of all times have been inspired by small, humble, kitchen gardens, where more than in the gardens or fields, human skillfulness is associated with a love for nature. In antiquity, a great poet, sensitive to the grace and poetry of a humble kitchen garden, was Virgil, whose poem we reprint here (Georgicae, Liber IV, 125-138):

... memini me ...
Corycium vidisse senem, cui pauca relicti
iugera ruris erant, nec fertilis illa juvencis
nec Cereri opportuna seges nec commoda Baccho.
Hic rarum tamen in dumis olus albaque circum
lilia verbenasque premens vescumque papaver,
regum aequabat opes animis, seraque revertens
nocte domum dapibus mensas onerabat inemptis.
Primus vere rosam atque autumno carpere poma,
et cum tristis hiems etiammum frigore saxa
rumperet et glacie cursus frenaret aquarum,
ille comam mollis iam tondebat hyacinthi,
aestatem increpitans seram Zephyrosque morantes.

Another proof of this sensitivity is given by a passage by Oscar Wilde in 'the devoted friend', ... 'In all the country-side there was no garden so lovely as his. Sweet-Williams grew there, and Gilly-flowers, and Shepherds' purses, and Fair-maids of France. There were damask Roses, and yellow Roses, lilac Crocuses and gold, purple Violets and white. Columbine and Ladysmock, Marjoram and Wild Basil, the Cowslip and the Flower-de-luce, the Daffodil and the Clove-Pink bloomed or blossomed in their proper order as the months went by, one flower taking another flower's place, so that there were always beautiful things to look at, and pleasant odours to smell'.

BIBLIOGRAPHY

Bodenheimer F.S., Materialien zur Geschichte der Entomologie zur Linné, Junk, Berlin, 1928-29 (2 vol.).

Geiger R., Das Klima der bodennahen Luftschicht, Vieweg, Braunschweig, 1961.

Marcuzzi G., Aspetti naturalistici della Dalmazia, *L'Universo,* 48, 1968, 567.

Marcuzzi G., Aspetti naturalistici delle Isole Eolie, *L'Universo,* 50, 1970, 915.

Marcuzzi G., Il saccheggio dei Colli Euganei: proposta per un piano di conservazione, *Acqua & Aria,* IV, N.16, 1971, 51.

Marcuzzi G., Ecologia umana, ed. Patron, Bologna, 1976.
Marcuzzi, G. & A. Camuffo, Prima applicazione della teoria dell' informazione allo studio dell' ecologia umana (First application of information theory to the study of human ecology), *Riv. Biol.*, 61, 1968, 293.
Moltoni E., Gli uccelli della città di Milano, *Atti Soc. It. Sc. Nat.*, 92, 1953, 153.
Nestman L., Oekologie und Humanoekologie. Das räumlich-funktionale System menschlicher Umwelten, *Verh. Gesell. Oekol., Göttingen,* 1976, 491.
Paul H.A., Der Begriff der 'Territorialität' in der allgemeinen Oekologie und in der Humanoekologie, *Verh. Gesell. Oekol., Göttingen,* 1976, 499.
Schweiger H., Das Tierleben in der Grossstadt, 1960.
Smailes A.E., The geography of towns, 1962 (Ital. transl. ed. Marsilio, Padova 1964).
Strawinski S., Die Vogelverstädterung vom ökologischen Standpunkt, *Ornithol. Mitt.*, 18, 1966, 72.
Tischler W., Ecology of arthropod fauna in man-made habitats: the problem of synanthropy, *Zool. Anz.*, 191, 1973, 157.
Zangheri P., La provincia di Forlì nei suoi aspetti naturali, Forlì, 1961.

ARTIFICIAL ECOSYSTEMS: THE CULTIVATED AREAS

οἷον τὸ γλυκύμαλον ἐρεύθεται ἄκρωι ἐπ' ὔσδωι
ἄκρον ἐπ' ἀκροτάτωι.
λελάθοντο δὲ μαλοδρόπηες,
οὐ μὰν ἐκλελάθοντ;
ἀλλ' οὐκ ἐδύναντ'
ἐπίκεσθαι, τοῖον . . .

(As a sweet apple which reddens on a high branch, high
on the highest ones;
it has been forgotten by the harvesters.
No. It has not been forgotten;
in vain they try to reach it . . . . .)

Sappho

The environment of the agricultural landscape is a peculiar terrestrial environment, characterized by annual crops, arboreal cultivations, etc. Earlier tree felling and water regulation brought about a uniform and stereotyped vegetation that has, in turn, caused a reduction of natural animal life. The latter has been replaced by a specialized fauna, made up of a small number of domesticated animal species more directly associated to Man's presence, plus cultivated plants. This artificial ecosystem is characterized by the parasites of plants and house animals, by phytophages, saprophages, coprophages and only to a lesser degree by their predators. As a matter of fact, the fauna of cultivated land comprises partly inhabitants of forest edges (birds, annelids, etc.), partly inhabitants of river banks (Carabids, e.g.) and partly species typical of saline areas. The retreat of woodlands northwards has permitted many steppic elements to move progressively into Central and Western Europe (Fig. 471).

Between the initial conditions and the cultivated ones (fields, etc.) there is a transitional series represented by uncut grass meadows, cut hay meadows, fertilized (or manured) pasture, cultivated perennial crops, cereal crops and finally intensive crops which require hoeing.

Among various problems of the cultivated landscape let us mention the presence of some birds particularly restricted to fields, the distribution of insects in fields, and the influence of the protective shelter belts and forest edges on the distribution of insects.

A human influenced, although uncultivated environment widespread in Europe is made up of the meadows of temperate climates. In Spain, for example, this environment represents 40% of all cultivated land. In Europe meadows can be subdivided into dry (subdivided into Corynephoretalia, Brometalia and Nardetalia) and wet meadows. The latter are further subdivided into fertile pasture (Arrhenatereta in the valleys, Lolieto-Cynosureta in the lowlands, Triseteta and Festuceto-Cynosureta in montane areas), into Molinieta, into swampy meadows (Caricetalia) and finally into coastal meadows. In each category a group of harmful insects is present. For instance, in Schleswig-Holstein the Molinietalia are characterized by the presence of *Charaeas graminis* and *Tipula czizeki.*

The environment of meadows and grasslands (secondary grasslands) represents ideal conditions for development of parasites restricted to soil (at least during a part of their life cycle) such as slugs, wireworms, grubs and Tipulid larvae. The fauna of this environment is greatly influenced by manuring, which selects the species present in a biocenosis and influences the distribution range of various species. Manure acts particularly upon the fauna of the vegetation cover (planticolous fauna) in which a reduction of the specific diversity is caused (*Dolichopodidae*, *Empididae*, etc.).

Still more effective is the action of man in fields in which each cultivated species is characterized by a peculiar zoocenosis. As in natural cenosis, these artificial cenosis (Ghilarov's 'agrarbiocenosis') also display a regular seasonal succession both in the insect fauna and in the weeds. Cereal fields are also characterized by a peculiar entomofauna, which, however, can also live on weeds. Other crops whose entomofauna is well known are beets and *Cruciferae* (*Brassica* and *Sinapis*) fields. Many species exhibit a quiescent stage (diapause) adapted to the

Fig. 471. Net primary productivity in Europe, according to the Miami Model (Lieth & Box, 21.II.1977). Going from north to south we find: arctic deserts and arctic tundra; tundra; forested tundra; northern taiga belt; middle taiga belt; southern taiga belt; mixed forest (or 'Mischwald'), deciduous forests, sclerophyllous forests, and finally forest-steppe or meadow-steppe, as well as temperate steppe; the last two are limited to European Russia.

life cycle of the different cultivated plants. A peculiar aspect of the fauna of cultivated areas is represented by the origin of several species: here the orientation of insects, the active establishment of several animal groups, the transport of others by water, winds, or through phoresy are very important. The transmission of viruses by insects which feed by sucking plant sap (*Rhynchota, Thysanoptera*) is also an interesting aspect in the study of pests.

Perhaps more so than in the natural ecosystems we have examined so far, the environmental factors have a very profound effect on the biota – both cultivated plants, domestic animals and their parasites. Indeed, the effect is in general more evident because of the simplicity of artificial ecosystems when compared with natural ones, and still more because the former are not buffered against various external factors as natural systems are in general. The instability of the former is

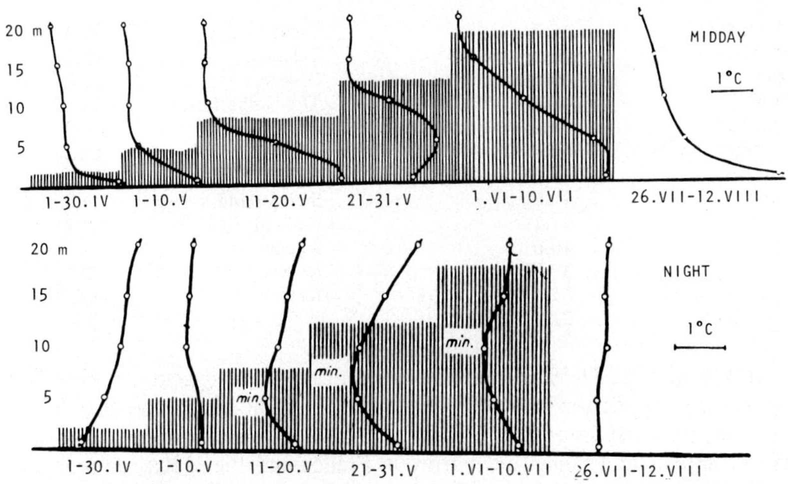

Fig. 472. Temperature profile from 0 tot 2.0 m at midday and during the night in a field of winter rye near Munich (from Geiger) (reproduced with the permission of Vieweg Verlag).

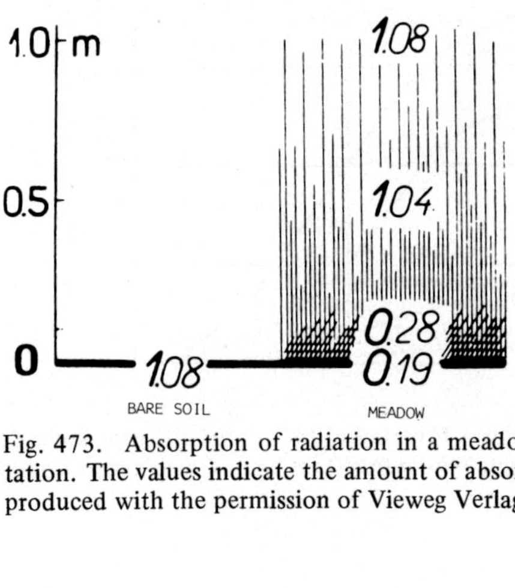

Fig. 473. Absorption of radiation in a meadow: left: bare soil; right: soil covered with vegetation. The values indicate the amount of absorbed radiation (cal/cm² /min) (from Geiger) (reproduced with the permission of Vieweg Verlag).

well recognized and measurable if we apply Information Theory methods to the study of its communities.

As examples of the main environmental factors acting upon cultivated plants and their parasites we may mention temperature (Fig. 472), solar radiation (Fig. 473-474), humidity and wind, climatic patterns, seasonality, soils and natural vegetation.

There is today a huge amount of information (especially in Central Europe and the U.K.) about the effects of various pesticides on crops and their biota, but it is not the aim of this work to describe them.

Below is given the proportion of each European country (with the exception of the smaller ones) given over to agriculture (ploughed areas plus arboreal cultivations), expressed as a percentage of the total surface of the territory. The data refer to 1967 and they are given in descending order:

Denmark	62.7	Great Britain	30.4
Hungary	60.5	Belgium	29.3
Italy	50.0	Greece	29.2
Poland	49.7	The Netherlands	25.5
Portugal	49.2	Austria	19.9
Eastern Germany	46.1	Albania	17.8
Romania	44.3	Ireland	17.0
Czechoslovakia	42.0	Switzerland	9.8
Bulgaria	41.1	Finland	8.2
Spain	40.6	Sweden	7.0
France	37.0	Norway	2.6
West Germany	33.1	Iceland	0.01
Jugoslavia	32.3		

The average value is 31.5%. Of course, levels of intensive agriculture vary from one country to another; the ratio ploughed areas/arboreal cultivation also differs.

Let us list the most important cultivated plants on our Continent. Wheat is most widely spread in France and barley in France and the U.K.; rye is mostly cultivated in Poland, to a lesser extent in West Germany but is absent in Northern

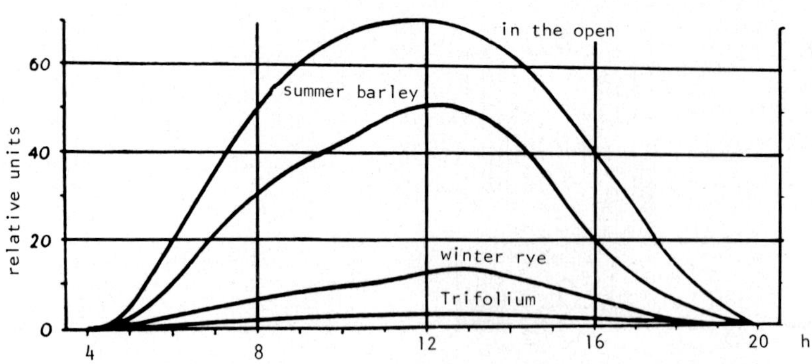

Fig. 474. Light intensity (relative units) during a May day at the soil in different kinds of cultures: barley, 12-15 cm high; rye, 80 cm high; *Trifolium*, 30 cm high (from Geiger) (reproduced with the permission of Vieweg Verlag).

Europe; millet is extensively cultivated only in Russia, maize has a maximum area in Jugoslavia and Romania, and is absent in Northern Europe; oats are mainly cultivated in Poland, France and West Germany. Rice is a typically Italian crop, absent in Northern Europe. Potato crops are commonest in Poland and secondarily in West Germany; sugar beet is cultivated mainly in West Germany and Poland, and to a lesser degree in France and Italy. The grape shows a maximum in Italy, followed by France and is absent from Northern Europe, although in this connection we must not forget the Rhine valley. Among industrial annuals we should mention the soya bean (*Glycine soia*) originally from China and Japan, cultivated also in Europe as forage, a green manure plant, or even as a plant for oil produc-

Fig. 475. Some examples of cultivated plants.

Fig. 476. Aspects of land use in the region of Sesto (Sexten) in Pusteria (Alto Adige or Süd Tirol). a. Hay stack, with highly characteristic, nordic features; b. hay cocks.

tion (soya bean oil). Tobacco (*Nicotiana tabacum*) shows a maximum in Bulgaria and Greece, i.e., in Southern Europe, and is absent in Northern Europe; sesamus (*Sesamum indicum*), originally from tropical Asia, is cultivated in Greece and to a much lesser extent in Southern Italy (i.e. in the sclerophyllous forest), cotton (*Gossypium herbaceum*), introduced by Arabs into Sicily in the 9th century, is

Fig. 477. Sheep grazing in high mountains on the Southeastern Alps (Marmolada, 2450 m). It is the highest altitude at which sheep can be found on a northern slope.

Fig. 478. Aspects of animal breeding on the Southeastern Alps. a. A summer abode, half-timber, for shepherds (tabià) at almost 1100 m. b. A very poor, raw summer abode on the Belluno Dolomites; the roof is made of spruce wood.

today cultivated in Greece, Spain and probably southern Russia, and in the past in Italy also. The hemp, *Cannabis sativa*, is cultivated in Jugoslavia, and secondarily in Hungary, Romania and Poland (i.e. in Eastern Europe) in areas characterized by a sub-continental climate; finally, lint, utilized both for fibres and for oil, is cultivated in Poland and secondarily in France.

Limited to some parts of Europe, there are other cultivated plants used either for food such as buck wheat *Fagopyrum esculentum*, belonging to the family *Polygonaceae,* which makes a type of grey-coloured 'polenta', although more often used for industrial purposes, as is the castor, *Ricinus communis*, which is utilized for the production of castor oil. Hops, *Humulus lupulus*, used in brewing beer and finally two trees of the family *Moraceae*, *Morus nigra* or red mulberry, with dark red, juicey edible fruits, and, more important, *Morus alba*, or white mulberry, cultivated for production of leaves upon which the silkworm feeds.

Among arboreal cultures we will mention here only some Mediterranean elements which play an important role in the economy of some countries, such as tea (*Camelia sinensis* or *Thea sinensis*) a plant of the family *Ternstroemiaceae*, growing in Russia between Batum and Suchum, near the Black Sea as well as on the southern slopes of the Caucasus.

Olive trees are most widely spread in Italy and secondarily in Spain and Greece (1967). They are present exclusively within the sclerophyllous forest zone. Limes are found in Italy and Spain in equal amounts, and much less in Portugal and Greece. They are limited to the Mediterranean climate and to the Atlantic climate prevailing along the coasts of Portugal.

Another tree cultivated all through the Mediterranean parts of Europe is the fig *Ficus carica* The method by which the fruit of this plant is produced has been known since antiquity. Aristotle describes in his *περὶ τα ξωα ἱστορίαι* the habits of the small Hymenopteron *Blastophaga psenes* (*ψηνες*) whose importance in ripening the fruits was clearly understood by the philosopher. The agricultural practice used today to achieve this aim ('caprification') is also much the same as that described several centuries B.C. by Aristotle. This process is due to the symbiosis between the fig and the *Chalcidoidei Hymenoptera Agaonidi*, through which the fertilisation of the flowers of a cycle with the pollen of the preceding cycle occurs, a phenomenon unique in the plant kingdom.

Finally we should mention a tree largely cultivated in Europe as a source of cellulose, namely the Canadian poplar, *Populus euramericana*, which exhibits a maximum spread in Sweden and Finland, followed by West Germany and the U.K., i.e. in the Fagetum zone. It grows also in Southern Europe in the Quercetum zone, although here maybe its utilization is not so commercialized. It is beyond the scope of this book to illustrate the main animal and plant groups harmful, in one way or another, to cultivated crops in Europe. Very few zoological and plant terrestrial groups do not possess at least a few representatives that are actual or potential pests. It is sufficient to mention nematodes, Acari, Apterygota and Pterygota insects, Isopods, Diplopods, Molluscs and Vertebrates (mammals and birds) among animals, viruses, bacteria, fungi and several flowering plants belonging to several families (*Rafflesiaceae, Convolvulaceae, Orobancaceae, Loranthaceae*, etc.) among vegetation. Also some weeds may in some instances cause serious damage to crops (Fig. 479).

Let us mention only a few insects noxious to European cultivated plants whose life history is generally well known.

Among beetles the most important families are without any doubt *Chrysomelidae, Cerambycidae* and *Curculionidae* or 'weevils'. The olive tree is attacked by the larva of the *Cerambycidae Macrotoma scutellaris*, typical of the sclerophyllous forest zone. An enemy of the potato, as well as of other *Solanaceae* (*Solanum melongena*, or aubergine), is the *Chrysomelidae Doryphora decemlineata* (in its original country known as the 'Colorado beetle') which penetrated from France into the Piemonte in 1944 and then in 4-5 years invaded all the Po lowlands,

Fig. 479. A plant of *Orobanche*, parasite of clover (*Trifolium*), alfalfa (*Medicago*), broad bean (*Vicia faba*), ivy, etc. sometimes very harmful to crops.

Fig. 480. Example of an insect noxious to sugar beet: the weevil *Temnorrhinus mendicus*.

reaching in 1950 the shores of the Isonzo. Successively it made its appearance near Trieste and in the meantime it arrived in the South (Pisino) in 1950. Today the Colorado beetle is present throughout Italy. It can be fought with lead arseniate or DDT containing insecticides such as Cytox (Fig. 480-482).

Wheat is attacked by the *Chrysomelidae Lema melanopa*, which in Eastern Europe is a major pest. Some ornamental plants such as *Chrysanthemum parthenium* and *Pyrethrum cinerariaefolium* are attacked by the *Cerambycidae Phytoecia pustulata*: the female lays eggs upon the stem or inside it, and the larva develops inside the plant until metamorphosis.

Cultivated species of the *Brassica* are victims of several beetles, among which are the *Bruchidae Entomoscelis adonis*, the *Chrysomelidae* genera *Psylliodes* and *Chaetocnema concinna*, the latter belonging to *Halticini*. Hemp can be attacked by another *Halticini, Psylliodes attenuata* and an enemy of carrots is the *Cerambycidae Phytoecia icterica.*

The common asparagus (*Asparagus officinalis*) is heavily damaged by the *Chrysomelidae Crioceris asparagi*, a strictly stenophagous species whose larva is characterized by the curious habit of covering itself with its own excrement. *Gramineae* are also attacked by some *Chrysomelidae*, such as *Lema melanopa* and *L. cyanella*, whose larva covers itself with a mucilage mixed with faeces. The rosemary, *Rosmarinus officinalis*, in South Europe both wild and cultivated (the leaves are used in the local, savoury, cooking), is attacked by an extremely beautiful, polychromous small beetle, *Chrysomela americana,* iridescent red and green and also a monophagous species.

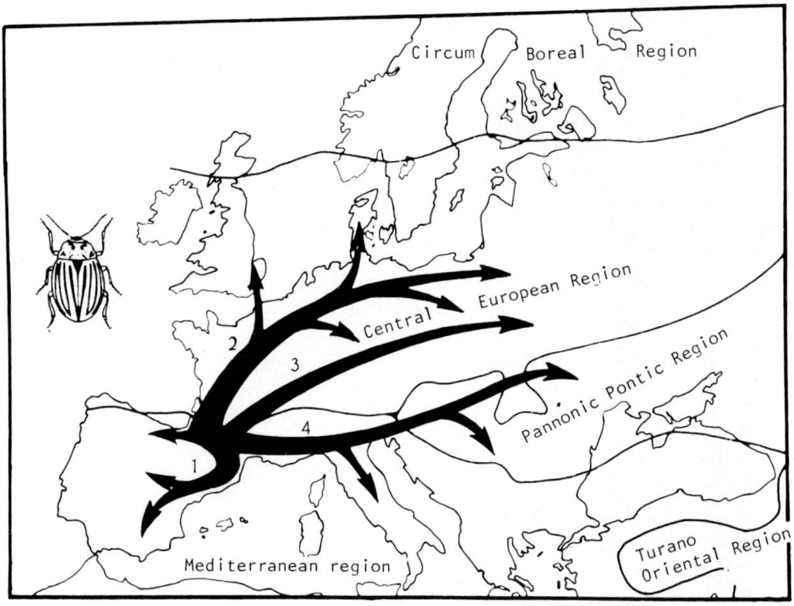

Fig. 481. Immigration routes in Europe of the Colorado beetle (*Leptinotarsa decemlineata*) (according to Wegorek, from Tischler).

The beets, which in some regions assume a remarkable economic importance, are damaged by several beetles, such as the *Chrysomelidae Chaetocnema tibialis, Cassida* sp. plur., the *Bruchidae Entomoscelis adonis*, etc.

Very many beetles exist at the expense of *Leguminosae* both wild and cultivated, living both on plants and stored seeds. Probably the most dangerous family is *Bruchidae*: the pea is attacked by *Bruchus pisorum*; the bean is attacked by *Acanthoscelides obsoletus* and *A. mimosae* (an imported species) and by *Callosobruchus sinensis*, also an imported species. The latter has been the object of some interesting research concerning its parasites, *Heterospilus prosopidis* and *Neocatolaccus mamezophagus*, which could possibly be used in the biological control of the *Bruchidae*. The variation in numbers of the host and its parasites closely recalls that of prey and predators. A mathematical formulation of the fluctuations of the three species is also possible (Utida's simultaneous equation).

The lentil *Lens esculenta* is attacked by *Bruchus signaticornis* (a pest in France and Italy), *B. ervi, B. lentis* and *B. luteicornis*; the *Vicia faba*, widespread in Southern Europe, where it is eaten both as a fresh vegetable and as a dry seed, is attacked by *Bruchus tristis, Bruchidius nanus, B. murinus, B. bimaculatus* and *Acanthoscelides mimosae* (imported species); the chick pea (*Cicer arietinum*), in

Fig. 482. Example of a harmful Carabid beetle, *Zabrus tenebrioides*, a species common to Europe. Asia Minor and Transcaspia, present in cereal fields.

Southern Italy consumed both as a fresh vegetable and as a dry seed, is attacked by *Acanthoscelides mimosae* and *Bruchus tristiculus*; the lupin (*Lupinus albus*) is attacked by *Bruchus tristiculus*; the arachis *Arachis hypogaea* (from the Greek αρακις = chick pea) is the victim of *Caryoborus acaciae* which should receive great attention in view of the economic importance of this plant for oil production. The forage *Leguminosae* are also attacked by some *Bruchidae*: let us mention *Bruchus brachialis, B. viciae, B. rufipes* and *B. luteicornis* on *Vicia* sp. plur., *Bruchidius varius* and *B. perparvulus* on clover.

Finally, some *Bruchidae* are polyphagous, and attack several *Leguminosae*, such as *Vicia, Lathyrus sativus, Lens esculenta* and *Pisum sativum.* These include *Bruchus rufimanus* and *B. atomarius.*

We could not finish this argument, however, without mentioning one of the most famous pests of all time: locusts, belonging to *Orthoptera Acrididae* (or *Chelifera*) which in Southern Europe make occasional appearances. Locusts have been known since ancient times; for example in the Bible, we find some very realistic and truthful descriptions of locust migrations, since locusts were, and are still today, very common in Palestine and Egypt. In Exodus X, 13-15 we read 'when it was morning, the east wind brought the locusts. They covered the face of the whole earth, so that the land was darkened; and they did eat every herb of the land, and all the fruit of the trees which the hail had left: and there remained not any green thing in the trees or in the herbs of the field, through all the land of Egypt.'

In England there is today an Anti-locust centre dealing with locusts and their control. It would be outside the aims of this book, although very interesting, to enter into further details on this subject.

The enemies of *Citraceae* (limes) have been very thoroughly investigated in Palestine (present-day Israel) by Bodenheimer (Fig. 484-485).

Birds also often prove to be very harmful to crops: in some instances, as in the newly cultivated areas of Poland, the greatest damage is inflicted on newly sown or ripening crops. From the agricultural point of view, the rook *Corvus frugilegus* is the most important species which, with the exception of the northern region of Poland, is an important breeding species. What proportion of individual rooks are resident is still a matter of conjecture. In Spring and Autumn large flocks of thousands of individuals migrate over Poland, and a large number of them winter in this country (Fig. 486).

Jackdaws *Corvus monedula* usually forage together with rooks and are similarly abundant in the south-eastern parts of Poland. Fields under spring crops, such as oats, barley and wheat, are the main feeding ground of rooks and jackdaws in Spring, as they supply them with food. The spring migration of rooks and jackdaws over Poland takes place, as a rule, earlier than the spring crops are sown. However, maize suffers most from breeding rooks and jackdaws: larger and larger areas are sown every year with maize, which in Poland is mainly used as a fodder for cattle. Rooks and jackdaws effectively prevent a more intensive spread of this crop. The sunflower (*Helianthus annuus*) which is also increasing its area and is a valuable fodder crop for cattle is favoured by these two species.

In contrast to the *Corvidae*, starlings in Poland cause a damage only to cherry orchards, but in some places they attack strawberries and currants. Inasmuch as the tree sparrow *P. montanus* is neutral or even useful for farming, the house sparrow causes considerable damage to ripening crops of wheat, barley and sunflowers. For instance, in the Zulawy region winter barley is no longer sown in view of the considerable damage caused by house sparrows. The damage caused to vegetable gardens and orchards by house sparrows may also be considerable. We must not forget that sparrows are carriers of diseases, mainly avituberculosis, which economically is extremely harmful.

Fig. 483. The moths *Hyponomeuta padellus*: a, adult; b, eggs and newly hatched larvae; c, eggs; d, young larvae on an apple leaf; e, nest with cocoons.

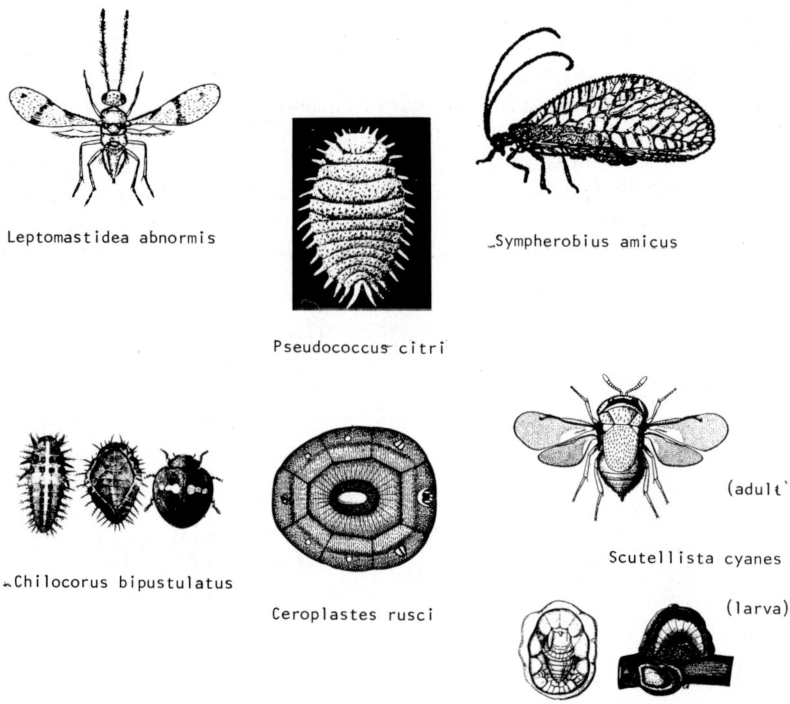

Fig. 484. Some European scale-insects and their parasites (*Pseudococcus citri* and *Ceroplastes rusci* from Silvestri, 1939; *Scutellista c.* from Grandi, 1911).

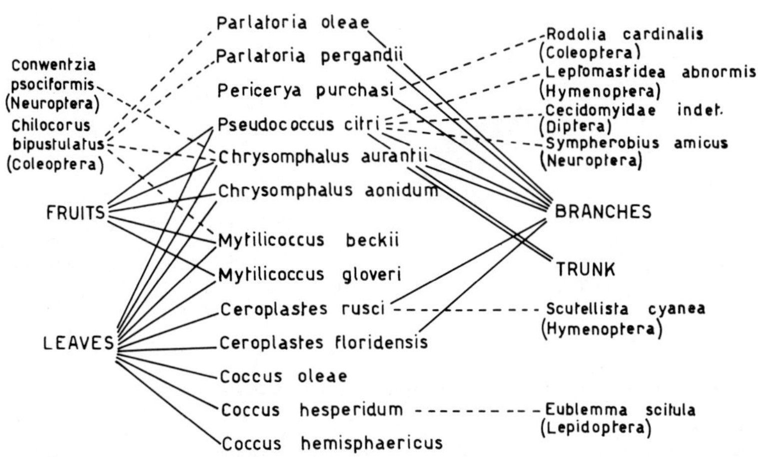

Fig. 485. The scale insects (*Coccini, Homoptera*) which attack *Citraceae* (citrus fruits) in Southern Europe, the parts of the tree attacked and the main parasites which attack the scale insects (constructed on the basis of Bodenheimer's data).

Great damage to crops of peas, wheat, maize and others used as fodder for cattle is caused by domestic pigeons within a radius of 50 km of towns. Until now, the damage caused by wood pigeons (*Columba palumbus*) in Poland has not been as great as in Western Europe. Besides sparrows and domesticated pigeons some migrating birds may also be harmful to crops.

Although less important compared to the planticolous fauna, which includes of course the true pests, the soil fauna of cultivated areas also offers many interesting aspects. The soil fauna of crops has little in common with that of prairies or grasslands, which is easily understandable if we consider the climatic peculiarities

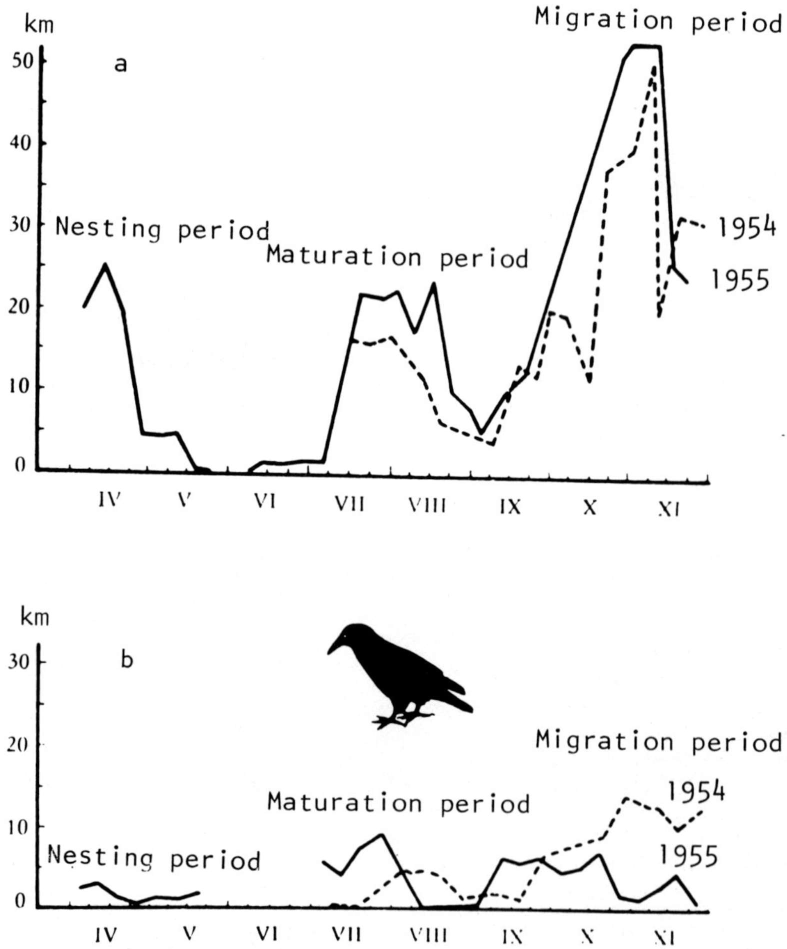

Fig. 486. Variation of numbers of the rook *Corvus frugilegus* in some Polish fields according to the growth of vegetation, corresponding to the different seasons: in correspondence to the minima of the curve the birds stay in meadows and mown meadows. a; cultivated areas poor in trees; b. cultivated areas rich in trees (according to Pinowski, from Tischler).

of a natural (or even secondary) grassland. The most striking differences between the fauna of cultivated areas and that of permanent meadows derives from the constant action of man. Some groups are very sensitive to agricultural practices, so that they are completely absent in cultivated soils: let us mention Gastropods and most isopods, i.e. mainly the surface living animals, unable to dig and requiring either an elevated humidity or, at least the chance of concealing themselves below the surface. Also animals unable to move easily and to find shelter in hedges or ditches (for instance as Carabids can) are very scarce in cultivated territories. On the whole, earthworms prevail in normal or heavy soils, small arthropods such as mites and springtails in light, sandy soils, but most animal groups are present in cultivated areas, although in lesser numbers, both of individuals and of species, when compared with natural areas.

The main agricultural practices which influence soil fauna are without any doubt soil working (more effective when mechanized), type of crop, harvesting and finally manuring. The plant cover is important for the pedofauna, first of all because the microclimate depends on vegetation, second because the litter serve as food both directly, or through the intermediary of microorganisms, for soil animals. Also root secretions of cultivated species act upon the microflora and therefore upon all soil life.

A general feature of soil zoocenosis is the character of instability compared with those of more natural environments. In general only the most euryoikous, or less exigent, species are able to thrive in fields, hence the monotonous character of their soil fauna.

Harvesting is perhaps the most catastrophic event in agricultural practice; the sudden deprivation of roots, litter, leaves, etc., together with the sudden changes in the microclimate, bring about a radical reduction or at least a radical change in the soil fauna. The phenomenon is readily apparent and well-known in the case of a complete felling of trees.

When we suddenly introduce manure to a soil which has been subjected to continuous cultivation for several years, some unexpected events may occur: animals previously established in the soil are destroyed by the substances produced by manure. On the other hand, the animals introduced into the soil with the manure are not able to penetrate it, so that the first application of manure behaves as an extraneous body. Only slowly does it allow introduction of a series of soil animals representing a true succession (Fig. 487). The manure generally causes an increase in the number of earthworms, *Enchytraeidae*, Nematodes, *Diptera*-larvae, springtails and mites (Fig. 488).

The effect of mineral fertilizers is quite different; it may even bring about a total momentary destruction of the fauna. Only successive development of the root systems and slow increase in micro-organisms allow the development of a good soil fauna. Apparently combined fertilization with manure and inorganic fertilizers is favourable to earthworms. Often discussed today are the effects on soil fauna of the various biocides (molluscicides, rodenticides, insecticides, herbicides, anti-cryptogamic substances, etc.) which are used more and more in Europe. We can only say that effects vary both according to chemicals and to animal groups, so that any generalization today is very difficult and inadvisable.

As a consequence of all these kinds of soil working and treatments, the soil

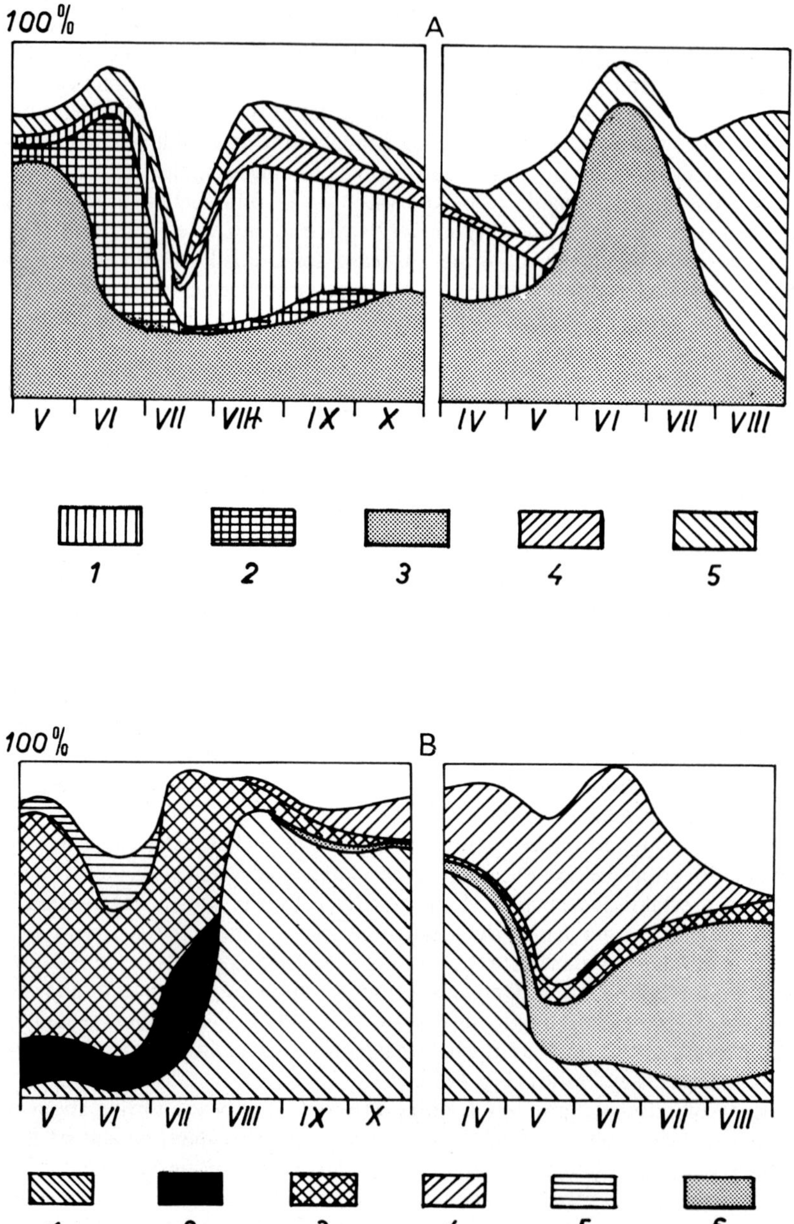

Fig. 487. A true 'succession' in soil micro-arthropods present in manure. A. Acari: 1. *Digamasidae*; 2. *Pyemotidae*; 3. *Oribatei*; 4. *Macrochelidae*; 5. *Parasitidae*; white: remaining groups. B. Collembols: 1. *Proisotoma minuta*; 2. *Hypogastrura manubrialis*; 3. *Lepidocyrtus cyaneus*; 4. *Isotoma notabilis*; 5. *Isotoma olivacea*; 6. *Onychiurus armatus*; white: remaining species (form Černova, modified).

fauna of cultivated areas is in general much poorer than that of natural ecosystems or even meadows. For instance, in a wheat field near Padua (Granze di Frassenelle) the present author has found the following (No. of specimens/500 cm^3 of soil):

	maximum	minimum
total fauna	301 (surf.)	10.1 (4-10 cm)
mites	176 (surf. May)	9.3 (4-10 cm, November)
springtails	82 (surf. June)	0

In a sugar beet crop near Perugia (Papiano) the present author has found a maximum value of 18.4/500 cc of soil, one of the lowest values in the world (cf. Fig. 489-491).

Closely associated with agriculture, and partly dependent on it, is the breeding of house or domesticated animals which occur practically throughout Europe, often modifying the environment. We have already seen that domesticated animals have been bred by man since pre-historic times (Neolithic), so that their impact upon the natural environment is very old and in some countries it has been of paramount importance in the destruction of the original climax vegetation and successive substitution with some disclimatic vegetational associations.

Sheep are today bred particularly in the U.K., but also in Spain and Romania. They are absent in Northern Europe, although several races are present in Europe. Goats are bred in Greece, Spain and Italy, i.e. in the Mediterranean countries where this animal probably originated. Cattle are raised mainly in France, only secondarily in West Germany and the U.K. They are absent in Northern Europe where they can be partially substituted by the reindeer (see the chapter on the

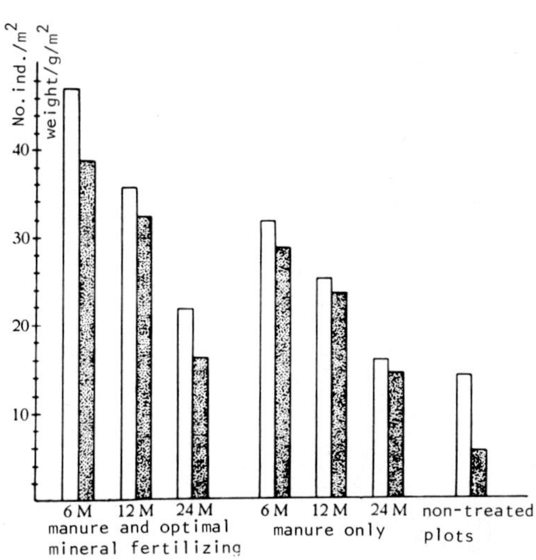

Fig. 488a. Effect of manure on the number of individuals (white column) and biomass (gray column) of earthworms several months (M) after the treatment.

tundra). There is an enormous variety of breeds of cows, whose origin is not clear. The horse is today less important than it was in the past; it is bred principally in Poland, to a lesser extent in Jugoslavia and France. In Northern Europe it is present only in Iceland.

The pig is bred mainly in West Germany and also in Poland; it is absent in Northern Europe, but its distribution depends much more on the eating habits of European people than on its ecological requirements, since pigs can also live in the Tropics.

Finally we should mention here a rather uncommon domesticated animal, the water buffalo, probably belonging to the Indian species *Bubalus bubalus* or 'arni'.

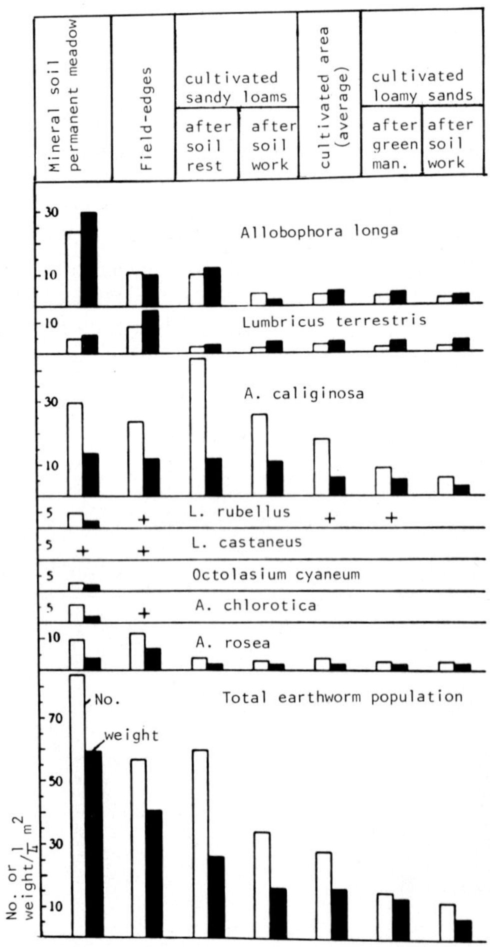

Fig. 488b. A striking difference between the earthworm populations of a permanent meadow and cultivated areas in Schleswig-Holstein (according to Finck, from Tischler).

It is represented by an undetermined number of specimens in Italy, more exactly in Lazio, Campania, Basilicata and Puglia, where it was introduced by Longobardi in the 6th century.

Animal breeding and husbandry is very ancient; in Neolithic times the farmers cleared trees and let cows graze in the clearings. In the Middle Ages oak forests were altered by pigs, looking for food. Since the 18th century, when the cultivation of potatoes began, pigs have been kept in stables or pig sties, so that their foraging on oak forests was stopped. The presence of the stable furthermore enhanced the formation and accentuation of the genetic characteristics in present-day domestic animals. In the late Middle Ages animal breeding was already developed: people operated selection upon several species, such as the Winscombe sheep, the Einsiedeln horses, the dogs named after St. Ubertus and St. Bernard and other races. The domestication of animals began in the Neolithic. The dog was already a domestic animal since the Mesolithic, probably because of its social behaviour.

The ox began to be domesticated in the Neolithic, but the origin of the present races is still a matter of controversy. In Denmark during the period 5500 to 3500 B.C., within a culture known as Kjökkenmöddinger, the wild ox *Bos primigenius* was disappearing, and *Bos longifrons* or 'domestic ox' makes it appearance, as is demonstrated by bone remains. In 2500 B.C., in the Neolithic of Switzerland, some remains of a 'bog ox' are found, attributed to a small domestic ox, whereas in the Bronze Age (1500 B.C.) a large domestic ox was present.

However, for a long period of time the wild ox lived together with the domestic one: in the Po valley (Northern Italy) the former is present up to the beginning of the 1st Century A.D. (Mantua: Virgilius); in 1176 it lived in Sweden and Russia. In the 13th century Siegfried – as the Nibelung saga (Nibelungen-Lied) tells us – hunts bison and wild oxen in the Vosges. In Braunschweig both wild and domestic oxen were living in 1240. In 1400 *Bos primigenius* was still present in the region today corresponding to Eastern Prussia. In 1501 Maximilian the First finds 5 'Ur' (Auroch) at Nürnberg; this animal was present in Lithuania in 1549

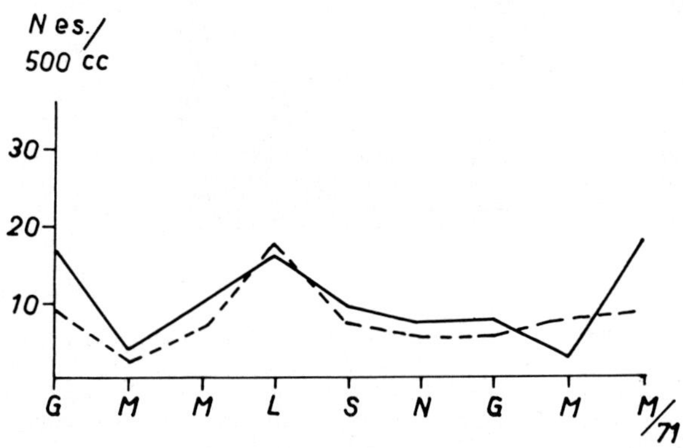

Fig. 489. N° of individuals/500 cc of soil in the surface (0-4 cm) (———) and in the depth (4-10 cm) (– – – –) at Papiano, near Perugia (Marcuzzi, 1973).

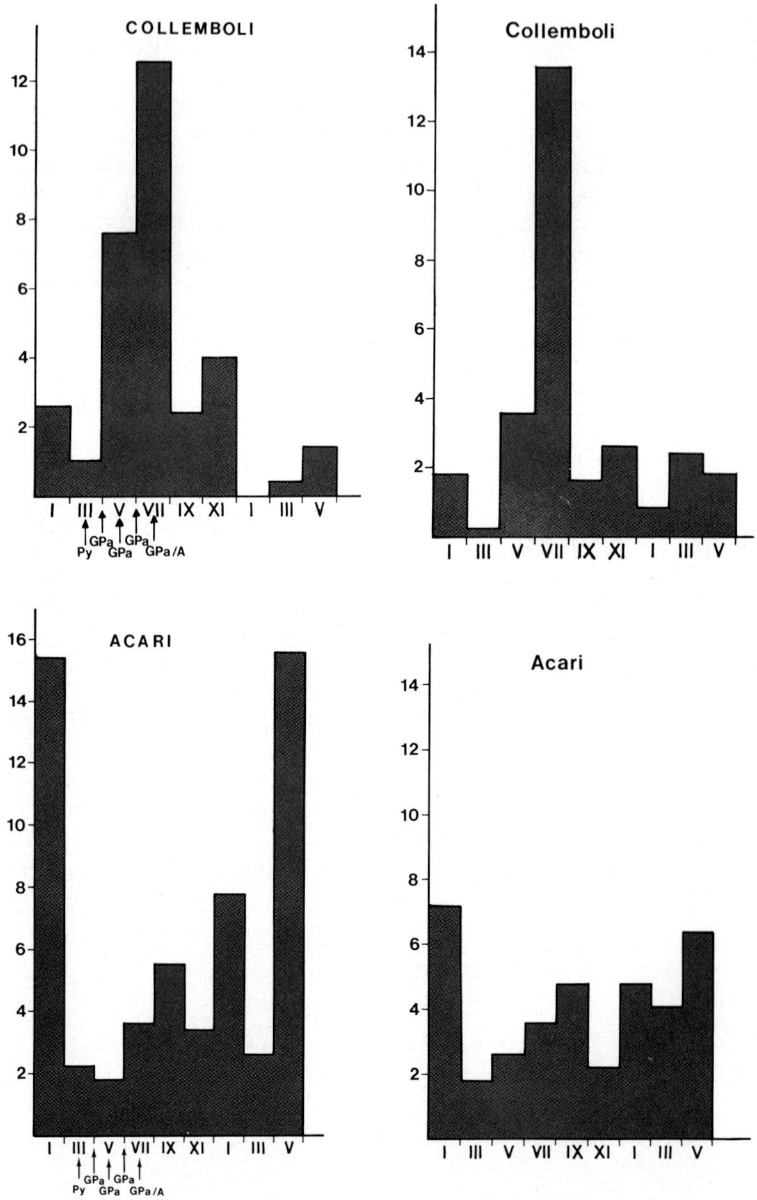

Fig. 490. Number of specimens of Collembols and Acari per 500 cc of soil in the upper layer (A, 0-4 cm) and in the lower layer (B, 4-10 cm) in a cultivated area near Perugia. The symbols at the bottom of the figures indicate the dates of the application of some biocides, i.e. Py = Pyramin; G = Gammasol DDT; Pa = Parathion; A = an anticercospora substance. The number of individuals is one of the lowest known so far in the world (from Marcuzzi, 1973).

together with bison and wild horses. The last wild oxen lived in a forest near Jaktorov (SW of Warsaw) and the last cow died in 1627.

Some think that 'blood' of the wild ox is still present in some domestic races, such as in fighting bulls of Andalucia, in the steppe oxen of Hungary, in the English Park ox or in the Scottish Highland cow.

In the Neolithic, beside oxen, sheep, goats, pigs, asses and lastly horses had been domesticated.

The origin of the domestic horse is obscure; during the Paleolithic the wild horse, *Equus przewalski*, was living in many European countries such as Spain and France, as is documented by the graffiti we find today in several caves (e.g. at Lascaux), but we cannot exclude the contemporaneous presence of some other species of horse, and today it is difficult to say from which wild species the domestic horse originated.

Today *Equus przewalski* lives in a small territory between Mongolia and China, more or less halfway between the two areas of the wild *Camelus bactrianus*, in high steppes (1000-1400 m), sand dunes, and arid soils covered with *Salsola* and 'saxaul' (*Haloxylon ammodendron*). It is known as 'tachi' while other recent Mongolian names are také or statur. It was known also as surtake (by Kirghizes) or even as kertag (by Kirghizes and Tartars in 1896). The name 'tarpan' used in Europe is probably due to an initial mistake.

The origin of the European domestic pig is also very obscure. According to Elsa Mohr *Sus scrofa vittatus* seems to be the original form, while *S.s. scrofa* is the historically younger form. Domestication occurred at different times at different places, starting from distinct races of *S. scrofa.* Since in the past a movement of pigs, both from the East towards the West and vice-versa occurred, today the relationships between the two extreme forms, *vittatus* and *scrofa*, are obscure. It must be remarked in this connection that the pig is the most recently domesticated animal.

In order to have an idea of the impact house animals have had on natural environment, we want to mention two insect species, the silk worm *Bombyx mori* and the bee *Apis mellifica*. In Europe the silk worm has been bred since the 12th

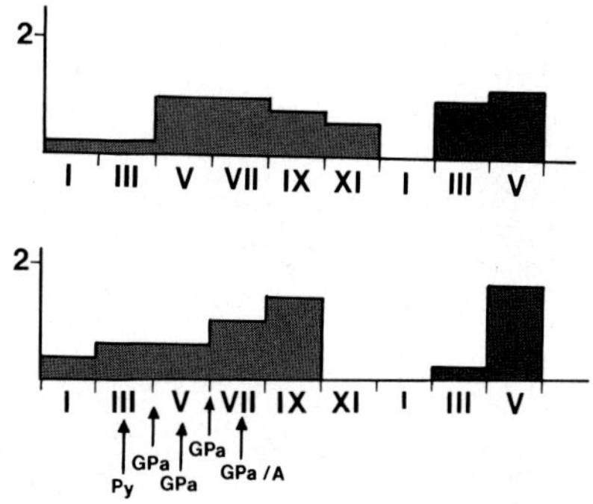

Fig. 491. The same as in the preceding figure for the remaining groups of the soil fauna (from Marcuzzi 1973).

century, although they say that in Italy it was bred before 700 A.D., at least if we believe a poem by Etilwaldus, King of Mercia (a kingdom of Anglo-Saxon England corresponding to present-day Derbyshire and Staffordshire). In the Far East, however, the first attempts to breed the silkworm date from 2698 B.C. and are attributed to a Chinese Empress, Si-Ling-Hsi, residing at Huangti, but others say that *Bombyx* was commercially produced in China before 2900 B.C. and possibly even earlier in India.

Today the silkworm is bred mainly in Southern Europe, with a peak in Italy, followed by Bulgaria. It is limited to the area of cultivation of the mulberry *Morus alba*, on whose leaves this animal feeds.

In biblical times honey was already known and consumed, although the honey was exclusively produced by wild bees. Greeks and Romans on the other hand bred bees, more precisely *Apis mellifica.* In Europe the following subspecies of *Apis mellifica* can be recognized: *m. mellifica*, in Central Europe; *m. ligustica*, in Italy, today almost cosmopolitan (imported); *m. carnica*, Krajn, Eastern Alps, the Carpathians and Balkans; *m. cypria*, Cyprus; *m. syrica*, Syria; *m. caucasica*, the Caucasus and *m. remipes* on the southern shores of the Caspian Sea. There are several more races in the different parts of the world (Vecchi). The best known races in Southern Europe are *m. mellifica*, the German bee, and *m. ligustica*, the Italian bee. Very often hybrids of the two races are present, for instance in North-eastern Italy. The German bee has a brown-black colour and is covered with thick grey hairs; the Italian bee has the first three abdominal segments with golden-yellow or orange-yellow bands, and is covered with fulvous grey hairs. Furthermore, the northern race has a shorter ligula, the southern a longer ligula, which makes it possible to visit also flowers with a long corolla such as red clover (Fig. 492). The Italian race is particularly important to apiarists since it is tame, very active, prolific, not so disposed to swarming and more resistant to diseases.

The bee is the theme of a famous passage from Georgicae (Liber IV, v. 538-558); Aristeus finds all his bees dead, and he goes to see his mother Cyrene to ask advice about what to do, and these are Cyrene's words:

> 'quattuor eximios praestanti corpore tauros
> qui tibi nunc viridis depascunt summa Lycaei,
> delige, et intacta totidem cervice iuvencas.
> Quattuor his aras alta ad delubra dearum
> constitue, et sacrum iugulus demitte cuorem,
> corporaque ipsa boum frondoso desere luco.
> post ubi nona suos Aurora ostenderit ortus,
> inferias Orphei Lethaea papavera mittes
> et nigram mactabis ovem lucumque revises:
> placatam Eurydicen vitula venerabere caesa.
> Haut mora: continuo matris praecepta facessit;
> ad delubra venit, monstratas excitat aras,
> quattuor eximios praestanti corpore tauros
> ducit et intacta totidem cervice iuvencas.
> post ubi none suous Aurora induxerat ortus,
> inferias Orphei mittit lucumque revisit.
> hic vero subitum ac dictu mirabile monstrum
> aspiciunt, liquefacta boum per viscera toto
> stridere apes utero et ruptis effervere costis,
> immensasque trahi nubes, iamque arbore summa
> confluere et lentis uvam demittere ramis.'

Needless to say Virgil falls into two errors: he believes in spontaneous generation and makes a confusion between bees and flies. But it is not at all strange, when we realize that the incorrectness of spontaneous generation was only proved by Redi (1625-1698) and that there are flies (*Eristalis, Eristalomyia*) that even a rather well-trained zoologist may confuse with a bee! During the Middle Ages scientific knowledge about bees did not progress much after the Persian Zakarīyā ben Muhammad ben Mahmūd al Kamūni Al Qazwīnī (known also as Al-Quazīwinī or Al-Kaswīni, a Moslem Arab born at Qazwīn in Iran in 1203 and who died in 1283 in Iraq) described the bee hierarchy, although he did mistake the queen for a king.

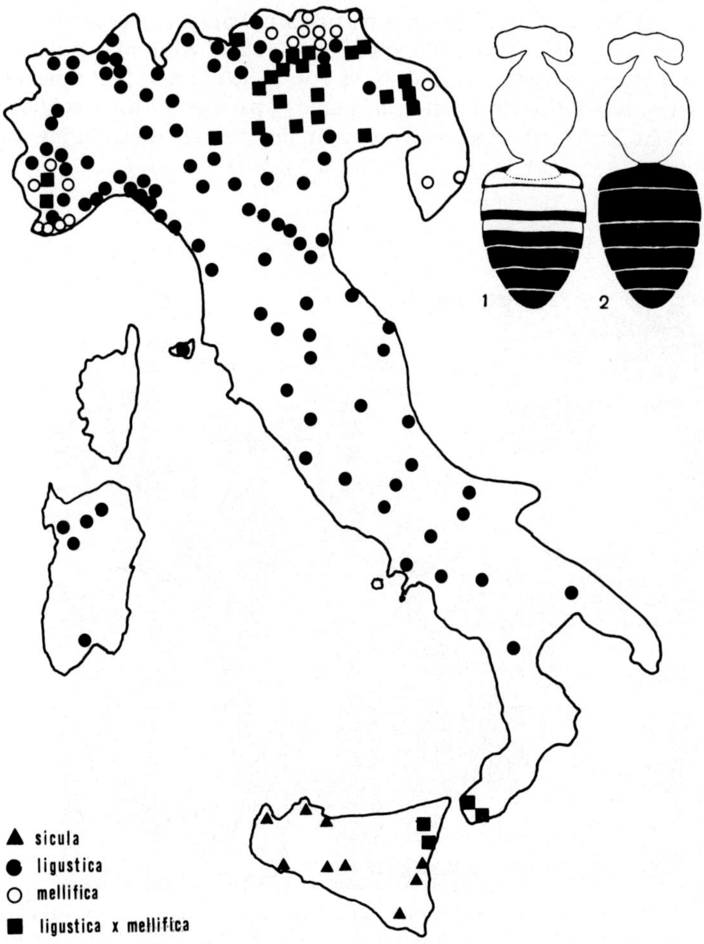

Fig. 492. Distribution in Italy of the different 'subspecies' of the bee, *Apis mellifica* (according to Vecchi, redrawn). 1. *A.m. ligustica;* 2. *A. mellifica.*

Besides its economic importance, the bee has a very peculiar role in artificial ecosystems, because it is one of the most effective pollenators. Many European crops need insect pollenation: *Cruciferae, Malvaceae, Linaceae, Rutaceae, Vitaceae, Papilionaceae* (alfalfa, clover, bean, *Vicia* etc.), *Rosaceae* (*Pyrus* and *Prunus*, among fruit trees, blackberry, strawberry, etc.), *Myrtaceae, Passifloraceae, Cucurbitaceae, Umbelliferae, Rubiaceae, Compositae, Vacciniaceae, Solanaceae, Chenopodiaceae, Polygonaceae, Lauraceae, Euphorbiaceae, Moraceae, Liliaceae* and other families. For this reason the life history and the social behaviour of the bee has been the object of much research, including experiments.

Agriculture in Europe has been developed considerably since the Neolithic and very early some sound knowledge about agriculture practice were acquired by European peoples. In Roman times the high standard of agricultural techniques is clearly revealed in several poetic or literary works, for instance those of Rutilius Taurus Aemilianus Palladius, Cato, Varo, but particularly Lucius Iunius Moderatus Columella (1st century A.D.). Also in Virgil's poetry – albeit without any technical aim – we find some passages which show us how agriculture of that time had made great progress. It is sufficient to quote a couple of passages of the Georgicae. In the 1st book (v. 259-310) Virgil tells us what the peasant does during the winter.

Fig. 493. The Bishop of Lausanne exorcizes cockchafers (from a 15th century drawing, according to Bodenheimer). The last exorcism was performed in Denmark in 1830, but this religious practice is possible even today. It is called 'benedictio deprecatoria contra mures, locustas, bruchos, et alia animalia nociva' (Rituale Romanum).

But the peasant of that epoch, like the farmer of today, has something to do all the year (Liber II, v. 516-526):

'Nec requies, quin aut pomis exuberet annus
aut fetu pecorum aut Cerealis mergite culmi
proventuque oneret sulcos atque horrea vincat.
Venit hiemps: territur Sicyonia baca trapetis,
glandes sues laeti redeunt, dant arbuta silvae;
et varios ponit fetus autumnus et alte
litis in apricis coquitur vindemia saxis.
Interea dulces pendent circum oscula nati,
casta pudicitiam servat domus, ubera vaccae
lactea demittunt pinguesque in gramine laeto
inter se adversis luctantur cornibus haedi.'

A little more advanced was the Arab knowledge of agriculture; in 825 under the dynasty of Abassides (Bagdad) Abu Bakr ben Wahasci (known also as Wahasci-

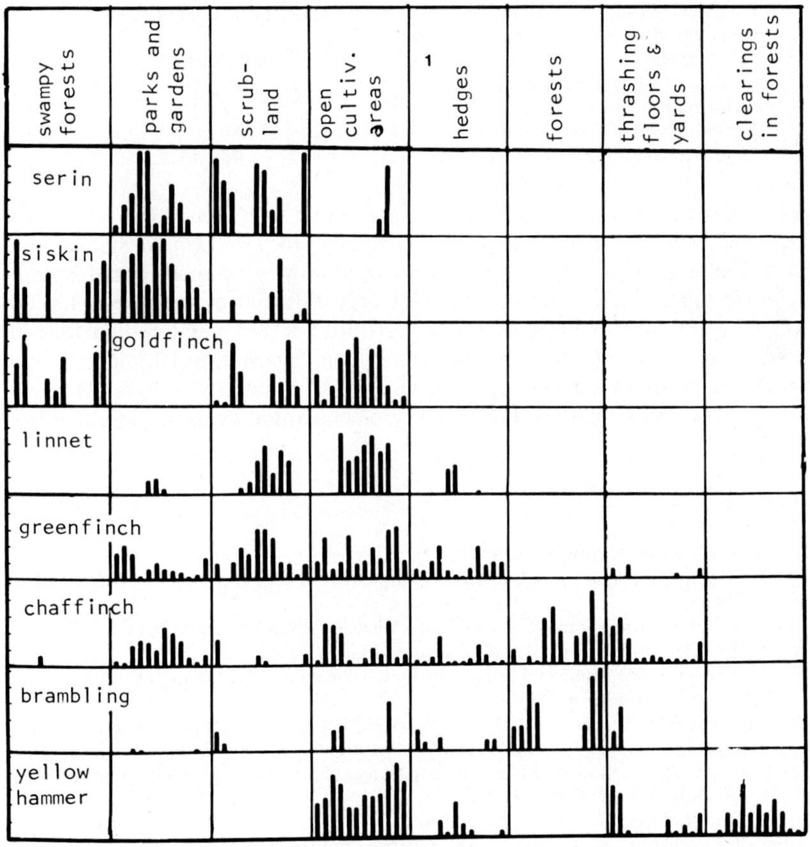

Fig. 494. Significance of the type of cultivation for the presence and abundance of *Fringillidae* during the year (according to Eber, from Tischler).

ja) translated into Arabic from Nabathaean (the language spoken by a people inhabiting the area south-east of the Dead Sea) the so-called 'Agricultura Nabathaea'. It is necessary to reach late Renaissance times to find deeper knowledge about animals harmful to agriculture. This period, from 1640 to 1750, has been called by Bodenheimer 'the bionomic period', and includes Bacon from Verulamius, Redi and Vallisneri. Indeed, the Middle Ages, together with the first part of the Renaissance, has represented for all natural sciences a period of obscurity so that even in the 15th century philosophy, theology and science were very much intermingled and people resorted to the power of the bishops to fight against cockchafers (Fig. 493). Aldrovandi also (1522-1605) put theology among the instruments of struggle against pests.

An applied aspect of ecology of cultivated areas is that of productivity. Without entering into details – which would be outside the aims of this book – we may mention some data on the secondary productivity of the bird populations of some European cultivated areas (Fig. 494):

Environment	No. of birds (pairs/km²)
Small woodlands surrounded by fields in Rhenania	442-400
Kitchen gardens in Rhenania	384-334
Fields with hedges in Rhenania	168-112
Agricultural land near Berlin	69
Field without hedges in Rhenania	96- 40

The number of pairs of birds in uncultivated areas falls from 2386 (old forest with oak and beech near a town in NW Germany) to 55-40 (*Pinus* stands with *Calluna* in Finland). It is easy to see how the productivity of uncultivated areas is much greater than that of cultivations. Not only this, but as we have noted on page 703, the number of birds of urban territories is also greater than that in cultivated areas. When all the data gathered by the International Biological Program (IBP) are published, a much greater amount of knowledge will be available on both primary and secondary productivity of cultivated areas (cf. also p. 724).

BIBLIOGRAPHY

Azzi, G. Ecologia agraria. Patron, Bologna (1967).

Černova, N.M. Zoologičeskaia charakteristika kompostov. *Akad. Nauk, SSR, Izd. 'Nauka'*, Moscow (1966).

Geiger, R. Das Klima der bodennahen Luftschicht. Vieweg, Braunschweig (1961).

Lengerkern, H. v. Der Ur. Die Neue Brehm Bücherei, Leipzig (1953).

Mohr, E. Das Urwildpferde. Die Neue Brehm Bücherei. Wittenberg-Lutherstadt (1959).

Mohr, E. Wildschweine. *Ibidem* (1960).

Pauer R., Zur Ausbreitung der Carabiden in der Agrarlandschaft, unter besonderer Berücksichtigung der Grenzbereiche verschiedener Feldkulturen, *Z. angew. Zool.*, 62, 1975, 457.

Schreiber K.F., Zur Sukzession und Flachenfreihaltung auf Brachland in Baden-Wurttemberg, *Verh. Gesell, Oekol., Göttingen,* 1976, 251.

Tischler, W. Agrarökologie. G. Fischer, Jena (1965).

World Atlas of Agriculture. Ist. Geografico De Agostini, Novara (1969).

ADDENDA

After the manuscript had been submitted for printing, it appeared that some observations were necessary for a better understanding of some specific aspects of the various European ecosystems, often based on works that have been recently published. These observations – which follow the order of the book – have been condensed in these 'Addenda'. For practical reasons, the most recent titles of papers and books have been included in the Bibliography of the relevant chapter.

The sclerophyllous forest

The presence of a dense evergreen vegetation is the main characteristic of the sclerophyllous forest, typical of all regions with heavy winter rainfalls and a dry summer, with an average annual temperature lower than 20°C (California, Chile, South Western Australia etc.). In the past, a forest with Aleppo pine (*Pinus halepensis*) and evergreen, sclerophyllous, oaks covered most of the shores of the Mediterranean basin: of this primordial forest there are today a few remains in Palestine and Transjordania.

Where the soil is arid and shallow, or where the original, climax forests have been degraded by human agency, a peculiar vegetation has substituted the sclerophyllous forest, the macchia (or 'maquis'), in which, together with *Pinus halepensis* and sclerophyllous oaks, some shrubs grow, such as the rosemary *Rosmarinus officinalis*, *Pistacia* and *Myrtus*. When old fires have destroyed the forest or a heavy deforestation occurred, the macchia has been replaced in turn by a new secondary plant formation, the so-called 'garigue' (from the Provençal 'garroulia' the name employed for *Quercus coccifera*, a shrub often dominant in this kind of vegetation), where only evergreen shrubs are present.

The mild, rainy winter associated with a hot, dry summer, induces in plants a twofold period of rest, with a vital cycle characterized by two stages. The flowering and fruitbearing at the end of the winter preceded by an autumnal leafage, causes olive trees and citraceae (limons, oranges etc.) to have their branches always covered with flowers, fruits or leaves. When the winds blowing from the deserts reach the sclerophyllous forest or the macchia, the leaves of a fig or an almond, which during the summer transpire plenty of water, lose it completely within one hour, but their roots and stems are able to substitute these losses very quickly. Other species living in the same environment exhibit a different water balance: the Aleppo pine or the olive tree reduce transpiration, so that they conserve their water. Carob (*Ceratonia siliqua*), arbutus (*Arbutus unedo*) or laurel (*Laurus nobilis*) on the contrary – more typically sclerophyllous species – exhibit a real period of rest or diapausa, a kind of drought dormancy, which it is not possible to interrupt even with heavy sprinkling. A diapausa is also present in many animals living under severe conditions.

There is a close relation between rainfall and soil, also disregarding evaporation. As an example of that it suffices to quote the following data:

Soil	Rainfalls	SiO_2/Fe_2O_3	Colour
Serosem	400 mm	12.4	yellowish brown
Terra rossa	500	12.9	brown-red
,, ,,	665	11	red
,, ,,	700	9.7	,,
,, ,,	735	9.1	,,
,, ,,	882	8.1	,,

These elements serve to explain plasticity, colour, field capacity (capacity of water retention) and the agricultural aptitudes of a given soil.

Let us mention some examples taken from Bodenheimer:

Rainfactor > 50: terra rossa, yellow soils. When the parent rock is basalt, the colour is a chocolate-brown; CaO = 3-18%, less upon basalt; organic substance = 0,5%; pH = 7.2-8.0. Soil good for cereals, olive trees, figs and carobs.

Rainfactor 30-50: on calcareous sandstones there is a sandy red soil; often the soil is poor in calcium (less than 0,1% CaO); pH on the surface 7-7.4. Good for citraceae. In the case of azonal soils (alluvional soils etc.) the soil can be dark (looking like the black soils of other regions) clayish, little permeability, with a great water capacity; pH 7.4.

Rainfactor 15-30: brown-gray steppe soil, partially formed by silt coming from degraded limestone, partially by small particles of the nearby deserts transported by the wind; rich in $CaCO_3$,pH = 8-9.2, often salty; good for cereals, particularly barley; underground rather impermeable.

Rainfactor < 15: arid soils, constituted by loess at the border of semiarid soils (in Europe extremely rare); yellowish, calcareous, slightly alkaline (pH 7.5-8.0). They can be utilized for the cultivation of barley.

Another group of Mediterranean plants with a very peculiar area of distribution is constituted by *Viola scorpiuroides, Sedum creticum, Teucrium brevifolium, Nepeta scordotis, Androcymbium rechingeri* etc., present in the South Aegean district and in Cyrenaica. In the past they may have been present all along the shores of the Eastern Mediterranean, and the climatic changes which have occurred since 5000 years or more have eliminated them in the greatest part of their area of distribution, so that today they exhibit a disjunct distribution. An analogous phenomenon is offered by an insect genus, *Polycoelogastridium*, belonging to the Tenebrionids.

Among the first descriptions of this kind of vegetation we would like to mention 'Noticia de las plantas y arboles de España. Comentario a los anécdotas Filológicas de Ábu Alí al-Qalí', by al-Becrí, or better Ábu Obáid al-Becrí, who died in 1094 (litt. News about the plants and trees of Spain. Commentary to Ábu Alí al-Qalí's philological anecdotes). Further descriptions of Sicily and other regions belonging to the dominion of the sclerophyllous forest are due to Ábu Hámid al-Garnatí (1080-1169). Famous is his 'Regalo a los amigos' (Gift to the friends) in which he describes inhabitants, peoples, the marvels of some countries and their noticeable buildings, the seas and the islands, with news about extraordinary animals which live in them, caves, sepulchres, fossil animals etc. Another description of Spain and Sicily, both belonging to, at least partially, the sclerophyllous forest, is due to al-Idrisí (1100-1169?) an Arab geographer living first in Spain (Córdoba) and then in Sicily. He was the great-grandson of Idris II, hamudí monarch of Málaga and since the latter wished 'to have a description of the world carried on according to direct observations and not according to books', he did a very serious geographical book (1154) with the title 'Recreo de quien desee recorrer el mundo' or 'Libro de Roger' (Recreation of anybody who wishes to go through the world, or Roger's book) known in non-Moslem Europe thanks to a compendium, published in Rome in 1592, of which a Latin translation has been done by the Maronites Gabriel Sionita and Juan Hesronita, who gave the book the erroneous title 'Geographia Nubiensis' (Paris, 1619). Al-Idrisí is considered today as the principal geographer of the Middle Ages (Gonzalez Palencia). He fixes the situation of Spain in the fourth climate and in the neighbourhood of 'Mar Tenebroso' (or Murky Sea), probably the Atlantic Ocean.

We should, however, point out that in approx 2500 B.C. in Creta there were some forests which made possible the existence of the deer *Cervus elaphus* and of a species of monkey, probably imported from Asia Minor or Egypt (*Cercopithecus*?).

Some bats should also be mentioned as characteristic though not quite exclusive of the sclerophyllous forest. Let us mention only *Rhinolophus euryale*, or Mediterranean horse-shoe bat, south-European caucasian-mediterranean element. It thrives in caves from which it comes our rather early in the evening; it is more gregarious than other *Rhinolophus*. It sleeps in groups not very compact and it is not enveloped by its wings, different from what is usually observed in bats. *Pipistrellus kuhli* lives in towns and villages; its winter and summer quarters are practically the same as those of *P. pipistrellus*, the common pipistrelle. It comes out from its living place rather early in the evening; it flies at low altitude in a straight line. *Pipistrellus savii* lives particularly on mountains, sometimes near houses. It hibernates in crevices of rocks, in hollow trees, garretts, granaries, small constructions and sometimes under stones. It comes out rather early in the evening and flies maybe all night long, and in some instances also during the day. It is a migratory species.

Finally we wish to mention the long-fingered bat, *Leuconoë capaccinii* living in caves during both winter and summer. It is a gregarious species, flying often above the water. It sleeps sometimes in very large groups. It flies for a great part of the night, generally in groups; it can probably swim as its kin *L. daubentonii*.

The Tyrrhenian islands (Neaples bay, Ponza, Zannone etc.) have been very thoroughly investigated from the ornithological point of view, so that we may now recognize the following ecological categories: a) resident species; b) migratory species; c) winter visitors and d) nesting or summer species.

Resident species are the hobby *Falco peregrinus*, the blue rock thrush *Monticola solitarius*, the blackcap *Sylvia atricapilla*, the Sardinian warbler *Sylvia melano-*

cephala, the wren *Troglodytes troglodytes*, the crag martin *Ptyonoprogne rupestris* (Capri!), the great tit *Parus major*, the Italian house sparrow *Passer italiae* (which on the Eolian Islands hybridizes with *P. hispaniolensis*!), the tree sparrow *Passer montanus*, the greenfinch *Chloris chloris*, the goldfinch *Carduelis carduelis*, the serin *Serinus canaria* and the finch *Fringilla coelebs*. Resident on the Eolian Islands, equidistant from Sicily (the Northeastern extremity) and Calabria, are the dartford warbler *Sylvia undata* (in Italian 'magnanina'), the rock sparrow *Petronia petronia* (in Italian 'passera lagia') and finally the cirl bunting *Emberiza cirlus* (in Italian 'zigolo nero'). The rock partridge *Alectoris graeca* has disappeared from the Eolians at least since 1913: it is mentioned for these islands by Spallanzani in his book 'Viaggi alle due Sicilie e in alcune parti dell'Appennino' of 1793.

Migratory birds are the honey buzzard *Pernis apivorus*, the pallid swift *Apus pallidus*, the alpine swift *Apus melba*, the red-rumped swallow *Hirundo daurica* (Zannone!), the Spanish sparrow *Passer hispaniolensis*. At the Eolian Islands the following species have been observed: Eleonora's falcon (the so-called 'falcone della regina') *Falco eleonorae*, the lesser kestrel *Falco naumanni*, the spotted eagle *Aquila clanga* (at Ustica!), the black eared wheatear *Oenanthe hispanica*, the subalpine warbler *Sylvia cantillans* (very common), the rare fan-tailed warbler *Cisticola juncidis* and not less than four subspecies or races of the wagtail *Motacilla flava*, namely *flava* (in German known as Schafstelze), *thunbergi* (in German Nordische Schafstelze), *feldeggi* (in German Maskenstelze) and *cinereocapilla*.

Hibernating species (migratory birds) are the white wagtail *Motacilla alba*, the grey wagtail *Motacilla cinera*, several thrushes, such as fieldfare *Turdus pilaris*, song thrush *Turdus philomelos* and the common blackbird *Turdus merula*; the beautiful back redstart *Phoenicurus ochruros*, the robin *Erithacus rubecula* (a species object of very remarkable experimental researches on behaviour), the hedgesparrow *Prunella modularis*, the alpine accentor *Prunella collaris*, the chiffchaff *Phylloscopus collybita* and the wall creeper *Tichodroma muraria*, well known from our mountains and during the winter present at Ischia, Ponza, Zannone and Capri.

Nesting birds (or summer visitors) are finally the alpine swift *Apus melba*, the orphean warbler *Sylvia hortensis*, the subalpine warbler *S. cantillans*, the marsh tit *Parus palustris*, the blue tit *Parus caeruleus* and the greenfinch *Chloris chloris*.

An extremely interesting ecological feature is the close relationship between the number of bird species observed on a given island (whether resident or migratory) and the surface of the same, as can be clearly seen for the Eolian islands from the valuable data gathered for this archipelago by two Italian students, Moltoni and Frugis. There is probably no other demonstration so clear for any part of the Mediterranean basin, as a consequence of the scarcity of thorough faunistic researches in this area. The dependence of the number of species on the size of the island shows that among this animal group there exists an evident competition for space, besides obviously for food. Even more interesting is the abundance of species registered at Ustica (186 species on only 8.6 km^2), which not only contrasts strongly with the data obtained for the Eolian archipelago, but probably represents a veritable 'record'. This fact could be explained if we assume that

Ustica is situated in an obligatory route for most migrants. So far this phenomenon has not been described.

On the Eolian islands, the extension of present day and past cultivations, the roads and the abodes have contributed to rarify, if not to destroy, the spontaneous vegetation. We must not forget that the proximity to the mainland and to Sicily already at the end of the III millennium B.C. made possible a numerous human settlement. Still more important to explain the serious impact man has had on this archipelago is the presence on Lipari of obsidian, the volcanic glass formed by the quick cooling of the lava erupted by the craters of Forgia Vecchia and Monte Pilato. Neolithic man used the obsidian to prepare his amygdaloid dagger, in Italian amigdala (from the Greek ἀμυγδάλη, the almond) the raw dagger with a form which recalls an almond, and the tools necessary to till the soil where he cultivated olive trees, grape and wheat.

The sclerophyllous forest, already changed into a 'maquis', retreats, restricts itself, is obliged to limit its vital space (or 'Lebensraum') to smaller and smaller areas, the sources of freshwater (documented by some actual toponyms), no more protected by the dense vegetation, exhaust themselves and the primigenious animal life will leave the place for the few species able to withstand severe conditions. In the same time the facility of communications with the Eolian Islands will permit the introduction of numerous cultures through the oriental navigators: the Neolithic is represented here by several stages and the successive Bronze Age will leave its traces at Panarea, Salina and Lipari. The liparitis found in Creta is exported from the archipelago, and obsidian has been found in some Ligurian neolithic caves.

In Italy many areas are protected, though in very different ways. First of all there are very wide zones subjected to the so-called 'hydrogeological bond' established on the ground of an old law of Dec. 1923; this bond is very loose. Other areas are subjected to the 'landscape bond', considered as 'natural beauties', established on the basis of a law of June 1939. The last criterion has been surpassed by the modern conceptions of town-planning of the so-called 'cultural commodities' (or goods) and 'environmental resources'. Unfortunately people in Italy are not very sensitive to the philosophy of nature and landscape.

Other forms of bond in Italy are the National Parks: in the case of the sclerophyllous forest only one such park exists, on the Circeo (Latium), but recently this Park has been degraded (or 'declassed') so that to-day there is practically no true National Park in this environment. There are, furthermore, the Natural Preserves, of which several exist in the sclerophyllous forest, but they are not much controlled by public authorities: there are many of which even the local population has no knowledge, as for instance near Trieste, in the Friuli-Venezia Giulia, not far from the Adriatic Sea (Gulf of Trieste).

The 'Preserves for refuge and repopulating' belong to the State: these are generally controlled by public officers, so that the fauna is relatively well protected. Finally, there are the Natural and Regional Parks, institutions of a very vague nature, theoretically useful, but practically very little effective.

The deciduous forest

A very peculiar kind of degradation of the heliophilous forest in Italy, limited to the Appennines, is the so called gariga, a translation of the French garrigue, of which we have talked when discussing the dominion of the sclerophyllous forest. Characteristic plants are many graminoid species, first of all *Bromus erectus*, associated to the bramble *Rubus ulmifolius*, the so-called 'agazzino' *Cotoneaster piracantha*, a thorny shrub with white flowers and scarlet red fruits, the hawthorn *Crataegus monogyna*, the broom *Spartium junceum*, often the so-called Reno-cane (from the River Reno running from the Northern Appennines to the Adriatic Sea), *Arundo plinii*, the common juniper *Juniperus communis*, some *Leguminosae* such as *Dorycnium hirsutum*, *D. herbaceum* and *Astragalus monspessulanus*, the elegant *Globularia elongata*, with pale bluish flowers, and several other herbs, both perennial and annual.

A completely different vegetation present on the Appennines in the dominion of the heliophilous forest (and in the Southern Appennines also in that of the sclerophyllous forest) is that thriving on the so-called 'calanchi', a morphological feature exclusive of the blue pliocenic clays of the Appennines. The calanchi are a type of narrow, deeply incised, valleys due to erosion – mainly determined by rain – in a peculiar substrate (particular nature of the clay, constituted by very fine elements, much colloidal, stability of the same etc.). The main factor is the soil, clayish and very compact, subjected to heavy sun drying and to a pronounced summer drought. The vegetation consists of trivial species adapted to endure the prevailing physical factors. Characteristic plants are *Artemisia cretacea* (almost exclusive of this geomorphological phenomenon) and a few 'marino-litoraee' as Prof. P. Zangheri calls them (personal communication), namely *Hordeum marinum*, *Atriplex hastatum angustifolium* and *Agropyrum litorale*. During winter a calanco is changed into a mudflow, during summer into 'an infernal pit' as Zangheri wrote me. Only two invertebrates seem to be closely bound to this environment, a beetle, *Pterostichus pantanelli*, and a woodlouse, *Armadillidium zangheri*. A thorough ecological (also quantitative) investigation in this South European environment would prove to be very interesting, particularly as far as soil fauna is concerned.

We must not forget a quite peculiar ecological category of plants present often in the European deciduous forests, though not exclusive to this biome, the epiphytes, i.e. the lower and higher plants living permanently, or almost so, upon other plants, generally trees. On the maple *Acer pseudoplatanus*, a beautiful tree common throughout Central Europe and the Alps, the following species have been observed: a) flowering plants: *Minuartia* sp., *Geranium robertianum*, *Thymus 'serpyllum'*, young plants of some trees such as *Picea excelsa*, *Abies alba*, *Sorbus aucuparia* and *Betula verrucosa*; b) ferns: the polypody, *Polypodium vulgare*; c) lichens: *Alectoria jubata, A. implexa, Usnea barbata, Cetraria glauca, Peltigera polydactyla, Parmelia physodes, P. furfuracea*, *Lobaria pulmonaria* and *Santoria parietina*; d) mosses: a great number of species (not less than 16) belonging to several genera such as *Abietina, Brachythecium, Dicranum, Ditrichum*, etc.

Lichens living on trees have also been described for the beech in Northern Italy (Cansiglio).

The Italian name of the little owl, 'civetta' would, according to some, be an onomatopoeia. The little owl has always stimulated and struck people's fantasy. Already in the Bible (Psalms, 102,4) we read of its mournful song. In Palestine Arabs consider it as a good omen, while in Europe even today people consider it as ominuous and a carrier of disgrace, because of its monotonous song: probably at the bottom of this belief there is some magic, and more precisely figurative magic, according to which 'each thing recalls a thing similar to it'. In ancient Athens the little owl was the symbol of wisdom and therefore it was the official symbol of Athens. Here a magic explanation is perhaps more difficult.

The hoopoe has always stimulated the fantasy of primitive peoples: in Palestine the Bedouins' wifes are still very anxious to possess some parts of an hoopoe, since they consider it as an amulet, which will ensure them of their husband's love. But even among more advanced peoples, the hoopoe has something mysterious: Foscolo in his 'Sepolcri' will say 'uscir dal teschio ove fuggìa la luna l'upupa e svolazzar su per le croci sparse per la funerea campagna'.

The raven has such a proud and majestic appearance – thanks to which it can keep at bay even several of the diurnal prey birds, like the kestrel (*Falco tinnunculus* – that it has inspired one of the most famous and beautiful of Poe's poems, 'The raven'. Its black colour, on the other hand, has induced in man the idea of funereal attributes (probably by means of a form of figurative magic) so that in Italy and perhaps also in other countries there is the expression 'raven of ill omen'. In Italian also the word 'corvaccio' exists, employed by Manzoni in his 'Promessi Sposi' to define the General Vicar of Milano.

The common wryneck *Iynx torquilla*, from the Greek Ἴυγξ, a mythological figure, daughter of Pan and Echo, is endowed with some magic powers (as an intermediary) in Teocritus, in Pindarus' Odes and in Xenophon's Memorabilia. It is easy to understand that when we consider that the mythological Iynx had been changed into a bird, the wryneck, because she made Zeus fall in love with Io. At the same time Iynx was one of the Greek names to designate the top, a toy bound to magic and mystery among the ancient Greeks.

The white willow *Salix alba* in Moravia is attacked by two series of animals, named by Křístek as primary and secondary noxious insects. Among the former there are a *Cecidomyidae* Dipteron, *Helicomyia saliciperda*, which causes damage comparable to that produced by the beetle *Aromia moschata*: it acts upon wounds of the tree, is very difficult to cicatrize and causes some damage which can be the start of fungal infections. These can also destroy the tree. The 'mining' Dipteron *Dizygomyza* sp., on the other hand, produces only slight effects. Secondary animals are the larvae of the weevil *Cryptorhynchus lapathi*, which prefers dying trees and, more precisely, the already decaying parts of them, the beetles *Aromia moschata*, present also on perfectly healthy willows, *Agrilus viridis, Xyphidria prolongata* and, lastly, the lepidopteron *Cossus cossus*.

Many elements of the soil fauna participate in the demolitive attack of dead wood, stumps etc. both in the deciduous and in the mixed forests of the middle latitudes of the European continent. Their rôle is extremely important, since they contribute to the formation of humus and indirectly to pedogenesis. At each stage of the decomposition of wood there is a predominant insect family, so that from the examination of insect coenosis we may determine the stage of the decomposition. Mamaev recognized: 1) a stage with *Cerambycidae*, 2) a stage with *Lucanidae*, 3) a stage with *Formicidae* and, finally, 4) a stage with earthworms. These animals, beside their taxonomical position, are different because of locomotion, body shape, alimentary habits and intestinal enzymes. According to Mamaev & Sokolov, the polysaccharides are digested by the different animals which attack dead wood in a different way, as one can see from the following scheme:

Animal group	Protease	Amylase	Lichenase	Cellulase
Ipidae (*Phloeosinus bicolor, Hylesinus fraxini*)	+	+	+	–
Cerambycidae (*Acanthocinus aedilis, Cerambyx cerdo, Ergates faber, Hylotrupes bajulus, Isotomus speciosus* etc.	+	+	+	+
Buprestidae (*Chalcophora mariana*)	+	+	+	+
Anobiidae (*Anobium punctatum, Xestobium rufovillosum, Ernobius mollis, Ptilinus pectinicornis*)	+	+	+	+
Lamellicornia (*Cetonia aurata, Potosia cuprea, Dorcus parallelopipedus*)	+	+	–	–

The composition of the animal community which attacks dead wood varies also according to locality and type of forest, as we may see from the following scheme (from Mamaev):

	European part of the USSR	Central Europe
1.	Ipidae Lymexilonidae Cerambycidae Buprestidae Siricidae Mordellidae Melandryidae	Ipidae Lymexilonidae
2.	Lucanidae Tipulidae Elateridae Scarabaeidae	Cerambycidae Elateridae
3.	Formicidae	Cerambycidae Syrphidae Lucanidae Tipulidae Elateridae
4.	Lumbricidae Enchytraeidae Diplopoda Gastropoda	Lucanidae Tipulidae Elateridae Formicidae Diplopoda Isopoda Gastropoda
5.	–	Annelida Diplopoda Gastropoda Elateridae (Lucanidae) (Formicidae)

In mixed forests, where oaks are present, a difference in the fauna is found, also according to the type of oak:

Animal group	Tula-oak	Tellerman-oak
Cerambycidae	*Leptura sexguttata, Hypulus quercinus*	*Strangalia* sp., *Xestobium rufovillosum*
Lucanidae	*Ceruchus chrysomelinus*	*Dorcus parallelopipedus, Lucanus cervus*
Formicidae	*Lasius niger, Myrmica ruginodis*	*Lasius niger*
Lumbricidae	*Bimastus tenuis*	*Eisenia foetida*
Diplopoda	*Proteroiulus fuscus, Polydesmus complanatus*	

Let us – very briefly – look at the main alterations of dead wood occurring in a forest of middle latitudes (deciduous or mixed forest). The alteration of the cambium is determined by *Cerambycidae* such as *Tetropium castaneum* and *T. fuscum*, by some *Ipidae* and *Bostricidae*. On the decomposing organic substance produced by the lignivorous beetles and on the fungi upon which *Scolitidae* feed, the following groups can be found (secondary degraders): larvae of *Diptera* (*Mycetophagidae*, *Fungivoridae* and *Cecidomyidae*), *Staphylinidae* and *Nitidulidae* (adults). The separation of the bark and the decay of wood are brought about by mites (*Phthyracaridae*, young stages of *Parasitiformes* and Oribatids), Enchytraeids, lignivorous nematodes (*Dorylaimus*, *Diphtherophora*) upon which some predatory mites feed (*Gamasidae*). In rotten wood one can find wireworms (*Dolopius* and *Athous*-larvae), larvae of Tenebrionids (mainly *Cylindronotus* and other *Helopini*, *Tenebrionini*, etc.) and larvae of *Diptera* (*Sciaridae* and *Sapromyziidae*). The fungal mycelium which penetrates the wood constitutes the food for the larvae of *Mordellidae* (beetles) which can destroy up a quarter of all the wood and the larvae of some moths. The rotten wood covered by mosses is destroyed mainly by springtails, mites and *Tipulidae*-larvae. In the debris so produced some earthworms live, such as *Bimastus tenuis, Eiseniella tetraedra* and *Allolobophora chlorotica*, which can already bring about the formation of the clay-humus compound ('Ton-Humus Komplex' of German authors). In the triturated wood (the so-called 'Mulm' of German authors) a rich community is now present: nematodes, enchytraeids, small snails (*Punctum pygmaeum*, only 1.5 mm wide), Scarab larvae (*Dorcus parallelopipedus, Oryctes nasicornis, Cetonia aurata*) feeding directly on the Mulm, larvae of *Cerambycidae* (*Prionus coriarius*, a beautiful insect up 40 mm long) and *Elateridae* (*Melanotus rufipes*), besides Tenebrionids (both larvae and adults) such as *Helops, Cylindronotus, Menephilus*, etc.). Also some earthworms live in Mulm (*Dendrobaena*, literally walking on the trees, from the Greek δένδρον tree and βαίνω to walk) and *Lumbricus rubellus*. At this stage the wood is attacked also by many animals coming from the soil, together with fungi and bacteria coming from the roots. In the lower part of the stump *Dorcus parallelopipedus* (adults) and *Athous rufus* (larvae) thrive, until earthworms, collembols and Oribatid mites arrive from the soil determining a complete mixing of Mulm with earth, so that a true clay-humus complex is formed. Nowadays the pedogenesis is complete. It is worth while mentioning that this process we have described in one page takes some six-seven years.

Besides the enzymes produced by animals, in the soil there are of course also enzymes produced by microorganisms, so that in a normal soil rich of life, there is always a given amount of free enzymes, which can attack some organic substances. Thus, e.g., in a soil belonging to the brown earths present on the Colli Euganei near Padova (M. Rua) investigated by the author, coming from the desintegration of trachyte, the following enzymes have been detected so far (unpublished data): α-glucosidases, able to split trehalose and melezitose, β-glucosidases, able to split arbutine and salycine, α-galactosidase plus β-fructosidase, able to split rafinose, α-glucosidase plus β-fructosidase, able to attack sucrose, besides several esterases. Whereas the latter are also present in ranker, the saccharases have been detected exclusively in the brown earth, i.e. in a much more mature soil. The ecological meaning of these findings is very evident.

As an appendix to soil fauna of the deciduous forest (and practically also of the coniferous forest) we should mention the nidicolous fauna, i.e. the animals more or less closely bound to the nests of birds and mammals. It should take too long to mention here even the most important and common species. An interesting work is due to Tischler, who has shown the stratification of the animals inside a small nest of a passerine bird, the *Parus major*: *(See table page 759).*

The blind, wingless endogeous insects to which Holdhaus refers, are mainly represented in Europe by beetles, and more precisely by Carabids (*Trechinae* etc.), *Silphidae* (*Bathysciinae*), *Staphylinidae* (many tribes), Pselaphids and weevils (*Troglorrhynchus*, *Raymondionymus*, etc.). According to the degree of dependence of these animals on soil or hypogeous dominion, Coiffait recognizes edaphobious, edaphophilous and edaphoxenous elements, going from the closest to the least dependence from soil. Recently very interesting papers have been devoted to this life-form. However, several invertebrate groups (particularly arthropods) belong to the true endogeous dominion, besides the beetles.

Enormous is the number of animals living permanently or at least during some periods of their life in caves, so that here is not even a case to indicate the most interesting. Sufficient to say that all terrestrial and freshwater groups (and also a few marine groups as, for instance, Polichetes) going from Protozoa to mammals can be found in caves. The most interesting groups, however, are Rhabdocoela, Triclades, Nematodes, Polichetes, Oligochetes, Hirudinea (or leeches), molluscs, Arthropods (including most of the hypogeous fauna) and vertebrates. Many of them are phoretic or parasitic species. The arthropods are represented by Pseudoscorpions, Opilionids, spiders, mites, millipedes, centipedes, springtails, Diplura, Thysanura, Orthoptera, Diptera and Coleoptera. According to Vandel, to the category of true troglobious belong essentially Isopods *Trichoniscidae*, the spiders belonging to *Leptonetidae*, diplopods, collembols belonging to *Entomobryidae*, Diplura *Campodeidae* and Coleoptera (*Trechinae* and *Bathysciinae*).

As far as aquatic animals are concerned, the true cavernicolous species are called stygobious (from the Greek Στύξ meaning the infernal or the river of life to come) and should correspond to what troglobious animals are within the terrestrial fauna. Here again we find an enormous number of species and groups, going

Species	Internal layer 0.5 cm under the cavity of nest t = 37-38°C	Middle layer 5 cm under the cavity of nest t = 29-30°C	External l. 7 cm under the the cavity t = 22-25°C	Thermic preferendum (in lab.)
Atheta nigricornis (a staphylinid)	80*			36.6
Ceratophyllus gallinae (a flea)	30	4		40
Eugamasus lunulatus (Gamasidae mites)	170			
Trachytes pyriformis Gamasidae	130	20		
Hister merdarius (Histeridae beetles)	10	96		30
Dendrophilus pygmaeus (Histeridae)	10	41		
Gnathoncus rotundatus (Histeridae)		157		30
Ceratophyllus gallinae	40	844	36	36.6
Phaulodynchus nidicola (mites)		580	12	30
Phaulodynchus sp.	10	268	92	30
Pteronyssus parinus (mites)		36		
Haffenreffia gilvipes (mites)		64	2	
Liacarus coracinus (mites)		88		
Galumna longiplumus (mites)	10	92		
Galumna dorsalis		80	16	23.3
Xenilla maritima (Coll.)		12	390	23.3
Entomobryia marginata (Collembol)		16	64	26.6
E. nivalis		13	72	
Phaonia querceti (fly)		12	18	
Total	490	2277	702	

* The number includes all life stages of the various species

from Protozoa to vertebrates (in Europe the salamander-like *Proteus anguinus*). Even more than for terrestrial fauna, it is easy for aquatic animals to pass from the subterranean water bodies to freatic waters. A peculiar subterranean aquatic fauna is present not only in the great caverns of subterranean lakes and rivers, but also in the interstices and fissures in which the subterranean water circulates. Evidence of this is, for instance, found in the presence of the amphibian *Proteus anguinus* in some wells near Gradisca (Gorizia, in the Isonzo basin) or even by the presence of some amphipods of the genus *Niphargus* (e.g. *N. puteanus elegans* in some very small rivulets at the foot of the Venetian Prealps, or *N. longicaudatus* in some wells of the outskirts of Padova).

We must not forget that whereas the terrestrial fauna of endogeous environment generally does not penetrate the freatic environment, several aquatic animals of freatic waters penetrate the endogeous milieu. For this reason *Niphargus* have been found under stones buried in the soil near a source, humid but never flooded. Also, experimental researches (Ginet) have shown that *Niphargus* can live out of their environment provided the atmosphere is saturated with water vapour.

A very peculiar subterranean environment, intermediary between soil and the true hypogeous dominion, is constituted by what Jeannel calls 'microcaverns', i.e. the holes, burrows etc. digged by micro-mammals in the soil, and where a whole fauna is living (foleophilous animals). This fauna is mainly constituted by small saprophagous insects and mites, together with their predators (generally insects and arachnids) and with some parasites of the host-mammal. This fauna has been studied very intensively.

Some nidicolous animals can sometimes be found in the cavernicolous environment (*Leptinus testaceus, Catops, Choleva*), either because they enter this environment following their host (*Leptinus*) or because they go there to reproduce (*Choleva*). However, in general there is no mixing of microcavernicolous and cavernicolous elements.

Actually, the distinction between troglobious and edaphobious species is very arbitrary: many are indeed the elements that are simultaneously cavernicolous and edaphobious, belonging to several zoological groups. Let us quote among Diplura *Campodea* (*Plusiocampa*) *caprai, C.* (*Paurocampa*) gr. *suensoni*, to which *P. suensoni*, both cavernicolous and edaphobious, and *P. spelaea*, exclusively cavernicolous, belong; among Collembols *Hypogastrura crassaegranulata, Onychiurus alticola* and *O. prolatus*; among Pseudoscorpions *Roncus assimilis*; among terrestrial Isopods or woodlice *Mesoniscus alpicola* and all the species of the genus *Spelaeonethes*; among Carabid beetles *Speluncarius stefani* or *Laemostenus schreibersi.*

The steppe

The first general description of steppes of Eastern Europe, Caspian Sea and the shores of the Volga River has probably been given by Abu Hamid al-Garnati (1080-1169) to which the 'Regalo a los amigos' (lit. A gift to the friends) is due. In this book, the author describes inhabitants, peoples, the marvels of some countries and their noticeable buildings, the seas and the islands, with some information about the extraordinary animals which live in caves, sepulchres, fossil animals, etc. He also visited the lands of the 'Kjazaros' and Bulgarians.

A very peculiar salty steppe is found near the Neusiedler See, more precisely at Seewinkel, belonging to the pannonic steppe and to steppe lakes. It is constituted both by an edaphic steppe, i.e. a steppe developed upon a salty soil, and by a steppe or a secondary prairie, due to deforestation (more or less as in Hungary). The steppe of Seewinkel includes some 80 small or very small lakes locally known as 'Lacken', which can be considered as true 'Zicklacken', since they contain sodium carbonate, sodium sulphate and sodium chloride. The solonetz are covered by a Statico-Artemisietum monogynae.

According to Wendelberger, in some places it is possible to speak of 'Rasensteppe' and 'Felssteppe', i.e. prairie-steppe and stoney or rocky steppe. As far as the salty soils of Neusiedler See are concerned, they belong to both solontchak and solonetz. Among the vegetation of these soils, some species come from the East, more precisely from the Aralo-Caspian region which during Miocene and to a lesser degree during Pliocene was practically united with Central, or better, Centro-Eastern Europe. Some species, on the other hand, are endemic: *Suaeda pannonica, Aster tripolium pannonicum* and *Cirsium brachycephalus*.

In Hungary (puszta) where the soil is denuded and there is a certain amount of calcium carbonate the following plants can be found: *Corispermum nitidum, Panicum crus-galli* and *P. glaucum, Eragrostis minor, Bromus tectorum, Tragus racemosus, Populus nigra, Chenopodium, Polygonum* etc. Steppic species highly characteristic for Hungary are *Aster cinereus* (sometimes also halophilous), *Centaurea orientalis* with large, magnificent, citrine-yellow flowers, *Jurinea arachnoides*, with some small basket-like spider-webs (hence its latin name), *Phlomis pungens* with red-lilac flowers, *Iris sintenisii*, characterized by some thin delicate leaves, *Rumex tuberosum*, with a tuber-like root, etc. Finally, in Moldavia characteristic steppic species are *Crambe tatarica* and *Phlomis pungens*.

In the nests of *Citellus* and other steppe rodents some *Gamasidae* mites are found, such as *Haemogamasus citelli* and *H. nidi*; some of these mites can transmit Rickettsiae. In the area of Stavropol, Reit'lat has found 22 species of *Gamasidae* on the following species of mammals: *Citellus pygmaeus, Mus musculus, Apodemus sylvaticus, Cricetulus migratorius, Cricetus raddei, Microtus arvalis, M. socialis, Lagurus lagurus, Ellobius talpinus, Spalax microphthalmus, Alactagulus acontion*, and *Dyromys nitedula*, as well as in the carnivor *Mustela nivalis. Gamasidae* are suspected to transmit tularemia.

In the USSR the following species of mammals are considered as bearers of the agent of tularemia (*Bacillus tularensis*): *Sciurus vulgaris, Eutamias sibiricus, Citellus pygmaeus, C. eversmanni, Marmota bobac, Alactaga elater, Arvicola terrestris, Ondatra zibethica, Lemmus lemmus, Lagurus lagurus, Microtus arvalis, M. ratticeps, M. gregalis, Clethrionomys glareolus, Mus musculus, Apodemus agrarius, A. sylvaticus, Micromys minutus, Rattus norvegicus* (limited to human settlements), *Cricetus cricetus, Cricetulus migratorius, Mesocricetus raddei, Lepus timidus* and *L. europaeus*.

Very numerous are the fleas that live on steppe mammals: in the Stavropol region, 84 species have been found, of which 39 belong to *Ceratophyllidae*, 25 to *Ctenophthalmidae* and 7 to *Pulicidae*. Fleas can transmit numerous bacterial diseases, for instance, the plague.

The ecology of *Cricetulus migratorius* has been studied on the European steppes by Bašenina (1951). The northern limit of distribution, reaching only a little south of Moscow, coincides with the July isotherme 20°C and the August isohyetal 60 mm. It passes through Kiev, Kursk and Kazan, bending towards Ufa. Its diet consists mainly of *Melilotus, Convolvulus arvensis, Reseda lutea*, and secondarily of *Agropyrum, Bromus, Medicago falcata, Atriplex* etc. It digs branched galleries, which in some burrows can meet those of *Microtus socialis*.

Some steppe mammals retreat when the steppe is cultivated: this is the case for *Lagurus lagurus, Ellobius talpinus, Microtus gregalis*. On the other hand, rodents feeding on cereals are favoured: e.g. *Mus musculus spicilegus* and *Cricetulus ever-*

smanni (*M.m. spicilegus* is substituted in the west by other subspecies such as *m. domesticus*.).

Some steppe rodents have invaded the zones of Central Europe that have been converted into herbaceous cultures: *Cricetus cricetus* and *Citellus citellus*. *Citellus pygmaeus* of the steppes and semi-deserts of Caucasus avoids cultivated areas. This proves once again that each species has its own auto-ecology, and that in ecology any generalization is extremely dangerous. At the same time it is evident that many more observations, both in the field and in the laboratory should be necessary. The same holds of course for invertebrates.

Almost all steppe animals have nocturnal habits; others are endogeous (rodents, insectivores). A characteristic feature is the presence of organs of sight that are much developed and, as a consequence, a reduction of the olfactory sense, contrary to what is the case with animals living in forests. Some animals exhibit a dormancy or an 'estivation'. Birds are generally very numerous: generally they join in herds and lay eggs in a rudimentary nest in the soil. The number of eggs is usually higher than that of species nesting upon trees or in the crevices of rocks, where the eggs or the nestlings are much more protected against their predators and enemies. Mammals and birds exhibit often a social behaviour, living in herds or groups where a leader which imposes its superiority is always recognizable. Herbivores or grazing mammals (ruminants, *Equidae*, etc.), rodents living in dens dug in the soil and carnivores (wolf) have always a leader. They move in groups from a region to another and they can even migrate very far when the alimentary supplies are exhausted (herbivores). Hunger has often compelled rodents and birds to migrate from oriental steppes to Europe in search of food. In Europe these animals have left some persisting colonies, e.g., some *Spermophilus* in Bohemia, in Central Europe *Otis tarda*, in Elsas the hamster and, according to some, in Champagne the Norway rat *Rattus norvegicus*.

The reptiles of the steppe near the Neusiedler See are numerous: let us remember the blindworm *Anguis fragilis*, the lizard *Lacerta agilis* (diffused from Central Europe as far North as Finland to Central Asia) the green lizard *Lacerta viridis*, the snake *Elaphe longissima* etc.

Tischler mentions that at Orsk, on the Ural River, 142 out of 330 steppe species have adapted themselves to cultivated areas. The number of specimens/m^2 in the steppe is 199, whereas in cultivated areas it rises to 351. Some species, however, show a decrease, for instance *Sminthurus viridis*, the cockroach *Ectobius duskei*, the ant *Leptothorax nassonovi*, the Homopteron *Graphacraerus ventralis*. Other species become more abundant, such as *Haplothrips tritici, Psammotettix collinus, Phyllotreta vittula* and *Chaetocnema aridula*, a species diffused in Europe, Turkestan and Siberia, living mainly on cereals such as oat, wheat and barley. Other steppe animals which adapt themselves very easily to life in the cultivated areas are, in the USSR, the wireworms *Agriotes sputator* and *A. gurgistanus*, the Tenebrionid *Opatrum sabulosum* and the Scarabeids *Anisoplia segetum* and *A. austriaca*. The latter became a pest in the Ukraine in the second half of the last century. *Zabrus tenebrioides* and *Loxostege sticticalis* also have a steppe origin.

In the USSR harmful insects are the Carabid beetles *Zabrus tenebrioides*, extending into Europe, Asia Minor and Transcaspia; *Z. morio, Z. blapoides*; other

Carabids harmful to crops are *Harpalus* and *Amara* (on maize). Dangerous insects are furthermore the Chrysomelid beetles *Phyllotreta vittula*, extended in Europe, Asia, Algeria and Tunisia, bound to several cultivated *Cruciferae* such as *Brassica, Sinapis* and *Raphanus*, besides some cereals, where some claim that the larvae live on the roots. In Crimea the *Aphididae Cuernavaca noxiae* is a pest. Let us also remember the mole cricket, *Gryllotalpa gryllotalpa* (dangerous to maize) and the *Diptera Mayetiola destructor*, in steppes and arbored steppes of the European part of the USSR. Lastly, harmful to maize are the *Diptera Oscinella frit* and *O. pusilla*, the Chrysomelid beetles *Chaetocnema* and *Lema* (the latter has already been mentioned) and the Lepidopteron *Loxostege sticticalis*.

Of course, many phytophagous insects prove to be true pests in several cultivated areas. Let us mention here only the Hymenopteron *Cephus pygmaeus*, harmful to wheat and rye, *Harmolita erectum*, to rye, the Heteroptera *Eurygaster integriceps* and *E. austriacum*, harmful to all cereals. Insects which attack all kinds of Graminoid plants are the *Diptera Oscinella pusilla* and *Meromyza saltatrix*, the Chrysomelid beetles *Chaetocnema aridula* and *C. hortensis*, as well as the Thysanopteron *Haplothrips tritici*. The Lepidopteron *Hadena sordida* is harmful to wheat. In Southeastern Europe among the insects the most dangerous to crops we find the Heteroptera *Pentatomidae*, the so-called 'stink bugs'. An Aphis or leaf-louse harmful to both herbs and cultivations in the Mediterranean Basin, southern USSR and USA is *Schizaphis graminum*. Subterranean aphids, finally, such as *Anoecia* and *Tetraneura*, diffused to Mediterranean Basin, Bulgaria and Ucraina, are harmful to wheat and corn.

The invertebrates of the steppe near the Neusiedler See are *Solenopsis, Tetramorium* (ants), *Helicella obvia, H. striata* and *Trochoidea geyeri* (snails), some of which are meso- or even xerophilous. Subfossil snails, characteristic of a warmer period (Bronze Age), are *Cyclostoma elegans, Acme sublineata, Graciliaria filograna* and *Clausilia pumila*. The arthropods of the dry prairies ('Trockenrasen') belong to Astragalo-Stipetum and are represented by *Solenopsis fugax, Tetramorium caespitum* and *Lasius alienus* (ants), together with some drought resistant, sometimes oriental species such as the millipede *Chromatoiulus unilineatus*. In salty soils (solonetz) covered by Statico-Artemisietum monogynae Kühnelt observed the Lepidopteron *Fidonia tessularia*, the Homopteron *Empoasca mendax*, the beetle *Chaetocnema scheffleri* (a species diffused to the Mediterranean Region, Hungary, the Balkans, Asia Minor and Caucasus; in Algeria on *Rumex*) and finally the Lepidoptera *Coleophora hungariae* and *Mamestra dianthi*.

Some elements of the salty steppes of South-Eastern Europe live also in the salty meadows of Central Europe, e.g. *Bledius* (algivorous) and its predator *Dyschirius*; *Dichirotrichus pubescens*, harmful to sugar beets etc. Some of them are only hygrophilous species, e.g. *Bembidion* and *Pogonus*. An interesting element is also *Dichirotrichus obsoletus*, extended to Central Europe, Siberia, on the shores of salty lakes and the shores of the Mediterranean Sea and the Atlantic Ocean. The species *lacustris* is a pontic form, today exclusively present at Neusiedler See and in Macedonia (Saloniki).

Most beetles belong to the so-called black beetles or Tenebrionids (in Russian černotelki). Along the shores of some steppe lakes *Penthicus dilectans* is abundant, together with other species (unfortunately the European steppes have not been studied as deeply as those of Asia). Particularly bound to solontchaks are *Clitobius salinicola* (limited to the Western Mediterranean basin) and *C. oblongiusculus* (the latter present under *Salicornia herbacea*), *Scleropatrum hirtulum* and *Pedinus volgensis*. In the more or less dry steppes of Eastern Europe the following species can be present: *Psammestes dilatatus, Scleropatrum hirtulum, Gonocephalum pusillum, G. setulosum* and *G. rusticum, Opatroides punctulatus, Penticus dilectans* and *Opatrum sabulosum* common also in other environments, but as a whole rather xerophilous. *Opatroides punctulatus*, which is not exclusive to steppes, exhibits sometimes some myrmecophilous habits, bound to the ant genus *Messor*.

As the most important elements, however, can be considered today the Tenebrionids of the European steppes, harmful to crops, in some instances veritable 'pests'. Let us remember *Dendarus punctatus*, in the Crimea harmful to vineyards; *Penthicus dilectans*, in Caucasia harmful to young cotton plants; *Dendarus crenulatus*, in Azherbaidjan a pest for several crops; *Pedinus strigicollis*, in the Crimea harmful to tobacco; *P. femoralis*, called in Russian 'the corn-Tenebrionid', present all over the European part of Southeastern Russia and Caucasia; *Opatrum sabulosum* (in Russian 'pesčanii medlyak') harmful to a number of cultivations. This species is a 'Rassenkreis' but unfortunately very little is known of the auto-ecology of each subspecies. It attacks barley, oat, corn, millet, sorgum, beans, lentils, potatoes, tomato, tobacco etc. Experimental research would prove very useful to find out more about the exact ecological requirements of this species. It can be easily bred in laboratories. *Opatrum triste* is harmful to tobacco in the Crimea; *Gonocephalum pusillum* proves to be a true pest for several crops: corn, sorgum, castorbean, fennel, anis, cauliflower, tomato, tobacco etc. *Opatroides punctulatus* in Transcaucasia is harmful to water melons and other cucurbitaceae, tobacco etc. and finally in Transcaucasia *Penthicus dilectans* is dangerous to the young cotton plantations.

As fas as the conservation of European steppes is concerned, we must mention Hortobágy National Park, a part of the puszta extending over some 120 000 ha between the Tibiscus and the town of Debrecen. Because of its natural and anthropic features, this territory has been a centre of touristic attraction for more than fifty years. Its horse raising ('nonius' race), the flora, the fauna and several anthropic characterictics are very attractive. We could also add the so-called 'fata morgana' (or mirage) which sometimes occurs here.

In the steppe of Bugac between the Danube and the Tibiscus the old, gray, Hungarian race of oxen is still preserved. Near Jasi, finally, there is a large steppe 'Naturschutzgebiet' with the features of a 'Heuwiese', or, as it is called in Hungarian, a 'Szénafüvek'. This steppe belongs, from a phytosociological standpoint, to a Salvio-Festucetum sulcatae moldavicum.

The montane environment

Among the organisms of the nival zone we can mention a micro-coleopterous insect, wingless, belonging to the soil fauna, coming from the nival zone of Dauphiné, Western Alps, *Chionostiba janetscheki*.

The local names of the common juniper *Juniperus communis* in many parts of the Alps, inhabited by German speaking populations, have been the origin of several toponyms and hydronyms, as well as of family names. Thus, we find near Innsbruck a Kranebitten, near Aibling a Kronwitt, near Inzell a Kranawitt, in the region of Laakirchen (Upper Austria) a Kranabeth, etc. From the dialect name Kranzen the toponym Kramsack in the lower valley of the Inn near Brixlegg is found. Also several family names of Southern Germany and the Eastern Alps take their origin from juniper: Kranewitter, Kronawitter, etc.

This abundance of traces of the name of a plant is due to the numerous tales and myths about the juniper: particularly important has been the presumed influence of this plant against several diseases, first of all the plague. As an example, we quote the following lines: 'Kranawet brennt im Haus, treibt Doktor und Bader aus', or 'Kranawetber muasst fruatlings (frish, munter) koin (kauen), aft (dann) kon di (ch) di Pest nöt hoin (hauen)'.

As for the upper limit to which the different vertebrates can reach on the Alps we may quote Erhard's data on the National Park of Gran Paradiso:

Amphibians		Mammals	
Rana temporaria	2667	*Apodemus sylvaticus*	2300
Reptiles		*Microtus nivalis*	3100
Vipera aspis	2250	*Eliomys quercinus*	1559
Birds		*Sciurus vulgaris*	2100
Lagopus mutus	3200	*Rupicapra rupicapra*	3200
Dryobates major	2100	*Capra ibex*	3600
Regulus regulus	2100		
Riparia rupestris	2300		
Delichon urbica	1600		
Oenanthe oenanthe	2500		

Humid or alluvial environments

Some birds that are found in the tundra of northern Europe during the summer are e.g., *Haematopus ostralegus, Somateria mollissima, Ardea cinerea, Anas platyrhynchos* and *Charadrius hiaticula*.

South of the Alps mires are often the remains of lakes from glacial times, surrounded by old morains. Typical examples are those of the Lake of Garda, with its morainic amphitheatre, that of the Fiavè mire, recently studied from the palethnological standpoint, and several others. Perhaps also the lake of Fimon in the Colli Berici, near Vicenza, where, however, there are traces of glacial phenomena, is responsible for the formation of some peat. The filling-up with earth of these lakes has determined, at least in cool and wet sites, the formation of mires. Here we constantly find the remains of some cultures going from the Neolithic up to Bronze Age. To the latter belong the remains found at Mercurago, not far from Arona, investigated in the last century by Gastaldi. The lake of Varese and the neighbouring ones have been inhabited by a numerous population if we judge from the numerous villages built upon palisades. In Switzerland (Neuweiler), in the palisades some remains of Neolithic legumes and the seeds of the fruits consumed by man (cherries, prunes) have been found.

Reptiles can also thrive on the shores of rivers and other water bodies, though they are not very common there. Famous is the pass of Georgicae (Liber IV, 455-460) where the poisonous snake biting Eurydice, Orpheus' woman, is described with insuperable words: 'illa quidem, dum te fugeret per flumina praeceps, immanem ante pedes hydrum moritura puella servantem ripas alta non vidit in herba'. The hydrum is probably a species of *Vipera, Vipera berus* which can reach a length of 80 cm. The river of which Virgil speaks is the Ebrus (today Maritza, in Thracia).

In the case of sandy shores, inside the sand where the pores are filled with water (also through capillary forces) we find an enormous amount of microscopic animals which can be easily collected simply by digging a hole in the sand (and fine gravel) in which immediately the water containing these microscopic creatures filters. To these microscopic animals the collective name of psammon (from the Greek *Ψάμμον*) is given, and, more precisely, limnopsammon, to distinguish it from the similar life-form present on the shores of rivers and sea.

The limnopsammon is constituted by very unconspicuous, often microscopical, animals, not very different from those of hyporrheic and littoral fauna. It can be in contact with hyporrheic fauna through the intermediary of the water table, which, running through alluvial deposits, discharges into lakes and rivers. Among the constituents of limnopsammon let us mention *Archianellidae, Protooligochaeta*, Rotifers, Syncarids, Amphipods and Hydracari. The limnopsammon of Lake Maggiore (in Lombardia) includes the following groups: Ciliates (Protozoa), Rotifers, Oligochaeta, secondarily flatworms and nematodes, besides a few Harpacticida, Tardigrades, Gastrotricha, mites and larvae of Chironomids and Ephemerids. The most interesting species is without any doubt *Rheomorpha neiwestnovae*.

In the case of sandy shores, inside the sand constantly wetted by the water (also through capillary forces) there is an enormous quantity of microscopic animals living in the small pores existing among the particles constituting the sand, the so called fluviatil psammon or hyporrheic fauna (from *ὑπό*, under and *ῥέω* to flow).

The fluviatil psammon or hyporrheic fauna includes Protozoa, flatworms, Celenterata, Nemertina, Rotifers, Gastrotricha, Nematodes, Oligochaeta, Archianellida, Tardigrada, Crustaceans (Cladocers, Ostracods, Copepods, *Bathynellaceae*, Isopods and Amphipods), arachnids (*Hydrachnellae* and *Halacarida*), Insects (preimaginal stages of Ephemerids, stone-flies, diptera and beetles) and finally Gastropods. Often hyporrheic fauna is constituted by some elements of the river which can penetrate the sand of the bed or of the shores. This fauna is closely correlated with the fauna living in the freatic spaces (water table) which, according to Ruffo, is constituted by freatoxenous, freatophilous and freatobious elements, according to the degree of dependence from this environment. Fretaophilous species belong to flatworms, nematodes, oligochaeta, tardigrades, ostracods, copepods and mites.

Cormorants are very able to dive to fish, but when they emerge, they have some difficulties in swallowing the fish: therefore, they throw their prey in the air in order to catch it again: the pelicans profit of this moment to steal the fish from the cormorant, since they are stronger than the latter.

The grebes or *Podiceps* dive, suddenly disappearing from the surface to re-emerge at a great distance. Sometimes they also fly in order to change their hunting territory.

Very interesting and well known are the *Alcidae*, as for instance *Alca torda*, in the Färöer svartføtti or even stórfuglur (both names employed also for *Uria aalge*), in Swedish tordmule, in Sicily carcarozza di mari, in Sardinia Dottori, at Malta mustal bahàr, etc. In English it is known as razorbill, i.e. bird whose beak has the shape of a razor. It nests upon rocks, often together with *Uria aalge*. In some periods it is extremely abundant in the Mediterranean Sea. Generally it lays only one egg on the bare rock, or in some hole or fissure. In Northern Iceland, on rocks and walls vertically cut upon the sea, the small auk or *Plautus alle* nests and lives. In Sweden it is known as alkekung, in Iceland as haftyrthill, in the Färöer as fulkobbi: it is a strictly marine bird, limited to the Arctic.

A bird very similar to the razorbill as distribution and habits are concerned is the guillemot *Uria aalge*, in Sweden known as sillgrissla, in Iceland as langnefja (literally 'long neck'), in the Färöer as lomwigi, langvíi, or even bakfuglur (the brooding female, literally, the bird of which only the back is visible'). The variety of names of this bird (as well as of *Alca torda*) in the Färöer, possibly unparalleled by other European birds with exception of the eider duck, is understandable when we think of the economic value this animal and its eggs have, or at least have had in the past, in this remote part of the Atlantic Ocean. It nests often together with the razorbill, but it appears only rarely in the Mediterranean Sea during its migrations.

Characteristic of the rocky shores of Scandinavia, the Färöer and Iceland are two other birds, *Cepphus grylle*, in English black guillemot, in Swedish tobrisgrissla, in Iceland theista, and *Fratercula arctica* or puffin, in the Färöer known as liri (the young bird), skrápur (the adult) or even smáfuglur (literally 'small bird', which is the name hunters gave it). Both are very frequent on rocks, rocky islands, etc., sometimes also far from the sea, particularly the latter. This is present also in Southern Europe as a winter visitor, a double-pass, irregular, bird, but almost annual, passing in any month of the year. The puffin has been seen also in the Po valley (Mantova and the Euganean Hills near Padova).

We may quote at this point a few marine birds practically indifferent to substrate, and therefore present on rocky, sandy or even muddy shores. They are the ruddy turnstone *Arenaria interpres*, in Iceland tildra, on the Färöer tjaldursgraelingur, frequent also in Southern Europe. Also some gulls prove to be indifferent to substrate: *Larus audouinii*, or Corsican gull, eupelagic, nesting on small islands and rocks of the Mediterranean Sea, characteristic for the shining, coral red colour of the beak, with a black strife and a yellow point, what has been responsible for the name the Germans have given this animal (Korallenmöwe). *Larus genei*, is similar to the former for its habits, nesting in Sardinia, where it is known as 'aipa cu pettu di rosa', and finally *Larus melanocephalus*, present in Northern Italy (Liguria), Central and Southern Italy, but in the past also in the Gulf of Venice.

Finally a bird not bound to substrate is the beautiful *Sterna sandwicensis*.

Similar to what occurs on the shores of inland or freshwaters (rivers and lakes), near the sandy shores of the sea we also find a 'psammon' and more precisely a litoral psammon. The litoral psammon, present practically all along the sandy shores of Europe and the world in general, is constituted by Protozoa, Celenterata (*Protohydra, Halammohydra*), Echinoderma (*Leptosynapta minuta*), Polichetes, Archianellida, Oligochaeta and Protooligochaeta (*Reomorpha neiwestnovae*), Bdelloid Rotifers, Gastrotricha, Nematodes, Echinoderes, flatworms, Gnathostomulida, Bryozoa, Nemertina, Crustaceans (Mystacocarids, Ostracods, Copepods, Syncarids, Thermosbaenacea, Mysidacea, Isopods and Amphipods), Tardigrada, Pycnogonida (very rare), mites (particularly the genus *Nematalycus*), collembols and Gastropods. According to Delamare Deboutteville, many marine interstitial phyletic lines have colonized the continental freatic waters passing through the subterranean litoral waters, and more precisely Foraminifera, Nematodes, Gastrotricha, Rotifers, Archianellida, Copepods, Syncarids, Thermosbaenacea, Isopods, Amphipods and Porohalacarida.

Freshwater environments

Many lakes are confined to high, sometimes very high, altitudes, more especially on the Pyrenees and the Alps. Here life – both animal and vegetal – is particularly well known, but a classification of the water bodies on the ground of biological data would be difficult today, and in any case would go outside the limits of this book. Let it be sufficient to mention a few lakes of the Dolomites (Eastern Alps) such as the lake of Landro (or Dürrenstein), at 1410 m, a pond in view of its little depth. It is rich in algae (especially *Zygnemales*) and in *Cladocera* (*Daphnia longispina, Scapholeberis mucronata* and *Bosmina longirostris* var. *cornuta*). An Ephemeropteron, *Cloëon dipterum* is extremely abundant, at least in some years. Much more interesting is the Misurina lake, at 1735 m, characterized by rich zooplankton (*Diaptomus denticornis, Alona rectangula, Daphnia longispina, Daphnia hyalina, Chydorus sphaericus*, the rotifers *Keratella quadrata, K. cochlearis, Filinia longiseta, Brachionus angularis* and *Polyarthra trigla*), besides amphipods (*Gammarus lacustris*), snails (*Armiger crista*), caddish flies (several species) and particularly dragon flies (some years ago no less than 14 or 15 species lived here). Some trouts also live in this beautiful, today highly polluted, lake.

A glacial lake, practically suspended in a glacial circle, is that of Sorapiss, near Cortina, at 1925 m, with a milky glaucous colour practically unique throughout the Dolomites. Several animals live in its extremely cool water: the copepod *Arctodiaptomus bacillifer*, larvae of the dipteron genus *Coretra* (*Chaoborus*) and the beetles *Gaurodytes congener, Hydroporus foveolatus* and *Helophorus glacialis*. Finally the lake of Federa (or Lago da Lago), near Cortina, at 2042 m, one of the most interesting and probably one of the richest dolomitic lakes at least as far as its fauna is concerned. The lake is certainly eutrophic, as its green colour testifies. Diptera (*Corethra plumicornis*), Cladocers (*Daphnia longispina, Ceriodaphnia reticulata, Chydorus sphaericus* and *Acroperus harpae*), copepods (*Cyclops strenuus*) and Rotifers (*Polyarthra trigla*) constitute the zooplanktion. On the bottom several Rhizopods thrive, such as *Arcella vulgaris, Difflugia globulosa, D. piriformis* and *D. acuminata*. The macrofauna is represented by gastropods (*Lymnaeidae*), stone flies, dragon flies, beetles, leeches and the alpine newt, *Triturus alpestris* (Marcuzzi, 1975).

A very interesting category of lakes because of its ecological implications is that of the man-make, or artificial lakes. Today the creation of artificial lakes occurs quite frequently, starting with rivers bridled for hydroelectric purposes, or lakes or estuaries closed up for domestic or irrigation purposes. Famous are the investigations on the effects of the construction of hydroelectric basins on the salmon fishing in Scotland.

We should remember that in a small region of Northern Italy (Trentino) there are at least 15 man-made lakes, going from a volume of 300,000 m^3 at 720 m, producing 27,4 millions kWh/year, to 172 millions m^3, at 530 m altitude, producing 280 millions kWh/year (Lago di S. Giustina). The latter is in harmony with the environment, so that no damage to it is determined, and it has been gradually and regularly colonized by plankton. The fishes living in this lake today are *Salmo irideus, Salmo fario* and *Salmo marmoratus* (the latter an endemic to Italy), *Coregonus lavaretus* (a kind of white-fish), *Phoxinus phoxinus, Rutilus rubilio, Tinca tinca, Alburnus albidus alborella, Cyprinus carpio* and *Eupomotis gibbosus* (the common 'sunfish'). The lake was made in 1951. In 1953 the zooplankton was composed already of at least 9 species, namely the Copepod *Cyclops strenuus*, the Cladocers *Daphnia longispina* and *Bosmina longirostris* and the Rotifers *Keratella cochlearis, K. quadrata, Testudinella patina, Asplanchna* sp. *Polyarthra trigla* and *Pedalion mirum*. The phytoplankton was constituted of at least 7 species, namely the Diatoms *Asterionella gracillima, Fragilaria capucina*, the Chlorophyta *Gloeococcus schroeteri*, the Conjugatae *Scenedesmus bijugatus* and *Pediastrum boryanum*, the Dinoflagellate *Ceratium hirundinella* and the Bacillariacea *Synedra ulna*.

A very peculiar artificial lake (or better a pond) is the so-called agricultural or colline-lake, present in some parts of the Appennines in Central Italy. The shape is variable (round, oval, rectangular etc.), the size is generally less than 1 ha (2.4 acres) and the depth rarely higher than 10-12 m. The average altitude is 200-450 m; in Umbria alone there are no less than 600 colline lakes. They are excavated in hilly regions on little permeable substrates (sandstone, clay, etc.); they are fed by rain and, secondarily, by some small brooks or springs. The use is exclusively for irrigation. Charasteristic are the high amount of clay, the shores built with earth, the turbidity of the water, the rather conspicuous variations in level, temperature, oxygen amount and pH. The oxygen can vary from almost nil to supersaturation (206.5%); the pH can oscillate between 5.5 and 7.5. The hardness is also rather variable. The organic substance is, in general, scarce; however, during the summer a remarkable amount of H_2S is present. Sometimes a thermocline is present between 1 and 3 m of depth.

The agricultural lakes are constantly eutrophic, inhabited by ubiquitary elements, many of which characteristic of temporary waters (or asthatic basins).

The phytoplankton is mainly constituted by the Chlorophyceae *Chlamydomonas braunii, Sphaerocystis schröteri, Spirogyra* sp., *Cosmarium botrytis*, the Diatoms *Fragilaria crotonensis, Synedra ulna, Gomphonema* sp., *Nitzschia acicula-*

ris, Surirella sp., the Dinoflagellates *Peridinium* sp., *Ceratium hirundinella*, the Euglenophyceae *Euglena* sp. and *Trachelomonas* sp. The zooplankton is formed by some Rotifers such as *Synchaeta* sp., *Polyarthra* sp., *Filinia longiseta, Keratella quadrata, K. cochlearis, Brachionus* sp., the Cladocers *Daphnia* gr. *pulex, Bosmina longirostris, Chydorus sphaericus*, the Copepods *Diaptomus* sp., *Cyclops* sp. and Dipteron larvae of genus *Chaoborus*.

On the shores or on the bottom a very rich plant and animal community is found, constituted of several *Chlorophyceae*, Diatoms, *Cryptophyceae, Eugleno-phyceae* and *Myxophyceae*. Practically all animal groups, from protozoa to fish, are present; the latter are, of course, introduced by man. Fishes are mainly represented by eel (rare), carp, tench and the blackbass *Micropterus salmoides*.

Interesting beetles common to swamps and mires are the *Elodidae Eucinetus*, *Cyphon, Microcara* and *Hydrocyphon*. We should also mention here the beetles that live in phytothelma, i.e., hollow trees, such as the *Elodidae Scirtes* and *Prionocyphon*.

Though not strictly pertinent to ecology, we should point out the enormous zoogeographical importance of Caspian Sea, as a consequence of its various and successive connections with several seas, such as the Arctic Ocean and the Black Sea (and also in some periods with the Mediterranean Sea). Many fish genera are relicts of the Upper Tertiary (Sarmatian, Maeotic and Pontic) fauna. The sea pikeperch *Lucioperca marina* and, among anadromous fish, the sturgeons are apparently relict forms. In addition to these relict forms, the Caspian Sea is inhabited also by migrants from the Arctic Ocean such as the salmon *Salmo salar caspius*, the inconnu *Stenodus leucichthys*, the seal *Phoca caspica* and the crustaceans *Limnocalanus grimaldii, Mesidothea entomon glacialis, Gammaracanthus loricatus caspius, Mysis caspia* and *M. microphthalma*: during the Quaternary, these species penetrated the Caspian Sea through the Pechora, Kama and Volga rivers, and partially from the Baltic Sea. Finally there are many purely freshwater fishes which arrived from the deltas of the inlowing rivers, such as the wild carp, the bream, the common pikeperch and the roach.

Little is known about life in the upper course of brooks. We wish to give here two examples taken from the Dolomites, one from the glacier of the Marmolada, the other from a source situated in the volcanic tuffs emerging north of the Marmolada at nearly 2300 m, and belonging to the chain of Mésola and Padón. In the first at 0.2°C no animal is present; at a temperature of 2.2°C the larvae of the Chironomid *Diamesa steinböcki* makes its appearance; at 7,2°C there are plenty of larvae of *Diamesa, Simuliidae* and stone flies, besides the flat worm *Dugesia alpina*. At 2200 m the larvae of the caddish fly *Cryptothrix nebulicola* (which literally means 'living on the clouds', referring to the high altitude!) thrives. In the second brooklet, at a temperature of 6.8°C a lotic community has been found by the author, constituted of larvae of *Orthocladiinae* (Chironomids), stone flies and a beetle (*Coelambus*?), besides *Dugesia alpina*. At 8°C larvae of *Orthocladiinae, Simuliidae, Chloroperla* (a stone-fly), *Rhithrogena nivata* and *Baëtis* sp. (*Ephemeroptera*), caddish flies, besides *Dugesia* were present. At 8.6°, finally, young individuals of the montane frog *Rana temporaria* make their first appearance. When the bottom of the brook is covered with mosses the animal life is much richer (Marcuzzi, 1975).

Though not strictly ecological, the subdivision of all European rivers based on the zoogeography of fishes proposed by Berg is of enormous interest. We may indeed recognize the following provinces and districts, each characterized by a given ichthic fauna (The reader interested particularly in this argument can find a huge amount of data in Berg's work):

1) Circumpolar section

Mediterranean subregion

a) Baltic province: α, Western or Rhein district (England is here included); β Eastern or Neva district

b) Mediterranean province

c) West Balcanian province

d) Ponto-Caspian-Aral province:

α) Black Sea district: 1) North Aegean subdistrict; 2) Danube-Kuban subdistrict

β) Caspian district: 1) Volga subdistrict; 2) Kura-Persian subdistrict.

A peculiar kind of geological formation bound to the presence of sources whose waters are very rich in calcium is constituted by the calcareous tuffs. The mechanism of tuff formation is very simple, and consists mainly of the precipitation of calcium carbonate (insoluble in water) from a solution of calcium bicarbonate, according to the formula

$$CaCO_3 + CO_2 + H_2O \rightleftarrows Ca(HCO_3)_2$$

When there is a high amount of CO_2 in the water, the reaction runs leftwards. When this CO_2 is produced by photosynthetic plants one can speak of phytogenic calcareous deposition. This is the course of the reaction due to plants:

$$Ca(HCO_3)_2 \rightarrow Ca(OH)_2 + 2\,CO_2 \qquad Ca(OH)_2 + Ca(HCO_3)_2 \rightarrow 2\,H_2O + 2\,CaCO_3$$

Numerous are the lower plants living in sources and able to determine the precipitation of calcium carbonate and automatically to construct some calcareous tuff. Let us mention, with Voigtländer (1967), the blue algae *Scytonema myochrous*, *Croococcus* sp. and *Lyngbya* sp., the desmidiacea *Oocardium stratum*, the Chlorophycea *Cladophora* sp., the Characea *Chara foetida*, besides the liverwort *Conocephalum conicum* and two mosses, *Eucladium verticillatum* and *Cratoneuron commutatum*.

A particularly well-developed calcareous tuff is present at Baun-Alm (m 864) near Bad Tölz.

Resultant waters exhibit some very peculiar physical and biotic characteristics through which they deserve a separate description. There is a constant flow which depends only very little on the general rhythm or conditions of surface waters. Temperature and chemical characteristics of the water are also very constant. The fauna of resultant waters comes, however, as a whole from surface waters: Protozoa, flatworms, rotifers, nematodes, oligochaeta, Gastrotricha, crustaceans (Copepods, Cladocers, Ostracods, Isopoda and Amphipods), Hydracari, Insects (Collembols, Ephemeroptera, Hemiptera, Coleoptera, Diptera, Trichoptera and Lepidoptera) and finally gastropods. The latter are sometimes very abundant. Highly characteristic for this type of water course is the continuity in the development cycle of their animals, the lack of ephippium-eggs ('Dauereiern') in Cladocers, the presence of some winged insects which are generally not present in other water courses, such as *Cloëon dipterum*. During the winter we may observe a migration of some insects towards the resultant waters from the nearby canals.

Fountains are extremely diffused and sometimes they can represent a highly interesting biotope for several organisms, both plants and animals, according to continuity of flow, physicochemical characteristics of the water, size and shape of the fountain, etc. In some instances we may have a remarkable depth (some 1.5 m), a continuous renewal of water, the walls completely covered with a rich vegetation formed by mosses (*Fontinalis antipyretica*, e.g.) and algae (particularly filamentous ones) so that a very rich aquatic fauna can develop. We have in mind with this description a fountain at the feet of the Dolomites, near Feltre, which offers a sight that we will rarely find. The temperature is practically constant throughout the year (in spring 15°C) and an almost neutral pH(6,8) allows almost perennial vegetation and animal life, making allowance for the difference in the life cycle of the different species. The animals which thrive in this biotope, several of which we cannot easily find in other water bodies or water courses of this region, are essentially larvae of Diptera (*Chironomidae* without haemoglobin as a sign of the elevated amount of oxygen, and a species of *Cylindrotomidae, Triogma trisulcata*), Amphipods (*Rivulogammarus balcanicus*), gastropods (*Lymnaea peregra* and *L. truncatula*, both very abundant) and beetles (*Gaurodytes paludosus*, known from rather low altitudes, *Protohaliplus lineatocollis*, known from slightly running waters and low altitudes, and *Laccobius alutaceus*). Fountains are, unfortunately, very scarcely known from the biológical point of view.

Springs coming from karstic subterranean waters contain often several representatives of the true cavernicolous fauna, both as permanent inhabitants and as extraneous elements transported by water in the epochs of flooding. Examples are the amphipod *Gammarus puteanus* and several species of the gastropod genus *Lartetia*, present often in the cold waters of some montane sources, as well as *Proteus anguinus*, whose discovery is due to the accidental appearance at the open in the resultant waters of Krajn (Slovenia).

Among fish living in brackish waters a great part reach this environment only occasionally or with seasonal migrations; scarcer are instead the species which thrive permanently in this so peculiar environment. For instance, in the Lagoon of Venice, stable species are some *Gobius, Aphanius fasciatus*, the stickleback *Gasterosteus aculeatus* and some *Syngnathidae*, whereas all valuable fishes such as bass, dory, mullets, soles, plaices, eels etc. enter only to feed but reproduce in the sea.

Already Aristotle (περί τα ζωα ὶστορίαι) knew that two euryhaline fishes, the eel and the mullet pass from one type of water to another. Of mullets he says 'they rise from sea towards freshwaters, whereas the eel leaves freshwaters for the sea'. According to this great philosopher, who was contemporaneously a zoologist, the reproduction of the eel is not known and the fish neither pairs nor lays eggs (Liber, VI, c. 15). As is well known, only in 1874 the male of the eel was discovered and today we still do not exactly know where reproduction occurs.

Out of the Mediterranean Basin, because of the relatively lower temperature and the scarce evaporation, brackish waters are much rarer, apart from, of course, the Baltic Sea, which is a brackish water 'per se'. Small brackish water ponds have been noticed along the River Thames, near the mouth, and the fauna of these waters has proved to be no less interesting than that of Mediterranean brackish waters, or even of internal salt waters of Algeria. Among bugs there is a terrestrial species, *Saldula palustris*, widespread throughout Centro-southern Europe, together with some aquatic forms, such as *Gerris thoracicus, Notonecta marmorea, Sigara stagnalis* and *S. selecta*, all widely distributed species. Among aquatic beetles living near the River Thames there is *Gyrinus caspius* (a southeastern element, in England reaching its northwestern limit of distribution); *Enochrus halophilus*, characteristic because its larvae fills the gut with air bubbles which allows the animal to displace very swiftly in the water where it lives, among filamentous algae, a little below the surface; two species of *Ochthebius*, a genus constantly linked to brackish or even marine environments and more precisely *O. marinus* and *O. pusillus*. The former, much more studied is diffused and is known from the shores of Arresee (a lake of Själland just in front of Kattegat) as well as from salty meadows of Nordwald, near Jaegerspris, on the shores of Själland. The adult can live in water, covered with an air bubble from which it slowly subtracts oxygen and into which it discharges carbonic anhydride. In these waters also a Chrysomelid beetle, belonging to *Donaciini, Haemonia* (= *Macroplea*) *mutica* var. *curtisi* lives.

These brackish waters of Southern England exhibit also a parasitological interest, since they offer a suitable environment to the larvae of mosquitos such as *Aedes detritus* and *A. caspius*, locally called salty meadow or salt-marsh mosquitos, as well as *Anopheles maculipennis* var. *atroparvus*, a transmitter of malaria in all temperate-warm parts of Europe. Since this is a race linked to cattle, i.e. a zoophilous form, it is less dangerous to man than the other races of *A. maculipennis*, often anthropophilous.

The environment of human settlement

Air pollution in towns consists essentially of small particles (diametre $< 100\ \mu$), coarse particles (diametre $> 100\ \mu$), sulphur compounds, organic compounds, nitrogen compounds, oxygen compounds, alogenated compounds and finally radioactive substances. Particularly dangerous are CO_2, SO_2, PAN (or peroxy-acetilnitrates), fluorine and ozone. Air pollution has a negative influence on the health of the inhabitants of a town: it determines specific diseases, prolongs the course of many others and increases the mortality rate.

Noise pollution is also very serious today, disregarding factories, where it represents a serious menace to workers. Let us remember some quantitative data about noise in towns:

Nature of noise	Decibels
hearing threshold	0
registration hall	20
quiet apartment	50
conversation	60
normal town road	80
fast train at 4 m	100
pain threshold	110-135
automobile horn at 1 m	110-120
pneumatic hammer	120-130
landing jets	130-170

We wish to add some quantitative data registered for some Italian towns in recent times about noise in general:

Town	year	decibels (mean value)
Milano	1959	90-100
Brescia	,,	90-100
Varese	,,	80-90
Mantova	,,	80-90
Cremona	,,	70-80
Bergamo	,,	70-80
Sondrio	,,	70-80
Napoli	1958-60	80
,,	1966	88
Sestri Levante		65

The first serious and precise description of a town, which we may consider 'ecological', is probably due to al-Idrisí (1100-1169?), an Arab geographer living first in Spain and later in Sicily. To him a perfect description of Toledo is due, which we wish to report completely (translation from the Spanish) 'The town of Toledo, situated east of the Talavera, is an important capital city, both because of its extension and the number of its inhabitants; already strong by nature, it is surrounded by beautiful walls and defended by an efficiently fortified town. It is situated upon a hill and few towns can be compared with it because of the height of the buildings, the beauty of the environs and the fertility of the fields, irrigated by the large Tajo river. Here a very curious aqueduct can be seen, constituted by an only arch under which the water throws itself with a great violence, and makes at the extremity of the aqueduct a hydraulic machine to move. The latter makes the water to raise up to 90 cubits (c. 40 m) of height: after having reached the upper part of the aqueduct, the water runs along the same direction and then penetrates into the town. ... The gardens which surround Toledo are crossed by canals above which some water-wheels have been constructed for the irrigation of the kitchen gardens. The latter produce a prodigious amount of fruits of an inexplicable beauty and savour. Everywhere beautiful estates and well fortified castles can be seen' (from Gonzales Palencia).

According to Strawinski (1962/63) the following habitats are generally present in towns: 1) territories of railway station and factories; 2) densely inhabited

areas, historic centre etc.; uncultivated areas; outskirts with desert, almost treeless gardens; recreational areas; new building areas; green places without shadow and meadows; 3) waters; 4) typical outskirts with houses and gardens; modern quarters with planted trees; 5) parks of recent constitution, cemeteries and small parks; old, large parks; and, finally 6) coastal areas and harbours.
The following scheme can give a clear idea of bird life in towns. Two Southern European towns are considered, Milano (Northern Italy) and Rome (Central Italy). These are the birds checked: *(See table page 776).*

Birds living in towns are, according to Strawinski, synanthropic species which easily adapt themselves to life in towns and which are generally present in urban biotopes. In Central Europe there are 28 species.

Among the first descriptions of European gardens is the 'Jardin aromático sobre las indicaciones de las ciudades y las comarcas' (Aromatic garden according to the indications of the towns and the country-sides) by Abd al-Monim, living in Spain at the beginning of the XIth century.

Besides several mammals living in the towns, two species are characteristic of the gardens, the Insectivores *Crocidura russula* (during winter in deposits of food) and *Crocidura leucodon*, sometimes also in abodes.

Artificial ecosystems: the cultivated areas

Among the first observations on insects harmful to agriculture we must mention those made by Reaumur (or better R.A. Ferchauld, Seigneur de Reaumur, des Alpes et de la Bermondière). He studied the animals by a method we could define as 'experimental', he investigated their control and the effects of temperature upon their development, namely *Plusia gamma, Apanteles, Tineola bisselliella*, *Cassida viridis, Merodon equestris* and *Xylocopa* among harmful insects, and *Ephialtes* among the useful ones. Other interesting observations on etho-ecology of harmful insects are due to three more authors of the 18th century, namely J.L. Frisch, A.J. Roese and C. Bonnet. John Ray (or Wray), finally, has done a classification of insects based on an essentially ecological criterium.

Two other insect groups deserve our attention because of their enormous economic importance to the Mediterranean basin, diptera and homoptera *Coccini* (or scale insects). Among diptera it is sufficient to quote *Dacus oleae* or the olive fly, found throughout the Mediterranean Basin, the Canary Islands, Eastern and Southern Africa, Asia Minor and from Western Asia to India, i.e. in every region where some species of *Olea* grows.

The Homoptera are the most noxious insects attacking Citraceae, i.e., lemons, oranges, grapefruits etc., a family which in some countries (e.g., Israel, Spain and Italy) has a very great economic meaning. (The production has been in Spain (1967) 21,970,000 q, in Israel (1967) 10.820,000 q and in Italy (1968) 25,097,000 q.)

We cannot even enter in the enormous amount of data referring to these

Milano	Roma
1) resident birds	1) resident birds
Columbia livia	idem
Streptopelia decaocto	–
Falco tinnunculus	idem
Carine noctua	idem
Tyto alba	idem
Turdus merula	idem
Sylvia atricapilla	idem
Parus major	idem
Motacilla alba	idem
Passer italiae	idem
Passer montanus	idem
Fringilla coelebs	idem
Carduelis carduelis	idem
Coccothraustes coccothraustes	winter bird!
Chloris chloris	idem
Sturnus vulgaris	winter bird!
	Monticola solitarius
	Saxicola torquata
	Cettia cetti
idem (winter bird!)	*Troglodytes troglodytes*
idem (winter bird)	*Parus coeruleus*
	Aegithalus caudatus italiae
	Regulus ignicapillus
	Corvus cornix
	Coleus monedula
	Certhia brachydactyla
	Serinus canaria
2) summer visitors	2) summer visitors
Columba palumbus	–
Apus apus	idem
Hirundo rustica	–
Chelidon urbica	idem
Muscicapa striata	idem
Phoenicurus phoenicurus	–
Luscinia megarhynchos	idem
Lanius collurio	idem
	Otus scops
	Dryobates major
	Jynx torquilla
	Anthus trivialis
	Hippolais icterica
	Oriolus oriolus
3) winter visitors	3) winter visitors
Troglodytes troglodytes	resident bird!
Phoenicurus ochruros	–
Erithacus rubecula	idem
Phylloscopus collibita	idem
Parus coeruleus	resident bird!
Aegithalus caudatus europaeus	–
Regulus regulus	idem
Motacilla cinerea	idem
Carduelis spinus	–
Carduelis cannabina	–
	Turdus philomelos
	Prunella modularis

animals, most of which are exclusive to the sclerophyllous forest both in Europe and in the Near East. It is enough to say that each part of a plant (trunk, leaves, flowers, fruits, seedlings etc.) is attacked by one or more species of scale-insects. Let us mention only a few of them, such as *Pseudococcus citri, Coccus hesperidum, Ceroplastes rusci, Mytilococcus becki* (named after the form of the female, recalling a minute mussel of genus *Mytilus*), *Parlatoria zizyphi* etc. Most of them have several parasites belonging to beetles (*Rodolia cardinalis* and *Chilocorus*, two kinds of lady-beetles), Neuroptera (*Sympherobius amicus*), Hymenoptera (*Scutellista cyanea*) and some Lepidopteron-larvae (*Eublemma scitula*), which can keep the scale insect under a certain control. The so-called biological contro[1] consists, in fact, of favouring or even introducing these parasites of Citraceae parasites. Most of the work on these pests has been done in Israel (at that time 'Palestine') by Bodenheimer.

Each day new findings show the paramount importance of the organic substance in the soil, and each day people adopt a stronger position against the use of chemical inorganic fertilizers. It has been shown – inter alia – that the organic substance of the soil, while favouring porosity, at the same time favours the amount of oxygen, so that plants are assisted in their active work of absorption of inorganic nitrogen. The enrichment of soil with organic substances is effectuated in several ways, namely with manure, with cultural residuals (like depuration muds), but more particularly with the so-called 'Kompost' (from the French 'compost') a mixture of manure and earth, torf or other substances. Recently the solid wastes of a big town have proved to be convertible to compost for nearly 50%. Another technique, employed in the USA since long, is sod-seeding (i.e. seeding without previous ploughing of the soil in order to keep intact all cultural residuals and in general the soil biological equilibrium), together with the so-called 'stubble mulch farming' and 'subtilling', in which the cultural residuals are mixed with the soil in order to increase its content in organic substances. Herbicides are also considered with suspicion nowadays. Of course, the organic substance is attacked partly by animals (e.g. earthworms), partly by microorganisms (bacteria and fungi), so that a slow, normal, process of mineralization occurs which furnishes the soil with all necessary nutrients (P,N,K etc.). The most important soil animals are without any doubt the earthworms, but also other groups, such as diptera-larvae, beetles, collembola, millipedes, mites, etc. are very useful, so that they have been defined as 'Bodenbildner', literally soil constructors. The bibliography on this argument is so rich that it would be impossible to quote even the most interesting and complete works.

SUBJECT INDEX